RSMeans

Mechanical Cost Data

25th Annual Edition

2002

Senior Editor

Melville J. Mossman, PE

Contributing Editors

Barbara Balboni
Robert A. Bastoni
Howard M. Chandler
John H. Chiang, PE
Jennifer L. Curran
Stephen E. Donnelly
J. Robert Lang
Robert C. McNichols
Robert W. Mewis, CCC
Melville J. Mossman, PE
John J. Moylan
Jeannene D. Murphy
Peter T. Nightingale
Stephen C. Plotner
Michael J. Regan
Marshall J. Stetson
Phillip R. Waier, PE

Senior Engineering Operations Manager

John H. Ferguson, PE

Senior Vice President

Charles Spahr

Vice President and General Manager

Roger J. Grant

Vice President, Sales and Marketing

John M. Shea

Production Manager

Michael Kokernak

Production Coordinator

Marion E. Schofield

Technical Support

Thomas J. Dion
Jonathan Forgit
Mary Lou Geary
Gary L. Hoitt
Paula Reale-Camelio
Robin Richardson
Kathryn S. Rodriguez
Sheryl A. Rose
Elizabeth Testa

Book & Cover Design

Norman R. Forgit

First Printing

Foreword

R.S. Means Co., Inc. is a subsidiary of CMD (Construction Market Data), a leading provider of construction information, products, and services in North America and globally. CMD's project information products include more than 100 regional editions, national construction data, sales leads, and over 70 local plan rooms in major business centers. CMD PlansDirect provides surveys, plans and specifications. The First Source suite of products consists of *First Source for Products,* SPEC-DATA™, MANU-SPEC™, CADBlocks, Manufacturer Catalogs and First Source Exchange (www.firstsourceexchange.com) for the selection of nationally available building products. CMD also publishes ProFile, a database of more then 20,000 U.S. architectural firms. R.S. Means provides construction cost data, training, and consulting services in print, CD-ROM and online. CMD is headquartered in Atlanta and has 1,400 employees worldwide. CMD is owned by Cahners Business Information (www.cahners.com), a leading provider of critical information and marketing solutions to business professionals in the media, manufacturing, electronics, construction and retail industries. Its market-leading properties include more than 135 business-to-business publications, over 125 Webzines and Web portals, as well as online services, custom publishing, directories, research and direct-marketing lists. Cahners is a member of the Reed Elsevier plc group (NYSE: RUK and ENL)—a world-leading publisher and information provider operating in the science and medical, legal, education and business-to-business industry sectors.

Our Mission

Since 1942, R.S. Means Company, Inc. has been actively engaged in construction cost publishing and consulting throughout North America.

Today, over 50 years after the company began, our primary objective remains the same: to provide you, the construction and facilities professional, with the most current and comprehensive construction cost data possible.

Whether you are a contractor, an owner, an architect, an engineer, a facilities manager, or anyone else who needs a fast and reliable construction cost estimate, you'll find this publication to be a highly useful and necessary tool.

Today, with the constant flow of new construction methods and materials, it's difficult to find the time to look at and evaluate all the different construction cost possibilities. In addition, because labor and material costs keep changing, last year's cost information is not a reliable basis for today's estimate or budget.

That's why so many construction professionals turn to R.S. Means. We keep track of the costs for you, along with a wide range of other key information, from city cost indexes . . . to productivity rates . . . to crew composition . . . to contractor's overhead and profit rates.

R.S. Means performs these functions by collecting data from all facets of the industry, and organizing it in a format that is instantly accessible to you. From the preliminary budget to the detailed unit price estimate, you'll find the data in this book useful for all phases of construction cost determination.

The Staff, the Organization, and Our Services

When you purchase one of R.S. Means' publications, you are in effect hiring the services of a full-time staff of construction and engineering professionals.

Our thoroughly experienced and highly qualified staff works daily at collecting, analyzing, and disseminating comprehensive cost information for your needs. These staff members have years of practical construction experience and engineering training prior to joining the firm. As a result, you can count on them not only for the cost figures, but also for additional background reference information that will help you create a realistic estimate.

The Means organization is always prepared to help you solve construction problems through its five major divisions: Construction and Cost Data Publishing, Electronic Products and Services, Consulting Services, Insurance Services, and Educational Services.

Besides a full array of construction cost estimating books, Means also publishes a number of other reference works for the construction industry. Subjects include construction estimating and project and business management; special topics such as HVAC, roofing, plumbing, and hazardous waste remediation; and a library of facility management references.

In addition, you can access all of our construction cost data through your computer with Means CostWorks 2002 CD-ROM, an electronic tool that offers over 50,000 lines of Means detailed construction cost data, along with assembly and whole building cost data. You can also access Means cost information from our Web site at www.rsmeans.com

What's more, you can increase your knowledge and improve your construction estimating and management performance with a Means Construction Seminar or In-House Training Program. These two-day seminar programs offer unparalleled opportunities for everyone in your organization to get updated on a wide variety of construction-related issues.

Means also is a worldwide provider of construction cost management and analysis services for commercial and government owners and of claims and valuation services for insurers.

In short, R.S. Means can provide you with the tools and expertise for constructing accurate and dependable construction estimates and budgets in a variety of ways.

Robert Snow Means Established a Tradition of Quality That Continues Today

Robert Snow Means spent years building his company, making certain he always delivered a quality product.

Today, at R.S. Means, we do more than talk about the quality of our data and the usefulness of our books. We stand behind all of our data, from historical cost indexes... to construction materials and techniques... to current costs.

If you have any questions about our products or services, please call us toll-free at 1-800-334-3509. Our customer service representatives will be happy to assist you or visit our Web site at www.rsmeans.com

Table of Contents

Foreword	ii
How the Book Is Built: An Overview	iv
Quick Start	v
How To Use the Book: The Details	vi
Unit Price Section	1
Assemblies Section	407
Reference Section	511
Reference Numbers	512
Change Orders	545
Crew Listings	549
Historical Cost Indexes	574
City Cost Indexes	577
Location Factors	595
Abbreviations	601
Index	604
Other Means Publications and Reference Works	Yellow Pages
Installing Contractor's Overhead & Profit	Inside Back Cover

UNIT PRICES

GENERAL REQUIREMENTS	1
SITE CONSTRUCTION	2
CONCRETE	3
MASONRY	4
METALS	5
WOOD & PLASTICS	6
THERMAL & MOISTURE PROTECTION	7
DOORS & WINDOWS	8
FINISHES	9
SPECIALTIES	10
EQUIPMENT	11
SPECIAL CONSTRUCTION	13
CONVEYING SYSTEMS	14
MECHANICAL	15
ELECTRICAL	16
SQUARE FOOT	17

ASSEMBLIES

SUBSTRUCTURE	A
SERVICES	D
BUILDING SITEWORK	G

REFERENCE INFORMATION

REFERENCE NUMBERS
CREWS
COST INDEXES
INDEX

How the Book Is Built: An Overview

A Powerful Construction Tool

You have in your hands one of the most powerful construction tools available today. A successful project is built on the foundation of an accurate and dependable estimate. This book will enable you to construct just such an estimate.

For the casual user the book is designed to be:

- quickly and easily understood so you can get right to your estimate
- filled with valuable information so you can understand the necessary factors that go into the cost estimate

For the regular user, the book is designed to be:

- a handy desk reference that can be quickly referred to for key costs
- a comprehensive, fully reliable source of current construction costs and productivity rates, so you'll be prepared to estimate any project
- a source book for preliminary project cost, product selections, and alternate materials and methods

To meet all of these requirements we have organized the book into the following clearly defined sections.

Quick Start

This one-page section (see following page) can quickly get you started on your estimate.

How To Use the Book: The Details

This section contains an in-depth explanation of how the book is arranged . . . and how you can use it to determine a reliable construction cost estimate. It includes information about how we develop our cost figures and how to completely prepare your estimate.

Unit Price Section

All cost data has been divided into the 16 divisions according to the MasterFormat system of classification and numbering as developed by the Construction Specifications Institute (CSI) and Construction Specifications Canada (CSC). For a listing of these divisions and an outline of their subdivisions, see the Unit Price Section Table of Contents.

Estimating tips are included at the beginning of each division.

Division 17: Quick Project Estimates

In addition to the 16 Unit Price Divisions there is a S.F. (Square Foot) and C.F. (Cubic Foot) Cost Division, Division 17. It contains costs for 59 different building types that allow you to quickly make a rough estimate for the overall cost of a project or its major components.

Assemblies Section

The cost data in this section has been organized in an "Assemblies" format. These assemblies are the functional elements of a building and are arranged according to the 7 divisions of the UNIFORMAT II classification system. For a complete explanation of a typical "Assemblies" page, see "How To Use the Assemblies Cost Tables."

Reference Section

This section includes information on Reference Numbers, Change Orders, Crew Listings, Historical Cost Indexes, City Cost Indexes, Location Factors and a listing of Abbreviations. It is visually identified by a vertical gray bar on the edge of pages.

Reference Numbers: Following selected major classifications throughout the book are "reference numbers" shown in bold squares. These numbers refer you to related information in other sections.

In these other sections, you'll find related tables, explanations, and estimating information. Also included are alternate pricing methods and technical data, along with information on design and economy in construction. You'll also find helpful tips on estimating and construction.

Change Orders: This section includes information on the factors that influence the pricing of change orders.

Crew Listings: This section lists all the crews referenced. For the purposes of this book, a crew is composed of more than one trade classification and/or the addition of equipment to any trade classification. Power equipment is included in the cost of the crew. Costs are shown both with bare labor rates and with the installing contractor's overhead and profit added. For each, the total crew cost per eight-hour day and the composite cost per labor-hour are listed.

Historical Cost Indexes: These indexes provide you with data to adjust construction costs over time. If you know costs for a past project, you can use these indexes to estimate the cost to construct the same project today.

City Cost Indexes: Obviously, costs vary depending on the regional economy. You can adjust the "national average" costs in this book to 316 major cities throughout the U.S. and Canada by using the data in this section.

Location Factors, to quickly adjust the data to over 930 zip code areas, are included .

Abbreviations: A listing of the abbreviations and the terms they represent, is included.

Index

A comprehensive listing of all terms and subjects in this book to help you find what you need quickly.

The Scope of This Book

This book is designed to be comprehensive and easy to use. To that end we have made certain assumptions:

1. We have established material prices based on a "national average."
2. We have computed labor costs based on a 30-city "national average" of union wage rates.
3. We have targeted the data for projects of a certain size range.

For a more detailed explanation of how the cost data is developed, see "How To Use the Book: The Details."

Project Size

This book is aimed primarily at commercial and industrial projects costing $1,000,000 and up, or large multi-family housing projects. Costs are primarily for new construction or major renovation of buildings rather than repairs or minor alterations.

With reasonable exercise of judgment the figures can be used for any building work. *However, for civil engineering structures such as bridges, dams, highways, or the like, please refer to **Means Heavy Construction Cost Data.***

Quick Start

If you feel you are ready to use this book and don't think you need the detailed instructions that begin on the following page, this Quick Start section is for you.

These steps will allow you to get started estimating in a matter of minutes.

1 First, decide whether you require a Unit Price or Assemblies type estimate. Unit price estimating requires a breakdown of the work to individual items. Assemblies estimates combine individual items or components into building systems.

If you need to estimate each line item separately, follow the instructions for **Unit Prices.**

If you can use an estimate for the entire assembly or system, follow the instructions for **Assemblies.**

2 Find each cost data section you need in the Table of Contents (either for Unit Prices or Assemblies).

Unit Prices: The cost data for Unit Prices has been divided into 16 divisions according to the CSI MasterFormat.

Assemblies: The cost data for Assemblies has been divided into 7 divisions according to the UNIFORMAT II.

3 Turn to the indicated section and locate the line item or assemblies table you need for your estimate. Portions of a sample page layout from both the Unit Price Listings and the Assemblies Cost Tables appear below.

Unit Prices: If there is a reference number listed at the beginning of the section, it refers to additional information you may find useful. See the referenced section for additional information.

- Note the crew code designation. You'll find full descriptions of crews in the Crew Listings including labor-hour and equipment costs.

Assemblies: The Assemblies (*not* shown in full here) are generally separated into three parts: 1) an illustration of the system to be estimated; 2) the components and related costs of a typical system; and 3) the costs for similar systems with dimensional and/or size variations. The Assemblies Section also contains reference numbers for additional useful information.

4 Determine the total number of units your job will require.

Unit Prices: Note the unit of measure for the material you're using is listed under "Unit."

- Bare Costs: These figures show unit costs for materials and installation. Labor and equipment costs are calculated according to crew costs and average daily output. Bare costs do not contain allowances for overhead, profit or taxes.
- "Labor-hours" allows you to calculate the total labor-hours to complete that task. Just multiply the quantity of work by this figure for an estimate of activity duration.

Assemblies: Note the unit of measure for the assembly or system you're estimating is listed in the Assemblies Table.

5 Then multiply the total units by . . .

Unit Prices: "Total Incl. O&P" which stands for the total cost including the installing contractor's overhead and profit. (See the "How To Use the Unit Price Pages" for a complete explanation.)

Assemblies: The "Total" in the right-hand column, which is the total cost **including the installing contractor's overhead and profit.** (See the "How To Use the Assemblies Cost Tables" section for a complete explanation.)

Material and equipment cost figures include a 10% markup. For labor markups, see the inside back cover of this book. If the work is to be subcontracted, add the general contractor's markup, approximately 10%.

6 The price you calculate will be an estimate for either an individual item of work or a completed assembly or *system.*

7 Compile a list of all items or assemblies included in the total project. Summarize cost information, and add project overhead.

Localize costs by using the City Cost Indexes or Location Factors found in the Reference Section.

For a more complete explanation of the way costs are derived, please see the following sections.

Editors' Note: We urge you to spend time reading and understanding all of the supporting material and to take into consideration the reference material such as Crews Listing and the "reference numbers."

Unit Price Pages

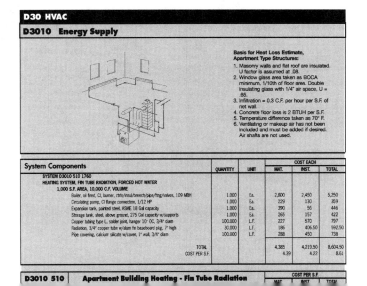

Assemblies Pages

How to Use the Book: The Details

What's Behind the Numbers? The Development of Cost Data

The staff at R.S. Means continuously monitors developments in the construction industry in order to ensure reliable, thorough and up-to-date cost information.

While *overall* construction costs may vary relative to general economic conditions, price fluctuations within the industry are dependent upon many factors. Individual price variations may, in fact, be opposite to overall economic trends. Therefore, costs are continually monitored and complete updates are published yearly. Also, new items are frequently added in response to changes in materials and methods.

Costs—$ (U.S.)

All costs represent U.S. national averages and are given in U.S. dollars. The Means City Cost Indexes can be used to adjust costs to a particular location. The City Cost Indexes for Canada can be used to adjust U.S. national averages to local costs in Canadian dollars.

Material Costs

The R.S. Means staff contacts manufacturers, dealers, distributors, and contractors all across the U.S. and Canada to determine national average material costs. If you have access to current material costs for your specific location, you may wish to make adjustments to reflect differences from the national average. Included within material costs are fasteners for a normal installation. R.S. Means engineers use manufacturers' recommendations, written specifications and/or standard construction practice for size and spacing of fasteners. Adjustments to material costs may be required for your specific application or location. Material costs do not include sales tax.

Labor Costs

Labor costs are based on the average of wage rates from 30 major U.S. cities. Rates are determined from labor union agreements or prevailing wages for construction trades for the current year. Rates along with overhead and profit markups are listed on the inside back cover of this book.

- If wage rates in your area vary from those used in this book, or if rate increases are expected within a given year, labor costs should be adjusted accordingly.

Labor costs reflect productivity based on actual working conditions. These figures include time spent during a normal workday on tasks other than actual installation, such as material receiving and handling, mobilization at site, site movement, breaks, and cleanup.

Productivity data is developed over an extended period so as not to be influenced by abnormal variations and reflects a typical average.

Equipment Costs

Equipment costs include not only rental, but also operating costs for equipment under normal use. The operating costs include parts and labor for routine servicing such as repair and replacement of pumps, filters and worn lines. Normal operating expendables such as fuel, lubricants, tires and electricity (where applicable) are also included. Extraordinary operating expendables with highly variable wear patterns such as diamond bits and blades are excluded. These costs are included under materials. Equipment rental rates are obtained from industry sources throughout North America—contractors, suppliers, dealers, manufacturers, and distributors.

Crew Equipment Cost/Day—The power equipment required for each crew is included in the crew cost. The daily cost for crew equipment is based on dividing the weekly bare rental rate by 5 (number of working days per week), and then adding the hourly operating cost times 8 (hours per day). This "Crew Equipment Cost/Day" is listed in Subdivision 01590.

General Conditions

Cost data in this book is presented in two ways: Bare Costs and Total Cost including O&P (Overhead and Profit). General Conditions, when applicable, should also be added to the Total Cost including O&P. The costs for General Conditions are listed in Division 1 of the Unit Price Section and the Reference Section of this book. General Conditions for the *Installing Contractor* may range from 0% to 10% of the Total Cost including O&P. For the *General* or *Prime Contractor*, costs for General Conditions may range from 5% to 15% of the Total Cost including O&P, with a figure of 10% as the most typical allowance.

Overhead and Profit

Total Cost including O&P for the *Installing Contractor* is shown in the last column on both the Unit Price and the Assemblies pages of this book. This figure is the sum of the bare material cost plus 10% for profit, the base labor cost plus total overhead and profit, and the bare equipment cost plus 10% for profit. Details for the calculation of Overhead and Profit on labor are shown on the inside back cover and in the Reference Section of this book. (See the "How to Use the Unit Price Pages" for an example of this calculation.)

Factors Affecting Costs

Costs can vary depending upon a number of variables. Here's how we have handled the main factors affecting costs.

Quality—The prices for materials and the workmanship upon which productivity is based represent sound construction work. They are also in line with U.S. government specifications.

Overtime—We have made no allowance for overtime. If you anticipate premium time or work beyond normal working hours, be sure to make an appropriate adjustment to your labor costs.

Productivity—The productivity, daily output, and labor-hour figures for each line item are based on working an eight-hour day in daylight hours in moderate temperatures. For work that extends beyond normal work hours or is performed under adverse conditions, productivity may decrease. (See the section in "How To Use the Unit Price Pages" for more on productivity.)

Size of Project—The size, scope of work, and type of construction project will have a significant impact on cost. Economies of scale can reduce costs for large projects. Unit costs can often run higher for small projects. Costs in this book are intended for the size and type of project as previously described in "How the Book Is Built: An Overview." Costs for projects of a significantly different size or type should be adjusted accordingly.

Location—Material prices in this book are for metropolitan areas. However, in dense urban areas, traffic and site storage limitations may increase costs. Beyond a 20-mile radius of large cities, extra trucking or transportation charges may also increase the material costs slightly. On the other hand, lower wage rates may be in effect. Be sure to consider both these factors when preparing an estimate, particularly if the job site is located in a central city or remote rural location.

In addition, highly specialized subcontract items may require travel and per diem expenses for mechanics.

Other factors —
- season of year
- contractor management
- weather conditions
- local union restrictions
- building code requirements
- availability of:
  - adequate energy
  - skilled labor
  - building materials
- owner's special requirements/restrictions
- safety requirements
- environmental considerations

Unpredictable Factors—General business conditions influence "in-place" costs of all items. Substitute materials and construction methods may have to be employed. These may affect the installed cost and/or life cycle costs. Such factors may be difficult to evaluate and cannot necessarily be predicted on the basis of the job's location in a particular section of the country. Thus, where these factors apply, you may find significant, but unavoidable cost variations for which you will have to apply a measure of judgment to your estimate.

Rounding of Costs

In general, all unit prices in excess of $5.00 have been rounded to make them easier to use and still maintain adequate precision of the results. The rounding rules we have chosen are in the following table.

Prices from . . .	Rounded to the nearest . . .
$.01 to $5.00	$.01
$5.01 to $20.00	$.05
$20.01 to $100.00	$.50
$100.01 to $300.00	$1.00
$300.01 to $1,000.00	$5.00
$1,000.01 to $10,000.00	$25.00
$10,000.01 to $50,000.00	$100.00
$50,000.01 and above	$500.00

Contingencies

Contingencies: The allowance for contingencies generally is to provide for unforeseen construction difficulties. On alterations or repair jobs 20% is none too much. If drawings are final and only field contingencies are being considered, 2% or 3% is probably sufficient and often nothing need be added. As far as the contract is concerned, future changes in plans will be covered by extras. The contractor should consider inflationary price trends and possible material shortages during the course of the job. If drawings are not complete or approved or a budget cost wanted, it is wise to add 5% to 10%. Contingencies, then, are a matter of judgment. Additional allowances are shown in Division 01250-200 for contingencies and job conditions and Division 01250-400 for factors to convert prices for repair and remodeling jobs.

Note: For an explanation of "Selective Demolition" versus "Removal for Replacement" see Reference number R15050-720.

Important Estimating Considerations

One reason for listing a job size "minimum" is to ensure that the construction craftsmen are productive for 8 hours a day on a continuous basis. Otherwise, tasks that only require 6 to 7 hours could be billed as an 8-hour day, thus providing an erroneous productivity estimate. The "productivity," or daily output of each craftsman, includes mobilization and cleanup time, break time, plan layout time, as well as an allowance to carry stock from the storage trailer or location on the job site up to 200′ into the building and on the first or second floor. If material has to be transported over greater distances or to higher floors, an additional allowance should be considered by the estimator. An allowance has also been included in the piping and fittings installation time for a leak check and minor tightening.

Equipment installation time includes all applicable items following: positioning, leveling and securing the unit in place, connecting all associated piping, ducts, vents, etc., which shall have been estimated separately, connecting to an adjacent power source, filling/bleeding, startup, adjusting the controls up and down to assure proper response, setting the integral controls/valves/ regulators/thermostats for proper operation (does not include external building type control systems, DDC systems, etc.), explaining/training owner's operator, and warranty. A reasonable breakdown of the labor costs would be as follows:

1. Movement into building, installation/setting of equipment — 35%
2. Connecting to piping/duct/power, etc. — 25%
3. Filling/flushing/cleaning/touchup, etc. — 15%
4. Startup/running adjustments — 5%
5. Training owner's rep. — 5%
6. Warranty/call back/service — 15%

Note that cranes or other lifting equipment are not included on any lines in Division 15. For example, if a crane is required to lift a heavy piece of pipe into place high above a gym floor, or to put a rooftop unit on the roof of a four-story building, etc., it must be added. Due to the potential for extreme variation—from nothing additional required, to a major crane or helicopter—we feel that including a nominal amount for "lifting contingency" would be useless and detract from the accuracy of the estimate. When using equipment rental from Means do not forget to include the cost of the operator(s).

Estimating Labor-Hours

The labor-hours expressed in this publication are based on Average Installation time, using an efficiency level of approximately 60-65% (see item 7) which has been found reasonable and acceptable by many contractors.

The book uses this national efficiency average to establish a consistent benchmark.

For bid situations, adjustments to this efficiency level should be the responsibility of the contractor bidding the project.

The unit labor-hour is divided in the following manner. A typical day for a journeyman might be:

1.	Study Plans	3%	14.4 min.
2.	Material Procurement	3%	14.4 min.
3.	Receiving and Storing	3%	14.4 min
4.	Mobilization	5%	24.0 min.
5.	Site Movement	5%	24.0 min.
6.	Layout and Marking	8%	38.4 min.
7.	Actual Installation	64%	307.2 min.
8.	Cleanup	3%	14.4 min.
9.	Breaks, Non-Productive	6%	28.8 min.
		100%	480.0 min.

If any of the percentages expressed in this breakdown do not apply to the particular work or project situation, then that percentage or a portion of it may be deducted from or added to labor hours.

Final Checklist

Estimating can be a straightforward process provided you remember the basics. Here's a checklist of some of the items you should remember to do before completing your estimate.

Did you remember to . . .

- factor in the City Cost Index for your locale
- take into consideration which items have been marked up and by how much
- mark up the entire estimate sufficiently for your purposes
- read the background information on techniques and technical matters that could impact your project time span and cost
- include all components of your project in the final estimate
- double check your figures to be sure of your accuracy
- for more information, please see "Tips for Accurate Estimating," R01100-005 in the Reference Section.
- call R.S. Means if you have any questions about your estimate or the data you've found in our publications

Remember, R.S. Means stands behind its publications. If you have any questions about your estimate . . . about the costs you've used from our books . . . or even about the technical aspects of the job that may affect your estimate, feel free to call the R.S. Means editors at 1-800-334-3509.

Unit Price Section

Table of Contents

Div No.		Page
	General Requirements	**5**
01100	Summary	6
01200	Price & Payment Procedures	6
01300	Administrative Requirements	7
01500	Temporary Facilities & Controls	8
01700	Execution Requirements	18
01800	Facility Operation	18
	Site Construction	**19**
02050	Basic Site Materials & Methods	20
02100	Site Remediation	21
02200	Site Preparation	22
02300	Earthwork	25
02400	Tunneling, Boring & Jacking	28
02500	Utility Services	28
02600	Drainage & Containment	40
02700	Bases, Ballasts, Pavements, & Appurtenances	43
02800	Site Improvements & Amenities	44
02950	Site Restoration & Rehab.	45
	Concrete	**49**
03100	Concrete Forms & Accessories	50
03200	Concrete Reinforcement	50
03300	Cast-In-Place Concrete	50
	Masonry	**53**
04050	Basic Masonry Mat. & Methods	54
04800	Masonry Assemblies	54
	Metals	**55**
05050	Basic Materials & Methods	56
05100	Structural Metal Framing	58
	Wood & Plastics	**61**
06050	Basic Wd. & Plastic Mat. & Meth.	62
06100	Rough Carpentry	63
06400	Architectural Woodwork	63
	Thermal & Moisture Protection	**65**
07600	Flashing & Sheet Metal	66
07700	Roof Specialties & Accessories	66
07800	Fire & Smoke Protection	67
07900	Joint Sealers	68

Div. No.		Page
	Doors & Windows	**71**
08300	Specialty Doors	72
	Finishes	**75**
09900	Paints & Coatings	76
	Specialties	**77**
10200	Louvers & Vents	78
10270	Access Flooring	78
10300	Fireplaces & Stoves	78
	Equipment	**81**
11010	Maintenance Equipment	82
11100	Mercantile Equipment	82
11110	Coml. Laun. & Dry Cln. Eq.	82
11140	Vehicle Service Equip.	83
11170	Solid Waste Handling Equip.	83
11300	Fluid Waste Treat. & Disp. Equip.	84
11400	Food Service Equipment	85
11450	Residential Equipment	86
11470	Darkroom Equipment	87
11500	Industrial & Process Equipment	88
11600	Laboratory Equipment	90
11700	Medical Equipment	90
	Special Construction	**93**
13030	Special Purpose Rooms	94
13120	Pre-Engineered Structures	95
13150	Swimming Pools	96
13175	Ice Rinks	96
13200	Storage Tanks	97
13280	Hazardous Mat. Remediation	100
13600	Solar & Wind Energy Equipment	103
13800	Building Automation & Control	106
13850	Detection & Alarm	115
	Conveying Systems	**117**
14500	Material Handling	118

Div. No.		Page
	Mechanical	**119**
15050	Basic Materials & Methods	120
15100	Building Services Piping	141
15200	Process Piping	276
15400	Plumbing Fixtures & Equipment	282
15500	Heat Generation Equipment	286
15600	Refrigeration Equipment	312
15700	Heating, Ventilation, & A/C Equip.	320
15800	Air Distribution	347
15900	HVAC Instrumentation & Controls	383
15950	Testing/Adjusting/Balancing	384
	Electrical	**387**
16100	Wiring Methods	388
16200	Electrical Power	396
16400	Low-Voltage Distribution	396
	Square Foot	**399**
17100	S.F., C.F. & % of Total Costs	400

1

How to Use the Unit Price Pages

The following is a detailed explanation of a sample entry in the Unit Price Section. Next to each bold number below is the item being described with appropriate component of the sample entry following in parenthesis. Some prices are listed as bare costs, others as costs that include overhead and profit of the installing contractor. In most cases, if the work is to be subcontracted, the general contractor will need to add an additional markup (R.S. Means suggests using 10%) to the figures in the column "Total Incl. O&P."

1 Division Number/Title (03300/Cast-In-Place Concrete)

Use the Unit Price Section Table of Contents to locate specific items. The sections are classified according to the CSI MasterFormat (1995 Edition).

2 Line Numbers (03310 240 3900)

Each unit price line item has been assigned a unique 12-digit code based on the CSI MasterFormat classification.

```
                          ┌──── Level One - CSI-MasterFormat Division
                     ┌──── Level Two - CSI
      03300
      03310-240-3900
                 └───── Means 12-digit
                        Line Number
              └───── Level Four - Means
         └───── Level Three - CSI
```

3 Description (Concrete-In-Place, etc.)

Each line item is described in detail. Sub-items and additional sizes are indented beneath the appropriate line items. The first line or two after the main item (in boldface) may contain descriptive information that pertains to all line items beneath this boldface listing.

4 Reference Number Information

R03310 -010 You'll see reference numbers shown in bold rectangles at the beginning of some sections. These refer to related items in the Reference Section, visually identified by a vertical gray bar on the edge of pages.

The relation may be: (1) an estimating procedure that should be read before estimating, (2) an alternate pricing method, or (3) technical information.

The "R" designates the Reference Section. The numbers refer to the MasterFormat classification system.

It is strongly recommended that you review all reference numbers that appear within the section in which you are working.

Note: Not all reference numbers appear in all Means publications.

03300 | Cast-In-Place Concrete

		03310 Structural Concrete	CREW	DAILY OUTPUT	LABOR-HOURS	UNIT	MAT.	LABOR	EQUIP.	TOTAL	TOTAL INCL O&P	
240	0010	**CONCRETE IN PLACE** including forms (4 uses), reinforcing										240
	0050	steel, including finishing unless otherwise indicated										
	3800	Footings, spread under 1 C.Y.	C-14C	38.07	2.942	C.Y.	100	84.50	.94	185.44	244	
	3850	Over 5 C.Y.		81.04	1.382		92	40	.44	132.44	164	
	3900	Footings, strip, 18" x 9", plain		41.04	2.729		91	78.50	.87	170.37	225	
	3950	36" x 12", reinforced		61.55	1.820		92.50	52.50	.58	145.58	185	
	4000	Foundation mat, under 10 C.Y.		38.67	2.896		124	83.50	.93	208.43	269	
	4050	Over 20 C.Y.		56	1.986		110	57	.64	167.64	210	
	4200	Grade walls, 8" thick, 8' high	C-14D	45	3.564		126	130	15.65	271.65	360	
	4250	14' high		27.26	7.337		149	219	26.50	394.50	540	
	4260	12" thick, 8' high		64.32	3.109		111	93	11.15	215.15	279	
	4270	14' high		40.01	4.999		119	149	17.95	285.95	385	

Crew (C-14C)

The "Crew" column designates the typical trade or crew used to install the item. If an installation can be accomplished by one trade and requires no power equipment, that trade and the number of workers are listed (for example, "2 Carpenters"). If an installation requires a composite crew, a crew code designation is listed (for example, "C-14C"). You'll find full details on all composite crews in the Crew Listings.
* For a complete list of all trades utilized in this book and their abbreviations, see the inside back cover.

Crews

Crew No.	Bare Costs		Incl. Subs O & P		Cost Per Labor-Hour	
Crew C-14C	Hr.	Daily	Hr.	Daily	Bare Costs	Incl. O&P
1 Carpenter Foreman (out)	$32.00	$256.00	$49.80	$398.40	$28.79	$45.29
6 Carpenters	30.00	1440.00	46.70	2241.60		
2 Rodmen (reinf.)	34.25	548.00	57.80	924.80		
4 Laborers	23.45	750.40	36.50	1168.00		
1 Cement Finisher	28.70	229.60	42.50	340.00		
1 Gas Engine Vibrator		35.60		39.15	.32	.35
112 L.H., Daily Totals		$3259.60		$5111.95	$29.11	$45.64

Productivity: Daily Output (41.04)/Labor-Hours (2.729)

The "Daily Output" represents the typical number of units the designated crew will install in a normal 8-hour day. To find out the number of days the given crew would require to complete the installation, divide your quantity by the daily output. For example:

Quantity	÷	Daily Output	=	Duration
100 C.Y.	÷	41.04/ Crew Day	=	2.44 Crew Days

The "Labor-Hours" figure represents the number of labor-hours required to install one unit of work. To find out the number of labor-hours required for your particular task, multiply the quantity of the item times the number of labor-hours shown. For example:

Quantity	x	Productivity Rate	=	Duration
100 C.Y.	x	2.729 Labor-Hours/ C.Y.	=	272.9 Labor-Hours

Unit (C.Y.)

The abbreviated designation indicates the unit of measure upon which the price, production, and crew are based (C.Y. = Cubic Yard). For a complete listing of abbreviations refer to the Abbreviations Listing in the Reference Section of this book.

Bare Costs:

Mat. (Bare Material Cost) (91)

The unit material cost is the "bare" material cost with no overhead and profit included. *Costs shown reflect national average material prices for January of the current year and include delivery to the job site. No sales taxes are included.*

Labor (78.50)

The unit labor cost is derived by multiplying bare labor-hour costs for Crew C-14C by labor-hour units. The bare labor-hour cost is found in the Crew Section under C-14C. (If a trade is listed, the hourly labor cost—the wage rate—is found on the inside back cover.)

Labor-Hour Cost Crew C-14C	x	Labor-Hour Units	=	Labor
$28.79	x	2.729	=	$78.50

Equip. (Equipment) (.87)

Equipment costs for each crew are listed in the description of each crew. Tools or equipment whose value justifies purchase or ownership by a contractor are considered overhead as shown on the inside back cover. The unit equipment cost is derived by multiplying the bare equipment hourly cost by the labor-hour units.

Equipment Cost Crew C-14C	x	Labor-Hour Units	=	Equip.
.32	x	2.729	=	.87

Total (170.37)

The total of the bare costs is the arithmetic total of the three previous columns: mat., labor, and equip.

Material	+	Labor	+	Equip.	=	Total
$91	+	$78.50	+	$.87	=	$170.37

Total Costs Including O&P

This figure is the sum of the bare material cost plus 10% for profit; the bare labor cost plus total overhead and profit (per the inside back cover or, if a crew is listed, from the crew listings); and the bare equipment cost plus 10% for profit.

Material is Bare Material Cost + 10% = 91 + 9.10	=	$100.10
Labor for Crew C-14C = Labor-Hour Cost (45.29) x Labor-Hour Units (2.729)	=	$123.60
Equip. is Bare Equip. Cost + 10% = .87 + .09	=	$ .96
Total (Rounded)	=	$225

Division 1
General Requirements

Estimating Tips

The General Requirements of any contract are very important to both the bidder and the owner. These lay the ground rules under which the contract will be executed and have a significant influence on the cost of operations. Therefore, it is extremely important to thoroughly read and understand the General Requirements both before preparing an estimate and when the estimate is complete, to ascertain that nothing in the contract is overlooked. Caution should be exercised when applying items listed in Division 1 to an estimate. Many of the items are included in the unit prices listed in the other divisions such as mark-ups on labor and company overhead.

01200 Price & Payment Procedures

- When estimating historic preservation projects (depending on the condition of the existing structure and the owner's requirements), a 15-20% contingency or allowance is recommended, regardless of the stage of the drawings.

01300 Administrative Requirements

- Before determining a final cost estimate, it is a good practice to review all the items listed in subdivision 01300 to make final adjustments for items that may need customizing to specific job conditions.
- Historic preservation projects may require specialty labor and methods, as well as extra time to protect existing materials that must be preserved and/or restored. Some additional expenses may be incurred in architectural fees for facility surveys and other special inspections and analyses.

01330 Submittal Procedures

- Requirements for initial and periodic submittals can represent a significant cost to the General Requirements of a job. Thoroughly check the submittal specifications when estimating a project to determine any costs that should be included.

01400 Quality Requirements

- All projects will require some degree of Quality Control. This cost is not included in the unit cost of construction listed in each division. Depending upon the terms of the contract, the various costs of inspection and testing can be the responsibility of either the owner or the contractor. Be sure to include the required costs in your estimate.

01500 Temporary Facilities & Controls

- Barricades, access roads, safety nets, scaffolding, security and many more requirements for the execution of a safe project are elements of direct cost. These costs can easily be overlooked when preparing an estimate. When looking through the major classifications of this subdivision, determine which items apply to each division in your estimate.

01590 Equipment Rental

- This subdivision contains transportation, handling, storage, protection and product options and substitutions. Listed in this cost manual are average equipment rental rates for all types of equipment. This is useful information when estimating the time and materials requirement of any particular operation in order to establish a unit or total cost.
- A good rule of thumb is that weekly rental is 3 times daily rental and that monthly rental is 3 times weekly rental.

- The figures in the column for Crew Equipment Cost represent the rental rate used in determining the daily cost of equipment in a crew. It is calculated by dividing the weekly rate by 5 days and adding the hourly operating cost times 8 hours.

01770 Closeout Procedures

- When preparing an estimate, read the specifications to determine the requirements for Contract Closeout thoroughly. Final cleaning, record documentation, operation and maintenance data, warranties and bonds, and spare parts and maintenance materials can all be elements of cost for the completion of a contract. Do not overlook these in your estimate.

01830 Operations & Maintenance

- If maintenance and repair are included in your contract, they require special attention. To estimate the cost to remove and replace any unit usually requires a site visit to determine the accessibility and the specific difficulty at that location. Obstructions, dust control, safety, and often overtime hours must be considered when preparing your estimate.

Reference Numbers

Reference numbers are shown in bold squares at the beginning of some major classifications. These numbers refer to related items in the Reference Section. The reference information may be an estimating procedure, an alternate pricing method or technical information.

Note: Not all subdivisions listed here necessarily appear in this publication.

01100 | Summary

01107	Professional Consultant		CREW	DAILY OUTPUT	LABOR-HOURS	UNIT	2002 BARE COSTS				TOTAL INCL O&P	
							MAT.	LABOR	EQUIP.	TOTAL		
300	0010	**ENGINEERING FEES**	R01107 -030										**300**
	0020	Educational planning consultant, minimum					Project					.50%	
	0100	Maximum					"					2.50%	
	0200	Electrical, minimum					Contrct					4.10%	
	0300	Maximum										10.10%	
	0400	Elevator & conveying systems, minimum										2.50%	
	0500	Maximum										5%	
	0600	Food service & kitchen equipment, minimum										8%	
	0700	Maximum										12%	
	1000	Mechanical (plumbing & HVAC), minimum										4.10%	
	1100	Maximum					▼					10.10%	

01200 | Price & Payment Procedures

01250	Contract Modification Procedures		CREW	DAILY OUTPUT	LABOR-HOURS	UNIT	2002 BARE COSTS				TOTAL INCL O&P	
							MAT.	LABOR	EQUIP.	TOTAL		
200	0010	**CONTINGENCIES** for estimate at conceptual stage					Project					20%	**200**
	0050	Schematic stage										15%	
	0100	Preliminary working drawing stage (Design Dev.)										10%	
	0150	Final working drawing stage					▼					3%	
300	0010	**CREWS** For building construction, see How To Use This Book											**300**
400	0010	**FACTORS** Cost adjustments	R01250 -010										**400**
	0100	Add to construction costs for particular job requirements											
	0500	Cut & patch to match existing construction, add, minimum					Costs	2%	3%				
	0550	Maximum						5%	9%				
	0800	Dust protection, add, minimum						1%	2%				
	0850	Maximum						4%	11%				
	1100	Equipment usage curtailment, add, minimum						1%	1%				
	1150	Maximum						3%	10%				
	1400	Material handling & storage limitation, add, minimum						1%	1%				
	1450	Maximum						6%	7%				
	1700	Protection of existing work, add, minimum						2%	2%				
	1750	Maximum						5%	7%				
	2000	Shift work requirements, add, minimum							5%				
	2050	Maximum							30%				
	2300	Temporary shoring and bracing, add, minimum						2%	5%				
	2350	Maximum						5%	12%				
	2400	Work inside prisons and high security areas, add, minimum							30%				
	2450	Maximum					▼		50%				
500	0010	**JOB CONDITIONS** Modifications to total											**500**
	0020	project cost summaries											
	0100	Economic conditions, favorable, deduct					Project					2%	
	0200	Unfavorable, add										5%	
	0300	Hoisting conditions, favorable, deduct										2%	
	0400	Unfavorable, add										5%	
	0700	Labor availability, surplus, deduct										1%	
	0800	Shortage, add										10%	
	0900	Material storage area, available, deduct										1%	
	1000	Not available, add					▼					2%	

Important: See the Reference Section for critical supporting data - Reference Nos., Crews, & City Cost Indexes

01200 | Price & Payment Procedures

		01250	Contract Modification Procedures	CREW	DAILY OUTPUT	LABOR-HOURS	UNIT	2002 BARE COSTS MAT.	LABOR	EQUIP.	TOTAL	TOTAL INCL O&P	
500	1100		Subcontractor availability, surplus, deduct				Project					5%	500
	1200		Shortage, add									12%	
	1300		Work space, available, deduct									2%	
	1400		Not available, add				↓					5%	
600	0010		**OVERTIME** For early completion of projects or where	R01100 -110									600
	0020		labor shortages exist, add to usual labor, up to				Costs		100%				

		01255	Cost Indexes										
200	0010		**CONSTRUCTION COST INDEX** (Reference) over 930 zip code locations in										200
	0020		The U.S. and Canada, total bldg cost, min. (Clarksdale, MS)				%					66.20%	
	0050		Average									100%	
	0100		Maximum (New York, NY)				↓					134.60%	
400	0010		**HISTORICAL COST INDEXES** (Reference) Back to 1952										400
500	0010		**LABOR INDEX** (Reference) For over 930 zip code locations in										500
	0020		the U.S. and Canada, minimum (Clarksdale, MS)				%		33.80%				
	0050		Average						100%				
	0100		Maximum (New York, NY)				↓		163.30%				
600	0011		**MATERIAL INDEX** For over 930 zip code locations in										600
	0020		the U.S. and Canada, minimum (Elizabethtown, KY)				%	92.10%					
	0040		Average					100%					
	0060		Maximum (Ketchikan, AK)				↓	143.90%					

		01290	Payment Procedures										
800	0010		**TAXES** Sales tax, State, average	R01100 -090			%	4.70%					800
	0050		Maximum					7%					
	0200		Social Security, on first $80,400 of wages	R01100 -100					7.65%				
	0300		Unemployment, MA, combined Federal and State, minimum						2.10%				
	0350		Average						7%				
	0400		Maximum	↓			↓		8%				

01300 | Administrative Requirements

		01310	Project Management/Coordination	CREW	DAILY OUTPUT	LABOR-HOURS	UNIT	2002 BARE COSTS MAT.	LABOR	EQUIP.	TOTAL	TOTAL INCL O&P	
150	0010		**PERMITS** Rule of thumb, most cities, minimum				Job					.50%	150
	0100		Maximum				"					2%	
200	0010		**PERFORMANCE BOND** For buildings, minimum	R01100 -080			Job					.60%	200
	0100		Maximum				"					2.50%	
350	0010		**INSURANCE** Builders risk, standard, minimum	R01100 -040			Job					.22%	350
	0050		Maximum									.59%	
	0200		All-risk type, minimum	R01100 -060								.25%	
	0250		Maximum				↓					.62%	
	0400		Contractor's equipment floater, minimum				Value					.50%	
	0450		Maximum				"					1.50%	
	0600		Public liability, average				Job					1.55%	
	0800		Workers' compensation & employer's liability, average										
	0850		by trade, carpentry, general				Payroll		18%				
	1000		Electrical						6.66%				
	1150		Insulation						16.17%				
	1450		Plumbing	↓			↓		8.25%				

01300 | Administrative Requirements

01310 | Project Management/Coordination

			CREW	DAILY OUTPUT	LABOR-HOURS	UNIT	2002 BARE COSTS				TOTAL INCL O&P	
							MAT.	LABOR	EQUIP.	TOTAL		
350	1550	Sheet metal work (HVAC) [R01100 -040]				Payroll		11.61%				350
500	0010	**MARK-UP** For General Contractors for change [R01100 -070]										500
	0100	of scope of job as bid										
	0200	Extra work, by subcontractors, add				%					10%	
	0250	By General Contractor, add									15%	
	0400	Omitted work, by subcontractors, deduct all but									5%	
	0450	By General Contractor, deduct all but									7.50%	
	0600	Overtime work, by subcontractors, add									15%	
	0650	By General Contractor, add									10%	
	1000	Installing contractors, on his own labor, minimum						47.30%				
	1100	Maximum						85.90%				
600	0010	**OVERHEAD** As percent of direct costs, minimum [R01100 -070]				%				5%		600
	0050	Average								13%		
	0100	Maximum								30%		
620	0010	**OVERHEAD & PROFIT** Allowance to add to items in this [R01100 -070]										620
	0020	book that do not include Subs O&P, average				%				25%		
	0100	Allowance to add to items in this book that										
	0110	do include Subs O&P, minimum				%					5%	
	0150	Average									10%	
	0200	Maximum									15%	
	0300	Typical, by size of project, under $100,000								30%		
	0350	$500,000 project								25%		
	0400	$2,000,000 project								20%		
	0450	Over $10,000,000 project								15%		

01321 | Construction Photos

			CREW	DAILY OUTPUT	LABOR-HOURS	UNIT	2002 BARE COSTS				TOTAL INCL O&P	
							MAT.	LABOR	EQUIP.	TOTAL		
500	0010	**PHOTOGRAPHS** 8″ x 10″, 4 shots, 2 prints ea., std. mounting				Set	276			276	305	500
	0100	Hinged linen mounts					300			300	330	
	0200	8″ x 10″, 4 shots, 2 prints each, in color					320			320	355	
	0300	For I.D. slugs, add to all above					3.89			3.89	4.28	
	1500	Time lapse equipment, camera and projector, buy					3,575			3,575	3,925	
	1550	Rent per month					530			530	585	
	1700	Cameraman and film, including processing, B.&W.				Day	1,250			1,250	1,375	
	1720	Color				″	1,250			1,250	1,375	

01500 | Temporary Facilities & Controls

01510 | Temporary Utilities

			CREW	DAILY OUTPUT	LABOR-HOURS	UNIT	2002 BARE COSTS				TOTAL INCL O&P	
							MAT.	LABOR	EQUIP.	TOTAL		
800	0010	**TEMPORARY UTILITIES**										800
	0100	Heat, incl. fuel and operation, per week, 12 hrs. per day	1 Skwk	100	.080	CSF Flr	5.20	2.48		7.68	9.60	
	0200	24 hrs. per day	″	60	.133		7.85	4.13		11.98	15.05	
	0350	Lighting, incl. service lamps, wiring & outlets, minimum	1 Elec	34	.235		2.11	8.35		10.46	14.75	
	0360	Maximum	″	17	.471		4.60	16.70		21.30	30	
	0400	Power for temp lighting only, per month, min/month 6.6 KWH								.75	1.18	
	0450	Maximum/month 23.6 KWH								2.85	3.14	
	0600	Power for job duration incl. elevator, etc., minimum								47	51.70	
	0650	Maximum								110	121	
	1000	Toilet, portable, see division 01590-400										

1 GENERAL REQUIREMENTS

8

Important: See the Reference Section for critical supporting data - Reference Nos., Crews, & City Cost Indexes

01520 | Construction Facilities

		CREW	DAILY OUTPUT	LABOR-HOURS	UNIT	2002 BARE COSTS MAT.	LABOR	EQUIP.	TOTAL	TOTAL INCL O&P	
500	**0010**	**OFFICE** Trailer, furnished, no hookups, 20' x 8', buy	2 Skwk	1	16	Ea.	5,450	495		5,945	6,750	**500**
	0250	Rent per month					153			153	169	
	0300	32' x 8', buy	2 Skwk	.70	22.857		8,000	705		8,705	9,900	
	0350	Rent per month					168			168	185	
	0400	50' x 10', buy	2 Skwk	.60	26.667		13,600	825		14,425	16,300	
	0450	Rent per month					300			300	330	
	0500	50' x 12', buy	2 Skwk	.50	32		16,100	990		17,090	19,300	
	0550	Rent per month					350			350	385	
	0700	For air conditioning, rent per month, add				↓	36.50			36.50	40	
	0800	For delivery, add per mile				Mile	1.53			1.53	1.68	
	1000	Portable buildings, prefab, on skids, economy, 8' x 8'	2 Carp	265	.060	S.F.	81.50	1.81		83.31	93	
	1100	Deluxe, 8' x 12'	"	150	.107	"	88.50	3.20		91.70	102	
	1200	Storage boxes, 20' x 8', buy	2 Skwk	1.80	8.889	Ea.	3,375	275		3,650	4,125	
	1250	Rent per month					74			74	81.50	
	1300	40' x 8', buy	2 Skwk	1.40	11.429		4,025	355		4,380	4,975	
	1350	Rent per month					105			105	115	
550	**0010**	**FIELD OFFICE EXPENSE**										**550**
	0100	Field office expense, office equipment rental average				Month	139			139	153	
	0120	Office supplies, average				"	83.50			83.50	92	
	0125	Office trailer rental, see division 01520-500										
	0140	Telephone bill; avg. bill/month incl. long dist.				Month	204			204	224	
	0160	Field office lights & HVAC				"	94			94	103	

01530 | Temporary Construction

		CREW	DAILY OUTPUT	LABOR-HOURS	UNIT	MAT.	LABOR	EQUIP.	TOTAL	TOTAL INCL O&P	
700	**0010**	**PROTECTION** Stair tread, 2" x 12" planks, 1 use	1 Carp	75	.107	Tread	5.10	3.20		8.30	10.60	**700**
	0100	Exterior plywood, 1/2" thick, 1 use	↓	65	.123	↓	1.47	3.69		5.16	7.35	
	0200	3/4" thick, 1 use	↓	60	.133	↓	1.99	4		5.99	8.45	
900	**0010**	**WINTER PROTECTION** Reinforced plastic on wood										**900**
	0100	framing to close openings	2 Clab	750	.021	S.F.	.37	.50		.87	1.19	
	0200	Tarpaulins hung over scaffolding, 8 uses, not incl. scaffolding		1,500	.011		.18	.25		.43	.59	
	0250	Tarpaulin polyester reinf. w/ integral fastening system 11 mils thick		1,600	.010		.76	.23		.99	1.21	
	0300	Prefab fiberglass panels, steel frame, 8 uses	↓	1,200	.013	↓	.72	.31		1.03	1.28	

01540 | Construction Aids

		CREW	DAILY OUTPUT	LABOR-HOURS	UNIT	MAT.	LABOR	EQUIP.	TOTAL	TOTAL INCL O&P	
750	**0010**	**SCAFFOLDING** R01540 -100										**750**
	0015	Steel tubular, reg, rent/mo, no plank, incl erect & dismantle										
	0090	Building exterior, wall face, 1 to 5 stories, 6'-4" x 5' frames	3 Carp	24	1	C.S.F.	24.50	30		54.50	73	
	0200	6 to 12 stories	4 Carp	21.20	1.509		24.50	45.50		70	97	
	0310	13 to 20 stories	5 Carp	20	2		24.50	60		84.50	120	
	0460	Building interior, wall face area, up to 16' high	3 Carp	25	.960	↓	24.50	29		53.50	71.50	
	0560	16' to 40' high		23	1.043	↓	24.50	31.50		56	75	
	0800	Building interior floor area, up to 30' high	↓	312	.077	C.C.F.	2.57	2.31		4.88	6.40	
	0900	Over 30' high	4 Carp	275	.116	"	2.57	3.49		6.06	8.30	
	0910	Steel tubular, heavy duty shoring, buy										
	0920	Frames 5' high 2' wide				Ea.	75			75	82.50	
	0925	5' high 4' wide					85			85	93.50	
	0930	6' high 2' wide					86			86	94.50	
	0935	6' high 4' wide				↓	101			101	111	
	0940	Accessories										
	0945	Cross braces				Ea.	16			16	17.60	
	0950	U-head, 8" x 8"					17.50			17.50	19.25	
	0955	J-head, 4" x 8"					12.80			12.80	14.10	
	0960	Base plate, 8" x 8"					14.20			14.20	15.60	
	0965	Leveling jack	↓			↓	30.50			30.50	33.50	

GENERAL REQUIREMENTS

1

	01540	Construction Aids		CREW	DAILY OUTPUT	LABOR-HOURS	UNIT	2002 BARE COSTS				TOTAL INCL O&P	
								MAT.	LABOR	EQUIP.	TOTAL		
750	1000	Steel tubular, regular, buy	R01540 -100										750
	1100	Frames 3' high 5' wide					Ea.	58			58	64	
	1150	5' high 5' wide						67			67	73.50	
	1200	6'-4" high 5' wide						84			84	92.50	
	1350	7'-6" high 6' wide						145			145	160	
	1500	Accessories cross braces						15			15	16.50	
	1550	Guardrail post						15			15	16.50	
	1600	Guardrail 7' section						7.25			7.25	8	
	1650	Screw jacks & plates						24			24	26.50	
	1700	Sidearm brackets						28			28	31	
	1750	8" casters						33			33	36.50	
	1800	Plank 2" x 10" x 16'-0"						42.50			42.50	47	
	1900	Stairway section						245			245	270	
	1910	Stairway starter bar						29			29	32	
	1920	Stairway inside handrail						53			53	58.50	
	1930	Stairway outside handrail						73			73	80.50	
	1940	Walk-thru frame guardrail					▼	37			37	40.50	
	2000	Steel tubular, regular, rent/mo.											
	2100	Frames 3' high 5' wide					Ea.	3.75			3.75	4.13	
	2150	5' high 5' wide						3.75			3.75	4.13	
	2200	6'-4" high 5' wide						3.75			3.75	4.13	
	2250	7'-6" high 6' wide						7			7	7.70	
	2500	Accessories, cross braces						.60			.60	.66	
	2550	Guardrail post						1			1	1.10	
	2600	Guardrail 7' section						.75			.75	.83	
	2650	Screw jacks & plates						1.50			1.50	1.65	
	2700	Sidearm brackets						1.50			1.50	1.65	
	2750	8" casters						6			6	6.60	
	2800	Outrigger for rolling tower						3			3	3.30	
	2850	Plank 2" x 10" x 16'-0"						5			5	5.50	
	2900	Stairway section						10			10	11	
	2910	Stairway starter bar						.10			.10	.11	
	2920	Stairway inside handrail						5			5	5.50	
	2930	Stairway outside handrail						5			5	5.50	
	2940	Walk-thru frame guardrail					▼	2			2	2.20	
	3000	Steel tubular, heavy duty shoring, rent/mo.											
	3250	5' high 2' & 4' wide					Ea.	5			5	5.50	
	3300	6' high 2' & 4' wide						5			5	5.50	
	3500	Accessories, cross braces						1			1	1.10	
	3600	U - head, 8" x 8"						1			1	1.10	
	3650	J - head, 4" x 8"						1			1	1.10	
	3700	Base plate, 8" x 8"						1			1	1.10	
	3750	Leveling jack						2			2	2.20	
	5700	Planks, 2"x10"x16'-0" in place, up to 50' high		3 Carp	144	.167			5		5	7.80	
	5800	In place over 50' high		4 Carp	160	.200	▼		6		6	9.35	
	6000	Heavy duty shoring for elevated slab forms to 8'-2" high, floor area											
	6010	incl erection & dismantle labor											
	6100	1 use/month		4 Carp	36	.889	C.S.F.	29.50	26.50		56	74	
	6150	2 uses/month		"	36	.889	"	14.80	26.50		41.30	58	
	6500	To 14'-8" high											
	6600	1 use/month		4 Carp	18	1.778	C.S.F.	43	53.50		96.50	131	
	6650	2 uses/month		"	18	1.778	"	21.50	53.50		75	107	
755	0010	**SCAFFOLDING SPECIALTIES**											755
	1200	Sidewalk bridge, heavy duty steel posts & beams, including											
	1210	parapet protection & waterproofing											
	1220	8' to 10' wide, 2 posts		3 Carp	15	1.600	L.F.	30.50	48		78.50	108	

Important: See the Reference Section for critical supporting data - Reference Nos., Crews, & City Cost Indexes

	01540	Construction Aids	CREW	DAILY OUTPUT	LABOR-HOURS	UNIT	2002 BARE COSTS				TOTAL INCL O&P	
							MAT.	LABOR	EQUIP.	TOTAL		
755	1230	3 posts	3 Carp	10	2.400	L.F.	47	72		119	164	**755**
	1500	Sidewalk bridge using tubular steel										
	1510	scaffold frames, including planking	3 Carp	45	.533	L.F.	4.72	16		20.72	30	
	1600	For 2 uses per month, deduct from all above					50%					
	1700	For 1 use every 2 months, add to all above					100%					
	1900	Catwalks, 20" wide, no guardrails, 7' span, buy				Ea.	125			125	138	
	2000	10' span, buy				"	160			160	176	
	2800	Hand winch-operated masons scaffolding, no plank										
	2810	plank moving not required										
	2900	98' long, 10'-6" high, buy				Ea.	25,900			25,900	28,400	
	3000	Rent per month					1,025			1,025	1,125	
	3100	28'-6" high, buy					31,800			31,800	35,000	
	3200	Rent per month					1,275			1,275	1,400	
	3400	196' long, 28'-6" high, buy					61,500			61,500	67,500	
	3500	Rent per month					2,450			2,450	2,700	
	3600	64'-6" high, buy					84,500			84,500	93,000	
	3700	Rent per month					3,375			3,375	3,725	
	3720	Putlog, standard, 8' span, with hangers, buy					61			61	67	
	3730	Rent per month					10			10	11	
	3750	12' span, buy					92			92	101	
	3755	Rent per month					15			15	16.50	
	3760	Trussed type, 16' span, buy					210			210	231	
	3770	Rent per month					20			20	22	
	3790	22' span, buy					252			252	277	
	3795	Rent per month					30			30	33	
	3800	Rolling ladders with handrails, 30" wide, buy, 2 step					146			146	161	
	4000	7 step					440			440	480	
	4050	10 step					660			660	725	
	4100	Rolling towers, buy, 5' wide, 7' long, 10' high					1,150			1,150	1,250	
	4200	For 5' high added sections, to buy, add					190			190	209	
	4300	Complete incl. wheels, railings, outriggers,										
	4350	21' high, to buy				Ea.	1,925			1,925	2,125	
	4400	Rent/month				"	145			145	160	
800	0010	**TARPAULINS** Cotton duck, 10 oz. to 13.13 oz. per S.Y., minimum				S.F.	.48			.48	.53	**800**
	0050	Maximum					.57			.57	.63	
	0100	Polyvinyl coated nylon, 14 oz. to 18 oz., minimum					.47			.47	.52	
	0150	Maximum					.67			.67	.74	
	0200	Reinforced polyethylene 3 mils thick, white					.11			.11	.12	
	0300	4 mils thick, white, clear or black					.14			.14	.15	
	0400	5.5 mils thick, clear					.19			.19	.21	
	0500	White, fire retardant					.18			.18	.20	
	0600	7.5 mils, oil resistant, fire retardant					.19			.19	.21	
	0700	8.5 mils, black					.24			.24	.26	
	0710	Woven polyethylene, 6 mils thick					.47			.47	.52	
	0730	Polyester reinforced w/ integral fastening system 11 mils thick					1.05			1.05	1.16	
	0740	Mylar polyester, non-reinforced, 7 mils thick					1.15			1.15	1.27	
820	0010	**SMALL TOOLS** As % of contractor's work, minimum				Total					.50%	**820**
	0100	Maximum				"					2%	

01550 | Vehicular Access & Parking

	0010	**ROADS AND SIDEWALKS** Temporary				S.Y.						**700**
700	0050	Roads, gravel fill, no surfacing, 4" gravel depth	B-14	715	.067	S.Y.	2.88	1.67	.25	4.80	6	
	0100	8" gravel depth	"	615	.078	"	5.75	1.94	.29	7.98	9.65	
	1000	Ramp, 3/4" plywood on 2" x 6" joists, 16" O.C.	2 Carp	300	.053	S.F.	1.23	1.60		2.83	3.84	

01550		Vehicular Access & Parking	CREW	DAILY OUTPUT	LABOR-HOURS	UNIT	2002 BARE COSTS				TOTAL INCL O&P	
							MAT.	LABOR	EQUIP.	TOTAL		
700	1100	On 2″ x 10″ joists, 16″ O.C.	2 Carp	275	.058	S.F.	1.90	1.75		3.65	4.81	**700**
	2200	Sidewalks, 2″ x 12″ planks, 2 uses	1 Carp	350	.023		.85	.69		1.54	2	
	2300	Exterior plywood, 2 uses, 1/2″ thick		750	.011		.25	.32		.57	.77	
	2400	5/8″ thick		650	.012		.28	.37		.65	.88	
	2500	3/4″ thick		600	.013		.33	.40		.73	.99	

01560		Barriers & Enclosures										
100	0010	**BARRICADES** 5′ high, 3 rail @ 2″ x 8″, fixed	2 Carp	30	.533	L.F.	11	16		27	37	**100**
	0150	Movable		20	.800		11	24		35	49.50	
	1000	Guardrail, wooden, 3′ high, 1″ x 6″, on 2″ x 4″ posts		200	.080		.98	2.40		3.38	4.82	
	1100	2″ x 6″, on 4″ x 4″ posts		165	.097		1.95	2.91		4.86	6.65	
	1200	Portable metal with base pads, buy					15.50			15.50	17.05	
	1250	Typical installation, assume 10 reuses	2 Carp	600	.027		1.60	.80		2.40	3.01	
	1300	Barricade tape, polyethelyne, 7 mil, 3″ wide x 500′ long roll				Ea.	25			25	27.50	
250	0010	**FENCING** Chain link, 11 ga, 5′ high	2 Clab	100	.160	L.F.	3.58	3.75		7.33	9.80	**250**
	0100	6′ high		75	.213		3.28	5		8.28	11.40	
	0200	Rented chain link, 6′ high, to 500′ (up to 12 mo.)		100	.160		2.26	3.75		6.01	8.35	
	0250	Over 1000′ (up to 12 mo.)		110	.145		1.64	3.41		5.05	7.10	
	0350	Plywood, painted, 2″ x 4″ frame, 4′ high	A-4	135	.178		4.58	5.15		9.73	13	
	0400	4″ x 4″ frame, 8′ high	″	110	.218		9.10	6.35		15.45	19.80	
	0500	Wire mesh on 4″ x 4″ posts, 4′ high	2 Carp	100	.160		7.25	4.80		12.05	15.40	
	0550	8′ high	″	80	.200		10.65	6		16.65	21	
400	0010	**TEMPORARY CONSTRUCTION** See also division 01530										**400**

01580		Project Signs										
700	0010	**SIGNS** Hi-intensity reflectorized, no posts, buy				S.F.	14			14	15.40	**700**

Important: See the Reference Section for critical supporting data - Reference Nos., Crews, & City Cost Indexes

01590 | Equipment Rental

			UNIT	HOURLY OPER. COST	RENT PER DAY	RENT PER WEEK	RENT PER MONTH	CREW EQUIPMENT COST/DAY	
200	0010	**EARTHWORK EQUIPMENT RENTAL** Without operators	R01590 -100						**200**
	0050	Augers for truck or trailer mounting, vertical drilling							
	0060	4″ to 36″ diam., 54 H.P., gas, 10' spindle travel	Ea.	19.05	665	2,000	6,000	552.40	
	0070	14' spindle travel		21.75	830	2,495	7,475	673	
	0100	Excavator, diesel hydraulic, crawler mounted, 1/2 C.Y. cap.		10.60	390	1,175	3,525	319.80	
	0120	5/8 C.Y. capacity		12.90	485	1,450	4,350	393.20	
	0140	3/4 C.Y. capacity		15.50	555	1,665	5,000	457	
	0150	1 C.Y. capacity		20.05	670	2,015	6,050	563.40	
	0200	1-1/2 C.Y. capacity		24.75	845	2,535	7,600	705	
	0300	2 C.Y. capacity		33.90	1,150	3,455	10,400	962.20	
	0320	2-1/2 C.Y. capacity		61.30	2,025	6,050	18,200	1,700	
	0340	3-1/2 C.Y. capacity		75.70	2,475	7,440	22,300	2,094	
	0341	Attachments							
	0342	Bucket thumbs		1.25	205	615	1,850	133	
	0345	Grapples		1	202	605	1,825	129	
	0350	Gradall type, truck mounted, 3 ton @ 15' radius, 5/8 C.Y.		24.80	895	2,690	8,075	736.40	
	0370	1 C.Y. capacity		28.95	1,050	3,185	9,550	868.60	
	0400	Backhoe-loader, 40 to 45 H.P., 5/8 C.Y. capacity		5.35	188	565	1,700	155.80	
	0450	45 H.P. to 60 H.P., 3/4 C.Y. capacity		6.95	205	615	1,850	178.60	
	0460	80 H.P., 1-1/4 C.Y. capacity		9.20	237	710	2,125	215.60	
	0470	112 H.P., 1-1/2 C.Y. capacity		12.80	415	1,240	3,725	350.40	
	0480	Attachments							
	0482	Compactor, 20,000 lb		1.95	173	520	1,550	119.60	
	0485	Hydraulic hammer, 750 ft-lbs		1.40	202	605	1,825	132.20	
	0486	Hydraulic hammer, 1200 ft-lbs		1.90	245	735	2,200	162.20	
	0750	Bucket, clamshell, general purpose, 3/8 C.Y.		.70	53.50	160	480	37.60	
	0800	1/2 C.Y.		.80	60	180	540	42.40	
	0850	3/4 C.Y.		.90	66.50	200	600	47.20	
	0900	1 C.Y.		.95	90	270	810	61.60	
	0950	1-1/2 C.Y.		1.45	115	345	1,025	80.60	
	1000	2 C.Y.		1.60	140	420	1,250	96.80	
	1200	Compactor, roller, 2 drum, 2000 lb., operator walking		4.65	143	430	1,300	123.20	
	1250	Rammer compactor, gas, 1000 lb. blow		.95	40	120	360	31.60	
	1300	Vibratory plate, gas, 13″ plate, 1000 lb. blow		.75	38.50	115	345	29	
	1350	24″ plate, 5000 lb. blow		1.95	66.50	200	600	55.60	
	3710	Screening plant 110 hp. w / 5' x 10'screen		15.85	370	1,115	3,350	349.80	
	3720	5' x 16' screen	▼	17.20	475	1,420	4,250	421.60	
	4891	Attachments for all skid steer loaders							
	4892	Auger	Ea.	.46	77.50	232	695	50.10	
	4893	Backhoe		.64	107	320	960	69.10	
	4894	Broom		.66	110	331	995	71.50	
	4895	Forks		.21	34.50	104	310	22.50	
	4896	Grapple		.54	90	270	810	58.30	
	4897	Concrete hammer		1.03	171	513	1,550	110.85	
	4898	Tree spade		.94	157	471	1,425	101.70	
	4899	Trencher		.71	119	356	1,075	76.90	
	4900	Trencher, chain, boom type, gas, operator walking, 12 H.P.		2	117	350	1,050	86	
	4910	Operator riding, 40 H.P.		5.75	245	735	2,200	193	
	5000	Wheel type, diesel, 4' deep, 12″ wide		12.80	645	1,930	5,800	488.40	
	5100	Diesel, 6' deep, 20″ wide		23.70	940	2,820	8,450	753.60	
	5150	Ladder type, diesel, 5' deep, 8″ wide		14.80	505	1,510	4,525	420.40	
	5200	Diesel, 8' deep, 16″ wide		32.45	985	2,950	8,850	849.60	
	5250	Truck, dump, tandem, 12 ton payload		19.90	325	970	2,900	353.20	
	5300	Three axle dump, 16 ton payload		27.20	420	1,265	3,800	470.60	
	5350	Dump trailer only, rear dump, 16-1/2 C.Y.		4.05	145	435	1,300	119.40	
	5400	20 C.Y.		4.50	155	465	1,400	129	
	5450	Flatbed, single axle, 1-1/2 ton rating		11.45	107	320	960	155.60	
	5500	3 ton rating	▼	14.55	110	330	990	182.40	

01590 | Equipment Rental

				UNIT	HOURLY OPER. COST	RENT PER DAY	RENT PER WEEK	RENT PER MONTH	CREW EQUIPMENT COST/DAY	
200	5550	Off highway rear dump, 25 ton capacity	R01590 -100	Ea.	40.45	1,000	3,020	9,050	927.60	200
	5600	35 ton capacity			41.50	1,050	3,165	9,500	965	
400	0010	**GENERAL EQUIPMENT RENTAL** Without operators								400
	0150	Aerial lift, scissor type, to 15' high, 1000 lb. cap., electric		Ea.	1.20	71.50	215	645	52.60	
	0160	To 25' high, 2000 lb. capacity			1.50	107	320	960	76	
	0170	Telescoping boom to 40' high, 750 lb. capacity, gas			7	263	790	2,375	214	
	0180	1000 lb. capacity			8.75	390	1,170	3,500	304	
	0190	To 60' high, 750 lb. capacity			8.85	400	1,200	3,600	310.80	
	0200	Air compressor, portable, gas engine, 60 C.F.M.			4.20	50	150	450	63.60	
	0300	160 C.F.M.			7.40	63.50	190	570	97.20	
	0400	Diesel engine, rotary screw, 250 C.F.M.			8.75	108	325	975	135	
	0500	365 C.F.M.			10.25	135	405	1,225	163	
	0600	600 C.F.M.			16.95	213	640	1,925	263.60	
	0700	750 C.F.M.			22	250	750	2,250	326	
	0800	For silenced models, small sizes, add			3%	5%	5%	5%		
	0900	Large sizes, add			5%	7%	7%	7%		
	0920	Air tools and accessories								
	0930	Breaker, pavement, 60 lb.		Ea.	.20	25	75	225	16.60	
	0940	80 lb.			.20	33.50	100	300	21.60	
	0980	Dust control per drill			1.24	15	45	135	18.90	
	1000	Hose, air with couplings, 50' long, 3/4" diameter			.02	4	12	36	2.55	
	1100	1" diameter			.03	5.35	16	48	3.45	
	1200	1-1/2" diameter			.04	7.35	22	66	4.70	
	1300	2" diameter			.10	16.35	49	147	10.60	
	1400	2-1/2" diameter			.11	19	57	171	12.30	
	1410	3" diameter			.16	27.50	82	246	17.70	
	1560	Tamper, single, 35 lb.			.44	29.50	88	264	21.10	
	1570	Triple, 140 lb.			.88	44	132	395	33.45	
	1580	Wrenches, impact, air powered, up to 3/4" bolt			.10	18.35	55	165	11.80	
	1590	Up to 1-1/4" bolt			.15	33.50	100	300	21.20	
	2100	Generator, electric, gas engine, 1.5 KW to 3 KW			1.30	33.50	100	300	30.40	
	2200	5 KW			1.90	50	150	450	45.20	
	2300	10 KW			2.35	112	335	1,000	85.80	
	2400	25 KW			6.70	133	400	1,200	133.60	
	2500	Diesel engine, 20 KW			4.40	96.50	290	870	93.20	
	2600	50 KW			9.55	113	340	1,025	144.40	
	2700	100 KW			14	157	470	1,400	206	
	2800	250 KW			43.45	277	830	2,500	513.60	
	2900	Heaters, space, oil or electric, 50 MBH			.93	22.50	67	201	20.85	
	3000	100 MBH			1.66	29	87	261	30.70	
	3100	300 MBH			5.33	43.50	130	390	68.65	
	3150	500 MBH			10.68	58.50	175	525	120.45	
	3200	Hose, water, suction with coupling, 20' long, 2" diameter			.01	8.35	25	75	5.10	
	3210	3" diameter			.02	12.65	38	114	7.75	
	3220	4" diameter			.03	16.65	50	150	10.25	
	3230	6" diameter			.06	31.50	95	285	19.50	
	3240	8" diameter			.31	51.50	154	460	33.30	
	3250	Discharge hose with coupling, 50' long, 2" diameter			.01	7	21	63	4.30	
	3260	3" diameter			.02	8.65	26	78	5.35	
	3270	4" diameter			.03	11.65	35	105	7.25	
	3280	6" diameter			.05	28.50	85	255	17.40	
	3290	8" diameter			.34	56.50	169	505	36.50	
	3300	Ladders, extension type, 16' to 36' long			.14	21	63	189	13.70	
	3400	40' to 60' long			.19	31	93	279	20.10	
	3500	Light towers, towable, with diesel generator, 2000 watt			1.75	93.50	280	840	70	
	3600	4000 watt			1.90	108	325	975	80.20	
	4100	Pump, centrifugal gas pump, 1-1/2", 4 MGPH			1	33.50	100	300	28	
	4200	2", 8 MGPH			1.55	38.50	115	345	35.40	

Important: See the Reference Section for critical supporting data - Reference Nos., Crews, & City Cost Indexes

01590 | Equipment Rental

		UNIT	HOURLY OPER. COST	RENT PER DAY	RENT PER WEEK	RENT PER MONTH	CREW EQUIPMENT COST/DAY	
400	4300	3", 15 MGPH	Ea.	1.75	41.50	125	375	39	**400**
	4400	6", 90 MGPH		11.90	172	515	1,550	198.20	
	4500	Submersible electric pump, 1-1/4", 55 GPM		.35	29.50	88	264	20.40	
	4600	1-1/2", 83 GPM		.41	33.50	100	300	23.30	
	4700	2", 120 GPM		.57	41.50	125	375	29.55	
	4800	3", 300 GPM		.97	51.50	155	465	38.75	
	4900	4", 560 GPM		6.26	96.50	290	870	108.10	
	5000	6", 1590 GPM		9.07	187	560	1,675	184.55	
	5100	Diaphragm pump, gas, single, 1-1/2" diameter		.70	33.50	100	300	25.60	
	5200	2" diameter		.95	41.50	125	375	32.60	
	5300	3" diameter		.90	45	135	405	34.20	
	5400	Double, 4" diameter		2.10	86.50	260	780	68.80	
	5500	Trash pump, self-priming, gas, 2" diameter		2.10	33	99	297	36.60	
	5600	Diesel, 4" diameter		3.40	76.50	230	690	73.20	
	5650	Diesel, 6" diameter		8.40	125	375	1,125	142.20	
	5700	Salamanders, L.P. gas fired, 100,000 B.T.U.		1.66	17	51	153	23.50	
	5800	Saw, chain, gas engine, 18" long		.48	50	150	450	33.85	
	5900	36" long		.20	63.50	190	570	39.60	
	5950	60" long		.20	72.50	218	655	45.20	
	6000	Masonry, table mounted, 14" diameter, 5 H.P.		1.31	56.50	170	510	44.50	
	6100	Circular, hand held, electric, 7-1/4" diameter		.18	18.65	56	168	12.65	
	6200	12" diameter		.24	26.50	80	240	17.90	
	6350	Torch, cutting, acetylene-oxygen, 150' hose		1.50	26.50	80	240	28	
	6360	Hourly operating cost includes tips and gas		4.25				34	
	6410	Toilet, portable chemical		.10	16.65	50	150	10.80	
	6420	Recycle flush type		.12	20.50	62	186	13.35	
	6430	Toilet, fresh water flush, garden hose,		.14	23.50	70	210	15.10	
	6440	Hoisted, non-flush, for high rise		.12	20.50	61	183	13.15	
	6450	Toilet, trailers, minimum		.21	34.50	104	310	22.50	
	6460	Maximum		.62	104	312	935	67.35	
	6500	Trailers, platform, flush deck, 2 axle, 25 ton capacity		3.40	105	315	945	90.20	
	6600	40 ton capacity		4.60	190	570	1,700	150.80	
	6700	3 axle, 50 ton capacity		5.05	210	630	1,900	166.40	
	6800	75 ton capacity		6.50	277	830	2,500	218	
	7100	Truck, pickup, 3/4 ton, 2 wheel drive		5.50	66.50	200	600	84	
	7200	4 wheel drive		5.65	75	225	675	90.20	
	7300	Tractor, 4 x 2, 30 ton capacity, 195 H.P.		13.85	300	905	2,725	291.80	
	7410	250 H.P.		18.85	380	1,140	3,425	378.80	
	7620	Vacuum truck, hazardous material, 2500 gallon		6.60	305	919	2,750	236.60	
	7625	5,000 gallon		7.15	410	1,225	3,675	302.20	
	7700	Welder, electric, 200 amp		3.73	50	150	450	59.85	
	7800	300 amp		5.27	63	189	565	79.95	
	7900	Gas engine, 200 amp		5.50	48.50	145	435	73	
	8000	300 amp		6.45	61.50	185	555	88.60	
	8100	Wheelbarrow, any size		.06	10.35	31	93	6.70	
600	0010	**LIFTING AND HOISTING EQUIPMENT RENTAL** R01590-100							**600**
	0100	without operators							
	0200	Crane, climbing, 106' jib, 6000 lb. capacity, 410 FPM	Ea.	41.15	1,275	3,830	11,500	1,095	
	0300	101' jib, 10,250 lb. capacity, 270 FPM	"	46.25	1,625	4,850	14,600	1,340	
	0400	Tower, static, 130' high, 106' jib,							
	0500	6200 lb. capacity at 400 FPM	Ea.	44.15	1,475	4,430	13,300	1,239	
	0600	Crawler, cable, 1/2 C.Y., 15 tons at 12' radius		4.97	455	1,360	4,075	311.75	
	0700	3/4 C.Y., 20 tons at 12' radius		22.76	535	1,600	4,800	502.10	
	0800	1 C.Y., 25 tons at 12' radius		30.35	665	1,990	5,975	640.80	
	0900	Crawler mounted, lattice boom, 1-1/2 C.Y., 40 tons at 12' radius		31.15	935	2,800	8,400	809.20	
	1000	2 C.Y., 50 tons at 12' radius		41.90	1,150	3,460	10,400	1,027	
	1100	3 C.Y., 75 tons at 12' radius		45.10	1,225	3,710	11,100	1,103	

GENERAL REQUIREMENTS 1

01590 | Equipment Rental

			UNIT	HOURLY OPER. COST	RENT PER DAY	RENT PER WEEK	RENT PER MONTH	CREW EQUIPMENT COST/DAY	
600	1200	100 ton capacity, standard boom	Ea.	54.40	1,575	4,740	14,200	1,383	600
	1300	165 ton capacity, standard boom		79.70	2,700	8,115	24,300	2,261	
	1400	200 ton capacity, 150' boom		81.45	2,800	8,400	25,200	2,332	
	1500	450' boom		119.60	3,825	11,440	34,300	3,245	
	1600	Truck mounted, lattice boom, 6 x 4, 20 tons at 10' radius		19.32	725	2,170	6,500	588.55	
	1700	25 tons at 10' radius		24.63	965	2,890	8,675	775.05	
	1800	8 x 4, 30 tons at 10' radius		29.10	760	2,280	6,850	688.80	
	1900	40 tons at 12' radius		34.95	945	2,840	8,525	847.60	
	2000	8 x 4, 60 tons at 15' radius		39.56	1,175	3,540	10,600	1,024	
	2050	82 tons at 15' radius		42.83	1,650	4,940	14,800	1,331	
	2100	90 tons at 15' radius		46.38	1,750	5,250	15,800	1,421	
	2200	115 tons at 15' radius		51.75	1,875	5,590	16,800	1,532	
	2300	150 tons at 18' radius		56.90	2,050	6,175	18,500	1,690	
	2350	165 tons at 18' radius		64.15	2,400	7,225	21,700	1,958	
	2400	Truck mounted, hydraulic, 12 ton capacity		31.70	590	1,770	5,300	607.60	
	2500	25 ton capacity		33.10	765	2,300	6,900	724.80	
	2550	33 ton capacity		35.80	950	2,855	8,575	857.40	
	2600	55 ton capacity		41.55	1,075	3,200	9,600	972.40	
	2700	80 ton capacity		55.70	1,525	4,550	13,700	1,356	
	2720	100 ton capacity		68.15	2,125	6,410	19,200	1,827	
	2740	120 ton capacity		72.05	2,300	6,870	20,600	1,950	
	2760	150 ton capacity		76.35	2,625	7,890	23,700	2,189	
	2900	12-1/2 ton capacity		20.15	520	1,560	4,675	473.20	
	3050	20 ton capacity		22.80	645	1,930	5,800	568.40	
	3100	25 ton capacity		24.80	720	2,155	6,475	629.40	
	3150	40 ton capacity		44.10	1,050	3,140	9,425	980.80	
	3200	Derricks, guy, 20 ton capacity, 60' boom, 75' mast		12.07	315	938	2,825	284.15	
	3300	100' boom, 115' mast		19.59	540	1,620	4,850	480.70	
	3400	Stiffleg, 20 ton capacity, 70' boom, 37' mast		13.97	405	1,210	3,625	353.75	
	3500	100' boom, 47' mast		21.97	655	1,960	5,875	567.75	
	3550	Helicopter, small, lift to 1250 lbs. maximum, w/pilot		64.13	2,525	7,590	22,800	2,031	
	3600	Hoists, chain type, overhead, manual, 3/4 ton		.05	5.35	16	48	3.60	
	3900	10 ton		.25	24.50	73	219	16.60	
	4000	Hoist and tower, 5000 lb. cap., portable electric, 40' high		4.03	180	541	1,625	140.45	
	4100	For each added 10' section, add		.08	14	42	126	9.05	
	4200	Hoist and single tubular tower, 5000 lb. electric, 100' high		5.43	252	755	2,275	194.45	
	4300	For each added 6'-6" section, add		.14	24	72	216	15.50	
	4400	Hoist and double tubular tower, 5000 lb., 100' high		5.81	277	831	2,500	212.70	
	4500	For each added 6'-6" section, add		.16	26.50	79	237	17.10	
	4550	Hoist and tower, mast type, 6000 lb., 100' high		6.29	287	862	2,575	222.70	
	4570	For each added 10' section, add		.10	17.35	52	156	11.20	
	4600	Hoist and tower, personnel, electric, 2000 lb., 100' @ 125 FPM		12.77	765	2,290	6,875	560.15	
	4700	3000 lb., 100' @ 200 FPM		14.65	865	2,600	7,800	637.20	
	4800	3000 lb., 150' @ 300 FPM		16.20	970	2,910	8,725	711.60	
	4900	4000 lb., 100' @ 300 FPM		16.83	990	2,970	8,900	728.65	
	5000	6000 lb., 100' @ 275 FPM		18.24	1,050	3,120	9,350	769.90	
	5100	For added heights up to 500', add	L.F.	.01	1.67	5	15	1.10	
	5200	Jacks, hydraulic, 20 ton	Ea.	.05	12.35	37	111	7.80	
	5500	100 ton	"	.15	35	105	315	22.20	
	6000	Jacks, hydraulic, climbing with 50' jackrods							
	6010	and control consoles, minimum 3 mo. rental							
	6100	30 ton capacity	Ea.	1.56	104	311	935	74.70	
	6150	For each added 10' jackrod section, add		.05	3.33	10	30	2.40	
	6300	50 ton capacity		2.50	167	500	1,500	120	
	6350	For each added 10' jackrod section, add		.06	4	12	36	2.90	
	6500	125 ton capacity		6.50	435	1,300	3,900	312	
	6550	For each added 10' jackrod section, add		.45	29.50	89	267	21.40	
	6600	Cable jack, 10 ton capacity with 200' cable		1.30	86.50	260	780	62.40	

R01590 -100

Important: See the Reference Section for critical supporting data - Reference Nos., Crews, & City Cost Indexes

01590 | Equipment Rental

			UNIT	HOURLY OPER. COST	RENT PER DAY	RENT PER WEEK	RENT PER MONTH	CREW EQUIPMENT COST/DAY	
600	6650	For each added 50' of cable, add R01590 -100	▼	.14	9	27	81	6.50	600
700	0010	**WELLPOINT EQUIPMENT RENTAL** See also division 02240							700
	0020	Based on 2 months rental							
	0100	Combination jetting & wellpoint pump, 60 H.P. diesel	Ea.	8.69	257	770	2,300	223.50	
	0200	High pressure gas jet pump, 200 H.P., 300 psi	"	15.61	219	658	1,975	256.50	
	0300	Discharge pipe, 8" diameter	L.F.	.01	.41	1.24	3.72	.35	
	0350	12" diameter		.01	.61	1.84	5.50	.45	
	0400	Header pipe, flows up to 150 G.P.M., 4" diameter		.01	.38	1.14	3.42	.30	
	0500	400 G.P.M., 6" diameter		.01	.44	1.33	3.99	.35	
	0600	800 G.P.M., 8" diameter		.01	.61	1.84	5.50	.45	
	0700	1500 G.P.M., 10" diameter		.01	.65	1.94	5.80	.45	
	0800	2500 G.P.M., 12" diameter		.02	1.22	3.67	11	.90	
	0900	4500 G.P.M., 16" diameter		.02	1.56	4.69	14.05	1.10	
	0950	For quick coupling aluminum and plastic pipe, add	▼	.02	1.62	4.85	14.55	1.15	
	1100	Wellpoint, 25' long, with fittings & riser pipe, 1-1/2" or 2" diameter	Ea.	.05	3.23	9.69	29	2.35	
	1200	Wellpoint pump, diesel powered, 4" diameter, 20 H.P.		4.21	148	444	1,325	122.50	
	1300	6" diameter, 30 H.P.		5.51	184	551	1,650	154.30	
	1400	8" suction, 40 H.P.		7.49	252	755	2,275	210.90	
	1500	10" suction, 75 H.P.		10.30	294	882	2,650	258.80	
	1600	12" suction, 100 H.P.		15.37	470	1,410	4,225	404.95	
	1700	12" suction, 175 H.P.	▼	20.55	520	1,560	4,675	476.40	

GENERAL REQUIREMENTS 1

01740	Cleaning	CREW	DAILY OUTPUT	LABOR-HOURS	UNIT	2002 BARE COSTS				TOTAL INCL O&P	
						MAT.	LABOR	EQUIP.	TOTAL		
500	0010 **CLEANING UP** After job completion, allow, minimum				Job					.30%	500
	0040 Maximum				"					1%	
	0050 Cleanup of floor area, continuous, per day, during const.	A-5	24	.750	M.S.F.	1.70	17.65	1.62	20.97	31	
	0100 Final by GC at end of job	"	11.50	1.565	"	2.65	37	3.38	43.03	63.50	
	1000 Mechanical demolition, see division 02225										

01810	Commissioning	CREW	DAILY OUTPUT	LABOR-HOURS	UNIT	2002 BARE COSTS				TOTAL INCL O&P	
						MAT.	LABOR	EQUIP.	TOTAL		
100	0010 **COMMISSIONING** Including documentation of design intent										100
	0100 performance verification, O&M, training, min				Project					.50%	
	0150 Maximum				"					.75%	

For information about Means Estimating Seminars, see yellow pages 11 and 12 in back of book

Important: See the Reference Section for critical supporting data - Reference Nos., Crews, & City Cost Indexes

Division 2
Site Construction

Estimating Tips

02200 Site Preparation

- If possible visit the site and take an inventory of the type, quantity and size of the trees. Certain trees may have a landscape resale value or firewood value. Stump disposal can be very expensive, particularly if they cannot be buried at the site. Consider using a bulldozer in lieu of hand cutting trees.

- Estimators should visit the site to determine the need for haul road, access, storage of materials, and security considerations. When estimating for access roads on unstable soil, consider using a geotextile stabilization fabric. It can greatly reduce the quantity of crushed stone or gravel. Sites of limited size and access can cause cost overruns due to lost productivity. Theft and damage is another consideration if the location is isolated. A temporary fence or security guards may be required. Investigate the site thoroughly.

02210 Subsurface Investigation

In preparing estimates on structures involving earthwork or foundations, all information concerning soil characteristics should be obtained. Look particularly for hazardous waste, evidence of prior dumping of debris, and previous stream beds.

02220 & 02225 Selective Demolition

The costs shown for selective demolition do not include rubbish handling or disposal. These items should be estimated separately using Means data or other sources.

- Historic preservation often requires that the contractor remove materials from the existing structure, rehab them and replace them. The estimator must be aware of any related measures and precautions that must be taken when doing selective demolition, and cutting and patching. Requirements may include special handling and storage, as well as security.

02300 Earthwork

- Estimating the actual cost of performing earthwork requires careful consideration of the variables involved. This includes items such as type of soil, whether or not water will be encountered, dewatering, whether or not banks need bracing, disposal of excavated earth, length of haul to fill or spoil sites, etc. If the project has large quantities of cut or fill, consider raising or lowering the site to reduce costs while paying close attention to the effect on site drainage and utilities if doing this.

- If the project has large quantities of fill, creating a borrow pit on the site can significantly lower the costs.

- It is very important to consider what time of year the project is scheduled for completion. Bad weather can create large cost overruns from dewatering, site repair and lost productivity from cold weather.

02500 Utility Services
02600 Drainage & Containment

- Never assume that the water, sewer and drainage lines will go in at the early stages of the project. Consider the site access needs before dividing the site in half with open trenches, loose pipe, and machinery obstructions. Always inspect the site to establish that the site drawings are complete. Check off all existing utilities on your drawing as you locate them. If you find any discrepancies, mark up the site plan for further research. Differing site conditions can be very costly if discovered later in the project.

02700 Bases, Ballasts, Pavements/Appurtenances

- When estimating paving, keep in mind the project schedule. If an asphaltic paving project is in a colder climate and runs through to the spring, consider placing the base course in the autumn, then topping it in the spring just prior to completion. This could save considerable costs in spring repair. Keep in mind that prices for asphalt and concrete are generally higher in the cold seasons.

02900 Planting

- The timing of planting and guarantee specifications often dictate the costs for establishing tree and shrub growth and a stand of grass or ground cover. Establish the work performance schedule to coincide with the local planting season. Maintenance and growth guarantees can add from 20% to 100% to the total landscaping cost. The cost to replace trees and shrubs can be as high as 5% of the total cost depending on the planting zone, soil conditions and time of year.

Reference Numbers

Reference numbers are shown in bold squares at the beginning of some major classifications. These numbers refer to related items in the Reference Section. The reference information may be an estimating procedure, an alternate pricing method or technical information.

Note: Not all subdivisions listed here necessarily appear in this publication.

2 SITE CONSTRUCTION

02065		Cement & Concrete	CREW	DAILY OUTPUT	LABOR-HOURS	UNIT	2002 BARE COSTS MAT.	LABOR	EQUIP.	TOTAL	TOTAL INCL O&P	
300	0010	**ASPHALTIC CONCRETE** plant mix (145 lb. per C.F.)				Ton	35			35	38.50	300
	0200	All weather patching mix, hot					34			34	37.50	
	0250	Cold patch					38			38	42	
	0300	Berm mix					34			34	37.50	

02080		Utility Materials										
790	0010	**UNDERGROUND MARKING TAPE** G1030 805										790
	0400	Underground tape, detectable, reinforced, alum. foil core, 2"	1 Clab	150	.053	C.L.F.	1.40	1.25		2.65	3.49	
	0500	6"	"	140	.057	"	3.50	1.34		4.84	5.95	
900	0010	**PIPING VALVES** Water distribution, see also div. 15110										900
	3000	Butterfly valves with boxes, cast iron, mech. jt.										
	3100	4" diameter	B-20	6	4	Ea.	228	106		334	415	
	3140	6" diameter	"	5	4.800		262	128		390	485	
	3180	8" diameter	B-21	4	7		315	192	39.50	546.50	690	
	3300	10" diameter		3.50	8		465	220	45	730	905	
	3340	12" diameter		3	9.333		610	256	52.50	918.50	1,125	
	3400	14" diameter		2	14		1,050	385	79	1,514	1,825	
	3440	16" diameter		2	14		1,400	385	79	1,864	2,225	
	3460	18" diameter		1.50	18.667		1,750	510	105	2,365	2,825	
	3480	20" diameter		1	28		2,200	770	158	3,128	3,800	
	3500	24" diameter		.50	56		3,075	1,525	315	4,915	6,100	
	3600	With lever operator										
	3610	4" diameter	B-20	6	4	Ea.	121	106		227	299	
	3614	6" diameter	"	5	4.800		169	128		297	385	
	3616	8" diameter	B-21	4	7		245	192	39.50	476.50	610	
	3618	10" diameter		3.50	8		395	220	45	660	820	
	3620	12" diameter		3	9.333		730	256	52.50	1,038.50	1,250	
	3622	14" diameter		2	14		1,175	385	79	1,639	1,950	
	3624	16" diameter		2	14		1,575	385	79	2,039	2,425	
	3626	18" diameter		1.50	18.667		1,825	510	105	2,440	2,900	
	3628	20" diameter		1	28		2,425	770	158	3,353	4,025	
	3630	24" diameter		.50	56		2,600	1,525	315	4,440	5,600	
	3700	Check valves, flanged										
	3710	4" diameter	B-20	6	4	Ea.	395	106		501	600	
	3714	6" diameter	"	5	4.800		665	128		793	935	
	3716	8" diameter	B-21	4	7		1,350	192	39.50	1,581.50	1,850	
	3718	10" diameter		3.50	8		2,175	220	45	2,440	2,800	
	3720	12" diameter		3	9.333		3,325	256	52.50	3,633.50	4,125	
	3722	14" diameter		2	14		5,325	385	79	5,789	6,525	
	3724	16" diameter		2	14		7,325	385	79	7,789	8,750	
	3726	18" diameter		1.50	18.667		10,900	510	105	11,515	12,800	
	3728	20" diameter		1	28		12,600	770	158	13,528	15,300	
	3730	24" diameter		.50	56		18,200	1,525	315	20,040	22,700	
	3800	Gate valves, C.I., 250 PSI, mechanical joint, w/boxes										
	3810	4" diameter	B-21A	8	5	Ea.	165	146	59	370	470	
	3814	6" diameter		6.80	5.882		199	172	69.50	440.50	560	
	3816	8" diameter		5.60	7.143		293	209	84.50	586.50	740	
	3818	10" diameter		4.80	8.333		475	243	98.50	816.50	1,000	
	3820	12" diameter		4.08	9.804		645	286	116	1,047	1,275	
	3822	14" diameter	B-21	2	14		755	385	79	1,219	1,500	
	3824	16" diameter		1	28		895	770	158	1,823	2,350	
	3826	18" diameter		.80	35		1,100	960	197	2,257	2,925	
	3828	20" diameter		.80	35		1,400	960	197	2,557	3,275	
	3830	24" diameter		.50	56		1,725	1,525	315	3,565	4,625	
	3831	30" diameter		.35	80		2,075	2,200	450	4,725	6,175	

Important: See the Reference Section for critical supporting data - Reference Nos., Crews, & City Cost Indexes

02080 | Utility Materials

		CREW	DAILY OUTPUT	LABOR-HOURS	UNIT	2002 BARE COSTS MAT.	LABOR	EQUIP.	TOTAL	TOTAL INCL O&P	
900	3832	36" diameter	B-21	.30	93.333	Ea.	2,475	2,550	525	5,550	7,275	**900**
	3900	Globe valves, flanged, iron body, class 125										
	3910	4" diameter	B-20	10	2.400	Ea.	865	64		929	1,050	
	3914	6" diameter	"	9	2.667		1,625	71		1,696	1,875	
	3916	8" diameter	B-21	6	4.667		3,000	128	26.50	3,154.50	3,525	
	3918	10" diameter		5	5.600		4,725	154	31.50	4,910.50	5,475	
	3920	12" diameter		4	7		6,625	192	39.50	6,856.50	7,625	
	3922	14" diameter		3	9.333		7,100	256	52.50	7,408.50	8,250	
	3924	16" diameter		2	14		7,000	385	79	7,464	8,375	
	3926	18" diameter		.80	35		7,700	960	197	8,857	10,200	
	3928	20" diameter		.60	46.667		7,600	1,275	263	9,138	10,600	
	3930	24" diameter	↓	.50	56	↓	9,000	1,525	315	10,840	12,600	

02100 | Site Remediation

02115 | Underground Tank Removal

			CREW	DAILY OUTPUT	LABOR-HOURS	UNIT	2002 BARE COSTS MAT.	LABOR	EQUIP.	TOTAL	TOTAL INCL O&P	
200	0010	**REMOVAL OF UNDERGROUND STORAGE TANKS** R02115 -200										**200**
	0011	Petroleum storage tanks, non-leaking										
	0100	Excavate & load onto trailer										
	0110	3000 gal. to 5000 gal. tank	B-14	4	12	Ea.		298	44.50	342.50	510	
	0120	6000 gal to 8000 gal tank	B-3A	3	13.333			335	235	570	775	
	0130	9000 gal to 12000 gal tank	"	2	20	↓		500	355	855	1,175	
	0190	Known leaking tank add				%				100%	100%	
	0200	Remove sludge, water and remaining product from tank bottom										
	0201	of tank with vacuum truck										
	0300	3000 gal to 5000 gal tank	A-13	5	1.600	Ea.		47.50	98	145.50	180	
	0310	6000 gal to 8000 gal tank		4	2			59.50	123	182.50	225	
	0320	9000 gal to 12000 gal tank	↓	3	2.667	↓		79.50	163	242.50	300	
	0390	Dispose of sludge off-site, average				Gal.					4	
	0400	Insert inert solid CO2 "dry ice" into tank										
	0401	For cleaning/transporting tanks (1.5 lbs./100 gal. cap)	1 Clab	500	.016	Lb.	1.40	.38		1.78	2.12	
	1020	Haul tank to certified salvage dump, 100 miles round trip										
	1023	3000 gal. to 5000 gal. tank				Ea.				550	630	
	1026	6000 gal. to 8000 gal. tank								650	750	
	1029	9,000 gal. to 12,000 gal. tank				↓				875	1,000	
	1100	Disposal of contaminated soil to landfill										
	1110	Minimum				C.Y.					100	
	1111	Maximum				"					300	
	1120	Disposal of contaminated soil to										
	1121	bituminous concrete batch plant										
	1130	Minimum				C.Y.					50	
	1131	Maximum				"					100	
	2010	Decontamination of soil on site incl poly tarp on top/bottom										
	2011	Soil containment berm, and chemical treatment										
	2020	Minimum	B-11C	100	.160	C.Y.	5.60	4.37	1.79	11.76	14.80	
	2021	Maximum	"	100	.160		7.25	4.37	1.79	13.41	16.65	
	2050	Disposal of decontaminated soil, minimum									60	
	2055	Maximum	↓			↓					125	

		02210	Subsurface Investigation	CREW	DAILY OUTPUT	LABOR-HOURS	UNIT	MAT.	LABOR	EQUIP.	TOTAL	TOTAL INCL O&P	
									2002 BARE COSTS				
320	0010		**DRILLING, CORE** Reinforced concrete slab, up to 6" thick slab										320
	0020		Including bit, layout and set up										
	0100		1" diameter core	B-89A	28	.571	Ea.	2.35	15.55	3.33	21.23	30.50	
	0150		Each added inch thick, add		300	.053		.42	1.45	.31	2.18	3.06	
	0300		3" diameter core		23	.696		5.25	18.90	4.05	28.20	39.50	
	0350		Each added inch thick, add		186	.086		.94	2.34	.50	3.78	5.25	
	0500		4" diameter core		19	.842		5.25	23	4.90	33.15	46.50	
	0550		Each added inch thick, add		170	.094		1.19	2.56	.55	4.30	5.90	
	0700		6" diameter core		14	1.143		8.60	31	6.65	46.25	65.50	
	0750		Each added inch thick, add		140	.114		1.46	3.11	.67	5.24	7.20	
	0900		8" diameter core		11	1.455		11.75	39.50	8.45	59.70	84	
	0950		Each added inch thick, add		95	.168		1.98	4.58	.98	7.54	10.40	
	1100		10" diameter core		10	1.600		15.70	43.50	9.30	68.50	95.50	
	1150		Each added inch thick, add		80	.200		2.60	5.45	1.16	9.21	12.65	
	1300		12" diameter core		9	1.778		18.85	48.50	10.35	77.70	107	
	1350		Each added inch thick, add		68	.235		3.12	6.40	1.37	10.89	14.95	
	1500		14" diameter core		7	2.286		23	62	13.30	98.30	137	
	1550		Each added inch thick, add		55	.291		3.96	7.90	1.69	13.55	18.55	
	1700		18" diameter core		4	4		29.50	109	23.50	162	228	
	1750		Each added inch thick, add	▼	28	.571		5.20	15.55	3.33	24.08	33.50	
	1760		For horizontal holes, add to above				▼				30%	30%	
	1770		Prestressed hollow core plank, 6" thick										
	1780		1" diameter core	B-89A	52	.308	Ea.	1.56	8.35	1.79	11.70	16.75	
	1790		Each added inch thick, add		350	.046		.27	1.24	.27	1.78	2.53	
	1800		3" diameter core		50	.320		3.44	8.70	1.86	14	19.40	
	1810		Each added inch thick, add		240	.067		.57	1.81	.39	2.77	3.89	
	1820		4" diameter core		48	.333		4.58	9.05	1.94	15.57	21.50	
	1830		Each added inch thick, add		216	.074		.79	2.01	.43	3.23	4.48	
	1840		6" diameter core		44	.364		5.65	9.90	2.12	17.67	24	
	1850		Each added inch thick, add		175	.091		.94	2.49	.53	3.96	5.50	
	1860		8" diameter core		32	.500		7.60	13.60	2.91	24.11	32.50	
	1870		Each added inch thick, add		118	.136		1.31	3.69	.79	5.79	8.05	
	1880		10" diameter core		28	.571		10.25	15.55	3.33	29.13	39	
	1890		Each added inch thick, add		99	.162		1.41	4.40	.94	6.75	9.45	
	1900		12" diameter core		22	.727		12.50	19.80	4.23	36.53	49.50	
	1910		Each added inch thick, add	▼	85	.188	▼	2.08	5.10	1.10	8.28	11.50	
	1950		Minimum charge for above, 3" diameter core		7	2.286	Total		62	13.30	75.30	112	
	2000		4" diameter core		6.80	2.353			64	13.70	77.70	115	
	2050		6" diameter core		6	2.667			72.50	15.50	88	130	
	2100		8" diameter core		5.50	2.909			79	16.95	95.95	142	
	2150		10" diameter core		4.75	3.368			91.50	19.60	111.10	165	
	2200		12" diameter core		3.90	4.103			112	24	136	201	
	2250		14" diameter core		3.38	4.734			129	27.50	156.50	232	
	2300		18" diameter core	▼	3.15	5.079	▼		138	29.50	167.50	248	

		02220	Site Demolition										
875	0010		**SITE DEMOLITION** No hauling, abandon catch basin or manhole	B-6	7	3.429	Ea.		87.50	25.50	113	163	875
	0020		Remove existing catch basin or manhole, masonry		4	6			153	44.50	197.50	285	
	0030		Catch basin or manhole frames and covers, stored		13	1.846			47	13.75	60.75	87.50	
	0040		Remove and reset	▼	7	3.429			87.50	25.50	113	163	
	0900		Hydrants, fire, remove only	B-21A	5	8			234	94.50	328.50	460	
	0950		Remove and reset	"	2	20	▼		585	237	822	1,150	
	1710		Pavement removal, bituminous roads, 3" thick	B-38	690	.058	S.Y.		1.55	1.26	2.81	3.76	
	1750		4" to 6" thick		420	.095			2.54	2.07	4.61	6.20	
	1800		Bituminous driveways		640	.063			1.67	1.36	3.03	4.06	
	1900		Concrete to 6" thick, hydraulic hammer, mesh reinforced	▼	255	.157	▼		4.18	3.41	7.59	10.20	

Important: See the Reference Section for critical supporting data - Reference Nos., Crews, & City Cost Indexes

2 SITE CONSTRUCTION

			CREW	DAILY OUTPUT	LABOR-HOURS	UNIT	MAT.	2002 BARE COSTS LABOR	EQUIP.	TOTAL	TOTAL INCL O&P	
02220		**Site Demolition**										
875	2000	Rod reinforced	B-38	200	.200	S.Y.		5.35	4.35	9.70	13	875
	2100	Concrete, 7" to 24" thick, plain		33	1.212	C.Y.		32.50	26.50	59	78.50	
	2200	Reinforced	↓	24	1.667	"		44.50	36.50	81	109	
	2300	With hand held air equipment, bituminous, to 6" thick	B-39	1,900	.025	S.F.		.63	.09	.72	1.07	
	2320	Concrete to 6" thick, no reinforcing		1,600	.030			.75	.11	.86	1.27	
	2340	Mesh reinforced		1,400	.034			.85	.13	.98	1.46	
	2360	Rod reinforced	↓	765	.063	↓		1.56	.23	1.79	2.67	
	4000	Sidewalk removal, bituminous, 2-1/2" thick	B-6	325	.074	S.Y.		1.89	.55	2.44	3.51	
	4050	Brick, set in mortar		185	.130			3.32	.97	4.29	6.15	
	4100	Concrete, plain, 4"		160	.150			3.84	1.12	4.96	7.15	
	4200	Mesh reinforced	↓	150	.160	↓		4.09	1.19	5.28	7.60	
02225		**Selective Demolition**										
730	0010	**RUBBISH HANDLING** The following are to be added to the										730
	0020	demolition prices										
	0400	Chute, circular, prefabricated steel, 18" diameter	B-1	40	.600	L.F.	26	14.45		40.45	51	
	0440	30" diameter	"	30	.800	"	35	19.30		54.30	68.50	
	0725	Dumpster, weekly rental, 1 dump/week, 20 C.Y. capacity (8 Tons)				Week					425	
	0800	30 C.Y. capacity (10 Tons)									640	
	0840	40 C.Y. capacity (13 Tons)				↓					775	
	1000	Dust partition, 6 mil polyethylene, 4' x 8' panels, 1" x 3" frame	2 Carp	2,000	.008	S.F.	.17	.24		.41	.55	
	1080	2" x 4" frame	"	2,000	.008	"	.30	.24		.54	.70	
	2000	Load, haul to chute & dumping into chute, 50' haul	2 Clab	24	.667	C.Y.		15.65		15.65	24.50	
	2040	100' haul		16.50	.970			22.50		22.50	35.50	
	2080	Over 100' haul, add per 100 L.F.		35.50	.451			10.55		10.55	16.45	
	2120	In elevators, per 10 floors, add	↓	140	.114			2.68		2.68	4.17	
	3000	Loading & trucking, including 2 mile haul, chute loaded	B-16	45	.711			17.30	10.45	27.75	38.50	
	3040	Hand loading truck, 50' haul	"	48	.667			16.25	9.80	26.05	36	
	3080	Machine loading truck	B-17	120	.267			6.80	4.43	11.23	15.25	
	5000	Haul, per mile, up to 8 C.Y. truck	B-34B	1,165	.007			.17	.40	.57	.70	
	5100	Over 8 C.Y. truck	"	1,550	.005	↓		.13	.30	.43	.53	
740	0010	**DUMP CHARGES** Typical urban city, tipping fees only										740
	0100	Building construction materials				Ton					55	
	0200	Trees, brush, lumber									45	
	0300	Rubbish only									50	
	0500	Reclamation station, usual charge				↓					80	
790	0010	**TORCH CUTTING** Steel, 1" thick plate	1 Clab	32	.250	L.F.	.18	5.85		6.03	9.35	790
	0040	1" diameter bar	"	210	.038	Ea.		.89		.89	1.39	
	1000	Oxygen lance cutting, reinforced concrete walls										
	1040	12" to 16" thick walls	1 Clab	10	.800	L.F.		18.75		18.75	29	
	1080	24" thick walls	"	6	1.333	"		31.50		31.50	48.50	
02240		**Dewatering**										
500	0010	**DEWATERING** Excavate drainage trench, 2' wide, 2' deep	B-11C	90	.178	C.Y.		4.86	1.98	6.84	9.65	500
	0100	2' wide, 3' deep, with backhoe loader	"	135	.119			3.24	1.32	4.56	6.40	
	0200	Excavate sump pits by hand, light soil	1 Clab	7.10	1.127			26.50		26.50	41	
	0300	Heavy soil	"	3.50	2.286	↓		53.50		53.50	83.50	
	0500	Pumping 8 hr., attended 2 hrs. per day, including 20 L.F.										
	0550	of suction hose & 100 L.F. discharge hose										
	0600	2" diaphragm pump used for 8 hours	B-10H	4	3	Day		86	11.60	97.60	144	
	0650	4" diaphragm pump used for 8 hours	B-10I	4	3			86	23.50	109.50	157	
	0800	8 hrs. attended, 2" diaphragm pump	B-10H	1	12			345	46.50	391.50	575	
	0900	3" centrifugal pump	B-10J	1	12	↓		345	57.50	402.50	590	

		02240	Dewatering	CREW	DAILY OUTPUT	LABOR-HOURS	UNIT	MAT.	2002 BARE COSTS LABOR	EQUIP.	TOTAL	TOTAL INCL O&P	
500	1000		4" diaphragm pump	B-10I	1	12	Day		345	93.50	438.50	630	500
	1100		6" centrifugal pump	B-10K	1	12	↓		345	252	597	805	
	1300		CMP, incl. excavation 3' deep, 12" diameter	B-6	115	.209	L.F.	6.35	5.35	1.55	13.25	16.90	
	1400		18" diameter		100	.240	"	7.90	6.15	1.79	15.84	20	
	1600		Sump hole construction, incl. excavation and gravel, pit		1,250	.019	C.F.	.67	.49	.14	1.30	1.65	
	1700		With 12" gravel collar, 12" pipe, corrugated, 16 ga.		70	.343	L.F.	12.10	8.75	2.55	23.40	29.50	
	1800		15" pipe, corrugated, 16 ga.		55	.436		15.45	11.15	3.25	29.85	37.50	
	1900		18" pipe, corrugated, 16 ga.		50	.480		18.15	12.25	3.57	33.97	43	
	2000		24" pipe, corrugated, 14 ga.		40	.600	↓	22	15.35	4.46	41.81	52.50	
	2200		Wood lining, up to 4' x 4', add	↓	300	.080	SFCA	12.30	2.05	.60	14.95	17.30	
	9950		See div. 02240-900 for wellpoints										
	9960		See div. 02240-700 for deep well systems										
700	0010	**WELLS** For dewatering 10' to 20' deep, 2' diameter											700
	0020		with steel casing, minimum	B-6	165	.145	V.L.F.	2.25	3.72	1.08	7.05	9.35	
	0050		Average		98	.245	↓	4.50	6.25	1.82	12.57	16.60	
	0100		Maximum	↓	49	.490	↓	12	12.50	3.64	28.14	36.50	
	0300		For pumps for dewatering, see division 01590-400-4100 to 4400										
	0500		For domestic water wells, see division 02520-900										
900	0010	**WELLPOINTS** For wellpoint equipment rental, see div. 01590-700	R02240 -900										900
	0100		Installation and removal of single stage system										
	0110		Labor only, .75 labor-hours per L.F., minimum	1 Clab	10.70	.748	LF Hdr		17.55		17.55	27.50	
	0200		2.0 labor-hours per L.F., maximum	"	4	2	"		47		47	73	
	0400		Pump operation, 4 @ 6 hr. shifts										
	0410		Per 24 hour day	4 Eqlt	1.27	25.197	Day		750		750	1,125	
	0500		Per 168 hour week, 160 hr. straight, 8 hr. double time		.18	177	Week		5,300		5,300	8,000	
	0550		Per 4.3 week month	↓	.04	800	Month		23,800		23,800	36,000	
	0600		Complete installation, operation, equipment rental, fuel &										
	0610		removal of system with 2" wellpoints 5' O.C.										
	0700		100' long header, 6" diameter, first month	4 Eqlt	3.23	9.907	LF Hdr	119	295		414	575	
	0800		Thereafter, per month		4.13	7.748		95	231		326	455	
	1000		200' long header, 8" diameter, first month		6	5.333		105	159		264	355	
	1100		Thereafter, per month		8.39	3.814		53.50	114		167.50	231	
	1300		500' long header, 8" diameter, first month		10.63	3.010		41.50	89.50		131	182	
	1400		Thereafter, per month		20.91	1.530		29.50	45.50		75	102	
	1600		1,000' long header, 10" diameter, first month		11.62	2.754		35.50	82		117.50	163	
	1700		Thereafter, per month	↓	41.81	.765	↓	17.80	23		40.80	54	
	1900		Note: above figures include pumping 168 hrs. per week										
	1910		and include the pump operator and one stand-by pump.										

		02250	Shoring & Underpinning										
400	0010	**SHEET PILING** Steel, not incl. wales, 22 psf, 15' excav., left in place		B-40	10.81	5.920	Ton	845	179	204	1,228	1,450	400
	0100		Drive, extract & salvage		6	10.667		355	325	365	1,045	1,325	
	0300		20' deep excavation, 27 psf, left in place	R02250 -450	12.95	4.942		845	150	170	1,165	1,350	
	0400		Drive, extract & salvage		6.55	9.771	↓	355	296	335	986	1,225	
	1200		15' deep excavation, 22 psf, left in place		983	.065	S.F.	9.80	1.97	2.24	14.01	16.45	
	1300		Drive, extract & salvage		545	.117		3.99	3.56	4.04	11.59	14.60	
	1500		20' deep excavation, 27 psf, left in place		960	.067		12.35	2.02	2.29	16.66	19.35	
	1600		Drive, extract & salvage		485	.132	↓	5.20	4	4.54	13.74	17.15	
	2100		Rent steel sheet piling and wales, first month				Ton	187			187	206	
	2200		Per added month					18.70			18.70	20.50	
	2300		Rental piling left in place, add to rental					600			600	660	
	2500		Wales, connections & struts, 2/3 salvage				↓	194			194	214	
	3900		Wood, solid sheeting, incl. wales, braces and spacers,										
	3910		drive, extract & salvage, 8' deep excavation	B-31	330	.121	S.F.	1.81	3.05	.47	5.33	7.25	
	4000		10' deep, 50 S.F./hr. in & 150 S.F./hr. out		300	.133		1.86	3.35	.52	5.73	7.80	
	4100		12' deep, 45 S.F./hr. in & 135 S.F./hr. out	↓	270	.148	↓	1.91	3.73	.57	6.21	8.55	

Important: See the Reference Section for critical supporting data - Reference Nos., Crews, & City Cost Indexes

2 SITE CONSTRUCTION

02200 | Site Preparation

02250 | Shoring & Underpinning

		CREW	DAILY OUTPUT	LABOR-HOURS	UNIT	2002 BARE COSTS				TOTAL INCL O&P
						MAT.	LABOR	EQUIP.	TOTAL	
400	**4200**	14' deep, 42 S.F./hr. in & 126 S.F./hr. out R02250-450	B-31	250	.160	S.F.	1.97	4.03	.62	6.62	9.10	**400**
4300	16' deep, 40 S.F./hr. in & 120 S.F./hr. out		240	.167		2.03	4.19	.65	6.87	9.50
4400	18' deep, 38 S.F./hr. in & 114 S.F./hr. out		230	.174		2.10	4.38	.67	7.15	9.85
4520	Left in place, 8' deep, 55 S.F./hr.		440	.091		3.26	2.29	.35	5.90	7.55
4540	10' deep, 50 S.F./hr.		400	.100		3.43	2.52	.39	6.34	8.10
4560	12' deep, 45 S.F./hr.		360	.111		3.62	2.80	.43	6.85	8.80
4565	14' deep, 42 S.F./hr.		335	.119		3.83	3	.46	7.29	9.40
4570	16' deep, 40 S.F./hr.		320	.125		4.07	3.15	.48	7.70	9.90
4700	Alternate pricing, left in place, 8' deep		1.76	22.727	M.B.F.	730	570	88	1,388	1,800
4800	Drive, extract and salvage, 8' deep		1.32	30.303	"	650	760	117	1,527	2,025

02260 | Excavation Support/Protection

		CREW	DAILY OUTPUT	LABOR-HOURS	UNIT	2002 BARE COSTS				TOTAL INCL O&P
						MAT.	LABOR	EQUIP.	TOTAL	
850	**0010**	**SOLDIER BEAMS & LAGGING** H piles with 3" wood sheeting									**850**
0020	horizontal between piles, including removal of wales & braces									
0100	No hydrostatic head, 15' deep, 1 line of braces, minimum	B-50	545	.205	S.F.	8.50	5.95	3.53	17.98	23
0200	Maximum		495	.226		9.40	6.55	3.89	19.84	25
1300	No hydrostatic head, left in place, 15' dp., 1 line of braces, min.		635	.176		11.30	5.10	3.03	19.43	24
1400	Maximum		575	.195		12.10	5.60	3.35	21.05	26
2350	Lagging only, 3" thick wood between piles 8' O.C., minimum	B-46	400	.120		1.88	3.24	.10	5.22	7.45
2370	Maximum		250	.192		2.83	5.20	.16	8.19	11.70
2400	Open sheeting no bracing, for trenches to 10' deep, min.		1,736	.028		.85	.75	.02	1.62	2.17
2450	Maximum		1,510	.032		.94	.86	.03	1.83	2.46
2500	Tie-back method, add to open sheeting, add, minimum								20%	20%
2550	Maximum								60%	60%
2700	Tie-backs only, based on tie-backs total length, minimum	B-46	86.80	.553	L.F.	11.20	14.90	.46	26.56	37
2750	Maximum		38.50	1.247	"	19.70	33.50	1.03	54.23	77
3500	Tie-backs only, typical average, 25' long		2	24	Ea.	495	645	19.90	1,159.90	1,600
3600	35' long		1.58	30.380	"	655	820	25	1,500	2,075

02300 | Earthwork

02315 | Excavation and Fill

		CREW	DAILY OUTPUT	LABOR-HOURS	UNIT	2002 BARE COSTS				TOTAL INCL O&P
						MAT.	LABOR	EQUIP.	TOTAL	
100	**0010**	**BACKFILL** By hand, no compaction, light soil G1030-805	1 Clab	14	.571	C.Y.		13.40		13.40	21	**100**
0100	Heavy soil		11	.727			17.05		17.05	26.50
0300	Compaction in 6" layers, hand tamp, add to above		20.60	.388			9.10		9.10	14.15
0400	Roller compaction operator walking, add	B-10A	100	.120			3.43	1.23	4.66	6.60
0500	Air tamp, add	B-9C	190	.211			5	.93	5.93	8.85
0600	Vibrating plate, add	A-1	60	.133			3.13	1.01	4.14	6
0800	Compaction in 12" layers, hand tamp, add to above	1 Clab	34	.235			5.50		5.50	8.60
0900	Roller compaction operator walking, add	B-10A	150	.080			2.29	.82	3.11	4.39
1000	Air tamp, add	B-9	285	.140			3.35	.62	3.97	5.90
1100	Vibrating plate, add	A-1	90	.089			2.08	.67	2.75	3.98
1300	Dozer backfilling, bulk, up to 300' haul, no compaction	B-10B	1,200	.010			.29	.69	.98	1.19
1400	Air tamped	B-11B	240	.067			1.82	4.24	6.06	7.45
1600	Compacting backfill, 6" to 12" lifts, vibrating roller	B-10C	800	.015			.43	1.25	1.68	2.02
1700	Sheepsfoot roller	B-10D	750	.016			.46	1.19	1.65	2.01
1900	Dozer backfilling, trench, up to 300' haul, no compaction	B-10B	900	.013			.38	.91	1.29	1.59
2000	Air tamped	B-11B	235	.068			1.86	4.33	6.19	7.60
2200	Compacting backfill, 6" to 12" lifts, vibrating roller	B-10C	700	.017			.49	1.43	1.92	2.32
2300	Sheepsfoot roller	B-10D	650	.018			.53	1.38	1.91	2.32

2 SITE CONSTRUCTION

02315 | Excavation and Fill

			CREW	DAILY OUTPUT	LABOR-HOURS	UNIT	MAT.	LABOR	EQUIP.	TOTAL	TOTAL INCL O&P	
130	0010	**BEDDING** For pipe and conduit, not incl. compaction	B-6	150	.160	C.Y.	18	4.09	1.19	23.28	27.50	130
	0050	Crushed or screened bank run gravel		150	.160		14.15	4.09	1.19	19.43	23	
	0100	Crushed stone 3/4" to 1/2"		150	.160		3.88	4.09	1.19	9.16	11.90	
	0200	Sand, dead or bank										
	0500	Compacting bedding in trench	A-1	90	.089			2.08	.67	2.75	3.98	
340	0010	**DRILLING AND BLASTING** Only, rock, open face, under 1500 C.Y.	B-47	225	.107	C.Y.	1.80	2.80	3.69	8.29	10.35	340
	0100	Over 1500 C.Y.		300	.080		1.80	2.10	2.76	6.66	8.25	
	0200	Areas where blasting mats are required, under 1500 C.Y.		175	.137		1.80	3.60	4.74	10.14	12.75	
	0250	Over 1500 C.Y.		250	.096		1.80	2.52	3.32	7.64	9.50	
	2200	Trenches, up to 1500 C.Y.		22	1.091		5.20	28.50	37.50	71.20	91.50	
	2300	Over 1500 C.Y.		26	.923		5.20	24	32	61.20	78.50	
440	0010	**EXCAVATING, STRUCTURAL** Hand, pits to 6' deep, sandy soil	1 Clab	8	1	C.Y.		23.50		23.50	36.50	440
	0100	Heavy soil or clay		4	2			47		47	73	
	0300	Pits 6' to 12' deep, sandy soil		5	1.600			37.50		37.50	58.50	
	0500	Heavy soil or clay		3	2.667			62.50		62.50	97.50	
	0700	Pits 12' to 18' deep, sandy soil		4	2			47		47	73	
	0900	Heavy soil or clay		2	4			94		94	146	
	1500	For wet or muck hand excavation, add to above				%				50%	50%	
505	0010	**FILL** Spread dumped material, by dozer, no compaction	B-10B	1,000	.012	C.Y.		.34	.82	1.16	1.43	505
	0100	By hand	1 Clab	12	.667	"		15.65		15.65	24.50	
	0500	Gravel fill, compacted, under floor slabs, 4" deep	B-37	10,000	.005	S.F.	.16	.12	.01	.29	.36	
	0600	6" deep		8,600	.006		.24	.14	.01	.39	.49	
	0700	9" deep		7,200	.007		.40	.17	.02	.59	.72	
	0800	12" deep		6,000	.008		.55	.20	.02	.77	.94	
	1000	Alternate pricing method, 4" deep		120	.400	C.Y.	11.90	9.95	.98	22.83	29.50	
	1100	6" deep		160	.300		11.90	7.45	.73	20.08	25.50	
	1200	9" deep		200	.240		11.90	5.95	.59	18.44	23	
	1300	12" deep		220	.218		11.90	5.40	.53	17.83	22	
	1400	Granular fill					7.90			7.90	8.70	
900	0010	**EXCAVATING, TRENCH** or continuous footing, common earth G1030 805										900
	0020	No sheeting or dewatering included										
	0050	1' to 4' deep, 3/8 C.Y. tractor loader/backhoe	B-11C	150	.107	C.Y.		2.92	1.19	4.11	5.75	
	0060	1/2 C.Y. tractor loader/backhoe	B-11M	200	.080			2.19	1.08	3.27	4.54	
	0090	4' to 6' deep, 1/2 C.Y. tractor loader/backhoe	"	200	.080			2.19	1.08	3.27	4.54	
	0100	5/8 C.Y. hydraulic backhoe	B-12Q	250	.064			1.89	1.57	3.46	4.58	
	0110	3/4 C.Y. hydraulic backhoe	B-12F	300	.053			1.57	1.52	3.09	4.06	
	0300	1/2 C.Y. hydraulic excavator, truck mounted	B-12J	200	.080			2.36	3.68	6.04	7.60	
	0500	6' to 10' deep, 3/4 C.Y. hydraulic backhoe, 6' to 10' deep	B-12F	225	.071			2.10	2.03	4.13	5.40	
	0600	1 C.Y. hydraulic excavator, truck mounted	B-12K	400	.040			1.18	2.17	3.35	4.17	
	0900	10' to 14' deep, 3/4 C.Y. hydraulic backhoe	B-12F	200	.080			2.36	2.28	4.64	6.10	
	1000	1-1/2 C.Y. hydraulic backhoe	B-12B	540	.030			.87	1.31	2.18	2.76	
	1300	14' to 20' deep, 1 C.Y. hydraulic backhoe	B-12A	320	.050			1.48	1.76	3.24	4.17	
	1400	By hand with pick and shovel 2' to 6' deep, light soil	1 Clab	8	1			23.50		23.50	36.50	
	1500	Heavy soil	"	4	2			47		47	73	
	1700	For tamping backfilled trenches, air tamp, add	A-1	100	.080			1.88	.61	2.49	3.59	
	1900	Vibrating plate, add	B-18	230	.104			2.52	.24	2.76	4.19	
	2100	Trim sides and bottom for concrete pours, common earth		1,500	.016	S.F.		.39	.04	.43	.64	
	2300	Hardpan		600	.040	"		.96	.09	1.05	1.60	
940	0010	**EXCAVATING, UTILITY TRENCH** Common earth										940
	0050	Trenching with chain trencher, 12 H.P., operator walking										
	0100	4" wide trench, 12" deep	B-53	800	.010	L.F.		.30	.11	.41	.57	
	0150	18" deep		750	.011			.32	.11	.43	.61	

26   **Important: See the Reference Section for critical supporting data - Reference Nos., Crews, & City Cost Indexes**

02315	Excavation and Fill	CREW	DAILY OUTPUT	LABOR-HOURS	UNIT	2002 BARE COSTS				TOTAL INCL O&P
						MAT.	LABOR	EQUIP.	TOTAL	
940 0200	24″ deep	B-53	700	.011	L.F.		.34	.12	.46	.65 **940**
0300	6″ wide trench, 12″ deep		650	.012			.37	.13	.50	.70
0350	18″ deep		600	.013			.40	.14	.54	.76
0400	24″ deep		550	.015			.43	.16	.59	.83
0450	36″ deep		450	.018			.53	.19	.72	1.01
0600	8″ wide trench, 12″ deep		475	.017			.50	.18	.68	.96
0650	18″ deep		400	.020			.60	.22	.82	1.14
0700	24″ deep		350	.023			.68	.25	.93	1.30
0750	36″ deep	▼	300	.027	▼		.79	.29	1.08	1.52
1000	Backfill by hand including compaction, add									
1050	4″ wide trench, 12″ deep	A-1	800	.010	L.F.		.23	.08	.31	.45
1100	18″ deep		530	.015			.35	.11	.46	.68
1150	24″ deep		400	.020			.47	.15	.62	.90
1300	6″ wide trench, 12″ deep		540	.015			.35	.11	.46	.66
1350	18″ deep		405	.020			.46	.15	.61	.88
1400	24″ deep		270	.030			.69	.22	.91	1.33
1450	36″ deep		180	.044			1.04	.34	1.38	1.99
1600	8″ wide trench, 12″ deep		400	.020			.47	.15	.62	.90
1650	18″ deep		265	.030			.71	.23	.94	1.35
1700	24″ deep		200	.040			.94	.30	1.24	1.79
1750	36″ deep	▼	135	.059	▼		1.39	.45	1.84	2.65
2000	Chain trencher, 40 H.P. operator riding									
2050	6″ wide trench and backfill, 12″ deep	B-54	1,200	.007	L.F.		.20	.16	.36	.48
2100	18″ deep		1,000	.008			.24	.19	.43	.57
2150	24″ deep		975	.008			.24	.20	.44	.59
2200	36″ deep		900	.009			.26	.21	.47	.64
2250	48″ deep		750	.011			.32	.26	.58	.76
2300	60″ deep		650	.012			.37	.30	.67	.88
2400	8″ wide trench and backfill, 12″ deep		1,000	.008			.24	.19	.43	.57
2450	18″ deep		950	.008			.25	.20	.45	.60
2500	24″ deep		900	.009			.26	.21	.47	.64
2550	36″ deep		800	.010			.30	.24	.54	.72
2600	48″ deep		650	.012			.37	.30	.67	.88
2700	12″ wide trench and backfill, 12″ deep		975	.008			.24	.20	.44	.59
2750	18″ deep		860	.009			.28	.22	.50	.67
2800	24″ deep		800	.010			.30	.24	.54	.72
2850	36″ deep		725	.011			.33	.27	.60	.79
3000	16″ wide trench and backfill, 12″ deep		835	.010			.29	.23	.52	.68
3050	18″ deep	▼	750	.011			.32	.26	.58	.76
3100	24″ deep	▼	700	.011	▼		.34	.28	.62	.81
3200	Compaction with vibratory plate, add								50%	50%
5100	Hand excavate and trim for pipe bells after trench excavation									
5200	8″ pipe	1 Clab	155	.052	L.F.		1.21		1.21	1.88
5300	18″ pipe	″	130	.062	″		1.44		1.44	2.25

02320	Hauling									
200 0011	**HAULING** Excavated or borrow material, loose cubic yards									**200**
0015	no loading included, highway haulers									
0020	6 C.Y. dump truck, 1/4 mile round trip, 5.0 loads/hr.	B-34A	195	.041	C.Y.		1.03	1.81	2.84	3.55
0030	1/2 mile round trip, 4.1 loads/hr.		160	.050			1.25	2.21	3.46	4.34
0040	1 mile round trip, 3.3 loads/hr.	R01590 -100	130	.062			1.54	2.72	4.26	5.35
0100	2 mile round trip, 2.6 loads/hr.		100	.080			2	3.53	5.53	6.95
0150	3 mile round trip, 2.1 loads/hr.		80	.100			2.50	4.42	6.92	8.65
0200	4 mile round trip, 1.8 loads/hr.		70	.114			2.86	5.05	7.91	9.90
0310	12 C.Y. dump truck, 1/4 mile round trip 3.7 loads/hr.	B-34B	288	.028			.69	1.63	2.32	2.86
0400	2 mile round trip, 2.2 loads/hr.	▼ ▼	180	.044	▼		1.11	2.61	3.72	4.57

SITE CONSTRUCTION **2**

02300 | Earthwork

02320 | Hauling

			CREW	DAILY OUTPUT	LABOR-HOURS	UNIT	\multicolumn{4}{c}{2002 BARE COSTS}	TOTAL INCL O&P	
							MAT.	LABOR	EQUIP.	TOTAL		
200	0450	3 mile round trip, 1.9 loads/hr.	B-34B	170	.047	C.Y.		1.18	2.77	3.95	4.84	200
	0500	4 mile round trip, 1.6 loads/hr.	↓	125	.064			1.60	3.77	5.37	6.60	
	1300	Hauling in medium traffic, add								20%	20%	
	1400	Heavy traffic, add								30%	30%	
	1600	Grading at dump, or embankment if required, by dozer	B-10B	1,000	.012	↓		.34	.82	1.16	1.43	
	1800	Spotter at fill or cut, if required	1 Clab	8	1	Hr.		23.50		23.50	36.50	

(R01590-100 reference at 0450/0500)

02400 | Tunneling, Boring & Jacking

02441 | Microtunneling

			CREW	DAILY OUTPUT	LABOR-HOURS	UNIT	\multicolumn{4}{c}{2002 BARE COSTS}	TOTAL INCL O&P	
							MAT.	LABOR	EQUIP.	TOTAL		
400	0010	**MICROTUNNELING** Not including excavation, backfill, shoring,										400
	0020	or dewatering, average 50'/day, slurry method										
	0100	24" to 48" outside diameter, minimum				L.F.					600	
	0110	Adverse conditions, add				%					50%	
	1000	Rent microtunneling machine, average monthly lease				Month					80,000	
	1010	Operating technician				Day					600	
	1100	Mobilization and demobilization, minimum				Job					40,000	
	1110	Maximum				"					400,000	

02445 | Boring or Jacking Conduits

			CREW	DAILY OUTPUT	LABOR-HOURS	UNIT	\multicolumn{4}{c}{2002 BARE COSTS}	TOTAL INCL O&P	
							MAT.	LABOR	EQUIP.	TOTAL		
300	0010	**HORIZONTAL BORING** Casing only, 100' minimum,										300
	0020	not incl. jacking pits or dewatering										
	0100	Roadwork, 1/2" thick wall, 24" diameter casing	B-42	20	3.200	L.F.	49	85	64	198	260	
	0200	36" diameter		16	4		78	106	80	264	340	
	0300	48" diameter		15	4.267		115	113	85.50	313.50	400	
	0500	Railroad work, 24" diameter		15	4.267		49	113	85.50	247.50	325	
	0600	36" diameter		14	4.571		78	121	91.50	290.50	380	
	0700	48" diameter	↓	12	5.333		115	142	107	364	465	
	0900	For ledge, add								145	175	
	1000	Small diameter boring, 3", sandy soil	B-82	900	.018		13.50	.47	.06	14.03	15.65	
	1040	Rocky soil	"	500	.032	↓	13.50	.85	.10	14.45	16.25	
	1100	Prepare jacking pits, incl. mobilization & demobilization, minimum				Ea.				2,650	3,150	
	1101	Maximum				"				15,000	18,000	

02500 | Utility Services

02510 | Water Distribution

			CREW	DAILY OUTPUT	LABOR-HOURS	UNIT	\multicolumn{4}{c}{2002 BARE COSTS}	TOTAL INCL O&P	
							MAT.	LABOR	EQUIP.	TOTAL		
350	0010	**DISTRIBUTION CONNECTION**										350
	7020	Crosses, 4" x 4"	B-21	37	.757	Ea.	380	21	4.26	405.26	455	
	7040	6" x 6"		25	1.120		455	30.50	6.30	491.80	555	
	7060	8" x 6"		21	1.333		555	36.50	7.50	599	675	
	7080	8" x 8"		21	1.333		605	36.50	7.50	649	730	
	7100	10" x 6"	↓	21	1.333	↓	1,100	36.50	7.50	1,144	1,275	

Important: See the Reference Section for critical supporting data - Reference Nos., Crews, & City Cost Indexes

02510	Water Distribution	CREW	DAILY OUTPUT	LABOR-HOURS	UNIT	2002 BARE COSTS				TOTAL INCL O&P
						MAT.	LABOR	EQUIP.	TOTAL	
350 7120	10" x 10"	B-21	21	1.333	Ea.	1,200	36.50	7.50	1,244	1,375 **350**
7140	12" x 6"		18	1.556		1,100	42.50	8.75	1,151.25	1,275
7160	12" x 12"		18	1.556		1,400	42.50	8.75	1,451.25	1,600
7180	14" x 6"		16	1.750		2,775	48	9.85	2,832.85	3,125
7200	14" x 14"		16	1.750		2,900	48	9.85	2,957.85	3,275
7220	16" x 6"		14	2		2,950	55	11.25	3,016.25	3,350
7240	16" x 10"		14	2		3,000	55	11.25	3,066.25	3,375
7260	16" x 16"		14	2		3,150	55	11.25	3,216.25	3,550
7280	18" x 6"		10	2.800		4,400	77	15.75	4,492.75	4,975
7300	18" x 12"		10	2.800		4,425	77	15.75	4,517.75	5,000
7320	18" x 18"		10	2.800		4,550	77	15.75	4,642.75	5,150
7340	20" x 6"		8	3.500		3,575	96	19.70	3,690.70	4,100
7360	20" x 12"		8	3.500		3,825	96	19.70	3,940.70	4,375
7380	20" x 20"		8	3.500		5,575	96	19.70	5,690.70	6,325
7400	24" x 6"		6	4.667		4,600	128	26.50	4,754.50	5,300
7420	24" x 12"		6	4.667		4,675	128	26.50	4,829.50	5,375
7440	24" x 18"		6	4.667		6,975	128	26.50	7,129.50	7,900
7460	24" x 24"		6	4.667		7,250	128	26.50	7,404.50	8,200
7600	Cut-in sleeves with rubber gaskets, 4"		18	1.556		127	42.50	8.75	178.25	215
7620	6"		12	2.333		161	64	13.15	238.15	291
7640	8"		10	2.800		219	77	15.75	311.75	375
7660	10"		10	2.800		305	77	15.75	397.75	470
7680	12"		9	3.111		360	85.50	17.50	463	545
7800	Cut-in valves with rubber gaskets, 4"		18	1.556		330	42.50	8.75	381.25	440
7820	6"		12	2.333		445	64	13.15	522.15	605
7840	8"		10	2.800		690	77	15.75	782.75	895
7860	10"		10	2.800		1,050	77	15.75	1,142.75	1,300
7880	12"		9	3.111		1,325	85.50	17.50	1,428	1,625
8000	Sleeves with rubber gaskets, 4" x 4"		37	.757		305	21	4.26	330.26	370
8020	6" x 6"		25	1.120		365	30.50	6.30	401.80	455
8040	8" x 6"		21	1.333		455	36.50	7.50	499	565
8060	8" x 8"		21	1.333		480	36.50	7.50	524	595
8080	10" x 6"		21	1.333		880	36.50	7.50	924	1,025
8100	10" x 10"		21	1.333		955	36.50	7.50	999	1,125
8120	12" x 6"		18	1.556		880	42.50	8.75	931.25	1,050
8140	12" x 12"		18	1.556		1,125	42.50	8.75	1,176.25	1,300
8160	14" x 6"		16	1.750		2,200	48	9.85	2,257.85	2,500
8180	14" x 14"		16	1.750		2,325	48	9.85	2,382.85	2,625
8200	16" x 6"		14	2		2,350	55	11.25	2,416.25	2,700
8220	16" x 10"		14	2		2,400	55	11.25	2,466.25	2,725
8240	16" x 16"		14	2		2,525	55	11.25	2,591.25	2,875
8260	18" x 6"		10	2.800		3,525	77	15.75	3,617.75	4,000
8280	18" x 12"		10	2.800		3,550	77	15.75	3,642.75	4,025
8300	18" x 18"		10	2.800		3,650	77	15.75	3,742.75	4,150
8320	20" x 6"		8	3.500		2,850	96	19.70	2,965.70	3,300
8340	20" x 12"		8	3.500		3,050	96	19.70	3,165.70	3,525
8360	20" x 20"		8	3.500		4,475	96	19.70	4,590.70	5,100
8380	24" x 6"		6	4.667		3,675	128	26.50	3,829.50	4,275
8400	24" x 12"		6	4.667		3,725	128	26.50	3,879.50	4,325
8420	24" x 18"		6	4.667		5,575	128	26.50	5,729.50	6,350
8440	24" x 24"	▼	6	4.667		5,800	128	26.50	5,954.50	6,600
8800	Curb box, 6' long	B-20	20	1.200		132	32		164	195
8820	8' long	"	18	1.333		138	35.50		173.50	208
9000	Valves, gate valve, N.R.S. post type, 4" diameter	B-21	32	.875		266	24	4.93	294.93	335
9020	6" diameter		20	1.400		340	38.50	7.90	386.40	445
9040	8" diameter	▼	16	1.750	▼	540	48	9.85	597.85	680

SITE CONSTRUCTION 2

02510 | Water Distribution

		CREW	DAILY OUTPUT	LABOR-HOURS	UNIT	2002 BARE COSTS				TOTAL INCL O&P	
						MAT.	LABOR	EQUIP.	TOTAL		
350											**350**
9060	10″ diameter	B-21	16	1.750	Ea.	840	48	9.85	897.85	1,000	
9080	12″ diameter		13	2.154		1,075	59	12.15	1,146.15	1,275	
9100	14″ diameter		11	2.545		2,525	70	14.35	2,609.35	2,900	
9120	O.S.&Y., 4″ diameter		32	.875		375	24	4.93	403.93	455	
9140	6″ diameter		20	1.400		485	38.50	7.90	531.40	605	
9160	8″ diameter		16	1.750		745	48	9.85	802.85	905	
9180	10″ diameter		16	1.750		1,075	48	9.85	1,132.85	1,250	
9200	12″ diameter		13	2.154		1,425	59	12.15	1,496.15	1,650	
9220	14″ diameter	↓	11	2.545		2,525	70	14.35	2,609.35	2,900	
9400	Check valves, rubber disc, 2-1/2″ diameter	B-20	44	.545		295	14.50		309.50	350	
9420	3″ diameter	"	38	.632		295	16.80		311.80	350	
9440	4″ diameter	B-21	32	.875		365	24	4.93	393.93	440	
9480	6″ diameter		20	1.400		545	38.50	7.90	591.40	670	
9500	8″ diameter		16	1.750		800	48	9.85	857.85	965	
9520	10″ diameter		16	1.750		1,425	48	9.85	1,482.85	1,650	
9540	12″ diameter		13	2.154		2,250	59	12.15	2,321.15	2,575	
9542	14″ diameter		11	2.545		3,175	70	14.35	3,259.35	3,625	
9700	Detector check valves, reducing, 4″ diameter		32	.875		640	24	4.93	668.93	745	
9720	6″ diameter		20	1.400		985	38.50	7.90	1,031.40	1,150	
9740	8″ diameter		16	1.750		1,600	48	9.85	1,657.85	1,850	
9760	10″ diameter		16	1.750		3,175	48	9.85	3,232.85	3,575	
9800	Galvanized, 4″ diameter		32	.875		835	24	4.93	863.93	960	
9820	6″ diameter		20	1.400		1,225	38.50	7.90	1,271.40	1,400	
9840	8″ diameter		16	1.750		1,950	48	9.85	2,007.85	2,225	
9860	10″ diameter	↓	16	1.750	↓	3,850	48	9.85	3,907.85	4,325	
800	0010 **PIPING, WATER DISTRIBUTION SYSTEMS** Pipe, laid											**800**
	0020 in trench, excavation and backfill not included											
1400	Ductile Iron, cement lined, class 50 water pipe, 18′ lengths											
1410	Mechanical joint, 4″ diameter	B-20	144	.167	L.F.	9.25	4.44		13.69	17.10	
1420	6″ diameter		126	.190		9.55	5.05		14.60	18.40	
1430	8″ diameter		108	.222		12.25	5.90		18.15	22.50	
1440	10″ diameter	↓	90	.267		15.90	7.10		23	28.50	
1450	12″ diameter	B-21	72	.389		20.50	10.65	2.19	33.34	41.50	
1460	14″ diameter		54	.519		25	14.25	2.92	42.17	52.50	
1470	16″ diameter		46	.609		27.50	16.70	3.43	47.63	60	
1480	18″ diameter		42	.667		36	18.30	3.75	58.05	72	
1490	24″ diameter	↓	35	.800		54	22	4.50	80.50	98	
1550	Push on joint, 4″ diameter	B-20	155	.155		6.80	4.12		10.92	13.85	
1560	6″ diameter		135	.178		7.70	4.73		12.43	15.85	
1570	8″ diameter		115	.209		11	5.55		16.55	21	
1580	10″ diameter		98	.245		16.70	6.50		23.20	28.50	
1590	12″ diameter	↓	78	.308		17.60	8.20		25.80	32	
1600	14″ diameter	B-21	58	.483		19.35	13.25	2.72	35.32	45	
1610	16″ diameter		52	.538		27	14.80	3.03	44.83	56	
1620	18″ diameter		43	.651		30	17.85	3.67	51.52	64.50	
1630	20″ diameter		41	.683		33	18.75	3.84	55.59	69	
1640	24″ diameter	↓	40	.700	↓	43	19.20	3.94	66.14	82	
1950	Butterfly valves with boxes, cast iron											
1970	4″ diameter	B-20	6	4	Ea.	360	106		466	560	
1990	6″ diameter	"	5	4.800		485	128		613	735	
2010	8″ diameter	B-21	4	7		725	192	39.50	956.50	1,150	
2030	10″ diameter		3.50	8		955	220	45	1,220	1,450	
2050	12″ diameter		3	9.333		1,325	256	52.50	1,633.50	1,900	
2070	14″ diameter		2	14		1,800	385	79	2,264	2,650	
2090	16″ diameter	↓ ↓	2	14	↓	2,225	385	79	2,689	3,125	

R02510 -800

Important: See the Reference Section for critical supporting data - Reference Nos., Crews, & City Cost Indexes

02500 | Utility Services

02510 | Water Distribution

			CREW	DAILY OUTPUT	LABOR-HOURS	UNIT	MAT.	LABOR	EQUIP.	TOTAL	TOTAL INCL O&P	
800	2650	Polyvinyl chloride pipe, class 160, S.D.R.-26, 1-1/2" diameter R02510-800	B-20	300	.080	L.F.	.59	2.13		2.72	3.97	800
	2700	2" diameter		250	.096		.89	2.56		3.45	4.96	
	2750	2-1/2" diameter		250	.096		1.32	2.56		3.88	5.45	
	2800	3" diameter		200	.120		1.89	3.19		5.08	7.05	
	2850	4" diameter		200	.120		3.09	3.19		6.28	8.40	
	2900	6" diameter		180	.133		6.65	3.55		10.20	12.90	
	2950	8" diameter	B-21	160	.175		11.30	4.80	.99	17.09	21	
	8000	Fittings, ductile iron, mechanical joint										
	8010	90° bend 4" diameter	B-20	37	.649	Ea.	153	17.25		170.25	195	
	8020	6" diameter		25	.960		205	25.50		230.50	265	
	8040	8" diameter		21	1.143		289	30.50		319.50	370	
	8060	10" diameter	B-21	21	1.333		455	36.50	7.50	499	565	
	8080	12" diameter		18	1.556		625	42.50	8.75	676.25	760	
	8100	14" diameter		16	1.750		640	48	9.85	697.85	790	
	8120	16" diameter		14	2		690	55	11.25	756.25	850	
	8140	18" diameter		10	2.800		1,600	77	15.75	1,692.75	1,875	
	8160	20" diameter		8	3.500		2,375	96	19.70	2,490.70	2,800	
	8180	24" diameter		6	4.667		2,725	128	26.50	2,879.50	3,225	
	8200	Wye or tee, 4" diameter	B-20	25	.960		128	25.50		153.50	181	
	8220	6" diameter		17	1.412		168	37.50		205.50	244	
	8240	8" diameter		14	1.714		246	45.50		291.50	340	
	8260	10" diameter	B-21	14	2		840	55	11.25	906.25	1,025	
	8280	12" diameter		12	2.333		1,025	64	13.15	1,102.15	1,250	
	8300	14" diameter		10	2.800		1,050	77	15.75	1,142.75	1,300	
	8320	16" diameter		8	3.500		1,100	96	19.70	1,215.70	1,375	
	8340	18" diameter		6	4.667		2,075	128	26.50	2,229.50	2,500	
	8360	20" diameter		4	7		2,725	192	39.50	2,956.50	3,350	
	8380	24" diameter		3	9.333		3,175	256	52.50	3,483.50	3,950	
820	0010	**PIPING, WATER DISTRIBUTION, DUCTILE IRON** cement lined										820
	4000	Drill and tap pressurized main (labor only)										
	4100	6" main, 1" to 2" service	Q-1	3	5.333	Ea.		173		173	260	
	4150	8" main, 1" to 2" service	"	2.75	5.818	"		188		188	284	
	4500	Tap and insert gate valve										
	4600	8" main, 4" branch	B-21	3.20	8.750	Ea.		240	49.50	289.50	425	
	4650	6" branch		2.70	10.370			285	58.50	343.50	505	
	4700	10" main, 4" branch		2.70	10.370			285	58.50	343.50	505	
	4750	6" branch		2.35	11.915			325	67	392	580	
	4800	12" main, 6" branch		2.35	11.915			325	67	392	580	
	4850	8" branch		2.35	11.915			325	67	392	580	
850	0010	**PIPING, HDPE BUTT FUSION JOINTS**, SDR 21, 40' lengths										850
	0100	4" diameter	B-22A	400	.095	L.F.	1.92	2.55	1.61	6.08	7.85	
	0200	6" diameter		380	.100		3.07	2.69	1.69	7.45	9.40	
	0300	8" diameter		320	.119		5.20	3.19	2.01	10.40	12.85	
	0400	10" diameter		300	.127		8.10	3.40	2.14	13.64	16.50	
	0500	12" diameter		260	.146		11.35	3.93	2.47	17.75	21.50	
	0600	14" diameter		220	.173		13.70	4.64	2.92	21.26	25.50	
	0700	16" diameter		180	.211		17.90	5.65	3.57	27.12	32.50	
	0800	18" diameter		140	.271		22.50	7.30	4.59	34.39	41.50	
	0900	24" diameter		100	.380		40.50	10.20	6.45	57.15	67.50	
	1000	Fittings										
	1100	Elbows, 90 degrees										
	1200	4" diameter	B-22B	32	.500	Ea.	29	13.60	4.72	47.32	58	
	1300	6" diameter		28	.571		75	15.55	5.40	95.95	112	
	1400	8" diameter		24	.667		159	18.15	6.30	183.45	210	
	1500	10" diameter		18	.889		275	24	8.40	307.40	345	

SITE CONSTRUCTION

SITE CONSTRUCTION | **2**

02510	Water Distribution	CREW	DAILY OUTPUT	LABOR-HOURS	UNIT	2002 BARE COSTS				TOTAL INCL O&P
						MAT.	LABOR	EQUIP.	TOTAL	
1600	12" diameter	B-22B	12	1.333	Ea.	465	36.50	12.60	514.10	585
1700	14" diameter		9	1.778		465	48.50	16.80	530.30	605
1800	16" diameter		6	2.667		560	72.50	25	657.50	755
1900	18" diameter		4	4		755	109	38	902	1,050
2000	24" diameter	▼	3	5.333	▼	1,525	145	50.50	1,720.50	1,950
2100	Tees									
2200	4" diameter	B-22B	30	.533	Ea.	33.50	14.50	5.05	53.05	65
2300	6" diameter		26	.615		88.50	16.75	5.80	111.05	129
2400	8" diameter		22	.727		224	19.80	6.85	250.65	285
2500	10" diameter		15	1.067		295	29	10.05	334.05	380
2600	12" diameter		10	1.600		405	43.50	15.10	463.60	530
2700	14" diameter		8	2		475	54.50	18.90	548.40	630
2800	16" diameter		6	2.667		560	72.50	25	657.50	755
2900	18" diameter		4	4		755	109	38	902	1,050
3000	24" diameter	▼	2	8	▼	1,575	218	75.50	1,868.50	2,150

02520	Wells	CREW	DAILY OUTPUT	LABOR-HOURS	UNIT	MAT.	LABOR	EQUIP.	TOTAL	TOTAL INCL O&P
0010	**WELLS** Domestic water									
0100	Drilled, 4" to 6" diameter	B-23	120	.333	L.F.		7.95	18.15	26.10	32.50
0200	8" diameter	"	95.20	.420	"		10	23	33	40.50
0400	Gravel pack well, 40' deep, incl. gravel & casing, complete									
0500	24" diameter casing x 18" diameter screen	B-23	.13	307	Total	21,000	7,350	16,700	45,050	53,000
0600	36" diameter casing x 18" diameter screen		.12	333	"	22,600	7,950	18,100	48,650	57,000
0800	Observation wells, 1-1/4" riser pipe	▼	163	.245	V.L.F.	11.55	5.85	13.35	30.75	36.50
0900	For flush Buffalo roadway box, add	1 Skwk	16.60	.482	Ea.	31.50	14.90		46.40	58
1200	Test well, 2-1/2" diameter, up to 50' deep (15 to 50 GPM)	B-23	1.51	26.490	"	475	630	1,450	2,555	3,075
1300	Over 50' deep, add	"	121.80	.328	L.F.	12.60	7.85	17.85	38.30	45.50
1500	Pumps, installed in wells to 100' deep, 4" submersible									
1510	1/2 H.P.	Q-1	3.22	4.969	Ea.	355	161		516	630
1520	3/4 H.P.		2.66	6.015		425	195		620	765
1600	1 H.P.	▼	2.29	6.987		480	226		706	865
1700	1-1/2 H.P.	Q-22	1.60	10		620	325	380	1,325	1,600
1800	2 H.P.		1.33	12.030		840	390	455	1,685	2,025
1900	3 H.P.		1.14	14.035		1,050	455	535	2,040	2,425
2000	5 H.P.		1.14	14.035		1,700	455	535	2,690	3,125
2050	Remove and install motor only, 4 H.P.		1.14	14.035		640	455	535	1,630	1,975
3000	Pump, 6" submersible, 25' to 150' deep, 25 H.P., 249 to 297 GPM		.89	17.978		4,075	580	685	5,340	6,125
3100	25' to 500' deep, 30 H.P., 100 to 300 GPM	▼	.73	21.918	▼	4,650	710	830	6,190	7,125
8110	Well screen assembly, stainless steel, 2" diameter	B-23A	273	.088	L.F.	35.50	2.35	7.60	45.45	51
8120	3" diameter		253	.095		49.50	2.53	8.20	60.23	67.50
8130	4" diameter		200	.120		56.50	3.20	10.40	70.10	78.50
8140	5" diameter		168	.143		66	3.81	12.35	82.16	92
8150	6" diameter		126	.190		79	5.10	16.50	100.60	113
8160	8" diameter		98.50	.244		103	6.50	21	130.50	146
8170	10" diameter		73	.329		129	8.80	28.50	166.30	187
8180	12" diameter		62.50	.384		153	10.25	33	196.25	220
8190	14" diameter		54.30	.442		173	11.80	38.50	223.30	250
8200	16" diameter		48.30	.497		191	13.25	43	247.25	278
8210	18" diameter		39.20	.612		237	16.35	53	306.35	345
8220	20" diameter		31.20	.769		270	20.50	66.50	357	400
8230	24" diameter		23.80	1.008		335	27	87.50	449.50	510
8240	26" diameter		21	1.143		370	30.50	99	499.50	560
8300	Slotted PVC, 1-1/4" diameter		521	.046		1.37	1.23	3.99	6.59	7.80
8310	1-1/2" diameter		488	.049		1.99	1.31	4.26	7.56	8.90
8320	2" diameter	▼	273	.088	▼	2.75	2.35	7.60	12.70	15

Important: See the Reference Section for critical supporting data - Reference Nos., Crews, & City Cost Indexes

02520 | Wells

		CREW	DAILY OUTPUT	LABOR-HOURS	UNIT	MAT.	LABOR	EQUIP.	TOTAL	TOTAL INCL O&P	
900	8330	3" diameter	B-23A	253	.095	L.F.	2.92	2.53	8.20	13.65	16.15	**900**
	8340	4" diameter		200	.120		3.39	3.20	10.40	16.99	20	
	8350	5" diameter		168	.143		3.56	3.81	12.35	19.72	23.50	
	8360	6" diameter		126	.190		4.94	5.10	16.50	26.54	31.50	
	8370	8" diameter	▼	98.50	.244		7.35	6.50	21	34.85	41	
	8400	Artificial gravel pack, 2" screen, 6" casing	B-23B	174	.138		1.97	3.68	13.10	18.75	22	
	8405	8" casing		111	.216		2.69	5.75	20.50	28.94	34.50	
	8410	10" casing		74.50	.322		3.38	8.60	30.50	42.48	50.50	
	8415	12" casing		60	.400		4.56	10.70	38	53.26	63	
	8420	14" casing		50.20	.478		5.90	12.75	45.50	64.15	76	
	8425	16" casing		40.70	.590		8.15	15.75	56	79.90	94.50	
	8430	18" casing		36	.667		9.45	17.80	63	90.25	107	
	8435	20" casing		29.50	.814		10.90	21.50	77	109.40	130	
	8440	24" casing		25.70	.934		12.05	25	88.50	125.55	149	
	8445	26" casing		24.60	.976		13.50	26	92.50	132	157	
	8450	30" casing		20	1.200		15.45	32	114	161.45	192	
	8455	36" casing		16.40	1.463	▼	16.65	39	139	194.65	231	
	8500	Develop well		8	3	Hr.	175	80	284	539	630	
	8550	Pump test well	▼	8	3		46.50	80	284	410.50	490	
	8560	Standby well	B-23A	8	3		45	80	260	385	460	
	8570	Standby, drill rig		8	3	▼		80	260	340	410	
	8580	Surface seal well, concrete filled	▼	1	24	Ea.	430	640	2,075	3,145	3,725	
	8590	Well test pump, install & remove	B-23	1	40			955	2,175	3,130	3,875	
	8600	Well sterilization, chlorine	2 Clab	1	16	▼	370	375		745	995	
	9950	See div. 02240-900 for wellpoints										
	9960	See div. 02240-700 for drainage wells										
910	0010	**PUMPS, WELL** Water system, with pressure control										**910**
	1000	Deep well, jet, 42 gal. galvanized tank										
	1040	3/4 HP	1 Plum	.80	10	Ea.	545	360		905	1,150	
	3000	Shallow well, jet, 30 gal. galvanized tank										
	3040	1/2 HP	1 Plum	2	4	Ea.	550	144		694	820	

02530 | Sanitary Sewerage

		CREW	DAILY OUTPUT	LABOR-HOURS	UNIT	MAT.	LABOR	EQUIP.	TOTAL	TOTAL INCL O&P	
730	0010	**PIPING, DRAINAGE & SEWAGE, CONCRETE**										**730**
	0020	Not including excavation or backfill										
	1000	Non-reinforced pipe, extra strength, B&S or T&G joints										
	1010	6" diameter	B-14	265.04	.181	L.F.	3.50	4.50	.67	8.67	11.55	
	1020	8" diameter		224	.214		3.85	5.30	.80	9.95	13.35	
	1030	10" diameter		216	.222		4.27	5.50	.83	10.60	14.15	
	1040	12" diameter		200	.240		5.25	5.95	.89	12.09	16.05	
	1050	15" diameter		180	.267		6.15	6.60	.99	13.74	18.10	
	1060	18" diameter		144	.333		7.55	8.30	1.24	17.09	22.50	
	1070	21" diameter		112	.429		9.30	10.65	1.59	21.54	28.50	
	1080	24" diameter	▼	100	.480	▼	11.40	11.90	1.79	25.09	33	
	2000	Reinforced culvert, class 3, no gaskets										
	2010	12" diameter	B-14	210	.229	L.F.	10.25	5.70	.85	16.80	21	
	2020	15" diameter		175	.274		12.65	6.80	1.02	20.47	25.50	
	2030	18" diameter		130	.369		12.75	9.15	1.37	23.27	30	
	2035	21" diameter		120	.400		15.55	9.95	1.49	26.99	34	
	2040	24" diameter	▼	100	.480		18.65	11.90	1.79	32.34	41	
	2045	27" diameter	B-13	92	.609		22.50	15.50	7.90	45.90	57.50	
	2050	30" diameter		88	.636		26	16.20	8.25	50.45	62.50	
	2060	36" diameter	▼	72	.778		36.50	19.80	10.05	66.35	82	
	2070	42" diameter	B-13B	72	.778		50	19.80	13.50	83.30	100	
	2080	48" diameter	▼	64	.875	▼	62	22.50	15.20	99.70	120	

SITE CONSTRUCTION 2

2 SITE CONSTRUCTION

		02530	Sanitary Sewerage	CREW	DAILY OUTPUT	LABOR-HOURS	UNIT	2002 BARE COSTS				TOTAL INCL O&P	
								MAT.	LABOR	EQUIP.	TOTAL		
730	2090		60" diameter	B-13B	48	1.167	L.F.	99.50	29.50	20.50	149.50	179	730
	2100		72" diameter		40	1.400		141	35.50	24.50	201	237	
	2120		84" diameter		32	1.750		257	44.50	30.50	332	385	
	2140		96" diameter		24	2.333		310	59.50	40.50	410	475	
	2200		With gaskets, class 3, 12" diameter	B-21	168	.167		11.30	4.57	.94	16.81	20.50	
	2220		15" diameter		160	.175		13.55	4.80	.99	19.34	23.50	
	2230		18" diameter		152	.184		16.95	5.05	1.04	23.04	27.50	
	2240		24" diameter		136	.206		25.50	5.65	1.16	32.31	38	
	2260		30" diameter	B-13	88	.636		34	16.20	8.25	58.45	71.50	
	2270		36" diameter	"	72	.778		51	19.80	10.05	80.85	97.50	
	2290		48" diameter	B-13B	64	.875		83	22.50	15.20	120.70	143	
	2310		72" diameter	"	40	1.400		215	35.50	24.50	275	320	
	2330		Flared ends, 6'-1" long, 12" diameter	B-21	190	.147		38.50	4.04	.83	43.37	49	
	2340		15" diameter		155	.181		44.50	4.96	1.02	50.48	58	
	2400		6'-2" long, 18" diameter		122	.230		47	6.30	1.29	54.59	63	
	2420		24" diameter		88	.318		57	8.75	1.79	67.54	78	
	2440		36" diameter	B-13	60	.933		87.50	24	12.10	123.60	146	
	3040		Vitrified plate lined, add to above, 30" to 36" diameter				SFCA	3.40			3.40	3.74	
	3050		42" to 54" diameter, add					3.64			3.64	4	
	3060		60" to 72" diameter, add					4.26			4.26	4.69	
	3070		Over 72" diameter, add					4.54			4.54	4.99	
	3080		Radius pipe, add to pipe prices, 12" to 60" diameter				L.F.	50%					
	3090		Over 60" diameter, add				"	20%					
	3500		Reinforced elliptical, 8' lengths, C507 class 3										
	3520		14" x 23" inside, round equivalent 18" diameter	B-21	82	.341	L.F.	23.50	9.35	1.92	34.77	42	
	3530		24" x 38" inside, round equivalent 30" diameter	B-13	58	.966		41.50	24.50	12.50	78.50	97.50	
	3540		29" x 45" inside, round equivalent 36" diameter		52	1.077		52.50	27.50	13.95	93.95	115	
	3550		38" x 60" inside, round equivalent 48" diameter		38	1.474		80	37.50	19.05	136.55	167	
	3560		48" x 76" inside, round equivalent 60" diameter		26	2.154		144	55	28	227	273	
	3570		58" x 91" inside, round equivalent 72" diameter		22	2.545		194	65	33	292	350	
	3780		Concrete slotted pipe, class 4 mortar joint										
	3800		12" diameter	B-21	168	.167	L.F.	11.90	4.57	.94	17.41	21	
	3840		18" diameter	"	152	.184	"	18.45	5.05	1.04	24.54	29.50	
	3900		Class 4 O-ring										
	3940		12" diameter	B-21	168	.167	L.F.	12.30	4.57	.94	17.81	21.50	
	3960		18" diameter	"	152	.184	"	17.25	5.05	1.04	23.34	28	
780	0010	**PIPING, DRAINAGE & SEWAGE, POLYVINYL CHLORIDE**											780
	0020		Not including excavation or backfill										
	2000		10' lengths, S.D.R. 35, B&S, 4" diameter	B-20	375	.064	L.F.	1.43	1.70		3.13	4.22	
	2040		6" diameter		350	.069		2.48	1.83		4.31	5.55	
	2080		8" diameter		335	.072		4.40	1.91		6.31	7.80	
	2120		10" diameter	B-21	330	.085		6.55	2.33	.48	9.36	11.35	
	2160		12" diameter		320	.087		7.35	2.40	.49	10.24	12.35	
	2200		15" diameter		190	.147		11.15	4.04	.83	16.02	19.45	
790	0010	**PIPING, DRAINAGE & SEWAGE, VITRIFIED CLAY** C700											790
	0020		Not including excavation or backfill,										
	4030		Extra strength, compression joints, C425										
	5000		4" diameter x 4' long	B-20	265	.091	L.F.	1.66	2.41		4.07	5.60	
	5020		6" diameter x 5' long	"	200	.120		2.78	3.19		5.97	8.05	
	5040		8" diameter x 5' long	B-21	200	.140		3.96	3.84	.79	8.59	11.20	
	5060		10" diameter x 5' long		190	.147		6.65	4.04	.83	11.52	14.45	
	5080		12" diameter x 6' long		150	.187		8.80	5.10	1.05	14.95	18.80	
	5100		15" diameter x 7' long		110	.255		15.45	7	1.43	23.88	29.50	
	5120		18" diameter x 7' long		88	.318		23	8.75	1.79	33.54	41	
	5140		24" diameter x 7' long		45	.622		42	17.05	3.50	62.55	77	
	5160		30" diameter x 7' long	B-22	31	.968		74.50	27	7.65	109.15	132	

Important: See the Reference Section for critical supporting data - Reference Nos., Crews, & City Cost Indexes

		02530 | **Sanitary Sewerage**	CREW	DAILY OUTPUT	LABOR-HOURS	UNIT	2002 BARE COSTS MAT.	LABOR	EQUIP.	TOTAL	TOTAL INCL O&P	
790	5180	36" diameter x 7' long	B-22	20	1.500	L.F.	123	41.50	11.85	176.35	213	**790**
	6000	For 3' lengths, add					30%	30%				
	6020	For 2' lengths, add					40%	60%				
	6060	For plain joints, deduct					25%					
	7060	2' lengths, add to above				↓	40%					

		02550 | **Piped Energy Distribution**										
450	0010	**GAS STATION PRODUCT LINE**										**450**
	0020	Primary containment pipe, fiberglass-reinforced										
	0030	Plastic pipe 15' & 30' lengths										
	0040	2" diameter	Q-6	425	.056	L.F.	2.98	1.91		4.89	6.15	
	0050	3" diameter		400	.060		3.92	2.03		5.95	7.35	
	0060	4" diameter	↓	375	.064	↓	5.05	2.16		7.21	8.80	
	0100	Fittings										
	0110	Elbows, 90° & 45°, bell-ends, 2"	Q-6	24	1	Ea.	30	34		64	84	
	0120	3" diameter		22	1.091		31.50	37		68.50	90.50	
	0130	4" diameter		20	1.200		42	40.50		82.50	107	
	0200	Tees, bell ends, 2"		21	1.143		36.50	38.50		75	98.50	
	0210	3" diameter		18	1.333		37	45		82	109	
	0220	4" diameter		15	1.600		50.50	54		104.50	137	
	0230	Flanges bell ends, 2"		24	1		12.10	34		46.10	64.50	
	0240	3" diameter		22	1.091		15.25	37		52.25	72.50	
	0250	4" diameter		20	1.200		21	40.50		61.50	84	
	0260	Sleeve couplings, 2"		21	1.143		7.70	38.50		46.20	66.50	
	0270	3" diameter		18	1.333		10.95	45		55.95	80	
	0280	4" diameter		15	1.600		15.10	54		69.10	98	
	0290	Threaded adapters 2"		21	1.143		10.15	38.50		48.65	69	
	0300	3" diameter		18	1.333		17.80	45		62.80	87.50	
	0310	4" diameter		15	1.600		24	54		78	108	
	0320	Reducers, 2"		27	.889		13.50	30		43.50	60.50	
	0330	3" diameter		22	1.091		15.60	37		52.60	72.50	
	0340	4" diameter	↓	20	1.200	↓	20	40.50		60.50	83	
	1010	Gas station product line for secondary containment (double wall)										
	1100	Fiberglass reinforced plastic pipe 25' lengths										
	1120	Pipe, plain end, 3"	Q-6	375	.064	L.F.	5.15	2.16		7.31	8.90	
	1130	4" diameter		350	.069		8.40	2.32		10.72	12.75	
	1140	5" diameter		325	.074		10.95	2.49		13.44	15.75	
	1150	6" diameter	↓	300	.080	↓	11.25	2.70		13.95	16.50	
	1200	Fittings										
	1230	Elbows, 90° & 45°, 3"	Q-6	18	1.333	Ea.	37	45		82	109	
	1240	4" diameter		16	1.500		64.50	50.50		115	148	
	1250	5" diameter		14	1.714		149	58		207	252	
	1260	6" diameter		12	2		151	67.50		218.50	268	
	1270	Tees, 3"		15	1.600		54.50	54		108.50	142	
	1280	4" diameter		12	2		80.50	67.50		148	191	
	1290	5" diameter		9	2.667		163	90		253	315	
	1300	6" diameter		6	4		170	135		305	390	
	1310	Couplings, 3"		18	1.333		26	45		71	96.50	
	1320	4" diameter		16	1.500		67	50.50		117.50	150	
	1330	5" diameter		14	1.714		139	58		197	241	
	1340	6" diameter		12	2		144	67.50		211.50	261	
	1350	Cross-over nipples, 3"		18	1.333		5.90	45		50.90	74.50	
	1360	4" diameter		16	1.500		6.95	50.50		57.45	84	
	1370	5" diameter		14	1.714		10.30	58		68.30	99	
	1380	6" diameter	↓	12	2	↓	10.80	67.50		78.30	114	

2

SITE CONSTRUCTION

2

SITE CONSTRUCTION

		02550 | **Piped Energy Distribution**	CREW	DAILY OUTPUT	LABOR-HOURS	UNIT	2002 BARE COSTS				TOTAL INCL O&P	
							MAT.	LABOR	EQUIP.	TOTAL		
450	1400	Telescoping, reducers, concentric 4" x 3"	Q-6	18	1.333	Ea.	19.75	45		64.75	89.50	**450**
	1410	5" x 4"		17	1.412		51.50	47.50		99	129	
	1420	6" x 5"	↓	16	1.500	↓	124	50.50		174.50	213	
464	0010	**PIPING, GAS SERVICE & DISTRIBUTION, POLYETHYLENE**										**464**
	0020	not including excavation or backfill										
	1000	60 psi coils, comp cplg @ 100', 1/2" diameter, SDR 9.3	B-20A	608	.053	L.F.	.38	1.49		1.87	2.70	
	1040	1-1/4" diameter, SDR 11		544	.059		.69	1.67		2.36	3.31	
	1100	2" diameter, SDR 11		488	.066		.86	1.86		2.72	3.79	
	1160	3" diameter, SDR 11	↓	408	.078		1.80	2.23		4.03	5.40	
	1500	60 PSI 40' joints with coupling, 3" diameter, SDR 11	B-21A	408	.098		1.80	2.86	1.16	5.82	7.60	
	1540	4" diameter, SDR 11		352	.114		4.14	3.32	1.34	8.80	11.10	
	1600	6" diameter, SDR 11		328	.122		12.85	3.56	1.44	17.85	21	
	1640	8" diameter, SDR 11	↓	272	.147	↓	17.55	4.29	1.74	23.58	28	
550	0010	**PIPE CONDUIT, PREFABRICATED**										**550**
	0020	Does not include trenching, fittings or crane.										
	0300	For cathodic protection, add 12 to 14%										
	0310	of total built-up price (casing plus service pipe)										
	0580	Polyurethane insulated system, 250°F. max. temp.										
	0620	Black steel service pipe, standard wt., 1/2" insulation										
	0660	3/4" diam. pipe size	Q-17	54	.296	L.F.	16.55	9.65	1.48	27.68	34.50	
	0670	1" diam. pipe size		50	.320		18.20	10.45	1.60	30.25	37.50	
	0680	1-1/4" diam. pipe size		47	.340		20.50	11.10	1.70	33.30	41	
	0690	1-1/2" diam. pipe size		45	.356		22	11.60	1.78	35.38	44	
	0700	2" diam. pipe size		42	.381		23	12.40	1.90	37.30	46	
	0710	2-1/2" diam. pipe size		34	.471		23	15.35	2.35	40.70	51	
	0720	3" diam. pipe size		28	.571		27	18.60	2.86	48.46	60.50	
	0730	4" diam. pipe size		22	.727		34	23.50	3.64	61.14	77	
	0740	5" diam. pipe size	↓	18	.889		43.50	29	4.44	76.94	96	
	0750	6" diam. pipe size	Q-18	23	1.043		50.50	35.50	3.47	89.47	112	
	0760	8" diam. pipe size		19	1.263		73.50	42.50	4.21	120.21	150	
	0770	10" diam. pipe size		16	1.500		97	50.50	5	152.50	189	
	0780	12" diam. pipe size		13	1.846		121	62.50	6.15	189.65	234	
	0790	14" diam. pipe size		11	2.182		135	73.50	7.25	215.75	267	
	0800	16" diam. pipe size		10	2.400		154	81	8	243	300	
	0810	18" diam. pipe size		8	3		178	101	10	289	360	
	0820	20" diam. pipe size		7	3.429		198	116	11.40	325.40	405	
	0830	24" diam. pipe size	↓	6	4		242	135	13.30	390.30	485	
	0900	For 1" thick insulation, add					10%					
	0940	For 1-1/2" thick insulation, add					13%					
	0980	For 2" thick insulation, add				↓	20%					
	1500	Gland seal for system, 3/4" diam. pipe size	Q-17	32	.500	Ea.	269	16.30	2.50	287.80	325	
	1510	1" diam. pipe size		32	.500		269	16.30	2.50	287.80	325	
	1540	1-1/4" diam. pipe size		30	.533		288	17.40	2.67	308.07	345	
	1550	1-1/2" diam. pipe size		30	.533		288	17.40	2.67	308.07	345	
	1560	2" diam. pipe size		28	.571		340	18.60	2.86	361.46	405	
	1570	2-1/2" diam. pipe size		26	.615		370	20	3.08	393.08	440	
	1580	3" diam. pipe size		24	.667		395	21.50	3.33	419.83	470	
	1590	4" diam. pipe size		22	.727		470	23.50	3.64	497.14	555	
	1600	5" diam. pipe size	↓	19	.842		575	27.50	4.21	606.71	680	
	1610	6" diam. pipe size	Q-18	26	.923		610	31	3.07	644.07	725	
	1620	8" diam. pipe size		25	.960		710	32.50	3.20	745.70	835	
	1630	10" diam. pipe size		23	1.043		865	35.50	3.47	903.97	1,000	
	1640	12" diam. pipe size		21	1.143		955	38.50	3.81	997.31	1,100	
	1650	14" diam. pipe size		19	1.263		1,075	42.50	4.21	1,121.71	1,250	
	1660	16" diam. pipe size	↓	18	1.333	↓	1,250	45	4.44	1,299.44	1,450	

Important: See the Reference Section for critical supporting data - Reference Nos., Crews, & City Cost Indexes

02550	Piped Energy Distribution	CREW	DAILY OUTPUT	LABOR-HOURS	UNIT	2002 BARE COSTS				TOTAL INCL O&P
						MAT.	LABOR	EQUIP.	TOTAL	
550 1670	18" diam. pipe size	Q-18	16	1.500	Ea.	1,350	50.50	5	1,405.50	1,550 **550**
1680	20" diam. pipe size		14	1.714		1,525	58	5.70	1,588.70	1,775
1690	24" diam. pipe size	▼	12	2	▼	1,700	67.50	6.65	1,774.15	1,950
2000	Elbow, 45° for system									
2020	3/4" diam. pipe size	Q-17	14	1.143	Ea.	170	37	5.70	212.70	249
2040	1" diam. pipe size		13	1.231		174	40	6.15	220.15	259
2050	1-1/4" diam. pipe size		11	1.455		198	47.50	7.25	252.75	298
2060	1-1/2" diam. pipe size		9	1.778		225	58	8.90	291.90	345
2070	2" diam. pipe size		6	2.667		216	87	13.35	316.35	385
2080	2-1/2" diam. pipe size		4	4		234	130	20	384	475
2090	3" diam. pipe size		3.50	4.571		270	149	23	442	545
2100	4" diam. pipe size		3	5.333		315	174	26.50	515.50	635
2110	5" diam. pipe size	▼	2.80	5.714		405	186	28.50	619.50	760
2120	6" diam. pipe size	Q-18	4	6		460	203	20	683	830
2130	8" diam. pipe size		3	8		665	270	26.50	961.50	1,175
2140	10" diam. pipe size		2.40	10		850	340	33.50	1,223.50	1,475
2150	12" diam. pipe size		2	12		1,125	405	40	1,570	1,875
2160	14" diam. pipe size		1.80	13.333		1,375	450	44.50	1,869.50	2,250
2170	16" diam. pipe size		1.60	15		1,650	505	50	2,205	2,650
2180	18" diam. pipe size		1.30	18.462		2,075	625	61.50	2,761.50	3,275
2190	20" diam. pipe size		1	24		2,600	810	80	3,490	4,200
2200	24" diam. pipe size	▼	.70	34.286		3,275	1,150	114	4,539	5,475
2260	For elbow, 90°, add					25%				
2300	For tee, straight, add					85%	30%			
2340	For tee, reducing, add					170%	30%			
2380	For weldolet, straight, add				▼	50%				
2400	Polyurethane insulation, 1"									
2410	FRP carrier and casing									
2420	4"	Q-5	18	.889	L.F.	34.50	29		63.50	81.50
2422	6"	Q-6	23	1.043		60	35.50		95.50	120
2424	8"		19	1.263		98	42.50		140.50	173
2426	10"		16	1.500		132	50.50		182.50	223
2428	12"	▼	13	1.846	▼	172	62.50		234.50	284
2430	FRP carrier and PVC casing									
2440	4"	Q-5	18	.889	L.F.	13.70	29		42.70	58.50
2444	8"	Q-6	19	1.263		28.50	42.50		71	96
2446	10"		16	1.500		41	50.50		91.50	122
2448	12"	▼	13	1.846	▼	52.50	62.50		115	152
2450	PVC carrier and casing									
2460	4"	Q-1	36	.444	L.F.	6.75	14.40		21.15	29
2462	6"	"	29	.552		9.45	17.85		27.30	37.50
2464	8"	Q-2	36	.667		13.15	22.50		35.65	48
2466	10"		32	.750		18	25		43	58
2468	12"	▼	31	.774	▼	21	26		47	62
2800	Calcium silicate insulated system, high temp. (1200°F)									
2840	Steel casing with protective exterior coating									
2850	6-5/8" diameter	Q-18	52	.462	L.F.	31	15.60	1.54	48.14	59
2860	8-5/8" diameter		50	.480		34	16.20	1.60	51.80	63.50
2870	10-3/4" diameter		47	.511		39.50	17.25	1.70	58.45	71.50
2880	12-3/4" diameter		44	.545		43	18.45	1.82	63.27	77.50
2890	14" diameter		41	.585		48.50	19.75	1.95	70.20	85.50
2900	16" diameter		39	.615		52	21	2.05	75.05	91.50
2910	18" diameter		36	.667		57.50	22.50	2.22	82.22	100
2920	20" diameter		34	.706		65	24	2.35	91.35	110
2930	22" diameter		32	.750		90	25.50	2.50	118	140
2940	24" diameter	▼	29	.828	▼	103	28	2.76	133.76	158

SITE CONSTRUCTION **2**

SITE CONSTRUCTION

2

02550	Piped Energy Distribution	CREW	DAILY OUTPUT	LABOR-HOURS	UNIT	2002 BARE COSTS				TOTAL INCL O&P
						MAT.	LABOR	EQUIP.	TOTAL	
2950	26" diameter	Q-18	26	.923	L.F.	117	31	3.07	151.07	179
2960	28" diameter		23	1.043		146	35.50	3.47	184.97	217
2970	30" diameter		21	1.143		155	38.50	3.81	197.31	232
2980	32" diameter		19	1.263		174	42.50	4.21	220.71	261
2990	34" diameter		18	1.333		176	45	4.44	225.44	267
3000	36" diameter		16	1.500		189	50.50	5	244.50	290
3040	For multi-pipe casings, add					10%				
3060	For oversize casings, add					2%				
3400	Steel casing gland seal, single pipe									
3420	6-5/8" diameter	Q-18	25	.960	Ea.	430	32.50	3.20	465.70	530
3440	8-5/8" diameter		23	1.043		505	35.50	3.47	543.97	610
3450	10-3/4" diameter		21	1.143		565	38.50	3.81	607.31	685
3460	12-3/4" diameter		19	1.263		675	42.50	4.21	721.71	810
3470	14" diameter		17	1.412		735	47.50	4.70	787.20	885
3480	16" diameter		16	1.500		865	50.50	5	920.50	1,025
3490	18" diameter		15	1.600		960	54	5.35	1,019.35	1,125
3500	20" diameter		13	1.846		1,050	62.50	6.15	1,118.65	1,275
3510	22" diameter		12	2		1,200	67.50	6.65	1,274.15	1,425
3520	24" diameter		11	2.182		1,350	73.50	7.25	1,430.75	1,600
3530	26" diameter		10	2.400		1,525	81	8	1,614	1,800
3540	28" diameter		9.50	2.526		1,725	85.50	8.40	1,818.90	2,050
3550	30" diameter		9	2.667		1,775	90	8.90	1,873.90	2,100
3560	32" diameter		8.50	2.824		1,975	95.50	9.40	2,079.90	2,325
3570	34" diameter		8	3		2,150	101	10	2,261	2,550
3580	36" diameter		7	3.429		2,300	116	11.40	2,427.40	2,725
3620	For multi-pipe casings, add					5%				
4000	Steel casing anchors, single pipe									
4020	6-5/8" diameter	Q-18	8	3	Ea.	385	101	10	496	590
4040	8-5/8" diameter		7.50	3.200		405	108	10.65	523.65	620
4050	10-3/4" diameter		7	3.429		530	116	11.40	657.40	775
4060	12-3/4" diameter		6.50	3.692		565	125	12.30	702.30	825
4070	14" diameter		6	4		665	135	13.30	813.30	950
4080	16" diameter		5.50	4.364		775	147	14.55	936.55	1,100
4090	18" diameter		5	4.800		870	162	16	1,048	1,225
4100	20" diameter		4.50	5.333		960	180	17.75	1,157.75	1,350
4110	22" diameter		4	6		1,075	203	20	1,298	1,500
4120	24" diameter		3.50	6.857		1,175	232	23	1,430	1,650
4130	26" diameter		3	8		1,325	270	26.50	1,621.50	1,900
4140	28" diameter		2.50	9.600		1,450	325	32	1,807	2,125
4150	30" diameter		2	12		1,550	405	40	1,995	2,350
4160	32" diameter		1.50	16		1,850	540	53.50	2,443.50	2,900
4170	34" diameter		1	24		2,075	810	80	2,965	3,600
4180	36" diameter		1	24		2,250	810	80	3,140	3,800
4220	For multi-pipe, add					5%	20%			
4800	Steel casing elbow									
4820	6-5/8" diameter	Q-18	15	1.600	Ea.	530	54	5.35	589.35	670
4830	8-5/8" diameter		15	1.600		565	54	5.35	624.35	710
4850	10-3/4" diameter		14	1.714		685	58	5.70	748.70	845
4860	12-3/4" diameter		13	1.846		810	62.50	6.15	878.65	990
4870	14" diameter		12	2		865	67.50	6.65	939.15	1,050
4880	16" diameter		11	2.182		945	73.50	7.25	1,025.75	1,175
4890	18" diameter		10	2.400		1,075	81	8	1,164	1,300
4900	20" diameter		9	2.667		1,150	90	8.90	1,248.90	1,425
4910	22" diameter		8	3		1,250	101	10	1,361	1,550
4920	24" diameter		7	3.429		1,375	116	11.40	1,502.40	1,700
4930	26" diameter		6	4		1,500	135	13.30	1,648.30	1,875

Important: See the Reference Section for critical supporting data - Reference Nos., Crews, & City Cost Indexes

02550	Piped Energy Distribution	CREW	DAILY OUTPUT	LABOR-HOURS	UNIT	2002 BARE COSTS				TOTAL INCL O&P
						MAT.	LABOR	EQUIP.	TOTAL	
4940	28" diameter	Q-18	5	4.800	Ea.	1,600	162	16	1,778	2,050
4950	30" diameter		4	6		1,625	203	20	1,848	2,100
4960	32" diameter		3	8		1,850	270	26.50	2,146.50	2,475
4970	34" diameter		2	12		2,025	405	40	2,470	2,875
4980	36" diameter	▼	2	12	▼	2,150	405	40	2,595	3,025
5500	Black steel service pipe, std. wt., 1" thick insulation									
5510	3/4" diameter pipe size	Q-17	54	.296	L.F.	13.50	9.65	1.48	24.63	31
5540	1" diameter pipe size		50	.320		14.05	10.45	1.60	26.10	33
5550	1-1/4" diameter pipe size		47	.340		15.75	11.10	1.70	28.55	36
5560	1-1/2" diameter pipe size		45	.356		17.15	11.60	1.78	30.53	38.50
5570	2" diameter pipe size		42	.381		19.05	12.40	1.90	33.35	42
5580	2-1/2" diameter pipe size		34	.471		20	15.35	2.35	37.70	47.50
5590	3" diameter pipe size		28	.571		23	18.60	2.86	44.46	56
5600	4" diameter pipe size		22	.727		29	23.50	3.64	56.14	71.50
5610	5" diameter pipe size	▼	18	.889		39.50	29	4.44	72.94	92
5620	6" diameter pipe size	Q-18	23	1.043	▼	43	35.50	3.47	81.97	104
6000	Black steel service pipe, std. wt., 1-1/2" thick insul.									
6010	3/4" diameter pipe size	Q-17	54	.296	L.F.	13.70	9.65	1.48	24.83	31
6040	1" diameter pipe size		50	.320		15.30	10.45	1.60	27.35	34.50
6050	1-1/4" diameter pipe size		47	.340		17.15	11.10	1.70	29.95	37.50
6060	1-1/2" diameter pipe size		45	.356		18.70	11.60	1.78	32.08	40
6070	2" diameter pipe size		42	.381		20.50	12.40	1.90	34.80	43.50
6080	2-1/2" diameter pipe size		34	.471		21.50	15.35	2.35	39.20	49.50
6090	3" diameter pipe size		28	.571		24.50	18.60	2.86	45.96	57.50
6100	4" diameter pipe size		22	.727		31	23.50	3.64	58.14	73.50
6110	5" diameter pipe size	▼	18	.889		39.50	29	4.44	72.94	92
6120	6" diameter pipe size	Q-18	23	1.043		45	35.50	3.47	83.97	106
6130	8" diameter pipe size		19	1.263		65	42.50	4.21	111.71	140
6140	10" diameter pipe size		16	1.500		84.50	50.50	5	140	175
6150	12" diameter pipe size	▼	13	1.846		101	62.50	6.15	169.65	212
6190	For 2" thick insulation, add					15%				
6220	For 2-1/2" thick insulation, add					25%				
6260	For 3" thick insulation, add				▼	30%				
6800	Black steel service pipe, ex. hvy. wt., 1" thick insul.									
6820	3/4" diameter pipe size	Q-17	50	.320	L.F.	14.20	10.45	1.60	26.25	33
6840	1" diameter pipe size		47	.340		15.10	11.10	1.70	27.90	35
6850	1-1/4" diameter pipe size		44	.364		17.45	11.85	1.82	31.12	39
6860	1-1/2" diameter pipe size		42	.381		18.55	12.40	1.90	32.85	41.50
6870	2" diameter pipe size		40	.400		19.60	13.05	2	34.65	43.50
6880	2-1/2" diameter pipe size		31	.516		24.50	16.80	2.58	43.88	55
6890	3" diameter pipe size		27	.593		27.50	19.30	2.96	49.76	62.50
6900	4" diameter pipe size		21	.762		36	25	3.81	64.81	81
6910	5" diameter pipe size	▼	17	.941		50.50	30.50	4.71	85.71	107
6920	6" diameter pipe size	Q-18	22	1.091	▼	56	37	3.63	96.63	121
7400	Black steel service pipe, ex. hvy. wt., 1-1/2" thick insul.									
7420	3/4" diameter pipe size	Q-17	50	.320	L.F.	14.20	10.45	1.60	26.25	33
7440	1" diameter pipe size		47	.340		16.20	11.10	1.70	29	36.50
7450	1-1/4" diameter pipe size		44	.364		18.80	11.85	1.82	32.47	40.50
7460	1-1/2" diameter pipe size		42	.381		20.50	12.40	1.90	34.80	43.50
7470	2" diameter pipe size		40	.400		21	13.05	2	36.05	45
7480	2-1/2" diameter pipe size		31	.516		24.50	16.80	2.58	43.88	55.50
7490	3" diameter pipe size		27	.593		29	19.30	2.96	51.26	64.50
7500	4" diameter pipe size		21	.762		38	25	3.81	66.81	83
7510	5" diameter pipe size	▼	17	.941		52	30.50	4.71	87.21	109
7520	6" diameter pipe size	Q-18	22	1.091		57.50	37	3.63	98.13	123
7530	8" diameter pipe size	▼	18	1.333	▼	86.50	45	4.44	135.94	168

SITE CONSTRUCTION **2**

02500 | Utility Services

02550 | Piped Energy Distribution

		CREW	DAILY OUTPUT	LABOR-HOURS	UNIT	2002 BARE COSTS MAT.	LABOR	EQUIP.	TOTAL	TOTAL INCL O&P	
550	7540	10" diameter pipe size	Q-18	15	1.600	L.F.	103	54	5.35	162.35	200	**550**
	7550	12" diameter pipe size	↓	13	1.846		126	62.50	6.15	194.65	239	
	7590	For 2" thick insulation, add					13%					
	7640	For 2-1/2" thick insulation, add					18%					
	7680	For 3" thick insulation, add				↓	24%					
	9000	Combined steam pipe with condensate return										
	9010	8" and 4" in 24" case	Q-18	100	.240	L.F.	185	8.10	.80	193.90	217	
	9020	6" and 3" in 20" case		104	.231		127	7.80	.77	135.57	152	
	9030	3" and 1-1/2" in 16" case		110	.218		95	7.35	.73	103.08	116	
	9040	2" and 1-1/4" in 12-3/4" case		114	.211		82.50	7.10	.70	90.30	102	
	9050	1-1/2" and 1-1/4" in 10-3/4" case	↓	116	.207	↓	72	7	.69	79.69	91	
	9100	Steam pipe only (no return)										
	9110	6" in 18" case	Q-18	108	.222	L.F.	104	7.50	.74	112.24	126	
	9120	4" in 14" case		112	.214		82.50	7.25	.71	90.46	102	
	9130	3" in 14" case		114	.211		76	7.10	.70	83.80	95.50	
	9140	2-1/2" in 12-3/4" case		118	.203		71	6.85	.68	78.53	89	
	9150	2" in 12-3/4" case	↓	122	.197	↓	68	6.65	.66	75.31	85.50	

02600 | Drainage & Containment

02620 | Subdrainage

		CREW	DAILY OUTPUT	LABOR-HOURS	UNIT	2002 BARE COSTS MAT.	LABOR	EQUIP.	TOTAL	TOTAL INCL O&P	
210	0010	**PIPING, SUBDRAINAGE, CONCRETE**										**210**
	0021	Not including excavation and backfill										
	3000	Porous wall concrete underdrain, std. strength, 4" diameter	B-20	335	.072	L.F.	1.84	1.91		3.75	4.99	
	3020	6" diameter	"	315	.076		2.39	2.03		4.42	5.80	
	3040	8" diameter	B-21	310	.090		2.95	2.48	.51	5.94	7.65	
	3100	18" diameter	"	165	.170		9.50	4.66	.96	15.12	18.65	
	4000	Extra strength, 6" diameter	B-20	315	.076		2.36	2.03		4.39	5.75	
	4020	8" diameter	B-21	310	.090		3.54	2.48	.51	6.53	8.30	
	4040	10" diameter		285	.098		7.10	2.70	.55	10.35	12.60	
	4060	12" diameter		230	.122		7.65	3.34	.69	11.68	14.40	
	4080	15" diameter		200	.140		8.50	3.84	.79	13.13	16.15	
	4100	18" diameter	↓	165	.170	↓	12.40	4.66	.96	18.02	22	
240	0010	**PIPING, SUBDRAINAGE, CORRUGATED METAL**										**240**
	0021	Not including excavation and backfill										
	2010	Aluminum, perforated										
	2020	6" diameter, 18 ga.	B-14	380	.126	L.F.	2.62	3.14	.47	6.23	8.25	
	2200	8" diameter, 16 ga.		370	.130		3.81	3.22	.48	7.51	9.70	
	2220	10" diameter, 16 ga.		360	.133		4.76	3.31	.50	8.57	10.95	
	2240	12" diameter, 16 ga.		285	.168		5.35	4.18	.63	10.16	13	
	2260	18" diameter, 16 ga.	↓	205	.234	↓	8	5.80	.87	14.67	18.75	
	3000	Uncoated galvanized, perforated										
	3020	6" diameter, 18 ga.	B-20	380	.063	L.F.	4.08	1.68		5.76	7.10	
	3200	8" diameter, 16 ga.	"	370	.065		5.60	1.73		7.33	8.85	
	3220	10" diameter, 16 ga.	B-21	360	.078		8.40	2.13	.44	10.97	13.05	
	3240	12" diameter, 16 ga.		285	.098		8.80	2.70	.55	12.05	14.50	
	3260	18" diameter, 16 ga.	↓	205	.137	↓	13.45	3.75	.77	17.97	21.50	
	4000	Steel, perforated, asphalt coated										
	4020	6" diameter 18 ga.	B-20	380	.063	L.F.	3.26	1.68		4.94	6.20	

Important: See the Reference Section for critical supporting data - Reference Nos., Crews, & City Cost Indexes

			CREW	DAILY OUTPUT	LABOR-HOURS	UNIT	2002 BARE COSTS				TOTAL INCL O&P	
							MAT.	LABOR	EQUIP.	TOTAL		
		**02620	Subdrainage**										
240	4030	8" diameter 18 ga	B-20	370	.065	L.F.	5.10	1.73		6.83	8.30	**240**
	4040	10" diameter 16 ga	B-21	360	.078		5.85	2.13	.44	8.42	10.25	
	4050	12" diameter 16 ga		285	.098		6.75	2.70	.55	10	12.20	
	4060	18" diameter 16 ga	▼	205	.137	▼	9.20	3.75	.77	13.72	16.75	
250	0010	**PIPING, SUBDRAINAGE, PLASTIC**										**250**
	0020	Not including excavation and backfill										
	1110	10"				Ea.	3.86			3.86	4.25	
	2100	Perforated PVC, 4" diameter	B-14	314	.153	L.F.	1.40	3.80	.57	5.77	8.05	
	2110	6" diameter		300	.160		2.59	3.97	.60	7.16	9.65	
	2120	8" diameter		290	.166		2.94	4.11	.62	7.67	10.25	
	2130	10" diameter		280	.171		4.55	4.26	.64	9.45	12.30	
	2140	12" diameter	▼	270	.178	▼	6.30	4.42	.66	11.38	14.55	
280	0010	**PIPING, SUBDRAINAGE, VITRIFIED CLAY**										**280**
	0020	Not including excavation and backfill										
	3000	Perforated, 5' lengths, C700, 4" diameter	B-14	400	.120	L.F.	2	2.98	.45	5.43	7.30	
	3020	6" diameter		315	.152		3.31	3.79	.57	7.67	10.10	
	3040	8" diameter		290	.166		4.71	4.11	.62	9.44	12.25	
	3060	12" diameter	▼	275	.175		9.95	4.34	.65	14.94	18.30	
	4000	Channel pipe, 4" diameter	B-20	430	.056		1.89	1.49		3.38	4.40	
	4020	6" diameter		335	.072		3.32	1.91		5.23	6.60	
	4060	8" diameter	▼	295	.081		5.05	2.17		7.22	8.95	
	4080	12" diameter	B-21	280	.100	▼	11.10	2.74	.56	14.40	17.05	
		**02630	Storm Drainage**										
100	0010	**PIPING, STORM DRAINAGE, CORRUGATED METAL**										**100**
	0020	Not including excavation or backfill										
	2000	Corrugated metal pipe, galvanized and coated										
	2020	Bituminous coated with paved invert, 20' lengths										
	2040	8" diameter, 16 ga.	B-14	330	.145	L.F.	5.85	3.61	.54	10	12.65	
	2060	10" diameter, 16 ga.		260	.185		7.05	4.59	.69	12.33	15.60	
	2080	12" diameter, 16 ga.		210	.229		8.50	5.70	.85	15.05	19.10	
	2100	15" diameter, 16 ga.		200	.240		10.25	5.95	.89	17.09	21.50	
	2120	18" diameter, 16 ga.		190	.253		13.20	6.30	.94	20.44	25	
	2140	24" diameter, 14 ga.	▼	160	.300		16.20	7.45	1.12	24.77	30.50	
	2160	30" diameter, 14 ga.	B-13	120	.467		21	11.90	6.05	38.95	48.50	
	2180	36" diameter, 12 ga.		120	.467		31	11.90	6.05	48.95	59.50	
	2200	48" diameter, 12 ga.	▼	100	.560		47.50	14.25	7.25	69	82	
	2220	60" diameter, 10 ga.	B-13B	75	.747		61.50	19	12.95	93.45	111	
	2240	72" diameter, 8 ga.	"	45	1.244	▼	93.50	31.50	21.50	146.50	176	
	2500	Galvanized, uncoated, 20' lengths										
	2520	8" diameter, 16 ga.	B-14	355	.135	L.F.	4.63	3.36	.50	8.49	10.85	
	2540	10" diameter, 16 ga.		280	.171		5.10	4.26	.64	10	12.90	
	2560	12" diameter, 16 ga.		220	.218		5.80	5.40	.81	12.01	15.65	
	2580	15" diameter, 16 ga.		220	.218		7.80	5.40	.81	14.01	17.90	
	2600	18" diameter, 16 ga.		205	.234		10.30	5.80	.87	16.97	21.50	
	2620	24" diameter, 14 ga.	▼	175	.274		14.15	6.80	1.02	21.97	27	
	2640	30" diameter, 14 ga.	B-13	130	.431		18	10.95	5.55	34.50	43	
	2660	36" diameter, 12 ga.		130	.431		29.50	10.95	5.55	46	55.50	
	2680	48" diameter, 12 ga.	▼	110	.509		39.50	12.95	6.60	59.05	71	
	2690	60" diameter, 10 ga.	B-13B	78	.718	▼	62	18.30	12.45	92.75	110	
	2780	End sections, 8" diameter	B-14	24	2	Ea.	55	49.50	7.45	111.95	146	
	2785	10" diameter		22	2.182		56.50	54	8.10	118.60	155	
	2790	12" diameter		35	1.371		42.50	34	5.10	81.60	105	
	2800	18" diameter	▼	30	1.600		65.50	39.50	5.95	110.95	140	
	2810	24" diameter	B-13	25	2.240		95.50	57	29	181.50	225	
	2820	30" diameter	▼	25	2.240		182	57	29	268	320	

For expanded coverage of these items see *Means Heavy Construction Cost Data 2002*

2 SITE CONSTRUCTION

02630	Storm Drainage	CREW	DAILY OUTPUT	LABOR-HOURS	UNIT	MAT.	LABOR	EQUIP.	TOTAL	TOTAL INCL O&P	
							2002 BARE COSTS				
100	2825	36" diameter	B-13	20	2.800	Ea.	241	71.50	36	348.50	415	100
	2830	48" diameter	↓	10	5.600		500	143	72.50	715.50	845	
	2835	60" diameter	B-13B	5	11.200		650	285	194	1,129	1,375	
	2840	72" diameter	"	4	14	↓	1,175	355	243	1,773	2,100	
	3000	Corrugated galvanized or alum. oval arch culverts, coated & paved										
	3020	17" x 13", 16 ga., 15" equivalent	B-14	200	.240	L.F.	17.40	5.95	.89	24.24	29.50	
	3040	21" x 15", 16 ga., 18" equivalent		150	.320		22.50	7.95	1.19	31.64	38	
	3060	28" x 20", 14 ga., 24" equivalent		125	.384		32	9.55	1.43	42.98	52	
	3080	35" x 24", 14 ga., 30" equivalent	↓	100	.480		40	11.90	1.79	53.69	64.50	
	3100	42" x 29", 12 ga., 36" equivalent	B-13	100	.560		58.50	14.25	7.25	80	94.50	
	3120	49" x 33", 12 ga., 42" equivalent		90	.622		70.50	15.85	8.05	94.40	111	
	3140	57" x 38", 12 ga., 48" equivalent	↓	75	.747	↓	79.50	19	9.65	108.15	128	
	3160	Steel, plain oval arch culverts, plain										
	3180	17" x 13", 16 ga., 15" equivalent	B-14	225	.213	L.F.	9.55	5.30	.79	15.64	19.55	
	3200	21" x 15", 16 ga., 18" equivalent		175	.274		11.35	6.80	1.02	19.17	24	
	3220	28" x 20", 14 ga., 24" equivalent	↓	150	.320		18.20	7.95	1.19	27.34	33.50	
	3240	35" x 24", 14 ga., 30" equivalent	B-13	108	.519		23	13.20	6.70	42.90	53	
	3260	42" x 29", 12 ga., 36" equivalent		108	.519		37.50	13.20	6.70	57.40	69.50	
	3280	49" x 33", 12 ga., 42" equivalent		92	.609		44.50	15.50	7.90	67.90	81.50	
	3300	57" x 38", 12 ga., 48" equivalent	↓	75	.747	↓	39.50	19	9.65	68.15	83	
	3320	End sections, 17" x 13"		22	2.545	Ea.	52.50	65	33	150.50	194	
	3340	42" x 29"	↓	17	3.294	"	215	84	42.50	341.50	410	
	3360	Multi-plate arch, steel	B-20	1,690	.014	Lb.	.60	.38		.98	1.25	
200	0010	**CATCH BASINS OR MANHOLES** not including footing, excavation,										200
	0020	backfill, frame and cover										
	0050	Brick, 4' inside diameter, 4' deep	D-1	1	16	Ea.	285	430		715	980	
	0100	6' deep		.70	22.857		400	615		1,015	1,375	
	0150	8' deep		.50	32	↓	510	865		1,375	1,875	
	0200	For depths over 8', add		4	4	V.L.F.	105	108		213	282	
	0400	Concrete blocks (radial), 4' I.D., 4' deep		1.50	10.667	Ea.	233	288		521	695	
	0500	6' deep		1	16		310	430		740	1,000	
	0600	8' deep		.70	22.857	↓	385	615		1,000	1,375	
	0700	For depths over 8', add	↓	5.50	2.909	V.L.F.	39	78.50		117.50	163	
	0800	Concrete, cast in place, 4' x 4', 8" thick, 4' deep	C-14H	2	24	Ea.	355	715	17.75	1,087.75	1,525	
	0900	6' deep		1.50	32		515	950	23.50	1,488.50	2,100	
	1000	8' deep		1	48	↓	735	1,425	35.50	2,195.50	3,100	
	1100	For depths over 8', add		8	6	V.L.F.	85.50	178	4.44	267.94	380	
	1110	Precast, 4' I.D., 4' deep	B-22	4.10	7.317	Ea.	420	203	57.50	680.50	840	
	1120	6' deep		3	10		540	278	79	897	1,100	
	1130	8' deep		2	15	↓	625	415	118	1,158	1,475	
	1140	For depths over 8', add		16	1.875	V.L.F.	89	52	14.80	155.80	195	
	1150	5' I.D., 4' deep	B-6	3	8	Ea.	445	205	59.50	709.50	870	
	1160	6' deep		2	12		605	305	89.50	999.50	1,225	
	1170	8' deep		1.50	16	↓	760	410	119	1,289	1,600	
	1180	For depths over 8', add		12	2	V.L.F.	99	51	14.90	164.90	204	
	1190	6' I.D., 4' deep		2	12	Ea.	730	305	89.50	1,124.50	1,375	
	1200	6' deep		1.50	16		950	410	119	1,479	1,800	
	1210	8' deep		1	24	↓	1,175	615	179	1,969	2,425	
	1220	For depths over 8', add		8	3	V.L.F.	153	76.50	22.50	252	310	
	1250	Slab tops, precast, 8" thick										
	1300	4' diameter manhole	B-6	8	3	Ea.	156	76.50	22.50	255	315	
	1400	5' diameter manhole		7.50	3.200		297	82	24	403	475	
	1500	6' diameter manhole		7	3.429		355	87.50	25.50	468	555	
	1600	Frames & covers, C.I., 24" square, 500 lb.		7.80	3.077		215	78.50	23	316.50	380	
	1700	26" D shape, 600 lb.	↓	7	3.429	↓	350	87.50	25.50	463	550	

Important: See the Reference Section for critical supporting data - Reference Nos., Crews, & City Cost Indexes

02630	Storm Drainage	CREW	DAILY OUTPUT	LABOR-HOURS	UNIT	2002 BARE COSTS				TOTAL INCL O&P
						MAT.	LABOR	EQUIP.	TOTAL	
200 1800	Light traffic, 18″ diameter, 100 lb.	B-6	10	2.400	Ea.	116	61.50	17.85	195.35	242 **200**
1900	24″ diameter, 300 lb.		8.70	2.759		158	70.50	20.50	249	305
2000	36″ diameter, 900 lb.		5.80	4.138		385	106	31	522	615
2100	Heavy traffic, 24″ diameter, 400 lb.		7.80	3.077		161	78.50	23	262.50	325
2200	36″ diameter, 1150 lb.		3	8		470	205	59.50	734.50	900
2300	Mass. State standard, 26″ diameter, 475 lb.		7	3.429		360	87.50	25.50	473	560
2400	30″ diameter, 620 lb.		7	3.429		300	87.50	25.50	413	495
2500	Watertight, 24″ diameter, 350 lb.		7.80	3.077		254	78.50	23	355.50	425
2600	26″ diameter, 500 lb.		7	3.429		249	87.50	25.50	362	435
2700	32″ diameter, 575 lb.	↓	6	4	↓	515	102	30	647	755
2800	3 piece cover & frame, 10″ deep,									
2900	1200 lbs., for heavy equipment	B-6	3	8	Ea.	800	205	59.50	1,064.50	1,250
3000	Raised for paving 1-1/4″ to 2″ high,									
3100	4 piece expansion ring									
3200	20″ to 26″ diameter	1 Clab	3	2.667	Ea.	105	62.50		167.50	214
3300	30″ to 36″ diameter	″	3	2.667	″	147	62.50		209.50	260
3320	Frames and covers, existing, raised for paving, 2″, including									
3340	row of brick, concrete collar, up to 12″ wide frame	B-6	18	1.333	Ea.	32	34	9.90	75.90	99
3360	20″ to 26″ wide frame		11	2.182		42	56	16.25	114.25	150
3380	30″ to 36″ wide frame	↓	9	2.667		52.50	68	19.85	140.35	185
3400	Inverts, single channel brick	D-1	3	5.333		59	144		203	286
3500	Concrete		5	3.200		46	86.50		132.50	184
3600	Triple channel, brick		2	8		89.50	216		305.50	430
3700	Concrete	↓	3	5.333		78	144		222	305
3800	Steps, heavyweight cast iron, 7″ x 9″	1 Bric	40	.200		9.45	6.10		15.55	19.75
3900	8″ x 9″		40	.200		14.20	6.10		20.30	25
3928	12″ x 10-1/2″		40	.200		13.05	6.10		19.15	23.50
4000	Standard sizes, galvanized steel		40	.200		11.75	6.10		17.85	22.50
4100	Aluminum	↓	40	.200	↓	15	6.10		21.10	26

SITE CONSTRUCTION **2**

02740	Flexible Pavement	CREW	DAILY OUTPUT	LABOR-HOURS	UNIT	2002 BARE COSTS				TOTAL INCL O&P
						MAT.	LABOR	EQUIP.	TOTAL	
300 0010	**ASPHALTIC CONCRETE PAVEMENT** for highways									**300**
0020	and large paved areas									
0080	Binder course, 1-1/2″ thick	B-25	7,725	.011	S.Y.	2.07	.29	.22	2.58	2.96
0120	2″ thick		6,345	.014		2.75	.36	.27	3.38	3.87
0160	3″ thick		4,905	.018		4.09	.46	.35	4.90	5.60
0200	4″ thick	↓	4,140	.021		5.45	.55	.41	6.41	7.30
0300	Wearing course, 1″ thick	B-25B	10,575	.009		1.52	.24	.18	1.94	2.24
0340	1-1/2″ thick		7,725	.012		2.32	.33	.24	2.89	3.32
0380	2″ thick		6,345	.015		3.12	.40	.29	3.81	4.36
0420	2-1/2″ thick		5,480	.018		3.84	.46	.34	4.64	5.30
0460	3″ thick	↓	4,900	.020	↓	4.58	.51	.38	5.47	6.25
0800	Alternate method of figuring paving costs									
0810	Binder course, 1-1/2″ thick	B-25	630	.140	Ton	27	3.60	2.69	33.29	38.50
0811	2″ thick		690	.128		27	3.28	2.46	32.74	38
0812	3″ thick		800	.110		27	2.83	2.12	31.95	36.50
0813	4″ thick	↓	850	.104	↓	27	2.67	2	31.67	36.50

2 SITE CONSTRUCTION

02740 | Flexible Pavement

		CREW	DAILY OUTPUT	LABOR-HOURS	UNIT	2002 BARE COSTS				TOTAL INCL O&P
						MAT.	LABOR	EQUIP.	TOTAL	
300	0850	Wearing course, 1" thick	B-25B	575	.167	Ton	28.50	4.37	3.24	36.11	41.50
	0851	1-1/2" thick		630	.152		28.50	3.99	2.96	35.45	40.50
	0852	2" thick		690	.139		28.50	3.65	2.70	34.85	39.50
	0853	2-1/2" thick		745	.129		28.50	3.38	2.50	34.38	39
	0854	3" thick	↓	800	.120	↓	28.50	3.14	2.33	33.97	38.50
	1000	Pavement replacement over trench, 2" thick	B-37	90	.533	S.Y.	3.17	13.25	1.30	17.72	25.50
	1050	4" thick		70	.686		6.30	17.05	1.67	25.02	35
	1080	6" thick	↓	55	.873	↓	10	21.50	2.13	33.63	47
315	0011	**PAVING** Asphaltic concrete, parking lots & driveways									
	0020	6" stone base, 2" binder course, 1" topping	B-25C	9,000	.005	S.F.	1.21	.14	.16	1.51	1.73
	0300	Binder course, 1-1/2" thick		35,000	.001		.32	.04	.04	.40	.46
	0400	2" thick		25,000	.002		.41	.05	.06	.52	.59
	0500	3" thick		15,000	.003		.64	.08	.10	.82	.94
	0600	4" thick		10,800	.004		.83	.12	.13	1.08	1.25
	0800	Sand finish course, 3/4" thick		41,000	.001		.18	.03	.04	.25	.29
	0900	1" thick	↓	34,000	.001		.22	.04	.04	.30	.35
	1000	Fill pot holes, hot mix, 2" thick	B-16	4,200	.008		.42	.19	.11	.72	.87
	1100	4" thick		3,500	.009		.61	.22	.13	.96	1.17
	1120	6" thick	↓	3,100	.010		.83	.25	.15	1.23	1.47
	1140	Cold patch, 2" thick	B-51	3,000	.016		.48	.38	.05	.91	1.17
	1160	4" thick		2,700	.018		.90	.43	.06	1.39	1.72
	1180	6" thick	↓	1,900	.025	↓	1.41	.60	.08	2.09	2.58

02800 | Site Improvements and Amenities

02810 | Irrigation System

		CREW	DAILY OUTPUT	LABOR-HOURS	UNIT	2002 BARE COSTS				TOTAL INCL O&P
						MAT.	LABOR	EQUIP.	TOTAL	
800	0010	**SPRINKLER IRRIGATION SYSTEM** For lawns									
	0100	Golf course with fully automatic system	C-17	.05	1,600	9 holes	79,000	50,000		129,000	165,000
	0200	24' diam. head at 15' O.C incl. piping, auto oper., minimum	B-20	70	.343	Head	17.35	9.15		26.50	33.50
	0300	Maximum		40	.600		40	15.95		55.95	69
	0500	60' diam. head at 40' O.C. incl. piping, auto oper., minimum		28	.857		52.50	23		75.50	93.50
	0600	Maximum		23	1.043	↓	147	28		175	206
	0800	Residential system, custom, 1" supply		2,000	.012	S.F.	.26	.32		.58	.79
	0900	1-1/2" supply	↓	1,800	.013	"	.29	.35		.64	.87

02815 | Fountains

		CREW	DAILY OUTPUT	LABOR-HOURS	UNIT	2002 BARE COSTS				TOTAL INCL O&P
						MAT.	LABOR	EQUIP.	TOTAL	
225	0010	**FOUNTAINS/AERATORS**									
	0100	Pump w/controls									
	0200	Single phase, 100' chord, 1/2 H.P. pump	2 Skwk	4.40	3.636	Ea.	2,550	113		2,663	3,000
	0300	3/4 H.P. pump		4.30	3.721		2,925	115		3,040	3,375
	0400	1 H.P. pump		4.20	3.810		2,950	118		3,068	3,425
	0500	1-1/2 H.P. pump		4.10	3.902		3,050	121		3,171	3,550
	0600	2 H.P. pump		4	4		3,100	124		3,224	3,600
	0700	Three phase, 200' chord, 5 H.P. pump		3.90	4.103		7,475	127		7,602	8,425
	0800	7-1/2 H.P. pump		3.80	4.211		8,450	130		8,580	9,500
	0900	10 H.P. pump		3.70	4.324		9,425	134		9,559	10,600
	1000	15 H.P. pump		3.60	4.444		11,200	138		11,338	12,500
	1100	Nozzles, minimum		8	2		187	62		249	305
	1200	Maximum		8	2		246	62		308	370
	1300	Lights w/mounting kits, 200 watt	↓	18	.889		310	27.50		337.50	385

Important: See the Reference Section for critical supporting data - Reference Nos., Crews, & City Cost Indexes

02800 | Site Improvements and Amenities

			DAILY	LABOR-		2002 BARE COSTS				TOTAL	
	02815	**Fountains**	CREW	OUTPUT	HOURS	UNIT	MAT.	LABOR	EQUIP.	TOTAL	INCL O&P
225	1400	300 watt	2 Skwk	18	.889	Ea.	345	27.50		372.50	420	**225**
	1500	500 watt		18	.889		375	27.50		402.50	460	
	1600	Color blender	↓	12	1.333	↓	292	41.50		333.50	385	

02950 | Site Restoration & Rehabilitation

			DAILY	LABOR-		2002 BARE COSTS				TOTAL	
	02955	**Restoration of Underground Piping**	CREW	OUTPUT	HOURS	UNIT	MAT.	LABOR	EQUIP.	TOTAL	INCL O&P
100	0010	**PIPE INTERNAL CLEANING & INSPECTION**										**100**
	0100	Cleaning, pressure pipe systems										
	0120	Pig method, lengths 1000' to 10,000'										
	0140	4" diameter thru 24" diameter, minimum				L.F.				2	2.30	
	0160	Maximum				"				5	6	
	6000	Sewage/sanitary systems										
	6100	Power rodder with header & cutters										
	6110	Mobilization charge, minimum				Total				300	350	
	6120	Mobilization charge, maximum				"				700	800	
	6140	4" diameter				L.F.				1	1.15	
	6150	6" diameter								1.25	1.45	
	6160	8" diameter								1.50	1.75	
	6170	10" diameter								1.75	2	
	6180	12" diameter								1.90	2.20	
	6190	14" diameter								2	2.30	
	6200	16" diameter								2.25	2.60	
	6210	18" diameter								2.50	2.85	
	6220	20" diameter								2.60	3	
	6230	24" diameter								2.90	3.35	
	6240	30" diameter								3.50	4	
	6250	36" diameter								4	4.60	
	6260	48" diameter								4.30	4.95	
	6270	60" diameter								5.60	6.45	
	6280	72" diameter				↓				6.20	7.15	
	9000	Inspection, television camera with film										
	9060	500 linear feet				Total				1,000	1,150	
680	0010	**PIPE REPAIR**										**680**
	0020	Not including excavation or backfill										
	0100	Clamp, stainless steel, lightweight, for steel pipe										
	0110	3" long, 1/2" diameter pipe	1 Plum	34	.235	Ea.	7.80	8.45		16.25	21.50	
	0120	3/4" diameter pipe		32	.250		8.05	9		17.05	22.50	
	0130	1" diameter pipe		30	.267		8.65	9.60		18.25	24	
	0140	1-1/4" diameter pipe		28	.286		9.20	10.25		19.45	25.50	
	0150	1-1/2" diameter pipe		26	.308		9.50	11.05		20.55	27	
	0160	2" diameter pipe		24	.333		10.35	12		22.35	29.50	
	0170	2-1/2" diameter pipe		23	.348		11.25	12.50		23.75	31.50	
	0180	3" diameter pipe		22	.364		12.45	13.05		25.50	33.50	
	0190	3-1/2" diameter pipe	↓	21	.381		13.35	13.70		27.05	35	
	0200	4" diameter pipe	B-20	56	.429		13.90	11.40		25.30	33	
	0210	5" diameter pipe		53	.453		16	12.05		28.05	36.50	
	0220	6" diameter pipe		48	.500		17.75	13.30		31.05	40	
	0230	8" diameter pipe	↓	30	.800	↓	20.50	21.50		42	56	

02955	Restoration of Underground Piping	CREW	DAILY OUTPUT	LABOR- HOURS	UNIT	2002 BARE COSTS				TOTAL INCL O&P	
						MAT.	LABOR	EQUIP.	TOTAL		
0240	10" diameter pipe	B-20	28	.857	Ea.	53.50	23		76.50	94	680
0250	12" diameter pipe		24	1		57	26.50		83.50	104	
0260	14" diameter pipe		22	1.091		59	29		88	111	
0270	16" diameter pipe		20	1.200		64	32		96	121	
0280	18" diameter pipe		18	1.333		67.50	35.50		103	130	
0290	20" diameter pipe		16	1.500		74.50	40		114.50	144	
0300	24" diameter pipe	↓	14	1.714		83	45.50		128.50	162	
0360	For 6" long, add					100%	40%				
0370	For 9" long, add					200%	100%				
0380	For 12" long, add					300%	150%				
0390	For 18" long, add				↓	500%	200%				
1000	Clamp, stainless steel, with threaded service tap										
1040	Full seal for iron, steel, PVC pipe										
1100	6" long, 2" diameter pipe	1 Plum	17	.471	Ea.	101	16.90		117.90	137	
1110	2-1/2" diameter pipe		16	.500		107	18		125	144	
1120	3" diameter pipe		15.60	.513		114	18.45		132.45	153	
1130	3-1/2" diameter pipe	↓	15	.533		118	19.15		137.15	159	
1140	4" diameter pipe	B-20	40	.600		130	15.95		145.95	168	
1150	6" diameter pipe		34	.706		150	18.80		168.80	195	
1160	8" diameter pipe		21	1.143		187	30.50		217.50	254	
1170	10" diameter pipe		20	1.200		207	32		239	278	
1180	12" diameter pipe	↓	17	1.412		237	37.50		274.50	320	
1205	For 9" long, add					20%	45%				
1210	For 12" long, add					40%	80%				
1220	For 18" long, add				↓	70%	110%				
1600	Clamp, stainless steel, single section										
1640	Full seal for iron, steel, PVC pipe										
1700	6" long, 2" diameter pipe	1 Plum	17	.471	Ea.	52	16.90		68.90	83	
1710	2-1/2" diameter pipe		16	.500		57	18		75	89.50	
1720	3" diameter pipe		15.60	.513		59	18.45		77.45	93	
1730	3-1/2" diameter pipe	↓	15	.533		58.50	19.15		77.65	93.50	
1740	4" diameter pipe	B-20	40	.600		71	15.95		86.95	103	
1750	6" diameter pipe		34	.706		89	18.80		107.80	127	
1760	8" diameter pipe		21	1.143		118	30.50		148.50	178	
1770	10" diameter pipe		20	1.200		136	32		168	200	
1780	12" diameter pipe	↓	17	1.412		148	37.50		185.50	222	
1800	For 9" long, add					40%	45%				
1810	For 12" long, add					60%	80%				
1820	For 18" long, add				↓	120%	110%				
2000	Clamp, stainless steel, two section										
2040	Full seal, for iron, steel, PVC pipe										
2100	6" long, 4" diameter pipe	B-20	24	1	Ea.	101	26.50		127.50	153	
2110	6" diameter pipe		20	1.200		118	32		150	180	
2120	8" diameter pipe		13	1.846		130	49		179	220	
2130	10" diameter pipe		12	2		190	53		243	291	
2140	12" diameter pipe		10	2.400		207	64		271	330	
2200	9" long, 4" diameter pipe		16	1.500		130	40		170	205	
2210	6" diameter pipe		13	1.846		148	49		197	240	
2220	8" diameter pipe		9	2.667		154	71		225	280	
2230	10" diameter pipe		8	3		213	80		293	360	
2240	12" diameter pipe		7	3.429		237	91.50		328.50	405	
2250	14" diameter pipe		6.40	3.750		355	100		455	545	
2260	16" diameter pipe		6	4		375	106		481	575	
2270	18" diameter pipe		5	4.800		385	128		513	625	
2280	20" diameter pipe		4.60	5.217		415	139		554	670	
2290	24" diameter pipe	↓	4	6	↓	710	160		870	1,025	

Important: See the Reference Section for critical supporting data - Reference Nos., Crews, & City Cost Indexes

02955	Restoration of Underground Piping	CREW	DAILY OUTPUT	LABOR-HOURS	UNIT	2002 BARE COSTS MAT.	LABOR	EQUIP.	TOTAL	TOTAL INCL O&P	
680											**680**
2320	For 12" long, add to 9"				Ea.	15%	25%				
2330	For 18" long, add to 9"				↓	70%	55%				
8000	For internal cleaning and inspection, see Div. 02955-100										
8100	For pipe testing, see Div. 15955-700										
800 0010	**CORROSION RESISTANCE** Wrap & coat, add to pipe, 4" dia.				L.F.	1.35			1.35	1.49	**800**
0020	5" diameter					1.42			1.42	1.56	
0040	6" diameter					1.42			1.42	1.56	
0060	8" diameter					2.19			2.19	2.41	
0080	10" diameter					3.27			3.27	3.60	
0100	12" diameter					3.27			3.27	3.60	
0120	14" diameter					4.79			4.79	5.25	
0140	16" diameter					4.79			4.79	5.25	
0160	18" diameter					4.84			4.84	5.30	
0180	20" diameter					5.05			5.05	5.60	
0200	24" diameter					6.20			6.20	6.80	
0220	Small diameter pipe, 1" diameter, add					1.02			1.02	1.12	
0240	2" diameter					1.12			1.12	1.23	
0260	2-1/2" diameter					1.23			1.23	1.35	
0280	3" diameter				↓	1.23			1.23	1.35	
0300	Fittings, field covered, add				S.F.	6.50			6.50	7.15	
0500	Coating, bituminous, per diameter inch, 1 coat, add				L.F.	.25			.25	.28	
0540	3 coat					.39			.39	.43	
0560	Coal tar epoxy, per diameter inch, 1 coat, add					.15			.15	.17	
0600	3 coat					.31			.31	.34	
1000	Polyethylene H.D. extruded, .025" thk., 1/2" diameter add					.22			.22	.24	
1020	3/4" diameter					.23			.23	.25	
1040	1" diameter					.26			.26	.29	
1060	1-1/4" diameter					.31			.31	.34	
1080	1-1/2" diameter					.35			.35	.39	
1100	.030" thk., 2" diameter					.38			.38	.42	
1120	2-1/2" diameter					.45			.45	.50	
1140	.035" thk., 3" diameter					.55			.55	.61	
1160	3-1/2" diameter					.63			.63	.69	
1180	4" diameter					.70			.70	.77	
1200	.040" thk, 5" diameter					.97			.97	1.07	
1220	6" diameter					1.01			1.01	1.11	
1240	8" diameter					1.32			1.32	1.45	
1260	10" diameter					1.61			1.61	1.77	
1280	12" diameter					1.92			1.92	2.11	
1300	.060" thk., 14" diameter					2.65			2.65	2.92	
1320	16" diameter					3.11			3.11	3.42	
1340	18" diameter					3.67			3.67	4.04	
1360	20" diameter				↓	4.13			4.13	4.54	
1380	Fittings, field wrapped, add				S.F.	4.26			4.26	4.69	

For information about Means Estimating Seminars, see yellow pages 11 and 12 in back of book

SITE CONSTRUCTION **2**

Division Notes

	CREW	DAILY OUTPUT	LABOR-HOURS	UNIT	2002 BARE COSTS				TOTAL INCL O&P
					MAT.	LABOR	EQUIP.	TOTAL	

Division 3
Concrete

Estimating Tips

General
- Carefully check all the plans and specifications. Concrete often appears on drawings other than structural drawings, including mechanical and electrical drawings for equipment pads. The cost of cutting and patching is often difficult to estimate. See Subdivision 02225 for demolition costs.
- Always obtain concrete prices from suppliers near the job site. A volume discount can often be negotiated depending upon competition in the area. Remember to add for waste, particularly for slabs and footings on grade.

03100 Concrete Forms & Accessories
- A primary cost for concrete construction is forming. Most jobs today are constructed with prefabricated forms. The selection of the forms best suited for the job and the total square feet of forms required for efficient concrete forming and placing are key elements in estimating concrete construction. Enough forms must be available for erection to make efficient use of the concrete placing equipment and crew.
- Concrete accessories for forming and placing depend upon the systems used. Study the plans and specifications to assure that all special accessory requirements have been included in the cost estimate such as anchor bolts, inserts and hangers.

03200 Concrete Reinforcement
- Ascertain that the reinforcing steel supplier has included all accessories, cutting, bending and an allowance for lapping, splicing and waste. A good rule of thumb is 10% for lapping, splicing and waste. Also, 10% waste should be allowed for welded wire fabric.

03300 Cast-in-Place Concrete
- When estimating structural concrete, pay particular attention to requirements for concrete additives, curing methods and surface treatments. Special consideration for climate, hot or cold, must be included in your estimate. Be sure to include requirements for concrete placing equipment and concrete finishing.

03400 Precast Concrete
03500 Cementitious Decks & Toppings
- The cost of hauling precast concrete structural members is often an important factor. For this reason, it is important to get a quote from the nearest supplier. It may become economically feasible to set up precasting beds on the site if the hauling costs are prohibitive.

Reference Numbers
Reference numbers are shown in bold squares at the beginning of some major classifications. These numbers refer to related items in the Reference Section. The reference information may be an estimating procedure, an alternate pricing method or technical information.

Note: Not all subdivisions listed here necessarily appear in this publication.

03100 | Concrete Forms & Accessories

03110 | Structural C.I.P. Forms

			CREW	DAILY OUTPUT	LABOR-HOURS	UNIT	2002 BARE COSTS				TOTAL INCL O&P	
							MAT.	LABOR	EQUIP.	TOTAL		
425	0010	FORMS IN PLACE, EQUIPMENT FOUNDATIONS 1 use	C-2	160	.300	SFCA	2.62	8.75		11.37	16.55	425
	0050	2 use		190	.253		1.44	7.40		8.84	13.10	
	0100	3 use		200	.240		1.05	7		8.05	12.05	
	0150	4 use		205	.234		.86	6.85		7.71	11.60	
620	0010	ACCESSORIES, SLEEVES AND CHASES										620
	0100	Plastic, 1 use, 9" long, 2" diameter	1 Carp	100	.080	Ea.	.53	2.40		2.93	4.32	
	0150	4" diameter		90	.089		1.56	2.67		4.23	5.85	
	0200	6" diameter		75	.107		2.75	3.20		5.95	8	
	0250	12" diameter		60	.133		18.05	4		22.05	26	
	5000	Sheet metal, 2" diameter		100	.080		.68	2.40		3.08	4.49	
	5100	4" diameter		90	.089		.85	2.67		3.52	5.10	
	5150	6" diameter		75	.107		1.23	3.20		4.43	6.35	
	5200	12" diameter		60	.133		2.46	4		6.46	8.95	
	6000	Steel pipe, 2" diameter		100	.080		1.80	2.40		4.20	5.70	
	6100	4" diameter		90	.089		5.30	2.67		7.97	10	
	6150	6" diameter		75	.107		12.85	3.20		16.05	19.15	
	6200	12" diameter		60	.133		36.50	4		40.50	47	

03200 | Concrete Reinforcement

03210 | Reinforcing Steel

			CREW	DAILY OUTPUT	LABOR-HOURS	UNIT	2002 BARE COSTS				TOTAL INCL O&P	
							MAT.	LABOR	EQUIP.	TOTAL		
600	0010	REINFORCING IN PLACE A615 Grade 60										600
	0602	Slab on grade, #3 to #7	4 Rodm	4,200	.008	Lb.	.28	.26		.54	.75	

03300 | Cast-In-Place Concrete

03310 | Structural Concrete

			CREW	DAILY OUTPUT	LABOR-HOURS	UNIT	2002 BARE COSTS				TOTAL INCL O&P	
							MAT.	LABOR	EQUIP.	TOTAL		
220	0010	CONCRETE, READY MIX Regular weight										220
	0020	2000 psi				C.Y.	65			65	71.50	
	0100	2500 psi					58			58	64	
	0150	3000 psi					69			69	76	
	0200	3500 psi					71			71	78	
	0300	4000 psi					74			74	81.50	
	0350	4500 psi					75.50			75.50	83	
	0400	5000 psi					77			77	84.50	
	0411	6000 psi					88			88	96.50	
	0412	8000 psi					143			143	158	
	0413	10,000 psi					203			203	224	
	0414	12,000 psi					246			246	270	
	1000	For high early strength cement, add					10%					
	2000	For all lightweight aggregate, add					45%					
240	0010	CONCRETE IN PLACE Including forms (4 uses), reinforcing										240
	0050	steel, including finishing unless otherwise indicated										

Important: See the Reference Section for critical supporting data - Reference Nos., Crews, & City Cost Indexes

03310 | Structural Concrete

			CREW	DAILY OUTPUT	LABOR-HOURS	UNIT	MAT.	LABOR	EQUIP.	TOTAL	TOTAL INCL O&P	
							2002 BARE COSTS					
240	0500	Chimney foundations, industrial, minimum	C-14C	32.22	3.476	C.Y.	140	100	1.11	241.11	310	**240**
	0510	Maximum		23.71	4.724		163	136	1.51	300.51	395	
	3800	Footings, spread under 1 C.Y.		38.07	2.942		100	84.50	.94	185.44	244	
	3850	Over 5 C.Y.		81.04	1.382		92	40	.44	132.44	164	
	3900	Footings, strip, 18" x 9", plain		41.04	2.729		91	78.50	.87	170.37	225	
	3950	36" x 12", reinforced		61.55	1.820		92.50	52.50	.58	145.58	185	
	4000	Foundation mat, under 10 C.Y.		38.67	2.896		124	83.50	.93	208.43	268	
	4050	Over 20 C.Y.		56.40	1.986		110	57	.64	167.64	212	
	4650	Slab on grade, not including finish, 4" thick	C-14E	60.75	1.449		84.50	43	.58	128.08	163	
	4700	6" thick	"	92	.957		81	28.50	.38	109.88	136	
	4751	Slab on grade, incl. troweled finish, not incl. forms										
	4760	or reinforcing, over 10,000 S.F., 4" thick slab	C-14F	3,425	.021	S.F.	.92	.57	.01	1.50	1.89	
	4820	6" thick slab		3,350	.021		1.35	.58	.01	1.94	2.37	
	4840	8" thick slab		3,184	.023		1.85	.61	.01	2.47	2.96	
	4900	12" thick slab		2,734	.026		2.77	.72	.01	3.50	4.14	
	4950	15" thick slab		2,505	.029		3.48	.78	.01	4.27	5	
700	0010	**PLACING CONCRETE** and vibrating, including labor & equipment										**700**
	1900	Footings, continuous, shallow, direct chute	C-6	120	.400	C.Y.		9.85	.59	10.44	15.85	
	1950	Pumped	C-20	150	.427			10.80	5	15.80	22	
	2000	With crane and bucket	C-7	90	.800			20.50	11.85	32.35	44.50	
	2100	Footings, continuous, deep, direct chute	C-6	140	.343			8.45	.51	8.96	13.60	
	2150	Pumped	C-20	160	.400			10.15	4.70	14.85	21	
	2200	With crane and bucket	C-7	110	.655			16.65	9.70	26.35	36	
	2900	Foundation mats, over 20 C.Y., direct chute	C-6	350	.137			3.38	.20	3.58	5.40	
	2950	Pumped	C-20	400	.160			4.05	1.88	5.93	8.30	
	3000	With crane and bucket	C-7	300	.240			6.10	3.55	9.65	13.30	

03350 | Concrete Finishing

			CREW	DAILY OUTPUT	LABOR-HOURS	UNIT	MAT.	LABOR	EQUIP.	TOTAL	TOTAL INCL O&P	
300	0010	**FINISHING FLOORS** Monolithic, screed finish	1 Cefi	900	.009	S.F.		.26		.26	.38	**300**
	0100	Screed and bull float (darby) finish		725	.011			.32		.32	.47	
	0150	Screed, float, and broom finish		630	.013			.36		.36	.54	
600	0010	**PATCHING CONCRETE**										**600**
	0100	Floors, 1/4" thick, small areas, regular grout	1 Cefi	170	.047	S.F.	.07	1.35		1.42	2.07	
	0150	Epoxy grout	"	100	.080	"	3.19	2.30		5.49	6.90	
	0300	Slab on Grade, cut outs, up to 50 C.F.	2 Cefi	50	.320	C.F.	5.10	9.20		14.30	19.20	
	2000	Walls, including chipping, cleaning and epoxy grout										
	2100	Minimum	1 Cefi	65	.123	S.F.	3.21	3.53		6.74	8.80	
	2150	Average		50	.160		6.40	4.59		10.99	13.85	
	2200	Maximum		40	.200		12.85	5.75		18.60	22.50	
	2510	Underlayment, P.C. based self-leveling, 4100 psi, pumped, 1/4"	C-8	20,000	.003		1.27	.07	.03	1.37	1.55	
	2520	1/2"		19,000	.003		2.29	.08	.04	2.41	2.67	
	2530	3/4"		18,000	.003		3.56	.08	.04	3.68	4.08	
	2540	1"		17,000	.003		4.83	.09	.04	4.96	5.45	
	2550	1-1/2"		15,000	.004		7.35	.10	.05	7.50	8.30	
	2560	Hand mix, 1/2"	C-18	4,000	.002		2.29	.05	.01	2.35	2.60	
	2610	Topping, P.C. based self-level/dry 6100 psi, pumped, 1/4"	C-8	20,000	.003		1.82	.07	.03	1.92	2.15	
	2620	1/2"		19,000	.003		3.28	.08	.04	3.40	3.76	
	2630	3/4"		18,000	.003		5.10	.08	.04	5.22	5.75	
	2660	1"		17,000	.003		6.90	.09	.04	7.03	7.75	
	2670	1-1/2"		15,000	.004		10.55	.10	.05	10.70	11.80	
	2680	Hand mix, 1/2"	C-18	4,000	.002		3.28	.05	.01	3.34	3.69	

CONCRETE 3

For information about Means Estimating Seminars, see yellow pages 11 and 12 in back of book

Division Notes

	CREW	DAILY OUTPUT	LABOR-HOURS	UNIT	2002 BARE COSTS				TOTAL INCL O&P
					MAT.	LABOR	EQUIP.	TOTAL	

Division 4
Masonry

Estimating Tips

04050 Basic Masonry Materials & Methods

- The terms *mortar* and *grout* are often used interchangeably, and incorrectly. Mortar is used to bed masonry units, seal the entry of air and moisture, provide architectural appearance, and allow for size variations in the units. Grout is used primarily in reinforced masonry construction and is used to bond the masonry to the reinforcing steel. Common mortar types are M(2500 psi), S(1800 psi), N(750 psi), and O(350 psi), and conform to ASTM C270. Grout is either fine or coarse, conforms to ASTM C476, and in-place strengths generally exceed 2500 psi. Mortar and grout are different components of masonry construction and are placed by entirely different methods. An estimator should be aware of their unique uses and costs.

- Waste, specifically the loss/droppings of mortar and the breakage of brick and block, is included in all masonry assemblies in this division. A factor of 25% is added for mortar and 3% for brick and concrete masonry units.

04200 Masonry Units

- The most common types of unit masonry are brick and concrete masonry. The major classifications of brick are building brick (ASTM C62), facing brick (ASTM C216) and glazed brick, fire brick and pavers. Many varieties of texture and appearance can exist within these classifications, and the estimator would be wise to check local custom and availability within the project area. On repair and remodeling jobs, matching the existing brick may be the most important criteria.

- Brick and concrete block are priced by the piece and then converted into a price per square foot of wall. Openings less than two square feet are generally ignored by the estimator because any savings in units used is offset by the cutting and trimming required.

- It is often difficult and expensive to find and purchase small lots of historic brick. Costs can vary widely. Many design issues affect costs, selection of mortar mix, and repairs or replacement of masonry materials. Cleaning techniques must be reflected in the estimate.

- All masonry walls, whether interior or exterior, require bracing. The cost of bracing walls during construction should be included by the estimator and this bracing must remain in place until permanent bracing is complete. Permanent bracing of masonry walls is accomplished by masonry itself, in the form of pilasters or abutting wall corners, or by anchoring the walls to the structural frame. Accessories in the form of anchors, anchor slots and ties are used, but their supply and installation can be by different trades. For instance, anchor slots on spandrel beams and columns are supplied and welded in place by the steel fabricator, but the ties from the slots into the masonry are installed by the bricklayer. Regardless of the installation method the estimator must be certain that these accessories are accounted for in pricing.

Reference Numbers

Reference numbers are shown in bold squares at the beginning of some major classifications. These numbers refer to related items in the Reference Section. The reference information may be an estimating procedure, an alternate pricing method or technical information.

Note: Not all subdivisions listed here necessarily appear in this publication.

04090	Masonry Accessories	CREW	DAILY OUTPUT	LABOR-HOURS	UNIT	2002 BARE COSTS				TOTAL INCL O&P	
						MAT.	LABOR	EQUIP.	TOTAL		
860	**0010**	**VENT BOX** Extruded aluminum, 4" deep, 2-3/8" x 8-1/8"	1 Bric	30	.267	Ea.	30.50	8.15		38.65	46	**860**
	0050	5" x 8-1/8"		25	.320		42.50	9.75		52.25	61.50	
	0100	2-1/4" x 25"		25	.320		85	9.75		94.75	109	
	0150	5" x 16-1/2"		22	.364		44	11.10		55.10	65.50	
	0200	6" x 16-1/2"		22	.364		85	11.10		96.10	111	
	0250	7-3/4" x 16-1/2"		20	.400		80	12.20		92.20	107	
	0400	For baked enamel finish, add					35%					
	0500	For cast aluminum, painted, add					60%					
	1000	Stainless steel ventilators, 6" x 6"	1 Bric	25	.320		91	9.75		100.75	115	
	1050	8" x 8"		24	.333		96	10.15		106.15	122	
	1100	12" x 12"		23	.348		112	10.60		122.60	139	
	1150	12" x 6"		24	.333		96	10.15		106.15	122	
	1200	Foundation block vent, galv., 1-1/4" thk, 8" high, 16" long, no damper		30	.267		18	8.15		26.15	32.50	
	1250	For damper, add					6			6	6.60	

04800 | Masonry Assemblies

04810	Unit Masonry Assemblies	CREW	DAILY OUTPUT	LABOR-HOURS	UNIT	2002 BARE COSTS				TOTAL INCL O&P	
						MAT.	LABOR	EQUIP.	TOTAL		
160	**0010**	**CHIMNEY** See Div. 03310 for foundation, add to prices below										**160**
	1800	Metal, high temp. steel jacket, factory lining, 24" diam.	E-2	65	.862	V.L.F.	174	28.50	22	224.50	265	
	1900	60" diameter	"	30	1.867		630	62	47.50	739.50	855	
	2100	Poured concrete, brick lining, 200' high x 10' diam.					5,500			5,500	6,050	
	2200	200' high x 18' diameter									4,995	
	2400	250' high x 9' diameter									3,920	
	2500	300' x 14' diameter									4,690	
	2700	400' high x 18' diameter									5,255	
	2800	500' x 20' diameter					9,700			9,700	10,700	

For information about Means Estimating Seminars, see yellow pages 11 and 12 in back of book

Important: See the Reference Section for critical supporting data - Reference Nos., Crews, & City Cost Indexes

Division 5
Metals

Estimating Tips

05050 Basic Metal Materials & Methods

- Nuts, bolts, washers, connection angles and plates can add a significant amount to both the tonnage of a structural steel job as well as the estimated cost. As a rule of thumb add 10% to the total weight to account for these accessories.

- Type 2 steel construction, commonly referred to as "simple construction," consists generally of field bolted connections with lateral bracing supplied by other elements of the building, such as masonry walls or x-bracing. The estimator should be aware, however, that shop connections may be accomplished by welding or bolting. The method may be particular to the fabrication shop and may have an impact on the estimated cost.

05200 Metal Joists

- In any given project the total weight of open web steel joists is determined by the loads to be supported and the design. However, economies can be realized in minimizing the amount of labor used to place the joists. This is done by maximizing the joist spacing and therefore minimizing the number of joists required to be installed on the job. Certain spacings and locations may be required by the design, but in other cases maximizing the spacing and keeping it as uniform as possible will keep the costs down.

05300 Metal Deck

- The takeoff and estimating of metal deck involves more than simply the area of the floor or roof and the type of deck specified or shown on the drawings. Many different sizes and types of openings may exist. Small openings for individual pipes or conduits may be drilled after the floor/roof is installed, but larger openings may require special deck lengths as well as reinforcing or structural support. The estimator should determine who will be supplying this reinforcing. Additionally, some deck terminations are part of the deck package, such as screed angles and pour stops, and others will be part of the steel contract, such as angles attached to structural members and cast-in-place angles and plates. The estimator must ensure that all pieces are accounted for in the complete estimate.

05500 Metal Fabrications

- The most economical steel stairs are those that use common materials, standard details and most importantly, a uniform and relatively simple method of field assembly. Commonly available A36 channels and plates are very good choices for the main stringers of the stairs, as are angles and tees for the carrier members. Risers and treads are usually made by specialty shops, and it is most economical to use a typical detail in as many places as possible. The stairs should be pre-assembled and shipped directly to the site. The field connections should be simple and straightforward to be accomplished efficiently and with a minimum of equipment and labor.

Reference Numbers

Reference numbers are shown in bold squares at the beginning of some major classifications. These numbers refer to related items in the Reference Section. The reference information may be an estimating procedure, an alternate pricing method or technical information.

Note: Not all subdivisions listed here necessarily appear in this publication.

05090	Metal Fastenings	CREW	DAILY OUTPUT	LABOR-HOURS	UNIT	2002 BARE COSTS				TOTAL INCL O&P
						MAT.	LABOR	EQUIP.	TOTAL	
340	**0010** **DRILLING** For anchors, up to 4" deep, incl. bit and layout									**340**
	0050 in concrete or brick walls and floors, no anchor									
	0100 Holes, 1/4" diameter	1 Carp	75	.107	Ea.	.09	3.20		3.29	5.10
	0150 For each additional inch of depth, add		430	.019		.02	.56		.58	.89
	0200 3/8" diameter		63	.127		.08	3.81		3.89	6.05
	0250 For each additional inch of depth, add		340	.024		.02	.71		.73	1.12
	0300 1/2" diameter		50	.160		.08	4.80		4.88	7.55
	0350 For each additional inch of depth, add		250	.032		.02	.96		.98	1.51
	0400 5/8" diameter		48	.167		.14	5		5.14	7.95
	0450 For each additional inch of depth, add		240	.033		.03	1		1.03	1.60
	0500 3/4" diameter		45	.178		.14	5.35		5.49	8.45
	0550 For each additional inch of depth, add		220	.036		.04	1.09		1.13	1.74
	0600 7/8" diameter		43	.186		.18	5.60		5.78	8.90
	0650 For each additional inch of depth, add		210	.038		.04	1.14		1.18	1.83
	0700 1" diameter		40	.200		.21	6		6.21	9.60
	0750 For each additional inch of depth, add		190	.042		.05	1.26		1.31	2.03
	0800 1-1/4" diameter		38	.211		.30	6.30		6.60	10.20
	0850 For each additional inch of depth, add		180	.044		.08	1.33		1.41	2.16
	0900 1-1/2" diameter		35	.229		.49	6.85		7.34	11.20
	0950 For each additional inch of depth, add	▼	165	.048	▼	.12	1.45		1.57	2.39
	1000 For ceiling installations, add						40%			
	1100 Drilling & layout for drywall or plaster walls, no anchor									
	1200 Holes, 1/4" diameter	1 Carp	150	.053	Ea.	.01	1.60		1.61	2.50
	1300 3/8" diameter		140	.057		.01	1.71		1.72	2.68
	1400 1/2" diameter		130	.062		.01	1.85		1.86	2.88
	1500 3/4" diameter		120	.067		.02	2		2.02	3.13
	1600 1" diameter		110	.073		.03	2.18		2.21	3.43
	1700 1-1/4" diameter		100	.080		.04	2.40		2.44	3.78
	1800 1-1/2" diameter	▼	90	.089		.06	2.67		2.73	4.22
	1900 For ceiling installations, add				▼		40%			
380	**0010** **EXPANSION ANCHORS** & shields									**380**
	0100 Bolt anchors for concrete, brick or stone, no layout and drilling									
	0200 Expansion shields, zinc, 1/4" diameter, 1-5/16" long, single	1 Carp	90	.089	Ea.	.95	2.67		3.62	5.20
	0300 1-3/8" long, double		85	.094		1.04	2.82		3.86	5.55
	0400 3/8" diameter, 1-1/2" long, single		85	.094		1.56	2.82		4.38	6.10
	0500 2" long, double		80	.100		1.93	3		4.93	6.80
	0600 1/2" diameter, 2-1/16" long, single		80	.100		2.59	3		5.59	7.50
	0700 2-1/2" long, double		75	.107		2.50	3.20		5.70	7.75
	0800 5/8" diameter, 2-5/8" long, single		75	.107		3.70	3.20		6.90	9.05
	0900 2-3/4" long, double		70	.114		3.70	3.43		7.13	9.40
	1000 3/4" diameter, 2-3/4" long, single		70	.114		5.50	3.43		8.93	11.40
	1100 3-15/16" long, double	▼	65	.123		7.35	3.69		11.04	13.80
	1410 Concrete anchor, w/rod & epoxy cartridge, 1-3/4" diameter x 15" long	E-22	20	1.200		71	38		109	138
	1415 18" long		17	1.412		85	44.50		129.50	163
	1420 2" diameter x 18" long		16	1.500		109	47.50		156.50	193
	1425 24" long		15	1.600		142	50.50		192.50	235
	1430 Chemical anchor, w/rod & epoxy cartridge, 3/4" diam. x 9-1/2" long		27	.889		10.40	28		38.40	55.50
	1435 1" diameter x 11-3/4" long		24	1		19.80	31.50		51.30	71.50
	1440 1-1/4" diameter x 14" long	▼	21	1.143		37.50	36		73.50	98
	1500 Self drilling anchor, snap-off, for 1/4" diameter bolt	1 Carp	26	.308		.92	9.25		10.17	15.35
	1600 3/8" diameter bolt		23	.348		1.36	10.45		11.81	17.75
	1700 1/2" diameter bolt		20	.400		2.06	12		14.06	21
	1800 5/8" diameter bolt		18	.444		3.42	13.35		16.77	25
	1900 3/4" diameter bolt	▼	16	.500	▼	6.10	15		21.10	30
	2100 Hollow wall anchors for gypsum wall board, plaster or tile									
	2300 1/8" diameter, short	1 Carp	160	.050	Ea.	.27	1.50		1.77	2.64

Important: See the Reference Section for critical supporting data - Reference Nos., Crews, & City Cost Indexes

05090	Metal Fastenings	CREW	DAILY OUTPUT	LABOR-HOURS	UNIT	2002 BARE COSTS				TOTAL INCL O&P
						MAT.	LABOR	EQUIP.	TOTAL	
2400	Long	1 Carp	160	.050	Ea.	.30	1.50		1.80	2.67
2500	3/16" diameter, short		150	.053		.60	1.60		2.20	3.15
2600	Long		150	.053		.68	1.60		2.28	3.24
2700	1/4" diameter, short		140	.057		.81	1.71		2.52	3.56
2800	Long		140	.057		.91	1.71		2.62	3.67
3000	Toggle bolts, bright steel, 1/8" diameter, 2" long		85	.094		.21	2.82		3.03	4.63
3100	4" long		80	.100		.34	3		3.34	5.05
3200	3/16" diameter, 3" long		80	.100		.34	3		3.34	5.05
3300	6" long		75	.107		.49	3.20		3.69	5.50
3400	1/4" diameter, 3" long		75	.107		.38	3.20		3.58	5.40
3500	6" long		70	.114		.54	3.43		3.97	5.95
3600	3/8" diameter, 3" long		70	.114		.67	3.43		4.10	6.10
3700	6" long		60	.133		1.22	4		5.22	7.60
3800	1/2" diameter, 4" long		60	.133		1.94	4		5.94	8.40
3900	6" long		50	.160		3.26	4.80		8.06	11.05
5000	Screw anchors for concrete, masonry,									
5100	stone & tile, no layout or drilling included									
5200	Jute fiber, #6, #8, & #10, 1" long	1 Carp	240	.033	Ea.	.22	1		1.22	1.80
5300	#12, 1-1/2" long		200	.040		.31	1.20		1.51	2.21
5400	#14, 2" long		160	.050		.49	1.50		1.99	2.88
5500	#16, 2" long		150	.053		.51	1.60		2.11	3.05
5600	#20, 2" long		140	.057		.82	1.71		2.53	3.57
5700	Lag screw shields, 1/4" diameter, short		90	.089		.40	2.67		3.07	4.59
5800	Long		85	.094		.46	2.82		3.28	4.91
5900	3/8" diameter, short		85	.094		.65	2.82		3.47	5.10
6000	Long		80	.100		.81	3		3.81	5.55
6100	1/2" diameter, short		80	.100		1.12	3		4.12	5.90
6200	Long		75	.107		1.27	3.20		4.47	6.40
6300	3/4" diameter, short		70	.114		2.69	3.43		6.12	8.30
6400	Long		65	.123		3.25	3.69		6.94	9.35
6600	Lead, #6 & #8, 3/4" long		260	.031		.19	.92		1.11	1.65
6700	#10 - #14, 1-1/2" long		200	.040		.28	1.20		1.48	2.18
6800	#16 & #18, 1-1/2" long		160	.050		.39	1.50		1.89	2.77
6900	Plastic, #6 & #8, 3/4" long		260	.031		.04	.92		.96	1.48
7000	#8 & #10, 7/8" long		240	.033		.04	1		1.04	1.60
7100	#10 & #12, 1" long		220	.036		.05	1.09		1.14	1.76
7200	#14 & #16, 1-1/2" long		160	.050		.07	1.50		1.57	2.42
8000	Wedge anchors, not including layout or drilling									
8050	Carbon steel, 1/4" diameter, 1-3/4" long	1 Carp	150	.053	Ea.	.32	1.60		1.92	2.84
8100	3 1/4" long		145	.055		.42	1.66		2.08	3.04
8150	3/8" diameter, 2-1/4" long		150	.053		.61	1.60		2.21	3.16
8200	5" long		145	.055		1.07	1.66		2.73	3.76
8250	1/2" diameter, 2-3/4" long		140	.057		.85	1.71		2.56	3.60
8300	7" long		130	.062		1.45	1.85		3.30	4.47
8350	5/8" diameter, 3-1/2" long		130	.062		1.57	1.85		3.42	4.60
8400	8-1/2" long		115	.070		3.34	2.09		5.43	6.90
8450	3/4" diameter, 4-1/4" long		115	.070		2	2.09		4.09	5.45
8500	10" long		100	.080		4.55	2.40		6.95	8.75
8550	1" diameter, 6" long		100	.080		6.85	2.40		9.25	11.25
8575	9" long		80	.100		8.90	3		11.90	14.40
8600	12" long		80	.100		9.60	3		12.60	15.20
8650	1-1/4" diameter, 9" long		70	.114		13.40	3.43		16.83	20
8700	12" long		60	.133		17.15	4		21.15	25
8750	For type 303 stainless steel, add					350%				
8800	For type 316 stainless steel, add					450%				

380

5

METALS

		05090	**Metal Fastenings**	CREW	DAILY OUTPUT	LABOR-HOURS	UNIT	2002 BARE COSTS				TOTAL INCL O&P	
								MAT.	LABOR	EQUIP.	TOTAL		
460	0010		**LAG SCREWS**										**460**
	0020		Steel, 1/4" diameter, 2" long	1 Carp	200	.040	Ea.	.05	1.20		1.25	1.93	
	0100		3/8" diameter, 3" long		150	.053		.15	1.60		1.75	2.66	
	0200		1/2" diameter, 3" long		130	.062		.25	1.85		2.10	3.15	
	0300		5/8" diameter, 3" long		120	.067		.62	2		2.62	3.79	
500	0010		**MACHINE SCREWS**										**500**
	0020		Steel, round head, #8 x 1" long				C	1.85			1.85	2.04	
	0110		#8 x 2" long					4.39			4.39	4.83	
	0200		#10 x 1" long					2.79			2.79	3.07	
	0300		#10 x 2" long					5.30			5.30	5.85	
540	0010		**MACHINERY ANCHORS**, heavy duty, incl. sleeve, floating base nut,										**540**
	0020		lower stud & coupling nut, fiber plug, connecting stud, washer & nut.										
	0030		For flush mounted embedment in poured concrete heavy equip. pads.										
	0200		Material only, 1/2" diameter stud & bolt				Ea.	45			45	49.50	
	0300		5/8" diameter					50			50	55	
	0500		3/4" diameter					57.50			57.50	63	
	0600		7/8" diameter					62.50			62.50	69	
	0800		1" diameter					66			66	72.50	
	0900		1-1/4" diameter					87.50			87.50	96.50	
580	0010		**POWDER ACTUATED** Tools & fasteners										**580**
	0020		Stud driver, .22 caliber, buy, minimum				Ea.	278			278	305	
	0100		Maximum				"	515			515	570	
	0300		Powder charges for above, low velocity				C	16.65			16.65	18.35	
	0400		Standard velocity					25			25	27.50	
	0600		Drive pins & studs, 1/4" & 3/8" diam., to 3" long, minimum					21.50			21.50	24	
	0700		Maximum					53.50			53.50	59	
	0800		Pneumatic stud driver for 1/8" diameter studs				Ea.	1,875			1,875	2,050	
	0900		Drive pins for above, 1/2" to 3/4" long				M	410			410	450	
600	0010		**RIVETS**										**600**
	0100		Aluminum rivet & mandrel, 1/2" grip length x 1/8" diameter				C	4.53			4.53	4.98	
	0200		3/16" diameter					7			7	7.70	
	0300		Aluminum rivet, steel mandrel, 1/8" diameter					4.23			4.23	4.65	
	0400		3/16" diameter					8.90			8.90	9.80	
	0500		Copper rivet, steel mandrel, 1/8" diameter					5.50			5.50	6.05	
	1200		Steel rivet and mandrel, 1/8" diameter					4.74			4.74	5.20	
	1300		3/16" diameter					7.15			7.15	7.85	
	1400		Hand riveting tool, minimum				Ea.	88.50			88.50	97.50	
	1500		Maximum					465			465	510	
	1600		Power riveting tool, minimum					880			880	965	
	1700		Maximum					2,000			2,000	2,200	

05100 | Structural Metal Framing

		05120	**Structural Steel**	CREW	DAILY OUTPUT	LABOR-HOURS	UNIT	2002 BARE COSTS				TOTAL INCL O&P	
								MAT.	LABOR	EQUIP.	TOTAL		
440	0010		**LIGHTWEIGHT FRAMING**										**440**
	0400		Angle framing, field fabricated, 4" and larger	E-3	440	.055	Lb.	.36	1.90	.20	2.46	4.04	

Important: See the Reference Section for critical supporting data - Reference Nos., Crews, & City Cost Indexes

5 METALS

05120	Structural Steel	CREW	DAILY OUTPUT	LABOR-HOURS	UNIT	2002 BARE COSTS				TOTAL INCL O&P	
						MAT.	LABOR	EQUIP.	TOTAL		
440	0450	Less than 4" angles	E-3	265	.091	Lb.	.37	3.16	.33	3.86	6.45	440
	0600	Channel framing, field fabricated, 8" and larger		500	.048		.37	1.68	.18	2.23	3.61	
	0650	Less than 8" channels	↓	335	.072		.37	2.50	.26	3.13	5.20	
	1000	Continuous slotted channel framing system, shop fabricated, min	2 Sswk	2,400	.007		1.90	.23		2.13	2.50	
	1200	Maximum	"	1,600	.010		2.14	.34		2.48	2.98	
	1250	Plate & bar stock for reinforcing beams and trusses					.67			.67	.74	
	1300	Cross bracing, rods, shop fabricated, 3/4" diameter	E-3	700	.034		.74	1.20	.13	2.07	3.11	
	1310	7/8" diameter		850	.028		.74	.99	.10	1.83	2.70	
	1320	1" diameter		1,000	.024		.74	.84	.09	1.67	2.42	
	1330	Angle, 5" x 5" x 3/8"		2,800	.009		.74	.30	.03	1.07	1.38	
	1350	Hanging lintels, shop fabricated, average	↓	850	.028		.74	.99	.10	1.83	2.70	
	1380	Roof frames, shop fabricated, 3'-0" square, 5' span	E-2	4,200	.013		.74	.44	.34	1.52	1.94	
	1400	Tie rod, not upset, 1-1/2" to 4" diameter, with turnbuckle	2 Sswk	800	.020		.80	.69		1.49	2.12	
	1420	No turnbuckle		700	.023		.77	.78		1.55	2.25	
	1500	Upset, 1-3/4" to 4" diameter, with turnbuckle		800	.020		.80	.69		1.49	2.12	
	1520	No turnbuckle	↓	700	.023	↓	.77	.78		1.55	2.25	
520	0010	**PIPE SUPPORT FRAMING**										520
	0020	Under 10#/L.F.	E-4	3,900	.008	Lb.	.82	.29	.02	1.13	1.44	
	0200	10.1 to 15#/L.F.		4,300	.007		.81	.26	.02	1.09	1.38	
	0400	15.1 to 20#/L.F.		4,800	.007		.80	.23	.02	1.05	1.32	
	0600	Over 20#/L.F.	↓	5,400	.006	↓	.78	.21	.02	1.01	1.25	

For information about Means Estimating Seminars, see yellow pages 11 and 12 in back of book

METALS 5

		CREW	DAILY OUTPUT	LABOR-HOURS	UNIT	2002 BARE COSTS				TOTAL INCL O&P
						MAT.	LABOR	EQUIP.	TOTAL	

Division 6
Wood & Plastics

Estimating Tips

06050 Basic Wood & Plastic Materials & Methods

- Common to any wood framed structure are the accessory connector items such as screws, nails, adhesives, hangers, connector plates, straps, angles and holdowns. For typical wood framed buildings, such as residential projects, the aggregate total for these items can be significant, especially in areas where seismic loading is a concern. For floor and wall framing, the material cost is based on 10 to 25 lbs. per MBF. Holdowns, hangers and other connectors should be taken off by the piece.

06100 Rough Carpentry

- Lumber is a traded commodity and therefore sensitive to supply and demand in the marketplace. Even in "budgetary" estimating of wood framed projects, it is advisable to call local suppliers for the latest market pricing.
- Common quantity units for wood framed projects are "thousand board feet" (MBF). A board foot is a volume of wood, 1″ x 1′ x 1′, or 144 cubic inches. Board foot quantities are generally calculated using nominal material dimensions—dressed sizes are ignored. Board foot per lineal foot of any stick of lumber can be calculated by dividing the nominal cross sectional area by 12. As an example, 2,000 lineal feet of 2 x 12 equates to 4 MBF by dividing the nominal area, 2 x 12, by 12, which equals 2, and multiplying by 2,000 to give 4,000 board feet. This simple rule applies to all nominal dimensioned lumber.
- Waste is an issue of concern at the quantity takeoff for any area of construction. Framing lumber is sold in even foot lengths, i.e., 10′, 12′, 14′, 16′, and depending on spans, wall heights and the grade of lumber, waste is inevitable. A rule of thumb for lumber waste is 5% to 10% depending on material quality and the complexity of the framing.
- Wood in various forms and shapes is used in many projects, even where the main structural framing is steel, concrete or masonry. Plywood as a back-up partition material and 2x boards used as blocking and cant strips around roof edges are two common examples. The estimator should ensure that the costs of all wood materials are included in the final estimate.

06200 Finish Carpentry

- It is necessary to consider the grade of workmanship when estimating labor costs for erecting millwork and interior finish. In practice, there are three grades: premium, custom and economy. The Means daily output for base and case moldings is in the range of 200 to 250 L.F. per carpenter per day. This is appropriate for most average custom grade projects. For premium projects an adjustment to productivity of 25% to 50% should be made depending on the complexity of the job.

Reference Numbers

Reference numbers are shown in bold squares at the beginning of some major classifications. These numbers refer to related items in the Reference Section. The reference information may be an estimating procedure, an alternate pricing method or technical information.

Note: Not all subdivisions listed here necessarily appear in this publication.

		06055	**Wood & Plastic Laminate**	CREW	DAILY OUTPUT	LABOR-HOURS	UNIT	2002 BARE COSTS				TOTAL INCL O&P	
								MAT.	LABOR	EQUIP.	TOTAL		
720	0010		**CONVECTOR COVERS** Laminated plastic on 3/4"		16	.500	L.F.	20	15		35	45.50	**720**
	0020		thick particle board, 12" wide, minimum	1 Carp									
	0050		Average		13	.615		25.50	18.45		43.95	56.50	
	0100		Maximum		13	.615		35	18.45		53.45	67	
	0300		Add to above for grille, minimum		150	.053	S.F.	1.60	1.60		3.20	4.25	
	0400		Maximum		75	.107	"	4.50	3.20		7.70	9.95	

		06090	**Wood & Plastic Fastenings**										
600	0010		**NAILS** Prices of material only, based on 50# box purchase, copper, plain				Lb.	4.50			4.50	4.95	**600**
	0400		Stainless steel, plain					4.87			4.87	5.35	
	0500		Box, 3d to 20d, bright					1.17			1.17	1.29	
	0520		Galvanized					1.34			1.34	1.47	
	0600		Common, 3d to 60d, plain					.77			.77	.85	
	0700		Galvanized					1.18			1.18	1.30	
	0800		Aluminum					3.31			3.31	3.64	
	1000		Annular or spiral thread, 4d to 60d, plain					1.74			1.74	1.91	
	1200		Galvanized					1.67			1.67	1.84	
	1400		Drywall nails, plain					1.33			1.33	1.46	
	1600		Galvanized					1.33			1.33	1.46	
	1800		Finish nails, 4d to 10d, plain					.94			.94	1.03	
	2000		Galvanized					1.04			1.04	1.14	
	2100		Aluminum					4.85			4.85	5.35	
	2300		Flooring nails, hardened steel, 2d to 10d, plain					1.22			1.22	1.34	
	2400		Galvanized					2.17			2.17	2.39	
	2500		Gypsum lath nails, 1-1/8", 13 ga. flathead, blued					1.33			1.33	1.46	
	2600		Masonry nails, hardened steel, 3/4" to 3" long, plain					1.71			1.71	1.88	
	2700		Galvanized					1.71			1.71	1.88	
	5000		Add to prices above for cement coating					.07			.07	.08	
	5200		Zinc or tin plating					.12			.12	.13	
	5500		Vinyl coated sinkers, 8d to 16d					.55			.55	.61	
650	0010		**NAILS** mat. only, for pneumatic tools, framing, per carton of 5000, 2"				Ea.	36.50			36.50	40	**650**
	0100		2-3/8"					41.50			41.50	45.50	
	0200		Per carton of 4000, 3"					37			37	40.50	
	0300		3-1/4"					39			39	43	
	0400		Per carton of 5000, 2-3/8", galv.					56.50			56.50	62	
	0500		Per carton of 4000, 3", galv.					63.50			63.50	70	
	0600		3-1/4", galv.					79			79	87	
	0700		Roofing, per carton of 7200, 1"					34.50			34.50	38	
	0800		1-1/4"					32			32	35.50	
	0900		1-1/2"					37			37	40.50	
	1000		1-3/4"					44.50			44.50	49	
700	0010		**SHEET METAL SCREWS** Steel, standard, #8 x 3/4", plain				C	2.69			2.69	2.96	**700**
	0100		Galvanized					3.37			3.37	3.71	
	0300		#10 x 1", plain					3.69			3.69	4.06	
	0400		Galvanized					4.26			4.26	4.69	
	0600		With washers, #14 x 1", plain					10			10	11	
	0700		Galvanized					10			10	11	
	0900		#14 x 2", plain					10			10	11	
	1000		Galvanized					17.80			17.80	19.55	
	1500		Self-drilling, with washers, (pinch point) #8 x 3/4", plain					6			6	6.60	
	1600		Galvanized					6			6	6.60	
	1800		#10 x 3/4", plain					6			6	6.60	
	1900		Galvanized					6			6	6.60	
	3000		Stainless steel w/aluminum or neoprene washers, #14 x 1", plain					18			18	19.80	
	3100		#14 x 2", plain					24.50			24.50	27	

Important: See the Reference Section for critical supporting data - Reference Nos., Crews, & City Cost Indexes

06090	Wood & Plastic Fastenings	CREW	DAILY OUTPUT	LABOR-HOURS	UNIT	2002 BARE COSTS				TOTAL INCL O&P	
						MAT.	LABOR	EQUIP.	TOTAL		
750	0010	**WOOD SCREWS** #8, 1" long, steel				C	3.29			3.29	3.62	750
	0100	Brass					11			11	12.10	
	0200	#8, 2" long, steel					3.69			3.69	4.06	
	0300	Brass					11.50			11.50	12.65	
	0400	#10, 1" long, steel					4.22			4.22	4.64	
	0500	Brass					22.50			22.50	24.50	
	0600	#10, 2" long, steel					7.50			7.50	8.25	
	0700	Brass					39.50			39.50	43.50	
	0800	#10, 3" long, steel					11.75			11.75	12.90	
	1000	#12, 2" long, steel					4.79			4.79	5.25	
	1100	Brass					16			16	17.60	
	1500	#12, 3" long, steel					15.90			15.90	17.50	
	2000	#12, 4" long, steel					28.50			28.50	31	

6 | **WOOD & PLASTICS**

06100 | Rough Carpentry

06160	Sheathing	CREW	DAILY OUTPUT	LABOR-HOURS	UNIT	2002 BARE COSTS				TOTAL INCL O&P	
						MAT.	LABOR	EQUIP.	TOTAL		
800	0010	**SHEATHING** Plywood on roof, CDX										800
	0030	5/16" thick	2 Carp	1,600	.010	S.F.	.43	.30		.73	.95	
	0035	Pneumatic nailed		1,952	.008		.43	.25		.68	.86	
	0050	3/8" thick		1,525	.010		.44	.31		.75	.97	
	0055	Pneumatic nailed		1,860	.009		.44	.26		.70	.88	
	0100	1/2" thick		1,400	.011		.49	.34		.83	1.07	
	0105	Pneumatic nailed		1,708	.009		.49	.28		.77	.98	
	0200	5/8" thick		1,300	.012		.57	.37		.94	1.19	
	0205	Pneumatic nailed		1,586	.010		.57	.30		.87	1.09	
	0300	3/4" thick		1,200	.013		.66	.40		1.06	1.35	
	0305	Pneumatic nailed		1,464	.011		.66	.33		.99	1.24	
	0500	Plywood on walls with exterior CDX, 3/8" thick		1,200	.013		.44	.40		.84	1.10	
	0505	Pneumatic nailed		1,488	.011		.44	.32		.76	.98	
	0600	1/2" thick		1,125	.014		.49	.43		.92	1.20	
	0605	Pneumatic nailed		1,395	.011		.49	.34		.83	1.08	
	0700	5/8" thick		1,050	.015		.57	.46		1.03	1.33	
	0705	Pneumatic nailed		1,302	.012		.57	.37		.94	1.19	
	0800	3/4" thick		975	.016		.66	.49		1.15	1.50	
	0805	Pneumatic nailed		1,209	.013		.66	.40		1.06	1.35	

06400 | Architectural Woodwork

06440	Wood Ornaments	CREW	DAILY OUTPUT	LABOR-HOURS	UNIT	2002 BARE COSTS				TOTAL INCL O&P	
						MAT.	LABOR	EQUIP.	TOTAL		
400	0010	**LOUVERS** Redwood, 2'-0" diameter, full circle	1 Carp	16	.500	Ea.	105	15		120	140	400
	0100	Half circle		16	.500		104	15		119	138	
	0200	Octagonal		16	.500		86	15		101	118	
	0300	Triangular, 5/12 pitch, 5'-0" at base		16	.500		180	15		195	222	

		CREW	DAILY OUTPUT	LABOR-HOURS	UNIT	2002 BARE COSTS				TOTAL INCL O&P
						MAT.	LABOR	EQUIP.	TOTAL	

Division 7
Thermal & Moisture Protection

Estimating Tips

07100 Dampproofing & Waterproofing

- Be sure of the job specifications before pricing this subdivision. The difference in cost between waterproofing and dampproofing can be great. Waterproofing will hold back standing water. Dampproofing prevents the transmission of water vapor. Also included in this section are vapor retarding membranes.

07200 Thermal Protection

- Insulation and fireproofing products are measured by area, thickness, volume or R value. Specifications may only give what the specific R value should be in a certain situation. The estimator may need to choose the type of insulation to meet that R value.

07300 Shingles, Roof Tiles & Roof Coverings
07400 Roofing & Siding Panels

- Many roofing and siding products are bought and sold by the square. One square is equal to an area that measures 100 square feet.

This simple change in unit of measure could create a large error if the estimator is not observant. Accessories and fasteners necessary for a complete installation must be figured into any calculations for both material and labor.

07500 Membrane Roofing
07600 Flashing & Sheet Metal
07700 Roof Specialties & Accessories

- The items in these subdivisions compose a roofing system. No one component completes the installation and all must be estimated. Built-up or single ply membrane roofing systems are made up of many products and installation trades. Wood blocking at roof perimeters or penetrations, parapet coverings, reglets, roof drains, gutters, downspouts, sheet metal flashing, skylights, smoke vents or roof hatches all need to be considered along with the roofing material. Several different installation trades will need to work together on the roofing system. Inherent difficulties in the scheduling and coordination of various trades must be accounted for when estimating labor costs.

07900 Joint Sealers

- To complete the weather-tight shell the sealants and caulkings must be estimated. Where different materials meet—at expansion joints, at flashing penetrations, and at hundreds of other locations throughout a construction project—they provide another line of defense against water penetration. Often, an entire system is based on the proper location and placement of caulking or sealants. The detail drawings that are included as part of a set of architectural plans, show typical locations for these materials. When caulking or sealants are shown at typical locations, this means the estimator must include them for all the locations where this detail is applicable. Be careful to keep different types of sealants separate, and remember to consider backer rods and primers if necessary.

Reference Numbers

Reference numbers are shown in bold squares at the beginning of some major classifications. These numbers refer to related items in the Reference Section. The reference information may be an estimating procedure, an alternate pricing method or technical information.

Note: Not all subdivisions listed here necessarily appear in this publication.

		07650	**Flexible Flashing**	CREW	DAILY OUTPUT	LABOR-HOURS	UNIT	2002 BARE COSTS				TOTAL INCL O&P	
								MAT.	LABOR	EQUIP.	TOTAL		
600	0010		**FLASHING** Aluminum, mill finish, .013" thick	1 Rofc	145	.055	S.F.	.35	1.47		1.82	2.89	600
	0030		.016" thick		145	.055		.51	1.47		1.98	3.06	
	0060		.019" thick		145	.055		.77	1.47		2.24	3.35	
	0100		.032" thick		145	.055		1.04	1.47		2.51	3.64	
	0200		.040" thick		145	.055		1.45	1.47		2.92	4.10	
	0300		.050" thick		145	.055		1.84	1.47		3.31	4.52	
	0400		Painted finish, add					.24			.24	.26	
	1300		Asphalt flashing cement, 5 gallon				Gal.	3.90			3.90	4.29	
	1600		Copper, 16 oz, sheets, under 1000 lbs.	1 Rofc	115	.070	S.F.	2.30	1.85		4.15	5.70	
	1700		Over 4000 lbs.		155	.052		2.91	1.37		4.28	5.55	
	1900		20 oz sheets, under 1000 lbs.		110	.073		3.90	1.93		5.83	7.60	
	2000		Over 4000 lbs.		145	.055		3.62	1.47		5.09	6.50	
	2200		24 oz sheets, under 1000 lbs.		105	.076		4.69	2.03		6.72	8.60	
	2300		Over 4000 lbs.		135	.059		4.34	1.58		5.92	7.45	
	2500		32 oz sheets, under 1000 lbs.		100	.080		6.20	2.13		8.33	10.45	
	2600		Over 4000 lbs.		130	.062		5.80	1.64		7.44	9.20	
	4300		Copper-clad stainless steel, .015" thick, under 500 lbs.		115	.070		3.16	1.85		5.01	6.65	
	4400		Over 2000 lbs.		155	.052		3.05	1.37		4.42	5.70	
	4600		.018" thick, under 500 lbs.		100	.080		4.18	2.13		6.31	8.20	
	4700		Over 2000 lbs.		145	.055		3.06	1.47		4.53	5.85	
	5800		Lead, 2.5 lb. per SF, up to 12" wide		135	.059		2.86	1.58		4.44	5.85	
	5900		Over 12" wide		135	.059		3.25	1.58		4.83	6.25	
	8500		Shower pan, bituminous membrane, 7 oz		155	.052		1.08	1.37		2.45	3.53	
	8550		3 ply copper and fabric, 3 oz		155	.052		1.63	1.37		3	4.13	
	8600		7 oz		155	.052		3.37	1.37		4.74	6.05	
	8650		Copper, 16 oz		100	.080		3.11	2.13		5.24	7.05	
	8700		Lead on copper and fabric, 5 oz		155	.052		1.62	1.37		2.99	4.12	
	8800		7 oz		155	.052		2.93	1.37		4.30	5.55	
	8850		Polyvinyl chloride, .030" thick		160	.050		.30	1.33		1.63	2.59	
	8900		Stainless steel sheets, 32 ga, .010" thick		155	.052		2.16	1.37		3.53	4.72	
	9000		28 ga, .015" thick		155	.052		2.55	1.37		3.92	5.15	
	9100		26 ga, .018" thick		155	.052		3.25	1.37		4.62	5.90	
	9200		24 ga, .025" thick		155	.052		4.22	1.37		5.59	7	
	9400		Terne coated stainless steel, .015" thick, 28 ga		155	.052		4.05	1.37		5.42	6.80	
	9500		.018" thick, 26 ga		155	.052		4.57	1.37		5.94	7.40	
	9600		Zinc and copper alloy (brass), .020" thick		155	.052		3.26	1.37		4.63	5.95	
	9700		.027" thick		155	.052		4.37	1.37		5.74	7.15	
	9800		.032" thick		155	.052		5.10	1.37		6.47	7.95	
	9900		.040" thick		155	.052		6.20	1.37		7.57	9.20	

07700 | Roof Specialties & Accessories

		07710	**Manufactured Roof Specialties**	CREW	DAILY OUTPUT	LABOR-HOURS	UNIT	2002 BARE COSTS				TOTAL INCL O&P	
								MAT.	LABOR	EQUIP.	TOTAL		
400	0010		**DOWNSPOUTS** Aluminum 2" x 3", .020" thick, embossed	1 Shee	190	.042	L.F.	.74	1.48		2.22	3.09	400
	0100		Enameled		190	.042		.91	1.48		2.39	3.28	
	4800		Steel, galvanized, round, corrugated, 2" or 3" diam, 28 ga		190	.042		.73	1.48		2.21	3.08	
	4900		4" diameter, 28 gauge		145	.055		1.07	1.94		3.01	4.16	
	5100		5" diameter, 28 gauge		130	.062		1.40	2.16		3.56	4.87	
	5200		26 gauge		130	.062		1.36	2.16		3.52	4.83	

7

THERMAL & MOISTURE PROTECTION

Important: See the Reference Section for critical supporting data - Reference Nos., Crews, & City Cost Indexes

07700 | Roof Specialties & Accessories

07710	Manufactured Roof Specialties	CREW	DAILY OUTPUT	LABOR-HOURS	UNIT	2002 BARE COSTS				TOTAL INCL O&P	
						MAT.	LABOR	EQUIP.	TOTAL		
400	5400	6" diameter, 28 gauge	1 Shee	105	.076	L.F.	1.94	2.67		4.61	6.25	**400**
	5500	26 gauge		105	.076		1.42	2.67		4.09	5.70	
	7500	Steel pipe, black, extra heavy, 4" diameter		20	.400		5.30	14.05		19.35	27.50	
	7600	6" diameter		18	.444		12.85	15.60		28.45	38	

07800 | Fire & Smoke Protection

07840	Firestopping		CREW	DAILY OUTPUT	LABOR-HOURS	UNIT	2002 BARE COSTS				TOTAL INCL O&P	
							MAT.	LABOR	EQUIP.	TOTAL		
100	0010	**FIRESTOPPING**	R07800 -030									**100**
	0100	Metallic piping, non insulated										
	0110	Through walls, 2" diameter	1 Carp	16	.500	Ea.	9.60	15		24.60	34	
	0120	4" diameter		14	.571		14.65	17.15		31.80	42.50	
	0130	6" diameter		12	.667		19.70	20		39.70	52.50	
	0140	12" diameter		10	.800		35	24		59	76	
	0150	Through floors, 2" diameter		32	.250		5.80	7.50		13.30	18.10	
	0160	4" diameter		28	.286		8.35	8.55		16.90	22.50	
	0170	6" diameter		24	.333		11	10		21	27.50	
	0180	12" diameter		20	.400		18.50	12		30.50	39	
	0190	Metallic piping, insulated										
	0200	Through walls, 2" diameter	1 Carp	16	.500	Ea.	13.60	15		28.60	38.50	
	0210	4" diameter		14	.571		18.65	17.15		35.80	47	
	0220	6" diameter		12	.667		23.50	20		43.50	57	
	0230	12" diameter		10	.800		39	24		63	80	
	0240	Through floors, 2" diameter		32	.250		9.80	7.50		17.30	22.50	
	0250	4" diameter		28	.286		12.35	8.55		20.90	27	
	0260	6" diameter		24	.333		15	10		25	32	
	0270	12" diameter		20	.400		18.50	12		30.50	39	
	0280	Non metallic piping, non insulated										
	0290	Through walls, 2" diameter	1 Carp	12	.667	Ea.	39.50	20		59.50	74.50	
	0300	4" diameter		10	.800		49.50	24		73.50	92	
	0310	6" diameter		8	1		69	30		99	123	
	0330	Through floors, 2" diameter		16	.500		31	15		46	57.50	
	0340	4" diameter		6	1.333		38.50	40		78.50	105	
	0350	6" diameter		6	1.333		46	40		86	113	
	0370	Ductwork, insulated & non insulated, round										
	0380	Through walls, 6" diameter	1 Carp	12	.667	Ea.	20	20		40	53	
	0390	12" diameter		10	.800		40	24		64	81.50	
	0400	18" diameter		8	1		65	30		95	118	
	0410	Through floors, 6" diameter		16	.500		11	15		26	35.50	
	0420	12" diameter		14	.571		20	17.15		37.15	48.50	
	0430	18" diameter		12	.667		35	20		55	69.50	
	0440	Ductwork, insulated & non insulated, rectangular										
	0450	With stiffener/closure angle, through walls, 6" x 12"	1 Carp	8	1	Ea.	16.65	30		46.65	65	
	0460	12" x 24"		6	1.333		22	40		62	87	
	0470	24" x 48"		4	2		63	60		123	163	
	0480	With stiffener/closure angle, through floors, 6" x 12"		10	.800		9	24		33	47.50	
	0490	12" x 24"		8	1		16.20	30		46.20	64.50	
	0500	24" x 48"		6	1.333		32	40		72	97.50	
	0510	Multi trade openings										
	0520	Through walls, 6" x 12"	1 Carp	2	4	Ea.	35	120		155	226	

THERMAL & MOISTURE PROTECTION

7

		07840	**Firestopping**		CREW	DAILY OUTPUT	LABOR-HOURS	UNIT	2002 BARE COSTS				TOTAL INCL O&P	
									MAT.	LABOR	EQUIP.	TOTAL		
100	0530		12" x 24"		1 Carp	1	8	Ea.	141	240		381	530	100
	0540		24" x 48"	R07800-030	2 Carp	1	16		565	480		1,045	1,375	
	0550		48" x 96"		"	.75	21.333		2,275	640		2,915	3,500	
	0560		Through floors, 6" x 12"		1 Carp	2	4		35	120		155	226	
	0570		12" x 24"		"	1	8		141	240		381	530	
	0580		24" x 48"		2 Carp	.75	21.333		565	640		1,205	1,625	
	0590		48" x 96"		"	.50	32		2,275	960		3,235	4,000	
	0600		Structural penetrations, through walls											
	0610		Steel beams, W8 x 10		1 Carp	8	1	Ea.	22	30		52	70.50	
	0620		W12 x 14			6	1.333		35	40		75	101	
	0630		W21 x 44			5	1.600		70	48		118	152	
	0640		W36 x 135			3	2.667		170	80		250	310	
	0650		Bar joists, 18" deep			6	1.333		32	40		72	97.50	
	0660		24" deep			6	1.333		40	40		80	107	
	0670		36" deep			5	1.600		60	48		108	141	
	0680		48" deep			4	2		70	60		130	171	
	0690		Construction joints, floor slab at exterior wall											
	0700		Precast, brick, block or drywall exterior											
	0710		2" wide joint		1 Carp	125	.064	L.F.	5	1.92		6.92	8.50	
	0720		4" wide joint		"	75	.107	"	10	3.20		13.20	16	
	0730		Metal panel, glass or curtain wall exterior											
	0740		2" wide joint		1 Carp	40	.200	L.F.	11.85	6		17.85	22.50	
	0750		4" wide joint		"	25	.320	"	16.15	9.60		25.75	33	
	0760		Floor slab to drywall partition											
	0770		Flat joint		1 Carp	100	.080	L.F.	4.90	2.40		7.30	9.15	
	0780		Fluted joint			50	.160		10	4.80		14.80	18.45	
	0790		Etched fluted joint			75	.107		6.50	3.20		9.70	12.15	
	0800		Floor slab to concrete/masonry partition											
	0810		Flat joint		1 Carp	75	.107	L.F.	11	3.20		14.20	17.10	
	0820		Fluted joint		"	50	.160	"	13	4.80		17.80	22	
	0830		Concrete/CMU wall joints											
	0840		1" wide		1 Carp	100	.080	L.F.	6	2.40		8.40	10.35	
	0850		2" wide			75	.107		11	3.20		14.20	17.10	
	0860		4" wide			50	.160		21	4.80		25.80	30.50	
	0870		Concrete/CMU floor joints											
	0880		1" wide		1 Carp	200	.040	L.F.	3	1.20		4.20	5.15	
	0890		2" wide			150	.053		5.50	1.60		7.10	8.55	
	0900		4" wide			100	.080		10.50	2.40		12.90	15.30	

		07920	**Joint Sealants**	CREW	DAILY OUTPUT	LABOR-HOURS	UNIT	2002 BARE COSTS				TOTAL INCL O&P	
								MAT.	LABOR	EQUIP.	TOTAL		
800	0010		**CAULKING AND SEALANTS**										800
	0020		Acoustical sealant, elastomeric, cartridges				Ea.	2.05			2.05	2.26	
	0030		Backer rod, polyethylene, 1/4" diameter	1 Bric	4.60	1.739	C.L.F.	1.29	53		54.29	83	
	0050		1/2" diameter		4.60	1.739		2.64	53		55.64	84.50	
	0070		3/4" diameter		4.60	1.739		5.10	53		58.10	87	
	0090		1" diameter		4.60	1.739		8.10	53		61.10	90.50	
	0100		Acrylic latex caulk, white										
	0200		11 fl. oz cartridge				Ea.	1.83			1.83	2.01	

07920	Joint Sealants	CREW	DAILY OUTPUT	LABOR-HOURS	UNIT	MAT.	LABOR	EQUIP.	TOTAL	TOTAL INCL O&P	
800						2002 BARE COSTS					800
0500	1/4" x 1/2"	1 Bric	248	.032	L.F.	.15	.98		1.13	1.67	
0600	1/2" x 1/2"		250	.032		.30	.98		1.28	1.83	
0800	3/4" x 3/4"		230	.035		.67	1.06		1.73	2.37	
0900	3/4" x 1"		200	.040		.90	1.22		2.12	2.86	
1000	1" x 1"	▼	180	.044	▼	1.12	1.36		2.48	3.31	
1400	Butyl based, bulk				Gal.	22			22	24.50	
1500	Cartridges				"	27			27	29.50	
1700	Bulk, in place 1/4" x 1/2", 154 L.F./gal.	1 Bric	230	.035	L.F.	.14	1.06		1.20	1.79	
1800	1/2" x 1/2", 77 L.F./gal.	"	180	.044	"	.29	1.36		1.65	2.40	
2000	Latex acrylic based, bulk				Gal.	23			23	25.50	
2100	Cartridges					26			26	28.50	
2300	Polysulfide compounds, 1 component, bulk				▼	44			44	48	
2600	1 or 2 component, in place, 1/4" x 1/4", 308 L.F./gal.	1 Bric	145	.055	L.F.	.14	1.68		1.82	2.74	
2700	1/2" x 1/4", 154 L.F./gal.		135	.059		.28	1.81		2.09	3.08	
2900	3/4" x 3/8", 68 L.F./gal.		130	.062		.64	1.88		2.52	3.59	
3000	1" x 1/2", 38 L.F./gal.	▼	130	.062	▼	1.15	1.88		3.03	4.15	
3200	Polyurethane, 1 or 2 component				Gal.	47.50			47.50	52	
3300	Cartridges				"	46.50			46.50	51.50	
3500	Bulk, in place, 1/4" x 1/4"	1 Bric	150	.053	L.F.	.15	1.63		1.78	2.67	
3600	1/2" x 1/4"		145	.055		.31	1.68		1.99	2.92	
3800	3/4" x 3/8", 68 L.F./gal.		130	.062		.70	1.88		2.58	3.65	
3900	1" x 1/2"	▼	110	.073	▼	1.23	2.22		3.45	4.75	
4100	Silicone rubber, bulk				Gal.	34.50			34.50	38	
4200	Cartridges				"	40.50			40.50	44.50	
4400	Neoprene gaskets, closed cell, adhesive, 1/8" x 3/8"	1 Bric	240	.033	L.F.	.20	1.02		1.22	1.78	
4500	1/4" x 3/4"		215	.037		.48	1.13		1.61	2.27	
4700	1/2" x 1"		200	.040		1.40	1.22		2.62	3.41	
4800	3/4" x 1-1/2"	▼	165	.048	▼	2.91	1.48		4.39	5.45	
5500	Resin epoxy coating, 2 component, heavy duty				Gal.	26			26	28.50	
5800	Tapes, sealant, P.V.C. foam adhesive, 1/16" x 1/4"				C.L.F.	4.60			4.60	5.05	
5900	1/16" x 1/2"					6.80			6.80	7.50	
5950	1/16" x 1"					11.30			11.30	12.45	
6000	1/8" x 1/2"				▼	7.65			7.65	8.40	
6200	Urethane foam, 2 component, handy pack, 1 C.F.				Ea.	27.50			27.50	30.50	
6300	50.0 C.F. pack				C.F.	14.05			14.05	15.45	

For information about Means Estimating Seminars, see yellow pages 11 and 12 in back of book

THERMAL & MOISTURE PROTECTION **7**

	CREW	DAILY OUTPUT	LABOR-HOURS	UNIT	2002 BARE COSTS				TOTAL INCL O&P
					MAT.	LABOR	EQUIP.	TOTAL	

Division 8
Doors & Windows

Estimating Tips

08100 Metal Doors & Frames
- Most metal doors and frames look alike, but there may be significant differences among them. When estimating these items be sure to choose the line item that most closely compares to the specification or door schedule requirements regarding:
  - type of metal
  - metal gauge
  - door core material
  - fire rating
  - finish

08200 Wood & Plastic Doors
- Wood and plastic doors vary considerably in price. The primary determinant is the veneer material. Lauan, birch and oak are the most common veneers. Other variables include the following:
  - hollow or solid core
  - fire rating
  - flush or raised panel
  - finish
- If the specifications require compliance with AWI (Architectural Woodwork Institute) standards or acoustical standards, the cost of the door may increase substantially. All wood doors are priced pre-mortised for hinges and predrilled for cylindrical locksets.

- Frequently doors, frames, and windows are unique in old buildings. Specified replacement units could be stock, custom (similar to the original) or exact reproduction. The estimator should work closely with a window consultant to determine any extra costs that may be associated with the unusual installation requirements.

08300 Specialty Doors
- There are many varieties of special doors, and they are usually priced per each. Add frames, hardware or operators required for a complete installation.

08510 Steel Windows
- Most metal windows are delivered preglazed. However, some metal windows are priced without glass. Refer to 08800 Glazing for glass pricing. The grade C indicates commercial grade windows, usually ASTM C-35.

08550 Wood Windows
- All wood windows are priced preglazed. The two glazing options priced are single pane float glass and insulating glass 1/2″ thick. Add the cost of screens and grills if required.

08700 Hardware
- Hardware costs add considerably to the cost of a door. The most efficient method to determine the hardware requirements for a project is to review the door schedule. This schedule, in conjunction with the specifications, is all you should need to take off the door hardware.

- Door hinges are priced by the pair, with most doors requiring 1-1/2 pairs per door. The hinge prices do not include installation labor because it is included in door installation. Hinges are classified according to the frequency of use.

08800 Glazing
- Different openings require different types of glass. The three most common types are:
  - float
  - tempered
  - insulating
- Most exterior windows are glazed with insulating glass. Entrance doors and window walls, where the glass is less than 18″ from the floor, are generally glazed with tempered glass. Interior windows and some residential windows are glazed with float glass.

08900 Glazed Curtain Wall
- Glazed curtain walls consist of the metal tube framing and the glazing material. The cost data in this subdivision is presented for the metal tube framing alone or the composite wall. If your estimate requires a detailed takeoff of the framing, be sure to add the glazing cost.

Reference Numbers
Reference numbers are shown in bold squares at the beginning of some major classifications. These numbers refer to related items in the Reference Section. The reference information may be an estimating procedure, an alternate pricing method or technical information.

Note: Not all subdivisions listed here necessarily appear in this publication.

8

DOORS & WINDOWS

08341 | Cold Storage Doors

		CREW	DAILY OUTPUT	LABOR-HOURS	UNIT	2002 BARE COSTS MAT.	LABOR	EQUIP.	TOTAL	TOTAL INCL O&P	
200	0010 **COLD STORAGE**										200
	0020 Single, 20 ga. galvanized steel										
	0300 Horizontal sliding, 5' x 7', manual operation, 2" thick	2 Carp	2	8	Ea.	2,300	240		2,540	2,925	
	0400 4" thick		2	8		2,775	240		3,015	3,450	
	0500 6" thick		2	8		2,700	240		2,940	3,350	
	0800 5' x 7', power operation, 2" thick		1.90	8.421		4,500	253		4,753	5,350	
	0900 4" thick		1.90	8.421		4,575	253		4,828	5,425	
	1000 6" thick		1.90	8.421		5,200	253		5,453	6,125	
	1300 9' x 10', manual operation, 2" insulation		1.70	9.412		3,625	282		3,907	4,450	
	1400 4" insulation		1.70	9.412		3,750	282		4,032	4,575	
	1500 6" insulation		1.70	9.412		4,525	282		4,807	5,425	
	1800 Power operation, 2" insulation		1.60	10		6,250	300		6,550	7,350	
	1900 4" insulation		1.60	10		6,400	300		6,700	7,525	
	2000 6" insulation	↓	1.70	9.412	↓	7,250	282		7,532	8,425	
	2300 For stainless steel face, add					20%					
	3000 Hinged, lightweight, 3' x 7'-0", galvanized 1 face, 2" thick	2 Carp	2	8	Ea.	1,050	240		1,290	1,525	
	3050 4" thick		1.90	8.421		1,200	253		1,453	1,725	
	3300 Aluminum doors, 3' x 7'-0", 4" thick		1.90	8.421		1,075	253		1,328	1,575	
	3350 6" thick		1.40	11.429		1,900	345		2,245	2,625	
	3600 Stainless steel, 3' x 7'-0", 4" thick		1.90	8.421		1,375	253		1,628	1,925	
	3650 6" thick		1.40	11.429		2,300	345		2,645	3,075	
	3900 Painted, 3' x 7'-0", 4" thick		1.90	8.421		985	253		1,238	1,475	
	3950 6" thick	↓	1.40	11.429	↓	1,875	345		2,220	2,600	
	5000 Bi-parting, electric operated										
	5010 6' x 8' opening, galv. faces, 4" thick for cooler	2 Carp	.80	20	Opng.	5,875	600		6,475	7,400	
	5050 For freezer, 4" thick		.80	20		6,475	600		7,075	8,050	
	5300 For door buck framing and door protection, add		2.50	6.400		480	192		672	830	
	6000 Galvanized batten door, galvanized hinges, 4' x 7'		2	8		1,450	240		1,690	1,975	
	6050 6' x 8'		1.80	8.889		1,975	267		2,242	2,600	
	6500 Fire door, 3 hr., 6' x 8', single slide		.80	20		6,800	600		7,400	8,400	
	6550 Double, bi-parting	↓	.70	22.857	↓	8,825	685		9,510	10,800	

08344 | Industrial Doors

		CREW	DAILY OUTPUT	LABOR-HOURS	UNIT	2002 BARE COSTS MAT.	LABOR	EQUIP.	TOTAL	TOTAL INCL O&P	
120	0010 **AIR CURTAINS** Incl. motor starters, transformers,										120
	0050 door switches & temperature controls										
	0100 Shipping and receiving doors, unheated, minimal wind stoppage										
	0150 8' high, multiples of 3' wide	2 Shee	6	2.667	L.F.	250	93.50		343.50	420	
	0160 5' wide		10	1.600		277	56		333	390	
	0210 10' high, multiples of 4' wide		8	2		199	70		269	325	
	0250 12' high, 3'-6" wide		7	2.286		206	80		286	350	
	0260 12' wide		6	2.667		208	93.50		301.50	375	
	0350 16' high, 3'-6" wide		7	2.286		284	80		364	440	
	0360 12' wide	↓	6	2.667	↓	325	93.50		418.50	500	
	1500 Customer entrance doors, unheated, minimal wind stoppage										
	1550 10' high, multiples of 3' wide	2 Shee	6	2.667	L.F.	250	93.50		343.50	420	
	1560 5' wide		10	1.600		272	56		328	385	
	1650 Maximum wind stoppage, 12' high, multiples of 4' wide	↓	8	2	↓	269	70		339	405	
	1700 Heated, minimal wind stoppage, electric heat										
	1750 8' high, multiples of 3' wide	2 Shee	6	2.667	L.F.	515	93.50		608.50	710	
	1850 10' high, multiples of 3' wide		6	2.667		515	93.50		608.50	710	
	1860 Multiples of 5' wide	↓	10	1.600	↓	355	56		411	475	
	1950 Maximum wind stoppage, steam heat										
	1960 12' high, multiples of 4' wide	2 Shee	8	2	L.F.	405	70		475	555	
	2000 Walk-in coolers and freezers, ambient air, minimal wind stoppage										
	2050 8' high, multiples of 3' wide	2 Shee	6	2.667	L.F.	325	93.50		418.50	505	

Important: See the Reference Section for critical supporting data - Reference Nos., Crews, & City Cost Indexes

08344	Industrial Doors	CREW	DAILY OUTPUT	LABOR-HOURS	UNIT	2002 BARE COSTS				TOTAL INCL O&P	
						MAT.	LABOR	EQUIP.	TOTAL		
120 2060	Multiples of 5' wide	2 Shee	10	1.600	L.F.	340	56		396	460	120
2250	Maximum wind stoppage, 12' high, multiples of 3' wide		6	2.667		335	93.50		428.50	515	
2450	Conveyor openings or service windows, unheated, 5' high		5	3.200		250	112		362	450	
2460	Heated, electric, 5' high, 2'-6" wide		5	3.200		250	112		362	450	

For information about Means Estimating Seminars, see yellow pages 11 and 12 in back of book

DOORS & WINDOWS **8**

For expanded coverage of these items see *Means Interior Cost Data 2002*

	CREW	DAILY OUTPUT	LABOR-HOURS	UNIT	2002 BARE COSTS				TOTAL INCL O&P
					MAT.	LABOR	EQUIP.	TOTAL	

Division 9
Finishes

Estimating Tips
General
- Room Finish Schedule: A complete set of plans should contain a room finish schedule. If one is not available, it would be well worth the time and effort to put one together. A room finish schedule should contain the room number, room name (for clarity), floor materials, base materials, wainscot materials, wainscot height, wall materials (for each wall), ceiling materials, ceiling height and special instructions.
- Surplus Finishes: Review the specifications to determine if there is any requirement to provide certain amounts of extra materials for the owner's maintenance department. In some cases the owner may require a substantial amount of materials, especially when it is a special order item or long lead time item.

09200 Plaster & Gypsum Board
- Lath is estimated by the square yard for both gypsum and metal lath, plus usually 5% allowance for waste. Furring, channels and accessories are measured by the linear foot. An extra foot should be allowed for each accessory miter or stop.
- Plaster is also estimated by the square yard. Deductions for openings vary by preference, from zero deduction to 50% of all openings over 2 feet in width. Some estimators deduct a percentage of the total yardage for openings. The estimator should allow one extra square foot for each linear foot of horizontal interior or exterior angle located below the ceiling level. Also, double the areas of small radius work.
- Each room should be measured, perimeter times maximum wall height. Ceiling areas are equal to length times width.
- Drywall accessories, studs, track, and acoustical caulking are all measured by the linear foot. Drywall taping is figured by the square foot. Gypsum wallboard is estimated by the square foot. No material deductions should be made for door or window openings under 32 S.F. Coreboard can be obtained in a 1" thickness for solid wall and shaft work. Additions should be made to price out the inside or outside corners.
- Different types of partition construction should be listed separately on the quantity sheets. There may be walls with studs of various widths, double studded, and similar or dissimilar surface materials. Shaft work is usually different construction from surrounding partitions requiring separate quantities and pricing of the work.

09300 Tile
09400 Terrazzo
- Tile and terrazzo areas are taken off on a square foot basis. Trim and base materials are measured by the linear foot. Accent tiles are listed per each. Two basic methods of installation are used. Mud set is approximately 30% more expensive than the thin set. In terrazzo work, be sure to include the linear footage of embedded decorative strips, grounds, machine rubbing and power cleanup.

09600 Flooring
- Wood flooring is available in strip, parquet, or block configuration. The latter two types are set in adhesives with quantities estimated by the square foot. The laying pattern will influence labor costs and material waste. In addition to the material and labor for laying wood floors, the estimator must make allowances for sanding and finishing these areas unless the flooring is prefinished.
- Most of the various types of flooring are all measured on a square foot basis. Base is measured by the linear foot. If adhesive materials are to be quantified, they are estimated at a specified coverage rate by the gallon depending upon the specified type and the manufacturer's recommendations.
- Sheet flooring is measured by the square yard. Roll widths vary, so consideration should be given to use the most economical width, as waste must be figured into the total quantity. Consider also the installation methods available, direct glue down or stretched .

09700 Wall Finishes
- Wall coverings are estimated by the square foot. The area to be covered is measured, length by height of wall above baseboards, to calculate the square footage of each wall. This figure is divided by the number of square feet in the single roll which is being used. Deduct, in full, the areas of openings such as doors and windows. Where a pattern match is required allow 25%-30% waste. One gallon of paste should be sufficient to hang 12 single rolls of light to medium weight paper.

09800 Acoustical Treatment
- Acoustical systems fall into several categories. The takeoff of these materials is by the square foot of area with a 5% allowance for waste. Do not forget about scaffolding, if applicable, when estimating these systems.

09900 Paints & Coatings
- Painting is one area where bids vary to a greater extent than almost any other section of a project. This arises from the many methods of measuring surfaces to be painted. The estimator should check the plans and specifications carefully to be sure of the required number of coats.
- Protection of adjacent surfaces is not included in painting costs. When considering the method of paint application, an important factor is the amount of protection and masking required. These must be estimated separately and may be the determining factor in choosing the method of application.

Reference Numbers
Reference numbers are shown in bold squares at the beginning of some major classifications. These numbers refer to related items in the Reference Section. The reference information may be an estimating procedure, an alternate pricing method or technical information.

Note: Not all subdivisions listed here necessarily appear in this publication.

09910	Paints & Coatings	CREW	DAILY OUTPUT	LABOR-HOURS	UNIT	2002 BARE COSTS				TOTAL INCL O&P
						MAT.	LABOR	EQUIP.	TOTAL	
630	0010	**MISCELLANEOUS, INTERIOR**									630
	3800	Grilles, per side, oil base, primer coat, brushwork	1 Pord	520	.015	S.F.	.09	.42		.51	.73
	3850	Spray		1,140	.007		.10	.19		.29	.40
	3880	Paint 1 coat, brushwork		520	.015		.10	.42		.52	.74
	3900	Spray		1,140	.007		.12	.19		.31	.42
	3920	Paint 2 coats, brushwork		325	.025		.20	.67		.87	1.23
	3940	Spray		650	.012		.23	.33		.56	.76
	3950	Prime & paint 1 coat		325	.025		.20	.67		.87	1.23
	3960	Prime & paint 2 coats		270	.030		.19	.80		.99	1.43
	4250	Paint 1 coat, brushwork	2 Pord	1,300	.012	L.F.	.11	.33		.44	.63
	4500	Louvers, one side, primer, brushwork	1 Pord	524	.015	S.F.	.06	.41		.47	.69
	4520	Paint one coat, brushwork		520	.015		.05	.42		.47	.69
	4530	Spray		1,140	.007		.06	.19		.25	.36
	4540	Paint two coats, brushwork		325	.025		.10	.67		.77	1.13
	4550	Spray		650	.012		.12	.33		.45	.64
	4560	Paint three coats, brushwork		270	.030		.16	.80		.96	1.39
	4570	Spray		500	.016		.17	.43		.60	.85
	5000	Pipe, to 4" diameter, primer or sealer coat, oil base, brushwork	2 Pord	1,250	.013	L.F.	.06	.35		.41	.59
	5100	Spray		2,165	.007		.05	.20		.25	.36
	5200	Paint 1 coat, brushwork		1,250	.013		.06	.35		.41	.60
	5300	Spray		2,165	.007		.05	.20		.25	.36
	5350	Paint 2 coats, brushwork		775	.021		.11	.56		.67	.97
	5400	Spray		1,240	.013		.12	.35		.47	.66
	5420	Paint 3 coats, brushwork		775	.021		.16	.56		.72	1.03
	5450	To 8" diameter, primer or sealer coat, brushwork		620	.026		.12	.70		.82	1.19
	5500	Spray		1,085	.015		.19	.40		.59	.82
	5550	Paint 1 coat, brushwork		620	.026		.17	.70		.87	1.24
	5600	Spray		1,085	.015		.19	.40		.59	.81
	5650	Paint 2 coats, brushwork		385	.042		.22	1.13		1.35	1.95
	5700	Spray		620	.026		.24	.70		.94	1.33
	5720	Paint 3 coats, brushwork		385	.042		.33	1.13		1.46	2.07
	5750	To 12" diameter, primer or sealer coat, brushwork		415	.039		.17	1.05		1.22	1.77
	5800	Spray		725	.022		.20	.60		.80	1.13
	5850	Paint 1 coat, brushwork		415	.039		.17	1.05		1.22	1.77
	6000	Spray		725	.022		.19	.60		.79	1.12
	6200	Paint 2 coats, brushwork		260	.062		.33	1.67		2	2.89
	6250	Spray		415	.039		.36	1.05		1.41	1.98
	6270	Paint 3 coats, brushwork		260	.062		.49	1.67		2.16	3.07
	6300	To 16" diameter, primer or sealer coat, brushwork		310	.052		.23	1.40		1.63	2.38
	6350	Spray		540	.030		.26	.80		1.06	1.51
	6400	Paint 1 coat, brushwork		310	.052		.23	1.40		1.63	2.37
	6450	Spray		540	.030		.25	.80		1.05	1.50
	6500	Paint 2 coats, brushwork		195	.082		.44	2.23		2.67	3.85
	6550	Spray		310	.052		.49	1.40		1.89	2.66
	6600	Radiators, per side, primer, brushwork	1 Pord	520	.015	S.F.	.06	.42		.48	.69
	6620	Paint one coat, brushwork		520	.015		.05	.42		.47	.69
	6640	Paint two coats, brushwork		340	.024		.10	.64		.74	1.09
	6660	Paint three coats, brushwork		283	.028		.16	.77		.93	1.33

For information about Means Estimating Seminars, see yellow pages 11 and 12 in back of book

Important: See the Reference Section for critical supporting data - Reference Nos., Crews, & City Cost Indexes

9 FINISHES

Division 10
Specialties

Estimating Tips

General

- The items in this division are usually priced per square foot or each.
- Many items in Division 10 require some type of support system or special anchors that are not usually furnished with the item. The required anchors must be added to the estimate in the appropriate division.
- Some items in Division 10, such as lockers, may require assembly before installation. Verify the amount of assembly required. Assembly can often exceed installation time.

10100 Visual Display Boards

- Toilet partitions are priced by the stall. A stall consists of a side wall, pilaster and door with hardware. Toilet tissue holders and grab bars are extra.

10150 Compartments & Cubicles

- Support angles and blocking are not included in the installation of toilet compartments, shower/dressing compartments or cubicles. Appropriate line items from Divisions 5 or 6 may need to be added to support the installations.

10600 Partitions

- The required acoustical rating of a folding partition can have a significant impact on costs. Verify the sound transmission coefficient rating of the panel priced to the specification requirements.

10800 Toilet/Bath/Laundry Accessories

- Grab bar installation does not include supplemental blocking or backing to support the required load. When grab bars are installed at an existing facility provisions must be made to attach the grab bars to solid structure.

Reference Numbers

Reference numbers are shown in bold squares at the beginning of some major classifications. These numbers refer to related items in the Reference Section. The reference information may be an estimating procedure, an alternate pricing method or technical information.

Note: Not all subdivisions listed here necessarily appear in this publication.

10210 | Wall Louvers

		CREW	DAILY OUTPUT	LABOR-HOURS	UNIT	2002 BARE COSTS MAT.	LABOR	EQUIP.	TOTAL	TOTAL INCL O&P	
800	0010	**LOUVERS** Aluminum with screen, residential, 8" x 8"	1 Carp	38	.211	Ea.	6.45	6.30		12.75	16.95	**800**
	0100	12" x 12"		38	.211		8.45	6.30		14.75	19.15	
	0200	12" x 18"		35	.229		12.30	6.85		19.15	24	
	0250	14" x 24"		30	.267		15.95	8		23.95	30	
	0300	18" x 24"		27	.296		18.55	8.90		27.45	34.50	
	0500	24" x 30"		24	.333		23.50	10		33.50	41.50	
	0700	Triangle, adjustable, small		20	.400		22	12		34	42.50	
	0800	Large	↓	15	.533	↓	39.50	16		55.50	68.50	
	1200	Extruded aluminum, see division 15850-600										
	2100	Midget, aluminum, 3/4" deep, 1" diameter	1 Carp	85	.094	Ea.	.66	2.82		3.48	5.15	
	2150	3" diameter		60	.133		1.37	4		5.37	7.75	
	2200	4" diameter		50	.160		1.89	4.80		6.69	9.55	
	2250	6" diameter	↓	30	.267	↓	2.50	8		10.50	15.20	
	2300	Ridge vent strip, mill finish	1 Shee	155	.052	L.F.	2.14	1.81		3.95	5.15	
	2400	Under eaves vent, aluminum, mill finish, 16" x 4"	1 Carp	48	.167	Ea.	1.24	5		6.24	9.15	
	2500	16" x 8"	"	48	.167	"	1.69	5		6.69	9.65	

10275 | Access Flooring

		CREW	DAILY OUTPUT	LABOR-HOURS	UNIT	2002 BARE COSTS MAT.	LABOR	EQUIP.	TOTAL	TOTAL INCL O&P	
150	0010	**PEDESTAL ACCESS FLOORS** Computer room application, metal										**150**
	0020	Particle board or steel panels, no covering, under 6,000 S.F.	2 Carp	400	.040	S.F.	9.95	1.20		11.15	12.80	
	0300	Metal covered, over 6,000 S.F.		450	.036		7.20	1.07		8.27	9.55	
	0400	Aluminum, 24" panels	↓	500	.032		24.50	.96		25.46	28.50	
	0600	For carpet covering, add					5.15			5.15	5.70	
	0700	For vinyl floor covering, add					5.25			5.25	5.75	
	0900	For high pressure laminate covering, add					3.49			3.49	3.84	
	0910	For snap on stringer system, add	2 Carp	1,000	.016	↓	1.19	.48		1.67	2.06	
	0950	Office applications, to 8" high, steel panels,										
	0960	no covering, over 6,000 S.F.	2 Carp	500	.032	S.F.	7.65	.96		8.61	9.90	
	1000	Machine cutouts after initial installation	1 Carp	10	.800	Ea.	39	24		63	80.50	
	1050	Pedestals, 6" to 12"	2 Carp	85	.188		7.35	5.65		13	16.85	
	1100	Air conditioning grilles, 4" x 12"	1 Carp	17	.471		46	14.10		60.10	72.50	
	1150	4" x 18"	"	14	.571	↓	63	17.15		80.15	96	
	1200	Approach ramps, minimum	2 Carp	85	.188	S.F.	23	5.65		28.65	34.50	
	1300	Maximum	"	60	.267	"	28	8		36	43	
	1500	Handrail, 2 rail, aluminum	1 Carp	15	.533	L.F.	82.50	16		98.50	116	

10305 | Manufactured Fireplaces

		CREW	DAILY OUTPUT	LABOR-HOURS	UNIT	2002 BARE COSTS MAT.	LABOR	EQUIP.	TOTAL	TOTAL INCL O&P	
100	0010	**FIREPLACE, PREFABRICATED** Free standing or wall hung										**100**
	0100	with hood & screen, minimum	1 Carp	1.30	6.154	Ea.	985	185		1,170	1,350	

Important: See the Reference Section for critical supporting data - Reference Nos., Crews, & City Cost Indexes

		10305	**Manufactured Fireplaces**	CREW	DAILY OUTPUT	LABOR-HOURS	UNIT	2002 BARE COSTS				TOTAL INCL O&P	
								MAT.	LABOR	EQUIP.	TOTAL		
100	0150		Average	1 Carp	1	8	Ea.	1,200	240		1,440	1,675	100
	0200		Maximum		.90	8.889	↓	2,850	267		3,117	3,550	
	0500		Chimney dbl. wall, all stainless, over 8'-6", 7" diam., add		33	.242	V.L.F.	46.50	7.25		53.75	63	
	0600		10" diameter, add		32	.250		47.50	7.50		55	64	
	0700		12" diameter, add		31	.258		62	7.75		69.75	80.50	
	0800		14" diameter, add		30	.267	↓	78.50	8		86.50	99	
	1000		Simulated brick chimney top, 4' high, 16" x 16"		10	.800	Ea.	169	24		193	224	
	1100		24" x 24"		7	1.143	"	315	34.50		349.50	405	
	1500		Simulated logs, gas fired, 40,000 BTU, 2' long, minimum		7	1.143	Set	425	34.50		459.50	525	
	1600		Maximum		6	1.333		600	40		640	725	
	1700		Electric, 1,500 BTU, 1'-6" long, minimum		7	1.143		117	34.50		151.50	183	
	1800		11,500 BTU, maximum	↓	6	1.333	↓	254	40		294	340	
		10320	**Stoves**										
100	0010		**WOODBURNING STOVES** Cast iron, minimum	2 Carp	1.30	12.308	Ea.	740	370		1,110	1,400	100
	0020		Average		1	16		1,175	480		1,655	2,050	
	0030		Maximum	↓	.80	20		2,200	600		2,800	3,350	
	0050		For gas log lighter, add				↓	37			37	40.50	

For information about Means Estimating Seminars, see yellow pages 11 and 12 in back of book

SPECIALTIES 10

		CREW	DAILY OUTPUT	LABOR-HOURS	UNIT	2002 BARE COSTS				TOTAL INCL O&P
						MAT.	LABOR	EQUIP.	TOTAL	

Division 11
Equipment

Estimating Tips
General
- The items in this division are usually priced per square foot or each. Many of these items are purchased by the owner for installation by the contractor. Check the specifications for responsibilities, and include time for receiving, storage, installation and mechanical and electrical hook-ups in the appropriate divisions.

- Many items in Division 11 require some type of support system that is not usually furnished with the item. Examples of these systems include blocking for the attachment of casework and support angles for ceiling hung projection screens. The required blocking or supports must be added to the estimate in the appropriate division.
- Some items in Division 11 may require assembly or electrical hook-ups. Verify the amount of assembly required or the need for a hard electrical connection and add the appropriate costs.

Reference Numbers
Reference numbers are shown in bold squares at the beginning of some major classifications. These numbers refer to related items in the Reference Section. The reference information may be an estimating procedure, an alternate pricing method or technical information.

Note: Not all subdivisions listed here necessarily appear in this publication.

11010 | Maintenance Equipment

		11013	Floor/Wall Cleaning Equipment	CREW	DAILY OUTPUT	LABOR-HOURS	UNIT	2002 BARE COSTS MAT.	LABOR	EQUIP.	TOTAL	TOTAL INCL O&P	
800	0010		**VACUUM CLEANING**	1 Skwk	.90	8.889	Total	570	275		845	1,050	800
	0020		Central, 3 inlet, residential	1 Skwk	.90	8.889	Total	570	275		845	1,050	
	0200		Commercial		.70	11.429		970	355		1,325	1,625	
	0400		5 inlet system, residential		.50	16		860	495		1,355	1,725	
	0600		7 inlet system, commercial		.40	20		1,075	620		1,695	2,150	
	0800		9 inlet system, residential		.30	26.667		1,200	825		2,025	2,625	
	4010		Rule of thumb: First 1200 S.F., installed									1,044	
	4020		For each additional S.F., add				S.F.					.17	

11100 | Mercantile Equipment

		11104	Display Cases & Systems	CREW	DAILY OUTPUT	LABOR-HOURS	UNIT	2002 BARE COSTS MAT.	LABOR	EQUIP.	TOTAL	TOTAL INCL O&P	
700	0010		**REFRIGERATED FOOD CASES**										700
	0030		Dairy, multi-deck, 12' long	Q-5	3	5.333	Ea.	7,800	174		7,974	8,825	
	0100		For rear sliding doors, add					1,075			1,075	1,175	
	0200		Delicatessen case, service deli, 12' long, single deck	Q-5	3.90	4.103		5,250	134		5,384	5,975	
	0300		Multi-deck, 18 S.F. shelf display		3	5.333		6,425	174		6,599	7,325	
	0400		Freezer, self-contained, chest-type, 30 C.F.		3.90	4.103		3,825	134		3,959	4,400	
	0500		Glass door, upright, 78 C.F.		3.30	4.848		7,300	158		7,458	8,275	
	0600		Frozen food, chest type, 12' long		3.30	4.848		5,275	158		5,433	6,050	
	0700		Glass door, reach-in, 5 door		3	5.333		10,100	174		10,274	11,400	
	0800		Island case, 12' long, single deck		3.30	4.848		5,975	158		6,133	6,800	
	0900		Multi-deck		3	5.333		12,600	174		12,774	14,200	
	1000		Meat case, 12' long, single deck		3.30	4.848		4,325	158		4,483	5,025	
	1050		Multi-deck		3.10	5.161		7,450	168		7,618	8,450	
	1100		Produce, 12' long, single deck		3.30	4.848		5,725	158		5,883	6,550	
	1200		Multi-deck		3.10	5.161		6,375	168		6,543	7,275	

11110 | Commercial Laundry & Dry Cleaning Equipment

		11119	Laundry Cleaning	CREW	DAILY OUTPUT	LABOR-HOURS	UNIT	2002 BARE COSTS MAT.	LABOR	EQUIP.	TOTAL	TOTAL INCL O&P	
450	0010		**LAUNDRY EQUIPMENT** Not incl. rough-in										450
	0500		Dryers, gas fired residential, 16 lb. capacity, average	1 Plum	3	2.667	Ea.	530	96		626	725	
	1000		Commercial, 30 lb. capacity, coin operated, single		3	2.667		2,425	96		2,521	2,800	
	1100		Double stacked		2	4		5,175	144		5,319	5,925	
	1500		Industrial, 30 lb. capacity		2	4		2,050	144		2,194	2,475	
	1600		50 lb. capacity		1.70	4.706		2,700	169		2,869	3,225	
	5000		Washers, residential, 4 cycle, average		3	2.667		605	96		701	810	
	5300		Commercial, coin operated, average		3	2.667		970	96		1,066	1,225	
	6000		Combination washer/extractor, 20 lb. capacity	L-6	1.50	8		3,475	286		3,761	4,250	
	6100		30 lb. capacity		.80	15		6,725	535		7,260	8,175	
	6200		50 lb. capacity		.68	17.647		8,375	630		9,005	10,200	
	6300		75 lb. capacity		.30	40		17,300	1,425		18,725	21,200	

Important: See the Reference Section for critical supporting data - Reference Nos., Crews, & City Cost Indexes

11110 | Commercial Laundry & Dry Cleaning Equipment

	11119	Laundry Cleaning	CREW	DAILY OUTPUT	LABOR-HOURS	UNIT	2002 BARE COSTS				TOTAL INCL O&P	
							MAT.	LABOR	EQUIP.	TOTAL		
450	6350	125 lb. capacity	L-6	.16	75	Ea.	20,900	2,675		23,575	27,000	450

11140 | Vehicle Service Equipment

	11141	Service Station Equipment	CREW	DAILY OUTPUT	LABOR-HOURS	UNIT	2002 BARE COSTS				TOTAL INCL O&P	
							MAT.	LABOR	EQUIP.	TOTAL		
150	0010	**AUTOMOTIVE**										150
	0030	Compressors, electric, 1-1/2 H.P., standard controls	L-4	1.50	16	Ea.	320	450		770	1,050	
	0550	Dual controls		1.50	16		470	450		920	1,225	
	0600	5 H.P., 115/230 volt, standard controls		1	24		1,525	675		2,200	2,750	
	0650	Dual controls		1	24		1,625	675		2,300	2,850	
	1100	Product dispenser with vapor recovery for 6 nozzles, installed, not										
	1110	including piping to storage tanks				Ea.	15,000			15,000	16,500	
	2200	Hoists, single post, 8,000# capacity, swivel arms	L-4	.40	60		3,500	1,700		5,200	6,500	
	2400	Two posts, adjustable frames, 11,000# capacity		.25	96		4,500	2,700		7,200	9,175	
	2500	24,000# capacity		.15	160		6,000	4,525		10,525	13,700	
	2700	7,500# capacity, frame supports		.50	48		5,000	1,350		6,350	7,625	
	2800	Four post, roll on ramp		.50	48		4,500	1,350		5,850	7,075	
	2810	Hydraulic lifts, above ground, 2 post, clear floor, 6000 lb cap		2.67	8.989		4,700	254		4,954	5,575	
	2815	9000 lb capacity		2.29	10.480		11,200	296		11,496	12,800	
	2820	15,000 lb capacity		2	12		12,600	340		12,940	14,400	
	2825	30,000 lb capacity		1.60	15		27,900	425		28,325	31,300	
	2830	4 post, ramp style, 25,000 lb capacity		2	12		10,900	340		11,240	12,500	
	2835	35,000 lb capacity		1	24		51,000	675		51,675	57,000	
	2840	50,000 lb capacity		1	24		57,000	675		57,675	63,500	
	2845	75,000 lb capacity		1	24		66,000	675		66,675	73,500	
	2850	For drive thru tracks, add, minimum					700			700	770	
	2855	Maximum					1,200			1,200	1,325	
	2860	Ramp extensions, 3'(set of 2)					575			575	635	
	2865	Rolling jack platform					2,000			2,000	2,200	
	2870	Elec/hyd jacking beam					5,350			5,350	5,875	
	2880	Scissor lift, portable, 6000 lb capacity					5,250			5,250	5,775	
	3000	Lube equipment, 3 reel type, with pumps, not including piping	L-4	.50	48	Set	6,000	1,350		7,350	8,725	
	3700	Pump lubrication, pneumatic, not incl. air compressor										
	3710	Oil/gear lube	Q-1	9.60	1.667	Ea.	680	54		734	830	
	3720	Grease	"	9.60	1.667		630	54		684	770	
	4000	Spray painting booth, 26' long, complete	L-4	.40	60		12,400	1,700		14,100	16,300	

11170 | Solid Waste Handling Equipment

	11179	Waste Handling Equipment	CREW	DAILY OUTPUT	LABOR-HOURS	UNIT	2002 BARE COSTS				TOTAL INCL O&P	
							MAT.	LABOR	EQUIP.	TOTAL		
150	0010	**WASTE HANDLING**										150
	0020	Compactors, 115 volt, 250#/hr., chute fed	L-4	1	24	Ea.	8,600	675		9,275	10,500	

EQUIPMENT **11**

83

		11179	Waste Handling Equipment	CREW	DAILY OUTPUT	LABOR-HOURS	UNIT	2002 BARE COSTS				TOTAL INCL O&P	
								MAT.	LABOR	EQUIP.	TOTAL		
150	0100		Hand fed	L-4	2.40	10	Ea.	6,200	282		6,482	7,275	150
	0300		Multi-bag, 230 volt, 600#/hr, chute fed		1	24		7,575	675		8,250	9,375	
	0400		Hand fed		1	24		6,500	675		7,175	8,200	
	0500		Containerized, hand fed, 2 to 6 C.Y. containers, 250#/hr.		1	24		8,175	675		8,850	10,000	
	0550		For chute fed, add per floor		1	24		940	675		1,615	2,075	
	1000		Heavy duty industrial compactor, 0.5 C.Y. capacity		1	24		5,525	675		6,200	7,125	
	1050		1.0 C.Y. capacity		1	24		8,150	675		8,825	10,000	
	1100		3 C.Y. capacity		.50	48		11,100	1,350		12,450	14,300	
	1150		5.0 C.Y. capacity		.50	48		14,500	1,350		15,850	18,000	
	1200		Combination shredder/compactor (5,000 lbs./hr.)		.50	48		27,500	1,350		28,850	32,400	
	1400		For handling hazardous waste materials, 55 gallon drum packer, std.					12,800			12,800	14,100	
	1410		55 gallon drum packer w/HEPA filter					16,300			16,300	17,900	
	1420		55 gallon drum packer w/charcoal & HEPA filter					21,700			21,700	23,900	
	1430		All of the above made explosion proof, add					9,925			9,925	10,900	
	1500		Crematory, not including building, 1 place	Q-3	.20	160		46,300	5,475		51,775	59,500	
	1750		2 place		.10	320		66,000	11,000		77,000	89,000	
	4400		Incinerator, gas, not incl. chimney, elec. or pipe, 50#/hr., minimum		.80	40		17,700	1,375		19,075	21,600	
	4420		Maximum		.70	45.714		23,100	1,575		24,675	27,800	
	4440		200 lb. per hr., minimum (batch type)		.60	53.333		23,100	1,825		24,925	28,200	
	4460		Maximum (with feeder)		.50	64		44,900	2,200		47,100	52,500	
	4480		400 lb. per hr., minimum (batch type)		.30	106		27,300	3,650		30,950	35,500	
	4500		Maximum (with feeder)		.25	128		51,500	4,400		55,900	63,000	
	4520		800 lb. per hr., with feeder, minimum		.20	160		67,000	5,475		72,475	82,500	
	4540		Maximum		.17	188		92,000	6,450		98,450	110,500	
	4560		1,200 lb. per hr., with feeder, minimum		.15	213		97,000	7,325		104,325	118,000	
	4580		Maximum		.11	290		117,000	9,975		126,975	143,500	
	4600		2,000 lb. per hr., with feeder, minimum		.10	320		170,000	11,000		181,000	203,500	
	4620		Maximum		.05	640		285,500	21,900		307,400	347,000	
	4700		For heat recovery system, add, minimum		.25	128		55,000	4,400		59,400	67,000	
	4710		Add, maximum		.11	290		176,500	9,975		186,475	209,000	
	4720		For automatic ash conveyer, add		.50	64		23,200	2,200		25,400	28,800	
	4750		Large municipal incinerators, incl. stack, minimum		.25	128	Ton/day	14,100	4,400		18,500	22,100	
	4850		Maximum		.10	320	"	37,500	11,000		48,500	57,500	
	5800		Shredder, industrial, minimum				Ea.	16,500			16,500	18,100	
	5850		Maximum					88,500			88,500	97,500	
	5900		Baler, industrial, minimum					6,600			6,600	7,275	
	5950		Maximum					386,000			386,000	425,000	

		11310	Sewage & Sludge Pumps	CREW	DAILY OUTPUT	LABOR-HOURS	UNIT	2002 BARE COSTS				TOTAL INCL O&P	
								MAT.	LABOR	EQUIP.	TOTAL		
350	0010		PUMPS, PNEUMATIC EJECTOR										350
	0020		With steel receiver, level controls, inlet/outlet gate/check valves										
	0030		Cross connect. not incl. compressor, fittings or piping										
	0040		Duplex										
	0050		30 GPM	Q-2	1.70	14.118	Ea.	14,400	475		14,875	16,500	
	0060		50 GPM		1.56	15.385		20,300	515		20,815	23,100	
	0070		100 GPM		1.30	18.462		24,500	620		25,120	27,900	
	0080		150 GPM		1.10	21.818		47,300	730		48,030	53,000	

11300 | Fluid Waste Treatment & Disposal Equipment

11310	Sewage & Sludge Pumps	CREW	DAILY OUTPUT	LABOR-HOURS	UNIT	2002 BARE COSTS MAT.	LABOR	EQUIP.	TOTAL	TOTAL INCL O&P	
350	0090	200 GPM	Q-2	.85	28.235	Ea.	60,000	945		60,945	67,500	**350**
	0100	250 GPM	Q-3	.91	35.165	↓	77,500	1,200		78,700	87,000	
	0110	300 GPM	"	.57	56.140	↓	94,500	1,925		96,425	107,000	

11400 | Food Service Equipment

11410	Food Prep Equipment	CREW	DAILY OUTPUT	LABOR-HOURS	UNIT	2002 BARE COSTS MAT.	LABOR	EQUIP.	TOTAL	TOTAL INCL O&P	
150	0010	**COMMERCIAL KITCHEN EQUIPMENT**										**150**
	0020	Bake oven, gas, one section	Q-1	8	2	Ea.	3,950	64.50		4,014.50	4,450	
	0300	Two sections		7	2.286		7,900	74		7,974	8,775	
	0600	Three sections	↓	6	2.667		11,300	86.50		11,386.50	12,500	
	0900	Electric convection, single deck	L-7	4	7		5,350	202		5,552	6,225	
	2350	Cooler, reach-in, beverage, 6' long	Q-1	6	2.667	↓	2,950	86.50		3,036.50	3,375	
	2700	Dishwasher, commercial, rack type										
	2720	10 to 12 racks per hour	Q-1	3.20	5	Ea.	2,700	162		2,862	3,225	
	2750	Semi-automatic 38 to 50 racks per hour	"	1.30	12.308		5,675	400		6,075	6,825	
	2800	Automatic, 190 to 230 racks per hour	L-6	.35	34.286		11,600	1,225		12,825	14,700	
	2820	235 to 275 racks per hour		.25	48		23,400	1,725		25,125	28,300	
	2840	8,750 to 12,500 dishes per hour	↓	.10	120		44,000	4,300		48,300	55,000	
	4300	Freezers, reach-in, 44 C.F.	Q-1	4	4		7,800	129		7,929	8,775	
	4500	68 C.F.		3	5.333		9,225	173		9,398	10,500	
	5800	Ice cube maker, 50 pounds per day		6	2.667		1,625	86.50		1,711.50	1,900	
	6050	500 pounds per day	↓	4	4		2,975	129		3,104	3,475	
	6350	20 Gal kettle w/steam jacket, tilting w/positive lock, S/S	L-7	7	4		5,900	116		6,016	6,650	
	6600	60 gallons	"	6	4.667		6,700	135		6,835	7,550	
	7950	Hood fire protection system, minimum	Q-1	3	5.333		3,050	173		3,223	3,600	
	8050	Maximum		1	16		22,100	520		22,620	25,100	
	8300	Refrigerators, reach-in type, 44 C.F.		5	3.200		5,325	104		5,429	6,025	
	8310	With glass doors, 68 C.F.	↓	4	4	↓	6,825	129		6,954	7,700	

11420	Food Cooking Equipment	CREW	DAILY OUTPUT	LABOR-HOURS	UNIT	2002 BARE COSTS MAT.	LABOR	EQUIP.	TOTAL	TOTAL INCL O&P	
550	0010	**KITCHEN VENTILATION,** Commercial										**550**
	1010	Heat reclaim unit, air to air										
	1100	Heat pipe exchanger										
	1110	Combined supply/exhaust air volume										
	1120	2.5 to 6.0 MCFM	Q-10	2.80	8.571	MCFM	4,725	281		5,006	5,625	
	1130	6 to 16 MCFM		5	4.800		2,900	157		3,057	3,425	
	1140	16 to 22 MCFM	↓	6	4	↓	2,250	131		2,381	2,700	
	2010	Packaged ventilation										
	2100	Combined supply/exhaust air volume										
	2110	With ambient supply										
	2120	3 to 7 MCFM	Q-10	5	4.800	MCFM	855	157		1,012	1,175	
	2130	7 to 12 MCFM		7.10	3.380		570	111		681	795	
	2140	12 to 22 MCFM	↓	10.70	2.243	↓	400	73.50		473.50	555	
	2400	With tempered supply										
	2410	3 to 7 MCFM	Q-10	5	4.800	MCFM	1,475	157		1,632	1,875	
	2420	7 to 12 MCFM		7.10	3.380		955	111		1,066	1,225	
	2430	12 to 22 MCFM	↓	10.70	2.243	↓	700	73.50		773.50	885	
	4000	Electrostatic precipitators										

EQUIPMENT **11**

85

	11420	Food Cooking Equipment	CREW	DAILY OUTPUT	LABOR-HOURS	UNIT	2002 BARE COSTS				TOTAL INCL O&P	
							MAT.	LABOR	EQUIP.	TOTAL		
550	4100	Smoke pollution control										550
	4110	Exhaust air volume										
	4120	1 to 2.4 MCFM	Q-10	2.15	11.163	MCFM	5,525	365		5,890	6,650	
	4130	2.4 to 4.8 MCFM		2.70	8.889		6,300	291		6,591	7,375	
	4140	4.8 to 7.2 MCFM		3.60	6.667		6,400	218		6,618	7,375	
	4150	7.2 to 12 MCFM	↓	5.40	4.444	↓	6,700	146		6,846	7,600	
	4400	Electrostatic precipitators with built-in exhaust fans										
	4410	Smoke pollution control										
	4420	Exhaust air volume										
	4440	1 to 2.4 MCFM	Q-10	2.15	11.163	MCFM	6,650	365		7,015	7,900	
	4450	2.4 to 4.8 MCFM		2.70	8.889		5,275	291		5,566	6,250	
	4460	4.8 to 7.2 MCFM		3.60	6.667		4,525	218		4,743	5,300	
	4470	7.2 to 12 MCFM	↓	5.40	4.444	↓	3,900	146		4,046	4,500	
	6000	Exhaust hoods										
	6100	Centrifugal grease extraction										
	6110	Water wash type										
	6120	Per foot of length, minimum	Q-10	7.10	3.380	L.F.	1,125	111		1,236	1,425	
	6130	Maximum	"	7.10	3.380	"	1,925	111		2,036	2,300	
	6200	Non water wash type										
	6210	Per foot of length	Q-10	13.40	1.791	L.F.	825	58.50		883.50	1,000	
	6400	Island style ventilator										
	6410	Water wash type										
	6420	Per foot of length, minimum	Q-10	7.10	3.380	L.F.	1,500	111		1,611	1,825	
	6430	Maximum	"	7.10	3.380	"	2,500	111		2,611	2,925	
	6460	Non water wash type										
	6470	Per foot of length, minimum	Q-10	13.40	1.791	L.F.	650	58.50		708.50	805	
	6480	Maximum	"	13.40	1.791	"	1,325	58.50		1,383.50	1,550	

	11454	Residential Appliances	CREW	DAILY OUTPUT	LABOR-HOURS	UNIT	2002 BARE COSTS				TOTAL INCL O&P	
							MAT.	LABOR	EQUIP.	TOTAL		
500	0010	**RESIDENTIAL APPLIANCES**										500
	0020	Cooking range, 30" free standing, 1 oven, minimum	2 Clab	10	1.600	Ea.	254	37.50		291.50	340	
	0050	Maximum		4	4		1,200	94		1,294	1,475	
	0150	2 oven, minimum		10	1.600		1,300	37.50		1,337.50	1,475	
	0200	Maximum	↓	10	1.600		1,450	37.50		1,487.50	1,650	
	1500	Combination range, refrigerator and sink, 30" wide, minimum	L-1	2	8		730	286		1,016	1,225	
	1550	Maximum		1	16		1,450	570		2,020	2,450	
	1570	60" wide, average		1.40	11.429		2,400	410		2,810	3,250	
	1590	72" wide, average		1.20	13.333		2,725	475		3,200	3,725	
	1600	Office model, 48" wide		2	8		2,050	286		2,336	2,700	
	1620	Refrigerator and sink only	↓	2.40	6.667	↓	2,050	238		2,288	2,625	
	1640	Combination range, refrigerator, sink, microwave										
	1660	oven and ice maker	L-1	.80	20	Ea.	4,025	715		4,740	5,500	
	2450	Dehumidifier, portable, automatic, 15 pint					203			203	223	
	2550	40 pint					225			225	247	
	2750	Dishwasher, built-in, 2 cycles, minimum	L-1	4	4		254	143		397	495	
	2800	Maximum		2	8		282	286		568	740	
	2950	4 or more cycles, minimum	↓	4	4		286	143		429	530	

Important: See the Reference Section for critical supporting data - Reference Nos., Crews, & City Cost Indexes

11454 | Residential Appliances

		CREW	DAILY OUTPUT	LABOR-HOURS	UNIT	2002 BARE COSTS				TOTAL INCL O&P	
						MAT.	LABOR	EQUIP.	TOTAL		
500	2960	Average	L-1	4	4	Ea.	380	143		523	635	500
	3000	Maximum	↓	2	8		520	286		806	1,000	
	3200	Dryer, automatic, minimum	L-2	3	5.333		251	141		392	495	
	3250	Maximum	"	2	8		560	211		771	950	
	3300	Garbage disposer, sink type, minimum	L-1	10	1.600		38.50	57		95.50	128	
	3350	Maximum	"	10	1.600		183	57		240	287	
	3550	Heater, electric, built-in, 1250 watt, ceiling type, minimum	1 Elec	4	2		70	71		141	183	
	3600	Maximum		3	2.667		114	94.50		208.50	267	
	3700	Wall type, minimum		4	2		99.50	71		170.50	215	
	3750	Maximum		3	2.667		132	94.50		226.50	286	
	3900	1500 watt wall type, with blower		4	2		123	71		194	241	
	3950	3000 watt	↓	3	2.667		251	94.50		345.50	415	
	4150	Hood for range, 2 speed, vented, 30" wide, minimum	L-3	5	3.200		35.50	104		139.50	199	
	4200	Maximum		3	5.333		450	174		624	760	
	4300	42" wide, minimum		5	3.200		172	104		276	350	
	4330	Custom		5	3.200		620	104		724	840	
	4350	Maximum	↓	3	5.333		755	174		929	1,100	
	4500	For ventless hood, 2 speed, add					14.65			14.65	16.10	
	4650	For vented 1 speed, deduct from maximum					28			28	31	
	4850	Humidifier, portable, 8 gallons per day					101			101	111	
	5000	15 gallons per day					163			163	180	
	5200	Icemaker, automatic, 20 lb. per day	1 Plum	7	1.143		550	41		591	665	
	5350	51 lb. per day		2	4		805	144		949	1,100	
	6400	Sump pump cellar drainer, pedestal, 1/3 H.P., molded PVC base		3	2.667		89	96		185	243	
	6450	Solid brass	↓	2	4	↓	185	144		329	420	
	6460	Sump pump, see also division 15440-940										
	6650	Washing machine, automatic, minimum	1 Plum	3	2.667	Ea.	330	96		426	510	
	6700	Maximum	"	1	8		650	288		938	1,150	
	6900	Water heater, electric, glass lined, 30 gallon, minimum	L-1	5	3.200		227	114		341	420	
	6950	Maximum		3	5.333		315	190		505	630	
	7100	80 gallon, minimum		2	8		440	286		726	915	
	7150	Maximum	↓	1	16		610	570		1,180	1,525	
	7180	Water heater, gas, glass lined, 30 gallon, minimum	2 Plum	5	3.200		298	115		413	505	
	7220	Maximum		3	5.333		415	192		607	745	
	7260	50 gallon, minimum		2.50	6.400		395	230		625	780	
	7300	Maximum	↓	1.50	10.667	↓	550	385		935	1,175	
	7310	Water heater, see also division 15480-200										
	7350	Water softener, automatic, to 30 grains per gallon	2 Plum	5	3.200	Ea.	420	115		535	635	
	7400	To 100 grains per gallon	"	4	4		550	144		694	820	
	7450	Vent kits for dryers	1 Carp	10	.800	↓	10.50	24		34.50	49	

11471 | Darkroom Processing

		CREW	DAILY OUTPUT	LABOR-HOURS	UNIT	2002 BARE COSTS				TOTAL INCL O&P	
						MAT.	LABOR	EQUIP.	TOTAL		
700	0010	**DARKROOM EQUIPMENT**										700
	0020	Developing sink, 5" deep, 24" x 48"	Q-1	2	8	Ea.	4,000	259		4,259	4,775	
	0050	48" x 52"		1.70	9.412		4,050	305		4,355	4,900	
	0200	10" deep, 24" x 48"		1.70	9.412		4,975	305		5,280	5,925	
	0250	24" x 108"		1.50	10.667		1,800	345		2,145	2,500	
	3500	Washers, round, minimum sheet 11" x 14"	↓	2	8	↓	2,450	259		2,709	3,100	

11470 | Darkroom Equipment

			DAILY	LABOR-			2002 BARE COSTS			TOTAL	
	11471	**Darkroom Processing**	CREW	OUTPUT	HOURS	UNIT	MAT.	LABOR	EQUIP.	TOTAL	INCL O&P
700	3550	Maximum sheet 20" x 24"	Q-1	1	16	Ea.	2,875	520		3,395	3,950	**700**
	3800	Square, minimum sheet 20" x 24"		1	16		2,575	520		3,095	3,625	
	3900	Maximum sheet 50" x 56"	▼	.80	20	▼	4,025	645		4,670	5,400	
	4500	Combination tank sink, tray sink, washers, with										
	4510	dry side tables, average	Q-1	.45	35.556	Ea.	7,650	1,150		8,800	10,100	

11500 | Industrial & Process Equipment

			DAILY	LABOR-			2002 BARE COSTS			TOTAL	
	11520	**Industrial Equipment**	CREW	OUTPUT	HOURS	UNIT	MAT.	LABOR	EQUIP.	TOTAL	INCL O&P
250	0010	**DUST COLLECTION SYSTEMS** Commercial / industrial										**250**
	0120	Central vacuum units										
	0130	Includes stand, filters and motorized shaker										
	0200	500 CFM, 10" inlet, 2 HP	Q-20	2.40	8.333	Ea.	3,250	270		3,520	4,000	
	0220	1000 CFM, 10" inlet, 3 HP		2.20	9.091		3,400	294		3,694	4,175	
	0240	1500 CFM, 10" inlet, 5 HP		2	10		3,475	325		3,800	4,300	
	0260	3000 CFM, 13" inlet, 10 HP		1.50	13.333		9,050	430		9,480	10,600	
	0280	5000 CFM, 16" inlet, 2 @ 10 HP	▼	1	20	▼	9,575	645		10,220	11,500	
	1000	Vacuum tubing, galvanized										
	1100	2-1/8" OD, 16 ga.	Q-9	440	.036	L.F.	1.76	1.15		2.91	3.71	
	1110	2-1/2" OD, 16 ga.		420	.038		2.04	1.20		3.24	4.10	
	1120	3" OD, 16 ga.		400	.040		2.59	1.26		3.85	4.80	
	1130	3-1/2" OD, 16 ga.		380	.042		3.68	1.33		5.01	6.10	
	1140	4" OD, 16 ga.		360	.044		3.76	1.40		5.16	6.30	
	1150	5" OD, 14 ga.		320	.050		6.95	1.58		8.53	10.10	
	1160	6" OD, 14 ga.		280	.057		7.80	1.81		9.61	11.40	
	1170	8" OD, 14 ga.		200	.080		11.80	2.53		14.33	16.90	
	1180	10" OD, 12 ga.		160	.100		11.80	3.16		14.96	17.85	
	1190	12" OD, 12 ga.		120	.133		34	4.21		38.21	43.50	
	1200	14" OD, 12 ga.	▼	80	.200	▼	37.50	6.30		43.80	51	
	1940	Hose, flexible wire reinforced rubber										
	1956	3" dia.	Q-9	400	.040	L.F.	5.55	1.26		6.81	8.05	
	1960	4" dia.		360	.044		7.50	1.40		8.90	10.40	
	1970	5" dia.		320	.050		8.65	1.58		10.23	11.95	
	1980	6" dia.	▼	280	.057	▼	10.10	1.81		11.91	13.90	
	2000	90° Elbow, slip fit										
	2110	2-1/8" dia.	Q-9	70	.229	Ea.	7.35	7.20		14.55	19.25	
	2120	2-1/2" dia.		65	.246		10.30	7.80		18.10	23.50	
	2130	3" dia.		60	.267		13.95	8.45		22.40	28.50	
	2140	3-1/2" dia.		55	.291		17.25	9.20		26.45	33	
	2150	4" dia.		50	.320		21.50	10.10		31.60	39	
	2160	5" dia.		45	.356		40.50	11.25		51.75	62	
	2170	6" dia.		40	.400		56	12.65		68.65	81	
	2180	8" dia.	▼	30	.533	▼	106	16.85		122.85	143	
	2400	45° Elbow, slip fit										
	2410	2-1/8" dia.	Q-9	70	.229	Ea.	6.45	7.20		13.65	18.25	
	2420	2-1/2" dia.		65	.246		9.50	7.80		17.30	22.50	
	2430	3" dia.		60	.267		11.45	8.45		19.90	25.50	
	2440	3-1/2" dia.		55	.291		14.40	9.20		23.60	30	
	2450	4" dia.	▼	50	.320	▼	18.70	10.10		28.80	36	

Important: See the Reference Section for critical supporting data - Reference Nos., Crews, & City Cost Indexes

11520	Industrial Equipment	CREW	DAILY OUTPUT	LABOR-HOURS	UNIT	2002 BARE COSTS				TOTAL INCL O&P
						MAT.	LABOR	EQUIP.	TOTAL	
250										**250**
2460	5″ dia.	Q-9	45	.356	Ea.	31.50	11.25		42.75	52
2470	6″ dia.		40	.400		42.50	12.65		55.15	66.50
2480	8″ dia.	▼	35	.457	▼	83.50	14.45		97.95	115
2800	90° TY, slip fit thru 6″ dia									
2810	2-1/8″ dia.	Q-9	42	.381	Ea.	14.40	12.05		26.45	34.50
2820	2-1/2″ dia.		39	.410		18.70	12.95		31.65	40.50
2830	3″ dia.		36	.444		24.50	14.05		38.55	48.50
2840	3-1/2″ dia.		33	.485		31.50	15.30		46.80	58
2850	4″ dia.		30	.533		45.50	16.85		62.35	76
2860	5″ dia.		27	.593		86	18.75		104.75	124
2870	6″ dia.	▼	24	.667		117	21		138	162
2880	8″ dia., butt end					249			249	274
2890	10″ dia., butt end					450			450	495
2900	12″ dia., butt end					450			450	495
2910	14″ dia., butt end					815			815	895
2920	6 x 4″ dia., butt end					71.50			71.50	78.50
2930	8″ x 4″ dia., butt end					135			135	149
2940	10″ x 4″ dia., butt end					177			177	195
2950	12″ x 4″ dia., butt end				▼	204			204	224
3100	90° Elbow, butt end segmented									
3110	8″ dia., butt end, segmented				Ea.	161			161	177
3120	10″ dia., butt end, segmented					264			264	291
3130	12″ dia., butt end, segmented					335			335	370
3140	14″ dia., butt end, segmented				▼	420			420	460
3200	45° Elbow, butt end segmented									
3210	8″ dia., butt end, segmented				Ea.	137			137	151
3220	10″ dia., butt end, segmented					191			191	210
3230	12″ dia., butt end, segmented					195			195	215
3240	14″ dia., butt end, segmented				▼	240			240	264
3400	All butt end fittings require one coupling per joint.									
3410	Labor for fitting included with couplings.									
3460	Compression coupling, galvanized, neoprene gasket									
3470	2-1/8″ dia.	Q-9	44	.364	Ea.	11.20	11.50		22.70	30
3480	2-1/2″ dia.		44	.364		11.20	11.50		22.70	30
3490	3″ dia.		38	.421		14.70	13.30		28	36.50
3500	3-1/2″ dia.		35	.457		16.45	14.45		30.90	40.50
3510	4″ dia.		33	.485		17.90	15.30		33.20	43
3520	5″ dia.		29	.552		20.50	17.45		37.95	49.50
3530	6″ dia.		26	.615		24	19.45		43.45	56
3540	8″ dia.		22	.727		43.50	23		66.50	83
3550	10″ dia.		20	.800		63.50	25.50		89	109
3560	12″ dia.		18	.889		79	28		107	131
3570	14″ dia.	▼	16	1	▼	112	31.50		143.50	172
3800	Air gate valves, galvanized									
3810	2-1/8″ dia.	Q-9	30	.533	Ea.	85	16.85		101.85	120
3820	2-1/2″ dia.		28	.571		89.50	18.05		107.55	127
3830	3″ dia.		26	.615		104	19.45		123.45	144
3840	4″ dia.		23	.696		136	22		158	184
3850	6″ dia.	▼	18	.889	▼	188	28		216	250

EQUIPMENT 11

11620	Laboratory Equipment	CREW	DAILY OUTPUT	LABOR-HOURS	UNIT	2002 BARE COSTS				TOTAL INCL O&P
						MAT.	LABOR	EQUIP.	TOTAL	
350	0010 **LABORATORY EQUIPMENT**									350
	0020 Cabinets, base, door units, metal	2 Carp	18	.889	L.F.	134	26.50		160.50	189
	0300 Drawer units	"	18	.889	"	289	26.50		315.50	360
	2000 Fume hood, with countertop & base, not including HVAC									
	2020 Simple, minimum	2 Carp	5.40	2.963	L.F.	505	89		594	695
	2050 Complex, including fixtures		2.40	6.667		1,200	200		1,400	1,600
	2100 Special, maximum	↓	1.70	9.412	↓	1,250	282		1,532	1,825
	2200 Ductwork, minimum	2 Shee	1	16	Hood	835	560		1,395	1,775
	2250 Maximum	"	.50	32	"	4,225	1,125		5,350	6,375
	2300 Service fixtures, average				Ea.	43			43	47
	2500 For sink assembly with hot and cold water, add	1 Plum	1.40	5.714		525	205		730	885
	2550 Glassware washer, undercounter, minimum	L-1	1.80	8.889		5,975	315		6,290	7,050
	2600 Maximum	"	1	16		7,675	570		8,245	9,300
	2800 Sink, one piece plastic, flask wash, hose, free standing	1 Plum	1.60	5		1,325	180		1,505	1,725
	2850 Epoxy resin sink, 25" x 16" x 10"	"	2	4	↓	147	144		291	380
	4200 Alternate pricing method: as percent of lab furniture									
	4400 Installation, not incl. plumbing & duct work				% Furn.					22%
	4800 Plumbing, final connections, simple system									10%
	5000 Moderately complex system									15%
	5200 Complex system									20%
	5400 Electrical, simple system									10%
	5600 Moderately complex system									20%
	5800 Complex system				↓					35%
	6000 Safety equipment, eye wash, hand held				Ea.	285			285	315
	6200 Deluge shower				"	550			550	605
	6300 Rule of thumb: lab furniture including installation & connection									
	6320 High school				S.F.					29
	6340 College									42.75
	6360 Clinical, health care									36.75
	6380 Industrial				↓					59.25

11710	Medical Equipment	CREW	DAILY OUTPUT	LABOR-HOURS	UNIT	2002 BARE COSTS				TOTAL INCL O&P
						MAT.	LABOR	EQUIP.	TOTAL	
500	0010 **MEDICAL EQUIPMENT**									500
	0015 Autopsy table, standard	1 Plum	1	8	Ea.	7,075	288		7,363	8,200
	0020 Deluxe	↓	.60	13.333		10,500	480		10,980	12,300
	0700 Distiller, water, steam heated, 50 gal. capacity	↓	1.40	5.714		14,700	205		14,905	16,500
	1800 Heat therapy unit, humidified, 26" x 78" x 28"					2,700			2,700	2,975
	3200 Mortuary refrigerator, end operated, 2 capacity					10,900			10,900	12,000
	3300 6 capacity					20,000			20,000	22,000
	5600 Sterilizers, floor loading, 26" x 62" x 42", single door, steam					128,500			128,500	141,500
	5650 Double door, steam					165,000			165,000	181,500
	5800 General purpose, 20" x 20" x 38", single door					12,400			12,400	13,600
	6000 Portable, counter top, steam, minimum					3,075			3,075	3,400
	6020 Maximum					4,825			4,825	5,300
	6050 Portable, counter top, gas, 17"x15"x32-1/2"					31,800			31,800	35,000
	6150 Manual washer/sterilizer, 16"x16"x26"	1 Plum	2	4	↓	43,600	144		43,744	48,200
	6200 Steam generators, electric 10 KW to 180 KW, freestanding									
	6250 Minimum	1 Elec	3	2.667	Ea.	6,475	94.50		6,569.50	7,275

Important: See the Reference Section for critical supporting data - Reference Nos., Crews, & City Cost Indexes

11700 | Medical Equipment

11710 | Medical Equipment

		CREW	DAILY OUTPUT	LABOR-HOURS	UNIT	2002 BARE COSTS MAT.	LABOR	EQUIP.	TOTAL	TOTAL INCL O&P	
500	6300	Maximum	1 Elec	.70	11.429	Ea.	23,300	405		23,705	26,300	**500**
	8100	Utensil washer-sanitizer	1 Plum	2	4		8,850	144		8,994	9,950	
	8200	Bed pan washer-sanitizer	"	2	4		5,775	144		5,919	6,575	
	8400	Whirlpool bath, mobile, sst, 18" x 24" x 60"					3,650			3,650	4,025	
	8450	Fixed, incl. mixing valves	1 Plum	2	4	↓	7,450	144		7,594	8,425	

11740 | Dental Equipment

		CREW	DAILY OUTPUT	LABOR-HOURS	UNIT	MAT.	LABOR	EQUIP.	TOTAL	TOTAL INCL O&P	
200	0010	**DENTAL EQUIPMENT**										**200**
	0020	Central suction system, minimum	1 Plum	1.20	6.667	Ea.	2,850	240		3,090	3,475	
	0100	Maximum	"	.90	8.889		21,900	320		22,220	24,600	
	0600	Chair, electric or hydraulic, minimum	1 Skwk	.50	16		3,575	495		4,070	4,700	
	0700	Maximum		.25	32		7,575	990		8,565	9,875	
	2000	Light, ceiling mounted, minimum		8	1		1,850	31		1,881	2,100	
	2100	Maximum	↓	8	1		3,750	31		3,781	4,150	
	2200	Unit light, minimum	2 Skwk	5.33	3.002		1,375	93		1,468	1,675	
	2210	Maximum		5.33	3.002		4,550	93		4,643	5,150	
	2220	Track light, minimum		3.20	5		2,450	155		2,605	2,950	
	2230	Maximum	↓	3.20	5		5,400	155		5,555	6,200	
	2300	Sterilizers, steam portable, minimum					3,125			3,125	3,450	
	2350	Maximum					10,800			10,800	11,900	
	2600	Steam, institutional					13,500			13,500	14,800	
	2650	Dry heat, electric, portable, 3 trays				↓	1,050			1,050	1,150	

For information about Means Estimating Seminars, see yellow pages 11 and 12 in back of book

EQUIPMENT 11

91

Division Notes

		CREW	DAILY OUTPUT	LABOR-HOURS	UNIT	2002 BARE COSTS				TOTAL INCL O&P
						MAT.	LABOR	EQUIP.	TOTAL	

Division 13
Special Construction

Estimating Tips

General

- The items and systems in this division are usually estimated, purchased, supplied and installed as a unit by one or more subcontractors. The estimator must ensure that all parties are operating from the same set of specifications and assumptions and that all necessary items are estimated and will be provided. Many times the complex items and systems are covered but the more common ones such as excavation or a crane are overlooked for the very reason that everyone assumes nobody could miss them. The estimator should be the central focus and be able to ensure that all systems are complete.

- Another area where problems can develop in this division is at the interface between systems. The estimator must ensure, for instance, that anchor bolts, nuts and washers are estimated and included for the air-supported structures and pre-engineered buildings to be bolted to their foundations.

Utility supply is a common area where essential items or pieces of equipment can be missed or overlooked due to the fact that each subcontractor may feel it is the others' responsibility. The estimator should also be aware of certain items which may be supplied as part of a package but installed by others, and ensure that the installing contractor's estimate includes the cost of installation. Conversely, the estimator must also ensure that items are not costed by two different subcontractors, resulting in an inflated overall estimate.

13120 Pre-Engineered Structures

- The foundations and floor slab, as well as rough mechanical and electrical, should be estimated, as this work is required for the assembly and erection of the structure. Generally, as noted in the book, the pre-engineered building comes as a shell and additional features must be included by the estimator. Here again, the estimator must have a clear understanding of the scope of each portion of the work and all the necessary interfaces.

13200 Storage Tanks

- The prices in this subdivision for above and below ground storage tanks do not include foundations or hold-down slabs. The estimator should refer to Divisions 2 and 3 for foundation system pricing. In addition to the foundations, required tank accessories such as tank gauges, leak detection devices, and additional manholes and piping must be added to the tank prices.

Reference Numbers

Reference numbers are shown in bold squares at the beginning of some major classifications. These numbers refer to related items in the Reference Section. The reference information may be an estimating procedure, an alternate pricing method or technical information.

Note: Not all subdivisions listed here necessarily appear in this publication.

13035	Special Purpose Rooms	CREW	DAILY OUTPUT	LABOR-HOURS	UNIT	2002 BARE COSTS				TOTAL INCL O&P
						MAT.	LABOR	EQUIP.	TOTAL	
200	0010	**CLEAN ROOMS**									**200**
	1100	Clean room, soft wall, 12' x 12', Class 100	1 Carp	.18	44.444	Ea.	12,200	1,325		13,525	15,500
	1110	Class 1,000		.18	44.444		9,450	1,325		10,775	12,500
	1120	Class 10,000		.21	38.095		8,050	1,150		9,200	10,600
	1130	Class 100,000	▼	.21	38.095	▼	7,450	1,150		8,600	9,975
	2800	Ceiling grid support, slotted channel struts 4'-0" O.C., ea. way				S.F.					6.50
	3000	Ceiling panel, vinyl coated foil on mineral substrate									
	3020	Sealed, non-perforated				S.F.					1.40
	4000	Ceiling panel seal, silicone sealant, 150 L.F./gal.	1 Carp	150	.053	L.F.	.22	1.60		1.82	2.73
	4100	Two sided adhesive tape	"	240	.033	"	.10	1		1.10	1.67
	4200	Clips, one per panel				Ea.	.85			.85	.94
	6000	HEPA filter,2'x4',99.97% eff.,3" dp beveled frame(silicone seal)					260			260	286
	6040	6" deep skirted frame (channel seal)					300			300	330
	6100	99.99% efficient, 3" deep beveled frame (silicone seal)					280			280	310
	6140	6" deep skirted frame (channel seal)					320			320	350
	6200	99.999% efficient, 3" deep beveled frame (silicone seal)					320			320	350
	6240	6" deep skirted frame (channel seal)				▼	360			360	395
	7000	Wall panel systems, including channel strut framing									
	7020	Polyester coated aluminum, particle board				S.F.					20
	7100	Porcelain coated aluminum, particle board									35
	7400	Wall panel support, slotted channel struts, to 12' high				▼					18
700	0010	**REFRIGERATION** Curbs, 12" high, 4" thick, concrete	2 Carp	58	.276	L.F.	3.02	8.30		11.32	16.20	**700**
	1000	Doors, see division 08341-200									
	2400	Finishes, 2 coat portland cement plaster, 1/2" thick	1 Plas	48	.167	S.F.	.83	4.68		5.51	8.10
	2500	For galvanized reinforcing mesh, add	1 Lath	335	.024		.56	.69		1.25	1.64
	2700	3/16" thick latex cement	1 Plas	88	.091		1.42	2.55		3.97	5.50
	2900	For glass cloth reinforced ceilings, add	"	450	.018		.34	.50		.84	1.14
	3100	Fiberglass panels, 1/8" thick	1 Carp	149.45	.054		1.88	1.61		3.49	4.57
	3200	Polystyrene, plastic finish ceiling, 1" thick		274	.029		1.74	.88		2.62	3.27
	3400	2" thick		274	.029		1.97	.88		2.85	3.53
	3500	4" thick	▼	219	.037		2.20	1.10		3.30	4.13
	3800	Floors, concrete, 4" thick	1 Cefi	93	.086		.82	2.47		3.29	4.56
	3900	6" thick	"	85	.094	▼	1.24	2.70		3.94	5.35
	4000	Insulation, 1" to 6" thick, cork				B.F.	.80			.80	.88
	4100	Urethane					.79			.79	.87
	4300	Polystyrene, regular					.54			.54	.59
	4400	Bead board				▼	.41			.41	.45
	4600	Installation of above, add per layer	2 Carp	657.60	.024	S.F.	.25	.73		.98	1.42
	4700	Wall and ceiling juncture		298.90	.054	L.F.	1.28	1.61		2.89	3.91
	4900	Partitions, galvanized sandwich panels, 4" thick, stock		219.20	.073	S.F.	5.40	2.19		7.59	9.35
	5000	Aluminum or fiberglass	▼	219.20	.073	"	5.95	2.19		8.14	9.90
	5200	Prefab walk-in, 7'-6" high, aluminum, incl. door & floors,									
	5210	not incl. partitions or refrigeration, 6' x 6' O.D. nominal	2 Carp	54.80	.292	SF Flr.	97	8.75		105.75	121
	5500	10' x 10' O.D. nominal		82.20	.195		78	5.85		83.85	95
	5700	12' x 14' O.D. nominal		109.60	.146		70	4.38		74.38	84
	5800	12' x 20' O.D. nominal	▼	109.60	.146		61	4.38		65.38	74
	6100	For 8'-6" high, add					5%				
	6300	Rule of thumb for complete units, w/o doors & refrigeration, cooler	2 Carp	146	.110		88	3.29		91.29	102
	6400	Freezer		109.60	.146	▼	104	4.38		108.38	121
	6600	Shelving, plated or galvanized, steel wire type		360	.044	SF Hor.	7.70	1.33		9.03	10.55
	6700	Slat shelf type	▼	375	.043		9.50	1.28		10.78	12.45
	6900	For stainless steel shelving, add				▼	300%				
	7000	Vapor barrier, on wood walls	2 Carp	1,644	.010	S.F.	.11	.29		.40	.57
	7200	On masonry walls	"	1,315	.012	"	.29	.37		.66	.89
	7500	For air curtain doors, see division 08344-120									

Important: See the Reference Section for critical supporting data - Reference Nos., Crews, & City Cost Indexes

13030 | Special Purpose Rooms

13035 | Special Purpose Rooms

		CREW	DAILY OUTPUT	LABOR-HOURS	UNIT	2002 BARE COSTS				TOTAL INCL O&P	
						MAT.	LABOR	EQUIP.	TOTAL		
800	0010	**SAUNA** Prefabricated, incl. heater & controls, 7' high, 6' x 4', C/C	L-7	2.20	12.727	Ea.	3,850	370		4,220	4,825	800
	0050	6' x 4', C/P		2	14		2,625	405		3,030	3,500	
	0400	6' x 5', C/C		2	14		3,350	405		3,755	4,300	
	0450	6' x 5', C/P		2	14		2,975	405		3,380	3,900	
	0600	6' x 6', C/C		1.80	15.556		4,475	450		4,925	5,625	
	0650	6' x 6', C/P		1.80	15.556		3,125	450		3,575	4,150	
	0800	6' x 9', C/C		1.60	17.500		4,325	505		4,830	5,550	
	0850	6' x 9', C/P		1.60	17.500		4,325	505		4,830	5,550	
	1000	8' x 12', C/C		1.10	25.455		6,875	735		7,610	8,675	
	1050	8' x 12', C/P		1.10	25.455		6,000	735		6,735	7,725	
	1200	8' x 8', C/C		1.40	20		5,125	580		5,705	6,525	
	1250	8' x 8', C/P		1.40	20		4,525	580		5,105	5,875	
	1400	8' x 10', C/C		1.20	23.333		5,600	675		6,275	7,225	
	1450	8' x 10', C/P		1.20	23.333		4,925	675		5,600	6,450	
	1600	10' x 12', C/C		1	28		7,475	810		8,285	9,475	
	1650	10' x 12', C/P		1	28		6,625	810		7,435	8,550	
	2500	Heaters only (incl. above), wall mounted, to 200 C.F.					600			600	660	
	2750	To 300 C.F.					625			625	690	
	3000	Floor standing, to 720 C.F., 10,000 watts, w/controls	1 Elec	3	2.667		1,125	94.50		1,219.50	1,375	
	3250	To 1,000 C.F., 16,000 watts	"	3	2.667		1,650	94.50		1,744.50	1,950	
940	0010	**STEAM BATH** Heater, timer & head, single, to 140 C.F.	1 Plum	1.20	6.667	Ea.	875	240		1,115	1,325	940
	0500	To 300 C.F.		1.10	7.273		1,000	261		1,261	1,500	
	1000	Commercial size, with blow-down assembly, to 800 C.F.		.90	8.889		3,950	320		4,270	4,825	
	1500	To 2500 C.F.		.80	10		4,150	360		4,510	5,125	
	2000	Multiple, motels, apts., 2 baths, w/ blow-down assm., 500 C.F.	Q-1	1.30	12.308		3,150	400		3,550	4,050	
	2500	4 baths	"	.70	22.857		3,825	740		4,565	5,325	
	2700	Conversion unit for residential tub, including door					1,850			1,850	2,025	

13120 | Pre-Engineered Structures

13128 | Pre-Eng. Structures

		CREW	DAILY OUTPUT	LABOR-HOURS	UNIT	2002 BARE COSTS				TOTAL INCL O&P	
						MAT.	LABOR	EQUIP.	TOTAL		
200	0010	**COMFORT STATIONS** Prefab., stock, w/doors, windows & fixt.										200
	0100	Not incl. interior finish or electrical										
	0300	Mobile, on steel frame, minimum				S.F.	36.50			36.50	40	
	0350	Maximum					57.50			57.50	63.50	
	0400	Permanent, including concrete slab, minimum	B-12J	50	.320		142	9.45	14.75	166.20	186	
	0500	Maximum	"	43	.372		206	11	17.15	234.15	262	
	0600	Alternate pricing method, mobile, minimum				Fixture	1,775			1,775	1,950	
	0650	Maximum					2,650			2,650	2,925	
	0700	Permanent, minimum	B-12J	.70	22.857		10,200	675	1,050	11,925	13,400	
	0750	Maximum	"	.50	32		17,000	945	1,475	19,420	21,800	

		CREW	DAILY OUTPUT	LABOR-HOURS	UNIT	2002 BARE COSTS				TOTAL INCL O&P	
13151	**Swimming Pools**					MAT.	LABOR	EQUIP.	TOTAL		
200	0010	**SWIMMING POOLS** Residential in-ground, vinyl lined, concrete sides										**200**
	0020	Sides including equipment, sand bottom	B-52	300	.187	SF Surf	10.60	5.15	1.15	16.90	21	
	0100	Metal or polystyrene sides	B-14	410	.117		8.85	2.91	.44	12.20	14.75	
	0200	Add for vermiculite bottom				↓	.68			.68	.75	
	0500	Gunite bottom and sides, white plaster finish										
	0600	12' x 30' pool	B-52	145	.386	SF Surf	17.55	10.70	2.38	30.63	38.50	
	0720	16' x 32' pool		155	.361		15.85	10	2.23	28.08	35.50	
	0750	20' x 40' pool	↓	250	.224	↓	14.15	6.20	1.38	21.73	26.50	
	0810	Concrete bottom and sides, tile finish										
	0820	12' x 30' pool	B-52	80	.700	SF Surf	17.75	19.40	4.32	41.47	54.50	
	0830	16' x 32' pool		95	.589		14.65	16.30	3.64	34.59	45.50	
	0840	20' x 40' pool	↓	130	.431	↓	11.65	11.90	2.66	26.21	34.50	
	1100	Motel, gunite with plaster finish, incl. medium										
	1150	capacity filtration & chlorination	B-52	115	.487	SF Surf	21.50	13.50	3	38	48.50	
	1200	Municipal, gunite with plaster finish, incl. high										
	1250	capacity filtration & chlorination	B-52	100	.560	SF Surf	28	15.50	3.46	46.96	59	
	1350	Add for formed gutters				L.F.	48.50			48.50	53.50	
	1360	Add for stainless steel gutters				"	144			144	158	
	1700	Filtration and deck equipment only, as % of total				Total				20%	20%	
	1800	Deck equipment, rule of thumb, 20' x 40' pool				SF Pool					1.30	
	1900	5000 S.F. pool				"					1.90	
	3000	Painting pools, preparation + 3 coats, 20' x 40' pool, epoxy	2 Pord	.33	48.485	Total	605	1,325		1,930	2,675	
	3100	Rubber base paint, 18 gallons	"	.33	48.485		460	1,325		1,785	2,500	
	3500	42' x 82' pool, 75 gallons, epoxy paint	3 Pord	.14	171		2,550	4,650		7,200	9,875	
	3600	Rubber base paint	"	.14	171	↓	1,975	4,650		6,625	9,225	
700	0010	**SWIMMING POOL EQUIPMENT** Diving stand, stainless steel, 3 meter	2 Carp	.40	40	Ea.	4,650	1,200		5,850	6,975	**700**
	0300	1 meter	"	2.70	5.926	"	3,775	178		3,953	4,425	
	0900	Filter system, sand or diatomite type, incl. pump, 6,000 gal./hr.	2 Plum	1.80	8.889	Total	1,050	320		1,370	1,625	
	1020	Add for chlorination system, 800 S.F. pool		3	5.333	Ea.	199	192		391	510	
	1040	5,000 S.F. pool	↓	3	5.333	"	1,200	192		1,392	1,600	
	1100	Gutter system, stainless steel, with grating, stock,										
	1110	contains supply and drainage system	E-1	20	1.200	L.F.	160	40	4.43	204.43	250	
	1120	Integral gutter and 5' high wall system, stainless steel	"	10	2.400	"	240	80	8.85	328.85	410	
	2100	Lights, underwater, 12 volt, with transformer, 300 watt	1 Elec	.40	20	Ea.	116	710		826	1,175	
	2200	110 volt, 500 watt, standard		.40	20		126	710		836	1,200	
	2400	Low water cutoff type	↓	.40	20	↓	130	710		840	1,200	
	2800	Heaters, see division 15510-880										

		CREW	DAILY OUTPUT	LABOR-HOURS	UNIT	2002 BARE COSTS				TOTAL INCL O&P	
13176	**Ice Rinks**					MAT.	LABOR	EQUIP.	TOTAL		
500	0010	**ICE SKATING** Equipment incl. refrigeration, plumbing & cooling										**500**
	0020	coils & concrete slab, 85' x 200' rink										
	0300	55° system, 5 mos., 100 ton				Total					451,000	
	0700	90° system, 12 mos., 135 ton				"					500,500	
	1000	Dasher boards, 1/2" H.D. polyethylene faced steel frame, 3' acrylic										
	1020	screen at sides, 5' acrylic ends, 85' x 200'	F-5	.06	533	Ea.	115,000	16,300		131,300	152,000	
	1100	Fiberglass & aluminum construction, same sides and ends	"	.06	533	↓	160,000	16,300		176,300	201,500	
	1200	Subsoil heating system (recycled from compressor), 85' x 200'	Q-7	.27	118		25,000	4,100		29,100	33,700	

Important: See the Reference Section for critical supporting data - Reference Nos., Crews, & City Cost Indexes

13175 | Ice Rinks

	13176	Ice Rinks	CREW	DAILY OUTPUT	LABOR-HOURS	UNIT	2002 BARE COSTS MAT.	LABOR	EQUIP.	TOTAL	TOTAL INCL O&P	
500	1300	Subsoil insulation, 2 lb. polystyrene with vapor barrier, 85' x 200'	2 Carp	.14	114	Ea.	30,000	3,425		33,425	38,300	**500**

13200 | Storage Tanks

	13201	Storage Tanks	CREW	DAILY OUTPUT	LABOR-HOURS	UNIT	2002 BARE COSTS MAT.	LABOR	EQUIP.	TOTAL	TOTAL INCL O&P	
200	0010	**ELEVATED STORAGE TANKS** Not incl. pipe or pumps										**200**
	3000	Elevated water tanks, 100' to bottom capacity line, incl painting										
	3010	50,000 gallons				Ea.					166,000	
	3300	100,000 gallons									229,000	
	3400	250,000 gallons									316,000	
	3600	500,000 gallons									510,000	
	3700	750,000 gallons									703,000	
	3900	1,000,000 gallons									812,000	
300	0010	**GROUND TANKS** Not incl. pipe or pumps, prestress conc., 250,000 gal.				Ea.					280,000	**300**
	0100	500,000 gallons									380,000	
	0300	1,000,000 gallons									540,000	
	0400	2,000,000 gallons									816,000	
	0600	4,000,000 gallons									1,290,000	
	0700	6,000,000 gallons									1,760,000	
	0750	8,000,000 gallons									2,240,000	
	0800	10,000,000 gallons									2,700,000	
	0900	Steel, ground level, ht/dia less than 1, not incl. fdn, 100,000 gallons									99,500	
	1000	250,000 gallons									137,000	
	1200	500,000 gallons									209,000	
	1250	750,000 gallons									249,000	
	1300	1,000,000 gallons									336,000	
	1500	2,000,000 gallons									615,000	
	1600	4,000,000 gallons									955,000	
	1800	6,000,000 gallons									1,325,000	
	1850	8,000,000 gallons									1,735,000	
	1900	10,000,000 gallons									1,962,000	
	2100	Steel standpipes, hgt/dia more than 1, 100' to overflow, no fdn										
	2200	500,000 gallons				Ea.					275,000	
	2400	750,000 gallons									335,000	
	2500	1,000,000 gallons									408,000	
	2700	1,500,000 gallons									570,000	
	2800	2,000,000 gallons									700,000	
	3000	Steel, storage, above ground, including cradles, coating,										
	3020	fittings, not including fdn, pumps or piping										
	3040	Single wall, interior, 275 gallon	Q-5	5	3.200	Ea.	241	104		345	420	
	3060	550 gallon	"	2.70	5.926		1,150	193		1,343	1,550	
	3080	1,000 gallon	Q-7	5	6.400		1,800	221		2,021	2,300	
	3100	1,500 gallon		4.75	6.737		2,700	232		2,932	3,325	
	3120	2,000 gallon		4.60	6.957		3,025	240		3,265	3,675	
	3140	5,000 gallon		3.20	10		4,200	345		4,545	5,150	
	3150	10,000 gallon		2	16		11,300	550		11,850	13,300	
	3160	15,000 gallon		1.70	18.824		13,700	650		14,350	16,100	
	3170	20,000 gallon		1.45	22.069		17,300	760		18,060	20,200	
	3180	25,000 gallon capacity		1.30	24.615		19,200	850		20,050	22,400	

SPECIAL CONSTRUCTION 13

13201	Storage Tanks	CREW	DAILY OUTPUT	LABOR-HOURS	UNIT	2002 BARE COSTS				TOTAL INCL O&P
						MAT.	LABOR	EQUIP.	TOTAL	
300 3190	30,000 gallon	Q-7	1.10	29.091	Ea.	23,900	1,000		24,900	27,800 **300**
3320	Double wall, 500 gallon capacity	Q-5	2.40	6.667		2,725	217		2,942	3,300
3330	2000 gallon capacity	Q-7	4.15	7.711		5,075	266		5,341	6,000
3340	4000 gallon capacity		3.60	8.889		9,025	305		9,330	10,400
3350	6000 gallon capacity		2.40	13.333		10,800	460		11,260	12,500
3360	8000 gallon capacity		2	16		12,300	550		12,850	14,400
3370	10000 gallon capacity		1.80	17.778		13,600	615		14,215	15,800
3380	15000 gallon capacity		1.50	21.333		20,200	735		20,935	23,300
3390	20000 gallon capacity		1.30	24.615		23,000	850		23,850	26,600
3400	25000 gallon capacity		1.15	27.826		28,000	960		28,960	32,300
3410	30000 gallon capacity	▼	1	32	▼	30,700	1,100		31,800	35,400
4000	Fixed roof oil storage tanks, steel, (1 BBL=42 GAL w/ fdn 3'D x 1'W)									
4200	5,000 barrels				Ea.					144,000
4300	25,000 barrels									202,000
4500	55,000 barrels									342,000
4600	100,000 barrels									550,000
4800	150,000 barrels									775,000
4900	225,000 barrels									1,168,000
5100	Floating roof gasoline tanks, steel, 5,000 barrels									122,000
5200	25,000 barrels									273,000
5400	55,000 barrels									413,000
5500	100,000 barrels									625,000
5700	150,000 barrels									846,000
5800	225,000 barrels									1,173,000
6000	Wood tanks, ground level, 2" cypress, 3,000 gallons	C-1	.19	168		5,775	4,775		10,550	13,800
6100	2-1/2" cypress, 10,000 gallons		.12	266		15,600	7,575		23,175	29,000
6300	3" redwood or 3" fir, 20,000 gallons		.10	320		23,800	9,075		32,875	40,300
6400	30,000 gallons		.08	400		29,500	11,300		40,800	50,000
6600	45,000 gallons	▼	.07	457	▼	44,600	13,000		57,600	69,500
6700	Larger sizes, minimum				Gal.					.66
6900	Maximum				"					.81
7000	Vinyl coated fabric pillow tanks, freestanding, 5,000 gallons	4 Clab	4	8	Ea.	3,525	188		3,713	4,175
7100	Supporting embankment not included, 25,000 gallons	6 Clab	2	24		6,700	565		7,265	8,250
7200	50,000 gallons	8 Clab	1.50	42.667		12,700	1,000		13,700	15,500
7300	100,000 gallons	9 Clab	.90	80		25,900	1,875		27,775	31,400
7400	150,000 gallons		.50	144		31,600	3,375		34,975	40,000
7500	200,000 gallons		.40	180		37,600	4,225		41,825	48,000
7600	250,000 gallons	▼	.30	240	▼	41,700	5,625		47,325	54,500
800 0010	**UNDERGROUND STORAGE TANKS**									**800**
0210	Fiberglass, underground, single wall, U.L. listed, not including									
0220	manway or hold-down strap									
0225	550 gallon capacity	Q-5	2.67	5.993	Ea.	1,450	195		1,645	1,900
0230	1,000 gallon capacity	"	2.46	6.504		1,825	212		2,037	2,325
0240	2,000 gallon capacity	Q-7	4.57	7.002		2,375	242		2,617	2,975
0250	4,000 gallon capacity		3.55	9.014		3,425	310		3,735	4,250
0260	6,000 gallon capacity		2.67	11.985		3,825	415		4,240	4,825
0270	8,000 gallon capacity		2.29	13.974		4,800	480		5,280	6,000
0280	10,000 gallon capacity		2	16		5,200	550		5,750	6,550
0282	12,000 gallon capacity		1.88	17.021		7,100	585		7,685	8,675
0284	15,000 gallon capacity		1.68	19.048		7,575	655		8,230	9,350
0290	20,000 gallon capacity		1.45	22.069		10,500	760		11,260	12,700
0300	25,000 gallon capacity		1.28	25		17,400	865		18,265	20,400
0320	30,000 gallon capacity		1.14	28.070		21,700	970		22,670	25,400
0340	40,000 gallon capacity		.89	35.955		30,500	1,250		31,750	35,400
0360	48,000 gallon capacity	▼	.81	39.506		50,000	1,375		51,375	57,000
0500	For manway, fittings and hold-downs, add				▼	20%	15%			

Important: See the Reference Section for critical supporting data - Reference Nos., Crews, & City Cost Indexes

13 SPECIAL CONSTRUCTION

13201	Storage Tanks	CREW	DAILY OUTPUT	LABOR-HOURS	UNIT	2002 BARE COSTS				TOTAL INCL O&P
						MAT.	LABOR	EQUIP.	TOTAL	
800 0600	For manways, add				Ea.	840			840	925 **800**
1000	For helical heating coil, add	Q-5	2.50	6.400	↓	2,125	209		2,334	2,675
1020	Fiberglass, underground, double wall, U.L. listed									
1030	includes manways, not incl. hold-down straps									
1040	600 gallon capacity	Q-5	2.42	6.612	Ea.	3,200	215		3,415	3,825
1050	1,000 gallon capacity	"	2.25	7.111		4,250	232		4,482	5,025
1060	2,500 gallon capacity	Q-7	4.16	7.692		6,050	265		6,315	7,050
1070	3,000 gallon capacity		3.90	8.205		6,500	283		6,783	7,575
1080	4,000 gallon capacity		3.64	8.791		7,100	305		7,405	8,250
1090	6,000 gallon capacity		2.42	13.223		8,225	455		8,680	9,750
1100	8,000 gallon capacity		2.08	15.385		10,400	530		10,930	12,200
1110	10,000 gallon capacity		1.82	17.582		10,800	605		11,405	12,800
1120	12,000 gallon capacity		1.70	18.824		13,200	650		13,850	15,500
1122	15,000 gallon capacity		1.52	21.053		18,600	725		19,325	21,600
1124	20,000 gallon capacity		1.33	24.060		22,600	830		23,430	26,200
1126	25,000 gallon capacity		1.16	27.586		29,600	950		30,550	34,000
1128	30,000 gallon capacity	↓	1.03	31.068		35,400	1,075		36,475	40,600
1140	For hold-down straps, add				↓	2%	10%			
1150	For hold-downs 500-4000 gal, add	Q-7	16	2	Set	162	69		231	282
1160	For hold-downs 5000-15000 gal, add		8	4		325	138		463	565
1170	For hold-downs 20,000 gal, add		5.33	6.004		485	207		692	850
1180	For hold-downs 25,000 gal, add		4	8		650	276		926	1,125
1190	For hold-downs 30,000 gal, add	↓	2.60	12.308	↓	975	425		1,400	1,725
2210	Fiberglass, underground, single wall, U.L. listed, including									
2220	hold-down straps, no manways									
2225	550 gallon capacity	Q-5	2	8	Ea.	1,625	261		1,886	2,175
2230	1,000 gallon capacity	"	1.88	8.511		1,975	277		2,252	2,600
2240	2,000 gallon capacity	Q-7	3.55	9.014		2,525	310		2,835	3,250
2250	4,000 gallon capacity		2.90	11.034		3,600	380		3,980	4,525
2260	6,000 gallon capacity		2	16		4,150	550		4,700	5,375
2270	8,000 gallon capacity		1.78	17.978		5,125	620		5,745	6,550
2280	10,000 gallon capacity		1.60	20		5,525	690		6,215	7,125
2282	12,000 gallon capacity		1.52	21.053		7,425	725		8,150	9,250
2284	15,000 gallon capacity		1.39	23.022		7,900	795		8,695	9,900
2290	20,000 gallon capacity		1.14	28.070		11,000	970		11,970	13,600
2300	25,000 gallon capacity		.96	33.333		18,000	1,150		19,150	21,500
2320	30,000 gallon capacity	↓	.80	40	↓	22,700	1,375		24,075	27,100
3020	Fiberglass, underground, double wall, U.L. listed									
3030	includes manways and hold-down straps									
3040	600 gallon capacity	Q-5	1.86	8.602	Ea.	3,350	280		3,630	4,100
3050	1,000 gallon capacity	"	1.70	9.412		4,425	305		4,730	5,300
3060	2,500 gallon capacity	Q-7	3.29	9.726		6,225	335		6,560	7,325
3070	3,000 gallon capacity		3.13	10.224		6,675	355		7,030	7,875
3080	4,000 gallon capacity		2.93	10.921		7,275	375		7,650	8,575
3090	6,000 gallon capacity		1.86	17.204		8,550	595		9,145	10,300
3100	8,000 gallon capacity		1.65	19.394		10,700	670		11,370	12,800
3110	10,000 gallon capacity		1.48	21.622		11,100	745		11,845	13,300
3120	12,000 gallon capacity		1.40	22.857		13,500	790		14,290	16,100
3122	15,000 gallon capacity		1.28	25		18,900	865		19,765	22,100
3124	20,000 gallon capacity		1.06	30.189		23,100	1,050		24,150	27,000
3126	25,000 gallon capacity		.90	35.556		30,200	1,225		31,425	35,200
3128	30,000 gallon capacity	↓	.74	43.243	↓	36,400	1,500		37,900	42,300
5000	Steel underground, sti-P3, set in place, not incl. hold-down bars.									
5500	Excavation, pad, pumps and piping not included									
5510	Single wall, 500 gallon capacity, 7 gauge shell	Q-5	2.70	5.926	Ea.	820	193		1,013	1,200
5520	1,000 gallon capacity, 7 gauge shell	"	2.50	6.400	↓	1,175	209		1,384	1,600

SPECIAL CONSTRUCTION 13

13201	Storage Tanks	CREW	DAILY OUTPUT	LABOR-HOURS	UNIT	2002 BARE COSTS				TOTAL INCL O&P	
						MAT.	LABOR	EQUIP.	TOTAL		
800											800
5530	2,000 gallon capacity, 1/4" thick shell	Q-7	4.60	6.957	Ea.	2,425	240		2,665	3,025	
5535	2,500 gallon capacity, 7 gauge shell	Q-5	3	5.333		2,500	174		2,674	3,000	
5540	5,000 gallon capacity, 1/4" thick shell	Q-7	3.20	10		4,675	345		5,020	5,650	
5560	10,000 gallon capacity, 1/4" thick shell		2	16		7,600	550		8,150	9,175	
5580	15,000 gallon capacity, 5/16" thick shell		1.70	18.824		10,900	650		11,550	12,900	
5600	20,000 gallon capacity, 5/16" thick shell		1.50	21.333		13,500	735		14,235	16,000	
5610	25,000 gallon capacity, 3/8" thick shell		1.30	24.615		17,600	850		18,450	20,700	
5620	30,000 gallon capacity, 3/8" thick shell		1.10	29.091		22,300	1,000		23,300	26,000	
5630	40,000 gallon capacity, 3/8" thick shell		.90	35.556		30,700	1,225		31,925	35,600	
5640	50,000 gallon capacity, 3/8" thick shell	↓	.80	40	↓	38,300	1,375		39,675	44,300	
6200	Steel, underground, 360°, double wall, U.L. listed,										
6210	with sti-P3 corrosion protection,										
6220	(dielectric coating, cathodic protection, electrical										
6230	isolation) 30 year warranty,										
6240	not incl. manholes or hold-downs.										
6250	500 gallon capacity	Q-5	2.40	6.667	Ea.	2,425	217		2,642	3,000	
6260	1,000 gallon capactiy	"	2.25	7.111		3,550	232		3,782	4,250	
6270	2,000 gallon capacity	Q-7	4.16	7.692		4,650	265		4,915	5,525	
6280	3,000 gallon capacity		3.90	8.205		5,250	283		5,533	6,200	
6290	4,000 gallon capacity		3.64	8.791		6,625	305		6,930	7,725	
6300	5,000 gallon capacity		2.91	10.997		8,575	380		8,955	10,000	
6310	6,000 gallon capacity		2.42	13.223		10,200	455		10,655	11,900	
6320	8,000 gallon capacity		2.08	15.385		10,600	530		11,130	12,500	
6330	10,000 gallon capacity		1.82	17.582		12,000	605		12,605	14,100	
6340	12,000 gallon capacity		1.70	18.824		14,400	650		15,050	16,900	
6350	15,000 gallon capacity		1.33	24.060		16,300	830		17,130	19,200	
6360	20,000 gallon capacity		1.33	24.060		19,600	830		20,430	22,900	
6370	25,000 gallon capacity		1.16	27.586		32,200	950		33,150	36,800	
6380	30,000 gallon capacity		1.03	31.068		38,800	1,075		39,875	44,300	
6390	40,000 gallon capacity		.80	40		49,900	1,375		51,275	57,000	
6395	50,000 gallon capacity		.73	43.836	↓	60,500	1,525		62,025	69,000	
6400	For hold-downs 500-2000 gal, add		16	2	Set	284	69		353	415	
6410	For hold-downs 3000-6000 gal, add		12	2.667		370	92		462	545	
6420	For hold-downs 8000-12,000 gal, add		11	2.909		905	100		1,005	1,150	
6430	For hold-downs 15,000 gal, add		9	3.556		1,125	123		1,248	1,425	
6440	For hold-downs 20,000 gal, add		8	4		1,250	138		1,388	1,575	
6450	For hold-downs 20,000 gal plus, add	↓	6	5.333	↓	2,675	184		2,859	3,200	
6500	For manways, add				Ea.	590			590	650	
6600	In place with hold-downs										
6652	550 gallon capacity	Q-5	1.84	8.696	Ea.	2,725	283		3,008	3,400	

13281	Hazardous Material Remediation	CREW	DAILY OUTPUT	LABOR-HOURS	UNIT	2002 BARE COSTS				TOTAL INCL O&P	
						MAT.	LABOR	EQUIP.	TOTAL		
120											120
0010	BULK ASBESTOS REMOVAL										
0020	Includes disposable tools and 2 suits and 1respirator filter/day/worker										
0200	Boiler insulation	A-9	480	.133	S.F.	.46	4.47		4.93	7.60	
0210	With metal lath add				%				50%		
0300	Boiler breeching or flue insulation	A-9	520	.123	S.F.	.37	4.12		4.49	6.95	
0310	For active boiler, add				%				100%		

Important: See the Reference Section for critical supporting data - Reference Nos., Crews, & City Cost Indexes

13 SPECIAL CONSTRUCTION

			DAILY	LABOR-		\multicolumn{4}{c}{2002 BARE COSTS}				TOTAL	
13281	**Hazardous Material Remediation**	CREW	OUTPUT	HOURS	UNIT	MAT.	LABOR	EQUIP.	TOTAL	INCL O&P	
120	0400	Duct or AHU insulation	A-10B	440	.073	S.F.	.22	2.44		2.66	4.12	**120**
	0500	Duct vibration isolation joints, up to 24 Sq. In. duct	A-9	56	1.143	Ea.	3.42	38.50		41.92	65	
	0520	25 Sq. In. to 48 Sq. In. duct		48	1.333		3.99	44.50		48.49	75.50	
	0530	49 Sq. In. to 76 Sq. In. duct		40	1.600		4.79	53.50		58.29	90.50	
	0600	Pipe insulation, air cell type, up to 4" diameter pipe		900	.071	L.F.	.21	2.38		2.59	4.01	
	0610	4" to 8" diameter pipe		800	.080		.24	2.68		2.92	4.52	
	0620	10" to 12" diameter pipe		700	.091		.27	3.06		3.33	5.15	
	0630	14" to 16" diameter pipe		550	.116		.35	3.90		4.25	6.60	
	0650	Over 16" diameter pipe		650	.098	S.F.	.29	3.30		3.59	5.55	
	0700	With glove bag up to 3" diameter pipe		100	.640	L.F.	3.15	21.50		24.65	37.50	
	1000	Pipe fitting insulation up to 4" diameter pipe		320	.200	Ea.	.60	6.70		7.30	11.30	
	1100	6" to 8" diameter pipe		304	.211		.63	7.05		7.68	11.90	
	1110	10" to 12" diameter pipe		192	.333		1	11.15		12.15	18.85	
	1120	14" to 16" diameter pipe		128	.500		1.50	16.75		18.25	28	
	1130	Over 16" diameter pipe		176	.364	S.F.	1.09	12.20		13.29	20.50	
	1200	With glove bag, up to 8" diameter pipe		40	1.600	Ea.	6.55	53.50		60.05	92	
	2000	Scrape foam fireproofing from flat surface		2,400	.027	S.F.	5.20	.89		6.09	7.10	
	2100	Irregular surfaces		1,200	.053		.16	1.79		1.95	3.02	
	3000	Remove cementitious material from flat surface		1,800	.036		.11	1.19		1.30	2.01	
	3100	Irregular surface		1,400	.046		.14	1.53		1.67	2.58	
	6000	Remove contaminated soil from crawl space by hand		400	.160	C.F.	.48	5.35		5.83	9.05	
	6100	With large production vacuum loader	A-12	700	.091	"	.27	3.06	.70	4.03	5.95	
	7000	Radiator backing, not including radiator removal	A-9	1,200	.053	S.F.	.16	1.79		1.95	3.02	
	9000	For type C (supplied air) respirator equipment, add				%					10%	
125	0010	**ASBESTOS ABATEMENT WORK AREA** Containment and preparation.										**125**
	0100	Pre-cleaning, HEPA vacuum and wet wipe, flat surfaces	A-10	12,000	.005	S.F.	.02	.18		.20	.30	
	0200	Protect carpeted area, 2 layers 6 mil poly on 3/4" plywood	"	1,000	.064		1.50	2.14		3.64	5.05	
	0300	Separation barrier, 2" x 4" @ 16", 1/2" plywood ea. side, 8' high	2 Carp	400	.040		1.25	1.20		2.45	3.25	
	0310	12' high		320	.050		1.40	1.50		2.90	3.88	
	0320	16' high	R13281-120	200	.080		1.50	2.40		3.90	5.40	
	0400	Personnel decontam. chamber, 2" x 4" @ 16", 3/4" ply ea. side		280	.057		2.50	1.71		4.21	5.40	
	0450	Waste decontam. chamber, 2" x 4" studs @ 16", 3/4" ply ea. side		360	.044		3	1.33		4.33	5.40	
	0500	Cover surfaces with polyethelene sheeting										
	0501	Including glue and tape										
	0550	Floors, each layer, 6 mil	A-10	8,000	.008	S.F.	.10	.27		.37	.54	
	0551	4 mil		9,000	.007		.06	.24		.30	.45	
	0560	Walls, each layer, 6 mil		6,000	.011		.09	.36		.45	.67	
	0561	4 mil		7,000	.009		.07	.31		.38	.57	
	0570	For heights above 12', add						20%				
	0575	For heights above 20', add						30%				
	0580	For fire retardant poly, add					100%					
	0590	For large open areas, deduct					10%	20%				
	0600	Seal floor penetrations with foam firestop to 36 Sq. In.	2 Carp	200	.080	Ea.	6.25	2.40		8.65	10.65	
	0610	36 Sq. In. to 72 Sq. In.		125	.128		12.50	3.84		16.34	19.75	
	0615	72 Sq. In. to 144 Sq. In.		80	.200		25	6		31	37	
	0620	Wall penetrations, to 36 square inches		180	.089		6.25	2.67		8.92	11.05	
	0630	36 Sq. In. to 72 Sq. In.		100	.160		12.50	4.80		17.30	21	
	0640	72 Sq. In. to 144 Sq. In.		60	.267		25	8		33	40	
	0800	Caulk seams with latex	1 Carp	230	.035	L.F.	.15	1.04		1.19	1.79	
	0900	Set up neg. air machine, 1-2k C.F.M. /25 M.C.F. volume	1 Asbe	4.30	1.860	Ea.		62		62	99	
130	0010	**DEMOLITION IN ASBESTOS CONTAMINATED AREA**										**130**
	0200	Ceiling, including suspension system, plaster and lath	A-9	2,100	.030	S.F.	.09	1.02		1.11	1.72	
	0210	Finished plaster, leaving wire lath		585	.109		.33	3.67		4	6.15	
	0220	Suspended acoustical tile		3,500	.018		.05	.61		.66	1.03	
	0230	Concealed tile grid system		3,000	.021		.06	.71		.77	1.20	
	0240	Metal pan grid system		1,500	.043		.13	1.43		1.56	2.41	

SPECIAL CONSTRUCTION 13

		13281 **Hazardous Material Remediation**	CREW	DAILY OUTPUT	LABOR-HOURS	UNIT	2002 BARE COSTS				TOTAL INCL O&P	
							MAT.	LABOR	EQUIP.	TOTAL		
130	0250	Gypsum board	A-9	2,500	.026	S.F.	.08	.86		.94	1.44	**130**
	0260	Lighting fixtures up to 2' x 4'	↓	72	.889	Ea.	2.66	30		32.66	50.50	
	0400	Partitions, non load bearing										
	0410	Plaster, lath, and studs	A-9	690	.093	S.F.	.83	3.11		3.94	5.85	
	0450	Gypsum board and studs	"	1,390	.046	"	.14	1.54		1.68	2.60	
	9000	For type C (supplied air) respirator equipment, add				%					10%	
135	0010	**ASBESTOS ABATEMENT EQUIPMENT** and supplies, buy R13281-120										**135**
	0200	Air filtration device, 2000 C.F.M.				Ea.	2,500			2,500	2,750	
	0250	Large volume air sampling pump, minimum					450			450	495	
	0260	Maximum					1,000			1,000	1,100	
	0300	Airless sprayer unit, 2 gun					2,000			2,000	2,200	
	0350	Light stand, 500 watt				↓	250			250	275	
	0400	Personal respirators										
	0410	Negative pressure, 1/2 face, dual operation, min.				Ea.	22.50			22.50	25	
	0420	Maximum					24			24	26.50	
	0450	P.A.P.R., full face, minimum					400			400	440	
	0460	Maximum					700			700	770	
	0470	Supplied air, full face, incl. air line, minimum					450			450	495	
	0480	Maximum					600			600	660	
	0500	Personnel sampling pump, minimum					450			450	495	
	0510	Maximum					750			750	825	
	1500	Power panel, 20 unit, incl. G.F.I.					1,800			1,800	1,975	
	1600	Shower unit, including pump and filters					1,125			1,125	1,250	
	1700	Supplied air system (type C)					10,000			10,000	11,000	
	1750	Vacuum cleaner, HEPA, 16 gal., stainless steel, wet/dry					1,000			1,000	1,100	
	1760	55 gallon					2,200			2,200	2,425	
	1800	Vacuum loader, 9-18 ton/hr					90,000			90,000	99,000	
	1900	Water atomizer unit, including 55 gal. drum					230			230	253	
	2000	Worker protection, whole body, foot, head cover & gloves, plastic					5.15			5.15	5.65	
	2500	Respirator, single use					10.50			10.50	11.55	
	2550	Cartridge for respirator					13.65			13.65	15	
	2570	Glove bag, 7 mil, 50" x 64"					8.50			8.50	9.35	
	2580	10 mil, 44" x 60"					8.40			8.40	9.25	
	3000	HEPA vacuum for work area, minimun					1,000			1,000	1,100	
	3050	Maximum					2,775			2,775	3,050	
	6000	Disposable polyethelene bags, 6 mil, 3 C.F.					1.15			1.15	1.27	
	6300	Disposable fiber drums, 3 C.F.					6.50			6.50	7.15	
	6400	Pressure sensitive caution lables, 3" x 5"					21.50			21.50	23.50	
	6450	11" x 17"					6.50			6.50	7.15	
	6500	Negative air machine, 1800 C.F.M.	↓		↓		775			775	855	
140	0010	**DECONTAMINATION CONTAINMENT AREA DEMOLITION** and clean-up										**140**
	0100	Spray exposed substrate with surfactant (bridging)										
	0200	Flat surfaces	A-9	6,000	.011	S.F.	.35	.36		.71	.96	
	0250	Irregular surfaces		4,000	.016	"	.30	.54		.84	1.18	
	0300	Pipes, beams, and columns		2,000	.032	L.F.	.55	1.07		1.62	2.31	
	1000	Spray encapsulate polyethelene sheeting		8,000	.008	S.F.	.30	.27		.57	.76	
	1100	Roll down polyethelene sheeting		8,000	.008	"		.27		.27	.43	
	1500	Bag polyethelene sheeting		400	.160	Ea.	.75	5.35		6.10	9.35	
	2000	Fine clean exposed substrate, with nylon brush		2,400	.027	S.F.		.89		.89	1.42	
	2500	Wet wipe substrate		4,800	.013			.45		.45	.71	
	2600	Vacuum surfaces, fine brush	↓	6,400	.010	↓		.34		.34	.53	
	3000	Structural demolition										
	3100	Wood stud walls	A-9	2,800	.023	S.F.		.77		.77	1.22	
	3500	Window manifolds, not incl. window replacement		4,200	.015			.51		.51	.81	
	3600	Plywood carpet protection	↓	2,000	.032	↓		1.07		1.07	1.70	
	4000	Remove custom decontamination facility	A-10A	8	3	Ea.	15	101		116	177	

13   **SPECIAL CONSTRUCTION**

13280 | Hazardous Material Remediation

		13281	Hazardous Material Remediation	CREW	DAILY OUTPUT	LABOR-HOURS	UNIT	2002 BARE COSTS MAT.	LABOR	EQUIP.	TOTAL	TOTAL INCL O&P	
140	4100		Remove portable decontamination facility	3 Asbe	12	2	Ea.	12.50	67		79.50	120	140
	5000		HEPA vacuum, shampoo carpeting	A-9	4,800	.013	S.F.	.05	.45		.50	.77	
	9000		Final cleaning of protected surfaces	A-10A	8,000	.003	"		.10		.10	.16	
145	0010		**OSHA TESTING**										145
	0100		Certified technician, minimum				Day					300	
	0110		Maximum				"					500	
	0200		Personal sampling, PCM analysis, minimum	1 Asbe	8	1	Ea.	2.75	33.50		36.25	56	
	0210		Maximum	"	4	2	"	3	67		70	109	
	0300		Industrial hygenist, minimum				Day					400	
	0310		Maximum				"					550	
	1000		Cleaned area samples	1 Asbe	8	1	Ea.	2.63	33.50		36.13	56	
	1100		PCM air sample analysis, minimum		8	1		30	33.50		63.50	86	
	1110		Maximum		4	2		3.09	67		70.09	109	
	1200		TEM air sample analysis, minimum								100	125	
	1210		Maximum								400	500	
150	0010		**ENCAPSULATION WITH SEALANTS**										150
	0100		Ceilings and walls, minimum	A-9	21,000	.003	S.F.	.26	.10		.36	.45	
	0110		Maximum		10,600	.006	"	.41	.20		.61	.77	
	0300		Pipes to 12" diameter including minor repairs, minimum		800	.080	L.F.	.36	2.68		3.04	4.66	
	0310		Maximum		400	.160	"	1.02	5.35		6.37	9.60	
155	0010		**WASTE PACKAGING, HANDLING, & DISPOSAL**										155
	0100		Collect and bag bulk material, 3 C.F. bags, by hand	A-9	400	.160	Ea.	1.15	5.35		6.50	9.75	
	0200		Large production vacuum loader	A-12	880	.073		.80	2.44	.56	3.80	5.35	
	1000		Double bag and decontaminate	A-9	960	.067		2.30	2.23		4.53	6.10	
	2000		Containerize bagged material in drums, per 3 C.F. drum	"	800	.080		6.50	2.68		9.18	11.40	
	3000		Cart bags 50' to dumpster	2 Asbe	400	.040			1.34		1.34	2.12	
	5000		Disposal charges, not including haul, minimum				C.Y.					50	
	5020		Maximum				"					175	
	5100		Remove refrigerant from system	1 Plum	40	.200	Lb.		7.20		7.20	10.85	
	9000		For type C (supplied air) respirator equipment, add				%					10%	

13600 | Solar and Wind Energy Equipment

		13630	Solar Collector Components		CREW	DAILY OUTPUT	LABOR-HOURS	UNIT	2002 BARE COSTS MAT.	LABOR	EQUIP.	TOTAL	TOTAL INCL O&P	
200	0010		**SOLAR ENERGY**	D3010 715										200
	0020		System/Package prices, not including connecting											
	0030		pipe, insulation, or special heating/plumbing fixtures	D3010 720										
	0150		For solar ultraviolet pipe insulation see Div. 15086											
	0500		Hot water, standard package, low temperature											
	0540		1 collector, circulator, fittings, 65 gal. tank	D3010 730	Q-1	.50	32	Ea.	1,350	1,025		2,375	3,050	
	0580		2 collectors, circulator, fittings, 120 gal. tank			.40	40		1,925	1,300		3,225	4,075	
	0620		3 collectors, circulator, fittings, 120 gal. tank	D3010 735		.34	47.059		2,350	1,525		3,875	4,875	
	0700		Medium temperature package											
	0720		1 collector, circulator, fittings, 80 gal. tank	D3010 745	Q-1	.50	32	Ea.	1,450	1,025		2,475	3,150	
	0740		2 collectors, circulator, fittings, 120 gal. tank	D3010 750		.40	40		2,075	1,300		3,375	4,250	
	0780		3 collectors, circulator, fittings, 120 gal. tank			.30	53.333		3,775	1,725		5,500	6,750	
	0980		For each additional 120 gal. tank, add	D3010 760					660			660	730	
	2250		Controller, liquid temperature		1 Plum	5	1.600		92	57.50		149.50	188	

			CREW	DAILY OUTPUT	LABOR-HOURS	UNIT	2002 BARE COSTS				TOTAL INCL O&P
	13630	**Solar Collector Components**					MAT.	LABOR	EQUIP.	TOTAL	
200	2300	Circulators, air									200
	2310	Blowers									
	2330	100-300 S.F. system, 1/10 HP	Q-9	16	1	Ea.	128	31.50		159.50	190
	2340	300-500 S.F. system, 1/5 HP		15	1.067		177	33.50		210.50	246
	2350	Two speed, 100-300 S.F., 1/10 HP		14	1.143		128	36		164	197
	2400	Reversible fan, 20" diameter, 2 speed		18	.889		102	28		130	156
	2520	Space & DHW system, less duct work		.50	32		1,400	1,000		2,400	3,100
	2550	Booster fan 6" diameter, 120 CFM		16	1		33	31.50		64.50	85
	2570	6" diameter, 225 CFM		16	1		40.50	31.50		72	93
	2580	8" diameter, 150 CFM		16	1		37	31.50		68.50	89.50
	2590	8" diameter, 310 CFM		14	1.143		57	36		93	119
	2600	8" diameter, 425 CFM		14	1.143		64	36		100	126
	2650	Rheostat		32	.500		13.95	15.80		29.75	40
	2660	Shutter/damper		12	1.333		45	42		87	115
	2670	Shutter motor		16	1		97	31.50		128.50	156
	2800	Circulators, liquid, 1/25 HP, 5.3 GPM	Q-1	14	1.143		139	37		176	209
	2820	1/20 HP, 17 GPM		12	1.333		146	43		189	226
	2850	1/20 HP, 17 GPM, stainless steel		12	1.333		230	43		273	320
	2870	1/12 HP, 30 GPM		10	1.600		315	52		367	425
	3000	Collector panels, air with aluminum absorber plate									
	3010	Wall or roof mount									
	3040	Flat black, plastic glazing									
	3080	4' x 8'	Q-9	6	2.667	Ea.	600	84.50		684.50	790
	3100	4' x 10'	"	5	3.200	"	740	101		841	970
	3200	Flush roof mount, 10' to 16' x 22" wide		96	.167	L.F.	146	5.25		151.25	169
	3210	Manifold, by L.F. width of collectors		160	.100	"	27	3.16		30.16	34.50
	3300	Collector panels, liquid with copper absorber plate									
	3320	Black chrome, tempered glass glazing									
	3330	Alum. frame, 4' x 8', 5/32" single glazing	Q-1	9.50	1.684	Ea.	545	54.50		599.50	680
	3390	Alum. frame, 4' x 10', 5/32" single glazing	"	6	2.667	"	650	86.50		736.50	840
	3440	Flat black									
	3450	Alum. frame, 3' x 8'	Q-1	9	1.778	Ea.	430	57.50		487.50	560
	3500	Alum. frame, 4' x 8.5'		5.50	2.909		490	94		584	680
	3520	Alum. frame, 4' x 10.5'		10	1.600		595	52		647	735
	3540	Alum. frame, 4' x 12.5'		5	3.200		735	104		839	965
	3550	Liquid with fin tube absorber plate									
	3560	Alum. frame 4' x 8' tempered glass	Q-1	10	1.600	Ea.	495	52		547	625
	3580	Liquid with vacuum tubes, 4' x 6'-10"		9	1.778		845	57.50		902.50	1,025
	3600	Liquid, full wetted, plastic, alum. frame, 3' x 10'		5	3.200		195	104		299	370
	3650	Collector panel mounting, flat roof or ground rack		7	2.286		64.50	74		138.50	183
	3670	Roof clamps		70	.229	Set	1.61	7.40		9.01	12.90
	3700	Roof strap, teflon	1 Plum	205	.039	L.F.	9.50	1.40		10.90	12.55
	3900	Differential controller with two sensors									
	3930	Thermostat, hard wired	1 Plum	8	1	Ea.	92	36		128	155
	3950	Line cord and receptacle		12	.667		77.50	24		101.50	122
	4050	Pool valve system		2.50	3.200		199	115		314	390
	4070	With 12 VAC actuator		2	4		256	144		400	500
	4080	Pool pump system, 2" pipe size		6	1.333		160	48		208	249
	4100	Five station with digital read-out		3	2.667		202	96		298	365
	4150	Sensors									
	4200	Brass plug, 1/2" MPT	1 Plum	32	.250	Ea.	16.60	9		25.60	32
	4210	Brass plug, reversed		32	.250		22.50	9		31.50	38.50
	4220	Freeze prevention		32	.250		20	9		29	35.50
	4240	Screw attached		32	.250		10.15	9		19.15	24.50
	4250	Brass, immersion		32	.250		24.50	9		33.50	40.50
	4300	Heat exchanger									

Reference boxes: D3010 775, D3010 780, R13600 -610

13 SPECIAL CONSTRUCTION

13630	Solar Collector Components		CREW	DAILY OUTPUT	LABOR-HOURS	UNIT	2002 BARE COSTS				TOTAL INCL O&P	
							MAT.	LABOR	EQUIP.	TOTAL		
200	4330	Fluid to air coil, up flow, 45 MBH	D3010 715	Q-1	4	4	Ea.	279	129		408	500	200
	4380	70 MBH			3.50	4.571		315	148		463	570	
	4400	80 MBH	D3010 720	▼	3	5.333	▼	420	173		593	720	
	4580	Fluid to fluid package includes two circulating pumps											
	4590	expansion tank, check valve, relief valve	D3010 730										
	4600	controller, high temperature cutoff and sensors		Q-1	2.50	6.400	Ea.	695	207		902	1,075	
	4650	Heat transfer fluid	D3010 735										
	4700	Propylene glycol, inhibited anti-freeze		1 Plum	28	.286	Gal.	8.80	10.25		19.05	25	
	4800	Solar storage tanks, knocked down	D3010 745										
	4810	Air, galvanized steel clad, double wall, 4″ fiberglass											
	4820	insulation, 20 Mil PVC lining	D3010 750										
	4870	4′ high, 4′ x 4′, = 64 C.F./450 gallons		Q-9	2	8	Ea.	1,750	253		2,003	2,325	
	4880	4′ x 8′ = 128 C.F./900 gallons	D3010 760		1.50	10.667		2,250	335		2,585	3,000	
	4890	4′ x 12′ = 190 C.F./1300 gallons			1.30	12.308		3,000	390		3,390	3,900	
	4900	8′ x 8′ = 250 C.F./1700 gallons	D3010 775	▼	1	16		3,350	505		3,855	4,450	
	5010	6′-3″ high, 7′ x 7′ = 306 C.F./2000 gallons		Q-10	1.20	20		8,150	655		8,805	9,950	
	5020	7′ x 10′-6″ = 459 C.F./3000 gallons	D3010 780		.80	30		10,300	985		11,285	12,800	
	5030	7′ x 14′ = 613 C.F./4000 gallons			.60	40		12,000	1,300		13,300	15,200	
	5040	10′-6″ x 10′-6″ = 689 C.F./4500 gallons	R13600 -610		.50	48		12,400	1,575		13,975	16,000	
	5050	10′-6″ x 14′ = 919 C.F./6000 gallons		▼	.40	60		14,600	1,975		16,575	19,000	
	5060	14′ x 14′ = 1225 C.F./8000 gallons		Q-11	.40	80		17,100	2,675		19,775	22,900	
	5070	14′ x 17′-6″ = 1531 C.F./10,000 gallons			.30	106		19,200	3,575		22,775	26,600	
	5080	17′-6″ x 17′-6″ = 1914 C.F./12,500 gallons			.25	128		22,600	4,275		26,875	31,500	
	5090	17′-6″ x 21′ = 2297 C.F./15,000 gallons			.20	160		25,200	5,350		30,550	36,100	
	5100	21′ x 21′ = 2756 C.F./18,000 gallons		▼	.18	177	▼	28,900	5,950		34,850	41,000	
	5120	30 Mil reinforced Chemflex lining,											
	5140	4′ high, 4′ x 4′ = 64 C.F./450 gallons		Q-9	2	8	Ea.	1,750	253		2,003	2,325	
	5150	4′ x 8′ = 128 C.F./900 gallons			1.50	10.667		2,650	335		2,985	3,450	
	5160	4′ x 12′ = 190 C.F./1300 gallons			1.30	12.308		3,525	390		3,915	4,475	
	5170	8′ x 8′ = 250 C.F./1700 gallons		▼	1	16		3,925	505		4,430	5,100	
	5190	6′-3″ high, 7′ x 7′ = 306 C.F./2000 gallons		Q-10	1.20	20		8,700	655		9,355	10,600	
	5200	7′ x 10′-6″ = 459 C.F./3000 gallons			.80	30		10,600	985		11,585	13,200	
	5210	7′ x 14′ = 613 C.F./4000 gallons			.60	40		12,300	1,300		13,600	15,600	
	5220	10′-6″ x 10′-6″ = 689 C.F./4500 gallons			.50	48		12,700	1,575		14,275	16,400	
	5230	10′-6″ x 14′ = 919 C.F./6000 gallons		▼	.40	60		15,000	1,975		16,975	19,500	
	5240	14′ x 14′ = 1225 C.F./8000 gallons		Q-11	.40	80		17,500	2,675		20,175	23,400	
	5250	14′ x 17′-6″ = 1531 C.F./10,000 gallons			.30	106		19,600	3,575		23,175	27,100	
	5260	17′-6″ x 17′-6″ = 1914 C.F./12,500 gallons			.25	128		23,100	4,275		27,375	32,000	
	5270	17′-6″ x 21′ = 2297 C.F./15,000 gallons			.20	160		26,000	5,350		31,350	36,900	
	5280	21′ x 21′ = 2756 C.F./18,000 gallons		▼	.18	177	▼	29,800	5,950		35,750	42,000	
	5290	30 Mil reinforced Hypalon lining, add						.02%					
	7000	Solar control valves and vents											
	7050	Air purger, 1″ pipe size		1 Plum	12	.667	Ea.	41	24		65	81	
	7070	Air eliminator, automatic 3/4″ size			32	.250		27	9		36	43	
	7090	Air vent, automatic, 1/8″ fitting			32	.250		8.80	9		17.80	23.50	
	7100	Manual, 1/8″ NPT			32	.250		1.94	9		10.94	15.70	
	7120	Backflow preventer, 1/2″ pipe size			16	.500		58.50	18		76.50	91	
	7130	3/4″ pipe size			16	.500		59	18		77	92	
	7150	Balancing valve, 3/4″ pipe size			20	.400		22.50	14.40		36.90	46	
	7180	Draindown valve, 1/2″ copper tube			9	.889		187	32		219	254	
	7200	Flow control valve, 1/2″ pipe size			22	.364		49.50	13.05		62.55	74	
	7220	Expansion tank, up to 5 gal.			32	.250		58	9		67	77.50	
	7250	Hydronic controller (aquastat)			8	1		41.50	36		77.50	99.50	
	7400	Pressure gauge, 2″ dial			32	.250		21	9		30	36.50	
	7450	Relief valve, temp. and pressure 3/4″ pipe size		▼	30	.267	▼	7.90	9.60		17.50	23	
	7500	Solenoid valve, normally closed											

13630 | Solar Collector Components

			CREW	DAILY OUTPUT	LABOR-HOURS	UNIT	2002 BARE COSTS MAT.	LABOR	EQUIP.	TOTAL	TOTAL INCL O&P	
200	7520	Brass, 3/4" NPT, 24V	D3010 715	1 Plum	9	.889	Ea.	212	32		244	281	200
	7530	1" NPT, 24V			9	.889		212	32		244	281	
	7750	Vacuum relief valve, 3/4" pipe size	D3010 720	▼	32	.250	▼	25	9		34	41	
	7800	Thermometers											
	7820	Digital temperature monitoring, 4 locations	D3010 730	1 Plum	2.50	3.200	Ea.	135	115		250	320	
	7900	Upright, 1/2" NPT			8	1		25.50	36		61.50	82	
	7970	Remote probe, 2" dial	D3010 735		8	1		29	36		65	86	
	7990	Stem, 2" dial, 9" stem		▼	16	.500	▼	19	18		37	48	
	8250	Water storage tank with heat exchanger and electric element	D3010 745										
	8270	66 gal. with 2" x 2 lb. density insulation		1 Plum	1.60	5	Ea.	650	180		830	985	
	8300	80 gal. with 2" x 2 lb. density insulation	D3010 750		1.60	5		745	180		925	1,100	
	8380	120 gal. with 2" x 2 lb. density insulation			1.40	5.714		840	205		1,045	1,225	
	8400	120 gal. with 2" x 2 lb. density insul., 40 S.F. heat coil	D3010 760	▼	1.40	5.714	▼	965	205		1,170	1,375	
	8500	Water storage module, plastic											
	8600	Tubular, 12" diameter, 4' high	D3010 775	1 Carp	48	.167	Ea.	76	5		81	91.50	
	8610	12" diameter, 8' high			40	.200		116	6		122	137	
	8620	18" diameter, 5' high	D3010 780		38	.211		127	6.30		133.30	150	
	8630	18" diameter, 10' high		▼	32	.250		167	7.50		174.50	196	
	8640	58" diameter, 5' high	R13600 -610	2 Carp	32	.500	▼	460	15		475	535	
	8650	Cap, 12" diameter						12			12	13.20	
	8660	18" diameter		▼			▼	15			15	16.50	

13832 | Direct Digital Controls

			CREW	DAILY OUTPUT	LABOR-HOURS	UNIT	2002 BARE COSTS MAT.	LABOR	EQUIP.	TOTAL	TOTAL INCL O&P	
200	0010	**CONTROL COMPONENTS / DDC SYSTEMS** (Sub's quote incl. M & L)										200
	0100	Analog inputs											
	0110	Sensors (avg. 50' run in 1/2" EMT)											
	0120	Duct temperature					Ea.					346.80	
	0130	Space temperature										622.20	
	0140	Duct humidity, +/- 3%										652.80	
	0150	Space humidity, +/- 2%										999.60	
	0160	Duct static pressure										530.40	
	0170	C.F.M./Transducer										714	
	0172	Water temp. (see 15120 for well tap add)										612	
	0174	Water flow (see 15120 for circuit sensor add)										2,244	
	0176	Water pressure differential (see 15120 for tap add)										918	
	0177	Steam flow (see 15120 for circuit sensor add)										2,244	
	0178	Steam pressure (see 15120 for tap add)										958.80	
	0180	K.W./Transducer										1,275	
	0182	K.W.H. totalization (not incl. elec. meter pulse xmtr.)										586	
	0190	Space static pressure					▼					1,000	
	1000	Analog outputs (avg. 50' run in 1/2" EMT)											
	1010	P/I Transducer					Ea.					591.60	
	1020	Analog output, matl. in MUX										285.60	
	1030	Pneumatic (not incl. control device)										601.80	
	1040	Electric (not incl control device)					▼					357	
	2000	Status (Alarms)											
	2100	Digital inputs (avg. 50' run in 1/2" EMT)											

T3 SPECIAL CONSTRUCTION

Important: See the Reference Section for critical supporting data - Reference Nos., Crews, & City Cost Indexes

13832	Direct Digital Controls	CREW	DAILY OUTPUT	LABOR-HOURS	UNIT	2002 BARE COSTS				TOTAL INCL O&P	
						MAT.	LABOR	EQUIP.	TOTAL		
200											200
2110	Freeze				Ea.					408	
2120	Fire									367.20	
2130	Differential pressure, (air)									561	
2140	Differential pressure, (water)									795.60	
2150	Current sensor									408	
2160	Duct high temperature thermostat									535.50	
2170	Duct smoke detector				▼					663	
2200	Digital output (avg. 50' run in 1/2" EMT)										
2210	Start/stop				Ea.					326.40	
2220	On/off (maintained contact)				"					561	
3000	Controller M.U.X. panel, incl. function boards										
3100	48 point				Ea.					4,947	
3110	128 point				"					6,783	
3200	D.D.C. controller (avg. 50' run in conduit)										
3210	Mechanical room										
3214	16 point controller (incl. 120v/1ph power supply)				Ea.					3,060	
3229	32 point controller (incl. 120v/1ph power supply)				"					5,100	
3230	Includes software programming and checkout										
3260	Space										
3266	V.A.V. terminal box (incl. space temp. sensor)				Ea.					790.50	
3280	Host computer (avg. 50' run in conduit)										
3281	Package complete with PC, keyboard,										
3282	printer, color CRT, modem, basic software				Ea.					9,180	
4000	Front end costs										
4100	Computer (P.C.)/software program				Ea.					6,120	
4200	Color graphics software									3,672	
4300	Color graphics slides									459	
4350	Additional dot matrix printer				▼					918	
4400	Communications trunk cable				L.F.					3.57	
4500	Engineering labor, (not incl. dftg.)				Point					77.52	
4600	Calibration labor									77.52	
4700	Start-up, checkout labor				▼					117.30	
4800	Drafting labor, as req'd										
5000	Communications bus (data transmission cable)										
5010	#18 twisted shielded pair in 1/2" EMT conduit				C.L.F.					357	
8000	Applications software										
8050	Basic maintenance manager software (not incl. data base entry)				Ea.					1,836	
8100	Time program				Point					6.43	
8120	Duty cycle									12.80	
8140	Optimum start/stop									38.76	
8160	Demand limiting									19.18	
8180	Enthalpy program				▼					38.76	
8200	Boiler optimization				Ea.					1,147.50	
8220	Chiller optimization				"					1,530	
8240	Custom applications										
8260	Cost varies with complexity										

13834	Electric/Electronic Control										
200											200
0010	CONTROL SYSTEMS, ELECTRONIC R15050 -710										
0020	For electronic costs, add to division 13836-200				Ea.					15%	

13836	Pneumatic Controls										
200											200
0010	CONTROL SYSTEMS, PNEUMATIC (Sub's quote incl. mat. & labor)										
0011	and nominal 50 Ft. of tubing. Add control panelboard if req'd.										
0100	Heating and Ventilating, split system R15050 -710										
0200	Mixed air control, economizer cycle,panel readout,tubing										

SPECIAL CONSTRUCTION 13

		13836	**Pneumatic Controls**		CREW	DAILY OUTPUT	LABOR-HOURS	UNIT	2002 BARE COSTS				TOTAL INCL O&P	
									MAT.	LABOR	EQUIP.	TOTAL		
200	0220		Up to 10 tons	R15050 -710	Q-19	.68	35.294	Ea.	2,825	1,175		4,000	4,900	200
	0240		For 10 to 20 tons			.63	37.915		3,025	1,275		4,300	5,225	
	0260		For over 20 tons			.58	41.096		3,275	1,375		4,650	5,675	
	0270		Enthalpy cycle, up to 10 tons			.50	48.387		3,125	1,625		4,750	5,900	
	0280		For 10 to 20 tons			.46	52.174		3,375	1,750		5,125	6,350	
	0290		For over 20 tons			.42	56.604		3,650	1,900		5,550	6,875	
	0300		Heating coil, hot water, 3 way valve,											
	0320		Freezestat, limit control on discharge, readout		Q-5	.69	23.088	Ea.	2,100	750		2,850	3,425	
	0500		Cooling coil, chilled water, room											
	0520		Thermostat, 3 way valve		Q-5	2	8	Ea.	935	261		1,196	1,425	
	0600		Cooling tower, fan cycle, damper control,											
	0620		Control system including water readout in/out at panel		Q-19	.67	35.821	Ea.	3,725	1,200		4,925	5,900	
	1000		Unit ventilator, day/night operation,											
	1100		freezestat, ASHRAE, cycle 2		Q-19	.91	26.374	Ea.	2,050	885		2,935	3,600	
	2000		Compensated hot water from boiler, valve control,											
	2100		readout and reset at panel, up to 60 GPM		Q-19	.55	43.956	Ea.	3,850	1,475		5,325	6,475	
	2120		For 120 GPM			.51	47.059		4,125	1,575		5,700	6,900	
	2140		For 240 GPM			.49	49.180		4,325	1,650		5,975	7,225	
	3000		Boiler room combustion air, damper to 5 SF, controls			1.36	17.582		1,850	590		2,440	2,925	
	3500		Fan coil, heating and cooling valves, 4 pipe control system			3	8		840	268		1,108	1,325	
	3600		Heat exchanger system controls			.86	27.907		1,800	935		2,735	3,400	
	3900		Multizone control (one per zone), includes thermostat, damper											
	3910		motor and reset of discharge temperature		Q-5	.51	31.373	Ea.	1,850	1,025		2,875	3,600	
	4000		Pneumatic thermostat, including controlling room radiator valve		"	2.43	6.593		560	215		775	940	
	4040		Program energy saving optimizer		Q-19	1.21	19.786		4,650	665		5,315	6,125	
	4060		Pump control system		"	3	8		860	268		1,128	1,350	
	4080		Reheat coil control system, not incl coil		Q-5	2.43	6.593		730	215		945	1,125	
	4500		Air supply for pneumatic control system											
	4600		Tank mounted duplex compressor, starter, alternator,											
	4620		piping, dryer, PRV station and filter											
	4630		1/2 HP		Q-19	.68	35.139	Ea.	6,925	1,175		8,100	9,375	
	4640		3/4 HP			.64	37.383		7,250	1,250		8,500	9,850	
	4650		1 HP			.61	39.539		7,925	1,325		9,250	10,700	
	4660		1-1/2 HP			.57	41.739		8,450	1,400		9,850	11,400	
	4680		3 HP			.55	43.956		11,500	1,475		12,975	14,900	
	4690		5 HP			.42	57.143		20,100	1,925		22,025	25,000	
	4800		Main air supply, includes 3/8" copper main and labor		Q-5	1.82	8.791	C.L.F.	233	286		519	685	
	4810		If poly tubing used, deduct										30%	
	7000		Static pressure control for air handling unit, includes pressure											
	7010		sensor, receiver controller, readout and damper motors		Q-19	.64	37.383	Ea.	5,500	1,250		6,750	7,925	
	7020		If return air fan requires control, add										70%	
	8600		VAV boxes, incl. thermostat, damper motor, reheat coil & tubing		Q-5	1.46	10.989		865	360		1,225	1,500	
	8610		If no reheat coil, deduct										204	
	9400		Sub assemblies for assembly systems											

		13838	**Pneumatic/Electric Controls**											
200	0010		**CONTROL COMPONENTS**											200
	0500		Aquastats											
	0508		Immersion type											
	0510		High/low limit, breaks contact w/temp rise		1 Stpi	4	2	Ea.	107	72.50		179.50	227	
	0514		Sequencing, break 2 switches w/temp rise			4	2		87.50	72.50		160	206	
	0518		Circulating, makes contact w/temp rise			4	2		79.50	72.50		152	196	
	0600		Carbon monoxide detector system											
	0606		Panel		1 Stpi	4	2	Ea.	2,350	72.50		2,422.50	2,700	
	0610		Sensor		"	7.30	1.096	"	320	39.50		359.50	410	
	0700		Controller, receiver											

T3 SPECIAL CONSTRUCTION

13838 | Pneumatic/Electric Controls

		CREW	DAILY OUTPUT	LABOR-HOURS	UNIT	2002 BARE COSTS MAT.	LABOR	EQUIP.	TOTAL	TOTAL INCL O&P	
200	0730	Pneumatic, panel mount, single input	1 Plum	8	1	Ea.	268	36		304	350	**200**
	0740	With conversion mounting bracket		8	1		277	36		313	360	
	0750	Dual input, with control point adjustment	↓	7	1.143		370	41		411	465	
	0850	Electric, single snap switch	1 Elec	4	2		340	71		411	475	
	0860	Dual snap switches		3	2.667		430	94.50		524.50	615	
	0870	Humidity controller		8	1		201	35.50		236.50	274	
	0880	Load limiting controller		8	1		440	35.50		475.50	540	
	0890	Temperature controller	↓	8	1	↓	251	35.50		286.50	330	
	0900	Control panel readout										
	0910	Panel up to 12 indicators	1 Stpi	1.20	6.667	Ea.	134	241		375	510	
	0914	Panel up to 24 indicators		.86	9.302		167	335		502	695	
	0918	Panel up to 48 indicators	↓	.50	16	↓	395	580		975	1,300	
	1000	Enthalpy control, boiler water temperature control										
	1010	governed by outdoor temperature, with timer	1 Elec	3	2.667	Ea.	199	94.50		293.50	360	
	1300	Energy control/monitor										
	1320	BTU computer meter and controller	1 Stpi	1	8	Ea.	1,425	290		1,715	2,000	
	1600	Flow meters										
	1610	Gas	1 Stpi	4	2	Ea.	79.50	72.50		152	197	
	1620	Liquid	"	4	2	"	79.50	72.50		152	197	
	1640	Freezestat										
	1644	20' sensing element, adjustable	1 Stpi	3.50	2.286	Ea.	87.50	82.50		170	221	
	2000	Gauges, pressure or vacuum										
	2100	2" diameter dial	1 Stpi	32	.250	Ea.	14.55	9.05		23.60	29.50	
	2200	2-1/2" diameter dial		32	.250		16.95	9.05		26	32.50	
	2300	3-1/2" diameter dial		32	.250		21	9.05		30.05	36.50	
	2400	4-1/2" diameter dial	↓	32	.250	↓	31.50	9.05		40.55	48	
	2700	Flanged iron case, black ring										
	2800	3-1/2" diameter dial	1 Stpi	32	.250	Ea.	66.50	9.05		75.55	87	
	2900	4-1/2" diameter dial		32	.250		91	9.05		100.05	114	
	3000	6" diameter dial	↓	32	.250	↓	72.50	9.05		81.55	93	
	3010	Steel case, 0 - 300 psi										
	3012	2" diam. dial	1 Stpi	16	.500	Ea.	21	18.10		39.10	50.50	
	3014	4" diam. dial	"	16	.500	"	35	18.10		53.10	66	
	3020	Aluminum case, 0 - 300 psi										
	3022	3-1/2" diam. dial	1 Stpi	16	.500	Ea.	48	18.10		66.10	80.50	
	3024	4-1/2" diam. dial		16	.500		71	18.10		89.10	106	
	3026	6" diam. dial		16	.500		118	18.10		136.10	158	
	3028	8-1/2" diam. dial	↓	16	.500	↓	281	18.10		299.10	340	
	3030	Brass case, 0 - 300 psi										
	3032	2" diam. dial	1 Stpi	16	.500	Ea.	23.50	18.10		41.60	53.50	
	3034	4-1/2" diam. dial	"	16	.500	"	53.50	18.10		71.60	86.50	
	3040	Steel case, high pressure, 0 -10,000 psi										
	3042	4-1/2" diam. dial	1 Stpi	16	.500	Ea.	134	18.10		152.10	175	
	3044	6-1/2" diam. dial		16	.500		215	18.10		233.10	264	
	3046	8-1/2" diam. dial	↓	16	.500	↓	390	18.10		408.10	460	
	3080	Pressure gauge, differential, magnehelic										
	3084	0 - 2" W.C., with air filter kit	1 Stpi	6	1.333	Ea.	74	48.50		122.50	154	
	3300	For compound pressure-vacuum, add					18%					
	3350	Humidistat										
	3360	Pneumatic operation										
	3361	Room humidistat, direct acting	1 Stpi	12	.667	Ea.	158	24		182	210	
	3362	Room humidistat, reverse acting		12	.667		158	24		182	210	
	3363	Room humidity transmitter		17	.471		170	17.05		187.05	213	
	3364	Duct mounted controller		12	.667		208	24		232	266	
	3365	Duct mounted transmitter		12	.667		187	24		211	243	
	3366	Humidity indicator, 3-1/2"	↓	28	.286	↓	75.50	10.35		85.85	98.50	

		CREW	DAILY OUTPUT	LABOR-HOURS	UNIT	MAT.	LABOR	EQUIP.	TOTAL	TOTAL INCL O&P	
13838	**Pneumatic/Electric Controls**					2002 BARE COSTS					
200	3390	Electric operated	1 Shee	8	1	Ea.	30	35		65	87	200
	3400	Relays									
	3430	Pneumatic/electric	1 Plum	16	.500	Ea.	184	18		202	229
	3440	Pneumatic proportioning		8	1		100	36		136	164
	3450	Pneumatic switching		12	.667		88.50	24		112.50	134
	3460	Selector, 3 point		6	1.333		63.50	48		111.50	143
	3470	Pneumatic time delay	▼	8	1	▼	235	36		271	310
	3500	Sensor, air operated									
	3520	Humidity	1 Plum	16	.500	Ea.	208	18		226	256
	3540	Pressure		16	.500		47	18		65	78.50
	3560	Temperature	▼	12	.667	▼	99.50	24		123.50	145
	3600	Electric operated									
	3620	Humidity	1 Elec	8	1	Ea.	26	35.50		61.50	82
	3650	Pressure		8	1		830	35.50		865.50	970
	3680	Temperature	▼	10	.800	▼	96	28.50		124.50	148
	3700	Switches									
	3710	Minimum position									
	3720	Electrical	1 Stpi	4	2	Ea.	25.50	72.50		98	137
	3730	Pneumatic	"	4	2	"	36.50	72.50		109	149
	4000	Thermometers									
	4100	Dial type, 3-1/2" diameter, vapor type, union connection	1 Stpi	32	.250	Ea.	97.50	9.05		106.55	121
	4120	Liquid type, union connection		32	.250		136	9.05		145.05	164
	4130	Remote reading, 15' capillary		32	.250		107	9.05		116.05	131
	4500	Stem type, 6-1/2" case, 2" stem, 1/2" NPT		32	.250		55.50	9.05		64.55	74.50
	4520	4" stem, 1/2" NPT		32	.250		76	9.05		85.05	97
	4600	9" case, 3-1/2" stem, 3/4" NPT		28	.286		92.50	10.35		102.85	117
	4620	6" stem, 3/4" NPT		28	.286		98	10.35		108.35	124
	4640	8" stem, 3/4" NPT		28	.286		121	10.35		131.35	149
	4660	12" stem, 1" NPT	▼	26	.308	▼	133	11.15		144.15	164
	4670	Bi-metal, dial type, steel case brass stem									
	4672	2" dial, 4" - 9" stem	1 Stpi	16	.500	Ea.	30	18.10		48.10	60.50
	4673	2-1/2" dial, 4" - 9" stem		16	.500		32.50	18.10		50.60	63.50
	4674	3-1/2" dial, 4" - 9" stem	▼	16	.500	▼	32.50	18.10		50.60	63.50
	4680	Mercury filled, industrial, union connection type									
	4682	Angle stem, 7" scale	1 Stpi	16	.500	Ea.	76.50	18.10		94.60	112
	4683	9" scale		16	.500		112	18.10		130.10	151
	4684	12" scale		16	.500		124	18.10		142.10	164
	4686	Straight stem, 7" scale		16	.500		98	18.10		116.10	136
	4687	9" scale		16	.500		105	18.10		123.10	143
	4688	12" scale	▼	16	.500	▼	125	18.10		143.10	166
	4690	Mercury filled, industrial, separable socket type									
	4692	Angle stem, with socket, 7" scale	1 Stpi	16	.500	Ea.	64.50	18.10		82.60	98
	4693	9" scale		16	.500		77	18.10		95.10	112
	4694	12" scale		16	.500		156	18.10		174.10	199
	4696	Straight stem, with socket, 7" scale		16	.500		64.50	18.10		82.60	98
	4697	9" scale		16	.500		77	18.10		95.10	112
	4698	12" scale	▼	16	.500	▼	90.50	18.10		108.60	128
	5000	Thermostats									
	5030	Manual	1 Shee	8	1	Ea.	27	35		62	84
	5040	1 set back, electric, timed		8	1		81	35		116	143
	5050	2 set back, electric, timed	▼	8	1		111	35		146	176
	5100	Locking cover					14.05			14.05	15.45
	5200	24 hour, automatic, clock	1 Shee	8	1		109	35		144	174
	5220	Electric, low voltage, 2 wire	1 Elec	13	.615		18	22		40	52.50
	5230	3 wire		10	.800		22	28.50		50.50	67
	5236	Heating/cooling, low voltage, with clock	▼	8	1	▼	105	35.50		140.50	169

Important: See the Reference Section for critical supporting data - Reference Nos., Crews, & City Cost Indexes

13838	Pneumatic/Electric Controls	CREW	DAILY OUTPUT	LABOR-HOURS	UNIT	2002 BARE COSTS				TOTAL INCL O&P
						MAT.	LABOR	EQUIP.	TOTAL	
5240	Pneumatic									
5250	Single temp., single pressure	1 Stpi	8	1	Ea.	115	36		151	182
5251	Dual pressure		8	1		170	36		206	242
5252	Dual temp., dual pressure		8	1		161	36		197	232
5253	Reverse acting w/averaging element		8	1		147	36		183	217
5254	Heating-cooling w/deadband		8	1		160	36		196	231
5255	Integral w/piston top valve actuator		8	1		157	36		193	228
5256	Dual temp., dual pressure		8	1		161	36		197	232
5257	Low limit, 8' averaging element		8	1		112	36		148	178
5258	Room single temp. proportional		8	1		46	36		82	106
5260	Dual temp, direct acting for VAV fan		8	1		146	36		182	216
5262	Capillary tube type, 20', -30 to +100°F	↓	8	1	↓	95.50	36		131.50	160
5300	Transmitter, pneumatic									
5320	Temperature averaging element	Q-1	8	2	Ea.	95.50	64.50		160	203
5350	Pressure differential	1 Plum	7	1.143		560	41		601	675
5370	Humidity, duct		8	1		187	36		223	260
5380	Room		12	.667		170	24		194	223
5390	Temperature, with averaging element	↓	6	1.333		107	48		155	191
5420	Electric operated, humidity	1 Elec	8	1		50.50	35.50		86	109
5430	DPST	"	8	1	↓	71	35.50		106.50	131
5500	Timer/time clocks									
5510	7 day, 12 hr. battery	1 Stpi	2.60	3.077	Ea.	226	111		337	415
5600	Recorders									
5610	Hydrograph humidity, wall mtd., aluminum case, grad. chart									
5612	8"	1 Stpi	1.60	5	Ea.	545	181		726	875
5614	10"	"	1.60	5	"	980	181		1,161	1,350
5620	Thermo-hydrograph, wall mtd., grad. chart, 5' SS probe									
5622	8"	1 Stpi	1.60	5	Ea.	475	181		656	800
5624	10"	"	1.60	5	"	790	181		971	1,150
5630	Time of operation, wall mtd., 1 pen cap, 24 hr. clock									
5632	6"	1 Stpi	1.60	5	Ea.	400	181		581	715
5634	8"	"	1.60	5	"	400	181		581	715
5640	Pressure & vacuum, bourdon tube type									
5642	Flush mount, 1 pen									
5644	6"	1 Stpi	1.60	5	Ea.	365	181		546	675
5646	8"	"	1.60	5	"	720	181		901	1,075
6000	Valves, motorized zone									
6100	Sweat connections, 1/2" C x C	1 Stpi	20	.400	Ea.	67.50	14.50		82	96.50
6110	3/4" C x C		20	.400		72.50	14.50		87	102
6120	1" C x C		19	.421		84.50	15.25		99.75	116
6140	1/2" C x C, with end switch, 2 wire		20	.400		71.50	14.50		86	101
6150	3/4" C x C, with end switch, 2 wire		20	.400		76	14.50		90.50	106
6160	1" C x C, with end switch, 2 wire		19	.421		86	15.25		101.25	118
6180	1-1/4" C x C, w/end switch, 2 wire	↓	15	.533	↓	103	19.30		122.30	143
7090	Valves, motor controlled, including actuator									
7100	Electric motor actuated									
7200	Brass, two way, screwed									
7210	1/2" pipe size	L-6	36	.333	Ea.	182	11.95		193.95	218
7220	3/4" pipe size		30	.400		276	14.30		290.30	325
7230	1" pipe size		28	.429		330	15.35		345.35	390
7240	1-1/2" pipe size		19	.632		355	22.50		377.50	425
7250	2" pipe size	↓	16	.750	↓	530	27		557	620
7350	Brass, three way, screwed									
7360	1/2" pipe size	L-6	33	.364	Ea.	212	13		225	253
7370	3/4" pipe size		27	.444		238	15.90		253.90	285
7380	1" pipe size	↓	25.50	.471	↓	210	16.85		226.85	257

13838 | Pneumatic/Electric Controls

		CREW	DAILY OUTPUT	LABOR-HOURS	UNIT	MAT.	LABOR	EQUIP.	TOTAL	TOTAL INCL O&P	
200	7384	1-1/4" pipe size	L-6	21	.571	Ea.	266	20.50		286.50	325	200
	7390	1-1/2" pipe size		17	.706		300	25.50		325.50	370	
	7400	2" pipe size	↓	14	.857	↓	465	30.50		495.50	560	
	7550	Iron body, two way, flanged										
	7560	2-1/2" pipe size	L-6	4	3	Ea.	355	107		462	550	
	7570	3" pipe size		3	4		470	143		613	730	
	7580	4" pipe size	↓	2	6	↓	655	215		870	1,050	
	7850	Iron body, three way, flanged										
	7860	2-1/2" pipe size	L-6	3	4	Ea.	770	143		913	1,075	
	7870	3" pipe size		2.50	4.800		930	172		1,102	1,275	
	7880	4" pipe size	↓	2	6	↓	1,200	215		1,415	1,650	
	8000	Pneumatic, air operated										
	8050	Brass, two way, screwed										
	8060	1/2" pipe size, class 250	1 Plum	24	.333	Ea.	100	12		112	128	
	8070	3/4" pipe size, class 250		20	.400		133	14.40		147.40	168	
	8080	1" pipe size, class 250		19	.421		151	15.15		166.15	189	
	8090	1-1/4" pipe size, class 125		15	.533		217	19.15		236.15	267	
	8100	1-1/2" pipe size, class 125		13	.615		251	22		273	310	
	8110	2" pipe size, class 125	↓	11	.727	↓	285	26		311	355	
	8180	Brass, three way, screwed										
	8190	1/2" pipe size, class 250	1 Plum	22	.364	Ea.	110	13.05		123.05	141	
	8200	3/4" pipe size, class 250		18	.444		134	16		150	171	
	8210	1" pipe size, class 250		17	.471		153	16.90		169.90	195	
	8214	1-1/4" pipe size, class 250		14	.571		210	20.50		230.50	262	
	8220	1-1/2" pipe size, class 125		11	.727		246	26		272	310	
	8230	2" pipe size, class 125	↓	9	.889	↓	310	32		342	390	
	8250	Iron body, two way, screwed										
	8260	1/2" pipe size, class 300	1 Plum	24	.333	Ea.	277	12		289	325	
	8270	3/4" pipe size, class 300		20	.400		305	14.40		319.40	355	
	8280	1" pipe size, class 300		19	.421		495	15.15		510.15	570	
	8290	1-1/4" pipe size, class 300		15	.533		525	19.15		544.15	610	
	8300	1-1/2" pipe size, class 300		13	.615		695	22		717	800	
	8310	2" pipe size, class 300		11	.727		765	26		791	885	
	8320	1-1/4" pipe size, class 125		15	.533		415	19.15		434.15	490	
	8330	1-1/2" pipe size, class 125		13	.615		375	22		397	445	
	8340	2" pipe size, class 125	↓	11	.727	↓	435	26		461	520	
	8450	Iron body, two way, flanged										
	8460	2-1/2" pipe size, 250 lb. flanges	Q-1	5	3.200	Ea.	2,775	104		2,879	3,200	
	8470	3" pipe size, 250 lb. flanges		4.50	3.556		2,950	115		3,065	3,425	
	8480	4" pipe size, 250 lb. flanges		3	5.333		3,600	173		3,773	4,200	
	8510	2-1/2" pipe size, class 125		5	3.200		595	104		699	810	
	8520	3" pipe size, class 125		4.50	3.556		710	115		825	960	
	8530	4" pipe size, class 125	↓	3	5.333		915	173		1,088	1,250	
	8540	5" pipe size, class 125	Q-2	3.40	7.059		1,900	237		2,137	2,425	
	8550	6" pipe size, class 125	"	3	8	↓	2,250	268		2,518	2,875	
	8560	Iron body, three way, flanged										
	8570	2-1/2" pipe size, class 125	Q-1	4.50	3.556	Ea.	680	115		795	920	
	8580	3" pipe size, class 125		4	4		845	129		974	1,125	
	8590	4" pipe size, class 125	↓	2.50	6.400		1,700	207		1,907	2,175	
	8600	6" pipe size, class 125	Q-2	3	8	↓	2,325	268		2,593	2,950	
	9005	Pneumatic system misc. components										
	9010	Adding/subtracting repeater	1 Stpi	20	.400	Ea.	98.50	14.50		113	130	
	9020	Adjustable ratio network		16	.500		92	18.10		110.10	129	
	9030	Comparator	↓	20	.400	↓	188	14.50		202.50	229	
	9040	Cumulator										
	9041	Air switching	1 Stpi	20	.400	Ea.	88.50	14.50		103	120	

Important: See the Reference Section for critical supporting data - Reference Nos., Crews, & City Cost Indexes

13838	Pneumatic/Electric Controls	CREW	DAILY OUTPUT	LABOR-HOURS	UNIT	MAT.	LABOR	EQUIP.	TOTAL	TOTAL INCL O&P	
							2002 BARE COSTS				
200	9042	Averaging	1 Stpi	20	.400	Ea.	92	14.50		106.50	123	200
9043	2:1 ratio		20	.400		83	14.50		97.50	113	
9044	Sequencing		20	.400		56.50	14.50		71	84	
9045	Two-position		16	.500		102	18.10		120.10	140	
9046	Two-position pilot	▼	16	.500	▼	95.50	18.10		113.60	133	
9050	Damper actuator										
9051	For smoke control	1 Stpi	8	1	Ea.	128	36		164	196	
9052	For ventilation	"	8	1	"	140	36		176	209	
9053	Series duplex pedestal mounted										
9054	For inlet vanes on fans, compressors	1 Stpi	6	1.333	Ea.	640	48.50		688.50	780	
9060	Enthalpy logic center		16	.500		199	18.10		217.10	247	
9070	High/low pressure selector		22	.364		35.50	13.15		48.65	59	
9080	Signal limiter		20	.400		56.50	14.50		71	84	
9081	Transmitter	▼	28	.286	▼	21.50	10.35		31.85	39	
9090	Optimal start										
9091	Warmup or cooldown programmer	1 Stpi	10	.800	Ea.	213	29		242	278	
9092	Mass temperature transmitter	"	18	.444	"	92.50	16.10		108.60	127	
9100	Step controller with positioner, time delay restrictor										
9110	recycler air valve, switches and cam settings										
9112	With cabinet										
9113	6 points	1 Stpi	4	2	Ea.	435	72.50		507.50	590	
9114	6 points, 6 manual sequences		2	4		460	145		605	730	
9115	8 points	▼	3	2.667	▼	555	96.50		651.50	755	
9200	Pressure controller and switches										
9210	High static pressure limit	1 Stpi	8	1	Ea.	129	36		165	196	
9220	Pressure transmitter		8	1		139	36		175	208	
9230	Differential pressure transmitter		8	1		560	36		596	670	
9240	Static pressure transmitter		8	1		325	36		361	410	
9250	Proportional-only control, single input		6	1.333		268	48.50		316.50	370	
9260	Proportional plus integral, single input		6	1.333		400	48.50		448.50	515	
9270	Proportional-only, dual input		6	1.333		370	48.50		418.50	480	
9280	Proportional plus integral, dual input		6	1.333		485	48.50		533.50	610	
9281	Time delay for above proportional units		16	.500		106	18.10		124.10	144	
9290	Proportional/2 position controller units		6	1.333		283	48.50		331.50	385	
9300	Differential press. control dir./rev.	▼	6	1.333	▼	190	48.50		238.50	282	
9310	Air pressure reducing valve										
9311	1/8" size	1 Stpi	18	.444	Ea.	20.50	16.10		36.60	47	
9315	Precision valve, 1/4" size		17	.471		109	17.05		126.05	146	
9320	Air flow controller		12	.667		110	24		134	158	
9330	Booster relay volume amplifier		18	.444		60.50	16.10		76.60	91	
9340	Series restrictor, straight		60	.133		3.24	4.83		8.07	10.85	
9341	T-fitting		40	.200		4.75	7.25		12	16.15	
9342	In-line adjustable		48	.167		20.50	6.05		26.55	31.50	
9350	Diode tee		40	.200		10.90	7.25		18.15	23	
9351	Restrictor tee		40	.200		10.90	7.25		18.15	23	
9360	Pneumatic gradual switch		8	1		145	36		181	214	
9361	Selector switch		8	1		63.50	36		99.50	125	
9370	Electro-pneumatic motor driven servo		10	.800		685	29		714	795	
9380	Regulated power supply		10	.800		475	29		504	565	
9390	Fan control switch and mounting base		16	.500		64	18.10		82.10	98	
9400	Circulating pump sequencer	▼	14	.571	▼	660	20.50		680.50	755	
9410	Pneumatic tubing, fittings and accessories										
9414	Tubing, urethane										
9415	1/8" OD x 1/16" ID	1 Stpi	120	.067	L.F.	.43	2.41		2.84	4.11	
9416	1/4" OD x 1/8" ID		115	.070		.85	2.52		3.37	4.74	
9417	5/32" OD x 3/32" ID	▼	110	.073	▼	.48	2.63		3.11	4.50	

SPECIAL CONSTRUCTION **13**

13838	Pneumatic/Electric Controls	CREW	DAILY OUTPUT	LABOR-HOURS	UNIT	2002 BARE COSTS				TOTAL INCL O&P
						MAT.	LABOR	EQUIP.	TOTAL	
9420	Coupling, straight									
9422	Barb x barb									
9423	1/4" x 5/32"	1 Stpi	160	.050	Ea.	.48	1.81		2.29	3.26
9424	1/4" x 1/4"		158	.051		.25	1.83		2.08	3.04
9425	3/8" x 3/8"		154	.052		.33	1.88		2.21	3.20
9426	3/8" x 1/4"		150	.053		.33	1.93		2.26	3.27
9427	1/2" x 1/4"		148	.054		.71	1.96		2.67	3.73
9428	1/2" x 3/8"		144	.056		.65	2.01		2.66	3.75
9429	1/2" x 1/2"		140	.057		.48	2.07		2.55	3.65
9440	Tube x tube									
9441	1/4" x 1/4"	1 Stpi	100	.080	Ea.	1.34	2.90		4.24	5.85
9442	3/8" x 1/4"		96	.083		1.65	3.02		4.67	6.35
9443	3/8" x 3/8"		92	.087		1.65	3.15		4.80	6.55
9444	1/2" x 3/8"		88	.091		2.33	3.29		5.62	7.50
9445	1/2" x 1/2"		84	.095		3.03	3.45		6.48	8.55
9450	Elbow coupling									
9452	Barb x barb									
9454	1/4" x 1/4"	1 Stpi	158	.051	Ea.	.53	1.83		2.36	3.34
9455	3/8" x 3/8"		154	.052		.71	1.88		2.59	3.62
9456	1/2" x 1/2"		140	.057		.95	2.07		3.02	4.17
9460	Tube x tube									
9462	1/2" x 1/2"	1 Stpi	84	.095	Ea.	3.71	3.45		7.16	9.30
9470	Tee coupling									
9472	Barb x barb x barb									
9474	1/4" x 1/4" x 5/32"	1 Stpi	108	.074	Ea.	.86	2.68		3.54	4.99
9475	1/4" x 1/4" x 1/4"		104	.077		.58	2.78		3.36	4.84
9476	3/8" x 3/8" x 5/32"		102	.078		1.85	2.84		4.69	6.30
9477	3/8" x 3/8" x 1/4"		99	.081		.86	2.93		3.79	5.35
9478	3/8" x 3/8" x 3/8"		98	.082		.92	2.96		3.88	5.45
9479	1/2" x 1/2" x 1/4"		96	.083		1.78	3.02		4.80	6.50
9480	1/2" x 1/2" x 3/8"		95	.084		1	3.05		4.05	5.70
9481	1/2" x 1/2" x 1/2"		92	.087		1.04	3.15		4.19	5.90
9484	Tube x tube x tube									
9485	1/4" x 1/4" x 1/4"	1 Stpi	66	.121	Ea.	2.57	4.39		6.96	9.45
9486	3/8" x 1/4" x 1/4"		64	.125		4.87	4.53		9.40	12.20
9487	3/8" x 3/8" x 1/4"		61	.131		3.82	4.75		8.57	11.35
9488	3/8" x 3/8" x 3/8"		60	.133		4.01	4.83		8.84	11.70
9489	1/2" x 1/2" x 3/8"		58	.138		6.20	4.99		11.19	14.35
9490	1/2" x 1/2" x 1/2"		56	.143		5.50	5.15		10.65	13.85
9492	Needle valve									
9494	Tube x tube									
9495	1/4" x 1/4"	1 Stpi	86	.093	Ea.	4.68	3.37		8.05	10.25
9496	3/8" x 3/8"	"	82	.098	"	5.20	3.53		8.73	11.10
9600	Electronic system misc. components									
9610	Electric motor damper actuator	1 Elec	8	1	Ea.	243	35.50		278.50	320
9620	Damper position indicator		10	.800		77	28.50		105.50	128
9630	Pneumatic-electronic transducer		12	.667		283	23.50		306.50	345
9640	Electro-pneumatic transducer		12	.667		283	23.50		306.50	345
9650	Remote reset control with motor drive		6	1.333		325	47.50		372.50	425
9660	Manual remote set point control		16	.500		62	17.75		79.75	95
9670	Alarm unit with two adjustable points		12	.667		299	23.50		322.50	365
9700	Modulating step controller									
9701	2-5 steps	1 Elec	5.30	1.509	Ea.	1,325	53.50		1,378.50	1,550
9702	6-10 steps		3.20	2.500		1,650	88.50		1,738.50	1,950
9703	11-15 steps		2.70	2.963		2,050	105		2,155	2,400
9704	16-20 steps		1.60	5		2,425	177		2,602	2,925

200

13838	Pneumatic/Electric Controls	CREW	DAILY OUTPUT	LABOR-HOURS	UNIT	2002 BARE COSTS				TOTAL INCL O&P	
						MAT.	LABOR	EQUIP.	TOTAL		
200	9710	Staging network										200
	9711	Master one stage	1 Elec	10	.800	Ea.	390	28.50		418.50	475
	9712	Two stage		9	.889		515	31.50		546.50	610
	9720	Sequencing network		8	1		590	35.50		625.50	700
	9730	Regulated DC power supply	↓	16	.500	↓	360	17.75		377.75	420
	9740	Accessory unit regulator									
	9741	Three function	1 Elec	9	.889	Ea.	194	31.50		225.50	260
	9742	Five function		8	1		375	35.50		410.50	465
	9750	Accessory power supply		18	.444		74.50	15.75		90.25	106
	9760	Step down transformer	↓	16	.500	↓	126	17.75		143.75	165

13851	Detection & Alarm	CREW	DAILY OUTPUT	LABOR-HOURS	UNIT	2002 BARE COSTS				TOTAL INCL O&P	
						MAT.	LABOR	EQUIP.	TOTAL		
065	0010	**DETECTION SYSTEMS**, not including wires & conduits										065
	0100	Burglar alarm, battery operated, mechanical trigger	1 Elec	4	2	Ea.	249	71		320	380
	0200	Electrical trigger		4	2		297	71		368	430
	0400	For outside key control, add		8	1		70.50	35.50		106	131
	0600	For remote signaling circuitry, add		8	1		112	35.50		147.50	176
	0800	Card reader, flush type, standard		2.70	2.963		835	105		940	1,075
	1000	Multi-code	↓	2.70	2.963	↓	1,075	105		1,180	1,325
	3594	Fire, alarm control panel									
	3600	4 zone	2 Elec	2	8	Ea.	920	284		1,204	1,425
	3800	8 zone		1	16		1,400	565		1,965	2,400
	4000	12 zone	↓	.67	23.988		1,825	850		2,675	3,275
	4020	Alarm device	1 Elec	8	1		122	35.50		157.50	187
	4050	Actuating device		8	1		292	35.50		327.50	375
	4200	Battery and rack		4	2		690	71		761	865
	4400	Automatic charger		8	1		445	35.50		480.50	545
	4600	Signal bell		8	1		49.50	35.50		85	108
	4800	Trouble buzzer or manual station		8	1		37	35.50		72.50	93.50
	5000	Detector, rate of rise		8	1		33.50	35.50		69	90
	5100	Fixed temperature		8	1		28	35.50		63.50	84
	5200	Smoke detector, ceiling type		6.20	1.290		75	45.50		120.50	151
	5400	Duct type		3.20	2.500		250	88.50		338.50	405
	5600	Strobe and horn		5.30	1.509		95	53.50		148.50	185
	5800	Fire alarm horn		6.70	1.194		36.50	42.50		79	103
	6000	Door holder, electro-magnetic		4	2		77.50	71		148.50	192
	6200	Combination holder and closer		3.20	2.500		430	88.50		518.50	605
	6400	Code transmitter		4	2		690	71		761	865
	6600	Drill switch		8	1		86.50	35.50		122	148
	6800	Master box		2.70	2.963		3,100	105		3,205	3,550
	7000	Break glass station		8	1		50	35.50		85.50	108
	7800	Remote annunciator, 8 zone lamp	↓	1.80	4.444		175	158		333	430
	8000	12 zone lamp	2 Elec	2.60	6.154		300	218		518	655
	8200	16 zone lamp	"	2.20	7.273	↓	300	258		558	715
350	0010	**TANK LEAK DETECTION SYSTEMS** Liquid and vapor										350
	0100	For hydrocarbons and hazardous liquids/vapors									
	0120	Controller, data acquisition, incl. printer, modem, RS232 port									
	0140	24 channel, for use with all probes				Ea.	2,050			2,050	2,250

SPECIAL CONSTRUCTION 13

	13851	Detection & Alarm	CREW	DAILY OUTPUT	LABOR-HOURS	UNIT	2002 BARE COSTS				TOTAL INCL O&P	
							MAT.	LABOR	EQUIP.	TOTAL		
350	0160	9 channel, for external monitoring				Ea.	1,575			1,575	1,750	350
	0200	Probes										
	0210	Well monitoring										
	0220	Liquid phase detection				Ea.	710			710	785	
	0230	Hydrocarbon vapor, fixed position					655			655	720	
	0240	Hydrocarbon vapor, float mounted					580			580	640	
	0250	Both liquid and vapor hydrocarbon				↓	880			880	970	
	0300	Secondary containment, liquid phase										
	0310	Pipe trench/manway sump				Ea.	320			320	350	
	0320	Double wall pipe and manual sump					310			310	340	
	0330	Double wall fiberglass annular space					286			286	315	
	0340	Double wall steel tank annular space				↓	286			286	315	
	0500	Accessories										
	0510	Modem, non-dedicated phone line				Ea.	256			256	282	
	0600	Monitoring, internal										
	0610	Automatic tank gauge, incl. overfill				Ea.	1,025			1,025	1,125	
	0620	Product line				"	1,100			1,100	1,200	
	0700	Monitoring, special										
	0710	Cathodic protection				Ea.	610			610	670	
	0720	Annular space chemical monitor				"	830			830	915	

For information about Means Estimating Seminars, see yellow pages 11 and 12 in back of book

Important: See the Reference Section for critical supporting data - Reference Nos., Crews, & City Cost Indexes

Division 14
Conveying Systems

Estimating Tips

General

• Many products in Division 14 will require some type of support or blocking for installation not included with the item itself. Examples are supports for conveyors or tube systems, attachment points for lifts, and footings for hoists or cranes. Add these supports in the appropriate division.

14100 Dumbwaiters
14200 Elevators

• Dumbwaiters and elevators are estimated and purchased in a method similar to buying a car. The manufacturer has a base unit with standard features. Added to this base unit price will be whatever options the owner or specifications require. Increased load capacity, additional vertical travel, additional stops, higher speed, and cab finish options are items to be considered. When developing an estimate for dumbwaiters and elevators, remember that some items needed by the installers may have to be included as part of the general contract.

Examples are:
- shaftway
- rail support brackets
- machine room
- electrical supply
- sill angles
- electrical connections
- pits
- roof penthouses
- pit ladders

Check the job specifications and drawings before pricing.

• Installation of elevators and handicapped lifts in historic structures can require significant additional costs. The associated structural requirements may involve cutting into and repairing finishes, mouldings, flooring, etc. The estimator must account for these special conditions.

14300 Escalators & Moving Walks

• Escalators and moving walks are specialty items installed by specialty contractors. There are numerous options associated with these items. For specific options contact a manufacturer or contractor. In a method similar to estimating dumbwaiters and elevators, you should verify the extent of general contract work and add items as necessary.

14400 Lifts
14500 Material Handling
14600 Hoists & Cranes

• Products such as correspondence lifts, conveyors, chutes, pneumatic tube systems, material handling cranes and hoists as well as other items specified in this subdivision may require trained installers. The general contractor might not have any choice as to who will perform the installation or when it will be performed. Long lead times are often required for these products, making early decisions in scheduling necessary.

Reference Numbers

Reference numbers are shown in bold squares at the beginning of some major classifications. These numbers refer to related items in the Reference Section. The reference information may be an estimating procedure, an alternate pricing method or technical information.

Note: Not all subdivisions listed here necessarily appear in this publication.

		14560	Chutes	CREW	DAILY OUTPUT	LABOR-HOURS	UNIT	2002 BARE COSTS				TOTAL INCL O&P	
								MAT.	LABOR	EQUIP.	TOTAL		
250	0010	**CHUTES** Linen or refuse, incl. sprinklers, 12' floor height											250
	0050	Aluminized steel, 16 ga., 18" diameter	2 Shee	3.50	4.571	Floor	735	160		895	1,050	
	0100	24" diameter		3.20	5		800	176		976	1,150	
	0200	30" diameter		3	5.333		880	187		1,067	1,250	
	0300	36" diameter		2.80	5.714		1,000	201		1,201	1,400	
	0400	Galvanized steel, 16 ga., 18" diameter		3.50	4.571		700	160		860	1,025	
	0500	24" diameter		3.20	5		790	176		966	1,125	
	0600	30" diameter		3	5.333		885	187		1,072	1,275	
	0700	36" diameter		2.80	5.714		1,050	201		1,251	1,450	
	0800	Stainless steel, 18" diameter		3.50	4.571		1,175	160		1,335	1,550	
	0900	24" diameter		3.20	5		1,250	176		1,426	1,650	
	1000	30" diameter		3	5.333		1,500	187		1,687	1,975	
	1005	36" diameter		2.80	5.714	▼	1,825	201		2,026	2,300	
	1200	Linen chute bottom collector, aluminized steel		4	4	Ea.	980	140		1,120	1,300	
	1300	Stainless steel		4	4		1,250	140		1,390	1,600	
	1500	Refuse, bottom hopper, aluminized steel, 18" diameter		3	5.333		615	187		802	965	
	1600	24" diameter		3	5.333		680	187		867	1,025	
	1800	36" diameter		3	5.333	▼	1,125	187		1,312	1,550	
	2900	Package chutes, spiral type, minimum		4.50	3.556	Floor	1,700	125		1,825	2,075	
	3000	Maximum	▼	1.50	10.667	"	4,425	375		4,800	5,450	

14580 | Pneumatic Tube Systems

		Description	CREW	DAILY OUTPUT	LABOR-HOURS	UNIT	MAT.	LABOR	EQUIP.	TOTAL	TOTAL INCL O&P	
800	0010	**PNEUMATIC TUBE SYSTEM** Single tube, 2 stations, blower										800
	0020	100' long, stock										
	0100	3" diameter	2 Stpi	.12	133	Total	4,800	4,825		9,625	12,600	
	0300	4" diameter	"	.09	177	"	5,400	6,425		11,825	15,700	
	0400	Twin tube, two stations or more, conventional system										
	0600	2-1/2" round	2 Stpi	62.50	.256	L.F.	9.65	9.25		18.90	24.50	
	0700	3" round		46	.348		11.10	12.60		23.70	31	
	0900	4" round		49.60	.323		12.30	11.70		24	31	
	1000	4" x 7" oval		37.60	.426	▼	17.90	15.40		33.30	42.50	
	1050	Add for blower		2	8	System	3,775	290		4,065	4,600	
	1110	Plus for each round station, add		7.50	2.133	Ea.	425	77		502	585	
	1150	Plus for each oval station, add		7.50	2.133	"	425	77		502	585	
	1200	Alternate pricing method: base cost, minimum		.75	21.333	Total	4,650	770		5,420	6,275	
	1300	Maximum		.25	64	"	9,200	2,325		11,525	13,600	
	1500	Plus total system length, add, minimum		93.40	.171	L.F.	5.95	6.20		12.15	15.90	
	1600	Maximum		37.60	.426	"	17.65	15.40		33.05	42.50	
	1800	Completely automatic system, 4" round, 15 to 50 stations		.29	55.172	Station	14,600	2,000		16,600	19,100	
	2200	51 to 144 stations		.32	50		11,300	1,800		13,100	15,200	
	2400	6" round or 4" x 7" oval, 15 to 50 stations		.24	66.667		18,300	2,425		20,725	23,800	
	2800	51 to 144 stations	▼	.23	69.565	▼	15,400	2,525		17,925	20,700	

14 CONVEYING SYSTEMS

For information about Means Estimating Seminars, see yellow pages 11 and 12 in back of book

Important: See the Reference Section for critical supporting data - Reference Nos., Crews, & City Cost Indexes

Division 15
Mechanical

Estimating Tips

15100 Building Services Piping
This subdivision is primarily basic pipe and related materials. The pipe may be used by any of the mechanical disciplines, i.e., plumbing, fire protection, heating, and air conditioning.

- The piping section lists the add to labor for elevated pipe installation. These adds apply to all elevated pipe, fittings, valves, insulation, etc., that are placed above 10' high. CAUTION: the correct percentage may vary for the same pipe. For example, the percentage add for the basic pipe installation should be based on the maximum height that the craftsman must install for that particular section. If the pipe is to be located 14' above the floor but it is suspended on threaded rod from beams, the bottom flange of which is 18' high (4' rods), then the height is actually 18' and the add is 20%. The pipe coverer, however, does not have to go above the 14' and so his add should be 10%.

- Most pipe is priced first as straight pipe with a joint (coupling, weld, etc.) every 10' and a hanger usually every 10'. There are exceptions with hanger spacing such as: for cast iron pipe (5') and plastic pipe (3 per 10'). Following each type of pipe there are several lines listing sizes and the amount to be subtracted to delete couplings and hangers. This is for pipe that is to be buried or supported together on trapeze hangers. The reason that the couplings are deleted is that these runs are usually long and frequently longer lengths of pipe are used. By deleting the couplings the estimator is expected to look up and add back the correct reduced number of couplings.

- When preparing an estimate it may be necessary to approximate the fittings. Fittings usually run between 25% and 50% of the cost of the pipe. The lower percentage is for simpler runs, and the higher number is for complex areas like mechanical rooms.

- For historic restoration projects, the systems must be as invisible as possible, and pathways must be sought for pipes, conduits, and ductwork. While installations in accessible spaces (such as basements and attics) are relatively straightforward to estimate, labor costs may be more difficult to determine when delivery systems must be concealed.

15400 Plumbing Fixtures & Equipment
- Plumbing fixture costs usually require two lines, the fixture itself and its "rough-in, supply and waste".

- In the Assemblies Section (Plumbing D2010) for the desired fixture, the System Components Group in the center of the page shows the fixture itself on the first line while the rest of the list (fittings, pipe, tubing, etc.) will total up to what we refer to in the Unit Price section as "Rough-in, supply, waste and vent". Note that for most fixtures we allow a nominal 5' of tubing to reach from the fixture to a main or riser.

- Remember that gas and oil fired units need venting.

15500 Heat Generation Equipment
- When estimating the cost of an HVAC system, check to see who is responsible for providing and installing the temperature control system. It is possible to overlook controls, assuming that they would be included in the electrical estimate.

- When looking up a boiler be careful on specified capacity. Some manufacturers rate their products on output while others use input.

- Include HVAC insulation for pipe, boiler and duct (wrap and liner).

- Be careful when looking up mechanical items to get the correct pressure rating and connection type (thread, weld, flange).

15700 Heating/Ventilation/ Air Conditioning Equipment
- Combination heating and cooling units are sized by the air conditioning requirements. (See Reference No. R15710-020 for preliminary sizing guide.)

- A ton of air conditioning is nominally 400 CFM.

- Rectangular duct is taken off by the linear foot for each size, but its cost is usually estimated by the pound. Remember that SMACNA standards now base duct on internal pressure.

- Prefabricated duct is estimated and purchased like pipe: straight sections and fittings.

- Note that cranes or other lifting equipment are not included on any lines in Division 15. For example, if a crane is required to lift a heavy piece of pipe into place high above a gym floor, or to put a rooftop unit on the roof of a four-story building, etc., it must be added. Due to the potential for extreme variation—from nothing additional required, to a major crane or helicopter—we feel that including a nominal amount for "lifting contingency" would be useless and detract from the accuracy of the estimate. When using equipment rental from Means do not forget to include the cost of the operator(s).

Reference Numbers
Reference numbers are shown in bold squares at the beginning of some major classifications. These numbers refer to related items in the Reference Section. The reference information may be an estimating procedure, an alternate pricing method or technical information.

Note: Not all subdivisions listed here necessarily appear in this publication.

15051 | Mechanical General

			CREW	DAILY OUTPUT	LABOR-HOURS	UNIT	2002 BARE COSTS				TOTAL INCL O&P	
							MAT.	LABOR	EQUIP.	TOTAL		
100	0010	**BOILERS, GENERAL** Prices do not include flue piping, elec. wiring,										100
	0020	gas or oil piping, boiler base, pad, or tankless unless noted										
	0100	Boiler H.P.: 10 KW = 34 lbs/steam/hr = 33,475 BTU/hr.	R15500 -020									
	0120											
	0150	To convert SFR to BTU rating: Hot water, 150 x SFR;										
	0160	Forced hot water, 180 x SFR; steam, 240 x SFR										
700	0010	**PIPING** See also divisions 02500 & 02600 for site work	R15100 -070									700
	1000	Add to labor for elevated installation										
	1080	10' to 15' high							10%			
	1100	15' to 20' high							20%			
	1120	20' to 25' high							25%			
	1140	25' to 30' high							35%			
	1160	30' to 35' high							40%			
	1180	35' to 40' high							50%			
	1200	Over 40' high							55%			

15055 | Selective Mech Demolition

			CREW	DAILY OUTPUT	LABOR-HOURS	UNIT	MAT.	LABOR	EQUIP.	TOTAL	TOTAL INCL O&P	
300	0010	**HVAC DEMOLITION**	R15050 -720									300
	0100	Air conditioner, split unit, 3 ton	Q-5	2	8	Ea.		261		261	395	
	0150	Package unit, 3 ton	Q-6	3	8			270		270	410	
	0300	Boiler, electric	Q-19	2	12			400		400	605	
	0340	Gas or oil, steel, under 150 MBH	Q-6	3	8			270		270	410	
	0380	Over 150 MBH	"	2	12			405		405	610	
	1000	Ductwork, 4" high, 8" wide	1 Clab	200	.040	L.F.		.94		.94	1.46	
	1020	10" wide		190	.042			.99		.99	1.54	
	1040	14" wide		180	.044			1.04		1.04	1.62	
	1100	6" high, 8" wide		165	.048			1.14		1.14	1.77	
	1120	12" wide		150	.053			1.25		1.25	1.95	
	1140	18" wide		135	.059			1.39		1.39	2.16	
	1200	10" high, 12" wide		125	.064			1.50		1.50	2.34	
	1220	18" wide		115	.070			1.63		1.63	2.54	
	1240	24" wide		110	.073			1.71		1.71	2.65	
	1300	12"-14" high, 16"-18" wide		85	.094			2.21		2.21	3.44	
	1320	24" wide		75	.107			2.50		2.50	3.89	
	1340	48" wide		71	.113			2.64		2.64	4.11	
	1400	18" high, 24" wide		67	.119			2.80		2.80	4.36	
	1420	36" wide		63	.127			2.98		2.98	4.63	
	1440	48" wide		59	.136			3.18		3.18	4.95	
	1500	30" high, 36" wide		56	.143			3.35		3.35	5.20	
	1520	48" wide		53	.151			3.54		3.54	5.50	
	1540	72" wide		50	.160			3.75		3.75	5.85	
	1550	Duct heater, electric strip	1 Elec	8	1	Ea.		35.50		35.50	53	
	2200	Furnace, electric	Q-20	2	10			325		325	495	
	2300	Gas or oil, under 120 MBH	Q-9	4	4			126		126	195	
	2340	Over 120 MBH	"	3	5.333			169		169	260	
	2800	Heat pump, package unit, 3 ton	Q-5	2.40	6.667			217		217	330	
	2840	Split unit, 3 ton		2	8			261		261	395	
	3000	Mechanical equipment, light items. Unit is weight, not cooling.		.90	17.778	Ton		580		580	875	
	3600	Heavy items		1.10	14.545	"		475		475	715	
	3700	Deduct for salvage (when applicable), minimum				Job					52	
	3750	Maximum				"					420	
600	0010	**PLUMBING DEMOLITION**	R15050 -720									600
	1020	Fixtures, including 10' piping										
	1100	Bath tubs, cast iron	1 Plum	4	2	Ea.		72		72	108	
	1120	Fiberglass		6	1.333			48		48	72.50	

15 MECHANICAL

Important: See the Reference Section for critical supporting data - Reference Nos., Crews, & City Cost Indexes

15055 | Selective Mech Demolition

			CREW	DAILY OUTPUT	LABOR-HOURS	UNIT	MAT.	2002 BARE COSTS LABOR	EQUIP.	TOTAL	TOTAL INCL O&P	
600	1140	Steel	1 Plum	5	1.600	Ea.		57.50		57.50	86.50	**600**
	1200	Lavatory, wall hung		10	.800			29		29	43.50	
	1220	Counter top		8	1			36		36	54	
	1300	Sink, steel or cast iron, single		8	1			36		36	54	
	1320	Double		7	1.143			41		41	62	
	1400	Water closet, floor mounted		8	1			36		36	54	
	1420	Wall mounted		7	1.143			41		41	62	
	1500	Urinal, floor mounted		4	2			72		72	108	
	1520	Wall mounted		7	1.143			41		41	62	
	1600	Water fountains, free standing		8	1			36		36	54	
	1620	Recessed		6	1.333	▼		48		48	72.50	
	2000	Piping, metal, to 1-1/2" diameter		200	.040	L.F.		1.44		1.44	2.17	
	2050	2" to 3-1/2" diameter	▼	150	.053			1.92		1.92	2.89	
	2100	4" to 6" diameter	2 Plum	100	.160			5.75		5.75	8.65	
	2150	8" to 14" diameter		60	.267			9.60		9.60	14.45	
	2153	16" diameter	▼	55	.291	▼		10.45		10.45	15.75	
	2160	Deduct for salvage, aluminum scrap				Ton					525	
	2170	Brass scrap									420	
	2180	Copper scrap									1,025	
	2200	Lead scrap									235	
	2220	Steel scrap				▼					65	
	2250	Water heater, 40 gal.	1 Plum	6	1.333	Ea.		48		48	72.50	
	6000	Remove and reset fixtures, minimum		6	1.333			48		48	72.50	
	6100	Maximum	▼	4	2	▼		72		72	108	

15060 | Hangers & Supports

			CREW	DAILY OUTPUT	LABOR-HOURS	UNIT	MAT.	2002 BARE COSTS LABOR	EQUIP.	TOTAL	TOTAL INCL O&P	
300	0010	**PIPE HANGERS AND SUPPORTS**, TYPE numbers per MSS-SP58										**300**
	0050	Brackets										
	0060	Beam side or wall, malleable iron, TYPE 34										
	0070	3/8" threaded rod size	1 Plum	48	.167	Ea.	2.74	6		8.74	12.05	
	0080	1/2" threaded rod size		48	.167		5.30	6		11.30	14.85	
	0090	5/8" threaded rod size		48	.167		7.05	6		13.05	16.80	
	0100	3/4" threaded rod size		48	.167		8.45	6		14.45	18.35	
	0110	7/8" threaded rod size	▼	48	.167		10.15	6		16.15	20	
	0120	For concrete installation, add				▼		30%				
	0150	Wall, welded steel, medium, TYPE 32										
	0160	0 size, 12" wide, 18" deep	1 Plum	34	.235	Ea.	206	8.45		214.45	239	
	0170	1 size, 18" wide 24" deep		34	.235		244	8.45		252.45	282	
	0180	2 size, 24" wide, 30" deep	▼	34	.235	▼	325	8.45		333.45	370	
	0200	Beam attachment, welded, TYPE 22										
	0202	3/8"	Q-15	80	.200	Ea.	12.40	6.45	1	19.85	24.50	
	0203	1/2"		76	.211		13.20	6.80	1.05	21.05	26	
	0204	5/8"		72	.222		14	7.20	1.11	22.31	27.50	
	0205	3/4"		68	.235		19.45	7.60	1.18	28.23	34.50	
	0206	7/8"		64	.250		24.50	8.10	1.25	33.85	40.50	
	0207	1"	▼	56	.286	▼	37	9.25	1.43	47.68	56	
	0300	Clamps										
	0310	C-clamp, for mounting on steel beam flange, w/locknut, TYPE 23										
	0320	3/8" threaded rod size	1 Plum	160	.050	Ea.	1.35	1.80		3.15	4.20	
	0330	1/2" threaded rod size		160	.050		1.57	1.80		3.37	4.44	
	0340	5/8" threaded rod size		160	.050		2.28	1.80		4.08	5.20	
	0350	3/4" threaded rod size		160	.050		3.29	1.80		5.09	6.35	
	0352	7/8" threaded rod size	▼	140	.057	▼	14	2.05		16.05	18.50	
	0400	High temperature to 1050°F, alloy steel										
	0410	4" pipe size	Q-1	106	.151	Ea.	33	4.88		37.88	44	
	0420	6" pipe size	▼	106	.151	▼	55.50	4.88		60.38	69	

R15050 -720

MECHANICAL 15

	15060	Hangers & Supports	CREW	DAILY OUTPUT	LABOR-HOURS	UNIT	2002 BARE COSTS				TOTAL INCL O&P	
							MAT.	LABOR	EQUIP.	TOTAL		
300	0430	8" pipe size	Q-1	97	.165	Ea.	63	5.35		68.35	77	**300**
	0440	10" pipe size		84	.190		85	6.15		91.15	103	
	0450	12" pipe size		72	.222		99	7.20		106.20	120	
	0460	14" pipe size		64	.250		219	8.10		227.10	253	
	0470	16" pipe size		56	.286		253	9.25		262.25	292	
	0480	Beam clamp, flange type, TYPE 25										
	0482	For 3/8" bolt	1 Plum	48	.167	Ea.	4.09	6		10.09	13.55	
	0483	For 1/2" bolt		44	.182		4.30	6.55		10.85	14.60	
	0484	For 5/8" bolt		40	.200		12.10	7.20		19.30	24	
	0485	For 3/4" bolt		36	.222		14.65	8		22.65	28	
	0486	For 1" bolt		32	.250		71	9		80	91.50	
	0500	I-beam, for mounting on bottom flange, strap iron, TYPE 21										
	0510	2" flange size	1 Plum	96	.083	Ea.	4.09	3		7.09	9	
	0520	3" flange size		95	.084		4.22	3.03		7.25	9.20	
	0530	4" flange size		93	.086		4.32	3.09		7.41	9.40	
	0540	5" flange size		92	.087		4.48	3.13		7.61	9.65	
	0550	6" flange size		90	.089		4.59	3.20		7.79	9.85	
	0560	7" flange size		88	.091		4.74	3.27		8.01	10.15	
	0570	8" flange size		86	.093		4.90	3.34		8.24	10.45	
	0600	One hole, vertical mounting, malleable iron										
	0610	1/2" pipe size	1 Plum	160	.050	Ea.	.71	1.80		2.51	3.49	
	0620	3/4" pipe size		145	.055		.88	1.98		2.86	3.96	
	0630	1" pipe size		136	.059		1.33	2.11		3.44	4.65	
	0640	1-1/4" pipe size		128	.063		1.66	2.25		3.91	5.20	
	0650	1-1/2" pipe size		120	.067		1.87	2.40		4.27	5.65	
	0660	2" pipe size		112	.071		2.82	2.57		5.39	6.95	
	0670	2-1/2" pipe size		104	.077		7.70	2.77		10.47	12.60	
	0680	3" pipe size		96	.083		10.65	3		13.65	16.25	
	0690	3-1/2" pipe size		90	.089		10.65	3.20		13.85	16.55	
	0700	4" pipe size		84	.095		41.50	3.42		44.92	50.50	
	0750	Riser or extension pipe, carbon steel, TYPE 8										
	0756	1/2" pipe size	1 Plum	52	.154	Ea.	2.51	5.55		8.06	11.10	
	0760	3/4" pipe size		48	.167		2.61	6		8.61	11.90	
	0770	1" pipe size		47	.170		2.71	6.10		8.81	12.25	
	0780	1-1/4" pipe size		46	.174		3.31	6.25		9.56	13.10	
	0790	1-1/2" pipe size		45	.178		3.47	6.40		9.87	13.45	
	0800	2" pipe size		43	.186		3.65	6.70		10.35	14.10	
	0810	2-1/2" pipe size		41	.195		3.85	7		10.85	14.85	
	0820	3" pipe size		40	.200		4.20	7.20		11.40	15.45	
	0830	3-1/2" pipe size		39	.205		5.15	7.35		12.50	16.75	
	0840	4" pipe size		38	.211		5.30	7.55		12.85	17.25	
	0850	5" pipe size		37	.216		7.50	7.75		15.25	19.95	
	0860	6" pipe size		36	.222		8.80	8		16.80	21.50	
	0870	8" pipe size		34	.235		14.25	8.45		22.70	28.50	
	0880	10" pipe size		32	.250		22.50	9		31.50	38.50	
	0890	12" pipe size		28	.286		27	10.25		37.25	45	
	0900	For plastic coating 3/4" to 4", add					190%					
	0910	For copper plating 3/4" to 4", add					58%					
	0950	Two piece, complete, carbon steel, medium weight, TYPE 4										
	0960	1/2" pipe size	Q-1	137	.117	Ea.	2.10	3.78		5.88	8	
	0970	3/4" pipe size		134	.119		2.10	3.86		5.96	8.10	
	0980	1" pipe size		132	.121		2.22	3.92		6.14	8.35	
	0990	1-1/4" pipe size		130	.123		2.76	3.98		6.74	9.05	
	1000	1-1/2" pipe size		126	.127		2.76	4.11		6.87	9.25	
	1010	2" pipe size		124	.129		3.41	4.17		7.58	10.05	
	1020	2-1/2" pipe size		120	.133		3.67	4.31		7.98	10.55	

15 MECHANICAL

Important: See the Reference Section for critical supporting data - Reference Nos., Crews, & City Cost Indexes

			DAILY	LABOR-		2002 BARE COSTS				TOTAL	
15060	**Hangers & Supports**	CREW	OUTPUT	HOURS	UNIT	MAT.	LABOR	EQUIP.	TOTAL	INCL O&P	
300	1030	3" pipe size	Q-1	117	.137	Ea.	4.09	4.42		8.51	11.15	300
	1040	3-1/2" pipe size		114	.140		5.80	4.54		10.34	13.25	
	1050	4" pipe size		110	.145		5.80	4.71		10.51	13.50	
	1060	5" pipe size		106	.151		8.40	4.88		13.28	16.60	
	1070	6" pipe size		104	.154		14.15	4.98		19.13	23	
	1080	8" pipe size		100	.160		16.90	5.20		22.10	26.50	
	1090	10" pipe size		96	.167		30.50	5.40		35.90	41.50	
	1100	12" pipe size		89	.180		35.50	5.80		41.30	48	
	1110	14" pipe size		82	.195		61.50	6.30		67.80	77	
	1120	16" pipe size	↓	68	.235		66.50	7.60		74.10	84.50	
	1130	For galvanized, add				↓	45%					
	1150	Insert, concrete										
	1160	Wedge type, carbon steel body, malleable iron nut										
	1170	1/4" threaded rod size	1 Plum	96	.083	Ea.	4.75	3		7.75	9.75	
	1180	3/8" threaded rod size		96	.083		4.99	3		7.99	10	
	1190	1/2" threaded rod size		96	.083		5.15	3		8.15	10.20	
	1200	5/8" threaded rod size		96	.083		5.25	3		8.25	10.25	
	1210	3/4" threaded rod size		96	.083		5.75	3		8.75	10.85	
	1220	7/8" threaded rod size	↓	96	.083		6.70	3		9.70	11.85	
	1230	For galvanized, add				↓	1.07			1.07	1.18	
	1250	Pipe guide sized for insulation										
	1260	No. 1, 1" pipe size, 1" thick insulation	1 Stpi	26	.308	Ea.	125	11.15		136.15	155	
	1270	No. 2, 1-1/4"-2" pipe size, 1" thick insulation		23	.348		148	12.60		160.60	182	
	1280	No. 3, 1-1/4"-2" pipe size, 1-1/2" thick insulation		21	.381		148	13.80		161.80	184	
	1290	No. 4, 2-1/2"-3-1/2" pipe size, 1-1/2" thick insulation		18	.444		148	16.10		164.10	188	
	1300	No. 5, 4"-5" pipe size, 1-1/2" thick insulation	↓	16	.500		164	18.10		182.10	209	
	1310	No. 6, 5"-6" pipe size, 2" thick insulation	Q-5	21	.762		182	25		207	239	
	1320	No. 7, 8" pipe size, 2" thick insulation		16	1		320	32.50		352.50	400	
	1330	No. 8, 10" pipe size, 2" thick insulation	↓	12	1.333		605	43.50		648.50	730	
	1340	No. 9, 12" pipe size, 2" thick insulation	Q-6	17	1.412		605	47.50		652.50	735	
	1350	No. 10, 12"-14" pipe size, 2-1/2" thick insulation		16	1.500		785	50.50		835.50	940	
	1360	No. 11, 16" pipe size, 2-1/2" thick insulation		10.50	2.286		785	77		862	980	
	1370	No. 12, 16"-18" pipe size, 3" thick insulation		9	2.667		830	90		920	1,050	
	1380	No. 13, 20" pipe size, 3" thick insulation		7.50	3.200		830	108		938	1,075	
	1390	No. 14, 24" pipe size, 3" thick insulation	↓	7	3.429	↓	1,050	116		1,166	1,350	
	1400	Bands										
	1410	Adjustable band, carbon steel, for non-insulated pipe, TYPE 7										
	1420	1/2" pipe size	Q-1	142	.113	Ea.	.82	3.65		4.47	6.40	
	1430	3/4" pipe size		140	.114		.82	3.70		4.52	6.50	
	1440	1" pipe size		137	.117		.93	3.78		4.71	6.70	
	1450	1-1/4" pipe size		134	.119		.97	3.86		4.83	6.85	
	1460	1-1/2" pipe size		131	.122		1.09	3.95		5.04	7.15	
	1470	2" pipe size		129	.124		1.50	4.01		5.51	7.70	
	1480	2-1/2" pipe size		125	.128		1.58	4.14		5.72	8	
	1490	3" pipe size		122	.131		1.76	4.24		6	8.35	
	1500	3-1/2" pipe size		119	.134		2.73	4.35		7.08	9.55	
	1510	4" pipe size		114	.140		2.73	4.54		7.27	9.85	
	1520	5" pipe size		110	.145		2.98	4.71		7.69	10.40	
	1530	6" pipe size		108	.148		4	4.79		8.79	11.65	
	1540	8" pipe size	↓	104	.154	↓	6.25	4.98		11.23	14.35	
	1550	For copper plated, add					50%					
	1560	For galvanized, add					30%					
	1570	For plastic coating, add					30%					
	1600	Adjusting nut malleable iron, steel band, TYPE 9										
	1610	1/2" pipe size, galvanized band	Q-1	137	.117	Ea.	.49	3.78		4.27	6.25	
	1620	3/4" pipe size, galvanized band	↓	135	.119		.49	3.83		4.32	6.35	

			CREW	DAILY OUTPUT	LABOR-HOURS	UNIT	2002 BARE COSTS				TOTAL INCL O&P	
	15060	**Hangers & Supports**					MAT.	LABOR	EQUIP.	TOTAL		
300	1630	1" pipe size, galvanized band	Q-1	132	.121	Ea.	.49	3.92		4.41	6.45	**300**
	1640	1-1/4" pipe size, galvanized band		129	.124		.54	4.01		4.55	6.65	
	1650	1-1/2" pipe size, galvanized band		126	.127		.54	4.11		4.65	6.80	
	1660	2" pipe size, galvanized band		124	.129		.57	4.17		4.74	6.95	
	1670	2-1/2" pipe size, galvanized band		120	.133		1.08	4.31		5.39	7.70	
	1680	3" pipe size, galvanized band		117	.137		1.17	4.42		5.59	7.95	
	1690	3-1/2" pipe size, galvanized band		114	.140		1.45	4.54		5.99	8.45	
	1700	4" pipe size, cadmium plated band		110	.145		2.17	4.71		6.88	9.50	
	1710	5" pipe size, cadmium plated band		106	.151		2.94	4.88		7.82	10.60	
	1720	6" pipe size, cadmium plated band		104	.154		3.29	4.98		8.27	11.10	
	1730	8" pipe size, cadmium plated band		100	.160		3.70	5.20		8.90	11.85	
	1740	For plastic coated band, add					35%					
	1750	For completely copper coated, add					45%					
	1800	Clevis, adjustable, carbon steel, for non-insulated pipe, TYPE 1										
	1810	1/2" pipe size	Q-1	137	.117	Ea.	.86	3.78		4.64	6.65	
	1820	3/4" pipe size		135	.119		.86	3.83		4.69	6.75	
	1830	1" pipe size		132	.121		.91	3.92		4.83	6.90	
	1840	1-1/4" pipe size		129	.124		1.04	4.01		5.05	7.20	
	1850	1-1/2" pipe size		126	.127		1.14	4.11		5.25	7.45	
	1860	2" pipe size		124	.129		1.27	4.17		5.44	7.70	
	1870	2-1/2" pipe size		120	.133		2.05	4.31		6.36	8.75	
	1880	3" pipe size		117	.137		2.59	4.42		7.01	9.50	
	1890	3-1/2" pipe size		114	.140		2.71	4.54		7.25	9.85	
	1900	4" pipe size		110	.145		3.12	4.71		7.83	10.55	
	1910	5" pipe size		106	.151		4.31	4.88		9.19	12.10	
	1920	6" pipe size		104	.154		4.97	4.98		9.95	12.95	
	1930	8" pipe size		100	.160		8.20	5.20		13.40	16.80	
	1940	10" pipe size		96	.167		22	5.40		27.40	32	
	1950	12" pipe size		89	.180		27.50	5.80		33.30	39.50	
	1960	14" pipe size		82	.195		40	6.30		46.30	53.50	
	1970	16" pipe size		68	.235		56.50	7.60		64.10	74	
	1971	18" pipe size		54	.296		67.50	9.60		77.10	89	
	1972	20" pipe size		38	.421		144	13.60		157.60	180	
	1980	For galvanized, add					66%					
	1990	For copper plated 1/2" to 4", add					77%					
	2000	For light weight 1/2" to 4", deduct					13%					
	2010	Insulated pipe type, 3/4" to 12" pipe, add					180%					
	2020	Insulated pipe type, chrome-moly U-strap, add					530%					
	2250	Split ring, malleable iron, for non-insulated pipe, TYPE 11										
	2260	1/2" pipe size	Q-1	137	.117	Ea.	3.15	3.78		6.93	9.15	
	2270	3/4" pipe size		135	.119		3.17	3.83		7	9.30	
	2280	1" pipe size		132	.121		3.35	3.92		7.27	9.60	
	2290	1-1/4" pipe size		129	.124		3.88	4.01		7.89	10.30	
	2300	1-1/2" pipe size		126	.127		5.05	4.11		9.16	11.75	
	2310	2" pipe size		124	.129		5.75	4.17		9.92	12.60	
	2320	2-1/2" pipe size		120	.133		8.25	4.31		12.56	15.60	
	2330	3" pipe size		117	.137		10.30	4.42		14.72	18	
	2340	3-1/2" pipe size		114	.140		14.30	4.54		18.84	22.50	
	2350	4" pipe size		110	.145		14.30	4.71		19.01	23	
	2360	5" pipe size		106	.151		32	4.88		36.88	42.50	
	2370	6" pipe size		104	.154		32	4.98		36.98	42.50	
	2380	8" pipe size		100	.160		56.50	5.20		61.70	70	
	2390	For copper plated, add					8%					
	2500	Washer, flat steel										
	2502	3/8"	1 Plum	240	.033	Ea.	.05	1.20		1.25	1.87	
	2503	1/2"		220	.036		.07	1.31		1.38	2.05	

Important: See the Reference Section for critical supporting data - Reference Nos., Crews, & City Cost Indexes

15 MECHANICAL

15060	Hangers & Supports	CREW	DAILY OUTPUT	LABOR-HOURS	UNIT	2002 BARE COSTS				TOTAL INCL O&P
						MAT.	LABOR	EQUIP.	TOTAL	
300 2504	5/8"	1 Plum	200	.040	Ea.	.15	1.44		1.59	2.34 **300**
2505	3/4"		180	.044		.11	1.60		1.71	2.53
2506	7/8"		160	.050		.16	1.80		1.96	2.89
2507	1"		140	.057		.16	2.05		2.21	3.28
2508	1-1/8"		120	.067		.48	2.40		2.88	4.14
2520	Nut, steel, hex									
2522	3/8"	1 Plum	200	.040	Ea.	.22	1.44		1.66	2.41
2523	1/2"		180	.044		.39	1.60		1.99	2.84
2524	5/8"		160	.050		.82	1.80		2.62	3.61
2525	3/4"		140	.057		1.10	2.05		3.15	4.31
2526	7/8"		120	.067		1.85	2.40		4.25	5.65
2527	1"		100	.080		2	2.88		4.88	6.55
2528	1-1/8"		80	.100		3.80	3.60		7.40	9.60
2532	Turnbuckle, TYPE 13									
2534	3/8"	1 Plum	80	.100	Ea.	14.90	3.60		18.50	22
2535	1/2"		72	.111		18.50	3.99		22.49	26.50
2536	5/8"		64	.125		19.30	4.49		23.79	28
2538	7/8"		48	.167		35.50	6		41.50	48
2539	1"		40	.200		43	7.20		50.20	58.50
2540	1-1/8"		32	.250		70	9		79	90.50
2544	Eye rod, welded									
2546	3/8"	1 Plum	56	.143	Ea.	17.95	5.15		23.10	27.50
2547	1/2"		48	.167		26.50	6		32.50	38.50
2548	5/8"		40	.200		40	7.20		47.20	55.50
2549	3/4"		32	.250		51	9		60	69.50
2550	7/8"		24	.333		62.50	12		74.50	87
2551	1"		16	.500		79	18		97	114
2556	Eye rod, welded, linked									
2558	3/8"	1 Plum	56	.143	Ea.	55.50	5.15		60.65	69
2559	1/2"		48	.167		75.50	6		81.50	92
2560	5/8"		40	.200		108	7.20		115.20	130
2561	3/4"		32	.250		140	9		149	168
2562	7/8"		24	.333		171	12		183	206
2563	1"		16	.500		220	18		238	268
2650	Rods, carbon steel									
2660	Continuous thread									
2670	1/4" thread size	1 Plum	144	.056	L.F.	.48	2		2.48	3.54
2680	3/8" thread size		144	.056		.53	2		2.53	3.59
2690	1/2" thread size		144	.056		.84	2		2.84	3.93
2700	5/8" thread size		144	.056		1.19	2		3.19	4.32
2710	3/4" thread size		144	.056		2.11	2		4.11	5.35
2720	7/8" thread size		144	.056		2.64	2		4.64	5.90
2721	1" thread size	Q-1	160	.100		4.49	3.24		7.73	9.80
2722	1-1/8" thread size	"	120	.133		5.60	4.31		9.91	12.65
2725	1/4" thread size, bright finish	1 Plum	144	.056		.68	2		2.68	3.76
2726	1/2" thread size, bright finish	"	144	.056		1.17	2		3.17	4.30
2730	For galvanized, add					40%				
2750	Both ends machine threaded 18" length									
2760	3/8" thread size	1 Plum	240	.033	Ea.	3.94	1.20		5.14	6.15
2770	1/2" thread size		240	.033		6.30	1.20		7.50	8.75
2780	5/8" thread size		240	.033		9.05	1.20		10.25	11.80
2790	3/4" thread size		240	.033		13.90	1.20		15.10	17.10
2800	7/8" thread size		240	.033		23.50	1.20		24.70	27.50
2810	1" thread size		240	.033		31.50	1.20		32.70	36.50
2820	Rod couplings									
2821	3/8"	1 Plum	60	.133	Ea.	1.26	4.79		6.05	8.65

MECHANICAL **15**

	15060	Hangers & Supports	CREW	DAILY OUTPUT	LABOR-HOURS	UNIT	2002 BARE COSTS				TOTAL INCL O&P	
							MAT.	LABOR	EQUIP.	TOTAL		
300	2822	1/2"	1 Plum	54	.148	Ea.	1.60	5.35		6.95	9.80	300
	2823	5/8"		48	.167		2.46	6		8.46	11.75	
	2824	3/4"		44	.182		3.59	6.55		10.14	13.80	
	2825	7/8"		40	.200		5.45	7.20		12.65	16.85	
	2826	1"		34	.235		7.35	8.45		15.80	21	
	2827	1-1/8"		30	.267		16.55	9.60		26.15	32.50	
	2860	Pipe hanger assy, adj. clevis, saddle, rod, clamp, insul. allowance										
	2864	1/2" pipe size	Q-5	35	.457	Ea.	24	14.90		38.90	48.50	
	2868	1" pipe size		34.60	.462		33	15.05		48.05	58.50	
	2872	2" pipe size		33.30	.480		36	15.65		51.65	63	
	2876	3" pipe size		31.20	.513		50.50	16.70		67.20	80.50	
	2880	4" pipe size		30.70	.521		53.50	17		70.50	84.50	
	2884	6" pipe size		29.80	.537		66.50	17.50		84	99.50	
	2888	8" pipe size		28	.571		104	18.60		122.60	143	
	2892	10" pipe size		25.20	.635		172	20.50		192.50	221	
	2896	12" pipe size		23.20	.690		193	22.50		215.50	247	
	2900	Rolls										
	2910	Adjustable yoke, carbon steel with CI roll, TYPE 43										
	2918	2" pipe size	Q-1	140	.114	Ea.	11.75	3.70		15.45	18.55	
	2920	2-1/2" pipe size		137	.117		11.75	3.78		15.53	18.65	
	2930	3" pipe size		131	.122		13.10	3.95		17.05	20.50	
	2940	3-1/2" pipe size		124	.129		17.40	4.17		21.57	25.50	
	2950	4" pipe size		117	.137		17.65	4.42		22.07	26	
	2960	5" pipe size		110	.145		20.50	4.71		25.21	29.50	
	2970	6" pipe size		104	.154		25.50	4.98		30.48	36	
	2980	8" pipe size		96	.167		38.50	5.40		43.90	50	
	2990	10" pipe size		80	.200		45.50	6.45		51.95	60	
	3000	12" pipe size		68	.235		67	7.60		74.60	85.50	
	3010	14" pipe size		56	.286		180	9.25		189.25	212	
	3020	16" pipe size		48	.333		231	10.80		241.80	270	
	3050	Chair, carbon steel with CI roll										
	3060	2" pipe size	1 Plum	68	.118	Ea.	15.90	4.23		20.13	24	
	3070	2-1/2" pipe size		65	.123		17.20	4.42		21.62	25.50	
	3080	3" pipe size		62	.129		18.60	4.64		23.24	27.50	
	3090	3-1/2" pipe size		60	.133		22.50	4.79		27.29	32	
	3100	4" pipe size		58	.138		25	4.96		29.96	35	
	3110	5" pipe size		56	.143		27.50	5.15		32.65	38	
	3120	6" pipe size		53	.151		38	5.45		43.45	50	
	3130	8" pipe size		50	.160		51.50	5.75		57.25	65	
	3140	10" pipe size		48	.167		64.50	6		70.50	80	
	3150	12" pipe size		46	.174		94	6.25		100.25	112	
	3170	Single pipe roll, (see line 2650 for rods), TYPE 41, 1" pipe size	Q-1	137	.117		9.35	3.78		13.13	16	
	3180	1-1/4" pipe size		131	.122		9.35	3.95		13.30	16.25	
	3190	1-1/2" pipe size		129	.124		9.50	4.01		13.51	16.50	
	3200	2" pipe size		124	.129		9.95	4.17		14.12	17.25	
	3210	2-1/2" pipe size		118	.136		10.70	4.39		15.09	18.35	
	3220	3" pipe size		115	.139		11.30	4.50		15.80	19.20	
	3230	3-1/2" pipe size		113	.142		13.10	4.58		17.68	21.50	
	3240	4" pipe size		112	.143		14.35	4.62		18.97	22.50	
	3250	5" pipe size		110	.145		15.90	4.71		20.61	24.50	
	3260	6" pipe size		101	.158		20.50	5.10		25.60	30.50	
	3270	8" pipe size		90	.178		32	5.75		37.75	43.50	
	3280	10" pipe size		80	.200		38	6.45		44.45	51.50	
	3290	12" pipe size		68	.235		55	7.60		62.60	72	
	3291	14" pipe size		56	.286		90.50	9.25		99.75	113	
	3292	16" pipe size		48	.333		123	10.80		133.80	151	

15 MECHANICAL

Important: See the Reference Section for critical supporting data - Reference Nos., Crews, & City Cost Indexes

15060	Hangers & Supports	CREW	DAILY OUTPUT	LABOR-HOURS	UNIT	2002 BARE COSTS				TOTAL INCL O&P
						MAT.	LABOR	EQUIP.	TOTAL	
300 3293	18" pipe size	Q-1	40	.400	Ea.	132	12.95		144.95	166
3294	20" pipe size		35	.457		181	14.80		195.80	222
3296	24" pipe size		30	.533		287	17.25		304.25	340
3297	30" pipe size		25	.640		565	20.50		585.50	650
3298	36" pipe size	▼	20	.800	▼	580	26		606	675
3300	Saddles (add vertical pipe riser, usually 3" diameter)									
3310	Pipe support, complete, adjust., CI saddle, TYPE 36									
3320	2-1/2" pipe size	1 Plum	96	.083	Ea.	147	3		150	167
3330	3" pipe size		88	.091		149	3.27		152.27	169
3340	3-1/2" pipe size		79	.101		150	3.64		153.64	171
3350	4" pipe size		68	.118		195	4.23		199.23	220
3360	5" pipe size		64	.125		198	4.49		202.49	225
3370	6" pipe size		59	.136		201	4.87		205.87	228
3380	8" pipe size		53	.151		213	5.45		218.45	242
3390	10" pipe size		50	.160		235	5.75		240.75	268
3400	12" pipe size	▼	48	.167	▼	222	6		228	253
3450	For standard pipe support, one piece, CI deduct					34%				
3460	For stanchion support, CI with steel yoke, deduct					60%				
3500	Multiple pipe support 2' long strut	1 Plum	30	.267	Ea.	4.40	9.60		14	19.30
3510	3' long strut		20	.400		6.60	14.40		21	29
3520	4' long strut	▼	17	.471		8.80	16.90		25.70	35
3550	Insulation shield 1" thick, 1/2" pipe size, TYPE 40	1 Asbe	100	.080		5.95	2.68		8.63	10.80
3560	3/4" pipe size		100	.080		5.95	2.68		8.63	10.80
3570	1" pipe size		98	.082		7.40	2.73		10.13	12.50
3580	1-1/4" pipe size		98	.082		7.40	2.73		10.13	12.50
3590	1-1/2" pipe size		96	.083		7.85	2.79		10.64	13
3600	2" pipe size		96	.083		8.15	2.79		10.94	13.40
3610	2-1/2" pipe size		94	.085		8.55	2.85		11.40	13.90
3620	3" pipe size		94	.085		9	2.85		11.85	14.40
3630	2" thick, 3-1/2" pipe size		92	.087		11.80	2.91		14.71	17.55
3640	4" pipe size		92	.087		11.80	2.91		14.71	17.55
3650	5" pipe size		90	.089		17.45	2.97		20.42	24
3660	6" pipe size		90	.089		19.90	2.97		22.87	26.50
3670	8" pipe size		88	.091		30	3.04		33.04	37.50
3680	10" pipe size		88	.091		44	3.04		47.04	53.50
3690	12" pipe size		86	.093		50.50	3.11		53.61	60.50
3700	14" pipe size		86	.093		59	3.11		62.11	70
3710	16" pipe size		84	.095		67.50	3.19		70.69	79.50
3720	18" pipe size		84	.095		74.50	3.19		77.69	86.50
3730	20" pipe size		82	.098		77.50	3.26		80.76	90.50
3732	24" pipe size		80	.100		88	3.35		91.35	102
3733	30" pipe size		78	.103		101	3.43		104.43	116
3735	36" pipe size	▼	75	.107	▼	108	3.57		111.57	125
3750	Covering protection saddle, TYPE 39									
3760	1" covering size									
3770	3/4" pipe size	1 Plum	68	.118	Ea.	10	4.23		14.23	17.40
3780	1" pipe size		68	.118		10	4.23		14.23	17.40
3790	1-1/4" pipe size		68	.118		10	4.23		14.23	17.40
3800	1-1/2" pipe size		66	.121		10.85	4.36		15.21	18.50
3810	2" pipe size		66	.121		10.85	4.36		15.21	18.50
3820	2-1/2" pipe size		64	.125		10.85	4.49		15.34	18.75
3830	3" pipe size		64	.125		15.65	4.49		20.14	24
3840	3-1/2" pipe size		62	.129		17.85	4.64		22.49	26.50
3850	4" pipe size		62	.129		16.70	4.64		21.34	25.50
3860	5" pipe size		60	.133		17.85	4.79		22.64	27
3870	6" pipe size	▼	60	.133	▼	21.50	4.79		26.29	31

MECHANICAL 15

15060	Hangers & Supports	CREW	DAILY OUTPUT	LABOR-HOURS	UNIT	MAT.	LABOR	EQUIP.	TOTAL	TOTAL INCL O&P
3900	1-1/2" covering size									
3910	3/4" pipe size	1 Plum	68	.118	Ea.	12.50	4.23		16.73	20
3920	1" pipe size		68	.118		12.50	4.23		16.73	20
3930	1-1/4" pipe size		68	.118		12.50	4.23		16.73	20
3940	1-1/2" pipe size		66	.121		12.50	4.36		16.86	20.50
3950	2" pipe size		66	.121		13.55	4.36		17.91	21.50
3960	2-1/2" pipe size		64	.125		16	4.49		20.49	24.50
3970	3" pipe size		64	.125		16	4.49		20.49	24.50
3980	3-1/2" pipe size		62	.129		16.25	4.64		20.89	25
3990	4" pipe size		62	.129		16.25	4.64		20.89	25
4000	5" pipe size		60	.133		16.25	4.79		21.04	25
4010	6" pipe size		60	.133		23.50	4.79		28.29	33
4020	8" pipe size		58	.138		29.50	4.96		34.46	39.50
4022	10" pipe size		56	.143		29.50	5.15		34.65	40
4024	12" pipe size	▼	54	.148	▼	43	5.35		48.35	55
4028	2" covering size									
4029	2-1/2" pipe size	1 Plum	62	.129	Ea.	15.35	4.64		19.99	24
4032	3" pipe size		60	.133		18.05	4.79		22.84	27
4033	4" pipe size		58	.138		18.05	4.96		23.01	27.50
4034	6" pipe size		56	.143		26	5.15		31.15	36.50
4035	8" pipe size		54	.148		30.50	5.35		35.85	41.50
4080	10" pipe size		58	.138		32.50	4.96		37.46	43
4090	12" pipe size		56	.143		48	5.15		53.15	61
4100	14" pipe size		56	.143		48	5.15		53.15	61
4110	16" pipe size		54	.148		67.50	5.35		72.85	82
4120	18" pipe size		54	.148		67.50	5.35		72.85	82
4130	20" pipe size		52	.154		75	5.55		80.55	91
4150	24" pipe size		50	.160		87.50	5.75		93.25	105
4160	30" pipe size		48	.167		96	6		102	115
4180	36" pipe size	▼	45	.178	▼	106	6.40		112.40	127
4186	2-1/2" covering size									
4187	3" pipe size	1 Plum	58	.138	Ea.	17.60	4.96		22.56	27
4188	4" pipe size		56	.143		19.10	5.15		24.25	29
4189	6" pipe size		52	.154		29	5.55		34.55	40.50
4190	8" pipe size		48	.167		35.50	6		41.50	48.50
4191	10" pipe size		44	.182		39	6.55		45.55	53
4192	12" pipe size		40	.200		52	7.20		59.20	68.50
4193	14" pipe size		36	.222		52	8		60	69.50
4194	16" pipe size		32	.250		77	9		86	98
4195	18" pipe size	▼	28	.286	▼	82	10.25		92.25	106
4200	Sockets									
4210	Rod end, malleable iron, TYPE 16									
4220	1/4" thread size	1 Plum	240	.033	Ea.	.99	1.20		2.19	2.90
4230	3/8" thread size		240	.033		1.59	1.20		2.79	3.56
4240	1/2" thread size		230	.035		1.90	1.25		3.15	3.98
4250	5/8" thread size		225	.036		3.76	1.28		5.04	6.05
4260	3/4" thread size		220	.036		6.10	1.31		7.41	8.70
4270	7/8" thread size		210	.038		12	1.37		13.37	15.25
4290	Strap, 1/2" pipe size, TYPE 26	Q-1	142	.113		2.39	3.65		6.04	8.15
4300	3/4" pipe size		140	.114		2.43	3.70		6.13	8.25
4310	1" pipe size		137	.117		2.48	3.78		6.26	8.45
4320	1-1/4" pipe size		134	.119		2.79	3.86		6.65	8.85
4330	1-1/2" pipe size		131	.122		3.50	3.95		7.45	9.80
4340	2" pipe size		129	.124		3.60	4.01		7.61	10
4350	2-1/2" pipe size		125	.128		6.30	4.14		10.44	13.15
4360	3" pipe size	▼	122	.131	▼	6.90	4.24		11.14	14

Important: See the Reference Section for critical supporting data - Reference Nos., Crews, & City Cost Indexes

15060	Hangers & Supports	CREW	DAILY OUTPUT	LABOR-HOURS	UNIT	2002 BARE COSTS				TOTAL INCL O&P	
						MAT.	LABOR	EQUIP.	TOTAL		
300	4370	3-1/2" pipe size	Q-1	119	.134	Ea.	11.55	4.35		15.90	19.25	300
	4380	4" pipe size	↓	114	.140	↓	12.90	4.54		17.44	21	
	4400	U-bolt, carbon steel										
	4410	Standard, with nuts, TYPE 42										
	4420	1/2" pipe size	1 Plum	160	.050	Ea.	1.12	1.80		2.92	3.94	
	4430	3/4" pipe size		158	.051		1.12	1.82		2.94	3.97	
	4450	1" pipe size		152	.053		1.22	1.89		3.11	4.19	
	4460	1-1/4" pipe size		148	.054		1.35	1.94		3.29	4.42	
	4470	1-1/2" pipe size		143	.056		1.40	2.01		3.41	4.57	
	4480	2" pipe size		139	.058		1.44	2.07		3.51	4.70	
	4490	2-1/2" pipe size		134	.060		2	2.15		4.15	5.45	
	4500	3" pipe size		128	.063		3.42	2.25		5.67	7.15	
	4510	3-1/2" pipe size		122	.066		3.54	2.36		5.90	7.45	
	4520	4" pipe size		117	.068		3.66	2.46		6.12	7.75	
	4530	5" pipe size		114	.070		3.87	2.52		6.39	8.05	
	4540	6" pipe size		111	.072		4.23	2.59		6.82	8.55	
	4550	8" pipe size		109	.073		5.05	2.64		7.69	9.60	
	4560	10" pipe size		107	.075		8.85	2.69		11.54	13.80	
	4570	12" pipe size	↓	104	.077	↓	11.35	2.77		14.12	16.65	
	4580	For plastic coating on 1/2" thru 6" size, add					150%					
	4700	U-hook, carbon steel, requires mounting screws or bolts										
	4710	3/4" thru 2" pipe size										
	4720	6" long	1 Plum	96	.083	Ea.	3.20	3		6.20	8.05	
	4730	8" long		96	.083		3.42	3		6.42	8.30	
	4740	10" long		96	.083		4.32	3		7.32	9.25	
	4750	12" long	↓	96	.083		4.79	3		7.79	9.75	
	4760	For copper plated, add					50%					
	8000	Pipe clamp, plastic, 1/2" CTS	1 Plum	80	.100		.19	3.60		3.79	5.60	
	8010	3/4" CTS		73	.110		.20	3.94		4.14	6.15	
	8020	1" CTS		68	.118		.46	4.23		4.69	6.90	
	8080	Economy clamp, 1/4" CTS		175	.046		.05	1.64		1.69	2.54	
	8090	3/8" CTS		168	.048		.06	1.71		1.77	2.65	
	8100	1/2" CTS		160	.050		.06	1.80		1.86	2.78	
	8110	3/4" CTS		145	.055		.12	1.98		2.10	3.12	
	8200	Half clamp, 1/2" CTS		80	.100		.06	3.60		3.66	5.45	
	8210	3/4" CTS		73	.110		.09	3.94		4.03	6.05	
	8300	Suspension clamp, 1/2" CTS		80	.100		.19	3.60		3.79	5.60	
	8310	3/4" CTS		73	.110		.20	3.94		4.14	6.15	
	8320	1" CTS		68	.118		.56	4.23		4.79	7	
	8400	Insulator, 1/2" CTS		80	.100		.33	3.60		3.93	5.75	
	8410	3/4" CTS		73	.110		.34	3.94		4.28	6.30	
	8420	1" CTS		68	.118		.36	4.23		4.59	6.80	
	8500	J hook clamp with nail 1/2" CTS		240	.033		.10	1.20		1.30	1.92	
	8501	3/4" CTS	↓	240	.033	↓	.12	1.20		1.32	1.94	

15070	Mech Sound/Vibration/Seismic Ctrl										
800	0010	VIBRATION ABSORBERS										800
	0100	Hangers, neoprene flex										
	0200	10 - 120 lb. capacity				Ea.	13.75			13.75	15.10	
	0220	75 - 550 lb. capacity					17.90			17.90	19.70	
	0240	250 - 1100 lb. capacity					36			36	40	
	0260	1000 - 4000 lb. capacity					60.50			60.50	66.50	
	0500	Spring flex, 60 lb. capacity					19.25			19.25	21	
	0520	450 lb. capacity					24			24	26.50	
	0540	900 lb. capacity					31.50			31.50	35	
	0560	1100-1300 lb. capacity	↓			↓	31.50			31.50	35	

MECHANICAL 15

15070	Mech Sound/Vibration/Seismic Ctrl	CREW	DAILY OUTPUT	LABOR-HOURS	UNIT	MAT.	LABOR	EQUIP.	TOTAL	TOTAL INCL O&P	
800							**2002 BARE COSTS**				**800**
0600	Rubber in shear										
0610	45 - 340 lb., up to 1/2" rod size	1 Stpi	22	.364	Ea.	12	13.15		25.15	33	
0620	130 - 700 lb., up to 3/4" rod size	↓	20	.400		27	14.50		41.50	51.50	
0630	50 - 1000 lb., up to 3/4" rod size	↓	18	.444		31	16.10		47.10	58.50	
1000	Mounts, neoprene, 45 - 380 lb. capacity					10.60			10.60	11.65	
1020	250 - 1100 lb. capacity					27.50			27.50	30.50	
1040	1000 - 4000 lb. capacity					62			62	68	
1100	Spring flex, 60 lb. capacity					37.50			37.50	41	
1120	165 lb. capacity					37			37	41	
1140	260 lb. capacity					37.50			37.50	41	
1160	450 lb. capacity					49			49	54	
1180	600 lb. capacity					49			49	54	
1200	750 lb. capacity					49			49	54	
1220	900 lb. capacity					49			49	54	
1240	1100 lb. capacity					54			54	59.50	
1260	1300 lb. capacity					54			54	59.50	
1280	1500 lb. capacity					79.50			79.50	87.50	
1300	1800 lb. capacity					79.50			79.50	87.50	
1320	2200 lb. capacity					85			85	93.50	
1340	2600 lb. capacity			↓		85			85	93.50	
1399	Spring type										
1400	2 Piece										
1410	50 - 1000 lb.	1 Stpi	12	.667	Ea.	102	24		126	149	
1420	1100 - 1600 lb.	"	12	.667	"	110	24		134	158	
1500	Double spring open										
1510	150 - 450 lb.	1 Stpi	24	.333	Ea.	74	12.05		86.05	99.50	
1520	500 - 1000 lb.		24	.333		74	12.05		86.05	99.50	
1530	1100 - 1600 lb.		24	.333		74	12.05		86.05	99.50	
1540	1700 - 2400 lb.		24	.333		80	12.05		92.05	106	
1550	2500 - 3400 lb.	↓	24	.333		136	12.05		148.05	168	
2000	Pads, cork rib, 18" x 18" x 1", 10-50 psi					102			102	112	
2020	18" x 36" x 1", 10-50 psi					204			204	224	
2100	Shear flexible pads, 18" x 18" x 3/8", 20-70 psi					48			48	53	
2120	18" x 36" x 3/8", 20-70 psi			↓		96.50			96.50	106	
2150	Laminated neoprene and cork										
2160	1" thick	1 Stpi	16	.500	S.F.	47	18.10		65.10	79	
3000	Note overlap in capacities due to deflections										

15075	Mech Identification										
400	0010	**PIPING SYSTEM IDENTIFICATION LABELS**									**400**
	0100	Indicate contents and flow direction									
	0106	Pipe markers									
	0110	Plastic snap around									
	0114	1/2" pipe	1 Plum	80	.100	Ea.	3	3.60		6.60	8.70
	0116	3/4" pipe		80	.100		3	3.60		6.60	8.70
	0118	1" pipe		80	.100		3	3.60		6.60	8.70
	0120	2" pipe		75	.107		3.95	3.83		7.78	10.15
	0122	3" pipe		70	.114		7.40	4.11		11.51	14.35
	0124	4" pipe		60	.133		7.40	4.79		12.19	15.40
	0126	6" pipe		60	.133		8	4.79		12.79	16.05
	0128	8" pipe		56	.143		11	5.15		16.15	19.85
	0130	10" pipe		56	.143		11	5.15		16.15	19.85
	0200	Over 10" pipe size	↓	50	.160	↓	13.50	5.75		19.25	23.50
	1110	Self adhesive									
	1114	1" pipe	1 Plum	80	.100	Ea.	2.10	3.60		5.70	7.70

15 MECHANICAL

15050 | Basic Materials & Methods

15075 | Mech Identification

		CREW	DAILY OUTPUT	LABOR-HOURS	UNIT	2002 BARE COSTS MAT.	LABOR	EQUIP.	TOTAL	TOTAL INCL O&P	
400	1116	2" pipe	1 Plum	75	.107	Ea.	2.10	3.83		5.93	8.10	**400**
	1118	3" pipe		70	.114		3.40	4.11		7.51	9.95	
	1120	4" pipe		60	.133		3.40	4.79		8.19	11	
	1122	6" pipe		60	.133		3.40	4.79		8.19	11	
	1124	8" pipe		56	.143		4.70	5.15		9.85	12.90	
	1126	10" pipe		56	.143		4.70	5.15		9.85	12.90	
	1200	Over 10" pipe size	▼	50	.160	▼	6.55	5.75		12.30	15.85	
	2000	Valve tags										
	2010	Numbered plus identifying legend										
	2100	Brass, 2" diameter	1 Plum	40	.200	Ea.	1.75	7.20		8.95	12.80	
	2200	Plastic, 1-1/2" Dia.	"	40	.200	"	1.60	7.20		8.80	12.60	

15082 | Mech Insulation

			CREW	DAILY OUTPUT	LABOR-HOURS	UNIT	2002 BARE COSTS MAT.	LABOR	EQUIP.	TOTAL	TOTAL INCL O&P	
200	0010	**INSULATION**										**200**
	0100	Rule of thumb, as a percentage of total mechanical costs				Job				10%		
	0110	Insulation req'd is based on the surface size/area to be covered										
	1000	Boiler, 1-1/2" calcium silicate, 1/2" cement finish	Q-14	50	.320	S.F.	2.81	9.65		12.46	18.40	
	1020	2" fiberglass	"	80	.200	"	1.65	6		7.65	11.35	
	2000	Breeching, 2" calcium silicate with 1/2" cement finish, no lath										
	2020	Rectangular	Q-14	42	.381	S.F.	3.17	11.45		14.62	21.50	
	2040	Round	"	38.70	.413	"	3.56	12.45		16.01	23.50	
	2300	Calcium silicate block, + 200°F to + 1200°F										
	2310	On irregular surfaces, valves and fittings										
	2340	1" thick	Q-14	30	.533	S.F.	2.22	16.05		18.27	28	
	2360	1-1/2" thick		25	.640		2.60	19.25		21.85	33.50	
	2380	2" thick		22	.727		3.57	22		25.57	39	
	2400	3" thick	▼	18	.889	▼	5.25	27		32.25	48.50	
	2410	On plane surfaces										
	2420	1" thick	Q-14	126	.127	S.F.	2.22	3.82		6.04	8.50	
	2430	1-1/2" thick		120	.133		2.60	4.01		6.61	9.20	
	2440	2" thick		100	.160		3.57	4.82		8.39	11.60	
	2450	3" thick	▼	70	.229	▼	5.25	6.90		12.15	16.65	
	2900	Domestic water heater wrap kit										
	2920	1-1/2" with vinyl jacket, 20-60 gal.	1 Plum	8	1	Ea.	16.25	36		52.25	72	
	2930	Insulated protectors, (ADA)										
	2935	For exposed piping under sinks or lavatories. ♿										
	2940	Vinyl coated foam, velcro tabs										
	2945	P Trap, 1-1/4" or 1-1/2"	1 Plum	32	.250	Ea.	16.80	9		25.80	32	
	2960	Valve and supply cover										
	2965	1/2", 3/8", and 7/16" pipe size	1 Plum	32	.250	Ea.	16.80	9		25.80	32	
	2970	Extension drain cover										
	2975	1-1/4", or 1-1/2" pipe size	1 Plum	32	.250	Ea.	17.20	9		26.20	32.50	
	2980	Tailpiece offset (wheelchair)										
	2985	1-1/4" pipe size	1 Plum	32	.250	Ea.	19.60	9		28.60	35	
	3000	Ductwork										
	3020	Blanket type, fiberglass, flexible										
	3030	Fire resistant liner, black coating one side										
	3050	1/2" thick, 2 lb. density	Q-14	380	.042	S.F.	.32	1.27		1.59	2.36	
	3060	1" thick, 1-1/2 lb. density		350	.046		.43	1.38		1.81	2.65	
	3070	1-1/2" thick, 1-1/2 lb. density		320	.050		.75	1.51		2.26	3.22	
	3080	2" thick, 1-1/2 lb. density	▼	300	.053	▼	.89	1.61		2.50	3.53	
	3140	FRK vapor barrier wrap, .75 lb. density										
	3160	1" thick	Q-14	350	.046	S.F.	.30	1.38		1.68	2.51	
	3170	1-1/2" thick		320	.050		.27	1.51		1.78	2.69	
	3180	2" thick	▼	300	.053	▼	.32	1.61		1.93	2.90	

MECHANICAL 15

131

			DAILY	LABOR-			2002 BARE COSTS				TOTAL
15082		**Mech Insulation**									
			CREW	OUTPUT	HOURS	UNIT	MAT.	LABOR	EQUIP.	TOTAL	INCL O&P
200	3190	3" thick	Q-14	260	.062	S.F.	.48	1.85		2.33	3.47	200
	3200	4" thick	↓	242	.066	↓	.64	1.99		2.63	3.86
	3210	Vinyl jacket, same as FRK									
	3280	Unfaced, 1 lb. density									
	3310	1" thick	Q-14	360	.044	S.F.	.27	1.34		1.61	2.42
	3320	1-1/2" thick		330	.048		.37	1.46		1.83	2.73
	3330	2" thick	↓	310	.052	↓	.43	1.55		1.98	2.94
	3340	Board type fiberglass liner, 1-1/2 lb. density									
	3344	1" thick	Q-14	150	.107	S.F.	.48	3.21		3.69	5.65
	3345	1-1/2" thick		130	.123		.63	3.70		4.33	6.60
	3346	2" thick	↓	120	.133	↓	.93	4.01		4.94	7.35
	3490	Board type, fiberglass liner, 3 lb. density									
	3500	Fire resistant, black pigmented, 1 side									
	3520	1" thick	Q-14	150	.107	S.F.	1.29	3.21		4.50	6.50
	3540	1-1/2" thick		130	.123		1.59	3.70		5.29	7.65
	3560	2" thick	↓	120	.133		1.91	4.01		5.92	8.45
	3600	FRK vapor barrier									
	3620	1" thick	Q-14	150	.107	S.F.	1.41	3.21		4.62	6.65
	3630	1-1/2" thick		130	.123		1.82	3.70		5.52	7.90
	3640	2" thick	↓	120	.133		2.21	4.01		6.22	8.80
	3680	No finish									
	3700	1" thick	Q-14	170	.094	S.F.	.75	2.83		3.58	5.35
	3710	1-1/2" thick		140	.114		1.13	3.44		4.57	6.70
	3720	2" thick	↓	130	.123	↓	1.44	3.70		5.14	7.50
	3730	Sheet insulation									
	3760	Polyethylene foam, closed cell, UV resistant									
	3770	Standard temperature (-90° to +212° F)									
	3771	1/4" thick	Q-14	450	.036	S.F.	1.10	1.07		2.17	2.91
	3772	3/8" thick		440	.036		1.62	1.09		2.71	3.52
	3773	1/2" thick		420	.038		2.02	1.15		3.17	4.04
	3774	3/4" thick		400	.040		2.95	1.20		4.15	5.15
	3775	1" thick	↓	380	.042	↓	3.75	1.27		5.02	6.15
	3779	Adhesive (see line 7878)									
	3780	Foam, rubber									
	3782	1" thick	1 Stpi	50	.160	S.F.	2.50	5.80		8.30	11.50
	3795	Finishes									
	3800	1/2" cement over 1" wire mesh, incl. corner bead	Q-14	100	.160	S.F.	1.37	4.82		6.19	9.15
	3820	Glass cloth, pasted on		170	.094		2.22	2.83		5.05	6.95
	3900	Weatherproof, non-metallic, 2 lb. per S.F.	↓	100	.160	↓	3.17	4.82		7.99	11.15
	4000	Pipe covering (price copper tube one size less than IPS)									
	4040	Air cell, asbestos free corrugated felt, with cover									
	4100	3 ply, 1/2" iron pipe size	Q-14	145	.110	L.F.	.75	3.32		4.07	6.10
	4130	3/4" iron pipe size		145	.110		.85	3.32		4.17	6.20
	4140	1" iron pipe size		145	.110		.90	3.32		4.22	6.25
	4150	1-1/4" iron pipe size		140	.114		1	3.44		4.44	6.55
	4160	1-1/2" iron pipe size		140	.114		1.10	3.44		4.54	6.65
	4170	2" iron pipe size		135	.119		1.20	3.57		4.77	6.95
	4180	2-1/2" iron pipe size		135	.119		1.35	3.57		4.92	7.15
	4190	3" iron pipe size		130	.123		1.50	3.70		5.20	7.55
	4200	3-1/2" iron pipe size		130	.123		1.76	3.70		5.46	7.85
	4210	4" iron pipe size		125	.128		1.95	3.85		5.80	8.25
	4220	5" iron pipe size		120	.133		2.25	4.01		6.26	8.85
	4230	6" iron pipe size		115	.139		2.40	4.19		6.59	9.30
	4240	8" iron pipe size		105	.152		3.30	4.59		7.89	10.95
	4250	10" iron pipe size		95	.168		3.90	5.05		8.95	12.35
	4260	12" iron pipe size	↓	90	.178	↓	4.50	5.35		9.85	13.45

Important: See the Reference Section for critical supporting data - Reference Nos., Crews, & City Cost Indexes

15082	Mech Insulation	CREW	DAILY OUTPUT	LABOR-HOURS	UNIT	2002 BARE COSTS				TOTAL INCL O&P
						MAT.	LABOR	EQUIP.	TOTAL	
4280	Cellular glass, closed cell foam, all service jacket, sealant,									
4281	working temp. (-450°F to +900°F), 0 water vapor xmission									
4284	1" wall,									
4286	1/2" iron pipe size	Q-14	120	.133	L.F.	1.73	4.01		5.74	8.25
4300	1-1/2" wall,									
4301	1" iron pipe size	Q-14	105	.152	L.F.	2.78	4.59		7.37	10.35
4304	2-1/2" iron pipe size		90	.178		4.68	5.35		10.03	13.65
4306	3" iron pipe size		85	.188		4.73	5.65		10.38	14.20
4308	4" iron pipe size		70	.229		6.30	6.90		13.20	17.80
4310	5" iron pipe size		65	.246		7.25	7.40		14.65	19.70
4320	2" wall,									
4322	1" iron pipe size	Q-14	100	.160	L.F.	4.13	4.82		8.95	12.20
4324	2-1/2" iron pipe size		85	.188		6.15	5.65		11.80	15.75
4326	3" iron pipe size		80	.200		6.30	6		12.30	16.45
4328	4" iron pipe size		65	.246		7.05	7.40		14.45	19.50
4330	5" iron pipe size		60	.267		8.45	8.05		16.50	22
4332	6" iron pipe size		50	.320		10	9.65		19.65	26.50
4336	8" iron pipe size		40	.400		12.95	12.05		25	33.50
4338	10" iron pipe size		35	.457		13.35	13.75		27.10	36.50
4350	2-1/2" wall,									
4360	12" iron pipe size	Q-14	32	.500	L.F.	21	15.05		36.05	47.50
4362	14" iron pipe size	"	28	.571	"	22.50	17.20		39.70	52.50
4370	3" wall,									
4378	6" iron pipe size	Q-14	48	.333	L.F.	13.35	10.05		23.40	30.50
4380	8" iron pipe size		38	.421		15.60	12.65		28.25	37
4382	10" iron pipe size		33	.485		20.50	14.60		35.10	45.50
4384	16" iron pipe size		25	.640		26.50	19.25		45.75	60
4386	18" iron pipe size		22	.727		29	22		51	67
4388	20" iron pipe size		20	.800		34.50	24		58.50	75.50
4400	3-1/2" wall,									
4412	12" iron pipe size	Q-14	27	.593	L.F.	30	17.85		47.85	61.50
4414	14" iron pipe size	"	25	.640	"	32	19.25		51.25	65.50
4430	4" wall,									
4446	16" iron pipe size	Q-14	22	.727	L.F.	32.50	22		54.50	70.50
4448	18" iron pipe size		20	.800		42	24		66	84
4450	20" iron pipe size		18	.889		43.50	27		70.50	90.50
4480	Fittings, average with fabric and mastic									
4484	1" wall,									
4486	1/2" iron pipe size	1 Asbe	40	.200	Ea.	3.07	6.70		9.77	14
4500	1-1/2" wall,									
4502	1" iron pipe size	1 Asbe	38	.211	Ea.	5.30	7.05		12.35	17
4504	2-1/2" iron pipe size		32	.250		6.60	8.35		14.95	20.50
4506	3" iron pipe size		30	.267		7.25	8.90		16.15	22
4508	4" iron pipe size		28	.286		11.15	9.55		20.70	27.50
4510	5" iron pipe size		24	.333		12.25	11.15		23.40	31
4520	2" wall,									
4522	1" iron pipe size	1 Asbe	36	.222	Ea.	6.15	7.45		13.60	18.55
4524	2-1/2" iron pipe size		30	.267		7.25	8.90		16.15	22
4526	3" iron pipe size		28	.286		8.70	9.55		18.25	24.50
4528	4" iron pipe size		24	.333		11.70	11.15		22.85	30.50
4530	5" iron pipe size		22	.364		14.75	12.15		26.90	35.50
4532	6" iron pipe size		20	.400		17.25	13.40		30.65	40
4536	8" iron pipe size		12	.667		26	22.50		48.50	64.50
4538	10" iron pipe size		8	1		32.50	33.50		66	88.50
4550	2-1/2" wall,									
4560	12" iron pipe size	1 Asbe	6	1.333	Ea.	55.50	44.50		100	133

200

200

			DAILY	LABOR-			2002 BARE COSTS				TOTAL	
15082	**Mech Insulation**	CREW	OUTPUT	HOURS	UNIT	MAT.	LABOR	EQUIP.	TOTAL	INCL O&P	
200	4562	14" iron pipe size	1 Asbe	4	2	Ea.	64.50	67		131.50	177	200
	4570	3" wall,										
	4578	6" iron pipe size	1 Asbe	16	.500	Ea.	20	16.75		36.75	48.50	
	4580	8" iron pipe size		10	.800		29.50	27		56.50	75	
	4582	10" iron pipe size		6	1.333		42	44.50		86.50	117	
	4584	16" iron pipe size		2	4		100	134		234	320	
	4586	18" iron pipe size		2	4		129	134		263	355	
	4588	20" iron pipe size	↓	2	4	↓	153	134		287	380	
	4600	3-1/2" wall,										
	4612	12" iron pipe size	1 Asbe	4	2	Ea.	72.50	67		139.50	186	
	4614	14" iron pipe size	"	2	4	"	80	134		214	300	
	4630	4" wall,										
	4646	16" iron pipe size	1 Asbe	2	4	Ea.	120	134		254	345	
	4648	18" iron pipe size		2	4		147	134		281	375	
	4650	20" iron pipe size	↓	2	4	↓	172	134		306	400	
	4900	Calcium silicate, with cover										
	5100	1" wall, 1/2" iron pipe size	Q-14	170	.094	L.F.	2.62	2.83		5.45	7.40	
	5130	3/4" iron pipe size		170	.094		2.63	2.83		5.46	7.40	
	5140	1" iron pipe size		170	.094		2.48	2.83		5.31	7.25	
	5150	1-1/4" iron pipe size		165	.097		2.51	2.92		5.43	7.40	
	5160	1-1/2" iron pipe size		165	.097		2.56	2.92		5.48	7.45	
	5170	2" iron pipe size		160	.100		3.03	3.01		6.04	8.10	
	5180	2-1/2" iron pipe size		160	.100		3.24	3.01		6.25	8.35	
	5190	3" iron pipe size		150	.107		3.43	3.21		6.64	8.85	
	5200	4" iron pipe size		140	.114		3.99	3.44		7.43	9.85	
	5210	5" iron pipe size		135	.119		4.67	3.57		8.24	10.80	
	5220	6" iron pipe size		130	.123		4.77	3.70		8.47	11.15	
	5280	1-1/2" wall, 1/2" iron pipe size		150	.107		2.86	3.21		6.07	8.25	
	5310	3/4" iron pipe size		150	.107		2.87	3.21		6.08	8.25	
	5320	1" iron pipe size		150	.107		3.13	3.21		6.34	8.55	
	5330	1-1/4" iron pipe size		145	.110		3.37	3.32		6.69	8.95	
	5340	1-1/2" iron pipe size		145	.110		3.57	3.32		6.89	9.20	
	5350	2" iron pipe size		140	.114		3.86	3.44		7.30	9.70	
	5360	2-1/2" iron pipe size		140	.114		4.24	3.44		7.68	10.10	
	5370	3" iron pipe size		135	.119		4.43	3.57		8	10.50	
	5380	4" iron pipe size		125	.128		5.15	3.85		9	11.75	
	5390	5" iron pipe size		120	.133		5.80	4.01		9.81	12.70	
	5400	6" iron pipe size		110	.145		5.95	4.38		10.33	13.50	
	5402	8" iron pipe size		95	.168		8	5.05		13.05	16.85	
	5404	10" iron pipe size		85	.188		10.60	5.65		16.25	20.50	
	5406	12" iron pipe size		80	.200		12.45	6		18.45	23.50	
	5408	14" iron pipe size		75	.213		14.10	6.40		20.50	25.50	
	5410	16" iron pipe size		70	.229		15.80	6.90		22.70	28.50	
	5412	18" iron pipe size		65	.246		18	7.40		25.40	31.50	
	5460	2" wall, 1/2" iron pipe size		135	.119		4.59	3.57		8.16	10.70	
	5490	3/4" iron pipe size		135	.119		4.60	3.57		8.17	10.70	
	5500	1" iron pipe size		135	.119		4.83	3.57		8.40	10.95	
	5510	1-1/4" iron pipe size		130	.123		5.40	3.70		9.10	11.80	
	5520	1-1/2" iron pipe size		130	.123		5.45	3.70		9.15	11.90	
	5530	2" iron pipe size		125	.128		5.75	3.85		9.60	12.40	
	5540	2-1/2" iron pipe size		125	.128		6.90	3.85		10.75	13.70	
	5550	3" iron pipe size		120	.133		7.60	4.01		11.61	14.70	
	5560	4" iron pipe size		115	.139		7.75	4.19		11.94	15.15	
	5570	5" iron pipe size		110	.145		8.80	4.38		13.18	16.65	
	5580	6" iron pipe size		105	.152		9.90	4.59		14.49	18.20	
	5581	8" iron pipe size	↓	95	.168	↓	11.80	5.05		16.85	21	

15 MECHANICAL

Important: See the Reference Section for critical supporting data - Reference Nos., Crews, & City Cost Indexes

15082	Mech Insulation	CREW	DAILY OUTPUT	LABOR-HOURS	UNIT	2002 BARE COSTS				TOTAL INCL O&P
						MAT.	LABOR	EQUIP.	TOTAL	
200 5582	10" iron pipe size	Q-14	85	.188	L.F.	14.70	5.65		20.35	25 **200**
5583	12" iron pipe size		80	.200		16.45	6		22.45	27.50
5584	14" iron pipe size		75	.213		18.10	6.40		24.50	30
5585	16" iron pipe size		70	.229		20.50	6.90		27.40	33.50
5586	18" iron pipe size		65	.246		22.50	7.40		29.90	36.50
5587	3" wall, 1-1/4" iron pipe size		100	.160		7.75	4.82		12.57	16.20
5588	1-1/2" iron pipe size		100	.160		8.25	4.82		13.07	16.70
5589	2" iron pipe size		95	.168		8.75	5.05		13.80	17.65
5590	2-1/2" iron pipe size		95	.168		10.40	5.05		15.45	19.50
5591	3" iron pipe size		90	.178		11.10	5.35		16.45	20.50
5592	4" iron pipe size		85	.188		13.15	5.65		18.80	23.50
5593	6" iron pipe size		80	.200		16.05	6		22.05	27
5594	8" iron pipe size		75	.213		19	6.40		25.40	31
5595	10" iron pipe size		65	.246		23	7.40		30.40	37
5596	12" iron pipe size		60	.267		25.50	8.05		33.55	41
5597	14" iron pipe size		55	.291		28	8.75		36.75	45
5598	16" iron pipe size		50	.320		31.50	9.65		41.15	50
5599	18" iron pipe size	▼	45	.356	▼	34.50	10.70		45.20	55
5600	Calcium silicate, no cover									
5720	1" wall, 1/2" iron pipe size	Q-14	180	.089	L.F.	2.47	2.68		5.15	6.95
5740	3/4" iron pipe size		180	.089		2.47	2.68		5.15	6.95
5750	1" iron pipe size		180	.089		2.31	2.68		4.99	6.80
5760	1-1/4" iron pipe size		175	.091		2.32	2.75		5.07	6.90
5770	1-1/2" iron pipe size		175	.091		2.37	2.75		5.12	7
5780	2" iron pipe size		170	.094		2.81	2.83		5.64	7.60
5790	2-1/2" iron pipe size		170	.094		3	2.83		5.83	7.80
5800	3" iron pipe size		160	.100		3.16	3.01		6.17	8.25
5810	4" iron pipe size		150	.107		3.67	3.21		6.88	9.15
5820	5" iron pipe size		145	.110		4.03	3.32		7.35	9.70
5830	6" iron pipe size		140	.114		4.35	3.44		7.79	10.25
5900	1-1/2" wall, 1/2" iron pipe size		160	.100		2.67	3.01		5.68	7.70
5920	3/4" iron pipe size		160	.100		2.67	3.01		5.68	7.70
5930	1" iron pipe size		160	.100		2.91	3.01		5.92	8
5940	1-1/4" iron pipe size		155	.103		3.14	3.11		6.25	8.40
5950	1-1/2" iron pipe size		155	.103		3.29	3.11		6.40	8.55
5960	2" iron pipe size		150	.107		3.59	3.21		6.80	9.05
5970	2-1/2" iron pipe size		150	.107		3.96	3.21		7.17	9.45
5980	3" iron pipe size		145	.110		4.12	3.32		7.44	9.80
5990	4" iron pipe size		135	.119		4.79	3.57		8.36	10.90
6000	5" iron pipe size		130	.123		5.35	3.70		9.05	11.80
6010	6" iron pipe size		120	.133		5.50	4.01		9.51	12.40
6020	7" iron pipe size		115	.139		6.60	4.19		10.79	13.90
6030	8" iron pipe size		105	.152		7.45	4.59		12.04	15.50
6040	9" iron pipe size		100	.160		8.95	4.82		13.77	17.50
6050	10" iron pipe size		95	.168		9.90	5.05		14.95	18.95
6060	12" iron pipe size		90	.178		11.70	5.35		17.05	21.50
6070	14" iron pipe size		85	.188		13.25	5.65		18.90	23.50
6080	16" iron pipe size		80	.200		14.90	6		20.90	26
6090	18" iron pipe size		75	.213		17	6.40		23.40	29
6120	2" wall, 1/2" iron pipe size		145	.110		4.35	3.32		7.67	10.05
6140	3/4" iron pipe size		145	.110		4.35	3.32		7.67	10.05
6150	1" iron pipe size		145	.110		4.57	3.32		7.89	10.30
6160	1-1/4" iron pipe size		140	.114		5.10	3.44		8.54	11.05
6170	1-1/2" iron pipe size		140	.114		5.15	3.44		8.59	11.15
6180	2" iron pipe size		135	.119		5.40	3.57		8.97	11.60
6190	2-1/2" iron pipe size	▼	135	.119		6.55	3.57		10.12	12.85

MECHANICAL 15

	15082	Mech Insulation	CREW	DAILY OUTPUT	LABOR- HOURS	UNIT	MAT.	LABOR	EQUIP.	TOTAL	TOTAL INCL O&P	
200	6200	3" iron pipe size	Q-14	130	.123	L.F.	6.70	3.70		10.40	13.30	200
	6210	4" iron pipe size		125	.128		7.30	3.85		11.15	14.15	
	6220	5" iron pipe size		120	.133		8.35	4.01		12.36	15.55	
	6230	6" iron pipe size		115	.139		9.40	4.19		13.59	16.95	
	6240	7" iron pipe size		110	.145		10.20	4.38		14.58	18.20	
	6250	8" iron pipe size		105	.152		11.20	4.59		15.79	19.65	
	6260	9" iron pipe size		100	.160		12.65	4.82		17.47	21.50	
	6270	10" iron pipe size		95	.168		14	5.05		19.05	23.50	
	6280	12" iron pipe size		90	.178		15.65	5.35		21	25.50	
	6290	14" iron pipe size		85	.188		17.25	5.65		22.90	28	
	6300	16" iron pipe size		80	.200		19.65	6		25.65	31	
	6310	18" iron pipe size		75	.213		21.50	6.40		27.90	33.50	
	6320	20" iron pipe size		65	.246		25	7.40		32.40	39.50	
	6330	22" iron pipe size		60	.267		27	8.05		35.05	42.50	
	6340	24" iron pipe size		55	.291		28	8.75		36.75	44.50	
	6360	3" wall, 1/2" iron pipe size		115	.139		7.90	4.19		12.09	15.35	
	6380	3/4" iron pipe size		115	.139		7.90	4.19		12.09	15.35	
	6390	1" iron pipe size		115	.139		7.15	4.19		11.34	14.50	
	6400	1-1/4" iron pipe size		110	.145		7.35	4.38		11.73	15.05	
	6410	1-1/2" iron pipe size		110	.145		7.85	4.38		12.23	15.60	
	6420	2" iron pipe size		105	.152		8.35	4.59		12.94	16.45	
	6430	2-1/2" iron pipe size		105	.152		10	4.59		14.59	18.30	
	6440	3" iron pipe size		100	.160		10.15	4.82		14.97	18.80	
	6450	4" iron pipe size		95	.168		12.60	5.05		17.65	22	
	6460	5" iron pipe size		90	.178		13.70	5.35		19.05	23.50	
	6470	6" iron pipe size		90	.178		15.45	5.35		20.80	25.50	
	6480	7" iron pipe size		85	.188		17.65	5.65		23.30	28.50	
	6490	8" iron pipe size		85	.188		18.35	5.65		24	29	
	6500	9" iron pipe size		80	.200		20.50	6		26.50	32.50	
	6510	10" iron pipe size		75	.213		22	6.40		28.40	34.50	
	6520	12" iron pipe size		70	.229		24.50	6.90		31.40	38	
	6530	14" iron pipe size		65	.246		27.50	7.40		34.90	42	
	6540	16" iron pipe size		60	.267		30.50	8.05		38.55	46.50	
	6550	18" iron pipe size		55	.291		33.50	8.75		42.25	50.50	
	6560	20" iron pipe size		50	.320		37.50	9.65		47.15	56.50	
	6570	22" iron pipe size		45	.356		40.50	10.70		51.20	61.50	
	6580	24" iron pipe size	↓	40	.400	↓	43.50	12.05		55.55	67	
	6600	Fiberglass, with all service jacket										
	6640	1/2" wall, 1/2" iron pipe size	Q-14	250	.064	L.F.	.98	1.93		2.91	4.14	
	6660	3/4" iron pipe size		240	.067		1.09	2.01		3.10	4.39	
	6670	1" iron pipe size		230	.070		1.17	2.09		3.26	4.61	
	6680	1-1/4" iron pipe size		220	.073		1.28	2.19		3.47	4.89	
	6690	1-1/2" iron pipe size		220	.073		1.36	2.19		3.55	4.98	
	6700	2" iron pipe size		210	.076		1.47	2.29		3.76	5.25	
	6710	2-1/2" iron pipe size		200	.080		1.68	2.41		4.09	5.65	
	6720	3" iron pipe size		190	.084		1.88	2.53		4.41	6.10	
	6730	3-1/2" iron pipe size		180	.089		2.01	2.68		4.69	6.45	
	6740	4" iron pipe size		160	.100		2.40	3.01		5.41	7.40	
	6750	5" iron pipe size		150	.107		2.76	3.21		5.97	8.15	
	6760	6" iron pipe size		120	.133		2.91	4.01		6.92	9.55	
	6840	1" wall, 1/2" iron pipe size		240	.067		1.17	2.01		3.18	4.48	
	6860	3/4" iron pipe size		230	.070		1.34	2.09		3.43	4.79	
	6870	1" iron pipe size		220	.073		1.36	2.19		3.55	4.98	
	6880	1-1/4" iron pipe size		210	.076		1.55	2.29		3.84	5.35	
	6890	1-1/2" iron pipe size		210	.076		1.66	2.29		3.95	5.45	
	6900	2" iron pipe size	↓	200	.080	↓	1.81	2.41		4.22	5.80	

		CREW	DAILY OUTPUT	LABOR-HOURS	UNIT	2002 BARE COSTS				TOTAL INCL O&P	
15082	**Mech Insulation**					MAT.	LABOR	EQUIP.	TOTAL		
200 6910	2-1/2" iron pipe size	Q-14	190	.084	L.F.	2	2.53		4.53	6.20	**200**
6920	3" iron pipe size		180	.089		2.26	2.68		4.94	6.75	
6930	3-1/2" iron pipe size		170	.094		2.48	2.83		5.31	7.25	
6940	4" iron pipe size		150	.107		2.91	3.21		6.12	8.30	
6950	5" iron pipe size		140	.114		3.35	3.44		6.79	9.15	
6960	6" iron pipe size		120	.133		3.56	4.01		7.57	10.25	
6970	7" iron pipe size		110	.145		4.38	4.38		8.76	11.75	
6980	8" iron pipe size		100	.160		5.05	4.82		9.87	13.20	
6990	9" iron pipe size		90	.178		5.55	5.35		10.90	14.60	
7000	10" iron pipe size		90	.178		5.95	5.35		11.30	15.05	
7010	12" iron pipe size		80	.200		6.75	6		12.75	17	
7020	14" iron pipe size		80	.200		7.80	6		13.80	18.15	
7030	16" iron pipe size		70	.229		9.30	6.90		16.20	21	
7040	18" iron pipe size		70	.229		11.30	6.90		18.20	23.50	
7050	20" iron pipe size		60	.267		12.65	8.05		20.70	26.50	
7060	24" iron pipe size		60	.267		15.45	8.05		23.50	29.50	
7080	1-1/2" wall, 1/2" iron pipe size		230	.070		2.38	2.09		4.47	5.95	
7100	3/4" iron pipe size		220	.073		2.48	2.19		4.67	6.20	
7110	1" iron pipe size		210	.076		2.61	2.29		4.90	6.50	
7120	1-1/4" iron pipe size		200	.080		2.80	2.41		5.21	6.90	
7130	1-1/2" iron pipe size		200	.080		2.95	2.41		5.36	7.05	
7140	2" iron pipe size		190	.084		3.22	2.53		5.75	7.55	
7150	2-1/2" iron pipe size		180	.089		3.55	2.68		6.23	8.15	
7160	3" iron pipe size		170	.094		3.65	2.83		6.48	8.50	
7170	3-1/2" iron pipe size		160	.100		3.99	3.01		7	9.15	
7180	4" iron pipe size		140	.114		4.18	3.44		7.62	10.05	
7190	5" iron pipe size		130	.123		4.65	3.70		8.35	11	
7200	6" iron pipe size		110	.145		4.78	4.38		9.16	12.20	
7210	7" iron pipe size		100	.160		5.55	4.82		10.37	13.75	
7220	8" iron pipe size		90	.178		6	5.35		11.35	15.10	
7230	9" iron pipe size		85	.188		6.85	5.65		12.50	16.50	
7240	10" iron pipe size		80	.200		7.55	6		13.55	17.85	
7250	12" iron pipe size		75	.213		8.25	6.40		14.65	19.25	
7260	14" iron pipe size		70	.229		9.50	6.90		16.40	21.50	
7270	16" iron pipe size		65	.246		10.70	7.40		18.10	23.50	
7280	18" iron pipe size		60	.267		13.80	8.05		21.85	28	
7290	20" iron pipe size		55	.291		14.65	8.75		23.40	30	
7300	24" iron pipe size		50	.320		17.15	9.65		26.80	34	
7320	2" wall, 1/2" iron pipe size		220	.073		3.72	2.19		5.91	7.55	
7340	3/4" iron pipe size		210	.076		3.85	2.29		6.14	7.90	
7350	1" iron pipe size		200	.080		4.11	2.41		6.52	8.35	
7360	1-1/4" iron pipe size		190	.084		4.35	2.53		6.88	8.80	
7370	1-1/2" iron pipe size		190	.084		4.60	2.53		7.13	9.05	
7380	2" iron pipe size		180	.089		4.77	2.68		7.45	9.50	
7390	2-1/2" iron pipe size		170	.094		5.15	2.83		7.98	10.15	
7400	3" iron pipe size		160	.100		5.55	3.01		8.56	10.90	
7410	3-1/2" iron pipe size		150	.107		5.95	3.21		9.16	11.65	
7420	4" iron pipe size		130	.123		6.35	3.70		10.05	12.85	
7430	5" iron pipe size		120	.133		7.15	4.01		11.16	14.25	
7440	6" iron pipe size		100	.160		7.35	4.82		12.17	15.75	
7450	7" iron pipe size		90	.178		8.45	5.35		13.80	17.75	
7460	8" iron pipe size		80	.200		9.10	6		15.10	19.55	
7470	9" iron pipe size		75	.213		10.05	6.40		16.45	21.50	
7480	10" iron pipe size		70	.229		10.85	6.90		17.75	23	
7490	12" iron pipe size		65	.246		12	7.40		19.40	25	
7500	14" iron pipe size		60	.267		13.55	8.05		21.60	27.50	

MECHANICAL 15

15082	Mech Insulation	CREW	DAILY OUTPUT	LABOR-HOURS	UNIT	2002 BARE COSTS				TOTAL INCL O&P
						MAT.	LABOR	EQUIP.	TOTAL	
200	7510	16" iron pipe size	Q-14	55	.291	L.F.	15.40	8.75		24.15	31	200
7520	18" iron pipe size		50	.320		18.40	9.65		28.05	35.50
7530	20" iron pipe size		45	.356		20.50	10.70		31.20	39.50
7540	24" iron pipe size		40	.400		23	12.05		35.05	44
7600	For fiberglass with standard canvas jacket, deduct					5%				
7660	For fittings, add 3 L.F. for each fitting									
7680	plus 4 L.F. for each flange of the fitting									
7700	For equipment or congested areas, add						20%			
7720	Finishes, for .010" aluminum jacket, add	Q-14	200	.080	S.F.	.32	2.41		2.73	4.17
7740	For .016" aluminum jacket, add		200	.080		.41	2.41		2.82	4.27
7760	For .010" stainless steel, add		160	.100		1.22	3.01		4.23	6.10
7780	For single layer of felt, add					10%	10%			
7782	For roofing paper, 45 lb to 55 lb, add					25%	10%			
7784	Polyethylene tubing flexible closed cell foam, UV resistant									
7828	Standard temperature (-90° to +212° F)									
7830	3/8" wall, 1/8" iron pipe size	1 Asbe	130	.062	L.F.	.25	2.06		2.31	3.55
7831	1/4" iron pipe size		130	.062		.28	2.06		2.34	3.58
7832	3/8" iron pipe size		130	.062		.29	2.06		2.35	3.59
7833	1/2" iron pipe size		126	.063		.33	2.12		2.45	3.73
7834	3/4" iron pipe size		122	.066		.36	2.19		2.55	3.88
7835	1" iron pipe size		120	.067		.46	2.23		2.69	4.05
7836	1-1/4" iron pipe size		118	.068		.59	2.27		2.86	4.25
7837	1-1/2" iron pipe size		118	.068		.66	2.27		2.93	4.33
7838	2" iron pipe size		116	.069		.88	2.31		3.19	4.63
7839	2-1/2" iron pipe size		114	.070		1.20	2.35		3.55	5.05
7840	3" iron pipe size		112	.071		1.34	2.39		3.73	5.25
7842	1/2" wall, 1/8" iron pipe size		120	.067		.39	2.23		2.62	3.97
7843	1/4" iron pipe size		120	.067		.41	2.23		2.64	3.99
7844	3/8" iron pipe size		120	.067		.42	2.23		2.65	4
7845	1/2" iron pipe size		118	.068		.51	2.27		2.78	4.16
7846	3/4" iron pipe size		116	.069		.53	2.31		2.84	4.24
7847	1" iron pipe size		114	.070		.68	2.35		3.03	4.48
7848	1-1/4" iron pipe size		112	.071		.83	2.39		3.22	4.70
7849	1-1/2" iron pipe size		110	.073		1.01	2.43		3.44	4.97
7850	2" iron pipe size		108	.074		1.25	2.48		3.73	5.30
7851	2-1/2" iron pipe size		106	.075		1.62	2.52		4.14	5.80
7852	3" iron pipe size		104	.077		1.76	2.57		4.33	6
7853	3-1/2" iron pipe size		102	.078		2.55	2.62		5.17	6.95
7854	4" iron pipe size		100	.080		2.79	2.68		5.47	7.30
7855	3/4" wall, 1/8" iron pipe size		110	.073		.91	2.43		3.34	4.86
7856	1/4" iron pipe size		110	.073		.91	2.43		3.34	4.86
7857	3/8" iron pipe size		108	.074		1.05	2.48		3.53	5.10
7858	1/2" iron pipe size		106	.075		1.26	2.52		3.78	5.40
7859	3/4" iron pipe size		104	.077		1.50	2.57		4.07	5.75
7860	1" iron pipe size		102	.078		1.75	2.62		4.37	6.10
7861	1-1/4" iron pipe size		100	.080		2.37	2.68		5.05	6.85
7862	1-1/2" iron pipe size		100	.080		2.70	2.68		5.38	7.20
7863	2" iron pipe size		98	.082		3.22	2.73		5.95	7.85
7864	2-1/2" iron pipe size		96	.083		3.99	2.79		6.78	8.80
7865	3" iron pipe size		94	.085		5.10	2.85		7.95	10.15
7866	3-1/2" iron pipe size		92	.087		6.30	2.91		9.21	11.55
7867	4" iron pipe size		90	.089		6.70	2.97		9.67	12.10
7868	1" wall, 1/4" iron pipe size		100	.080		1.96	2.68		4.64	6.40
7869	3/8" iron pipe size		98	.082		1.96	2.73		4.69	6.50
7870	1/2" iron pipe size		96	.083		2.45	2.79		5.24	7.10
7871	3/4" iron pipe size		94	.085		2.48	2.85		5.33	7.25

Important: See the Reference Section for critical supporting data - Reference Nos., Crews, & City Cost Indexes

		DAILY	LABOR-			2002 BARE COSTS				TOTAL	
15082	**Mech Insulation**	CREW	OUTPUT	HOURS	UNIT	MAT.	LABOR	EQUIP.	TOTAL	INCL O&P	
200 7872	1" iron pipe size	1 Asbe	92	.087	L.F.	2.87	2.91		5.78	7.80	**200**
7873	1-1/4" iron pipe size		90	.089		4.02	2.97		6.99	9.15	
7874	1-1/2" iron pipe size		90	.089		4.31	2.97		7.28	9.45	
7875	2" iron pipe size		88	.091		5.65	3.04		8.69	11.10	
7876	2-1/2" iron pipe size		86	.093		7.85	3.11		10.96	13.60	
7877	3" iron pipe size	▼	84	.095	▼	8.20	3.19		11.39	14.10	
7878	Contact cement, quart can				Ea.	18.55			18.55	20.50	
7879	Rubber tubing, flexible closed cell foam										
7880	3/8" wall, 1/4" iron pipe size	1 Asbe	120	.067	L.F.	.28	2.23		2.51	3.85	
7900	3/8" iron pipe size		120	.067		.31	2.23		2.54	3.88	
7910	1/2" iron pipe size		115	.070		.35	2.33		2.68	4.08	
7920	3/4" iron pipe size		115	.070		.40	2.33		2.73	4.13	
7930	1" iron pipe size		110	.073		.49	2.43		2.92	4.40	
7940	1-1/4" iron pipe size		110	.073		.60	2.43		3.03	4.52	
7950	1-1/2" iron pipe size		110	.073		.72	2.43		3.15	4.65	
8100	1/2" wall, 1/4" iron pipe size		90	.089		.60	2.97		3.57	5.40	
8120	3/8" iron pipe size		90	.089		.92	2.97		3.89	5.75	
8130	1/2" iron pipe size		89	.090		.48	3.01		3.49	5.30	
8140	3/4" iron pipe size		89	.090		.53	3.01		3.54	5.35	
8150	1" iron pipe size		88	.091		.55	3.04		3.59	5.45	
8160	1-1/4" iron pipe size		87	.092		.71	3.08		3.79	5.65	
8170	1-1/2" iron pipe size		87	.092		.86	3.08		3.94	5.85	
8180	2" iron pipe size		86	.093		1.59	3.11		4.70	6.70	
8190	2-1/2" iron pipe size		86	.093		1.58	3.11		4.69	6.70	
8200	3" iron pipe size		85	.094		1.66	3.15		4.81	6.85	
8210	3-1/2" iron pipe size		85	.094		2.22	3.15		5.37	7.45	
8220	4" iron pipe size		80	.100		2.62	3.35		5.97	8.20	
8230	5" iron pipe size		80	.100		3.61	3.35		6.96	9.25	
8300	3/4" wall, 1/4" iron pipe size		90	.089		.65	2.97		3.62	5.45	
8320	3/8" iron pipe size		90	.089		.72	2.97		3.69	5.50	
8330	1/2" iron pipe size		89	.090		.78	3.01		3.79	5.65	
8340	3/4" iron pipe size		89	.090		.86	3.01		3.87	5.70	
8350	1" iron pipe size		88	.091		1.14	3.04		4.18	6.10	
8360	1-1/4" iron pipe size		87	.092		1.30	3.08		4.38	6.30	
8370	1-1/2" iron pipe size		87	.092		1.76	3.08		4.84	6.80	
8380	2" iron pipe size		86	.093		2.34	3.11		5.45	7.50	
8390	2-1/2" iron pipe size		86	.093		3.12	3.11		6.23	8.35	
8400	3" iron pipe size		85	.094		3.19	3.15		6.34	8.50	
8410	3-1/2" iron pipe size		85	.094		3.90	3.15		7.05	9.30	
8420	4" iron pipe size		80	.100		4.51	3.35		7.86	10.25	
8430	5" iron pipe size		80	.100		5.45	3.35		8.80	11.30	
8440	6" iron pipe size		80	.100		6.55	3.35		9.90	12.55	
8444	1" wall, 1/2" iron pipe size		86	.093		1.63	3.11		4.74	6.75	
8445	3/4" iron pipe size		84	.095		1.82	3.19		5.01	7.05	
8446	1" iron pipe size		84	.095		2.21	3.19		5.40	7.50	
8447	1-1/4" iron pipe size		82	.098		2.57	3.26		5.83	8.05	
8448	1-1/2" iron pipe size		82	.098		3.38	3.26		6.64	8.90	
8449	2" iron pipe size		80	.100		4.51	3.35		7.86	10.25	
8450	2-1/2" iron pipe size	▼	80	.100	▼	5.90	3.35		9.25	11.75	
8456	Rubber insulation tape, 1/8" x 2" x 30'				Ea.	10.90			10.90	12	
8460	Polyolefin tubing, flexible closed cell foam, UV stabilized, work										
8462	temp.-165° F to +210° F, 0 water vapor transmission.										
8464	3/8" wall, 1/8" iron pipe size	1 Asbe	140	.057	L.F.	.24	1.91		2.15	3.29	
8466	1/4" iron pipe size		140	.057		.26	1.91		2.17	3.32	
8468	3/8" iron pipe size		140	.057		.29	1.91		2.20	3.35	
8470	1/2" iron pipe size	▼	136	.059	▼	.32	1.97		2.29	3.47	

	15082	Mech Insulation	CREW	DAILY OUTPUT	LABOR-HOURS	UNIT	2002 BARE COSTS				TOTAL INCL O&P	
							MAT.	LABOR	EQUIP.	TOTAL		
200	8472	3/4" iron pipe size	1 Asbe	132	.061	L.F.	.37	2.03		2.40	3.63	200
	8474	1" iron pipe size		130	.062		.40	2.06		2.46	3.71	
	8476	1-1/4" iron pipe size		128	.063		.53	2.09		2.62	3.90	
	8478	1-1/2" iron pipe size		128	.063		.64	2.09		2.73	4.02	
	8480	2" iron pipe size		126	.063		.72	2.12		2.84	4.16	
	8482	2-1/2" iron pipe size		123	.065		1.04	2.18		3.22	4.59	
	8484	3" iron pipe size		121	.066		1.36	2.21		3.57	5	
	8486	4" iron pipe size		118	.068		2.25	2.27		4.52	6.10	
	8500	1/2" wall, 1/8" iron pipe size		130	.062		.37	2.06		2.43	3.68	
	8502	1/4" iron pipe size		130	.062		.39	2.06		2.45	3.70	
	8504	3/8" iron pipe size		130	.062		.42	2.06		2.48	3.73	
	8506	1/2" iron pipe size		128	.063		.48	2.09		2.57	3.85	
	8508	3/4" iron pipe size		126	.063		.53	2.12		2.65	3.95	
	8510	1" iron pipe size		123	.065		.58	2.18		2.76	4.09	
	8512	1-1/4" iron pipe size		121	.066		.72	2.21		2.93	4.30	
	8514	1-1/2" iron pipe size		119	.067		.86	2.25		3.11	4.52	
	8516	2" iron pipe size		117	.068		1.01	2.29		3.30	4.74	
	8518	2-1/2" iron pipe size		114	.070		1.36	2.35		3.71	5.25	
	8520	3" iron pipe size		112	.071		1.71	2.39		4.10	5.65	
	8522	4" iron pipe size		110	.073		2.36	2.43		4.79	6.45	
	8534	3/4" wall, 1/8" iron pipe size		120	.067		.55	2.23		2.78	4.15	
	8536	1/4" iron pipe size		120	.067		.59	2.23		2.82	4.19	
	8538	3/8" iron pipe size		117	.068		.67	2.29		2.96	4.37	
	8540	1/2" iron pipe size		114	.070		.81	2.35		3.16	4.62	
	8542	3/4" iron pipe size		112	.071		.97	2.39		3.36	4.86	
	8544	1" iron pipe size		110	.073		1.12	2.43		3.55	5.10	
	8546	1-1/4" iron pipe size		108	.074		1.51	2.48		3.99	5.60	
	8548	1-1/2" iron pipe size		108	.074		1.77	2.48		4.25	5.90	
	8550	2" iron pipe size		106	.075		2.07	2.52		4.59	6.30	
	8552	2-1/2" iron pipe size		104	.077		2.61	2.57		5.18	6.95	
	8554	3" iron pipe size		102	.078		3.14	2.62		5.76	7.60	
	8556	4" iron pipe size		100	.080		4.14	2.68		6.82	8.80	
	8570	1" wall, 1/8" iron pipe size		110	.073		1.01	2.43		3.44	4.97	
	8572	1/4" iron pipe size		108	.074		1.03	2.48		3.51	5.05	
	8574	3/8" iron pipe size		106	.075		1.05	2.52		3.57	5.15	
	8576	1/2" iron pipe size		104	.077		1.31	2.57		3.88	5.50	
	8578	3/4" iron pipe size		102	.078		1.46	2.62		4.08	5.75	
	8580	1" iron pipe size		100	.080		1.68	2.68		4.36	6.10	
	8582	1-1/4" iron pipe size		97	.082		1.92	2.76		4.68	6.50	
	8584	1-1/2" iron pipe size		97	.082		2.19	2.76		4.95	6.80	
	8586	2" iron pipe size		95	.084		2.88	2.82		5.70	7.65	
	8588	2-1/2" iron pipe size		93	.086		3.74	2.88		6.62	8.70	
	8590	3" iron pipe size	▼	91	.088	▼	4.40	2.94		7.34	9.50	
	8600	Seam Roller				Ea.	6.60			6.60	7.25	
	8602	Fuse Seal applicator (Butane)				"	176			176	193	
	8604	Fuse Seal 1/2" sticks				Lb.	12.90			12.90	14.20	
	8606	Leaktite adhesive				Qt.	21			21	23.50	
	8608	Leaktite adhesive				Gal.	59.50			59.50	65	
	8610	NOTE: Preslit/preglued vs unslit, same price										

15 MECHANICAL

Important: See the Reference Section for critical supporting data - Reference Nos., Crews, & City Cost Indexes

15107	Metal Pipe & Fittings	CREW	DAILY OUTPUT	LABOR-HOURS	UNIT	2002 BARE COSTS				TOTAL INCL O&P	
						MAT.	LABOR	EQUIP.	TOTAL		
220	0010	**PIPE, BRASS** Plain end,										**220**
	0900	Field threaded, coupling & clevis hanger 10' O.C.									
	0920	Regular weight									
	0980	1/8" diameter	1 Plum	62	.129	L.F.	2.58	4.64		7.22	9.85
	1000	1/4" diameter		57	.140		3.21	5.05		8.26	11.15
	1100	3/8" diameter		52	.154		3.82	5.55		9.37	12.55
	1120	1/2" diameter		48	.167		4.55	6		10.55	14.05
	1140	3/4" diameter		46	.174		6.05	6.25		12.30	16.10
	1160	1" diameter	▼	43	.186		8.60	6.70		15.30	19.55
	1180	1-1/4" diameter	Q-1	72	.222		12.85	7.20		20.05	25
	1200	1-1/2" diameter		65	.246		15.10	7.95		23.05	28.50
	1220	2" diameter		53	.302		20.50	9.75		30.25	37.50
	1240	2-1/2" diameter		41	.390		31	12.60		43.60	53
	1260	3" diameter	▼	31	.516		44	16.70		60.70	73
	1280	3-1/2" diameter	Q-2	39	.615		72.50	20.50		93	111
	1300	4" diameter	"	37	.649	▼	73	22		95	113
	1930	To delete coupling & hanger, subtract									
	1940	1/8" diam. to 1/2" diam.					14%	46%			
	1950	3/4" diam. to 1-1/2" diam.					8%	47%			
	1960	2" diam. to 4" diam.					10%	37%			
260	0010	**PIPE, BRASS, FITTINGS** Rough bronze, threaded										**260**
	1000	Standard wt., 90° Elbow									
	1040	1/8"	1 Plum	13	.615	Ea.	7.35	22		29.35	41.50
	1060	1/4"		13	.615		7.35	22		29.35	41.50
	1080	3/8"		13	.615		7.35	22		29.35	41.50
	1100	1/2"		12	.667		7.35	24		31.35	44
	1120	3/4"		11	.727		9.80	26		35.80	50.50
	1140	1"	▼	10	.800		15.95	29		44.95	61
	1160	1-1/4"	Q-1	17	.941		26	30.50		56.50	74.50
	1180	1-1/2"		16	1		32	32.50		64.50	84
	1200	2"		14	1.143		51.50	37		88.50	113
	1220	2-1/2"		11	1.455		125	47		172	208
	1240	3"	▼	8	2		190	64.50		254.50	305
	1260	4"	Q-2	11	2.182		385	73		458	535
	1280	5"		8	3		1,075	101		1,176	1,325
	1300	6"	▼	7	3.429		1,575	115		1,690	1,900
	1500	45° Elbow, 1/8"	1 Plum	13	.615		9	22		31	43.50
	1540	1/4"		13	.615		9	22		31	43.50
	1560	3/8"		13	.615		9	22		31	43.50
	1580	1/2"		12	.667		9	24		33	46
	1600	3/4"		11	.727		12.80	26		38.80	53.50
	1620	1"	▼	10	.800		22	29		51	67.50
	1640	1-1/4"	Q-1	17	.941		43.50	30.50		74	94
	1660	1-1/2"		16	1		56.50	32.50		89	111
	1680	2"		14	1.143		101	37		138	168
	1700	2-1/2"		11	1.455		128	47		175	212
	1720	3"	▼	8	2		198	64.50		262.50	315
	1740	4"	Q-2	11	2.182		475	73		548	635
	1760	5"		8	3		880	101		981	1,125
	1780	6"	▼	7	3.429		1,225	115		1,340	1,500
	2000	Tee, 1/8"	1 Plum	9	.889		8.60	32		40.60	57.50
	2040	1/4"		9	.889		8.60	32		40.60	57.50
	2060	3/8"		9	.889		8.60	32		40.60	57.50
	2080	1/2"		8	1		8.60	36		44.60	63.50
	2100	3/4"		7	1.143		12.25	41		53.25	75.50
	2120	1"	▼	6	1.333	▼	22	48		70	97

MECHANICAL 15

			CREW	DAILY OUTPUT	LABOR-HOURS	UNIT	2002 BARE COSTS				TOTAL INCL O&P
							MAT.	LABOR	EQUIP.	TOTAL	
260	2140	1-1/4"	Q-1	10	1.600	Ea.	38	52		90	120
	2160	1-1/2"		9	1.778		43	57.50		100.50	134
	2180	2"		8	2		71	64.50		135.50	176
	2200	2-1/2"		7	2.286		170	74		244	298
	2220	3"		5	3.200		259	104		363	440
	2240	4"	Q-2	7	3.429		645	115		760	880
	2260	5"		5	4.800		1,350	161		1,511	1,725
	2280	6"		4	6		2,100	201		2,301	2,600
	2500	Coupling, 1/8"	1 Plum	26	.308		6.15	11.05		17.20	23.50
	2540	1/4"		22	.364		6.15	13.05		19.20	26.50
	2560	3/8"		18	.444		6.15	16		22.15	31
	2580	1/2"		15	.533		6.15	19.15		25.30	36
	2600	3/4"		14	.571		8.60	20.50		29.10	40.50
	2620	1"		13	.615		14.70	22		36.70	49.50
	2640	1-1/4"	Q-1	22	.727		24.50	23.50		48	62.50
	2660	1-1/2"		20	.800		32	26		58	74
	2680	2"		18	.889		53	29		82	102
	2700	2-1/2"		14	1.143		90	37		127	155
	2720	3"		10	1.600		125	52		177	215
	2740	4"	Q-2	12	2		257	67		324	385
	2760	5"		10	2.400		480	80.50		560.50	650
	2780	6"		9	2.667		685	89.50		774.50	885
	3000	Union, 125 lb									
	3020	1/8"	1 Plum	12	.667	Ea.	21	24		45	59
	3040	1/4"		12	.667		21	24		45	59
	3060	3/8"		12	.667		21	24		45	59
	3080	1/2"		11	.727		21	26		47	62.50
	3100	3/4"		10	.800		28.50	29		57.50	75
	3120	1"		9	.889		38	32		70	89.50
	3140	1-1/4"	Q-1	16	1		55	32.50		87.50	110
	3160	1-1/2"		15	1.067		65.50	34.50		100	124
	3180	2"		13	1.231		101	40		141	171
	3200	2-1/2"		10	1.600		262	52		314	365
	3220	3"		7	2.286		405	74		479	555
	3240	4"	Q-2	10	2.400		1,025	80.50		1,105.50	1,250
	3320	For 250 lb. (navy pattern), add						100%			
420	0010	**PIPE, COPPER** Solder joints									
	0100	Solder									
	0120	Solder, lead free, roll				Lb.	8.70			8.70	9.55
	1000	Type K tubing, couplings & clevis hangers 10' O.C.									
	1100	1/4" diameter	1 Plum	84	.095	L.F.	1.43	3.42		4.85	6.75
	1120	3/8" diameter		82	.098		1.57	3.51		5.08	7.05
	1140	1/2" diameter		78	.103		1.70	3.69		5.39	7.40
	1160	5/8" diameter		77	.104		1.97	3.74		5.71	7.80
	1180	3/4" diameter		74	.108		2.64	3.89		6.53	8.75
	1200	1" diameter		66	.121		3.26	4.36		7.62	10.15
	1220	1-1/4" diameter		56	.143		3.97	5.15		9.12	12.10
	1240	1-1/2" diameter		50	.160		4.92	5.75		10.67	14.05
	1260	2" diameter		40	.200		7.50	7.20		14.70	19.10
	1280	2-1/2" diameter	Q-1	60	.267		10.95	8.65		19.60	25
	1300	3" diameter		54	.296		15.05	9.60		24.65	31
	1320	3-1/2" diameter		42	.381		19.10	12.30		31.40	39.50
	1330	4" diameter		38	.421		23.50	13.60		37.10	46.50
	1340	5" diameter		32	.500		62	16.20		78.20	93
	1360	6" diameter	Q-2	38	.632		90	21		111	131
	1380	8" diameter	"	34	.706		159	23.50		182.50	210

15107 | Metal Pipe & Fittings

R15100 -050

Important: See the Reference Section for critical supporting data - Reference Nos., Crews, & City Cost Indexes

15107 | Metal Pipe & Fittings

		DAILY OUTPUT	LABOR-HOURS	UNIT	2002 BARE COSTS				TOTAL INCL O&P
					MAT.	LABOR	EQUIP.	TOTAL	
420									**420**
1390	For other than full hard temper, add [R15100-050]			L.F.	13%				
1440	For silver solder, add					15%			
1800	For medical clean, (oxygen class), add				12%				
1950	To delete cplgs. & hngrs., 1/4"-1" pipe, subtract				27%	60%			
1960	1-1/4" -3" pipe, subtract				14%	52%			
1970	3-1/2"-5" pipe, subtract				10%	60%			
1980	6"-8" pipe, subtract				19%	53%			
2000	Type L tubing, couplings & hangers 10' O.C.								
2100	1/4" diameter	1 Plum	88	.091	L.F.	1.13	3.27		4.40	6.15
2120	3/8" diameter		84	.095		1.35	3.42		4.77	6.65
2140	1/2" diameter		81	.099		1.54	3.55		5.09	7.05
2160	5/8" diameter		79	.101		1.91	3.64		5.55	7.60
2180	3/4" diameter		76	.105		2.07	3.78		5.85	7.95
2200	1" diameter		68	.118		2.70	4.23		6.93	9.35
2220	1-1/4" diameter		58	.138		3.47	4.96		8.43	11.30
2240	1-1/2" diameter		52	.154		4.25	5.55		9.80	13.05
2260	2" diameter	▼	42	.190		6.35	6.85		13.20	17.30
2280	2-1/2" diameter	Q-1	62	.258		9.40	8.35		17.75	23
2300	3" diameter		56	.286		12.65	9.25		21.90	28
2320	3-1/2" diameter		43	.372		16.35	12.05		28.40	36
2340	4" diameter		39	.410		20	13.25		33.25	42
2360	5" diameter	▼	34	.471		56.50	15.20		71.70	85
2380	6" diameter	Q-2	40	.600		71	20		91	109
2400	8" diameter	"	36	.667		122	22.50		144.50	168
2410	For other than full hard temper, add				▼	21%			
2590	For silver solder, add					15%			
2900	For medical clean, (oxygen class), add				12%				
2940	To delete cplgs. & hngrs., 1/4"-1" pipe, subtract				37%	63%			
2960	1-1/4"-3" pipe, subtract				12%	53%			
2970	3-1/2"-5" pipe, subtract				12%	63%			
2980	6"-8" pipe, subtract				24%	55%			
3000	Type M tubing, couplings & hangers 10' O.C.								
3100	1/4" diameter	1 Plum	90	.089	L.F.	1.19	3.20		4.39	6.15
3120	3/8" diameter		87	.092		1.20	3.31		4.51	6.30
3140	1/2" diameter		84	.095		1.26	3.42		4.68	6.55
3160	5/8" diameter		81	.099		1.58	3.55		5.13	7.10
3180	3/4" diameter		78	.103		1.68	3.69		5.37	7.40
3200	1" diameter		70	.114		2.23	4.11		6.34	8.65
3220	1-1/4" diameter		60	.133		2.92	4.79		7.71	10.45
3240	1-1/2" diameter		54	.148		3.83	5.35		9.18	12.25
3260	2" diameter	▼	44	.182		5.90	6.55		12.45	16.35
3280	2-1/2" diameter	Q-1	64	.250		8.30	8.10		16.40	21.50
3300	3" diameter		58	.276		10.65	8.90		19.55	25
3320	3-1/2" diameter		45	.356		13.85	11.50		25.35	32.50
3340	4" diameter		40	.400		18	12.95		30.95	39.50
3360	5" diameter	▼	36	.444		54.50	14.40		68.90	81.50
3370	6" diameter	Q-2	42	.571		69.50	19.15		88.65	106
3380	8" diameter	"	38	.632		115	21		136	158
3440	For silver solder, add					15%			
3960	To delete cplgs. & hngrs., 1/4"-1" pipe, subtract				35%	65%			
3970	1-1/4"-3" pipe, subtract				19%	56%			
3980	3-1/2"-5" pipe, subtract				13%	65%			
3990	6"-8" pipe, subtract				28%	58%			
4000	Type DWV tubing, couplings & hangers 10' O.C.								
4100	1-1/4" diameter	1 Plum	60	.133	L.F.	2.88	4.79		7.67	10.40
4120	1-1/2" diameter	▼	54	.148	▼	3.47	5.35		8.82	11.85

MECHANICAL 15

15107	Metal Pipe & Fittings		CREW	DAILY OUTPUT	LABOR-HOURS	UNIT	2002 BARE COSTS				TOTAL INCL O&P	
							MAT.	LABOR	EQUIP.	TOTAL		
420	4140	2" diameter	R15100	1 Plum	44	.182	L.F.	4.47	6.55		11.02	14.75	420
	4160	3" diameter	-050	Q-1	58	.276		7.55	8.90		16.45	22	
	4180	4" diameter			40	.400		13.25	12.95		26.20	34	
	4200	5" diameter			36	.444		39	14.40		53.40	64.50	
	4220	6" diameter		Q-2	42	.571		55	19.15		74.15	89.50	
	4240	8" diameter		"	38	.632		124	21		145	168	
	4730	To delete cplgs. & hngrs., 1-1/4"-2" pipe, subtract						16%	53%				
	4740	2-1/2"-4" pipe, subtract						13%	60%				
	4750	5"-8" pipe, subtract						23%	58%				
	5200	ACR tubing, type L, hard temper, cleaned and											
	5220	capped, no couplings or hangers											
	5240	3/8" OD					L.F.	.58			.58	.64	
	5250	1/2" OD						.84			.84	.92	
	5260	5/8" OD						1.07			1.07	1.18	
	5270	3/4" OD						1.42			1.42	1.56	
	5280	7/8" OD						1.68			1.68	1.85	
	5290	1-1/8" OD						2.39			2.39	2.63	
	5300	1-3/8" OD						3.30			3.30	3.63	
	5310	1-5/8" OD						4.16			4.16	4.58	
	5320	2-1/8" OD						6.70			6.70	7.35	
	5330	2-5/8" OD						9.50			9.50	10.45	
	5340	3-1/8" OD						13.15			13.15	14.45	
	5350	3-5/8" OD						16.65			16.65	18.30	
	5360	4-1/8" OD						21.50			21.50	23.50	
	5380	ACR tubing, type L, annealed, cleaned and capped											
	5381	No couplings or hangers											
	5384	3/8"		1 Stpi	160	.050	L.F.	.58	1.81		2.39	3.37	
	5385	1/2"			160	.050		.84	1.81		2.65	3.65	
	5386	5/8"			160	.050		1.07	1.81		2.88	3.91	
	5387	3/4"			130	.062		1.42	2.23		3.65	4.92	
	5388	7/8"			130	.062		1.68	2.23		3.91	5.20	
	5389	1-1/8"			115	.070		2.39	2.52		4.91	6.45	
	5390	1-3/8"			100	.080		3.30	2.90		6.20	8	
	5391	1-5/8"			90	.089		4.16	3.22		7.38	9.45	
	5392	2-1/8"			80	.100		6.70	3.62		10.32	12.80	
	5393	2-5/8"		Q-5	125	.128		9.50	4.17		13.67	16.75	
	5394	3-1/8"			105	.152		13.15	4.96		18.11	22	
	5395	4-1/8"			95	.168		21.50	5.50		27	32	
	5800	Refrigeration tubing, dryseal, 50' coils											
	5840	1/8" OD					Coil	14.20			14.20	15.60	
	5850	3/16" OD						15.55			15.55	17.10	
	5860	1/4" OD						19.15			19.15	21	
	5870	5/16" OD						24.50			24.50	27	
	5880	3/8" OD						28.50			28.50	31.50	
	5890	1/2" OD						36.50			36.50	40	
	5900	5/8" OD						49			49	54	
	5910	3/4" OD						57.50			57.50	63	
	5920	7/8" OD						85			85	93.50	
	5930	1-1/8" OD						122			122	134	
	5940	1-3/8" OD						169			169	186	
	5950	1-5/8" OD						218			218	240	
	9400	Sub assemblies used in assembly systems											
	9410	Chilled water unit, coil connections per unit under 10 ton		Q-5	.80	20	System	500	650		1,150	1,525	
	9420	Chilled water unit, coil connections per unit 10 ton and up			1	16		935	520		1,455	1,800	
	9430	Chilled water dist. piping per ton, less than 61 ton systems			26	.615		9.60	20		29.60	40.50	
	9440	Chilled water dist. piping per ton, 61 through 120 ton systems		Q-6	31	.774		21	26		47	63	

Important: See the Reference Section for critical supporting data - Reference Nos., Crews, & City Cost Indexes

15107	Metal Pipe & Fittings		CREW	DAILY OUTPUT	LABOR-HOURS	UNIT	2002 BARE COSTS				TOTAL INCL O&P	
							MAT.	LABOR	EQUIP.	TOTAL		
420	9450	Chilled water dist. piping/ton, 135 ton systems and up R15100 -050	Q-8	25.40	1.260	System	28	43.50	3.15	74.65	99.50	**420**
	9510	Refrigerant piping/ton of cooling for remote condensers	Q-5	2	8		127	261		388	535	
	9520	Refrigerant piping per ton up to 10 ton w/remote condensing unit		2.40	6.667		71	217		288	410	
	9530	Refrigerant piping per ton, 20 ton w/remote condensing unit		2	8		90	261		351	495	
	9540	Refrigerant piping per ton, 40 ton w/remote condensing unit		1.90	8.421		115	274		389	540	
	9550	Refrigerant piping per ton, 75-80 ton w/remote condensing unit	Q-6	2.40	10		161	340		501	685	
	9560	Refrigerant piping per ton, 100 ton w/remote condensing unit	"	2.20	10.909		211	370		581	785	
460	0010	**PIPE, COPPER, FITTINGS** Wrought unless otherwise noted										**460**
	0040	Solder joints, copper x copper										
	0070	90° elbow, 1/4"	1 Plum	22	.364	Ea.	1.09	13.05		14.14	21	
	0090	3/8"		22	.364		1.09	13.05		14.14	21	
	0100	1/2"		20	.400		.37	14.40		14.77	22	
	0110	5/8"		19	.421		2.24	15.15		17.39	25.50	
	0120	3/4"		19	.421		.81	15.15		15.96	24	
	0130	1"		16	.500		2	18		20	29	
	0140	1-1/4"		15	.533		3.03	19.15		22.18	32.50	
	0150	1-1/2"		13	.615		4.73	22		26.73	38.50	
	0160	2"		11	.727		8.60	26		34.60	49	
	0170	2-1/2"	Q-1	13	1.231		16.40	40		56.40	78	
	0180	3"		11	1.455		23	47		70	96.50	
	0190	3-1/2"		10	1.600		76.50	52		128.50	162	
	0200	4"		9	1.778		59	57.50		116.50	151	
	0210	5"		6	2.667		217	86.50		303.50	370	
	0220	6"	Q-2	9	2.667		300	89.50		389.50	470	
	0230	8"	"	8	3		1,225	101		1,326	1,500	
	0250	45° elbow, 1/4"	1 Plum	22	.364		2.28	13.05		15.33	22	
	0270	3/8"		22	.364		1.85	13.05		14.90	21.50	
	0280	1/2"		20	.400		.71	14.40		15.11	22.50	
	0290	5/8"		19	.421		3.55	15.15		18.70	27	
	0300	3/4"		19	.421		1.24	15.15		16.39	24.50	
	0310	1"		16	.500		3.12	18		21.12	30.50	
	0320	1-1/4"		15	.533		4.22	19.15		23.37	33.50	
	0330	1-1/2"		13	.615		5.05	22		27.05	39	
	0340	2"		11	.727		8.50	26		34.50	49	
	0350	2-1/2"	Q-1	13	1.231		18.10	40		58.10	80	
	0360	3"		13	1.231		27	40		67	89.50	
	0370	3-1/2"		10	1.600		44	52		96	127	
	0380	4"		9	1.778		57	57.50		114.50	150	
	0390	5"		6	2.667		221	86.50		307.50	375	
	0400	6"	Q-2	9	2.667		330	89.50		419.50	500	
	0410	8"	"	8	3		1,175	101		1,276	1,450	
	0450	Tee, 1/4"	1 Plum	14	.571		2.41	20.50		22.91	33.50	
	0470	3/8"		14	.571		1.85	20.50		22.35	33	
	0480	1/2"		13	.615		.61	22		22.61	34	
	0490	5/8"		12	.667		4	24		28	40.50	
	0500	3/4"		12	.667		1.50	24		25.50	37.50	
	0510	1"		10	.800		4.63	29		33.63	48.50	
	0520	1-1/4"		9	.889		7	32		39	55.50	
	0530	1-1/2"		8	1		9.75	36		45.75	65	
	0540	2"		7	1.143		15.20	41		56.20	78.50	
	0550	2-1/2"	Q-1	8	2		30.50	64.50		95	131	
	0560	3"		7	2.286		47	74		121	164	
	0570	3-1/2"		6	2.667		135	86.50		221.50	279	
	0580	4"		5	3.200		113	104		217	280	
	0590	5"		4	4		350	129		479	580	

MECHANICAL 15

		CREW	DAILY OUTPUT	LABOR-HOURS	UNIT	2002 BARE COSTS				TOTAL INCL O&P	
	**15107	Metal Pipe & Fittings**					MAT.	LABOR	EQUIP.	TOTAL		
460	0600	6"	Q-2	6	4	Ea.	480	134		614	730	**460**
	0610	8"	"	5	4.800		1,950	161		2,111	2,400	
	0612	Tee, reducing on the outlet, 1/4"	1 Plum	15	.533		3.88	19.15		23.03	33.50	
	0613	3/8"		15	.533		3.66	19.15		22.81	33	
	0614	1/2"		14	.571		3.39	20.50		23.89	34.50	
	0615	5/8"		13	.615		3.19	22		25.19	37	
	0616	3/4"		12	.667		1.45	24		25.45	37.50	
	0617	1"		11	.727		4.63	26		30.63	44.50	
	0618	1-1/4"		10	.800		6.85	29		35.85	51	
	0619	1-1/2"		9	.889		7.25	32		39.25	56	
	0620	2"	▼	8	1		11.45	36		47.45	66.50	
	0621	2-1/2"	Q-1	9	1.778		35.50	57.50		93	126	
	0622	3"		8	2		39.50	64.50		104	141	
	0623	4"		6	2.667		76.50	86.50		163	214	
	0624	5"	▼	5	3.200		360	104		464	550	
	0625	6"	Q-2	7	3.429		480	115		595	705	
	0626	8"	"	6	4		1,950	134		2,084	2,350	
	0630	Tee, reducing on the run, 1/4"	1 Plum	15	.533		4.85	19.15		24	34.50	
	0631	3/8"		15	.533		5.50	19.15		24.65	35	
	0632	1/2"		14	.571		4.39	20.50		24.89	36	
	0633	5/8"		13	.615		6.50	22		28.50	40.50	
	0634	3/4"		12	.667		1.77	24		25.77	38	
	0635	1"		11	.727		5.50	26		31.50	45.50	
	0636	1-1/4"		10	.800		8.95	29		37.95	53.50	
	0637	1-1/2"	▼	9	.889		14.55	32		46.55	64	
	0638	2"		8	1		19.75	36		55.75	76	
	0639	2-1/2"	Q-1	9	1.778		49	57.50		106.50	141	
	0640	3"		8	2		66	64.50		130.50	170	
	0641	4"		6	2.667		151	86.50		237.50	296	
	0642	5"	▼	5	3.200		350	104		454	540	
	0643	6"	Q-2	7	3.429		480	115		595	705	
	0644	8"	"	6	4		1,950	134		2,084	2,350	
	0650	Coupling, 1/4"	1 Plum	24	.333		.27	12		12.27	18.35	
	0670	3/8"		24	.333		.36	12		12.36	18.45	
	0680	1/2"		22	.364		.27	13.05		13.32	20	
	0690	5/8"		21	.381		.82	13.70		14.52	21.50	
	0700	3/4"		21	.381		.54	13.70		14.24	21	
	0710	1"		18	.444		1.12	16		17.12	25	
	0715	1-1/4"		17	.471		1.97	16.90		18.87	27.50	
	0716	1-1/2"		15	.533		2.59	19.15		21.74	32	
	0718	2"	▼	13	.615		4.31	22		26.31	38	
	0721	2-1/2"	Q-1	15	1.067		9.20	34.50		43.70	62	
	0722	3"		13	1.231		14.05	40		54.05	75.50	
	0724	3-1/2"		8	2		26.50	64.50		91	127	
	0726	4"		7	2.286		27.50	74		101.50	143	
	0728	5"	▼	6	2.667		64.50	86.50		151	201	
	0731	6"	Q-2	8	3		101	101		202	263	
	0732	8"	"	7	3.429		345	115		460	555	
	0741	Coupling, reducing, concentric										
	0743	1/2"	1 Plum	23	.348	Ea.	.65	12.50		13.15	19.55	
	0745	3/4"		21.50	.372		1.14	13.40		14.54	21.50	
	0747	1"		19.50	.410		2.12	14.75		16.87	24.50	
	0749	1-1/2"		16	.500		2.64	18		20.64	30	
	0751	2"	▼	14	.571		6.95	20.50		27.45	38.50	
	0753	3"	Q-1	14	1.143		21	37		58	79	
	0755	4"	"	8	2	▼	44.50	64.50		109	147	

Important: See the Reference Section for critical supporting data - Reference Nos., Crews, & City Cost Indexes

15107	Metal Pipe & Fittings	CREW	DAILY OUTPUT	LABOR-HOURS	UNIT	2002 BARE COSTS				TOTAL INCL O&P
						MAT.	LABOR	EQUIP.	TOTAL	
0757	5"	Q-2	7.50	3.200	Ea.	139	107		246	315
0759	6"		7	3.429		239	115		354	435
0761	8"	▼	6.50	3.692	▼	395	124		519	620
0771	Cap, sweat									
0773	1/2"	1 Plum	40	.200	Ea.	.27	7.20		7.47	11.15
0775	3/4"		38	.211		.46	7.55		8.01	11.90
0777	1"		32	.250		1.11	9		10.11	14.75
0779	1-1/2"		26	.308		2.29	11.05		13.34	19.20
0781	2"	▼	22	.364	▼	3.95	13.05		17	24
0791	Flange, sweat									
0793	3"	Q-1	22	.727	Ea.	40.50	23.50		64	80
0795	4"		18	.889		57.50	29		86.50	107
0797	5"	▼	12	1.333		111	43		154	187
0799	6"	Q-2	18	1.333		117	44.50		161.50	196
0801	8"	"	16	1.500		345	50.50		395.50	455
0850	Unions, 1/4"	1 Plum	21	.381		6.85	13.70		20.55	28
0870	3/8"		21	.381		6.55	13.70		20.25	27.50
0880	1/2"		19	.421		3.77	15.15		18.92	27
0890	5/8"		18	.444		15.65	16		31.65	41.50
0900	3/4"		18	.444		4.72	16		20.72	29
0910	1"		15	.533		8.35	19.15		27.50	38
0920	1-1/4"		14	.571		14.50	20.50		35	47
0930	1-1/2"		12	.667		19.05	24		43.05	57
0940	2"	▼	10	.800		32.50	29		61.50	79.50
0950	2-1/2"	Q-1	12	1.333		67.50	43		110.50	140
0960	3"	"	10	1.600		174	52		226	270
0980	Adapter, copper x male IPS, 1/4"	1 Plum	20	.400		3.78	14.40		18.18	25.50
0990	3/8"		20	.400		1.88	14.40		16.28	23.50
1000	1/2"		18	.444		.77	16		16.77	25
1010	3/4"		17	.471		1.27	16.90		18.17	27
1020	1"		15	.533		3.31	19.15		22.46	32.50
1030	1-1/4"		13	.615		4.44	22		26.44	38.50
1040	1-1/2"		12	.667		5.60	24		29.60	42
1050	2"	▼	11	.727		10.05	26		36.05	50.50
1060	2-1/2"	Q-1	10.50	1.524		30	49.50		79.50	108
1070	3"		10	1.600		41	52		93	124
1080	3-1/2"		9	1.778		68	57.50		125.50	162
1090	4"		8	2		68	64.50		132.50	173
1200	5"	▼	6	2.667		325	86.50		411.50	490
1210	6"	Q-2	8.50	2.824	▼	360	94.50		454.50	545
1214	Adapter, copper x female IPS									
1216	1/2"	1 Plum	18	.444	Ea.	1.20	16		17.20	25.50
1218	3/4"		17	.471		1.66	16.90		18.56	27.50
1220	1"		15	.533		3.83	19.15		22.98	33
1222	1-1/2"		12	.667		8.75	24		32.75	45.50
1224	2"		11	.727		11.90	26		37.90	52.50
1250	Cross, 1/2"		10	.800		4.98	29		33.98	49
1260	3/4"		9.50	.842		9.65	30.50		40.15	56
1270	1"		8	1		16.40	36		52.40	72
1280	1-1/4"		7.50	1.067		22.50	38.50		61	82.50
1290	1-1/2"		6.50	1.231		33.50	44.50		78	104
1300	2"	▼	5.50	1.455		63.50	52.50		116	149
1310	2-1/2"	Q-1	6.50	2.462		146	79.50		225.50	281
1320	3"		5.50	2.909		192	94		286	355
1340	4"		4.70	3.404		269	110		379	460
1350	5"	▼	3.20	5	▼	380	162		542	665

MECHANICAL 15

15107	Metal Pipe & Fittings	CREW	DAILY OUTPUT	LABOR-HOURS	UNIT	2002 BARE COSTS				TOTAL INCL O&P
						MAT.	LABOR	EQUIP.	TOTAL	
460 1360	6"	Q-2	4.70	5.106	Ea.	535	171		706	850	460
1500	Tee fitting, mechanically formed, (Type 1).									
1520	1/2" run size, 3/8" to 1/2" branch size	1 Plum	80	.100	Ea.		3.60		3.60	5.40
1530	3/4" run size, 3/8" to 3/4" branch size		60	.133			4.79		4.79	7.25
1540	1" run size, 3/8" to 1" branch size		54	.148			5.35		5.35	8.05
1550	1-1/4" run size, 3/8" to 1-1/4" branch size		48	.167			6		6	9.05
1560	1-1/2" run size, 3/8" to 1-1/2" branch size		40	.200			7.20		7.20	10.85
1570	2" run size, 3/8" to 2" branch size		35	.229			8.20		8.20	12.40
1580	2-1/2" run size, 1/2" to 2" branch size		32	.250			9		9	13.55
1590	3" run size, 1" to 2" branch size		26	.308			11.05		11.05	16.70
1600	4" run size, 1" to 2" branch size	↓	24	.333	↓		12		12	18.05
1640	Tee fitting, mechanically formed, (Type 2)									
1650	2-1/2" run size, 2-1/2" branch size	1 Plum	12.50	.640	Ea.		23		23	34.50
1660	3" run size, 2-1/2" to 3" branch size		12	.667			24		24	36
1670	3-1/2" run size, 2-1/2" to 3-1/2" branch size		11	.727			26		26	39.50
1680	4" run size, 2-1/2" to 4" branch size		10.50	.762			27.50		27.50	41.50
1698	5" run size, 2" to 4" branch size		9.50	.842			30.50		30.50	45.50
1700	6" run size, 2" to 4" branch size		8.50	.941			34		34	51
1710	8" run size, 2" to 4" branch size	↓	7	1.143	↓		41		41	62
1800	ACR fittings									
1802	Tee, straight									
1808	5/8"	1 Stpi	12	.667	Ea.	4	24		28	41
1810	3/4"		12	.667		1.50	24		25.50	38
1812	7/8"		10	.800		10.85	29		39.85	55.50
1814	1-1/8"		10	.800		4.63	29		33.63	48.50
1816	1-3/8"		9	.889		7	32		39	56
1818	1-5/8"		8	1		9.75	36		45.75	65.50
1820	2-1/8"	↓	7	1.143		15.20	41.50		56.70	79
1822	2-5/8"	Q-5	8	2		30.50	65		95.50	132
1824	3-1/8"		7	2.286		47	74.50		121.50	164
1826	4-1/8"	↓	5	3.200	↓	113	104		217	281
1830	90° elbow									
1836	5/8"	1 Stpi	19	.421	Ea.	1.21	15.25		16.46	24.50
1838	3/4"		19	.421		2.21	15.25		17.46	25.50
1840	7/8"		16	.500		2.20	18.10		20.30	30
1842	1-1/8"		16	.500		2.94	18.10		21.04	30.50
1844	1-3/8"		15	.533		3.03	19.30		22.33	32.50
1846	1-5/8"		13	.615		4.73	22.50		27.23	38.50
1848	2-1/8"	↓	11	.727		8.60	26.50		35.10	49
1850	2-5/8"	Q-5	13	1.231		16.40	40		56.40	78.50
1852	3-1/8"		11	1.455		23	47.50		70.50	97
1854	4-1/8"	↓	9	1.778	↓	59	58		117	152
1860	Coupling									
1866	5/8"	1 Stpi	21	.381	Ea.	.27	13.80		14.07	21.50
1868	3/4"		21	.381		.82	13.80		14.62	22
1870	7/8"		18	.444		.54	16.10		16.64	25
1872	1-1/8"		18	.444		1.12	16.10		17.22	25.50
1874	1-3/8"		17	.471		1.97	17.05		19.02	27.50
1876	1-5/8"		15	.533		2.59	19.30		21.89	32
1878	2-1/8"	↓	13	.615		4.31	22.50		26.81	38
1880	2-5/8"	Q-5	15	1.067		9.20	35		44.20	62.50
1882	3-1/8"		13	1.231		13.80	40		53.80	75.50
1884	4-1/8"	↓	7	2.286	↓	27.50	74.50		102	143
2000	DWV, solder joints, copper x copper									
2030	90° Elbow, 1-1/4"	1 Plum	13	.615	Ea.	3.53	22		25.53	37.50
2050	1-1/2"	↓	12	.667	↓	4.73	24		28.73	41

Important: See the Reference Section for critical supporting data - Reference Nos., Crews, & City Cost Indexes

		CREW	DAILY OUTPUT	LABOR-HOURS	UNIT	2002 BARE COSTS				TOTAL INCL O&P	
15107	**Metal Pipe & Fittings**					MAT.	LABOR	EQUIP.	TOTAL		
460 2070	2"	1 Plum	10	.800	Ea.	6.85	29		35.85	51	**460**
2090	3"	Q-1	10	1.600		17	52		69	96.50	
2100	4"	"	9	1.778		79	57.50		136.50	174	
2150	45° Elbow, 1-1/4"	1 Plum	13	.615		4.98	22		26.98	39	
2170	1-1/2"		12	.667		3.76	24		27.76	40	
2180	2"		10	.800		7.60	29		36.60	52	
2190	3"	Q-1	10	1.600		18.40	52		70.40	98	
2200	4"	"	9	1.778		55	57.50		112.50	147	
2250	Tee, Sanitary, 1-1/4"	1 Plum	9	.889		6.95	32		38.95	55.50	
2270	1-1/2"		8	1		8.65	36		44.65	63.50	
2290	2"		7	1.143		10.10	41		51.10	73	
2310	3"	Q-1	7	2.286		36	74		110	152	
2330	4"	"	6	2.667		91.50	86.50		178	231	
2400	Coupling, 1-1/4"	1 Plum	14	.571		1.64	20.50		22.14	33	
2420	1-1/2"		13	.615		2.05	22		24.05	36	
2440	2"		11	.727		2.83	26		28.83	42.50	
2460	3"	Q-1	11	1.455		5.50	47		52.50	77	
2480	4"	"	10	1.600		17.50	52		69.50	97.50	
2600	Traps, see division 15155										
3500	Compression joint fittings										
3510	As used for plumbing and oil burner work										
3520	Fitting price includes nuts and sleeves										
3540	Sleeve, 1/8"				Ea.	.09			.09	.10	
3550	3/16"					.09			.09	.10	
3560	1/4"					.05			.05	.06	
3570	5/16"					.10			.10	.11	
3580	3/8"					.21			.21	.23	
3600	1/2"					.28			.28	.31	
3620	Nut, 1/8"					.15			.15	.17	
3630	3/16"					.17			.17	.19	
3640	1/4"					.16			.16	.18	
3650	5/16"					.22			.22	.24	
3660	3/8"					.24			.24	.26	
3670	1/2"					.44			.44	.48	
3710	Union, 1/8"	1 Plum	26	.308		.80	11.05		11.85	17.60	
3720	3/16"		24	.333		.88	12		12.88	19	
3730	1/4"		24	.333		.92	12		12.92	19.05	
3740	5/16"		23	.348		1.17	12.50		13.67	20	
3750	3/8"		22	.364		1.19	13.05		14.24	21	
3760	1/2"		22	.364		1.82	13.05		14.87	21.50	
3780	5/8"		21	.381		2.02	13.70		15.72	22.50	
3820	Union tee, 1/8"		17	.471		3.47	16.90		20.37	29.50	
3830	3/16"		16	.500		2.01	18		20.01	29	
3840	1/4"		15	.533		1.82	19.15		20.97	31	
3850	5/16"		15	.533		2.56	19.15		21.71	32	
3860	3/8"		15	.533		2.61	19.15		21.76	32	
3870	1/2"		15	.533		4.39	19.15		23.54	34	
3910	Union elbow, 1/4"		24	.333		1.12	12		13.12	19.30	
3920	5/16"		23	.348		1.97	12.50		14.47	21	
3930	3/8"		22	.364		1.91	13.05		14.96	22	
3940	1/2"		22	.364		3.11	13.05		16.16	23	
3980	Female connector, 1/8"		26	.308		.74	11.05		11.79	17.50	
4000	3/16" x 1/8"		24	.333		.90	12		12.90	19.05	
4010	1/4" x 1/8"		24	.333		.80	12		12.80	18.95	
4020	1/4"		24	.333		1.07	12		13.07	19.25	
4030	3/8" x 1/4"		22	.364		1.07	13.05		14.12	21	

MECHANICAL 15

	15107	Metal Pipe & Fittings	CREW	DAILY OUTPUT	LABOR-HOURS	UNIT	2002 BARE COSTS				TOTAL INCL O&P	
							MAT.	LABOR	EQUIP.	TOTAL		
460	4040	1/2" x 3/8"	1 Plum	22	.364	Ea.	1.70	13.05		14.75	21.50	460
	4050	5/8" x 1/2"		21	.381		2.15	13.70		15.85	23	
	4090	Male connector, 1/8"		26	.308		.57	11.05		11.62	17.35	
	4100	3/16" x 1/8"		24	.333		.59	12		12.59	18.70	
	4110	1/4" x 1/8"		24	.333		.61	12		12.61	18.70	
	4120	1/4"		24	.333		.61	12		12.61	18.70	
	4130	5/16" x 1/8"		23	.348		.71	12.50		13.21	19.65	
	4140	5/16" x 1/4"		23	.348		.82	12.50		13.32	19.75	
	4150	3/8" x 1/8"		22	.364		.85	13.05		13.90	20.50	
	4160	3/8" x 1/4"		22	.364		.98	13.05		14.03	21	
	4170	3/8"		22	.364		.98	13.05		14.03	21	
	4180	3/8" x 1/2"		22	.364		1.34	13.05		14.39	21	
	4190	1/2" x 3/8"		22	.364		1.30	13.05		14.35	21	
	4200	1/2"		22	.364		1.47	13.05		14.52	21.50	
	4210	5/8" x 1/2"		21	.381		1.69	13.70		15.39	22.50	
	4240	Male elbow, 1/8"		26	.308		1.17	11.05		12.22	18	
	4250	3/16" x 1/8"		24	.333		1.29	12		13.29	19.45	
	4260	1/4" x 1/8"		24	.333		.98	12		12.98	19.15	
	4270	1/4"		24	.333		1.11	12		13.11	19.25	
	4280	3/8" x 1/4"		22	.364		1.41	13.05		14.46	21.50	
	4290	3/8"		22	.364		2.42	13.05		15.47	22.50	
	4300	1/2" x 1/4"		22	.364		2.35	13.05		15.40	22.50	
	4310	1/2" x 3/8"		22	.364		3.68	13.05		16.73	24	
	4340	Female elbow, 1/8"		26	.308		2.51	11.05		13.56	19.45	
	4350	1/4" x 1/8"		24	.333		1.43	12		13.43	19.60	
	4360	1/4"		24	.333		1.88	12		13.88	20	
	4370	3/8" x 1/4"		22	.364		1.85	13.05		14.90	21.50	
	4380	1/2" x 3/8"		22	.364		3.16	13.05		16.21	23	
	4390	1/2"		22	.364		4.25	13.05		17.30	24.50	
	4420	Male run tee, 1/4" x 1/8"		15	.533		1.76	19.15		20.91	31	
	4430	5/16" x 1/8"		15	.533		2.62	19.15		21.77	32	
	4440	3/8" x 1/4"		15	.533		2.71	19.15		21.86	32	
	4480	1/4" x 1/8"		15	.533		2.03	19.15		21.18	31	
	4490	1/4"		15	.533		2.19	19.15		21.34	31.50	
	4500	3/8" x 1/4"		15	.533		2.54	19.15		21.69	32	
	4510	1/2" x 3/8"		15	.533		3.95	19.15		23.10	33.50	
	4520	1/2"	▼	15	.533	▼	6.35	19.15		25.50	36	
	4800	Flare joint fittings										
	4810	Refrigeration fittings										
	4820	Flare joint nuts and labor not incl. in price. Add 1 nut per jnt.										
	4830	90° Elbow, 1/4"				Ea.	1.73			1.73	1.90	
	4840	3/8"					2.70			2.70	2.97	
	4850	1/2"					2.82			2.82	3.10	
	4860	5/8"					4.13			4.13	4.54	
	4870	3/4"					11.30			11.30	12.45	
	5030	Tee, 1/4"					3.15			3.15	3.47	
	5040	5/16"					3.32			3.32	3.65	
	5050	3/8"					4.48			4.48	4.93	
	5060	1/2"					7.20			7.20	7.90	
	5070	5/8"					9.25			9.25	10.20	
	5080	3/4"					11.20			11.20	12.35	
	5140	Union, 3/16"					1.97			1.97	2.17	
	5150	1/4"					.60			.60	.66	
	5160	5/16"					.67			.67	.74	
	5170	3/8"					1.73			1.73	1.90	
	5180	1/2"				▼	2.95			2.95	3.25	

Important: See the Reference Section for critical supporting data - Reference Nos., Crews, & City Cost Indexes

15 MECHANICAL

	15107	Metal Pipe & Fittings	CREW	DAILY OUTPUT	LABOR-HOURS	UNIT	2002 BARE COSTS				TOTAL INCL O&P	
							MAT.	LABOR	EQUIP.	TOTAL		
460	5190	5/8"				Ea.	6			6	6.60	460
	5200	3/4"					5.20			5.20	5.70	
	5260	Long flare nut, 3/16"	1 Stpi	42	.190		.42	6.90		7.32	10.85	
	5270	1/4"		41	.195		.48	7.05		7.53	11.20	
	5280	5/16"		40	.200		.73	7.25		7.98	11.70	
	5290	3/8"		39	.205		1.13	7.45		8.58	12.45	
	5300	1/2"		38	.211		1.70	7.60		9.30	13.35	
	5310	5/8"		37	.216		3.42	7.85		11.27	15.55	
	5320	3/4"		34	.235		5.80	8.50		14.30	19.20	
	5380	Short flare nut, 3/16"		42	.190		.29	6.90		7.19	10.70	
	5390	1/4"		41	.195		.39	7.05		7.44	11.10	
	5400	5/16"		40	.200		.57	7.25		7.82	11.55	
	5410	3/8"		39	.205		.49	7.45		7.94	11.75	
	5420	1/2"		38	.211		.71	7.60		8.31	12.30	
	5430	5/8"		36	.222		1.01	8.05		9.06	13.25	
	5440	3/4"		34	.235		3.36	8.50		11.86	16.55	
	5500	90° Elbow flare by MIPS, 1/4"					1.38			1.38	1.52	
	5510	3/8"					1.92			1.92	2.11	
	5520	1/2"					3.08			3.08	3.39	
	5530	5/8"					3.91			3.91	4.30	
	5540	3/4"					5.60			5.60	6.15	
	5600	Flare by FIPS, 1/4"					4.09			4.09	4.50	
	5610	3/8"					3.57			3.57	3.93	
	5620	1/2"					6.75			6.75	7.45	
	5670	Flare by sweat, 1/4"					1.42			1.42	1.56	
	5680	3/8"					1.38			1.38	1.52	
	5690	1/2"					2.51			2.51	2.76	
	5700	5/8"					3.47			3.47	3.82	
	5760	Tee flare by IPS, 1/4"					2.19			2.19	2.41	
	5770	3/8"					2.16			2.16	2.38	
	5780	1/2"					6.30			6.30	6.95	
	5790	5/8"					6.40			6.40	7.05	
	5850	Connector, 1/4"					.43			.43	.47	
	5860	3/8"					1.13			1.13	1.24	
	5870	1/2"					1.16			1.16	1.28	
	5880	5/8"					1.33			1.33	1.46	
	5890	3/4"					7.55			7.55	8.30	
	5950	Seal cap, 1/4"					.29			.29	.32	
	5960	3/8"					.37			.37	.41	
	5970	1/2"					.59			.59	.65	
	5980	5/8"					1.79			1.79	1.97	
	5990	3/4"					6.20			6.20	6.85	
	6000	Water service fittings										
	6010	Flare joints nut and labor are included in the fitting price.										
	6020	90° Elbow, cxc, 3/8"	1 Plum	19	.421	Ea.	10.25	15.15		25.40	34.50	
	6030	1/2"		18	.444		10.25	16		26.25	35.50	
	6040	3/4"		16	.500		14.50	18		32.50	43	
	6050	1"		15	.533		27.50	19.15		46.65	59	
	6060	1-1/4"		12	.667		94.50	24		118.50	140	
	6070	1-1/2"		12	.667		94.50	24		118.50	140	
	6080	2"		10	.800		94.50	29		123.50	148	
	6090	C x MPT, 3/8"		19	.421		7.40	15.15		22.55	31	
	6100	1/2"		18	.444		7.40	16		23.40	32	
	6110	3/4"		16	.500		8.80	18		26.80	36.50	
	6120	1"		15	.533		24.50	19.15		43.65	56	
	6130	1-1/4"		13	.615		53.50	22		75.50	92	

MECHANICAL 15

15107	Metal Pipe & Fittings	CREW	DAILY OUTPUT	LABOR-HOURS	UNIT	2002 BARE COSTS				TOTAL INCL O&P	
						MAT.	LABOR	EQUIP.	TOTAL		
460 6140	1-1/2"	1 Plum	12	.667	Ea.	53.50	24		77.50	94.50	**460**
6150	2"		10	.800		94.50	29		123.50	148	
6160	C x FPT, 3/8"		19	.421		7.95	15.15		23.10	31.50	
6170	1/2"		18	.444		7.95	16		23.95	32.50	
6180	3/4"		16	.500		10.90	18		28.90	39	
6190	1"		15	.533		24	19.15		43.15	55	
6200	1-1/4"		13	.615		63.50	22		85.50	104	
6210	1-1/2"		12	.667		146	24		170	197	
6220	2"		10	.800		223	29		252	290	
6230	Tee, cxcxc, 3/8"		13	.615		16.60	22		38.60	52	
6240	1/2"		12	.667		16.60	24		40.60	54.50	
6250	3/4"		11	.727		24	26		50	65.50	
6260	1"		10	.800		34.50	29		63.50	81.50	
6270	1-1/4"		8	1		115	36		151	180	
6280	1-1/2"		8	1		138	36		174	205	
6290	2"		7	1.143		246	41		287	335	
6300	Tee, cxcxFPT, 3/8"		13	.615		12.95	22		34.95	48	
6310	1/2"		12	.667		19.20	24		43.20	57	
6320	3/4"		11	.727		27	26		53	69.50	
6330	Tube nut, cxnut seat, 3/8"		40	.200		1.71	7.20		8.91	12.75	
6340	1/2"		38	.211		1.71	7.55		9.26	13.30	
6350	3/4"		34	.235		3.38	8.45		11.83	16.45	
6360	1"		32	.250		6.10	9		15.10	20.50	
6370	1-1/4"		25	.320		31	11.50		42.50	51.50	
6380	Coupling, cxc, 3/8"		19	.421		10.20	15.15		25.35	34.50	
6390	1/2"		18	.444		10.20	16		26.20	35.50	
6400	3/4"		16	.500		12	18		30	40	
6410	1"		15	.533		23	19.15		42.15	54.50	
6420	1-1/4"		12	.667		37.50	24		61.50	77	
6430	1-1/2"		12	.667		58.50	24		82.50	101	
6440	2"		10	.800		88	29		117	141	
6450	Adapter, cxFPT, 3/8"		19	.421		7.45	15.15		22.60	31	
6460	1/2"		18	.444		7.45	16		23.45	32	
6470	3/4"		16	.500		8.85	18		26.85	37	
6480	1"		15	.533		19.85	19.15		39	51	
6490	1-1/4"		13	.615		38	22		60	75.50	
6500	1-1/2"		12	.667		38	24		62	78	
6510	2"		10	.800		53.50	29		82.50	102	
6520	Adapter, cxMPT, 3/8"		19	.421		6.40	15.15		21.55	30	
6530	1/2"		18	.444		6.40	16		22.40	31	
6540	3/4"		16	.500		9.05	18		27.05	37	
6550	1"		15	.533		15.95	19.15		35.10	46.50	
6560	1-1/4"		13	.615		42	22		64	79.50	
6570	1-1/2"		12	.667		42	24		66	82	
6580	2"	▼	10	.800	▼	57	29		86	106	
6990	Polybutylene/polyethylene pipe, See 15108 for plastic ftng.										
7000	Insert type Brass/copper, 100 psi @ 180°F, CTS										
7010	Adapter MPT 3/8" x 3/8" CTS	1 Plum	29	.276	Ea.	.97	9.90		10.87	16	
7020	1/2" x 1/2"		26	.308		.97	11.05		12.02	17.75	
7030	3/4" x 1/2"		26	.308		1.44	11.05		12.49	18.30	
7040	3/4" x 3/4"		25	.320		1.42	11.50		12.92	18.90	
7050	Adapter CTS 1/2" x 1/2" sweat		24	.333		.38	12		12.38	18.45	
7060	3/4" x 3/4" sweat		22	.364		.47	13.05		13.52	20	
7070	Coupler center set 3/8" CTS		25	.320		.49	11.50		11.99	17.90	
7080	1/2" CTS		23	.348		.38	12.50		12.88	19.25	
7090	3/4" CTS	▼	22	.364	▼	.47	13.05		13.52	20	

Important: See the Reference Section for critical supporting data - Reference Nos., Crews, & City Cost Indexes

		15107	Metal Pipe & Fittings	CREW	DAILY OUTPUT	LABOR-HOURS	UNIT	2002 BARE COSTS				TOTAL INCL O&P	
								MAT.	LABOR	EQUIP.	TOTAL		
460	7100		Elbow 90°, copper 3/8"	1 Plum	25	.320	Ea.	.65	11.50		12.15	18.05	**460**
	7110		1/2" CTS		23	.348		.69	12.50		13.19	19.60	
	7120		3/4" CTS		22	.364		.84	13.05		13.89	20.50	
	7130		Tee copper 3/8" CTS		17	.471		1.28	16.90		18.18	27	
	7140		1/2" CTS		15	.533		.88	19.15		20.03	30	
	7150		3/4" CTS		14	.571		1.35	20.50		21.85	32.50	
	7160		3/8" x 3/8" x 1/2"		16	.500		1.35	18		19.35	28.50	
	7170		1/2" x 3/8" x 1/2"		15	.533		1.30	19.15		20.45	30.50	
	7180		3/4" x 1/2" x 3/4"		14	.571		1.30	20.50		21.80	32.50	
480	0010		**PIPE/TUBE, GROOVED-JOINT FOR COPPER**										**480**
	4000		Fittings: coupling and labor required at joints not incl.										
	4001		in fitting price. add 1 per joint for installed price.										
	4010		Coupling, rigid style										
	4018		2" diameter	1 Plum	50	.160	Ea.	7.90	5.75		13.65	17.35	
	4020		2-1/2" diameter	Q-1	80	.200		8.95	6.45		15.40	19.60	
	4022		3" diameter		67	.239		10	7.75		17.75	22.50	
	4024		4" diameter		50	.320		15.40	10.35		25.75	32.50	
	4026		5" diameter		40	.400		26.50	12.95		39.45	48.50	
	4028		6" diameter	Q-2	50	.480		35.50	16.10		51.60	63.50	
	4100		Elbow, 90° or 45°										
	4108		2" diameter	1 Plum	25	.320	Ea.	12.05	11.50		23.55	30.50	
	4110		2-1/2" diameter	Q-1	40	.400		13.15	12.95		26.10	34	
	4112		3" diameter		33	.485		18.60	15.70		34.30	44	
	4114		4" diameter		25	.640		43.50	20.50		64	79	
	4116		5" diameter		20	.800		124	26		150	176	
	4118		6" diameter	Q-2	25	.960		199	32		231	268	
	4200		Tee										
	4208		2" diameter	1 Plum	17	.471	Ea.	20	16.90		36.90	47.50	
	4210		2-1/2" diameter	Q-1	27	.593		21.50	19.15		40.65	52.50	
	4212		3" diameter		22	.727		32	23.50		55.50	71	
	4214		4" diameter		17	.941		71	30.50		101.50	125	
	4216		5" diameter		13	1.231		201	40		241	281	
	4218		6" diameter	Q-2	17	1.412		248	47.50		295.50	345	
	4300		Reducer, concentric										
	4310		3" x 2-1/2" diameter	Q-1	35	.457	Ea.	17.60	14.80		32.40	42	
	4312		4" x 2-1/2" diameter		32	.500		36.50	16.20		52.70	65	
	4314		4" x 3" diameter		29	.552		36.50	17.85		54.35	67.50	
	4316		5" x 3" diameter		25	.640		104	20.50		124.50	145	
	4318		5" x 4" diameter		22	.727		104	23.50		127.50	150	
	4320		6" x 3" diameter	Q-2	28	.857		113	29		142	168	
	4322		6" x 4" diameter		26	.923		113	31		144	171	
	4324		6" x 5" diameter		24	1		113	33.50		146.50	175	
	4350		Flange, w/groove gasket										
	4351		ANSI class 125 and 150										
	4355		2" diameter	1 Plum	23	.348	Ea.	64	12.50		76.50	89.50	
	4356		2-1/2" diameter	Q-1	37	.432		67	14		81	94.50	
	4358		3" diameter		31	.516		70	16.70		86.70	102	
	4360		4" diameter		23	.696		77	22.50		99.50	119	
	4362		5" diameter		19	.842		104	27		131	155	
	4364		6" diameter	Q-2	23	1.043		113	35		148	177	
620	0010		**PIPE, STEEL**										**620**
	0020		All pipe sizes are to Spec. A-53 unless noted otherwise	R15100 -050									
	0030		Schedule 10, see 15107-690-0500										
	0050		Schedule 40, threaded, with couplings, and clevis type										

MECHANICAL 15

		15107	Metal Pipe & Fittings		CREW	DAILY OUTPUT	LABOR-HOURS	UNIT	2002 BARE COSTS				TOTAL INCL O&P
									MAT.	LABOR	EQUIP.	TOTAL	
620	0060		hangers sized for covering, 10' O.C.											**620**
	0540		Black, 1/4" diameter	R15100 -050	1 Plum	66	.121	L.F.	1.40	4.36		5.76	8.10
	0550		3/8" diameter			65	.123		1.53	4.42		5.95	8.35
	0560		1/2" diameter			63	.127		1.32	4.56		5.88	8.35
	0570		3/4" diameter			61	.131		1.47	4.71		6.18	8.70
	0580		1" diameter			53	.151		1.88	5.45		7.33	10.25
	0590		1-1/4" diameter		Q-1	89	.180		2.23	5.80		8.03	11.20
	0600		1-1/2" diameter			80	.200		2.51	6.45		8.96	12.50
	0610		2" diameter			64	.250		3.18	8.10		11.28	15.70
	0620		2-1/2" diameter			50	.320		5.05	10.35		15.40	21
	0630		3" diameter			43	.372		6.35	12.05		18.40	25
	0640		3-1/2" diameter			40	.400		8.30	12.95		21.25	28.50
	0650		4" diameter			36	.444		9.35	14.40		23.75	32
	0660		5" diameter			26	.615		17.60	19.90		37.50	49.50
	0670		6" diameter		Q-2	31	.774		21.50	26		47.50	63
	0680		8" diameter			27	.889		31	30		61	79
	0690		10" diameter			23	1.043		47	35		82	105
	0700		12" diameter			18	1.333		59	44.50		103.50	133
	1220		To delete coupling & hanger, subtract										
	1230		1/4" diam. to 3/4" diam.						31%	56%			
	1240		1" diam. to 1-1/2" diam.						23%	51%			
	1250		2" diam. to 4" diam.						23%	41%			
	1260		5" diam. to 12" diam.						21%	45%			
	1290		Galvanized, 1/4" diameter		1 Plum	66	.121	L.F.	1.73	4.36		6.09	8.45
	1300		3/8" diameter			65	.123		1.86	4.42		6.28	8.70
	1310		1/2" diameter			63	.127		1.51	4.56		6.07	8.55
	1320		3/4" diameter			61	.131		1.93	4.71		6.64	9.20
	1330		1" diameter			53	.151		2.29	5.45		7.74	10.70
	1340		1-1/4" diameter		Q-1	89	.180		2.84	5.80		8.64	11.85
	1350		1-1/2" diameter			80	.200		3.06	6.45		9.51	13.10
	1360		2" diameter			64	.250		4.11	8.10		12.21	16.70
	1370		2-1/2" diameter			50	.320		5.35	10.35		15.70	21.50
	1380		3" diameter			43	.372		6.70	12.05		18.75	25.50
	1390		3-1/2" diameter			40	.400		8.60	12.95		21.55	29
	1400		4" diameter			36	.444		9.75	14.40		24.15	32
	1410		5" diameter			26	.615		20	19.90		39.90	52
	1420		6" diameter		Q-2	31	.774		25	26		51	66.50
	1430		8" diameter			27	.889		38.50	30		68.50	87.50
	1440		10" diameter			23	1.043		58.50	35		93.50	118
	1450		12" diameter			18	1.333		73.50	44.50		118	149
	1750		To delete coupling & hanger, subtract										
	1760		1/4" diam. to 3/4" diam.						31%	56%			
	1770		1" diam. to 1-1/2" diam.						23%	51%			
	1780		2" diam. to 4" diam.						23%	41%			
	1790		5" diam. to 12" diam.						21%	45%			
	2000		Welded, sch. 40, on yoke & roll hangers, sized for covering,										
	2010		10' O.C. (no hangers incl. for 14" diam. and up)										
	2040		Black, 1" diameter		Q-15	93	.172	L.F.	2.73	5.55	.86	9.14	12.35
	2050		1-1/4" diameter			84	.190		3.44	6.15	.95	10.54	14.15
	2060		1-1/2" diameter			76	.211		3.65	6.80	1.05	11.50	15.45
	2070		2" diameter			61	.262		4.17	8.50	1.31	13.98	18.80
	2080		2-1/2" diameter			47	.340		5.15	11	1.70	17.85	24
	2090		3" diameter			43	.372		6	12.05	1.86	19.91	27
	2100		3-1/2" diameter			39	.410		7.50	13.25	2.05	22.80	30.50
	2110		4" diameter			37	.432		9	14	2.16	25.16	33.50
	2120		5" diameter			32	.500		13.35	16.20	2.50	32.05	42

15107 | Metal Pipe & Fittings

		CREW	DAILY OUTPUT	LABOR-HOURS	UNIT	MAT.	LABOR	EQUIP.	TOTAL	TOTAL INCL O&P	
								2002 BARE COSTS			
620	2130	6" diameter R15100-050	Q-16	36	.667	L.F.	15.70	22.50	2.22	40.42	53	**620**
	2140	8" diameter		29	.828		23.50	28	2.76	54.26	71	
	2150	10" diameter		24	1		43	33.50	3.33	79.83	102	
	2160	12" diameter		19	1.263		48	42.50	4.21	94.71	122	
	2170	14" diameter		15	1.600		40	53.50	5.35	98.85	131	
	2180	16" diameter		13	1.846		41.50	62	6.15	109.65	146	
	2190	18" diameter		11	2.182		55.50	73	7.25	135.75	179	
	2200	20" diameter		9	2.667		61	89.50	8.90	159.40	212	
	2220	24" diameter		8	3		76	101	10	187	247	
	2230	26" diameter		7.50	3.200		145	107	10.65	262.65	335	
	2240	30" diameter		6	4		158	134	13.30	305.30	390	
	2260	36" diameter		4.50	5.333		181	179	17.75	377.75	490	
	2323	.375" wall, A-53, gr. A/B, seamless, welded w/hngrs.										
	2324	12" diameter	Q-16	20	1.200	L.F.	26	40.50	4	70.50	93.50	
	2325	14" diameter		17	1.412		31.50	47.50	4.70	83.70	112	
	2326	16" diameter		14	1.714		47	57.50	5.70	110.20	144	
	2334	.50" wall, 8" diameter		25	.960		17.10	32	3.20	52.30	71	
	2335	10" diameter		20	1.200		27	40.50	4	71.50	94.50	
	2336	12" diameter		16	1.500		32.50	50.50	5	88	118	
	2337	14" diameter		13	1.846		48	62	6.15	116.15	153	
	2345	Sch. 40, A-53, gr. A/B, ERW, welded w/hngrs.										
	2346	2" diameter	Q-15	61	.262	L.F.	2.59	8.50	1.31	12.40	17.10	
	2347	2-1/2" diameter		47	.340		3.74	11	1.70	16.44	22.50	
	2348	3" diameter		43	.372		4.72	12.05	1.86	18.63	25.50	
	2349	4" diameter		38	.421		6.45	13.60	2.11	22.16	30	
	2350	6" diameter	Q-16	37	.649		8.25	22	2.16	32.41	44.50	
	2351	8" diameter		29	.828		12.40	28	2.76	43.16	58.50	
	2352	10" diameter		24	1		22	33.50	3.33	58.83	78.50	
	2363	.375" wall, A-53, gr. A/B, ERW, welded w/hngrs.										
	2364	12" diameter	Q-16	20	1.200	L.F.	26	40.50	4	70.50	93.50	
	2365	14" diameter		17	1.412		31.50	47.50	4.70	83.70	112	
	2366	16" diameter		14	1.714		34.50	57.50	5.70	97.70	131	
	2560	To delete hanger, subtract										
	2570	1" diam. to 1-1/2" diam.					15%	34%				
	2580	2" diam. to 3-1/2" diam.					9%	21%				
	2590	4" diam. to 12" diam.					5%	12%				
	2640	Galvanized, 1" diameter	Q-15	93	.172	L.F.	2.90	5.55	.86	9.31	12.55	
	2650	1-1/4" diameter		84	.190		3.65	6.15	.95	10.75	14.35	
	2660	1-1/2" diameter		76	.211		3.91	6.80	1.05	11.76	15.70	
	2670	2" diameter		61	.262		4.52	8.50	1.31	14.33	19.20	
	2680	2-1/2" diameter		47	.340		5.75	11	1.70	18.45	25	
	2690	3" diameter		43	.372		6.75	12.05	1.86	20.66	27.50	
	2700	3-1/2" diameter		39	.410		8.40	13.25	2.05	23.70	31.50	
	2710	4" diameter		37	.432		8.55	14	2.16	24.71	33	
	2720	5" diameter		32	.500		16.40	16.20	2.50	35.10	45.50	
	2730	6" diameter	Q-16	36	.667		19.30	22.50	2.22	44.02	57	
	2740	8" diameter		29	.828		29	28	2.76	59.76	77	
	2750	10" diameter		24	1		51.50	33.50	3.33	88.33	111	
	2760	12" diameter		19	1.263		58.50	42.50	4.21	105.21	133	
	3160	To delete hanger, subtract										
	3170	1" diam. to 1-1/2" diam.					30%	34%				
	3180	2" diam. to 3-1/2" diam.					19%	21%				
	3190	4" diam. to 12" diam.					10%	12%				
	3250	Flanged, 150 lb. weld neck, on yoke & roll hangers										
	3260	sized for covering, 10' O.C.										
	3290	Black, 1" diameter	Q-15	70	.229	L.F.	6.30	7.40	1.14	14.84	19.35	

MECHANICAL 15

			CREW	DAILY OUTPUT	LABOR-HOURS	UNIT	\multicolumn{4}{c}{2002 BARE COSTS}	TOTAL INCL O&P	
15107		**Metal Pipe & Fittings**					MAT.	LABOR	EQUIP.	TOTAL		
620	3300	1-1/4" diameter	Q-15	64	.250	L.F.	7	8.10	1.25	16.35	21.50	620
	3310	1-1/2" diameter	R15100 -050	58	.276		7.20	8.90	1.38	17.48	23	
	3320	2" diameter		45	.356		8.20	11.50	1.78	21.48	28.50	
	3330	2-1/2" diameter		36	.444		9.85	14.40	2.22	26.47	34.50	
	3340	3" diameter		32	.500		11.10	16.20	2.50	29.80	39.50	
	3350	3-1/2" diameter		29	.552		13.50	17.85	2.76	34.11	45	
	3360	4" diameter		26	.615		14.85	19.90	3.08	37.83	49.50	
	3370	5" diameter		21	.762		22.50	24.50	3.81	50.81	66	
	3380	6" diameter	Q-16	25	.960		25	32	3.20	60.20	79.50	
	3390	8" diameter		19	1.263		38	42.50	4.21	84.71	110	
	3400	10" diameter		16	1.500		68.50	50.50	5	124	157	
	3410	12" diameter		14	1.714		89.50	57.50	5.70	152.70	191	
	3470	For 300 lb. flanges, add					63%					
	3480	For 600 lb. flanges, add					310%					
	3960	To delete flanges & hanger, subtract										
	3970	1" diam. to 2" diam.					76%	65%				
	3980	2-1/2" diam. to 4" diam.					62%	59%				
	3990	5" diam. to 12" diam.					60%	46%				
	4040	Galvanized, 1" diameter	Q-15	70	.229	L.F.	6.45	7.40	1.14	14.99	19.50	
	4050	1-1/4" diameter		64	.250		7.20	8.10	1.25	16.55	21.50	
	4060	1-1/2" diameter		58	.276		7.50	8.90	1.38	17.78	23	
	4070	2" diameter		45	.356		8.20	11.50	1.78	21.48	28.50	
	4080	2-1/2" diameter		36	.444		10.15	14.40	2.22	26.77	35	
	4090	3" diameter		32	.500		11.40	16.20	2.50	30.10	40	
	4100	3-1/2" diameter		29	.552		13.55	17.85	2.76	34.16	45	
	4110	4" diameter		26	.615		14.45	19.90	3.08	37.43	49.50	
	4120	5" diameter		21	.762		25.50	24.50	3.81	53.81	69.50	
	4130	6" diameter	Q-16	25	.960		29	32	3.20	64.20	84	
	4140	8" diameter		19	1.263		38	42.50	4.21	84.71	111	
	4150	10" diameter		16	1.500		67.50	50.50	5	123	156	
	4160	12" diameter		14	1.714		80	57.50	5.70	143.20	181	
	4220	For 300 lb. flanges, add					65%					
	4240	For 600 lb. flanges, add					350%					
	4660	To delete flanges & hanger, subtract										
	4670	1" diam. to 2" diam.					73%	65%				
	4680	2-1/2" diam. to 4" diam.					58%	59%				
	4690	5" diam. to 12" diam.					43%	46%				
	4750	Schedule 80, threaded, with couplings, and clevis type hangers										
	4760	sized for covering, 10' O.C.										
	4790	Black, 1/4" diameter	1 Plum	54	.148	L.F.	1.71	5.35		7.06	9.95	
	4800	3/8" diameter		53	.151		1.69	5.45		7.14	10.05	
	4810	1/2" diameter		52	.154		1.83	5.55		7.38	10.35	
	4820	3/4" diameter		50	.160		2.08	5.75		7.83	10.95	
	4830	1" diameter		45	.178		2.52	6.40		8.92	12.40	
	4840	1-1/4" diameter	Q-1	75	.213		3.31	6.90		10.21	14.05	
	4850	1-1/2" diameter		69	.232		3.64	7.50		11.14	15.30	
	4860	2" diameter		56	.286		4.88	9.25		14.13	19.30	
	4870	2-1/2" diameter		44	.364		7.10	11.75		18.85	25.50	
	4880	3" diameter		38	.421		8.80	13.60		22.40	30	
	4890	3-1/2" diameter		35	.457		10.90	14.80		25.70	34.50	
	4900	4" diameter		32	.500		12.85	16.20		29.05	38.50	
	4910	5" diameter		23	.696		22.50	22.50		45	59	
	4920	6" diameter	Q-2	28	.857		31.50	29		60.50	78	
	4930	8" diameter		23	1.043		52.50	35		87.50	111	
	4940	10" diameter		20	1.200		81	40.50		121.50	150	
	4950	12" diameter		15	1.600		93	53.50		146.50	183	

Important: See the Reference Section for critical supporting data - Reference Nos., Crews, & City Cost Indexes

				DAILY	LABOR-		2002 BARE COSTS				TOTAL	
15107		**Metal Pipe & Fittings**	CREW	OUTPUT	HOURS	UNIT	MAT.	LABOR	EQUIP.	TOTAL	INCL O&P	
620	5061	A-106, gr. A/B seamless with cplgs. & hgrs,1/4" dia. R15100-050	1 Plum	63	.127	L.F.	2.03	4.56		6.59	9.15	**620**
	5062	3/8" diameter		62	.129		2.18	4.64		6.82	9.40	
	5063	1/2" diameter		61	.131		2.74	4.71		7.45	10.10	
	5064	3/4" diameter		57	.140		2.53	5.05		7.58	10.40	
	5065	1" diameter	▼	51	.157		3.45	5.65		9.10	12.30	
	5066	1-1/4" diameter	Q-1	85	.188		5	6.10		11.10	14.70	
	5067	1-1/2" diameter		77	.208		5.45	6.70		12.15	16.15	
	5071	A-53 2" diameter		61	.262		6.10	8.50		14.60	19.55	
	5072	2-1/2" diameter		48	.333		6.20	10.80		17	23	
	5073	3" diameter		41	.390		7.65	12.60		20.25	27.50	
	5074	4" diameter	▼	35	.457	▼	13.50	14.80		28.30	37.50	
	5430	To delete coupling & hanger, subtract										
	5440	1/4" diam. to 1/2" diam.					31%	54%				
	5450	3/4" diam. to 1-1/2" diam.					28%	49%				
	5460	2" diam. to 4" diam.					21%	40%				
	5470	5" diam. to 12" diam.					17%	43%				
	5510	Galvanized, 1/4" diameter	1 Plum	54	.148	L.F.	1.91	5.35		7.26	10.15	
	5520	3/8" diameter		53	.151		1.94	5.45		7.39	10.35	
	5530	1/2" diameter		52	.154		2.08	5.55		7.63	10.65	
	5540	3/4" diameter		50	.160		2.39	5.75		8.14	11.25	
	5550	1" diameter	▼	45	.178		2.75	6.40		9.15	12.70	
	5560	1-1/4" diameter	Q-1	75	.213		3.63	6.90		10.53	14.40	
	5570	1-1/2" diameter		69	.232		4.03	7.50		11.53	15.75	
	5580	2" diameter		56	.286		5.40	9.25		14.65	19.90	
	5590	2-1/2" diameter		44	.364		7.90	11.75		19.65	26.50	
	5600	3" diameter		38	.421		9.90	13.60		23.50	31.50	
	5610	3-1/2" diameter		35	.457		12.25	14.80		27.05	36	
	5620	4" diameter		32	.500		14.45	16.20		30.65	40.50	
	5630	5" diameter	▼	23	.696		23	22.50		45.50	59.50	
	5640	6" diameter	Q-2	28	.857		30.50	29		59.50	77	
	5650	8" diameter		23	1.043		51.50	35		86.50	110	
	5660	10" diameter		20	1.200		67.50	40.50		108	135	
	5670	12" diameter	▼	15	1.600	▼	91.50	53.50		145	181	
	5930	To delete coupling & hanger, subtract										
	5940	1/4" diam. to 1/2" diam.					31%	54%				
	5950	3/4" diam. to 1-1/2" diam.					28%	49%				
	5960	2" diam. to 4" diam.					21%	40%				
	5970	5" diam. to 12" diam.					17%	43%				
	6000	Welded, on yoke & roller hangers										
	6010	sized for covering, 10' O.C.										
	6040	Black, 1" diameter	Q-15	85	.188	L.F.	3	6.10	.94	10.04	13.55	
	6050	1-1/4" diameter		79	.203		3.86	6.55	1.01	11.42	15.25	
	6060	1-1/2" diameter		72	.222		4.19	7.20	1.11	12.50	16.70	
	6070	2" diameter		57	.281		4.93	9.10	1.40	15.43	20.50	
	6080	2-1/2" diameter		44	.364		6.65	11.75	1.82	20.22	27	
	6090	3" diameter		40	.400		8	12.95	2	22.95	30.50	
	6100	3-1/2" diameter		34	.471		9.65	15.20	2.35	27.20	36	
	6110	4" diameter		33	.485		11	15.70	2.42	29.12	38.50	
	6120	5" diameter	▼	26	.615		20	19.90	3.08	42.98	55.50	
	6130	6" diameter	Q-16	30	.800		29	27	2.66	58.66	75	
	6140	8" diameter		25	.960		51	32	3.20	86.20	108	
	6150	10" diameter		20	1.200		83	40.50	4	127.50	156	
	6160	12" diameter	▼	15	1.600	▼	94.50	53.50	5.35	153.35	191	
	6540	To delete hanger, subtract										
	6550	1" diam. to 1-1/2" diam.					30%	14%				
	6560	2" diam. to 3" diam.					23%	9%				

MECHANICAL 15

15107	Metal Pipe & Fittings		CREW	DAILY OUTPUT	LABOR-HOURS	UNIT	2002 BARE COSTS				TOTAL INCL O&P	
							MAT.	LABOR	EQUIP.	TOTAL		
620	6570	3-1/2″ diam. to 5″ diam.	R15100 -050					12%	6%			**620**
	6580	6″ diam. to 12″ diam.						10%	4%			
	6610	Galvanized, 1″ diameter		Q-15	85	.188	L.F.	3.23	6.10	.94	10.27	13.80
	6650	1-1/4″ diameter			79	.203		4.18	6.55	1.01	11.74	15.60
	6660	1-1/2″ diameter			72	.222		4.58	7.20	1.11	12.89	17.10
	6670	2″ diameter			57	.281		5.45	9.10	1.40	15.95	21
	6680	2-1/2″ diameter			44	.364		7.50	11.75	1.82	21.07	28
	6690	3″ diameter			40	.400		9.10	12.95	2	24.05	31.50
	6700	3-1/2″ diameter			34	.471		11	15.20	2.35	28.55	37.50
	6710	4″ diameter			33	.485		12.65	15.70	2.42	30.77	40
	6720	5″ diameter			26	.615		20.50	19.90	3.08	43.48	56
	6730	6″ diameter		Q-16	30	.800		28	27	2.66	57.66	74
	6740	8″ diameter			25	.960		50	32	3.20	85.20	107
	6750	10″ diameter			20	1.200		70	40.50	4	114.50	142
	6760	12″ diameter			15	1.600		93	53.50	5.35	151.85	189
	7150	To delete hanger, subtract										
	7160	1″ diam. to 1-1/2″ diam.						26%	14%			
	7170	2″ diam. to 3″ diam.						16%	9%			
	7180	3-1/2″ diam. to 5″ diam.						10%	6%			
	7190	6″ diam. to 12″ diam.						7%	4%			
	7250	Flanged, 300 lb. weld neck, on yoke & roll hangers										
	7260	sized for covering, 10' O.C.										
	7290	Black, 1″ diameter		Q-15	66	.242	L.F.	7.50	7.85	1.21	16.56	21.50
	7300	1-1/4″ diameter			61	.262		8.55	8.50	1.31	18.36	23.50
	7310	1-1/2″ diameter			54	.296		8.70	9.60	1.48	19.78	25.50
	7320	2″ diameter			42	.381		10.30	12.30	1.90	24.50	32
	7330	2-1/2″ diameter			33	.485		12.70	15.70	2.42	30.82	40
	7340	3″ diameter			29	.552		14.05	17.85	2.76	34.66	45.50
	7350	3-1/2″ diameter			24	.667		19.60	21.50	3.33	44.43	57.50
	7360	4″ diameter			23	.696		21	22.50	3.48	46.98	61
	7370	5″ diameter			19	.842		35.50	27	4.21	66.71	84.50
	7380	6″ diameter		Q-16	23	1.043		41.50	35	3.47	79.97	103
	7390	8″ diameter			17	1.412		73.50	47.50	4.70	125.70	157
	7400	10″ diameter			14	1.714		130	57.50	5.70	193.20	236
	7410	12″ diameter			12	2		155	67	6.65	228.65	278
	7470	For 600 lb. flanges, add						100%				
	7940	To delete flanges & hanger, subtract										
	7950	1″ diam. to 1-1/2″ diam.						75%	66%			
	7960	2″ diam. to 3″ diam.						62%	60%			
	7970	3-1/2″ diam. to 5″ diam.						54%	66%			
	7980	6″ diam. to 12″ diam.						55%	62%			
	8040	Galvanized, 1″ diameter		Q-15	66	.242	L.F.	7.75	7.85	1.21	16.81	21.50
	8050	1-1/4″ diameter			61	.262		8.90	8.50	1.31	18.71	24
	8060	1-1/2″ diameter			54	.296		9.10	9.60	1.48	20.18	26
	8070	2″ diameter			42	.381		10.85	12.30	1.90	25.05	32.50
	8080	2-1/2″ diameter			33	.485		13.50	15.70	2.42	31.62	41
	8090	3″ diameter			29	.552		15.15	17.85	2.76	35.76	46.50
	8100	3-1/2″ diameter			24	.667		21	21.50	3.33	45.83	59
	8110	4″ diameter			23	.696		22.50	22.50	3.48	48.48	63
	8120	5″ diameter			19	.842		36	27	4.21	67.21	85
	8130	6″ diameter		Q-16	23	1.043		40.50	35	3.47	78.97	101
	8140	8″ diameter			17	1.412		72.50	47.50	4.70	124.70	156
	8150	10″ diameter			14	1.714		117	57.50	5.70	180.20	222
	8160	12″ diameter			12	2		153	67	6.65	226.65	276
	8240	For 600 lb. flanges, add						100%				
	8900	To delete flanges & hangers, subtract										

15 MECHANICAL

Important: See the Reference Section for critical supporting data - Reference Nos., Crews, & City Cost Indexes

15107	Metal Pipe & Fittings	CREW	DAILY OUTPUT	LABOR-HOURS	UNIT	2002 BARE COSTS				TOTAL INCL O&P
						MAT.	LABOR	EQUIP.	TOTAL	
620	8910	1" diam. to 1-1/2" diam.	R15100				72%	66%				620
8920	2" diam. to 3" diam.	-050				59%	60%			
8930	3-1/2" diam. to 5" diam.					51%	66%			
8940	6" diam. to 12" diam.					49%	62%			
9000	Threading pipe labor, one end, all schedules through 80									
9010	1/4" through 3/4" pipe size	1 Plum	80	.100	Ea.		3.60		3.60	5.40
9020	1" through 2" pipe size		73	.110			3.94		3.94	5.95
9030	2-1/2" pipe size		53	.151			5.45		5.45	8.20
9040	3" pipe size	▼	50	.160			5.75		5.75	8.65
9050	3-1/2" pipe size	Q-1	89	.180			5.80		5.80	8.75
9060	4" pipe size		73	.219			7.10		7.10	10.70
9070	5" pipe size		53	.302			9.75		9.75	14.75
9080	6" pipe size		46	.348			11.25		11.25	16.95
9090	8" pipe size		29	.552			17.85		17.85	27
9100	10" pipe size		21	.762			24.50		24.50	37
9110	12" pipe size	▼	13	1.231	▼		40		40	60
9200	Welding labor per joint									
9210	Schedule 40,									
9230	1/2" pipe size	Q-15	32	.500	Ea.		16.20	2.50	18.70	27.50
9240	3/4" pipe size		27	.593			19.15	2.96	22.11	32.50
9250	1" pipe size		23	.696			22.50	3.48	25.98	38
9260	1-1/4" pipe size		20	.800			26	4	30	43.50
9270	1-1/2" pipe size		19	.842			27	4.21	31.21	45.50
9280	2" pipe size		16	1			32.50	5	37.50	54.50
9290	2-1/2" pipe size		13	1.231			40	6.15	46.15	67
9300	3" pipe size		12	1.333			43	6.65	49.65	72.50
9310	4" pipe size		10	1.600			52	8	60	87
9320	5" pipe size		9	1.778			57.50	8.90	66.40	96.50
9330	6" pipe size		8	2			64.50	10	74.50	109
9340	8" pipe size		5	3.200			104	16	120	174
9350	10" pipe size		4	4			129	20	149	217
9360	12" pipe size		3	5.333			173	26.50	199.50	290
9370	14" pipe size		2.60	6.154			199	31	230	335
9380	16" pipe size		2.20	7.273			235	36.50	271.50	395
9390	18" pipe size		2	8			259	40	299	435
9400	20" pipe size		1.80	8.889			288	44.50	332.50	485
9410	22" pipe size		1.70	9.412			305	47	352	510
9420	24" pipe size	▼	1.50	10.667	▼		345	53.50	398.50	580
9450	Schedule 80,									
9460	1/2" pipe size	Q-15	27	.593	Ea.		19.15	2.96	22.11	32.50
9470	3/4" pipe size		23	.696			22.50	3.48	25.98	38
9480	1" pipe size		20	.800			26	4	30	43.50
9490	1-1/4" pipe size		19	.842			27	4.21	31.21	45.50
9500	1-1/2" pipe size		18	.889			29	4.44	33.44	48.50
9510	2" pipe size		15	1.067			34.50	5.35	39.85	58
9520	2-1/2" pipe size		12	1.333			43	6.65	49.65	72.50
9530	3" pipe size		11	1.455			47	7.25	54.25	79
9540	4" pipe size		8	2			64.50	10	74.50	109
9550	5" pipe size		6	2.667			86.50	13.35	99.85	145
9560	6" pipe size		5	3.200			104	16	120	174
9570	8" pipe size		4	4			129	20	149	217
9580	10" pipe size		3	5.333			173	26.50	199.50	290
9590	12" pipe size	▼	2	8			259	40	299	435
9600	14" pipe size	Q-16	2.60	9.231			310	30.50	340.50	500
9610	16" pipe size		2.30	10.435			350	35	385	570
9620	18" pipe size	▼	2	12	▼		405	40	445	650

MECHANICAL 15

15107 | Metal Pipe & Fittings

		CREW	DAILY OUTPUT	LABOR-HOURS	UNIT	2002 BARE COSTS				TOTAL INCL O&P	
						MAT.	LABOR	EQUIP.	TOTAL		
620	9630	20" pipe size	Q-16	1.80	13.333	Ea.		445	44.50	489.50	725	**620**
	9640	22" pipe size	↓	1.60	15	↓		505	50	555	815	
	9650	24" pipe size		1.50	16			535	53.50	588.50	870	
640	0010	**PIPE, STEEL, FITTINGS** Threaded										**640**
	0020	Cast Iron										
	0040	Standard weight, black										
	0060	90° Elbow, straight										
	0070	1/4"	1 Plum	16	.500	Ea.	3.17	18		21.17	30.50	
	0080	3/8"		16	.500		4.47	18		22.47	32	
	0090	1/2"		15	.533		2	19.15		21.15	31	
	0100	3/4"		14	.571		2.09	20.50		22.59	33.50	
	0110	1"	↓	13	.615		2.48	22		24.48	36	
	0120	1-1/4"	Q-1	22	.727		3.52	23.50		27.02	39.50	
	0130	1-1/2"		20	.800		4.87	26		30.87	44.50	
	0140	2"		18	.889		7.60	29		36.60	52	
	0150	2-1/2"		14	1.143		18.25	37		55.25	76	
	0160	3"		10	1.600		30	52		82	111	
	0170	3-1/2"		8	2		61.50	64.50		126	165	
	0180	4"		6	2.667		55.50	86.50		142	191	
	0190	5"	↓	5	3.200		108	104		212	274	
	0200	6"	Q-2	7	3.429		136	115		251	325	
	0210	8"	"	6	4	↓	284	134		418	510	
	0250	45° Elbow, straight										
	0260	1/4"	1 Plum	16	.500	Ea.	3.38	18		21.38	30.50	
	0270	3/8"		16	.500		3.64	18		21.64	31	
	0280	1/2"		15	.533		3.07	19.15		22.22	32.50	
	0300	3/4"		14	.571		3.09	20.50		23.59	34.50	
	0320	1"	↓	13	.615		3.62	22		25.62	37.50	
	0330	1-1/4"	Q-1	22	.727		4.88	23.50		28.38	41	
	0340	1-1/2"		20	.800		8.10	26		34.10	48	
	0350	2"		18	.889		9.30	29		38.30	54	
	0360	2-1/2"		14	1.143		23.50	37		60.50	82	
	0370	3"		10	1.600		38.50	52		90.50	120	
	0380	3-1/2"		8	2		78	64.50		142.50	184	
	0400	4"		6	2.667		78	86.50		164.50	216	
	0420	5"	↓	5	3.200		133	104		237	300	
	0440	6"	Q-2	7	3.429		171	115		286	360	
	0460	8"	"	6	4	↓	330	134		464	565	
	0500	Tee, straight										
	0510	1/4"	1 Plum	10	.800	Ea.	4.42	29		33.42	48.50	
	0520	3/8"		10	.800		4.70	29		33.70	48.50	
	0530	1/2"		9	.889		3.13	32		35.13	51.50	
	0540	3/4"		9	.889		3.65	32		35.65	52	
	0550	1"	↓	8	1		3.24	36		39.24	57.50	
	0560	1-1/4"	Q-1	14	1.143		5.95	37		42.95	62.50	
	0570	1-1/2"		13	1.231		7.70	40		47.70	68.50	
	0580	2"		11	1.455		10.75	47		57.75	83	
	0590	2-1/2"		9	1.778		28	57.50		85.50	117	
	0600	3"		6	2.667		43	86.50		129.50	177	
	0610	3-1/2"		5	3.200		83.50	104		187.50	248	
	0620	4"		4	4		83.50	129		212.50	287	
	0630	5"	↓	3	5.333		169	173		342	445	
	0640	6"	Q-2	4	6		192	201		393	515	
	0650	8"	"	3	8	↓	460	268		728	910	
	0660	Tee, reducing, run or outlet										

R15100 -050

Important: See the Reference Section for critical supporting data - Reference Nos., Crews, & City Cost Indexes

15107	Metal Pipe & Fittings	CREW	DAILY OUTPUT	LABOR-HOURS	UNIT	2002 BARE COSTS				TOTAL INCL O&P
						MAT.	LABOR	EQUIP.	TOTAL	
640 0661	1/2"	1 Plum	9	.889	Ea.	6.75	32		38.75	55.50	**640**
0662	3/4"		9	.889		5.75	32		37.75	54.50
0663	1"		8	1		5.95	36		41.95	60.50
0664	1-1/4"	Q-1	14	1.143		5.85	37		42.85	62.50
0665	1-1/2"		13	1.231		10.85	40		50.85	72
0666	2"		11	1.455		16.65	47		63.65	89.50
0667	2-1/2"		9	1.778		31.50	57.50		89	122
0668	3"		6	2.667		54	86.50		140.50	189
0669	3-1/2"		5	3.200		97.50	104		201.50	263
0670	4"		4	4		97.50	129		226.50	300
0671	5"		3	5.333		125	173		298	400
0672	6"	Q-2	4	6		230	201		431	560
0673	8"	"	3	8		263	268		531	695
0674	Reducer, concentric									
0675	3/4"	1 Plum	18	.444	Ea.	5.15	16		21.15	29.50
0676	1"	"	15	.533		3.80	19.15		22.95	33
0677	1-1/4"	Q-1	26	.615		9.85	19.90		29.75	41
0678	1-1/2"		24	.667		13.95	21.50		35.45	48
0679	2"		21	.762		19.10	24.50		43.60	58
0680	2-1/2"		18	.889		26	29		55	72
0681	3"		14	1.143		35.50	37		72.50	95.50
0682	3-1/2"		12	1.333		62.50	43		105.50	134
0683	4"		10	1.600		62.50	52		114.50	147
0684	5"		6	2.667		113	86.50		199.50	254
0685	6"	Q-2	8	3		113	101		214	276
0686	8"	"	7	3.429		148	115		263	335
0687	Reducer, eccentric									
0688	3/4"	1 Plum	16	.500	Ea.	11.90	18		29.90	40
0689	1"	"	14	.571		11.65	20.50		32.15	44
0690	1-1/4"	Q-1	25	.640		16.85	20.50		37.35	49.50
0691	1-1/2"		22	.727		25	23.50		48.50	63
0692	2"		20	.800		31.50	26		57.50	73.50
0693	2-1/2"		16	1		47	32.50		79.50	101
0694	3"		12	1.333		58	43		101	129
0695	3-1/2"		10	1.600		113	52		165	203
0696	4"		9	1.778		113	57.50		170.50	212
0697	5"		5	3.200		161	104		265	335
0698	6"	Q-2	8	3		218	101		319	390
0699	8"	"	7	3.429		825	115		940	1,075
0700	Standard weight, galvanized cast iron									
0720	90° Elbow, straight									
0730	1/4"	1 Plum	16	.500	Ea.	6.10	18		24.10	33.50
0740	3/8"		16	.500		6.10	18		24.10	33.50
0750	1/2"		15	.533		6.10	19.15		25.25	36
0760	3/4"		14	.571		6.85	20.50		27.35	38.50
0770	1"		13	.615		7.90	22		29.90	42
0780	1-1/4"	Q-1	22	.727		12.30	23.50		35.80	49
0790	1-1/2"		20	.800		16.85	26		42.85	57.50
0800	2"		18	.889		25	29		54	71
0810	2-1/2"		14	1.143		51	37		88	113
0820	3"		10	1.600		78	52		130	164
0830	3-1/2"		8	2		124	64.50		188.50	235
0840	4"		6	2.667		143	86.50		229.50	287
0850	5"		5	3.200		144	104		248	315
0860	6"	Q-2	7	3.429		320	115		435	525
0870	8"	"	6	4		650	134		784	915

MECHANICAL 15

15107 | Metal Pipe & Fittings

		DAILY	LABOR-		2002 BARE COSTS				TOTAL
		CREW	OUTPUT	HOURS	UNIT	MAT.	LABOR	EQUIP.	TOTAL	INCL O&P

		CREW	OUTPUT	HOURS	UNIT	MAT.	LABOR	EQUIP.	TOTAL	INCL O&P
0900	45° Elbow, straight									
0910	1/4"	1 Plum	16	.500	Ea.	6.95	18		24.95	34.50
0920	3/8"		16	.500		6.95	18		24.95	34.50
0930	1/2"		15	.533		6.95	19.15		26.10	36.50
0940	3/4"		14	.571		8	20.50		28.50	40
0950	1"		13	.615		10.10	22		32.10	44.50
0960	1-1/4"	Q-1	22	.727		13.95	23.50		37.45	51
0970	1-1/2"		20	.800		20.50	26		46.50	62
0980	2"		18	.889		29	29		58	75.50
0990	2-1/2"		14	1.143		57	37		94	119
1000	3"		10	1.600		91.50	52		143.50	178
1010	3-1/2"		8	2		168	64.50		232.50	282
1020	4"		6	2.667		168	86.50		254.50	315
1030	5"		5	3.200		207	104		311	385
1040	6"	Q-2	7	3.429		266	115		381	465
1050	8"	"	6	4		605	134		739	865
1100	Tee, straight									
1110	1/4"	1 Plum	10	.800	Ea.	7.80	29		36.80	52
1120	3/8"		10	.800		7.80	29		36.80	52
1130	1/2"		9	.889		6.75	32		38.75	55.50
1140	3/4"		9	.889		9.85	32		41.85	59
1150	1"		8	1		10.65	36		46.65	66
1160	1-1/4"	Q-1	14	1.143		18.65	37		55.65	76.50
1170	1-1/2"		13	1.231		24.50	40		64.50	87
1180	2"		11	1.455		31	47		78	105
1190	2-1/2"		9	1.778		66.50	57.50		124	160
1200	3"		6	2.667		172	86.50		258.50	320
1210	3-1/2"		5	3.200		176	104		280	350
1220	4"		4	4		198	129		327	410
1230	5"		3	5.333		249	173		422	535
1240	6"	Q-2	4	6		299	201		500	635
1250	8"	"	3	8		680	268		948	1,150
1300	Extra heavy weight, black									
1310	Couplings, steel straight									
1320	1/4"	1 Plum	19	.421	Ea.	2.86	15.15		18.01	26
1330	3/8"		19	.421		3.12	15.15		18.27	26.50
1340	1/2"		19	.421		4.22	15.15		19.37	27.50
1350	3/4"		18	.444		4.50	16		20.50	29
1360	1"		15	.533		5.75	19.15		24.90	35.50
1370	1-1/4"	Q-1	26	.615		9.20	19.90		29.10	40
1380	1-1/2"		24	.667		9.20	21.50		30.70	42.50
1390	2"		21	.762		14	24.50		38.50	52.50
1400	2-1/2"		18	.889		21	29		50	66.50
1410	3"		14	1.143		24.50	37		61.50	83
1420	3-1/2"		12	1.333		33	43		76	102
1430	4"		10	1.600		39	52		91	121
1440	5"		6	2.667		55.50	86.50		142	192
1450	6"	Q-2	8	3		68.50	101		169.50	227
1460	8"		7	3.429		74	115		189	255
1470	10"		6	4		116	134		250	330
1480	12"		4	6		155	201		356	475
1510	90° Elbow, straight									
1520	1/2"	1 Plum	15	.533	Ea.	10.60	19.15		29.75	40.50
1530	3/4"		14	.571		11.20	20.50		31.70	43.50
1540	1"		13	.615		13.55	22		35.55	48.50
1550	1-1/4"	Q-1	22	.727		20	23.50		43.50	57.50

Important: See the Reference Section for critical supporting data - Reference Nos., Crews, & City Cost Indexes

15107	Metal Pipe & Fittings	CREW	DAILY OUTPUT	LABOR-HOURS	UNIT	2002 BARE COSTS				TOTAL INCL O&P	
						MAT.	LABOR	EQUIP.	TOTAL		
640	1560	1-1/2"	Q-1	20	.800	Ea.	25	26		51	66.50	**640**
1580	2"		18	.889		31	29		60	77.50	
1590	2-1/2"		14	1.143		60	37		97	122	
1600	3"		10	1.600		100	52		152	188	
1610	4"		6	2.667		167	86.50		253.50	315	
1620	6"	Q-2	7	3.429		435	115		550	650	
1650	45° Elbow, straight										
1660	1/2"	1 Plum	15	.533	Ea.	12.60	19.15		31.75	43	
1670	3/4"		14	.571		14.25	20.50		34.75	46.50	
1680	1"		13	.615		17.10	22		39.10	52.50	
1690	1-1/4"	Q-1	22	.727		24	23.50		47.50	62	
1700	1-1/2"		20	.800		31	26		57	73	
1710	2"		18	.889		44	29		73	92	
1720	2-1/2"		14	1.143		66.50	37		103.50	129	
1800	Tee, straight										
1810	1/2"	1 Plum	9	.889	Ea.	16.35	32		48.35	66	
1820	3/4"		9	.889		16.70	32		48.70	66.50	
1830	1"		8	1		20	36		56	76	
1840	1-1/4"	Q-1	14	1.143		30	37		67	89	
1850	1-1/2"		13	1.231		38.50	40		78.50	103	
1860	2"		11	1.455		48	47		95	124	
1870	2-1/2"		9	1.778		85.50	57.50		143	181	
1880	3"		6	2.667		148	86.50		234.50	293	
1890	4"		4	4		233	129		362	450	
1900	6"	Q-2	4	6		515	201		716	870	
4000	Standard weight, black										
4010	Couplings, steel straight, merchants										
4030	1/4"	1 Plum	19	.421	Ea.	.70	15.15		15.85	24	
4040	3/8"		19	.421		.84	15.15		15.99	24	
4050	1/2"		19	.421		.90	15.15		16.05	24	
4060	3/4"		18	.444		1.14	16		17.14	25.50	
4070	1"		15	.533		1.60	19.15		20.75	31	
4080	1-1/4"	Q-1	26	.615		2.09	19.90		21.99	32.50	
4090	1-1/2"		24	.667		2.59	21.50		24.09	35.50	
4100	2"		21	.762		3.70	24.50		28.20	41	
4110	2-1/2"		18	.889		10.55	29		39.55	55	
4120	3"		14	1.143		14.80	37		51.80	72.50	
4130	3-1/2"		12	1.333		26	43		69	94	
4140	4"		10	1.600		26	52		78	107	
4150	5"		6	2.667		48	86.50		134.50	183	
4160	6"	Q-2	8	3		57.50	101		158.50	216	
4166	Plug, 1/4"	1 Plum	38	.211		1.15	7.55		8.70	12.65	
4167	3/8"		38	.211		1.07	7.55		8.62	12.60	
4168	1/2"		38	.211		.93	7.55		8.48	12.40	
4169	3/4"		32	.250		1.18	9		10.18	14.85	
4170	1"		30	.267		1.88	9.60		11.48	16.50	
4171	1-1/4"	Q-1	52	.308		2.64	9.95		12.59	17.90	
4172	1-1/2"		48	.333		3.10	10.80		13.90	19.65	
4173	2"		42	.381		3.10	12.30		15.40	22	
4176	2-1/2"		36	.444		7.55	14.40		21.95	30	
4200	Standard weight, galvanized										
4210	Couplings, steel straight, merchants										
4230	1/4"	1 Plum	19	.421	Ea.	1.46	15.15		16.61	24.50	
4240	3/8"		19	.421		1.61	15.15		16.76	25	
4250	1/2"		19	.421		3.21	15.15		18.36	26.50	
4260	3/4"		18	.444		3.44	16		19.44	28	

MECHANICAL 15

			DAILY	LABOR-			2002 BARE COSTS				TOTAL	
15107	**Metal Pipe & Fittings**	CREW	OUTPUT	HOURS	UNIT	MAT.	LABOR	EQUIP.	TOTAL	INCL O&P	
640	4270	1"	1 Plum	15	.533	Ea.	4.43	19.15		23.58	34	**640**
	4280	1-1/4"	Q-1	26	.615		6.15	19.90		26.05	37	
	4290	1-1/2"		24	.667		6.15	21.50		27.65	39.50	
	4300	2"		21	.762		10.45	24.50		34.95	48.50	
	4310	2-1/2"		18	.889		9.70	29		38.70	54	
	4320	3"		14	1.143		12.90	37		49.90	70	
	4330	3-1/2"		12	1.333		22.50	43		65.50	90	
	4340	4"		10	1.600		22.50	52		74.50	103	
	4350	5"	▼	6	2.667		42.50	86.50		129	177	
	4360	6"	Q-2	8	3	▼	52	101		153	210	
	4370	Plug, galvanized, square head										
	4374	1/2"	1 Plum	38	.211	Ea.	2	7.55		9.55	13.60	
	4375	3/4"		32	.250		2.17	9		11.17	15.95	
	4376	1"	▼	30	.267		2.17	9.60		11.77	16.85	
	4377	1-1/4"	Q-1	52	.308		3.62	9.95		13.57	19	
	4378	1-1/2"		48	.333		4.89	10.80		15.69	21.50	
	4379	2"		42	.381		6.05	12.30		18.35	25.50	
	4380	2-1/2"		36	.444		12.85	14.40		27.25	35.50	
	4381	3"		28	.571		16.45	18.50		34.95	46	
	4382	4"		20	.800		40	26		66	83	
	4384	6"	▼	8	2	▼	75.50	64.50		140	181	
	5000	Malleable iron, 150 lb.										
	5020	Black										
	5040	90° elbow, straight										
	5060	1/4"	1 Plum	16	.500	Ea.	1.79	18		19.79	29	
	5070	3/8"		16	.500		1.79	18		19.79	29	
	5080	1/2"		15	.533		1.16	19.15		20.31	30.50	
	5090	3/4"		14	.571		1.38	20.50		21.88	32.50	
	5100	1"	▼	13	.615		2.45	22		24.45	36	
	5110	1-1/4"	Q-1	22	.727		4.02	23.50		27.52	40	
	5120	1-1/2"		20	.800		5.30	26		31.30	45	
	5130	2"		18	.889		9.15	29		38.15	53.50	
	5140	2-1/2"		14	1.143		20.50	37		57.50	78.50	
	5150	3"		10	1.600		30	52		82	111	
	5160	3-1/2"		8	2		64	64.50		128.50	168	
	5170	4"		6	2.667		64	86.50		150.50	201	
	5180	5"	▼	5	3.200		197	104		301	375	
	5190	6"	Q-2	7	3.429	▼	197	115		312	390	
	5250	45° elbow, straight										
	5270	1/4"	1 Plum	16	.500	Ea.	2.53	18		20.53	30	
	5280	3/8"		16	.500		2.53	18		20.53	30	
	5290	1/2"		15	.533		1.92	19.15		21.07	31	
	5300	3/4"		14	.571		2.37	20.50		22.87	33.50	
	5310	1"	▼	13	.615		3	22		25	37	
	5320	1-1/4"	Q-1	22	.727		5.30	23.50		28.80	41.50	
	5330	1-1/2"		20	.800		6.55	26		32.55	46	
	5340	2"		18	.889		9.90	29		38.90	54.50	
	5350	2-1/2"		14	1.143		28.50	37		65.50	87.50	
	5360	3"		10	1.600		37.50	52		89.50	119	
	5370	3-1/2"		8	2		73	64.50		137.50	178	
	5380	4"		6	2.667		73	86.50		159.50	211	
	5390	5"	▼	5	3.200		262	104		366	445	
	5400	6"	Q-2	7	3.429	▼	262	115		377	460	
	5450	Tee, straight										
	5470	1/4"	1 Plum	10	.800	Ea.	2.45	29		31.45	46	
	5480	3/8"	▼	10	.800	▼	2.45	29		31.45	46	

Important: See the Reference Section for critical supporting data - Reference Nos., Crews, & City Cost Indexes

640

15107	Metal Pipe & Fittings	CREW	DAILY OUTPUT	LABOR-HOURS	UNIT	2002 BARE COSTS				TOTAL INCL O&P
						MAT.	LABOR	EQUIP.	TOTAL	
5490	1/2"	1 Plum	9	.889	Ea.	1.56	32		33.56	49.50
5500	3/4"		9	.889		2.24	32		34.24	50.50
5510	1"		8	1		3.82	36		39.82	58
5520	1-1/4"	Q-1	14	1.143		6.20	37		43.20	63
5530	1-1/2"		13	1.231		7.70	40		47.70	68.50
5540	2"		11	1.455		13.15	47		60.15	85.50
5550	2-1/2"		9	1.778		28.50	57.50		86	118
5560	3"		6	2.667		41.50	86.50		128	176
5570	3-1/2"		5	3.200		101	104		205	267
5580	4"		4	4		101	129		230	305
5590	5"		3	5.333		225	173		398	505
5600	6"	Q-2	4	6		291	201		492	625
5601	Tee, reducing, on outlet									
5602	1/2"	1 Plum	9	.889	Ea.	3.40	32		35.40	51.50
5603	3/4"		9	.889		4.25	32		36.25	52.50
5604	1"		8	1		4.36	36		40.36	59
5605	1-1/4"	Q-1	14	1.143		7.55	37		44.55	64.50
5606	1-1/2"		13	1.231		10.85	40		50.85	72
5607	2"		11	1.455		15.80	47		62.80	88.50
5608	2-1/2"		9	1.778		31.50	57.50		89	122
5609	3"		6	2.667		62	86.50		148.50	199
5610	3-1/2"		5	3.200		113	104		217	280
5611	4"		4	4		110	129		239	315
5612	5"		3	5.333		137	173		310	410
5650	Coupling									
5670	1/4"	1 Plum	19	.421	Ea.	2.08	15.15		17.23	25.50
5680	3/8"		19	.421		2.08	15.15		17.23	25.50
5690	1/2"		19	.421		1.61	15.15		16.76	25
5700	3/4"		18	.444		2	16		18	26
5710	1"		15	.533		2.82	19.15		21.97	32
5720	1-1/4"	Q-1	26	.615		3.66	19.90		23.56	34
5730	1-1/2"		24	.667		4.94	21.50		26.44	38
5740	2"		21	.762		7.30	24.50		31.80	45
5750	2-1/2"		18	.889		20	29		49	65.50
5760	3"		14	1.143		27.50	37		64.50	86
5770	3-1/2"		12	1.333		55	43		98	126
5780	4"		10	1.600		55	52		107	139
5790	5"		6	2.667		114	86.50		200.50	255
5800	6"	Q-2	8	3		114	101		215	277
5840	Reducer, concentric, 1/4"	1 Plum	19	.421		2.20	15.15		17.35	25.50
5850	3/8"		19	.421		2.20	15.15		17.35	25.50
5860	1/2"		19	.421		2	15.15		17.15	25
5870	3/4"		16	.500		2.12	18		20.12	29.50
5880	1"		15	.533		3.37	19.15		22.52	32.50
5890	1-1/4"	Q-1	26	.615		4.97	19.90		24.87	35.50
5900	1-1/2"		24	.667		6.20	21.50		27.70	39.50
5910	2"		21	.762		8.90	24.50		33.40	47
5911	2-1/2"		18	.889		21	29		50	66.50
5912	3"		14	1.143		28	37		65	86.50
5913	3-1/2"		12	1.333		71	43		114	143
5914	4"		10	1.600		56	52		108	140
5915	5"		6	2.667		112	86.50		198.50	253
5916	6"	Q-2	8	3		143	101		244	310
5981	Bushing, 1/4"	1 Plum	19	.421		2.09	15.15		17.24	25.50
5982	3/8"		19	.421		2.09	15.15		17.24	25.50
5983	1/2"		19	.421		2.09	15.15		17.24	25.50

640

MECHANICAL **15**

15107	Metal Pipe & Fittings	CREW	DAILY OUTPUT	LABOR-HOURS	UNIT	2002 BARE COSTS				TOTAL INCL O&P
						MAT.	LABOR	EQUIP.	TOTAL	
640 5984	3/4"	1 Plum	16	.500	Ea.	2.09	18		20.09	29.50 **640**
5985	1"	↓	15	.533		2.73	19.15		21.88	32
5986	1-1/4"	Q-1	26	.615		3.81	19.90		23.71	34
5987	1-1/2"		24	.667		4.56	21.50		26.06	37.50
5988	2"	↓	21	.762		4.95	24.50		29.45	42.50
5989	Cap, 1/4"	1 Plum	38	.211		1.54	7.55		9.09	13.10
5991	3/8"		38	.211		1.30	7.55		8.85	12.85
5992	1/2"		38	.211		1.17	7.55		8.72	12.70
5993	3/4"		32	.250		1.59	9		10.59	15.30
5994	1"	↓	30	.267		1.93	9.60		11.53	16.55
5995	1-1/4"	Q-1	52	.308		2.54	9.95		12.49	17.80
5996	1-1/2"		48	.333		3.49	10.80		14.29	20
5997	2"	↓	42	.381		5.10	12.30		17.40	24
6000	For galvanized elbows, tees, and couplings add				↓	20%				
6058	For galvanized reducers, caps and plugs add					20%				
6100	90° elbow, galvanized, 150 lb., reducing									
6110	3/4" x 1/2"	1 Plum	15.40	.519	Ea.	3.37	18.70		22.07	31.50
6112	1" x 3/4"		14	.571		4.32	20.50		24.82	36
6114	1" x 1/2"	↓	14.50	.552		4.58	19.85		24.43	35
6116	1-1/4" x 1"	Q-1	24.20	.661		7.25	21.50		28.75	40.50
6118	1-1/4" x 3/4"		25.40	.630		8.70	20.50		29.20	40
6120	1-1/4" x 1/2"		26.20	.611		9.25	19.75		29	40
6122	1-1/2" x 1-1/4"		21.60	.741		11.60	24		35.60	49
6124	1-1/2" x 1"		23.50	.681		11.60	22		33.60	46
6126	1-1/2" x 3/4"		24.60	.650		11.60	21		32.60	44.50
6128	2" x 1-1/2"		20.50	.780		12.10	25.50		37.60	51.50
6130	2" x 1-1/4"		21	.762		15.50	24.50		40	54
6132	2" x 1"		22.80	.702		16.10	22.50		38.60	51.50
6134	2" x 3/4"		23.90	.669		16.40	21.50		37.90	50.50
6136	2-1/2" x 2"		12.30	1.301		46	42		88	114
6138	2-1/2" x 1-1/2"		12.50	1.280		46	41.50		87.50	113
6140	3" x 2-1/2"		8.60	1.860		74	60		134	173
6142	3" x 2"		11.80	1.356		72.50	44		116.50	146
6144	4" x 3"	↓	8.20	1.951	↓	143	63		206	252
6160	90° elbow, black, 150 lb., reducing									
6170	1" x 3/4"	1 Plum	14	.571	Ea.	3.12	20.50		23.62	34.50
6174	1-1/2" x 1"	Q-1	23.50	.681		7.05	22		29.05	41
6178	1-1/2" x 3/4"		24.60	.650		8.45	21		29.45	41
6182	2" x 1-1/2"		20.50	.780		10.70	25.50		36.20	50
6186	2" x 1"		22.80	.702		12.20	22.50		34.70	47.50
6190	2" x 3/4"		23.90	.669		12.90	21.50		34.40	46.50
6194	2-1/2" x 2"	↓	12.30	1.301	↓	29.50	42		71.50	96
7000	Union, with brass seat									
7010	1/4"	1 Plum	15	.533	Ea.	9.60	19.15		28.75	39.50
7020	3/8"		15	.533		9.60	19.15		28.75	39.50
7030	1/2"		14	.571		6	20.50		26.50	37.50
7040	3/4"		13	.615		6.30	22		28.30	40.50
7050	1"	↓	12	.667		9	24		33	46
7060	1-1/4"	Q-1	21	.762		12	24.50		36.50	50
7070	1-1/2"		19	.842		16.10	27		43.10	58.50
7080	2"		17	.941		18.70	30.50		49.20	66.50
7090	2-1/2"		13	1.231		55.50	40		95.50	122
7100	3"	↓	9	1.778	↓	67	57.50		124.50	161
7120	Union, galvanized									
7124	1/2"	1 Plum	14	.571	Ea.	6.85	20.50		27.35	38.50
7125	3/4"	↓	13	.615	↓	7.90	22		29.90	42

		CREW	DAILY OUTPUT	LABOR-HOURS	UNIT	2002 BARE COSTS				TOTAL INCL O&P	
15107	**Metal Pipe & Fittings**					MAT.	LABOR	EQUIP.	TOTAL		
640 7126	1"	1 Plum	12	.667	Ea.	10.35	24		34.35	47.50	**640**
7127	1-1/4"	Q-1	21	.762		14.95	24.50		39.45	53.50	
7128	1-1/2"		19	.842		18.10	27		45.10	61	
7129	2"		17	.941		21	30.50		51.50	69	
7130	2-1/2"		13	1.231		72.50	40		112.50	140	
7131	3"		9	1.778		101	57.50		158.50	198	
7500	Malleable iron, 300 lb										
7520	Black										
7540	90° Elbow, straight, 1/4"	1 Plum	16	.500	Ea.	6.90	18		24.90	34.50	
7560	3/8"		16	.500		6.10	18		24.10	33.50	
7570	1/2"		15	.533		7.90	19.15		27.05	37.50	
7580	3/4"		14	.571		8.95	20.50		29.45	41	
7590	1"		13	.615		11.45	22		33.45	46	
7600	1-1/4"	Q-1	22	.727		16.70	23.50		40.20	54	
7610	1-1/2"		20	.800		19.70	26		45.70	60.50	
7620	2"		18	.889		28.50	29		57.50	74.50	
7630	2-1/2"		14	1.143		67	37		104	130	
7640	3"		10	1.600		84.50	52		136.50	171	
7650	4"		6	2.667		176	86.50		262.50	325	
7700	45° Elbow, straight, 1/4"	1 Plum	16	.500		10.25	18		28.25	38.50	
7720	3/8"		16	.500		10.25	18		28.25	38.50	
7730	1/2"		15	.533		11.40	19.15		30.55	41.50	
7740	3/4"		14	.571		12.70	20.50		33.20	45	
7750	1"		13	.615		14.05	22		36.05	49	
7760	1-1/4"	Q-1	22	.727		22.50	23.50		46	60	
7770	1-1/2"		20	.800		29.50	26		55.50	71	
7780	2"		18	.889		44.50	29		73.50	92	
7790	2-1/2"		14	1.143		88	37		125	153	
7800	3"		10	1.600		115	52		167	205	
7810	4"		6	2.667		223	86.50		309.50	375	
7850	Tee, straight, 1/4"	1 Plum	10	.800		8.65	29		37.65	53	
7870	3/8"		10	.800		9.20	29		38.20	53.50	
7880	1/2"		9	.889		11.70	32		43.70	61	
7890	3/4"		9	.889		12.60	32		44.60	62	
7900	1"		8	1		15.15	36		51.15	70.50	
7910	1-1/4"	Q-1	14	1.143		21.50	37		58.50	80	
7920	1-1/2"		13	1.231		25.50	40		65.50	88	
7930	2"		11	1.455		37.50	47		84.50	112	
7940	2-1/2"		9	1.778		81	57.50		138.50	176	
7950	3"		6	2.667		130	86.50		216.50	273	
7960	4"		4	4		279	129		408	500	
7980	6"	Q-2	4	6		296	201		497	630	
7990	8"	"	3	8		465	268		733	915	
8050	Couplings, straight, 1/4"	1 Plum	19	.421		6.20	15.15		21.35	30	
8070	3/8"		19	.421		6.20	15.15		21.35	30	
8080	1/2"		19	.421		6.95	15.15		22.10	30.50	
8090	3/4"		18	.444		7.95	16		23.95	33	
8100	1"		15	.533		9.10	19.15		28.25	39	
8110	1-1/4"	Q-1	26	.615		10.85	19.90		30.75	42	
8120	1-1/2"		24	.667		16.20	21.50		37.70	50.50	
8130	2"		21	.762		23.50	24.50		48	62.50	
8140	2-1/2"		18	.889		38	29		67	85.50	
8150	3"		14	1.143		55	37		92	117	
8160	4"		10	1.600		69	52		121	154	
8200	Galvanized										
8220	90° Elbow, straight, 1/4"	1 Plum	16	.500	Ea.	11.65	18		29.65	40	

MECHANICAL **15**

640

15107	Metal Pipe & Fittings	CREW	DAILY OUTPUT	LABOR-HOURS	UNIT	MAT.	LABOR	EQUIP.	TOTAL	TOTAL INCL O&P
						2002 BARE COSTS				
8222	3/8"	1 Plum	16	.500	Ea.	11.90	18		29.90	40
8224	1/2"		15	.533		14.15	19.15		33.30	44.50
8226	3/4"		14	.571		15.90	20.50		36.40	48.50
8228	1"		13	.615		20.50	22		42.50	56
8230	1-1/4"	Q-1	22	.727		32	23.50		55.50	70.50
8232	1-1/2"		20	.800		34.50	26		60.50	77
8234	2"		18	.889		58.50	29		87.50	108
8236	2-1/2"		14	1.143		114	37		151	181
8238	3"		10	1.600		153	52		205	246
8240	4"		6	2.667		345	86.50		431.50	510
8280	45° Elbow, straight									
8282	1/2"	1 Plum	15	.533	Ea.	21	19.15		40.15	52
8284	3/4"		14	.571		24.50	20.50		45	57.50
8286	1"		13	.615		26.50	22		48.50	62.50
8288	1-1/4"	Q-1	22	.727		41	23.50		64.50	81
8290	1-1/2"		20	.800		53.50	26		79.50	97.50
8292	2"		18	.889		74	29		103	125
8310	Tee, straight, 1/4"	1 Plum	10	.800		16.55	29		45.55	61.50
8312	3/8"		10	.800		16.90	29		45.90	62
8314	1/2"		9	.889		21	32		53	71
8316	3/4"		9	.889		23.50	32		55.50	73.50
8318	1"		8	1		28.50	36		64.50	85.50
8320	1-1/4"	Q-1	14	1.143		41	37		78	101
8322	1-1/2"		13	1.231		42.50	40		82.50	107
8324	2"		11	1.455		68	47		115	146
8326	2-1/2"		9	1.778		163	57.50		220.50	266
8328	3"		6	2.667		244	86.50		330.50	400
8330	4"		4	4		475	129		604	715
8380	Couplings, straight, 1/4"	1 Plum	19	.421		8.90	15.15		24.05	33
8382	3/8"		19	.421		8.90	15.15		24.05	33
8384	1/2"		19	.421		9.15	15.15		24.30	33
8386	3/4"		18	.444		10.35	16		26.35	35.50
8388	1"		15	.533		13.65	19.15		32.80	44
8390	1-1/4"	Q-1	26	.615		17.80	19.90		37.70	49.50
8392	1-1/2"		24	.667		28.50	21.50		50	64
8394	2"		21	.762		34.50	24.50		59	75
8396	2-1/2"		18	.889		68.50	29		97.50	119
8398	3"		14	1.143		96	37		133	161
8399	4"		10	1.600		102	52		154	190
8529	Black									
8530	Reducer, concentric, 1/4"	1 Plum	19	.421	Ea.	7.55	15.15		22.70	31.50
8531	3/8"		19	.421		7.55	15.15		22.70	31.50
8532	1/2"		17	.471		9.10	16.90		26	35.50
8533	3/4"		16	.500		9.30	18		27.30	37
8534	1"		15	.533		11.10	19.15		30.25	41
8535	1-1/4"	Q-1	26	.615		15.15	19.90		35.05	46.50
8536	1-1/2"		24	.667		21	21.50		42.50	55.50
8537	2"		21	.762		29	24.50		53.50	69
8550	Cap, 1/4"	1 Plum	38	.211		5.85	7.55		13.40	17.80
8551	3/8"		38	.211		5.85	7.55		13.40	17.80
8552	1/2"		34	.235		5.80	8.45		14.25	19.15
8553	3/4"		32	.250		7.30	9		16.30	21.50
8554	1"		30	.267		9.60	9.60		19.20	25
8555	1-1/4"	Q-1	52	.308		10.65	9.95		20.60	26.50
8556	1-1/2"		48	.333		15.55	10.80		26.35	33.50
8557	2"		42	.381		21.50	12.30		33.80	42.50

640

15 MECHANICAL

Important: See the Reference Section for critical supporting data - Reference Nos., Crews, & City Cost Indexes

			DAILY	LABOR-		2002 BARE COSTS				TOTAL	
15107	**Metal Pipe & Fittings**	CREW	OUTPUT	HOURS	UNIT	MAT.	LABOR	EQUIP.	TOTAL	INCL O&P	
640	8570	Plug, 1/4"	1 Plum	38	.211	Ea.	.12	7.55		7.67	11.55	**640**
	8571	3/8"		38	.211		.15	7.55		7.70	11.55	
	8572	1/2"		34	.235		.23	8.45		8.68	13	
	8573	3/4"		32	.250		.37	9		9.37	13.95	
	8574	1"	↓	30	.267		.77	9.60		10.37	15.30	
	8575	1-1/4"	Q-1	52	.308		1.52	9.95		11.47	16.65	
	8576	1-1/2"		48	.333		2.77	10.80		13.57	19.30	
	8577	2"	↓	42	.381		3.88	12.30		16.18	23	
	9500	Union with brass seat, 1/4"	1 Plum	15	.533		12.80	19.15		31.95	43	
	9530	3/8"		15	.533		13	19.15		32.15	43.50	
	9540	1/2"		14	.571		10.50	20.50		31	42.50	
	9550	3/4"		13	.615		11.60	22		33.60	46.50	
	9560	1"	↓	12	.667		15.20	24		39.20	52.50	
	9570	1-1/4"	Q-1	21	.762		25	24.50		49.50	64.50	
	9580	1-1/2"		19	.842		26	27		53	69.50	
	9590	2"		17	.941		32	30.50		62.50	81.50	
	9600	2-1/2"		13	1.231		104	40		144	174	
	9610	3"		9	1.778		134	57.50		191.50	234	
	9620	4"	↓	5	3.200		330	104		434	520	
	9630	Union, all iron, 1/4"	1 Plum	15	.533		22	19.15		41.15	53	
	9650	3/8"		15	.533		23	19.15		42.15	54	
	9660	1/2"		14	.571		23	20.50		43.50	56	
	9670	3/4"		13	.615		24.50	22		46.50	60.50	
	9680	1"	↓	12	.667		30.50	24		54.50	69.50	
	9690	1-1/4"	Q-1	21	.762		46.50	24.50		71	88	
	9700	1-1/2"		19	.842		56.50	27		83.50	103	
	9710	2"		17	.941		71.50	30.50		102	125	
	9720	2-1/2"		13	1.231		261	40		301	345	
	9730	3"	↓	9	1.778		261	57.50		318.50	375	
	9750	For galvanized unions, add				↓	15%					
	9757	Forged steel, 3000 lb.										
	9758	Black										
	9760	90° Elbow, 1/4"	1 Plum	16	.500	Ea.	8.10	18		26.10	36	
	9761	3/8"		16	.500		11.15	18		29.15	39.50	
	9762	1/2"		15	.533		8.25	19.15		27.40	38	
	9763	3/4"		14	.571		9.90	20.50		30.40	42	
	9764	1"	↓	13	.615		14.35	22		36.35	49.50	
	9765	1-1/4"	Q-1	22	.727		28	23.50		51.50	66.50	
	9766	1-1/2"		20	.800		36.50	26		62.50	79	
	9767	2"	↓	18	.889		44	29		73	92	
	9780	45° Elbow 1/4"	1 Plum	16	.500		12.20	18		30.20	40.50	
	9781	3/8"		16	.500		12.20	18		30.20	40.50	
	9782	1/2"		15	.533		12.20	19.15		31.35	42.50	
	9783	3/4"		14	.571		14.10	20.50		34.60	46.50	
	9784	1"	↓	13	.615		19.25	22		41.25	54.50	
	9785	1-1/4"	Q-1	22	.727		26.50	23.50		50	65	
	9786	1-1/2"		20	.800		40	26		66	83	
	9787	2"	↓	18	.889		55	29		84	104	
	9800	Tee, 1/4"	1 Plum	10	.800		10.90	29		39.90	55.50	
	9801	3/8"		10	.800		12.60	29		41.60	57.50	
	9802	1/2"		9	.889		10.90	32		42.90	60	
	9803	3/4"		9	.889		15.55	32		47.55	65	
	9804	1"	↓	8	1		20	36		56	76	
	9805	1-1/4"	Q-1	14	1.143		38	37		75	98	
	9806	1-1/2"		13	1.231		45	40		85	110	
	9807	2"	↓	11	1.455	↓	60	47		107	137	

MECHANICAL 15

			DAILY	LABOR-			2002 BARE COSTS				TOTAL	
15107	**Metal Pipe & Fittings**	CREW	OUTPUT	HOURS	UNIT	MAT.	LABOR	EQUIP.	TOTAL	INCL O&P	
640	9820	Reducer, concentric, 1/4"	1 Plum	19	.421	Ea.	6.60	15.15		21.75	30.50	**640**
	9821	3/8"		19	.421		6.60	15.15		21.75	30.50	
	9822	1/2"		17	.471		6.60	16.90		23.50	33	
	9823	3/4"		16	.500		7.40	18		25.40	35	
	9824	1"		15	.533		10.15	19.15		29.30	40	
	9825	1-1/4"	Q-1	26	.615		15.90	19.90		35.80	47.50	
	9826	1-1/2"		24	.667		18.50	21.50		40	53	
	9827	2"		21	.762		26.50	24.50		51	66.50	
	9840	Cap, 1/4"	1 Plum	38	.211		4.07	7.55		11.62	15.90	
	9841	3/8"		38	.211		4.07	7.55		11.62	15.90	
	9842	1/2"		34	.235		4.07	8.45		12.52	17.25	
	9843	3/4"		32	.250		5.65	9		14.65	19.80	
	9844	1"		30	.267		8.75	9.60		18.35	24	
	9845	1-1/4"	Q-1	52	.308		13.10	9.95		23.05	29.50	
	9846	1-1/2"		48	.333		15.15	10.80		25.95	33	
	9847	2"		42	.381		23.50	12.30		35.80	44	
	9860	Plug, 1/4"	1 Plum	38	.211		1.27	7.55		8.82	12.80	
	9861	3/8"		38	.211		1.58	7.55		9.13	13.15	
	9862	1/2"		34	.235		1.58	8.45		10.03	14.50	
	9863	3/4"		32	.250		2.75	9		11.75	16.60	
	9864	1"		30	.267		3.61	9.60		13.21	18.40	
	9865	1-1/4"	Q-1	52	.308		6.25	9.95		16.20	22	
	9866	1-1/2"		48	.333		8.10	10.80		18.90	25	
	9867	2"		42	.381		15.15	12.30		27.45	35.50	
	9880	Union, bronze seat, 1/4"	1 Plum	15	.533		14.60	19.15		33.75	45	
	9881	3/8"		15	.533		14.60	19.15		33.75	45	
	9882	1/2"		14	.571		14.60	20.50		35.10	47	
	9883	3/4"		13	.615		17.55	22		39.55	53	
	9884	1"		12	.667		23	24		47	61	
	9885	1-1/4"	Q-1	21	.762		38.50	24.50		63	79.50	
	9886	1-1/2"		19	.842		40	27		67	85	
	9887	2"		17	.941		51.50	30.50		82	103	
	9900	Coupling, 1/4"	1 Plum	19	.421		3.03	15.15		18.18	26.50	
	9901	3/8"		19	.421		3.37	15.15		18.52	26.50	
	9902	1/2"		17	.471		2.87	16.90		19.77	28.50	
	9903	3/4"		16	.500		4.07	18		22.07	31.50	
	9904	1"		15	.533		5.95	19.15		25.10	35.50	
	9905	1-1/4"	Q-1	26	.615		11.05	19.90		30.95	42	
	9906	1-1/2"		24	.667		15.15	21.50		36.65	49	
	9907	2"		21	.762		18.85	24.50		43.35	58	
660	0010	**PIPE, STEEL, FITTINGS** Flanged, welded and special type										**660**
	0020	flanged joints, C.I., standard weight, black. One gasket & bolt										
	0040	set, mat'l only, required at each joint, not included (see line 0620)										
	0060	90° Elbow, straight, 1-1/2" pipe size	Q-1	14	1.143	Ea.	143	37		180	214	
	0080	2" pipe size		13	1.231		112	40		152	184	
	0090	2-1/2" pipe size		12	1.333		122	43		165	199	
	0100	3" pipe size		11	1.455		101	47		148	183	
	0110	4" pipe size		8	2		126	64.50		190.50	236	
	0120	5" pipe size		7	2.286		231	74		305	365	
	0130	6" pipe size	Q-2	9	2.667		197	89.50		286.50	350	
	0140	8" pipe size		8	3		310	101		411	490	
	0150	10" pipe size		7	3.429		680	115		795	920	
	0160	12" pipe size		6	4		1,175	134		1,309	1,475	
	0171	90° Elbow, reducing										
	0172	2-1/2" by 2" pipe size	Q-1	12	1.333	Ea.	278	43		321	370	
	0173	3" by 2-1/2" pipe size		11	1.455		292	47		339	390	

Important: See the Reference Section for critical supporting data - Reference Nos., Crews, & City Cost Indexes

660

15107	Metal Pipe & Fittings	CREW	DAILY OUTPUT	LABOR-HOURS	UNIT	2002 BARE COSTS MAT.	LABOR	EQUIP.	TOTAL	TOTAL INCL O&P
0174	4" by 3" pipe size	Q-1	8	2	Ea.	330	64.50		394.50	465
0175	5" by 3" pipe size		7	2.286		425	74		499	575
0176	6" by 4" pipe size	Q-2	9	2.667		475	89.50		564.50	660
0177	8" by 6" pipe size		8	3		700	101		801	920
0178	10" by 8" pipe size		7	3.429		1,100	115		1,215	1,375
0179	12" by 10" pipe size		6	4		1,900	134		2,034	2,275
0200	45° Elbow, straight, 1-1/2" pipe size	Q-1	14	1.143		173	37		210	246
0220	2" pipe size		13	1.231		172	40		212	249
0230	2-1/2" pipe size		12	1.333		173	43		216	256
0240	3" pipe size		11	1.455		169	47		216	257
0250	4" pipe size		8	2		190	64.50		254.50	305
0260	5" pipe size		7	2.286		355	74		429	500
0270	6" pipe size	Q-2	9	2.667		310	89.50		399.50	475
0280	8" pipe size		8	3		455	101		556	650
0290	10" pipe size		7	3.429		960	115		1,075	1,225
0300	12" pipe size		6	4		1,475	134		1,609	1,825
0310	Cross, straight									
0311	2-1/2" pipe size	Q-1	6	2.667	Ea.	375	86.50		461.50	545
0312	3" pipe size		5	3.200		375	104		479	570
0313	4" pipe size		4	4		490	129		619	735
0314	5" pipe size		3	5.333		800	173		973	1,150
0315	6" pipe size	Q-2	5	4.800		800	161		961	1,125
0316	8" pipe size		4	6		1,275	201		1,476	1,725
0317	10" pipe size		3	8		1,975	268		2,243	2,550
0318	12" pipe size		2	12		2,975	405		3,380	3,875
0350	Tee, straight, 1-1/2" pipe size	Q-1	10	1.600		123	52		175	213
0370	2" pipe size		9	1.778		123	57.50		180.50	222
0380	2-1/2" pipe size		8	2		180	64.50		244.50	296
0390	3" pipe size		7	2.286		126	74		200	250
0400	4" pipe size		5	3.200		192	104		296	370
0410	5" pipe size		4	4		400	129		529	635
0420	6" pipe size	Q-2	6	4		277	134		411	505
0430	8" pipe size		5	4.800		435	161		596	720
0440	10" pipe size		4	6		990	201		1,191	1,400
0450	12" pipe size		3	8		1,550	268		1,818	2,125
0459	Tee, reducing on outlet									
0460	2-1/2" by 2" pipe size	Q-1	8	2	Ea.	265	64.50		329.50	390
0461	3" by 2-1/2" pipe size		7	2.286		305	74		379	445
0462	4" by 3" pipe size		5	3.200		745	104		849	975
0463	5" by 4" pipe size		4	4		940	129		1,069	1,225
0464	6" by 4" pipe size	Q-2	6	4		940	134		1,074	1,225
0465	8" by 6" pipe size		5	4.800		940	161		1,101	1,275
0466	10" by 8" pipe size		4	6		1,025	201		1,226	1,425
0467	12" by 10" pipe size		3	8		1,600	268		1,868	2,175
0476	Reducer, concentric									
0477	3" by 2-1/2"	Q-1	12	1.333	Ea.	281	43		324	375
0478	4" by 3"		9	1.778		300	57.50		357.50	415
0479	5" by 4"		8	2		475	64.50		539.50	620
0480	6" by 4"	Q-2	10	2.400		530	80.50		610.50	705
0481	8" by 6"		9	2.667		600	89.50		689.50	795
0482	10" by 8"		8	3		960	101		1,061	1,200
0483	12" by 10"		7	3.429		1,750	115		1,865	2,100
0492	Reducer, eccentric									
0493	4" by 3"	Q-1	8	2	Ea.	370	64.50		434.50	505
0494	5" by 4"	"	7	2.286		475	74		549	630
0495	6" by 4"	Q-2	9	2.667		505	89.50		594.50	690

660

MECHANICAL **15**

15107	Metal Pipe & Fittings	CREW	DAILY OUTPUT	LABOR-HOURS	UNIT	2002 BARE COSTS				TOTAL INCL O&P
						MAT.	LABOR	EQUIP.	TOTAL	
660 0496	8" by 6"	Q-2	8	3	Ea.	665	101		766	885 **660**
0497	10" by 8"		7	3.429		1,150	115		1,265	1,425
0498	12" by 10"	↓	6	4		1,750	134		1,884	2,125
0500	For galvanized elbows and tees, add					100%				
0520	For extra heavy weight elbows and tees, add					140%				
0620	Gasket and bolt set, 150#, 1/2" pipe size	1 Plum	20	.400		2.36	14.40		16.76	24
0622	3/4" pipe size		19	.421		2.36	15.15		17.51	25.50
0624	1" pipe size		18	.444		2.42	16		18.42	26.50
0626	1-1/4" pipe size		17	.471		2.54	16.90		19.44	28.50
0628	1-1/2" pipe size		15	.533		2.71	19.15		21.86	32
0630	2" pipe size		13	.615		4.33	22		26.33	38.50
0640	2-1/2" pipe size		12	.667		4.58	24		28.58	41
0650	3" pipe size		11	.727		4.80	26		30.80	45
0660	3-1/2" pipe size		9	.889		8.70	32		40.70	57.50
0670	4" pipe size		8	1		8.70	36		44.70	63.50
0680	5" pipe size		7	1.143		12.90	41		53.90	76
0690	6" pipe size		6	1.333		13.55	48		61.55	87.50
0700	8" pipe size		5	1.600		14.45	57.50		71.95	102
0710	10" pipe size		4.50	1.778		30	64		94	130
0720	12" pipe size		4.20	1.905		32	68.50		100.50	139
0730	14" pipe size		4	2		45.50	72		117.50	158
0740	16" pipe size		3	2.667		59.50	96		155.50	211
0750	18" pipe size		2.70	2.963		83	107		190	253
0760	20" pipe size		2.30	3.478		111	125		236	310
0780	24" pipe size		1.90	4.211		149	151		300	390
0790	26" pipe size		1.60	5		168	180		348	455
0810	30" pipe size		1.40	5.714		208	205		413	540
0830	36" pipe size	↓	1.10	7.273	↓	256	261		517	675
0850	For 300 lb gasket set, add					40%				
2000	Flanged unions, 125 lb., black, 1/2" pipe size	1 Plum	17	.471	Ea.	28	16.90		44.90	56.50
2040	3/4" pipe size		17	.471		37	16.90		53.90	66
2050	1" pipe size	↓	16	.500		36	18		54	66.50
2060	1-1/4" pipe size	Q-1	28	.571		42.50	18.50		61	74.50
2070	1-1/2" pipe size		27	.593		39	19.15		58.15	72
2080	2" pipe size		26	.615		47	19.90		66.90	82
2090	2-1/2" pipe size		24	.667		63.50	21.50		85	103
2100	3" pipe size		22	.727		72	23.50		95.50	115
2110	3-1/2" pipe size		18	.889		150	29		179	209
2120	4" pipe size		16	1		98	32.50		130.50	157
2130	5" pipe size	↓	14	1.143		174	37		211	247
2140	6" pipe size	Q-2	19	1.263		210	42.50		252.50	294
2150	8" pipe size	"	16	1.500		360	50.50		410.50	475
2200	For galvanized unions, add				↓	150%				
2290	Threaded flange									
2300	Cast iron									
2310	Black, 125 lb., per flange									
2320	1" pipe size	1 Plum	27	.296	Ea.	16.40	10.65		27.05	34
2330	1-1/4" pipe size	Q-1	44	.364		19.60	11.75		31.35	39.50
2340	1-1/2" pipe size		40	.400		18.15	12.95		31.10	39.50
2350	2" pipe size		36	.444		18.10	14.40		32.50	41.50
2360	2-1/2" pipe size		28	.571		21	18.50		39.50	51.50
2370	3" pipe size		20	.800		27.50	26		53.50	69
2380	3-1/2" pipe size		16	1		36.50	32.50		69	89.50
2390	4" pipe size		12	1.333		36.50	43		79.50	106
2400	5" pipe size	↓	10	1.600		52	52		104	135
2410	6" pipe size	Q-2	14	1.714	↓	58.50	57.50		116	151

Important: See the Reference Section for critical supporting data - Reference Nos., Crews, & City Cost Indexes

15107 | Metal Pipe & Fittings

		DAILY OUTPUT	LABOR-HOURS	UNIT	2002 BARE COSTS				TOTAL INCL O&P
					MAT.	LABOR	EQUIP.	TOTAL	
660									**660**
2420	8" pipe size	Q-2 12	2	Ea.	92.50	67		159.50	203
2430	10" pipe size	10	2.400		161	80.50		241.50	298
2440	12" pipe size	8	3		267	101		368	445
2460	For galvanized flanges, add				95%				
2580	Forged steel,								
2590	Black 150 lb., per flange								
2600	1/2" pipe size	1 Plum 30	.267	Ea.	17.30	9.60		26.90	33.50
2610	3/4" pipe size	28	.286		17.30	10.25		27.55	34.50
2620	1" pipe size	27	.296		17.30	10.65		27.95	35
2630	1-1/4" pipe size	Q-1 44	.364		17.30	11.75		29.05	37
2640	1-1/2" pipe size	40	.400		17.30	12.95		30.25	38.50
2650	2" pipe size	36	.444		17.30	14.40		31.70	40.50
2660	2-1/2" pipe size	28	.571		21	18.50		39.50	51
2670	3" pipe size	20	.800		21	26		47	62
2690	4" pipe size	12	1.333		24.50	43		67.50	92
2700	5" pipe size	10	1.600		36	52		88	118
2710	6" pipe size	Q-2 14	1.714		40.50	57.50		98	131
2720	8" pipe size	12	2		69	67		136	177
2730	10" pipe size	10	2.400		125	80.50		205.50	258
2860	Black 300 lb., per flange								
2870	1/2" pipe size	1 Plum 30	.267	Ea.	19.20	9.60		28.80	35.50
2880	3/4" pipe size	28	.286		19.20	10.25		29.45	36.50
2890	1" pipe size	27	.296		19.20	10.65		29.85	37
2900	1-1/4" pipe size	Q-1 44	.364		19.20	11.75		30.95	39
2910	1-1/2" pipe size	40	.400		19.20	12.95		32.15	40.50
2920	2" pipe size	36	.444		20	14.40		34.40	43.50
2930	2-1/2" pipe size	28	.571		28	18.50		46.50	58.50
2940	3" pipe size	20	.800		30	26		56	72
2960	4" pipe size	12	1.333		42.50	43		85.50	112
2970	6" pipe size	Q-2 14	1.714		75	57.50		132.50	169
3000	Weld joint, butt, carbon steel, standard weight								
3040	90° elbow, long radius								
3050	1/2" pipe size	Q-15 16	1	Ea.	14.10	32.50	5	51.60	70
3060	3/4" pipe size	16	1		14.10	32.50	5	51.60	70
3070	1" pipe size	16	1		7.05	32.50	5	44.55	62.50
3080	1-1/4" pipe size	14	1.143		7.05	37	5.70	49.75	70
3090	1-1/2" pipe size	13	1.231		7.05	40	6.15	53.20	74.50
3100	2" pipe size	10	1.600		7.05	52	8	67.05	94.50
3110	2-1/2" pipe size	8	2		9.30	64.50	10	83.80	119
3120	3" pipe size	7	2.286		10.90	74	11.45	96.35	137
3130	4" pipe size	5	3.200		18.55	104	16	138.55	194
3136	5" pipe size	4	4		39.50	129	20	188.50	261
3140	6" pipe size	Q-16 5	4.800		44	161	16	221	310
3150	8" pipe size	3.75	6.400		82.50	215	21.50	319	440
3160	10" pipe size	3	8		151	268	26.50	445.50	600
3170	12" pipe size	2.50	9.600		217	320	32	569	760
3180	14" pipe size	2	12		299	405	40	744	980
3190	16" pipe size	1.50	16		525	535	53.50	1,113.50	1,450
3191	18" pipe size	1.25	19.200		525	645	64	1,234	1,625
3192	20" pipe size	1.15	20.870		775	700	69.50	1,544.50	1,975
3194	24" pipe size	1.02	23.529		1,025	790	78.50	1,893.50	2,400
3200	45° Elbow, long								
3210	1/2" pipe size	Q-15 16	1	Ea.	13.75	32.50	5	51.25	69.50
3220	3/4" pipe size	16	1		13.75	32.50	5	51.25	69.50
3230	1" pipe size	16	1		6.70	32.50	5	44.20	62
3240	1-1/4" pipe size	14	1.143		6.70	37	5.70	49.40	69.50

MECHANICAL 15

			DAILY	LABOR-		2002 BARE COSTS				TOTAL	
	15107	**Metal Pipe & Fittings**	CREW	OUTPUT	HOURS	UNIT	MAT.	LABOR	EQUIP.	TOTAL	INCL O&P
660	3250	1-1/2" pipe size	Q-15	13	1.231	Ea.	6.70	40	6.15	52.85	74
	3260	2" pipe size		10	1.600		6.70	52	8	66.70	94
	3270	2-1/2" pipe size		8	2		8	64.50	10	82.50	117
	3280	3" pipe size		7	2.286		9.30	74	11.45	94.75	135
	3290	4" pipe size		5	3.200		16.30	104	16	136.30	192
	3296	5" pipe size		4	4		25	129	20	174	245
	3300	6" pipe size	Q-16	5	4.800		35	161	16	212	299
	3310	8" pipe size		3.75	6.400		59	215	21.50	295.50	415
	3320	10" pipe size		3	8		107	268	26.50	401.50	550
	3330	12" pipe size		2.50	9.600		151	320	32	503	685
	3340	14" pipe size		2	12		205	405	40	650	875
	3341	16" pipe size		1.50	16		281	535	53.50	869.50	1,175
	3342	18" pipe size		1.25	19.200		375	645	64	1,084	1,450
	3343	20" pipe size		1.15	20.870		495	700	69.50	1,264.50	1,675
	3345	24" pipe size		1.05	22.857		715	765	76	1,556	2,025
	3346	26" pipe size		.85	28.235		1,475	945	94	2,514	3,150
	3347	30" pipe size		.45	53.333		1,625	1,800	178	3,603	4,700
	3349	36" pipe size		.38	63.158		2,700	2,125	210	5,035	6,400
	3350	Tee, straight									
	3352	For reducing tees and concentrics see starting line 4600									
	3360	1/2" pipe size	Q-15	10	1.600	Ea.	33	52	8	93	123
	3370	3/4" pipe size		10	1.600		33	52	8	93	123
	3380	1" pipe size		10	1.600		18.55	52	8	78.55	107
	3390	1-1/4" pipe size		9	1.778		21	57.50	8.90	87.40	119
	3400	1-1/2" pipe size		8	2		21	64.50	10	95.50	132
	3410	2" pipe size		6	2.667		18.55	86.50	13.35	118.40	165
	3420	2-1/2" pipe size		5	3.200		23.50	104	16	143.50	199
	3430	3" pipe size		4	4		25.50	129	20	174.50	245
	3440	4" pipe size		3	5.333		36.50	173	26.50	236	330
	3446	5" pipe size		2.50	6.400		66	207	32	305	420
	3450	6" pipe size	Q-16	3	8		66	268	26.50	360.50	510
	3460	8" pipe size		2.50	9.600		121	320	32	473	655
	3470	10" pipe size		2	12		202	405	40	647	870
	3480	12" pipe size		1.60	15		305	505	50	860	1,150
	3481	14" pipe size		1.30	18.462		510	620	61.50	1,191.50	1,575
	3482	16" pipe size		1	24		615	805	80	1,500	2,000
	3483	18" pipe size		.80	30		975	1,000	100	2,075	2,700
	3484	20" pipe size		.75	32		1,500	1,075	107	2,682	3,400
	3486	24" pipe size		.70	34.286		1,875	1,150	114	3,139	3,900
	3487	26" pipe size		.55	43.636		2,600	1,475	145	4,220	5,200
	3488	30" pipe size		.30	80		2,825	2,675	266	5,766	7,450
	3490	36" pipe size		.20	120		4,550	4,025	400	8,975	11,500
	3491	Eccentric reducer, 1-1/2" pipe size	Q-15	14	1.143		25.50	37	5.70	68.20	90.50
	3492	2" pipe size		11	1.455		19.50	47	7.25	73.75	101
	3493	2-1/2" pipe size		9	1.778		16.65	57.50	8.90	83.05	115
	3494	3" pipe size		8	2		18.90	64.50	10	93.40	130
	3495	4" pipe size		6	2.667		25	86.50	13.35	124.85	172
	3496	6" pipe size	Q-16	5	4.800		71	161	16	248	340
	3497	8" pipe size		4	6		99.50	201	20	320.50	435
	3498	10" pipe size		3	8		121	268	26.50	415.50	570
	3499	12" pipe size		2.50	9.600		180	320	32	532	720
	3501	Cap, 1-1/2" pipe size	Q-15	28	.571		7.70	18.50	2.86	29.06	39.50
	3502	2" pipe size		22	.727		7.05	23.50	3.64	34.19	47.50
	3503	2-1/2" pipe size		18	.889		7.05	29	4.44	40.49	56
	3504	3" pipe size		16	1		7.05	32.50	5	44.55	62.50
	3505	4" pipe size		12	1.333		8.95	43	6.65	58.60	82

660

15 MECHANICAL

Important: See the Reference Section for critical supporting data - Reference Nos., Crews, & City Cost Indexes

15107	Metal Pipe & Fittings	CREW	DAILY OUTPUT	LABOR-HOURS	UNIT	2002 BARE COSTS				TOTAL INCL O&P
						MAT.	LABOR	EQUIP.	TOTAL	
3506	6" pipe size	Q-16	10	2.400	Ea.	16.95	80.50	8	105.45	148
3507	8" pipe size		8	3		26	101	10	137	192
3508	10" pipe size		6	4		46	134	13.30	193.30	267
3509	12" pipe size		5	4.800		59	161	16	236	325
3511	14" pipe size		4	6		105	201	20	326	440
3512	16" pipe size		4	6		123	201	20	344	465
3513	18" pipe size	▼	3	8	▼	141	268	26.50	435.50	590
3517	Weld joint, butt, carbon steel, extra strong									
3519	90° elbow, long									
3520	1/2" pipe size	Q-15	13	1.231	Ea.	16.95	40	6.15	63.10	85.50
3530	3/4" pipe size		12	1.333		16.95	43	6.65	66.60	91
3540	1" pipe size		11	1.455		8.95	47	7.25	63.20	89
3550	1-1/4" pipe size		10	1.600		8.95	52	8	68.95	96.50
3560	1-1/2" pipe size		9	1.778		8.95	57.50	8.90	75.35	106
3570	2" pipe size		8	2		8.95	64.50	10	83.45	118
3580	2-1/2" pipe size		7	2.286		13.10	74	11.45	98.55	139
3590	3" pipe size		6	2.667		16.30	86.50	13.35	116.15	163
3600	4" pipe size		4	4		27.50	129	20	176.50	248
3606	5" pipe size	▼	3.50	4.571		59.50	148	23	230.50	315
3610	6" pipe size	Q-16	4.50	5.333		66	179	17.75	262.75	365
3620	8" pipe size		3.50	6.857		125	230	23	378	505
3630	10" pipe size		2.50	9.600		224	320	32	576	765
3640	12" pipe size	▼	2.25	10.667	▼	315	360	35.50	710.50	925
3650	45° Elbow, long									
3660	1/2" pipe size	Q-15	13	1.231	Ea.	19.50	40	6.15	65.65	88.50
3670	3/4" pipe size		12	1.333		19.50	43	6.65	69.15	94
3680	1" pipe size		11	1.455		9.30	47	7.25	63.55	89
3690	1-1/4" pipe size		10	1.600		9.30	52	8	69.30	97
3700	1-1/2" pipe size		9	1.778		9.30	57.50	8.90	75.70	107
3710	2" pipe size		8	2		8.30	64.50	10	82.80	118
3720	2-1/2" pipe size		7	2.286		21	74	11.45	106.45	148
3730	3" pipe size		6	2.667		12.80	86.50	13.35	112.65	159
3740	4" pipe size		4	4		19.50	129	20	168.50	239
3746	5" pipe size	▼	3.50	4.571		47.50	148	23	218.50	300
3750	6" pipe size	Q-16	4.50	5.333		51.50	179	17.75	248.25	345
3760	8" pipe size		3.50	6.857		88	230	23	341	465
3770	10" pipe size		2.50	9.600		142	320	32	494	675
3780	12" pipe size	▼	2.25	10.667	▼	211	360	35.50	606.50	810
3800	Tee, straight									
3810	1/2" pipe size	Q-15	9	1.778	Ea.	41.50	57.50	8.90	107.90	142
3820	3/4" pipe size		8.50	1.882		41.50	61	9.40	111.90	148
3830	1" pipe size		8	2		22	64.50	10	96.50	133
3840	1-1/4" pipe size		7	2.286		22	74	11.45	107.45	149
3850	1-1/2" pipe size		6	2.667		25.50	86.50	13.35	125.35	173
3860	2" pipe size		5	3.200		22	104	16	142	198
3870	2-1/2" pipe size		4	4		38.50	129	20	187.50	260
3880	3" pipe size		3.50	4.571		36	148	23	207	288
3890	4" pipe size		2.50	6.400		58.50	207	32	297.50	410
3896	5" pipe size	▼	2.25	7.111		119	230	35.50	384.50	515
3900	6" pipe size	Q-16	2.25	10.667		92	360	35.50	487.50	680
3910	8" pipe size		2	12		187	405	40	632	855
3920	10" pipe size		1.75	13.714		300	460	45.50	805.50	1,075
3930	12" pipe size	▼	1.50	16		440	535	53.50	1,028.50	1,350
4000	Eccentric reducer, 1-1/2" pipe size	Q-15	10	1.600		38	52	8	98	128
4010	2" pipe size		9	1.778		29	57.50	8.90	95.40	128
4020	2-1/2" pipe size	▼	8	2	▼	30.50	64.50	10	105	143

660

MECHANICAL 15

	15107	Metal Pipe & Fittings	CREW	DAILY OUTPUT	LABOR-HOURS	UNIT	MAT.	LABOR	EQUIP.	TOTAL	TOTAL INCL O&P	
								2002 BARE COSTS				
660	4030	3" pipe size	Q-15	7	2.286	Ea.	30.50	74	11.45	115.95	159	660
	4040	4" pipe size		5	3.200		30	104	16	150	207	
	4046	5" pipe size	▼	4.70	3.404		85	110	17	212	278	
	4050	6" pipe size	Q-16	4.50	5.333		74.50	179	17.75	271.25	370	
	4060	8" pipe size		3.50	6.857		214	230	23	467	605	
	4070	10" pipe size		2.50	9.600		223	320	32	575	765	
	4080	12" pipe size		2.25	10.667		221	360	35.50	616.50	820	
	4090	14" pipe size		2.10	11.429		500	385	38	923	1,175	
	4100	16" pipe size	▼	1.90	12.632		470	425	42	937	1,200	
	4151	Cap, 1-1/2" pipe size	Q-15	24	.667		8	21.50	3.33	32.83	45	
	4152	2" pipe size		18	.889		7.05	29	4.44	40.49	56	
	4153	2-1/2" pipe size		16	1		8	32.50	5	45.50	63.50	
	4154	3" pipe size		14	1.143		8.95	37	5.70	51.65	72	
	4155	4" pipe size	▼	10	1.600		12.50	52	8	72.50	101	
	4156	6" pipe size	Q-16	9	2.667		31.50	89.50	8.90	129.90	179	
	4157	8" pipe size		7	3.429		42	115	11.40	168.40	232	
	4158	10" pipe size		5	4.800		66	161	16	243	335	
	4159	12" pipe size	▼	4	6	▼	93.50	201	20	314.50	430	
	4190	Weld fittings, reducing, standard weight										
	4200	Welding ring w/spacer pins, 2" pipe size				Ea.	2.10			2.10	2.31	
	4210	2-1/2" pipe size					2.10			2.10	2.31	
	4220	3" pipe size					2.50			2.50	2.75	
	4230	4" pipe size					2.80			2.80	3.08	
	4236	5" pipe size					3			3	3.30	
	4240	6" pipe size					3.50			3.50	3.85	
	4250	8" pipe size					5.65			5.65	6.20	
	4260	10" pipe size					6.80			6.80	7.50	
	4270	12" pipe size					8			8	8.80	
	4280	14" pipe size					9			9	9.90	
	4290	16" pipe size					10			10	11	
	4300	18" pipe size					10.50			10.50	11.55	
	4310	20" pipe size					12			12	13.20	
	4330	24" pipe size					14			14	15.40	
	4340	26" pipe size					14			14	15.40	
	4350	30" pipe size					18			18	19.80	
	4370	36" pipe size				▼	25			25	27.50	
	4600	Tee, reducing on outlet										
	4601	2-1/2" x 2" pipe size	Q-15	5	3.200	Ea.	33	104	16	153	210	
	4602	3" x 2-1/2" pipe size		4	4		33.50	129	20	182.50	254	
	4603	3-1/2" x 3" pipe size		3.50	4.571		42	148	23	213	295	
	4604	4" x 3" pipe size		3	5.333		42	173	26.50	241.50	335	
	4605	5" x 4" pipe size	▼	2.50	6.400		91.50	207	32	330.50	445	
	4606	6" x 5" pipe size	Q-16	3	8		82.50	268	26.50	377	525	
	4607	8" x 6" pipe size		2.50	9.600		170	320	32	522	705	
	4608	10" x 8" pipe size		2	12		266	405	40	711	940	
	4609	12" x 10" pipe size		1.60	15		395	505	50	950	1,250	
	4610	16" x 12" pipe size		1.50	16		645	535	53.50	1,233.50	1,575	
	4611	14" x 12" pipe size	▼	1.52	15.789	▼	420	530	52.50	1,002.50	1,325	
	4618	Reducer, concentric										
	4619	2-1/2" by 2" pipe size	Q-15	10	1.600	Ea.	11.85	52	8	71.85	100	
	4620	3" by 2-1/2" pipe size		9	1.778		15.05	57.50	8.90	81.45	113	
	4621	3-1/2" by 3" pipe size		8	2		25.50	64.50	10	100	137	
	4622	4" by 2-1/2" pipe size		7	2.286		14.70	74	11.45	100.15	141	
	4623	5" by 3" pipe size	▼	7	2.286		35	74	11.45	120.45	163	
	4624	6" by 4" pipe size	Q-16	6	4		28.50	134	13.30	175.80	248	
	4625	8" by 6" pipe size	▼	5	4.800	▼	38	161	16	215	300	

Important: See the Reference Section for critical supporting data - Reference Nos., Crews, & City Cost Indexes

15107	Metal Pipe & Fittings	CREW	DAILY OUTPUT	LABOR-HOURS	UNIT	2002 BARE COSTS				TOTAL INCL O&P
						MAT.	LABOR	EQUIP.	TOTAL	
660 4626	10" by 8" pipe size	Q-16	4	6	Ea.	57.50	201	20	278.50	390 **660**
4627	12" by 10" pipe size	↓	3	8	↓	89	268	26.50	383.50	535
4660	Reducer, eccentric									
4662	3" x 2"	Q-15	8	2	Ea.	13.35	64.50	10	87.85	123
4664	4" x 3"		6	2.667		18.90	86.50	13.35	118.75	166
4666	4" x 2"	↓	6	2.667		25	86.50	13.35	124.85	172
4670	6" x 4"	Q-16	5	4.800		38	161	16	215	305
4672	6" x 3"		5	4.800		71	161	16	248	340
4676	8" x 6"		4	6		38	201	20	259	370
4678	8" x 4"		4	6		99.50	201	20	320.50	435
4682	10" x 8"		3	8		68	268	26.50	362.50	510
4684	10" x 6"		3	8		121	268	26.50	415.50	570
4688	12" x 10"		2.50	9.600		114	320	32	466	645
4690	12" x 8"		2.50	9.600		168	320	32	520	705
4691	14" x 12"		2.20	10.909		167	365	36.50	568.50	775
4693	16" x 14"		1.80	13.333		223	445	44.50	712.50	970
4694	16" x 12"		2	12		340	405	40	785	1,025
4696	18" x 16"	↓	1.60	15	↓	360	505	50	915	1,200
5000	Weld joint, socket, forged steel, 3000 lb., schedule 40 pipe									
5010	90° elbow, straight									
5020	1/4" pipe size	Q-15	22	.727	Ea.	7.45	23.50	3.64	34.59	47.50
5030	3/8" pipe size		22	.727		9.20	23.50	3.64	36.34	49.50
5040	1/2" pipe size		20	.800		7.45	26	4	37.45	51.50
5050	3/4" pipe size		20	.800		7.65	26	4	37.65	52
5060	1" pipe size		20	.800		9.90	26	4	39.90	54.50
5070	1-1/4" pipe size		18	.889		18.90	29	4.44	52.34	69.50
5080	1-1/2" pipe size		16	1		21	32.50	5	58.50	77.50
5090	2" pipe size		12	1.333		31.50	43	6.65	81.15	107
5100	2-1/2" pipe size		10	1.600		93	52	8	153	189
5110	3" pipe size		8	2		157	64.50	10	231.50	282
5120	4" pipe size	↓	6	2.667	↓	425	86.50	13.35	524.85	615
5130	45° Elbow, straight									
5134	1/4" pipe size	Q-15	22	.727	Ea.	12.10	23.50	3.64	39.24	53
5135	3/8" pipe size		22	.727		12.10	23.50	3.64	39.24	53
5136	1/2" pipe size		20	.800		9.40	26	4	39.40	53.50
5137	3/4" pipe size		20	.800		10.95	26	4	40.95	55.50
5140	1" pipe size		20	.800		14.50	26	4	44.50	59.50
5150	1-1/4" pipe size		18	.889		19.30	29	4.44	52.74	69.50
5160	1-1/2" pipe size		16	1		24	32.50	5	61.50	81
5170	2" pipe size		12	1.333		41	43	6.65	90.65	117
5180	2-1/2" pipe size		10	1.600		106	52	8	166	203
5190	3" pipe size		8	2		178	64.50	10	252.50	305
5200	4" pipe size	↓	6	2.667	↓	350	86.50	13.35	449.85	530
5250	Tee, straight									
5254	1/4" pipe size	Q-15	15	1.067	Ea.	8.55	34.50	5.35	48.40	67.50
5255	3/8" pipe size		15	1.067		10.55	34.50	5.35	50.40	69.50
5256	1/2" pipe size		13	1.231		8.55	40	6.15	54.70	76
5257	3/4" pipe size		13	1.231		10.80	40	6.15	56.95	78.50
5260	1" pipe size		13	1.231		14.85	40	6.15	61	83
5270	1-1/4" pipe size		12	1.333		23.50	43	6.65	73.15	98
5280	1-1/2" pipe size		11	1.455		31.50	47	7.25	85.75	114
5290	2" pipe size		8	2		44	64.50	10	118.50	157
5300	2-1/2" pipe size		6	2.667		130	86.50	13.35	229.85	288
5310	3" pipe size		5	3.200		315	104	16	435	520
5320	4" pipe size	↓	4	4	↓	500	129	20	649	765
5350	For reducing sizes, add					60%				

15107	Metal Pipe & Fittings	CREW	DAILY OUTPUT	LABOR-HOURS	UNIT	2002 BARE COSTS				TOTAL INCL O&P	
						MAT.	LABOR	EQUIP.	TOTAL		
660											**660**
5450	Couplings										
5451	1/4" pipe size	Q-15	23	.696	Ea.	6.55	22.50	3.48	32.53	45	
5452	3/8" pipe size		23	.696		6.55	22.50	3.48	32.53	45	
5453	1/2" pipe size		21	.762		3.31	24.50	3.81	31.62	45	
5454	3/4" pipe size		21	.762		4.41	24.50	3.81	32.72	46	
5460	1" pipe size		20	.800		5.15	26	4	35.15	49	
5470	1-1/4" pipe size		20	.800		10.10	26	4	40.10	54.50	
5480	1-1/2" pipe size		18	.889		11.15	29	4.44	44.59	60.50	
5490	2" pipe size		14	1.143		17.55	37	5.70	60.25	81.50	
5500	2-1/2" pipe size		12	1.333		37	43	6.65	86.65	113	
5510	3" pipe size		9	1.778		86.50	57.50	8.90	152.90	191	
5520	4" pipe size		7	2.286		130	74	11.45	215.45	268	
5570	Union, 1/4" pipe size		21	.762		14.35	24.50	3.81	42.66	57	
5571	3/8" pipe size		21	.762		14.35	24.50	3.81	42.66	57	
5572	1/2" pipe size		19	.842		14.35	27	4.21	45.56	61.50	
5573	3/4" pipe size		19	.842		16.30	27	4.21	47.51	63.50	
5574	1" pipe size		19	.842		21	27	4.21	52.21	68.50	
5575	1-1/4" pipe size		17	.941		34	30.50	4.71	69.21	88.50	
5576	1-1/2" pipe size		15	1.067		36	34.50	5.35	75.85	97.50	
5577	2" pipe size		11	1.455		50	47	7.25	104.25	134	
5600	Reducer, 1/4" pipe size		23	.696		17.65	22.50	3.48	43.63	57.50	
5601	3/8" pipe size		23	.696		17.65	22.50	3.48	43.63	57.50	
5602	1/2" pipe size		21	.762		12	24.50	3.81	40.31	54.50	
5603	3/4" pipe size		21	.762		13.95	24.50	3.81	42.26	56.50	
5604	1" pipe size		21	.762		15.30	24.50	3.81	43.61	58	
5605	1-1/4" pipe size		19	.842		27.50	27	4.21	58.71	76	
5607	1-1/2" pipe size		17	.941		29.50	30.50	4.71	64.71	83.50	
5608	2" pipe size		13	1.231		32.50	40	6.15	78.65	103	
5612	Cap, 1/4" pipe size		46	.348		6.95	11.25	1.74	19.94	26.50	
5613	3/8" pipe size		46	.348		6.95	11.25	1.74	19.94	26.50	
5614	1/2" pipe size		42	.381		4.91	12.30	1.90	19.11	26	
5615	3/4" pipe size		42	.381		5.70	12.30	1.90	19.90	27	
5616	1" pipe size		42	.381		8.80	12.30	1.90	23	30.50	
5617	1-1/4" pipe size		38	.421		10.25	13.60	2.11	25.96	34	
5618	1-1/2" pipe size		34	.471		13.95	15.20	2.35	31.50	41	
5619	2" pipe size		26	.615		22	19.90	3.08	44.98	57.50	
5630	T-O-L, 1/4" pipe size, nozzle		23	.696		4.31	22.50	3.48	30.29	42.50	
5631	3/8" pipe size, nozzle		23	.696		4.31	22.50	3.48	30.29	42.50	
5632	1/2" pipe size, nozzle		22	.727		4.31	23.50	3.64	31.45	44	
5633	3/4" pipe size, nozzle		21	.762		5.05	24.50	3.81	33.36	46.50	
5634	1" pipe size, nozzle		20	.800		5.70	26	4	35.70	49.50	
5635	1-1/4" pipe size, nozzle		18	.889		8.05	29	4.44	41.49	57	
5636	1-1/2" pipe size, nozzle		16	1		8.75	32.50	5	46.25	64	
5637	2" pipe size, nozzle		12	1.333		10.05	43	6.65	59.70	83.50	
5638	2-1/2" pipe size, nozzle		10	1.600		31.50	52	8	91.50	122	
5639	4" pipe size, nozzle		6	2.667		72	86.50	13.35	171.85	224	
5640	W-O-L, 1/4" pipe size, nozzle		23	.696		8.25	22.50	3.48	34.23	47	
5641	3/8" pipe size, nozzle		23	.696		8.25	22.50	3.48	34.23	47	
5642	1/2" pipe size, nozzle		22	.727		8.25	23.50	3.64	35.39	48.50	
5643	3/4" pipe size, nozzle		21	.762		8.45	24.50	3.81	36.76	50.50	
5644	1" pipe size, nozzle		20	.800		8.85	26	4	38.85	53	
5645	1-1/4" pipe size, nozzle		18	.889		10.75	29	4.44	44.19	60	
5646	1-1/2" pipe size, nozzle		16	1		10.75	32.50	5	48.25	66.50	
5647	2" pipe size, nozzle		12	1.333		11.20	43	6.65	60.85	84.50	
5648	2-1/2" pipe size, nozzle		10	1.600		25.50	52	8	85.50	115	
5649	3" pipe size, nozzle		8	2		29.50	64.50	10	104	141	

Important: See the Reference Section for critical supporting data - Reference Nos., Crews, & City Cost Indexes

			DAILY	LABOR-		2002 BARE COSTS				TOTAL	
15107	**Metal Pipe & Fittings**	CREW	OUTPUT	HOURS	UNIT	MAT.	LABOR	EQUIP.	TOTAL	INCL O&P	
660	5650	4" pipe size, nozzle	Q-15	6	2.667	Ea.	37.50	86.50	13.35	137.35	186	**660**
	5651	5" pipe size, nozzle		5	3.200		93.50	104	16	213.50	277	
	5652	6" pipe size, nozzle		4	4		119	129	20	268	350	
	5653	8" pipe size, nozzle		3	5.333		215	173	26.50	414.50	525	
	5654	10" pipe size, nozzle		2.60	6.154		310	199	31	540	680	
	5655	12" pipe size, nozzle		2.20	7.273		490	235	36.50	761.50	930	
	5656	14" pipe size, nozzle		2	8		905	259	40	1,204	1,425	
	5660	L-O-L, 1/4" pipe size, outlet		23	.696		47	22.50	3.48	72.98	89.50	
	5661	3/8" pipe size, outlet		23	.696		47	22.50	3.48	72.98	89.50	
	5662	1/2" pipe size, outlet		22	.727		47	23.50	3.64	74.14	91	
	5663	3/4" pipe size, outlet		21	.762		51.50	24.50	3.81	79.81	98	
	5664	1" pipe size, outlet		20	.800		69.50	26	4	99.50	120	
	5665	1-1/4" pipe size, outlet		18	.889		108	29	4.44	141.44	167	
	5666	1-1/2" pipe size, outlet		16	1		108	32.50	5	145.50	174	
	5667	2" pipe size, outlet		12	1.333		171	43	6.65	220.65	261	
	5674	S-O-L, 1/4" pipe size, outlet		23	.696		4.81	22.50	3.48	30.79	43	
	5675	3/8" pipe size, outlet		23	.696		4.81	22.50	3.48	30.79	43	
	5676	1/2" pipe size, outlet		22	.727		4.81	23.50	3.64	31.95	45	
	5677	3/4" pipe size, outlet		21	.762		4.88	24.50	3.81	33.19	46.50	
	5678	1" pipe size, outlet		20	.800		5.30	26	4	35.30	49.50	
	5679	1-1/4" pipe size, outlet		18	.889		8.20	29	4.44	41.64	57.50	
	5680	1-1/2" pipe size, outlet		16	1		8.60	32.50	5	46.10	64	
	5681	2" pipe size, outlet	▼	12	1.333	▼	9.65	43	6.65	59.30	83	
	6000	Weld-on flange, forged steel										
	6020	Slip-on, 150 lb. flange (welded front and back)										
	6050	1/2" pipe size	Q-15	18	.889	Ea.	11.85	29	4.44	45.29	61.50	
	6060	3/4" pipe size		18	.889		11.85	29	4.44	45.29	61.50	
	6070	1" pipe size		17	.941		11.85	30.50	4.71	47.06	64	
	6080	1-1/4" pipe size		16	1		11.85	32.50	5	49.35	67.50	
	6090	1-1/2" pipe size		15	1.067		11.85	34.50	5.35	51.70	71	
	6100	2" pipe size		12	1.333		11.85	43	6.65	61.50	85.50	
	6110	2-1/2" pipe size		10	1.600		21	52	8	81	110	
	6120	3" pipe size		9	1.778		16.65	57.50	8.90	83.05	115	
	6130	3-1/2" pipe size		7	2.286		19.50	74	11.45	104.95	146	
	6140	4" pipe size		6	2.667		19.50	86.50	13.35	119.35	166	
	6150	5" pipe size	▼	5	3.200		38	104	16	158	215	
	6160	6" pipe size	Q-16	6	4		34	134	13.30	181.30	254	
	6170	8" pipe size		5	4.800		51	161	16	228	315	
	6180	10" pipe size		4	6		89.50	201	20	310.50	425	
	6190	12" pipe size		3	8		132	268	26.50	426.50	580	
	6191	14" pipe size		2.50	9.600		175	320	32	527	715	
	6192	16" pipe size	▼	1.80	13.333	▼	290	445	44.50	779.50	1,050	
	6200	300 lb. flange										
	6210	1/2" pipe size	Q-15	17	.941	Ea.	17.90	30.50	4.71	53.11	71	
	6220	3/4" pipe size		17	.941		17.90	30.50	4.71	53.11	71	
	6230	1" pipe size		16	1		17.90	32.50	5	55.40	74	
	6240	1-1/4" pipe size		13	1.231		17.90	40	6.15	64.05	86.50	
	6250	1-1/2" pipe size		12	1.333		17.90	43	6.65	67.55	92	
	6260	2" pipe size		11	1.455		22	47	7.25	76.25	104	
	6270	2-1/2" pipe size		9	1.778		24.50	57.50	8.90	90.90	123	
	6280	3" pipe size		7	2.286		29	74	11.45	114.45	156	
	6290	4" pipe size		6	2.667		41	86.50	13.35	140.85	190	
	6300	5" pipe size	▼	4	4		67.50	129	20	216.50	292	
	6310	6" pipe size	Q-16	5	4.800		67	161	16	244	335	
	6320	8" pipe size		4	6		115	201	20	336	455	
	6330	10" pipe size	▼	3.40	7.059	▼	204	237	23.50	464.50	605	

MECHANICAL 15

15107	Metal Pipe & Fittings	CREW	DAILY OUTPUT	LABOR-HOURS	UNIT	2002 BARE COSTS				TOTAL INCL O&P
						MAT.	LABOR	EQUIP.	TOTAL	
660 6340	12" pipe size	Q-16	2.80	8.571	Ea.	222	288	28.50	538.50	710 **660**
6400	Welding neck, 150 lb. flange									
6410	1/2" pipe size	Q-15	40	.400	Ea.	16.95	12.95	2	31.90	40.50
6420	3/4" pipe size		36	.444		16.95	14.40	2.22	33.57	42.50
6430	1" pipe size		32	.500		16.95	16.20	2.50	35.65	46
6440	1-1/4" pipe size		29	.552		16.95	17.85	2.76	37.56	48.50
6450	1-1/2" pipe size		26	.615		16.95	19.90	3.08	39.93	52
6460	2" pipe size		20	.800		18.55	26	4	48.55	64
6470	2-1/2" pipe size		16	1		19.85	32.50	5	57.35	76.50
6480	3" pipe size		14	1.143		22	37	5.70	64.70	86.50
6500	4" pipe size		10	1.600		26	52	8	86	115
6510	5" pipe size	↓	8	2		41.50	64.50	10	116	154
6520	6" pipe size	Q-16	10	2.400		40	80.50	8	128.50	174
6530	8" pipe size		7	3.429		65.50	115	11.40	191.90	258
6540	10" pipe size		6	4		113	134	13.30	260.30	340
6550	12" pipe size		5	4.800		164	161	16	341	440
6551	14" pipe size		4.50	5.333		236	179	17.75	432.75	550
6552	16" pipe size		3	8		400	268	26.50	694.50	875
6553	18" pipe size		2.50	9.600		570	320	32	922	1,150
6554	20" pipe size		2.30	10.435		690	350	35	1,075	1,325
6556	24" pipe size		2	12		925	405	40	1,370	1,675
6557	26" pipe size		1.70	14.118		1,500	475	47	2,022	2,425
6558	30" pipe size		.90	26.667		2,000	895	89	2,984	3,650
6559	36" pipe size	↓	.75	32	↓	2,300	1,075	107	3,482	4,275
6560	300 lb. flange									
6570	1/2" pipe size	Q-15	36	.444	Ea.	21.50	14.40	2.22	38.12	47.50
6580	3/4" pipe size		34	.471		21.50	15.20	2.35	39.05	49
6590	1" pipe size		30	.533		21.50	17.25	2.67	41.42	52.50
6600	1-1/4" pipe size		28	.571		22.50	18.50	2.86	43.86	55.50
6610	1-1/2" pipe size		24	.667		21.50	21.50	3.33	46.33	59.50
6620	2" pipe size		18	.889		24.50	29	4.44	57.94	75.50
6630	2-1/2" pipe size		14	1.143		28	37	5.70	70.70	93
6640	3" pipe size		12	1.333		28	43	6.65	77.65	103
6650	4" pipe size		8	2		45.50	64.50	10	120	159
6660	5" pipe size	↓	7	2.286		70.50	74	11.45	155.95	202
6670	6" pipe size	Q-16	9	2.667		66	89.50	8.90	164.40	218
6680	8" pipe size		6	4		114	134	13.30	261.30	345
6690	10" pipe size		5	4.800		221	161	16	398	505
6700	12" pipe size		4	6		284	201	20	505	635
6710	14" pipe size		3.50	6.857		630	230	23	883	1,075
6720	16" pipe size	↓	2	12	↓	745	405	40	1,190	1,475
7740	Plain ends for plain end pipe, mechanically coupled									
7750	Cplg & labor required at joints not included add 1 per									
7760	joint for installed price, see line 9180									
7770	Malleable iron, painted, unless noted otherwise									
7800	90° Elbow 1"				Ea.	20.50			20.50	22.50
7810	1-1/2"					20.50			20.50	22.50
7820	2"					20.50			20.50	22.50
7830	2-1/2"					20.50			20.50	22.50
7840	3"					24.50			24.50	27
7860	4"					36			36	39.50
7870	5" welded steel					124			124	136
7880	6"					159			159	174
7890	8" welded steel					208			208	229
7900	10" welded steel					262			262	288
7910	12" welded steel				↓	335			335	365

Important: See the Reference Section for critical supporting data - Reference Nos., Crews, & City Cost Indexes

15107	Metal Pipe & Fittings	CREW	DAILY OUTPUT	LABOR-HOURS	UNIT	2002 BARE COSTS				TOTAL INCL O&P	
						MAT.	LABOR	EQUIP.	TOTAL		
660	7970	45° Elbow 1"				Ea.	20.50			20.50	22.50	660
	7980	1-1/2"					20.50			20.50	22.50
	7990	2"					20.50			20.50	22.50
	8000	2-1/2"					20.50			20.50	22.50
	8010	3"					24.50			24.50	27
	8030	4"					36			36	39.50
	8040	5" welded steel					124			124	136
	8050	6"					98.50			98.50	108
	8060	8"					114			114	126
	8070	10" welded steel					160			160	176
	8080	12" welded steel					206			206	227
	8140	Tee, straight 1"					27.50			27.50	30.50
	8150	1-1/2"					27.50			27.50	30.50
	8160	2"					27.50			27.50	30.50
	8170	2-1/2"					27.50			27.50	30.50
	8180	3"					39.50			39.50	43.50
	8200	4"					55			55	60.50
	8210	5" welded steel					121			121	133
	8220	6"					149			149	164
	8230	8" welded steel					207			207	228
	8240	10" welded steel					315			315	345
	8250	12" welded steel					425			425	470
	8340	Segmentally welded steel, painted									
	8390	Wye 2"				Ea.	80.50			80.50	88.50
	8400	2-1/2"					80.50			80.50	88.50
	8410	3"					90.50			90.50	99.50
	8430	4"					134			134	148
	8440	5"					164			164	180
	8450	6"					225			225	248
	8460	8"					292			292	320
	8470	10"					430			430	475
	8480	12"					665			665	730
	8540	Wye, lateral 2"					86.50			86.50	95
	8550	2-1/2"					86.50			86.50	95
	8560	3"					96.50			96.50	106
	8580	4"					145			145	160
	8590	5"					189			189	208
	8600	6"					242			242	266
	8610	8"					355			355	390
	8620	10"					500			500	550
	8630	12"					675			675	740
	8690	Cross, 2"					91.50			91.50	101
	8700	2-1/2"					91.50			91.50	101
	8710	3"					109			109	120
	8730	4"					149			149	164
	8740	5"					212			212	233
	8750	6"					280			280	310
	8760	8"					360			360	395
	8770	10"					525			525	575
	8780	12"					755			755	830
	8800	Tees, reducing 2" x 1"					54.50			54.50	60
	8810	2" x 1-1/2"					54.50			54.50	60
	8820	3" x 1"					54.50			54.50	60
	8830	3" x 1-1/2"					54.50			54.50	60
	8840	3" x 2"					54.50			54.50	60
	8850	4" x 1"					79			79	87

MECHANICAL 15

181

		CREW	DAILY OUTPUT	LABOR-HOURS	UNIT	2002 BARE COSTS				TOTAL INCL O&P	
15107 | Metal Pipe & Fittings						MAT.	LABOR	EQUIP.	TOTAL		
660											**660**
8860	4" x 1-1/2"				Ea.	79			79	87	
8870	4" x 2"					79			79	87	
8880	4" x 2-1/2"					81.50			81.50	89.50	
8890	4" x 3"					81.50			81.50	89.50	
8900	6" x 2"					132			132	145	
8910	6" x 3"					139			139	153	
8920	6" x 4"					139			139	153	
8930	8" x 2"					169			169	186	
8940	8" x 3"					177			177	195	
8950	8" x 4"					181			181	199	
8960	8" x 5"					186			186	204	
8970	8" x 6"					186			186	204	
8980	10" x 4"					266			266	293	
8990	10" x 6"					282			282	310	
9000	10" x 8"					282			282	310	
9010	12" x 6"					410			410	450	
9020	12" x 8"					410			410	455	
9030	12" x 10"					425			425	470	
9080	Adapter nipples 3" long										
9090	1"				Ea.	7.10			7.10	7.80	
9100	1-1/2"					7.10			7.10	7.80	
9110	2"					7.10			7.10	7.80	
9120	2-1/2"					8.30			8.30	9.15	
9130	3"					10.25			10.25	11.30	
9140	4"					17.15			17.15	18.85	
9150	6"					46.50			46.50	51	
9180	Coupling, mechanical, plain end pipe to plain end pipe or fitting										
9190	1"	Q-1	29	.552	Ea.	22	17.85		39.85	51	
9200	1-1/2"		28	.571		28	18.50		46.50	59	
9210	2"		27	.593		26.50	19.15		45.65	58	
9220	2-1/2"		26	.615		26.50	19.90		46.40	59	
9230	3"		25	.640		39	20.50		59.50	74	
9240	3-1/2"		24	.667		45	21.50		66.50	82	
9250	4"		22	.727		45	23.50		68.50	85	
9260	5"	Q-2	28	.857		64	29		93	114	
9270	6"		24	1		78.50	33.50		112	137	
9280	8"		19	1.263		139	42.50		181.50	217	
9290	10"		16	1.500		181	50.50		231.50	275	
9300	12"		12	2		227	67		294	350	
9310	Outlets for precut holes through pipe wall										
9331	Strapless type, with gasket										
9332	4" to 8" pipe x 1/2"	1 Plum	13	.615	Ea.	25.50	22		47.50	62	
9333	4" to 8" pipe x 3/4"		13	.615		27.50	22		49.50	64	
9334	10" pipe and larger x 1/2"		11	.727		25.50	26		51.50	68	
9335	10" pipe and larger x 3/4"		11	.727		27.50	26		53.50	70	
9341	Thermometer wells with gasket										
9342	4" to 8" pipe, 6" stem	1 Plum	14	.571	Ea.	39.50	20.50		60	74.50	
9343	8" pipe and larger, 6" stem	"	13	.615	"	39.50	22		61.50	77	
9400	Mechanical joint ends for plain end pipe										
9410	Malleable iron, black										
9420	90° Elbows, 1-1/4"	Q-1	29	.552	Ea.	9.30	17.85		27.15	37.50	
9430	1-1/2"		27	.593		10.20	19.15		29.35	40	
9440	2"		24	.667		12.05	21.50		33.55	46	
9490	Tee, reducing outlet										
9510	1-1/4" x 1/2"	Q-1	18	.889	Ea.	7.45	29		36.45	51.50	
9520	1-1/4" x 3/4"		18	.889		7.45	29		36.45	51.50	

15107	Metal Pipe & Fittings	CREW	DAILY OUTPUT	LABOR-HOURS	UNIT	2002 BARE COSTS				TOTAL INCL O&P	
						MAT.	LABOR	EQUIP.	TOTAL		
660	9530	1-1/4" x 1"	Q-1	18	.889	Ea.	7.45	29		36.45	51.50	**660**
	9540	1-1/2" x 1/2"		17	.941		7.90	30.50		38.40	54.50	
	9550	1-1/2" x 3/4"		17	.941		7.90	30.50		38.40	54.50	
	9560	1-1/2" x 1"		17	.941		7.90	30.50		38.40	54.50	
	9570	2" x 1/2"		15	1.067		9.65	34.50		44.15	62.50	
	9580	2" x 3/4"		15	1.067		9.65	34.50		44.15	62.50	
	9590	2" x 1"	▼	15	1.067	▼	9.65	34.50		44.15	62.50	
	9640	Tee, reducing run and outlet										
	9660	1-1/4" x 1" x 1/2"	Q-1	18	.889	Ea.	8	29		37	52.50	
	9670	1-1/4" x 1" x 3/4"		18	.889		8	29		37	52.50	
	9680	1-1/4" x 1" x 1"		18	.889		8	29		37	52.50	
	9690	1-1/2" x 1-1/4" x 1/2"		17	.941		8.45	30.50		38.95	55.50	
	9700	1-1/2" x 1-1/4" x 3/4"		17	.941		8.45	30.50		38.95	55.50	
	9710	1-1/2" x 1-1/4" x 1"		17	.941		8.45	30.50		38.95	55.50	
	9720	2" x 1-1/2" x 1/2"		15	1.067		10.25	34.50		44.75	63.50	
	9730	2" x 1-1/2" x 3/4"		15	1.067		10.25	34.50		44.75	63.50	
	9740	2" x 1-1/2" x 1"	▼	15	1.067	▼	10.25	34.50		44.75	63.50	
	9790	Tee, outlet										
	9810	3" x 1-1/4"	Q-1	29	.552	Ea.	18.05	17.85		35.90	47	
	9820	3" x 1-1/2"		28	.571		18.05	18.50		36.55	48	
	9830	3" x 2"		26	.615		20	19.90		39.90	52	
	9840	4" x 1-1/4"		28	.571		22.50	18.50		41	52.50	
	9850	4" x 1-1/2"		26	.615		23.50	19.90		43.40	56	
	9860	4" x 2"	▼	24	.667	▼	23.50	21.50		45	58.50	
	9940	For galvanized fittings for plain end pipe, add					20%					
690	0010	**PIPE, GROOVED-JOINT STEEL FITTINGS & VALVES**										**690**
	0020	Pipe includes coupling & clevis type hanger 10' O.C.										
	0500	Schedule 10, black										
	0550	2" diameter	1 Plum	43	.186	L.F.	3.07	6.70		9.77	13.50	
	0560	2-1/2" diameter	Q-1	61	.262		3.94	8.50		12.44	17.15	
	0570	3" diameter		55	.291		4.46	9.40		13.86	19.10	
	0580	3-1/2" diameter		53	.302		5.55	9.75		15.30	21	
	0590	4" diameter		49	.327		5.65	10.55		16.20	22	
	0600	5" diameter	▼	40	.400		9.40	12.95		22.35	30	
	0610	6" diameter	Q-2	46	.522		10.05	17.50		27.55	37.50	
	0620	8" diameter	"	41	.585	▼	20	19.65		39.65	51.50	
	0700	To delete couplings & hangers, subtract										
	0710	2" diam. to 5" diam.					25%	20%				
	0720	6" diam. to 8" diam.					27%	15%				
	1000	Schedule 40, black										
	1040	3/4" diameter	1 Plum	71	.113	L.F.	1.97	4.05		6.02	8.25	
	1050	1" diameter		63	.127		2.10	4.56		6.66	9.20	
	1060	1-1/4" diameter		58	.138		2.61	4.96		7.57	10.35	
	1070	1-1/2" diameter		51	.157		2.91	5.65		8.56	11.70	
	1080	2" diameter	▼	40	.200		3.41	7.20		10.61	14.60	
	1090	2-1/2" diameter	Q-1	57	.281		4.75	9.10		13.85	18.90	
	1100	3" diameter		50	.320		5.75	10.35		16.10	22	
	1110	4" diameter		45	.356		8.50	11.50		20	26.50	
	1120	5" diameter	▼	37	.432		12.75	14		26.75	35	
	1130	6" diameter	Q-2	42	.571		15.15	19.15		34.30	45.50	
	1140	8" diameter		37	.649		23	22		45	58.50	
	1150	10" diameter		31	.774		37	26		63	79.50	
	1160	12" diameter		27	.889		44	30		74	93	
	1170	14" diameter		20	1.200		58	40.50		98.50	125	
	1180	16" diameter	▼	17	1.412	▼	65.50	47.50		113	144	

MECHANICAL 15

15107	Metal Pipe & Fittings	CREW	DAILY OUTPUT	LABOR-HOURS	UNIT	2002 BARE COSTS				TOTAL INCL O&P
						MAT.	LABOR	EQUIP.	TOTAL	
690 1190	18" diameter	Q-2	14	1.714	L.F.	83.50	57.50		141	179	**690**
1200	20" diameter		12	2		108	67		175	220
1210	24" diameter	▼	10	2.400	▼	126	80.50		206.50	260
1740	To delete coupling & hanger, subtract									
1750	3/4" diam. to 2" diam.					65%	27%			
1760	2-1/2" diam. to 5" diam.					41%	18%			
1770	6" diam. to 12" diam.					31%	13%			
1780	14" diam. to 24" diam.					35%	10%			
1800	Galvanized									
1840	3/4" diameter	1 Plum	71	.113	L.F.	2.20	4.05		6.25	8.50
1850	1" diameter		63	.127		2.27	4.56		6.83	9.40
1860	1-1/4" diameter		58	.138		2.82	4.96		7.78	10.60
1870	1-1/2" diameter		51	.157		3.17	5.65		8.82	12
1880	2" diameter	▼	40	.200		3.76	7.20		10.96	15
1890	2-1/2" diameter	Q-1	57	.281		5.30	9.10		14.40	19.55
1900	3" diameter		50	.320		6.50	10.35		16.85	23
1910	4" diameter		45	.356		8.05	11.50		19.55	26
1920	5" diameter	▼	37	.432		15.75	14		29.75	38.50
1930	6" diameter	Q-2	42	.571		18.75	19.15		37.90	49.50
1940	8" diameter		37	.649		28.50	22		50.50	64.50
1950	10" diameter		31	.774		45	26		71	88.50
1960	12" diameter	▼	27	.889	▼	54	30		84	105
2540	To delete coupling & hanger, subtract									
2550	3/4" diam. to 2" diam.					36%	27%			
2560	2-1/2" diam. to 5" diam.					19%	18%			
2570	6" diam. to 12" diam.					14%	13%			
2600	Schedule 80, black									
2610	3/4" diameter	1 Plum	65	.123	L.F.	2.24	4.42		6.66	9.10
2650	1" diameter		61	.131		2.43	4.71		7.14	9.80
2660	1-1/4" diameter		55	.145		3.09	5.25		8.34	11.30
2670	1-1/2" diameter		49	.163		3.51	5.85		9.36	12.70
2680	2" diameter	▼	38	.211		4.30	7.55		11.85	16.15
2690	2-1/2" diameter	Q-1	54	.296		6	9.60		15.60	21
2700	3" diameter		48	.333		7.50	10.80		18.30	24.50
2710	4" diameter		44	.364		10.70	11.75		22.45	29.50
2720	5" diameter	▼	35	.457		19.70	14.80		34.50	44
2730	6" diameter	Q-2	40	.600		26.50	20		46.50	59.50
2740	8" diameter		35	.686		49	23		72	88
2750	10" diameter		29	.828		77	28		105	127
2760	12" diameter	▼	24	1	▼	85	33.50		118.50	144
3240	To delete coupling & hanger, subtract									
3250	3/4" diam. to 2" diam.					30%	25%			
3260	2-1/2" diam. to 5" diam.					14%	17%			
3270	6" diam. to 12" diam.					12%	12%			
3300	Galvanized									
3310	3/4" diameter	1 Plum	65	.123	L.F.	2.55	4.42		6.97	9.45
3350	1" diameter		61	.131		2.66	4.71		7.37	10.05
3360	1-1/4" diameter		55	.145		3.41	5.25		8.66	11.65
3370	1-1/2" diameter		46	.174		3.90	6.25		10.15	13.75
3380	2" diameter	▼	38	.211		4.74	7.55		12.29	16.60
3390	2-1/2" diameter	Q-1	54	.296		6.70	9.60		16.30	22
3400	3" diameter		48	.333		8.55	10.80		19.35	25.50
3410	4" diameter		44	.364		12.15	11.75		23.90	31
3420	5" diameter	▼	35	.457		19.55	14.80		34.35	44
3430	6" diameter	Q-2	40	.600		25	20		45	58
3440	8" diameter	▼	35	.686	▼	47	23		70	86.50

Important: See the Reference Section for critical supporting data - Reference Nos., Crews, & City Cost Indexes

		CREW	DAILY OUTPUT	LABOR-HOURS	UNIT	2002 BARE COSTS				TOTAL INCL O&P	
	15107 │ Metal Pipe & Fittings					MAT.	LABOR	EQUIP.	TOTAL		
690 3450	10" diameter	Q-2	29	.828	L.F.	62.50	28		90.50	111	**690**
3460	12" diameter	↓	24	1	↓	82.50	33.50		116	142	
3920	To delete coupling & hanger, subtract										
3930	3/4" diam. to 2" diam.					30%	25%				
3940	2-1/2" diam. to 5" diam.					15%	17%				
3950	6" diam. to 12" diam.					11%	12%				
3990	Fittings: cplg. & labor required at joints not incl. in fitting										
3994	price. Add 1 per joint for installed price.										
4000	Elbow, 90° or 45°, painted										
4030	3/4" diameter	1 Plum	50	.160	Ea.	12.35	5.75		18.10	22	
4040	1" diameter		50	.160		12.35	5.75		18.10	22	
4050	1-1/4" diameter		40	.200		12.35	7.20		19.55	24.50	
4060	1-1/2" diameter		33	.242		12.35	8.70		21.05	26.50	
4070	2" diameter	↓	25	.320		12.35	11.50		23.85	31	
4080	2-1/2" diameter	Q-1	40	.400		12.35	12.95		25.30	33	
4090	3" diameter		33	.485		22	15.70		37.70	47.50	
4100	4" diameter		25	.640		24	20.50		44.50	57.50	
4110	5" diameter	↓	20	.800		57.50	26		83.50	103	
4120	6" diameter	Q-2	25	.960		68	32		100	123	
4130	8" diameter		21	1.143		142	38.50		180.50	214	
4140	10" diameter		18	1.333		258	44.50		302.50	350	
4150	12" diameter		15	1.600		415	53.50		468.50	535	
4170	14" diameter		12	2		740	67		807	915	
4180	16" diameter	↓	11	2.182		960	73		1,033	1,150	
4190	18" diameter	Q-3	15	2.133		1,225	73		1,298	1,450	
4200	20" diameter		13	2.462		1,600	84.50		1,684.50	1,900	
4210	24" diameter	↓	11	2.909		2,325	99.50		2,424.50	2,725	
4250	For galvanized elbows, add				↓	26%					
4690	Tee, painted										
4700	3/4" diameter	1 Plum	38	.211	Ea.	19	7.55		26.55	32.50	
4740	1" diameter		33	.242		19	8.70		27.70	34	
4750	1-1/4" diameter		27	.296		19	10.65		29.65	37	
4760	1-1/2" diameter		22	.364		19	13.05		32.05	40.50	
4770	2" diameter	↓	17	.471		19	16.90		35.90	46.50	
4780	2-1/2" diameter	Q-1	27	.593		19	19.15		38.15	50	
4790	3" diameter		22	.727		26.50	23.50		50	65	
4800	4" diameter		17	.941		40.50	30.50		71	90.50	
4810	5" diameter	↓	13	1.231		95.50	40		135.50	165	
4820	6" diameter	Q-2	17	1.412		110	47.50		157.50	193	
4830	8" diameter		14	1.714		242	57.50		299.50	355	
4840	10" diameter		12	2		505	67		572	655	
4850	12" diameter		10	2.400		700	80.50		780.50	890	
4851	14" diameter		9	2.667		745	89.50		834.50	955	
4852	16" diameter	↓	8	3		840	101		941	1,075	
4853	18" diameter	Q-3	11	2.909		1,050	99.50		1,149.50	1,300	
4854	20" diameter		10	3.200		1,500	110		1,610	1,825	
4855	24" diameter	↓	8	4		2,300	137		2,437	2,725	
4900	For galvanized tees, add				↓	24%					
4906	Couplings, rigid style, painted										
4908	1" diameter	1 Plum	100	.080	Ea.	9.10	2.88		11.98	14.35	
4909	1-1/4" diameter		100	.080		9.10	2.88		11.98	14.35	
4910	1-1/2" diameter		67	.119		9.10	4.29		13.39	16.45	
4912	2" diameter	↓	50	.160		9.10	5.75		14.85	18.65	
4914	2-1/2" diameter	Q-1	80	.200		10.75	6.45		17.20	21.50	
4916	3" diameter		67	.239		12.55	7.75		20.30	25.50	
4918	4" diameter	↓	50	.320	↓	17.80	10.35		28.15	35	

MECHANICAL 15

	15107	Metal Pipe & Fittings	CREW	DAILY OUTPUT	LABOR-HOURS	UNIT	2002 BARE COSTS				TOTAL INCL O&P	
							MAT.	LABOR	EQUIP.	TOTAL		
690	4920	5" diameter	Q-1	40	.400	Ea.	23	12.95		35.95	45	690
	4922	6" diameter	Q-2	50	.480		31	16.10		47.10	58.50	
	4924	8" diameter		42	.571		48.50	19.15		67.65	82.50	
	4926	10" diameter		35	.686		87	23		110	130	
	4928	12" diameter		32	.750		97.50	25		122.50	145	
	4930	14" diameter		24	1		127	33.50		160.50	191	
	4931	16" diameter		20	1.200		166	40.50		206.50	243	
	4932	18" diameter		18	1.333		192	44.50		236.50	279	
	4933	20" diameter		16	1.500		228	50.50		278.50	325	
	4934	24" diameter	Q-9	13	1.231		335	39		374	430	
	4940	Flexible, standard, painted										
	4950	3/4" diameter	1 Plum	100	.080	Ea.	6.55	2.88		9.43	11.55	
	4960	1" diameter		100	.080		6.55	2.88		9.43	11.55	
	4970	1-1/4" diameter		80	.100		8.70	3.60		12.30	14.95	
	4980	1-1/2" diameter		67	.119		9.50	4.29		13.79	16.90	
	4990	2" diameter		50	.160		10.05	5.75		15.80	19.70	
	5000	2-1/2" diameter	Q-1	80	.200		12	6.45		18.45	23	
	5010	3" diameter		67	.239		13.30	7.75		21.05	26.50	
	5020	3-1/2" diameter		57	.281		19.35	9.10		28.45	35	
	5030	4" diameter		50	.320		19.35	10.35		29.70	37	
	5040	5" diameter		40	.400		29.50	12.95		42.45	52	
	5050	6" diameter	Q-2	50	.480		35	16.10		51.10	63	
	5070	8" diameter		42	.571		57	19.15		76.15	92	
	5090	10" diameter		35	.686		95	23		118	139	
	5110	12" diameter		32	.750		108	25		133	157	
	5120	14" diameter		24	1		131	33.50		164.50	195	
	5130	16" diameter		20	1.200		172	40.50		212.50	250	
	5140	18" diameter		18	1.333		201	44.50		245.50	290	
	5150	20" diameter		16	1.500		315	50.50		365.50	425	
	5160	24" diameter		13	1.846		345	62		407	475	
	5176	Lightweight style, painted										
	5178	1-1/2" diameter	1 Plum	67	.119	Ea.	8.30	4.29		12.59	15.55	
	5180	2" diameter	"	50	.160		8.45	5.75		14.20	17.95	
	5182	2-1/2" diameter	Q-1	80	.200		9.85	6.45		16.30	20.50	
	5184	3" diameter		67	.239		11.40	7.75		19.15	24	
	5186	3-1/2" diameter		57	.281		16.15	9.10		25.25	31.50	
	5188	4" diameter		50	.320		16.15	10.35		26.50	33.50	
	5190	5" diameter		40	.400		23.50	12.95		36.45	45	
	5192	6" diameter	Q-2	50	.480		28	16.10		44.10	55.50	
	5194	8" diameter		42	.571		44	19.15		63.15	77.50	
	5196	10" diameter		35	.686		118	23		141	165	
	5198	12" diameter		32	.750		132	25		157	184	
	5200	For galvanized couplings, add					33%					
	5220	Tee, reducing, painted										
	5225	2" x 1-1/2" diameter	Q-1	38	.421	Ea.	41	13.60		54.60	65.50	
	5226	2-1/2" x 2" diameter		28	.571		41	18.50		59.50	73	
	5227	3" x 2-1/2" diameter		23	.696		36	22.50		58.50	74	
	5228	4" x 3" diameter		18	.889		49	29		78	97.50	
	5229	5" x 4" diameter		15	1.067		105	34.50		139.50	168	
	5230	6" x 4" diameter	Q-2	18	1.333		116	44.50		160.50	196	
	5231	8" x 6" diameter		15	1.600		242	53.50		295.50	345	
	5232	10" x 8" diameter		13	1.846		320	62		382	445	
	5233	12" x 10" diameter		11	2.182		485	73		558	645	
	5234	14" x 12" diameter		10	2.400		540	80.50		620.50	715	
	5235	16" x 12" diameter		9	2.667		670	89.50		759.50	870	
	5236	18" x 12" diameter	Q-3	12	2.667		805	91.50		896.50	1,025	

15 MECHANICAL

15107	Metal Pipe & Fittings	CREW	DAILY OUTPUT	LABOR-HOURS	UNIT	2002 BARE COSTS				TOTAL INCL O&P
						MAT.	LABOR	EQUIP.	TOTAL	
5237	18" x 16" diameter	Q-3	11	2.909	Ea.	1,025	99.50		1,124.50	1,275
5238	20" x 16" diameter		10	3.200		1,325	110		1,435	1,625
5239	24" x 20" diameter	▼	9	3.556	▼	2,100	122		2,222	2,475
5240	Reducer, concentric, painted									
5241	2-1/2" x 2" diameter	Q-1	43	.372	Ea.	14.35	12.05		26.40	34
5242	3" x 2-1/2" diameter		35	.457		17.35	14.80		32.15	41.50
5243	4" x 3" diameter		29	.552		21	17.85		38.85	50
5244	5" x 4" diameter	▼	22	.727		29.50	23.50		53	68
5245	6" x 4" diameter	Q-2	26	.923		34	31		65	84
5246	8" x 6" diameter		23	1.043		89	35		124	151
5247	10" x 8" diameter		20	1.200		182	40.50		222.50	261
5248	12" x 10" diameter	▼	16	1.500	▼	325	50.50		375.50	435
5255	Eccentric, painted									
5256	2-1/2" x 2" diameter	Q-1	42	.381	Ea.	30.50	12.30		42.80	52
5257	3" x 2-1/2" diameter		34	.471		35	15.20		50.20	61.50
5258	4" x 3" diameter		28	.571		42.50	18.50		61	75
5259	5" x 4" diameter	▼	21	.762		58	24.50		82.50	101
5260	6" x 4" diameter	Q-2	25	.960		67.50	32		99.50	123
5261	8" x 6" diameter		22	1.091		137	36.50		173.50	206
5262	10" x 8" diameter		19	1.263		375	42.50		417.50	480
5263	12" x 10" diameter	▼	15	1.600	▼	520	53.50		573.50	650
5270	Coupling, reducing, painted									
5272	2" x 1-1/2" diameter	1 Plum	52	.154	Ea.	14.35	5.55		19.90	24
5274	2-1/2" x 2" diameter	Q-1	82	.195		18.70	6.30		25	30
5276	3" x 2" diameter		69	.232		21.50	7.50		29	35
5278	4" x 2" diameter		52	.308		34	9.95		43.95	52.50
5280	5" x 4" diameter	▼	42	.381		38.50	12.30		50.80	61
5282	6" x 4" diameter	Q-2	52	.462		58.50	15.50		74	88
5284	8" x 6" diameter	"	44	.545	▼	88	18.30		106.30	124
5290	Outlet coupling, painted									
5294	1-1/2" x 1" pipe size	1 Plum	65	.123	Ea.	20.50	4.42		24.92	29
5296	2" x 1" pipe size	"	48	.167		22	6		28	33.50
5298	2-1/2" x 1" pipe size	Q-1	78	.205		34.50	6.65		41.15	48
5300	2-1/2" x 1" pipe size	1 Plum	70	.114		40	4.11		44.11	50
5302	3" x 1" pipe size	Q-1	65	.246		32.50	7.95		40.45	48
5304	4" x 3/4" pipe size		48	.333		49	10.80		59.80	70.50
5306	4" x 1-1/2" pipe size	▼	46	.348		71.50	11.25		82.75	95.50
5308	6" x 1-1/2" pipe size	Q-2	44	.545	▼	103	18.30		121.30	142
5750	Flange, w/groove gasket, black steel									
5752	See 15107-660-0620 for gasket & bolt set									
5760	ANSI class 125 and 150, painted									
5780	2" pipe size	1 Plum	23	.348	Ea.	41.50	12.50		54	64.50
5790	2-1/2" pipe size	Q-1	37	.432		51.50	14		65.50	78
5800	3" pipe size		31	.516		56	16.70		72.70	86.50
5820	4" pipe size		23	.696		75	22.50		97.50	117
5830	5" pipe size	▼	19	.842		87	27		114	137
5840	6" pipe size	Q-2	23	1.043		95	35		130	157
5850	8" pipe size		17	1.412		107	47.50		154.50	190
5860	10" pipe size		14	1.714		170	57.50		227.50	274
5870	12" pipe size		12	2		222	67		289	345
5880	14" pipe size		10	2.400		510	80.50		590.50	680
5890	16" pipe size		9	2.667		590	89.50		679.50	785
5900	18" pipe size		6	4		730	134		864	1,000
5910	20" pipe size		5	4.800		875	161		1,036	1,200
5920	24" pipe size	▼	4.50	5.333	▼	1,125	179		1,304	1,500
5940	ANSI class 350, painted									

15107	Metal Pipe & Fittings	CREW	DAILY OUTPUT	LABOR-HOURS	UNIT	MAT.	LABOR	EQUIP.	TOTAL	TOTAL INCL O&P
5946	2" pipe size	1 Plum	23	.348	Ea.	52	12.50		64.50	76.50
5948	2-1/2" pipe size	Q-1	37	.432		60.50	14		74.50	87.50
5950	3" pipe size		31	.516		82.50	16.70		99.20	116
5952	4" pipe size		23	.696		110	22.50		132.50	155
5954	5" pipe size	▼	19	.842		125	27		152	179
5956	6" pipe size	Q-2	23	1.043		146	35		181	214
5958	8" pipe size		17	1.412		168	47.50		215.50	256
5960	10" pipe size	▼	14	1.714		267	57.50		324.50	380
5962	12" pipe size	1 Plum	12	.667	▼	285	24		309	350
6100	Coupling, for PVC plastic pipe									
6110	2" diameter	1 Plum	50	.160	Ea.	11.20	5.75		16.95	21
6112	2-1/2" diameter	Q-1	80	.200		15.70	6.45		22.15	27
6114	3" diameter		67	.239		19.05	7.75		26.80	32.50
6116	4" diameter		50	.320		25	10.35		35.35	43
6118	6" diameter	Q-2	50	.480		41	16.10		57.10	70
6120	8" diameter		42	.571		68	19.15		87.15	104
6122	10" diameter		35	.686		111	23		134	157
6124	12" diameter	▼	32	.750	▼	141	25		166	193
7400	Suction diffuser									
7402	Grooved end inlet x flanged outlet									
7410	3" x 3"	Q-1	50	.320	Ea.	435	10.35		445.35	490
7412	4" x 4"		38	.421		585	13.60		598.60	665
7414	5" x 5"	▼	30	.533		685	17.25		702.25	780
7416	6" x 6"	Q-2	38	.632		865	21		886	980
7418	8" x 8"		27	.889		1,600	30		1,630	1,825
7420	10" x 10"		20	1.200		2,200	40.50		2,240.50	2,450
7422	12" x 12"		16	1.500		3,600	50.50		3,650.50	4,050
7424	14" x 14"		15	1.600		4,550	53.50		4,603.50	5,075
7426	16" x 14"	▼	14	1.714	▼	4,700	57.50		4,757.50	5,225
7500	Strainer, tee type, painted									
7506	2" pipe size	1 Plum	38	.211	Ea.	270	7.55		277.55	310
7508	2-1/2" pipe size	Q-1	62	.258		283	8.35		291.35	325
7510	3" pipe size		50	.320		320	10.35		330.35	365
7512	4" pipe size		38	.421		360	13.60		373.60	415
7514	5" pipe size	▼	30	.533		520	17.25		537.25	600
7516	6" pipe size	Q-2	38	.632		565	21		586	655
7518	8" pipe size		27	.889		870	30		900	1,000
7520	10" pipe size		20	1.200		1,275	40.50		1,315.50	1,450
7522	12" pipe size		16	1.500		1,650	50.50		1,700.50	1,875
7524	14" pipe size		15	1.600		5,850	53.50		5,903.50	6,500
7526	16" pipe size	▼	14	1.714	▼	7,275	57.50		7,332.50	8,075
7570	Expansion joint, max. 3" travel									
7572	2" diameter	1 Plum	38	.211	Ea.	221	7.55		228.55	255
7574	3" diameter	Q-1	50	.320		260	10.35		270.35	300
7576	4" diameter	"	38	.421		325	13.60		338.60	380
7578	6" diameter	Q-2	38	.632	▼	565	21		586	655
7800	Ball valve w/handle, carbon steel trim									
7810	1-1/2" pipe size	1 Plum	50	.160	Ea.	71	5.75		76.75	87
7812	2" pipe size	"	38	.211		79	7.55		86.55	98.50
7814	2-1/2" pipe size	Q-1	62	.258		173	8.35		181.35	203
7816	3" pipe size		50	.320		280	10.35		290.35	325
7818	4" pipe size	▼	38	.421		435	13.60		448.60	495
7820	6" pipe size	Q-2	30	.800	▼	1,325	27		1,352	1,500
7830	With gear operator									
7834	2-1/2" pipe size	Q-1	62	.258	Ea.	330	8.35		338.35	380
7836	3" pipe size	▼	50	.320	▼	435	10.35		445.35	495

Important: See the Reference Section for critical supporting data - Reference Nos., Crews, & City Cost Indexes

15107	Metal Pipe & Fittings	CREW	DAILY OUTPUT	LABOR-HOURS	UNIT	2002 BARE COSTS				TOTAL INCL O&P
						MAT.	LABOR	EQUIP.	TOTAL	
7838	4" pipe size	Q-1	38	.421	Ea.	585	13.60		598.60	665
7840	6" pipe size	Q-2	30	.800	↓	1,500	27		1,527	1,700
7870	Check valve									
7874	2-1/2" pipe size	Q-1	62	.258	Ea.	118	8.35		126.35	143
7876	3" pipe size		50	.320		140	10.35		150.35	170
7878	4" pipe size		38	.421		147	13.60		160.60	183
7880	5" pipe size	↓	30	.533		246	17.25		263.25	296
7882	6" pipe size	Q-2	38	.632		291	21		312	350
7884	8" pipe size		27	.889		400	30		430	485
7886	10" pipe size		20	1.200		1,150	40.50		1,190.50	1,325
7888	12" pipe size	↓	16	1.500	↓	1,375	50.50		1,425.50	1,575
7900	Plug valve, balancing, w/lever operator									
7906	3" pipe size	Q-1	50	.320	Ea.	242	10.35		252.35	283
7908	4" pipe size	"	38	.421		267	13.60		280.60	315
7909	6" pipe size	Q-2	30	.800	↓	480	27		507	565
7916	With gear operator									
7920	3" pipe size	Q-1	50	.320	Ea.	440	10.35		450.35	495
7922	4" pipe size	"	38	.421		460	13.60		473.60	530
7924	6" pipe size	Q-2	38	.632		675	21		696	770
7926	8" pipe size		27	.889		900	30		930	1,025
7928	10" pipe size		20	1.200		1,425	40.50		1,465.50	1,600
7930	12" pipe size	↓	16	1.500	↓	2,075	50.50		2,125.50	2,350
8000	Butterfly valve, 2 position handle, with standard trim									
8010	1-1/2" pipe size	1 Plum	50	.160	Ea.	104	5.75		109.75	124
8020	2" pipe size	"	38	.211		104	7.55		111.55	126
8030	3" pipe size	Q-1	50	.320		150	10.35		160.35	181
8050	4" pipe size	"	38	.421		165	13.60		178.60	202
8070	6" pipe size	Q-2	38	.632		335	21		356	395
8080	8" pipe size		27	.889		470	30		500	565
8090	10" pipe size	↓	20	1.200	↓	680	40.50		720.50	810
8200	With stainless steel trim									
8240	1-1/2" pipe size	1 Plum	50	.160	Ea.	132	5.75		137.75	155
8250	2" pipe size	"	38	.211		132	7.55		139.55	157
8270	3" pipe size	Q-1	50	.320		178	10.35		188.35	212
8280	4" pipe size	"	38	.421		193	13.60		206.60	234
8300	6" pipe size	Q-2	38	.632		360	21		381	425
8310	8" pipe size		27	.889		665	30		695	775
8320	10" pipe size		20	1.200		1,100	40.50		1,140.50	1,275
8322	12" pipe size		16	1.500		1,350	50.50		1,400.50	1,550
8324	14" pipe size		15	1.600		1,625	53.50		1,678.50	1,875
8326	16" pipe size	↓	14	1.714		2,250	57.50		2,307.50	2,550
8328	18" pipe size	Q-3	12	2.667		2,750	91.50		2,841.50	3,175
8330	20" pipe size		11	2.909		3,900	99.50		3,999.50	4,450
8332	24" pipe size	↓	10	3.200	↓	4,750	110		4,860	5,400
8336	Note: sizes 8" up w/manual gear operator									
9000	Cut one groove, labor									
9010	3/4" pipe size	Q-1	152	.105	Ea.		3.41		3.41	5.15
9020	1" pipe size		140	.114			3.70		3.70	5.60
9030	1-1/4" pipe size		124	.129			4.17		4.17	6.30
9040	1-1/2" pipe size		114	.140			4.54		4.54	6.85
9050	2" pipe size		104	.154			4.98		4.98	7.50
9060	2-1/2" pipe size		96	.167			5.40		5.40	8.15
9070	3" pipe size		88	.182			5.90		5.90	8.85
9080	3-1/2" pipe size		83	.193			6.25		6.25	9.40
9090	4" pipe size	↓	78	.205			6.65		6.65	10
9100	5" pipe size	↓	72	.222	↓		7.20		7.20	10.85

690

690

15107	Metal Pipe & Fittings	CREW	DAILY OUTPUT	LABOR-HOURS	UNIT	2002 BARE COSTS				TOTAL INCL O&P	
						MAT.	LABOR	EQUIP.	TOTAL		
690	9110	6" pipe size	Q-1	70	.229	Ea.		7.40		7.40	11.15	**690**
	9120	8" pipe size		54	.296			9.60		9.60	14.45	
	9130	10" pipe size		38	.421			13.60		13.60	20.50	
	9140	12" pipe size		30	.533			17.25		17.25	26	
	9150	14" pipe size		20	.800			26		26	39	
	9160	16" pipe size		19	.842			27		27	41	
	9170	18" pipe size		18	.889			29		29	43.50	
	9180	20" pipe size		17	.941			30.50		30.50	46	
	9190	24" pipe size	▼	15	1.067	▼		34.50		34.50	52	
	9210	Roll one groove										
	9220	3/4" pipe size	Q-1	266	.060	Ea.		1.95		1.95	2.93	
	9230	1" pipe size		228	.070			2.27		2.27	3.42	
	9240	1-1/4" pipe size		200	.080			2.59		2.59	3.90	
	9250	1-1/2" pipe size		178	.090			2.91		2.91	4.38	
	9260	2" pipe size		116	.138			4.46		4.46	6.75	
	9270	2-1/2" pipe size		110	.145			4.71		4.71	7.10	
	9280	3" pipe size		100	.160			5.20		5.20	7.80	
	9290	3-1/2" pipe size		94	.170			5.50		5.50	8.30	
	9300	4" pipe size		86	.186			6		6	9.10	
	9310	5" pipe size		84	.190			6.15		6.15	9.30	
	9320	6" pipe size		80	.200			6.45		6.45	9.75	
	9330	8" pipe size		66	.242			7.85		7.85	11.85	
	9340	10" pipe size		58	.276			8.90		8.90	13.45	
	9350	12" pipe size		46	.348			11.25		11.25	16.95	
	9360	14" pipe size		30	.533			17.25		17.25	26	
	9370	16" pipe size		28	.571			18.50		18.50	28	
	9380	18" pipe size		27	.593			19.15		19.15	29	
	9390	20" pipe size		25	.640			20.50		20.50	31	
	9400	24" pipe size	▼	23	.696	▼		22.50		22.50	34	
920	0010	**PIPE, STAINLESS STEEL**										**920**
	0020	Welded, with clevis type hangers 10' O.C.										
	0500	Schedule 5, type 304										
	0540	1/2" diameter	Q-15	128	.125	L.F.	4.86	4.04	.63	9.53	12.15	
	0550	3/4" diameter		116	.138		5.50	4.46	.69	10.65	13.55	
	0560	1" diameter		103	.155		6.70	5.05	.78	12.53	15.85	
	0570	1-1/4" diameter		93	.172		7.65	5.55	.86	14.06	17.80	
	0580	1-1/2" diameter		85	.188		8.80	6.10	.94	15.84	19.90	
	0590	2" diameter		69	.232		10.55	7.50	1.16	19.21	24	
	0600	2-1/2" diameter		53	.302		14.70	9.75	1.51	25.96	32.50	
	0610	3" diameter		48	.333		17.70	10.80	1.67	30.17	37.50	
	0620	4" diameter		44	.364		22.50	11.75	1.82	36.07	44.50	
	0630	5" diameter	▼	36	.444		45	14.40	2.22	61.62	74	
	0640	6" diameter	Q-16	42	.571		42	19.15	1.90	63.05	77	
	0650	8" diameter		34	.706		64	23.50	2.35	89.85	109	
	0660	10" diameter		26	.923		89.50	31	3.07	123.57	148	
	0670	12" diameter	▼	21	1.143	▼	118	38.50	3.81	160.31	192	
	0700	To delete hangers, subtract										
	0710	1/2" diam. to 1-1/2" diam.					8%	19%				
	0720	2" diam. to 5" diam.					4%	9%				
	0730	6" diam. to 12" diam.					3%	4%				
	0750	For small quantities, add				L.F.	10%					
	1250	Schedule 5, type 316										
	1290	1/2" diameter	Q-15	128	.125	L.F.	6.10	4.04	.63	10.77	13.50	
	1300	3/4" diameter		116	.138		7.05	4.46	.69	12.20	15.25	
	1310	1" diameter	▼	103	.155	▼	8	5.05	.78	13.83	17.25	

Important: See the Reference Section for critical supporting data - Reference Nos., Crews, & City Cost Indexes

15107	Metal Pipe & Fittings	CREW	DAILY OUTPUT	LABOR-HOURS	UNIT	MAT.	LABOR	EQUIP.	TOTAL	TOTAL INCL O&P	
920	1320	1-1/4" diameter	Q-15	93	.172	L.F.	9.60	5.55	.86	16.01	19.90	920
1330	1-1/2" diameter		85	.188		10.90	6.10	.94	17.94	22
1340	2" diameter		69	.232		13.20	7.50	1.16	21.86	27
1350	2-1/2" diameter		53	.302		18.95	9.75	1.51	30.21	37.50
1360	3" diameter		48	.333		24	10.80	1.67	36.47	44.50
1370	4" diameter		44	.364		29.50	11.75	1.82	43.07	52.50
1380	5" diameter	▼	36	.444		56.50	14.40	2.22	73.12	86
1390	6" diameter	Q-16	42	.571		55	19.15	1.90	76.05	91.50
1400	8" diameter		34	.706		82.50	23.50	2.35	108.35	129
1410	10" diameter		26	.923		119	31	3.07	153.07	181
1420	12" diameter	▼	21	1.143		153	38.50	3.81	195.31	230
1490	For small quantities, add				▼	10%				
1940	To delete hanger, subtract									
1950	1/2" diam. to 1-1/2" diam.					5%	19%			
1960	2" diam. to 5" diam.					3%	9%			
1970	6" diam. to 12" diam.					2%	4%			
2000	Schedule 10, type 304									
2040	1/4" diameter	Q-15	131	.122	L.F.	3.83	3.95	.61	8.39	10.85
2050	3/8" diameter		128	.125		4.04	4.04	.63	8.71	11.25
2060	1/2" diameter		125	.128		4.97	4.14	.64	9.75	12.40
2070	3/4" diameter		113	.142		5.75	4.58	.71	11.04	14.05
2080	1" diameter		100	.160		8.25	5.20	.80	14.25	17.75
2090	1-1/4" diameter		91	.176		9.90	5.70	.88	16.48	20.50
2100	1-1/2" diameter		83	.193		10.90	6.25	.96	18.11	22.50
2110	2" diameter		67	.239		13.50	7.75	1.19	22.44	28
2120	2-1/2" diameter		51	.314		16.75	10.15	1.57	28.47	35.50
2130	3" diameter		46	.348		20.50	11.25	1.74	33.49	41.50
2140	4" diameter		42	.381		26.50	12.30	1.90	40.70	50
2150	5" diameter	▼	35	.457		36	14.80	2.29	53.09	64.50
2160	6" diameter	Q-16	40	.600		42	20	2	64	78.50
2170	8" diameter		33	.727		64	24.50	2.42	90.92	110
2180	10" diameter		25	.960		89.50	32	3.20	124.70	151
2190	12" diameter	▼	21	1.143	▼	112	38.50	3.81	154.31	185
2250	For small quantities, add				▼	10%				
2650	To delete hanger, subtract									
2660	1/4" diam. to 3/4" diam.					9%	22%			
2670	1" diam. to 2" diam.					4%	15%			
2680	2-1/2" diam. to 5" diam.					3%	8%			
2690	6" diam. to 12" diam.					3%	4%			
2750	Schedule 10, type 316									
2790	1/4" diameter	Q-15	131	.122	L.F.	4.31	3.95	.61	8.87	11.35
2800	3/8" diameter		128	.125		4.78	4.04	.63	9.45	12.05
2810	1/2" diameter		125	.128		6.10	4.14	.64	10.88	13.65
2820	3/4" diameter		113	.142		7.05	4.58	.71	12.34	15.50
2830	1" diameter		100	.160		9.30	5.20	.80	15.30	18.90
2840	1-1/4" diameter		91	.176		12.45	5.70	.88	19.03	23.50
2850	1-1/2" diameter		83	.193		13.90	6.25	.96	21.11	26
2860	2" diameter		67	.239		17.15	7.75	1.19	26.09	32
2870	2-1/2" diameter		51	.314		22.50	10.15	1.57	34.22	41.50
2880	3" diameter		46	.348		28	11.25	1.74	40.99	50
2890	4" diameter		42	.381		34	12.30	1.90	48.20	58
2900	5" diameter	▼	35	.457		50.50	14.80	2.29	67.59	81
2910	6" diameter	Q-16	40	.600		54.50	20	2	76.50	92.50
2920	8" diameter		33	.727		90	24.50	2.42	116.92	139
2930	10" diameter		25	.960		119	32	3.20	154.20	183
2940	12" diameter	▼	21	1.143	▼	149	38.50	3.81	191.31	226

MECHANICAL 15

15107	Metal Pipe & Fittings	CREW	DAILY OUTPUT	LABOR-HOURS	UNIT	2002 BARE COSTS				TOTAL INCL O&P	
						MAT.	LABOR	EQUIP.	TOTAL		
920											920
2990	For small quantities, add				L.F.	10%					
3430	To delete hanger, subtract										
3440	1/4" diam. to 3/4" diam.					6%	22%				
3450	1" diam. to 2" diam.					3%	15%				
3460	2-1/2" diam. to 5" diam.					2%	8%				
3470	6" diam. to 12" diam.					2%	4%				
3500	Threaded, couplings and hangers 10' O.C.										
3520	Schedule 40, type 304										
3540	1/4" diameter	1 Plum	54	.148	L.F.	5.05	5.35		10.40	13.60	
3550	3/8" diameter		53	.151		5.75	5.45		11.20	14.50	
3560	1/2" diameter		52	.154		6.60	5.55		12.15	15.60	
3570	3/4" diameter		51	.157		8.90	5.65		14.55	18.30	
3580	1" diameter		45	.178		11.20	6.40		17.60	22	
3590	1-1/4" diameter	Q-1	76	.211		13.20	6.80		20	25	
3600	1-1/2" diameter		69	.232		15.40	7.50		22.90	28.50	
3610	2" diameter		57	.281		20.50	9.10		29.60	36	
3620	2-1/2" diameter		44	.364		32.50	11.75		44.25	54	
3630	3" diameter		38	.421		42	13.60		55.60	67	
3640	4" diameter	Q-2	51	.471		58	15.80		73.80	88	
3740	For small quantities, add					10%					
4200	To delete couplings & hangers, subtract										
4210	1/4" diam. to 3/4" diam.					15%	56%				
4220	1" diam. to 2" diam.					18%	49%				
4230	2-1/2" diam. to 4" diam.					34%	40%				
4250	Schedule 40, type 316										
4290	1/4" diameter	1 Plum	54	.148	L.F.	5.90	5.35		11.25	14.50	
4300	3/8" diameter		53	.151		7.65	5.45		13.10	16.60	
4310	1/2" diameter		52	.154		8.60	5.55		14.15	17.85	
4320	3/4" diameter		51	.157		10.85	5.65		16.50	20.50	
4330	1" diameter		45	.178		14.25	6.40		20.65	25.50	
4340	1-1/4" diameter	Q-1	76	.211		16.80	6.80		23.60	29	
4350	1-1/2" diameter		69	.232		19.65	7.50		27.15	33	
4360	2" diameter		57	.281		26	9.10		35.10	42	
4370	2-1/2" diameter		44	.364		42.50	11.75		54.25	64.50	
4380	3" diameter		38	.421		57	13.60		70.60	83.50	
4390	4" diameter	Q-2	51	.471		77	15.80		92.80	109	
4490	For small quantities, add					10%					
4900	To delete couplings & hangers, subtract										
4910	1/4" diam. to 3/4" diam.					12%	56%				
4920	1" diam. to 2" diam.					14%	49%				
4930	2-1/2" diam. to 4" diam.					27%	40%				
5000	Schedule 80, type 304										
5040	1/4" diameter	1 Plum	53	.151	L.F.	6.65	5.45		12.10	15.55	
5050	3/8" diameter		52	.154		7.55	5.55		13.10	16.65	
5060	1/2" diameter		51	.157		9.45	5.65		15.10	18.90	
5070	3/4" diameter		48	.167		12.15	6		18.15	22.50	
5080	1" diameter		43	.186		16.60	6.70		23.30	28.50	
5090	1-1/4" diameter	Q-1	73	.219		23.50	7.10		30.60	36.50	
5100	1-1/2" diameter		67	.239		27.50	7.75		35.25	41.50	
5110	2" diameter		54	.296		34.50	9.60		44.10	52.50	
5190	For small quantities, add					10%					
5700	To delete couplings & hangers, subtract										
5710	1/4" diam. to 3/4" diam.					10%	53%				
5720	1" diam. to 2" diam.					14%	47%				
5750	Schedule 80, type 316										
5790	1/4" diameter	1 Plum	53	.151	L.F.	8.20	5.45		13.65	17.20	

Important: See the Reference Section for critical supporting data - Reference Nos., Crews, & City Cost Indexes

15 MECHANICAL

		CREW	DAILY OUTPUT	LABOR-HOURS	UNIT	2002 BARE COSTS				TOTAL INCL O&P	
	15107 | Metal Pipe & Fittings					MAT.	LABOR	EQUIP.	TOTAL		
920											**920**
5800	3/8" diameter	1 Plum	52	.154	L.F.	9.30	5.55		14.85	18.60	
5810	1/2" diameter		51	.157		13.30	5.65		18.95	23	
5820	3/4" diameter		48	.167		15.05	6		21.05	25.50	
5830	1" diameter		43	.186		20.50	6.70		27.20	33	
5840	1-1/4" diameter	Q-1	73	.219		30	7.10		37.10	43.50	
5850	1-1/2" diameter		67	.239		34.50	7.75		42.25	49.50	
5860	2" diameter		54	.296		44.50	9.60		54.10	63	
5950	For small quantities, add					10%					
7000	To delete couplings & hangers, subtract										
7010	1/4" diam. to 3/4" diam.					9%	53%				
7020	1" diam. to 2" diam.					14%	47%				
8000	Weld joints with clevis type hangers 10' O.C.										
8010	Schedule 40, type 304										
8050	1/8" pipe size	Q-15	126	.127	L.F.	3.57	4.11	.63	8.31	10.85	
8060	1/4" pipe size		125	.128		4.45	4.14	.64	9.23	11.85	
8070	3/8" pipe size		122	.131		5.05	4.24	.66	9.95	12.65	
8080	1/2" pipe size		118	.136		6.10	4.39	.68	11.17	14.05	
8090	3/4" pipe size		109	.147		7.60	4.75	.73	13.08	16.30	
8100	1" pipe size		95	.168		9.50	5.45	.84	15.79	19.60	
8110	1-1/4" pipe size		86	.186		11.70	6	.93	18.63	23	
8120	1-1/2" pipe size		78	.205		13.60	6.65	1.03	21.28	26	
8130	2" pipe size		62	.258		17.45	8.35	1.29	27.09	33	
8140	2-1/2" pipe size		49	.327		26	10.55	1.63	38.18	47	
8150	3" pipe size		44	.364		33.50	11.75	1.82	47.07	57	
8160	3-1/2" pipe size		44	.364		39.50	11.75	1.82	53.07	63.50	
8170	4" pipe size		39	.410		45	13.25	2.05	60.30	72	
8180	5" pipe size		32	.500		65.50	16.20	2.50	84.20	99.50	
8190	6" pipe size	Q-16	37	.649		78.50	22	2.16	102.66	122	
8200	8" pipe size		29	.828		127	28	2.76	157.76	184	
8210	10" pipe size		24	1		215	33.50	3.33	251.83	290	
8220	12" pipe size		20	1.200		243	40.50	4	287.50	330	
8300	Schedule 40, type 316										
8310	1/8" pipe size	Q-15	126	.127	L.F.	4.39	4.11	.63	9.13	11.75	
8320	1/4" pipe size		125	.128		5.20	4.14	.64	9.98	12.65	
8330	3/8" pipe size		122	.131		6.80	4.24	.66	11.70	14.60	
8340	1/2" pipe size		118	.136		8	4.39	.68	13.07	16.15	
8350	3/4" pipe size		109	.147		9.25	4.75	.73	14.73	18.15	
8360	1" pipe size		95	.168		12.15	5.45	.84	18.44	22.50	
8370	1-1/4" pipe size		86	.186		15	6	.93	21.93	26.50	
8380	1-1/2" pipe size		78	.205		17.50	6.65	1.03	25.18	30.50	
8390	2" pipe size		62	.258		22.50	8.35	1.29	32.14	39	
8400	2-1/2" pipe size		49	.327		35	10.55	1.63	47.18	56	
8410	3" pipe size		44	.364		47	11.75	1.82	60.57	71.50	
8420	3-1/2" pipe size		44	.364		54.50	11.75	1.82	68.07	79.50	
8430	4" pipe size		39	.410		61.50	13.25	2.05	76.80	90.50	
8440	5" pipe size		32	.500		93	16.20	2.50	111.70	129	
8450	6" pipe size	Q-16	37	.649		104	22	2.16	128.16	150	
8460	8" pipe size		29	.828		195	28	2.76	225.76	259	
8470	10" pipe size		24	1		261	33.50	3.33	297.83	340	
8480	12" pipe size		20	1.200		355	40.50	4	399.50	455	
8500	Schedule 80, type 304										
8510	1/4" pipe size	Q-15	110	.145	L.F.	6	4.71	.73	11.44	14.50	
8520	3/8" pipe size		109	.147		6.75	4.75	.73	12.23	15.40	
8530	1/2" pipe size		106	.151		8.50	4.88	.75	14.13	17.55	
8540	3/4" pipe size		96	.167		10.85	5.40	.83	17.08	21	
8550	1" pipe size		87	.184		14.45	5.95	.92	21.32	26	

MECHANICAL 15

			DAILY	LABOR-			2002 BARE COSTS				TOTAL	
15107	**Metal Pipe & Fittings**	CREW	OUTPUT	HOURS	UNIT	MAT.	LABOR	EQUIP.	TOTAL	INCL O&P	
920	8560	1-1/4" pipe size	Q-15	81	.198	L.F.	18.60	6.40	.99	25.99	31	920
	8570	1-1/2" pipe size		74	.216		21	7	1.08	29.08	35	
	8580	2" pipe size		58	.276		26.50	8.90	1.38	36.78	44	
	8590	2-1/2" pipe size		46	.348		54	11.25	1.74	66.99	78.50	
	8600	3" pipe size		41	.390		72.50	12.60	1.95	87.05	101	
	8610	4" pipe size		33	.485		113	15.70	2.42	131.12	151	
	8630	6" pipe size	Q-16	30	.800		249	27	2.66	278.66	315	
	8640	Schedule 80, type 316										
	8650	1/4" pipe size	Q-15	110	.145	L.F.	7.45	4.71	.73	12.89	16.10	
	8660	3/8" pipe size		109	.147		8.50	4.75	.73	13.98	17.30	
	8670	1/2" pipe size		106	.151		12.15	4.88	.75	17.78	21.50	
	8680	3/4" pipe size		96	.167		13.40	5.40	.83	19.63	24	
	8690	1" pipe size		87	.184		17.95	5.95	.92	24.82	29.50	
	8700	1-1/4" pipe size		81	.198		24.50	6.40	.99	31.89	37	
	8710	1-1/2" pipe size		74	.216		27.50	7	1.08	35.58	41.50	
	8720	2" pipe size		58	.276		34.50	8.90	1.38	44.78	52.50	
	8730	2-1/2" pipe size		46	.348		89	11.25	1.74	101.99	117	
	8740	3" pipe size		41	.390		104	12.60	1.95	118.55	135	
	8760	4" pipe size		33	.485		144	15.70	2.42	162.12	185	
	8770	6" pipe size	Q-16	30	.800		390	27	2.66	419.66	470	
	8790	Schedule 160, type 304										
	8800	1/2" pipe size	Q-15	96	.167	L.F.	11.95	5.40	.83	18.18	22	
	8810	3/4" pipe size		88	.182		14	5.90	.91	20.81	25.50	
	8820	1" pipe size		80	.200		23.50	6.45	1	30.95	37	
	8840	1-1/2" pipe size		67	.239		32.50	7.75	1.19	41.44	48.50	
	8850	2" pipe size		53	.302		46.50	9.75	1.51	57.76	68	
	8870	3" pipe size		37	.432		76	14	2.16	92.16	107	
	8900	Schedule 160, type 316										
	8910	1/2" pipe size	Q-15	96	.167	L.F.	17.65	5.40	.83	23.88	28.50	
	8920	3/4" pipe size		88	.182		18.55	5.90	.91	25.36	30.50	
	8930	1" pipe size		80	.200		26.50	6.45	1	33.95	40	
	8940	1-1/4" pipe size		73	.219		37.50	7.10	1.10	45.70	53	
	8950	1-1/2" pipe size		67	.239		43	7.75	1.19	51.94	60.50	
	8960	2" pipe size		53	.302		59.50	9.75	1.51	70.76	82	
	8990	4" pipe size		30	.533		150	17.25	2.67	169.92	194	
	9100	Threading pipe labor, sst, one end, schedules 40 & 80										
	9110	1/4" through 3/4" pipe size	1 Plum	61.50	.130	Ea.		4.68		4.68	7.05	
	9120	1" through 2" pipe size		55.90	.143			5.15		5.15	7.75	
	9130	2-1/2" pipe size		41.50	.193			6.95		6.95	10.45	
	9140	3" pipe size		38.50	.208			7.45		7.45	11.25	
	9150	3-1/2" pipe size	Q-1	68.40	.234			7.55		7.55	11.40	
	9160	4" pipe size		73	.219			7.10		7.10	10.70	
	9170	5" pipe size		40.70	.393			12.70		12.70	19.20	
	9180	6" pipe size		35.40	.452			14.60		14.60	22	
	9190	8" pipe size		22.30	.717			23		23	35	
	9200	10" pipe size		16.10	.994			32		32	48.50	
	9210	12" pipe size		12.30	1.301			42		42	63.50	
	9250	Welding labor per joint for stainless steel										
	9260	Schedule 5 and 10										
	9270	1/4" pipe size	Q-15	36	.444	Ea.		14.40	2.22	16.62	24	
	9280	3/8" pipe size		35	.457			14.80	2.29	17.09	25	
	9290	1/2" pipe size		35	.457			14.80	2.29	17.09	25	
	9300	3/4" pipe size		28	.571			18.50	2.86	21.36	31	
	9310	1" pipe size		25	.640			20.50	3.20	23.70	34.50	
	9320	1-1/4" pipe size		22	.727			23.50	3.64	27.14	39.50	
	9330	1-1/2" pipe size		21	.762			24.50	3.81	28.31	41	

15 MECHANICAL

Important: See the Reference Section for critical supporting data - Reference Nos., Crews, & City Cost Indexes

		CREW	DAILY OUTPUT	LABOR-HOURS	UNIT	MAT.	2002 BARE COSTS LABOR	EQUIP.	TOTAL	TOTAL INCL O&P	
15107	**Metal Pipe & Fittings**										
920	9340	2" pipe size	Q-15	18	.889	Ea.		29	4.44	33.44	48.50	**920**
	9350	2-1/2" pipe size		12	1.333			43	6.65	49.65	72.50	
	9360	3" pipe size		9.73	1.644			53	8.20	61.20	89	
	9370	4" pipe size		7.37	2.171			70	10.85	80.85	118	
	9380	5" pipe size		6.15	2.602			84	13	97	141	
	9390	6" pipe size		5.71	2.802			90.50	14	104.50	152	
	9400	8" pipe size		3.69	4.336			140	21.50	161.50	236	
	9410	10" pipe size		2.91	5.498			178	27.50	205.50	298	
	9420	12" pipe size	▼	2.31	6.926	▼		224	34.50	258.50	380	
	9500	Schedule 40										
	9510	1/4" pipe size	Q-15	28	.571	Ea.		18.50	2.86	21.36	31	
	9520	3/8" pipe size		27	.593			19.15	2.96	22.11	32.50	
	9530	1/2" pipe size		25.40	.630			20.50	3.15	23.65	34	
	9540	3/4" pipe size		22.22	.720			23.50	3.60	27.10	39	
	9550	1" pipe size		20.25	.790			25.50	3.95	29.45	43	
	9560	1-1/4" pipe size		18.82	.850			27.50	4.25	31.75	46	
	9570	1-1/2" pipe size		17.78	.900			29	4.50	33.50	49	
	9580	2" pipe size		15.09	1.060			34.50	5.30	39.80	57.50	
	9590	2-1/2" pipe size		7.96	2.010			65	10.05	75.05	109	
	9600	3" pipe size		6.43	2.488			80.50	12.45	92.95	135	
	9610	4" pipe size		4.88	3.279			106	16.40	122.40	178	
	9620	5" pipe size		4.26	3.756			122	18.80	140.80	204	
	9630	6" pipe size		3.77	4.244			137	21	158	231	
	9640	8" pipe size		2.44	6.557			212	33	245	355	
	9650	10" pipe size		1.92	8.333			270	41.50	311.50	450	
	9660	12" pipe size	▼	1.52	10.526	▼		340	52.50	392.50	575	
	9750	Schedule 80										
	9760	1/4" pipe size	Q-15	21.55	.742	Ea.		24	3.71	27.71	40	
	9770	3/8" pipe size		20.75	.771			25	3.86	28.86	41.50	
	9780	1/2" pipe size		19.54	.819			26.50	4.09	30.59	44.50	
	9790	3/4" pipe size		17.09	.936			30.50	4.68	35.18	50.50	
	9800	1" pipe size		15.58	1.027			33	5.15	38.15	55.50	
	9810	1-1/4" pipe size		14.48	1.105			36	5.50	41.50	60	
	9820	1-1/2" pipe size		13.68	1.170			38	5.85	43.85	63.50	
	9830	2" pipe size		11.61	1.378			44.50	6.90	51.40	74.50	
	9840	2-1/2" pipe size		6.12	2.614			84.50	13.05	97.55	142	
	9850	3" pipe size		4.94	3.239			105	16.20	121.20	176	
	9860	4" pipe size		3.75	4.267			138	21.50	159.50	232	
	9870	5" pipe size		3.27	4.893			158	24.50	182.50	266	
	9880	6" pipe size		2.90	5.517			178	27.50	205.50	300	
	9890	8" pipe size		1.87	8.556			277	43	320	460	
	9900	10" pipe size		1.48	10.811			350	54	404	585	
	9910	12" pipe size		1.17	13.675			440	68.50	508.50	740	
	9920	Schedule 160, 1/2" pipe size		17	.941			30.50	4.71	35.21	51	
	9930	3/4" pipe size		14.81	1.080			35	5.40	40.40	58.50	
	9940	1" pipe size		13.50	1.185			38.50	5.95	44.45	64.50	
	9950	1-1/4" pipe size		12.55	1.275			41	6.35	47.35	69	
	9960	1-1/2" pipe size		11.85	1.350			43.50	6.75	50.25	73.50	
	9970	2" pipe size		10	1.600			52	8	60	87	
	9980	3" pipe size		4.28	3.738			121	18.70	139.70	203	
	9990	4" pipe size	▼	3.25	4.923	▼		159	24.50	183.50	267	
960	0010	**PIPE, STAINLESS STEEL, FITTINGS**										**960**
	0100	Butt weld joint, schedule 5, type 304										
	0120	90° Elbow, long										
	0140	1/2"	Q-15	17.50	.914	Ea.	10.05	29.50	4.57	44.12	60.50	

MECHANICAL 15

		15107	Metal Pipe & Fittings	CREW	DAILY OUTPUT	LABOR-HOURS	UNIT	2002 BARE COSTS				TOTAL INCL O&P	
								MAT.	LABOR	EQUIP.	TOTAL		
960	0150		3/4″	Q-15	14	1.143	Ea.	15.75	37	5.70	58.45	79.50	960
	0160		1″		12.50	1.280		11.50	41.50	6.40	59.40	82	
	0170		1-1/4″		11	1.455		18.40	47	7.25	72.65	99	
	0180		1-1/2″		10.50	1.524		14.45	49.50	7.60	71.55	99	
	0190		2″		9	1.778		19.05	57.50	8.90	85.45	117	
	0200		2-1/2″		6	2.667		27.50	86.50	13.35	127.35	175	
	0210		3″		4.86	3.292		35	107	16.45	158.45	218	
	0220		3-1/2″		4.27	3.747		184	121	18.75	323.75	405	
	0230		4″		3.69	4.336		64.50	140	21.50	226	305	
	0240		5″		3.08	5.195		460	168	26	654	785	
	0250		6″	Q-16	4.29	5.594		184	188	18.65	390.65	505	
	0260		8″		2.76	8.696		310	292	29	631	815	
	0270		10″		2.18	11.009		645	370	36.50	1,051.50	1,300	
	0280		12″		1.73	13.873		1,025	465	46	1,536	1,875	
	0320		For schedule 5, type 316, add					30%					
	0600		45° Elbow, long										
	0620		1/2″	Q-15	17.50	.914	Ea.	14.45	29.50	4.57	48.52	65.50	
	0630		3/4″		14	1.143		14.45	37	5.70	57.15	78	
	0640		1″		12.50	1.280		16.30	41.50	6.40	64.20	87.50	
	0650		1-1/4″		11	1.455		27.50	47	7.25	81.75	110	
	0660		1-1/2″		10.50	1.524		14.80	49.50	7.60	71.90	99	
	0670		2″		9	1.778		17	57.50	8.90	83.40	115	
	0680		2-1/2″		6	2.667		28	86.50	13.35	127.85	176	
	0690		3″		4.86	3.292		29	107	16.45	152.45	211	
	0700		3-1/2″		4.27	3.747		184	121	18.75	323.75	405	
	0710		4″		3.69	4.336		51.50	140	21.50	213	293	
	0720		5″		3.08	5.195		320	168	26	514	635	
	0730		6″	Q-16	4.29	5.594		110	188	18.65	316.65	425	
	0740		8″		2.76	8.696		193	292	29	514	685	
	0750		10″		2.18	11.009		430	370	36.50	836.50	1,075	
	0760		12″		1.73	13.873		690	465	46	1,201	1,500	
	0800		For schedule 5, type 316, add					25%					
	1100		Tee, straight										
	1130		1/2″	Q-15	11.66	1.372	Ea.	40.50	44.50	6.85	91.85	119	
	1140		3/4″		9.33	1.715		40.50	55.50	8.55	104.55	137	
	1150		1″		8.33	1.921		38.50	62	9.60	110.10	147	
	1160		1-1/4″		7.33	2.183		66	70.50	10.90	147.40	191	
	1170		1-1/2″		7	2.286		38.50	74	11.45	123.95	167	
	1180		2″		6	2.667		40.50	86.50	13.35	140.35	189	
	1190		2-1/2″		4	4		116	129	20	265	345	
	1200		3″		3.24	4.938		75.50	160	24.50	260	350	
	1210		3-1/2″		2.85	5.614		202	182	28	412	525	
	1220		4″		2.46	6.504		107	210	32.50	349.50	470	
	1230		5″		2	8		430	259	40	729	910	
	1240		6″	Q-16	2.85	8.421		320	283	28	631	810	
	1250		8″		1.84	13.043		735	440	43.50	1,218.50	1,525	
	1260		10″		1.45	16.552		1,150	555	55	1,760	2,175	
	1270		12″		1.15	20.870		1,450	700	69.50	2,219.50	2,725	
	1320		For schedule 5, type 316, add					25%					
	2000		Butt weld joint, schedule 10, type 304										
	2020		90° elbow, long										
	2040		1/2″	Q-15	17	.941	Ea.	9.45	30.50	4.71	44.66	61.50	
	2050		3/4″		14	1.143		9.45	37	5.70	52.15	72.50	
	2060		1″		12.50	1.280		9.45	41.50	6.40	57.35	80	
	2070		1-1/4″		11	1.455		17.65	47	7.25	71.90	98.50	
	2080		1-1/2″		10.50	1.524		13.75	49.50	7.60	70.85	98	

15107	Metal Pipe & Fittings	CREW	DAILY OUTPUT	LABOR-HOURS	UNIT	2002 BARE COSTS				TOTAL INCL O&P
						MAT.	LABOR	EQUIP.	TOTAL	
2090	2"	Q-15	9	1.778	Ea.	18.40	57.50	8.90	84.80	116
2100	2-1/2"		6	2.667		40	86.50	13.35	139.85	188
2110	3"		4.86	3.292		47.50	107	16.45	170.95	231
2120	3-1/2"		4.27	3.747		202	121	18.75	341.75	425
2130	4"		3.69	4.336		77.50	140	21.50	239	320
2140	5"		3.08	5.195		415	168	26	609	735
2150	6"	Q-16	4.29	5.594		176	188	18.65	382.65	500
2160	8"		2.76	8.696		350	292	29	671	855
2170	10"		2.18	11.009		755	370	36.50	1,161.50	1,425
2180	12"		1.73	13.873		1,175	465	46	1,686	2,025
2500	45° elbow, long									
2520	1/2"	Q-15	17.50	.914	Ea.	16.50	29.50	4.57	50.57	67.50
2530	3/4"		14	1.143		16.50	37	5.70	59.20	80.50
2540	1"		12.50	1.280		15.45	41.50	6.40	63.35	86.50
2550	1-1/4"		11	1.455		31.50	47	7.25	85.75	114
2560	1-1/2"		10.50	1.524		13.75	49.50	7.60	70.85	98
2570	2"		9	1.778		15.80	57.50	8.90	82.20	114
2580	2-1/2"		6	2.667		36	86.50	13.35	135.85	184
2590	3"		4.86	3.292		40.50	107	16.45	163.95	224
2600	3-1/2"		4.27	3.747		202	121	18.75	341.75	425
2610	4"		3.69	4.336		48	140	21.50	209.50	289
2620	5"		3.08	5.195		315	168	26	509	625
2630	6"	Q-16	4.29	5.594		103	188	18.65	309.65	415
2640	8"		2.76	8.696		233	292	29	554	730
2650	10"		2.18	11.009		540	370	36.50	946.50	1,200
2660	12"		1.73	13.873		825	465	46	1,336	1,650
3000	Tee, straight									
3030	1/2"	Q-15	11.66	1.372	Ea.	41.50	44.50	6.85	92.85	120
3040	3/4"		9.33	1.715		41.50	55.50	8.55	105.55	138
3050	1"		8.33	1.921		33	62	9.60	104.60	141
3060	1-1/4"		7.33	2.183		63	70.50	10.90	144.40	188
3070	1-1/2"		7	2.286		36	74	11.45	121.45	165
3080	2"		6	2.667		37	86.50	13.35	136.85	185
3090	2-1/2"		4	4		116	129	20	265	345
3100	3"		3.24	4.938		81	160	24.50	265.50	355
3110	3-1/2"		2.85	5.614		122	182	28	332	440
3120	4"		2.46	6.504		101	210	32.50	343.50	460
3130	5"		2	8		390	259	40	689	860
3140	6"	Q-16	2.85	8.421		305	283	28	616	790
3150	8"		1.84	13.043		705	440	43.50	1,188.50	1,475
3151	10"		1.45	16.552		1,175	555	55	1,785	2,175
3152	12"		1.15	20.870		1,625	700	69.50	2,394.50	2,900
3154	For schedule 10, type 316, add					25%				
3281	Butt weld joint, schedule 40, type 304									
3284	90° Elbow, long, 1/2"	Q-15	12.70	1.260	Ea.	15.75	41	6.30	63.05	86
3288	3/4"		11.10	1.441		15.75	46.50	7.20	69.45	96
3289	1"		10.13	1.579		17.25	51	7.90	76.15	105
3290	1-1/4"		9.40	1.702		25.50	55	8.50	89	120
3300	1-1/2"		8.89	1.800		20.50	58	9	87.50	120
3310	2"		7.55	2.119		28.50	68.50	10.60	107.60	146
3320	2-1/2"		3.98	4.020		59.50	130	20	209.50	283
3330	3"		3.21	4.984		76	161	25	262	355
3340	3-1/2"		2.83	5.654		315	183	28.50	526.50	655
3350	4"		2.44	6.557		140	212	33	385	510
3360	5"		2.13	7.512		495	243	37.50	775.50	950
3370	6"	Q-16	2.83	8.481		440	285	28	753	945

960

MECHANICAL 15

		CREW	DAILY OUTPUT	LABOR-HOURS	UNIT	MAT.	LABOR	EQUIP.	TOTAL	TOTAL INCL O&P
15107	**Metal Pipe & Fittings**						2002 BARE COSTS			
3380	8″	Q-16	1.83	13.115	Ea.	740	440	43.50	1,223.50	1,525
3390	10″		1.44	16.667		1,550	560	55.50	2,165.50	2,600
3400	12″	↓	1.14	21.053	↓	2,175	705	70	2,950	3,550
3410	For schedule 40, type 316, add					25%				
3460	45° Elbow, long, 1/2″	Q-15	12.70	1.260	Ea.	21	41	6.30	68.30	91.50
3470	3/4″		11.10	1.441		21	46.50	7.20	74.70	101
3480	1″		10.13	1.579		21	51	7.90	79.90	109
3490	1-1/4″		9.40	1.702		31	55	8.50	94.50	126
3500	1-1/2″		8.89	1.800		24	58	9	91	124
3510	2″		7.55	2.119		29.50	68.50	10.60	108.60	147
3520	2-1/2″		3.98	4.020		59.50	130	20	209.50	283
3530	3″		3.21	4.984		55	161	25	241	330
3540	3-1/2″		2.83	5.654		320	183	28.50	531.50	660
3550	4″		2.44	6.557		99	212	33	344	465
3560	5″	↓	2.13	7.512		345	243	37.50	625.50	785
3570	6″	Q-16	2.83	8.481		258	285	28	571	745
3580	8″		1.83	13.115		490	440	43.50	973.50	1,250
3590	10″		1.44	16.667		920	560	55.50	1,535.50	1,900
3600	12″	↓	1.14	21.053	↓	1,275	705	70	2,050	2,550
3610	For schedule 40, type 316, add					25%				
3660	Tee, straight 1/2″	Q-15	8.46	1.891	Ea.	44.50	61	9.45	114.95	151
3670	3/4″		7.40	2.162		44.50	70	10.80	125.30	165
3680	1″		6.74	2.374		44.50	77	11.85	133.35	178
3690	1-1/4″		6.27	2.552		73.50	82.50	12.75	168.75	219
3700	1-1/2″		5.92	2.703		44.50	87.50	13.50	145.50	195
3710	2″		5.03	3.181		48	103	15.90	166.90	226
3720	2-1/2″		2.65	6.038		120	195	30	345	460
3730	3″		2.14	7.477		97	242	37.50	376.50	510
3740	3-1/2″		1.88	8.511		139	275	42.50	456.50	615
3750	4″		1.62	9.877		139	320	49.50	508.50	690
3760	5″	↓	1.42	11.268		515	365	56.50	936.50	1,175
3770	6″	Q-16	1.88	12.766		445	430	42.50	917.50	1,175
3780	8″		1.22	19.672		1,025	660	65.50	1,750.50	2,200
3790	10″		.96	25		2,175	840	83.50	3,098.50	3,750
3800	12″	↓	.76	31.579	↓	2,300	1,050	105	3,455	4,250
3810	For schedule 40, type 316, add					25%				
3820	Tee, reducing on outlet, 3/4″ x 1/2″	Q-15	7.73	2.070	Ea.	50.50	67	10.35	127.85	168
3822	1″ x 1/2″		7.24	2.210		70	71.50	11.05	152.55	197
3824	1″ x 3/4″		6.96	2.299		58.50	74.50	11.50	144.50	189
3826	1-1/4″ x 1″		6.43	2.488		185	80.50	12.45	277.95	340
3828	1-1/2″ x 1/2″		6.58	2.432		97	78.50	12.15	187.65	239
3830	1-1/2″ x 3/4″		6.35	2.520		97	81.50	12.60	191.10	244
3832	1-1/2″ x 1″		6.18	2.589		70	84	12.95	166.95	217
3834	2″ x 1″		5.50	2.909		68	94	14.55	176.55	233
3836	2″ x 1-1/2″		5.30	3.019		64	97.50	15.10	176.60	234
3838	2-1/2″ x 2″		3.15	5.079		165	164	25.50	354.50	455
3840	3″ x 1-1/2″		2.72	5.882		141	190	29.50	360.50	475
3842	3″ x 2″		2.65	6.038		119	195	30	344	460
3844	4″ x 2″		2.10	7.619		214	246	38	498	650
3846	4″ x 3″		1.77	9.040		175	292	45	512	680
3848	5″ x 4″	↓	1.48	10.811		550	350	54	954	1,200
3850	6″ x 3″	Q-16	2.19	10.959		440	370	36.50	846.50	1,075
3852	6″ x 4″		2.04	11.765		400	395	39	834	1,075
3854	8″ x 4″		1.46	16.438		735	550	54.50	1,339.50	1,700
3856	10″ x 8″		.69	34.783		2,025	1,175	116	3,316	4,125
3858	12″ x 10″	↓	.55	43.636	↓	2,525	1,475	145	4,145	5,125

Important: See the Reference Section for critical supporting data - Reference Nos., Crews, & City Cost Indexes

15107	Metal Pipe & Fittings	CREW	DAILY OUTPUT	LABOR-HOURS	UNIT	2002 BARE COSTS				TOTAL INCL O&P
						MAT.	LABOR	EQUIP.	TOTAL	
3950	Reducer, concentric, 3/4" x 1/2"	Q-15	11.85	1.350	Ea.	23.50	43.50	6.75	73.75	99
3952	1" x 3/4"		10.60	1.509		27	49	7.55	83.55	112
3954	1-1/4" x 3/4"		10.19	1.570		77.50	51	7.85	136.35	171
3956	1-1/4" x 1"		9.76	1.639		42	53	8.20	103.20	135
3958	1-1/2" x 3/4"		9.88	1.619		48.50	52.50	8.10	109.10	141
3960	1-1/2" x 1"		9.47	1.690		16.20	54.50	8.45	79.15	110
3962	2" x 1"		8.65	1.850		19.45	60	9.25	88.70	122
3964	2" x 1-1/2"		8.16	1.961		16.20	63.50	9.80	89.50	124
3966	2-1/2" x 1"		5.71	2.802		117	90.50	14	221.50	280
3968	2-1/2" x 2"		5.21	3.071		58.50	99.50	15.35	173.35	231
3970	3" x 1"		4.88	3.279		77.50	106	16.40	199.90	264
3972	3" x 1-1/2"		4.72	3.390		36	110	16.95	162.95	223
3974	3" x 2"		4.51	3.548		29	115	17.75	161.75	225
3976	4" x 2"		3.69	4.336		40	140	21.50	201.50	280
3978	4" x 3"		2.77	5.776		35	187	29	251	355
3980	5" x 3"		2.56	6.250		291	202	31.50	524.50	660
3982	5" x 4"	▼	2.27	7.048		233	228	35	496	640
3984	6" x 3"	Q-16	3.57	6.723		111	226	22.50	359.50	485
3986	6" x 4"		3.19	7.524		92.50	252	25	369.50	510
3988	8" x 4"		2.44	9.836		350	330	33	713	920
3990	8" x 6"		2.22	10.811		224	365	36	625	830
3992	10" x 6"		1.91	12.565		465	420	42	927	1,200
3994	10" x 8"		1.61	14.907		276	500	49.50	825.50	1,125
3995	12" x 6"		1.63	14.724		700	495	49	1,244	1,575
3996	12" x 8"		1.41	17.021		640	570	56.50	1,266.50	1,625
3997	12" x 10"	▼	1.27	18.898	▼	385	635	63	1,083	1,450
4000	Socket weld joint, 3000 lb., type 304									
4100	90° Elbow									
4140	1/4"	Q-15	13.47	1.188	Ea.	21.50	38.50	5.95	65.95	88
4150	3/8"		12.97	1.234		27.50	40	6.15	73.65	97.50
4160	1/2"		12.21	1.310		29	42.50	6.55	78.05	103
4170	3/4"		10.68	1.498		35	48.50	7.50	91	120
4180	1"		9.74	1.643		53	53	8.20	114.20	147
4190	1-1/4"		9.05	1.768		92	57	8.85	157.85	197
4200	1-1/2"		8.55	1.871		114	60.50	9.35	183.85	227
4210	2"	▼	7.26	2.204	▼	184	71.50	11	266.50	320
4300	45° Elbow									
4340	1/4"	Q-15	13.47	1.188	Ea.	41	38.50	5.95	85.45	110
4350	3/8"		12.97	1.234		41	40	6.15	87.15	112
4360	1/2"		12.21	1.310		41	42.50	6.55	90.05	116
4370	3/4"		10.68	1.498		47	48.50	7.50	103	133
4380	1"		9.74	1.643		68	53	8.20	129.20	164
4390	1-1/4"		9.05	1.768		110	57	8.85	175.85	217
4400	1-1/2"		8.55	1.871		111	60.50	9.35	180.85	224
4410	2"	▼	7.26	2.204	▼	201	71.50	11	283.50	340
4500	Tee									
4540	1/4"	Q-15	8.97	1.784	Ea.	26	57.50	8.90	92.40	126
4550	3/8"		8.64	1.852		31	60	9.25	100.25	135
4560	1/2"		8.13	1.968		41.50	63.50	9.85	114.85	152
4570	3/4"		7.12	2.247		48	72.50	11.25	131.75	175
4580	1"		6.48	2.469		64.50	80	12.35	156.85	205
4590	1-1/4"		6.03	2.653		114	86	13.25	213.25	270
4600	1-1/2"		5.69	2.812		167	91	14.05	272.05	335
4610	2"	▼	4.83	3.313	▼	253	107	16.55	376.55	460
5000	Socket weld joint, 3000 lb., type 316									
5100	90° Elbow									

15107	Metal Pipe & Fittings	CREW	DAILY OUTPUT	LABOR-HOURS	UNIT	2002 BARE COSTS				TOTAL INCL O&P
						MAT.	LABOR	EQUIP.	TOTAL	
5140	1/4"	Q-15	13.47	1.188	Ea.	25	38.50	5.95	69.45	92
5150	3/8"		12.97	1.234		29	40	6.15	75.15	99
5160	1/2"		12.21	1.310		35	42.50	6.55	84.05	110
5170	3/4"		10.68	1.498		46	48.50	7.50	102	132
5180	1"		9.74	1.643		68.50	53	8.20	129.70	165
5190	1-1/4"		9.05	1.768		122	57	8.85	187.85	231
5200	1-1/2"		8.55	1.871		141	60.50	9.35	210.85	257
5210	2"		7.26	2.204		240	71.50	11	322.50	385
5300	45° Elbow									
5340	1/4"	Q-15	13.47	1.188	Ea.	52.50	38.50	5.95	96.95	123
5350	3/8"		12.97	1.234		52.50	40	6.15	98.65	125
5360	1/2"		12.21	1.310		52.50	42.50	6.55	101.55	129
5370	3/4"		10.68	1.498		59.50	48.50	7.50	115.50	147
5380	1"		9.74	1.643		90	53	8.20	151.20	188
5390	1-1/4"		9.05	1.768		126	57	8.85	191.85	235
5400	1-1/2"		8.55	1.871		141	60.50	9.35	210.85	258
5410	2"		7.26	2.204		211	71.50	11	293.50	350
5500	Tee									
5540	1/4"	Q-15	8.97	1.784	Ea.	31.50	57.50	8.90	97.90	132
5550	3/8"		8.64	1.852		38.50	60	9.25	107.75	143
5560	1/2"		8.13	1.968		48	63.50	9.85	121.35	160
5570	3/4"		7.12	2.247		59.50	72.50	11.25	143.25	188
5580	1"		6.48	2.469		91	80	12.35	183.35	234
5590	1-1/4"		6.03	2.653		145	86	13.25	244.25	305
5600	1-1/2"		5.69	2.812		208	91	14.05	313.05	380
5610	2"		4.83	3.313		330	107	16.55	453.55	545
5700	For socket weld joint, 6000 lb., type 304 and 316, add					100%				
6000	Threaded companion flange									
6010	Stainless steel, 150 lb., type 304									
6020	1/2" diam.	1 Plum	30	.267	Ea.	45.50	9.60		55.10	64.50
6030	3/4" diam.		28	.286		50.50	10.25		60.75	71.50
6040	1" diam.		27	.296		56.50	10.65		67.15	78.50
6050	1-1/4" diam.	Q-1	44	.364		77	11.75		88.75	103
6060	1-1/2" diam.		40	.400		79	12.95		91.95	107
6070	2" diam.		36	.444		88	14.40		102.40	119
6080	2-1/2" diam.		28	.571		135	18.50		153.50	176
6090	3" diam.		20	.800		147	26		173	200
6110	4" diam.		12	1.333		231	43		274	320
6130	6" diam.	Q-2	14	1.714		440	57.50		497.50	570
6140	8" diam.	"	12	2		825	67		892	1,000
6150	For type 316 add					40%				
6260	Weld flanges, stainless steel, type 304									
6270	Slip on, 150 lb. (welded, front and back)									
6280	1/2" diam.	Q-15	12.70	1.260	Ea.	45.50	41	6.30	92.80	118
6290	3/4" diam.		11.11	1.440		50.50	46.50	7.20	104.20	134
6300	1" diam.		10.13	1.579		36	51	7.90	94.90	126
6310	1-1/4" diam.		9.41	1.700		77	55	8.50	140.50	177
6320	1-1/2" diam.		8.89	1.800		53.50	58	9	120.50	156
6330	2" diam.		7.55	2.119		70	68.50	10.60	149.10	192
6340	2-1/2" diam.		3.98	4.020		135	130	20	285	365
6350	3" diam.		3.21	4.984		147	161	25	333	430
6370	4" diam.		2.44	6.557		147	212	33	392	520
6390	6" diam.	Q-16	1.89	12.698		231	425	42.50	698.50	940
6400	8" diam.	"	1.22	19.672		825	660	65.50	1,550.50	1,975
6410	For type 316, add					40%				
6530	Weld neck 150 lb.									

960

Important: See the Reference Section for critical supporting data - Reference Nos., Crews, & City Cost Indexes

15107	Metal Pipe & Fittings	CREW	DAILY OUTPUT	LABOR-HOURS	UNIT	2002 BARE COSTS				TOTAL INCL O&P	
						MAT.	LABOR	EQUIP.	TOTAL		
960	6540	1/2" diam.	Q-15	25.40	.630	Ea.	34	20.50	3.15	57.65	71.50	**960**
	6550	3/4" diam.		22.22	.720		38.50	23.50	3.60	65.60	81.50	
	6560	1" diam.		20.25	.790		46	25.50	3.95	75.45	93.50	
	6570	1-1/4" diam.		18.82	.850		55	27.50	4.25	86.75	107	
	6580	1-1/2" diam.		17.78	.900		64.50	29	4.50	98	119	
	6590	2" diam.		15.09	1.060		82.50	34.50	5.30	122.30	148	
	6600	2-1/2" diam.		7.96	2.010		138	65	10.05	213.05	260	
	6610	3" diam.		6.43	2.488		156	80.50	12.45	248.95	305	
	6630	4" diam.		4.88	3.279		202	106	16.40	324.40	400	
	6640	5" diam.	▼	4.26	3.756		268	122	18.80	408.80	500	
	6650	6" diam.	Q-16	5.66	4.240		315	142	14.10	471.10	575	
	6652	8" diam.		3.65	6.575		440	221	22	683	845	
	6654	10" diam.		2.88	8.333		600	280	28	908	1,100	
	6656	12" diam.	▼	2.28	10.526	▼	955	355	35	1,345	1,625	
	6670	For type 316 add					23%					
	7000	Threaded joint, 150 lb., type 304										
	7030	90° elbow										
	7040	1/8"	1 Plum	13	.615	Ea.	12.80	22		34.80	47.50	
	7050	1/4"		13	.615		12.80	22		34.80	47.50	
	7070	3/8"		13	.615		15.45	22		37.45	50.50	
	7080	1/2"		12	.667		8.30	24		32.30	45	
	7090	3/4"		11	.727		9.40	26		35.40	50	
	7100	1"	▼	10	.800		12.50	29		41.50	57.50	
	7110	1-1/4"	Q-1	17	.941		33	30.50		63.50	82	
	7120	1-1/2"		16	1		21	32.50		53.50	72.50	
	7130	2"		14	1.143		30.50	37		67.50	89.50	
	7140	2-1/2"		11	1.455		144	47		191	230	
	7150	3"	▼	8	2		182	64.50		246.50	299	
	7160	4"	Q-2	11	2.182	▼	211	73		284	340	
	7180	45° elbow										
	7190	1/8"	1 Plum	13	.615	Ea.	17.60	22		39.60	53	
	7200	1/4"		13	.615		17.60	22		39.60	53	
	7210	3/8"		13	.615		19.60	22		41.60	55	
	7220	1/2"		12	.667		10	24		34	47	
	7230	3/4"		11	.727		12.25	26		38.25	53	
	7240	1"	▼	10	.800		13.50	29		42.50	58.50	
	7250	1-1/4"	Q-1	17	.941		31	30.50		61.50	80	
	7260	1-1/2"		16	1		20.50	32.50		53	71.50	
	7270	2"		14	1.143		32	37		69	91	
	7280	2-1/2"		11	1.455		204	47		251	295	
	7290	3"	▼	8	2		298	64.50		362.50	430	
	7300	4"	Q-2	11	2.182	▼	267	73		340	405	
	7320	Tee, straight										
	7330	1/8"	1 Plum	9	.889	Ea.	19.70	32		51.70	69.50	
	7340	1/4"		9	.889		19.70	32		51.70	69.50	
	7350	3/8"		9	.889		22.50	32		54.50	73	
	7360	1/2"		8	1		12.75	36		48.75	68	
	7370	3/4"		7	1.143		13.40	41		54.40	77	
	7380	1"	▼	6.50	1.231		16	44.50		60.50	84	
	7390	1-1/4"	Q-1	11	1.455		45	47		92	120	
	7400	1-1/2"		10	1.600		32.50	52		84.50	114	
	7410	2"		9	1.778		38.50	57.50		96	129	
	7420	2-1/2"		7	2.286		208	74		282	340	
	7430	3"	▼	5	3.200		315	104		419	505	
	7440	4"	Q-2	7	3.429	▼	315	115		430	520	
	7460	Coupling, straight										

MECHANICAL 15

15107 | Metal Pipe & Fittings

		CREW	DAILY OUTPUT	LABOR-HOURS	UNIT	MAT.	LABOR	EQUIP.	TOTAL	TOTAL INCL O&P	
960	7470	1/8"	1 Plum	19	.421	Ea.	4.86	15.15		20.01	28.50	960
	7480	1/4"		19	.421		5.85	15.15		21	29.50	
	7490	3/8"		19	.421		6.90	15.15		22.05	30.50	
	7500	1/2"		19	.421		5.05	15.15		20.20	28.50	
	7510	3/4"		18	.444		13.05	16		29.05	38.50	
	7520	1"		15	.533		17.35	19.15		36.50	48	
	7530	1-1/4"	Q-1	26	.615		14.85	19.90		34.75	46.50	
	7540	1-1/2"		24	.667		17.90	21.50		39.40	52	
	7550	2"		21	.762		28.50	24.50		53	68	
	7560	2-1/2"		18	.889		64.50	29		93.50	114	
	7570	3"		14	1.143		86.50	37		123.50	151	
	7580	4"	Q-2	16	1.500		129	50.50		179.50	217	
	7600	Reducer, concentric, 1/2"	1 Plum	12	.667		4.44	24		28.44	41	
	7610	3/4"		11	.727		5.80	26		31.80	46	
	7612	1"		10	.800		10.10	29		39.10	54.50	
	7614	1-1/4"	Q-1	17	.941		20.50	30.50		51	68.50	
	7616	1-1/2"		16	1		22.50	32.50		55	74	
	7618	2"		14	1.143		35.50	37		72.50	95	
	7620	2-1/2"		11	1.455		74.50	47		121.50	153	
	7622	3"		8	2		108	64.50		172.50	217	
	7624	4"	Q-2	11	2.182		155	73		228	280	
	7710	Union										
	7720	1/8"	1 Plum	12	.667	Ea.	23.50	24		47.50	62	
	7730	1/4"		12	.667		23.50	24		47.50	62	
	7740	3/8"		12	.667		27.50	24		51.50	66	
	7750	1/2"		11	.727		17.15	26		43.15	58.50	
	7760	3/4"		10	.800		24	29		53	70	
	7770	1"		9	.889		34.50	32		66.50	86	
	7780	1-1/4"	Q-1	16	1		151	32.50		183.50	215	
	7790	1-1/2"		15	1.067		168	34.50		202.50	237	
	7800	2"		13	1.231		212	40		252	294	
	7810	2-1/2"		10	1.600		435	52		487	555	
	7820	3"		7	2.286		575	74		649	745	
	7830	4"	Q-2	10	2.400		795	80.50		875.50	995	
	7838	Caps										
	7840	1/2"	1 Plum	24	.333	Ea.	4.61	12		16.61	23	
	7841	3/4"		22	.364		6.70	13.05		19.75	27	
	7842	1"		20	.400		10.70	14.40		25.10	33.50	
	7843	1-1/2"	Q-1	32	.500		28	16.20		44.20	55.50	
	7844	2"	"	28	.571		36	18.50		54.50	67.50	
	7845	4"	Q-2	22	1.091		141	36.50		177.50	210	
	7850	For 150 lb., type 316, add					25%					
	8750	Threaded joint, 2000 lb., type 304										
	8770	90° Elbow										
	8780	1/8"	1 Plum	13	.615	Ea.	15.40	22		37.40	50.50	
	8790	1/4"		13	.615		15.40	22		37.40	50.50	
	8800	3/8"		13	.615		19.05	22		41.05	54.50	
	8810	1/2"		12	.667		26	24		50	64.50	
	8820	3/4"		11	.727		30	26		56	72.50	
	8830	1"		10	.800		39.50	29		68.50	87	
	8840	1-1/4"	Q-1	17	.941		63.50	30.50		94	116	
	8850	1-1/2"		16	1		102	32.50		134.50	161	
	8860	2"		14	1.143		143	37		180	213	
	8880	45° Elbow										
	8890	1/8"	1 Plum	13	.615	Ea.	32	22		54	69	
	8900	1/4"		13	.615		32	22		54	69	

Important: See the Reference Section for critical supporting data - Reference Nos., Crews, & City Cost Indexes

15107	Metal Pipe & Fittings	CREW	DAILY OUTPUT	LABOR-HOURS	UNIT	MAT.	LABOR	EQUIP.	TOTAL	TOTAL INCL O&P	
						2002 BARE COSTS					
960 8910	3/8"	1 Plum	13	.615	Ea.	40	22		62	77.50	960
8920	1/2"		12	.667		40	24		64	80	
8930	3/4"		11	.727		44.50	26		70.50	88	
8940	1"		10	.800		52.50	29		81.50	102	
8950	1-1/4"	Q-1	17	.941		90.50	30.50		121	146	
8960	1-1/2"		16	1		121	32.50		153.50	182	
8970	2"		14	1.143		140	37		177	210	
8990	Tee, straight										
9000	1/8"	1 Plum	9	.889	Ea.	19.90	32		51.90	70	
9010	1/4"		9	.889		19.90	32		51.90	70	
9020	3/8"		9	.889		25.50	32		57.50	76	
9030	1/2"		8	1		33	36		69	90	
9040	3/4"		7	1.143		37	41		78	103	
9050	1"		6.50	1.231		51.50	44.50		96	123	
9060	1-1/4"	Q-1	11	1.455		85	47		132	165	
9070	1-1/2"		10	1.600		141	52		193	234	
9080	2"		9	1.778		197	57.50		254.50	305	
9100	For couplings and unions use 3000 lb., type 304										
9120	2000 lb., type 316										
9130	90° Elbow										
9140	1/8"	1 Plum	13	.615	Ea.	18.70	22		40.70	54	
9150	1/4"		13	.615		18.70	22		40.70	54	
9160	3/8"		13	.615		22	22		44	57.50	
9170	1/2"		12	.667		26.50	24		50.50	65	
9180	3/4"		11	.727		34.50	26		60.50	77.50	
9190	1"		10	.800		51.50	29		80.50	100	
9200	1-1/4"	Q-1	17	.941		92	30.50		122.50	147	
9210	1-1/2"		16	1		106	32.50		138.50	165	
9220	2"		14	1.143		180	37		217	254	
9240	45° Elbow										
9250	1/8"	1 Plum	13	.615	Ea.	39.50	22		61.50	77	
9260	1/4"		13	.615		39.50	22		61.50	77	
9270	3/8"		13	.615		39.50	22		61.50	77	
9280	1/2"		12	.667		39.50	24		63.50	79.50	
9300	3/4"		11	.727		44.50	26		70.50	88.50	
9310	1"		10	.800		67.50	29		96.50	118	
9320	1-1/4"	Q-1	17	.941		94.50	30.50		125	150	
9330	1-1/2"		16	1		129	32.50		161.50	191	
9340	2"		14	1.143		159	37		196	230	
9360	Tee, straight										
9370	1/8"	1 Plum	9	.889	Ea.	24	32		56	74	
9380	1/4"		9	.889		24	32		56	74	
9390	3/8"		9	.889		29	32		61	80	
9400	1/2"		8	1		36	36		72	93.50	
9410	3/4"		7	1.143		44.50	41		85.50	111	
9420	1"		6.50	1.231		68.50	44.50		113	142	
9430	1-1/4"	Q-1	11	1.455		109	47		156	190	
9440	1-1/2"		10	1.600		156	52		208	249	
9450	2"		9	1.778		249	57.50		306.50	360	
9470	For couplings and unions use 3000 lb., type 316										
9490	3000 lb., type 304										
9510	Coupling										
9520	1/8"	1 Plum	19	.421	Ea.	6.20	15.15		21.35	30	
9530	1/4"		19	.421		6.80	15.15		21.95	30.50	
9540	3/8"		19	.421		7.90	15.15		23.05	31.50	
9550	1/2"		19	.421		9.55	15.15		24.70	33.50	

MECHANICAL 15

15107 | Metal Pipe & Fittings

		CREW	DAILY OUTPUT	LABOR-HOURS	UNIT	2002 BARE COSTS				TOTAL INCL O&P
						MAT.	LABOR	EQUIP.	TOTAL	
9560	3/4"	1 Plum	18	.444	Ea.	12.70	16		28.70	38
9570	1"		15	.533		21.50	19.15		40.65	52.50
9580	1-1/4"	Q-1	26	.615		48.50	19.90		68.40	83.50
9590	1-1/2"		24	.667		61	21.50		82.50	100
9600	2"		21	.762		80.50	24.50		105	126
9620	Union									
9630	1/8"	1 Plum	12	.667	Ea.	48	24		72	88.50
9640	1/4"		12	.667		48	24		72	88.50
9650	3/8"		12	.667		51	24		75	92.50
9660	1/2"		11	.727		51	26		77	96
9670	3/4"		10	.800		62.50	29		91.50	113
9680	1"		9	.889		96.50	32		128.50	154
9690	1-1/4"	Q-1	16	1		162	32.50		194.50	227
9700	1-1/2"		15	1.067		182	34.50		216.50	253
9710	2"		13	1.231		244	40		284	330
9730	3000 lb., type 316									
9750	Coupling									
9770	1/8"	1 Plum	19	.421	Ea.	6.55	15.15		21.70	30
9780	1/4"		19	.421		7.40	15.15		22.55	31
9790	3/8"		19	.421		7.90	15.15		23.05	31.50
9800	1/2"		19	.421		11.20	15.15		26.35	35.50
9810	3/4"		18	.444		16.50	16		32.50	42
9820	1"		15	.533		27.50	19.15		46.65	59.50
9830	1-1/4"	Q-1	26	.615		57.50	19.90		77.40	93.50
9840	1-1/2"		24	.667		72	21.50		93.50	112
9850	2"		21	.762		100	24.50		124.50	147
9870	Union									
9880	1/8"	1 Plum	12	.667	Ea.	49	24		73	90
9890	1/4"		12	.667		49	24		73	90
9900	3/8"		12	.667		56.50	24		80.50	98.50
9910	1/2"		11	.727		56.50	26		82.50	102
9920	3/4"		10	.800		74	29		103	125
9930	1"		9	.889		117	32		149	176
9940	1-1/4"	Q-1	16	1		185	32.50		217.50	252
9950	1-1/2"		15	1.067		228	34.50		262.50	305
9960	2"		13	1.231		290	40		330	380

15108 | Plastic Pipe & Fittings

		CREW	DAILY OUTPUT	LABOR-HOURS	UNIT	2002 BARE COSTS				TOTAL INCL O&P
						MAT.	LABOR	EQUIP.	TOTAL	
0010	**PIPE, PLASTIC**									
0020	Fiberglass reinforced, couplings 10' O.C., hangers 3 per 10'									
0080	General service									
0120	2" diameter	Q-1	59	.271	L.F.	10.20	8.75		18.95	24.50
0140	3" diameter		52	.308		12.90	9.95		22.85	29
0150	4" diameter		48	.333		14.65	10.80		25.45	32.50
0160	6" diameter		39	.410		22	13.25		35.25	44.50
0170	8" diameter	Q-2	49	.490		34	16.45		50.45	62.50
0180	10" diameter		41	.585		51.50	19.65		71.15	86
0190	12" diameter		36	.667		65.50	22.50		88	106
0200	High strength									
0240	2" diameter	Q-1	58	.276	L.F.	10.50	8.90		19.40	25
0260	3" diameter		51	.314		14.75	10.15		24.90	31.50
0280	4" diameter		47	.340		17.45	11		28.45	36
0300	6" diameter		38	.421		26.50	13.60		40.10	49.50
0320	8" diameter	Q-2	48	.500		41.50	16.80		58.30	71
0340	10" diameter		40	.600		60.50	20		80.50	97.50
0360	12" diameter		36	.667		74.50	22.50		97	116

Important: See the Reference Section for critical supporting data - Reference Nos., Crews, & City Cost Indexes

		CREW	DAILY OUTPUT	LABOR-HOURS	UNIT	2002 BARE COSTS				TOTAL INCL O&P
15108	**Plastic Pipe & Fittings**					MAT.	LABOR	EQUIP.	TOTAL	
0550	To delete coupling & hangers, subtract									
0560	2" diam. to 6" diam.					33%	56%			
0570	8" diam. to 12" diam.					31%	52%			
0600	PVC, high impact/pressure, cplgs. 10' O.C., hangers 3 per 10'									
1020	Schedule 80									
1040	1/4" diameter	1 Plum	58	.138	L.F.	2.58	4.96		7.54	10.35
1060	3/8" diameter		55	.145		2.77	5.25		8.02	10.95
1070	1/2" diameter		50	.160		2.46	5.75		8.21	11.35
1080	3/4" diameter		47	.170		2.64	6.10		8.74	12.15
1090	1" diameter		43	.186		3.02	6.70		9.72	13.40
1100	1-1/4" diameter		39	.205		3.48	7.35		10.83	14.95
1110	1-1/2" diameter		34	.235		3.75	8.45		12.20	16.90
1120	2" diameter	Q-1	55	.291		4.36	9.40		13.76	19
1130	2-1/2" diameter		52	.308		5.75	9.95		15.70	21.50
1140	3" diameter		50	.320		7.55	10.35		17.90	24
1150	4" diameter		46	.348		9.90	11.25		21.15	28
1160	5" diameter		42	.381		13.30	12.30		25.60	33.50
1170	6" diameter		38	.421		17.25	13.60		30.85	39.50
1180	8" diameter	Q-2	47	.511		29	17.15		46.15	57.50
1190	10" diameter		42	.571		41	19.15		60.15	74
1200	12" diameter		38	.632		53	21		74	90
1730	To delete coupling & hangers, subtract									
1740	1/4" diam. to 1/2" diam.					62%	80%			
1750	3/4" diam. to 1-1/4" diam.					58%	73%			
1760	1-1/2" diam. to 6" diam.					40%	57%			
1770	8" diam. to 12" diam.					34%	50%			
1800	PVC, couplings 10' O.C., hangers 3 per 10'									
1820	Schedule 40									
1860	1/2" diameter	1 Plum	54	.148	L.F.	2.16	5.35		7.51	10.40
1870	3/4" diameter		51	.157		2.24	5.65		7.89	10.95
1880	1" diameter		46	.174		2.43	6.25		8.68	12.10
1890	1-1/4" diameter		42	.190		2.65	6.85		9.50	13.20
1900	1-1/2" diameter		36	.222		2.76	8		10.76	15.10
1910	2" diameter	Q-1	59	.271		3.10	8.75		11.85	16.65
1920	2-1/2" diameter		56	.286		3.86	9.25		13.11	18.20
1930	3" diameter		53	.302		4.70	9.75		14.45	19.90
1940	4" diameter		48	.333		5.70	10.80		16.50	22.50
1950	5" diameter		43	.372		7.20	12.05		19.25	26
1960	6" diameter		39	.410		9.35	13.25		22.60	30.50
1970	8" diameter	Q-2	48	.500		16.30	16.80		33.10	43.50
1980	10" diameter		43	.558		28	18.75		46.75	59
1990	12" diameter		42	.571		36.50	19.15		55.65	69
2000	14" diameter		31	.774		57.50	26		83.50	103
2010	16" diameter		23	1.043		90.50	35		125.50	153
2340	To delete coupling & hangers, subtract									
2360	1/2" diam. to 1-1/4" diam.					65%	74%			
2370	1-1/2" diam. to 6" diam.					44%	57%			
2380	8" diam. to 12" diam.					41%	53%			
2390	14" diam. to 16" diam.					48%	45%			
2420	Schedule 80									
2440	1/4" diameter	1 Plum	58	.138	L.F.	2.03	4.96		6.99	9.75
2450	3/8" diameter		55	.145		2.03	5.25		7.28	10.15
2460	1/2" diameter		50	.160		2.20	5.75		7.95	11.05
2470	3/4" diameter		47	.170		2.30	6.10		8.40	11.80
2480	1" diameter		43	.186		2.52	6.70		9.22	12.85
2490	1-1/4" diameter		39	.205		2.78	7.35		10.13	14.15

MECHANICAL 15

15108	Plastic Pipe & Fittings	CREW	DAILY OUTPUT	LABOR-HOURS	UNIT	2002 BARE COSTS				TOTAL INCL O&P
						MAT.	LABOR	EQUIP.	TOTAL	
520 2500	1-1/2" diameter	1 Plum	34	.235	L.F.	2.92	8.45		11.37	15.95 **520**
2510	2" diameter	Q-1	55	.291		3.22	9.40		12.62	17.75
2520	2-1/2" diameter		52	.308		3.93	9.95		13.88	19.30
2530	3" diameter		50	.320		5.20	10.35		15.55	21.50
2540	4" diameter		46	.348		6.45	11.25		17.70	24
2550	5" diameter		42	.381		9.50	12.30		21.80	29
2560	6" diameter		38	.421		11	13.60		24.60	32.50
2570	8" diameter	Q-2	47	.511		18.85	17.15		36	47
2580	10" diameter		42	.571		34	19.15		53.15	66
2590	12" diameter		38	.632		43	21		64	79.50
2830	To delete coupling & hangers, subtract									
2840	1/4" diam. to 1/2" diam.					66%	80%			
2850	3/4" diam. to 1-1/4" diam.					61%	73%			
2860	1-1/2" diam. to 6" diam.					41%	57%			
2870	8" diam. to 12" diam.					31%	50%			
2900	Schedule 120									
2910	1/2" diameter	1 Plum	50	.160	L.F.	2.59	5.75		8.34	11.50
2950	3/4" diameter		47	.170		2.80	6.10		8.90	12.35
2960	1" diameter		43	.186		3.26	6.70		9.96	13.70
2970	1-1/4" diameter		39	.205		3.82	7.35		11.17	15.30
2980	1-1/2" diameter		33	.242		4.17	8.70		12.87	17.75
2990	2" diameter	Q-1	54	.296		4.99	9.60		14.59	19.95
3000	2-1/2" diameter		52	.308		7.10	9.95		17.05	23
3010	3" diameter		49	.327		8.80	10.55		19.35	25.50
3020	4" diameter		45	.356		12.50	11.50		24	31
3030	6" diameter		37	.432		22.50	14		36.50	45.50
3240	To delete coupling & hangers, subtract									
3250	1/2" diam. to 1-1/4" diam.					52%	74%			
3260	1-1/2" diam. to 4" diam.					30%	57%			
3270	6" diam.					17%	50%			
3300	PVC, pressure, couplings 10' O.C., hangers 3 per 10'									
3310	SDR 26, 160 psi									
3350	1-1/4" diameter	1 Plum	42	.190	L.F.	2.48	6.85		9.33	13.05
3360	1-1/2" diameter	"	36	.222		2.57	8		10.57	14.90
3370	2" diameter	Q-1	59	.271		2.75	8.75		11.50	16.30
3380	2-1/2" diameter		56	.286		3.49	9.25		12.74	17.80
3390	3" diameter		53	.302		4.30	9.75		14.05	19.50
3400	4" diameter		48	.333		5.30	10.80		16.10	22
3420	6" diameter		39	.410		9.10	13.25		22.35	30
3430	8" diameter	Q-2	48	.500		16.30	16.80		33.10	43.50
3660	To delete coupling & hangers, subtract									
3670	1-1/4" diam.					63%	68%			
3680	1-1/2" diam. to 4" diam.					48%	57%			
3690	6" diam. to 8" diam.					60%	54%			
3720	SDR 21, 200 psi, 1/2" diameter	1 Plum	54	.148	L.F.	2.12	5.35		7.47	10.40
3740	3/4" diameter		51	.157		2.16	5.65		7.81	10.90
3750	1" diameter		46	.174		2.31	6.25		8.56	12
3760	1-1/4" diameter		42	.190		2.51	6.85		9.36	13.05
3770	1-1/2" diameter		36	.222		2.61	8		10.61	14.90
3780	2" diameter	Q-1	59	.271		2.83	8.75		11.58	16.35
3790	2-1/2" diameter		56	.286		4.02	9.25		13.27	18.35
3800	3" diameter		53	.302		4.46	9.75		14.21	19.65
3810	4" diameter		48	.333		5.60	10.80		16.40	22.50
3830	6" diameter		39	.410		9.65	13.25		22.90	30.50
3840	8" diameter	Q-2	48	.500		17.45	16.80		34.25	44.50
4000	To delete coupling & hangers, subtract									

Important: See the Reference Section for critical supporting data - Reference Nos., Crews, & City Cost Indexes

15108	Plastic Pipe & Fittings	CREW	DAILY OUTPUT	LABOR-HOURS	UNIT	2002 BARE COSTS				TOTAL INCL O&P
						MAT.	LABOR	EQUIP.	TOTAL	
520											**520**
4010	1/2" diam. to 3/4" diam.					71%	77%			
4020	1" diam. to 1-1/4" diam.					63%	70%			
4030	1-1/2" diam. to 6" diam.					44%	57%			
4040	8" diam.					46%	54%			
4100	DWV type, schedule 40, couplings 10' O.C., hangers 3 per 10'									
4120	ABS									
4140	1-1/4" diameter	1 Plum	42	.190	L.F.	2.37	6.85		9.22	12.90
4150	1-1/2" diameter	"	36	.222		2.40	8		10.40	14.70
4160	2" diameter	Q-1	59	.271		2.53	8.75		11.28	16.05
4170	3" diameter		53	.302		3.75	9.75		13.50	18.85
4180	4" diameter		48	.333		4.57	10.80		15.37	21.50
4190	6" diameter		39	.410		13.50	13.25		26.75	35
4360	To delete coupling & hangers, subtract									
4370	1-1/4" diam.					64%	68%			
4380	1-1/2" diam. to 6" diam.					54%	57%			
4400	PVC									
4410	1-1/4" diameter	1 Plum	42	.190	L.F.	2.37	6.85		9.22	12.90
4420	1-1/2" diameter	"	36	.222		2.48	8		10.48	14.80
4460	2" diameter	Q-1	59	.271		2.65	8.75		11.40	16.15
4470	3" diameter		53	.302		3.98	9.75		13.73	19.15
4480	4" diameter		48	.333		4.87	10.80		15.67	21.50
4490	6" diameter		39	.410		8.45	13.25		21.70	29.50
4500	8" diameter	Q-2	48	.500		17.90	16.80		34.70	45
4750	To delete coupling & hangers, subtract									
4760	1-1/4" diam. to 1-1/2" diam.					71%	64%			
4770	2" diam. to 8" diam.					60%	57%			
4800	PVC, clear pipe, cplgs. 10' O.C., hangers 3 per 10', Sched. 40									
4840	1/4" diameter	1 Plum	59	.136	L.F.	2.49	4.87		7.36	10.10
4850	3/8" diameter		56	.143		2.70	5.15		7.85	10.70
4860	1/2" diameter		54	.148		3.08	5.35		8.43	11.45
4870	3/4" diameter		51	.157		3.47	5.65		9.12	12.30
4880	1" diameter		46	.174		4.34	6.25		10.59	14.20
4890	1-1/4" diameter		42	.190		5.20	6.85		12.05	16.05
4900	1-1/2" diameter		36	.222		5.85	8		13.85	18.50
4910	2" diameter	Q-1	59	.271		7.25	8.75		16	21
4920	2-1/2" diameter		56	.286		10.70	9.25		19.95	25.50
4930	3" diameter		53	.302		13.40	9.75		23.15	29.50
4940	3-1/2" diameter		50	.320		17.75	10.35		28.10	35
4950	4" diameter		48	.333		19.15	10.80		29.95	37.50
5250	To delete coupling & hangers, subtract									
5260	1/4" diam. to 3/8" diam.					60%	81%			
5270	1/2" diam. to 3/4" diam.					41%	77%			
5280	1" diam. to 1-1/2" diam.					26%	67%			
5290	2" diam. to 4" diam.					16%	58%			
5360	CPVC, couplings 10' O.C., hangers 3 per 10'									
5380	Schedule 40									
5460	1/2" diameter	1 Plum	54	.148	L.F.	2.86	5.35		8.21	11.20
5470	3/4" diameter		51	.157		3.21	5.65		8.86	12.05
5480	1" diameter		46	.174		3.77	6.25		10.02	13.60
5490	1-1/4" diameter		42	.190		4.38	6.85		11.23	15.10
5500	1-1/2" diameter		36	.222		5.40	8		13.40	18
5510	2" diameter	Q-1	59	.271		6.25	8.75		15	20
5520	2-1/2" diameter		56	.286		10	9.25		19.25	25
5530	3" diameter		53	.302		12.10	9.75		21.85	28
5540	4" diameter		48	.333		17.45	10.80		28.25	35.50
5550	6" diameter		43	.372		24.50	12.05		36.55	45

15108	Plastic Pipe & Fittings	CREW	DAILY OUTPUT	LABOR-HOURS	UNIT	2002 BARE COSTS				TOTAL INCL O&P
						MAT.	LABOR	EQUIP.	TOTAL	
5730	To delete coupling & hangers, subtract									520
5740	1/2" diam. to 3/4" diam.					37%	77%			
5750	1" diam. to 1-1/4" diam.					27%	70%			
5760	1-1/2" diam. to 3" diam.					21%	57%			
5770	4" diam. to 6" diam.					16%	57%			
5800	Schedule 80									
5860	1/2" diameter	1 Plum	50	.160	L.F.	3.25	5.75		9	12.25
5870	3/4" diameter		47	.170		3.39	6.10		9.49	12.95
5880	1" diameter		43	.186		4.04	6.70		10.74	14.55
5890	1-1/4" diameter		39	.205		4.79	7.35		12.14	16.35
5900	1-1/2" diameter	↓	34	.235		5.95	8.45		14.40	19.25
5910	2" diameter	Q-1	55	.291		7.05	9.40		16.45	22
5920	2-1/2" diameter		52	.308		11.15	9.95		21.10	27.50
5930	3" diameter		50	.320		13.75	10.35		24.10	30.50
5940	4" diameter		46	.348		20	11.25		31.25	39
5950	6" diameter	↓	38	.421		30.50	13.60		44.10	54
5960	8" diameter	Q-2	47	.511	↓	66	17.15		83.15	98.50
6060	To delete couplings & hangers, subtract									
6070	1/2" diam. to 3/4" diam.					44%	77%			
6080	1" diam. to 1-1/4" diam.					32%	71%			
6090	1-1/2" diam. to 4" diam.					25%	58%			
6100	6" diam. to 8" diam.					20%	53%			
6240	CTS, 1/2" diameter	1 Plum	54	.148	L.F.	2.10	5.35		7.45	10.35
6250	3/4" diameter		51	.157		2.63	5.65		8.28	11.40
6260	1" diameter		46	.174		2.93	6.25		9.18	12.65
6270	1 1/4"		42	.190		3.44	6.85		10.29	14.10
6280	1 1/2" diameter	↓	36	.222		3.96	8		11.96	16.40
6290	2" diameter	Q-1	59	.271	↓	5.35	8.75		14.10	19.15
6370	To delete coupling & hangers, subtract									
6380	1/2" diam.					51%	79%			
6390	3/4" diam.					40%	76%			
6392	1" thru 2" diam.					72%	68%			
7280	Polyethylene, flexible, no couplings or hangers									
7282	Note: For labor costs add 25% to the couplings and fittings labor total.									
7300	SDR 15, 100 psi									
7310	3/4" diameter				L.F.	.22			.22	.24
7350	1" diameter					.29			.29	.32
7360	1-1/4" diameter					.53			.53	.58
7370	1-1/2" diameter					.64			.64	.70
7380	2" diameter				↓	1.05			1.05	1.16
7700	SIDR 9, 160 psi									
7710	1/2" diameter				L.F.	.28			.28	.31
7750	3/4" diameter					.29			.29	.32
7760	1" diameter					.45			.45	.50
7770	1-1/4" diameter					.79			.79	.87
7780	1-1/2" diameter					1.10			1.10	1.21
7790	2" diameter				↓	1.82			1.82	2
8120	SDR 9, 200 psi									
8150	3/4" diameter				L.F.	.26			.26	.29
8160	1" diameter					.45			.45	.50
8170	1-1/4" diameter				↓	.64			.64	.70
8420	SDR 7, 250 psi									
8440	3/4" diameter				L.F.	.37			.37	.41
8450	1" diameter					.60			.60	.66
8460	1-1/4" diameter				↓	1.05			1.05	1.16
8800	PVC, type PSP, drain & sewer, belled end gasket jnt., no hngr.									

Important: See the Reference Section for critical supporting data - Reference Nos., Crews, & City Cost Indexes

15108	Plastic Pipe & Fittings	CREW	DAILY OUTPUT	LABOR-HOURS	UNIT	2002 BARE COSTS				TOTAL INCL O&P	
						MAT.	LABOR	EQUIP.	TOTAL		
520	8840	3" diameter				L.F.	1.25			1.25	1.38	**520**
	8850	4" diameter					1.55			1.55	1.71	
	8860	6" diameter				↓	2.99			2.99	3.29	
	9000	Perforated										
	9040	4" diameter				L.F.	1.55			1.55	1.71	
560	0010	**PIPE, PLASTIC, FITTINGS**										**560**
	0030	Epoxy resin, fiberglass reinforced, general service										
	0090	Elbow, 90°, 2"	Q-1	33.10	.483	Ea.	45.50	15.65		61.15	73.50	
	0100	3"		20.80	.769		64.50	25		89.50	108	
	0110	4"		16.50	.970		88	31.50		119.50	145	
	0120	6"	↓	10.10	1.584		134	51.50		185.50	226	
	0130	8"	Q-2	9.30	2.581		248	86.50		334.50	405	
	0140	10"		8.50	2.824		310	94.50		404.50	485	
	0150	12"	↓	7.60	3.158	↓	445	106		551	650	
	0160	45° Elbow, same as 90°										
	0170	Elbow, 90°, flanged										
	0172	2"	Q-1	23	.696	Ea.	91	22.50		113.50	134	
	0173	3"		16	1		105	32.50		137.50	165	
	0174	4"		13	1.231		138	40		178	211	
	0176	6"	↓	8	2		249	64.50		313.50	370	
	0177	8"	Q-2	9	2.667		450	89.50		539.50	625	
	0178	10"		7	3.429		615	115		730	850	
	0179	12"	↓	5	4.800	↓	835	161		996	1,150	
	0186	Elbow, 45°, flanged										
	0188	2"	Q-1	23	.696	Ea.	91.50	22.50		114	135	
	0189	3"		16	1		105	32.50		137.50	165	
	0190	4"		13	1.231		137	40		177	211	
	0192	6"	↓	8	2		251	64.50		315.50	375	
	0193	8"	Q-2	9	2.667		450	89.50		539.50	630	
	0194	10"		7	3.429		615	115		730	850	
	0195	12"	↓	5	4.800		835	161		996	1,150	
	0290	Tee, 2"	Q-1	20	.800		55	26		81	99.50	
	0300	3"		13.90	1.151		64.50	37		101.50	127	
	0310	4"		11	1.455		76.50	47		123.50	155	
	0320	6"	↓	6.70	2.388		205	77.50		282.50	340	
	0330	8"	Q-2	6.20	3.871		237	130		367	455	
	0340	10"		5.70	4.211		380	141		521	630	
	0350	12"	↓	5.10	4.706	↓	460	158		618	745	
	0352	Tee, flanged										
	0354	2"	Q-1	17	.941	Ea.	124	30.50		154.50	182	
	0355	3"		10	1.600		166	52		218	261	
	0356	4"		8	2		185	64.50		249.50	300	
	0358	6"	↓	5	3.200		365	104		469	555	
	0359	8"	Q-2	6	4		625	134		759	885	
	0360	10"		5	4.800		905	161		1,066	1,250	
	0361	12"	↓	4	6	↓	1,150	201		1,351	1,575	
	0365	Wye, flanged										
	0367	2"	Q-1	17	.941	Ea.	320	30.50		350.50	395	
	0368	3"		10	1.600		355	52		407	470	
	0369	4"		8	2		440	64.50		504.50	585	
	0371	6"	↓	5	3.200		530	104		634	735	
	0372	8"	Q-2	6	4		690	134		824	955	
	0373	10"		5	4.800		1,175	161		1,336	1,525	
	0374	12"	↓	4	6	↓	1,425	201		1,626	1,850	
	0380	Couplings										

			DAILY	LABOR-		2002 BARE COSTS				TOTAL	
15108	**Plastic Pipe & Fittings**	CREW	OUTPUT	HOURS	UNIT	MAT.	LABOR	EQUIP.	TOTAL	INCL O&P	
560 0410	2″	Q-1	33.10	.483	Ea.	9.85	15.65		25.50	34.50	**560**
0420	3″		20.80	.769		16.95	25		41.95	56	
0430	4″		16.50	.970		20	31.50		51.50	69.50	
0440	6″	▼	10.10	1.584		38	51.50		89.50	120	
0450	8″	Q-2	9.30	2.581		58	86.50		144.50	195	
0460	10″		8.50	2.824		87.50	94.50		182	239	
0470	12″	▼	7.60	3.158	▼	117	106		223	289	
0473	High corrosion resistant couplings, add					30%					
0474	Reducer, concentric, flanged										
0475	2″ x 1-1/2″	Q-1	30	.533	Ea.	132	17.25		149.25	171	
0476	3″ x 2″		24	.667		150	21.50		171.50	198	
0477	4″ x 3″		19	.842		158	27		185	215	
0479	6″ x 4″	▼	15	1.067		206	34.50		240.50	279	
0480	8″ x 6″	Q-2	16	1.500		289	50.50		339.50	395	
0481	10″ x 8″		13	1.846		385	62		447	520	
0482	12″ x 10″	▼	11	2.182	▼	555	73		628	720	
0486	Adapter, bell x male or female										
0488	2″	Q-1	28	.571	Ea.	17.70	18.50		36.20	47.50	
0489	3″		20	.800		22.50	26		48.50	64	
0491	4″		17	.941		31	30.50		61.50	80	
0492	6″	▼	12	1.333		61.50	43		104.50	133	
0493	8″	Q-2	15	1.600		86	53.50		139.50	176	
0494	10″	″	11	2.182	▼	151	73		224	276	
0528	Flange										
0532	2″	Q-1	46	.348	Ea.	17.40	11.25		28.65	36	
0533	3″		32	.500		21	16.20		37.20	48	
0534	4″		26	.615		29.50	19.90		49.40	62.50	
0536	6″	▼	16	1		50	32.50		82.50	104	
0537	8″	Q-2	18	1.333		82	44.50		126.50	158	
0538	10″		14	1.714		114	57.50		171.50	213	
0539	12″	▼	10	2.400	▼	161	80.50		241.50	298	
2100	PVC schedule 80										
2110	90° elbow, 1/2″	1 Plum	30.30	.264	Ea.	.71	9.50		10.21	15.10	
2130	3/4″		26	.308		.91	11.05		11.96	17.70	
2140	1″		22.70	.352		1.47	12.65		14.12	20.50	
2150	1-1/4″		20.20	.396		1.96	14.25		16.21	23.50	
2160	1-1/2″	▼	18.20	.440		2.10	15.80		17.90	26.50	
2170	2″	Q-1	33.10	.483		2.55	15.65		18.20	26.50	
2180	3″		20.80	.769		6.70	25		31.70	45	
2190	4″		16.50	.970		7.65	31.50		39.15	56	
2200	6″	▼	10.10	1.584		57.50	51.50		109	141	
2210	8″	Q-2	9.30	2.581		158	86.50		244.50	305	
2250	45° elbow, 1/2″	1 Plum	30.30	.264		1.35	9.50		10.85	15.80	
2270	3/4″		26	.308		2.05	11.05		13.10	18.95	
2280	1″		22.70	.352		3.08	12.65		15.73	22.50	
2290	1-1/4″		20.20	.396		3.92	14.25		18.17	26	
2300	1-1/2″	▼	18.20	.440		4.63	15.80		20.43	29	
2310	2″	Q-1	33.10	.483		6	15.65		21.65	30	
2320	3″		20.80	.769		15.35	25		40.35	54.50	
2330	4″		16.50	.970		27.50	31.50		59	78	
2340	6″	▼	10.10	1.584		35	51.50		86.50	116	
2350	8″	Q-2	9.30	2.581		149	86.50		235.50	295	
2400	Tee, 1/2″	1 Plum	20.20	.396		2.21	14.25		16.46	24	
2420	3/4″		17.30	.462		2.32	16.60		18.92	27.50	
2430	1″		15.20	.526		2.89	18.90		21.79	31.50	
2440	1-1/4″	▼	13.50	.593	▼	5.90	21.50		27.40	38.50	

Important: See the Reference Section for critical supporting data - Reference Nos., Crews, & City Cost Indexes

15108	Plastic Pipe & Fittings	CREW	DAILY OUTPUT	LABOR-HOURS	UNIT	MAT.	LABOR	EQUIP.	TOTAL	TOTAL INCL O&P
						2002 BARE COSTS				
2450	1-1/2"	1 Plum	12.10	.661	Ea.	7.25	24		31.25	44
2460	2"	Q-1	20	.800		9.95	26		35.95	50
2470	3"		13.90	1.151		13.55	37		50.55	71
2480	4"		11	1.455		15.70	47		62.70	88.50
2490	6"		6.70	2.388		53.50	77.50		131	175
2500	8"	Q-2	6.20	3.871		213	130		343	430
2510	Flange, socket, 150 lb., 1/2"	1 Plum	55.60	.144		3.90	5.15		9.05	12.10
2514	3/4"		47.60	.168		4.17	6.05		10.22	13.70
2518	1"		41.70	.192		4.65	6.90		11.55	15.50
2522	1-1/2"		33.30	.240		4.90	8.65		13.55	18.40
2526	2"	Q-1	60.60	.264		6.50	8.55		15.05	20
2530	4"		30.30	.528		14.05	17.10		31.15	41.50
2534	6"		18.50	.865		22	28		50	66.50
2538	8"	Q-2	17.10	1.404		78.50	47		125.50	157
2550	Coupling, 1/2"	1 Plum	30.30	.264		1.29	9.50		10.79	15.70
2570	3/4"		26	.308		1.75	11.05		12.80	18.65
2580	1"		22.70	.352		1.71	12.65		14.36	21
2590	1-1/4"		20.20	.396		2.73	14.25		16.98	24.50
2600	1-1/2"		18.20	.440		2.85	15.80		18.65	27
2610	2"	Q-1	33.10	.483		3.16	15.65		18.81	27
2620	3"		20.80	.769		8.95	25		33.95	47.50
2630	4"		16.50	.970		11.20	31.50		42.70	60
2640	6"		10.10	1.584		24	51.50		75.50	104
2650	8"	Q-2	9.30	2.581		65	86.50		151.50	203
2660	10"		8.50	2.824		61	94.50		155.50	210
2670	12"		7.60	3.158		69	106		175	236
2700	PVC (white), schedule 40, socket joints									
2760	90° elbow, 1/2"	1 Plum	33.30	.240	Ea.	.26	8.65		8.91	13.30
2770	3/4"		28.60	.280		.29	10.05		10.34	15.45
2780	1"		25	.320		.53	11.50		12.03	17.95
2790	1-1/4"		22.20	.360		.94	12.95		13.89	20.50
2800	1-1/2"		20	.400		1	14.40		15.40	22.50
2810	2"	Q-1	36.40	.440		1.56	14.20		15.76	23
2820	2-1/2"		26.70	.599		4.75	19.40		24.15	34.50
2830	3"		22.90	.699		5.70	22.50		28.20	40.50
2840	4"		18.20	.879		10.20	28.50		38.70	54
2850	5"		12.10	1.322		26.50	43		69.50	93.50
2860	6"		11.10	1.441		32.50	46.50		79	106
2870	8"	Q-2	10.30	2.330		89	78		167	216
2980	45° elbow, 1/2"	1 Plum	33.30	.240		.43	8.65		9.08	13.45
2990	3/4"		28.60	.280		.66	10.05		10.71	15.90
3000	1"		25	.320		.80	11.50		12.30	18.25
3010	1-1/4"		22.20	.360		1.12	12.95		14.07	21
3020	1-1/2"		20	.400		.90	14.40		15.30	22.50
3030	2"	Q-1	36.40	.440		1.83	14.20		16.03	23.50
3040	2-1/2"		26.70	.599		4.76	19.40		24.16	34.50
3050	3"		22.90	.699		7.40	22.50		29.90	42
3060	4"		18.20	.879		13.25	28.50		41.75	57.50
3070	5"		12.10	1.322		26.50	43		69.50	93.50
3080	6"		11.10	1.441		32.50	46.50		79	107
3090	8"	Q-2	10.30	2.330		93	78		171	220
3180	Tee, 1/2"	1 Plum	22.20	.360		.32	12.95		13.27	19.90
3190	3/4"		19	.421		.36	15.15		15.51	23.50
3200	1"		16.70	.479		.70	17.20		17.90	27
3210	1-1/4"		14.80	.541		1.10	19.45		20.55	30.50
3220	1-1/2"		13.30	.601		1.33	21.50		22.83	34

560

560

MECHANICAL 15

	15108	Plastic Pipe & Fittings	CREW	DAILY OUTPUT	LABOR-HOURS	UNIT	2002 BARE COSTS				TOTAL INCL O&P	
							MAT.	LABOR	EQUIP.	TOTAL		
560	3230	2"	Q-1	24.20	.661	Ea.	1.93	21.50		23.43	34.50	560
	3240	2-1/2"		17.80	.899		6.35	29		35.35	51	
	3250	3"		15.20	1.053		10.20	34		44.20	62.50	
	3260	4"		12.10	1.322		15.10	43		58.10	81	
	3270	5"		8.10	1.975		36.50	64		100.50	137	
	3280	6"		7.40	2.162		51	70		121	161	
	3290	8"	Q-2	6.80	3.529		56.50	118		174.50	241	
	3380	Coupling, 1/2"	1 Plum	33.30	.240		.17	8.65		8.82	13.20	
	3390	3/4"		28.60	.280		.24	10.05		10.29	15.40	
	3400	1"		25	.320		.41	11.50		11.91	17.80	
	3410	1-1/4"		22.20	.360		.56	12.95		13.51	20	
	3420	1-1/2"		20	.400		.60	14.40		15	22	
	3430	2"	Q-1	36.40	.440		.94	14.20		15.14	22.50	
	3440	2-1/2"		26.70	.599		2.06	19.40		21.46	31.50	
	3450	3"		22.90	.699		3.24	22.50		25.74	37.50	
	3460	4"		18.20	.879		5.05	28.50		33.55	48.50	
	3470	5"		12.10	1.322		11.40	43		54.40	77	
	3480	6"		11.10	1.441		14.75	46.50		61.25	87	
	3490	8"	Q-2	10.30	2.330		28	78		106	149	
	3710	Reducing insert, schedule 40, socket weld										
	3712	3/4"	1 Plum	31.50	.254	Ea.	.29	9.15		9.44	14.05	
	3713	1"		27.50	.291		.53	10.45		10.98	16.35	
	3715	1-1/2"		22	.364		.74	13.05		13.79	20.50	
	3716	2"	Q-1	40	.400		1.24	12.95		14.19	21	
	3717	4"		20	.800		6.60	26		32.60	46.50	
	3718	6"		12.20	1.311		16.30	42.50		58.80	82	
	3719	8"	Q-2	11.30	2.124		57	71.50		128.50	170	
	3730	Reducing insert, socket weld x female/male thread										
	3732	1/2"	1 Plum	38.30	.209	Ea.	1.32	7.50		8.82	12.75	
	3733	3/4"		32.90	.243		.82	8.75		9.57	14.10	
	3734	1"		28.80	.278		1.14	10		11.14	16.30	
	3736	1-1/2"		23	.348		2.06	12.50		14.56	21	
	3737	2"	Q-1	41.90	.382		1.10	12.35		13.45	19.85	
	3738	4"	"	20.90	.766		6.15	25		31.15	44.50	
	3742	Male adapter, socket weld x male thread										
	3744	1/2"	1 Plum	38.30	.209	Ea.	.26	7.50		7.76	11.60	
	3745	3/4"		32.90	.243		.38	8.75		9.13	13.60	
	3746	1"		28.80	.278		.50	10		10.50	15.60	
	3748	1-1/2"		23	.348		.82	12.50		13.32	19.75	
	3749	2"	Q-1	41.90	.382		1.07	12.35		13.42	19.85	
	3750	4"	"	20.90	.766		5.90	25		30.90	44	
	3754	Female adapter, socket weld x female thread										
	3756	1/2"	1 Plum	38.30	.209	Ea.	.41	7.50		7.91	11.75	
	3757	3/4"		32.90	.243		.53	8.75		9.28	13.80	
	3758	1"		28.80	.278		.61	10		10.61	15.70	
	3760	1-1/2"		23	.348		.82	12.50		13.32	19.75	
	3761	2"	Q-1	41.90	.382		1.10	12.35		13.45	19.85	
	3762	4"	"	20.90	.766		6.15	25		31.15	44.50	
	3800	PVC, schedule 80, socket joints										
	3810	Reducing insert										
	3812	3/4"	1 Plum	28.60	.280	Ea.	.78	10.05		10.83	16	
	3813	1"		25	.320		3.03	11.50		14.53	20.50	
	3815	1-1/2"		20	.400		4.79	14.40		19.19	27	
	3816	2"	Q-1	36.40	.440		6.80	14.20		21	29	
	3817	4"		18.20	.879		26	28.50		54.50	71.50	
	3818	6"		11.10	1.441		36	46.50		82.50	111	

Important: See the Reference Section for critical supporting data - Reference Nos., Crews, & City Cost Indexes

15108	Plastic Pipe & Fittings	CREW	DAILY OUTPUT	LABOR-HOURS	UNIT	2002 BARE COSTS				TOTAL INCL O&P
						MAT.	LABOR	EQUIP.	TOTAL	
3819	8"	Q-2	10.20	2.353	Ea.	55.50	79		134.50	181
3830	Reducing insert, socket weld x female/male thread									
3832	1/2"	1 Plum	34.80	.230	Ea.	2.40	8.25		10.65	15.10
3833	3/4"		29.90	.268		1.98	9.60		11.58	16.70
3834	1"		26.10	.307		3.43	11		14.43	20.50
3836	1-1/2"	↓	20.90	.383		5.75	13.75		19.50	27.50
3837	2"	Q-1	38	.421		8.35	13.60		21.95	29.50
3838	4"	"	19	.842	↓	40	27		67	85
3844	Adapter, male socket x male thread									
3846	1/2"	1 Plum	34.80	.230	Ea.	1.80	8.25		10.05	14.45
3847	3/4"		29.90	.268		1.98	9.60		11.58	16.70
3848	1"		26.10	.307		3.43	11		14.43	20.50
3850	1-1/2"	↓	20.90	.383		5.75	13.75		19.50	27.50
3851	2"	Q-1	38	.421		8.35	13.60		21.95	29.50
3852	4"	"	19	.842	↓	18.70	27		45.70	61.50
3860	Adapter, female socket x female thread									
3862	1/2"	1 Plum	34.80	.230	Ea.	1.44	8.25		9.69	14.05
3863	3/4"		29.90	.268		3.37	9.60		12.97	18.20
3864	1"		26.10	.307		4.97	11		15.97	22
3866	1-1/2"	↓	20.90	.383		9.90	13.75		23.65	32
3867	2"	Q-1	38	.421		17.25	13.60		30.85	39.50
3868	4"	"	19	.842	↓	52.50	27		79.50	99
3872	Union, socket joints									
3874	1/2"	1 Plum	25.80	.310	Ea.	4.97	11.15		16.12	22.50
3875	3/4"		22.10	.362		7.20	13		20.20	27.50
3876	1"		19.30	.415		6.15	14.90		21.05	29.50
3878	1-1/2"	↓	15.50	.516		16.20	18.55		34.75	46
3879	2"	Q-1	28.10	.569	↓	22	18.40		40.40	52
3888	Cap									
3890	1/2"	1 Plum	54.50	.147	Ea.	2.38	5.30		7.68	10.55
3891	3/4"		46.70	.171		3.33	6.15		9.48	12.95
3892	1"		41	.195		4.44	7		11.44	15.50
3894	1-1/2"	↓	32.80	.244		5.35	8.75		14.10	19.10
3895	2"	Q-1	59.50	.269		14.10	8.70		22.80	28.50
3896	4"		30	.533		42.50	17.25		59.75	73
3897	6"	↓	18.20	.879		112	28.50		140.50	166
3898	8"	Q-2	16.70	1.437	↓	143	48		191	231
4500	DWV, ABS, non pressure, socket joints									
4540	1/4 Bend, 1-1/4"	1 Plum	20.20	.396	Ea.	2.12	14.25		16.37	24
4560	1-1/2"	"	18.20	.440		1.13	15.80		16.93	25
4570	2"	Q-1	33.10	.483		1.85	15.65		17.50	25.50
4580	3"		20.80	.769		4.39	25		29.39	42.50
4590	4"		16.50	.970		7.65	31.50		39.15	56
4600	6"	↓	10.10	1.584	↓	48	51.50		99.50	131
4650	1/8 Bend, same as 1/4 Bend									
4800	Tee, sanitary									
4820	1-1/4"	1 Plum	13.50	.593	Ea.	2.67	21.50		24.17	35
4830	1-1/2"	"	12.10	.661		1.66	24		25.66	38
4840	2"	Q-1	20	.800		2.35	26		28.35	41.50
4850	3"		13.90	1.151		5.95	37		42.95	62.50
4860	4"		11	1.455		14.80	47		61.80	87.50
4862	Tee, sanitary, reducing, 2" x 1-1/2"		22	.727		2.48	23.50		25.98	38
4864	3" x 2"		15.30	1.046		4.57	34		38.57	56
4868	4" x 3"	↓	12.10	1.322	↓	13.45	43		56.45	79.50
4870	Combination Y and 1/8 bend									
4872	1-1/2"	1 Plum	12.10	.661	Ea.	5.55	24		29.55	42

MECHANICAL 15

560

	15108	Plastic Pipe & Fittings	CREW	DAILY OUTPUT	LABOR-HOURS	UNIT	2002 BARE COSTS				TOTAL INCL O&P	
							MAT.	LABOR	EQUIP.	TOTAL		
560	4874	2"	Q-1	20	.800	Ea.	6.75	26		32.75	46.50	560
	4876	3"		13.90	1.151		11.95	37		48.95	69	
	4878	4"		11	1.455		23.50	47		70.50	97	
	4880	3" x 1-1/2"		15.50	1.032		13.90	33.50		47.40	66	
	4882	4" x 3"	▼	12.10	1.322		21	43		64	87.50	
	4900	Wye, 1-1/4"	1 Plum	13.50	.593		3.08	21.50		24.58	35.50	
	4902	1-1/2"	"	12.10	.661		3.15	24		27.15	39.50	
	4904	2"	Q-1	20	.800		3.22	26		29.22	42.50	
	4906	3"		13.90	1.151		7.40	37		44.40	64	
	4908	4"		11	1.455		17.60	47		64.60	90.50	
	4910	6"		6.70	2.388		66.50	77.50		144	190	
	4918	3" x 1-1/2"		15.50	1.032		7.35	33.50		40.85	58.50	
	4920	4" x 3"		12.10	1.322		12.15	43		55.15	78	
	4922	6" x 4"	▼	6.90	2.319		39.50	75		114.50	157	
	4930	Double Wye, 1-1/2"	1 Plum	9.10	.879		6.65	31.50		38.15	55	
	4932	2"	Q-1	16.60	.964		8.60	31		39.60	56.50	
	4934	3"		10.40	1.538		22	50		72	99.50	
	4936	4"		8.25	1.939		45	62.50		107.50	144	
	4940	2" x 1-1/2"		16.80	.952		8.60	31		39.60	56	
	4942	3" x 2"		10.60	1.509		16.50	49		65.50	91.50	
	4944	4" x 3"		8.45	1.893		35.50	61.50		97	132	
	4946	6" x 4"		7.25	2.207		74.50	71.50		146	190	
	4950	Reducer bushing, 2" x 1-1/2"		36.40	.440		1.20	14.20		15.40	23	
	4952	3" x 1-1/2"		27.30	.586		3.57	18.95		22.52	32.50	
	4954	4" x 2"		18.20	.879		10.15	28.50		38.65	54	
	4956	6" x 4"	▼	11.10	1.441		24.50	46.50		71	97.50	
	4960	Couplings, 1-1/2"	1 Plum	18.20	.440		2.32	15.80		18.12	26.50	
	4962	2"	Q-1	33.10	.483		.74	15.65		16.39	24.50	
	4963	3"		20.80	.769		2.48	25		27.48	40	
	4964	4"		16.50	.970		3.81	31.50		35.31	51.50	
	4966	6"		10.10	1.584		20	51.50		71.50	99.50	
	4970	2" x 1-1/2"		33.30	.480		2.17	15.55		17.72	26	
	4972	3" x 1-1/2"		21	.762		6.20	24.50		30.70	44	
	4974	4" x 3"	▼	16.70	.958		10.75	31		41.75	58.50	
	4978	Closet flange, 4"	1 Plum	32	.250		4.79	9		13.79	18.80	
	4980	4" x 3"	"	34	.235	▼	4.50	8.45		12.95	17.70	
	5000	DWV, PVC, schedule 40, socket joints										
	5040	1/4 bend, 1-1/4"	1 Plum	20.20	.396	Ea.	1.82	14.25		16.07	23.50	
	5060	1-1/2"	"	18.20	.440		.92	15.80		16.72	25	
	5070	2"	Q-1	33.10	.483		1.42	15.65		17.07	25	
	5080	3"		20.80	.769		3.54	25		28.54	41.50	
	5090	4"		16.50	.970		6.35	31.50		37.85	54.50	
	5100	6"	▼	10.10	1.584		38.50	51.50		90	120	
	5105	8"	Q-2	9.30	2.581		87	86.50		173.50	227	
	5106	10"	"	8.50	2.824		65.50	94.50		160	215	
	5110	1/4 bend, long sweep, 1-1/2"	1 Plum	18.20	.440		2.90	15.80		18.70	27	
	5112	2"	Q-1	33.10	.483		2.35	15.65		18	26	
	5114	3"		20.80	.769		6.15	25		31.15	44.50	
	5116	4"	▼	16.50	.970		12.15	31.50		43.65	61	
	5150	1/8 bend, 1-1/4"	1 Plum	20.20	.396		1.85	14.25		16.10	23.50	
	5170	1-1/2"	"	18.20	.440		1.03	15.80		16.83	25	
	5180	2"	Q-1	33.10	.483		1.50	15.65		17.15	25	
	5190	3"		20.80	.769		3.05	25		28.05	41	
	5200	4"		16.50	.970		5.05	31.50		36.55	53	
	5210	6"	▼	10.10	1.584		35.50	51.50		87	117	
	5215	8"	Q-2	9.30	2.581	▼	38.50	86.50		125	173	

Important: See the Reference Section for critical supporting data - Reference Nos., Crews, & City Cost Indexes

15108	Plastic Pipe & Fittings	CREW	DAILY OUTPUT	LABOR-HOURS	UNIT	MAT.	LABOR	EQUIP.	TOTAL	TOTAL INCL O&P	
560							2002 BARE COSTS					**560**
5216	10"	Q-2	8.50	2.824	Ea.	69	94.50		163.50	219	
5217	12"	↓	7.60	3.158		95	106		201	265	
5250	Tee, sanitary 1-1/4"	1 Plum	13.50	.593		3.27	21.50		24.77	35.50	
5254	1-1/2"	"	12.10	.661		1.38	24		25.38	37.50	
5255	2"	Q-1	20	.800		1.88	26		27.88	41	
5256	3"		13.90	1.151		5.15	37		42.15	61.50	
5257	4"		11	1.455		9.85	47		56.85	82	
5259	6"	↓	6.70	2.388		63.50	77.50		141	186	
5261	8"	Q-2	6.20	3.871		191	130		321	405	
5264	2" x 1-1/2"	Q-1	22	.727		7.35	23.50		30.85	43.50	
5266	3" x 1-1/2"		15.50	1.032		4.40	33.50		37.90	55.50	
5268	4" x 3"		12.10	1.322		20	43		63	86.50	
5271	6" x 4"	↓	6.90	2.319	↓	68.50	75		143.50	189	
5276	Tee, sanitary, reducing										
5281	2" x 1-1/2" x 1-1/2"	Q-1	23	.696	Ea.	3.03	22.50		25.53	37.50	
5282	2" x 1-1/2" x 2"		22	.727		4.19	23.50		27.69	40	
5283	2" x 2" x 1-1/2"		22	.727		2.41	23.50		25.91	38	
5284	3" x 3" x 1-1/2"		15.50	1.032		4.89	33.50		38.39	56	
5285	3" x 3" x 2"		15.30	1.046		5.20	34		39.20	57	
5286	4" x 4" x 1-1/2"		12.30	1.301		16.75	42		58.75	82	
5287	4" x 4" x 2"		12.20	1.311		16.50	42.50		59	82	
5288	4" x 4" x 3"		12.10	1.322		22.50	43		65.50	89	
5291	6" x 6" x 4"	↓	6.90	2.319	↓	68.50	75		143.50	189	
5294	Tee, double sanitary										
5295	1-1/2"	1 Plum	9.10	.879	Ea.	5.05	31.50		36.55	53	
5296	2"	Q-1	16.60	.964		7.40	31		38.40	55	
5297	3"		10.40	1.538		20.50	50		70.50	98	
5298	4"	↓	8.25	1.939	↓	33.50	62.50		96	131	
5303	Wye, reducing										
5304	2" x 1-1/2" x 1-1/2"	Q-1	23	.696	Ea.	6.25	22.50		28.75	41	
5305	2" x 2" x 1-1/2"		22	.727		5.95	23.50		29.45	42	
5306	3" x 3" x 2"		15.30	1.046		11	34		45	63	
5307	4" x 4" x 2"		12.20	1.311		11.40	42.50		53.90	76.50	
5309	4" x 4" x 3"	↓	12.10	1.322		13.35	43		56.35	79	
5314	Combination Y & 1/8 bend, 1-1/2"	1 Plum	12.10	.661		4.35	24		28.35	41	
5315	2"	Q-1	20	.800		6.25	26		32.25	46	
5317	3"		13.90	1.151		10.30	37		47.30	67.50	
5318	4"		11	1.455		20.50	47		67.50	93.50	
5319	6"	↓	6.70	2.388		66.50	77.50		144	190	
5320	8"	Q-2	6.20	3.871	↓	106	130		236	315	
5324	Combination Y & 1/8 bend, reducing										
5325	2" x 2" x 1-1/2"	Q-1	22	.727	Ea.	6.95	23.50		30.45	43	
5327	3" x 3" x 1-1/2"		15.50	1.032		11.60	33.50		45.10	63.50	
5328	3" x 3" x 2"		15.30	1.046		7.50	34		41.50	59.50	
5329	4" x 4" x 2"	↓	12.20	1.311		16.75	42.50		59.25	82.50	
5331	Wye, 1-1/4"	1 Plum	13.50	.593		3.13	21.50		24.63	35.50	
5332	1-1/2"	"	12.10	.661		3.35	24		27.35	39.50	
5333	2"	Q-1	20	.800		3.36	26		29.36	42.50	
5334	3"		13.90	1.151		7.65	37		44.65	64.50	
5335	4"		11	1.455		11.55	47		58.55	83.50	
5336	6"	↓	6.70	2.388		35.50	77.50		113	155	
5337	8"	Q-2	6.20	3.871		64	130		194	267	
5338	10"		5.70	4.211		124	141		265	350	
5339	12"	↓	5.10	4.706		202	158		360	460	
5341	2" x 1-1/2"	Q-1	22	.727		4.79	23.50		28.29	41	
5342	3" x 1-1/2"	↓	15.50	1.032	↓	6.40	33.50		39.90	57.50	

		CREW	DAILY OUTPUT	LABOR-HOURS	UNIT	2002 BARE COSTS				TOTAL INCL O&P	
15108	**Plastic Pipe & Fittings**					MAT.	LABOR	EQUIP.	TOTAL		
560 5343	4" x 3"	Q-1	12.10	1.322	Ea.	10.80	43		53.80	76.50	560
5344	6" x 4"	↓	6.90	2.319		42.50	75		117.50	160	
5345	8" x 6"	Q-2	6.40	3.750		48.50	126		174.50	244	
5347	Double wye, 1-1/2"	1 Plum	9.10	.879		5.80	31.50		37.30	54	
5348	2"	Q-1	16.60	.964		7.45	31		38.45	55	
5349	3"		10.40	1.538		19.20	50		69.20	96	
5350	4"		8.25	1.939		39	62.50		101.50	138	
5354	2" x 1-1/2"		16.80	.952		6.80	31		37.80	54	
5355	3" x 2"		10.60	1.509		14.30	49		63.30	89.50	
5356	4" x 3"		8.45	1.893		31	61.50		92.50	127	
5357	6" x 4"	↓	7.25	2.207		64	71.50		135.50	179	
5374	Coupling, 1-1/4"	1 Plum	20.20	.396		1.76	14.25		16.01	23.50	
5376	1-1/2"	"	18.20	.440		.59	15.80		16.39	24.50	
5378	2"	Q-1	33.10	.483		.61	15.65		16.26	24	
5380	3"		20.80	.769		1.76	25		26.76	39.50	
5390	4"		16.50	.970		2.90	31.50		34.40	50.50	
5400	6"	↓	10.10	1.584		15.45	51.50		66.95	94.50	
5402	8"	Q-2	9.30	2.581		35	86.50		121.50	170	
5404	2" x 1-1/2"	Q-1	33.30	.480		1.53	15.55		17.08	25	
5406	3" x 1-1/2"		21	.762		4.85	24.50		29.35	42.50	
5408	4" x 3"		16.70	.958		8.40	31		39.40	56	
5410	Reducer bushing, 2" x 1-1/4"		36.50	.438		1.03	14.20		15.23	22.50	
5411	2" x 1-1/2"		36.40	.440		1.69	14.20		15.89	23.50	
5412	3" x 1-1/2"		27.30	.586		4.52	18.95		23.47	33.50	
5413	3" x 2"		27.10	.590		5	19.10		24.10	34.50	
5414	4" x 2"		18.20	.879		9.75	28.50		38.25	54	
5415	4" x 3"		16.70	.958		5.40	31		36.40	52.50	
5416	6" x 4"	↓	11.10	1.441		28	46.50		74.50	102	
5418	8" x 6"	Q-2	10.20	2.353		56	79		135	181	
5425	Closet flange 4"	Q-1	32	.500		5.10	16.20		21.30	30	
5426	4" x 3"	"	34	.471	↓	4.51	15.20		19.71	28	
5450	Solvent cement for PVC, industrial grade, per quart				Qt.	11.30			11.30	12.45	
5500	CPVC, Schedule 80, threaded joints										
5540	90° Elbow, 1/4"	1 Plum	32	.250	Ea.	6.25	9		15.25	20.50	
5560	1/2"		30.30	.264		2.45	9.50		11.95	17	
5570	3/4"		26	.308		3.13	11.05		14.18	20	
5580	1"		22.70	.352		4.96	12.65		17.61	24.50	
5590	1-1/4"		20.20	.396		10.75	14.25		25	33.50	
5600	1-1/2"	↓	18.20	.440		12	15.80		27.80	37	
5610	2"	Q-1	33.10	.483		14.45	15.65		30.10	39.50	
5620	2-1/2"		24.20	.661		33.50	21.50		55	69	
5630	3"		20.80	.769		37.50	25		62.50	79	
5640	4"		16.50	.970		68	31.50		99.50	122	
5650	6"	↓	10.10	1.584	↓	136	51.50		187.50	228	
5700	45° Elbow same as 90° Elbow										
5850	Tee, 1/4"	1 Plum	22	.364	Ea.	5.75	13.05		18.80	26	
5870	1/2"		20.20	.396		5.75	14.25		20	28	
5880	3/4"		17.30	.462		5.85	16.60		22.45	31.50	
5890	1"		15.20	.526		7.15	18.90		26.05	36.50	
5900	1-1/4"		13.50	.593		15.10	21.50		36.60	48.50	
5910	1-1/2"	↓	12.10	.661		17.30	24		41.30	55	
5920	2"	Q-1	20	.800		19.25	26		45.25	60	
5930	2-1/2"		16.20	.988		49	32		81	102	
5940	3"		13.90	1.151		49	37		86	110	
5950	4"		11	1.455		65	47		112	143	
5960	6"	↓	6.70	2.388	↓	170	77.50		247.50	305	

Important: See the Reference Section for critical supporting data - Reference Nos., Crews, & City Cost Indexes

	15108	Plastic Pipe & Fittings	CREW	DAILY OUTPUT	LABOR-HOURS	UNIT	2002 BARE COSTS				TOTAL INCL O&P	
							MAT.	LABOR	EQUIP.	TOTAL		
560	6000	Coupling, 1/4"	1 Plum	32	.250	Ea.	3.88	9		12.88	17.80	**560**
	6020	1/2"		30.30	.264		2.72	9.50		12.22	17.30	
	6030	3/4"		26	.308		3.87	11.05		14.92	21	
	6040	1"		22.70	.352		4.16	12.65		16.81	23.50	
	6050	1-1/4"		20.20	.396		5.15	14.25		19.40	27	
	6060	1-1/2"		18.20	.440		11.45	15.80		27.25	36.50	
	6070	2"	Q-1	33.10	.483		12	15.65		27.65	36.50	
	6080	2-1/2"		24.20	.661		28.50	21.50		50	64	
	6090	3"		20.80	.769		33.50	25		58.50	74.50	
	6100	4"		16.50	.970		58	31.50		89.50	112	
	6110	6"		10.10	1.584		45.50	51.50		97	128	
	6120	8"	Q-2	9.30	2.581		227	86.50		313.50	380	
	6200	CTS, 100 psi at 180°F, hot and cold water										
	6230	90° Elbow, 1/2"	1 Plum	20	.400	Ea.	.08	14.40		14.48	21.50	
	6250	3/4"		19	.421		.18	15.15		15.33	23	
	6251	1"		16	.500		.58	18		18.58	27.50	
	6252	1-1/4"		15	.533		.99	19.15		20.14	30	
	6253	1-1/2"		14	.571		1.59	20.50		22.09	33	
	6254	2"	Q-1	23	.696		3.43	22.50		25.93	38	
	6260	45° Elbow, 1/2"	1 Plum	20	.400		.13	14.40		14.53	21.50	
	6280	3/4"		19	.421		.20	15.15		15.35	23	
	6281	1"		16	.500		.53	18		18.53	27.50	
	6282	1-1/4"		15	.533		1.04	19.15		20.19	30	
	6283	1-1/2"		14	.571		1.65	20.50		22.15	33	
	6284	2"	Q-1	23	.696		3.48	22.50		25.98	38	
	6290	Tee, 1/2"	1 Plum	13	.615		.12	22		22.12	33.50	
	6310	3/4"		12	.667		.26	24		24.26	36.50	
	6311	1"		11	.727		1.31	26		27.31	41	
	6312	1-1/4"		10	.800		2	29		31	45.50	
	6313	1-1/2"		10	.800		2.60	29		31.60	46.50	
	6314	2"	Q-1	17	.941		4.22	30.50		34.72	50.50	
	6320	Coupling, 1/2"	1 Plum	22	.364		.10	13.05		13.15	19.80	
	6340	3/4"		21	.381		.12	13.70		13.82	20.50	
	6341	1"		18	.444		.50	16		16.50	24.50	
	6342	1-1/4"		17	.471		.62	16.90		17.52	26	
	6343	1-1/2"		16	.500		.88	18		18.88	28	
	6344	2"	Q-1	28	.571		1.75	18.50		20.25	30	
	6360	Solvent cement for CPVC, commercial grade, per quart				Qt.	10.95			10.95	12.05	
	7990	Polybutyl/polyethyl pipe, for copper fittings see 15107-460-7000										
	8000	Compression type, PVC, 160 psi cold water										
	8010	Coupling, 3/4" CTS	1 Plum	21	.381	Ea.	2.11	13.70		15.81	23	
	8020	1" CTS		18	.444		2.61	16		18.61	27	
	8030	1-1/4" CTS		17	.471		3.63	16.90		20.53	29.50	
	8040	1-1/2" CTS		16	.500		5	18		23	32.50	
	8050	2" CTS		15	.533		6.95	19.15		26.10	36.50	
	8060	Female adapter, 3/4" FPT x 3/4" CTS		23	.348		3.35	12.50		15.85	22.50	
	8070	3/4" FPT x 1" CTS		21	.381		5	13.70		18.70	26	
	8080	1" FPT x 1" CTS		20	.400		5.05	14.40		19.45	27	
	8090	1-1/4" FPT x 1-1/4" CTS		18	.444		6.60	16		22.60	31.50	
	8100	1-1/2" FPT x 1-1/2" CTS		16	.500		7.70	18		25.70	35.50	
	8110	2" FPT x 2" CTS		13	.615		11.45	22		33.45	46	
	8130	Male adapter, 3/4" MPT x 3/4" CTS		23	.348		3.02	12.50		15.52	22	
	8140	3/4" MPT x 1" CTS		21	.381		3.71	13.70		17.41	24.50	
	8150	1" MPT x 1" CTS		20	.400		3.79	14.40		18.19	25.50	
	8160	1-1/4" MPT x 1-1/4" CTS		18	.444		5.80	16		21.80	30.50	
	8170	1-1/2" MPT x 1-1/2" CTS		16	.500		6.75	18		24.75	34.50	

MECHANICAL **15**

	15108	Plastic Pipe & Fittings	CREW	DAILY OUTPUT	LABOR-HOURS	UNIT	MAT.	LABOR	EQUIP.	TOTAL	TOTAL INCL O&P	
560	8180	2" MPT x 2" CTS	1 Plum	13	.615	Ea.	10.75	22		32.75	45.50	**560**
	8200	Spigot adapter, 3/4" IPS x 3/4" CTS		23	.348		3.02	12.50		15.52	22	
	8210	3/4" IPS x 1" CTS		21	.381		3.02	13.70		16.72	24	
	8220	1" IPS x 1" CTS		20	.400		3.26	14.40		17.66	25	
	8230	1-1/4" IPS x 1-1/4" CTS		18	.444		5.80	16		21.80	30.50	
	8240	1-1/2" IPS x 1-1/2" CTS		16	.500		6.75	18		24.75	34.50	
	8250	2" IPS x 2" CTS		13	.615		10.75	22		32.75	45.50	
	8270	Price includes insert stiffeners										
	8280	250 psi is same price as 160 psi										
	8300	Insert type, nylon, 160 & 250 psi, cold water										
	8310	Clamp ring stainless steel, 3/4" IPS	1 Plum	115	.070	Ea.	.63	2.50		3.13	4.46	
	8320	1" IPS		107	.075		.65	2.69		3.34	4.77	
	8330	1-1/4" IPS		101	.079		.65	2.85		3.50	5	
	8340	1-1/2" IPS		95	.084		.66	3.03		3.69	5.30	
	8350	2" IPS		85	.094		.66	3.38		4.04	5.85	
	8370	Coupling, 3/4" IPS		22	.364		.62	13.05		13.67	20.50	
	8390	1-1/4" IPS		18	.444		1.49	16		17.49	25.50	
	8400	1-1/2" IPS		17	.471		1.94	16.90		18.84	27.50	
	8410	2" IPS		16	.500		2.33	18		20.33	29.50	
	8430	Elbow, 90°, 3/4" IPS		22	.364		.91	13.05		13.96	20.50	
	8440	1" IPS		19	.421		1	15.15		16.15	24	
	8450	1-1/4" IPS		18	.444		1.12	16		17.12	25	
	8460	1-1/2" IPS		17	.471		1.33	16.90		18.23	27	
	8470	2" IPS		16	.500		1.85	18		19.85	29	
	8490	Male adapter, 3/4" IPS x 3/4" MPT		25	.320		.63	11.50		12.13	18.05	
	8500	1" IPS x 1" MPT		21	.381		.83	13.70		14.53	21.50	
	8510	1-1/4" IPS x 1-1/4" MPT		20	.400		1.37	14.40		15.77	23	
	8520	1-1/2" IPS x 1-1/2" MPT		18	.444		1.81	16		17.81	26	
	8530	2" IPS x 2" MPT		15	.533		2.33	19.15		21.48	31.50	
	8550	Tee, 3/4" IPS		14	.571		1.19	20.50		21.69	32.50	
	8560	1" IPS		13	.615		1.94	22		23.94	35.50	
	8570	1-1/4" IPS		12	.667		3.50	24		27.50	40	
	8580	1-1/2" IPS		11	.727		4.40	26		30.40	44.50	
	8590	2" IPS		10	.800		5.95	29		34.95	50	
	8610	Insert type, PVC, 100 psi @ 180°F, hot & cold water										
	8620	Coupler, male, 3/8" CTS x 3/8" MPT	1 Plum	29	.276	Ea.	.38	9.90		10.28	15.35	
	8630	3/8" CTS x 1/2" MPT		28	.286		.38	10.25		10.63	15.90	
	8640	1/2" CTS x 1/2" MPT		27	.296		.38	10.65		11.03	16.45	
	8650	1/2" CTS x 3/4" MPT		26	.308		1.07	11.05		12.12	17.90	
	8660	3/4" CTS x 1/2" MPT		25	.320		1.07	11.50		12.57	18.55	
	8670	3/4" CTS x 3/4" MPT		25	.320		.45	11.50		11.95	17.85	
	8700	Coupling, 3/8" CTS x 1/2" CTS		25	.320		2.15	11.50		13.65	19.70	
	8710	1/2" CTS		23	.348		2.50	12.50		15	21.50	
	8720	1/2" CTS x stub		23	.348		2.52	12.50		15.02	21.50	
	8730	3/4" CTS		22	.364		3.96	13.05		17.01	24	
	8750	Elbow 90°, 3/8" CTS		25	.320		2.55	11.50		14.05	20	
	8760	1/2" CTS		23	.348		3.01	12.50		15.51	22	
	8770	3/4" CTS		22	.364		4.61	13.05		17.66	25	
	8800	Rings, crimp, copper, 3/8" CTS		120	.067		.11	2.40		2.51	3.73	
	8810	1/2" CTS		117	.068		.12	2.46		2.58	3.84	
	8820	3/4" CTS		115	.070		.15	2.50		2.65	3.94	
	8850	Reducer tee, 3/8" x 3/8" x 1/2" CTS		17	.471		1.46	16.90		18.36	27	
	8860	1/2" x 3/8" x 1/2" CTS		15	.533		1.25	19.15		20.40	30.50	
	8870	3/4" x 1/2" x 1/2" CTS		14	.571		1.31	20.50		21.81	32.50	
	8890	3/4" x 3/4" x 1/2" CTS		14	.571		1.24	20.50		21.74	32.50	
	8900	3/4" x 1/2" x 3/8" CTS		14	.571		1.09	20.50		21.59	32	

Important: See the Reference Section for critical supporting data - Reference Nos., Crews, & City Cost Indexes

15108 | Plastic Pipe & Fittings

		CREW	DAILY OUTPUT	LABOR-HOURS	UNIT	MAT.	LABOR	EQUIP.	TOTAL	TOTAL INCL O&P	
560	8930	Tee, 3/8" CTS	1 Plum	17	.471	Ea.	1.05	16.90		17.95	26.50	**560**
	8940	1/2" CTS		15	.533		1.19	19.15		20.34	30.50	
	8950	3/4" CTS		14	.571		1.29	20.50		21.79	32.50	
	8960	Copper rings included in fitting price										
	9000	Flare type, assembled, acetal, hot & cold water										
	9010	Coupling, 1/4" & 3/8" CTS	1 Plum	24	.333	Ea.	2.14	12		14.14	20.50	
	9020	1/2" CTS		22	.364		2.45	13.05		15.50	22.50	
	9030	3/4" CTS		21	.381		3.96	13.70		17.66	25	
	9040	1" CTS		18	.444		3.96	16		19.96	28.50	
	9050	Elbow 90°, 1/4" CTS		26	.308		2.38	11.05		13.43	19.30	
	9060	3/8" CTS		24	.333		2.55	12		14.55	21	
	9070	1/2" CTS		22	.364		3.01	13.05		16.06	23	
	9080	3/4" CTS		21	.381		4.61	13.70		18.31	25.50	
	9090	1" CTS		18	.444		5.80	16		21.80	30.50	
	9110	Tee, 1/4" & 3/8" CTS		15	.533		2.62	19.15		21.77	32	
	9120	1/2" CTS		14	.571		3.44	20.50		23.94	35	
	9130	3/4" CTS		13	.615		5.25	22		27.25	39.50	
	9140	1" CTS		12	.667		7	24		31	43.50	
	9550	For plastic hangers see 15060-300-8000										
	9560	For copper/brass fittings see 15107-460-7000										
590	0010	**PIPE, HIGH DENSITY POLYETHYLENE PLASTIC (HDPE)**										**590**
	0020	Not incl. hangers, trenching, backfill, hoisting or digging equipment.										
	0030	Standard length is 40', add a weld for each joint.										
	0040	Single wall										
	0050	Straight										
	0054	1" diameter DR 11				L.F.	.28			.28	.31	
	0058	1-1/2" diameter DR 11					.62			.62	.68	
	0062	2" diameter DR 11					.79			.79	.87	
	0066	3" diameter DR 11					1.72			1.72	1.89	
	0070	3" diameter DR 17					1.15			1.15	1.27	
	0074	4" diameter DR 11					2.84			2.84	3.12	
	0078	4" diameter DR 17					1.91			1.91	2.10	
	0082	6" diameter DR 11					6.15			6.15	6.75	
	0086	6" diameter DR 17					4.13			4.13	4.54	
	0090	8" diameter DR 11					10.40			10.40	11.45	
	0094	8" diameter DR 26					4.69			4.69	5.15	
	0098	10" diameter DR 11					16.20			16.20	17.85	
	0102	10" diameter DR 26					7.25			7.25	8	
	0106	12" diameter DR 11					23			23	25	
	0110	12" diameter DR 26					10.25			10.25	11.25	
	0114	16" diameter DR 11					34			34	37	
	0118	16" diameter DR 26					15.20			15.20	16.70	
	0122	18" diameter DR 11					43			43	47	
	0126	18" diameter DR 26					19.20			19.20	21	
	0130	20" diameter DR 11					53			53	58	
	0134	20" diameter DR 26					23.50			23.50	26	
	0138	22" diameter DR 11					64			64	70.50	
	0142	22" diameter DR 26					28.50			28.50	31.50	
	0146	24" diameter DR 11					76			76	84	
	0150	24" diameter DR 26					34			34	37.50	
	0154	28" diameter DR 17					69.50			69.50	76.50	
	0158	28" diameter DR 26					46.50			46.50	51	
	0162	30" diameter DR 21					65.50			65.50	72	
	0166	30" diameter DR 26					43			43	47.50	
	0170	36" diameter DR 26					43			43	47.50	
	0174	42" diameter DR 26					105			105	115	

MECHANICAL 15

	15108	**Plastic Pipe & Fittings**	CREW	DAILY OUTPUT	LABOR-HOURS	UNIT	MAT.	LABOR	EQUIP.	TOTAL	TOTAL INCL O&P	
590	0178	48″ diameter DR 26				L.F.	132			132	146	**590**
	0182	54″ diameter DR 26				↓	173			173	190	
	0300	90° Elbow										
	0304	1″ diameter DR 11				Ea.	6.45			6.45	7.10	
	0308	1-1/2″ diameter DR 11					10.75			10.75	11.85	
	0312	2″ diameter DR 11					12			12	13.20	
	0316	3″ diameter DR 11					23			23	25.50	
	0320	3″ diameter DR 17					23			23	25.50	
	0324	4″ diameter DR 11					27			27	29.50	
	0328	4″ diameter DR 17					27			27	29.50	
	0332	6″ diameter DR 11					79			79	87	
	0336	6″ diameter DR 17					79			79	87	
	0340	8″ diameter DR 11					260			260	286	
	0344	8″ diameter DR 26					177			177	195	
	0348	10″ diameter DR 11					645			645	710	
	0352	10″ diameter DR 26					325			325	360	
	0356	12″ diameter DR 11					645			645	710	
	0360	12″ diameter DR 26					355			355	390	
	0364	16″ diameter DR 11					720			720	795	
	0368	16″ diameter DR 26					580			580	640	
	0372	18″ diameter DR 11					755			755	830	
	0376	18″ diameter DR 26					755			755	830	
	0380	20″ diameter DR 11					1,125			1,125	1,225	
	0384	20″ diameter DR 26					1,125			1,125	1,225	
	0388	22″ diameter DR 11					1,400			1,400	1,550	
	0392	22″ diameter DR 26					1,400			1,400	1,550	
	0396	24″ diameter DR 11					1,625			1,625	1,775	
	0400	24″ diameter DR 26					1,625			1,625	1,775	
	0404	28″ diameter DR 17					2,450			2,450	2,700	
	0408	28″ diameter DR 26					2,450			2,450	2,700	
	0412	30″ diameter DR 17					2,250			2,250	2,475	
	0416	30″ diameter DR 26					2,250			2,250	2,475	
	0420	36″ diameter DR 26					3,450			3,450	3,775	
	0424	42″ diameter DR 26					3,875			3,875	4,250	
	0428	48″ diameter DR 26					4,075			4,075	4,500	
	0432	54″ diameter DR 26				↓	9,475			9,475	10,400	
	0500	45° Elbow										
	0512	2″ diameter DR 11				Ea.	24			24	26.50	
	0516	3″ diameter DR 11					21			21	23	
	0520	3″ diameter DR 17					34			34	37.50	
	0524	4″ diameter DR 11					35			35	38.50	
	0528	4″ diameter DR 17					42			42	46	
	0532	6″ diameter DR 11					74			74	81.50	
	0536	6″ diameter DR 17					70			70	77	
	0540	8″ diameter DR 11					175			175	193	
	0544	8″ diameter DR 26					105			105	116	
	0548	10″ diameter DR 11					490			490	540	
	0552	10″ diameter DR 26					142			142	156	
	0556	12″ diameter DR 11					590			590	650	
	0560	12″ diameter DR 26					224			224	246	
	0564	16″ diameter DR 11					282			282	310	
	0568	16″ diameter DR 26					282			282	310	
	0572	18″ diameter DR 11					415			415	455	
	0576	18″ diameter DR 26					415			415	455	
	0580	20″ diameter DR 11					645			645	710	
	0584	20″ diameter DR 26				↓	645			645	710	

15 MECHANICAL

Important: See the Reference Section for critical supporting data - Reference Nos., Crews, & City Cost Indexes

15108	Plastic Pipe & Fittings	CREW	DAILY OUTPUT	LABOR-HOURS	UNIT	2002 BARE COSTS				TOTAL INCL O&P
						MAT.	LABOR	EQUIP.	TOTAL	
590 0588	22" diameter DR 11				Ea.	755			755	830
0592	22" diameter DR 26					755			755	830
0596	24" diameter DR 11					925			925	1,025
0600	24" diameter DR 26					925			925	1,025
0604	28" diameter DR 17					1,350			1,350	1,500
0608	28" diameter DR 26					1,350			1,350	1,500
0612	30" diameter DR 17					1,300			1,300	1,425
0616	30" diameter DR 26					1,300			1,300	1,425
0620	36" diameter DR 26					2,125			2,125	2,350
0624	42" diameter DR 26					2,600			2,600	2,850
0628	48" diameter DR 26					2,675			2,675	2,925
0632	54" diameter DR 26				▼	4,350			4,350	4,775
0700	Tee									
0704	1" diameter DR 11				Ea.	6.45			6.45	7.10
0708	1-1/2" diameter DR 11					10.75			10.75	11.85
0712	2" diameter DR 11					15			15	16.50
0716	3" diameter DR 11					24			24	26.50
0720	3" diameter DR 17					24			24	26.50
0724	4" diameter DR 11					38			38	42
0728	4" diameter DR 17					38			38	42
0732	6" diameter DR 11					94			94	103
0736	6" diameter DR 17					94			94	103
0740	8" diameter DR 11					217			217	239
0744	8" diameter DR 17					197			197	217
0748	10" diameter DR 11					675			675	745
0752	10" diameter DR 17					242			242	266
0756	12" diameter DR 11					975			975	1,075
0760	12" diameter DR 17					276			276	305
0764	16" diameter DR 11					645			645	710
0768	16" diameter DR 17					645			645	710
0772	18" diameter DR 11					685			685	755
0776	18" diameter DR 17					685			685	755
0780	20" diameter DR 11					850			850	935
0784	20" diameter DR 17					850			850	935
0788	22" diameter DR 11					1,175			1,175	1,300
0792	22" diameter DR 17					1,175			1,175	1,300
0796	24" diameter DR 11					1,175			1,175	1,300
0800	24" diameter DR 17					1,175			1,175	1,300
0804	28" diameter DR 17					2,100			2,100	2,300
0812	30" diameter DR 17					2,350			2,350	2,575
0820	36" diameter DR 17					4,675			4,675	5,125
0824	42" diameter DR 26					5,125			5,125	5,650
0828	48" diameter DR 26				▼	6,000			6,000	6,600
1000	Flange adptr, w/back-up ring and 1/2 cost of plated bolt set									
1004	1" diameter DR 11				Ea.	59			59	65
1008	1-1/2" diameter DR 11					60.50			60.50	66.50
1012	2" diameter DR 11					33.50			33.50	36.50
1016	3" diameter DR 11					41			41	45
1020	3" diameter DR 17					41			41	45
1024	4" diameter DR 11					54			54	59
1028	4" diameter DR 17					54			54	59
1032	6" diameter DR 11					78.50			78.50	86
1036	6" diameter DR 17					78.50			78.50	86
1040	8" diameter DR 11					112			112	123
1044	8" diameter DR 26					112			112	123
1048	10" diameter DR 11				▼	168			168	185

15108	Plastic Pipe & Fittings	CREW	DAILY OUTPUT	LABOR-HOURS	UNIT	2002 BARE COSTS				TOTAL INCL O&P
						MAT.	LABOR	EQUIP.	TOTAL	
590	1052	10" diameter DR 26				Ea.	168			168	185	590
	1056	12" diameter DR 11					243			243	267
	1060	12" diameter DR 26					243			243	267
	1064	16" diameter DR 11					520			520	570
	1068	16" diameter DR 26					520			520	570
	1072	18" diameter DR 11					665			665	730
	1076	18" diameter DR 26					665			665	730
	1080	20" diameter DR 11					950			950	1,050
	1084	20" diameter DR 26					950			950	1,050
	1088	22" diameter DR 11					1,125			1,125	1,225
	1092	22" diameter DR 26					1,125			1,125	1,225
	1096	24" diameter DR 17					1,175			1,175	1,300
	1100	24" diameter DR 32.5					1,175			1,175	1,300
	1104	28" diameter DR 15.5					2,025			2,025	2,225
	1108	28" diameter DR 32.5					1,975			1,975	2,175
	1112	30" diameter DR 11					2,175			2,175	2,375
	1116	30" diameter DR 21					1,875			1,875	2,075
	1120	36" diameter DR 26					2,800			2,800	3,075
	1124	42" diameter DR 26					3,150			3,150	3,475
	1128	48" diameter DR 26					4,450			4,450	4,900
	1132	54" diameter DR 26				↓	4,975			4,975	5,475
	1200	Reducer									
	1204	1-1/2" x 1" diameter DR 11				Ea.	6.90			6.90	7.55
	1208	2" x 1-1/2" diameter DR 11					10.75			10.75	11.85
	1212	3" x 2" diameter DR 11					14			14	15.40
	1216	4" x 2" diameter DR 11					21			21	23
	1220	4" x 3" diameter DR 11					21			21	23
	1224	6" x 4" diameter DR 11					49			49	54
	1228	8" x 6" diameter DR 11					86			86	94.50
	1232	10" x 8" diameter DR 11					115			115	127
	1236	12" x 8" diameter DR 11					215			215	237
	1240	12" x 10" diameter DR 11					154			154	169
	1244	14" x 12" diameter DR 11					173			173	190
	1248	16" x 14" diameter DR 11					208			208	229
	1252	18" x 16" diameter DR 11					415			415	460
	1256	20" x 18" diameter DR 11					565			565	620
	1260	22" x 20" diameter DR 11					1,150			1,150	1,275
	1264	24" x 22" diameter DR 11					740			740	810
	1268	26" x 24" diameter DR 11					865			865	950
	1272	28" x 24" diameter DR 11					995			995	1,100
	1276	32" x 28" diameter DR 17					1,200			1,200	1,325
	1280	36" x 32" diameter DR 17				↓	1,425			1,425	1,550
	4000	Welding labor per joint, not including welding machine									
	4010	Pipe joint size (cost based on thickest wall for each dia.)									
	4030	1" pipe size	4 Skwk	273	.117	Ea.		3.63		3.63	5.65
	4040	1-1/2" pipe size		175	.183			5.65		5.65	8.85
	4050	2" pipe size		128	.250			7.75		7.75	12.10
	4060	3" pipe size		100	.320			9.90		9.90	15.45
	4070	4" pipe size	↓	77	.416			12.85		12.85	20
	4080	6" pipe size	5 Skwk	63	.635			19.65		19.65	30.50
	4090	8" pipe size		48	.833			26		26	40.50
	4100	10" pipe size	↓	40	1			31		31	48.50
	4110	12" pipe size	6 Skwk	41	1.171			36		36	56.50
	4120	16" pipe size		34	1.412			43.50		43.50	68.50
	4130	18" pipe size	↓	32	1.500			46.50		46.50	72.50
	4140	20" pipe size	8 Skwk	37	1.730	↓		53.50		53.50	83.50

Important: See the Reference Section for critical supporting data - Reference Nos., Crews, & City Cost Indexes

15108	Plastic Pipe & Fittings	CREW	DAILY OUTPUT	LABOR-HOURS	UNIT	MAT.	2002 BARE COSTS LABOR	EQUIP.	TOTAL	TOTAL INCL O&P	
4150	22" pipe size	8 Skwk	35	1.829	Ea.		56.50		56.50	88.50	590
4160	24" pipe size		34	1.882			58.50		58.50	91	
4170	28" pipe size		33	1.939			60		60	94	
4180	30" pipe size		32	2			62		62	96.50	
4190	36" pipe size		31	2.065			64		64	100	
4200	42" pipe size		30	2.133			66		66	103	
4210	48" pipe size	9 Skwk	33	2.182			67.50		67.50	105	
4220	54" pipe size	"	31	2.323			72		72	112	
4300	Note: Cost for set up each time welder is moved.										
4301	Add 50% of a weld cost										
4310	Welder usually remains stationary with pipe moved through it.										
4340	Weld machine, rental per day based on dia. capacity										
4350	1" thru 2" diameter				Ea.			35	35	38.50	
4360	3" thru 4" diameter							40	40	44	
4370	6" thru 8" diameter							90	90	99	
4380	10" thru 12" diameter							155	155	170.50	
4390	16" thru 18" diameter							225	225	247.50	
4400	20" thru 24" diameter							435	435	478.50	
4410	28" thru 32" diameter							475	475	525	
4420	36" diameter							495	495	545	
4430	42" thru 54" diameter							775	775	855	
5000	Dual wall contained pipe										
5040	Straight										
5054	1" DR 11 x 3" DR 11				L.F.	5.05			5.05	5.55	
5058	1" DR 11 x 4" DR 11					5.55			5.55	6.10	
5062	1-1/2" DR 11 x 4" DR 17					5.85			5.85	6.45	
5066	2" DR 11 x 4" DR 17					5.90			5.90	6.50	
5070	2" DR 11 x 6" DR 17					9.25			9.25	10.15	
5074	3" DR 11 x 6" DR 17					10.15			10.15	11.20	
5078	3" DR 11 x 6" DR 26					8.15			8.15	8.95	
5082	3" DR 17 x 8" DR 11					16.80			16.80	18.45	
5086	3" DR 17 x 8" DR 17					13.25			13.25	14.60	
5090	4" DR 11 x 8" DR 17					15			15	16.50	
5094	4" DR 17 x 8" DR 26					11.85			11.85	13.05	
5098	6" DR 11 x 10" DR 17					24			24	26.50	
5102	6" DR 17 x 10" DR 26					18.25			18.25	20	
5106	6" DR 26 x 10" DR 26					16.85			16.85	18.50	
5110	8" DR 17 x 12" DR 26					26.50			26.50	29	
5114	8" DR 26 x 12" DR 32.5					22			22	24	
5118	10" DR 17 x 14" DR 26					34.50			34.50	38	
5122	10" DR 17 x 16" DR 26					39			39	43	
5126	10" DR 26 x 16" DR 26					35.50			35.50	39	
5130	12" DR 26 x 16" DR 26					40			40	44	
5134	12" DR 17 x 18" DR 26					51.50			51.50	56.50	
5138	12" DR 26 x 18" DR 26					46			46	50.50	
5142	14" DR 26 x 20" DR 32.5					50.50			50.50	55.50	
5146	16" DR 26 x 22" DR 32.5					76			76	83.50	
5150	18" DR 26 x 24" DR 32.5					70			70	77	
5154	20" DR 32.5 x 28" DR 32.5					90.50			90.50	99.50	
5158	22" DR 32.5 x 30" DR 32.5					90.50			90.50	99.50	
5162	24" DR 32.5 x 32" DR 32.5					115			115	127	
5166	36" DR 32.5 x 42" DR 32.5					195			195	215	
5300	Force transfer coupling										
5354	1" DR 11 x 3" DR 11				Ea.	158			158	174	
5358	1" DR 11 x 4" DR 17					166			166	183	
5362	1-1/2" DR 11 x 4" DR 17					174			174	192	

15108	Plastic Pipe & Fittings	CREW	DAILY OUTPUT	LABOR-HOURS	UNIT	2002 BARE COSTS				TOTAL INCL O&P
						MAT.	LABOR	EQUIP.	TOTAL	
590	5366	2" DR 11 x 4" DR 17				Ea.	182			182	200	590
5370	2" DR 11 x 6" DR 17					264			264	291
5374	3" DR 11 x 6" DR 17					264			264	291
5378	3" DR 11 x 6" DR 26					264			264	291
5382	3" DR 11 x 8" DR 11					279			279	305
5386	3" DR 11 x 8" DR 17					279			279	305
5390	4" DR 11 x 8" DR 17					279			279	305
5394	4" DR 17 x 8" DR 26					245			245	270
5398	6" DR 11 x 10" DR 17					355			355	390
5402	6" DR 17 x 10" DR 26					292			292	320
5406	6" DR 26 x 10" DR 26					292			292	320
5410	8" DR 17 x 12" DR 26					390			390	425
5414	8" DR 26 x 12" DR 32.5					315			315	345
5418	10" DR 17 x 14" DR 26					510			510	560
5422	10" DR 17 x 16" DR 26					510			510	560
5426	10" DR 26 x 16" DR 26					510			510	560
5430	12" DR 26 x 16" DR 26					635			635	700
5434	12" DR 17 x 18" DR 26					655			655	720
5438	12" DR 26 x 18" DR 26					655			655	720
5442	14" DR 26 x 20" DR 32.5					690			690	760
5446	16" DR 26 x 22" DR 32.5					710			710	780
5450	18" DR 26 x 24" DR 32.5					770			770	845
5454	20" DR 32.5 x 28" DR 32.5					940			940	1,025
5458	22" DR 32.5 x 30" DR 32.5					1,025			1,025	1,150
5462	24" DR 32.5 x 32" DR 32.5					1,125			1,125	1,225
5466	36" DR 32.5 x 42" DR 32.5				▼	1,850			1,850	2,050
5600	90° Elbow									
5654	1" DR 11 x 3" DR 11				Ea.	131			131	144
5658	1" DR 11 x 4" DR 17					125			125	138
5662	1-1/2" DR 11 x 4" DR 17					137			137	151
5666	2" DR 11 x 4" DR 17					145			145	160
5670	2" DR 11 x 6" DR 17					184			184	203
5674	3" DR 11 x 6" DR 17					211			211	232
5678	3" DR 17 x 6" DR 26					168			168	185
5682	3" DR 17 x 8" DR 11					340			340	375
5686	3" DR 17 x 8" DR 17					265			265	292
5690	4" DR 11 x 8" DR 17					310			310	345
5694	4" DR 17 x 8" DR 26					248			248	273
5698	6" DR 11 x 10" DR 17					415			415	455
5702	6" DR 17 x 10" DR 26					320			320	350
5706	6" DR 26 x 10" DR 26					298			298	330
5710	8" DR 17 x 12" DR 26					475			475	520
5714	8" DR 26 x 12" DR 32.5					435			435	480
5718	10" DR 17 x 14" DR 26					745			745	820
5722	10" DR 17 x 16" DR 26					710			710	785
5726	10" DR 26 x 16" DR 26					660			660	725
5730	12" DR 26 x 16" DR 26					730			730	805
5734	12" DR 17 x 18" DR 26					870			870	960
5738	12" DR 26 x 18" DR 26					800			800	880
5742	14" DR 26 x 20" DR 32.5					1,050			1,050	1,150
5746	16" DR 26 x 22" DR 32.5					1,075			1,075	1,175
5750	18" DR 26 x 24" DR 32.5					1,300			1,300	1,425
5754	20" DR 32.5 x 28" DR 32.5					1,500			1,500	1,650
5758	22" DR 32.5 x 30" DR 32.5					2,050			2,050	2,250
5762	24" DR 32.5 x 32" DR 32.5					2,000			2,000	2,200
5766	36" DR 32.5 x 42" DR 32.5				▼	3,250			3,250	3,550

T5 MECHANICAL

Important: See the Reference Section for critical supporting data - Reference Nos., Crews, & City Cost Indexes

15108	Plastic Pipe & Fittings	CREW	DAILY OUTPUT	LABOR-HOURS	UNIT	2002 BARE COSTS				TOTAL INCL O&P
						MAT.	LABOR	EQUIP.	TOTAL	
5800	45° Elbow									
5804	1" DR 11 x 3" DR 11				Ea.	78.50			78.50	86
5808	1" DR 11 x 4" DR 17					78			78	85.50
5812	1-1/2" DR 11 x 4" DR 17					83.50			83.50	92
5816	2" DR 11 x 4" DR 17					91			91	100
5820	2" DR 11 x 6" DR 17					114			114	126
5824	3" DR 11 x 6" DR 17					127			127	140
5828	3" DR 17 x 6" DR 26					105			105	116
5832	3" DR 17 x 8" DR 11					194			194	214
5836	3" DR 17 x 8" DR 17					155			155	171
5840	4" DR 11 x 8" DR 17					182			182	200
5844	4" DR 17 x 8" DR 26					148			148	163
5848	6" DR 11 x 10" DR 17					240			240	264
5852	6" DR 17 x 10" DR 26					188			188	207
5856	6" DR 26 x 10" DR 26					177			177	195
5860	8" DR 17 x 12" DR 26					286			286	315
5864	8" DR 26 x 12" DR 32.5					267			267	294
5868	10" DR 17 x 14" DR 26					450			450	495
5872	10" DR 17 x 16" DR 26					440			440	480
5876	10" DR 26 x 16" DR 26					410			410	450
5880	12" DR 26 x 16" DR 26					465			465	510
5884	12" DR 17 x 18" DR 26					545			545	600
5888	12" DR 26 x 18" DR 26					510			510	560
5892	14" DR 26 x 20" DR 32.5					655			655	720
5896	16" DR 26 x 22" DR 32.5					685			685	755
5900	18" DR 26 x 24" DR 32.5					810			810	895
5904	20" DR 32.5 x 28" DR 32.5					995			995	1,100
5908	22" DR 32.5 x 30" DR 32.5					1,275			1,275	1,400
5912	24" DR 32.5 x 32" DR 32.5					1,250			1,250	1,375
5916	36" DR 32.5 x 42" DR 32.5				▼	1,950			1,950	2,150
6000	Access port with 4" riser									
6050	1" DR 11 x 4" DR 17				Ea.	171			171	188
6054	1-1/2" DR 11 x 4" DR 17					175			175	192
6058	2" DR 11 x 6" DR 17					213			213	234
6062	3" DR 11 x 6" DR 17					214			214	235
6066	3" DR 17 x 6" DR 26					210			210	231
6070	3" DR 17 x 8" DR 11					240			240	264
6074	3" DR 17 x 8" DR 17					229			229	252
6078	4" DR 11 x 8" DR 17					237			237	261
6082	4" DR 17 x 8" DR 26					227			227	250
6086	6" DR 11 x 10" DR 17					297			297	325
6090	6" DR 17 x 10" DR 26					273			273	300
6094	6" DR 26 x 10" DR 26					268			268	295
6098	8" DR 17 x 12" DR 26					287			287	315
6102	8" DR 26 x 12" DR 32.5				▼	274			274	300
6200	End termination with vent plug									
6204	1" DR 11 x 3" DR 11				Ea.	142			142	156
6208	1" DR 11 x 4" DR 17					151			151	167
6212	1-1/2" DR 11 x 4" DR 17					151			151	167
6216	2" DR 11 x 4" DR 17					165			165	181
6220	2" DR 11 x 6" DR 17					188			188	206
6224	3" DR 11 x 6" DR 17					188			188	206
6228	3" DR 17 x 6" DR 26					188			188	206
6232	3" DR 17 x 8" DR 11					260			260	285
6236	3" DR 17 x 8" DR 17					260			260	285
6240	4" DR 11 x 8" DR 17				▼	260			260	285

590

MECHANICAL 15

15108	Plastic Pipe & Fittings	CREW	DAILY OUTPUT	LABOR-HOURS	UNIT	2002 BARE COSTS				TOTAL INCL O&P
						MAT.	LABOR	EQUIP.	TOTAL	
590											**590**
6244	4″ DR 17 x 8″ DR 26				Ea.	225			225	248
6248	6″ DR 11 x 10″ DR 17					310			310	340
6252	6″ DR 17 x 10″ DR 26					253			253	278
6256	6″ DR 26 x 10″ DR 26					253			253	278
6260	8″ DR 17 x 12″ DR 26					375			375	410
6264	8″ DR 26 x 12″ DR 32.5					300			300	330
6268	10″ DR 17 x 14″ DR 26					390			390	430
6272	10″ DR 17 x 16″ DR 26					390			390	430
6276	10″ DR 26 x 16″ DR 26					390			390	430
6280	12″ DR 26 x 16″ DR 26					390			390	430
6284	12″ DR 17 x 18″ DR 26					515			515	565
6288	12″ DR 26 x 18″ DR 26					515			515	565
6292	14″ DR 26 x 20″ DR 32.5					525			525	575
6296	16″ DR 26 x 22″ DR 32.5					605			605	665
6300	18″ DR 26 x 24″ DR 32.5					625			625	690
6304	20″ DR 32.5 x 28″ DR 32.5					800			800	880
6308	22″ DR 32.5 x 30″ DR 32.5					880			880	965
6312	24″ DR 32.5 x 32″ DR 32.5					970			970	1,075
6316	36″ DR 32.5 x 42″ DR 32.5				▼	1,425			1,425	1,550
6600	Tee									
6604	1″ DR 11 x 3″ DR 11				Ea.	144			144	159
6608	1″ DR 11 x 4″ DR 17					179			179	197
6612	1-1/2″ DR 11 x 4″ DR 17					199			199	219
6616	2″ DR 11 x 4″ DR 17					217			217	239
6620	2″ DR 11 x 6″ DR 17					264			264	290
6624	3″ DR 11 x 6″ DR 17					305			305	335
6628	3″ DR 17 x 6″ DR 26					296			296	325
6632	3″ DR 17 x 8″ DR 11					360			360	395
6636	3″ DR 17 x 8″ DR 17					340			340	375
6640	4″ DR 11 x 8″ DR 17					370			370	410
6644	4″ DR 17 x 8″ DR 26					355			355	390
6648	6″ DR 11 x 10″ DR 17					465			465	510
6652	6″ DR 17 x 10″ DR 26					440			440	485
6656	6″ DR 26 x 10″ DR 26					435			435	480
6660	8″ DR 17 x 12″ DR 26					435			435	480
6664	8″ DR 26 x 12″ DR 32.5					570			570	625
6668	10″ DR 17 x 14″ DR 26					655			655	725
6672	10″ DR 17 x 16″ DR 26					690			690	760
6676	10″ DR 26 x 16″ DR 26					690			690	760
6680	12″ DR 26 x 16″ DR 26					745			745	815
6684	12″ DR 17 x 18″ DR 26					745			745	815
6688	12″ DR 26 x 18″ DR 26					825			825	905
6692	14″ DR 26 x 20″ DR 32.5					1,100			1,100	1,200
6696	16″ DR 26 x 22″ DR 32.5					1,325			1,325	1,475
6700	18″ DR 26 x 24″ DR 32.5					1,450			1,450	1,600
6704	20″ DR 32.5 x 28″ DR 32.5					1,825			1,825	2,025
6708	22″ DR 32.5 x 30″ DR 32.5					2,175			2,175	2,400
6712	24″ DR 32.5 x 32″ DR 32.5					2,600			2,600	2,875
6716	36″ DR 32.5 x 42″ DR 32.5				▼	5,300			5,300	5,850
6800	Wye									
6816	2″ DR 11 x 4″ DR 17				Ea.	274			274	300
6820	2″ DR 11 x 6″ DR 17					310			310	340
6824	3″ DR 11 x 6″ DR 17					315			315	345
6828	3″ DR 17 x 6″ DR 26					305			305	335
6832	3″ DR 17 x 8″ DR 11					380			380	420
6836	3″ DR 17 x 8″ DR 17				▼	355			355	395

Important: See the Reference Section for critical supporting data - Reference Nos., Crews, & City Cost Indexes

15108 | Plastic Pipe & Fittings

		CREW	DAILY OUTPUT	LABOR-HOURS	UNIT	2002 BARE COSTS				TOTAL INCL O&P
						MAT.	LABOR	EQUIP.	TOTAL	
590	6840	4" DR 11 x 8" DR 17				Ea.	390			390	430	590
	6844	4" DR 17 x 8" DR 26					370			370	405
	6848	6" DR 11 x 10" DR 17					480			480	530
	6852	6" DR 17 x 10" DR 26					450			450	495
	6856	6" DR 26 x 10" DR 26					445			445	490
	6860	8" DR 17 x 12" DR 26					640			640	700
	6864	8" DR 26 x 12" DR 32.5					615			615	680
	6868	10" DR 17 x 14" DR 26					770			770	845
	6872	10" DR 17 x 16" DR 26					815			815	900
	6876	10" DR 26 x 16" DR 26					805			805	885
	6880	12" DR 26 x 16" DR 26					905			905	995
	6884	12" DR 17 x 18" DR 26					1,025			1,025	1,125
	6888	12" DR 26 x 18" DR 26					1,025			1,025	1,125
	6892	14" DR 26 x 20" DR 32.5					1,250			1,250	1,375
	6896	16" DR 26 x 22" DR 32.5					1,800			1,800	2,000
	6900	18" DR 26 x 24" DR 32.5					2,125			2,125	2,350
	6904	20" DR 32.5 x 28" DR 32.5					2,675			2,675	2,925
	6908	22" DR 32.5 x 30" DR 32.5					3,025			3,025	3,350
	6912	24" DR 32.5 x 32" DR 32.5				▼	3,450			3,450	3,800
	9000	Welding labor per joint, not including welding machine									
	9010	Pipe joint size, outer pipe (cost based on the thickest walls)									
	9020	Straight pipe									
	9050	3" pipe size	4 Skwk	96	.333	Ea.		10.30		10.30	16.10
	9060	4" pipe size	"	77	.416			12.85		12.85	20
	9070	6" pipe size	5 Skwk	60	.667			20.50		20.50	32
	9080	8" pipe size	"	40	1			31		31	48.50
	9090	10" pipe size	6 Skwk	41	1.171			36		36	56.50
	9100	12" pipe size		39	1.231			38		38	59.50
	9110	14" pipe size		38	1.263			39		39	61
	9120	16" pipe size	▼	35	1.371			42.50		42.50	66.50
	9130	18" pipe size	8 Skwk	45	1.422			44		44	69
	9140	20" pipe size		42	1.524			47		47	73.50
	9150	22" pipe size		40	1.600			49.50		49.50	77.50
	9160	24" pipe size		38	1.684			52		52	81.50
	9170	28" pipe size		37	1.730			53.50		53.50	83.50
	9180	30" pipe size		36	1.778			55		55	86
	9190	32" pipe size		35	1.829			56.50		56.50	88.50
	9200	42" pipe size	▼	32	2	▼		62		62	96.50
	9300	Note: Cost for set up each time welder is moved.									
	9301	Add 100% of weld labor cost									
	9310	For handling between fitting welds add 50% of weld labor cost									
	9320	Welder usually remains stationary with pipe moved through it									
	9360	Weld machine, rental per day based on dia. capacity									
	9380	3" thru 4" diameter				Ea.			55	55	60.50
	9390	6" thru 8" diameter							180	180	198
	9400	10" thru 12" diameter							270	270	297
	9410	14" thru 18" diameter							365	365	402
	9420	20" thru 24" diameter							650	650	715
	9430	28" thru 32" diameter							765	765	840
	9440	42" thru 58" diameter				▼			790	790	870

15110 | Valves

		CREW	DAILY OUTPUT	LABOR-HOURS	UNIT	MAT.	LABOR	EQUIP.	TOTAL	TOTAL INCL O&P
100	0010 **VALVES, BRASS**										100
	0030 For motorized valves, see Division 13838-200									
	0500 Gas cocks, threaded									
	0510 1/4"	1 Plum	26	.308	Ea.	7.80	11.05		18.85	25.50

			DAILY	LABOR-			2002 BARE COSTS				TOTAL	
15110		**Valves**	CREW	OUTPUT	HOURS	UNIT	MAT.	LABOR	EQUIP.	TOTAL	INCL O&P	
100	0520	3/8"	1 Plum	24	.333	Ea.	8.25	12		20.25	27	**100**
	0530	1/2"		24	.333		7.65	12		19.65	26.50	
	0540	3/4"		22	.364		10.30	13.05		23.35	31	
	0550	1"		19	.421		12.60	15.15		27.75	37	
	0560	1-1/4"		15	.533		36	19.15		55.15	68.50	
	0570	1-1/2"		13	.615		45	22		67	83	
	0580	2"	▼	11	.727	▼	65.50	26		91.50	112	
	0670	For larger sizes use lubricated plug valve, Section 15110-600										
160	0010	**VALVES, BRONZE**										**160**
	1020	Angle, 150 lb., rising stem, threaded R15100 -090										
	1030	1/8"	1 Plum	24	.333	Ea.	42	12		54	64	
	1040	1/4"		24	.333		42	12		54	64	
	1050	3/8"		24	.333		44.50	12		56.50	66.50	
	1060	1/2"		22	.364		44.50	13.05		57.55	68	
	1070	3/4"		20	.400		59.50	14.40		73.90	86.50	
	1080	1"		19	.421		85	15.15		100.15	116	
	1090	1-1/4"		15	.533		110	19.15		129.15	149	
	1100	1-1/2"		13	.615		142	22		164	190	
	1110	2"	▼	11	.727	▼	230	26		256	292	
	1380	Ball, 150 psi, threaded										
	1400	1/4"	1 Plum	24	.333	Ea.	6.75	12		18.75	25.50	
	1430	3/8"		24	.333		6.75	12		18.75	25.50	
	1450	1/2"		22	.364		6.75	13.05		19.80	27	
	1460	3/4"		20	.400		11.10	14.40		25.50	33.50	
	1470	1"		19	.421		13.95	15.15		29.10	38.50	
	1480	1-1/4"		15	.533		23.50	19.15		42.65	55	
	1490	1-1/2"		13	.615		30	22		52	66.50	
	1500	2"		11	.727		37.50	26		63.50	81	
	1510	2-1/2"		9	.889		177	32		209	243	
	1520	3"	▼	8	1	▼	289	36		325	375	
	1600	Butterfly, 175 psi, full port, solder or threaded ends										
	1610	Stainless steel disc and stem										
	1620	1/4"	1 Plum	24	.333	Ea.	5.75	12		17.75	24.50	
	1630	3/8"		24	.333		5.75	12		17.75	24.50	
	1640	1/2"		22	.364		6.10	13.05		19.15	26.50	
	1650	3/4"		20	.400		9.85	14.40		24.25	32.50	
	1660	1"		19	.421		12.25	15.15		27.40	36.50	
	1670	1-1/4"		15	.533		19.45	19.15		38.60	50.50	
	1680	1-1/2"		13	.615		25	22		47	61	
	1690	2"	▼	11	.727	▼	31.50	26		57.50	74	
	1750	Check, swing, class 150, regrinding disc, threaded										
	1800	1/8"	1 Plum	24	.333	Ea.	23.50	12		35.50	43.50	
	1830	1/4"		24	.333		23.50	12		35.50	43.50	
	1840	3/8"		24	.333		23.50	12		35.50	43.50	
	1850	1/2"		24	.333		23.50	12		35.50	43.50	
	1860	3/4"		20	.400		31.50	14.40		45.90	56	
	1870	1"		19	.421		47.50	15.15		62.65	75	
	1880	1-1/4"		15	.533		65.50	19.15		84.65	101	
	1890	1-1/2"		13	.615		77.50	22		99.50	119	
	1900	2"	▼	11	.727		113	26		139	165	
	1910	2-1/2"	Q-1	15	1.067		188	34.50		222.50	259	
	1920	3"	"	13	1.231	▼	278	40		318	365	
	2000	For 200 lb, add						5%	10%				
	2040	For 300 lb, add						15%	15%				
	2060	Check swing, 300#, sweat, 3/8" size	1 Plum	24	.333	Ea.	13.55	12		25.55	33	
	2070	1/2"	▼	24	.333	▼	13.55	12		25.55	33	

Important: See the Reference Section for critical supporting data - Reference Nos., Crews, & City Cost Indexes

15110	Valves		CREW	DAILY OUTPUT	LABOR-HOURS	UNIT	2002 BARE COSTS				TOTAL INCL O&P
							MAT.	LABOR	EQUIP.	TOTAL	
160 2080	3/4"	R15100 -090	1 Plum	20	.400	Ea.	17.30	14.40		31.70	40.50 **160**
2090	1"			19	.421		24	15.15		39.15	49.50
2100	1-1/4"			15	.533		31	19.15		50.15	63
2110	1-1/2 "			13	.615		41.50	22		63.50	79
2120	2"		▼	11	.727		62	26		88	108
2130	2-1/2"		Q-1	15	1.067		147	34.50		181.50	214
2140	3"		"	13	1.231	▼	197	40		237	277
2350	Check, lift, class 150, horizontal composition disc, threaded										
2430	1/4"		1 Plum	24	.333	Ea.	55.50	12		67.50	79
2440	3/8"			24	.333		55.50	12		67.50	79
2450	1/2"			24	.333		61.50	12		73.50	85.50
2460	3/4"			20	.400		75	14.40		89.40	104
2470	1"			19	.421		107	15.15		122.15	140
2480	1-1/4"			15	.533		143	19.15		162.15	187
2490	1-1/2"			13	.615		171	22		193	222
2500	2"		▼	11	.727	▼	285	26		311	355
2850	Gate, N.R.S., soldered, 125 psi										
2900	3/8"		1 Plum	24	.333	Ea.	17.70	12		29.70	37.50
2920	1/2"			24	.333		15.50	12		27.50	35
2940	3/4"			20	.400		17.70	14.40		32.10	41
2950	1"			19	.421		25.50	15.15		40.65	51
2960	1-1/4"			15	.533		40	19.15		59.15	72.50
2970	1-1/2"			13	.615		45	22		67	83
2980	2"		▼	11	.727		60.50	26		86.50	106
2990	2-1/2"		Q-1	15	1.067		152	34.50		186.50	219
3000	3"		"	13	1.231	▼	199	40		239	279
3350	Threaded, class 150										
3410	1/4"		1 Plum	24	.333	Ea.	26	12		38	47
3420	3/8"			24	.333		26	12		38	47
3430	1/2"			24	.333		24	12		36	44.50
3440	3/4"			20	.400		26.50	14.40		40.90	51
3450	1"			19	.421		35	15.15		50.15	61.50
3460	1-1/4" size			15	.533		47	19.15		66.15	80.50
3470	1-1/2"			13	.615		59	22		81	98.50
3480	2"		▼	11	.727		80.50	26		106.50	128
3490	2-1/2"		Q-1	15	1.067		186	34.50		220.50	257
3500	3" size		"	13	1.231	▼	264	40		304	350
3600	Gate, flanged, 150 lb.										
3610	1"		1 Plum	7	1.143	Ea.	455	41		496	565
3620	1-1/2"			6	1.333		570	48		618	700
3630	2"		▼	5	1.600		840	57.50		897.50	1,000
3640	3"		Q-1	4.50	3.556	▼	1,500	115		1,615	1,825
3850	Rising stem, soldered, 300 psi										
3900	3/8"		1 Plum	24	.333	Ea.	41.50	12		53.50	63.50
3920	1/2"			24	.333		38.50	12		50.50	60
3940	3/4"			20	.400		49	14.40		63.40	75
3950	1"			19	.421		59.50	15.15		74.65	88
3960	1-1/4"			15	.533		83.50	19.15		102.65	121
3970	1-1/2"			13	.615		102	22		124	146
3980	2"		▼	11	.727		151	26		177	206
3990	2-1/2"		Q-1	15	1.067		365	34.50		399.50	450
4000	3"		"	13	1.231	▼	425	40		465	530
4250	Threaded, class 150										
4310	1/4"		1 Plum	24	.333	Ea.	23	12		35	43
4320	3/8"			24	.333		23	12		35	43
4330	1/2"			24	.333		21.50	12		33.50	41.50

MECHANICAL 15

15110 | Valves

			CREW	DAILY OUTPUT	LABOR-HOURS	UNIT	MAT.	LABOR	EQUIP.	TOTAL	TOTAL INCL O&P	
160	4340	3/4"	1 Plum	20	.400	Ea.	25	14.40		39.40	49	160
	4350	1"		19	.421		33.50	15.15		48.65	60	
	4360	1-1/4"		15	.533		45.50	19.15		64.65	79	
	4370	1-1/2"		13	.615		57	22		79	96	
	4380	2"		11	.727		77	26		103	124	
	4390	2-1/2"	Q-1	15	1.067		179	34.50		213.50	249	
	4400	3"	"	13	1.231		251	40		291	335	
	4500	For 300 psi, threaded, add					100%	15%				
	4540	For chain operated type, add					15%					
	4850	Globe, class 150, rising stem, threaded										
	4920	1/4"	1 Plum	24	.333	Ea.	33	12		45	54	
	4940	3/8" size		24	.333		33	12		45	54	
	4950	1/2"		24	.333		33	12		45	54	
	4960	3/4"		20	.400		45	14.40		59.40	71	
	4970	1"		19	.421		70	15.15		85.15	99.50	
	4980	1-1/4"		15	.533		110	19.15		129.15	150	
	4990	1-1/2"		13	.615		134	22		156	181	
	5000	2"		11	.727		201	26		227	261	
	5010	2-1/2"	Q-1	15	1.067		405	34.50		439.50	495	
	5020	3"	"	13	1.231		575	40		615	690	
	5120	For 300 lb threaded, add					50%	15%				
	5130	Globe, 300 lb., sweat, 3/8" size	1 Plum	24	.333	Ea.	33.50	12		45.50	55	
	5140	1/2"		24	.333		34.50	12		46.50	56	
	5150	3/4"		20	.400		49.50	14.40		63.90	76	
	5160	1"		19	.421		75	15.15		90.15	106	
	5170	1-1/4"		15	.533		116	19.15		135.15	157	
	5180	1-1/2"		13	.615		138	22		160	185	
	5190	2"		11	.727		206	26		232	267	
	5200	2-1/2"	Q-1	15	1.067		455	34.50		489.50	550	
	5210	3"	"	13	1.231		665	40		705	790	
	5600	Relief, pressure & temperature, self-closing, ASME, threaded										
	5640	3/4"	1 Plum	28	.286	Ea.	65	10.25		75.25	87	
	5650	1"		24	.333		94.50	12		106.50	122	
	5660	1-1/4"		20	.400		189	14.40		203.40	230	
	5670	1-1/2"		18	.444		360	16		376	425	
	5680	2"		16	.500		395	18		413	460	
	5950	Pressure, poppet type, threaded										
	6000	1/2"	1 Plum	30	.267	Ea.	10.80	9.60		20.40	26.50	
	6040	3/4"	"	28	.286	"	45	10.25		55.25	65	
	6400	Pressure, water, ASME, threaded										
	6440	3/4"	1 Plum	28	.286	Ea.	43	10.25		53.25	63	
	6450	1"		24	.333		93.50	12		105.50	121	
	6460	1-1/4"		20	.400		153	14.40		167.40	190	
	6470	1-1/2"		18	.444		211	16		227	256	
	6480	2"		16	.500		305	18		323	360	
	6490	2-1/2" size		15	.533		1,225	19.15		1,244.15	1,375	
	6900	Reducing, water pressure										
	6920	300 psi to 25-75 psi, threaded or sweat										
	6940	1/2"	1 Plum	24	.333	Ea.	116	12		128	145	
	6950	3/4"		20	.400		116	14.40		130.40	149	
	6960	1"		19	.421		179	15.15		194.15	220	
	6970	1-1/4"		15	.533		300	19.15		319.15	360	
	6980	1-1/2"		13	.615		485	22		507	565	
	6990	2"		11	.727		655	26		681	760	
	7100	For built-in by-pass or 10-35 psi, add					7			7	7.70	
	7700	High capacity, 250 psi to 25-75 psi, threaded										

R15100 -090

Important: See the Reference Section for critical supporting data - Reference Nos., Crews, & City Cost Indexes

			CREW	DAILY OUTPUT	LABOR-HOURS	UNIT	2002 BARE COSTS				TOTAL INCL O&P	
15110	**Valves**						MAT.	LABOR	EQUIP.	TOTAL		
160	7740	1/2″	R15100 -090	1 Plum	24	.333	Ea.	116	12		128	145	**160**
	7780	3/4″			20	.400		116	14.40		130.40	149	
	7790	1″			19	.421		179	15.15		194.15	220	
	7800	1-1/4″			15	.533		320	19.15		339.15	385	
	7810	1-1/2″			13	.615		485	22		507	565	
	7820	2″			11	.727		700	26		726	810	
	7830	2-1/2″			9	.889		890	32		922	1,025	
	7840	3″			8	1		1,100	36		1,136	1,250	
	7850	3″ flanged (iron body)		Q-1	10	1.600		1,125	52		1,177	1,325	
	7860	4″ flanged (iron body)		″	8	2		1,850	64.50		1,914.50	2,150	
	7920	For higher pressure, add						25%					
	8000	Silent check, bronze trim											
	8010	Compact wafer type, for 125 or 150 lb. flanges											
	8020	1-1/2″		1 Plum	11	.727	Ea.	255	26		281	320	
	8021	2″		″	9	.889		274	32		306	350	
	8022	2-1/2″		Q-1	9	1.778		288	57.50		345.50	400	
	8023	3″			8	2		320	64.50		384.50	455	
	8024	4″			5	3.200		550	104		654	760	
	8025	5″		Q-2	6	4		760	134		894	1,025	
	8026	6″			5	4.800		865	161		1,026	1,200	
	8027	8″			4.50	5.333		1,850	179		2,029	2,300	
	8028	10″			4	6		3,250	201		3,451	3,875	
	8029	12″			3	8		7,600	268		7,868	8,750	
	8050	For 250 or 300 lb flanges, thru 6″ no change											
	8051	For 8″ and 10″, add					Ea.	40%	10%				
	8060	Full flange wafer type, 150 lb.											
	8061	1″		1 Plum	14	.571	Ea.	272	20.50		292.50	330	
	8062	1-1/4″			12	.667		284	24		308	345	
	8063	1-1/2″			11	.727		340	26		366	415	
	8064	2″			9	.889		460	32		492	555	
	8065	2-1/2″		Q-1	9	1.778		520	57.50		577.50	655	
	8066	3″			8	2		575	64.50		639.50	730	
	8067	4″			5	3.200		850	104		954	1,100	
	8068	5″		Q-2	6	4		1,100	134		1,234	1,400	
	8069	6″		″	5	4.800		1,450	161		1,611	1,825	
	8080	For 300 lb., add						40%	10%				
	8100	Globe type, 150 lb.											
	8110	2″		1 Plum	9	.889	Ea.	535	32		567	640	
	8111	2-1/2″		Q-1	9	1.778		660	57.50		717.50	815	
	8112	3″			8	2		790	64.50		854.50	970	
	8113	4″			5	3.200		1,100	104		1,204	1,350	
	8114	5″		Q-2	6	4		1,350	134		1,484	1,675	
	8115	6″		″	5	4.800		1,875	161		2,036	2,300	
	8130	For 300 lb., add						20%	10%				
	8140	Screwed end type, 250 lb.											
	8141	1/2″		1 Plum	24	.333	Ea.	48	12		60	71	
	8142	3/4″			20	.400		48	14.40		62.40	74.50	
	8143	1″			19	.421		55	15.15		70.15	83.50	
	8144	1-1/4″			15	.533		76	19.15		95.15	113	
	8145	1-1/2″			13	.615		85	22		107	127	
	8146	2″			11	.727		114	26		140	166	
	8350	Tempering, water, sweat connections											
	8400	1/2″		1 Plum	24	.333	Ea.	42	12		54	64.50	
	8440	3/4″		″	20	.400	″	51.50	14.40		65.90	78	
	8650	Threaded connections											
	8700	1/2″		1 Plum	24	.333	Ea.	51.50	12		63.50	74.50	

15110 | Valves

			CREW	DAILY OUTPUT	LABOR-HOURS	UNIT	2002 BARE COSTS				TOTAL INCL O&P	
							MAT.	LABOR	EQUIP.	TOTAL		
160	8740	3/4"	1 Plum	20	.400	Ea.	197	14.40		211.40	238	160
	8750	1"		19	.421		217	15.15		232.15	261	
	8760	1-1/4"		15	.533		345	19.15		364.15	410	
	8770	1-1/2"		13	.615		375	22		397	450	
	8780	2"		11	.727		565	26		591	660	
200	0010	**VALVES, IRON BODY** R15100-090										200
	0020	For grooved joint, see Division 15107-690										
	0100	Angle, 125 lb.										
	0110	Flanged										
	0116	2"	1 Plum	5	1.600	Ea.	650	57.50		707.50	800	
	0118	4"	Q-1	3	5.333		1,075	173		1,248	1,425	
	0120	6"	Q-2	3	8		2,100	268		2,368	2,700	
	0122	8"	"	2.50	9.600		3,750	320		4,070	4,600	
	0560	Butterfly, lug type, pneumatic operator, 2" size	1 Stpi	14	.571		221	20.50		241.50	274	
	0570	3"	Q-1	8	2		229	64.50		293.50	350	
	0580	4"	"	5	3.200		250	104		354	430	
	0590	6"	Q-2	5	4.800		335	161		496	615	
	0600	8"		4.50	5.333		405	179		584	715	
	0610	10"		4	6		510	201		711	865	
	0620	12"		3	8		685	268		953	1,150	
	0630	14"		2.30	10.435		985	350		1,335	1,600	
	0640	18"		1.50	16		1,700	535		2,235	2,675	
	0650	20"		1	24		2,050	805		2,855	3,475	
	0800	Butterfly, lug type, gear operated, 2-1/2" size, 200 lb. except noted	Q-1	9	1.778		179	57.50		236.50	283	
	0810	3"		8	2		188	64.50		252.50	305	
	0820	4"		5	3.200		212	104		316	390	
	0830	5"	Q-2	5	4.800		263	161		424	535	
	0840	6"		5	4.800		292	161		453	565	
	0850	8"		4.50	5.333		380	179		559	685	
	0860	10"		4	6		490	201		691	840	
	0870	12"		3	8		710	268		978	1,175	
	0880	14", 150 lb.		2.30	10.435		815	350		1,165	1,425	
	0890	16", 150 lb.		1.75	13.714		1,150	460		1,610	1,975	
	0900	18", 150 lb.		1.50	16		1,500	535		2,035	2,450	
	0910	20", 150 lb.		1	24		1,775	805		2,580	3,175	
	0930	24", 150 lb.		.75	32		2,475	1,075		3,550	4,350	
	1020	Butterfly, wafer type, gear actuator, 200 lb.										
	1030	2"	1 Plum	14	.571	Ea.	109	20.50		129.50	150	
	1040	2-1/2"	Q-1	9	1.778		112	57.50		169.50	210	
	1050	3"		8	2		116	64.50		180.50	226	
	1060	4"		5	3.200		145	104		249	315	
	1070	5"	Q-2	5	4.800		175	161		336	435	
	1080	6"		5	4.800		198	161		359	460	
	1090	8"		4.50	5.333		264	179		443	560	
	1100	10"		4	6		330	201		531	670	
	1110	12"		3	8		460	268		728	910	
	1200	Wafer type, lever actuator, 200 lb.										
	1220	2"	1 Plum	14	.571	Ea.	78	20.50		98.50	117	
	1230	2-1/2"	Q-1	9	1.778		81	57.50		138.50	176	
	1240	3"		8	2		85.50	64.50		150	192	
	1250	4"		5	3.200		104	104		208	271	
	1260	5"	Q-2	5	4.800		145	161		306	400	
	1270	6"		5	4.800		174	161		335	435	
	1280	8"		4.50	5.333		258	179		437	555	
	1290	10"		4	6		455	201		656	805	

Important: See the Reference Section for critical supporting data - Reference Nos., Crews, & City Cost Indexes

15110 | Valves

		CREW	DAILY OUTPUT	LABOR-HOURS	UNIT	MAT.	LABOR	EQUIP.	TOTAL	TOTAL INCL O&P	
200	1300	12"	R15100 -090	Q-2	3	8	Ea.	645	268		913	1,100	200
	1400	Diverter, 150 lb. flanged, bronze or iron plugs									
	1440	2"	Q-1	2	8	Ea.	2,725	259		2,984	3,400
	1450	3"	"	1.50	10.667	"	3,900	345		4,245	4,825
	1600	Gate, threaded, 125 lb.									
	1604	1-1/2"	1 Plum	13	.615	Ea.	200	22		222	254
	1605	2"		11	.727		530	26		556	625
	1606	3"		8	1		440	36		476	540
	1607	4"		5	1.600		955	57.50		1,012.50	1,125
	1650	Gate, 125 lb., N.R.S.									
	2150	Flanged									
	2200	2"	1 Plum	5	1.600	Ea.	284	57.50		341.50	395
	2240	2-1/2"	Q-1	5	3.200		291	104		395	475
	2260	3"		4.50	3.556		325	115		440	535
	2280	4"		3	5.333		465	173		638	775
	2290	5"	Q-2	3.40	7.059		795	237		1,032	1,225
	2300	6"		3	8		795	268		1,063	1,275
	2320	8"		2.50	9.600		1,375	320		1,695	1,975
	2340	10"		2.20	10.909		2,400	365		2,765	3,200
	2360	12"		1.70	14.118		3,300	475		3,775	4,350
	2370	14"		1.30	18.462		5,650	620		6,270	7,150
	2380	16"		1	24		7,900	805		8,705	9,900
	2420	For 250 lb flanged, add					200%	10%			
	3500	OS&Y, 125 lb., (225# noted), threaded									
	3504	3/4" (225 lb.)	1 Plum	18	.444	Ea.	385	16		401	450
	3505	1" (225 lb.)		16	.500		440	18		458	505
	3506	1-1/2" (225 lb.)		13	.615		640	22		662	735
	3507	2"		11	.727		500	26		526	590
	3508	2-1/2"		9	.889		515	32		547	615
	3510	4"		7	1.143		830	41		871	970
	3550	OS&Y, 125 lb., flanged									
	3600	2"	1 Plum	5	1.600	Ea.	215	57.50		272.50	325
	3640	2-1/2"	Q-1	5	3.200		223	104		327	400
	3660	3"		4.50	3.556		250	115		365	450
	3670	3-1/2"		3	5.333		420	173		593	720
	3680	4"		3	5.333		355	173		528	650
	3690	5"	Q-2	3.40	7.059		585	237		822	1,000
	3700	6"		3	8		585	268		853	1,050
	3720	8"		2.50	9.600		1,050	320		1,370	1,625
	3740	10"		2.20	10.909		1,900	365		2,265	2,650
	3760	12"		1.70	14.118		2,550	475		3,025	3,525
	3770	14"		1.30	18.462		4,850	620		5,470	6,275
	3780	16"		1	24		7,275	805		8,080	9,225
	3790	18"		.80	30		9,775	1,000		10,775	12,300
	3800	20"		.60	40		13,600	1,350		14,950	17,000
	3830	24"		.50	48		20,200	1,600		21,800	24,600
	3900	For 250 lb, flanged, add					200%	10%			
	4350	Globe, OS&Y									
	4540	Class 125, flanged									
	4550	2"	1 Plum	5	1.600	Ea.	440	57.50		497.50	570
	4560	2-1/2"	Q-1	5	3.200		625	104		729	840
	4570	3"		4.50	3.556		540	115		655	770
	4580	4"		3	5.333		770	173		943	1,100
	4590	5"	Q-2	3.40	7.059		1,400	237		1,637	1,900
	4600	6"		3	8		1,400	268		1,668	1,950
	4610	8"		2.50	9.600		2,750	320		3,070	3,500

MECHANICAL 15

15110 | Valves

			CREW	DAILY OUTPUT	LABOR-HOURS	UNIT	2002 BARE COSTS				TOTAL INCL O&P
							MAT.	LABOR	EQUIP.	TOTAL	
200	4612	10"	Q-2	2.20	10.909	Ea.	4,300	365		4,665	5,275	200
	4614	12" R15100 -090	↓	1.70	14.118	↓	4,850	475		5,325	6,050
	5040	Class 250, flanged									
	5050	2"	1 Plum	4.50	1.778	Ea.	630	64		694	790
	5060	2-1/2"	Q-1	4.50	3.556		820	115		935	1,075
	5070	3"		4	4		850	129		979	1,125
	5080	4"	↓	2.70	5.926		1,250	192		1,442	1,675
	5090	5"	Q-2	3	8		2,225	268		2,493	2,850
	5100	6"		2.70	8.889		2,225	298		2,523	2,900
	5110	8"		2.20	10.909		3,775	365		4,140	4,700
	5120	10"		2	12		5,375	405		5,780	6,525
	5130	12"	↓	1.60	15		8,400	505		8,905	9,975
	5240	Valve sprocket rim w/chain, for 2" valve	1 Stpi	30	.267		71.50	9.65		81.15	93
	5250	2-1/2" valve		27	.296		71.50	10.75		82.25	94.50
	5260	3-1/2" valve		25	.320		71.50	11.60		83.10	96
	5270	6" valve		20	.400		71.50	14.50		86	101
	5280	8" valve		18	.444		108	16.10		124.10	144
	5290	12" valve		16	.500		108	18.10		126.10	147
	5300	16" valve		12	.667		147	24		171	198
	5310	20" valve		10	.800		147	29		176	205
	5320	36" valve	↓	8	1	↓	272	36		308	355
	5450	Swing check, 125 lb., threaded									
	5500	2"	1 Plum	11	.727	Ea.	355	26		381	430
	5540	2-1/2"	Q-1	15	1.067		420	34.50		454.50	510
	5550	3"		13	1.231		470	40		510	575
	5560	4"	↓	10	1.600		750	52		802	905
	5950	Flanged									
	5994	1"	1 Plum	7	1.143	Ea.	83	41		124	154
	5998	1-1/2"		6	1.333		120	48		168	204
	6000	2"	↓	5	1.600		148	57.50		205.50	250
	6040	2-1/2"	Q-1	5	3.200		179	104		283	355
	6050	3"		4.50	3.556		192	115		307	385
	6060	4"	↓	3	5.333		300	173		473	595
	6070	6"	Q-2	3	8		515	268		783	975
	6080	8"		2.50	9.600		975	320		1,295	1,550
	6090	10"		2.20	10.909		1,650	365		2,015	2,375
	6100	12"		1.70	14.118		2,575	475		3,050	3,575
	6110	18"		1.30	18.462		17,400	620		18,020	20,000
	6114	24"	↓	.75	32	↓	32,300	1,075		33,375	37,100
	6160	For 250 lb flanged, add					200%	20%			
	6600	Silent check, bronze trim									
	6610	Compact wafer type, for 125 or 150 lb. flanges									
	6630	1-1/2"	1 Plum	11	.727	Ea.	118	26		144	169
	6640	2"	"	9	.889		140	32		172	202
	6650	2-1/2"	Q-1	9	1.778		154	57.50		211.50	256
	6660	3"		8	2		165	64.50		229.50	279
	6670	4"	↓	5	3.200		210	104		314	385
	6680	5"	Q-2	6	4		275	134		409	505
	6690	6"		6	4		370	134		504	605
	6700	8"		4.50	5.333		660	179		839	995
	6710	10"		4	6		1,150	201		1,351	1,550
	6720	12"	↓	3	8	↓	2,175	268		2,443	2,800
	6740	For 250 or 300 lb. flanges, thru 6" no change									
	6741	For 8" and 10", add				Ea.	11%	10%			
	6750	Twin disc									
	6752	2"	1 Plum	9	.889	Ea.	168	32		200	233

Important: See the Reference Section for critical supporting data - Reference Nos., Crews, & City Cost Indexes

15100 | Building Services Piping

15110 | Valves

		CREW	DAILY OUTPUT	LABOR-HOURS	UNIT	MAT.	LABOR	EQUIP.	TOTAL	TOTAL INCL O&P	
200	6754	4"	Q-1	5	3.200	Ea.	282	104		386	465	**200**
	6756	6"	Q-2	5	4.800		415	161		576	700	
	6758	8"		4.50	5.333		645	179		824	975	
	6760	10"		4	6		1,025	201		1,226	1,425	
	6762	12"		3	8		1,250	268		1,518	1,750	
	6764	18"		1.50	16		5,050	535		5,585	6,350	
	6766	24"		.75	32		7,475	1,075		8,550	9,850	
	6800	Full flange type, 150 lb.										
	6810	1"	1 Plum	14	.571	Ea.	138	20.50		158.50	182	
	6811	1-1/4"		12	.667		145	24		169	195	
	6812	1-1/2"		11	.727		168	26		194	225	
	6813	2"		9	.889		193	32		225	260	
	6814	2-1/2"	Q-1	9	1.778		234	57.50		291.50	345	
	6815	3"		8	2		242	64.50		306.50	365	
	6816	4"		5	3.200		299	104		403	485	
	6817	5"	Q-2	6	4		470	134		604	720	
	6818	6"		5	4.800		620	161		781	925	
	6819	8"		4.50	5.333		1,050	179		1,229	1,425	
	6820	10"		4	6		1,325	201		1,526	1,750	
	6840	For 250 lb., add					30%	10%				
	6900	Globe type, 125 lb.										
	6910	2"	1 Plum	9	.889	Ea.	267	32		299	340	
	6911	2-1/2"	Q-1	9	1.778		294	57.50		351.50	410	
	6912	3"		8	2		315	64.50		379.50	450	
	6913	4"		5	3.200		425	104		529	625	
	6914	5"	Q-2	6	4		550	134		684	805	
	6915	6"		5	4.800		685	161		846	1,000	
	6916	8"		4.50	5.333		1,250	179		1,429	1,650	
	6917	10"		4	6		1,550	201		1,751	2,000	
	6918	12"		3	8		2,625	268		2,893	3,300	
	6919	14"		2.30	10.435		3,600	350		3,950	4,475	
	6920	16"		1.75	13.714		5,175	460		5,635	6,400	
	6921	18"		1.50	16		7,475	535		8,010	9,025	
	6922	20"		1	24		9,075	805		9,880	11,200	
	6923	24"		.75	32		13,200	1,075		14,275	16,200	
	6940	For 250 lb., add					40%	10%				
	6980	Screwed end type, 125 lb.										
	6981	1"	1 Plum	19	.421	Ea.	50.50	15.15		65.65	78.50	
	6982	1-1/4"		15	.533		65.50	19.15		84.65	101	
	6983	1-1/2"		13	.615		81	22		103	123	
	6984	2"		11	.727		110	26		136	161	
400	0010	**MULTIPURPOSE VALVES**										**400**
	0100	Functions as a shut off, balancing, check & metering valve										
	1000	Cast iron body										
	1010	Threaded										
	1020	1-1/2" size	1 Stpi	11	.727	Ea.	288	26.50		314.50	355	
	1030	2" size	"	8	1		350	36		386	440	
	1040	2-1/2" size	Q-5	5	3.200		380	104		484	570	
	1050	3" size	"	4.50	3.556		400	116		516	615	
	1200	Flanged										
	1210	3" size	Q-5	4.20	3.810	Ea.	435	124		559	665	
	1220	4" size	"	3	5.333		880	174		1,054	1,225	
	1230	5" size	Q-6	3.80	6.316		1,050	213		1,263	1,475	
	1240	6" size		3	8		1,300	270		1,570	1,825	
	1250	8" size		2.50	9.600		1,900	325		2,225	2,600	

R15100 -090

MECHANICAL 15

15110	Valves	CREW	DAILY OUTPUT	LABOR-HOURS	UNIT	2002 BARE COSTS				TOTAL INCL O&P
						MAT.	LABOR	EQUIP.	TOTAL	
400 1260	10" size	Q-6	2.20	10.909	Ea.	2,725	370		3,095	3,550	**400**
1270	12" size		2.10	11.429		7,450	385		7,835	8,750
1280	14" size	▼	2	12	▼	10,500	405		10,905	12,200
500 0010	**VALVES, PLASTIC** R15100 -090										**500**
1100	Angle, PVC, threaded									
1110	1/4"	1 Plum	26	.308	Ea.	49.50	11.05		60.55	71
1120	1/2"		26	.308		49.50	11.05		60.55	71
1130	3/4"		25	.320		58.50	11.50		70	82
1140	1"	▼	23	.348	▼	70	12.50		82.50	96
1150	Ball, PVC, socket or threaded, single union									
1200	1/4"	1 Plum	26	.308	Ea.	32	11.05		43.05	51.50
1220	3/8"		26	.308		32	11.05		43.05	51.50
1230	1/2"		26	.308		32	11.05		43.05	51.50
1240	3/4"		25	.320		38	11.50		49.50	59.50
1250	1"		23	.348		45.50	12.50		58	69
1260	1-1/4"		21	.381		61	13.70		74.70	87.50
1270	1-1/2"		20	.400		76	14.40		90.40	105
1280	2"	▼	17	.471		109	16.90		125.90	146
1290	2-1/2"	Q-1	26	.615		275	19.90		294.90	330
1300	3"		24	.667		275	21.50		296.50	335
1310	4"	▼	20	.800		355	26		381	430
1360	For PVC, flanged, add					100%	15%			
1450	Double union 1/2"	1 Plum	26	.308		13.80	11.05		24.85	32
1460	3/4"		25	.320		16.20	11.50		27.70	35
1470	1"		23	.348		17.75	12.50		30.25	38.50
1480	1-1/4"		21	.381		27	13.70		40.70	50
1490	1-1/2"		20	.400		30	14.40		44.40	54.50
1500	2"	▼	17	.471	▼	42.50	16.90		59.40	72
1650	CPVC, socket or threaded, single union									
1700	1/2"	1 Plum	26	.308	Ea.	33.50	11.05		44.55	53
1720	3/4"		25	.320		42	11.50		53.50	63.50
1730	1"		23	.348		50	12.50		62.50	74
1750	1-1/4"		21	.381		83.50	13.70		97.20	113
1760	1-1/2"		20	.400		83.50	14.40		97.90	114
1770	2"	▼	17	.471		116	16.90		132.90	154
1780	3"	Q-1	24	.667		345	21.50		366.50	415
1840	For CPVC, flanged, add					65%	15%			
1880	For true union, socket or threaded, add				▼	50%	5%			
2050	Polypropylene, threaded									
2100	1/4"	1 Plum	26	.308	Ea.	31.50	11.05		42.55	51
2120	3/8"		26	.308		31.50	11.05		42.55	51
2130	1/2"		26	.308		31.50	11.05		42.55	51
2140	3/4"		25	.320		39	11.50		50.50	60.50
2150	1"		23	.348		46.50	12.50		59	70.50
2160	1-1/4"		21	.381		67.50	13.70		81.20	95
2170	1-1/2"		20	.400		78	14.40		92.40	107
2180	2"	▼	17	.471		106	16.90		122.90	142
2190	3"	Q-1	24	.667		280	21.50		301.50	345
2200	4"	"	20	.800	▼	465	26		491	550
2550	PVC, three way, socket or threaded									
2600	1/2"	1 Plum	26	.308	Ea.	59.50	11.05		70.55	82
2640	3/4"		25	.320		67.50	11.50		79	92
2650	1"		23	.348		73	12.50		85.50	99
2660	1-1/2"		20	.400		148	14.40		162.40	184
2670	2"	▼	17	.471		197	16.90		213.90	243

Important: See the Reference Section for critical supporting data - Reference Nos., Crews, & City Cost Indexes

15110 | Valves

			CREW	DAILY OUTPUT	LABOR-HOURS	UNIT	2002 BARE COSTS				TOTAL INCL O&P	
							MAT.	LABOR	EQUIP.	TOTAL		
500	2680	3"	Q-1	24	.667	Ea.	430	21.50		451.50	505	500
	2740	For flanged, add	↓			↓	60%	15%				
	3150	Ball check, PVC, socket or threaded										
	3200	1/4"	1 Plum	26	.308	Ea.	29	11.05		40.05	48.50	
	3220	3/8"		26	.308		29	11.05		40.05	48.50	
	3240	1/2"		26	.308		29	11.05		40.05	48.50	
	3250	3/4"		25	.320		32.50	11.50		44	53	
	3260	1"		23	.348		40.50	12.50		53	64	
	3270	1-1/4"		21	.381		68	13.70		81.70	95.50	
	3280	1-1/2"		20	.400		68	14.40		82.40	96.50	
	3290	2"	↓	17	.471		93	16.90		109.90	128	
	3310	3"	Q-1	24	.667		257	21.50		278.50	315	
	3320	4"	"	20	.800		365	26		391	440	
	3360	For PVC, flanged, add				↓	50%	15%				
	3750	CPVC, socket or threaded										
	3800	1/2"	1 Plum	26	.308	Ea.	42.50	11.05		53.55	63	
	3840	3/4"		25	.320		50.50	11.50		62	73	
	3850	1"		23	.348		60	12.50		72.50	84.50	
	3860	1-1/2"		20	.400		103	14.40		117.40	135	
	3870	2"	↓	17	.471		138	16.90		154.90	178	
	3880	3"	Q-1	24	.667		375	21.50		396.50	445	
	3920	4"	"	20	.800		505	26		531	595	
	3930	For CPVC, flanged, add				↓	40%	15%				
	4340	Polypropylene, threaded										
	4360	1/2"	1 Plum	26	.308	Ea.	26.50	11.05		37.55	45.50	
	4400	3/4"		25	.320		30.50	11.50		42	51	
	4440	1"		23	.348		39	12.50		51.50	62	
	4450	1-1/2"		20	.400		76	14.40		90.40	106	
	4460	2"	↓	17	.471		96	16.90		112.90	131	
	4500	For polypropylene flanged, add				↓	200%	15%				
	4850	Foot valve, PVC, socket or threaded										
	4900	1/2"	1 Plum	34	.235	Ea.	42.50	8.45		50.95	60	
	4930	3/4"		32	.250		48.50	9		57.50	67	
	4940	1"		28	.286		63	10.25		73.25	85	
	4950	1-1/4"		27	.296		121	10.65		131.65	149	
	4960	1-1/2"		26	.308		121	11.05		132.05	150	
	4970	2"		24	.333		140	12		152	172	
	4980	3"		20	.400		335	14.40		349.40	390	
	4990	4"	↓	18	.444		590	16		606	670	
	5000	For flanged, add				↓	25%	10%				
	5050	CPVC, socket or threaded										
	5060	1/2"	1 Plum	34	.235	Ea.	93.50	8.45		101.95	116	
	5070	3/4"		32	.250		114	9		123	139	
	5080	1"		28	.286		134	10.25		144.25	163	
	5090	1-1/4"		27	.296		218	10.65		228.65	256	
	5100	1-1/2"		26	.308		218	11.05		229.05	257	
	5110	2"		24	.333		280	12		292	330	
	5120	3"		20	.400		670	14.40		684.40	760	
	5130	4"	↓	18	.444		1,150	16		1,166	1,275	
	5140	For flanged, add				↓	25%	10%				
	5280	Needle valve, PVC, threaded										
	5300	1/4"	1 Plum	26	.308	Ea.	33.50	11.05		44.55	53.50	
	5340	3/8"		26	.308		39.50	11.05		50.55	59.50	
	5360	1/2"	↓	26	.308		39.50	11.05		50.55	59.50	
	5380	For polypropylene, add				↓	10%					
	5800	Y check, PVC, socket or threaded										

R15100 -090

MECHANICAL 15

15110 | Valves

			CREW	DAILY OUTPUT	LABOR-HOURS	UNIT	MAT.	LABOR	EQUIP.	TOTAL	TOTAL INCL O&P	
500	5820	1/2" R15100-090	1 Plum	26	.308	Ea.	55	11.05		66.05	77	**500**
	5840	3/4"		25	.320		59.50	11.50		71	83	
	5850	1"		23	.348		65	12.50		77.50	90.50	
	5860	1-1/4"		21	.381		102	13.70		115.70	133	
	5870	1-1/2"		20	.400		111	14.40		125.40	144	
	5880	2"		17	.471		138	16.90		154.90	177	
	5890	2-1/2"		15	.533		291	19.15		310.15	350	
	5900	3"	Q-1	24	.667		273	21.50		294.50	335	
	5910	4"	"	20	.800		475	26		501	565	
	5960	For PVC flanged, add					45%	15%				
	6350	Y sediment strainer, PVC, socket or threaded										
	6400	1/2"	1 Plum	26	.308	Ea.	31	11.05		42.05	50.50	
	6440	3/4"		24	.333		34	12		46	55	
	6450	1"		23	.348		41	12.50		53.50	64	
	6460	1-1/4"		21	.381		67.50	13.70		81.20	95	
	6470	1-1/2"		20	.400		67.50	14.40		81.90	96	
	6480	2"		17	.471		82	16.90		98.90	116	
	6490	2-1/2"		15	.533		202	19.15		221.15	251	
	6500	3"	Q-1	24	.667		202	21.50		223.50	255	
	6510	4"	"	20	.800		335	26		361	410	
	6560	For PVC, flanged, add					55%	15%				
600	0010	**VALVES, SEMI-STEEL** R15100-090										**600**
	1020	Lubricated plug valve, threaded, 200 psi										
	1030	1/2"	1 Plum	18	.444	Ea.	61.50	16		77.50	91.50	
	1040	3/4"		16	.500		61.50	18		79.50	94.50	
	1050	1"		14	.571		79	20.50		99.50	118	
	1060	1-1/4"		12	.667		94.50	24		118.50	140	
	1070	1-1/2"		11	.727		102	26		128	152	
	1080	2"		8	1		120	36		156	186	
	1090	2-1/2"	Q-1	5	3.200		186	104		290	360	
	1100	3"	"	4.50	3.556		228	115		343	425	
	6990	Flanged, 200 psi										
	7000	2"	1 Plum	8	1	Ea.	145	36		181	213	
	7010	2-1/2"	Q-1	5	3.200		218	104		322	395	
	7020	3"		4.50	3.556		263	115		378	465	
	7030	4"		3	5.333		335	173		508	625	
	7036	5"		2.50	6.400		500	207		707	860	
	7040	6"	Q-2	3	8		655	268		923	1,125	
	7050	8"		2.50	9.600		1,175	320		1,495	1,775	
	7060	10"		2.20	10.909		1,725	365		2,090	2,450	
	7070	12"		1.70	14.118		2,975	475		3,450	3,975	
700	0010	**VALVES, STEEL** R15100-090										**700**
	0800	Cast										
	1350	Check valve, swing type, 150 lb., flanged										
	1400	2"	1 Plum	8	1	Ea.	465	36		501	570	
	1440	2-1/2"	Q-1	5	3.200		505	104		609	710	
	1450	3"		4.50	3.556		595	115		710	830	
	1460	4"		3	5.333		875	173		1,048	1,225	
	1470	6"	Q-2	3	8		1,350	268		1,618	1,900	
	1480	8"		2.50	9.600		2,225	320		2,545	2,925	
	1490	10"		2.20	10.909		3,300	365		3,665	4,175	
	1500	12"		1.70	14.118		4,725	475		5,200	5,925	
	1510	14"		1.30	18.462		6,750	620		7,370	8,350	
	1520	16"		1	24		8,225	805		9,030	10,300	
	1548	For 600 lb., flanged, add					110%	20%				

Important: See the Reference Section for critical supporting data - Reference Nos., Crews, & City Cost Indexes

15110 | Valves

			CREW	DAILY OUTPUT	LABOR-HOURS	UNIT	MAT.	LABOR	EQUIP.	TOTAL	TOTAL INCL O&P	
700	1571	300 lb., 2"	R15100 -090	1 Plum	7.40	1.081	Ea.	545	39		584	660	**700**
	1572	2-1/2"		Q-1	4.20	3.810		825	123		948	1,100	
	1573	3"			4	4		825	129		954	1,100	
	1574	4"		↓	2.80	5.714		1,175	185		1,360	1,550	
	1575	6"		Q-2	2.90	8.276		2,100	278		2,378	2,750	
	1576	8"			2.40	10		3,475	335		3,810	4,300	
	1577	10"			2.10	11.429		4,400	385		4,785	5,425	
	1578	12"			1.60	15		5,975	505		6,480	7,325	
	1579	14"			1.20	20		9,725	670		10,395	11,700	
	1581	16"		↓	.90	26.667	↓	13,500	895		14,395	16,200	
	1700	Diverter, 150 lb., flanged, bronze or iron plugs											
	1720	1-1/2" pipe size		Q-1	3	5.333	Ea.	2,450	173		2,623	2,950	
	1730	2" pipe size			2	8		2,775	259		3,034	3,450	
	1740	3" pipe size			1.50	10.667		3,950	345		4,295	4,875	
	1750	4" pipe size		↓	1	16		6,025	520		6,545	7,400	
	1760	6" pipe size		Q-2	1	24		11,500	805		12,305	13,900	
	1790	For 300 lb., add					↓	20%	5%				
	1810	Diverter, 150 lb., flanged, stainless steel plugs											
	1830	1-1/2" pipe size		Q-1	3	5.333	Ea.	2,400	173		2,573	2,900	
	1840	2" pipe size			2	8		3,200	259		3,459	3,925	
	1850	3" pipe size			1.50	10.667		4,325	345		4,670	5,275	
	1860	4" pipe size		↓	1	16		6,525	520		7,045	7,950	
	1870	6" pipe size		Q-2	1	24		13,200	805		14,005	15,700	
	1890	For 300 lb., add					↓	10%	5%				
	1950	Gate valve, 150 lb., flanged											
	2000	2"		1 Plum	8	1	Ea.	540	36		576	650	
	2040	2-1/2"		Q-1	5	3.200		765	104		869	995	
	2050	3"			4.50	3.556		765	115		880	1,025	
	2060	4"		↓	3	5.333		930	173		1,103	1,275	
	2070	6"		Q-2	3	8		1,475	268		1,743	2,025	
	2080	8"			2.50	9.600		2,350	320		2,670	3,075	
	2090	10"			2.20	10.909		3,700	365		4,065	4,625	
	2100	12"			1.70	14.118		5,050	475		5,525	6,275	
	2110	14"			1.30	18.462		8,400	620		9,020	10,200	
	2120	16"			1	24		10,200	805		11,005	12,400	
	2130	18"			.80	30		11,300	1,000		12,300	13,900	
	2140	20"		↓	.60	40	↓	13,100	1,350		14,450	16,400	
	2650	300 lb., flanged											
	2700	2"		1 Plum	7.40	1.081	Ea.	740	39		779	870	
	2740	2-1/2"		Q-1	4.20	3.810		995	123		1,118	1,275	
	2750	3"			4	4		995	129		1,124	1,300	
	2760	4"		↓	2.80	5.714		1,375	185		1,560	1,800	
	2770	6"		Q-2	2.90	8.276		2,350	278		2,628	3,000	
	2780	8"			2.40	10		3,675	335		4,010	4,525	
	2790	10"			2.10	11.429		5,425	385		5,810	6,550	
	2800	12"			1.60	15		7,200	505		7,705	8,675	
	2810	14"			1.20	20		11,300	670		11,970	13,500	
	2820	16"			.90	26.667		14,200	895		15,095	17,000	
	2830	18"			.70	34.286		17,800	1,150		18,950	21,200	
	2840	20"		↓	.50	48	↓	21,500	1,600		23,100	26,100	
	3650	Globe valve, 150 lb., flanged											
	3700	2"		1 Plum	8	1	Ea.	675	36		711	800	
	3740	2-1/2"		Q-1	5	3.200		850	104		954	1,100	
	3750	3"			4.50	3.556		850	115		965	1,100	
	3760	4"		↓	3	5.333		1,250	173		1,423	1,625	
	3770	6"		Q-2	3	8	↓	1,950	268		2,218	2,550	

MECHANICAL 15

15110 | Valves

			CREW	DAILY OUTPUT	LABOR-HOURS	UNIT	MAT.	LABOR	EQUIP.	TOTAL	TOTAL INCL O&P	
700	3780	8"	R15100-090	Q-2	2.50	9.600	Ea.	3,625	320		3,945	4,450	700
	3790	10"			2.20	10.909		8,925	365		9,290	10,400	
	3800	12"			1.70	14.118		11,400	475		11,875	13,200	
	4080	300 lb., flanged											
	4100	2"		1 Plum	7.40	1.081	Ea.	935	39		974	1,075	
	4140	2-1/2"		Q-1	4.20	3.810		1,250	123		1,373	1,575	
	4150	3"			4	4		1,250	129		1,379	1,600	
	4160	4"			2.80	5.714		1,750	185		1,935	2,200	
	4170	6"		Q-2	2.90	8.276		3,150	278		3,428	3,875	
	4180	8"			2.40	10		5,175	335		5,510	6,200	
	4190	10"			2.10	11.429		12,700	385		13,085	14,500	
	4200	12"			1.60	15		14,900	505		15,405	17,200	
	4680	600 lb., flanged											
	4700	2"		1 Plum	7	1.143	Ea.	1,325	41		1,366	1,500	
	4740	2-1/2"		Q-1	4	4		1,975	129		2,104	2,375	
	4750	3"			3.60	4.444		1,975	144		2,119	2,400	
	4760	4"			2.50	6.400		3,200	207		3,407	3,800	
	4770	6"		Q-2	2.60	9.231		6,600	310		6,910	7,750	
	4780	8"		"	2.10	11.429		10,600	385		10,985	12,300	
	4800	Silent check, 316 S.S. trim											
	4810	Full flange type, 150 lb.											
	4811	1"		1 Plum	14	.571	Ea.	305	20.50		325.50	365	
	4812	1-1/4"			12	.667		305	24		329	370	
	4813	1-1/2"			11	.727		335	26		361	405	
	4814	2"			9	.889		400	32		432	495	
	4815	2-1/2"		Q-1	9	1.778		485	57.50		542.50	620	
	4816	3"			8	2		565	64.50		629.50	720	
	4817	4"			5	3.200		770	104		874	1,000	
	4818	5"		Q-2	6	4		965	134		1,099	1,250	
	4819	6"			5	4.800		1,250	161		1,411	1,625	
	4820	8"			4.50	5.333		1,800	179		1,979	2,250	
	4821	10"			4	6		2,550	201		2,751	3,125	
	4840	For 300 lb., add						20%	10%				
	4860	For 600 lb., add						50%	15%				
	4900	Globe type, 150 lb.											
	4910	2"		1 Plum	9	.889	Ea.	465	32		497	565	
	4911	2-1/2"		Q-1	9	1.778		575	57.50		632.50	715	
	4912	3"			8	2		640	64.50		704.50	805	
	4913	4"			5	3.200		850	104		954	1,100	
	4914	5"		Q-2	6	4		1,075	134		1,209	1,375	
	4915	6"			5	4.800		1,325	161		1,486	1,700	
	4916	8"			4.50	5.333		1,850	179		2,029	2,300	
	4917	10"			4	6		3,200	201		3,401	3,800	
	4918	12"			3	8		3,925	268		4,193	4,700	
	4919	14"			2.30	10.435		5,925	350		6,275	7,050	
	4920	16"			1.75	13.714		8,275	460		8,735	9,825	
	4921	18"			1.50	16		8,700	535		9,235	10,400	
	4940	For 300 lb., add						30%	10%				
	4960	For 600 lb., add						60%	15%				
	5150	Forged											
	5340	Ball valve, 800 psi, threaded, 1/4" size		1 Plum	24	.333	Ea.	35	12		47	56.50	
	5350	3/8"			24	.333		35	12		47	56.50	
	5360	1/2"			24	.333		43.50	12		55.50	66	
	5370	3/4"			20	.400		58	14.40		72.40	85	
	5380	1"			19	.421		72	15.15		87.15	102	
	5390	1-1/4"			15	.533		96	19.15		115.15	135	

T5 MECHANICAL

Important: See the Reference Section for critical supporting data - Reference Nos., Crews, & City Cost Indexes

15110 | Valves

		CREW	DAILY OUTPUT	LABOR-HOURS	UNIT	MAT.	LABOR	EQUIP.	TOTAL	TOTAL INCL O&P	
							2002 BARE COSTS				
700	5400	1-1/2" R15100 -090	1 Plum	13	.615	Ea.	125	22		147	172	**700**
	5410	2"	▼	11	.727		166	26		192	222	
	5460	Ball valve, 800 lb., socket weld, 1/4" size	Q-15	19	.842		36.50	27	4.21	67.71	85.50	
	5470	3/8"		19	.842		36.50	27	4.21	67.71	85.50	
	5480	1/2"		19	.842		43.50	27	4.21	74.71	93.50	
	5490	3/4"		19	.842		58	27	4.21	89.21	109	
	5500	1"		15	1.067		72	34.50	5.35	111.85	137	
	5510	1-1/4"		13	1.231		96	40	6.15	142.15	173	
	5520	1-1/2"		11	1.455		125	47	7.25	179.25	217	
	5530	2"	▼	8.50	1.882	▼	166	61	9.40	236.40	284	
	5550	Ball valve, 150 lb., flanged										
	5560	4"	Q-1	3	5.333	Ea.	415	173		588	720	
	5570	6"	Q-2	3	8		1,000	268		1,268	1,500	
	5580	8"	"	2.50	9.600	▼	1,800	320		2,120	2,450	
	5650	Check valve, class 800, horizontal, socket										
	5652	1/4"	Q-15	19	.842	Ea.	41	27	4.21	72.21	90.50	
	5654	3/8"		19	.842		41	27	4.21	72.21	90.50	
	5656	1/2"		19	.842		41	27	4.21	72.21	90.50	
	5658	3/4"		19	.842		44.50	27	4.21	75.71	94.50	
	5660	1"		15	1.067		52	34.50	5.35	91.85	115	
	5662	1-1/4"		13	1.231		99.50	40	6.15	145.65	176	
	5664	1-1/2"		11	1.455		99.50	47	7.25	153.75	188	
	5666	2"	▼	8.50	1.882	▼	139	61	9.40	209.40	255	
	5698	Threaded										
	5700	1/4"	1 Plum	24	.333	Ea.	41	12		53	63	
	5720	3/8"		24	.333		41	12		53	63	
	5730	1/2"		24	.333		41	12		53	63	
	5740	3/4"		20	.400		44.50	14.40		58.90	70.50	
	5750	1"		19	.421		52	15.15		67.15	80	
	5760	1-1/4"		15	.533		99.50	19.15		118.65	138	
	5770	1-1/2"		13	.615		99.50	22		121.50	143	
	5780	2"	▼	11	.727		139	26		165	193	
	5840	For class 150, flanged, add					100%	15%				
	5860	For class 300, flanged, add				▼	120%	20%				
	6100	Gate, class 800, inside screw stem, socket										
	6102	3/8"	Q-15	19	.842	Ea.	43	27	4.21	74.21	93	
	6103	1/2"		19	.842		43	27	4.21	74.21	93	
	6104	3/4"		19	.842		50.50	27	4.21	81.71	101	
	6105	1"		15	1.067		57.50	34.50	5.35	97.35	121	
	6106	1-1/4"		13	1.231		115	40	6.15	161.15	194	
	6107	1-1/2"		11	1.455		115	47	7.25	169.25	206	
	6108	2"	▼	8.50	1.882	▼	144	61	9.40	214.40	260	
	6110	For OS&Y, socket, add					50%					
	6118	Threaded										
	6120	3/8"	1 Plum	24	.333	Ea.	43	12		55	65.50	
	6130	1/2"		24	.333		43	12		55	65.50	
	6140	3/4"		20	.400		50.50	14.40		64.90	77	
	6150	1"		19	.421		57.50	15.15		72.65	86.50	
	6160	1-1/4"		15	.533		115	19.15		134.15	156	
	6170	1-1/2"		13	.615		115	22		137	161	
	6180	2"	▼	11	.727		144	26		170	198	
	6240	For OS&Y, threaded, add					50%					
	6260	For OS&Y, flanged, add				▼	100%	20%				
	6700	Globe, OS&Y, class 800, socket										
	6710	1/4"	Q-15	19	.842	Ea.	49.50	27	4.21	80.71	100	
	6720	3/8"	▼	19	.842	▼	49.50	27	4.21	80.71	100	

15110	Valves	CREW	DAILY OUTPUT	LABOR-HOURS	UNIT	2002 BARE COSTS				TOTAL INCL O&P	
						MAT.	LABOR	EQUIP.	TOTAL		
700											**700**
6730	1/2"	Q-15 R15100 -090	19	.842	Ea.	49.50	27	4.21	80.71	100	
6740	3/4"		19	.842		52.50	27	4.21	83.71	104	
6750	1"		15	1.067		66	34.50	5.35	105.85	131	
6760	1-1/4"		13	1.231		126	40	6.15	172.15	206	
6770	1-1/2"		11	1.455		126	47	7.25	180.25	218	
6780	2"		8.50	1.882		162	61	9.40	232.40	280	
6860	For OS&Y, flanged, add					300%	20%				
6880	Threaded										
6882	1/4"	1 Plum	24	.333	Ea.	49.50	12		61.50	72.50	
6884	3/8"		24	.333		49.50	12		61.50	72.50	
6886	1/2"		24	.333		49.50	12		61.50	72.50	
6888	3/4"		20	.400		52.50	14.40		66.90	79.50	
6890	1"		19	.421		66	15.15		81.15	96	
6892	1-1/4"		15	.533		126	19.15		145.15	168	
6894	1-1/2"		13	.615		126	22		148	173	
6896	2"		11	.727		162	26		188	218	
800	0010	**VALVES, STAINLESS STEEL**									**800**
1610	Ball, threaded 1/4"	1 Stpi R15100 -090	24	.333	Ea.	29.50	12.05		41.55	50	
1620	3/8"		24	.333		29.50	12.05		41.55	50	
1630	1/2"		22	.364		29.50	13.15		42.65	52	
1640	3/4"		20	.400		48.50	14.50		63	75.50	
1650	1"		19	.421		59	15.25		74.25	88	
1660	1-1/4		15	.533		113	19.30		132.30	153	
1670	1-1/2"		13	.615		118	22.50		140.50	164	
1680	2"		11	.727		153	26.50		179.50	208	
1700	Check, 200 lb., threaded										
1710	1/4"	1 Plum	24	.333	Ea.	72	12		84	97	
1720	1/2"		22	.364		72	13.05		85.05	98.50	
1730	3/4"		20	.400		79	14.40		93.40	109	
1750	1"		19	.421		101	15.15		116.15	134	
1760	1-1/2"		13	.615		191	22		213	244	
1770	2"		11	.727		320	26		346	390	
1800	150 lb., flanged										
1810	2-1/2"	Q-1	5	3.200	Ea.	765	104		869	995	
1820	3"		4.50	3.556		765	115		880	1,025	
1830	4"		3	5.333		1,150	173		1,323	1,525	
1840	6"	Q-2	3	8		2,050	268		2,318	2,650	
1850	8"	"	2.50	9.600		3,425	320		3,745	4,225	
1900	Diverter, 150 lb. flanged, stainless steel plugs										
1910	1-1/2" pipe size	Q-1	3	5.333	Ea.	4,750	173		4,923	5,475	
1920	2" pipe size		2	8		6,175	259		6,434	7,175	
1940	3" pipe size		1.50	10.667		7,850	345		8,195	9,150	
1950	4" pipe size		1	16		13,200	520		13,720	15,300	
2100	Gate, OS&Y, 150 lb., flanged										
2120	1/2"	1 Plum	18	.444	Ea.	310	16		326	365	
2140	3/4"		16	.500		335	18		353	395	
2150	1"		14	.571		375	20.50		395.50	440	
2155	1-1/4"		12	.667		470	24		494	550	
2160	1-1/2"		11	.727		470	26		496	555	
2170	2"		8	1		520	36		556	630	
2180	2-1/2"	Q-1	5	3.200		810	104		914	1,050	
2190	3"		4.50	3.556		885	115		1,000	1,150	
2200	4"		3	5.333		1,225	173		1,398	1,600	
2205	5"		2.80	5.714		2,200	185		2,385	2,700	
2210	6"	Q-2	3	8		2,200	268		2,468	2,825	
2220	8"		2.50	9.600		3,400	320		3,720	4,225	

Important: See the Reference Section for critical supporting data - Reference Nos., Crews, & City Cost Indexes

			DAILY	LABOR-		2002 BARE COSTS				TOTAL	
15110	**Valves**	CREW	OUTPUT	HOURS	UNIT	MAT.	LABOR	EQUIP.	TOTAL	INCL O&P	
800	2230	10"	R15100 -090	Q-2	2.30	10.435	Ea.	4,950	350		5,300	5,975	800
	2240	12"		↓	1.90	12.632	↓	6,900	425		7,325	8,250	
	2260	For 300 lb., flanged, add					↓	120%	15%				
	2600	600 lb., flanged											
	2620	1/2"		1 Plum	16	.500	Ea.	720	18		738	815	
	2640	3/4"			14	.571		880	20.50		900.50	995	
	2650	1"			12	.667		1,100	24		1,124	1,250	
	2660	1-1/2"			10	.800		1,550	29		1,579	1,750	
	2670	2"		↓	7	1.143		2,775	41		2,816	3,100	
	2680	2-1/2"		Q-1	4	4		7,975	129		8,104	8,975	
	2690	3"		"	3.60	4.444	↓	8,050	144		8,194	9,075	
	3100	Globe, OS&Y, 150 lb., flanged											
	3120	1/2"		1 Plum	18	.444	Ea.	345	16		361	405	
	3140	3/4"			16	.500		380	18		398	445	
	3150	1"			14	.571		470	20.50		490.50	550	
	3160	1-1/2"			11	.727		700	26		726	810	
	3170	2"		↓	8	1		855	36		891	995	
	3180	2-1/2"		Q-1	5	3.200		1,800	104		1,904	2,125	
	3190	3"			4.50	3.556		1,850	115		1,965	2,200	
	3200	4"		↓	3	5.333		2,150	173		2,323	2,625	
	3210	6"		Q-2	3	8	↓	4,075	268		4,343	4,875	
	5000	Silent check, 316 S.S. body and trim											
	5010	Compact wafer type, 300 lb.											
	5020	1"		1 Plum	14	.571	Ea.	555	20.50		575.50	640	
	5021	1-1/4"			12	.667		600	24		624	695	
	5022	1-1/2"			11	.727		620	26		646	725	
	5023	2"		↓	9	.889		685	32		717	800	
	5024	2-1/2"		Q-1	9	1.778		975	57.50		1,032.50	1,150	
	5025	3"		"	8	2	↓	1,075	64.50		1,139.50	1,275	
	5100	Full flange wafer type, 150 lb.											
	5110	1"		1 Plum	14	.571	Ea.	390	20.50		410.50	460	
	5111	1-1/4"			12	.667		420	24		444	495	
	5112	1-1/2"			11	.727		480	26		506	570	
	5113	2"		↓	9	.889		625	32		657	740	
	5114	2-1/2"		Q-1	9	1.778		750	57.50		807.50	905	
	5115	3"			8	2		845	64.50		909.50	1,025	
	5116	4"		↓	5	3.200		1,250	104		1,354	1,525	
	5117	5"		Q-2	6	4		1,475	134		1,609	1,800	
	5118	6"			5	4.800		1,825	161		1,986	2,275	
	5119	8"			4.50	5.333		3,050	179		3,229	3,625	
	5120	10"		↓	4	6	↓	4,225	201		4,426	4,950	
	5200	Globe type, 300 lb.											
	5210	2"		1 Plum	9	.889	Ea.	715	32		747	835	
	5211	2-1/2"		Q-1	9	1.778		880	57.50		937.50	1,050	
	5212	3"			8	2		1,100	64.50		1,164.50	1,300	
	5213	4"		↓	5	3.200		1,500	104		1,604	1,800	
	5214	5"		Q-2	6	4		2,125	134		2,259	2,525	
	5215	6"			5	4.800		2,400	161		2,561	2,875	
	5216	8"			4.50	5.333		3,375	179		3,554	4,000	
	5217	10"		↓	4	6	↓	4,775	201		4,976	5,550	
	5300	Screwed end type, 300 lb.											
	5310	1/2"		1 Plum	24	.333	Ea.	60.50	12		72.50	84.50	
	5311	3/4"			20	.400		72	14.40		86.40	101	
	5312	1"			19	.421		84	15.15		99.15	115	
	5313	1-1/4"			15	.533		111	19.15		130.15	151	
	5314	1-1/2"		↓	13	.615	↓	120	22		142	166	

			CREW	DAILY OUTPUT	LABOR-HOURS	UNIT	2002 BARE COSTS				TOTAL INCL O&P	
							MAT.	LABOR	EQUIP.	TOTAL		
15110		**Valves**										
800	5315	2" — R15100-090	1 Plum	11	.727	Ea.	149	26		175	204	**800**
15120		**Piping Specialties**										
120	0010	**AIR CONTROL**										**120**
	0030	Air separator, with strainer										
	0040	2" diameter	Q-5	6	2.667	Ea.	530	87		617	715	
	0080	2-1/2" diameter		5	3.200		605	104		709	820	
	0100	3" diameter		4	4		920	130		1,050	1,200	
	0120	4" diameter	▼	3	5.333		1,325	174		1,499	1,725	
	0130	5" diameter	Q-6	3.60	6.667		1,700	225		1,925	2,225	
	0140	6" diameter		3.40	7.059		2,025	238		2,263	2,575	
	0160	8" diameter		3	8		3,025	270		3,295	3,725	
	0180	10" diameter		2.20	10.909		4,700	370		5,070	5,725	
	0200	12" diameter		1.70	14.118		7,950	475		8,425	9,475	
	0210	14" diameter		1.30	18.462		9,475	625		10,100	11,300	
	0220	16" diameter		1	24		14,300	810		15,110	16,900	
	0230	18" diameter		.80	30		18,800	1,025		19,825	22,100	
	0240	20" diameter	▼	.60	40	▼	21,000	1,350		22,350	25,200	
	0300	Without strainer										
	0310	2" diameter	Q-5	6	2.667	Ea.	410	87		497	585	
	0320	2-1/2" diameter		5	3.200		490	104		594	695	
	0330	3" diameter		4	4		690	130		820	955	
	0340	4" diameter		3	5.333		1,075	174		1,249	1,450	
	0350	5" diameter	▼	2.40	6.667		1,475	217		1,692	1,950	
	0360	6" diameter	Q-6	3.40	7.059		1,700	238		1,938	2,225	
	0370	8" diameter		3	8		2,325	270		2,595	2,975	
	0380	10" diameter		2.20	10.909		3,425	370		3,795	4,300	
	0390	12" diameter		1.70	14.118		5,325	475		5,800	6,575	
	0400	14" diameter		1.30	18.462		7,950	625		8,575	9,700	
	0410	16" diameter		1	24		11,100	810		11,910	13,400	
	0420	18" diameter		.80	30		14,500	1,025		15,525	17,500	
	0430	20" diameter	▼	.60	40	▼	17,400	1,350		18,750	21,200	
	1000	Micro-bubble separator for total air removal, closed loop system										
	1010	Requires bladder type tank in system.										
	1020	Water (hot or chilled) or glycol system										
	1030	Threaded										
	1040	3/4" diameter	1 Stpi	20	.400	Ea.	67	14.50		81.50	95.50	
	1050	1" diameter		19	.421		75	15.25		90.25	106	
	1060	1-1/4" diameter		16	.500		103	18.10		121.10	142	
	1070	1-1/2" diameter		13	.615		135	22.50		157.50	182	
	1080	2" diameter	▼	11	.727		655	26.50		681.50	760	
	1090	2-1/2" diameter	Q-5	15	1.067		725	35		760	855	
	1100	3" diameter		13	1.231		1,050	40		1,090	1,200	
	1110	4" diameter	▼	10	1.600	▼	1,125	52		1,177	1,300	
	1230	Flanged										
	1250	2" diameter	1 Stpi	8	1	Ea.	835	36		871	970	
	1260	2-1/2" diameter	Q-5	5	3.200		905	104		1,009	1,150	
	1270	3" diameter		4.50	3.556		1,225	116		1,341	1,525	
	1280	4" diameter	▼	3	5.333		1,350	174		1,524	1,725	
	1290	5" diameter	Q-6	3.40	7.059		2,525	238		2,763	3,125	
	1300	6" diameter	"	3	8	▼	2,625	270		2,895	3,275	
	1400	Larger sizes available										
	1590	With extended tank dirt catcher										
	1600	Flanged										
	1620	2" diameter	1 Stpi	7.40	1.081	Ea.	1,100	39		1,139	1,275	

Important: See the Reference Section for critical supporting data - Reference Nos., Crews, & City Cost Indexes

T5 MECHANICAL

15120	Piping Specialties	CREW	DAILY OUTPUT	LABOR-HOURS	UNIT	2002 BARE COSTS				TOTAL INCL O&P
						MAT.	LABOR	EQUIP.	TOTAL	
120										**120**
1630	2-1/2" diameter	Q-5	4.20	3.810	Ea.	1,150	124		1,274	1,450
1640	3" diameter	↓	4	4		1,525	130		1,655	1,875
1650	4" diameter	↓	2.80	5.714		1,650	186		1,836	2,075
1660	5" diameter	Q-6	3.10	7.742		2,775	262		3,037	3,450
1670	6" diameter	"	2.90	8.276	↓	2,950	280		3,230	3,675
1800	With drain/dismantle/cleaning access flange									
1810	Flanged									
1820	2" diameter	1 Stpi	7	1.143	Ea.	2,100	41.50		2,141.50	2,400
1830	2-1/2" diameter	Q-5	4	4		2,175	130		2,305	2,600
1840	3" diameter		3.60	4.444		2,850	145		2,995	3,350
1850	4" diameter	↓	2.50	6.400		3,000	209		3,209	3,625
1860	5" diameter	Q-6	2.80	8.571		5,550	290		5,840	6,550
1870	6" diameter	"	2.60	9.231	↓	5,925	310		6,235	7,000
2000	Boiler air fitting, in line separator									
2010	1-1/4", 19 GPM	Q-5	22	.727	Ea.	80	23.50		103.50	124
2020	1-1/2", 29 GPM		20	.800		90.50	26		116.50	140
2021	2", 50 GPM	↓	18	.889	↓	410	29		439	495
2040	Boiler fitting, top outlet									
2050	2", 12 GPM	Q-5	18	.889	Ea.	119	29		148	175
2060	2", 19 GPM		17	.941		80	30.50		110.50	134
2070	2", 26 GPM		16	1		88	32.50		120.50	146
2100	3", 46 GPM		8	2		116	65		181	226
2110	3", 68 GPM		6	2.667		123	87		210	266
2130	4", 125 GPM	↓	4.50	3.556		144	116		260	335
2150	6", 220 GPM	Q-6	3.80	6.316		263	213		476	610
2160	6", 640 GPM	"	3.60	6.667	↓	375	225		600	750
2400	Compression tank air fitting									
2410	For tanks 9" to 24" diameter	1 Stpi	15	.533	Ea.	33.50	19.30		52.80	66
2420	For tanks 100 gallon or larger	"	12	.667	"	140	24		164	191
140 0010	**AIR PURGING SCOOP** with tappings									**140**
0020	for air vent and expansion tank connection									
0100	1" pipe size, threaded	1 Stpi	19	.421	Ea.	13.60	15.25		28.85	38
0110	1-1/4" pipe size, threaded		15	.533		13.70	19.30		33	44
0120	1-1/2" pipe size, threaded		13	.615		30	22.50		52.50	66.50
0130	2" pipe size, threaded	↓	11	.727		34	26.50		60.50	77
0140	2-1/2" pipe size, threaded	Q-5	15	1.067		76.50	35		111.50	137
0150	3" pipe size, threaded		13	1.231		95	40		135	166
0160	4" 150 lb. flanges	↓	3	5.333	↓	214	174		388	495
160 0010	**AUTOMATIC AIR VENT**									**160**
0020	Cast iron body, stainless steel internals, float type									
0060	1/2" NPT inlet, 300 psi	1 Stpi	12	.667	Ea.	66.50	24		90.50	110
0140	3/4" NPT inlet, 300 psi		12	.667		66.50	24		90.50	110
0180	1/2" NPT inlet, 250 psi		10	.800		212	29		241	277
0220	3/4" NPT inlet, 250 psi		10	.800		212	29		241	277
0260	1" NPT inlet, 250 psi	↓	10	.800		310	29		339	385
0340	1-1/2" NPT inlet, 250 psi	Q-5	12	1.333		660	43.50		703.50	790
0380	2" NPT inlet, 250 psi	"	12	1.333	↓	660	43.50		703.50	790
0600	Forged steel body, stainless steel internals, float type									
0640	1/2" NPT inlet, 750 psi	1 Stpi	12	.667	Ea.	655	24		679	760
0680	3/4" NPT inlet, 750 psi		12	.667		655	24		679	760
0760	3/4" NPT inlet, 1000 psi	↓	10	.800		985	29		1,014	1,125
0800	1" NPT inlet, 1000 psi	Q-5	12	1.333		985	43.50		1,028.50	1,150
0880	1-1/2" NPT inlet, 1000 psi		10	1.600		2,775	52		2,827	3,125
0920	2" NPT inlet, 1000 psi	↓	10	1.600	↓	2,775	52		2,827	3,125

MECHANICAL 15

15120	Piping Specialties	CREW	DAILY OUTPUT	LABOR-HOURS	UNIT	2002 BARE COSTS				TOTAL INCL O&P
						MAT.	LABOR	EQUIP.	TOTAL	
160	1100	Formed steel body, non corrosive									**160**
	1110	1/8" NPT inlet 150 psi	1 Stpi	32	.250	Ea.	6.90	9.05		15.95	21.50
	1120	1/4" NPT inlet 150 psi		32	.250		23	9.05		32.05	39
	1130	3/4" NPT inlet 150 psi	↓	32	.250	↓	23	9.05		32.05	39
	1300	Chrome plated brass, automatic/manual, for radiators									
	1310	1/8" NPT inlet, nickel plated brass	1 Stpi	32	.250	Ea.	4.03	9.05		13.08	18.10
180	0010	**CIRCUIT SENSOR** Flow meter									**180**
	0020	Metering stations									
	0040	Wafer orifice insert type									
	0060	2-1/2" pipe size	Q-5	12	1.333	Ea.	123	43.50		166.50	201
	0100	3" pipe size		11	1.455		137	47.50		184.50	222
	0140	4" pipe size		8	2		154	65		219	268
	0180	5" pipe size	↓	7.30	2.192		202	71.50		273.50	330
	0220	6" pipe size		6.40	2.500		235	81.50		316.50	380
	0260	8" pipe size	Q-6	5.30	4.528		330	153		483	590
	0280	10" pipe size		4.60	5.217		380	176		556	685
	0360	12" pipe size	↓	4.20	5.714	↓	630	193		823	985
	2000	In-line probe type									
	2200	Copper, soldered									
	2210	1/2" size	1 Stpi	24	.333	Ea.	107	12.05		119.05	135
	2220	3/4" size		20	.400		111	14.50		125.50	144
	2230	1" size		19	.421		126	15.25		141.25	161
	2240	1-1/4" size		15	.533		132	19.30		151.30	175
	2250	1-1/2" size		13	.615		155	22.50		177.50	205
	2260	2" size	↓	11	.727		174	26.50		200.50	231
	2270	2-1/2" size	Q-5	15	1.067		184	35		219	256
	2280	3" size	"	13	1.231	↓	200	40		240	281
	2500	Brass, threaded									
	2510	1" size	1 Stpi	19	.421	Ea.	132	15.25		147.25	169
	2520	1-1/4" size		15	.533		146	19.30		165.30	190
	2530	1-1/2" size		13	.615		175	22.50		197.50	227
	2540	2" size	↓	11	.727		184	26.50		210.50	243
	2550	2-1/2" size	Q-5	15	1.067		201	35		236	274
	2560	3" size	"	13	1.231	↓	205	40		245	286
	2700	Steel, threaded									
	2710	1" size	1 Stpi	19	.421	Ea.	127	15.25		142.25	162
	2720	1-1/4" size		15	.533		138	19.30		157.30	181
	2730	1-1/2" size		13	.615		156	22.50		178.50	206
	2740	2" size	↓	11	.727		174	26.50		200.50	231
	2750	2-1/2" size	Q-5	15	1.067		184	35		219	256
	2760	3" size	"	13	1.231	↓	204	40		244	286
	3000	Pitot tube probe type									
	3100	Weld-on mounting with probe									
	3110	2" size	Q-17	18	.889	Ea.	143	29	4.44	176.44	205
	3120	2-1/2" size		18	.889		146	29	4.44	179.44	209
	3130	3" size		18	.889		155	29	4.44	188.44	219
	3140	4" size		18	.889		205	29	4.44	238.44	273
	3150	5" size		17	.941		230	30.50	4.71	265.21	305
	3160	6" size		17	.941		297	30.50	4.71	332.21	375
	3170	8" size		16	1		425	32.50	5	462.50	520
	3180	10" size		15	1.067		460	35	5.35	500.35	565
	3190	12" size		15	1.067		490	35	5.35	530.35	600
	3200	14" size		14	1.143		560	37	5.70	602.70	680
	3210	16" size		14	1.143		665	37	5.70	707.70	790
	3220	18" size		13	1.231		710	40	6.15	756.15	845
	3230	20" size	↓	13	1.231	↓	765	40	6.15	811.15	910

15 MECHANICAL

Important: See the Reference Section for critical supporting data - Reference Nos., Crews, & City Cost Indexes

15120	Piping Specialties	CREW	DAILY OUTPUT	LABOR-HOURS	UNIT	2002 BARE COSTS				TOTAL INCL O&P
						MAT.	LABOR	EQUIP.	TOTAL	
180 3240	24" size	Q-17	12	1.333	Ea.	860	43.50	6.65	910.15	1,025 **180**
3400	Weld-on wet tap with probe									
3410	2" size, 5/16" probe dia.	Q-17	14	1.143	Ea.	345	37	5.70	387.70	440
3420	2-1/2" size, 5/16" probe dia.		14	1.143		355	37	5.70	397.70	450
3430	3" size, 5/16" probe dia.		14	1.143		375	37	5.70	417.70	470
3440	4" size, 5/16" probe dia.		14	1.143		385	37	5.70	427.70	480
3450	5" size, 3/8" probe dia.		13	1.231		400	40	6.15	446.15	505
3460	6" size, 3/8" probe dia.		13	1.231		425	40	6.15	471.15	535
3470	8" size, 1/2" probe dia.		12	1.333		445	43.50	6.65	495.15	560
3480	10" size, 3/4" probe dia.		11	1.455		470	47.50	7.25	524.75	600
3490	12" size, 3/4" probe dia.		11	1.455		515	47.50	7.25	569.75	645
3500	14" size, 1" probe dia.		10	1.600		580	52	8	640	725
3510	16" size, 1" probe dia.		10	1.600		695	52	8	755	850
3520	18" size, 1" probe dia.		9	1.778		745	58	8.90	811.90	915
3530	20" size, 1" probe dia.		8	2		800	65	10	875	990
3540	24" size, 1" probe dia.	▼	8	2	▼	895	65	10	970	1,100
3700	Clamp-on wet tap with probe									
3710	2" size, 5/16" probe dia.	Q-5	16	1	Ea.	405	32.50		437.50	495
3720	2-1/2" size, 5/16" probe dia.		16	1		420	32.50		452.50	515
3730	3" size, 5/16" probe dia.		15	1.067		435	35		470	530
3740	4" size, 5/16" probe dia.		15	1.067		500	35		535	605
3750	5" size, 3/8" probe dia.		14	1.143		525	37		562	635
3760	6" size, 3/8" probe dia.		14	1.143		560	37		597	670
3770	8" size, 1/2" probe dia.		13	1.231		615	40		655	740
3780	10" size, 3/4" probe dia.		12	1.333		645	43.50		688.50	775
3790	12" size, 3/4" probe dia.		12	1.333		720	43.50		763.50	860
3800	14" size, 1" probe dia.		11	1.455		975	47.50		1,022.50	1,150
3810	16" size, 1" probe dia.		11	1.455		1,125	47.50		1,172.50	1,325
3820	18" size, 1" probe dia.		10	1.600		1,200	52		1,252	1,400
3830	20" size, 1" probe dia		9	1.778		1,325	58		1,383	1,550
3840	24" size, 1" probe dia.	▼	9	1.778	▼	1,425	58		1,483	1,675
4000	Wet tap drills									
4010	3/8" dia. for 5/16" probe				Ea.	83			83	91.50
4020	1/2" dia. for 3/8" probe					86.50			86.50	95
4030	5/8" dia for 1/2" probe					103			103	113
4040	7/8" dia for 3/4" probe					134			134	147
4050	1-3/16" dia for 1" probe				▼	143			143	157
4100	Wet tap punch									
4110	5/16" dia for 5/16" probe				Ea.	86.50			86.50	95
4120	3/8" dia for 3/8" probe					93			93	102
4130	1/2" dia for 1/2" probe				▼	111			111	123
4150	Note: labor for wet tap drill or punch									
4160	is included with the mounting and probe assembly									
9000	Readout instruments									
9200	Gauges									
9220	GPM gauge				Ea.	1,400			1,400	1,525
9230	Dual gauge kit				"	1,450			1,450	1,600
200 0010	**CIRCUIT SETTER** Balance valve									**200**
0012	Bronze body, soldered									
0013	1/2" size	1 Stpi	24	.333	Ea.	43	12.05		55.05	65
0014	3/4" size	"	20	.400	"	45	14.50		59.50	71.50
0018	Threaded									
0019	1/2" pipe size	1 Stpi	22	.364	Ea.	44.50	13.15		57.65	69
0020	3/4" pipe size		20	.400		47.50	14.50		62	74
0040	1" pipe size	▼	18	.444	▼	61	16.10		77.10	92

MECHANICAL 15

15120	Piping Specialties	CREW	DAILY OUTPUT	LABOR-HOURS	UNIT	2002 BARE COSTS				TOTAL INCL O&P
						MAT.	LABOR	EQUIP.	TOTAL	
200										**200**
0050	1-1/4" pipe size	1 Stpi	15	.533	Ea.	90.50	19.30		109.80	129
0060	1-1/2" pipe size		12	.667		106	24		130	154
0080	2" pipe size	↓	10	.800		152	29		181	211
0100	2-1/2" pipe size	Q-5	15	1.067		345	35		380	435
0120	3" pipe size	"	10	1.600	↓	485	52		537	615
0130	Cast iron body, flanged									
0136	3" pipe size, flanged	Q-5	4	4	Ea.	510	130		640	760
0140	4" pipe size	"	3	5.333		725	174		899	1,050
0200	For differential meter, accurate to 1%, add					590			590	650
0300	For differential meter 400 psi/200° F continuous, add				↓	1,450			1,450	1,600
220	0010	**COCKS, DRAINS & SPECIALTIES**									**220**
	1000	Boiler drain									
	1010	Pipe thread to hose									
	1020	Bronze									
	1030	1/2" size	1 Stpi	36	.222	Ea.	4.10	8.05		12.15	16.65
	1040	3/4" size	"	34	.235	"	4.44	8.50		12.94	17.75
	1100	Solder to hose									
	1110	Bronze									
	1120	1/2" size	1 Stpi	46	.174	Ea.	4.10	6.30		10.40	14
	1130	3/4" size	"	44	.182	"	4.44	6.60		11.04	14.85
	1600	With built-in vacuum breaker									
	1610	1/2" IP or solder	1 Stpi	36	.222	Ea.	13.25	8.05		21.30	26.50
	1630	With tamper proof vacuum breaker									
	1640	1/2" IP or solder	1 Stpi	36	.222	Ea.	16.45	8.05		24.50	30.50
	1650	3/4" IP or solder	"	34	.235	"	16.75	8.50		25.25	31.50
	3000	Cocks									
	3010	Air, lever or tee handle									
	3020	Bronze, single thread									
	3030	1/8" size	1 Stpi	52	.154	Ea.	5.75	5.55		11.30	14.70
	3040	1/4" size		46	.174		6.20	6.30		12.50	16.35
	3050	3/8" size		40	.200		6.95	7.25		14.20	18.55
	3060	1/2" size	↓	36	.222	↓	7	8.05		15.05	19.85
	3100	Bronze, double thread									
	3110	1/8" size	1 Stpi	26	.308	Ea.	7.70	11.15		18.85	25.50
	3120	1/4" size		22	.364		7.45	13.15		20.60	28
	3130	3/8" size		18	.444		8.15	16.10		24.25	33.50
	3140	1/2" size	↓	15	.533	↓	9.75	19.30		29.05	39.50
	4000	Steam									
	4010	Bronze									
	4020	1/8" size	1 Stpi	26	.308	Ea.	14.25	11.15		25.40	32.50
	4030	1/4" size		22	.364		13.75	13.15		26.90	35
	4040	3/8" size		18	.444		13.45	16.10		29.55	39.50
	4050	1/2" size		15	.533		12.60	19.30		31.90	43
	4060	3/4" size		14	.571		13.90	20.50		34.40	46.50
	4070	1" size		13	.615		25.50	22.50		48	61.50
	4080	1-1/4" size		11	.727		37	26.50		63.50	80
	4090	1-1/2" size	↓	10	.800	↓	46.50	29		75.50	94.50
	4100	2" size		9	.889		68.50	32		100.50	124
	4300	Bronze, 3 way									
	4320	1/4" size	1 Stpi	15	.533	Ea.	27.50	19.30		46.80	59
	4330	3/8" size		12	.667		28.50	24		52.50	68
	4340	1/2" size		10	.800		16.85	29		45.85	62
	4350	3/4" size		9.50	.842		29	30.50		59.50	78
	4360	1" size		8.50	.941		35	34		69	90
	4370	1-1/4" size		7.50	1.067		52.50	38.50		91	116
	4380	1-1/2" size	↓	6.50	1.231	↓	61	44.50		105.50	134

15 MECHANICAL

Important: See the Reference Section for critical supporting data - Reference Nos., Crews, & City Cost Indexes

			CREW	DAILY OUTPUT	LABOR-HOURS	UNIT	2002 BARE COSTS				TOTAL INCL O&P	
		15120 Piping Specialties					MAT.	LABOR	EQUIP.	TOTAL		
220	4390	2" size	1 Stpi	6	1.333	Ea.	90	48.50		138.50	172	**220**
	4500	Gauge cock, brass										
	4510	1/4" FPT	1 Stpi	24	.333	Ea.	6.15	12.05		18.20	25	
	4512	1/4" MPT	"	24	.333	"	7.15	12.05		19.20	26	
	4600	Pigtail, steam syphon										
	4604	1/4"	1 Stpi	24	.333	Ea.	9.80	12.05		21.85	29	
	4650	Snubber valve										
	4654	1/4"	1 Stpi	22	.364	Ea.	9.40	13.15		22.55	30	
	4660	Nipple, black steel										
	4664	1/4" x 3"	1 Stpi	37	.216	Ea.	.41	7.85		8.26	12.25	
250	0010	**DIELECTRIC UNIONS** Standard gaskets for water and air										**250**
	0020	250 psi maximum pressure										
	0280	Female IPT to sweat, straight										
	0300	1/2" pipe size	1 Plum	24	.333	Ea.	3.15	12		15.15	21.50	
	0340	3/4" pipe size		20	.400		3.15	14.40		17.55	25	
	0360	1" pipe size		19	.421		6.10	15.15		21.25	29.50	
	0380	1-1/4" pipe size		15	.533		10	19.15		29.15	40	
	0400	1-1/2" pipe size		13	.615		15	22		37	50	
	0420	2" pipe size	▼	11	.727	▼	21	26		47	62.50	
	0580	Female IPT to brass pipe thread, straight										
	0600	1/2" pipe size	1 Plum	24	.333	Ea.	5.95	12		17.95	24.50	
	0640	3/4" pipe size		20	.400		7.60	14.40		22	30	
	0660	1" pipe size		19	.421		11.40	15.15		26.55	35.50	
	0680	1-1/4" pipe size		15	.533		17.05	19.15		36.20	48	
	0700	1-1/2" pipe size		13	.615		23.50	22		45.50	59.50	
	0720	2" pipe size	▼	11	.727	▼	30.50	26		56.50	73.50	
	0780	Female IPT to female IPT, straight										
	0800	1/2" pipe size	1 Plum	24	.333	Ea.	4.75	12		16.75	23.50	
	0840	3/4" pipe size		20	.400		5.30	14.40		19.70	27.50	
	0860	1" pipe size		19	.421		7	15.15		22.15	30.50	
	0880	1-1/4" pipe size		15	.533		9.75	19.15		28.90	40	
	0900	1-1/2" pipe size		13	.615		14.95	22		36.95	50	
	0920	2" pipe size	▼	11	.727	▼	21.50	26		47.50	63	
	2000	175 psi maximum pressure										
	2180	Female IPT to sweat										
	2200	1-1/2" pipe size	1 Plum	11	.727	Ea.	38.50	26		64.50	81.50	
	2240	2" pipe size	"	9	.889		44	32		76	96.50	
	2260	2-1/2" pipe size	Q-1	15	1.067		48	34.50		82.50	105	
	2280	3" pipe size		14	1.143		65.50	37		102.50	128	
	2300	4" pipe size	▼	11	1.455		174	47		221	263	
	2320	5" pipe size	Q-2	14	1.714		284	57.50		341.50	395	
	2340	6" pipe size		12	2		330	67		397	460	
	2360	8" pipe size	▼	10	2.400	▼	910	80.50		990.50	1,125	
	2480	Female IPT to brass pipe										
	2500	1-1/2" pipe size	1 Plum	11	.727	Ea.	58	26		84	103	
	2540	2" pipe size	"	9	.889		69	32		101	124	
	2560	2-1/2" pipe size	Q-1	15	1.067		100	34.50		134.50	162	
	2580	3" pipe size		14	1.143		116	37		153	184	
	2600	4" pipe size	▼	11	1.455		199	47		246	290	
	2620	5" pipe size	Q-2	14	1.714		224	57.50		281.50	335	
	2640	6" pipe size		12	2		267	67		334	395	
	2680	8" pipe size	▼	10	2.400	▼	435	80.50		515.50	600	
	2880	Female IPT to female IPT										
	2900	1-1/2" pipe size	1 Plum	11	.727	Ea.	36	26		62	79	
	2940	2" pipe size	"	9	.889		38.50	32		70.50	90.50	
	2960	2-1/2" pipe size	Q-1	15	1.067		48	34.50		82.50	105	

MECHANICAL 15

15120	Piping Specialties	CREW	DAILY OUTPUT	LABOR-HOURS	UNIT	2002 BARE COSTS				TOTAL INCL O&P
						MAT.	LABOR	EQUIP.	TOTAL	
250	**2980**	3" pipe size	Q-1	14	1.143	Ea.	60	37		97	123	**250**
	3000	4" pipe size	↓	11	1.455		91	47		138	171
	3020	5" pipe size	Q-2	14	1.714		105	57.50		162.50	203
	3040	6" pipe size		12	2		141	67		208	256
	3060	8" pipe size	↓	10	2.400	↓	192	80.50		272.50	330
	3380	Copper to copper									
	3400	1-1/2" pipe size	1 Plum	11	.727	Ea.	47.50	26		73.50	92
	3440	2" pipe size	"	9	.889		54	32		86	108
	3460	2-1/2" pipe size	Q-1	15	1.067		58	34.50		92.50	116
	3480	3" pipe size		14	1.143		72.50	37		109.50	136
	3500	4" pipe size	↓	11	1.455		230	47		277	325
	3520	5" pipe size	Q-2	14	1.714		370	57.50		427.50	490
	3540	6" pipe size		12	2		450	67		517	595
	3560	8" pipe size	↓	10	2.400		1,200	80.50		1,280.50	1,450
280	**0010**	**EXPANSION COUPLINGS** Hydronic										**280**
	0100	Copper to copper, sweat									
	0120	3/4" diameter	1 Stpi	20	.400	Ea.	28	14.50		42.50	53
	0140	1" diameter		19	.421		35.50	15.25		50.75	62
	0160	1-1/4" diameter	↓	15	.533	↓	48.50	19.30		67.80	82.50
	1000	Baseboard riser fitting, 5" stub by coupling 12" long									
	1020	1/2" diameter	1 Stpi	24	.333	Ea.	10.50	12.05		22.55	30
	1040	3/4" diameter		20	.400		14.90	14.50		29.40	38.50
	1060	1" diameter		19	.421		20.50	15.25		35.75	46
	1080	1-1/4" diameter	↓	15	.533	↓	27	19.30		46.30	59
	1180	9" Stub by tubing 8" long									
	1200	1/2" diameter	1 Stpi	24	.333	Ea.	9.40	12.05		21.45	28.50
	1220	3/4" diameter		20	.400		12.45	14.50		26.95	35.50
	1240	1" diameter		19	.421		17.30	15.25		32.55	42
	1260	1-1/4" diameter	↓	15	.533	↓	24	19.30		43.30	55.50
300	**0010**	**EXPANSION JOINTS**										**300**
	0100	Bellows type, neoprene cover, flanged spool									
	0120	6" face to face, 1" diameter	1 Stpi	13	.615	Ea.	158	22.50		180.50	208
	0140	1-1/4" diameter		11	.727		161	26.50		187.50	217
	0160	1-1/2" diameter	↓	10.60	.755		161	27.50		188.50	218
	0180	2" diameter	Q-5	13.30	1.203		163	39		202	238
	0190	2-1/2" diameter		12.40	1.290		170	42		212	251
	0200	3" diameter		11.40	1.404		193	45.50		238.50	281
	0210	4" diameter		8.40	1.905		206	62		268	320
	0220	5" diameter		7.60	2.105		254	68.50		322.50	380
	0230	6" diameter		6.80	2.353		261	76.50		337.50	405
	0240	8" diameter		5.40	2.963		299	96.50		395.50	475
	0250	10" diameter		5	3.200		400	104		504	595
	0260	12" diameter		4.60	3.478		455	113		568	670
	0480	10" face to face, 2" diameter		13	1.231		240	40		280	325
	0500	2-1/2" diameter		12	1.333		253	43.50		296.50	345
	0520	3" diameter		11	1.455		257	47.50		304.50	355
	0540	4" diameter		8	2		287	65		352	415
	0560	5" diameter		7	2.286		390	74.50		464.50	540
	0580	6" diameter		6	2.667		355	87		442	525
	0600	8" diameter		5	3.200		425	104		529	620
	0620	10" diameter		4.60	3.478		455	113		568	670
	0640	12" diameter		4	4		520	130		650	765
	0660	14" diameter		3.80	4.211		650	137		787	920
	0680	16" diameter		2.90	5.517		745	180		925	1,100
	0700	18" diameter	↓	2.50	6.400	↓	840	209		1,049	1,250

Important: See the Reference Section for critical supporting data - Reference Nos., Crews, & City Cost Indexes

15120	Piping Specialties	CREW	DAILY OUTPUT	LABOR-HOURS	UNIT	2002 BARE COSTS				TOTAL INCL O&P
						MAT.	LABOR	EQUIP.	TOTAL	
300	0720	20" diameter	Q-5	2.10	7.619	Ea.	880	248		1,128	1,350	300
	0740	24" diameter		1.80	8.889		1,000	290		1,290	1,550
	0760	26" diameter		1.40	11.429		1,125	370		1,495	1,775
	0780	30" diameter		1.20	13.333		1,225	435		1,660	2,000
	0800	36" diameter	▼	1	16	▼	1,525	520		2,045	2,450
	1000	Bellows with internal sleeves and external covers									
	1010	Stainless steel, 150 lb.									
	1020	With male threads									
	1030	1/2" diameter	1 Stpi	22	.364	Ea.	71.50	13.15		84.65	98.50
	1040	3/4" diameter		20	.400		71.50	14.50		86	101
	1050	1" diameter		19	.421		76	15.25		91.25	107
	1060	1-1/4" diameter		16	.500		79	18.10		97.10	114
	1070	1-1/2" diameter		13	.615		96	22.50		118.50	140
	1080	2" diameter	▼	11	.727	▼	122	26.50		148.50	174
	1110	With flanged ends									
	1120	1-1/4" diameter	1 Stpi	12	.667	Ea.	310	24		334	375
	1130	1-1/2" diameter		11	.727		310	26.50		336.50	380
	1140	2" diameter	▼	9	.889		495	32		527	595
	1150	3" diameter	Q-5	8	2		500	65		565	650
	1160	4" diameter	"	5	3.200		535	104		639	740
	1170	5" diameter	Q-6	6	4		570	135		705	830
	1180	6" diameter		5	4.800		625	162		787	930
	1190	8" diameter		4.50	5.333		750	180		930	1,100
	1200	10" diameter		4	6		915	203		1,118	1,300
	1210	12" diameter	▼	3.60	6.667	▼	1,175	225		1,400	1,625
320	0010	**EXPANSION TANKS**									320
	1400	Plastic, corrosion resistant, see Plumbing Cost Data									
	1410	Div. 15220									
	1505	Fiberglass and steel single / double wall storage, see Div 13201									
	1510	Tank leak detection systems, see Div 13851-350									
	2000	Steel, liquid expansion, ASME, painted, 15 gallon capacity	Q-5	17	.941	Ea.	350	30.50		380.50	430
	2020	24 gallon capacity		14	1.143		355	37		392	445
	2040	30 gallon capacity		12	1.333		385	43.50		428.50	490
	2060	40 gallon capacity		10	1.600		370	52		422	485
	2080	60 gallon capacity		8	2		530	65		595	680
	2100	80 gallon capacity		7	2.286		545	74.50		619.50	710
	2120	100 gallon capacity		6	2.667		700	87		787	900
	2130	120 gallon capacity		5	3.200		780	104		884	1,025
	2140	135 gallon capacity		4.50	3.556		815	116		931	1,075
	2150	175 gallon capacity		4	4		1,125	130		1,255	1,450
	2160	220 gallon capacity		3.60	4.444		1,300	145		1,445	1,650
	2170	240 gallon capacity		3.30	4.848		1,450	158		1,608	1,825
	2180	305 gallon capacity		3	5.333		1,950	174		2,124	2,400
	2190	400 gallon capacity	▼	2.80	5.714	▼	2,175	186		2,361	2,675
	2360	Galvanized									
	2370	15 gallon capacity	Q-5	17	.941	Ea.	510	30.50		540.50	605
	2380	24 gallon capacity		14	1.143		640	37		677	760
	2390	30 gallon capacity		12	1.333		635	43.50		678.50	765
	2400	40 gallon capacity		10	1.600		725	52		777	880
	2410	60 gallon capacity		8	2		835	65		900	1,025
	2420	80 gallon capacity		7	2.286		1,125	74.50		1,199.50	1,325
	2430	100 gallon capacity		6	2.667		1,375	87		1,462	1,625
	2440	120 gallon capacity		5	3.200		1,600	104		1,704	1,900
	2450	135 gallon capacity		4.50	3.556		1,700	116		1,816	2,050
	2460	175 gallon capacity	▼	4	4	▼	2,075	130		2,205	2,500

MECHANICAL 15

		15120 | **Piping Specialties**	CREW	DAILY OUTPUT	LABOR-HOURS	UNIT	2002 BARE COSTS				TOTAL INCL O&P	
							MAT.	LABOR	EQUIP.	TOTAL		
320	2470	220 gallon capacity	Q-5	3.60	4.444	Ea.	2,400	145		2,545	2,875	**320**
	2480	240 gallon capacity		3.30	4.848		2,500	158		2,658	3,000	
	2490	305 gallon capacity		3	5.333		3,400	174		3,574	3,975	
	2500	400 gallon capacity		2.80	5.714		4,125	186		4,311	4,825	
	3000	Steel ASME expansion, rubber diaphragm, 19 gal. cap. accept.		12	1.333		1,275	43.50		1,318.50	1,475	
	3020	31 gallon capacity		8	2		1,450	65		1,515	1,700	
	3040	61 gallon capacity		6	2.667		2,050	87		2,137	2,375	
	3060	79 gallon capacity		5	3.200		2,175	104		2,279	2,550	
	3080	119 gallon capacity		4	4		2,300	130		2,430	2,750	
	3100	158 gallon capacity		3.80	4.211		3,250	137		3,387	3,775	
	3120	211 gallon capacity		3.30	4.848		3,750	158		3,908	4,350	
	3140	317 gallon capacity		2.80	5.714		4,900	186		5,086	5,650	
	3160	422 gallon capacity		2.60	6.154		7,225	200		7,425	8,250	
	3180	528 gallon capacity	▼	2.40	6.667	▼	7,925	217		8,142	9,050	
350	0010	**FLEXIBLE CONNECTORS**, Corrugated, 7/8" O.D., 1/2" I.D.										**350**
	0050	Gas, seamless brass, steel fittings										
	0200	12" long	1 Plum	36	.222	Ea.	10.45	8		18.45	23.50	
	0220	18" long		36	.222		12.95	8		20.95	26.50	
	0240	24" long		34	.235		15.35	8.45		23.80	29.50	
	0260	30" long		34	.235		16.55	8.45		25	31	
	0280	36" long		32	.250		18.30	9		27.30	33.50	
	0320	48" long		30	.267		23	9.60		32.60	40	
	0340	60" long		30	.267		27.50	9.60		37.10	45	
	0360	72" long	▼	30	.267	▼	32	9.60		41.60	49.50	
	2000	Water, copper tubing, dielectric separators										
	2100	12" long	1 Plum	36	.222	Ea.	8.95	8		16.95	22	
	2220	15" long		36	.222		9.95	8		17.95	23	
	2240	18" long		36	.222		10.80	8		18.80	24	
	2260	24" long	▼	34	.235	▼	13.25	8.45		21.70	27.50	
370	0010	**FLEXIBLE METAL HOSE** Connectors, standard lengths										**370**
	0100	Bronze braided, bronze ends										
	0120	3/8" diameter x 12"	1 Stpi	26	.308	Ea.	15.40	11.15		26.55	34	
	0140	1/2" diameter x 12"		24	.333		14.65	12.05		26.70	34.50	
	0160	3/4" diameter x 12"		20	.400		21.50	14.50		36	45.50	
	0180	1" diameter x 18"		19	.421		27.50	15.25		42.75	53.50	
	0200	1-1/2" diameter x 18"		13	.615		44	22.50		66.50	82	
	0220	2" diameter x 18"	▼	11	.727	▼	53	26.50		79.50	97.50	
	1000	Carbon steel ends										
	1020	1/4" diameter x 12"	1 Stpi	28	.286	Ea.	12.65	10.35		23	29.50	
	1040	3/8" diameter x 12"		26	.308		13	11.15		24.15	31	
	1060	1/2" diameter x 12"		24	.333		12.65	12.05		24.70	32	
	1080	1/2" diameter x 24"		24	.333		16.20	12.05		28.25	36	
	1100	1/2" diameter x 36"		24	.333		19.80	12.05		31.85	40	
	1120	3/4" diameter x 12"		20	.400		16.80	14.50		31.30	40.50	
	1140	3/4" diameter x 24"		20	.400		20.50	14.50		35	44.50	
	1160	3/4" diameter x 36"		20	.400		24.50	14.50		39	49	
	1180	1" diameter x 18"		19	.421		22	15.25		37.25	47.50	
	1200	1" diameter x 30"		19	.421		26	15.25		41.25	51.50	
	1220	1" diameter x 36"		19	.421		28	15.25		43.25	54	
	1240	1-1/4" diameter x 18"		15	.533		22.50	19.30		41.80	54	
	1260	1-1/4" diameter x 36"		15	.533		30	19.30		49.30	62	
	1280	1-1/2" diameter x 18"		13	.615		24.50	22.50		47	60	
	1300	1-1/2" diameter x 36"		13	.615		32.50	22.50		55	69	
	1320	2" diameter x 24"		11	.727		32	26.50		58.50	74.50	
	1340	2" diameter x 36"	▼	11	.727		39.50	26.50		66	83	

15 MECHANICAL

	15120	Piping Specialties	CREW	DAILY OUTPUT	LABOR-HOURS	UNIT	2002 BARE COSTS				TOTAL INCL O&P	
							MAT.	LABOR	EQUIP.	TOTAL		
370	1360	2-1/2" diameter x 24"	1 Stpi	9	.889	Ea.	45	32		77	98	**370**
	1380	2-1/2" diameter x 36"		9	.889		53	32		85	107	
	1400	3" diameter x 24"		7	1.143		53.50	41.50		95	122	
	1420	3" diameter x 36"	↓	7	1.143	↓	62	41.50		103.50	131	
	2000	Carbon steel braid, carbon steel solid ends										
	2100	1/2" diameter x 12"	1 Stpi	24	.333	Ea.	12.65	12.05		24.70	32	
	2120	3/4" diameter x 12"		20	.400		16.80	14.50		31.30	40.50	
	2140	1" diameter x 12"		19	.421		20	15.25		35.25	45	
	2160	1-1/4" diameter x 12"		15	.533		20	19.30		39.30	51	
	2180	1-1/2" diameter x 12"	↓	13	.615	↓	21.50	22.50		44	57.50	
	3000	Stainless steel braid, welded on carbon steel ends										
	3100	1/2" diameter x 12"	1 Stpi	24	.333	Ea.	18.50	12.05		30.55	38.50	
	3120	3/4" diameter x 12"		20	.400		26.50	14.50		41	51	
	3140	3/4" diameter x 24"		20	.400		32	14.50		46.50	57	
	3160	3/4" diameter x 36"		20	.400		37.50	14.50		52	63	
	3180	1" diameter x 12"		19	.421		30	15.25		45.25	56	
	3200	1" diameter x 24"		19	.421		36	15.25		51.25	63	
	3220	1" diameter x 36"		19	.421		42.50	15.25		57.75	70	
	3240	1-1/4" diameter x 12"		15	.533		32.50	19.30		51.80	65	
	3260	1-1/4" diameter x 24"		15	.533		41.50	19.30		60.80	74.50	
	3280	1-1/4" diameter x 36"		15	.533		50	19.30		69.30	84	
	3300	1-1/2" diameter x 12"		13	.615		36	22.50		58.50	73	
	3320	1-1/2" diameter x 24"		13	.615		46	22.50		68.50	84	
	3340	1-1/2" diameter x 36"	↓	13	.615	↓	56	22.50		78.50	95	
	3400	Metal sst braid, over corrugated stainless steel, flanged ends										
	3410	150 PSI										
	3420	1/2" diameter x 12"	1 Stpi	24	.333	Ea.	20	12.05		32.05	40	
	3430	1" diameter x 12"		20	.400		32	14.50		46.50	57	
	3440	1-1/2" diameter x 12"		15	.533		37	19.30		56.30	69.50	
	3450	2-1/2" diameter x 9"		12	.667		54	24		78	96	
	3460	3" diameter x 9"		9	.889		65	32		97	120	
	3470	4" diameter x 9"		7	1.143		80	41.50		121.50	151	
	3480	4" diameter x 30"		5	1.600		170	58		228	275	
	3490	4" diameter x 36"		4.80	1.667		190	60.50		250.50	300	
	3500	6" diameter x 11"		5	1.600		132	58		190	233	
	3510	6" diameter x 36"		3.80	2.105		251	76		327	390	
	3520	8" diameter x 12"		4	2		260	72.50		332.50	395	
	3530	10" diameter x 13"	↓	3	2.667		330	96.50		426.50	510	
	3540	12" diameter x 14"	Q-5	4	4	↓	655	130		785	915	
	6000	Molded rubber with helical wire reinforcement										
	6010	150 PSI										
	6020	1-1/2" diameter x 12"	1 Stpi	15	.533	Ea.	39	19.30		58.30	71.50	
	6030	2" diameter x 12"		12	.667		145	24		169	197	
	6040	3" diameter x 12"		8	1		187	36		223	261	
	6050	4" diameter x 12"		6	1.333		218	48.50		266.50	315	
	6060	6" diameter x 18"		4	2		275	72.50		347.50	415	
	6070	8" diameter x 24"		3	2.667		355	96.50		451.50	535	
	6080	10" diameter x 24"	↓	2	4		400	145		545	660	
	6090	12" diameter x 24"	Q-5	3	5.333	↓	455	174		629	760	
	7000	Molded teflon with stainless steel flanges										
	7010	150 PSI										
	7020	2-1/2" diameter x 3-3/16"	Q-1	7.80	2.051	Ea.	282	66.50		348.50	410	
	7030	3" diameter x 3-5/8"		6.50	2.462		350	79.50		429.50	505	
	7040	4" diameter x 3-5/8"		5	3.200		445	104		549	645	
	7050	6" diameter x 4"		4.30	3.721		650	120		770	895	
	7060	8" diameter x 6"	↓	3.80	4.211	↓	860	136		996	1,150	

MECHANICAL 15

15120	Piping Specialties	CREW	DAILY OUTPUT	LABOR-HOURS	UNIT	2002 BARE COSTS				TOTAL INCL O&P
						MAT.	LABOR	EQUIP.	TOTAL	
400	0010	**FLOAT VALVES**									**400**
	0020	With ball and bracket								
	0030	Single seat, threaded								
	0040	Brass body								
	0050	1/2"	1 Stpi	11	.727	Ea.	38.50	26.50		65	81.50
	0060	3/4"		9	.889		44	32		76	97
	0070	1"		7	1.143		56.50	41.50		98	125
	0080	1-1/2"		4.50	1.778		84	64.50		148.50	190
	0090	2"	▼	3.60	2.222	▼	85.50	80.50		166	215
	0300	For condensate receivers, CI, in-line mount								
	0320	1" inlet	1 Stpi	7	1.143	Ea.	97.50	41.50		139	170
	0360	For condensate receiver, CI, external float, flanged tank mount								
	0370	3/4" inlet	1 Stpi	5	1.600	Ea.	82.50	58		140.50	179
420	0010	**FLOW CHECK CONTROL**									**420**
	0100	Bronze body, soldered								
	0110	3/4" size	1 Stpi	20	.400	Ea.	35.50	14.50		50	61
	0120	1" size	"	19	.421	"	42.50	15.25		57.75	70
	0200	Cast iron body, threaded								
	0210	3/4" size	1 Stpi	20	.400	Ea.	28.50	14.50		43	53
	0220	1" size		19	.421		32	15.25		47.25	58
	0230	1-1/4" size		15	.533		39	19.30		58.30	71.50
	0240	1-1/2" size		13	.615		59.50	22.50		82	98.50
	0250	2" size	▼	11	.727	▼	85	26.50		111.50	133
	0300	Flanged inlet, threaded outlet								
	0310	2-1/2" size	1 Stpi	8	1	Ea.	189	36		225	263
	0320	3" size	Q-5	9	1.778		225	58		283	335
	0330	4" size	"	7	2.286	▼	440	74.50		514.50	590
520	0010	**HYDRONIC HEATING CONTROL VALVES**									**520**
	0050	Hot water, nonelectric, thermostatic								
	0100	Radiator supply, 1/2" diameter	1 Stpi	24	.333	Ea.	36	12.05		48.05	57.50
	0120	3/4" diameter		20	.400		36.50	14.50		51	62.50
	0140	1" diameter		19	.421		47.50	15.25		62.75	75
	0160	1-1/4" diameter	▼	15	.533	▼	66.50	19.30		85.80	103
	0500	For low pressure steam, add					25%				
	1000	Manual, radiator supply								
	1010	1/2" pipe size, angle union	1 Stpi	24	.333	Ea.	28.50	12.05		40.55	49
	1020	3/4" pipe size, angle union		20	.400		26.50	14.50		41	51.50
	1030	1" pipe size, angle union	▼	19	.421	▼	34.50	15.25		49.75	61
	1100	Radiator, balancing, straight, sweat connections								
	1110	1/2" pipe size	1 Stpi	24	.333	Ea.	8.10	12.05		20.15	27
	1120	3/4" pipe size		20	.400		11.30	14.50		25.80	34.50
	1130	1" pipe size	▼	19	.421	▼	18.40	15.25		33.65	43.50
	1200	Steam, radiator, supply								
	1210	1/2" pipe size, angle union	1 Stpi	24	.333	Ea.	25.50	12.05		37.55	46
	1220	3/4" pipe size, angle union		20	.400		28	14.50		42.50	53
	1230	1" pipe size, angle union		19	.421		35.50	15.25		50.75	62
	1240	1-1/4" pipe size, angle union	▼	15	.533		44	19.30		63.30	77.50
	8000	System balancing and shut-off								
	8020	Butterfly, quarter turn, calibrated, threaded or solder								
	8040	Bronze, -30° F to +350° F, pressure to 175 psi								
	8060	1/2" size	1 Stpi	22	.364	Ea.	9.80	13.15		22.95	30.50
	8070	3/4" size		20	.400		10.55	14.50		25.05	33.50
	8080	1" size		19	.421		11.70	15.25		26.95	36
	8090	1-1/4" size		15	.533		19.25	19.30		38.55	50
	8100	1-1/2" size	▼	13	.615	▼	24	22.50		46.50	60

15 MECHANICAL

Important: See the Reference Section for critical supporting data - Reference Nos., Crews, & City Cost Indexes

15120	Piping Specialties	CREW	DAILY OUTPUT	LABOR-HOURS	UNIT	2002 BARE COSTS				TOTAL INCL O&P	
						MAT.	LABOR	EQUIP.	TOTAL		
520 8110	2" size	1 Stpi	11	.727	Ea.	28.50	26.50		55	71	**520**
550 0010	**LIQUID DRAINERS**										**550**
0100	Guided lever type										
0110	Cast iron body, threaded										
0120	1/2" pipe size	1 Stpi	22	.364	Ea.	66.50	13.15		79.65	93	
0130	3/4" pipe size		20	.400		177	14.50		191.50	216	
0140	1" pipe size		19	.421		283	15.25		298.25	335	
0150	1-1/2" pipe size		11	.727		615	26.50		641.50	715	
0160	2" pipe size	↓	8	1	↓	615	36		651	730	
0200	Forged steel body, threaded										
0210	1/2" pipe size	1 Stpi	18	.444	Ea.	620	16.10		636.10	705	
0220	3/4" pipe size		16	.500		620	18.10		638.10	710	
0230	1" pipe size	↓	14	.571		930	20.50		950.50	1,050	
0240	1-1/2" pipe size	Q-5	10	1.600		2,700	52		2,752	3,050	
0250	2" pipe size	"	8	2	↓	2,700	65		2,765	3,075	
0300	Socket weld										
0310	1/2" pipe size	Q-17	16	1	Ea.	715	32.50	5	752.50	840	
0320	3/4" pipe size		14	1.143		715	37	5.70	757.70	845	
0330	1" pipe size		12	1.333		1,050	43.50	6.65	1,100.15	1,225	
0340	1-1/2" pipe size		9	1.778		2,800	58	8.90	2,866.90	3,200	
0350	2" pipe size	↓	7	2.286		2,800	74.50	11.45	2,885.95	3,225	
0400	Flanged										
0410	1/2" pipe size	1 Stpi	12	.667	Ea.	1,325	24		1,349	1,475	
0420	3/4" pipe size		10	.800		1,325	29		1,354	1,500	
0430	1" pipe size	↓	7	1.143		1,725	41.50		1,766.50	1,975	
0440	1-1/2" pipe size	Q-5	9	1.778		3,775	58		3,833	4,275	
0450	2" pipe size	"	7	2.286	↓	3,775	74.50		3,849.50	4,275	
1000	Fixed lever and snap action type										
1100	Cast iron body, threaded										
1110	1/2" pipe size	1 Stpi	22	.364	Ea.	104	13.15		117.15	134	
1120	3/4" pipe size		20	.400		274	14.50		288.50	320	
1130	1" pipe size	↓	19	.421	↓	274	15.25		289.25	325	
1200	Forged steel body, threaded										
1210	1/2" pipe size	1 Stpi	18	.444	Ea.	670	16.10		686.10	765	
1220	3/4" pipe size		16	.500		670	18.10		688.10	770	
1230	1" pipe size		14	.571		1,650	20.50		1,670.50	1,850	
1240	1-1/4" pipe size	↓	12	.667	↓	1,650	24		1,674	1,850	
2000	High pressure high leverage type										
2100	Forged steel body, threaded										
2110	1/2" pipe size	1 Stpi	18	.444	Ea.	1,325	16.10		1,341.10	1,475	
2120	3/4" pipe size		16	.500		1,325	18.10		1,343.10	1,475	
2130	1" pipe size		14	.571		1,550	20.50		1,570.50	1,725	
2140	1-1/4" pipe size	↓	12	.667		2,200	24		2,224	2,450	
2150	1-1/2" pipe size	Q-5	10	1.600		3,350	52		3,402	3,750	
2160	2" pipe size	"	8	2	↓	3,350	65		3,415	3,775	
2200	Socket weld										
2210	1/2" pipe size	Q-17	16	1	Ea.	1,425	32.50	5	1,462.50	1,625	
2220	3/4" pipe size		14	1.143		1,425	37	5.70	1,467.70	1,625	
2230	1" pipe size		12	1.333		2,325	43.50	6.65	2,375.15	2,625	
2240	1-1/4" pipe size		11	1.455		2,325	47.50	7.25	2,379.75	2,625	
2250	1-1/2" pipe size		9	1.778		3,450	58	8.90	3,516.90	3,900	
2260	2" pipe size	↓	7	2.286	↓	3,450	74.50	11.45	3,535.95	3,925	
2300	Flanged										
2310	1/2" pipe size	1 Stpi	12	.667	Ea.	2,125	24		2,149	2,350	
2320	3/4" pipe size	↓	10	.800	↓	2,125	29		2,154	2,375	

MECHANICAL 15

				DAILY	LABOR-		2002 BARE COSTS				TOTAL	
	15120	**Piping Specialties**	CREW	OUTPUT	HOURS	UNIT	MAT.	LABOR	EQUIP.	TOTAL	INCL O&P	
550	2330	1" pipe size	1 Stpi	7	1.143	Ea.	3,250	41.50		3,291.50	3,675	550
	2340	1-1/4" pipe size	↓	6	1.333		3,250	48.50		3,298.50	3,675	
	2350	1-1/2" pipe size	Q-5	9	1.778		4,425	58		4,483	4,950	
	2360	2" pipe size	"	7	2.286	↓	4,425	74.50		4,499.50	4,950	
	3000	Guided lever dual gravity type										
	3100	Cast iron body, threaded										
	3110	1/2" pipe size	1 Stpi	22	.364	Ea.	255	13.15		268.15	300	
	3120	3/4" pipe size		20	.400		255	14.50		269.50	300	
	3130	1" pipe size		19	.421		370	15.25		385.25	435	
	3140	1-1/2" pipe size		11	.727		720	26.50		746.50	830	
	3150	2" pipe size	↓	8	1	↓	720	36		756	845	
	3200	Forged steel body, threaded										
	3210	1/2" pipe size	1 Stpi	18	.444	Ea.	720	16.10		736.10	820	
	3220	3/4" pipe size		16	.500		715	18.10		733.10	815	
	3230	1" pipe size	↓	14	.571		1,075	20.50		1,095.50	1,200	
	3240	1-1/2" pipe size	Q-5	10	1.600		2,850	52		2,902	3,200	
	3250	2" pipe size	"	8	2	↓	2,850	65		2,915	3,225	
	3300	Socket weld										
	3310	1/2" pipe size	Q-17	16	1	Ea.	815	32.50	5	852.50	950	
	3320	3/4" pipe size		14	1.143		815	37	5.70	857.70	955	
	3330	1" pipe size		12	1.333		1,175	43.50	6.65	1,225.15	1,375	
	3340	1-1/2" pipe size		9	1.778		2,950	58	8.90	3,016.90	3,350	
	3350	2" pipe size	↓	7	2.286	↓	2,950	74.50	11.45	3,035.95	3,375	
	3400	Flanged										
	3410	1/2" pipe size	1 Stpi	12	.667	Ea.	1,425	24		1,449	1,600	
	3420	3/4" pipe size		10	.800		1,425	29		1,454	1,625	
	3430	1" pipe size	↓	7	1.143		1,850	41.50		1,891.50	2,125	
	3440	1-1/2" pipe size	Q-5	9	1.778		3,925	58		3,983	4,425	
	3450	2" pipe size	"	7	2.286	↓	3,925	74.50		3,999.50	4,425	
580	0010	**MIXING VALVE** Automatic water tempering										580
	0040	1/2" size	1 Stpi	19	.421	Ea.	375	15.25		390.25	440	
	0050	3/4" size		18	.444		375	16.10		391.10	440	
	0100	1" size		16	.500		540	18.10		558.10	625	
	0120	1-1/4" size		13	.615		780	22.50		802.50	895	
	0140	1-1/2" size		10	.800		930	29		959	1,075	
	0160	2" size		8	1		1,175	36		1,211	1,350	
	0170	2-1/2" size		6	1.333		1,175	48.50		1,223.50	1,375	
	0180	3" size		4	2		2,700	72.50		2,772.50	3,075	
	0190	4" size	↓	3	2.667	↓	2,700	96.50		2,796.50	3,125	
610	0010	**MONOFLOW TEE FITTING**										610
	1100	For one pipe hydronic, supply and return										
	1110	Copper, soldered										
	1120	3/4" x 1/2" size	1 Stpi	13	.615	Ea.	12.40	22.50		34.90	47	
	1130	1" x 1/2" size		12	.667		12.40	24		36.40	50	
	1140	1" x 3/4" size		11	.727		12.40	26.50		38.90	53	
	1150	1-1/4" x 1/2" size		11	.727		13.15	26.50		39.65	54	
	1160	1-1/4" x 3/4" size		10	.800		13.15	29		42.15	58	
	1170	1-1/2" x 3/4" size		10	.800		18.40	29		47.40	63.50	
	1180	1-1/2" x 1" size		9	.889		18.40	32		50.40	68.50	
	1190	2" x 3/4" size		9	.889		28	32		60	79	
	1200	2" x 1" size	↓	8	1	↓	28	36		64	85	
640	0010	**PRESSURE REDUCING VALVE** Steam, pilot operated										640
	0100	Threaded, iron body										
	0120	1/2" size	1 Stpi	18	.444	Ea.	565	16.10		581.10	650	
	0140	3/4" size	↓	14	.571		475	20.50		495.50	550	

Important: See the Reference Section for critical supporting data - Reference Nos., Crews, & City Cost Indexes

T5
MECHANICAL

			DAILY	LABOR-		2002 BARE COSTS				TOTAL	
15120		**Piping Specialties**	CREW	OUTPUT	HOURS	UNIT	MAT.	LABOR	EQUIP.	TOTAL	INCL O&P
640	0160	1" size	1 Stpi	13	.615	Ea.	505	22.50		527.50	590	**640**
	0180	1-1/4" size		10	.800		590	29		619	695
	0200	1-1/2" size		8	1		640	36		676	760
	0220	2" size	▼	5	1.600	▼	835	58		893	1,000
	1000	Flanged, iron body, 125 lb. flanges									
	1020	2" size	1 Stpi	8	1	Ea.	880	36		916	1,025
	1040	2-1/2" size	"	4	2		1,100	72.50		1,172.50	1,325
	1060	3" size	Q-5	4.50	3.556		1,125	116		1,241	1,425
	1080	4" size	"	3	5.333		1,425	174		1,599	1,825
	1500	For 250 lb. flanges, add				▼	5%				
670	0010	**PRESSURE REGULATOR**										**670**
	0100	Gas appliance regulators									
	0106	Main burner and pilot applications									
	0108	Rubber seat poppet type									
	0109	1/8" pipe size	1 Stpi	24	.333	Ea.	10.60	12.05		22.65	30
	0110	1/4" pipe size		24	.333		12.15	12.05		24.20	31.50
	0112	3/8" pipe size		24	.333		13.10	12.05		25.15	32.50
	0113	1/2" pipe size		24	.333		14.55	12.05		26.60	34
	0114	3/4" pipe size	▼	20	.400	▼	17.45	14.50		31.95	41
	0122	Lever action type									
	0123	3/8" pipe size	1 Stpi	24	.333	Ea.	21.50	12.05		33.55	41.50
	0124	1/2" pipe size		24	.333		21.50	12.05		33.55	41.50
	0125	3/4" pipe size		20	.400		43	14.50		57.50	69.50
	0126	1" pipe size	▼	19	.421	▼	43	15.25		58.25	70.50
	0132	Double diaphragm type									
	0133	3/8" pipe size	1 Stpi	24	.333	Ea.	24.50	12.05		36.55	45
	0134	1/2" pipe size		24	.333		37.50	12.05		49.55	59
	0135	3/4" pipe size		20	.400		64	14.50		78.50	92.50
	0136	1" pipe size		19	.421		78.50	15.25		93.75	109
	0137	1-1/4" pipe size		15	.533		255	19.30		274.30	310
	0138	1-1/2" pipe size		13	.615		475	22.50		497.50	555
	0139	2" pipe size	▼	11	.727		475	26.50		501.50	560
	0140	2-1/2" pipe size	Q-5	15	1.067		925	35		960	1,075
	0141	3" pipe size		13	1.231		925	40		965	1,075
	0142	4" pipe size (flanged)	▼	8	2	▼	1,700	65		1,765	1,975
	0160	Main burner only									
	0162	Straight-thru-flow design									
	0163	1/2" pipe size	1 Stpi	24	.333	Ea.	27.50	12.05		39.55	48.50
	0164	3/4" pipe size		20	.400		37.50	14.50		52	63.50
	0165	1" pipe size		19	.421		56	15.25		71.25	84.50
	0166	1-1/4" pipe size	▼	15	.533	▼	56	19.30		75.30	90.50
	0200	Oil, light, hot water, ordinary steam, threaded									
	0220	Bronze body, 1/4" size	1 Stpi	24	.333	Ea.	100	12.05		112.05	128
	0230	3/8" size		24	.333		108	12.05		120.05	137
	0240	1/2" size		24	.333		134	12.05		146.05	165
	0250	3/4" size		20	.400		160	14.50		174.50	198
	0260	1" size		19	.421		243	15.25		258.25	291
	0270	1-1/4" size		15	.533		325	19.30		344.30	390
	0280	1-1/2" size		13	.615		350	22.50		372.50	425
	0290	2" size		11	.727		640	26.50		666.50	745
	0320	Iron body, 1/4" size		24	.333		78	12.05		90.05	104
	0330	3/8" size		24	.333		92	12.05		104.05	119
	0340	1/2" size		24	.333		97.50	12.05		109.55	125
	0350	3/4" size		20	.400		118	14.50		132.50	152
	0360	1" size		19	.421		154	15.25		169.25	193
	0370	1-1/4" size	▼	15	.533	▼	214	19.30		233.30	264

MECHANICAL 15

			DAILY	LABOR-		2002 BARE COSTS				TOTAL	
15120	**Piping Specialties**	CREW	OUTPUT	HOURS	UNIT	MAT.	LABOR	EQUIP.	TOTAL	INCL O&P	
670	0380	1-1/2″ size	1 Stpi	13	.615	Ea.	238	22.50		260.50	296	**670**
	0390	2″ size	↓	11	.727	↓	345	26.50		371.50	420	
	0500	Oil, heavy, viscous fluids, threaded										
	0520	Bronze body, 3/8″ size	1 Stpi	24	.333	Ea.	172	12.05		184.05	207	
	0530	1/2″ size		24	.333		221	12.05		233.05	261	
	0540	3/4″ size		20	.400		245	14.50		259.50	292	
	0550	1″ size		19	.421		305	15.25		320.25	365	
	0560	1-1/4″ size		15	.533		435	19.30		454.30	505	
	0570	1-1/2″ size		13	.615		495	22.50		517.50	580	
	0600	Iron body, 3/8″ size		24	.333		137	12.05		149.05	168	
	0620	1/2″ size		24	.333		166	12.05		178.05	200	
	0630	3/4″ size		20	.400		188	14.50		202.50	228	
	0640	1″ size		19	.421		222	15.25		237.25	267	
	0650	1-1/4″ size		15	.533		320	19.30		339.30	385	
	0660	1-1/2″ size	↓	13	.615	↓	345	22.50		367.50	415	
	0800	Process steam, wet or super heated, monel trim, threaded										
	0820	Bronze body, 1/4″ size	1 Stpi	24	.333	Ea.	287	12.05		299.05	335	
	0830	3/8″ size		24	.333		287	12.05		299.05	335	
	0840	1/2″ size		24	.333		315	12.05		327.05	365	
	0850	3/4″ size		20	.400		350	14.50		364.50	400	
	0860	1″ size		19	.421		445	15.25		460.25	515	
	0870	1-1/4″ size		15	.533		540	19.30		559.30	620	
	0880	1-1/2″ size		13	.615		645	22.50		667.50	745	
	0890	2″ size		11	.727		935	26.50		961.50	1,075	
	0920	Iron body, max 125 PSIG press out, 1/4″ size		24	.333		98	12.05		110.05	126	
	0930	3/8″ size		24	.333		116	12.05		128.05	145	
	0940	1/2″ size		24	.333		123	12.05		135.05	153	
	0950	3/4″ size		20	.400		148	14.50		162.50	185	
	0960	1″ size		19	.421		194	15.25		209.25	236	
	0970	1-1/4″ size		15	.533		269	19.30		288.30	325	
	0980	1-1/2″ size		13	.615		299	22.50		321.50	365	
	0990	2″ size	↓	11	.727	↓	435	26.50		461.50	515	
	1000	Flanged, Class 125										
	1006	2-1/2″ size	Q-5	10	1.600	Ea.	1,100	52		1,152	1,300	
	1008	3″ size		9	1.778		1,125	58		1,183	1,350	
	1010	4″ size		8	2		1,425	65		1,490	1,675	
	1020	6″ size	↓	7	2.286	↓	4,575	74.50		4,649.50	5,125	
	3000	Steam, high capacity, bronze body, stainless steel trim										
	3020	Threaded, 1/2″ diameter	1 Stpi	24	.333	Ea.	705	12.05		717.05	795	
	3030	3/4″ diameter		24	.333		705	12.05		717.05	795	
	3040	1″ diameter		19	.421		790	15.25		805.25	895	
	3060	1-1/4″ diameter		15	.533		870	19.30		889.30	990	
	3080	1-1/2″ diameter		13	.615		995	22.50		1,017.50	1,125	
	3100	2″ diameter	↓	11	.727		1,225	26.50		1,251.50	1,400	
	3120	2-1/2″ diameter	Q-5	12	1.333		1,525	43.50		1,568.50	1,750	
	3140	3″ diameter	″	11	1.455	↓	1,750	47.50		1,797.50	2,000	
	3500	Flanged connection, iron body, 125 lb. W.S.P.										
	3520	3″ diameter	Q-5	11	1.455	Ea.	1,900	47.50		1,947.50	2,175	
	3540	4″ diameter	″	5	3.200	″	2,400	104		2,504	2,800	
	9000	For water pressure regulators, see Div. 15110-160										
700	0010	**PRESSURE & TEMPERATURE SAFETY PLUG**										**700**
	1000	3/4″ external thread, 3/8″ diam. element										
	1020	Carbon steel										
	1050	7-1/2″ insertion	1 Stpi	32	.250	Ea.	36.50	9.05		45.55	53.50	
	1120	304 stainless steel										
	1150	7-1/2″ insertion	1 Stpi	32	.250	Ea.	36.50	9.05		45.55	53.50	

15 MECHANICAL

Important: See the Reference Section for critical supporting data - Reference Nos., Crews, & City Cost Indexes

15120	Piping Specialties	CREW	DAILY OUTPUT	LABOR-HOURS	UNIT	MAT.	LABOR	EQUIP.	TOTAL	TOTAL INCL O&P	
700 1220	316 Stainless steel										**700**
1250	7-1/2" insertion	1 Stpi	32	.250	Ea.	45	9.05		54.05	63	
730 0010	**SLEEVES AND ESCUTCHEONS**										**730**
0100	Pipe sleeve										
0110	Steel, w/water stop, 12" long, with link seal										
0120	2" diam. for 1/2" carrier pipe	1 Plum	8.40	.952	Ea.	25	34		59	79	
0130	2-1/2" diam. for 3/4" carrier pipe		8	1		30	36		66	87	
0140	2-1/2" diam. for 1" carrier pipe		8	1		27.50	36		63.50	84.50	
0150	3" diam. for 1-1/4" carrier pipe		7.20	1.111		39	40		79	103	
0160	3-1/2" diam. for 1-1/2" carrier pipe		6.80	1.176		40.50	42.50		83	109	
0170	4" diam. for 2" carrier pipe		6	1.333		45	48		93	122	
0180	4" diam. for 2-1/2" carrier pipe		6	1.333		44.50	48		92.50	122	
0190	5" diam. for 3" carrier pipe		5.40	1.481		56	53.50		109.50	142	
0200	6" diam. for 4" carrier pipe		4.80	1.667		65	60		125	162	
0210	10" diam. for 6" carrier pipe	Q-1	8	2		121	64.50		185.50	231	
0220	12" diam. for 8" carrier pipe		7.20	2.222		174	72		246	299	
0230	14" diam. for 10" carrier pipe		6.40	2.500		183	81		264	325	
0240	16" diam. for 12" carrier pipe		5.80	2.759		214	89		303	370	
0250	18" diam. for 14" carrier pipe		5.20	3.077		241	99.50		340.50	415	
0260	24" diam. for 18" carrier pipe		4	4		400	129		529	635	
0270	24" diam. for 20" carrier pipe		4	4		325	129		454	550	
0280	30" diam. for 24" carrier pipe		3.20	5		735	162		897	1,050	
0500	Wall sleeve										
0510	Ductile iron with rubber gasket seal										
0520	3"	1 Plum	8.40	.952	Ea.	96	34		130	158	
0530	4"		7.20	1.111		120	40		160	192	
0540	6"		6	1.333		161	48		209	250	
0550	8"		4	2		206	72		278	335	
0560	10"		3	2.667		231	96		327	400	
0570	12"		2.40	3.333		274	120		394	480	
5000	Escutcheon										
5100	Split ring, pipe										
5110	Chrome plated										
5120	1/2"	1 Plum	160	.050	Ea.	.95	1.80		2.75	3.76	
5130	3/4"		160	.050		.98	1.80		2.78	3.79	
5140	1"		135	.059		1.06	2.13		3.19	4.38	
5150	1-1/2"		115	.070		1.60	2.50		4.10	5.55	
5160	2"		100	.080		1.66	2.88		4.54	6.15	
5170	4"		80	.100		3.69	3.60		7.29	9.45	
5180	6"		68	.118		6.10	4.23		10.33	13.15	
760 0010	**STEAM TRAP**										**760**
0030	Cast iron body, threaded										
0040	Inverted bucket										
0050	1/2" pipe size	1 Stpi	12	.667	Ea.	84.50	24		108.50	130	
0070	3/4" pipe size		10	.800		140	29		169	198	
0100	1" pipe size		9	.889		224	32		256	296	
0120	1-1/4" pipe size		8	1		340	36		376	425	
0130	1-1/2" pipe size		7	1.143		375	41.50		416.50	480	
0140	2" pipe size		6	1.333		560	48.50		608.50	695	
0200	With thermic vent & check valve										
0202	1/2" pipe size	1 Stpi	12	.667	Ea.	138	24		162	189	
0204	3/4" pipe size		10	.800		204	29		233	269	
0206	1" pipe size		9	.889		299	32		331	380	
0208	1-1/4" pipe size		8	1		445	36		481	545	
0210	1-1/2" pipe size		7	1.143		605	41.50		646.50	730	
0212	2" pipe size		6	1.333		920	48.50		968.50	1,075	

MECHANICAL 15

15120	Piping Specialties	CREW	DAILY OUTPUT	LABOR-HOURS	UNIT	2002 BARE COSTS				TOTAL INCL O&P
						MAT.	LABOR	EQUIP.	TOTAL	
760											760
1000	Float & thermostatic, 15 psi									
1010	3/4" pipe size	1 Stpi	16	.500	Ea.	97.50	18.10		115.60	135
1020	1" pipe size		15	.533		97.50	19.30		116.80	136
1030	1-1/4" pipe size		13	.615		120	22.50		142.50	166
1040	1-1/2" pipe size		9	.889		174	32		206	241
1060	2" pipe size		6	1.333		320	48.50		368.50	430
1100	For vacuum breaker, add					43.50			43.50	48
1110	Differential condensate controller									
1120	With union and manual metering valve									
1130	1/2" pipe size	1 Stpi	24	.333	Ea.	350	12.05		362.05	405
1140	3/4" pipe size		20	.400		405	14.50		419.50	470
1150	1" pipe size		19	.421		550	15.25		565.25	630
1160	1-1/4" pipe size		15	.533		740	19.30		759.30	840
1170	1-1/2" pipe size		13	.615		890	22.50		912.50	1,025
1180	2" pipe size		11	.727		1,175	26.50		1,201.50	1,325
1290	Brass body, threaded									
1300	Thermostatic, angle union, 25 psi									
1310	1/2" pipe size	1 Stpi	24	.333	Ea.	38.50	12.05		50.55	60
1320	3/4" pipe size		20	.400		68.50	14.50		83	97
1330	1" pipe size		19	.421		99	15.25		114.25	132
1990	Carbon steel body, threaded									
2000	Controlled disc, integral strainer & blow down									
2010	Threaded									
2020	3/8" pipe size	1 Stpi	24	.333	Ea.	154	12.05		166.05	187
2030	1/2" pipe size		24	.333		154	12.05		166.05	187
2040	3/4" pipe size		20	.400		207	14.50		221.50	249
2050	1" pipe size		19	.421		265	15.25		280.25	315
2100	Socket weld									
2110	3/8" pipe size	Q-17	17	.941	Ea.	184	30.50	4.71	219.21	253
2120	1/2" pipe size		16	1		184	32.50	5	221.50	257
2130	3/4" pipe size		14	1.143		239	37	5.70	281.70	325
2140	1" pipe size		12	1.333		300	43.50	6.65	350.15	405
2170	Flanged, 600 lb. ASA									
2180	3/8" pipe size	1 Stpi	12	.667	Ea.	455	24		479	535
2190	1/2" pipe size		12	.667		455	24		479	535
2200	3/4" pipe size		10	.800		545	29		574	640
2210	1" pipe size		7	1.143		640	41.50		681.50	770
5000	Forged steel body									
5010	Inverted bucket									
5020	Threaded									
5030	1/2" pipe size	1 Stpi	12	.667	Ea.	274	24		298	335
5040	3/4" pipe size		10	.800		475	29		504	565
5050	1" pipe size		9	.889		710	32		742	830
5060	1-1/4" pipe size		8	1		1,025	36		1,061	1,175
5070	1-1/2" pipe size		7	1.143		1,475	41.50		1,516.50	1,675
5080	2" pipe size		6	1.333		2,575	48.50		2,623.50	2,900
5100	Socket weld									
5110	1/2" pipe size	Q-17	16	1	Ea.	330	32.50	5	367.50	420
5120	3/4" pipe size		14	1.143		530	37	5.70	572.70	645
5130	1" pipe size		12	1.333		770	43.50	6.65	820.15	920
5140	1-1/4" pipe size		10	1.600		1,100	52	8	1,160	1,275
5150	1-1/2" pipe size		9	1.778		1,525	58	8.90	1,591.90	1,775
5160	2" pipe size		7	2.286		2,650	74.50	11.45	2,735.95	3,025
5200	Flanged									
5210	1/2" pipe size	1 Stpi	12	.667	Ea.	555	24		579	645
5220	3/4" pipe size		10	.800		895	29		924	1,025

Important: See the Reference Section for critical supporting data - Reference Nos., Crews, & City Cost Indexes

15120	Piping Specialties	CREW	DAILY OUTPUT	LABOR-HOURS	UNIT	2002 BARE COSTS				TOTAL INCL O&P	
						MAT.	LABOR	EQUIP.	TOTAL		
760	5230	1" pipe size	1 Stpi	7	1.143	Ea.	1,275	41.50		1,316.50	1,475	**760**
	5240	1-1/4" pipe size	↓	6	1.333		1,750	48.50		1,798.50	2,025
	5250	1-1/2" pipe size	Q-5	9	1.778		2,200	58		2,258	2,525
	5260	2" pipe size	"	7	2.286	↓	3,350	74.50		3,424.50	3,800
	6000	Stainless steel body									
	6010	Inverted bucket									
	6020	Threaded									
	6030	1/2" pipe size	1 Stpi	12	.667	Ea.	91	24		115	137
	6040	3/4" pipe size		10	.800		119	29		148	175
	6050	1" pipe size	↓	9	.889	↓	425	32		457	515
	6100	Threaded with thermic vent									
	6110	1/2" pipe size	1 Stpi	12	.667	Ea.	104	24		128	151
	6120	3/4" pipe size		10	.800		134	29		163	191
	6130	1" pipe size	↓	9	.889	↓	440	32		472	535
	6200	Threaded with check valve									
	6210	1/2" pipe size	1 Stpi	12	.667	Ea.	130	24		154	180
	6220	3/4" pipe size	"	10	.800	"	177	29		206	238
	6300	Socket weld									
	6310	1/2" pipe size	Q-17	16	1	Ea.	106	32.50	5	143.50	171
	6320	3/4" pipe size	"	14	1.143	"	136	37	5.70	178.70	211
790	0010	**STRAINERS, BASKET TYPE** Perforated stainless steel basket										**790**
	0100	Brass or monel available									
	2000	Simplex style									
	2300	Bronze body									
	2320	Screwed, 3/8" pipe size	1 Stpi	22	.364	Ea.	98	13.15		111.15	128
	2340	1/2" pipe size		20	.400		100	14.50		114.50	132
	2360	3/4" pipe size		17	.471		103	17.05		120.05	139
	2380	1" pipe size		15	.533		134	19.30		153.30	176
	2400	1-1/4" pipe size		13	.615		201	22.50		223.50	255
	2420	1-1/2" pipe size		12	.667		204	24		228	261
	2440	2" pipe size	↓	10	.800		310	29		339	385
	2460	2-1/2" pipe size	Q-5	15	1.067		440	35		475	540
	2480	3" pipe size	"	14	1.143		635	37		672	755
	2600	Flanged, 2" pipe size	1 Stpi	6	1.333		465	48.50		513.50	590
	2620	2-1/2" pipe size	Q-5	4.50	3.556		730	116		846	980
	2640	3" pipe size		3.50	4.571		840	149		989	1,150
	2660	4" pipe size	↓	3	5.333		1,425	174		1,599	1,825
	2680	5" pipe size	Q-6	3.40	7.059		2,150	238		2,388	2,700
	2700	6" pipe size		3	8		2,725	270		2,995	3,375
	2710	8" pipe size	↓	2.50	9.600	↓	4,500	325		4,825	5,450
	3600	Iron body									
	3700	Screwed, 3/8" pipe size	1 Stpi	22	.364	Ea.	112	13.15		125.15	143
	3720	1/2" pipe size		20	.400		112	14.50		126.50	145
	3740	3/4" pipe size		17	.471		147	17.05		164.05	188
	3760	1" pipe size		15	.533		147	19.30		166.30	191
	3780	1-1/4" pipe size		13	.615		197	22.50		219.50	250
	3800	1-1/2" pipe size		12	.667		216	24		240	274
	3820	2" pipe size	↓	10	.800		258	29		287	325
	3840	2-1/2" pipe size	Q-5	15	1.067		370	35		405	460
	3860	3" pipe size	"	14	1.143		420	37		457	515
	4000	Flanged, 2" pipe size	1 Stpi	6	1.333		350	48.50		398.50	460
	4020	2-1/2" pipe size	Q-5	4.50	3.556		475	116		591	700
	4040	3" pipe size		3.50	4.571		505	149		654	780
	4060	4" pipe size	↓	3	5.333		760	174		934	1,100
	4080	5" pipe size	Q-6	3.40	7.059		1,200	238		1,438	1,675
	4100	6" pipe size		3	8	↓	1,475	270		1,745	2,025

MECHANICAL 15

15120	Piping Specialties	CREW	DAILY OUTPUT	LABOR-HOURS	UNIT	2002 BARE COSTS				TOTAL INCL O&P	
						MAT.	LABOR	EQUIP.	TOTAL		
790	4120	8″ pipe size	Q-6	2.50	9.600	Ea.	2,725	325		3,050	3,500	790
	4140	10″ pipe size		2.20	10.909		4,100	370		4,470	5,050	
	4160	12″ pipe size		1.70	14.118		4,875	475		5,350	6,100	
	4180	14″ pipe size		1.40	17.143		6,150	580		6,730	7,625	
	4200	16″ pipe size	▼	1	24	▼	6,425	810		7,235	8,300	
	6000	Cast steel body										
	6400	Screwed, 1″ pipe size	1 Stpi	15	.533	Ea.	241	19.30		260.30	294	
	6410	1-1/4″ pipe size		13	.615		365	22.50		387.50	435	
	6420	1-1/2″ pipe size		12	.667		365	24		389	435	
	6440	2″ pipe size	▼	10	.800		515	29		544	610	
	6460	2-1/2″ pipe size	Q-5	15	1.067		720	35		755	850	
	6480	3″ pipe size	″	14	1.143		945	37		982	1,100	
	6560	Flanged, 2″ pipe size	1 Stpi	6	1.333		750	48.50		798.50	900	
	6580	2-1/2″ pipe size	Q-5	4.50	3.556		1,100	116		1,216	1,375	
	6600	3″ pipe size		3.50	4.571		1,200	149		1,349	1,525	
	6620	4″ pipe size	▼	3	5.333		1,675	174		1,849	2,100	
	6640	6″ pipe size	Q-6	3	8		3,050	270		3,320	3,750	
	6660	8″ pipe size	″	2.50	9.600	▼	4,950	325		5,275	5,950	
	7000	Stainless steel body										
	7200	Screwed, 1″ pipe size	1 Stpi	15	.533	Ea.	395	19.30		414.30	465	
	7210	1-1/4″ pipe size		13	.615		575	22.50		597.50	670	
	7220	1-1/2″ pipe size		12	.667		575	24		599	670	
	7240	2″ pipe size	▼	10	.800		845	29		874	975	
	7260	2-1/2″ pipe size	Q-5	15	1.067		1,400	35		1,435	1,600	
	7280	3″ pipe size	″	14	1.143		1,850	37		1,887	2,100	
	7400	Flanged, 2″ pipe size	1 Stpi	6	1.333		1,150	48.50		1,198.50	1,350	
	7420	2-1/2″ pipe size	Q-5	4.50	3.556		2,100	116		2,216	2,500	
	7440	3″ pipe size		3.50	4.571		2,150	149		2,299	2,600	
	7460	4″ pipe size	▼	3	5.333		3,375	174		3,549	3,975	
	7480	6″ pipe size	Q-6	3	8		5,850	270		6,120	6,850	
	7500	8″ pipe size	″	2.50	9.600	▼	9,925	325		10,250	11,400	
	8100	Duplex style										
	8200	Bronze body										
	8240	Screwed, 3/4″ pipe size	1 Stpi	16	.500	Ea.	1,050	18.10		1,068.10	1,175	
	8260	1″ pipe size		14	.571		1,050	20.50		1,070.50	1,175	
	8280	1-1/4″ pipe size		12	.667		2,050	24		2,074	2,275	
	8300	1-1/2″ pipe size		11	.727		2,050	26.50		2,076.50	2,300	
	8320	2″ pipe size	▼	9	.889		3,250	32		3,282	3,625	
	8340	2-1/2″ pipe size	Q-5	14	1.143		3,250	37		3,287	3,625	
	8420	Flanged, 2″ pipe size	1 Stpi	6	1.333		3,675	48.50		3,723.50	4,125	
	8440	2-1/2″ pipe size	Q-5	4.50	3.556		3,775	116		3,891	4,325	
	8460	3″ pipe size		3.50	4.571		5,200	149		5,349	5,925	
	8480	4″ pipe size	▼	3	5.333		8,200	174		8,374	9,275	
	8500	5″ pipe size	Q-6	3.40	7.059		14,300	238		14,538	16,100	
	8520	6″ pipe size	″	3	8	▼	18,000	270		18,270	20,200	
	8700	Iron body										
	8740	Screwed, 3/4″ pipe size	1 Stpi	16	.500	Ea.	545	18.10		563.10	630	
	8760	1″ pipe size		14	.571		545	20.50		565.50	630	
	8780	1-1/4″ pipe size		12	.667		950	24		974	1,075	
	8800	1-1/2″ pipe size		11	.727		950	26.50		976.50	1,100	
	8820	2″ pipe size	▼	9	.889		1,625	32		1,657	1,825	
	8840	2-1/2″ pipe size	Q-5	14	1.143		1,625	37		1,662	1,825	
	9000	Flanged, 2″ pipe size	1 Stpi	6	1.333		1,700	48.50		1,748.50	1,950	
	9020	2-1/2″ pipe size	Q-5	4.50	3.556		1,775	116		1,891	2,125	
	9040	3″ pipe size		3.50	4.571		2,250	149		2,399	2,700	
	9060	4″ pipe size	▼	3	5.333		3,550	174		3,724	4,175	

Important: See the Reference Section for critical supporting data - Reference Nos., Crews, & City Cost Indexes

15120 | Piping Specialties

		CREW	DAILY OUTPUT	LABOR-HOURS	UNIT	2002 BARE COSTS				TOTAL INCL O&P
						MAT.	LABOR	EQUIP.	TOTAL	
790	9080	5" pipe size	Q-6	3.40	7.059	Ea.	7,150	238		7,388	8,225
	9100	6" pipe size		3	8		7,150	270		7,420	8,275
	9120	8" pipe size		2.50	9.600		13,700	325		14,025	15,600
	9140	10" pipe size		2.20	10.909		19,100	370		19,470	21,600
	9160	12" pipe size		1.70	14.118		21,200	475		21,675	24,000
	9170	14" pipe size		1.40	17.143		25,700	580		26,280	29,200
	9180	16" pipe size	▼	1	24	▼	32,200	810		33,010	36,700
	9300	Cast steel body									
	9340	Screwed, 1" pipe size	1 Stpi	14	.571	Ea.	1,525	20.50		1,545.50	1,700
	9360	1-1/2" pipe size		11	.727		2,250	26.50		2,276.50	2,525
	9380	2" pipe size		9	.889		3,825	32		3,857	4,250
	9460	Flanged, 2" pipe size	▼	6	1.333		3,825	48.50		3,873.50	4,300
	9480	2-1/2" pipe size	Q-5	4.50	3.556		4,225	116		4,341	4,825
	9500	3" pipe size		3.50	4.571		5,575	149		5,724	6,375
	9520	4" pipe size	▼	3	5.333		8,175	174		8,349	9,250
	9540	6" pipe size	Q-6	3	8		14,000	270		14,270	15,800
	9560	8" pipe size	"	2.50	9.600	▼	27,700	325		28,025	31,000
	9700	Stainless steel body									
	9740	Screwed, 1" pipe size	1 Stpi	14	.571	Ea.	2,775	20.50		2,795.50	3,075
	9760	1-1/2" pipe size		11	.727		4,200	26.50		4,226.50	4,675
	9780	2" pipe size		9	.889		6,775	32		6,807	7,500
	9860	Flanged, 2" pipe size	▼	6	1.333		7,050	48.50		7,098.50	7,825
	9880	2-1/2" pipe size	Q-5	4.50	3.556		7,400	116		7,516	8,325
	9900	3" pipe size		3.50	4.571		11,600	149		11,749	13,000
	9920	4" pipe size	▼	3	5.333		14,200	174		14,374	15,900
	9940	6" pipe size	Q-6	3	8		25,000	270		25,270	27,900
	9960	8" pipe size	"	2.50	9.600	▼	46,300	325		46,625	51,500
820	0010	**STRAINERS, Y TYPE** Bronze body									
	0050	Screwed, 150 lb., 1/4" pipe size	1 Stpi	24	.333	Ea.	11.40	12.05		23.45	31
	0070	3/8" pipe size		24	.333		15.90	12.05		27.95	35.50
	0100	1/2" pipe size		20	.400		15.90	14.50		30.40	39.50
	0120	3/4" pipe size		19	.421		17.40	15.25		32.65	42
	0140	1" pipe size		17	.471		18.70	17.05		35.75	46
	0150	1-1/4" pipe size		15	.533		42	19.30		61.30	75
	0160	1-1/2" pipe size		14	.571		40.50	20.50		61	75.50
	0180	2" pipe size		13	.615		53.50	22.50		76	92.50
	0182	3" pipe size		12	.667		286	24		310	350
	0200	300 lb., 2-1/2" pipe size	Q-5	17	.941		286	30.50		316.50	360
	0220	3" pipe size		16	1		565	32.50		597.50	675
	0240	4" pipe size	▼	15	1.067	▼	1,300	35		1,335	1,475
	0500	For 300 lb rating 1/4" thru 2", add					15%				
	1000	Flanged, 150 lb., 1-1/2" pipe size	1 Stpi	11	.727	Ea.	305	26.50		331.50	375
	1020	2" pipe size	"	8	1		390	36		426	485
	1030	2-1/2" pipe size	Q-5	5	3.200		505	104		609	710
	1040	3" pipe size		4.50	3.556		620	116		736	860
	1060	4" pipe size	▼	3	5.333		945	174		1,119	1,275
	1080	5" pipe size	Q-6	3.40	7.059		1,875	238		2,113	2,425
	1100	6" pipe size		3	8		2,500	270		2,770	3,150
	1106	8" pipe size	▼	2.60	9.231	▼	3,900	310		4,210	4,775
	1500	For 300 lb rating, add					40%				
840	0010	**STRAINERS, Y TYPE** Iron body									
	0050	Screwed, 250 lb., 1/4" pipe size	1 Stpi	20	.400	Ea.	6.65	14.50		21.15	29.50
	0070	3/8" pipe size		20	.400		6.30	14.50		20.80	29
	0100	1/2" pipe size	▼	20	.400	▼	6.65	14.50		21.15	29.50

15120	Piping Specialties	CREW	DAILY OUTPUT	LABOR-HOURS	UNIT	2002 BARE COSTS				TOTAL INCL O&P
						MAT.	LABOR	EQUIP.	TOTAL	
840	0120	3/4" pipe size	1 Stpi	18	.444	Ea.	7.95	16.10		24.05	33.50	**840**
	0140	1" pipe size		16	.500		10.75	18.10		28.85	39.50
	0150	1-1/4" pipe size		15	.533		14.75	19.30		34.05	45
	0160	1-1/2" pipe size		12	.667		17.75	24		41.75	56
	0180	2" pipe size		8	1		27.50	36		63.50	84.50
	0200	2-1/2" pipe size	Q-5	12	1.333		140	43.50		183.50	220
	0220	3" pipe size		11	1.455		151	47.50		198.50	238
	0240	4" pipe size		5	3.200		255	104		359	435
	0500	For galvanized body, add					50%				
	1000	Flanged, 125 lb., 1-1/2" pipe size	1 Stpi	11	.727	Ea.	86	26.50		112.50	134
	1020	2" pipe size	"	8	1		64	36		100	125
	1030	2-1/2" pipe size	Q-5	5	3.200		72.50	104		176.50	237
	1040	3" pipe size		4.50	3.556		84.50	116		200.50	268
	1060	4" pipe size		3	5.333		155	174		329	430
	1080	5" pipe size	Q-6	3.40	7.059		242	238		480	625
	1100	6" pipe size		3	8		295	270		565	735
	1120	8" pipe size		2.50	9.600		495	325		820	1,025
	1140	10" pipe size		2	12		950	405		1,355	1,650
	1160	12" pipe size		1.70	14.118		1,425	475		1,900	2,300
	1170	14" pipe size		1.30	18.462		2,625	625		3,250	3,850
	1180	16" pipe size		1	24		3,725	810		4,535	5,325
	1500	For 250 lb rating, add					20%				
	2000	For galvanized body, add					50%				
	2500	For steel body, add					40%				
870	0010	**SUCTION DIFFUSERS**										**870**
	0100	Cast iron body with integral straightening vanes, strainer									
	1000	Flanged									
	1010	2" inlet, 1-1/2" pump side	1 Stpi	6	1.333	Ea.	219	48.50		267.50	315
	1020	2" pump side	"	5	1.600		227	58		285	335
	1030	3" inlet, 2" pump side	Q-5	6.50	2.462		278	80		358	425
	1040	2-1/2" pump side		5.30	3.019		405	98.50		503.50	600
	1050	3" pump side		4.50	3.556		410	116		526	625
	1060	4" inlet, 3" pump side		3.50	4.571		480	149		629	755
	1070	4" pump side		3	5.333		560	174		734	875
	1080	5" inlet, 4" pump side	Q-6	3.80	6.316		635	213		848	1,025
	1090	5" pump side		3.40	7.059		760	238		998	1,200
	1100	6" inlet, 4" pump side		3.30	7.273		660	246		906	1,100
	1110	5" pump side		3.10	7.742		795	262		1,057	1,275
	1120	6" pump side		3	8		835	270		1,105	1,325
	1130	8" inlet, 6" pump side		2.70	8.889		900	300		1,200	1,450
	1140	8" pump side		2.50	9.600		1,550	325		1,875	2,225
	1150	10" inlet, 6" pump side		2.40	10		2,175	340		2,515	2,875
	1160	8" pump side		2.30	10.435		1,825	350		2,175	2,525
	1170	10" pump side		2.20	10.909		2,600	370		2,970	3,425
	1180	12" inlet, 8" pump side		2	12		2,750	405		3,155	3,625
	1190	10" pump side		1.80	13.333		3,050	450		3,500	4,025
	1200	12" pump side		1.70	14.118		3,425	475		3,900	4,500
	1210	14" inlet, 12" pump side		1.50	16		4,450	540		4,990	5,725
	1220	14" pump side		1.30	18.462		4,625	625		5,250	6,050
900	0010	**THERMOFLO INDICATOR** For balancing										**900**
	1000	Sweat connections, 1-1/4" pipe size	1 Stpi	12	.667	Ea.	355	24		379	430
	1020	1-1/2" pipe size		10	.800		365	29		394	445
	1040	2" pipe size		8	1		385	36		421	480
	1060	2-1/2" pipe size		7	1.143		585	41.50		626.50	710
	2000	Flange connections, 3" pipe size	Q-5	5	3.200		695	104		799	920

Important: See the Reference Section for critical supporting data - Reference Nos., Crews, & City Cost Indexes

		15120	Piping Specialties	CREW	DAILY OUTPUT	LABOR-HOURS	UNIT	2002 BARE COSTS				TOTAL INCL O&P	
								MAT.	LABOR	EQUIP.	TOTAL		
900	2020	4" pipe size		Q-5	4	4	Ea.	835	130		965	1,125	900
	2030	5" pipe size			3.50	4.571		1,050	149		1,199	1,375	
	2040	6" pipe size			3	5.333		1,125	174		1,299	1,500	
	2060	8" pipe size			2	8		1,375	261		1,636	1,900	
920	0010	**VENTURI FLOW** Measuring device											920
	0050	1/2" diameter		1 Stpi	24	.333	Ea.	175	12.05		187.05	211	
	0100	3/4" diameter			20	.400		163	14.50		177.50	201	
	0120	1" diameter			19	.421		169	15.25		184.25	209	
	0140	1-1/4" diameter			15	.533		203	19.30		222.30	252	
	0160	1-1/2" diameter			13	.615		206	22.50		228.50	261	
	0180	2" diameter			11	.727		222	26.50		248.50	284	
	0200	2-1/2" diameter		Q-5	16	1		330	32.50		362.50	415	
	0220	3" diameter			14	1.143		380	37		417	475	
	0240	4" diameter			11	1.455		595	47.50		642.50	725	
	0260	5" diameter		Q-6	4	6		720	203		923	1,100	
	0280	6" diameter			3.50	6.857		780	232		1,012	1,200	
	0300	8" diameter			3	8		1,075	270		1,345	1,575	
	0320	10" diameter			2	12		2,050	405		2,455	2,875	
	0330	12" diameter			1.80	13.333		3,400	450		3,850	4,400	
	0340	14" diameter			1.60	15		3,650	505		4,155	4,800	
	0350	16" diameter			1.40	17.143		3,950	580		4,530	5,225	
	0500	For meter, add						1,175			1,175	1,300	
940	0010	**WATER SUPPLY METERS**											940
	1000	Detector, serves dual systems such as fire and domestic or											
	1020	process water, wide range cap., UL and FM approved											
	1100	3" mainline x 2" by-pass, 400 GPM		Q-1	3.60	4.444	Ea.	5,150	144		5,294	5,875	
	1140	4" mainline x 2" by-pass, 700 GPM		"	2.50	6.400		5,150	207		5,357	5,950	
	1180	6" mainline x 3" by-pass, 1600 GPM		Q-2	2.60	9.231		7,775	310		8,085	9,025	
	1220	8" mainline x 4" by-pass, 2800 GPM			2.10	11.429		11,500	385		11,885	13,200	
	1260	10" mainline x 6" by-pass, 4400 GPM			2	12		15,700	405		16,105	17,900	
	1300	10"x12" mainlines x 6" by-pass, 5400 GPM			1.70	14.118		20,700	475		21,175	23,500	
	2000	Domestic/commercial, bronze											
	2020	Threaded											
	2060	5/8" diameter, to 20 GPM		1 Plum	16	.500	Ea.	40	18		58	71	
	2080	3/4" diameter, to 30 GPM			14	.571		67.50	20.50		88	106	
	2100	1" diameter, to 50 GPM			12	.667		94	24		118	139	
	2300	Threaded/flanged											
	2340	1-1/2" diameter, to 100 GPM		1 Plum	8	1	Ea.	330	36		366	420	
	2360	2" diameter, to 160 GPM		"	6	1.333	"	415	48		463	530	
	2600	Flanged, compound											
	2640	3" diameter, 320 GPM		Q-1	3	5.333	Ea.	1,950	173		2,123	2,375	
	2660	4" diameter, to 500 GPM			1.50	10.667		3,025	345		3,370	3,850	
	2680	6" diameter, to 1,000 GPM			1	16		4,350	520		4,870	5,550	
	2700	8" diameter, to 1,800 GPM			.80	20		8,600	645		9,245	10,400	
	7000	Turbine											
	7260	Flanged											
	7300	2" diameter, to 160 GPM		1 Plum	7	1.143	Ea.	455	41		496	560	
	7320	3" diameter, to 450 GPM		Q-1	3.60	4.444		745	144		889	1,025	
	7340	4" diameter, to 650 GPM		"	2.50	6.400		1,350	207		1,557	1,775	
	7360	6" diameter, to 1800 GPM		Q-2	2.60	9.231		2,550	310		2,860	3,275	
	7380	8" diameter, to 2500 GPM			2.10	11.429		4,050	385		4,435	5,025	
	7400	10" diameter, to 5500 GPM			1.70	14.118		5,400	475		5,875	6,675	
960	0010	**WELD END BALL JOINTS** Steel											960
	0050	2-1/2" diameter		Q-17	13	1.231	Ea.	455	40	6.15	501.15	565	
	0100	3" diameter			12	1.333		475	43.50	6.65	525.15	600	
	0120	4" diameter			11	1.455		765	47.50	7.25	819.75	925	

MECHANICAL 15

15120 | Piping Specialties

		CREW	DAILY OUTPUT	LABOR-HOURS	UNIT	2002 BARE COSTS				TOTAL INCL O&P
						MAT.	LABOR	EQUIP.	TOTAL	
960	0140	5" diameter	Q-18	14	1.714	Ea.	1,100	58	5.70	1,163.70	1,325	**960**
	0160	6" diameter		12	2		1,250	67.50	6.65	1,324.15	1,475
	0180	8" diameter		9	2.667		1,775	90	8.90	1,873.90	2,100
	0200	10" diameter		8	3		2,625	101	10	2,736	3,075
	0220	12" diameter		6	4		3,475	135	13.30	3,623.30	4,050
	0240	14" diameter	▼	4	6	▼	4,175	203	20	4,398	4,900
980	0010	**ZONE VALVES**										**980**
	1000	Bronze body									
	1010	2 way, 125 psi									
	1020	Standard, 65' pump head									
	1030	1/2" soldered	1 Stpi	24	.333	Ea.	71.50	12.05		83.55	96.50
	1040	3/4" soldered		20	.400		71.50	14.50		86	101
	1050	1" soldered		19	.421		76.50	15.25		91.75	108
	1060	1-1/4" soldered	▼	15	.533	▼	89.50	19.30		108.80	127
	1200	High head, 125' pump head									
	1210	1/2" soldered	1 Stpi	24	.333	Ea.	96.50	12.05		108.55	124
	1220	3/4" soldered		20	.400		100	14.50		114.50	132
	1230	1" soldered	▼	19	.421	▼	110	15.25		125.25	144
	1300	Geothermal, 150' pump head									
	1310	3/4" soldered	1 Stpi	20	.400	Ea.	115	14.50		129.50	149
	1320	1" soldered		19	.421		123	15.25		138.25	158
	1330	3/4" threaded	▼	20	.400	▼	115	14.50		129.50	149
	3000	3 way, 125 psi									
	3200	By-pass, 65' pump head									
	3210	1/2" soldered	1 Stpi	24	.333	Ea.	101	12.05		113.05	129
	3220	3/4" soldered		20	.400		105	14.50		119.50	137
	3230	1" soldered		19	.421		119	15.25		134.25	154
	9000	Transformer, for up to 3 zone valves	▼	20	.400	▼	14.70	14.50		29.20	38

15140 | Domestic Water Piping

		CREW	DAILY OUTPUT	LABOR-HOURS	UNIT	2002 BARE COSTS				TOTAL INCL O&P
						MAT.	LABOR	EQUIP.	TOTAL	
100	0010	**BACKFLOW PREVENTER** Includes valves										**100**
	0020	and four test cocks, corrosion resistant, automatic operation									
	1000	Double check principle									
	1010	Threaded, with ball valves									
	1020	3/4" pipe size	1 Plum	16	.500	Ea.	191	18		209	237
	1030	1" pipe size		14	.571		194	20.50		214.50	244
	1040	1-1/2" pipe size		10	.800		340	29		369	415
	1050	2" pipe size	▼	7	1.143		405	41		446	505
	1080	Threaded, with gate valves									
	1100	3/4" pipe size	1 Plum	16	.500	Ea.	620	18		638	705
	1120	1" pipe size		14	.571		645	20.50		665.50	735
	1140	1-1/2" pipe size		10	.800		735	29		764	855
	1160	2" pipe size	▼	7	1.143	▼	905	41		946	1,050
	1200	Flanged, valves are gate									
	1210	3" pipe size	Q-1	4.50	3.556	Ea.	1,425	115		1,540	1,750
	1220	4" pipe size	"	3	5.333		1,800	173		1,973	2,225
	1230	6" pipe size	Q-2	3	8		3,025	268		3,293	3,725
	1240	8" pipe size		2	12		5,625	405		6,030	6,800
	1250	10" pipe size	▼	1	24	▼	8,000	805		8,805	10,000
	1300	Flanged, valves are OS&Y									
	1370	1" pipe size	1 Plum	5	1.600	Ea.	675	57.50		732.50	830
	1374	1-1/2" pipe size		5	1.600		770	57.50		827.50	935
	1378	2" pipe size	▼	4.80	1.667		940	60		1,000	1,125
	1380	3" pipe size	Q-1	4.50	3.556		1,525	115		1,640	1,850
	1400	4" pipe size	"	3	5.333		2,175	173		2,348	2,650
	1420	6" pipe size	Q-2	3	8		3,425	268		3,693	4,175

Important: See the Reference Section for critical supporting data - Reference Nos., Crews, & City Cost Indexes

15140 | Domestic Water Piping

		CREW	DAILY OUTPUT	LABOR-HOURS	UNIT	MAT.	LABOR	EQUIP.	TOTAL	TOTAL INCL O&P	
100	1430	8" pipe size	Q-2	2	12	Ea.	6,350	405		6,755	7,600	**100**
	4000	Reduced pressure principle										
	4100	Threaded, valves are ball										
	4120	3/4" pipe size	1 Plum	16	.500	Ea.	715	18		733	810	
	4140	1" pipe size		14	.571		760	20.50		780.50	865	
	4150	1-1/4" pipe size		12	.667		1,050	24		1,074	1,175	
	4160	1-1/2" pipe size		10	.800		1,025	29		1,054	1,175	
	4180	2" pipe size		7	1.143		1,100	41		1,141	1,250	
	5000	Flanged, bronze, valves are OS&Y										
	5060	2-1/2" pipe size	Q-1	5	3.200	Ea.	2,750	104		2,854	3,175	
	5080	3" pipe size		4.50	3.556		3,425	115		3,540	3,950	
	5100	4" pipe size		3	5.333		4,850	173		5,023	5,600	
	5120	6" pipe size	Q-2	3	8		10,300	268		10,568	11,700	
	5200	Flanged, iron, valves are gate										
	5210	2-1/2" pipe size	Q-1	5	3.200	Ea.	1,775	104		1,879	2,100	
	5220	3" pipe size		4.50	3.556		1,975	115		2,090	2,350	
	5230	4" pipe size		3	5.333		2,525	173		2,698	3,025	
	5240	6" pipe size	Q-2	3	8		3,875	268		4,143	4,675	
	5250	8" pipe size		2	12		8,350	405		8,755	9,775	
	5260	10" pipe size		1	24		11,000	805		11,805	13,300	
	5600	Flanged, iron, valves are OS&Y										
	5660	2-1/2" pipe size	Q-1	5	3.200	Ea.	1,900	104		2,004	2,250	
	5680	3" pipe size		4.50	3.556		2,025	115		2,140	2,400	
	5700	4" pipe size		3	5.333		2,675	173		2,848	3,200	
	5720	6" pipe size	Q-2	3	8		4,200	268		4,468	5,025	
	5740	8" pipe size		2	12		8,500	405		8,905	9,950	
	5760	10" pipe size		1	24		11,700	805		12,505	14,000	
600	0010	**VACUUM BREAKERS** Hot or cold water										**600**
	1030	Anti-siphon, brass										
	1040	1/4" size	1 Plum	24	.333	Ea.	19.20	12		31.20	39	
	1050	3/8" size		24	.333		19.20	12		31.20	39	
	1060	1/2" size		24	.333		22	12		34	42	
	1080	3/4" size		20	.400		26	14.40		40.40	50	
	1100	1" size		19	.421		40	15.15		55.15	67.50	
	1120	1-1/4" size		15	.533		70.50	19.15		89.65	107	
	1140	1-1/2" size		13	.615		83	22		105	125	
	1160	2" size		11	.727		129	26		155	182	
	1180	2-1/2" size		9	.889		370	32		402	460	
	1200	3" size		7	1.143		495	41		536	605	
	1300	For polished chrome, (1/4" thru 1"), add					50%					
	1900	Vacuum relief, water service, bronze										
	2000	1/2" size	1 Plum	30	.267	Ea.	18.60	9.60		28.20	35	
	2040	3/4" size	"	28	.286	"	23.50	10.25		33.75	41	
700	0010	**VACUUM BREAKERS**										**700**
	0011	See also backflow preventers 15140-100										
	1000	Anti-siphon continuous pressure type										
	1010	Max. 150 PSI - 210°F										
	1020	Bronze body										
	1030	1/2" size	1 Stpi	24	.333	Ea.	84	12.05		96.05	110	
	1040	3/4" size		20	.400		84	14.50		98.50	114	
	1050	1" size		19	.421		86.50	15.25		101.75	119	
	1060	1-1/4" size		15	.533		171	19.30		190.30	217	
	1070	1-1/2" size		13	.615		210	22.50		232.50	266	
	1080	2" size		11	.727		217	26.50		243.50	279	
	1200	Max. 125 PSI with atmospheric vent										

MECHANICAL 15

267

15140 | Domestic Water Piping

			CREW	DAILY OUTPUT	LABOR-HOURS	UNIT	2002 BARE COSTS				TOTAL INCL O&P
							MAT.	LABOR	EQUIP.	TOTAL	
700	1210	Brass, in-line construction									
	1220	1/4" size	1 Stpi	24	.333	Ea.	32	12.05		44.05	53.50
	1230	3/8" size	"	24	.333		32	12.05		44.05	53.50
	1260	For polished chrome finish, add					13%				
	2000	Anti-siphon, non-continuous pressure type									
	2010	Hot or cold water 125 PSI - 210° F									
	2020	Bronze body									
	2030	1/4" size	1 Stpi	24	.333	Ea.	19.20	12.05		31.25	39
	2040	3/8" size		24	.333		19.20	12.05		31.25	39
	2050	1/2" size		24	.333		22	12.05		34.05	42
	2060	3/4" size		20	.400		26	14.50		40.50	50.50
	2070	1" size		19	.421		40	15.25		55.25	67.50
	2080	1-1/4" size		15	.533		70.50	19.30		89.80	107
	2090	1-1/2" size		13	.615		83	22.50		105.50	125
	2100	2" size		11	.727		129	26.50		155.50	182
	2110	2-1/2" size		8	1		370	36		406	465
	2120	3" size		6	1.333		495	48.50		543.50	620
	2150	For polished chrome finish, add					50%				
800	0010	**WATER HAMMER ARRESTORS / SHOCK ABSORBERS**									
	0490	Copper									
	0500	3/4" male I.P.S. For 1 to 11 fixtures	1 Plum	12	.667	Ea.	16.05	24		40.05	53.50
	0600	1" male I.P.S., For 12 to 32 fixtures		8	1		41	36		77	99
	0700	1-1/4" male I.P.S. For 33 to 60 fixtures		8	1		47.50	36		83.50	107
	0800	1-1/2" male I.P.S. For 61 to 113 fixtures		8	1		65	36		101	126
	0900	2" male I.P.S. For 114 to 154 fixtures		8	1		104	36		140	168
	1000	2-1/2" male I.P.S. For 155 to 330 fixtures		4	2		231	72		303	360
	4000	Bellows type									
	4010	3/4" FNPT, to 11 fixture units	1 Plum	10	.800	Ea.	68.50	29		97.50	119
	4020	1" FNPT, to 32 fixture units		8.80	.909		138	32.50		170.50	202
	4030	1" FNPT, to 60 fixture units		8.80	.909		206	32.50		238.50	277
	4040	1" FNPT, to 113 fixture units		8.80	.909		515	32.50		547.50	620
	4050	1" FNPT, to 154 fixture units		6.60	1.212		585	43.50		628.50	705
	4060	1-1/2" FNPT, to 300 fixture units		5	1.600		720	57.50		777.50	875

15155 | Drainage Specialties

			CREW	DAILY OUTPUT	LABOR-HOURS	UNIT	MAT.	LABOR	EQUIP.	TOTAL	TOTAL INCL O&P
680	0010	**SEPARATORS** Entrainment eliminator, steel body, 150 PSIG									
	0100	1/4" size	1 Stpi	24	.333	Ea.	137	12.05		149.05	168
	0120	1/2" size		24	.333		147	12.05		159.05	180
	0140	3/4" size		20	.400		158	14.50		172.50	196
	0160	1" size		19	.421		210	15.25		225.25	254
	0180	1-1/4" size		15	.533		242	19.30		261.30	295
	0200	1-1/2" size		13	.615		273	22.50		295.50	335
	0220	2" size		11	.727		305	26.50		331.50	375
	0240	2-1/2" size	Q-5	15	1.067		1,200	35		1,235	1,375
	0260	3" size		13	1.231		1,425	40		1,465	1,625
	0280	4" size		10	1.600		1,675	52		1,727	1,925
	0300	5" size		6	2.667		1,900	87		1,987	2,200
	0320	6" size		3	5.333		2,175	174		2,349	2,625
	0340	8" size	Q-6	4.40	5.455		2,600	184		2,784	3,125
	0360	10" size	"	4	6		3,900	203		4,103	4,575
	1000	For 300 PSIG, add					15%				
780	0010	**TRAPS**									
	4700	Copper, drainage, drum trap									
	4800	3" x 5" solid, 1-1/2" pipe size	1 Plum	16	.500	Ea.	27.50	18		45.50	57
	4840	3" x 6" swivel, 1-1/2" pipe size	"	16	.500	"	40.50	18		58.50	71.50

Important: See the Reference Section for critical supporting data - Reference Nos., Crews, & City Cost Indexes

15155	Drainage Specialties	CREW	DAILY OUTPUT	LABOR-HOURS	UNIT	MAT.	LABOR	EQUIP.	TOTAL	TOTAL INCL O&P	
780	5100	P trap, standard pattern										780
	5200	1-1/4" pipe size	1 Plum	18	.444	Ea.	19.65	16		35.65	45.50	
	5240	1-1/2" pipe size		17	.471		19.55	16.90		36.45	47	
	5260	2" pipe size		15	.533		30	19.15		49.15	62.50	
	5280	3" pipe size		11	.727		72.50	26		98.50	120	
	5340	With cleanout and slip joint										
	5360	1-1/4" pipe size	1 Plum	18	.444	Ea.	21	16		37	47	
	5400	1-1/2" pipe size		17	.471		21.50	16.90		38.40	49	
	5420	2" pipe size		15	.533		35	19.15		54.15	67.50	
	5460	For swivel, add					3.63			3.63	3.99	
	6710	ABS DWV P trap, solvent weld joint										
	6720	1-1/2" pipe size	1 Plum	18	.444	Ea.	3.82	16		19.82	28	
	6722	2" pipe size		17	.471		6.15	16.90		23.05	32.50	
	6724	3" pipe size		15	.533		28	19.15		47.15	60	
	6726	4" pipe size		14	.571		58	20.50		78.50	94.50	
	6732	PVC DWV P trap, solvent weld joint										
	6733	1-1/2" pipe size	1 Plum	18	.444	Ea.	3.29	16		19.29	27.50	
	6734	2" pipe size		17	.471		4.47	16.90		21.37	30.50	
	6735	3" pipe size		15	.533		22.50	19.15		41.65	53.50	
	6736	4" pipe size		14	.571		53.50	20.50		74	90	
	6760	PP DWV, dilution trap, 1-1/2" pipe size		16	.500		108	18		126	146	
	6770	P trap, 1-1/2" pipe size		17	.471		29	16.90		45.90	57.50	
	6780	2" pipe size		16	.500		39.50	18		57.50	70.50	
	6790	3" pipe size		14	.571		91.50	20.50		112	132	
	6800	4" pipe size		13	.615		117	22		139	162	
	6810	Running trap, 1-1/2" pipe size		16	.500		28.50	18		46.50	58.50	
	6820	2" pipe size		15	.533		40	19.15		59.15	73	
	6830	S trap, 1-1/2" pipe size		16	.500		29	18		47	59	
	6840	2" pipe size		15	.533		36.50	19.15		55.65	69	
	6850	Universal trap, 1-1/2" pipe size		14	.571		82.50	20.50		103	122	
	6860	PVC DWV hub x hub, basin trap, 1-1/4" pipe size		18	.444		5.30	16		21.30	30	
	6870	Sink P trap, 1-1/2" pipe size		18	.444		5.30	16		21.30	30	
	6880	Tubular S trap, 1-1/2" pipe size		17	.471		9.10	16.90		26	35.50	
	6890	PVC sch. 40 DWV, drum trap										
	6900	1-1/2" pipe size	1 Plum	16	.500	Ea.	17.05	18		35.05	46	
	6910	P trap, 1-1/2" pipe size		18	.444		3.29	16		19.29	27.50	
	6920	2" pipe size		17	.471		4.47	16.90		21.37	30.50	
	6930	3" pipe size		15	.533		22.50	19.15		41.65	53.50	
	6940	4" pipe size		14	.571		53.50	20.50		74	90	
	6950	P trap w/clean out, 1-1/2" pipe size		18	.444		7.95	16		23.95	33	
	6960	2" pipe size		17	.471		13.55	16.90		30.45	40.50	
	6970	P trap adjustable, 1-1/2" pipe size		17	.471		4.28	16.90		21.18	30	
	6980	P trap adj. w/union & cleanout, 1-1/2" pipe size		16	.500		13.50	18		31.50	42	
940	0010	**VENT FLASHING, CAPS**										940
	0120	Vent caps										
	0140	Cast iron										
	0180	2-1/2" - 3-5/8" pipe	1 Plum	21	.381	Ea.	33	13.70		46.70	57	
	0190	4" - 4-1/8" pipe	"	19	.421	"	39.50	15.15		54.65	66.50	
	0900	Vent flashing										
	1000	Aluminum with lead ring										
	1020	1-1/4" pipe	1 Plum	20	.400	Ea.	6.40	14.40		20.80	28.50	
	1030	1-1/2" pipe		20	.400		6.40	14.40		20.80	28.50	
	1040	2" pipe		18	.444		6.85	16		22.85	31.50	
	1050	3" pipe		17	.471		7.60	16.90		24.50	34	
	1060	4" pipe		16	.500		9.15	18		27.15	37	

MECHANICAL 15

269

15155	Drainage Specialties	CREW	DAILY OUTPUT	LABOR-HOURS	UNIT	2002 BARE COSTS				TOTAL INCL O&P	
						MAT.	LABOR	EQUIP.	TOTAL		
940 1350	Copper with neoprene ring										**940**
1400	1-1/4" pipe	1 Plum	20	.400	Ea.	14.05	14.40		28.45	37	
1430	1-1/2" pipe		20	.400		14.05	14.40		28.45	37	
1440	2" pipe		18	.444		14.85	16		30.85	40.50	
1450	3" pipe		17	.471		17.40	16.90		34.30	44.50	
1460	4" pipe	▼	16	.500	▼	19.20	18		37.20	48	
2000	Galvanized with neoprene ring										
2020	1-1/4" pipe	1 Plum	20	.400	Ea.	5.10	14.40		19.50	27	
2030	1-1/2" pipe		20	.400		5.10	14.40		19.50	27	
2040	2" pipe		18	.444		5.40	16		21.40	30	
2050	3" pipe		17	.471		5.85	16.90		22.75	32	
2060	4" pipe	▼	16	.500	▼	7.60	18		25.60	35.50	
2980	Neoprene, one piece										
3000	1-1/4" pipe	1 Plum	24	.333	Ea.	5.15	12		17.15	24	
3030	1-1/2" pipe		24	.333		4.92	12		16.92	23.50	
3040	2" pipe		23	.348		5.40	12.50		17.90	25	
3050	3" pipe		21	.381		6.55	13.70		20.25	28	
3060	4" pipe	▼	20	.400	▼	8.90	14.40		23.30	31.50	

15180	Heating and Cooling Piping										
100 0010	**ANTI-FREEZE** Inhibited										**100**
0900	Ethylene glycol concentrated										
1000	55 gallon drums, small quantities				Gal.	7.20			7.20	7.95	
1200	Large quantities					6.65			6.65	7.30	
2000	Propylene glycol, for solar heat, small quantities					7.95			7.95	8.75	
2100	Large quantities				▼	7.75			7.75	8.55	
200 0010	**PUMPS, CIRCULATING** Heated or chilled water application										**200**
0600	Bronze, sweat connections, 1/40 HP, in line										
0640	3/4" size	Q-1	16	1	Ea.	117	32.50		149.50	178	
1000	Flange connection, 3/4" to 1-1/2" size										
1040	1/12 HP	Q-1	6	2.667	Ea.	320	86.50		406.50	480	
1060	1/8 HP		6	2.667		550	86.50		636.50	735	
1100	1/3 HP		6	2.667		615	86.50		701.50	805	
1140	2" size, 1/6 HP		5	3.200		790	104		894	1,025	
1180	2-1/2" size, 1/4 HP		5	3.200		1,025	104		1,129	1,275	
1220	3" size, 1/4 HP		4	4		1,075	129		1,204	1,375	
1260	1/3 HP		4	4		1,300	129		1,429	1,650	
1300	1/2 HP		4	4		1,350	129		1,479	1,675	
1340	3/4 HP		4	4		1,450	129		1,579	1,800	
1380	1 HP	▼	4	4	▼	2,350	129		2,479	2,800	
2000	Cast iron, flange connection										
2040	3/4" to 1-1/2" size, in line, 1/12 HP	Q-1	6	2.667	Ea.	209	86.50		295.50	360	
2060	1/8 HP		6	2.667		350	86.50		436.50	515	
2100	1/3 HP		6	2.667		390	86.50		476.50	560	
2140	2" size, 1/6 HP		5	3.200		425	104		529	625	
2180	2-1/2" size, 1/4 HP		5	3.200		560	104		664	770	
2220	3" size, 1/4 HP		4	4		565	129		694	820	
2260	1/3 HP		4	4		765	129		894	1,025	
2300	1/2 HP		4	4		795	129		924	1,075	
2340	3/4 HP		4	4		920	129		1,049	1,200	
2380	1 HP	▼	4	4	▼	1,325	129		1,454	1,650	
2600	For non-ferrous impeller, add					3%					
3000	High head, bronze impeller										
3030	1-1/2" size 1/2 HP	Q-1	5	3.200	Ea.	680	104		784	900	
3040	1-1/2" size 3/4 HP		5	3.200		725	104		829	955	
3050	2" size 1 HP	▼	4	4	▼	870	129		999	1,150	

Important: See the Reference Section for critical supporting data - Reference Nos., Crews, & City Cost Indexes

15180	Heating and Cooling Piping	CREW	DAILY OUTPUT	LABOR-HOURS	UNIT	2002 BARE COSTS				TOTAL INCL O&P	
						MAT.	LABOR	EQUIP.	TOTAL		
200	3090	2" size 1-1/2 HP	Q-1	4	4	Ea.	1,050	129		1,179	1,350	**200**
	4000	Close coupled, end suction, bronze impeller										
	4040	1-1/2" size, 1-1/2 HP, to 40 GPM	Q-1	3	5.333	Ea.	1,100	173		1,273	1,475	
	4090	2" size, 2 HP, to 50 GPM		3	5.333		1,125	173		1,298	1,475	
	4100	2" size, 3 HP, to 90 GPM		2.30	6.957		1,225	225		1,450	1,700	
	4190	2-1/2" size, 3 HP, to 150 GPM		2	8		1,325	259		1,584	1,850	
	4300	3" size, 5 HP, to 225 GPM		1.80	8.889		1,425	288		1,713	2,000	
	4410	3" size, 10 HP, to 350 GPM		1.60	10		2,075	325		2,400	2,775	
	4420	4" size, 7-1/2 HP, to 350 GPM	▼	1.60	10		2,000	325		2,325	2,700	
	4520	4" size, 10 HP, to 600 GPM	Q-2	1.70	14.118		2,475	475		2,950	3,425	
	4530	5" size, 15 HP, to 1000 GPM		1.70	14.118		2,975	475		3,450	4,000	
	4610	5" size, 20 HP, to 1350 GPM		1.50	16		3,275	535		3,810	4,400	
	4620	5" size, 25 HP, to 1550 GPM	▼	1.50	16	▼	3,900	535		4,435	5,100	
	5000	Base mounted, bronze impeller, coupling guard										
	5040	1-1/2" size, 1-1/2 HP, to 40 GPM	Q-1	2.30	6.957	Ea.	1,575	225		1,800	2,075	
	5090	2" size, 2 HP, to 50 GPM		2.30	6.957		1,775	225		2,000	2,325	
	5100	2" size, 3 HP, to 90 GPM		2	8		1,800	259		2,059	2,375	
	5190	2-1/2" size, 3 HP, to 150 GPM		1.80	8.889		1,900	288		2,188	2,500	
	5300	3" size, 5 HP, to 225 GPM		1.60	10		1,950	325		2,275	2,650	
	5410	4" size, 5 HP, to 350 GPM		1.50	10.667		2,125	345		2,470	2,875	
	5420	4" size, 7-1/2 HP, to 350 GPM	▼	1.50	10.667		2,450	345		2,795	3,225	
	5520	5" size, 10 HP, to 600 GPM	Q-2	1.60	15		2,450	505		2,955	3,425	
	5530	5" size, 15 HP, to 1000 GPM		1.60	15		3,375	505		3,880	4,450	
	5610	6" size, 20 HP, to 1350 GPM		1.40	17.143		3,725	575		4,300	4,950	
	5620	6" size, 25 HP, to 1550 GPM	▼	1.40	17.143	▼	4,200	575		4,775	5,500	
	5800	The above pump capacities are based on 1800 RPM,										
	5810	at a 60 foot head. Increasing the RPM										
	5820	or decreasing the head will increase the GPM.										
300	0010	**PUMPS, CONDENSATE RETURN SYSTEM**										**300**
	2000	Simplex										
	2010	With pump, motor, CI receiver, float switch										
	2020	3/4 HP, 15 GPM	Q-1	1.80	8.889	Ea.	3,400	288		3,688	4,175	
	2100	Duplex										
	2110	With 2 pumps and motors, CI receiver, float switch, alternator										
	2120	3/4 HP, 15 GPM, 15 Gal CI rcvr	Q-1	1.40	11.429	Ea.	3,725	370		4,095	4,650	
	2130	1 HP, 25 GPM		1.20	13.333		4,075	430		4,505	5,125	
	2140	1-1/2 HP, 45 GPM		1	16		4,875	520		5,395	6,150	
	2150	1-1/2 HP, 60 GPM	▼	1	16	▼	3,875	520		4,395	5,050	
700	0010	**REFRIGERATION SPECIALTIES**										**700**
	0600	Accumulator										
	0610	3/4"	1 Stpi	8.80	.909	Ea.	34.50	33		67.50	87.50	
	0614	7/8"		6.40	1.250		41	45.50		86.50	114	
	0618	1-1/8"		4.80	1.667		50	60.50		110.50	146	
	0622	1-3/8"		4	2		68	72.50		140.50	184	
	0626	1-5/8"		3.20	2.500		83.50	90.50		174	229	
	0630	2-1/8"	▼	2.40	3.333	▼	278	121		399	485	
	1000	Filter dryer										
	1010	Replaceable core type, solder										
	1020	1/2"	1 Stpi	20	.400	Ea.	102	14.50		116.50	135	
	1030	5/8"		19	.421		107	15.25		122.25	141	
	1040	7/8"		18	.444		107	16.10		123.10	143	
	1050	1-1/8"		15	.533		112	19.30		131.30	152	
	1060	1-3/8"		14	.571		134	20.50		154.50	179	
	1070	1-5/8"		12	.667		134	24		158	185	
	1080	2-1/8"		10	.800		203	29		232	268	
	1090	2-5/8"	▼	9	.889	▼	236	32		268	310	

MECHANICAL 15

15180	Heating and Cooling Piping	CREW	DAILY OUTPUT	LABOR- HOURS	UNIT	2002 BARE COSTS				TOTAL INCL O&P
						MAT.	LABOR	EQUIP.	TOTAL	
700 1100	3-1/8"	1 Stpi	8	1	Ea.	230	36		266	305 **700**
1200	Sealed in-line, solder									
1210	1/4", 3 cubic inches	1 Stpi	22	.364	Ea.	8.10	13.15		21.25	29
1220	3/8", 5 cubic inches		21	.381		7.40	13.80		21.20	29
1230	1/2", 9 cubic inches		20	.400		12	14.50		26.50	35
1240	1/2", 16 cubic inches		20	.400		15.60	14.50		30.10	39
1250	5/8", 16 cubic inches		19	.421		16.35	15.25		31.60	41
1260	5/8", 30 cubic inches		19	.421		26.50	15.25		41.75	52
1270	7/8", 30 cubic inches		18	.444		28	16.10		44.10	55.50
1280	7/8", 41 cubic inches		17	.471		37.50	17.05		54.55	67
1290	1-1/8", 60 cubic inches	▼	15	.533	▼	52	19.30		71.30	86.50
4000	P-Trap, suction line, solder									
4010	5/8"	1 Stpi	19	.421	Ea.	9.30	15.25		24.55	33
4020	3/4"		19	.421		7.35	15.25		22.60	31
4030	7/8"		18	.444		7.35	16.10		23.45	32.50
4040	1-1/8"		15	.533		10.50	19.30		29.80	40.50
4050	1-3/8"		14	.571		16.40	20.50		36.90	49
4060	1-5/8"		12	.667		25	24		49	64
4070	2-1/8"	▼	10	.800	▼	51	29		80	99.50
4440	Refrigerant, R-507, 25 lb. disposable cylinder				Lb.	239			239	263
5000	Sightglass									
5010	Moisture and liquid indicator, solder									
5020	1/4"	1 Stpi	22	.364	Ea.	10	13.15		23.15	31
5030	3/8"		21	.381		10.60	13.80		24.40	32.50
5040	1/2"		20	.400		13.45	14.50		27.95	37
5050	5/8"		19	.421		13.75	15.25		29	38
5060	7/8"		18	.444		18	16.10		34.10	44.50
5070	1-1/8"		15	.533		18.75	19.30		38.05	49.50
5080	1-3/8"		14	.571		22.50	20.50		43	55.50
5090	1-5/8"		12	.667		26	24		50	65
5100	2-1/8"	▼	10	.800	▼	32.50	29		61.50	79
7400	Vacuum pump set									
7410	Two stage, high vacuum continuous duty	1 Stpi	8	1	Ea.	2,125	36		2,161	2,375
8000	Valves									
8100	Check valve, soldered									
8110	5/8"	1 Stpi	36	.222	Ea.	29	8.05		37.05	44
8114	7/8"		26	.308		50	11.15		61.15	72
8118	1-1/8"		18	.444		137	16.10		153.10	176
8122	1-3/8"		14	.571		187	20.50		207.50	237
8126	1-5/8"		13	.615		254	22.50		276.50	315
8130	2-1/8"	▼	12	.667		385	24		409	460
8134	2-5/8"	Q-5	22	.727		950	23.50		973.50	1,075
8138	3-1/8"	"	20	.800	▼	1,175	26		1,201	1,350
8500	Refrigeration valve, packless, soldered									
8510	1/2"	1 Stpi	38	.211	Ea.	24.50	7.60		32.10	38.50
8514	5/8"		36	.222		47	8.05		55.05	63.50
8518	7/8"	▼	26	.308	▼	74.50	11.15		85.65	99
8520	Packed, soldered									
8522	1-1/8"	1 Stpi	18	.444	Ea.	106	16.10		122.10	141
8526	1-3/8"		14	.571		172	20.50		192.50	220
8530	1-5/8"		13	.615		213	22.50		235.50	268
8534	2-1/8"	▼	12	.667		315	24		339	380
8538	2-5/8"	Q-5	22	.727		470	23.50		493.50	550
8542	3-1/8"		20	.800		625	26		651	725
8546	4-1/8"	▼	18	.889	▼	895	29		924	1,025
8600	Solenoid valve, flange/solder									

Important: See the Reference Section for critical supporting data - Reference Nos., Crews, & City Cost Indexes

			DAILY	LABOR-		2002 BARE COSTS				TOTAL	
15180	**Heating and Cooling Piping**	CREW	OUTPUT	HOURS	UNIT	MAT.	LABOR	EQUIP.	TOTAL	INCL O&P	
700	8610	1/2"	1 Stpi	38	.211	Ea.	122	7.60		129.60	146	**700**
	8614	5/8"		36	.222		152	8.05		160.05	179	
	8618	3/4"		30	.267		190	9.65		199.65	224	
	8622	7/8"		26	.308		282	11.15		293.15	325	
	8626	1-1/8"		18	.444		315	16.10		331.10	370	
	8630	1-3/8"		14	.571		380	20.50		400.50	450	
	8634	1-5/8"		13	.615		535	22.50		557.50	625	
	8638	2-1/8"		12	.667		590	24		614	680	
	8800	Thermostatic valve, flange/solder										
	8810	1/2 - 3 ton, 3/8" x 5/8"	1 Stpi	9	.889	Ea.	111	32		143	171	
	8814	4 - 5 ton, 1/2" x 7/8"		7	1.143		111	41.50		152.50	185	
	8818	6 - 8 ton, 5/8" x 7/8"		5	1.600		111	58		169	210	
	8822	7 - 12 ton, 7/8" x 1-1/8"		4	2		129	72.50		201.50	251	
	8826	15 - 20 ton, 7/8" x 1-3/8"		3.20	2.500		187	90.50		277.50	345	
800	0010	**STEAM CONDENSATE METER**										**800**
	0100	500 lb. per hour	1 Stpi	14	.571	Ea.	2,300	20.50		2,320.50	2,550	
	0140	1500 lb. per hour		7	1.143		2,475	41.50		2,516.50	2,800	
	0160	3000 lb. per hour		5	1.600		3,325	58		3,383	3,750	
	0200	12,000 lb. per hour	Q-5	3.50	4.571		4,450	149		4,599	5,100	

15195 | Fuel Systems

			CREW	DAILY OUTPUT	LABOR-HOURS	UNIT	MAT.	LABOR	EQUIP.	TOTAL	INCL O&P	
490	0010	**FUEL OIL SPECIALTIES**										**490**
	0020	Foot valve, single poppet, metal to metal construction										
	0040	Bevel seat, 1/2" diameter	1 Stpi	20	.400	Ea.	36	14.50		50.50	61.50	
	0060	3/4" diameter		18	.444		37.50	16.10		53.60	66	
	0080	1" diameter		16	.500		26.50	18.10		44.60	56.50	
	0100	1-1/4" diameter		15	.533		53	19.30		72.30	87	
	0120	1-1/2" diameter		13	.615		45	22.50		67.50	83	
	0140	2" diameter		11	.727		52.50	26.50		79	97.50	
	0160	Foot valve, double poppet, metal to metal construction										
	0164	1" diameter	1 Stpi	15	.533	Ea.	79.50	19.30		98.80	117	
	0166	1-1/2" diameter	"	12	.667	"	111	24		135	159	
	0400	Fuel fill box, flush type										
	0408	Nonlocking, watertight										
	0410	1-1/2" diameter	1 Stpi	12	.667	Ea.	13.60	24		37.60	51.50	
	0440	2" diameter		10	.800		14.65	29		43.65	59.50	
	0450	2-1/2" diameter		9	.889		49	32		81	103	
	0460	3" diameter		7	1.143		47.50	41.50		89	115	
	0470	4" diameter		5	1.600		65	58		123	159	
	0500	Locking inner cover										
	0510	2" diameter	1 Stpi	8	1	Ea.	71	36		107	133	
	0520	2-1/2" diameter		7	1.143		95.50	41.50		137	168	
	0530	3" diameter		5	1.600		102	58		160	200	
	0540	4" diameter		4	2		131	72.50		203.50	253	
	0600	Fuel system components										
	0620	Spill container	1 Stpi	4	2	Ea.	485	72.50		557.50	645	
	0640	Fill adapter, 4", straight drop		8	1		30	36		66	87.50	
	0680	Fill cap, 4"		30	.267		25.50	9.65		35.15	42.50	
	0700	Extractor fitting, 4" x 1-1/2"		8	1		340	36		376	425	
	0740	Vapor hose adapter, 4"		8	1		64.50	36		100.50	126	
	0760	Wood gage stick, 10'					10.40			10.40	11.45	
	1000	Oil filters, 3/8" IPT., 20 gal. per hour	1 Stpi	20	.400		19.25	14.50		33.75	43	
	1020	32 gal. per hour		18	.444		32.50	16.10		48.60	60	
	1040	40 gal. per hour		16	.500		37	18.10		55.10	68	
	1060	50 gal. per hour		14	.571		38	20.50		58.50	73	

MECHANICAL 15

			DAILY	LABOR-		2002 BARE COSTS				TOTAL	
	15195	**Fuel Systems**	CREW	OUTPUT	HOURS	UNIT	MAT.	LABOR	EQUIP.	TOTAL	INCL O&P
490	1800	Pump and motor sets									**490**
	1810	Light fuel and diesel oils, 100 PSI									
	1820	25 GPH 1/4 HP	Q-5	6	2.667	Ea.	795	87		882	1,000
	1830	45 GPH, 1/4 HP		6	2.667		800	87		887	1,000
	1840	90 GPH, 1/4 HP		5	3.200		860	104		964	1,100
	1850	160 GPH, 1/3 HP		4	4		885	130		1,015	1,175
	1860	325 GPH, 3/4 HP		4	4		970	130		1,100	1,275
	1870	700 GPH, 1-1/2 HP		3	5.333		1,750	174		1,924	2,175
	1880	1000 GPH, 2 HP		3	5.333		1,800	174		1,974	2,225
	1890	1800 GPH, 5 HP		1.80	8.889		2,750	290		3,040	3,475
	2000	Remote tank gauging system, self contained									
	2100	Gage and one probe	1 Stpi	2.50	3.200	Ea.	2,700	116		2,816	3,150
	2120	Gage with two probes		2	4		3,700	145		3,845	4,300
	3000	Valve, ball check, globe type, 3/8" diameter		24	.333		6.55	12.05		18.60	25.50
	3500	Fusible, 3/8" diameter		24	.333		5.20	12.05		17.25	24
	3600	1/2" diameter		24	.333		13.65	12.05		25.70	33
	3610	3/4" diameter		20	.400		29	14.50		43.50	54
	3620	1" diameter		19	.421		86.50	15.25		101.75	118
	4000	Nonfusible, 3/8" diameter		24	.333		9.95	12.05		22	29
	4500	Shutoff, gate type, lever handle, spring-fusible kit									
	4520	1/4" diameter	1 Stpi	14	.571	Ea.	15.45	20.50		35.95	48
	4540	3/8" diameter		12	.667		15.55	24		39.55	53.50
	4560	1/2" diameter		10	.800		20.50	29		49.50	66
	4570	3/4" diameter		8	1		34.50	36		70.50	92.50
	4580	Lever handle, requires weight and fusible kit									
	4600	1" diameter	1 Stpi	9	.889	Ea.	284	32		316	360
	4620	1-1/4" diameter		8	1		294	36		330	380
	4640	1-1/2" diameter		7	1.143		400	41.50		441.50	505
	4660	2" diameter		6	1.333		420	48.50		468.50	535
	4680	For fusible link, weight and braided wire, add					5%				
	5000	Vent alarm, whistling signal					26			26	28.50
	5500	Vent protector/breather, 1-1/4" diameter	1 Stpi	32	.250		7.20	9.05		16.25	21.50
	5520	1-1/2" diameter		32	.250		8	9.05		17.05	22.50
	5540	2" diameter		32	.250		12.80	9.05		21.85	28
	5560	3" diameter		28	.286		40	10.35		50.35	59.50
	5580	4" diameter		24	.333		47	12.05		59.05	70
	5600	Dust cap, breather, 2"		40	.200		10.20	7.25		17.45	22
	8000	Fuel oil and tank heaters									
	8020	Electric, capacity rated at 230 volts									
	8040	Immersion element in steel manifold									
	8060	96 GPH at 50°F rise	Q-5	6.40	2.500	Ea.	630	81.50		711.50	815
	8070	128 GPH at 50°F rise		6.20	2.581		665	84		749	855
	8080	160 GPH at 50°F rise		5.90	2.712		695	88.50		783.50	900
	8090	192 GPH at 50°F rise		5.50	2.909		740	95		835	960
	8100	240 GPH at 50°F rise		5.10	3.137		890	102		992	1,125
	8110	288 GPH at 50°F rise		4.60	3.478		1,050	113		1,163	1,350
	8120	384 GPH at 50°F rise		3.10	5.161		1,200	168		1,368	1,575
	8130	480 GPH at 50°F rise		2.30	6.957		1,500	227		1,727	2,000
	8140	576 GPH at 50°F rise		2.10	7.619		1,625	248		1,873	2,175
	8300	Suction stub, immersion type									
	8320	75" long, 750 watts	1 Stpi	14	.571	Ea.	530	20.50		550.50	615
	8330	99" long, 2000 watts		12	.667		745	24		769	855
	8340	123" long, 3000 watts		10	.800		805	29		834	930
	8400	Electro/steam, electric capacity									
	8420	rated at 230 volts/steam at 5 PSIG									
	8440	28/50 GPH	Q-5	3.50	4.571	Ea.	1,900	149		2,049	2,300

Important: See the Reference Section for critical supporting data - Reference Nos., Crews, & City Cost Indexes

15 MECHANICAL

15195 | Fuel Systems

		CREW	DAILY OUTPUT	LABOR-HOURS	UNIT	2002 BARE COSTS MAT.	LABOR	EQUIP.	TOTAL	TOTAL INCL O&P	
490	8450	64/135 GPH	Q-5	3	5.333	Ea.	2,250	174		2,424	2,725	**490**
	8460	80/257 GPH		2.50	6.400		2,975	209		3,184	3,600	
	8470	96/303 GPH		2.10	7.619		3,225	248		3,473	3,925	
	8480	120/384 GPH	Q-6	3	8		3,475	270		3,745	4,225	
	8490	144/720 GPH		1.80	13.333		5,875	450		6,325	7,125	
	8500	192/800 GPH		1.60	15		6,725	505		7,230	8,175	
	8510	240/1000 GPH		1.30	18.462		7,575	625		8,200	9,275	
	8660	Steam, cross flow, rated at 5 PSIG										
	8680	42 GPH	Q-5	7	2.286	Ea.	1,225	74.50		1,299.50	1,450	
	8690	73 GPH		6.70	2.388		1,425	78		1,503	1,700	
	8700	112 GPH		6.20	2.581		1,475	84		1,559	1,750	
	8710	158 GPH		5.80	2.759		1,625	90		1,715	1,900	
	8720	187 GPH		4	4		2,225	130		2,355	2,650	
	8730	270 GPH		3.60	4.444		2,400	145		2,545	2,850	
	8740	365 GPH	Q-6	4.90	4.898		2,800	165		2,965	3,350	
	8750	635 GPH		3.70	6.486		3,550	219		3,769	4,225	
	8760	845 GPH		2.50	9.600		5,200	325		5,525	6,225	
	8770	1420 GPH		1.60	15		8,475	505		8,980	10,100	
	8780	2100 GPH		1.10	21.818		11,600	735		12,335	13,900	
	8860	Steam or water, safety type										
	8880	Water capacity at 180°F/steam at 5 PSIG										
	8900	70/84 GPH	Q-5	3.70	4.324	Ea.	1,500	141		1,641	1,875	
	8910	90/108 GPH		3.10	5.161		1,775	168		1,943	2,225	
	8920	120/144 GPH		2.70	5.926		2,075	193		2,268	2,600	
	8930	150/180 GPH		2.30	6.957		2,450	227		2,677	3,050	
	8940	180/216 GPH		1.80	8.889		3,050	290		3,340	3,800	
	8950	250/300 GPH		1.40	11.429		3,850	370		4,220	4,800	
	8960	320/384 GPH	Q-6	1.90	12.632		4,625	425		5,050	5,725	
	8970	400/449 GPH		1.50	16		5,650	540		6,190	7,050	
	8980	480/536 GPH		1.30	18.462		6,625	625		7,250	8,225	
	8990	600/670 GPH		1.10	21.818		8,400	735		9,135	10,300	
	9100	Suction bell type										
	9120	160 GPH, 17" high	Q-5	8	2	Ea.	2,175	65		2,240	2,475	
	9140	240 GPH, 24" high	"	7	2.286	"	2,350	74.50		2,424.50	2,675	
	9200	Tank heating coil										
	9240	73" high by 15-3/8" diameter	Q-5	5	3.200	Ea.	3,450	104		3,554	3,950	
	9280	Tank outlet heater (large capacity tank)										
	9300	rated for steam at 5/125 PSIG										
	9320	95/200 GPH	Q-5	4	4	Ea.	1,375	130		1,505	1,725	
	9330	190/400 GPH	"	2.70	5.926		1,950	193		2,143	2,450	
	9340	360/810 GPH	Q-6	3.20	7.500		2,750	253		3,003	3,400	
	9350	490/1070 GPH		2.60	9.231		3,425	310		3,735	4,250	
	9360	830/1850 GPH		2	12		4,375	405		4,780	5,425	
	9370	1000/2200 GPH		1.70	14.118		4,950	475		5,425	6,175	
	9380	1310/2920 GPH		1.30	18.462		6,725	625		7,350	8,350	
	9390	1500/3300 GPH		1.10	21.818		7,400	735		8,135	9,225	
	9400	2280/5050 GPH		.83	28.916		10,000	975		10,975	12,500	
	9410	2500/5550 GPH		.80	30		10,800	1,025		11,825	13,400	
	9420	3000/6660 GPH		.72	33.333		12,100	1,125		13,225	15,000	
	9430	4100/9150 GPH		.60	40		14,400	1,350		15,750	17,900	
	9440	5450/12,100 GPH		.50	48		17,700	1,625		19,325	22,000	
	9450	7000/15,500 GPH		.40	60		21,900	2,025		23,925	27,100	
	9460	8700/19,300 GPH		.33	72.727		26,400	2,450		28,850	32,800	
	9470	10,400/23,200 GPH		.29	82.759		29,300	2,800		32,100	36,400	
605	0010	**METERS**										**605**
	4000	Residential										

15100 | Building Services Piping

				DAILY	LABOR-		2002 BARE COSTS				TOTAL	
	15195	**Fuel Systems**	CREW	OUTPUT	HOURS	UNIT	MAT.	LABOR	EQUIP.	TOTAL	INCL O&P	
605	4010	Gas meter, residential, 3/4" pipe size	1 Plum	14	.571	Ea.	116	20.50		136.50	159	**605**
	4020	Gas meter, residential, 1" pipe size	↓	12	.667		142	24		166	192	
	4030	Gas meter, residential, 1-1/4" pipe size	▼	10	.800	▼	142	29		171	200	

15200 | Process Piping

				DAILY	LABOR-		2002 BARE COSTS				TOTAL	
	15210	**Process Air/Gas Piping**	CREW	OUTPUT	HOURS	UNIT	MAT.	LABOR	EQUIP.	TOTAL	INCL O&P	
100	0010	**COMPRESSORS**										**100**
	0100	Refrigeration, hermetic, switches and protective devices										
	0200	High back pressure types for R-22										
	0210	1.25 ton	1 Stpi	3	2.667	Ea.	350	96.50		446.50	535	
	0220	1.42 ton		3	2.667		360	96.50		456.50	540	
	0230	1.68 ton		2.80	2.857		410	103		513	605	
	0240	2.00 ton		2.60	3.077		505	111		616	725	
	0250	2.37 ton		2.50	3.200		610	116		726	845	
	0260	2.67 ton		2.40	3.333		365	121		486	580	
	0270	3.53 ton		2.30	3.478		460	126		586	695	
	0280	4.43 ton		2.20	3.636		665	132		797	930	
	0290	5.08 ton		2.10	3.810		790	138		928	1,075	
	0300	5.22 ton	▼	2	4		790	145		935	1,075	
	0310	7.31 ton	Q-5	3	5.333		1,450	174		1,624	1,850	
	0320	9.95 ton	"	2.60	6.154	▼	1,575	200		1,775	2,050	
	0990	Refrigeration, recip. hermetic, switches & protective devices										
	1000	(Ratings are ARI standard 515 group IV using R-22) 2 ton	Q-5	1	16	Ea.	2,900	520		3,420	3,950	
	1100	15 ton	Q-6	.72	33.333		9,300	1,125		10,425	11,900	
	1200	20 ton		.64	37.500		11,200	1,275		12,475	14,200	
	1300	40 ton		.44	54.545		12,100	1,850		13,950	16,100	
	1400	50 ton	▼	.20	120		20,400	4,050		24,450	28,500	
	1500	75 ton	Q-7	.27	118		23,900	4,100		28,000	32,500	
	1600	130 ton	"	.21	152	▼	29,800	5,250		35,050	40,700	
	1800	Refrigeration, scroll type										
	1810	1.9 ton	1 Stpi	2.70	2.963	Ea.	455	107		562	665	
	1820	2.35 ton		2.50	3.200		500	116		616	725	
	1830	2.82 ton		2.35	3.404		535	123		658	775	
	1840	3.3 ton		2.30	3.478		555	126		681	800	
	1850	3.83 ton		2.25	3.556		580	129		709	835	
	1860	4.1 ton		2.20	3.636		730	132		862	1,000	
	1870	4.8 ton		2.10	3.810		765	138		903	1,050	
	1880	5 ton	▼	2.07	3.865	▼	780	140		920	1,075	
	5250	Air, reciprocating air cooled, splash lubricated, tank mounted										
	5300	Single stage, 1 phase, 140 psi										
	5303	1/2 HP, 30 gal tank	1 Stpi	3	2.667	Ea.	935	96.50		1,031.50	1,175	
	5305	3/4 HP, 30 gal tank		2.60	3.077		940	111		1,051	1,200	
	5307	1 HP, 30 gal tank	▼	2.20	3.636		1,200	132		1,332	1,525	
	5309	2 HP, 30 gal tank	Q-5	4	4		1,625	130		1,755	1,975	
	5310	3 HP, 30 gal tank		3.60	4.444		2,125	145		2,270	2,575	
	5314	3 HP, 60 gal tank		3.50	4.571		2,425	149		2,574	2,900	
	5320	5 HP, 60 gal tank		3.20	5		2,825	163		2,988	3,350	
	5330	5 HP, 80 gal tank	▼	3	5.333	▼	2,775	174		2,949	3,325	

Important: See the Reference Section for critical supporting data - Reference Nos., Crews, & City Cost Indexes

15210	Process Air/Gas Piping	CREW	DAILY OUTPUT	LABOR-HOURS	UNIT	2002 BARE COSTS				TOTAL INCL O&P
						MAT.	LABOR	EQUIP.	TOTAL	
100	5340	7.5 HP, 80 gal tank	Q-5	2.60	6.154	Ea.	4,525	200		4,725	5,275	**100**

15210	Process Air/Gas Piping	CREW	DAILY OUTPUT	LABOR-HOURS	UNIT	MAT.	LABOR	EQUIP.	TOTAL	TOTAL INCL O&P	
100	5340	7.5 HP, 80 gal tank	Q-5	2.60	6.154	Ea.	4,525	200		4,725	5,275	100
	5600	2 stage pkg., 3 phase										
	5650	6 CFM at 125 psi 1-1/2 HP, 60 gal tank	Q-5	3	5.333	Ea.	2,100	174		2,274	2,550	
	5670	10.9 CFM at 125 psi, 3 HP, 80 gal tank		1.50	10.667		2,500	350		2,850	3,275	
	5680	38.7 CFM at 125 psi, 10 HP, 120 gal tank	▼	.60	26.667		4,925	870		5,795	6,725	
	5690	105 CFM at 125 psi, 25 HP, 250 gal tank	Q-6	.60	40	▼	9,200	1,350		10,550	12,200	
	5800	With single stage pump										
	5850	8.3 CFM at 125 psi, 2 HP, 80 gal tank	Q-6	3.50	6.857	Ea.	2,575	232		2,807	3,200	
	5860	38.7 CFM at 125 psi, 10 HP, 120 gal tank	"	.90	26.667	"	5,450	900		6,350	7,350	
	6000	Reciprocating, 2 stage, tank mtd, 3 Ph., Cap. rated @175 PSIG										
	6050	Pressure lubcatd, hvy duty, 9.7 CFM, 3 HP, 120 Gal. tank	Q-5	1.30	12.308	Ea.	4,050	400		4,450	5,050	
	6054	5 CFM, 1-1/2 HP, 80 gal tank		2.80	5.714		2,450	186		2,636	2,975	
	6056	6.4 CFM, 2 HP, 80 gal tank		2	8		2,575	261		2,836	3,250	
	6058	8.1 CFM, 3 HP, 80 gal tank		1.70	9.412		2,750	305		3,055	3,475	
	6059	14.8 CFM, 5 HP, 80 gal tank		1	16		2,850	520		3,370	3,900	
	6060	16.5 CFM, 5 HP, 120 gal. tank		1	16		4,100	520		4,620	5,275	
	6063	13 CFM, 6 HP, 80 gal tank		.90	17.778		4,325	580		4,905	5,625	
	6066	19.8 CFM, 7.5 HP, 80 gal tank		.80	20		4,550	650		5,200	5,975	
	6070	25.8 CFM, 7-1/2 HP, 120 gal. tank		.80	20		6,100	650		6,750	7,700	
	6078	34.8 CFM, 10 HP, 80 gal tank		.70	22.857		5,775	745		6,520	7,475	
	6080	34.8 CFM, 10 HP, 120 gal. tank	▼	.60	26.667		6,075	870		6,945	7,975	
	6090	53.7 CFM, 15 HP, 120 gal. tank	Q-6	.80	30		6,300	1,025		7,325	8,450	
	6100	76.7 CFM, 20 HP, 120 gal. tank		.70	34.286		9,400	1,150		10,550	12,100	
	6104	76.7 CFM, 20 HP, 240 gal tank		.68	35.294		9,250	1,200		10,450	12,000	
	6110	90.1 CFM, 25 HP, 120 gal. tank		.63	38.095		9,300	1,275		10,575	12,200	
	6120	101 CFM, 30 HP, 120 gal. tank		.57	42.105		9,500	1,425		10,925	12,600	
	6130	101 CFM, 30 HP, 250 gal. tank	▼	.52	46.154		11,500	1,550		13,050	15,000	
	6200	Oil-less, 13.6 CFM, 5 HP, 120 gal. tank	Q-5	.88	18.182		12,600	590		13,190	14,800	
	6210	13.6 CFM, 5 HP, 250 gal. tank		.80	20		13,500	650		14,150	15,900	
	6220	18.2 CFM, 7.5 HP, 120 gal. tank		.73	21.918		12,800	715		13,515	15,200	
	6230	18.2 CFM, 7.5 HP, 250 gal. tank		.67	23.881		13,700	780		14,480	16,300	
	6250	30.5 CFM, 10 HP, 120 gal. tank		.57	28.070		15,000	915		15,915	17,800	
	6260	30.5 CFM, 10 HP, 250 gal. tank	▼	.53	30.189		15,800	985		16,785	18,900	
	6270	41.3 CFM, 15 HP, 120 gal. tank	Q-6	.70	34.286		16,200	1,150		17,350	19,600	
	6280	41.3 CFM, 15 HP, 250 gal. tank	"	.67	35.821	▼	17,100	1,200		18,300	20,600	
200	0010	**COMPRESSOR ACCESSORIES**										**200**
	0100	Aftercooler										
	0130	Air cooled, based on 100 psig and 250°F										
	0140	Tank mounted single stage, 20 SCFM	Q-5	7.50	2.133	Ea.	650	69.50		719.50	820	
	0150	46 SCFM		7	2.286		755	74.50		829.50	940	
	0160	77 SCFM		6.50	2.462		1,250	80		1,330	1,525	
	0170	Floor mounted horizontal draft, 36 SCFM		6.50	2.462		670	80		750	855	
	0180	78 SCFM		6	2.667		830	87		917	1,050	
	0190	190 SCFM		5.70	2.807		1,275	91.50		1,366.50	1,550	
	0200	260 SCFM		5.20	3.077		1,425	100		1,525	1,725	
	0210	440 SCFM		4	4		1,900	130		2,030	2,300	
	0220	650 SCFM		3.50	4.571		2,075	149		2,224	2,500	
	0230	Floor mounted vertical draft, 850 SCFM		2.50	6.400		3,575	209		3,784	4,250	
	0260	2060 SCFM		2	8		5,875	261		6,136	6,850	
	0280	3640 SCFM	▼	1.70	9.412	▼	14,400	305		14,705	16,300	
	0300	Separator with automatic float trap										
	0330	600 lb/hr drain cap. 125 psi 3/4" conn.	Q-5	7.50	2.133	Ea.	325	69.50		394.50	460	
	0340	1200 lb/hr drain cap. 125 psi 1" conn.		7	2.286		345	74.50		419.50	490	
	0350	1200 lb/hr drain cap. 125 psi 1-1/2" conn.		6	2.667		390	87		477	560	
	0360	1200 lb/hr drain cap. 125 psi 2" conn.	▼	5	3.200	▼	420	104		524	620	

MECHANICAL 15

277

			DAILY	LABOR-		2002 BARE COSTS				TOTAL
15210	**Process Air/Gas Piping**	CREW	OUTPUT	HOURS	UNIT	MAT.	LABOR	EQUIP.	TOTAL	INCL O&P
200 0370	1500 CFM at 200 psi 2-1/2" conn.	Q-5	4	4	Ea.	995	130		1,125	1,300 **200**
0380	4000 CFM at 200 psi 4" conn.	↓	3	5.333	↓	1,075	174		1,249	1,425
0400	Water cooled, fixed tube at 250°F									
0430	52 CFM	Q-5	5	3.200	Ea.	420	104		524	620
0440	104 CFM		4	4		490	130		620	730
0450	182 CFM		3.50	4.571		755	149		904	1,050
0460	312 CFM		3	5.333		840	174		1,014	1,175
0480	494 CFM		2.80	5.714		1,750	186		1,936	2,200
0490	702 CFM		2.40	6.667		2,350	217		2,567	2,900
0500	806 CFM		2.20	7.273		2,775	237		3,012	3,400
0510	1092 CFM		2	8		3,450	261		3,711	4,200
0520	1430 CFM	↓	1.50	10.667	↓	4,050	350		4,400	5,000
0600	Separator, horizontal									
0610	100 CFM	Q-5	15	1.067	Ea.	229	35		264	305
0620	350 CFM		10	1.600		330	52		382	445
0630	900 CFM		7	2.286		805	74.50		879.50	995
0640	1500 CFM		5	3.200		890	104		994	1,125
0650	2800 CFM	↓	3.50	4.571	↓	1,300	149		1,449	1,675
0700	Dryers, air, deliquescent at 90 psi									
0730	21 CFM	Q-5	14	1.143	Ea.	680	37		717	805
0740	63 CFM		12	1.333		1,250	43.50		1,293.50	1,450
0750	140 CFM		8	2		2,225	65		2,290	2,525
0760	335 CFM		5.50	2.909		3,000	95		3,095	3,450
0770	525 CFM		5	3.200		3,925	104		4,029	4,475
0780	760 CFM		3	5.333		5,475	174		5,649	6,275
0790	1050 CFM		2.50	6.400		6,825	209		7,034	7,825
0800	1340 CFM		2	8		8,025	261		8,286	9,225
0810	1700 CFM	↓	1.50	10.667		10,200	350		10,550	11,700
0820	2550 CFM	Q-6	2	12		13,200	405		13,605	15,200
0830	3550 CFM		1.50	16		16,600	540		17,140	19,100
0840	4700 CFM	↓	1	24		20,700	810		21,510	23,900
0850	6100 CFM	Q-7	1	32		24,200	1,100		25,300	28,300
0860	8400 CFM		.80	40		30,000	1,375		31,375	35,100
0870	10200 CFM		.60	53.333		32,700	1,850		34,550	38,700
0880	12100 CFM		.50	64		34,900	2,200		37,100	41,700
0890	14200 CFM	↓	.40	80		36,900	2,750		39,650	44,700
0950	Desiccant 33 lb. bag					31.50			31.50	35
0960	55 lb. bag					51.50			51.50	57
0980	2200 lb. bulk bag 1 or 2					2,075			2,075	2,275
0990	2200 lb. bulk bag 3 or 4				↓	2,025			2,025	2,225
1250	Refrigeration, 50°F pressure dewpoint									
1280	20 CFM	Q-5	7	2.286	Ea.	1,275	74.50		1,349.50	1,525
1290	60 CFM		6	2.667		1,850	87		1,937	2,150
1300	130 CFM		5.50	2.909		2,375	95		2,470	2,775
1310	150 CFM		5	3.200		5,000	104		5,104	5,650
1320	200 CFM		4.50	3.556		5,125	116		5,241	5,825
1330	280 CFM		4	4		5,600	130		5,730	6,350
1340	350 CFM		3.50	4.571		6,100	149		6,249	6,925
1350	450 CFM		3	5.333		7,025	174		7,199	7,975
1360	520 CFM		2.80	5.714		7,725	186		7,911	8,775
1370	650 CFM		2.40	6.667		8,500	217		8,717	9,700
1380	800 CFM		2.20	7.273		9,075	237		9,312	10,400
1390	1000 CFM		2	8		11,700	261		11,961	13,300
1400	1400 CFM		1.50	10.667		13,800	350		14,150	15,600
1410	2000 CFM		1.20	13.333		16,800	435		17,235	19,200
1420	2400 CFM	↓	1	16	↓	24,200	520		24,720	27,400

Important: See the Reference Section for critical supporting data - Reference Nos., Crews, & City Cost Indexes

15210	Process Air/Gas Piping	CREW	DAILY OUTPUT	LABOR-HOURS	UNIT	2002 BARE COSTS				TOTAL INCL O&P
						MAT.	LABOR	EQUIP.	TOTAL	
1430	3000 CFM	Q-6	1.30	18.462	Ea.	32,900	625		33,525	37,000
1440	3600 CFM		1.20	20		37,000	675		37,675	41,700
1450	4800 CFM		1	24		43,200	810		44,010	48,800
1460	6000 CFM		.80	30		53,000	1,025		54,025	60,000
1470	7200 CFM	▼	.50	48	▼	62,000	1,625		63,625	71,000
1500	Package includes dryer & filtration systems									
1530	150 CFM	Q-5	4.20	3.810	Ea.	7,300	124		7,424	8,200
1540	200 CFM		3.90	4.103		7,450	134		7,584	8,400
1550	280 CFM		3.50	4.571		8,550	149		8,699	9,625
1560	350 CFM		3.20	5		9,600	163		9,763	10,800
1570	450 CFM		3	5.333		10,600	174		10,774	12,000
1580	520 CFM		2.50	6.400		11,300	209		11,509	12,700
1590	650 CFM		2.20	7.273		11,700	237		11,937	13,300
1600	800 CFM		2	8		12,400	261		12,661	14,100
1610	1000 CFM		1.80	8.889		15,900	290		16,190	17,900
1620	1200 CFM		1.40	11.429		17,900	370		18,270	20,300
1630	1400 CFM		1.20	13.333		19,500	435		19,935	22,200
1640	2000 CFM		1	16		22,500	520		23,020	25,500
1650	2400 CFM	▼	.70	22.857	▼	29,000	745		29,745	33,000
2000	Regenerative adsorption with desiccant									
2030	50 CFM	Q-5	5	3.200	Ea.	4,400	104		4,504	5,000
2040	75 CFM		4.50	3.556		4,700	116		4,816	5,325
2050	100 CFM		4.20	3.810		5,150	124		5,274	5,825
2060	125 CFM		4	4		5,600	130		5,730	6,350
2070	170 CFM		3.70	4.324		6,050	141		6,191	6,850
2080	240 CFM		3.40	4.706		7,350	153		7,503	8,300
2090	350 CFM		3	5.333		7,875	174		8,049	8,900
2100	520 CFM		2.50	6.400		9,400	209		9,609	10,700
2110	650 CFM		2	8		10,800	261		11,061	12,300
2120	800 CFM		1.50	10.667		13,600	350		13,950	15,400
2130	1000 CFM		1.20	13.333		15,800	435		16,235	18,000
2140	1200 CFM		1	16		16,600	520		17,120	19,000
2150	1600 CFM		.80	20		20,700	650		21,350	23,800
2160	2000 CFM		.70	22.857		23,600	745		24,345	27,100
2170	2500 CFM	▼	.60	26.667		29,200	870		30,070	33,400
2180	3000 CFM	Q-6	.90	26.667		30,500	900		31,400	35,000
2190	3500 CFM		.70	34.286		44,200	1,150		45,350	50,500
2200	4000 CFM	▼	.50	48		45,500	1,625		47,125	52,500
2300	Desiccant, 50 lb. bag					126			126	138
2330	350 lb. drum				▼	875			875	965
3000	Filters Air									
3100	Adsorptive, odors by glass fiber & carbon, 90 psi									
3140	110 CFM	Q-5	8	2	Ea.	865	65		930	1,050
3150	190 CFM		7.50	2.133		985	69.50		1,054.50	1,175
3160	290 CFM		7	2.286		1,225	74.50		1,299.50	1,450
3170	380 CFM		6	2.667		1,650	87		1,737	1,950
3180	570 CFM		5	3.200		2,325	104		2,429	2,725
3190	870 CFM		4.50	3.556		2,650	116		2,766	3,100
3200	1160 CFM		4	4		3,400	130		3,530	3,925
3210	1450 CFM		3.50	4.571		3,975	149		4,124	4,600
3220	1740 CFM	▼	3	5.333	▼	4,450	174		4,624	5,125
3260	Coalescent system, liquids & solids 90 psi high purity									
3300	110 CFM	Q-5	7	2.286	Ea.	3,225	74.50		3,299.50	3,650
3310	190 CFM		6.70	2.388		3,500	78		3,578	3,975
3320	290 CFM		6.50	2.462		3,700	80		3,780	4,200
3330	380 CFM	▼	6	2.667	▼	4,350	87		4,437	4,925

200

MECHANICAL 15

15210	Process Air/Gas Piping	CREW	DAILY OUTPUT	LABOR-HOURS	UNIT	2002 BARE COSTS				TOTAL INCL O&P
						MAT.	LABOR	EQUIP.	TOTAL	
200											**200**
3340	570 CFM	Q-5	4.50	3.556	Ea.	5,450	116		5,566	6,175
3350	870 CFM		4	4		5,775	130		5,905	6,550
3360	1160 CFM		3.50	4.571		6,550	149		6,699	7,450
3370	1450 CFM		3	5.333		7,400	174		7,574	8,375
3380	1740 CFM		2.50	6.400		7,825	209		8,034	8,925
3390	2030 CFM	↓	2	8	↓	8,100	261		8,361	9,300
3500	Combination air line filter, regulator and lubricator									
3510	1/4" NPT	1 Stpi	16	.500	Ea.	121	18.10		139.10	161
3520	3/8" NPT		16	.500		137	18.10		155.10	178
3530	1/2" NPT		15	.533		137	19.30		156.30	179
3540	3/4" NPT		14	.571		290	20.50		310.50	350
3550	1" NPT	↓	12	.667		290	24		314	355
4000	Couplers, air line, sleeve type									
4010	Female, connection size NPT									
4020	1/4"	1 Stpi	38	.211	Ea.	4.30	7.60		11.90	16.25
4030	3/8"		36	.222		5.35	8.05		13.40	18.05
4040	1/2"		35	.229		11.25	8.25		19.50	25
4050	3/4"	↓	34	.235	↓	12.90	8.50		21.40	27
4100	Male									
4110	1/4"	1 Stpi	38	.211	Ea.	4.25	7.60		11.85	16.20
4120	3/8"		36	.222		5.35	8.05		13.40	18.05
4130	1/2"		35	.229		10.65	8.25		18.90	24
4140	3/4"	↓	34	.235	↓	12.60	8.50		21.10	26.50
4150	Coupler, combined male and female halves									
4160	1/2"	1 Stpi	17	.471	Ea.	22	17.05		39.05	49.50
4170	3/4"	"	15	.533	"	25.50	19.30		44.80	57
900	0010 **VACUUM PUMPS**										**900**
0100	Medical, with receiver									
0110	Duplex									
0120	20 SCFM	Q-1	1.14	14.035	Ea.	14,100	455		14,555	16,200
0130	60 SCFM, 10 HP	Q-2	1.20	20	"	24,000	670		24,670	27,300
0200	Triplex									
0220	180 SCFM	Q-2	.86	27.907	Ea.	46,400	935		47,335	52,500
0300	Dental oral									
0310	Duplex									
0330	165 SCFM with 77 Gal separator	Q-2	1.30	18.462	Ea.	42,000	620		42,620	47,100

15230	Industrial Process Piping									
500	0010 **PUMPS, GENERAL UTILITY** With motor										**500**
0200	Multi-stage, horizontal split, for boiler feed applications									
0300	Two stage, 3" discharge x 4" suction, 75 HP	Q-7	.30	106	Ea.	14,900	3,675		18,575	22,000
0340	Four stage, 3" discharge x 4" suction, 150 HP	"	.18	177	"	24,500	6,125		30,625	36,300
2000	Single stage									
2060	End suction, 1"D. x 2"S., 3 HP	Q-1	.50	32	Ea.	3,650	1,025		4,675	5,550
2100	1-1/2"D. x 3"S., 10 HP	"	.40	40		3,925	1,300		5,225	6,275
2140	2"D. x 3"S., 15 HP	Q-2	.60	40		4,275	1,350		5,625	6,725
2180	3"D. x 4"S., 20 HP		.50	48		4,550	1,600		6,150	7,425
2220	4"D. x 6"S., 30 HP	↓	.40	60		4,925	2,025		6,950	8,450
3000	Double suction, 2"D. x 2-1/2"S., 10 HP	Q-1	.30	53.333		5,050	1,725		6,775	8,150
3060	3"D. x 4"S., 15 HP	Q-2	.46	52.174		5,200	1,750		6,950	8,375
3100	4"D. x 5"S., 30 HP		.40	60		7,050	2,025		9,075	10,800
3140	5"D. x 6"S., 50 HP	↓	.33	72.727		7,575	2,450		10,025	12,000
3180	6"D. x 8"S., 60 HP	Q-3	.30	106		8,475	3,650		12,125	14,800
3190	75 HP, to 2500 GPM		.28	114		10,800	3,925		14,725	17,800
3220	100 HP, to 3000 GPM		.26	123		13,700	4,225		17,925	21,500
3240	150 HP, to 4000 GPM	↓	.24	133	↓	18,500	4,575		23,075	27,200

15 MECHANICAL

Important: See the Reference Section for critical supporting data - Reference Nos., Crews, & City Cost Indexes

15230	Industrial Process Piping	CREW	DAILY OUTPUT	LABOR-HOURS	UNIT	MAT.	LABOR	EQUIP.	TOTAL	TOTAL INCL O&P	
						2002 BARE COSTS					
500	4000	Centrifugal, end suction, mounted on base										500
4010	Horizontal mounted, with drip proof motor, rated @ 100' head										
4020	Vertical split case, single stage										
4040	100 GPM, 5 HP, 1-1/2" discharge	Q-1	1.70	9.412	Ea.	1,975	305		2,280	2,625	
4050	200 GPM, 10 HP, 2" discharge		1.30	12.308		2,400	400		2,800	3,225	
4060	250 GPM, 10 HP, 3" discharge		1.28	12.500		2,500	405		2,905	3,350	
4070	300 GPM, 15 HP, 2" discharge	Q-2	1.56	15.385		2,750	515		3,265	3,800	
4080	500 GPM, 20 HP, 4" discharge		1.44	16.667		3,200	560		3,760	4,350	
4090	750 GPM, 30 HP, 4" discharge		1.20	20		4,275	670		4,945	5,700	
4100	1050 GPM, 40 HP, 5" discharge		1	24		4,625	805		5,430	6,300	
4110	1500 GPM, 60 HP, 6" discharge		.60	40		6,725	1,350		8,075	9,425	
4120	2000 GPM, 75 HP, 6" discharge		.50	48		7,725	1,600		9,325	10,900	
4130	3000 GPM, 100 HP, 8" discharge		.40	60		11,600	2,025		13,625	15,800	
4200	Horizontal split case, single stage										
4210	100 GPM, 7.5 HP, 1-1/2" discharge	Q-1	1.70	9.412	Ea.	3,275	305		3,580	4,075	
4220	250 GPM, 15 HP, 2-1/2" discharge	"	1.30	12.308		4,550	400		4,950	5,600	
4230	500 GPM, 20 HP, 4" discharge	Q-2	1.60	15		5,200	505		5,705	6,475	
4240	750 GPM, 25 HP, 5" discharge		1.54	15.584		5,375	525		5,900	6,725	
4250	1000 GPM, 40 HP, 5" discharge		1.20	20		7,050	670		7,720	8,750	
4260	1500 GPM, 50 HP, 6" discharge	Q-3	1.42	22.535		8,850	775		9,625	10,900	
4270	2000 GPM, 75 HP, 8" discharge		1.14	28.070		12,400	960		13,360	15,200	
4280	3000 GPM, 100 HP, 10" discharge		.96	33.333		13,400	1,150		14,550	16,400	
4290	3500 GPM, 150 HP, 10" discharge		.86	37.209		16,700	1,275		17,975	20,300	
4300	4000 GPM, 200 HP, 10" discharge		.66	48.485		18,500	1,650		20,150	22,800	
4330	Horizontal split case, two stage, 500' head										
4340	100 GPM, 40 HP, 1-1/2" discharge	Q-2	1.70	14.118	Ea.	4,300	475		4,775	5,450	
4350	200 GPM, 50 HP, 1-1/2" discharge	"	1.44	16.667		4,800	560		5,360	6,125	
4360	300 GPM, 75 HP, 2" discharge	Q-3	1.57	20.382		5,100	700		5,800	6,650	
4370	400 GPM, 100 HP, 3" discharge		1.14	28.070		6,550	960		7,510	8,650	
4380	800 GPM, 200 HP, 4" discharge		.86	37.209		15,300	1,275		16,575	18,800	
5000	Centrifugal, in-line										
5006	Vertical mount, iron body, 125 lb. flgd, 3550 RPM TEFC mtr										
5010	Single stage										
5020	50 GPM, 3 HP, 1-1/2" discharge	Q-1	2.30	6.957	Ea.	525	225		750	915	
5030	75 GPM, 5 HP, 1-1/2" discharge		1.60	10		885	325		1,210	1,475	
5040	100 GPM, 7.5 HP, 1-1/2" discharge		1.30	12.308		955	400		1,355	1,650	
5050	125 GPM, 10 HP, 1-1/2" discharge	Q-2	1.70	14.118		1,175	475		1,650	2,025	
5060	150 GPM, 15 HP, 1-1/2" discharge		1.60	15		1,525	505		2,030	2,425	
5070	200 GPM, 30 HP, 1-1/2" discharge		1.50	16		2,475	535		3,010	3,525	
5080	250 GPM, 40 HP, 2" discharge		1.40	17.143		3,275	575		3,850	4,475	
5090	300 GPM, 50 HP, 3" discharge		1.30	18.462		3,975	620		4,595	5,300	
5100	400 GPM, 75 HP, 3" discharge	Q-3	.60	53.333		6,475	1,825		8,300	9,875	
5110	600 GPM, 100 HP, 3" discharge	"	.50	64		8,675	2,200		10,875	12,900	
8000	Vertical submerged, with non-submerged motor										
8060	1"D., 3 HP	Q-1	.50	32	Ea.	5,275	1,025		6,300	7,350	
8100	1-1/2"D., 10 HP	"	.30	53.333		5,525	1,725		7,250	8,675	
8140	2"D., 15 HP	Q-2	.40	60		5,675	2,025		7,700	9,275	
8180	3"D., 25 HP		.30	80		5,800	2,675		8,475	10,400	
8220	4"D., 30 HP		.20	120		6,500	4,025		10,525	13,200	
600	0010	**PUMPS MISCELLANEOUS**										600
0020	Water pump, portable, gasoline powered										
0100	6000 GPH, 2" discharge	Q-1	11	1.455	Ea.	800	47		847	950	
0110	8000 GPH, 2" discharge		10.50	1.524		800	49.50		849.50	955	
0120	10,000 GPH, 2" discharge		10	1.600		1,300	52		1,352	1,500	
0500	Pump, propylene body, housing and impeller										

MECHANICAL 15

281

15230	Industrial Process Piping	CREW	DAILY OUTPUT	LABOR-HOURS	UNIT	2002 BARE COSTS				TOTAL INCL O&P
						MAT.	LABOR	EQUIP.	TOTAL	
600 0510	22 GPM, 1/3 HP, 40' HD	1 Plum	5	1.600	Ea.	515	57.50		572.50	655	600
0520	33 GPM, 1/2 HP, 40' HD	Q-1	5	3.200		525	104		629	730
0530	53 GPM, 3/4 HP, 40' HD	"	4	4	↓	560	129		689	815
0600	Rotary pump, CI									
0610	10 GPM, 1/2 HP, 1" discharge	Q-1	5	3.200	Ea.	945	104		1,049	1,200
0614	10 GPM, 3/4 HP, 1" discharge		4.50	3.556		945	115		1,060	1,225
0618	10 GPM 1 HP, 1" discharge		4	4		980	129		1,109	1,275
0620	25 GPM, 1 HP, 1-1/4" discharge		3.80	4.211		980	136		1,116	1,275
0624	25 GPM, 1.5 HP, 1-1/4" discharge		3.60	4.444		1,075	144		1,219	1,400
0628	25 GPM, 2 HP, 1-1/4" discharge	↓	3.20	5	↓	1,075	162		1,237	1,425
0700	Fuel oil pump, 2 stage, 3450 RPM									
0710	1/4 HP, 100 PSI	Q-1	8	2	Ea.	1,600	64.50		1,664.50	1,875
0800	Hydraulic oil pump									
0810	75 HP, 3 PH, 480 V	Q-3	1.20	26.667	Ea.	5,175	915		6,090	7,050
1000	Turbine pump, CI									
1010	50 GPM, 2 HP, 3" discharge	Q-1	.80	20	Ea.	1,075	645		1,720	2,175
1020	100 GPM, 3 HP, 4" discharge	Q-2	.96	25		1,625	840		2,465	3,050
1030	250 GPM, 15 HP, 6" discharge	"	.94	25.532		3,250	855		4,105	4,875
1040	500 GPM, 25 HP, 6" discharge	Q-3	1.30	24.615		5,400	845		6,245	7,225
1050	1000 GPM, 50 HP, 8" discharge		1.14	28.070		8,650	960		9,610	11,000
1060	2000 GPM, 100 HP, 10" discharge		1	32		9,725	1,100		10,825	12,400
1070	3000 GPM, 150 HP, 10" discharge		.80	40		10,800	1,375		12,175	14,000
1080	4000 GPM, 200 HP, 12" discharge		.70	45.714		16,200	1,575		17,775	20,200
1090	6000 GPM, 300 HP, 14" discharge		.60	53.333		17,900	1,825		19,725	22,500
1100	10,000 GPM, 300 HP, 18" discharge	↓	.58	55.172	↓	37,900	1,900		39,800	44,500

15410	Plumbing Fixtures	CREW	DAILY OUTPUT	LABOR-HOURS	UNIT	2002 BARE COSTS				TOTAL INCL O&P
						MAT.	LABOR	EQUIP.	TOTAL	
300 0010	**FAUCETS/FITTINGS**										300
5000	Sillcock, compact, brass, IPS or copper to hose	1 Plum	24	.333	Ea.	4.94	12		16.94	23.50
6000	Stop and waste valves, bronze									
6100	Angle, solder end 1/2"	1 Plum	24	.333	Ea.	4.18	12		16.18	22.50
6110	3/4"		20	.400		4.59	14.40		18.99	26.50
6300	Straightway, solder end 3/8"		24	.333		3.59	12		15.59	22
6310	1/2"		24	.333		2.32	12		14.32	20.50
6320	3/4"		20	.400		2.68	14.40		17.08	24.50
6330	1"		19	.421		4	15.15		19.15	27.50
6400	Straightway, threaded 3/8"		24	.333		4.07	12		16.07	22.50
6410	1/2"		24	.333		3.13	12		15.13	21.50
6420	3/4"		20	.400		3.73	14.40		18.13	25.50
6430	1"	↓	19	.421	↓	10.50	15.15		25.65	34.50

15411	Commercial/Indust Fixtures	CREW	DAILY OUTPUT	LABOR-HOURS	UNIT	2002 BARE COSTS				TOTAL INCL O&P
						MAT.	LABOR	EQUIP.	TOTAL	
500 0010	**HYDRANTS**										500
0050	Wall type, moderate climate, bronze, encased									
0200	3/4" IPS connection	1 Plum	16	.500	Ea.	226	18		244	276
0300	1" IPS connection	"	14	.571	↓	226	20.50		246.50	280

15411	Commercial/Indust Fixtures	CREW	DAILY OUTPUT	LABOR-HOURS	UNIT	2002 BARE COSTS				TOTAL INCL O&P
						MAT.	LABOR	EQUIP.	TOTAL	
500 0500	Anti-siphon type				Ea.	30.50			30.50	33.50	500
1000	Non-freeze, bronze, exposed									
1100	3/4" IPS connection, 4" to 9" thick wall	1 Plum	14	.571	Ea.	153	20.50		173.50	200
1120	10" to 14" thick wall		12	.667		166	24		190	219
1140	15" to 19" thick wall		12	.667		185	24		209	240
1160	20" to 24" thick wall		10	.800		201	29		230	265
1200	For 1" IPS connection, add					15%	10%			
1240	For 3/4" adapter type vacuum breaker, add				Ea.	19.25			19.25	21
1280	For anti-siphon type				"	40.50			40.50	44.50
2000	Non-freeze bronze, encased									
2100	3/4" IPS connection, 5" to 9" thick wall	1 Plum	14	.571	Ea.	264	20.50		284.50	320
2120	10" to 14" thick wall		12	.667		277	24		301	340
2140	15" to 19" thick wall		12	.667		297	24		321	360
2160	20" to 24" thick wall		10	.800		310	29		339	385
2200	For 1" IPS connection, add					10%	10%			
2280	Anti-siphon type				Ea.	124			124	136
3000	Ground box type, bronze frame, 3/4" IPS connection									
3080	Non-freeze, all bronze, polished face, set flush									
3100	2 feet depth of bury	1 Plum	8	1	Ea.	282	36		318	365
3120	3 feet depth of bury		8	1		305	36		341	390
3140	4 feet depth of bury		8	1		325	36		361	410
3160	5 feet depth of bury		7	1.143		350	41		391	440
3180	6 feet depth of bury		7	1.143		370	41		411	465
3200	7 feet depth of bury		6	1.333		390	48		438	505
3220	8 feet depth of bury		5	1.600		415	57.50		472.50	540
3240	9 feet depth of bury		4	2		435	72		507	590
3260	10 feet depth of bury		4	2		455	72		527	610
3400	For 1" IPS connection, add					15%	10%			
3450	For 1-1/4" IPS connection, add					325%	14%			
3500	For 1-1/2" connection, add					370%	18%			
3550	For 2" connection, add					445%	24%			
3600	For tapped drain port in box, add					25.50			25.50	28.50
4000	Non-freeze, CI body, bronze frame & scoriated cover									
4010	with hose storage									
4100	2 feet depth of bury	1 Plum	7	1.143	Ea.	535	41		576	645
4120	3 feet depth of bury		7	1.143		555	41		596	670
4140	4 feet depth of bury		7	1.143		575	41		616	690
4160	5 feet depth of bury		6.50	1.231		585	44.50		629.50	710
4180	6 feet depth of bury		6	1.333		595	48		643	730
4200	7 feet depth of bury		5.50	1.455		615	52.50		667.50	760
4220	8 feet depth of bury		5	1.600		640	57.50		697.50	785
4240	9 feet depth of bury		4.50	1.778		660	64		724	820
4260	10 feet depth of bury		4	2		680	72		752	860
4280	For 1" IPS connection, add					117			117	129
4300	For tapped drain port in box, add					25.50			25.50	28.50
5000	Moderate climate, all bronze, polished face									
5020	and scoriated cover, set flush									
5100	3/4" IPS connection	1 Plum	16	.500	Ea.	201	18		219	248
5120	1" IPS connection	"	14	.571		200	20.50		220.50	251
5200	For tapped drain port in box, add					25.50			25.50	28.50
6000	Ground post type, all non-freeze, all bronze, aluminum casing									
6010	guard, exposed head, 3/4" IPS connection									
6100	2 feet depth of bury	1 Plum	8	1	Ea.	283	36		319	365
6120	3 feet depth of bury		8	1		305	36		341	390
6140	4 feet depth of bury		8	1		330	36		366	420
6160	5 feet depth of bury		7	1.143		355	41		396	450

15411	Commercial/Indust Fixtures	CREW	DAILY OUTPUT	LABOR-HOURS	UNIT	2002 BARE COSTS				TOTAL INCL O&P	
						MAT.	LABOR	EQUIP.	TOTAL		
500	**6180**	6 feet depth of bury	1 Plum	7	1.143	Ea.	375	41		416	475	**500**
	6200	7 feet depth of bury		6	1.333		405	48		453	520	
	6220	8 feet depth of bury		5	1.600		430	57.50		487.50	555	
	6240	9 feet depth of bury		4	2		455	72		527	610	
	6260	10 feet depth of bury	▼	4	2	▼	480	72		552	640	
	6300	For 1" IPS connection, add					40%	10%				
	6350	For 1-1/4" IPS connection, add					140%	14%				
	6400	For 1-1/2" IPS connection, add					225%	18%				
	6450	For 2" IPS connection, add					315%	24%				

15440	Plumbing Pumps	CREW	DAILY OUTPUT	LABOR-HOURS	UNIT	MAT.	LABOR	EQUIP.	TOTAL	TOTAL INCL O&P	
240	**0010**	**PUMPS, PRESSURE BOOSTER SYSTEM**										**240**
	0200	Pump system, with diaphragm tank, control, press. switch										
	0300	1 HP pump	Q-1	1.30	12.308	Ea.	3,275	400		3,675	4,200	
	0400	1-1/2 HP pump		1.25	12.800		3,325	415		3,740	4,275	
	0420	2 HP pump		1.20	13.333		3,400	430		3,830	4,375	
	0440	3 HP pump	▼	1.10	14.545		3,450	470		3,920	4,500	
	0460	5 HP pump	Q-2	1.50	16		3,800	535		4,335	5,000	
	0480	7-1/2 HP pump		1.42	16.901		4,250	565		4,815	5,525	
	0500	10 HP pump	▼	1.34	17.910	▼	4,450	600		5,050	5,775	
	1000	Pump/ energy storage system, diaphragm tank, 3 HP pump										
	1100	motor, PRV, switch, gauge, control center, flow switch										
	1200	125 lb. working pressure	Q-2	.70	34.286	Ea.	9,550	1,150		10,700	12,200	
	1300	250 lb. working pressure	"	.64	37.500	"	10,400	1,250		11,650	13,400	
400	**0010**	**PUMPS, GRINDER SYSTEM** Complete, incl. check valve, tank, std.										**400**
	0020	controls incl. alarm/disconnect panel w/wire. Excavation not included										
	0260	Simplex, 9 GPM at 60 PSIG, 70 gal. tank				Ea.	2,225			2,225	2,450	
	0300	For manway, 26" I.D., 18" high, add					420			420	460	
	0340	26" I.D., 36" high, add					490			490	540	
	0380	43" I.D., 4' high, add					540			540	590	
	0600	Simplex, 9 GPM at 60 PSIG, 150 gal. tank				▼	2,525			2,525	2,775	
	0660	For manway, add										
	0700	26" I.D., 36" high, add				Ea.	505			505	555	
	0740	26" I.D., 4' high, add					585			585	645	
	2000	Duplex, 18 GPM at 60 PSIG, 150 gal. tank					4,175			4,175	4,600	
	2060	For manway 43" I.D., 4' high, add					1,325			1,325	1,475	
	2400	For core only					1,650			1,650	1,825	
	2410	For salt water core, add				▼	161			161	177	
	2600	For addit. manway 5' to 10' high, 60 gal tank, per ft., add				V.L.F.	149			149	164	
	2620	For manways 5' to 10', 120 gal tank, add				Ea.	162			162	178	
800	**0010**	**PUMPS, SEWAGE EJECTOR** With operating and level controls										**800**
	0100	Simplex system incl. tank, cover, pump 15' head										
	0500	37 gal PE tank, 12 GPM, 1/2 HP, 2" discharge	Q-1	3.20	5	Ea.	375	162		537	660	
	0510	3" discharge		3.10	5.161		405	167		572	695	
	0530	87 GPM, .7 HP, 2" discharge		3.20	5		580	162		742	880	
	0540	3" discharge		3.10	5.161		625	167		792	940	
	0600	45 gal. coated stl tank, 12 GPM, 1/2 HP, 2" discharge		3	5.333		665	173		838	990	
	0610	3" discharge		2.90	5.517		695	178		873	1,025	
	0630	87 GPM, .7 HP, 2" discharge		3	5.333		855	173		1,028	1,200	
	0640	3" discharge		2.90	5.517		905	178		1,083	1,275	
	0660	134 GPM, 1 HP, 2" discharge		2.80	5.714		925	185		1,110	1,300	
	0680	3" discharge		2.70	5.926		975	192		1,167	1,375	
	0700	70 gal. PE tank, 12 GPM, 1/2 HP, 2" discharge		2.60	6.154		755	199		954	1,125	
	0710	3" discharge		2.40	6.667	▼	805	216		1,021	1,200	

Important: See the Reference Section for critical supporting data - Reference Nos., Crews, & City Cost Indexes

15440 | Plumbing Pumps

			CREW	DAILY OUTPUT	LABOR-HOURS	UNIT	2002 BARE COSTS				TOTAL INCL O&P	
							MAT.	LABOR	EQUIP.	TOTAL		
800	0730	87 GPM, 0.7 HP, 2" discharge	Q-1	2.50	6.400	Ea.	1,000	207		1,207	1,400	800
	0740	3" discharge		2.30	6.957		1,025	225		1,250	1,475	
	0760	134 GPM, 1 HP, 2" discharge		2.20	7.273		1,050	235		1,285	1,500	
	0770	3" discharge		2	8		1,125	259		1,384	1,625	
	0800	75 gal. coated stl. tank, 12 GPM, 1/2 HP, 2" discharge		2.40	6.667		755	216		971	1,150	
	0810	3" discharge		2.20	7.273		785	235		1,020	1,225	
	0830	87 GPM, .7 HP, 2" discharge		2.30	6.957		950	225		1,175	1,400	
	0840	3" discharge		2.10	7.619		1,000	246		1,246	1,475	
	0860	134 GPM, 1 HP, 2" discharge		2	8		1,025	259		1,284	1,525	
	0880	3" discharge	▼	1.80	8.889	▼	1,075	288		1,363	1,600	
	1040	Duplex system incl. tank, covers, pumps										
	1060	110 gal. fiberglass tank, 24 GPM, 1/2 HP, 2" discharge	Q-1	1.60	10	Ea.	1,475	325		1,800	2,125	
	1080	3" discharge		1.40	11.429		1,525	370		1,895	2,225	
	1100	174 GPM, .7 HP, 2" discharge		1.50	10.667		1,925	345		2,270	2,650	
	1120	3" discharge		1.30	12.308		2,000	400		2,400	2,800	
	1140	268 GPM, 1 HP, 2" discharge		1.20	13.333		2,100	430		2,530	2,950	
	1160	3" discharge	▼	1	16		2,175	520		2,695	3,150	
	1260	135 gal. coated stl. tank, 24 GPM, 1/2 HP, 2" discharge	Q-2	1.70	14.118		1,500	475		1,975	2,350	
	2000	3" discharge		1.60	15		1,575	505		2,080	2,475	
	2640	174 GPM, .7 HP, 2" discharge		1.60	15		1,950	505		2,455	2,900	
	2660	3" discharge		1.50	16		2,050	535		2,585	3,050	
	2700	268 GPM, 1 HP, 2" discharge		1.30	18.462		2,125	620		2,745	3,250	
	3040	3" discharge		1.10	21.818		2,225	730		2,955	3,550	
	3060	275 gal. coated stl. tank, 24 GPM, 1/2 HP, 2" discharge		1.50	16		1,850	535		2,385	2,850	
	3080	3" discharge		1.40	17.143		1,875	575		2,450	2,950	
	3100	174 GPM, .7 HP, 2" discharge		1.40	17.143		2,400	575		2,975	3,500	
	3120	3" discharge		1.30	18.462		2,425	620		3,045	3,575	
	3140	268 GPM, 1 HP, 2" discharge		1.10	21.818		2,625	730		3,355	4,000	
	3160	3" discharge	▼	.90	26.667	▼	2,725	895		3,620	4,350	
	3260	Pump system accessories, add										
	3300	Alarm horn and lights, 115V mercury switch	Q-1	8	2	Ea.	84	64.50		148.50	190	
	3340	Switch, mag. contactor, alarm bell, light, 3 level control		5	3.200		400	104		504	595	
	3380	Alternator, mercury switch activated	▼	4	4	▼	810	129		939	1,075	
900	0010	**PUMPS, PEDESTAL SUMP** With float control										900
	0400	Molded PVC base, 21 GPM at 15' head, 1/3 HP	1 Plum	5	1.600	Ea.	89	57.50		146.50	184	
	0800	Iron base, 21 GPM at 15' head, 1/3 HP		5	1.600		114	57.50		171.50	212	
	1200	Solid brass, 21 GPM at 15' head, 1/3 HP	▼	5	1.600	▼	185	57.50		242.50	290	
940	0010	**PUMPS, SUBMERSIBLE** Sump										940
	7000	Sump pump, automatic										
	7100	Plastic, 1-1/4" discharge, 1/4 HP	1 Plum	6	1.333	Ea.	105	48		153	189	
	7140	1/3 HP		5	1.600		127	57.50		184.50	227	
	7160	1/2 HP		5	1.600		153	57.50		210.50	255	
	7180	1-1/2" discharge, 1/2 HP		4	2		171	72		243	296	
	7500	Cast iron, 1-1/4" discharge, 1/4 HP		6	1.333		125	48		173	210	
	7540	1/3 HP		6	1.333		140	48		188	227	
	7560	1/2 HP	▼	5	1.600	▼	176	57.50		233.50	280	

15450 | Potable Water Tanks

			CREW	DAILY OUTPUT	LABOR-HOURS	UNIT	MAT.	LABOR	EQUIP.	TOTAL	TOTAL INCL O&P	
900	0010	**WATER HEATER STORAGE TANKS** 125 psi ASME										900
	2000	Galvanized steel, 15 gal., 14" diam., 26" LOA	1 Plum	12	.667	Ea.	505	24		529	590	
	2060	30 gal., 14" diam. x 49" LOA		11	.727		600	26		626	700	
	2080	80 gal., 20" diam. x 64" LOA		9	.889		930	32		962	1,075	
	2100	135 gal., 24" diam. x 75" LOA		6	1.333		1,375	48		1,423	1,575	
	2120	240 gal., 30" diam. x 86" LOA	▼	4	2	▼	2,375	72		2,447	2,700	

MECHANICAL 15

15400 | Plumbing Fixtures & Equipment

		15450	Potable Water Tanks	CREW	DAILY OUTPUT	LABOR-HOURS	UNIT	2002 BARE COSTS MAT.	LABOR	EQUIP.	TOTAL	TOTAL INCL O&P	
900	2140		300 gal., 36" diam. x 76" LOA	1 Plum	3	2.667	Ea.	3,850	96		3,946	4,375	900
	2160		400 gal., 36" diam. x 100" LOA	Q-1	4	4		4,750	129		4,879	5,425	
	2180		500 gal., 36" diam., x 126" LOA	"	3	5.333		5,850	173		6,023	6,700	
	3000		Glass lined, P.E., 80 gal., 20" diam. x 60" LOA	1 Plum	9	.889		1,600	32		1,632	1,800	
	3060		140 gal., 24" diam. x 80" LOA		6	1.333		2,275	48		2,323	2,575	
	3080		225 gal., 30" diam. x 78" LOA		4	2		2,650	72		2,722	3,025	
	3100		325 gal., 36" diam. x 81" LOA	▼	3	2.667		3,400	96		3,496	3,900	
	3120		460 gal., 42" diam. x 84" LOA	Q-1	4	4		3,575	129		3,704	4,150	
	3140		605 gal., 48" diam. x 87" LOA		3	5.333		6,850	173		7,023	7,775	
	3160		740 gal., 54" diam. x 91" LOA		3	5.333		7,700	173		7,873	8,725	
	3180		940 gal., 60" diam. x 93" LOA		2.50	6.400		8,900	207		9,107	10,100	
	3200		1330 gal., 66" diam. x 107" LOA		2	8		12,200	259		12,459	13,800	
	3220		1615 gal., 72" diam. x 110" LOA	▼	1.50	10.667		13,900	345		14,245	15,700	
	3240		2285 gal., 84" diam. x 128" LOA		1	16		17,100	520		17,620	19,600	
	3260		3440 gal., 96" diam. x 157" LOA	Q-2	1.50	16	▼	25,400	535		25,935	28,700	

		15480	Domestic Water Heaters										
900	0010	**HEAT TRANSFER PACKAGES** Complete, controls,											900
	0020	expansion tank, converter, air separator											
	1000	Hot water, 180°F enter, 200°F leaving, 15# steam											
	1010	One pump system, 28 GPM	Q-6	.75	32	Ea.	11,100	1,075		12,175	13,800	
	1020	35 GPM		.70	34.286		12,200	1,150		13,350	15,200	
	1040	55 GPM		.65	36.923		13,900	1,250		15,150	17,200	
	1060	130 GPM		.55	43.636		17,200	1,475		18,675	21,200	
	1080	255 GPM		.40	60		23,700	2,025		25,725	29,100	
	1100	550 GPM		.30	80		30,100	2,700		32,800	37,200	
	1120	800 GPM		.25	96		36,900	3,250		40,150	45,500	
	1220	Two pump system, 28 GPM		.70	34.286		14,700	1,150		15,850	18,000	
	1240	35 GPM		.65	36.923		17,600	1,250		18,850	21,300	
	1260	55 GPM		.60	40		18,000	1,350		19,350	21,900	
	1280	130 GPM		.50	48		23,100	1,625		24,725	27,900	
	1300	255 GPM		.35	68.571		31,700	2,325		34,025	38,400	
	1320	550 GPM		.25	96		40,000	3,250		43,250	48,900	
	1340	800 GPM	▼	.20	120	▼	48,900	4,050		52,950	60,000	

15500 | Heat Generation Equipment

		15510	Heating Boilers and Accessories	CREW	DAILY OUTPUT	LABOR-HOURS	UNIT	2002 BARE COSTS MAT.	LABOR	EQUIP.	TOTAL	TOTAL INCL O&P	
110	0010	**BOILER BLOWDOWN SYSTEMS**											110
	1010	Boiler blowdown, auto/manual to 2000 MBH	Q-5	3.75	4.267	Ea.	2,525	139		2,664	2,975	
	1020	7300 MBH	"	3	5.333	"	2,900	174		3,074	3,425	
120	0010	**BURNERS**											120
	0990	Residential, conversion, gas fired, LP or natural											
	1000	Gun type, atmospheric input 72 to 200 MBH	Q-1	2.50	6.400	Ea.	655	207		862	1,025	
	1020	120 to 360 MBH		2	8		725	259		984	1,200	
	1040	280 to 800 MBH	▼	1.70	9.412	▼	1,400	305		1,705	2,000	
	2000	Commercial and industrial, gas/oil, input											

Important: See the Reference Section for critical supporting data - Reference Nos., Crews, & City Cost Indexes

			DAILY	LABOR-		\multicolumn{4}{c}{2002 BARE COSTS}				TOTAL	
	15510	**Heating Boilers and Accessories**	CREW	OUTPUT	HOURS	UNIT	MAT.	LABOR	EQUIP.	TOTAL	INCL O&P	
120	2050	400 MBH	Q-1	1.50	10.667	Ea.	1,900	345		2,245	2,625	**120**
	2080	For mounting plate, add		35	.457		137	14.80		151.80	174	
	2090	670 MBH		1.40	11.429		2,350	370		2,720	3,125	
	2110	For mounting plate, add		35	.457		137	14.80		151.80	174	
	2140	1155 MBH		1.30	12.308		3,025	400		3,425	3,950	
	2170	For mounting plate, add		35	.457		137	14.80		151.80	174	
	2200	1800 MBH		1.20	13.333		3,300	430		3,730	4,275	
	2230	For mounting plate, add		35	.457		137	14.80		151.80	174	
	2260	3000 MBH		1.10	14.545		4,550	470		5,020	5,700	
	2290	For mounting plate, add		35	.457		137	14.80		151.80	174	
	2320	4100 MBH		1	16		4,700	520		5,220	5,925	
	2350	For mounting plate, add		35	.457		137	14.80		151.80	174	
	2500	Impinged jet, rectangular, burner only, 3-1/2" WC,										
	2520	Gas fired, input, 420 to 640 MBH	Q-1	2	8	Ea.	570	259		829	1,025	
	2540	560 to 860 MBH		2	8		820	259		1,079	1,300	
	2560	630 to 950 MBH		1.90	8.421		855	272		1,127	1,350	
	2580	950 to 1200 MBH		1.80	8.889		1,225	288		1,513	1,775	
	2600	1120 to 1700 MBH		1.80	8.889		1,625	288		1,913	2,225	
	2620	1700 to 1900 MBH		1.70	9.412		1,850	305		2,155	2,475	
	2640	1400 to 2100 MBH		1.60	10		2,050	325		2,375	2,750	
	2660	1600 to 2400 MBH		1.60	10		2,150	325		2,475	2,850	
	2680	1700 to 2500 MBH		1.50	10.667		2,450	345		2,795	3,225	
	2700	1880 to 2880 MBH		1.40	11.429		2,575	370		2,945	3,375	
	2720	2260 to 3680 MBH		1.40	11.429		3,000	370		3,370	3,850	
	2740	2590 to 4320 MBH		1.30	12.308		3,650	400		4,050	4,625	
	2760	3020 to 5040 MBH		1.20	13.333		4,275	430		4,705	5,350	
	2780	3450 to 5760 MBH		1.20	13.333		4,875	430		5,305	6,000	
	2850	Burner pilot		50	.320		15.25	10.35		25.60	32.50	
	2860	Thermocouple		50	.320		15.70	10.35		26.05	33	
	2870	Thermocouple & pilot bracket		45	.356		41.50	11.50		53	63.50	
	2900	For 7" W.C. pressure, increase MBH 40%										
	3000	Flame retention oil fired assembly, input										
	3020	.50 to 2.25 GPH	Q-1	2.40	6.667	Ea.	291	216		507	645	
	3040	2.0 to 5.0 GPH		2	8		525	259		784	970	
	3060	3.0 to 7.0 GPH		1.80	8.889		535	288		823	1,025	
	3080	6.0 to 12.0 GPH		1.60	10		835	325		1,160	1,400	
	4600	Gas safety, shut off valve, 3/4" threaded	1 Stpi	20	.400		276	14.50		290.50	325	
	4610	1" threaded		19	.421		276	15.25		291.25	330	
	4620	1-1/4" threaded		15	.533		298	19.30		317.30	360	
	4630	1-1/2" threaded		13	.615		325	22.50		347.50	395	
	4640	2" threaded		11	.727		365	26.50		391.50	445	
	4650	2-1/2" threaded	Q-1	15	1.067		420	34.50		454.50	510	
	4660	3" threaded		13	1.231		575	40		615	695	
	4670	4" flanged		3	5.333		2,200	173		2,373	2,650	
	4680	6" flanged	Q-2	3	8		5,075	268		5,343	5,975	
	8020	Coal stoker, automatic, complete system 1000 MBH boiler	Q-7	.33	96.970		14,200	3,350		17,550	20,700	
	8040	2000 MBH boiler		.33	96.970		15,300	3,350		18,650	21,900	
	8060	7300 MBH boiler		.30	106		17,000	3,675		20,675	24,300	
300	0010	**BOILERS, ELECTRIC, ASME** Standard controls and trim R15500-030										**300**
	1000	Steam, 6 KW, 20.5 MBH	Q-19	1.20	20	Ea.	7,900	670		8,570	9,700	
	1040	9 KW, 30.7 MBH		1.20	20		8,250	670		8,920	10,100	
	1060	18 KW, 61.4 MBH D3020 102		1.20	20		8,325	670		8,995	10,200	
	1080	24 KW, 81.8 MBH		1.10	21.818		8,400	730		9,130	10,400	
	1120	36 KW, 123 MBH D3020 104		1.10	21.818		8,575	730		9,305	10,500	
	1160	60 KW, 205 MBH		1	24		9,775	805		10,580	11,900	
	1220	112 KW, 382 MBH		.75	32		11,200	1,075		12,275	13,900	

MECHANICAL 15

	15510	Heating Boilers and Accessories		CREW	DAILY OUTPUT	LABOR-HOURS	UNIT	MAT.	LABOR	EQUIP.	TOTAL	TOTAL INCL O&P	
300	1240	148 KW, 505 MBH	R15500 -030	Q-19	.65	36.923	Ea.	12,600	1,250		13,850	15,800	300
	1260	168 KW, 573 MBH			.60	40		14,700	1,350		16,050	18,200	
	1280	222 KW, 758 MBH	D3020 102		.55	43.636		15,300	1,475		16,775	19,000	
	1300	296 KW, 1010 MBH			.45	53.333		15,400	1,800		17,200	19,700	
	1320	300 KW, 1023 MBH	D3020 104		.40	60		17,700	2,000		19,700	22,400	
	1340	370 KW, 1263 MBH			.35	68.571		19,100	2,300		21,400	24,500	
	1360	444 KW, 1515 MBH			.30	80		20,100	2,675		22,775	26,100	
	1380	518 KW, 1768 MBH		Q-21	.36	88.889		20,600	3,050		23,650	27,200	
	1400	592 KW, 2020 MBH			.34	94.118		23,800	3,225		27,025	31,100	
	1420	666 KW, 2273 MBH			.32	100		26,000	3,425		29,425	33,800	
	1460	740 KW, 2526 MBH			.28	114		29,700	3,900		33,600	38,500	
	1480	814 KW, 2778 MBH			.25	128		31,500	4,375		35,875	41,300	
	1500	962 KW, 3283 MBH			.22	145		39,600	4,975		44,575	51,000	
	1520	1036 KW, 3536 MBH			.20	160		45,300	5,475		50,775	58,000	
	1540	1110 KW, 3788 MBH			.19	168		52,500	5,750		58,250	66,500	
	1560	2070 KW, 7063 MBH			.18	177		56,500	6,075		62,575	71,000	
	1580	2250 KW, 7677 MBH			.17	188		58,500	6,450		64,950	73,500	
	1600	2,340 KW, 7984 MBH			.16	200		60,000	6,850		66,850	76,000	
	2000	Hot water, 6 KW, 20.5 MBH		Q-19	1.30	18.462		3,400	620		4,020	4,650	
	2020	9 KW, 30.7 MBH			1.30	18.462		3,425	620		4,045	4,675	
	2040	24 KW, 82 MBH			1.20	20		3,625	670		4,295	5,000	
	2060	30 KW, 103 MBH			1.20	20		3,725	670		4,395	5,100	
	2070	36 KW, 123 MBH			1.20	20		3,950	670		4,620	5,350	
	2080	48 KW, 164 MBH			1.10	21.818		3,975	730		4,705	5,475	
	2100	74 KW, 253 MBH			1.10	21.818		4,725	730		5,455	6,300	
	2120	93 KW, 317 MBH			1	24		5,550	805		6,355	7,300	
	2140	112 KW, 382 MBH			.90	26.667		6,025	895		6,920	7,975	
	2160	148 KW, 505 MBH			.75	32		8,550	1,075		9,625	11,000	
	2180	168 KW, 573 MBH			.65	36.923		9,150	1,250		10,400	12,000	
	2200	222 KW, 758 MBH			.60	40		10,000	1,350		11,350	13,000	
	2220	296 KW, 1010 MBH			.55	43.636		10,600	1,475		12,075	13,900	
	2280	370 KW, 1263 MBH			.40	60		14,600	2,000		16,600	19,100	
	2300	444 KW, 1515 MBH			.35	68.571		16,900	2,300		19,200	22,100	
	2340	518 KW, 1768 MBH		Q-21	.44	72.727		18,900	2,475		21,375	24,600	
	2360	592 KW, 2020 MBH			.43	74.419		20,100	2,550		22,650	25,900	
	2400	666 KW, 2273 MBH			.40	80		23,500	2,725		26,225	29,900	
	2420	740 KW, 2526 MBH			.39	82.051		23,800	2,800		26,600	30,400	
	2440	814 KW, 2778 MBH			.38	84.211		27,200	2,875		30,075	34,200	
	2460	888 KW, 3031 MBH			.37	86.486		29,000	2,950		31,950	36,400	
	2480	962 KW, 3283 MBH			.36	88.889		30,800	3,050		33,850	38,500	
	2500	1,036 KW, 3536 MBH			.34	94.118		33,600	3,225		36,825	41,900	
	2520	1110 KW, 3788 MBH			.33	96.970		38,800	3,325		42,125	47,600	
	2540	1440 KW, 4915 MBH			.32	100		40,900	3,425		44,325	50,000	
	2560	1560 KW, 5323 MBH			.31	103		43,100	3,525		46,625	52,500	
	2580	1680 KW, 5733 MBH			.30	106		44,800	3,650		48,450	55,000	
	2600	1800 KW, 6143 MBH			.29	110		46,800	3,775		50,575	57,000	
	2620	1980 KW, 6757 MBH			.28	114		51,000	3,900		54,900	62,000	
	2640	2100 KW, 7167 MBH			.27	118		52,500	4,050		56,550	64,000	
	2660	2220 KW, 7576 MBH			.26	123		55,000	4,200		59,200	67,000	
	2680	2,400 KW, 8191 MBH			.25	128		57,500	4,375		61,875	69,500	
	2700	2610 KW, 8905 MBH			.24	133		60,500	4,550		65,050	73,500	
	2720	2790 KW, 9519 MBH			.23	139		63,000	4,750		67,750	76,500	
	2740	2970 KW, 10133 MBH			.21	152		65,000	5,200		70,200	79,500	
	2760	3150 KW, 10748 MBH			.19	168		66,500	5,750		72,250	81,500	
	2780	3240 KW, 11055 MBH			.18	177		67,500	6,075		73,575	83,500	
	2800	3420 KW, 11669 MBH			.17	188		69,500	6,450		75,950	85,500	

Important: See the Reference Section for critical supporting data - Reference Nos., Crews, & City Cost Indexes

15500 | Heat Generation Equipment

		15510	**Heating Boilers and Accessories**	CREW	DAILY OUTPUT	LABOR-HOURS	UNIT	2002 BARE COSTS				TOTAL INCL O&P	
								MAT.	LABOR	EQUIP.	TOTAL		
300	2820		3,600 KW, 12,283 MBH R15500-030	Q-21	.16	200	Ea.	72,000	6,850		78,850	90,000	300
400	0010		**BOILERS, GAS FIRED** Natural or propane, standard controls										400
	1000		Cast iron, with insulated jacket										
	2000		Steam, gross output, 81 MBH	Q-7	1.40	22.857	Ea.	1,175	790		1,965	2,500	
	2020		102 MBH		1.30	24.615		1,350	850		2,200	2,750	
	2040		122 MBH		1	32		1,475	1,100		2,575	3,300	
	2060		163 MBH		.90	35.556		1,800	1,225		3,025	3,825	
	2080		203 MBH		.90	35.556		2,000	1,225		3,225	4,050	
	2100		240 MBH		.85	37.647		2,000	1,300		3,300	4,150	
	2120		280 MBH		.80	40		2,550	1,375		3,925	4,875	
	2140		320 MBH		.70	45.714		2,775	1,575		4,350	5,425	
	2160		360 MBH		.63	51.200		3,125	1,775		4,900	6,125	
	2180		400 MBH		.56	56.838		3,325	1,950		5,275	6,625	
	2200		440 MBH		.51	62.500		3,575	2,150		5,725	7,175	
	2220		544 MBH		.45	71.588		4,175	2,475		6,650	8,325	
	2240		765 MBH		.43	74.419		5,325	2,575		7,900	9,750	
	2260		892 MBH		.38	84.211		6,400	2,900		9,300	11,400	
	2280		1275 MBH		.34	94.118		8,175	3,250		11,425	13,900	
	2300		1530 MBH		.32	100		9,525	3,450		12,975	15,700	
	2320		1,875 MBH		.30	106		11,100	3,675		14,775	17,800	
	2340		2170 MBH		.26	122		12,700	4,225		16,925	20,300	
	2360		2675 MBH		.20	163		14,400	5,625		20,025	24,400	
	2380		3060 MBH		.19	172		15,700	5,925		21,625	26,200	
	2400		3570 MBH		.18	181		17,500	6,275		23,775	28,700	
	2420		4207 MBH		.16	205		19,300	7,075		26,375	31,900	
	2440		4,720 MBH		.15	207		21,200	7,175		28,375	34,100	
	2460		5660 MBH		.15	220		45,000	7,625		52,625	61,000	
	2480		6,100 MBH		.13	246		51,500	8,500		60,000	69,500	
	2500		6390 MBH		.12	266		54,500	9,200		63,700	74,000	
	2520		6680 MBH		.11	290		55,000	10,000		65,000	75,500	
	2540		6,970 MBH		.10	320		58,000	11,000		69,000	80,000	
	3000		Hot water, gross output, 80 MBH		1.46	21.918		1,025	755		1,780	2,275	
	3020		100 MBH		1.35	23.704		1,175	820		1,995	2,525	
	3040		122 MBH		1.10	29.091		1,300	1,000		2,300	2,950	
	3060		163 MBH		1	32		1,625	1,100		2,725	3,450	
	3080		203 MBH		1	32		1,800	1,100		2,900	3,650	
	3100		240 MBH		.95	33.684		1,825	1,150		2,975	3,750	
	3120		280 MBH		.90	35.556		2,275	1,225		3,500	4,350	
	3140		320 MBH		.80	40		2,500	1,375		3,875	4,800	
	3160		360 MBH		.71	45.070		2,825	1,550		4,375	5,475	
	3180		400 MBH		.64	50		3,025	1,725		4,750	5,925	
	3200		440 MBH		.58	54.983		3,250	1,900		5,150	6,425	
	3220		544 MBH		.51	62.992		3,975	2,175		6,150	7,650	
	3240		765 MBH		.46	70.022		5,125	2,425		7,550	9,275	
	3260		1,088 MBH		.40	80		6,700	2,750		9,450	11,600	
	3280		1,275 MBH		.36	89.888		7,875	3,100		10,975	13,400	
	3300		1,530 MBH		.31	104		9,325	3,625		12,950	15,800	
	3320		2,000 MBH		.26	125		11,500	4,325		15,825	19,200	
	3340		2,312 MBH		.22	148		13,400	5,125		18,525	22,400	
	3360		2,856 MBH		.20	160		15,700	5,525		21,225	25,500	
	3380		3,264 MBH		.18	179		16,600	6,200		22,800	27,700	
	3400		3,808 MBH		.16	195		18,300	6,725		25,025	30,400	
	3420		4,488 MBH		.15	210		27,900	7,275		35,175	41,600	
	3440		4,720 MBH		.15	220		46,100	7,625		53,725	62,000	
	3460		5,520 MBH		.14	228		52,000	7,900		59,900	69,000	

MECHANICAL 15

15510	Heating Boilers and Accessories	CREW	DAILY OUTPUT	LABOR-HOURS	UNIT	2002 BARE COSTS				TOTAL INCL O&P	
						MAT.	LABOR	EQUIP.	TOTAL		
400											**400**
3480	6,100 MBH	Q-7	.13	250	Ea.	52,500	8,625		61,125	71,000	
3500	6,390 MBH		.11	285		53,500	9,850		63,350	74,000	
3520	6,680 MBH		.10	310		57,000	10,700		67,700	78,500	
3540	6,970 MBH	↓	.09	359	↓	59,000	12,400		71,400	83,500	
4000	Steel, insulating jacket										
4500	Steam, not including burner, gross output										
5500	1,440 MBH	Q-6	.35	68.571	Ea.	16,100	2,325		18,425	21,200	
5520	1630 MBH		.30	80		16,700	2,700		19,400	22,500	
5540	1800 MBH		.26	92.308		17,300	3,125		20,425	23,800	
5560	2520 MBH		.20	120		21,500	4,050		25,550	29,700	
5580	3,065 MBH		.18	133		22,900	4,500		27,400	32,000	
5600	3600 MBH		.16	150		24,300	5,075		29,375	34,500	
5620	5400 MBH	↓	.14	171		33,900	5,800		39,700	46,000	
5640	7,200 MBH	Q-7	.18	177		42,100	6,125		48,225	55,500	
5660	10,800 MBH		.17	188		52,500	6,500		59,000	67,500	
5680	12,600 MBH		.15	213		58,000	7,350		65,350	75,000	
5700	16,190 MBH		.12	266		73,500	9,200		82,700	95,000	
5720	17,990 MBH	↓	.10	320	↓	78,500	11,000		89,500	103,000	
6000	Hot water, including burner & one zone valve, gross output										
6010	51.2 MBH	Q-6	2	12	Ea.	1,575	405		1,980	2,325	
6020	72 MBH		2	12		1,750	405		2,155	2,525	
6040	89 MBH		1.90	12.632		1,775	425		2,200	2,625	
6060	105 MBH		1.80	13.333		2,000	450		2,450	2,875	
6080	132 MBH		1.70	14.118		2,275	475		2,750	3,225	
6100	155 MBH		1.50	16		2,650	540		3,190	3,725	
6110	186 MBH		1.40	17.143		3,200	580		3,780	4,375	
6120	227 MBH		1.30	18.462		4,050	625		4,675	5,400	
6140	292 MBH		1.20	20		4,575	675		5,250	6,050	
6160	400 MBH		.80	30		5,775	1,025		6,800	7,875	
6180	480 MBH		.70	34.286		6,600	1,150		7,750	9,025	
6200	640 MBH		.60	40		7,850	1,350		9,200	10,700	
6220	800 MBH		.50	48		9,150	1,625		10,775	12,600	
6240	960 MBH		.45	53.333		11,400	1,800		13,200	15,200	
6260	1200 MBH		.40	60		14,000	2,025		16,025	18,400	
6280	1440 MBH		.35	68.571		15,800	2,325		18,125	20,900	
6300	1960 MBH		.30	80		20,900	2,700		23,600	27,100	
6320	2400 MBH		.20	120		25,500	4,050		29,550	34,200	
6340	3,000 MBH	↓	.15	160	↓	31,600	5,400		37,000	42,900	
7000	For tankless water heater on smaller gas units, add					10%					
7050	For additional zone valves up to 312 MBH add				Ea.	101			101	111	
7990	Special feature gas fired boilers										
8000	Pulse combustion, standard controls / trim, 44,000 BTU	Q-5	1.50	10.667	Ea.	2,275	350		2,625	3,025	
8050	88,000 BTU		1.40	11.429		2,700	370		3,070	3,500	
8080	134,000 BTU	↓	1.20	13.333	↓	3,500	435		3,935	4,475	
9000	Wall hung, C.I., sealed combustion, direct vent										
9010	Packaged, net output										
9020	45,000 BTUH	Q-6	1.33	18.045	Ea.	1,375	610		1,985	2,425	
9030	60,800 BTUH		1.33	18.045		1,525	610		2,135	2,600	
9040	77,000 BTUH	↓	1.33	18.045		1,650	610		2,260	2,750	
460	0010	**BOILERS, GAS/OIL** Combination with burners and controls										**460**
	1000	Cast iron with insulated jacket										
	2000	Steam, gross output, 720 MBH	Q-7	.43	74.074	Ea.	7,000	2,550		9,550	11,600	
	2020	810 MBH		.38	83.989		7,000	2,900		9,900	12,100	
	2040	1,084 MBH		.34	93.023		8,025	3,200		11,225	13,700	
	2060	1,360 MBH	↓	.33	98.160	↓	9,500	3,375		12,875	15,600	

Important: See the Reference Section for critical supporting data - Reference Nos., Crews, & City Cost Indexes

			DAILY	LABOR-			2002 BARE COSTS				TOTAL	
15510 | Heating Boilers and Accessories		CREW	OUTPUT	HOURS	UNIT	MAT.	LABOR	EQUIP.	TOTAL	INCL O&P	
460	2080	1,600 MBH	Q-7	.30	107	Ea.	10,700	3,700		14,400	17,400	**460**
	2100	2,040 MBH		.25	130		13,200	4,500		17,700	21,300	
	2120	2,450 MBH		.20	156		14,400	5,375		19,775	24,000	
	2140	2,700 MBH		.19	165		15,300	5,725		21,025	25,400	
	2160	3,000 MBH		.18	175		16,500	6,075		22,575	27,300	
	2180	3,270 MBH		.17	183		17,700	6,350		24,050	29,100	
	2200	3,770 MBH		.17	191		40,700	6,625		47,325	55,000	
	2220	4,070 MBH		.16	200		42,800	6,900		49,700	57,500	
	2240	4,650 MBH		.15	210		45,100	7,275		52,375	60,500	
	2260	5,230 MBH		.14	223		50,000	7,725		57,725	66,500	
	2280	5,520 MBH		.14	235		54,500	8,125		62,625	72,000	
	2300	5,810 MBH		.13	248		55,000	8,550		63,550	73,500	
	2320	6,100 MBH		.12	260		55,500	8,975		64,475	74,500	
	2340	6,390 MBH		.11	296		58,500	10,200		68,700	79,500	
	2360	6,680 MBH		.10	320		61,500	11,000		72,500	84,000	
	2380	6,970 MBH	↓	.09	372	↓	62,000	12,800		74,800	87,500	
	2900	Hot water, gross output										
	2910	200 MBH	Q-6	.61	39.024	Ea.	4,525	1,325		5,850	6,975	
	2920	300 MBH		.49	49.080		4,525	1,650		6,175	7,475	
	2930	400 MBH		.41	57.971		5,300	1,950		7,250	8,775	
	2940	500 MBH	↓	.36	67.039		5,725	2,275		8,000	9,700	
	3000	584 MBH	Q-7	.44	72.072		6,000	2,475		8,475	10,400	
	3020	876 MBH		.40	79.012		8,350	2,725		11,075	13,300	
	3040	1,168 MBH		.31	103		9,175	3,575		12,750	15,500	
	3060	1,460 MBH		.28	113		10,800	3,900		14,700	17,800	
	3080	2,044 MBH		.26	122		13,900	4,225		18,125	21,700	
	3100	2,628 MBH		.21	150		16,900	5,175		22,075	26,400	
	3120	3,210 MBH		.18	174		20,800	6,025		26,825	32,000	
	3140	3,796 MBH		.17	186		23,300	6,425		29,725	35,300	
	3160	4,088 MBH		.16	195		26,500	6,725		33,225	39,400	
	3180	4,672 MBH		.16	203		28,900	7,025		35,925	42,400	
	3200	5,256 MBH		.15	217		34,400	7,500		41,900	49,100	
	3220	6,000 MBH, 179 BHP		.13	256		46,400	8,825		55,225	64,500	
	3240	7,130 MBH, 213 BHP		.08	385		48,600	13,300		61,900	73,500	
	3260	9,800 MBH, 286 BHP		.06	533		56,000	18,400		74,400	89,500	
	3280	10,900 MBH, 325.6 BHP		.05	592		68,000	20,500		88,500	105,500	
	3290	12,200 MBH, 364.5 BHP		.05	666		74,000	23,000		97,000	116,000	
	3300	13,500 MBH, 403.3 BHP	↓	.04	727	↓	82,000	25,100		107,100	128,000	
	4000	Steel, insulated jacket, skid base, tubeless										
	4100	Steam, 15 psi, gross output 517 MBH, 15 BHP	Q-6	.41	57.971	Ea.	12,400	1,950		14,350	16,600	
	4110	690 MBH, 20 BHP		.38	62.992		12,700	2,125		14,825	17,200	
	4120	862 MBH, 25 BHP		.36	67.039		13,500	2,275		15,775	18,300	
	4130	1,004 MBH, 30 BHP		.32	74.074		15,900	2,500		18,400	21,300	
	4140	1,340 MBH, 40 BHP		.31	78.431		17,500	2,650		20,150	23,200	
	4150	1,675 MBH, 50 BHP		.28	85.714		18,100	2,900		21,000	24,400	
	4160	2,175 MBH, 65 BHP		.23	104		18,700	3,525		22,225	25,800	
	4170	2,510 MBH, 75 BHP		.19	125		20,100	4,225		24,325	28,500	
	4180	3,015 MBH, 90 BHP	↓	.17	140		22,800	4,750		27,550	32,300	
	4190	3,350 MBH, 100 BHP	Q-7	.22	146		26,000	5,075		31,075	36,300	
	4200	3,880 MBH, 116 BHP		.21	153		28,100	5,300		33,400	38,900	
	4210	4,685 MBH, 140 BHP		.19	168		30,000	5,800		35,800	41,700	
	4220	5,025 MBH, 150 BHP		.18	177		30,900	6,125		37,025	43,300	
	4230	5,550 MBH, 166 BHP		.17	189		40,100	6,525		46,625	54,000	
	4240	6,700 MBH, 200 BHP		.13	256		43,800	8,825		52,625	61,500	
	4250	7,760 MBH, 232 BHP		.11	296		49,000	10,200		59,200	69,500	
	4260	8,870 MBH, 265 BHP	↓	.09	336	↓	53,000	11,600		64,600	76,000	

MECHANICAL 15

		CREW	DAILY OUTPUT	LABOR-HOURS	UNIT	2002 BARE COSTS				TOTAL INCL O&P	
15510	**Heating Boilers and Accessories**					MAT.	LABOR	EQUIP.	TOTAL		
460	4270	11,100 MBH, 322 BHP	Q-7	.07	426	Ea.	57,000	14,700		71,700	84,500	**460**
	4280	13,400 MBH, 400 BHP		.06	516		65,000	17,800		82,800	98,500	
	4290	17,750 MBH, 530 BHP		.05	680		81,500	23,500		105,000	125,000	
	4300	20,750 MBH, 620 BHP	↓	.04	800		91,000	27,600		118,600	141,500	
	4500	Steam, 150 psi gross output, 335 MBH, 10 BHP	Q-6	.54	44.037		10,000	1,500		11,500	13,300	
	4520	502 MBH, 15 BHP		.40	60		11,400	2,025		13,425	15,600	
	4560	670 MBH, 20 BHP		.36	65.934		12,400	2,225		14,625	17,000	
	4600	1,005 MBH, 30 BHP		.31	77.922		16,200	2,625		18,825	21,800	
	4640	1,339 MBH, 40 BHP		.29	81.911		19,600	2,775		22,375	25,800	
	4660	1,674 MBH, 50 BHP		.24	100		21,900	3,375		25,275	29,200	
	4680	2,009 MBH, 60 BHP		.21	114		25,900	3,850		29,750	34,300	
	4720	2,511 MBH, 75 BHP		.17	141		26,700	4,775		31,475	36,600	
	5000	Hot water, gross output, 525 MBH		.45	52.980		9,850	1,800		11,650	13,500	
	5020	630 MBH		.43	55.944		9,925	1,900		11,825	13,800	
	5040	735 MBH		.41	58.968		12,300	2,000		14,300	16,500	
	5060	840 MBH		.38	64		13,600	2,150		15,750	18,200	
	5080	1,050 MBH		.35	67.989		15,900	2,300		18,200	21,000	
	5100	1,365 MBH		.33	72.727		19,800	2,450		22,250	25,500	
	5120	1,680 MBH		.29	82.759		22,700	2,800		25,500	29,100	
	5140	2,310 MBH		.20	120		29,500	4,050		33,550	38,600	
	5160	2,835 MBH		.18	129		35,600	4,375		39,975	45,700	
	5180	3,150 MBH	↓	.16	146	↓	39,300	4,950		44,250	50,500	
500	0010	**BOILERS, OIL FIRED** Standard controls, flame retention burner										**500**
	1000	Cast iron, with insulated flush jacket										
	2000	Steam, gross output, 109 MBH	Q-7	1.20	26.667	Ea.	1,450	920		2,370	3,000	
	2020	144 MBH		1.10	29.091		1,675	1,000		2,675	3,375	
	2040	173 MBH		1	32		1,825	1,100		2,925	3,675	
	2060	207 MBH		.90	35.556		1,975	1,225		3,200	4,025	
	2080	236 MBH		.85	37.647		2,825	1,300		4,125	5,075	
	2100	300 MBH		.70	45.714		2,800	1,575		4,375	5,450	
	2120	480 MBH		.50	64		3,525	2,200		5,725	7,225	
	2140	665 MBH		.45	71.111		4,925	2,450		7,375	9,125	
	2160	794 MBH		.41	78.049		5,750	2,700		8,450	10,400	
	2180	1,084 MBH		.38	85.106		6,800	2,925		9,725	11,900	
	2200	1,360 MBH		.33	98.160		7,950	3,375		11,325	13,900	
	2220	1,600 MBH		.26	122		9,025	4,225		13,250	16,300	
	2240	2,175 MBH		.24	133		10,900	4,625		15,525	19,000	
	2260	2,480 MBH		.20	156		12,600	5,375		17,975	21,900	
	2280	3,000 MBH		.19	170		14,500	5,875		20,375	24,800	
	2300	3,550 MBH		.17	187		16,200	6,450		22,650	27,600	
	2320	3,820 MBH		.16	200		17,600	6,900		24,500	29,700	
	2340	4,360 MBH		.15	214		19,500	7,400		26,900	32,700	
	2360	4,940 MBH		.14	225		38,900	7,775		46,675	54,500	
	2380	5,520 MBH		.14	235		45,600	8,125		53,725	62,000	
	2400	6,100 MBH		.13	256		52,500	8,825		61,325	71,000	
	2420	6,390 MBH		.11	290		55,000	10,000		65,000	76,000	
	2440	6,680 MBH		.10	313		56,500	10,800		67,300	79,000	
	2460	6,970 MBH	↓	.09	363	↓	59,000	12,500		71,500	83,500	
	3000	Hot water, same price as steam										
	4000	For tankless coil in smaller sizes, add					15%					
	5000	Steel, insulated jacket, burner										
	6000	Steam, full water leg construction, gross output										
	6020	144 MBH	Q-6	1.33	18.045	Ea.	3,425	610		4,035	4,700	
	6040	198 MBH		1.20	20		3,725	675		4,400	5,125	
	6060	252 MBH		1.13	21.295		4,650	720		5,370	6,200	
	6080	324 MBH	↓	1.04	23.011	↓	5,300	775		6,075	7,000	

Important: See the Reference Section for critical supporting data - Reference Nos., Crews, & City Cost Indexes

15 MECHANICAL

15510	Heating Boilers and Accessories	CREW	DAILY OUTPUT	LABOR-HOURS	UNIT	2002 BARE COSTS				TOTAL INCL O&P	
						MAT.	LABOR	EQUIP.	TOTAL		
500											**500**
6100	396 MBH	Q-6	.78	30.691	Ea.	5,750	1,025		6,775	7,900	
6120	468 MBH		.70	34.483		6,325	1,175		7,500	8,700	
6140	648 MBH		.52	45.977		6,725	1,550		8,275	9,750	
6160	792 MBH		.44	55.046		7,925	1,850		9,775	11,500	
6180	1,008 MBH		.39	61.381		9,800	2,075		11,875	13,900	
6200	1,260 MBH		.35	68.966		10,800	2,325		13,125	15,400	
6220	1,512 MBH		.31	78.689		13,000	2,650		15,650	18,300	
6240	1,800 MBH		.29	83.624		14,200	2,825		17,025	20,000	
6260	2,100 MBH		.23	106		16,800	3,575		20,375	23,900	
6280	2,400 MBH	▼	.19	125	▼	18,700	4,225		22,925	27,000	
6400	Larger sizes are same as steel, gas fired										
7000	Hot water, gross output, 103 MBH	Q-6	1.60	15	Ea.	1,275	505		1,780	2,175	
7020	122 MBH		1.45	16.506		2,200	560		2,760	3,275	
7040	137 MBH		1.36	17.595		2,450	595		3,045	3,600	
7060	168 MBH		1.30	18.405		2,775	620		3,395	4,000	
7080	225 MBH		1.22	19.704		2,850	665		3,515	4,125	
7100	315 MBH		.96	25.105		3,625	850		4,475	5,275	
7120	420 MBH		.70	34.483		4,375	1,175		5,550	6,575	
7140	525 MBH		.57	42.403		4,825	1,425		6,250	7,475	
7160	630 MBH		.52	45.977		5,725	1,550		7,275	8,650	
7180	735 MBH		.48	50.104		5,725	1,700		7,425	8,850	
7200	840 MBH		.43	55.172		9,150	1,875		11,025	12,900	
7220	1,050 MBH		.36	65.753		9,300	2,225		11,525	13,600	
7240	1,365 MBH		.32	74.534		10,800	2,525		13,325	15,700	
7260	1,680 MBH		.29	83.624		12,200	2,825		15,025	17,700	
7280	2,310 MBH		.21	114		13,100	3,875		16,975	20,300	
7300	2,835 MBH		.17	141		16,900	4,775		21,675	25,800	
7320	3,150 MBH	▼	.13	184	▼	17,700	6,225		23,925	28,900	
7340	For tankless coil in steam or hot water, add					7%					
700	0010	**BOILERS, PACKAGED SCOTCH MARINE** Steam or hot water										**700**
	1000	Packaged fire tube, #2 oil, gross output										
	1006	15 PSI steam										
	1010	1005 MBH, 30 HP	Q-6	.33	72.072	Ea.	23,200	2,425		25,625	29,200	
	1014	1675 MBH, 50 HP	"	.26	93.023		29,000	3,150		32,150	36,700	
	1020	3348 MBH, 100 HP	Q-7	.21	152		30,700	5,250		35,950	41,700	
	1030	5025 MBH, 150 HP		.17	192		38,800	6,650		45,450	52,500	
	1040	6696 MBH, 200 HP		.14	223		48,400	7,725		56,125	64,500	
	1050	8375 MBH, 250 HP		.14	231		49,500	8,000		57,500	66,500	
	1060	10,044 MBH, 300 HP		.13	251		65,500	8,700		74,200	85,000	
	1080	16,740 MBH, 500 HP		.08	380		95,500	13,100		108,600	125,000	
	1100	23,435 MBH, 700 HP		.07	484		117,000	16,700		133,700	153,500	
	1140	To fire #6, and gas, add	▼	.42	76.190	▼	10,400	2,625		13,025	15,400	
	1180	For duplex package feed system										
	1200	To 3348 MBH boiler, add	Q-7	.54	59.259	Ea.	5,600	2,050		7,650	9,250	
	1220	To 6696 MBH boiler, add		.41	78.049		6,650	2,700		9,350	11,400	
	1240	To 10,044 MBH boiler, add		.38	84.211		7,625	2,900		10,525	12,800	
	1260	To 16,740 MBH boiler, add		.28	114		8,675	3,950		12,625	15,500	
	1280	To 23,435 MBH boiler, add	▼	.25	128	▼	10,300	4,425		14,725	18,000	
	1300	150 PSI steam										
	1310	1005 MBH, 30 HP	Q-6	.33	72.072	Ea.	23,100	2,425		25,525	29,100	
	1320	1675 MBH, 50 HP	"	.26	93.023		28,900	3,150		32,050	36,600	
	1330	3350 MBH, 100 HP	Q-7	.21	152		39,400	5,250		44,650	51,000	
	1340	5025 MBH, 150 HP		.17	192		50,500	6,650		57,150	65,500	
	1350	6700 MBH, 200 HP		.14	223		56,000	7,725		63,725	73,000	
	1360	10,050 MBH, 300 HP	▼	.13	251	▼	82,000	8,700		90,700	103,500	

MECHANICAL 15

15510	Heating Boilers and Accessories	CREW	DAILY OUTPUT	LABOR-HOURS	UNIT	2002 BARE COSTS				TOTAL INCL O&P
						MAT.	LABOR	EQUIP.	TOTAL	
700 1400	Packaged scotch marine, #6 oil									**700**
1410	15 PSI steam									
1420	1675 MBH, 50 HP	Q-6	.25	97.166	Ea.	35,600	3,275		38,875	44,200
1430	3350 MBH, 100 HP	Q-7	.20	159		38,000	5,500		43,500	50,000
1440	5025 MBH, 150 HP		.16	202		47,700	7,000		54,700	63,000
1450	6700 MBH, 200 HP		.14	233		52,000	8,050		60,050	69,500
1460	8375 MBH, 250 HP		.13	244		66,500	8,425		74,925	85,500
1470	10,050 MBH, 300 HP		.12	262		78,000	9,050		87,050	99,500
1480	13,400 MBH, 400 HP	↓	.09	351	↓	89,500	12,100		101,600	117,000
1500	150 PSI steam									
1510	1675 MBH, 50 HP	Q-6	.25	97.166	Ea.	42,200	3,275		45,475	51,500
1520	3350 MBH, 100 HP	Q-7	.20	159		45,600	5,500		51,100	58,500
1530	5025 MBH, 150 HP		.16	202		57,000	7,000		64,000	73,000
1540	6700 MBH, 200 HP		.14	233		61,000	8,050		69,050	79,000
1550	8375 MBH, 250 HP		.13	244		77,500	8,425		85,925	97,500
1560	10,050 MBH, 300 HP		.12	262		87,500	9,050		96,550	110,000
1570	13,400 MBH, 400 HP	↓	.09	351	↓	101,500	12,100		113,600	130,500
2000	Packaged water tube, #2 oil, steam or hot water, gross output									
2010	200 MBH	Q-6	1.50	16	Ea.	7,425	540		7,965	9,000
2014	275 MBH		1.40	17.143		8,050	580		8,630	9,750
2018	360 MBH		1.10	21.818		9,025	735		9,760	11,000
2022	520 MBH		.65	36.923		9,250	1,250		10,500	12,100
2026	600 MBH		.60	40		9,400	1,350		10,750	12,400
2030	720 MBH		.55	43.636		9,800	1,475		11,275	13,000
2034	960 MBH	↓	.48	50		10,800	1,700		12,500	14,500
2040	1200 MBH	Q-7	.50	64		12,300	2,200		14,500	16,800
2044	1440 MBH		.45	71.111		15,200	2,450		17,650	20,400
2060	1600 MBH		.40	80		15,500	2,750		18,250	21,300
2068	1920 MBH		.35	91.429		16,900	3,150		20,050	23,300
2072	2160 MBH		.33	96.970		18,700	3,350		22,050	25,700
2080	2400 MBH		.30	106		20,900	3,675		24,575	28,600
2100	3200 MBH		.25	128		23,900	4,425		28,325	33,000
2120	4800 MBH	↓	.20	160	↓	31,200	5,525		36,725	42,700
2200	Gas fired									
2204	200 MBH	Q-6	1.50	16	Ea.	5,050	540		5,590	6,375
2208	275 MBH		1.40	17.143		5,250	580		5,830	6,650
2212	360 MBH		1.10	21.818		5,250	735		5,985	6,875
2216	520 MBH		.65	36.923		6,200	1,250		7,450	8,700
2220	600 MBH		.60	40		6,200	1,350		7,550	8,875
2224	720 MBH		.55	43.636		6,850	1,475		8,325	9,750
2228	960 MBH	↓	.48	50		8,050	1,700		9,750	11,400
2232	1220 MBH	Q-7	.50	64		8,675	2,200		10,875	12,900
2236	1440 MBH		.45	71.111		9,250	2,450		11,700	13,900
2240	1680 MBH		.40	80		10,300	2,750		13,050	15,600
2244	1920 MBH		.35	91.429		10,900	3,150		14,050	16,800
2248	2160 MBH		.33	96.970		11,800	3,350		15,150	18,100
2252	2400 MBH	↓	.30	106	↓	13,200	3,675		16,875	20,100
760 0010	**BOILERS, SOLID FUEL**									**760**
0020	Wood, coal, wood/paper waste, peat. Steel, 3″ diam. horiz.									
0040	fire tubes, low pressure steam or hot water, natural or									
0060	mechanical draft. Not including dry base.									
1000	Natural draft, gross output, 1445 MBH	Q-6	.45	53.333	Ea.	12,300	1,800		14,100	16,300
1050	1635 MBH		.43	55.814		12,900	1,875		14,775	17,100
1100	1805 MBH		.41	58.537		13,400	1,975		15,375	17,800
1150	2160 MBH	↓	.38	63.158		14,400	2,125		16,525	19,100

15 MECHANICAL

Important: See the Reference Section for critical supporting data - Reference Nos., Crews, & City Cost Indexes

15510	Heating Boilers and Accessories	CREW	DAILY OUTPUT	LABOR-HOURS	UNIT	2002 BARE COSTS				TOTAL INCL O&P
						MAT.	LABOR	EQUIP.	TOTAL	
760										**760**
1200	2520 MBH	Q-6	.32	75	Ea.	17,400	2,525		19,925	23,000
1250	3065 MBH	Q-7	.40	80		18,700	2,750		21,450	24,800
1300	3600 MBH		.38	84.211		20,100	2,900		23,000	26,500
1350	4500 MBH		.29	110		26,000	3,800		29,800	34,400
1400	5400 MBH		.26	123		27,900	4,250		32,150	37,100
1450	6295 MBH		.23	139		32,400	4,800		37,200	43,000
1500	7200 MBH		.21	152		36,700	5,250		41,950	48,200
1550	9000 MBH		.19	168		39,800	5,800		45,600	52,500
1600	10,800 MBH		.17	188		45,600	6,500		52,100	60,000
1650	12,600 MBH		.15	213		49,900	7,350		57,250	66,000
1700	14,390 MBH		.13	246		60,000	8,500		68,500	79,000
1750	16,190 MBH		.12	266		65,500	9,200		74,700	86,000
1800	17,990 MBH	▼	.10	320	▼	72,500	11,000		83,500	96,500
3000	Stoker fired (coal) cast iron with flush jacket and									
3080	insulation, steam or water, gross output, 148 MBH	Q-6	1.50	16	Ea.	3,650	540		4,190	4,850
3100	200 MBH		1.30	18.462		4,200	625		4,825	5,550
3120	247 MBH		1.20	20		4,675	675		5,350	6,175
3140	300 MBH		1	24		5,250	810		6,060	7,000
3160	410 MBH		.96	25		6,000	845		6,845	7,875
3180	495 MBH		.88	27.273		6,700	920		7,620	8,775
3200	580 MBH		.80	30		7,425	1,025		8,450	9,700
3400	1280 MBH		.36	66.667		18,300	2,250		20,550	23,500
3420	1460 MBH		.30	80		20,100	2,700		22,800	26,200
3440	1640 MBH		.28	85.714		21,500	2,900		24,400	28,000
3460	1820 MBH		.26	92.308		23,000	3,125		26,125	30,000
3480	2000 MBH		.25	96		24,500	3,250		27,750	31,800
3500	2360 MBH		.23	104		27,700	3,525		31,225	35,700
3540	2725 MBH	▼	.20	120		30,700	4,050		34,750	39,900
3800	2950 MBH	Q-7	.16	200		51,500	6,900		58,400	67,000
3820	3210 MBH		.15	213		54,000	7,350		61,350	70,500
3840	3480 MBH		.14	228		56,500	7,900		64,400	74,000
3860	3745 MBH		.14	228		59,000	7,900		66,900	77,000
3880	4000 MBH		.13	246		61,500	8,500		70,000	80,500
3900	4200 MBH		.13	246		64,500	8,500		73,000	83,500
3920	4400 MBH		.12	266		66,500	9,200		75,700	87,500
3940	4600 MBH	▼	.12	266	▼	69,500	9,200		78,700	90,000
880	**SWIMMING POOL HEATERS** Not including wiring, external									**880**
0020	piping, base or pad,									
0060	Gas fired, input, 115 MBH	Q-6	3	8	Ea.	1,575	270		1,845	2,125
0100	135 MBH		2	12		1,775	405		2,180	2,550
0160	155 MBH		1.50	16		1,875	540		2,415	2,875
0180	175 MBH		1.30	18.462		2,075	625		2,700	3,225
0200	190 MBH		1	24		2,425	810		3,235	3,900
0220	235 MBH		.70	34.286		2,975	1,150		4,125	5,000
0240	285 MBH		.60	40		3,800	1,350		5,150	6,250
0260	365 MBH		.50	48		4,300	1,625		5,925	7,175
0280	500 MBH		.40	60		5,875	2,025		7,900	9,500
0300	600 MBH		.35	68.571		6,750	2,325		9,075	10,900
0320	800 MBH		.33	72.727		8,050	2,450		10,500	12,600
0360	1000 MBH		.22	109		9,400	3,675		13,075	16,000
0370	1,200 MBH		.21	114		11,700	3,850		15,550	18,700
0380	1,500 MBH		.19	126		14,500	4,275		18,775	22,300
0400	1,800 MBH		.14	171		16,400	5,800		22,200	26,700
0410	2,450 MBH		.13	184		21,800	6,225		28,025	33,400
0420	3,000 MBH		.11	218		26,600	7,375		33,975	40,300
0440	3,750 MBH	▼	.09	266	▼	33,000	9,000		42,000	49,900

MECHANICAL 15

15510	Heating Boilers and Accessories	CREW	DAILY OUTPUT	LABOR-HOURS	UNIT	2002 BARE COSTS				TOTAL INCL O&P	
						MAT.	LABOR	EQUIP.	TOTAL		
880	**1000**	Oil fired, gross output, 97 MBH	Q-6	1.70	14.118	Ea.	1,800	475		2,275	2,700	**880**
1020	118 MBH		1.70	14.118		1,825	475		2,300	2,750	
1040	134 MBH		1.60	15		1,950	505		2,455	2,925	
1060	161 MBH		1.30	18.462		2,375	625		3,000	3,575	
1080	187 MBH		1.20	20		2,500	675		3,175	3,775	
1100	214 MBH		1.10	21.818		2,850	735		3,585	4,250	
1120	262 MBH		.89	26.966		3,225	910		4,135	4,925	
1140	340 MBH		.68	35.294		4,275	1,200		5,475	6,500	
1160	420 MBH		.58	41.379		4,900	1,400		6,300	7,475	
1180	525 MBH		.46	52.174		6,275	1,750		8,025	9,550	
1200	630 MBH		.41	58.537		7,000	1,975		8,975	10,700	
1220	735 MBH		.35	68.571		8,600	2,325		10,925	13,000	
1240	840 MBH		.33	72.727		9,850	2,450		12,300	14,500	
1260	1,050 MBH		.27	88.889		12,000	3,000		15,000	17,700	
1280	1,365 MBH		.21	114		15,400	3,850		19,250	22,700	
1300	1,680 MBH		.18	133		18,100	4,500		22,600	26,700	
1320	2,310 MBH		.13	184		23,400	6,225		29,625	35,100	
1340	2,835 MBH		.11	218		29,100	7,375		36,475	43,100	
1360	3,150 MBH		.09	266		32,600	9,000		41,600	49,500	
2000	Electric, 12 KW, 4,800 gallon pool	Q-19	3	8		1,550	268		1,818	2,125	
2020	15 KW, 7,200 gallon pool		2.80	8.571		1,550	287		1,837	2,150	
2040	24 KW, 9,600 gallon pool		2.40	10		2,100	335		2,435	2,825	
2060	30 KW, 12,000 gallon pool		2	12		2,175	400		2,575	3,000	
2080	35 KW, 14,400 gallon pool		1.60	15		2,175	505		2,680	3,150	
2100	55 KW, 24,000 gallon pool		1.20	20		3,000	670		3,670	4,300	
9000	To select pool heater: 12 BTUH x S.F. pool area										
9010	X temperature differential =required output										
9050	For electric, KW = gallons x 2.5 divided by 1000										
9100	For family home type pool, double the										
9110	Rated gallon capacity = 1/2°F rise per hour										

15520	Feedwater Equipment										
050	**0010**	**SHOT CHEMICAL FEEDER**										**050**
0200	Shot chem feeder, by pass, in line mount, 125 PSIG, 1.7 gallon	2 Stpi	1.80	8.889	Ea.	215	320		535	720	
0220	Floor mount, 175 PSIG, 5 gallon		1.24	12.903		285	465		750	1,025	
0230	12 gallon		.93	17.204		455	625		1,080	1,450	
0300	150 lb., ASME, 5 gallon		1.24	12.903		1,200	465		1,665	2,025	
0320	10 gallon		.93	17.204		1,300	625		1,925	2,375	
0400	300 lb., ASME, 5 gallon		1.24	12.903		3,225	465		3,690	4,225	
0420	10 gallon		.93	17.204		4,000	625		4,625	5,350	

15530	Furnaces										
200	**0010**	**FURNACE COMPONENTS AND COMBINATIONS**										**200**
0080	Coils, A/C evaporator, for gas or oil furnaces										
0090	Add-on, with holding charge										
0100	Upflow										
0120	1-1/2 ton cooling	Q-5	4	4	Ea.	127	130		257	335	
0130	2 ton cooling		3.70	4.324		153	141		294	380	
0140	3 ton cooling		3.30	4.848		191	158		349	450	
0150	4 ton cooling		3	5.333		270	174		444	560	
0160	5 ton cooling		2.70	5.926		345	193		538	670	
0300	Downflow										
0330	2-1/2 ton cooling	Q-5	3	5.333	Ea.	178	174		352	455	
0340	3-1/2 ton cooling		2.60	6.154		239	200		439	565	
0350	5 ton cooling		2.20	7.273		345	237		582	735	
0600	Horizontal										

			DAILY	LABOR-		2002 BARE COSTS				TOTAL	
15530	**Furnaces**	CREW	OUTPUT	HOURS	UNIT	MAT.	LABOR	EQUIP.	TOTAL	INCL O&P	
200	0630	2 ton cooling	Q-5	3.90	4.103	Ea.	180	134		314	400	**200**
	0640	3 ton cooling		3.50	4.571		206	149		355	450	
	0650	4 ton cooling		3.20	5		259	163		422	530	
	0660	5 ton cooling		2.90	5.517		345	180		525	650	
	2000	Cased evaporator coils for air handlers										
	2100	1-1/2 ton cooling	Q-5	4.40	3.636	Ea.	191	118		309	390	
	2110	2 ton cooling		4.10	3.902		194	127		321	405	
	2120	2-1/2 ton cooling		3.90	4.103		224	134		358	450	
	2130	3 ton cooling		3.70	4.324		265	141		406	505	
	2140	3-1/2 ton cooling		3.50	4.571		271	149		420	525	
	2150	4 ton cooling		3.20	5		297	163		460	570	
	2160	5 ton cooling		2.90	5.517		345	180		525	650	
	3010	Air handler, modular										
	3100	With cased evaporator cooling coil										
	3120	1-1/2 ton cooling	Q-5	3.80	4.211	Ea.	565	137		702	825	
	3130	2 ton cooling		3.50	4.571		590	149		739	875	
	3140	2-1/2 ton cooling		3.30	4.848		640	158		798	945	
	3150	3 ton cooling		3.10	5.161		700	168		868	1,025	
	3160	3-1/2 ton cooling		2.90	5.517		850	180		1,030	1,200	
	3170	4 ton cooling		2.50	6.400		955	209		1,164	1,375	
	3180	5 ton cooling		2.10	7.619		1,000	248		1,248	1,475	
	3500	With no cooling coil										
	3520	1-1/2 ton coil size	Q-5	12	1.333	Ea.	360	43.50		403.50	460	
	3530	2 ton coil size		10	1.600		385	52		437	505	
	3540	2-1/2 ton coil size		10	1.600		425	52		477	550	
	3554	3 ton coil size		9	1.778		475	58		533	615	
	3560	3-1/2 ton coil size		9	1.778		475	58		533	615	
	3570	4 ton coil size		8.50	1.882		680	61.50		741.50	845	
	3580	5 ton coil size		8	2		735	65		800	910	
	4000	With heater										
	4120	5 kW, 17.1 MBH	Q-5	16	1	Ea.	268	32.50		300.50	345	
	4130	7.5 kW, 25.6 MBH		15.60	1.026		268	33.50		301.50	345	
	4140	10 kW, 34.2 MBH		15.20	1.053		320	34.50		354.50	405	
	4150	12.5 KW, 42.7 MBH		14.80	1.081		320	35		355	410	
	4160	15 KW, 51.2 MBH		14.40	1.111		400	36		436	495	
	4170	25 KW, 85.4 MBH		14	1.143		400	37		437	495	
	4180	30 KW, 102 MBH		13	1.231		540	40		580	655	
400	0010	**FURNACES** Hot air heating, blowers, standard controls										**400**
	0020	not including gas, oil or flue piping										
	1000	Electric, UL listed										
	1020	10.2 MBH	Q-20	5	4	Ea.	315	129		444	545	
	1040	17.1 MBH		4.80	4.167		330	135		465	565	
	1060	27.3 MBH		4.60	4.348		395	141		536	650	
	1100	34.1 MBH		4.40	4.545		405	147		552	670	
	1120	51.6 MBH		4.20	4.762		505	154		659	790	
	1140	68.3 MBH		4	5		610	162		772	920	
	1160	85.3 MBH		3.80	5.263		670	170		840	995	
	3000	Gas, AGA certified, upflow, direct drive models										
	3020	45 MBH input	Q-9	4	4	Ea.	435	126		561	675	
	3040	60 MBH input		3.80	4.211		580	133		713	845	
	3060	75 MBH input		3.60	4.444		615	140		755	895	
	3100	100 MBH input		3.20	5		655	158		813	965	
	3120	125 MBH input		3	5.333		755	169		924	1,100	
	3130	150 MBH input		2.80	5.714		875	181		1,056	1,250	
	3140	200 MBH input		2.60	6.154		1,600	194		1,794	2,075	

MECHANICAL 15

		CREW	DAILY OUTPUT	LABOR-HOURS	UNIT	2002 BARE COSTS				TOTAL INCL O&P	
	15530	**Furnaces**				MAT.	LABOR	EQUIP.	TOTAL		
400	3160	300 MBH input	Q-9	2.30	6.957	Ea.	1,700	220		1,920	2,225	**400**
	3180	400 MBH input	↓	2	8	↓	2,200	253		2,453	2,825	
	3200	Gas fired wall furnace										
	3204	Horizontal flow										
	3208	7.7 MBH	Q-9	7	2.286	Ea.	400	72		472	550	
	3212	14 MBH		6.50	2.462		430	78		508	595	
	3216	24 MBH		5	3.200		520	101		621	725	
	3220	49 MBH		4	4		515	126		641	760	
	3224	65 MBH	↓	3.60	4.444	↓	695	140		835	975	
	3240	Up-flow										
	3244	7.7 MBH	Q-9	7	2.286	Ea.	254	72		326	390	
	3248	14 MBH		6.50	2.462		294	78		372	445	
	3252	24 MBH		5	3.200		360	101		461	555	
	3256	49 MBH		4	4		705	126		831	970	
	3260	65 MBH	↓	3.60	4.444	↓	730	140		870	1,025	
	3500	Gas furnace										
	3510	Up-flow										
	3514	1200 CFM, 45 MBH	Q-9	4	4	Ea.	685	126		811	950	
	3518	75 MBH		3.60	4.444		755	140		895	1,050	
	3522	100 MBH		3.20	5		865	158		1,023	1,200	
	3526	125 MBH		3	5.333		1,000	169		1,169	1,350	
	3530	150 MBH		2.80	5.714		1,225	181		1,406	1,625	
	3540	2000 CFM, 75 MBH		3.60	4.444		830	140		970	1,125	
	3544	100 MBH		3.20	5		940	158		1,098	1,275	
	3548	120 MBH		3	5.333		1,075	169		1,244	1,450	
	3552	150 MBH	↓	2.80	5.714	↓	1,300	181		1,481	1,700	
	3560	Horizontal flow										
	3564	1200 CFM, 45 MBH	Q-9	4	4	Ea.	690	126		816	950	
	3568	60 MBH		3.80	4.211		790	133		923	1,075	
	3572	70 MBH		3.60	4.444		800	140		940	1,100	
	3576	2000 CFM, 90 MBH		3.30	4.848		865	153		1,018	1,175	
	3580	110 MBH	↓	3.10	5.161	↓	905	163		1,068	1,250	
	6000	Oil, UL listed, atomizing gun type burner										
	6020	56 MBH output	Q-9	3.60	4.444	Ea.	745	140		885	1,025	
	6030	84 MBH output		3.50	4.571		775	144		919	1,075	
	6040	95 MBH output		3.40	4.706		790	149		939	1,100	
	6060	134 MBH output		3.20	5		1,100	158		1,258	1,450	
	6080	151 MBH output		3	5.333		1,225	169		1,394	1,575	
	6100	200 MBH input		2.60	6.154		1,650	194		1,844	2,125	
	6120	300 MBH input		2.30	6.957		1,925	220		2,145	2,475	
	6140	400 MBH input	↓	2	8	↓	2,275	253		2,528	2,900	
	6200	Up flow										
	6210	105 MBH	Q-9	3.40	4.706	Ea.	1,125	149		1,274	1,475	
	6220	140 MBH		3.30	4.848		1,300	153		1,453	1,650	
	6230	168 MBH	↓	2.90	5.517	↓	1,400	174		1,574	1,825	
	6260	Lo-boy style										
	6270	105 MBH	Q-9	3.40	4.706	Ea.	1,300	149		1,449	1,675	
	6280	140 MBH		3.30	4.848		1,425	153		1,578	1,775	
	6290	189 MBH	↓	2.80	5.714	↓	1,575	181		1,756	2,025	
	6320	Down flow										
	6330	105 MBH	Q-9	3.40	4.706	Ea.	1,250	149		1,399	1,600	
	6340	140 MBH		3.30	4.848		1,425	153		1,578	1,800	
	6350	168 MBH	↓	2.90	5.517	↓	1,475	174		1,649	1,900	
	6380	Horizontal										
	6390	105 MBH	Q-9	3.40	4.706	Ea.	1,225	149		1,374	1,575	
	6400	140 MBH	↓	3.30	4.848		1,350	153		1,503	1,725	

Important: See the Reference Section for critical supporting data - Reference Nos., Crews, & City Cost Indexes

15500 | Heat Generation Equipment

15530 | Furnaces

		CREW	DAILY OUTPUT	LABOR-HOURS	UNIT	MAT.	LABOR	EQUIP.	TOTAL	TOTAL INCL O&P	
400	6410	189 MBH	Q-9	2.80	5.714	Ea.	1,475	181		1,656	1,900	**400**
	8500	Wood and coal combination complete										
	8520	95 MBH output	Q-9	3	5.333	Ea.	4,425	169		4,594	5,125	
	8540	140 MBH output (oil)		3	5.333		4,425	169		4,594	5,125	
	8560	163 MBH output (oil)	↓	3	5.333	↓	4,425	169		4,594	5,125	
440	0010	**FURNACES, COMBINATION SYSTEMS** Heating, cooling, electric air										**440**
	0020	cleaner, humidification, dehumidification.										
	2000	Gas fired, 80 MBH heat output, 24 MBH cooling	Q-9	1.20	13.333	Ea.	2,850	420		3,270	3,800	
	2020	80 MBH heat output, 36 MBH cooling		1.20	13.333		3,125	420		3,545	4,100	
	2040	100 MBH heat output, 29 MBH cooling		1	16		3,200	505		3,705	4,300	
	2060	100 MBH heat output, 36 MBH cooling		1	16		3,350	505		3,855	4,475	
	2080	100 MBH heat output, 47 MBH cooling	↓	.90	17.778		3,975	560		4,535	5,250	
	2100	120 MBH heat output, 29 MBH cooling	Q-10	1.30	18.462		3,400	605		4,005	4,650	
	2120	120 MBH heat output, 42 MBH cooling		1.30	18.462		3,575	605		4,180	4,875	
	2140	120 MBH heat output, 47 MBH cooling		1.20	20		4,000	655		4,655	5,400	
	2160	120 MBH heat output, 55 MBH cooling		1.10	21.818		4,300	715		5,015	5,825	
	2180	144 MBH heat output, 42 MBH cooling		1.20	20		3,725	655		4,380	5,100	
	2200	144 MBH heat output, 47 MBH cooling		1.20	20		4,150	655		4,805	5,550	
	2220	144 MBH heat output, 58 MBH cooling	↓	1	24	↓	4,350	785		5,135	6,000	
	2500	For precharged tubing with connection										
	2520	15 feet				Ea.	119			119	131	
	2540	25 feet					159			159	175	
	2560	35 feet					205			205	226	
	3000	Oil fired, 84 MBH heat output, 24 MBH cooling	Q-9	1.20	13.333		3,200	420		3,620	4,175	
	3020	84 MBH heat output, 36 MBH cooling		1.20	13.333		3,525	420		3,945	4,525	
	3040	95.2 MBH heat output, 29 MBH cooling		1	16		3,450	505		3,955	4,550	
	3060	95.2 MBH heat output, 36 MBH cooling		1	16		3,575	505		4,080	4,725	
	3080	95.2 MBH heat output, 47 MBH cooling	↓	.90	17.778		4,150	560		4,710	5,425	
	3100	112 MBH heat output, 29 MBH cooling	Q-10	1.30	18.462		3,675	605		4,280	4,975	
	3120	112 MBH heat output, 42 MBH cooling		1.30	18.462		4,000	605		4,605	5,325	
	3140	112 MBH heat output, 47 MBH cooling		1.20	20		4,425	655		5,080	5,850	
	3160	112 MBH heat output, 58 MBH cooling		1.10	21.818		4,825	715		5,540	6,425	
	3180	151.2 MBH heat output, 42 MBH cooling		1.20	20		4,300	655		4,955	5,725	
	3200	151.2 MBH heat output, 47 MBH cooling		1.20	20		4,700	655		5,355	6,175	
	3240	151.2 MBH heat output, 58 MBH cooling	↓	1	24	↓	4,925	785		5,710	6,625	
	3500	For precharged tubing with connection, add										
	3520	15 feet				Ea.	119			119	131	
	3540	25 feet					159			159	175	
	3560	35 feet				↓	190			190	209	

15540 | Fuel-Fired Heaters

		CREW	DAILY OUTPUT	LABOR-HOURS	UNIT	MAT.	LABOR	EQUIP.	TOTAL	TOTAL INCL O&P	
300	0010	**DUCT FURNACES** Includes burner, controls, stainless steel										**300**
	0020	heat exchanger. Gas fired, electric ignition										
	0030	Indoor installation										
	0080	100 MBH output	Q-5	5	3.200	Ea.	1,050	104		1,154	1,300	
	0100	120 MBH output		4	4		1,200	130		1,330	1,500	
	0130	200 MBH output		2.70	5.926		1,650	193		1,843	2,125	
	0140	240 MBH output		2.30	6.957		1,675	227		1,902	2,200	
	0160	280 MBH output		2	8		1,925	261		2,186	2,525	
	0180	320 MBH output	↓	1.60	10	↓	2,100	325		2,425	2,825	
	0300	For powered venter and adapter, add					284			284	310	
	0500	For required flue pipe, see 15550										
	1000	Outdoor installation, with vent cap										
	1020	75 MBH output	Q-5	4	4	Ea.	1,650	130		1,780	2,000	
	1040	94 MBH output		4	4		1,775	130		1,905	2,150	

299

15540	Fuel-Fired Heaters	CREW	DAILY OUTPUT	LABOR-HOURS	UNIT	2002 BARE COSTS				TOTAL INCL O&P	
						MAT.	LABOR	EQUIP.	TOTAL		
300											**300**
1060	120 MBH output	Q-5	4	4	Ea.	2,025	130		2,155	2,425	
1080	157 MBH output		3.50	4.571		2,350	149		2,499	2,825	
1100	187 MBH output		3	5.333		2,900	174		3,074	3,450	
1120	225 MBH output		2.50	6.400		3,025	209		3,234	3,650	
1140	300 MBH output		1.80	8.889		3,800	290		4,090	4,600	
1160	375 MBH output		1.60	10		4,150	325		4,475	5,075	
1180	450 MBH output		1.40	11.429		4,350	370		4,720	5,350	
1200	600 MBH output		1	16		5,300	520		5,820	6,600	
1300	Aluminized exchanger, subtract					15%					
1400	For powered venter, add					10%					
1500	For two stage gas valve, add					160			160	176	
900	**0010** **SPACE HEATERS** Cabinet, grilles, fan, controls, burner,										**900**
0020	thermostat, no piping. For flue see 15550										
1000	Gas fired, floor mounted										
1100	60 MBH output	Q-5	10	1.600	Ea.	545	52		597	680	
1120	80 MBH output		9	1.778		555	58		613	700	
1140	100 MBH output		8	2		595	65		660	755	
1160	120 MBH output		7	2.286		670	74.50		744.50	850	
1180	180 MBH output		6	2.667		815	87		902	1,025	
1500	Rooftop mounted, gravity vent, stainless steel exchanger										
1520	75 MBH output	Q-6	4	6	Ea.	3,300	203		3,503	3,950	
1540	95 MBH output		3.60	6.667		3,375	225		3,600	4,050	
1560	120 MBH output		3.30	7.273		3,600	246		3,846	4,350	
1580	159 MBH output		3	8		3,725	270		3,995	4,500	
1600	190 MBH output		2.60	9.231		4,200	310		4,510	5,075	
1620	225 MBH output		2.30	10.435		4,250	350		4,600	5,200	
1640	300 MBH output		1.90	12.632		5,075	425		5,500	6,225	
1660	375 MBH output		1.40	17.143		5,750	580		6,330	7,200	
1680	450 MBH output		1.20	20		5,875	675		6,550	7,475	
1700	600 MBH output		1	24		7,075	810		7,885	9,025	
1720	750 MBH output		.80	30		12,100	1,025		13,125	14,800	
1740	900 MBH output		.60	40		13,600	1,350		14,950	17,000	
1760	1200 MBH output		.30	80		16,500	2,700		19,200	22,200	
1800	For power vent, add					10%					
1900	For aluminized steel exchanger, subtract					10%					
2000	Suspension mounted, propeller fan, 20 MBH output	Q-5	8.50	1.882		420	61.50		481.50	555	
2020	40 MBH output		7.50	2.133		460	69.50		529.50	610	
2040	60 MBH output		7	2.286		515	74.50		589.50	675	
2060	80 MBH output		6	2.667		575	87		662	760	
2080	100 MBH output		5.50	2.909		685	95		780	895	
2100	130 MBH output		5	3.200		765	104		869	995	
2120	140 MBH output		4.50	3.556		820	116		936	1,075	
2140	160 MBH output		4	4		850	130		980	1,125	
2160	180 MBH output		3.50	4.571		930	149		1,079	1,250	
2180	200 MBH output		3	5.333		990	174		1,164	1,350	
2200	240 MBH output		2.70	5.926		1,175	193		1,368	1,600	
2220	280 MBH output		2.30	6.957		1,375	227		1,602	1,875	
2240	320 MBH output		2	8		1,575	261		1,836	2,125	
2500	For powered venter and adapter, add					254			254	279	
3000	Suspension mounted, blower type, 40 MBH output	Q-5	6.80	2.353		880	76.50		956.50	1,075	
3020	60 MBH output		6.60	2.424		790	79		869	985	
3040	84 MBH output		5.80	2.759		865	90		955	1,075	
3060	104 MBH output		5.20	3.077		1,175	100		1,275	1,425	
3080	140 MBH output		4.30	3.721		1,400	121		1,521	1,725	
3100	180 MBH output		3.30	4.848		1,450	158		1,608	1,850	

15 MECHANICAL

			DAILY	LABOR-		2002 BARE COSTS				TOTAL	
15540	**Fuel-Fired Heaters**	CREW	OUTPUT	HOURS	UNIT	MAT.	LABOR	EQUIP.	TOTAL	INCL O&P	
900	3120	240 MBH output	Q-5	2.50	6.400	Ea.	2,125	209		2,334	2,650	900
	3140	280 MBH output	↓	2	8	↓	2,325	261		2,586	2,975	
	4000	Suspension mounted, sealed combustion system,										
	4020	Aluminized steel exchanger, powered vent										
	4040	100 MBH output	Q-5	5	3.200	Ea.	1,600	104		1,704	1,900	
	4060	120 MBH output		4.70	3.404		1,725	111		1,836	2,075	
	4080	160 MBH output		3.70	4.324		1,925	141		2,066	2,300	
	4100	200 MBH output		2.90	5.517		2,275	180		2,455	2,775	
	4120	240 MBH output		2.50	6.400		2,425	209		2,634	3,000	
	4140	320 MBH output		1.70	9.412		2,775	305		3,080	3,525	
	5000	Wall furnace, 17.5 MBH output		6	2.667		535	87		622	715	
	5020	24 MBH output		5	3.200		550	104		654	760	
	5040	35 MBH output		4	4		745	130		875	1,000	
	6000	Oil fired, suspension mounted, 94 MBH output		4	4		1,975	130		2,105	2,375	
	6040	140 MBH output		3	5.333		2,100	174		2,274	2,550	
	6060	184 MBH output	↓	3	5.333	↓	2,275	174		2,449	2,750	

15550	**Breechings, Chimneys & Stacks**										
200	0010	**DRAFT CONTROLS**										200
	1000	Barometric, gas fired system only, 6" size for 5" and 6" pipes	1 Shee	20	.400	Ea.	39.50	14.05		53.55	65	
	1020	7" size, for 6" and 7" pipes		19	.421		46	14.80		60.80	73.50	
	1040	8" size, for 7" and 8" pipes		18	.444		53.50	15.60		69.10	83	
	1060	9" size, for 8" and 9" pipes	↓	16	.500	↓	60.50	17.55		78.05	93.50	
	2000	All fuel, oil, oil/gas, coal										
	2020	10" for 9" and 10" pipes	1 Shee	15	.533	Ea.	91.50	18.70		110.20	130	
	2040	12" for 11" and 12" pipes		15	.533		122	18.70		140.70	163	
	2060	14" for 13" and 14" pipes		14	.571		152	20		172	199	
	2080	16" for 15" and 16" pipes		13	.615		213	21.50		234.50	269	
	2100	18" for 17" and 18" pipes		12	.667		278	23.50		301.50	340	
	2120	20" for 19" and 21" pipes	↓	10	.800		340	28		368	420	
	2140	24" for 22" and 25" pipes	Q-9	12	1.333		415	42		457	520	
	2160	28" for 26" and 30" pipes		10	1.600		495	50.50		545.50	625	
	2180	32" for 31" and 34" pipes	↓	8	2		635	63		698	800	
	3260	For thermal switch for above, add	1 Shee	24	.333		38.50	11.70		50.20	60.50	
	5000	Vent damper, bi-metal, gas, 3" diameter	Q-9	24	.667		22	21		43	56.50	
	5010	4" diameter		24	.667		22	21		43	56.50	
	5020	5" diameter		23	.696		22	22		44	58	
	5030	6" diameter		22	.727		22	23		45	59.50	
	5040	7" diameter		21	.762		23.50	24		47.50	63	
	5050	8" diameter		20	.800		25.50	25.50		51	67	
	5101	Electric, automatic, gas, 4" diameter		24	.667		137	21		158	183	
	5110	5" diameter		23	.696		139	22		161	187	
	5121	6" diameter		22	.727		142	23		165	192	
	5130	7" diameter		21	.762		144	24		168	196	
	5140	8" diameter		20	.800		147	25.50		172.50	201	
	5150	9" diameter		20	.800		150	25.50		175.50	204	
	5160	10" diameter		19	.842		155	26.50		181.50	211	
	5170	12" diameter		19	.842		157	26.50		183.50	214	
	5250	Automatic, oil, 4" diameter		24	.667		153	21		174	201	
	5260	5" diameter		23	.696		156	22		178	205	
	5270	6" diameter		22	.727		159	23		182	211	
	5280	7" diameter		21	.762		162	24		186	215	
	5290	8" diameter		20	.800		169	25.50		194.50	225	
	5300	9" diameter		20	.800		171	25.50		196.50	228	
	5310	10" diameter		19	.842		172	26.50		198.50	230	
	5320	12" diameter	↓	19	.842	↓	175	26.50		201.50	233	

MECHANICAL 15

15500 | Heat Generation Equipment

			CREW	DAILY OUTPUT	LABOR-HOURS	UNIT	MAT.	LABOR	EQUIP.	TOTAL	TOTAL INCL O&P	
15550		**Breechings, Chimneys & Stacks**						2002 BARE COSTS				
200	5330	14" diameter	Q-9	18	.889	Ea.	234	28		262	300	200
	5340	16" diameter		17	.941		271	29.50		300.50	345	
	5350	18" diameter	↓	16	1	↓	305	31.50		336.50	385	
440	0010	**VENT CHIMNEY** Prefab metal, U.L. listed										440
	0020	Gas, double wall, galvanized steel										
	0080	3" diameter	Q-9	72	.222	V.L.F.	3.48	7		10.48	14.70	
	0100	4" diameter		68	.235		4.35	7.45		11.80	16.25	
	0120	5" diameter		64	.250		5.05	7.90		12.95	17.80	
	0140	6" diameter		60	.267		6.25	8.45		14.70	19.85	
	0160	7" diameter		56	.286		8.65	9.05		17.70	23.50	
	0180	8" diameter		52	.308		9.65	9.70		19.35	25.50	
	0200	10" diameter		48	.333		20.50	10.55		31.05	39	
	0220	12" diameter		44	.364		27	11.50		38.50	47.50	
	0240	14" diameter		42	.381		46	12.05		58.05	69	
	0260	16" diameter		40	.400		62	12.65		74.65	87.50	
	0280	18" diameter	↓	38	.421		80	13.30		93.30	109	
	0300	20" diameter	Q-10	36	.667		94.50	22		116.50	138	
	0320	22" diameter		34	.706		120	23		143	168	
	0340	24" diameter		32	.750		147	24.50		171.50	200	
	0360	26" diameter		31	.774		177	25.50		202.50	234	
	0380	28" diameter		30	.800		187	26		213	246	
	0400	30" diameter		28	.857		197	28		225	261	
	0420	32" diameter		27	.889		229	29		258	297	
	0440	34" diameter		26	.923		261	30.50		291.50	335	
	0460	36" diameter		25	.960		275	31.50		306.50	355	
	0480	38" diameter		24	1		300	33		333	380	
	0500	40" diameter		23	1.043		335	34		369	425	
	0520	42" diameter		22	1.091		350	36		386	440	
	0540	44" diameter		21	1.143		390	37.50		427.50	490	
	0560	46" diameter		20	1.200		430	39.50		469.50	530	
	0580	48" diameter	↓	19	1.263	↓	475	41.50		516.50	585	
	0600	For 4", 5" and 6" oval, add						50%				
	0650	Gas, double wall, galvanized steel, fittings										
	0660	Elbow 45°, 3" diameter	Q-9	36	.444	Ea.	8.40	14.05		22.45	31	
	0670	4" diameter		34	.471		10.20	14.85		25.05	34.50	
	0680	5" diameter		32	.500		12	15.80		27.80	37.50	
	0690	6" diameter		30	.533		14.35	16.85		31.20	42	
	0700	7" diameter		28	.571		20.50	18.05		38.55	51	
	0710	8" diameter		26	.615		24	19.45		43.45	56	
	0720	10" diameter		24	.667		59	21		80	97.50	
	0730	12" diameter		22	.727		75	23		98	118	
	0740	14" diameter		21	.762		105	24		129	152	
	0750	16" diameter		20	.800		131	25.50		156.50	184	
	0760	18" diameter	↓	19	.842		177	26.50		203.50	235	
	0770	20" diameter	Q-10	18	1.333		194	43.50		237.50	281	
	0780	22" diameter		17	1.412		292	46.50		338.50	390	
	0790	24" diameter		16	1.500		380	49		429	495	
	0800	26" diameter		16	1.500		750	49		799	900	
	0810	28" diameter		15	1.600		770	52.50		822.50	930	
	0820	30" diameter		14	1.714		815	56		871	980	
	0830	32" diameter		14	1.714		915	56		971	1,075	
	0840	34" diameter		13	1.846		1,000	60.50		1,060.50	1,200	
	0850	36" diameter		12	2		1,025	65.50		1,090.50	1,225	
	0860	38" diameter		12	2		1,125	65.50		1,190.50	1,325	
	0870	40" diameter	↓	11	2.182	↓	1,175	71.50		1,246.50	1,400	

15550	Breechings, Chimneys & Stacks	CREW	DAILY OUTPUT	LABOR-HOURS	UNIT	2002 BARE COSTS				TOTAL INCL O&P
						MAT.	LABOR	EQUIP.	TOTAL	
0880	42" diameter	Q-10	11	2.182	Ea.	1,225	71.50		1,296.50	1,450
0890	44" diameter		10	2.400		1,325	78.50		1,403.50	1,575
0900	46" diameter		10	2.400		1,400	78.50		1,478.50	1,675
0910	48" diameter		10	2.400		1,475	78.50		1,553.50	1,750
0916	Adjustable length									
0918	3" diameter, to 12"	Q-9	36	.444	Ea.	7.05	14.05		21.10	29.50
0920	4" diameter, to 12"		34	.471		8.35	14.85		23.20	32
0924	6" diameter, to 12"		30	.533		10.80	16.85		27.65	38
0928	8" diameter, to 12"		26	.615		15.85	19.45		35.30	47.50
0930	10" diameter, to 18"		24	.667		59.50	21		80.50	98
0932	12" diameter, to 18"		22	.727		72.50	23		95.50	116
0936	16" diameter, to 18"		20	.800		140	25.50		165.50	193
0938	18" diameter, to 18"		19	.842		167	26.50		193.50	225
0944	24" diameter, to 18"	Q-10	16	1.500		299	49		348	405
0950	Elbow 90°, adjustable, 3" diameter	Q-9	36	.444		14.40	14.05		28.45	37.50
0960	4" diameter		34	.471		17.50	14.85		32.35	42.50
0970	5" diameter		32	.500		20.50	15.80		36.30	47
0980	6" diameter		30	.533		24.50	16.85		41.35	53
0990	7" diameter		28	.571		38.50	18.05		56.55	70.50
1010	8" diameter		26	.615		41	19.45		60.45	75
1020	Wall thimble, 4 to 7" adjustable, 3" diameter		36	.444		9.10	14.05		23.15	31.50
1022	4" diameter		34	.471		10.15	14.85		25	34
1024	5" diameter		32	.500		11.40	15.80		27.20	37
1026	6" diameter		30	.533		14	16.85		30.85	41.50
1028	7" diameter		28	.571		16.25	18.05		34.30	46
1030	8" diameter		26	.615		16.85	19.45		36.30	48.50
1040	Roof flashing, 3" diameter		36	.444		5.20	14.05		19.25	27
1050	4" diameter		34	.471		5.85	14.85		20.70	29.50
1060	5" diameter		32	.500		9.10	15.80		24.90	34.50
1070	6" diameter		30	.533		12.75	16.85		29.60	40
1080	7" diameter		28	.571		16.10	18.05		34.15	45.50
1090	8" diameter		26	.615		17.50	19.45		36.95	49.50
1100	10" diameter		24	.667		47.50	21		68.50	85
1110	12" diameter		22	.727		61	23		84	103
1120	14" diameter		20	.800		75	25.50		100.50	122
1130	16" diameter		18	.889		92.50	28		120.50	146
1140	18" diameter		16	1		125	31.50		156.50	187
1150	20" diameter	Q-10	18	1.333		163	43.50		206.50	247
1160	22" diameter		14	1.714		206	56		262	315
1170	24" diameter		12	2		247	65.50		312.50	375
1200	Tee, 3" diameter	Q-9	27	.593		21.50	18.75		40.25	52.50
1210	4" diameter		26	.615		22	19.45		41.45	54.50
1220	5" diameter		25	.640		24	20		44	57.50
1230	6" diameter		24	.667		26	21		47	61
1240	7" diameter		23	.696		28.50	22		50.50	65.50
1250	8" diameter		22	.727		39	23		62	78.50
1260	10" diameter		21	.762		106	24		130	154
1270	12" diameter		20	.800		128	25.50		153.50	179
1280	14" diameter		18	.889		208	28		236	273
1290	16" diameter		16	1		295	31.50		326.50	375
1300	18" diameter		14	1.143		370	36		406	460
1310	20" diameter	Q-10	17	1.412		500	46.50		546.50	620
1320	22" diameter		13	1.846		620	60.50		680.50	780
1330	24" diameter		12	2		705	65.50		770.50	880
1340	26" diameter		12	2		1,025	65.50		1,090.50	1,225
1350	28" diameter		11	2.182		1,150	71.50		1,221.50	1,375

440

MECHANICAL 15

15550	Breechings, Chimneys & Stacks	CREW	DAILY OUTPUT	LABOR-HOURS	UNIT	2002 BARE COSTS				TOTAL INCL O&P
						MAT.	LABOR	EQUIP.	TOTAL	
440 1360	30" diameter	Q-10	10	2.400	Ea.	1,250	78.50		1,328.50	1,500	440
1370	32" diameter		10	2.400		1,475	78.50		1,553.50	1,750
1380	34" diameter		9	2.667		1,475	87.50		1,562.50	1,750
1390	36" diameter		9	2.667		1,775	87.50		1,862.50	2,075
1400	38" diameter		8	3		1,775	98.50		1,873.50	2,100
1410	40" diameter		7	3.429		1,925	112		2,037	2,300
1420	42" diameter		6	4		2,100	131		2,231	2,525
1430	44" diameter		5	4.800		2,225	157		2,382	2,700
1440	46" diameter		5	4.800		2,450	157		2,607	2,925
1450	48" diameter		5	4.800		2,600	157		2,757	3,125
1460	Tee cap, 3" diameter	Q-9	45	.356		1.53	11.25		12.78	19
1470	4" diameter		42	.381		1.69	12.05		13.74	20.50
1480	5" diameter		40	.400		2.28	12.65		14.93	22
1490	6" diameter		37	.432		2.63	13.65		16.28	24
1500	7" diameter		35	.457		4.29	14.45		18.74	27
1510	8" diameter		34	.471		5	14.85		19.85	28.50
1520	10" diameter		32	.500		9.60	15.80		25.40	35
1530	12" diameter		30	.533		12.80	16.85		29.65	40
1540	14" diameter		28	.571		26.50	18.05		44.55	57
1550	16" diameter		25	.640		30.50	20		50.50	64.50
1560	18" diameter		24	.667		34.50	21		55.50	70.50
1570	20" diameter	Q-10	27	.889		37	29		66	85.50
1580	22" diameter		22	1.091		42.50	36		78.50	102
1590	24" diameter		21	1.143		48	37.50		85.50	110
1600	26" diameter		20	1.200		83	39.50		122.50	152
1610	28" diameter		19	1.263		103	41.50		144.50	177
1620	30" diameter		18	1.333		113	43.50		156.50	192
1630	32" diameter		17	1.412		131	46.50		177.50	216
1640	34" diameter		16	1.500		146	49		195	237
1650	36" diameter		15	1.600		163	52.50		215.50	260
1660	38" diameter		14	1.714		177	56		233	281
1670	40" diameter		14	1.714		193	56		249	299
1680	42" diameter		13	1.846		205	60.50		265.50	320
1690	44" diameter		12	2		222	65.50		287.50	345
1700	46" diameter		12	2		235	65.50		300.50	360
1710	48" diameter		11	2.182		265	71.50		336.50	400
1750	Top, 3" diameter	Q-9	46	.348		8	11		19	26
1760	4" diameter		44	.364		8.40	11.50		19.90	27
1770	5" diameter		42	.381		11.55	12.05		23.60	31.50
1780	6" diameter		40	.400		14.65	12.65		27.30	35.50
1790	7" diameter		38	.421		22	13.30		35.30	44.50
1800	8" diameter		36	.444		27.50	14.05		41.55	51.50
1810	10" diameter		34	.471		66	14.85		80.85	95.50
1820	12" diameter		32	.500		99.50	15.80		115.30	134
1830	14" diameter		30	.533		113	16.85		129.85	150
1840	16" diameter		28	.571		204	18.05		222.05	252
1850	18" diameter		26	.615		370	19.45		389.45	435
1860	20" diameter	Q-10	28	.857		400	28		428	485
1870	22" diameter		22	1.091		545	36		581	655
1880	24" diameter		20	1.200		680	39.50		719.50	805
1882	26" diameter		19	1.263		1,025	41.50		1,066.50	1,200
1884	28" diameter		18	1.333		1,175	43.50		1,218.50	1,350
1886	30" diameter		17	1.412		1,300	46.50		1,346.50	1,525
1888	32" diameter		16	1.500		1,375	49		1,424	1,600
1890	34" diameter		15	1.600		1,475	52.50		1,527.50	1,700
1892	36" diameter		14	1.714		1,575	56		1,631	1,825

15 MECHANICAL

Important: See the Reference Section for critical supporting data - Reference Nos., Crews, & City Cost Indexes

15550	Breechings, Chimneys & Stacks	CREW	DAILY OUTPUT	LABOR-HOURS	UNIT	2002 BARE COSTS				TOTAL INCL O&P
						MAT.	LABOR	EQUIP.	TOTAL	
1900	Gas, double wall, galvanized steel, oval									
1904	4" x 1'	Q-9	68	.235	V.L.F.	11.20	7.45		18.65	24
1906	5"		64	.250		23	7.90		30.90	37
1908	5"/6" x 1'		60	.267		23	8.45		31.45	38.50
1910	Oval fittings									
1912	Adjustable length									
1914	4" diameter to 12" lg	Q-9	34	.471	Ea.	11.85	14.85		26.70	36
1916	5" diameter to 12" lg		32	.500		33	15.80		48.80	60.50
1918	5"/6" diameter to 12"		30	.533		30	16.85		46.85	59
1920	Elbow 45°									
1922	4"	Q-9	34	.471	Ea.	21.50	14.85		36.35	47
1924	5"		32	.500		41	15.80		56.80	69.50
1926	5"/6"		30	.533		39	16.85		55.85	68.50
1930	Elbow 45°, flat									
1932	4"	Q-9	34	.471	Ea.	21.50	14.85		36.35	47
1934	5"		32	.500		41	15.80		56.80	69.50
1936	5"/6"		30	.533		41	16.85		57.85	71
1940	Top									
1942	4"	Q-9	44	.364	Ea.	16.10	11.50		27.60	35.50
1944	5"		42	.381		24.50	12.05		36.55	45.50
1946	5"/6"		40	.400		24.50	12.65		37.15	46.50
1950	Adjustable flashing									
1952	4"	Q-9	34	.471	Ea.	9.70	14.85		24.55	33.50
1954	5"		32	.500		28.50	15.80		44.30	56
1956	5"/6"		30	.533		19.20	16.85		36.05	47
1960	Tee									
1962	4"	Q-9	26	.615	Ea.	32	19.45		51.45	65.50
1964	5"		25	.640		61.50	20		81.50	99
1966	5"/6"		24	.667		62	21		83	101
1970	Tee with short snout									
1972	4"	Q-9	26	.615	Ea.	32	19.45		51.45	65.50
3200	All fuel, pressure tight, double wall, U.L. listed, 1400°F.									
3210	304 stainless steel liner, aluminized steel outer jacket									
3220	6" diameter	Q-9	60	.267	L.F.	35.50	8.45		43.95	52
3221	8" diameter		52	.308		43.50	9.70		53.20	63
3222	10" diameter		48	.333		48.50	10.55		59.05	70
3223	12" diameter		44	.364		55.50	11.50		67	78.50
3224	14" diameter		42	.381		63.50	12.05		75.55	88.50
3225	16" diameter		40	.400		67.50	12.65		80.15	93.50
3226	18" diameter		38	.421		76	13.30		89.30	104
3227	20" diameter	Q-10	36	.667		87	22		109	129
3228	24" diameter		32	.750		112	24.50		136.50	161
3229	28" diameter		30	.800		128	26		154	182
3230	32" diameter		27	.889		144	29		173	203
3231	36" diameter		25	.960		161	31.50		192.50	226
3232	42" diameter		22	1.091		184	36		220	258
3233	48" diameter		19	1.263		217	41.50		258.50	300
3260	For 316 stainless steel liner add					6%				
3280	All fuel, pressure tight, double wall fittings									
3284	304 stainless steel inner, aluminized steel jacket									
3288	Adjustable 18"/30" section									
3291	5" diameter	Q-9	32	.500	Ea.	91	15.80		106.80	125
3292	6" diameter		30	.533		115	16.85		131.85	153
3293	8" diameter		26	.615		133	19.45		152.45	176
3294	10" diameter		24	.667		149	21		170	197
3295	12" diameter		22	.727		170	23		193	223

440

MECHANICAL 15

		15550	Breechings, Chimneys & Stacks	CREW	DAILY OUTPUT	LABOR-HOURS	UNIT	2002 BARE COSTS				TOTAL INCL O&P	
								MAT.	LABOR	EQUIP.	TOTAL		
440	3296		14" diameter	Q-9	21	.762	Ea.	196	24		220	253	440
	3297		16" diameter		20	.800		204	25.50		229.50	263	
	3298		18" diameter	▼	19	.842		251	26.50		277.50	320	
	3299		20" diameter	Q-10	18	1.333		290	43.50		333.50	390	
	3300		24" diameter		16	1.500		375	49		424	485	
	3301		28" diameter		15	1.600		415	52.50		467.50	540	
	3302		32" diameter		14	1.714		430	56		486	560	
	3303		36" diameter		12	2		450	65.50		515.50	595	
	3304		42" diameter		11	2.182		515	71.50		586.50	680	
	3305		48" diameter	▼	10	2.400		605	78.50		683.50	785	
	3326		For 316 stainless steel liner, add				▼	15%					
	3350		Elbow 90° fixed										
	3353		5" diameter	Q-9	32	.500	Ea.	170	15.80		185.80	212	
	3354		6" diameter		30	.533		200	16.85		216.85	246	
	3355		8" diameter		26	.615		221	19.45		240.45	274	
	3356		10" diameter		24	.667		254	21		275	310	
	3357		12" diameter		22	.727		287	23		310	350	
	3358		14" diameter		21	.762		335	24		359	405	
	3359		16" diameter		20	.800		375	25.50		400.50	455	
	3360		18" diameter	▼	19	.842		440	26.50		466.50	520	
	3361		20" diameter	Q-10	18	1.333		490	43.50		533.50	605	
	3362		24" diameter		16	1.500		605	49		654	740	
	3363		28" diameter		15	1.600		755	52.50		807.50	910	
	3364		32" diameter		14	1.714		1,000	56		1,056	1,175	
	3365		36" diameter		12	2		1,125	65.50		1,190.50	1,350	
	3366		42" diameter		11	2.182		1,725	71.50		1,796.50	2,000	
	3367		48" diameter	▼	10	2.400		2,175	78.50		2,253.50	2,500	
	3380		For 316 stainless steel liner, add				▼	20%					
	3400		Elbow 45°										
	3404		6" diameter	Q-9	30	.533	Ea.	127	16.85		143.85	165	
	3405		8" diameter		26	.615		143	19.45		162.45	188	
	3406		10" diameter		24	.667		161	21		182	210	
	3407		12" diameter		22	.727		184	23		207	238	
	3408		14" diameter		21	.762		211	24		235	269	
	3409		16" diameter		20	.800		238	25.50		263.50	300	
	3410		18" diameter	▼	19	.842		269	26.50		295.50	335	
	3411		20" diameter	Q-10	18	1.333		310	43.50		353.50	410	
	3412		24" diameter		16	1.500		395	49		444	510	
	3413		28" diameter		15	1.600		445	52.50		497.50	570	
	3414		32" diameter		14	1.714		600	56		656	745	
	3415		36" diameter		12	2		805	65.50		870.50	985	
	3416		42" diameter		11	2.182		1,100	71.50		1,171.50	1,300	
	3417		48" diameter	▼	10	2.400		1,475	78.50		1,553.50	1,750	
	3430		For 316 stainless steel liner, add				▼	20%					
	3450		Tee 90°										
	3454		6" diameter	Q-9	24	.667	Ea.	132	21		153	178	
	3455		8" diameter		22	.727		152	23		175	203	
	3456		10" diameter		21	.762		178	24		202	232	
	3457		12" diameter		20	.800		208	25.50		233.50	267	
	3458		14" diameter		18	.889		245	28		273	315	
	3459		16" diameter		16	1		275	31.50		306.50	355	
	3460		18" diameter	▼	14	1.143		325	36		361	410	
	3461		20" diameter	Q-10	17	1.412		365	46.50		411.50	470	
	3462		24" diameter		12	2		450	65.50		515.50	595	
	3463		28" diameter		11	2.182		565	71.50		636.50	730	
	3464		32" diameter	▼	10	2.400		770	78.50		848.50	970	

Important: See the Reference Section for critical supporting data - Reference Nos., Crews, & City Cost Indexes

15550	Breechings, Chimneys & Stacks	CREW	DAILY OUTPUT	LABOR-HOURS	UNIT	2002 BARE COSTS				TOTAL INCL O&P
						MAT.	LABOR	EQUIP.	TOTAL	
3465	36" diameter	Q-10	9	2.667	Ea.	860	87.50		947.50	1,075
3466	42" diameter		6	4		1,375	131		1,506	1,700
3467	48" diameter	↓	5	4.800		1,750	157		1,907	2,175
3480	For Tee Cap, add					35%	20%			
3500	For 316 stainless steel liner, add				↓	22%				
3520	Plate support, galvanized									
3524	6" diameter	Q-9	26	.615	Ea.	83	19.45		102.45	121
3525	8" diameter		22	.727		93.50	23		116.50	139
3526	10" diameter		20	.800		102	25.50		127.50	151
3527	12" diameter		18	.889		106	28		134	161
3528	14" diameter		17	.941		128	29.50		157.50	187
3529	16" diameter		16	1		136	31.50		167.50	198
3530	18" diameter	↓	15	1.067		143	33.50		176.50	209
3531	20" diameter	Q-10	16	1.500		151	49		200	242
3532	24" diameter		14	1.714		159	56		215	262
3533	28" diameter		13	1.846		196	60.50		256.50	310
3534	32" diameter		12	2		254	65.50		319.50	380
3535	36" diameter		10	2.400		315	78.50		393.50	470
3536	42" diameter		9	2.667		560	87.50		647.50	750
3537	48" diameter	↓	8	3	↓	675	98.50		773.50	895
3570	Bellows, lined, 316 stainless steel only									
3574	6" diameter	Q-9	30	.533	Ea.	284	16.85		300.85	340
3575	8" diameter		26	.615		310	19.45		329.45	370
3576	10" diameter		24	.667		340	21		361	410
3577	12" diameter		22	.727		375	23		398	450
3578	14" diameter		21	.762		425	24		449	500
3579	16" diameter		20	.800		470	25.50		495.50	555
3580	18" diameter	↓	19	.842		525	26.50		551.50	615
3581	20" diameter	Q-10	18	1.333		590	43.50		633.50	720
3582	24" diameter		16	1.500		740	49		789	890
3583	28" diameter		15	1.600		830	52.50		882.50	995
3584	32" diameter		14	1.714		955	56		1,011	1,125
3585	36" diameter		12	2		1,100	65.50		1,165.50	1,300
3586	42" diameter		11	2.182		1,375	71.50		1,446.50	1,625
3587	48" diameter	↓	10	2.400	↓	1,700	78.50		1,778.50	2,000
3600	Ventilated roof thimble, 304 stainless steel									
3620	6" diameter	Q-9	26	.615	Ea.	191	19.45		210.45	240
3624	8" diameter		22	.727		205	23		228	261
3625	10" diameter		20	.800		206	25.50		231.50	266
3626	12" diameter		18	.889		222	28		250	288
3627	14" diameter		17	.941		239	29.50		268.50	310
3628	16" diameter		16	1		254	31.50		285.50	330
3629	18" diameter	↓	15	1.067		274	33.50		307.50	350
3630	20" diameter	Q-10	16	1.500		293	49		342	400
3631	24" diameter		14	1.714		330	56		386	450
3632	28" diameter		13	1.846		380	60.50		440.50	515
3633	32" diameter		12	2		425	65.50		490.50	570
3634	36" diameter		10	2.400		465	78.50		543.50	630
3635	42" diameter		9	2.667		620	87.50		707.50	815
3636	48" diameter	↓	8	3	↓	780	98.50		878.50	1,000
3650	For 316 stainless steel, add					10%				
3670	Exit cone, 316 stainless steel only									
3674	6" diameter	Q-9	46	.348	Ea.	125	11		136	155
3675	8" diameter		42	.381		131	12.05		143.05	164
3676	10" diameter		40	.400		142	12.65		154.65	176
3677	12" diameter	↓	38	.421		149	13.30		162.30	185

			DAILY	LABOR-		2002 BARE COSTS				TOTAL
	15550	**Breechings, Chimneys & Stacks**				MAT.	LABOR	EQUIP.	TOTAL	INCL O&P
			CREW	OUTPUT	HOURS	UNIT				
440	3678	14" diameter	Q-9	37	.432	Ea.	166	13.65		179.65	203	440
	3679	16" diameter		36	.444		182	14.05		196.05	223
	3680	18" diameter	↓	35	.457		224	14.45		238.45	270
	3681	20" diameter	Q-10	28	.857		267	28		295	340
	3682	24" diameter		26	.923		345	30.50		375.50	425
	3683	28" diameter		25	.960		415	31.50		446.50	510
	3684	32" diameter		24	1		520	33		553	620
	3685	36" diameter		22	1.091		595	36		631	710
	3686	42" diameter		21	1.143		760	37.50		797.50	895
	3687	48" diameter	↓	20	1.200	↓	980	39.50		1,019.50	1,125
	3720	Roof support assy., incl. 30" pipe sect, 304 st. st.									
	3724	6" diameter	Q-9	25	.640	Ea.	218	20		238	271
	3725	8" diameter		21	.762		238	24		262	299
	3726	10" diameter		19	.842		261	26.50		287.50	330
	3727	12" diameter		17	.941		278	29.50		307.50	350
	3728	14" diameter		16	1		310	31.50		341.50	390
	3729	16" diameter		15	1.067		330	33.50		363.50	415
	3730	18" diameter	↓	14	1.143		355	36		391	445
	3731	20" diameter	Q-10	15	1.600		380	52.50		432.50	500
	3732	24" diameter		13	1.846		425	60.50		485.50	565
	3733	28" diameter		12	2		500	65.50		565.50	650
	3734	32" diameter		11	2.182		600	71.50		671.50	770
	3735	36" diameter		9	2.667		690	87.50		777.50	895
	3736	42" diameter		8	3		1,075	98.50		1,173.50	1,325
	3737	48" diameter	↓	7	3.429	↓	1,325	112		1,437	1,625
	3750	For 316 stainless steel, add					20%				
	3770	Stack cap, 316 stainless steel only									
	3773	5" diameter	Q-9	48	.333	Ea.	119	10.55		129.55	147
	3774	6" diameter		46	.348		140	11		151	171
	3775	8" diameter		42	.381		162	12.05		174.05	197
	3776	10" diameter		40	.400		190	12.65		202.65	229
	3777	12" diameter		38	.421		221	13.30		234.30	264
	3778	14" diameter		37	.432		260	13.65		273.65	305
	3779	16" diameter		36	.444		297	14.05		311.05	345
	3780	18" diameter	↓	35	.457		335	14.45		349.45	395
	3781	20" diameter	Q-10	28	.857		385	28		413	465
	3782	24" diameter		26	.923		460	30.50		490.50	555
	3783	28" diameter		25	.960		505	31.50		536.50	605
	3784	32" diameter		24	1		580	33		613	685
	3785	36" diameter		22	1.091		675	36		711	800
	3786	42" diameter		21	1.143		805	37.50		842.50	945
	3787	48" diameter	↓	20	1.200	↓	975	39.50		1,014.50	1,125
	3800	Round storm collar									
	3806	5" diameter	Q-9	32	.500	Ea.	25	15.80		40.80	52
	7800	All fuel, double wall, stainless steel, 6" diameter		60	.267	V.L.F.	28.50	8.45		36.95	44.50
	7802	7" diameter		56	.286		37	9.05		46.05	54.50
	7804	8" diameter		52	.308		43	9.70		52.70	62.50
	7806	10" diameter		48	.333		63	10.55		73.55	86
	7808	12" diameter		44	.364		84.50	11.50		96	111
	7810	14" diameter	↓	42	.381	↓	111	12.05		123.05	141
	8000	All fuel, double wall, stainless steel fittings									
	8010	Roof support 6" diameter	Q-9	30	.533	Ea.	73.50	16.85		90.35	107
	8020	7" diameter		28	.571		83.50	18.05		101.55	120
	8030	8" diameter		26	.615		90	19.45		109.45	129
	8040	10" diameter		24	.667		119	21		140	164
	8050	12" diameter	↓	22	.727	↓	143	23		166	193

Important: See the Reference Section for critical supporting data - Reference Nos., Crews, & City Cost Indexes

15550	Breechings, Chimneys & Stacks	CREW	DAILY OUTPUT	LABOR-HOURS	UNIT	2002 BARE COSTS				TOTAL INCL O&P
						MAT.	LABOR	EQUIP.	TOTAL	
8060	14" diameter	Q-9	21	.762	Ea.	181	24		205	236
8100	Elbow 15°, 6" diameter		30	.533		64	16.85		80.85	96.50
8120	7" diameter		28	.571		83.50	18.05		101.55	120
8140	8" diameter		26	.615		82.50	19.45		101.95	121
8160	10" diameter		24	.667		108	21		129	151
8180	12" diameter		22	.727		131	23		154	180
8200	14" diameter		21	.762		156	24		180	209
8300	Insulated tee with insulated tee cap, 6" diameter		30	.533		121	16.85		137.85	159
8340	7" diameter		28	.571		159	18.05		177.05	203
8360	8" diameter		26	.615		179	19.45		198.45	227
8380	10" diameter		24	.667		252	21		273	310
8400	12" diameter		22	.727		355	23		378	425
8420	14" diameter		21	.762		465	24		489	545
8500	Joist shield, 6" diameter		30	.533		37	16.85		53.85	66.50
8510	7" diameter		28	.571		40	18.05		58.05	72
8520	8" diameter		26	.615		49.50	19.45		68.95	84.50
8530	10" diameter		24	.667		66.50	21		87.50	106
8540	12" diameter		22	.727		83	23		106	127
8550	14" diameter		21	.762		103	24		127	150
8600	Round top, 6" diameter		30	.533		41	16.85		57.85	71
8620	7" diameter		28	.571		56	18.05		74.05	89.50
8640	8" diameter		26	.615		75	19.45		94.45	113
8660	10" diameter		24	.667		135	21		156	182
8680	12" diameter		22	.727		193	23		216	248
8700	14" diameter		21	.762		258	24		282	320
8800	Adjustable roof flashing, 6" diameter		30	.533		49	16.85		65.85	79.50
8820	7" diameter		28	.571		56	18.05		74.05	89.50
8840	8" diameter		26	.615		60.50	19.45		79.95	96.50
8860	10" diameter		24	.667		78	21		99	118
8880	12" diameter		22	.727		100	23		123	147
8900	14" diameter		21	.762		125	24		149	175
9000	High temp. (2000° F), steel jacket, acid resist. refractory lining									
9010	11 ga. galvanized jacket, U.L. listed									
9020	Straight section, 48" long, 10" diameter	Q-10	13.30	1.805	Ea.	335	59		394	455
9030	12" diameter		11.20	2.143		360	70		430	510
9040	18" diameter		7.40	3.243		510	106		616	725
9050	24" diameter		4.60	5.217		695	171		866	1,025
9060	30" diameter		3.70	6.486		885	213		1,098	1,300
9070	36" diameter		2.70	8.889		1,075	291		1,366	1,650
9080	42" diameter	Q-11	3.10	10.323		1,350	345		1,695	2,000
9090	48" diameter		2.70	11.852		1,575	395		1,970	2,350
9100	54" diameter		2.20	14.545		1,925	485		2,410	2,875
9110	60" diameter		2	16		2,525	535		3,060	3,600
9120	Tee section, 10" diameter	Q-10	4.40	5.455		705	179		884	1,050
9130	12" diameter		3.70	6.486		745	213		958	1,150
9140	18" diameter		2.40	10		1,000	330		1,330	1,600
9150	24" diameter		1.50	16		1,400	525		1,925	2,325
9160	30" diameter		1.20	20		1,950	655		2,605	3,150
9170	36" diameter		.80	30		3,300	985		4,285	5,150
9180	42" diameter	Q-11	1	32		4,325	1,075		5,400	6,425
9190	48" diameter		.90	35.556		5,600	1,200		6,800	8,000
9200	54" diameter		.70	45.714		8,175	1,525		9,700	11,400
9210	60" diameter		.60	53.333		9,875	1,775		11,650	13,700
9220	Cleanout pier section, 10" diameter	Q-10	3.50	6.857		545	225		770	945
9230	12" diameter		2.50	9.600		570	315		885	1,100
9240	18" diameter		1.90	12.632		730	415		1,145	1,450

440

MECHANICAL 15

			DAILY	LABOR-		2002 BARE COSTS				TOTAL
15550	**Breechings, Chimneys & Stacks**	CREW	OUTPUT	HOURS	UNIT	MAT.	LABOR	EQUIP.	TOTAL	INCL O&P
440 9250	24" diameter	Q-10	1.30	18.462	Ea.	925	605		1,530	1,950	440
9260	30" diameter		1	24		1,125	785		1,910	2,450
9270	36" diameter	▼	.75	32		1,350	1,050		2,400	3,100
9280	42" diameter	Q-11	.90	35.556		1,700	1,200		2,900	3,700
9290	48" diameter		.75	42.667		2,050	1,425		3,475	4,450
9300	54" diameter		.60	53.333		2,450	1,775		4,225	5,450
9310	60" diameter	▼	.50	64		3,125	2,150		5,275	6,725
9320	For drain, add					53%				
9330	Elbow, 30° and 45°, 10" diameter	Q-10	6.60	3.636		460	119		579	690
9340	12" diameter		5.60	4.286		495	140		635	760
9350	18" diameter		3.70	6.486		745	213		958	1,150
9360	24" diameter		2.30	10.435		1,100	340		1,440	1,750
9370	30" diameter		1.85	12.973		1,525	425		1,950	2,325
9380	36" diameter	▼	1.30	18.462		1,850	605		2,455	2,950
9390	42" diameter	Q-11	1.60	20		2,525	670		3,195	3,800
9400	48" diameter		1.35	23.704		3,225	795		4,020	4,775
9410	54" diameter		1.10	29.091		4,175	975		5,150	6,100
9420	60" diameter	▼	1	32		5,075	1,075		6,150	7,250
9430	For 60° and 90° elbow, add					112%				
9440	End cap, 10" diameter	Q-10	26	.923		790	30.50		820.50	910
9450	12" diameter		22	1.091		900	36		936	1,050
9460	18" diameter		15	1.600		1,125	52.50		1,177.50	1,300
9470	24" diameter		9	2.667		1,375	87.50		1,462.50	1,625
9480	30" diameter		7.30	3.288		1,700	108		1,808	2,025
9490	36" diameter	▼	5	4.800		1,950	157		2,107	2,400
9500	42" diameter	Q-11	6.30	5.079		2,425	170		2,595	2,925
9510	48" diameter		5.30	6.038		2,875	202		3,077	3,450
9520	54" diameter		4.40	7.273		3,275	243		3,518	3,975
9530	60" diameter	▼	3.90	8.205		3,275	275		3,550	4,025
9540	Increaser (1 diameter), 10" diameter	Q-10	6.60	3.636		390	119		509	615
9550	12" diameter		5.60	4.286		420	140		560	675
9560	18" diameter		3.70	6.486		610	213		823	1,000
9570	24" diameter		2.30	10.435		880	340		1,220	1,500
9580	30" diameter		1.85	12.973		1,075	425		1,500	1,850
9590	36" diameter	▼	1.30	18.462		1,200	605		1,805	2,250
9592	42" diameter	Q-11	1.60	20		1,325	670		1,995	2,475
9594	48" diameter		1.35	23.704		1,575	795		2,370	2,975
9596	54" diameter	▼	1.10	29.091	▼	1,925	975		2,900	3,600
9600	For expansion joints, add to straight section					7%				
9610	For 1/4" hot rolled steel jacket, add					157%				
9620	For 2950°F very high temperature, add					89%				
9630	26 ga. aluminized jacket, straight section, 48" long									
9640	10" diameter	Q-10	15.30	1.569	V.L.F.	197	51.50		248.50	296
9650	12" diameter		12.90	1.860		208	61		269	325
9660	18" diameter		8.50	2.824		300	92.50		392.50	475
9670	24" diameter		5.30	4.528		430	148		578	705
9680	30" diameter		4.30	5.581		550	183		733	885
9690	36" diameter	▼	3.10	7.742	▼	695	254		949	1,150
9700	Accessories (all models)									
9710	Guy band, 10" diameter	Q-10	32	.750	Ea.	61	24.50		85.50	105
9720	12" diameter		30	.800		64.50	26		90.50	112
9730	18" diameter		26	.923		83.50	30.50		114	139
9740	24" diameter		24	1		124	33		157	187
9750	30" diameter		20	1.200		143	39.50		182.50	218
9760	36" diameter	▼	18	1.333		158	43.50		201.50	241
9770	42" diameter	Q-11	22	1.455	▼	287	48.50		335.50	390

Important: See the Reference Section for critical supporting data - Reference Nos., Crews, & City Cost Indexes

15550 | Breechings, Chimneys & Stacks

		CREW	DAILY OUTPUT	LABOR-HOURS	UNIT	2002 BARE COSTS				TOTAL INCL O&P
						MAT.	LABOR	EQUIP.	TOTAL	
440	9780	48" diameter	Q-11	20	1.600	Ea.	315	53.50		368.50	430
	9790	54" diameter		16	2		350	67		417	485
	9800	60" diameter	↓	12	2.667		375	89.50		464.50	555
	9810	Draw band, galv. stl., 11 gauge, 10" diameter	Q-10	32	.750		53	24.50		77.50	96.50
	9820	12" diameter		30	.800		57	26		83	103
	9830	18" diameter		26	.923		71.50	30.50		102	125
	9840	24" diameter		24	1		90.50	33		123.50	150
	9850	30" diameter		20	1.200		105	39.50		144.50	177
	9860	36" diameter	↓	18	1.333		121	43.50		164.50	201
	9870	42" diameter	Q-11	22	1.455		162	48.50		210.50	253
	9880	48" diameter		20	1.600		207	53.50		260.50	310
	9890	54" diameter		16	2		241	67		308	370
	9900	60" diameter	↓	12	2.667		305	89.50		394.50	475
	9910	Draw band, aluminized stl., 26 gauge, 10" diameter	Q-10	32	.750		15.20	24.50		39.70	54.50
	9920	12" diameter		30	.800		15.20	26		41.20	57
	9930	18" diameter		26	.923		19	30.50		49.50	67.50
	9940	24" diameter		24	1		23	33		56	75.50
	9950	30" diameter		20	1.200		26.50	39.50		66	90
	9960	36" diameter	↓	18	1.333		30.50	43.50		74	101
	9962	42" diameter	Q-11	22	1.455		34	48.50		82.50	113
	9964	48" diameter		20	1.600		42	53.50		95.50	129
	9966	54" diameter		16	2		42	67		109	149
	9968	60" diameter	↓	12	2.667	↓	45.50	89.50		135	188
600	0010	**INDUCED DRAFT FANS**									
	1000	Breeching installation									
	1800	Hot gas, 600°F, variable pitch pulley and motor									
	1840	6" diam. inlet, 1/4 H.P., 1 phase, 400 CFM	Q-9	6	2.667	Ea.	1,025	84.50		1,109.50	1,250
	1850	7" diam. inlet, 1/4 H.P., 1 phase, 800 CFM		5	3.200		1,100	101		1,201	1,350
	1860	8" diam. inlet, 1/4 H.P., 1phase, 1120 CFM		4	4		1,600	126		1,726	1,950
	1870	9" diam. inlet, 3/4 H.P., 1 phase, 1440 CFM		3.60	4.444		1,875	140		2,015	2,300
	1880	10" diam. inlet, 3/4 H.P., 1 phase, 2000 CFM		3.30	4.848		2,125	153		2,278	2,550
	1900	12" diam. inlet, 3/4 H.P., 3 phase, 2960 CFM		3	5.333		2,175	169		2,344	2,650
	1910	14" diam. inlet, 1 H.P., 3 phase, 4160 CFM		2.60	6.154		2,575	194		2,769	3,125
	1920	16" diam. inlet, 2 H.P., 3 phase, 6720 CFM		2.30	6.957		2,800	220		3,020	3,425
	1940	18" diam. inlet, 3 H.P., 3 phase, 9120 CFM		2	8		3,375	253		3,628	4,125
	1950	20" diam. inlet, 3 H.P., 3 phase, 9760 CFM		1.50	10.667		3,675	335		4,010	4,575
	1960	22" diam. inlet, 5 H.P., 3 phase, 13,360 CFM		1	16		4,325	505		4,830	5,525
	1980	24" diam. inlet, 7-1/2 H.P., 3 phase, 17,760 CFM	↓	.80	20	↓	5,300	630		5,930	6,825
	2300	For multi-blade damper at fan inlet, add					20%				
	3600	Chimneytop installation									
	3700	6" size	1 Shee	8	1	Ea.	440	35		475	540
	3740	8" size		7	1.143		440	40		480	545
	3780	13" size	↓	6	1.333		605	47		652	735
	3880	For speed control switch, add					39.50			39.50	43.50
	3920	For thermal fan control, add				↓	39.50			39.50	43.50
	5500	Flue shutter damper for draft control,									
	5510	parallel blades									
	5550	8" size	Q-9	8	2	Ea.	360	63		423	495
	5560	9" size		7.50	2.133		390	67.50		457.50	530
	5570	10" size		7	2.286		400	72		472	550
	5580	12" size		6.50	2.462		440	78		518	600
	5590	14" size		6	2.667		500	84.50		584.50	680
	5600	16" size		5.50	2.909		540	92		632	735
	5610	18" size		5	3.200		595	101		696	810
	5620	20" size	↓	4.50	3.556	↓	655	112		767	900

MECHANICAL 15

	15550	Breechings, Chimneys & Stacks	CREW	DAILY OUTPUT	LABOR-HOURS	UNIT	2002 BARE COSTS				TOTAL INCL O&P	
							MAT.	LABOR	EQUIP.	TOTAL		
600	5630	22" size	Q-9	4	4	Ea.	725	126		851	990	600
	5640	24" size		3.50	4.571		790	144		934	1,100	
	5650	27" size		3	5.333		790	169		959	1,125	
	5660	30" size		2.50	6.400		840	202		1,042	1,225	
	5670	32" size		2	8		880	253		1,133	1,350	
	5680	36" size		1.50	10.667		950	335		1,285	1,575	

	15620	Packaged Water Chillers	CREW	DAILY OUTPUT	LABOR-HOURS	UNIT	2002 BARE COSTS				TOTAL INCL O&P	
							MAT.	LABOR	EQUIP.	TOTAL		
100	0010	**ABSORPTION WATER CHILLERS**										100
	0020	Steam or hot water, water cooled										
	0050	100 ton	Q-7	.13	240	Ea.	114,000	8,275		122,275	138,000	
	0100	148 ton		.12	258		117,000	8,900		125,900	142,500	
	0200	200 ton		.12	275		130,500	9,525		140,025	158,000	
	0240	250 ton		.11	290		143,500	10,000		153,500	173,000	
	0300	354 ton		.10	304		179,500	10,500		190,000	213,000	
	0400	420 ton		.10	323		196,000	11,200		207,200	232,500	
	0500	665 ton		.09	340		275,000	11,700		286,700	320,000	
	0600	750 ton		.09	363		321,500	12,500		334,000	372,500	
	0700	850 ton		.08	385		343,000	13,300		356,300	397,500	
	0800	955 ton		.08	410		367,000	14,200		381,200	425,000	
	0900	1125 ton		.08	421		423,500	14,500		438,000	488,000	
	1000	1250 ton		.07	444		455,500	15,300		470,800	524,000	
	1100	1465 ton		.07	463		540,000	16,000		556,000	618,000	
	1200	1660 ton		.07	477		635,500	16,500		652,000	724,000	
	2000	For two stage unit, add					80%	25%				
	3000	Gas fired, air cooled										
	3180	3 ton	Q-5	1.30	12.308	Ea.	3,175	400		3,575	4,100	
	3200	4 ton		.90	17.778		3,750	580		4,330	5,000	
	3220	5 ton		.60	26.667		4,225	870		5,095	5,925	
	3250	8 ton		.50	32		9,250	1,050		10,300	11,800	
	3270	10 ton		.40	40		10,600	1,300		11,900	13,600	
	4000	Water cooled, duplex										
	4130	100 ton	Q-7	.13	246	Ea.	110,500	8,500		119,000	134,500	
	4140	200 ton		.11	283		163,500	9,775		173,275	194,500	
	4150	300 ton		.11	299		187,500	10,300		197,800	221,500	
	4160	400 ton		.10	316		250,000	10,900		260,900	291,500	
	4170	500 ton		.10	329		312,500	11,400		323,900	360,500	
	4180	600 ton		.09	340		375,000	11,700		386,700	430,000	
	4190	700 ton		.09	359		437,000	12,400		449,400	499,500	
	4200	800 ton		.08	380		499,500	13,100		512,600	569,500	
	4210	900 ton		.08	400		562,000	13,800		575,800	639,500	
	4220	1000 ton		.08	421		624,500	14,500		639,000	709,000	
600	0010	**CENTRIFUGAL/RECIP. WATER CHILLERS,** With standard controls										600
	0020	Centrifugal liquid chiller, water cooled										
	0030	not including water tower										
	0100	Open drive, 2000 ton	Q-7	.07	477	Ea.	521,500	16,500		538,000	598,500	
	0110	Centrifugal liquid chiller, air cooled, insulated evaporator										
	0120	130 ton	Q-7	.14	228	Ea.	56,000	7,900		63,900	73,500	

Important: See the Reference Section for critical supporting data - Reference Nos., Crews, & City Cost Indexes

The Cost Data you know and trust is just a click away

Means CostWorks 2002

Fax back the form below toll free: 1-800-632-6732 or order by phone: 1-800-334-3509

Select the individual digital cost books or packages you need.

Catalog No.	Title	Price	Cost
65062	Assemblies Cost Data	$175.95	
65012	Building Construction Cost Data	$122.95	
65112	Concrete & Masonry Cost Data	$122.95	
65032	Electrical Cost Data	$122.95	
65162	Heavy Construction Cost Data	$122.95	
65092	Interior Cost Data	$122.95	
65182	Light Commercial Cost Data	$142.95	
65022	Mechanical Cost Data	$122.95	
65152	Open Shop BCCD	$122.95	
65212	Plumbing Cost Data	$122.95	
65282	Site Work & Landscape Cost Data	$122.95	
65052	Square Foot Cost Data	$175.95	
65902	Wage Rate Utility (Available for above titles only)	$122.95	
65602	CostWorks Estimator	$299.95	
65202	Facilities Construction Cost Data	$285.95	
65042	Repair & Remodeling Cost Data	$122.95	
65312	Building Construction Cost Data/Metric	$122.95	
65362	Heavy Construction Cost Data/Metric	$122.95	
65172	Residential Cost Data	$142.95	
65302	Facilities Maintenance & Repair Cost Data	$285.95	

Packages

Catalog No.	Title	Price	Cost
65512	Builder's Package	$329.99	
65522	Building Professional's Package	$384.99	
65532	Facility Manager's Package	$495.99	
65542	Design Professional's Package	$384.99	

Name: _____

Company: _____

Street: _____

City: _____ State: _____ Zip: _____

Phone: (___) _____ Fax: (___) _____

E-Mail: _____

☐ Bill me   ☐ Please charge my credit card

☐ Visa   ☐ MasterCard   ☐ American Express   ☐ Discover

Card # _____ Expiration Date: _____

Cardholder Signature: _____

Subtotal	
MA residents please add 5% Sales Tax	
Shipping & Handling	$5.00
Grand Total	

All items shipped on a single CD with software code to unlock specific selections purchased.

Buy additional keys any time you need to unlock additional cost data books!

RSMeans CMD Understanding your craft. Advancing your business.

CWCW-2002

Check out all you get . . . see back cover!
Order on our secure Web site: www.rsmeans.com/cwcw2002.html

At the job site or in the office . . .

Means CostWorks gives you the cost data you need at the click of a mouse

- Search and organize detailed cost data in an instant to simplify estimating
- Test various options quickly and easily to determine best-cost solutions
- Make changes on the fly and calculate the results automatically to save time and headaches
- Localize costs to your geographic area for greater accuracy and more competitive pricing
- Stay current with rising costs thanks to online updates available every three months

Built-in tools and accessories that work for you!

Means Cost Books are a great reference when you need a quick price on an item. But for complex estimates involving numerous cost factors, you need a resource that can go to work in an instant — and produce the results that make for successful projects. That's what you get with the **Means CostWorks 2002 CD.**

This comprehensive digital library provides 18 detailed cost titles that cover every job costing requirement you can imagine — all at your fingertips using your desktop or laptop PC. Each "book" addresses a specific area of the building and maintenance industry, available separately or in specially priced packages. Plus, you have access to regular cost updates and unlimited access to additional valuable resources on the R.S. Means Web site included in the price of the CD.

New for 2002! Means CostWorks Estimator . . .
This powerful new utility helps you create customized reports and efficiently manage your estimates and data. Easily group information the way you want it. Develop detailed reports that include user defined columns, rows, and formulas. Add subtotals, create reusable estimates, and merge multiple CostLists. Quickly adjust or add your own costs. Create custom templates, XML documents, and more . . . Get the ultimate tool to harness the power of Means Data — CostWorks Estimator!

Job costing has never been easier . . . or more precise

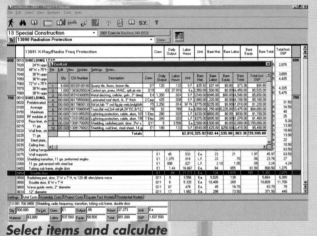

Select items and calculate project costs with the CostList

New!
30 Day Free Trial!

Design custom reports that meet your specifications

Just one CD is packed with extensive amounts of cost data covering every detail from the ground up

- 18 detailed cost data titles – available individually, in specially priced packages, or in total.
- CostWorks Estimator – Customize reports and manage your estimates and data more efficiently.

Hot Links to:
- Architect's First Source® – featuring thousands of products listed by manufacturer.
- The Construction Specifications Institute's SPEC-DATA® for technical product information.
- CSI's MANU-SPEC® for manufacturers' complete product specifications.

Plus the productivity-enhancing tools to really put this valuable information to work.

Bonus! Get the Means Construction Dictionary with over 17,000 construction terms – a $99.95 value – FREE on the Means CostWorks 2002 CD!

Save hundreds on specially priced packages! Choose the one that's right for you . . .

Builder's Package
Includes:
- Residential Cost Data
- Light Commercial Cost Data
- Site Work & Landscape Cost Data
- Concrete & Masonry Cost Data
- Open Shop Building Construction Cost Data

Just $329.99
Save over $300 off the price of buying each book separately!

Building Professional's Package
Includes:
- Building Construction Cost Data
- Mechanical Cost Data
- Plumbing Cost Data
- Electrical Cost Data
- Square Foot Model Costs

Just $384.99
Save over $250 off the price of buying each book separately!

Facility Manager's Package
Includes:
- Facilities Construction Cost Data
- Facilities Maintenance & Repair
- Repair & Remodeling Cost Data
- Building Construction Cost Data

Just $495.99
Save over $300 off the price of buying each book separately!

Design Professional's Package
Includes:
- Building Construction Cost Data
- Square Foot Cost Data
- Interior Cost Data
- Assemblies Cost Data

Just $384.99
Save over $200 off the price of buying each book separately!

Your 2002 Annual Means CostWorks license includes four free quarterly updates from the Means CostWorks Web site

15620	Packaged Water Chillers	CREW	DAILY OUTPUT	LABOR-HOURS	UNIT	2002 BARE COSTS				TOTAL INCL O&P
			CREW	DAILY OUTPUT	LABOR-HOURS	UNIT	MAT.	LABOR	EQUIP.	TOTAL	TOTAL INCL O&P
---	---	---	---	---	---	---	---	---	---	---	---
0124		160 ton	Q-7	.13	246	Ea.	67,500	8,500		76,000	87,000
0128		180 ton		.13	250		75,500	8,625		84,125	96,500
0132		210 ton		.12	258		88,500	8,900		97,400	110,500
0136		270 ton		.12	266		113,500	9,200		122,700	139,000
0140		320 ton		.12	275		134,500	9,525		144,025	162,500
0200		Packaged unit, water cooled, not incl. tower									
0210		80 ton	Q-7	.14	223	Ea.	26,400	7,725		34,125	40,700
0220		100 ton		.14	230		33,000	7,950		40,950	48,300
0230		150 ton		.13	240		45,800	8,300		54,100	63,000
0240		200 ton		.13	251		70,500	8,700		79,200	90,500
0250		250 ton		.12	260		76,500	8,975		85,475	97,500
0260		300 ton		.12	266		91,500	9,200		100,700	115,000
0270		350 ton		.12	275		107,000	9,525		116,525	132,000
0280		400 ton		.11	283		125,500	9,775		135,275	153,000
0282		450 ton		.11	290		137,500	10,000		147,500	166,000
0290		500 ton		.11	296		152,500	10,200		162,700	183,500
0292		550 ton		.10	304		168,000	10,500		178,500	200,500
0300		600 ton		.10	310		183,000	10,700		193,700	217,500
0302		650 ton		.10	320		198,500	11,000		209,500	235,000
0304		700 ton		.10	326		213,500	11,300		224,800	252,000
0306		750 ton		.10	333		229,000	11,500		240,500	269,500
0310		800 ton		.09	340		244,500	11,700		256,200	286,000
0312		850 ton		.09	351		259,500	12,100		271,600	304,000
0316		900 ton		.09	359		286,000	12,400		298,400	333,000
0318		950 ton		.09	363		302,000	12,500		314,500	351,000
0320		1000 ton		.09	372		323,000	12,800		335,800	374,500
0324		1100 ton		.08	385		349,500	13,300		362,800	404,500
0330		1200 ton		.08	395		381,500	13,600		395,100	440,000
0340		1300 ton		.08	410		415,000	14,200		429,200	478,000
0350		1400 ton		.08	421		445,000	14,500		459,500	511,500
0360		1500 ton		.07	426		476,500	14,700		491,200	546,000
0370		1600 ton		.07	438		508,500	15,100		523,600	582,500
0380		1700 ton		.07	450		540,500	15,600		556,100	618,000
0390		1800 ton		.07	457		572,000	15,800		587,800	653,500
0400		1900 ton		.07	470		604,000	16,200		620,200	689,000
0490		Reciprocating, packaged w/integral air cooled condenser, 15 ton cool		.37	86.486		15,000	2,975		17,975	21,000
0500		20 ton cooling		.34	94.955		19,100	3,275		22,375	26,000
0510		25 ton cooling		.33	97.859		21,700	3,375		25,075	28,900
0515		30 ton cooling		.31	101		23,400	3,525		26,925	31,000
0517		35 ton cooling		.31	104		26,000	3,625		29,625	34,100
0520		40 ton cooling		.30	108		32,100	3,725		35,825	41,000
0528		45 ton cooling		.29	109		36,300	3,800		40,100	45,600
0536		50 ton cooling		.28	113		40,900	3,900		44,800	51,000
0538		60 ton cooling		.28	115		47,000	3,975		50,975	57,500
0540		65 ton cooling		.27	118		41,100	4,075		45,175	51,500
0546		70 ton cooling		.27	119		54,000	4,125		58,125	65,500
0552		80 ton cooling		.26	123		58,000	4,250		62,250	70,000
0554		90 ton cooling		.25	125		62,500	4,350		66,850	75,500
0600		100 ton cooling		.25	129		70,000	4,450		74,450	83,500
0620		110 ton cooling		.24	132		76,000	4,575		80,575	91,000
0630		130 ton cooling		.24	135		84,000	4,650		88,650	99,500
0640		150 ton cooling		.23	137		97,500	4,750		102,250	114,000
0650		175 ton cooling		.23	140		103,500	4,875		108,375	121,500
0654		190 ton cooling		.22	144		117,500	4,975		122,475	136,500
0660		210 ton cooling		.22	148		128,500	5,125		133,625	148,500
0662		240 ton cooling		.21	151		154,000	5,225		159,225	177,500

15620	Packaged Water Chillers	CREW	DAILY OUTPUT	LABOR-HOURS	UNIT	2002 BARE COSTS				TOTAL INCL O&P
						MAT.	LABOR	EQUIP.	TOTAL	
0664	270 ton cooling	Q-7	.20	156	Ea.	173,500	5,375		178,875	199,000
0666	300 ton cooling		.20	160		205,000	5,525		210,525	234,000
0668	330 ton cooling		.19	164		207,500	5,675		213,175	237,000
0670	360 ton cooling		.19	168		235,000	5,800		240,800	267,500
0672	390 ton cooling		.18	172		246,500	5,975		252,475	280,000
0674	420 ton cooling		.18	177		258,000	6,125		264,125	293,000
0680	Water cooled, single compressor, semi-hermetic, tower not incl.									
0700	2 to 5 ton cooling	Q-5	.57	28.021	Ea.	6,025	915		6,940	8,000
0720	6 ton cooling		.42	38.005		9,150	1,250		10,400	12,000
0740	8 ton cooling		.31	52.117		9,825	1,700		11,525	13,400
0760	10 ton cooling	Q-6	.36	67.039		10,700	2,275		12,975	15,200
0780	15 ton cooling	"	.33	72.727		11,000	2,450		13,450	15,700
0800	20 ton cooling	Q-7	.38	83.989		12,800	2,900		15,700	18,400
0820	30 ton cooling		.33	96.096		15,400	3,325		18,725	21,900
0840	35 ton cooling		.33	97.859		16,100	3,375		19,475	22,800
0980	Water cooled, multiple compress., semi-hermetic, tower not incl.									
1000	15 ton cooling	Q-6	.36	65.934	Ea.	19,500	2,225		21,725	24,800
1020	20 ton cooling	Q-7	.41	78.049		21,300	2,700		24,000	27,500
1040	25 ton cooling		.36	89.888		23,900	3,100		27,000	31,000
1060	30 ton cooling		.31	101		24,400	3,525		27,925	32,100
1080	40 ton cooling		.30	108		30,600	3,725		34,325	39,300
1100	50 ton cooling		.28	113		34,600	3,925		38,525	43,900
1120	60 ton cooling		.25	125		38,000	4,350		42,350	48,500
1130	70 ton cooling		.23	139		51,500	4,800		56,300	64,000
1140	80 ton cooling		.21	151		53,500	5,225		58,725	67,000
1150	90 ton cooling		.19	164		57,500	5,700		63,200	72,000
1160	100 ton cooling		.18	179		61,500	6,200		67,700	77,000
1170	110 ton cooling		.17	190		65,000	6,575		71,575	81,500
1180	120 ton cooling		.16	196		70,000	6,775		76,775	87,000
1200	140 ton cooling		.16	202		83,000	7,000		90,000	101,500
1210	160 ton cooling		.15	210		87,500	7,275		94,775	107,000
1300	Water cooled, single compressor, direct drive, tower not incl.									
1320	40 ton cooling,	Q-7	.30	108	Ea.	24,000	3,725		27,725	32,000
1340	50 ton cooling		.28	113		28,000	3,925		31,925	36,700
1360	60 ton cooling		.25	125		30,400	4,350		34,750	40,100
1380	80 ton cooling		.21	151		40,200	5,225		45,425	52,000
1450	Water cooled, dual compressors, direct drive, tower not incl.									
1500	80 ton cooling	Q-7	.14	222	Ea.	46,400	7,675		54,075	62,500
1520	100 ton cooling		.14	228		54,000	7,900		61,900	71,500
1540	120 ton cooling		.14	231		59,500	8,000		67,500	77,500
1560	135 ton cooling		.14	235		68,000	8,125		76,125	86,500
1580	150 ton cooling		.13	240		72,500	8,300		80,800	92,000
1600	175 ton cooling		.13	244		83,000	8,425		91,425	103,500
1620	200 ton cooling		.13	250		88,500	8,625		97,125	110,000
1640	225 ton cooling		.13	256		96,000	8,825		104,825	119,500
1660	250 ton cooling		.12	260		102,000	8,975		110,975	125,500
4000	Packaged chiller, remote air cooled condensers not incl.									
4020	15 ton cooling	Q-7	.30	108	Ea.	21,300	3,725		25,025	29,000
4030	20 ton cooling		.28	115		23,200	4,000		27,200	31,500
4040	25 ton cooling		.25	125		26,200	4,350		30,550	35,400
4050	30 ton cooling		.24	133		27,600	4,625		32,225	37,300
4060	40 ton cooling		.22	144		32,000	4,975		36,975	42,700
4070	50 ton cooling		.21	153		33,200	5,300		38,500	44,600
4080	60 ton cooling		.19	164		36,200	5,675		41,875	48,400
4090	70 ton cooling		.18	173		51,500	6,000		57,500	65,500
4100	80 ton cooling		.17	183		53,500	6,350		59,850	68,500

15 MECHANICAL

Important: See the Reference Section for critical supporting data - Reference Nos., Crews, & City Cost Indexes

15620 | Packaged Water Chillers

		CREW	DAILY OUTPUT	LABOR-HOURS	UNIT	2002 BARE COSTS				TOTAL INCL O&P
						MAT.	LABOR	EQUIP.	TOTAL	
600	4110	90 ton cooling	Q-7	.16	193	Ea.	57,500	6,700		64,200	73,500	600
	4120	100 ton cooling		.16	203		61,500	7,025		68,525	78,000	
	4130	110 ton cooling		.15	213		65,000	7,350		72,350	82,500	
	4140	120 ton cooling		.14	223		70,000	7,725		77,725	88,500	
	4150	140 ton cooling		.14	233		83,000	8,050		91,050	103,000	
	8000	Direct expansion, shell and tube type, for built up systems										
	8020	1 ton	Q-5	2	8	Ea.	4,625	261		4,886	5,500	
	8030	5 ton		1.90	8.421		7,700	274		7,974	8,875	
	8040	10 ton		1.70	9.412		9,525	305		9,830	11,000	
	8050	20 ton		1.50	10.667		10,700	350		11,050	12,300	
	8060	30 ton		1	16		13,300	520		13,820	15,400	
	8070	50 ton		.90	17.778		20,700	580		21,280	23,700	
	8080	100 ton	Q-6	.90	26.667		38,300	900		39,200	43,500	

15640 | Packaged Cooling Towers

		CREW	DAILY OUTPUT	LABOR-HOURS	UNIT	MAT.	LABOR	EQUIP.	TOTAL	TOTAL INCL O&P
400	0010	**COOLING TOWERS** Packaged units D3030 115										400
	0070	Galvanized steel										
	0080	Draw thru, single flow										
	0100	Belt drive, 60 tons D3030 210	Q-6	90	.267	TonAC	80.50	9		89.50	102	
	0150	95 tons		100	.240		71	8.10		79.10	90.50	
	0200	110 tons		109	.220		69.50	7.45		76.95	87.50	
	0250	125 tons		120	.200		67.50	6.75		74.25	84.50	
	1000	For higher capacities, use multiples										
	1500	Induced air, double flow										
	1900	Gear drive, 150 ton	Q-6	126	.190	TonAC	94	6.45		100.45	113	
	2000	300 ton		129	.186		64	6.30		70.30	79.50	
	2100	600 ton		132	.182		60	6.15		66.15	75.50	
	2150	840 ton		142	.169		48	5.70		53.70	61.50	
	2200	Up to 1,000 tons		150	.160		45	5.40		50.40	57.50	
	3000	For higher capacities, use multiples										
	3500	For pumps and piping, add	Q-6	38	.632	TonAC	40.50	21.50		62	76.50	
	4000	For absorption systems, add				"	75%	75%				
	4100	Cooling water chemical feeder	Q-5	3	5.333	Ea.	276	174		450	565	
	4500	For rigging, see division 01559										
	5000	Fiberglass										
	5010	Draw thru										
	5100	60 tons	Q-6	1.50	16	Ea.	2,775	540		3,315	3,875	
	5120	125 tons		.99	24.242		6,250	820		7,070	8,100	
	5140	300 tons		.43	55.814		14,600	1,875		16,475	19,000	
	5160	600 tons		.22	109		26,600	3,675		30,275	34,800	
	5180	1000 tons		.15	160		45,600	5,400		51,000	58,000	
	6000	Stainless steel										
	6010	Draw thru										
	6100	60 tons	Q-6	1.50	16	Ea.	7,800	540		8,340	9,400	
	6120	110 tons		.99	24.242		12,500	820		13,320	14,900	
	6140	300 tons		.43	55.814		34,500	1,875		36,375	40,900	
	6160	600 tons		.22	109		53,500	3,675		57,175	64,500	
	6180	1000 tons		.15	160		87,000	5,400		92,400	103,500	

15660 | Liquid Coolers/Evap Condensers

		CREW	DAILY OUTPUT	LABOR-HOURS	UNIT	MAT.	LABOR	EQUIP.	TOTAL	TOTAL INCL O&P
100	0010	**CONDENSERS** Ratings are for 30°F TD, R-22 D3030 110										100
	0080	Air cooled, belt drive, propeller fan										
	0220	45 ton	Q-6	.70	34.286	Ea.	10,400	1,150		11,550	13,300	
	0240	50 ton		.69	34.985		10,700	1,175		11,875	13,600	

MECHANICAL 15

			CREW	DAILY OUTPUT	LABOR-HOURS	UNIT	2002 BARE COSTS				TOTAL INCL O&P	
15660	**Liquid Coolers/Evap Condensers**						MAT.	LABOR	EQUIP.	TOTAL		
100	0260	54 ton	D3030 110	Q-6	.63	37.795	Ea.	11,100	1,275		12,375	14,100	100
	0280	59 ton			.58	41.308		11,800	1,400		13,200	15,100	
	0300	65 ton			.53	45.541		12,300	1,550		13,850	15,800	
	0320	73 ton			.47	51.173		13,800	1,725		15,525	17,800	
	0340	81 ton			.42	56.738		16,600	1,925		18,525	21,100	
	0360	86 ton			.40	60.302		17,200	2,025		19,225	22,100	
	0380	88 ton			.39	61.697		18,300	2,075		20,375	23,400	
	0400	101 ton		Q-7	.45	70.640		21,200	2,450		23,650	27,100	
	0500	159 ton			.31	102		34,100	3,550		37,650	42,900	
	0600	228 ton			.22	148		49,200	5,125		54,325	61,500	
	0700	314 ton			.16	203		60,000	7,025		67,025	76,500	
	0800	471 ton			.11	283		70,000	9,775		79,775	91,500	
	0900	630 ton			.08	376		112,500	13,000		125,500	143,500	
	1000	840 ton			.06	507		151,000	17,500		168,500	192,500	
	1500	May be specified single or multi-circuit											
	1550	Air cooled, direct drive, propeller fan											
	1590	1 ton		Q-5	3.80	4.211	Ea.	505	137		642	760	
	1600	1-1/2 ton			3.60	4.444		600	145		745	885	
	1620	2 ton			3.20	5		665	163		828	975	
	1630	3 ton			2.40	6.667		740	217		957	1,150	
	1640	5 ton			2	8		1,350	261		1,611	1,900	
	1650	8 ton			1.80	8.889		1,700	290		1,990	2,300	
	1660	10 ton			1.40	11.429		2,100	370		2,470	2,850	
	1670	12 ton			1.30	12.308		2,350	400		2,750	3,175	
	1680	14 ton			1.20	13.333		2,675	435		3,110	3,600	
	1690	16 ton			1.10	14.545		2,950	475		3,425	3,950	
	1700	21 ton			1	16		3,400	520		3,920	4,500	
	1720	26 ton			.84	19.002		3,700	620		4,320	4,975	
	1740	30 ton			.70	22.792		6,325	745		7,070	8,075	
	1760	41 ton		Q-6	.77	31.008		6,975	1,050		8,025	9,250	
	1780	52 ton			.66	36.419		9,975	1,225		11,200	12,900	
	1800	63 ton			.54	44.037		11,000	1,500		12,500	14,400	
	1820	76 ton			.45	52.980		12,500	1,800		14,300	16,400	
	1840	86 ton			.40	60		15,400	2,025		17,425	20,100	
	1860	97 ton			.35	67.989		17,600	2,300		19,900	22,800	
	1880	105 ton		Q-7	.43	73.563		20,200	2,550		22,750	26,100	
	1890	118 ton			.39	82.687		21,800	2,850		24,650	28,300	
	1900	126 ton			.36	88.154		25,300	3,050		28,350	32,500	
	1910	136 ton			.34	95.238		28,300	3,275		31,575	36,200	
	1920	142 ton			.32	99.379		29,100	3,425		32,525	37,200	
	3400	Evaporative, copper coil, pump, fan motor											
	3440	10 ton		Q-5	.54	29.630	Ea.	4,225	965		5,190	6,100	
	3460	15 ton			.50	32		4,375	1,050		5,425	6,400	
	3480	20 ton			.47	34.043		4,575	1,100		5,675	6,700	
	3500	25 ton			.45	35.556		4,875	1,150		6,025	7,125	
	3520	30 ton			.42	38.095		5,200	1,250		6,450	7,600	
	3540	40 ton		Q-6	.49	48.980		6,875	1,650		8,525	10,100	
	3560	50 ton			.39	61.538		8,625	2,075		10,700	12,600	
	3580	65 ton			.35	68.571		9,575	2,325		11,900	14,000	
	3600	80 ton			.33	72.727		10,600	2,450		13,050	15,400	
	3620	90 ton			.29	82.759		11,400	2,800		14,200	16,700	
	3640	100 ton		Q-7	.36	88.889		12,400	3,075		15,475	18,300	
	3660	110 ton			.33	96.970		13,800	3,350		17,150	20,200	
	3680	125 ton			.30	106		14,900	3,675		18,575	22,000	
	3700	135 ton			.28	114		16,000	3,950		19,950	23,600	
	3720	150 ton			.25	128		17,600	4,425		22,025	26,100	

Important: See the Reference Section for critical supporting data - Reference Nos., Crews, & City Cost Indexes

15660 | Liquid Coolers/Evap Condensers

			CREW	DAILY OUTPUT	LABOR-HOURS	UNIT	MAT.	LABOR	EQUIP.	TOTAL	TOTAL INCL O&P	
100	3740	165 ton	Q-7	.23	139	Ea.	19,400	4,800		24,200	28,600	100
	3760	185 ton	▼	.22	145	▼	20,500	5,025		25,525	30,100	
	3860	For fan damper control, add	Q-5	2	8	▼	440	261		701	880	
	8000	Oils										
	8100	Lubricating										
	8120	Oil, lubricating				Oz.	.17			.17	.19	
	8500	Refrigeration										
	8520	Oil, refrigeration	▼			Gal.	10.25			10.25	11.30	

D3030 110

15670 | Refrig Condensing Units

			CREW	DAILY OUTPUT	LABOR-HOURS	UNIT	MAT.	LABOR	EQUIP.	TOTAL	TOTAL INCL O&P	
300	0010	**CONDENSING UNITS**										300
	0030	Air cooled, compressor, standard controls										
	0050	1.5 ton	Q-5	2.50	6.400	Ea.	790	209		999	1,175	
	0100	2 ton		2.10	7.619		835	248		1,083	1,300	
	0200	2.5 ton		1.70	9.412		1,050	305		1,355	1,600	
	0300	3 ton		1.30	12.308		1,200	400		1,600	1,925	
	0350	3.5 ton		1.10	14.545		1,300	475		1,775	2,150	
	0400	4 ton		.90	17.778		1,500	580		2,080	2,525	
	0500	5 ton		.60	26.667		1,750	870		2,620	3,225	
	0550	7.5 ton		.55	29.091		4,300	950		5,250	6,150	
	0560	8.5 ton		.53	30.189		4,525	985		5,510	6,475	
	0600	10 ton		.50	32		4,925	1,050		5,975	7,000	
	0620	11 ton		.48	33.333		6,825	1,075		7,900	9,150	
	0650	15 ton	▼	.40	40		7,775	1,300		9,075	10,500	
	0700	20 ton	Q-6	.40	60		13,900	2,025		15,925	18,400	
	0720	25 ton		.35	68.571		15,700	2,325		18,025	20,800	
	0750	30 ton		.30	80		17,900	2,700		20,600	23,800	
	0800	40 ton		.20	120		21,700	4,050		25,750	30,000	
	0840	50 ton		.18	133		25,900	4,500		30,400	35,300	
	0860	60 ton		.16	150		29,800	5,075		34,875	40,500	
	0900	75 ton		.14	171		33,400	5,800		39,200	45,400	
	1000	80 ton		.12	200		37,800	6,750		44,550	52,000	
	1100	100 ton	▼	.09	266	▼	46,300	9,000		55,300	64,500	
	2000	Water cooled, compressor, heat exchanger, controls										
	2100	5 ton	Q-5	.70	22.857	Ea.	5,150	745		5,895	6,775	
	2110	10 ton		.60	26.667		6,675	870		7,545	8,650	
	2200	15 ton	▼	.50	32		7,600	1,050		8,650	9,925	
	2300	20 ton	Q-6	.40	60		14,700	2,025		16,725	19,300	
	2310	30 ton		.30	80		16,600	2,700		19,300	22,300	
	2400	40 ton		.20	120		19,900	4,050		23,950	27,900	
	2410	60 ton		.18	133		29,900	4,500		34,400	39,700	
	2420	80 ton		.14	171		34,600	5,800		40,400	46,700	
	2500	100 ton		.11	218		42,300	7,375		49,675	57,500	
	2510	120 ton	▼	.10	240	▼	68,000	8,100		76,100	87,000	

15680 | Evaporators, DX Coils

			CREW	DAILY OUTPUT	LABOR-HOURS	UNIT	MAT.	LABOR	EQUIP.	TOTAL	TOTAL INCL O&P	
300	0010	**EVAPORATORS** DX coils, remote compressors not included										300
	1000	Coolers, reach-in type, above freezing temperatures										
	1300	Shallow depth, wall mount, 7 fins per inch, air defrost										
	1310	600 BTUH, 8" fan, rust-proof core	Q-5	3.80	4.211	Ea.	264	137		401	495	
	1320	900 BTUH, 8" fan, rust-proof core		3.30	4.848		325	158		483	600	
	1330	1200 BTUH, 8" fan, rust-proof core		3	5.333		385	174		559	685	
	1340	1800 BTUH, 8" fan, rust-proof core		2.40	6.667		445	217		662	820	
	1350	2500 BTUH, 10" fan, rust-proof core		2.20	7.273		485	237		722	885	
	1360	3500 BTUH, 10" fan		2	8		570	261		831	1,025	
	1370	4500 BTUH, 10" fan	▼	1.90	8.421	▼	600	274		874	1,075	

MECHANICAL 15

15680	Evaporators, DX Coils	CREW	DAILY OUTPUT	LABOR-HOURS	UNIT	2002 BARE COSTS				TOTAL INCL O&P	
						MAT.	LABOR	EQUIP.	TOTAL		
300 1390	For pan drain, add				Ea.	36			36	39.50	**300**
1600	Undercounter refrigerators, ceiling or wall mount,										
1610	8 fins per inch, air defrost, rust-proof core										
1630	800 BTUH, one 6" fan	Q-5	5	3.200	Ea.	165	104		269	340	
1640	1300 BTUH, two 6" fans		4	4		202	130		332	420	
1650	1700 BTUH, two 6" fans	↓	3.60	4.444	↓	226	145		371	465	
2000	Coolers, reach-in and walk-in types, above freezing temperatures										
2600	Two-way discharge, ceiling mount, 150-4100 CFM,										
2610	above 34°F applications, air defrost										
2630	900 BTUH, 7 fins per inch, 8" fan	Q-5	4	4	Ea.	188	130		318	405	
2660	2500 BTUH, 7 fins per inch, 10" fan		2.60	6.154		350	200		550	685	
2690	5500 BTUH, 7 fins per inch, 12" fan		1.70	9.412		705	305		1,010	1,225	
2720	8500 BTUH, 8 fins per inch, 16" fan		1.20	13.333		915	435		1,350	1,650	
2750	15,000 BTUH, 7 fins per inch, 18" fan		1.10	14.545		1,275	475		1,750	2,125	
2770	24,000 BTUH, 7 fins per inch, two 16" fans		1	16		1,900	520		2,420	2,875	
2790	30,000 BTUH, 7 fins per inch, two 18" fans	↓	.90	17.778	↓	2,375	580		2,955	3,475	
2850	Two-way discharge, low profile, ceiling mount,										
2860	8 fins per inch, 200-570 CFM, air defrost										
2880	800 BTUH, one 6" fan	Q-5	4	4	Ea.	214	130		344	430	
2890	1300 BTUH, two 6" fans		3.50	4.571		254	149		403	505	
2900	1800 BTUH, two 6" fans		3.30	4.848		268	158		426	535	
2910	2700 BTUH, three 6" fans	↓	2.70	5.926		315	193		508	635	
3000	Coolers, walk-in type, above freezing temperatures										
3300	General use, ceiling mount, 108-2080 CFM										
3320	600 BTUH, 7 fins per inch, 6" fan	Q-5	5.20	3.077	Ea.	168	100		268	335	
3340	1200 BTUH, 7 fins per inch, 8" fan		3.70	4.324		232	141		373	470	
3360	1800 BTUH, 7 fins per inch, 10" fan		2.70	5.926		325	193		518	650	
3380	3500 BTUH, 7 fins per inch, 12" fan		2.40	6.667		480	217		697	860	
3400	5500 BTUH, 8 fins per inch, 12" fan		2	8		535	261		796	985	
3430	8500 BTUH, 8 fins per inch, 16" fan		1.40	11.429		795	370		1,165	1,425	
3460	15,000 BTUH, 7 fins per inch, two 16" fans	↓	1.10	14.545	↓	1,275	475		1,750	2,125	
3640	Low velocity, high latent load, ceiling mount,										
3650	1050-2420 CFM, air defrost, 6 fins per inch										
3670	6700 BTUH, two 10" fans	Q-5	1.30	12.308	Ea.	1,050	400		1,450	1,750	
3680	10,000 BTUH, three 10" fans		1	16		1,225	520		1,745	2,125	
3690	13,500 BTUH, three 10" fans		1	16		1,475	520		1,995	2,400	
3700	18,000 BTUH, four 10" fans		1	16		1,950	520		2,470	2,925	
3710	26,500 BTUH, four 10" fans	↓	.80	20		2,400	650		3,050	3,625	
3730	For electric defrost, add				↓	17%					
5000	Freezers and coolers, reach-in type, above 34°F										
5030	to sub-freezing temperature range, low latent load,										
5050	air defrost, 7 fins per inch										
5070	1200 BTUH, 8" fan, rustproof core	Q-5	4.40	3.636	Ea.	209	118		327	410	
5080	1500 BTUH, 8" fan, rustproof core		4.20	3.810		215	124		339	425	
5090	1800 BTUH, 8" fan, rustproof core		3.70	4.324		222	141		363	455	
5100	2500 BTUH, 10" fan, rustproof core		2.90	5.517		253	180		433	550	
5110	3500 BTUH, 10" fan, rustproof core		2.50	6.400		350	209		559	700	
5120	4500 BTUH, 12" fan, rustproof core	↓	2.10	7.619	↓	455	248		703	875	
6000	Freezers and coolers, walk-in type										
6050	1960-18,000 CFM, medium profile										
6060	Standard motor, 6 fins per inch, to -30°F, all aluminum										
6080	10,500 BTUH, air defrost, one 18" fan	Q-5	1.40	11.429	Ea.	940	370		1,310	1,575	
6090	12,500 BTUH, air defrost, one 18" fan		1.40	11.429		1,025	370		1,395	1,675	
6100	16,400 BTUH, air defrost, one 20" fan		1.30	12.308		1,050	400		1,450	1,750	
6110	20,900 BTUH, air defrost, one 18" fans		1	16		1,225	520		1,745	2,125	
6120	27,000 BTUH, air defrost, two 18" fans	↓	1	16	↓	1,800	520		2,320	2,775	

Important: See the Reference Section for critical supporting data - Reference Nos., Crews, & City Cost Indexes

T5 MECHANICAL

		CREW	DAILY OUTPUT	LABOR-HOURS	UNIT	2002 BARE COSTS				TOTAL INCL O&P
15680	**Evaporators, DX Coils**					MAT.	LABOR	EQUIP.	TOTAL	
6130	32,900 BTUH, air defrost, two 20" fans	Q-6	1.30	18.462	Ea.	2,025	625		2,650	3,175
6140	39,000 BTUH, air defrost, two 18" fans		1.25	19.200		2,350	650		3,000	3,575
6150	44,100 BTUH, air defrost, three 20" fans		1	24		2,675	810		3,485	4,150
6155	53,000 BTUH, air defrost, three 20" fans		1	24		2,675	810		3,485	4,150
6160	66,200 BTUH, air defrost, four 20" fans		1	24		3,325	810		4,135	4,875
6170	78,000 BTUH, air defrost, four 20" fans		.80	30		3,900	1,025		4,925	5,800
6180	88,200 BTUH, air defrost, four 24" fans		.60	40		4,250	1,350		5,600	6,725
6190	110,000 BTUH, air defrost, four 24" fans	▼	.50	48		5,000	1,625		6,625	7,950
6330	Hot gas defrost, standard motor units, add					35%				
6370	Electric defrost, 230V, standard motor, add					16%				
6410	460V, standard or high capacity motor, add				▼	25%				
6800	Eight fins per inch, increases BTUH 75%									
6810	Standard capacity, add				Ea.	5%				
6830	Hot gas & electric defrost not recommended									
7000	800-4150 CFM, 12" fans									
7010	Four fins per inch									
7030	3400 BTUH, 1 fan	Q-5	2.80	5.714	Ea.	330	186		516	645
7040	4200 BTUH, 1 fan		2.40	6.667		335	217		552	700
7050	5300 BTUH, 1 fan		1.90	8.421		390	274		664	845
7060	6800 BTUH, 1 fan		1.70	9.412		440	305		745	940
7070	8400 BTUH, 2 fans		1.60	10		585	325		910	1,125
7080	10,500 BTUH, 2 fans		1.40	11.429		635	370		1,005	1,250
7090	13,000 BTUH, 3 fans		1.30	12.308		820	400		1,220	1,500
7100	17,000 BTUH, 4 fans		1.20	13.333		1,050	435		1,485	1,800
7110	21,500 BTUH, 5 fans	▼	1	16		1,275	520		1,795	2,200
7160	Six fins per inch increases BTUH 35%, add				▼	5%				
7180	Eight fins per inch increases BTUH 53%,									
7190	not recommended for below 35°F, add				Ea.	8%				
7210	For 12°F temperature differential, add					20%				
7230	For adjustable thermostat control, add				▼	67			67	73.50
7240	For hot gas defrost, except 8 fin models									
7250	on applications below 35°F, add				Ea.	58%				
7260	For electric defrost, except 8 fin models									
7270	on applications below 35°F, add				Ea.	36%				
8000	Freezers, pass-thru door uprights, walk-in storage									
8300	Low temperature, thin profile, electric defrost									
8310	138-1800 CFM, 5 fins per inch									
8320	900 BTUH, one 6" fan	Q-5	3.30	4.848	Ea.	248	158		406	510
8340	1500 BTUH, two 6" fans		2.20	7.273		370	237		607	765
8360	2600 BTUH, three 6" fans		1.90	8.421		535	274		809	1,000
8370	3300 BTUH, four 6" fans		1.70	9.412		600	305		905	1,125
8380	4400 BTUH, five 6" fans		1.40	11.429		840	370		1,210	1,475
8400	7000 BTUH, three 10" fans		1.30	12.308		1,000	400		1,400	1,700
8410	8700 BTUH, four 10" fans	▼	1.10	14.545		1,150	475		1,625	2,000
8450	For air defrost, deduct				▼	124			124	136

300

MECHANICAL 15

319

		15705 | **Curbs/Pads/Stands Prefab**	CREW	DAILY OUTPUT	LABOR-HOURS	UNIT	2002 BARE COSTS				TOTAL INCL O&P	
							MAT.	LABOR	EQUIP.	TOTAL		
600	0010	**CURBS/PADS PREFABRICATED**										**600**
	6000	Pad, fiberglass reinforced concrete with polystyrene foam core										
	6050	Condenser, 2" thick, 20" x 38"	1 Shee	8	1	Ea.	8.55	35		43.55	63.50	
	6070	24" x 24"		16	.500		7.60	17.55		25.15	35.50	
	6090	24" x 36"		12	.667		11.80	23.50		35.30	49	
	6110	24" x 42"		8	1		13.70	35		48.70	69	
	6150	26" x 36"	Q-9	8	2		12.90	63		75.90	112	
	6170	28" x 38"	1 Shee	8	1		14.15	35		49.15	69.50	
	6190	30" x 30"		12	.667		12.20	23.50		35.70	49.50	
	6220	30" x 36"		8	1		14.60	35		49.60	70	
	6240	30" x 40"	Q-9	8	2		15.90	63		78.90	115	
	6260	32" x 32"		7	2.286		14.20	72		86.20	127	
	6280	36" x 36"		8	2		17	63		80	116	
	6300	36" x 40"		7	2.286		19.30	72		91.30	132	
	6320	36" x 48"		7	2.286		22.50	72		94.50	136	
	6340	36" x 54"		6	2.667		27	84.50		111.50	160	
	6360	36" x 60"		7	2.286		29	72		101	143	
	6400	41" x 47"		6	2.667		27.50	84.50		112	161	
	6490	26" round		10	1.600		7.30	50.50		57.80	86	
	6550	30" round		9	1.778		13.70	56		69.70	102	
	6600	36" round		8	2		22.50	63		85.50	123	
	6700	For 3" thick pad, add					20%	5%				
	8800	Stand, corrosion-resistant plastic										
	8850	Heat pump, 6" high	1 Shee	32	.250	Ea.	6.80	8.80		15.60	21	
	8870	12" high	"	32	.250	"	9.20	8.80		18	23.50	

		15710 | **Heat Exchangers**	CREW	DAILY OUTPUT	LABOR-HOURS	UNIT	MAT.	LABOR	EQUIP.	TOTAL	TOTAL INCL O&P	
900	0010	**HEAT EXCHANGERS**										**900**
	0016	Shell & tube type, 2 or 4 pass, 3/4" O.D. copper tubes,										
	0020	C.I. heads, C.I. tube sheet, steel shell										
	0100	Hot water 40°F to 180°F, by steam at 10 PSI										
	0120	8 GPM	Q-5	6	2.667	Ea.	920	87		1,007	1,125	
	0140	10 GPM		5	3.200		1,375	104		1,479	1,675	
	0160	40 GPM		4	4		2,150	130		2,280	2,575	
	0180	64 GPM		2	8		3,300	261		3,561	4,025	
	0200	96 GPM		1	16		4,350	520		4,870	5,575	
	0220	120 GPM	Q-6	1.50	16		5,725	540		6,265	7,125	
	0240	168 GPM		1	24		7,025	810		7,835	8,950	
	0260	240 GPM		.80	30		11,000	1,025		12,025	13,600	
	0300	600 GPM		.70	34.286		23,500	1,150		24,650	27,700	
	0500	For bronze head and tube sheet, add					50%					
	1000	Hot water 40°F to 140°F, by water at 200°F										
	1020	7 GPM	Q-5	6	2.667	Ea.	1,125	87		1,212	1,375	
	1040	16 GPM		5	3.200		1,600	104		1,704	1,925	
	1060	34 GPM		4	4		2,425	130		2,555	2,875	
	1080	55 GPM		3	5.333		3,525	174		3,699	4,125	
	1100	74 GPM		1.50	10.667		4,350	350		4,700	5,300	
	1120	86 GPM		1.40	11.429		5,825	370		6,195	6,950	
	1140	112 GPM	Q-6	2	12		7,250	405		7,655	8,575	
	1160	126 GPM		1.80	13.333		8,975	450		9,425	10,600	
	1180	152 GPM		1	24		11,500	810		12,310	13,900	
	3000	Plate type,										
	3100	400 GPM	Q-6	.80	30	Ea.	16,000	1,025		17,025	19,100	
	3120	800 GPM	"	.50	48		27,500	1,625		29,125	32,800	
	3140	1200 GPM	Q-7	.34	94.118		33,500	3,250		36,750	41,800	

Important: See the Reference Section for critical supporting data - Reference Nos., Crews, & City Cost Indexes

			DAILY	LABOR-		2002 BARE COSTS				TOTAL	
	15710	**Heat Exchangers**	CREW	OUTPUT	HOURS	UNIT	MAT.	LABOR	EQUIP.	TOTAL	INCL O&P	
900	3160	1800 GPM	Q-7	.24	133	Ea.	50,000	4,600		54,600	62,000	**900**
	8000	Heat pipe type, glycol, 50% efficient										
	8010	100 MBH, 1700 CFM	1 Stpi	.80	10	Ea.	16,200	360		16,560	18,300	
	8020	160 MBH, 2700 CFM	↓	.60	13.333		20,400	485		20,885	23,100	
	8030	620 MBH, 4000 CFM	▼	.40	20	▼	27,700	725		28,425	31,600	

			DAILY	LABOR-		2002 BARE COSTS				TOTAL	
	15720	**Air Handling Units**	CREW	OUTPUT	HOURS	UNIT	MAT.	LABOR	EQUIP.	TOTAL	INCL O&P	
100	0010	**AIR HANDLING UNIT** Built-up										**100**
	0100	With cooling/heating coil section, filters, mixing box										
	0880	Single zone, horizontal / vertical										
	0890	Constant volume										
	0900	1600 CFM	Q-5	1.20	13.333	Ea.	750	435		1,185	1,475	
	0906	2000 CFM		1.10	14.545		940	475		1,415	1,750	
	0910	3000 CFM		1	16		1,400	520		1,920	2,325	
	0916	4000 CFM	▼	.95	16.842		1,875	550		2,425	2,900	
	0920	5000 CFM	Q-6	1.40	17.143		2,350	580		2,930	3,450	
	0926	6500 CFM		1.30	18.462		3,050	625		3,675	4,300	
	0930	7500 CFM		1.20	20		3,525	675		4,200	4,900	
	0936	9200 CFM		1.10	21.818		4,325	735		5,060	5,850	
	0940	11,500 CFM		1	24		5,400	810		6,210	7,175	
	0946	13,200 CFM		.90	26.667		6,200	900		7,100	8,175	
	0950	16,500 CFM		.80	30		7,750	1,025		8,775	10,100	
	0960	19,500 CFM		.70	34.286		9,150	1,150		10,300	11,900	
	0970	22,000 CFM		.60	40		10,300	1,350		11,650	13,500	
	0980	27,000 CFM		.50	48		12,700	1,625		14,325	16,400	
	0990	34,000 CFM		.40	60		16,000	2,025		18,025	20,700	
	1000	40,000 CFM		.30	80		18,500	2,700		21,200	24,500	
	1010	47,000 CFM	▼	.20	120		22,100	4,050		26,150	30,400	
	1140	60,000 CFM	Q-7	.18	177		49,300	6,125		55,425	64,000	
	1160	75,000 CFM	"	.16	200	▼	61,500	6,900		68,400	78,500	
	2000	Variable air volume										
	2100	75,000 CFM	Q-7	.12	266	Ea.	63,500	9,200		72,700	83,500	
	2140	100,000 CFM		.11	290		93,000	10,000		103,000	117,000	
	2160	150,000 CFM	▼	.10	320	▼	140,000	11,000		151,000	170,500	
	2300	Multi-zone, horizontal or vertical, blow-thru fan										
	2310	3000 CFM	Q-5	1	16	Ea.	1,425	520		1,945	2,350	
	2314	4000 CFM	"	.95	16.842		1,925	550		2,475	2,925	
	2318	5000 CFM	Q-6	1.40	17.143		2,400	580		2,980	3,500	
	2322	6500 CFM		1.30	18.462		3,100	625		3,725	4,375	
	2326	7500 CFM		1.20	20		3,600	675		4,275	4,975	
	2330	9200 CFM		1.10	21.818		4,400	735		5,135	5,950	
	2334	11,500 CFM		1	24		5,500	810		6,310	7,275	
	2336	13,200 CFM		.90	26.667		6,325	900		7,225	8,300	
	2340	16,500 CFM		.80	30		7,900	1,025		8,925	10,200	
	2344	19,500 CFM		.70	34.286		9,325	1,150		10,475	12,100	
	2348	22,000 CFM		.60	40		10,500	1,350		11,850	13,700	
	2352	27,000 CFM		.50	48		12,900	1,625		14,525	16,700	
	2356	34,000 CFM		.40	60		16,300	2,025		18,325	21,000	
	2360	40,000 CFM		.30	80		19,200	2,700		21,900	25,200	
	2364	47,000 CFM	▼	.20	120	▼	22,500	4,050		26,550	30,900	
	3000	Rooftop										
	3010	Constant volume										
	3100	2000 CFM	Q-5	1	16	Ea.	3,700	520		4,220	4,850	
	3140	5000 CFM	Q-6	.79	30.380		6,650	1,025		7,675	8,850	
	3150	10,000 CFM	↓	.52	46.154	▼	8,425	1,550		9,975	11,600	

MECHANICAL 15

15720	Air Handling Units	CREW	DAILY OUTPUT	LABOR-HOURS	UNIT	2002 BARE COSTS				TOTAL INCL O&P
						MAT.	LABOR	EQUIP.	TOTAL	
100 3160	15,000 CFM	Q-6	.39	61.538	Ea.	12,700	2,075		14,775	17,100 **100**
3170	20,000 CFM	↓	.30	80	↓	15,800	2,700		18,500	21,500
3200	Variable air volume									
3240	5000 CFM	Q-6	.70	34.286	Ea.	8,400	1,150		9,550	11,000
3250	10,000 CFM		.50	48		10,900	1,625		12,525	14,500
3260	15,000 CFM		.40	60		15,200	2,025		17,225	19,800
3270	20,000 CFM		.29	82.759		19,800	2,800		22,600	25,900
3280	30,000 CFM	↓	.20	120	↓	32,500	4,050		36,550	41,900
200 0010	**CENTRAL STATION AIR-HANDLING UNIT** Chilled water									**200**
1000	Modular, capacity at 700 FPM face velocity									
1090	Constant volume									
1100	1300 CFM	Q-5	1.20	13.333	Ea.	2,575	435		3,010	3,475
1200	1900 CFM		1.10	14.545		2,850	475		3,325	3,875
1300	3200 CFM	↓	.80	20		3,400	650		4,050	4,700
1400	5400 CFM	Q-6	.80	30		5,100	1,025		6,125	7,150
1500	8000 CFM		.60	40		5,950	1,350		7,300	8,600
1550	10,000 CFM		.54	44.444		6,475	1,500		7,975	9,400
1600	12,000 CFM		.46	52.174		7,375	1,750		9,125	10,800
1650	14,000 CFM		.42	57.143		8,725	1,925		10,650	12,500
1670	16,000 CFM		.38	63.158		9,775	2,125		11,900	14,000
1700	20,000 CFM		.31	77.419		12,200	2,625		14,825	17,400
1800	23,000 CFM		.28	85.714		14,000	2,900		16,900	19,800
1850	28,000 CFM		.24	100		17,700	3,375		21,075	24,600
1900	33,500 CFM		.19	126		21,200	4,275		25,475	29,700
1950	42,500 CFM	↓	.15	160		25,100	5,400		30,500	35,900
2000	52,500 CFM	Q-7	.16	200		34,600	6,900		41,500	48,500
2100	63,000 CFM	"	.13	246		41,100	8,500		49,600	58,000
2200	For hot water heating coil, add				↓	21%	5%			
2300	Variable air volume, also incl. heating coil									
2340	5000 CFM	Q-6	.70	34.286	Ea.	7,875	1,150		9,025	10,400
2350	10,000 CFM		.50	48		10,200	1,625		11,825	13,800
2360	15,000 CFM		.40	60		14,200	2,025		16,225	18,700
2370	20,000 CFM		.29	82.759		18,500	2,800		21,300	24,600
2380	30,000 CFM		.20	120		30,400	4,050		34,450	39,500
3000	Packaged, 3000 CFM, 7.5 ton		1.70	14.118		2,475	475		2,950	3,450
3050	3200 CFM, 8 ton		1.60	15		2,900	505		3,405	3,950
3100	4000 CFM, 10 ton		1.40	17.143		3,050	580		3,630	4,225
3150	4400 CFM, 11 ton		1.35	17.778		3,625	600		4,225	4,900
3200	6000 CFM, 15 ton		1	24		3,925	810		4,735	5,550
3250	7000 CFM, 17.5 ton		.90	26.667		4,750	900		5,650	6,575
3300	10,000 CFM, 25 ton		.65	36.923		7,250	1,250		8,500	9,850
3400	12,000 CFM, 30 ton	↓	.50	48	↓	7,775	1,625		9,400	11,000
3500	For hot water heat coil, add				↓	30%	5%			
300 0010	**EVAPORATIVE COOLERS** Ducted, not incl. duct.									**300**
0100	Side discharge style, capacities at .25" S.P.									
0120	1785 CFM, 1/3 HP, 115 V	Q-9	5	3.200	Ea.	340	101		441	530
0140	2740 CFM, 1/3 HP, 115 V		4.50	3.556		415	112		527	630
0160	3235 CFM, 1/2 HP, 115 V		4	4		425	126		551	665
0180	3615 CFM, 1/2 HP, 230 V		3.60	4.444		435	140		575	690
0200	4215 CFM, 3/4 HP, 230 V		3.20	5		550	158		708	850
0220	5255 CFM. 1 HP, 115/230 V		3	5.333		905	169		1,074	1,250
0240	6090 CFM, 1 HP, 230/460 V		2.80	5.714		1,275	181		1,456	1,675
0260	8300 CFM, 1-1/2 HP, 230/460 V		2.60	6.154		1,275	194		1,469	1,700
0280	8360 CFM, 1-1/2 HP, 230/460 V		2.20	7.273		1,700	230		1,930	2,225
0300	9725 CFM, 2 HP, 230/460 V	↓	1.80	8.889	↓	1,725	281		2,006	2,300

Important: See the Reference Section for critical supporting data - Reference Nos., Crews, & City Cost Indexes

15720	Air Handling Units	CREW	DAILY OUTPUT	LABOR-HOURS	UNIT	2002 BARE COSTS MAT.	LABOR	EQUIP.	TOTAL	TOTAL INCL O&P	
300	0320	11,715 CFM, 3 HP, 230/460 V	Q-9	1.40	11.429	Ea.	1,775	360		2,135	2,500	300
	0340	14,410 CFM, 5 HP, 230/460 V	↓	1	16	↓	1,875	505		2,380	2,850	
	0400	For two-speed motor, add					5%					
	0500	For down discharge style, add					10%					
	1000	High capacities at .4" S.P.										
	1040	15,000 CFM, 5 HP, 230/460 V	Q-9	.54	29.630	Ea.	5,575	935		6,510	7,575	
	1080	20,000 CFM, 10HP, 230/460 V	"	.48	33.333		7,150	1,050		8,200	9,475	
	1100	25,000 CFM, 20 HP, 230/460 V	Q-10	.54	44.444		7,875	1,450		9,325	10,900	
	1140	30,000 CFM, 20 HP, 230/460 V	↓	.46	52.174		9,075	1,700		10,775	12,600	
	1160	35,000 CFM, 20 HP, 230/460 V		.44	54.545		9,925	1,775		11,700	13,700	
	1180	40,000 CFM, 30 HP, 230/460 V	↓	.42	57.143	↓	10,400	1,875		12,275	14,300	
500	0010	**MAKE-UP AIR UNIT**										500
	0020	Indoor suspension, natural/LP gas, direct fired,										
	0030	standard control. For flue see division 15550-440										
	0040	70°F temperature rise, MBH is input										
	0100	2000 CFM, 168 MBH	Q-6	3	8	Ea.	5,725	270		5,995	6,700	
	0120	3000 CFM, 252 MBH		2	12		5,725	405		6,130	6,900	
	0140	4000 CFM, 336 MBH		1.80	13.333		5,975	450		6,425	7,250	
	0160	6000 CFM, 502 MBH		1.50	16		7,200	540		7,740	8,750	
	0180	8000 CFM, 670 MBH		1.40	17.143		7,900	580		8,480	9,550	
	0200	10,000 CFM, 838 MBH		1.20	20		8,900	675		9,575	10,800	
	0220	12,000 CFM, 1005 MBH		1	24		8,975	810		9,785	11,100	
	0240	14,000 CFM, 1180 MBH		.94	25.532		9,100	860		9,960	11,300	
	0260	18,000 CFM, 1340 MBH		.86	27.907		9,175	945		10,120	11,500	
	0280	20,000 CFM, 1675 MBH	↓	.76	31.579		11,300	1,075		12,375	14,000	
	0300	24,000 CFM, 2007 MBH	Q-7	1	32		11,700	1,100		12,800	14,600	
	0320	30,000 CFM, 2510 MBH		.96	33.333		12,000	1,150		13,150	14,900	
	0340	35,000 CFM, 2930 MBH		.92	34.783		15,100	1,200		16,300	18,400	
	0360	40,000 CFM, 3350 MBH		.88	36.364		15,500	1,250		16,750	18,900	
	0380	45,000 CFM, 3770 MBH		.84	38.095		15,900	1,325		17,225	19,500	
	0400	50,000 CFM, 4180 MBH		.80	40		16,100	1,375		17,475	19,800	
	0420	55,000 CFM, 4600 MBH		.75	42.667		20,500	1,475		21,975	24,700	
	0440	60,000 CFM, 5020 MBH		.70	45.714		21,300	1,575		22,875	25,800	
	0460	65,000 CFM, 5435 MBH		.60	53.333		21,400	1,850		23,250	26,400	
	0480	75,000 CFM, 6275 MBH	↓	.50	64		22,600	2,200		24,800	28,100	
	0600	For discharge louver assembly, add					10%					
	0700	For filters, add					20%					
	0800	For air shut-off damper section, add					10%					
	0900	For vertical unit, add				↓	10%					
	1000	Rooftop unit, natural gas, gravity vent, S.S. exchanger										
	1010	70°F temperature rise, MBH is input										
	1020	995 CFM, 75 MBH	Q-6	4	6	Ea.	3,850	203		4,053	4,525	
	1040	1245 CFM, 95 MBH		3.60	6.667		3,875	225		4,100	4,625	
	1060	1590 CFM, 120 MBH		3.30	7.273		4,175	246		4,421	4,975	
	1080	2090 CFM, 159 MBH		3	8		4,300	270		4,570	5,150	
	1100	2490 CFM, 190 MBH		2.60	9.231		4,925	310		5,235	5,900	
	1120	2985 CFM, 225 MBH		2.30	10.435		5,000	350		5,350	6,025	
	1140	3975 CFM, 300 MBH		1.90	12.632		5,925	425		6,350	7,175	
	1160	4970 CFM, 375 MBH		1.40	17.143		6,875	580		7,455	8,425	
	1180	5950 CFM, 450 MBH		1.20	20		7,250	675		7,925	9,000	
	1200	7940 CFM, 600 MBH		1	24		8,425	810		9,235	10,500	
	1220	9930 CFM, 750 MBH		.80	30		14,400	1,025		15,425	17,300	
	1240	11,900 CFM, 900 MBH		.60	40		16,100	1,350		17,450	19,800	
	1260	15,880 CFM, 1200 MBH	↓	.30	80	↓	19,300	2,700		22,000	25,300	
	1500	For power vent, add					8%					

15720 | Air Handling Units

Line	Description	CREW	DAILY OUTPUT	LABOR-HOURS	UNIT	MAT.	LABOR	EQUIP.	TOTAL	TOTAL INCL O&P
500 1600	For cleanable filters, add				Ea.	112			112	123 **500**
1700	For electric modulating gas control, add				"	1,200			1,200	1,325
1800	For aluminized steel exchanger, deduct					10%				
8000	Rooftop unit, electric, with curb, controls									
8010	33 kW	Q-5	4	4	Ea.	3,175	130		3,305	3,675
8020	100 kW		3.60	4.444		5,250	145		5,395	6,000
8030	150 kW		3.20	5		6,100	163		6,263	6,950
8040	250 kW		2.80	5.714		7,925	186		8,111	8,975
8050	400 kW		2.20	7.273		11,600	237		11,837	13,100

15730 | Unitary Air Conditioning Equip

Line	Description	CREW	DAILY OUTPUT	LABOR-HOURS	UNIT	MAT.	LABOR	EQUIP.	TOTAL	TOTAL INCL O&P
200 0010	COMPUTER ROOM UNITS (D3030 222)									**200**
1000	Air cooled, includes remote condenser but not									
1020	interconnecting tubing or refrigerant									
1080	3 ton	Q-5	.50	32	Ea.	7,525	1,050		8,575	9,850
1120	5 ton		.45	35.556		10,100	1,150		11,250	12,900
1160	6 ton		.30	53.333		18,500	1,750		20,250	23,000
1200	8 ton		.27	59.259		21,000	1,925		22,925	26,000
1240	10 ton		.25	64		21,800	2,075		23,875	27,200
1280	15 ton		.22	72.727		24,400	2,375		26,775	30,400
1290	18 ton		.20	80		27,800	2,600		30,400	34,400
1320	20 ton	Q-6	.29	82.759		29,100	2,800		31,900	36,200
1360	23 ton	"	.28	85.714		30,500	2,900		33,400	38,000
2200	Chilled water, for connection to									
2220	existing chiller system of adequate capacity									
2260	5 ton	Q-5	.74	21.622	Ea.	8,475	705		9,180	10,400
2280	6 ton		.52	30.769		11,300	1,000		12,300	13,900
2300	8 ton		.50	32		12,100	1,050		13,150	14,900
2320	10 ton		.49	32.653		13,500	1,075		14,575	16,400
2330	12 ton		.48	32.990		13,900	1,075		14,975	16,900
2360	15 ton		.48	33.333		14,400	1,075		15,475	17,600
2400	20 ton		.46	34.783		14,800	1,125		15,925	18,000
2440	23 ton		.42	38.095		15,500	1,250		16,750	19,000
4000	Glycol system, complete except for interconnecting tubing									
4060	3 ton	Q-5	.40	40	Ea.	9,225	1,300		10,525	12,100
4100	5 ton		.38	42.105		12,000	1,375		13,375	15,300
4120	6 ton		.25	64		24,000	2,075		26,075	29,600
4140	8 ton		.23	69.565		25,900	2,275		28,175	31,900
4160	10 ton		.21	76.190		28,100	2,475		30,575	34,700
4200	15 ton	Q-6	.26	92.308		34,100	3,125		37,225	42,200
4240	20 ton		.24	100		36,500	3,375		39,875	45,200
4280	23 ton		.22	109		39,800	3,675		43,475	49,300
8000	Water cooled system, not including condenser,									
8020	water supply or cooling tower									
8060	3 ton	Q-5	.62	25.806	Ea.	6,800	840		7,640	8,750
8100	5 ton		.54	29.630		9,575	965		10,540	12,000
8120	6 ton		.35	45.714		17,900	1,500		19,400	22,000
8140	8 ton		.33	48.485		19,800	1,575		21,375	24,100
8160	10 ton		.31	51.613		21,000	1,675		22,675	25,600
8200	15 ton		.27	59.259		24,100	1,925		26,025	29,400
8240	20 ton	Q-6	.38	63.158		28,500	2,125		30,625	34,500
8280	23 ton	"	.34	70.588		30,200	2,375		32,575	36,800
500 0010	PACKAGED TERMINAL AIR CONDITIONER Cabinet, wall sleeve,									**500**
0100	louver, electric heat, thermostat, manual changeover, 208 V									

324

15730	Unitary Air Conditioning Equip		CREW	DAILY OUTPUT	LABOR-HOURS	UNIT	2002 BARE COSTS				TOTAL INCL O&P	
							MAT.	LABOR	EQUIP.	TOTAL		
500	0200	6,000 BTUH cooling, 8800 BTU heat		Q-5	6	2.667	Ea.	1,025	87		1,112	1,275	500
	0220	9,000 BTUH cooling, 13,900 BTU heat			5	3.200		1,050	104		1,154	1,300	
	0240	12,000 BTUH cooling, 13,900 BTU heat			4	4		1,075	130		1,205	1,400	
	0260	15,000 BTUH cooling, 13,900 BTU heat			3	5.333		1,175	174		1,349	1,525	
	0280	18,000 BTUH cooling, 10 KW heat			2.40	6.667		1,350	217		1,567	1,825	
	0300	24,000 BTUH cooling, 10 KW heat			1.90	8.421		1,375	274		1,649	1,925	
	0320	30,000 BTUH cooling, 10 KW heat			1.40	11.429		1,550	370		1,920	2,250	
	0340	36,000 BTUH cooling, 10 KW heat			1.25	12.800		1,625	415		2,040	2,425	
	0360	42,000 BTUH cooling, 10 KW heat			1	16		2,050	520		2,570	3,050	
	0380	48,000 BTUH cooling, 10 KW heat			.90	17.778		2,225	580		2,805	3,325	
	0500	For hot water coil, increase heat by 10%, add						5%	10%				
	1000	For steam, increase heat output by 30%, add						8%	10%				
600	0010	**ROOF TOP AIR CONDITIONERS** Standard controls, curb, economizer											600
	1000	Single zone, electric cool, gas heat											
	1090	2 ton cooling, 55 MBH heating	D3030 202	Q-5	.93	17.204	Ea.	4,050	560		4,610	5,300	
	1100	3 ton cooling, 60 MBH heating			.70	22.857		4,075	745		4,820	5,600	
	1120	4 ton cooling, 95 MBH heating	D3030 206		.61	26.403		5,125	860		5,985	6,950	
	1140	5 ton cooling, 112 MBH heating			.56	28.520		5,800	930		6,730	7,800	
	1145	6 ton cooling, 140 MBH heating	R15700 -010		.52	30.769		7,075	1,000		8,075	9,275	
	1150	7.5 ton cooling, 170 MBH heating			.50	32.258		9,225	1,050		10,275	11,800	
	1160	10 ton cooling, 200 MBH heating	R15700 -020	Q-6	.67	35.982		12,500	1,225		13,725	15,500	
	1170	12.5 ton cooling, 230 MBH heating			.63	37.975		13,900	1,275		15,175	17,200	
	1180	15 ton cooling, 270 MBH heating			.57	42.032		13,600	1,425		15,025	17,200	
	1190	18 ton cooling, 330 MBH heating			.52	45.889		16,200	1,550		17,750	20,200	
	1200	20 ton cooling, 360 MBH heating		Q-7	.67	47.976		21,100	1,650		22,750	25,700	
	1210	25 ton cooling, 450 MBH heating			.56	57.554		25,000	1,975		26,975	30,500	
	1220	30 ton cooling, 540 MBH heating			.47	68.376		32,600	2,350		34,950	39,500	
	1240	40 ton cooling, 675 MBH heating			.35	91.168		42,000	3,150		45,150	51,000	
	1300	Electric cool, electric heat											
	1310	5 ton		Q-5	.63	25.600	Ea.	1,575	835		2,410	2,975	
	1314	10 ton		Q-6	.74	32.432		3,150	1,100		4,250	5,125	
	1318	15 ton		"	.63	37.795		4,725	1,275		6,000	7,125	
	1322	20 ton		Q-7	.74	43.185		6,300	1,500		7,800	9,175	
	1326	25 ton			.62	51.696		7,875	1,775		9,650	11,400	
	1330	30 ton			.52	61.657		9,450	2,125		11,575	13,600	
	1334	40 ton			.39	82.051		12,600	2,825		15,425	18,200	
	1338	50 ton			.31	102		15,800	3,550		19,350	22,700	
	1342	60 ton			.26	123		18,900	4,250		23,150	27,200	
	1346	75 ton			.21	153		23,600	5,300		28,900	34,000	
	1700	Gas cool, gas heat											
	1720	3 ton cooling, 90 MBH heating		Q-5	.93	17.204	Ea.	4,525	560		5,085	5,825	
	1730	4 ton cooling, 112.5 MBH heating			.61	26.403		5,300	860		6,160	7,125	
	1750	5 ton cooling, 135 MBH heating			.56	28.571		5,825	930		6,755	7,800	
	2000	Multizone, electric cool, gas heat, economizer											
	2100	15 ton cooling, 360 MBH heating		Q-7	.61	52.545	Ea.	49,500	1,825		51,325	57,000	
	2120	20 ton cooling, 360 MBH heating			.53	60.038		57,000	2,075		59,075	65,500	
	2140	25 ton cooling, 450 MBH heating			.44	71.910		64,500	2,475		66,975	75,000	
	2160	28 ton cooling, 450 MBH heating			.40	79.012		67,500	2,725		70,225	78,500	
	2180	30 ton cooling, 540 MBH heating			.37	85.562		73,500	2,950		76,450	85,500	
	2200	40 ton cooling, 540 MBH heating			.28	113		90,500	3,925		94,425	106,000	
	2210	50 ton cooling, 540 MBH heating			.22	142		87,500	4,900		92,400	104,000	
	2220	70 ton cooling, 1500 MBH heating			.16	198		108,000	6,850		114,850	129,500	
	2240	80 ton cooling, 1500 MBH heating			.14	228		123,500	7,900		131,400	148,000	
	2260	90 ton cooling, 1500 MBH heating			.13	256		129,500	8,825		138,325	156,000	
	2280	105 ton cooling, 1500 MBH heating			.11	290		151,000	10,000		161,000	181,500	
	2400	For hot water heat coil, deduct						5%					

MECHANICAL 15

15730	Unitary Air Conditioning Equip		CREW	DAILY OUTPUT	LABOR-HOURS	UNIT	2002 BARE COSTS				TOTAL INCL O&P
							MAT.	LABOR	EQUIP.	TOTAL	
600	2500	For steam heat coil, deduct	D3030 202				Ea.	2%					600
	2600	For electric heat, deduct					↓	3%	5%			
	4000	Single zone, electric cool, gas heat, variable air volume	D3030 206									
	4100	12.5 ton cooling, 230 MBH heating		Q-6	.55	43.716	Ea.	16,400	1,475		17,875	20,200
	4120	18 ton cooling, 330 MBH heating	R15700 -010	"	.45	52.747		18,900	1,775		20,675	23,500
	4140	25 ton cooling, 450 MBH heating		Q-7	.48	66.116		29,000	2,275		31,275	35,400
	4160	40 ton cooling, 675 MBH heating	R15700 -020		.31	104		49,300	3,625		52,925	60,000
	4180	60 ton cooling, 900 MBH heating			.20	157		69,500	5,450		74,950	84,500
	4200	80 ton cooling, 1000 MBH heating		↓	.15	209	↓	109,500	7,225		116,725	131,000
	5000	Single zone, electric cool only										
	5050	3 ton cooling		Q-5	.88	18.203	Ea.	3,200	595		3,795	4,425
	5060	4 ton cooling			.76	21.108		3,775	690		4,465	5,175
	5070	5 ton cooling			.70	22.792		4,450	745		5,195	6,025
	5080	6 ton cooling			.65	24.502		5,450	800		6,250	7,200
	5090	7.5 ton cooling			.62	25.806		7,450	840		8,290	9,475
	5100	8.5 ton cooling		↓	.59	26.981		8,450	880		9,330	10,600
	5110	10 ton cooling		Q-6	.83	28.812		9,775	975		10,750	12,300
	5120	12 ton cooling			.80	30		12,700	1,025		13,725	15,400
	5130	15 ton cooling			.71	33.613		11,900	1,125		13,025	14,700
	5140	18 ton cooling		↓	.65	36.697		14,500	1,250		15,750	17,900
	5150	20 ton cooling		Q-7	.83	38.415		19,800	1,325		21,125	23,800
	5160	25 ton cooling			.70	45.977		24,200	1,575		25,775	29,000
	5170	30 ton cooling			.58	54.701		31,400	1,900		33,300	37,500
	5180	40 ton cooling		↓	.44	73.059		40,700	2,525		43,225	48,500
	5400	For low heat, add						7%				
	5410	For high heat, add					↓	10%				
	6000	Single zone electric cooling with variable volume distribution										
	6020	20 ton cooling		Q-7	.72	44.199	Ea.	23,200	1,525		24,725	27,800
	6030	25 ton cooling			.60	52.893		27,500	1,825		29,325	33,100
	6040	30 ton cooling			.51	62.868		30,800	2,175		32,975	37,200
	6050	40 ton cooling			.38	83.989		39,200	2,900		42,100	47,600
	6060	50 ton cooling			.31	104		47,300	3,625		50,925	57,500
	6070	60 ton cooling		↓	.25	125		53,000	4,350		57,350	64,500
	6200	For low heat, add						7%				
	6210	For high heat, add					↓	10%				
	7000	Multizone, cool/heat, variable volume distribution										
	7100	50 ton cooling		Q-7	.19	164	Ea.	62,000	5,675		67,675	76,500
	7110	70 ton cooling			.14	228		87,000	7,900		94,900	107,500
	7120	90 ton cooling			.11	296		94,500	10,200		104,700	119,500
	7130	105 ton cooling			.10	333		110,500	11,500		122,000	139,000
	7140	120 ton cooling			.08	380		126,000	13,100		139,100	158,500
	7150	140 ton cooling		↓	.07	444	↓	145,500	15,300		160,800	183,000
	7300	Penthouse unit, cool/heat, variable volume distribution										
	7400	150 ton cooling		Q-7	.10	310	Ea.	225,000	10,700		235,700	263,500
	7410	170 ton cooling			.10	320		245,000	11,000		256,000	286,000
	7420	200 ton cooling			.10	326		286,000	11,300		297,300	331,500
	7430	225 ton cooling			.09	340		322,000	11,700		333,700	371,500
	7440	250 ton cooling			.09	359		322,500	12,400		334,900	373,500
	7450	270 ton cooling			.09	363		348,500	12,500		361,000	402,000
	7460	300 ton cooling		↓	.09	372	↓	387,000	12,800		399,800	445,000
	9400	Sub assemblies for assembly systems										
	9410	Duct work per ton roof top 1 zone units		Q-9	.96	16.667	Ton	226	525		751	1,050
	9420	Ductwork per ton rooftop multizone units			.50	32		298	1,000		1,298	1,875
	9430	Ductwork, VAV cooling/ton rooftop multizone, 1 zone/ton			.32	50		390	1,575		1,965	2,850
	9440	Duct work per ton packaged water & air cooled units			1.26	12.698		49.50	400		449.50	675
	9450	Duct work per ton split system remote condensing units		↓	1.32	12.121	↓	46.50	385		431.50	640

T5 MECHANICAL

Important: See the Reference Section for critical supporting data - Reference Nos., Crews, & City Cost Indexes

15730	Unitary Air Conditioning Equip	CREW	DAILY OUTPUT	LABOR-HOURS	UNIT	2002 BARE COSTS				TOTAL INCL O&P
						MAT.	LABOR	EQUIP.	TOTAL	
800	**0010**	**WINDOW UNIT AIR CONDITIONERS**										**800**
4000	Portable/window, 15 amp 125V grounded receptacle required									
4060	5000 BTUH	1 Carp	8	1	Ea.	315	30		345	390
4340	6000 BTUH		8	1		370	30		400	450
4480	8000 BTUH		6	1.333		420	40		460	530
4500	10,000 BTUH		6	1.333		500	40		540	615
4520	12,000 BTUH	L-2	8	2		545	53		598	680
4600	Window/thru-the-wall, 15 amp 230V grounded receptacle required									
4780	17,000 BTUH	L-2	6	2.667	Ea.	785	70.50		855.50	970
4940	25,000 BTUH		4	4		1,125	106		1,231	1,400
4960	29,000 BTUH		4	4		1,325	106		1,431	1,625
840	**0010**	**SELF-CONTAINED SINGLE PACKAGE** D3030 210										**840**
0100	Air cooled, for free blow or duct, not incl. remote condenser									
0110	Constant volume									
0200	3 ton cooling D3030 214	Q-5	1	16	Ea.	4,825	520		5,345	6,075
0210	4 ton cooling	"	.80	20		5,250	650		5,900	6,750
0220	5 ton cooling	Q-6	1.20	20		6,075	675		6,750	7,700
0230	7.5 ton cooling	"	.90	26.667		8,250	900		9,150	10,400
0240	10 ton cooling	Q-7	1	32		10,300	1,100		11,400	13,100
0250	15 ton cooling		.95	33.684		14,600	1,150		15,750	17,900
0260	20 ton cooling		.90	35.556		16,800	1,225		18,025	20,400
0270	25 ton cooling		.85	37.647		20,300	1,300		21,600	24,300
0280	30 ton cooling		.80	40		21,200	1,375		22,575	25,500
0300	40 ton cooling		.60	53.333		29,400	1,850		31,250	35,100
0320	50 ton cooling		.50	64		36,100	2,200		38,300	43,000
0340	60 ton cooling	Q-8	.40	80		42,200	2,750	200	45,150	51,000
0380	With hot water heat, variable air volume									
0390	10 ton cooling	Q-7	.96	33.333	Ea.	13,800	1,150		14,950	16,900
0400	20 ton cooling		.86	37.209		22,400	1,275		23,675	26,600
0410	30 ton cooling		.76	42.105		28,400	1,450		29,850	33,400
0420	40 ton cooling		.58	55.172		39,700	1,900		41,600	46,500
0430	50 ton cooling		.48	66.667		47,800	2,300		50,100	56,000
0440	60 ton cooling	Q-8	.38	84.211		58,500	2,900	211	61,611	69,000
0490	For duct mounting no price change									
0500	For hot water or steam heating coils, add				Ea.	10%	10%			
0550	Packaged, with electric heat									
0552	Constant volume									
0560	5 ton cooling	Q-6	1.20	20	Ea.	8,475	675		9,150	10,400
0570	10 ton cooling	Q-7	1	32		14,000	1,100		15,100	17,100
0580	20 ton cooling		.90	35.556		22,100	1,225		23,325	26,200
0590	30 ton cooling		.80	40		30,100	1,375		31,475	35,200
0600	40 ton cooling		.60	53.333		38,900	1,850		40,750	45,600
0610	50 ton cooling		.50	64		49,400	2,200		51,600	58,000
0700	Variable air volume									
0710	10 ton cooling	Q-7	1	32	Ea.	16,500	1,100		17,600	19,900
0720	20 ton cooling		.90	35.556		26,000	1,225		27,225	30,500
0730	30 ton cooling		.80	40		35,100	1,375		36,475	40,700
0740	40 ton cooling		.60	53.333		46,300	1,850		48,150	54,000
0750	50 ton cooling		.50	64		57,500	2,200		59,700	66,500
0760	60 ton cooling	Q-8	.40	80		66,500	2,750	200	69,450	77,500
1000	Water cooled for free blow or duct, not including tower									
1010	Constant volume									
1100	3 ton cooling	Q-6	1	24	Ea.	3,375	810		4,185	4,950
1120	5 ton cooling		1	24		4,300	810		5,110	5,975
1130	7.5 ton cooling		.80	30		6,125	1,025		7,150	8,275

		15730 | **Unitary Air Conditioning Equip**	CREW	DAILY OUTPUT	LABOR-HOURS	UNIT	2002 BARE COSTS				TOTAL INCL O&P	
							MAT.	LABOR	EQUIP.	TOTAL		
840	1140	10 ton cooling D3030 210	Q-7	.90	35.556	Ea.	8,950	1,225		10,175	11,700	840
	1150	15 ton cooling		.85	37.647		12,200	1,300		13,500	15,400	
	1160	20 ton cooling D3030 214		.80	40		14,600	1,375		15,975	18,200	
	1170	25 ton cooling		.75	42.667		17,200	1,475		18,675	21,200	
	1180	30 ton cooling		.70	45.714		19,400	1,575		20,975	23,700	
	1200	40 ton cooling		.40	80		24,500	2,750		27,250	31,200	
	1220	50 ton cooling	↓	.30	106		30,500	3,675		34,175	39,200	
	1240	60 ton cooling	Q-8	.30	106		35,600	3,675	267	39,542	45,000	
	1300	For hot water or steam heat coils, add				↓	12%	10%				
	2000	Water cooled with electric heat, not including tower										
	2010	Constant volume										
	2100	5 ton cooling	Q-6	1.20	20	Ea.	6,725	675		7,400	8,425	
	2120	10 ton cooling	Q-7	1	32		12,700	1,100		13,800	15,600	
	2130	15 ton cooling	"	.90	35.556		17,500	1,225		18,725	21,100	
	2400	Variable air volume										
	2440	10 ton cooling	Q-7	1	32		15,100	1,100		16,200	18,300	
	2450	15 ton cooling	"	.90	35.556	"	21,400	1,225		22,625	25,500	
	2600	Water cooled, hot water coils, not including tower										
	2610	Variable air volume										
	2640	10 ton cooling	Q-7	1	32	Ea.	12,300	1,100		13,400	15,300	
	2650	15 ton cooling	"	.80	40	"	17,400	1,375		18,775	21,200	
900	0010	**SPLIT DUCTLESS SYSTEM**										900
	0100	Cooling only, single zone										
	0110	Wall mount										
	0120	3/4 ton cooling	Q-5	2	8	Ea.	815	261		1,076	1,300	
	0130	1 ton cooling		1.80	8.889		965	290		1,255	1,475	
	0140	1-1/2 ton cooling		1.60	10		1,250	325		1,575	1,875	
	0150	2 ton cooling	↓	1.40	11.429	↓	1,450	370		1,820	2,150	
	1000	Ceiling mount										
	1020	2 ton cooling	Q-5	1.40	11.429	Ea.	1,800	370		2,170	2,550	
	1030	3 ton cooling	"	1.20	13.333	"	2,850	435		3,285	3,800	
	2000	T-Bar mount										
	2010	2 ton cooling	Q-5	1.40	11.429	Ea.	2,075	370		2,445	2,825	
	2020	3 ton cooling		1.20	13.333		2,550	435		2,985	3,450	
	2030	3-1/2 ton cooling	↓	1.10	14.545	↓	3,075	475		3,550	4,100	
	3000	Multizone										
	3010	Wall mount										
	3020	2 @ 3/4 ton cooling	Q-5	1.80	8.889	Ea.	2,550	290		2,840	3,225	
	5000	Cooling / Heating										
	5010	Wall mount										
	5110	1 ton cooling	Q-5	1.70	9.412	Ea.	1,250	305		1,555	1,825	
	5120	1-1/2 ton cooling	"	1.50	10.667	"	1,700	350		2,050	2,375	
	5300	Ceiling mount										
	5310	3 ton cooling	Q-5	1	16	Ea.	3,300	520		3,820	4,425	
	7000	Accessories for all split ductless systems										
	7010	Add for ambient frost control	Q-5	8	2	Ea.	140	65		205	253	
	7020	Add for tube / wiring kit										
	7030	15' kit	Q-5	32	.500	Ea.	52	16.30		68.30	82	
	7040	35' kit	"	24	.667	"	104	21.50		125.50	147	

		15740 | **Heat Pumps**										
100	0010	**HEAT PUMPS** (Not including interconnecting tubing)										100
	1000	Air to air, split system, not including curbs, pads, or ductwork										
	1010	For curbs/pads see division 15705										
	1015	1.5 ton cooling, 7 MBH heat @ 0°F	Q-5	1.22	13.115	Ea.	1,225	425		1,650	2,000	

Important: See the Reference Section for critical supporting data - Reference Nos., Crews, & City Cost Indexes

15 MECHANICAL

15740	Heat Pumps	CREW	DAILY OUTPUT	LABOR-HOURS	UNIT	2002 BARE COSTS				TOTAL INCL O&P
						MAT.	LABOR	EQUIP.	TOTAL	
1020	2 ton cooling, 8.5 MBH heat @ 0°F	Q-5	1.20	13.333	Ea.	1,275	435		1,710	2,050
1030	2.5 ton cooling, 10 MBH heat @ 0°F		1	16		1,475	520		1,995	2,400
1040	3 ton cooling, 13 MBH heat @ 0°F		.80	20		1,625	650		2,275	2,775
1050	3.5 ton cooling, 18 MBH heat @ 0°F		.75	21.333		1,825	695		2,520	3,075
1054	4 ton cooling, 24 MBH heat @ 0°F		.60	26.667		2,050	870		2,920	3,550
1060	5 ton cooling, 27 MBH heat @ 0°F		.50	32		2,350	1,050		3,400	4,175
1080	7.5 ton cooling, 33 MBH heat @ 0°F		.30	53.333		8,100	1,750		9,850	11,600
1100	10 ton cooling, 50 MBH heat @ 0°F	Q-6	.38	63.158		11,100	2,125		13,225	15,500
1120	15 ton cooling, 64 MBH heat @ 0°F		.26	92.308		14,500	3,125		17,625	20,700
1130	20 ton cooling, 85 MBH heat @ 0°F		.20	120		19,700	4,050		23,750	27,800
1140	25 ton cooling, 119 MBH heat @ 0°F		.20	120		23,400	4,050		27,450	31,800
1300	Supplementary electric heat coil, included									
1500	Single package, not including curbs, pads, or plenums									
1502	1/2 ton cooling, supplementary heat not incl.	Q-5	8	2	Ea.	1,325	65		1,390	1,550
1504	3/4 ton cooling, supplementary heat not incl.		6	2.667		1,375	87		1,462	1,625
1506	1 ton cooling, supplementary heat not incl.		4	4		1,475	130		1,605	1,825
1510	1.5 ton cooling, 5 MBH heat @ 0°F		1.55	10.323		3,100	335		3,435	3,900
1520	2 ton cooling, 6.5 MBH heat @ 0°F		1.50	10.667		3,175	350		3,525	4,000
1540	2.5 ton cooling, 8 MBH heat @ 0°F		1.40	11.429		3,400	370		3,770	4,275
1560	3 ton cooling, 10 MBH heat @ 0°F		1.20	13.333		3,800	435		4,235	4,825
1570	3.5 ton cooling, 11 MBH heat @ 0°F		1	16		4,125	520		4,645	5,325
1580	4 ton cooling, 13 MBH heat @ 0°F		.96	16.667		4,475	545		5,020	5,750
1620	5 ton cooling, 27 MBH heat @ 0°F		.65	24.615		5,125	800		5,925	6,825
1640	7.5 ton cooling, 35 MBH heat @ 0°F		.40	40		8,225	1,300		9,525	11,000
1648	10 ton cooling, 45 MBH heat @ 0°F	Q-6	.40	60		10,900	2,025		12,925	15,100
1652	12 ton cooling, 50 MBH heat @ 0°F	"	.36	66.667		13,300	2,250		15,550	18,000
1696	Supplementary electric heat coil incl., except as noted									
1800	Rooftop, with curb, economizer and supplementary electric heat									
1820	10 ton cooling	Q-6	.40	60	Ea.	36,100	2,025		38,125	42,900
1830	20 ton cooling	"	.20	120		51,500	4,050		55,550	63,000
1840	30 ton cooling	Q-7	.19	168		69,500	5,800		75,300	85,500
1850	40 ton cooling		.17	188		77,500	6,500		84,000	95,000
1860	50 ton cooling		.15	213		90,500	7,350		97,850	110,500
1870	60 ton cooling		.13	246		95,500	8,500		104,000	118,000
2000	Water source to air, single package									
2100	1 ton cooling, 13 MBH heat @ 75°F	Q-5	2	8	Ea.	975	261		1,236	1,475
2120	1.5 ton cooling, 17 MBH heat @ 75°F		1.80	8.889		1,175	290		1,465	1,700
2140	2 ton cooling, 19 MBH heat @ 75°F		1.70	9.412		1,225	305		1,530	1,800
2160	2.5 ton cooling, 25 MBH heat @ 75°F		1.60	10		1,300	325		1,625	1,925
2180	3 ton cooling, 27 MBH heat @ 75°F		1.40	11.429		1,350	370		1,720	2,050
2190	3.5 ton cooling, 29 MBH heat @ 75°F		1.30	12.308		1,450	400		1,850	2,200
2200	4 ton cooling, 31 MBH heat @ 75°F		1.20	13.333		1,775	435		2,210	2,625
2220	5 ton cooling, 29 MBH heat @ 75°F		.90	17.778		2,150	580		2,730	3,250
2240	7.5 ton cooling, 35 MBH heat @ 75°F		.60	26.667		2,450	870		3,320	3,975
2250	8.5 ton cooling, 40 MBH heat @ 75°F		.58	27.586		6,975	900		7,875	9,025
2260	10 ton cooling, 50 MBH heat @ 75°F		.53	30.189		7,675	985		8,660	9,900
2280	15 ton cooling, 64 MBH heat @ 75°F	Q-6	.47	51.064		8,000	1,725		9,725	11,400
2300	20 ton cooling, 100 MBH heat @ 75°F		.41	58.537		8,250	1,975		10,225	12,100
2320	30 ton cooling, (twin 15 ton units)		.23	102		16,000	3,450		19,450	22,800
2340	40 ton cooling, (twin 20 ton units)		.20	117		16,500	3,950		20,450	24,100
2360	50 ton cooling, (twin 15 + 20 ton unit)		.15	160		24,300	5,400		29,700	34,900
3960	For supplementary heat coil, add					10%				
4000	For increase in capacity thru use									
4020	of solar collector, size boiler at 60%									

15750	Humidity Control Equipment	CREW	DAILY OUTPUT	LABOR-HOURS	UNIT	2002 BARE COSTS				TOTAL INCL O&P
						MAT.	LABOR	EQUIP.	TOTAL	
300	0010 **DEHUMIDIFIERS**										300
	6000 Selfcontained with filters and standard controls									
	6040 1.5 lb/hr, 50 cfm	1 Plum	8	1	Ea.	2,950	36		2,986	3,300
	6060 3 lb/hr, 150 cfm	Q-1	12	1.333		3,350	43		3,393	3,775
	6065 6 lb/hr, 150 cfm		9	1.778		6,200	57.50		6,257.50	6,900
	6070 16 to 20 lb/hr, 600 cfm		5	3.200		12,400	104		12,504	13,800
	6080 30 to 40 lb/hr, 1125 cfm		4	4		20,400	129		20,529	22,700
	6090 60 to 75 lb/hr, 2250 cfm		3	5.333		26,800	173		26,973	29,800
	6100 120 to 155 lb/hr, 4500 cfm		2	8		49,000	259		49,259	54,500
	6110 240 to 310 lb/hr, 9000 cfm		1.50	10.667		69,000	345		69,345	76,500
	6120 400 to 515 lb/hr, 15,000 cfm	Q-2	1.60	15		94,000	505		94,505	104,000
	6130 530 to 690 lb/hr, 20,000 cfm		1.40	17.143		101,500	575		102,075	113,000
	6140 800 to 1030 lb/hr, 30,000 cfm		1.20	20		116,000	670		116,670	128,500
	6150 1060 to 1375 lb/hr, 40,000 cfm		1	24		161,500	805		162,305	179,000
500	0010 **HUMIDIFIERS**										500
	0520 Steam, room or duct, filter, regulators, auto. controls, 220 V									
	0540 11 lb. per hour	Q-5	6	2.667	Ea.	2,050	87		2,137	2,375
	0560 22 lb. per hour		5	3.200		2,250	104		2,354	2,625
	0580 33 lb. per hour		4	4		2,325	130		2,455	2,750
	0600 50 lb. per hour		4	4		2,850	130		2,980	3,325
	0620 100 lb. per hour		3	5.333		3,400	174		3,574	4,000
	0640 150 lb. per hour		2.50	6.400		4,500	209		4,709	5,275
	0660 200 lb. per hour		2	8		5,575	261		5,836	6,550
	0700 With blower									
	0720 11 lb. per hour	Q-5	5.50	2.909	Ea.	2,850	95		2,945	3,300
	0740 22 lb. per hour		4.75	3.368		3,075	110		3,185	3,550
	0760 33 lb. per hour		3.75	4.267		3,125	139		3,264	3,650
	0780 50 lb. per hour		3.50	4.571		3,775	149		3,924	4,375
	0800 100 lb. per hour		2.75	5.818		4,250	190		4,440	4,950
	0820 150 lb. per hour		2	8		6,300	261		6,561	7,325
	0840 200 lb. per hour		1.50	10.667		7,375	350		7,725	8,650
	1000 Steam for duct installation, manifold with pneum. controls									
	1010 3 - 191 lb/hr, 24" x 24"	Q-5	5	3.200	Ea.	1,725	104		1,829	2,050
	1020 3 - 191 lb/hr, 48" x 48"		4.60	3.478		2,225	113		2,338	2,600
	1030 65 - 334 lb/hr, 36" x 36"		4.80	3.333		1,750	109		1,859	2,100
	1040 65 - 334 lb/hr, 72" x 72"		4	4		2,975	130		3,105	3,475
	1050 100 - 690 lb/hr, 48" x 48"		4	4		3,050	130		3,180	3,550
	1060 100 - 690 lb/hr, 84" x 84"		3.60	4.444		5,475	145		5,620	6,250
	1070 220 - 2000 lb/hr, 72" x 72"		3.80	4.211		4,050	137		4,187	4,675
	1080 220 - 2000 lb/hr, 144" x 144"		3.40	4.706		8,450	153		8,603	9,500
	1090 For electrically controlled unit, add					697			697	767
	1092 For additional manifold, add					25%	15%			
	1200 Electric steam, self generating, with manifold									
	1210 1 - 96 lb/hr, 12" x 72"	Q-5	3.60	4.444	Ea.	4,900	145		5,045	5,625
	1220 20 - 168 lb/hr, 12" x 72"		3.60	4.444		4,900	145		5,045	5,625
	1290 For room mounting, add		16	1		142	32.50		174.50	205
	5000 Furnace type, wheel bypass.									
	5020 10 GPD	1 Stpi	4	2	Ea.	136	72.50		208.50	259
	5040 14 GPD		3.80	2.105		181	76		257	315
	5060 19 GPD		3.60	2.222		210	80.50		290.50	350

15761	Air Coils									
200	0010 **DUCT HEATERS** Electric, 480 V, 3 Ph										200
	0020 Finned tubular insert, 500°F									
	0100 8" wide x 6" high, 4.0 kW	Q-20	16	1.250	Ea.	525	40.50		565.50	640
	0120 12" high, 8.0 kW		15	1.333		875	43		918	1,025

Important: See the Reference Section for critical supporting data - Reference Nos., Crews, & City Cost Indexes

15761	Air Coils	CREW	DAILY OUTPUT	LABOR-HOURS	UNIT	MAT.	LABOR	EQUIP.	TOTAL	TOTAL INCL O&P	
						2002 BARE COSTS					
200											200
0140	18" high, 12.0 kW	Q-20	14	1.429	Ea.	1,225	46		1,271	1,425	
0160	24" high, 16.0 kW		13	1.538		1,575	50		1,625	1,800	
0180	30" high, 20.0 kW		12	1.667		1,925	54		1,979	2,200	
0300	12" wide x 6" high, 6.7 kW		15	1.333		560	43		603	680	
0320	12" high, 13.3 kW		14	1.429		905	46		951	1,075	
0340	18" high, 20.0 kW		13	1.538		1,275	50		1,325	1,475	
0360	24" high, 26.7 kW		12	1.667		1,625	54		1,679	1,875	
0380	30" high, 33.3 kW		11	1.818		2,000	59		2,059	2,300	
0500	18" wide x 6" high, 13.3 kW		14	1.429		600	46		646	730	
0520	12" high, 26.7 kW		13	1.538		1,025	50		1,075	1,225	
0540	18" high, 40.0 kW		12	1.667		1,375	54		1,429	1,600	
0560	24" high, 53.3 kW		11	1.818		1,825	59		1,884	2,125	
0580	30" high, 66.7 kW		10	2		2,300	64.50		2,364.50	2,625	
0700	24" wide x 6" high, 17.8 kW		13	1.538		660	50		710	800	
0720	12" high, 35.6 kW		12	1.667		1,125	54		1,179	1,300	
0740	18" high, 53.3 kW		11	1.818		1,575	59		1,634	1,825	
0760	24" high, 71.1 kW		10	2		2,025	64.50		2,089.50	2,325	
0780	30" high, 88.9 kW		9	2.222		2,500	72		2,572	2,850	
0900	30" wide x 6" high, 22.2 kW		12	1.667		700	54		754	855	
0920	12" high, 44.4 kW		11	1.818		1,175	59		1,234	1,400	
0940	18" high, 66.7 kW		10	2		1,675	64.50		1,739.50	1,925	
0960	24" high, 88.9 kW		9	2.222		2,150	72		2,222	2,475	
0980	30" high, 111.0 kW		8	2.500		2,625	81		2,706	3,025	
1400	Note decreased kW available for										
1410	each duct size at same cost										
1420	See line 5000 for modifications and accessories										
2000	Finned tubular flange with insulated										
2020	terminal box, 500°F										
2100	12" wide x 36" high, 54 kW	Q-20	10	2	Ea.	2,600	64.50		2,664.50	2,950	
2120	40" high, 60 kW		9	2.222		2,975	72		3,047	3,375	
2200	24" wide x 36" high, 118.8 kW		9	2.222		2,875	72		2,947	3,275	
2220	40" high, 132 kW		8	2.500		3,300	81		3,381	3,750	
2400	36" wide x 8" high, 40 kW		11	1.818		1,325	59		1,384	1,550	
2420	16" high, 80 kW		10	2		1,775	64.50		1,839.50	2,050	
2440	24" high, 120 kW		9	2.222		2,300	72		2,372	2,650	
2460	32" high, 160 kW		8	2.500		2,950	81		3,031	3,375	
2480	36" high, 180 kW		7	2.857		3,650	92.50		3,742.50	4,150	
2500	40" high, 200 kW		6	3.333		4,025	108		4,133	4,625	
2600	40" wide x 8" high, 45 kW		11	1.818		1,375	59		1,434	1,625	
2620	16" high, 90 kW		10	2		1,900	64.50		1,964.50	2,200	
2640	24" high, 135 kW		9	2.222		2,400	72		2,472	2,750	
2660	32" high, 180 kW		8	2.500		3,200	81		3,281	3,650	
2680	36" high, 202.5 kW		7	2.857		3,700	92.50		3,792.50	4,225	
2700	40" high, 225 kW		6	3.333		4,200	108		4,308	4,800	
2800	48" wide x 8" high, 54.8 kW		10	2		1,450	64.50		1,514.50	1,700	
2820	16" high, 109.8 kW		9	2.222		2,000	72		2,072	2,300	
2840	24" high, 164.4 kW		8	2.500		2,550	81		2,631	2,925	
2860	32" high, 219.2 kW		7	2.857		3,400	92.50		3,492.50	3,900	
2880	36" high, 246.6 kW		6	3.333		3,925	108		4,033	4,475	
2900	40" high, 274 kW		5	4		4,425	129		4,554	5,075	
3000	56" wide x 8" high, 64 kW		9	2.222		1,650	72		1,722	1,900	
3020	16" high, 128 kW		8	2.500		2,275	81		2,356	2,625	
3040	24" high, 192 kW		7	2.857		2,775	92.50		2,867.50	3,200	
3060	32" high, 256 kW		6	3.333		3,875	108		3,983	4,425	
3080	36" high, 288kW		5	4		4,425	129		4,554	5,075	
3100	40" high, 320kW		4	5		4,900	162		5,062	5,650	

MECHANICAL 15

	15761	Air Coils	CREW	DAILY OUTPUT	LABOR-HOURS	UNIT	2002 BARE COSTS				TOTAL INCL O&P	
							MAT.	LABOR	EQUIP.	TOTAL		
200	3200	64″ wide x 8″ high, 74kW	Q-20	8	2.500	Ea.	1,700	81		1,781	2,000	200
	3220	16″ high, 148kW		7	2.857		2,350	92.50		2,442.50	2,725	
	3240	24″ high, 222kW		6	3.333		3,000	108		3,108	3,450	
	3260	32″ high, 296kW		5	4		4,000	129		4,129	4,600	
	3280	36″ high, 333kW		4	5		4,775	162		4,937	5,500	
	3300	40″ high, 370kW		3	6.667		5,275	216		5,491	6,125	
	3800	Note decreased kW available for										
	3820	each duct size at same cost										
	5000	Duct heater modifications and accessories										
	5120	T.C.O. limit auto or manual reset	Q-20	42	.476	Ea.	86.50	15.40		101.90	119	
	5140	Thermostat		28	.714		370	23		393	440	
	5160	Overheat thermocouple (removable)		7	2.857		520	92.50		612.50	715	
	5180	Fan interlock relay		18	1.111		126	36		162	194	
	5200	Air flow switch		20	1		109	32.50		141.50	170	
	5220	Split terminal box cover		100	.200		36	6.45		42.45	49.50	
	8000	To obtain BTU multiply kW by 3413										
700	0010	**COILS, FLANGED**										700
	0100	Basic water, DX, or condenser coils										
	0110	Copper tubes, alum. fins, galv. end sheets										
	0112	H is finned height, L is finned length										
	0120	3/8″ x .016 tube, .0065 aluminum fins										
	0130	2 row, 8 fins per inch										
	0140	4″ H x 12″ L	Q-5	48	.333	Ea.	510	10.85		520.85	575	
	0150	4″ H x 24″ L		38.38	.417		515	13.60		528.60	590	
	0160	4″ H x 48″ L		19.25	.831		530	27		557	625	
	0170	4″ H x 72″ L		12.80	1.250		650	40.50		690.50	775	
	0180	6″ H x 12″ L		48	.333		520	10.85		530.85	590	
	0190	6″ H x 24″ L		25.60	.625		530	20.50		550.50	615	
	0200	6″ H x 48″ L		12.80	1.250		550	40.50		590.50	665	
	0210	6″ H x 72″ L		8.53	1.876		675	61		736	835	
	0220	10″ H x 12″ L		30.73	.521		550	16.95		566.95	630	
	0230	10″ H x 24″ L		15.33	1.044		560	34		594	670	
	0240	10″ H x 48″ L		7.69	2.081		590	68		658	745	
	0250	10″ H x 72″ L		5.12	3.125		730	102		832	955	
	0260	12″ H x 12″ L		25.60	.625		560	20.50		580.50	650	
	0270	12″ H x 24″ L		12.80	1.250		575	40.50		615.50	695	
	0280	12″ H x 48″ L		6.40	2.500		660	81.50		741.50	850	
	0290	12″ H x 72″ L		4.27	3.747		755	122		877	1,025	
	0300	16″ H x 16″ L		14.38	1.113		595	36.50		631.50	710	
	0310	16″ H x 24″ L		9.59	1.668		605	54.50		659.50	750	
	0320	16″ H x 48″ L		4.80	3.333		700	109		809	935	
	0330	16″ H x 72″ L		3.20	5		805	163		968	1,125	
	0340	20″ H x 24″ L		7.69	2.081		640	68		708	800	
	0350	20″ H x 48″ L		3.84	4.167		740	136		876	1,025	
	0360	20″ H x 72″ L		2.56	6.250		860	204		1,064	1,250	
	0370	24″ H x 24″ L		6.40	2.500		710	81.50		791.50	905	
	0380	24″ H x 48″ L		3.20	5		785	163		948	1,100	
	0390	24″ H x 72″ L		2.13	7.512		910	245		1,155	1,375	
	0400	30″ H x 28″ L		4.39	3.645		775	119		894	1,025	
	0410	30″ H x 48″ L		2.56	6.250		850	204		1,054	1,250	
	0420	30″ H x 72″ L		1.71	9.357		990	305		1,295	1,550	
	0430	30″ H x 84″ L		1.46	10.959		1,075	355		1,430	1,725	
	0440	34″ H x 32″ L		3.39	4.720		830	154		984	1,150	
	0450	34″ H x 48″ L		2.26	7.080		890	231		1,121	1,325	
	0460	34″ H x 72″ L		1.51	10.596		1,050	345		1,395	1,675	
	0470	34″ H x 84″ L		1.29	12.403		1,125	405		1,530	1,850	

Important: See the Reference Section for critical supporting data - Reference Nos., Crews, & City Cost Indexes

15761	Air Coils	CREW	DAILY OUTPUT	LABOR-HOURS	UNIT	2002 BARE COSTS				TOTAL INCL O&P
						MAT.	LABOR	EQUIP.	TOTAL	
700										**700**
0600	For 10 fins per inch, add					1%				
0610	For 12 fins per inch, add					3%				
0620	For 4 row, add				Ea.	30%	50%			
0630	For 6 row, add					60%	100%			
0640	For 8 row, add				↓	80%	150%			
1100	1/2" x .017 tube, .0065 aluminum fins									
1110	2 row, 8 fins per inch									
1120	5" H x 5" L	Q-5	40	.400	Ea.	510	13.05		523.05	585
1130	5" H x 20" L		37.10	.431		525	14.05		539.05	595
1140	5" H x 45" L		16.41	.975		545	32		577	650
1150	5" H x 90" L		8.21	1.949		745	63.50		808.50	915
1160	7.5" H x 5" L		38.50	.416		525	13.55		538.55	600
1170	7.5" H x 20" L		24.62	.650		545	21		566	630
1180	7.5" H x 45" L		10.94	1.463		575	47.50		622.50	700
1190	7.5" H x 90" L		5.46	2.930		790	95.50		885.50	1,000
1200	12.5" H x 10" L		29.49	.543		565	17.70		582.70	650
1210	12.5" H x 20" L		14.75	1.085		585	35.50		620.50	700
1220	12.5" H x 45" L		6.55	2.443		675	79.50		754.50	865
1230	12.5" H x 90" L		3.28	4.878		880	159		1,039	1,200
1240	15" H x 15" L		16.41	.975		595	32		627	705
1250	15" H x 30" L		8.21	1.949		625	63.50		688.50	780
1260	15" H x 45" L		5.46	2.930		705	95.50		800.50	920
1270	15" H x 90" L		2.73	5.861		920	191		1,111	1,325
1280	20" H x 20" L		9.22	1.735		645	56.50		701.50	795
1290	20" H x 45" L		4.10	3.902		765	127		892	1,025
1300	20" H x 90" L		2.05	7.805		1,000	254		1,254	1,500
1310	25" H x 25" L		5.90	2.712		745	88.50		833.50	955
1320	25" H x 45" L		3.28	4.878		825	159		984	1,150
1330	25" H x 90" L		1.64	9.756		1,100	320		1,420	1,675
1340	30" H x 30" L		4.10	3.902		820	127		947	1,100
1350	30" H x 45" L		2.73	5.861		885	191		1,076	1,275
1360	30" H x 90" L		1.37	11.679		1,200	380		1,580	1,900
1370	35" H x 35" L		3.01	5.316		895	173		1,068	1,250
1380	35" H x 45" L		2.34	6.838		950	223		1,173	1,375
1390	35" H x 90" L		1.17	13.675		1,300	445		1,745	2,100
1400	42.5" H x 45" L		1.93	8.290		1,050	270		1,320	1,550
1410	42.5" H x 75" L		1.16	13.793		1,300	450		1,750	2,100
1420	42.5" H x 105" L	↓	.83	19.277		1,525	630		2,155	2,625
1600	For 10 fins per inch, add					2%				
1610	For 12 fins per inch, add					5%				
1620	For 4 row, add					35%	50%			
1630	For 6 row, add					65%	100%			
1640	For 8 row, add				↓	90%	150%			
2010	5/8" x .020" tube, .0065 aluminum fins									
2020	2 row, 8 fins per inch									
2030	6" H x 6" L	Q-5	38	.421	Ea.	575	13.70		588.70	650
2040	6" H x 24" L		24.96	.641		605	21		626	695
2050	6" H x 48" L		12.48	1.282		645	42		687	775
2060	6" H x 96" L		6.24	2.564		1,100	83.50		1,183.50	1,325
2070	12" H x 12" L		24.96	.641		294	21		315	355
2080	12" H x 24" L		12.48	1.282		320	42		362	415
2090	12" H x 48" L		6.24	2.564		815	83.50		898.50	1,025
2100	12" H x 96" L		3.12	5.128		1,325	167		1,492	1,700
2110	18" H x 24" L		8.32	1.923		3,575	62.50		3,637.50	4,025
2120	18" H x 48" L		4.16	3.846		940	125		1,065	1,225
2130	18" H x 96" L	↓	2.08	7.692	↓	1,550	251		1,801	2,075

MECHANICAL 15

15761 | Air Coils

		CREW	DAILY OUTPUT	LABOR-HOURS	UNIT	2002 BARE COSTS				TOTAL INCL O&P
						MAT.	LABOR	EQUIP.	TOTAL	
700	2140	24" H x 24" L	Q-5	6.24	2.564	Ea.	905	83.50		988.50	1,125
	2150	24" H x 48" L		3.12	5.128		510	167		677	810
	2160	24" H x 96" L		1.56	10.256		1,775	335		2,110	2,450
	2170	30" H x 30" L		3.99	4.010		1,050	131		1,181	1,350
	2180	30" H x 48" L		2.50	6.400		570	209		779	940
	2190	30" H x 96" L		1.25	12.800		2,000	415		2,415	2,825
	2200	30" H x 120" L		1	16		2,450	520		2,970	3,475
	2210	36" H x 36" L		2.77	5.776		1,200	188		1,388	1,600
	2220	36" H x 48" L		2.08	7.692		1,325	251		1,576	1,825
	2230	36" H x 90" L		1.11	14.414		2,175	470		2,645	3,100
	2240	36" H x 120" L		.83	19.277		2,750	630		3,380	3,975
	2250	42" H x 48" L		1.78	8.989		1,450	293		1,743	2,050
	2260	42" H x 90" L		.95	16.842		2,400	550		2,950	3,450
	2270	42" H x 120" L		.71	22.535		3,050	735		3,785	4,475
	2280	51" H x 48" L		1.47	10.884		1,675	355		2,030	2,350
	2290	51" H x 96" L		.73	21.918		2,825	715		3,540	4,175
	2300	51" H x 120" L		.59	27.119		3,525	885		4,410	5,200
	2500	For 10 fins per inch, add					3%				
	2510	For 12 fins per inch, add					6%				
	2520	For 4 row, add				Ea.	40%	50%			
	2530	For 6 row, add					80%	100%			
	2540	For 8 row, add					110%	150%			
	3000	Hot water booster coils									
	3010	Copper tubes, alum. fins, galv. end sheets									
	3012	H is finned height, L is finned length									
	3020	1/2" x .017" tube, .0065 Al fins									
	3030	1 row, 10 fins per inch									
	3040	5"H x 10"L	Q-5	40	.400	Ea.	202	13.05		215.05	242
	3050	5"H x 25"L		33.18	.482		208	15.70		223.70	253
	3060	10"H x 10"L		39	.410		210	13.35		223.35	251
	3070	10"H x 20"L		20.73	.772		217	25		242	277
	3080	10"H x 30"L		13.83	1.157		225	37.50		262.50	305
	3090	15"H x 15"L		18.44	.868		223	28.50		251.50	289
	3110	15"H x 20"L		13.83	1.157		228	37.50		265.50	310
	3120	15"H x 30"L		9.22	1.735		239	56.50		295.50	350
	3130	20"H x 20"L		10.37	1.543		240	50.50		290.50	340
	3140	20"H x 30"L		6.91	2.315		253	75.50		328.50	395
	3150	20"H x 40"L		5.18	3.089		315	101		416	500
	3160	25"H x 25"L		6.64	2.410		300	78.50		378.50	450
	3170	25"H x 35"L		4.74	3.376		330	110		440	525
	3180	25"H x 45"L		3.69	4.336		350	141		491	600
	3190	30"H x 30"L		4.61	3.471		335	113		448	535
	3200	30"H x 40"L		3.46	4.624		360	151		511	625
	3210	30"H x 50"L		2.76	5.797		390	189		579	715
	3300	For 2 row, add					20%	50%			
	3400	1/2" x .020" tube, .008 Al fins									
	3410	2 row, 10 fins per inch									
	3420	20" H x 27" L	Q-5	7.27	2.201	Ea.	300	71.50		371.50	440
	3430	20" H x 36" L		5.44	2.941		370	96		466	555
	3440	20" H x 45" L		4.35	3.678		400	120		520	620
	3450	25" H x 45" L		3.49	4.585		445	149		594	715
	3460	25" H x 54" L		2.90	5.517		530	180		710	850
	3470	30" H x 54" L		2.42	6.612		575	215		790	960
	3480	35" H x 54" L		2.07	7.729		625	252		877	1,075
	3490	35" H x 63" L		1.78	8.989		675	293		968	1,175
	3500	40" H x 72" L		1.36	11.765		780	385		1,165	1,450

Important: See the Reference Section for critical supporting data - Reference Nos., Crews, & City Cost Indexes

		CREW	DAILY OUTPUT	LABOR-HOURS	UNIT	2002 BARE COSTS				TOTAL INCL O&P
15761	**Air Coils**					MAT.	LABOR	EQUIP.	TOTAL	
3510	55" H x 63" L	Q-5	1.13	14.159	Ea.	785	460		1,245	1,550
3520	55" H x 72" L		.99	16.162		975	525		1,500	1,875
3530	55" H x 81" L		.88	18.182		1,100	590		1,690	2,100
3540	75" H x 81" L		.64	25		1,250	815		2,065	2,600
3550	75" H x 90" L		.58	27.586		1,450	900		2,350	2,950
3560	75" H x 108" L		.48	33.333		1,700	1,075		2,775	3,525
3800	5/8" x .020 tubing, .0065 aluminum fins									
3810	1 row, 10 fins per inch									
3820	6" H x 12" L	Q-5	38	.421	Ea.	205	13.70		218.70	246
3830	6" H x 24" L		28.80	.556		210	18.10		228.10	259
3840	6" H x 30" L		23.04	.694		213	22.50		235.50	268
3850	12" H x 12" L		28.80	.556		214	18.10		232.10	263
3860	12" H x 24" L		14.40	1.111		224	36		260	300
3870	12" H x 30" L		11.52	1.389		228	45.50		273.50	320
3880	18" H x 18" L		12.80	1.250		230	40.50		270.50	315
3890	18" H x 24" L		9.60	1.667		237	54.50		291.50	345
3900	18" H x 36" L		6.40	2.500		251	81.50		332.50	400
3910	24" H x 24" L		7.20	2.222		251	72.50		323.50	385
3920	24" H x 30" L		4.80	3.333		310	109		419	505
3930	24" H x 36" L		5.05	3.168		320	103		423	510
3940	30" H x 30" L		4.61	3.471		330	113		443	530
3950	30" H x 36" L		3.85	4.159		345	136		481	585
3960	30" H x 42" L		3.29	4.863		360	158		518	635
3970	36" H x 36" L		3.20	5		330	163		493	610
3980	36" H x 42" L		2.74	5.839		385	190		575	710
4100	For 2 row, add					30%	50%			
4400	Steam and hot water heating coils									
4410	Copper tubes, alum. fins. galv. end sheets									
4412	H is finned height, L is finned length									
4420	5/8" tubing, to 175 psig working pressure									
4430	1 row, 8 fins per inch									
4440	6"H x 6"L	Q-5	38	.421	Ea.	560	13.70		573.70	635
4450	6"H x 12"L		37	.432		565	14.10		579.10	640
4460	6"H x 24"L		24.48	.654		575	21.50		596.50	660
4470	6"H x 36"L		16.32	.980		585	32		617	695
4480	6"H x 48"L		12.24	1.307		600	42.50		642.50	725
4490	6"H x 84"L		6.99	2.289		975	74.50		1,049.50	1,175
4500	6" H x 108" L		5.44	2.941		1,075	96		1,171	1,325
4510	12" H x 12" L		24.48	.654		610	21.50		631.50	705
4520	12" H x 24" L		12.24	1.307		630	42.50		672.50	760
4530	12" H x 36" L		8.16	1.961		650	64		714	810
4540	12" H x 48" L		6.12	2.614		720	85		805	925
4550	12" H x 84" L		3.50	4.571		1,100	149		1,249	1,450
4560	12" H x 108" L		2.72	5.882		1,250	192		1,442	1,675
4570	18" H x18" L		10.88	1.471		675	48		723	815
4580	18" H x 24" L		8.16	1.961		690	64		754	850
4590	18" H x 48" L		4.08	3.922		800	128		928	1,075
4600	18" H x 84" L		2.33	6.867		1,250	224		1,474	1,700
4610	18" H x 108" L		1.81	8.840		1,425	288		1,713	1,975
4620	24" H x 24" L		6.12	2.614		785	85		870	995
4630	24" H x 48" L		3.06	5.229		880	170		1,050	1,225
4640	24" H x 84" L		1.75	9.143		1,375	298		1,673	1,950
4650	24" H x 108" L		1.36	11.765		1,575	385		1,960	2,300
4660	36" H x 36" L		2.72	5.882		980	192		1,172	1,375
4670	36" H x 48" L		2.04	7.843		1,050	256		1,306	1,525
4680	36" H x 84" L		1.17	13.675		1,650	445		2,095	2,475

700

MECHANICAL 15

15761 | Air Coils

			DAILY OUTPUT	LABOR-HOURS	UNIT	2002 BARE COSTS				TOTAL INCL O&P	
			CREW			MAT.	LABOR	EQUIP.	TOTAL		
700	4690	36"H x 108"L	Q-5	.91	17.582	Ea.	1,950	575		2,525	3,000	700
	4700	48"H x 48"L		1.53	10.458		1,200	340		1,540	1,850	
	4710	48"H x 84"L		.87	18.391		1,925	600		2,525	3,000	
	4720	48"H x 108"L		.68	23.529		2,300	765		3,065	3,675	
	4730	48"H x 126"L		.58	27.586		2,425	900		3,325	4,025	
	4910	For 10 fins per inch, add					2%					
	4920	For 12 fins per inch, add					5%					
	4930	For 2 row, add					30%	50%				
	4940	For 3 row, add					50%	100%				
	4950	For 4 row, add					80%	123%				
	4960	For 230 PSIG heavy duty, add					20%	30%				
	6000	Steam, heavy duty, 200 PSIG										
	6010	Copper tubes, alum. fins, galv. end sheets										
	6012	H is finned height, L is finned length										
	6020	1" x .035" tubing, .010 aluminum fins										
	6030	1 row, 8 fins per inch										
	6040	12"H x 12"L	Q-5	24.48	.654	Ea.	625	21.50		646.50	720	
	6050	12"H x 24"L		12.24	1.307		665	42.50		707.50	800	
	6060	12"H x 36"L		8.16	1.961		705	64		769	870	
	6070	12"H x 48"L		6.12	2.614		795	85		880	1,000	
	6080	12"H x 84"L		3.50	4.571		1,025	149		1,174	1,350	
	6090	12"H x 108"L		2.72	5.882		1,175	192		1,367	1,575	
	6100	18"H x 24"L		8.16	1.961		740	64		804	910	
	6110	18"H x 36"L		5.44	2.941		845	96		941	1,075	
	6120	18"H x 48"L		4.08	3.922		910	128		1,038	1,200	
	6130	18"H x 84"L		2.33	6.867		1,200	224		1,424	1,650	
	6140	18"H x 108"L		1.81	8.840		1,375	288		1,663	1,925	
	6150	24"H x 24"L		6.12	2.614		860	85		945	1,075	
	6160	24"H x 36"L		4.08	3.922		945	128		1,073	1,250	
	6170	24"H x 48"L		3.06	5.229		1,025	170		1,195	1,375	
	6180	24"H x 84"L		1.75	9.143		1,375	298		1,673	1,975	
	6190	24"H x 108"L		1.36	11.765		1,600	385		1,985	2,350	
	6200	30"H x 36"L		3.26	4.908		1,050	160		1,210	1,400	
	6210	30"H x 48"L		2.45	6.531		1,125	213		1,338	1,550	
	6220	30"H x 84"L		1.40	11.429		1,525	370		1,895	2,225	
	6230	30"H x 108"L		1.09	14.679		1,775	480		2,255	2,675	
	6240	36"H x 36"L		2.72	5.882		1,125	192		1,317	1,525	
	6250	36"H x 48"L		2.04	7.843		1,225	256		1,481	1,725	
	6260	36"H x 84"L		1.17	13.675		1,675	445		2,120	2,525	
	6270	36"H x 108"L		.91	17.582		1,975	575		2,550	3,050	
	6280	36"H x 120"L		.82	19.512		2,100	635		2,735	3,275	
	6290	48"H x 48"L		1.53	10.458		1,475	340		1,815	2,150	
	6300	48"H x 84"L		.87	18.391		2,025	600		2,625	3,125	
	6310	48"H x 108"L		.68	23.529		2,425	765		3,190	3,800	
	6320	48"H x 120"L		.61	26.230		2,575	855		3,430	4,125	
	6400	For 10 fins per inch, add					4%					
	6410	For 12 fins per inch, add					10%					

15765 | Fan Coil Unit/Unit Ventilators

			CREW	DAILY OUTPUT	LABOR-HOURS	UNIT	MAT.	LABOR	EQUIP.	TOTAL	TOTAL INCL O&P	
200	0010	**FAN COIL AIR CONDITIONING** Cabinet mounted, filters, controls										200
	0100	Chilled water, 1/2 ton cooling	Q-5	8	2	Ea.	765	65		830	940	
	0110	3/4 ton cooling		7	2.286		805	74.50		879.50	1,000	
	0120	1 ton cooling		6	2.667		920	87		1,007	1,125	
	0140	1.5 ton cooling		5.50	2.909		1,050	95		1,145	1,300	
	0150	2 ton cooling		5.25	3.048		1,350	99.50		1,449.50	1,625	

15 MECHANICAL

15765 | Fan Coil Unit/Unit Ventilators

		CREW	DAILY OUTPUT	LABOR-HOURS	UNIT	2002 BARE COSTS MAT.	LABOR	EQUIP.	TOTAL	TOTAL INCL O&P	
200	0160	2.5 ton cooling	Q-5	5	3.200	Ea.	1,975	104		2,079	2,325	**200**
	0180	3 ton cooling		4	4		2,175	130		2,305	2,600	
	0190	7.5 ton cooling	↓	2.70	5.926	↓	2,500	193		2,693	3,050	
	0262	For hot water coil, add					40%	10%				
	0300	Console, 2 pipe with electric heat										
	0310	1/2 ton cooling	Q-5	8	2	Ea.	925	65		990	1,125	
	0315	3/4 ton cooling		7	2.286		1,000	74.50		1,074.50	1,200	
	0320	1 ton cooling		6	2.667		1,100	87		1,187	1,325	
	0330	1.5 ton cooling		5.50	2.909		1,350	95		1,445	1,625	
	0340	2 ton cooling		4.70	3.404		1,775	111		1,886	2,125	
	0345	2-1/2 ton cooling		4.70	3.404		2,575	111		2,686	3,000	
	0350	3 ton cooling		4	4		3,450	130		3,580	3,975	
	0360	4 ton cooling		4	4		3,575	130		3,705	4,150	
	0940	Direct expansion, for use w/air cooled condensing, 1.5 ton cooling		5	3.200		575	104		679	785	
	0950	2 ton cooling		4.80	3.333		610	109		719	835	
	0960	2.5 ton cooling		4.40	3.636		715	118		833	965	
	0970	3 ton cooling		3.80	4.211		745	137		882	1,025	
	0980	3.5 ton cooling		3.60	4.444		910	145		1,055	1,225	
	0990	4 ton cooling		3.40	4.706		965	153		1,118	1,275	
	1000	5 ton cooling		3	5.333		1,025	174		1,199	1,375	
	1020	7.5 ton cooling	↓	3	5.333		2,400	174		2,574	2,900	
	1040	10 ton cooling	Q-6	2.60	9.231		2,900	310		3,210	3,650	
	1042	11 ton cooling		2.30	10.435		3,725	350		4,075	4,625	
	1050	15 ton cooling		1.60	15		3,850	505		4,355	5,000	
	1060	20 ton cooling		.70	34.286		5,100	1,150		6,250	7,350	
	1070	25 ton cooling		.65	36.923		6,375	1,250		7,625	8,900	
	1080	30 ton cooling	↓	.60	40		7,050	1,350		8,400	9,800	
	1500	For hot water coil, add				↓	40%	10%				
	1510	For condensing unit add see division 15670.										
	3000	Chilled water, horizontal unit, housing, 2 pipe, controls										
	3100	1/2 ton cooling	Q-5	8	2	Ea.	860	65		925	1,050	
	3110	1 ton cooling		6	2.667		1,050	87		1,137	1,275	
	3120	1.5 ton cooling		5.50	2.909		1,250	95		1,345	1,525	
	3130	2 ton cooling		5.25	3.048		1,500	99.50		1,599.50	1,800	
	3140	3 ton cooling		4	4		1,725	130		1,855	2,100	
	3150	3.5 ton cooling		3.80	4.211		1,900	137		2,037	2,300	
	3160	4 ton cooling		3.80	4.211		1,900	137		2,037	2,300	
	3170	5 ton cooling		3.40	4.706		2,200	153		2,353	2,650	
	3180	6 ton cooling		3.40	4.706		2,200	153		2,353	2,650	
	3190	7 ton cooling		2.90	5.517		2,400	180		2,580	2,925	
	3200	8 ton cooling	↓	2.90	5.517	↓	2,400	180		2,580	2,925	
	3210	10 ton cooling	Q-6	2.80	8.571		2,650	290		2,940	3,350	
	3240	For hot water coil, add					40%	10%				
	4000	With electric heat, 2 pipe										
	4100	1/2 ton cooling	Q-5	8	2	Ea.	1,050	65		1,115	1,250	
	4105	3/4 ton cooling		7	2.286		1,150	74.50		1,224.50	1,375	
	4110	1 ton cooling		6	2.667		1,275	87		1,362	1,525	
	4120	1.5 ton cooling		5.50	2.909		1,425	95		1,520	1,700	
	4130	2 ton cooling		5.25	3.048		1,775	99.50		1,874.50	2,100	
	4135	2.5 ton cooling		4.80	3.333		2,450	109		2,559	2,875	
	4140	3 ton cooling		4	4		3,300	130		3,430	3,825	
	4150	3.5 ton cooling		3.80	4.211		3,925	137		4,062	4,525	
	4160	4 ton cooling		3.60	4.444		4,550	145		4,695	5,225	
	4170	5 ton cooling		3.40	4.706		5,100	153		5,253	5,825	
	4180	6 ton cooling		3.20	5		5,100	163		5,263	5,850	
	4190	7 ton cooling	↓	2.90	5.517	↓	5,350	180		5,530	6,175	

MECHANICAL 15

15765 | Fan Coil Unit/Unit Ventilators

			DAILY OUTPUT	LABOR-HOURS	UNIT	2002 BARE COSTS				TOTAL INCL O&P	
			CREW			MAT.	LABOR	EQUIP.	TOTAL		
200	4200	8 ton cooling	Q-5	2.20	7.273	Ea.	6,050	237		6,287	7,000	**200**
	4210	10 ton cooling	Q-6	2.80	8.571	↓	6,400	290		6,690	7,475	
600	0010	**HEATING & VENTILATING UNITS** Classroom										**600**
	0020	Includes filter, heating/cooling coils, standard controls										
	0080	750 CFM, 2 tons cooling	Q-6	2	12	Ea.	3,050	405		3,455	3,950	
	0100	1000 CFM, 2-1/2 tons cooling		1.60	15		3,600	505		4,105	4,750	
	0120	1250 CFM, 3 tons cooling		1.40	17.143		3,725	580		4,305	4,975	
	0140	1500 CFM, 4 tons cooling		.80	30		3,975	1,025		5,000	5,900	
	0160	2000 CFM 5 tons cooling	↓	.50	48		4,675	1,625		6,300	7,600	
	0500	For electric heat, add					35%					
	1000	For no cooling, deduct				↓	25%	10%				

15766 | Fin Tube Radiation

			DAILY OUTPUT	LABOR-HOURS	UNIT	2002 BARE COSTS				TOTAL INCL O&P	
			CREW			MAT.	LABOR	EQUIP.	TOTAL		
190	0010	**HYDRONIC HEATING** Terminal units, not incl. main supply pipe										**190**
	1000	Radiation										
	1100	Panel, baseboard, C.I., including supports, no covers	Q-5	46	.348	L.F.	21	11.35		32.35	40	
	1150	Fin tube, wall hung, 14" slope top cover, with damper										
	1200	1-1/4" copper tube, 4-1/4" alum. fin	Q-5	38	.421	L.F.	27	13.70		40.70	50.50	
	1250	1-1/4" steel tube, 4-1/4" steel fin		36	.444		24.50	14.50		39	49	
	1255	2" steel tube, 4-1/4" steel fin	↓	32	.500	↓	26.50	16.30		42.80	53.50	
	1260	21", two tier, slope top, w/damper										
	1262	1-1/4" copper tube, 4-1/4" alum. fin	Q-5	30	.533	L.F.	32	17.40		49.40	61	
	1264	1-1/4" steel tube, 4-1/4" steel fin		28	.571		29.50	18.60		48.10	60	
	1266	2" steel tube, 4-1/4" steel fin	↓	25	.640	↓	31	21		52	66	
	1270	10", flat top, wtih damper										
	1272	1-1/4" copper tube, 4-1/4" alum. fin	Q-5	38	.421	L.F.	26	13.70		39.70	49	
	1274	1-1/4" steel tube, 4-1/4" steel fin		36	.444		24.50	14.50		39	49	
	1276	2" steel tube, 4-1/4" steel fin	↓	32	.500	↓	26.50	16.30		42.80	53.50	
	1280	17", two tier, flat top, w/damper										
	1282	1-1/4" copper tube, 4-1/4" alum. fin	Q-5	30	.533	L.F.	30.50	17.40		47.90	59.50	
	1284	1-1/4" steel tube, 4-1/4" steel fin		28	.571		29.50	18.60		48.10	60	
	1286	2" steel tube, 4-1/4" steel fin	↓	25	.640	↓	31	21		52	66	
	1301	Rough in wall hung, steel fin w/supply & balance valves	1 Stpi	.91	8.791	Ea.	143	320		463	635	
	1310	Baseboard, pkgd, 1/2" copper tube, alum. fin, 7" high	Q-5	60	.267	L.F.	5.35	8.70		14.05	18.95	
	1320	3/4" copper tube, alum. fin, 7" high		58	.276		5.65	9		14.65	19.75	
	1340	1" copper tube, alum. fin, 8-7/8" high		56	.286		11.65	9.30		20.95	27	
	1360	1-1/4" copper tube, alum. fin, 8-7/8" high		54	.296		17.25	9.65		26.90	33.50	
	1380	1-1/4" IPS steel tube with steel fins	↓	52	.308	↓	16.60	10		26.60	33.50	
	1381	Rough in baseboard panel & fin tube, supply & balance valves	1 Stpi	1.06	7.547	Ea.	104	273		377	525	
	1500	Note: fin tube may also require corners, caps, etc.										
	1990	Convector unit, floor recessed, flush, with trim										
	2000	for under large glass wall areas, no damper	Q-5	20	.800	L.F.	24.50	26		50.50	66.50	
	2100	For unit with damper, add				"	9.95			9.95	10.95	
	3000	Radiators, cast iron										
	3100	Free standing or wall hung, 6 tube, 25" high	Q-5	96	.167	Section	21.50	5.45		26.95	31.50	
	3150	4 tube 25" high		96	.167		15.90	5.45		21.35	25.50	
	3200	4 tube, 19" high	↓	96	.167	↓	14.80	5.45		20.25	24.50	
	3250	Adj. brackets, 2 per wall radiator up to 30 sections	1 Stpi	32	.250	Ea.	32	9.05		41.05	48.50	
	3500	Recessed, 20" high x 5" deep, without grille	Q-5	60	.267	Section	17.60	8.70		26.30	32.50	
	3600	For inlet grille, add				"	1.25			1.25	1.38	
	3950	Unit heaters, propeller, 115 V 2 psi steam, 60°F entering air										
	4000	Horizontal, 12 MBH	Q-5	12	1.333	Ea.	247	43.50		290.50	340	
	4020	28.2 MBH		10	1.600		340	52		392	455	
	4040	36.5 MBH		8	2		400	65		465	540	
	4060	43.9 MBH	↓	8	2	↓	405	65		470	545	

Important: See the Reference Section for critical supporting data - Reference Nos., Crews, & City Cost Indexes

			DAILY	LABOR-			2002 BARE COSTS				TOTAL
15766	**Fin Tube Radiation**	CREW	OUTPUT	HOURS	UNIT	MAT.	LABOR	EQUIP.	TOTAL		INCL O&P
190	4080	56.5 MBH	Q-5	7.50	2.133	Ea.	440	69.50		509.50	590	190
	4100	65.6 MBH		7	2.286		445	74.50		519.50	600
	4120	87.6 MBH		6.50	2.462		470	80		550	635
	4140	96.8 MBH		6	2.667		570	87		657	760
	4160	133.3 MBH		5	3.200		590	104		694	805
	4180	157.6 MBH		4	4		740	130		870	1,000
	4200	197.7 MBH		3	5.333		875	174		1,049	1,225
	4220	257.2 MBH		2.50	6.400		1,125	209		1,334	1,575
	4240	286.9 MBH		2	8		1,225	261		1,486	1,750
	4250	326.0 MBH		1.90	8.421		1,350	274		1,624	1,925
	4260	364 MBH		1.80	8.889		1,450	290		1,740	2,000
	4270	404 MBH		1.60	10		1,975	325		2,300	2,675
	4280	Rough in unit heater, 72.7 MBH, piping, supports & controls	1 Stpi	.47	17.021	Set	330	615		945	1,300
	4300	For vertical diffuser, add				Ea.	139			139	153
	4310	Vertical flow, 40 MBH	Q-5	11	1.455		395	47.50		442.50	500
	4314	58.5 MBH		8	2		445	65		510	590
	4318	92 MBH		7	2.286		535	74.50		609.50	700
	4322	109.7 MBH		6	2.667		625	87		712	815
	4326	131.0 MBH		4	4		625	130		755	880
	4330	160.0 MBH		3	5.333		745	174		919	1,075
	4334	194.0 MBH		2.20	7.273		805	237		1,042	1,250
	4338	212.0 MBH		2.10	7.619		1,250	248		1,498	1,750
	4342	247.0 MBH		1.96	8.163		1,300	266		1,566	1,850
	4346	297.0 MBH		1.80	8.889		1,450	290		1,740	2,025
	4350	333.0 MBH	Q-6	1.90	12.632		1,800	425		2,225	2,650
	4354	420 MBH,(460 V)		1.80	13.333		1,975	450		2,425	2,850
	4358	500 MBH, (460 V)		1.71	14.035		2,450	475		2,925	3,400
	4362	570 MBH, (460 V)		1.40	17.143		2,925	580		3,505	4,100
	4366	620 MBH, (460 V)		1.30	18.462		3,375	625		4,000	4,650
	4370	960 MBH, (460 V)		1.10	21.818		5,900	735		6,635	7,600
	4400	Unit heaters, propeller, hot water, 115 V									
	4404	Horizontal									
	4408	6 MBH	Q-5	11	1.455	Ea.	247	47.50		294.50	345
	4412	12 MBH		11	1.455		247	47.50		294.50	345
	4416	16 MBH		10	1.600		267	52		319	375
	4420	24 MBH		9	1.778		340	58		398	465
	4424	29 MBH		8	2		400	65		465	540
	4428	47 MBH		7	2.286		445	74.50		519.50	600
	4432	63 MBH		6	2.667		470	87		557	645
	4436	81 MBH		5.50	2.909		570	95		665	775
	4440	90 MBH		5	3.200		590	104		694	805
	4460	133 MBH		4	4		740	130		870	1,000
	4464	139 MBH		3.50	4.571		875	149		1,024	1,200
	4468	198 MBH		2.50	6.400		1,200	209		1,409	1,650
	4472	224 MBH		2	8		1,250	261		1,511	1,775
	4476	273 MBH		1.50	10.667		1,325	350		1,675	1,975
	4500	Vertical									
	4504	8 MBH	Q-5	13	1.231	Ea.	395	40		435	490
	4508	12 MBH		13	1.231		395	40		435	490
	4512	16 MBH		13	1.231		395	40		435	490
	4516	30 MBH		12	1.333		395	43.50		438.50	495
	4520	43 MBH		11	1.455		445	47.50		492.50	560
	4524	57 MBH		8	2		500	65		565	650
	4528	68 MBH		7	2.286		535	74.50		609.50	700
	4540	105 MBH		6	2.667		625	87		712	815
	4544	123 MBH		5	3.200		740	104		844	970

MECHANICAL 15

			DAILY	LABOR-		2002 BARE COSTS				TOTAL	
	15766	**Fin Tube Radiation**	CREW	OUTPUT	HOURS	UNIT	MAT.	LABOR	EQUIP.	TOTAL	INCL O&P
190	4550	140 MBH	Q-5	5	3.200	Ea.	805	104		909	1,050	**190**
	4554	156 MBH		4	4		910	130		1,040	1,200	
	4560	210 MBH		4	4		995	130		1,125	1,300	
	4564	223 MBH		3	5.333		1,475	174		1,649	1,875	
	4568	257 MBH	▼	2	8	▼	1,475	261		1,736	2,025	
	5400	Cabinet, horizontal, hot water, blower type										
	5420	20 MBH	Q-5	10	1.600	Ea.	675	52		727	820	
	5430	60 MBH		7	2.286		1,175	74.50		1,249.50	1,400	
	5440	100 MBH		5	3.200		1,200	104		1,304	1,475	
	5450	120 MBH	▼	4	4	▼	1,275	130		1,405	1,600	
	6000	Valance units, complete with 1/2" cooling coil, enclosure										
	6020	2 tube	Q-5	18	.889	L.F.	26	29		55	72	
	6040	3 tube		16	1		31	32.50		63.50	83	
	6060	4 tube		16	1		35.50	32.50		68	88	
	6080	5 tube		15	1.067		42.50	35		77.50	99.50	
	6100	6 tube		15	1.067		46	35		81	104	
	6120	8 tube	▼	14	1.143	▼	68.50	37		105.50	132	
	6200	For 3/4" cooling coil, add					10%					
	9500	To convert SFR to BTU rating: Hot water, 150 x SFR										
	9510	Forced hot water, 180 x SFR; steam, 240 x SFR										

	15768	**Infrared Heaters**									
600	0010	**INFRA-RED UNIT**										**600**
	0020	Gas fired, unvented, electric ignition, 100% shutoff. Piping and										
	0030	wiring not included										
	0060	Input, 15 MBH	Q-5	7	2.286	Ea.	350	74.50		424.50	495	
	0100	30 MBH		6	2.667		370	87		457	540	
	0120	45 MBH		5	3.200		430	104		534	630	
	0140	50 MBH		4.50	3.556		470	116		586	690	
	0160	60 MBH		4	4		485	130		615	730	
	0180	75 MBH		3	5.333		540	174		714	855	
	0200	90 MBH		2.50	6.400		555	209		764	925	
	0220	105 MBH		2	8		720	261		981	1,200	
	0240	120 MBH	▼	2	8	▼	895	261		1,156	1,375	
	2000	Electric, single or three phase										
	2050	6 kW, 20,478 BTU	1 Elec	2.30	3.478	Ea.	385	123		508	610	
	2100	13.5 KW, 40,956 BTU		2.20	3.636		600	129		729	850	
	2150	24 KW, 81,912 BTU	▼	2	4	▼	1,375	142		1,517	1,700	
	3000	Oil fired, pump, controls, fusible valve, oil supply tank										
	3050	91,000 BTU	Q-5	2.50	6.400	Ea.	5,100	209		5,309	5,950	
	3080	105.000 BTU		2.25	7.111		5,350	232		5,582	6,225	
	3110	119,000 BTU	▼	2	8	▼	5,625	261		5,886	6,600	

	15770	**Floor-Heating & Snow-Melting Eq.**									
200	0010	**ELECTRIC HEATING**, equipments are not including wires & conduits										**200**
	0200	Snow melting for paved surface embedded mat heaters & controls	1 Elec	130	.062	S.F.	6.25	2.18		8.43	10.10	
	0400	Cable heating, radiant heat plaster, no controls, in South		130	.062		7.45	2.18		9.63	11.40	
	0600	In North		90	.089		7.45	3.15		10.60	12.85	
	0800	Cable on 1/2" board not incl. controls, tract housing		90	.089		6.25	3.15		9.40	11.55	
	1000	Custom housing		80	.100	▼	7.30	3.55		10.85	13.30	
	1100	Rule of thumb: Baseboard units, including control		4.40	1.818	kW	77	64.50		141.50	181	
	1200	Duct heaters, including controls		5.30	1.509	"	63	53.50		116.50	150	
	1300	Baseboard heaters, 2' long, 375 watt		8	1	Ea.	34.50	35.50		70	91	
	1400	3' long, 500 watt		8	1		42	35.50		77.50	99	
	1600	4' long, 750 watt		6.70	1.194		49	42.50		91.50	117	
	1800	5' long, 935 watt	▼	5.70	1.404	▼	58	50		108	139	

15 MECHANICAL

Important: See the Reference Section for critical supporting data - Reference Nos., Crews, & City Cost Indexes

15770	Floor-Heating & Snow-Melting Eq.	CREW	DAILY OUTPUT	LABOR-HOURS	UNIT	2002 BARE COSTS				TOTAL INCL O&P
						MAT.	LABOR	EQUIP.	TOTAL	
200 2000	6' long, 1125 watt	1 Elec	5	1.600	Ea.	66.50	56.50		123	158	200
2200	7' long, 1310 watt		4.40	1.818		77.50	64.50		142	182
2400	8' long, 1500 watt		4	2		82.50	71		153.50	197
2600	9' long, 1680 watt		3.60	2.222		137	79		216	268
2800	10' long, 1875 watt		3.30	2.424		137	86		223	278
2950	Wall heaters with fan, 120 to 277 volt									
2970	surface mounted, residential, 750 watt	1 Elec	7	1.143	Ea.	107	40.50		147.50	179
2980	1000 watt		7	1.143		109	40.50		149.50	181
2990	1250 watt		6	1.333		122	47.50		169.50	206
3000	1500 watt		5	1.600		146	56.50		202.50	246
3010	2000 watt		5	1.600		148	56.50		204.50	248
3040	2250 watt		4	2		153	71		224	274
3050	2500 watt		4	2		153	71		224	274
3070	4000 watt		3.50	2.286		179	81		260	320
3080	Commercial, 750 watt		7	1.143		173	40.50		213.50	251
3090	1000 watt		7	1.143		173	40.50		213.50	251
3100	1250 watt		6	1.333		173	47.50		220.50	261
3110	1500 watt		5	1.600		173	56.50		229.50	275
3120	2000 watt		5	1.600		173	56.50		229.50	275
3130	2500 watt		4.50	1.778		335	63		398	465
3140	3000 watt		4	2		335	71		406	475
3150	4000 watt		3.50	2.286		360	81		441	515
3160	Recessed, residential, 750 watt		6	1.333		85	47.50		132.50	164
3170	1000 watt		6	1.333		86	47.50		133.50	165
3180	1250 watt		5	1.600		95	56.50		151.50	189
3190	1500 watt		4	2		95	71		166	210
3210	2000 watt		4	2		106	71		177	223
3230	2500 watt		3.50	2.286		229	81		310	370
3240	3000 watt		3	2.667		229	94.50		323.50	390
3250	4000 watt		2.70	2.963		234	105		339	415
3260	Commercial, 750 watt		6	1.333		173	47.50		220.50	261
3270	1000 watt		6	1.333		173	47.50		220.50	261
3280	1250 watt		5	1.600		173	56.50		229.50	275
3290	1500 watt		4	2		173	71		244	296
3300	2000 watt		4	2		173	71		244	296
3310	2500 watt		3.50	2.286		173	81		254	310
3320	3000 watt		3	2.667		310	94.50		404.50	480
3330	4000 watt		2.70	2.963		310	105		415	495
3600	Thermostats, integral		16	.500		25.50	17.75		43.25	54.50
3800	Line voltage, 1 pole		8	1		25.50	35.50		61	81
3810	2 pole		8	1		25.50	35.50		61	81
3820	Low voltage, 1 pole		8	1		25.50	35.50		61	81
4000	Heat trace system, 400 degree									
4020	115V, 2.5 watts per L.F.	1 Elec	530	.015	L.F.	5.95	.53		6.48	7.35
4030	5 watts per L.F.		530	.015		5.90	.53		6.43	7.30
4050	10 watts per L.F.		530	.015		7.55	.53		8.08	9.15
4060	220V, 4 watts per L.F.		530	.015		5.95	.53		6.48	7.35
4080	480V, 8 watts per L.F.		530	.015		5.95	.53		6.48	7.35
4200	Heater raceway									
4220	5/8" W x 3/8" H	1 Elec	200	.040	L.F.	4.07	1.42		5.49	6.60
4240	5/8" W x 1/2" H	"	190	.042	"	5.65	1.49		7.14	8.50
4260	Heat transfer cement									
4280	1 gallon				Ea.	51.50			51.50	56.50
4300	5 gallon				"	211			211	232
4320	Snap band, clamp									
4340	3/4" pipe size	1 Elec	470	.017	Ea.		.60		.60	.90

15770	Floor-Heating & Snow-Melting Eq.	CREW	DAILY OUTPUT	LABOR-HOURS	UNIT	2002 BARE COSTS				TOTAL INCL O&P
						MAT.	LABOR	EQUIP.	TOTAL	
200 4360	1" pipe size	1 Elec	444	.018	Ea.		.64		.64	.95 **200**
4380	1-1/4" pipe size		400	.020			.71		.71	1.06
4400	1-1/2" pipe size		355	.023			.80		.80	1.19
4420	2" pipe size		320	.025			.89		.89	1.32
4440	3" pipe size		160	.050			1.77		1.77	2.65
4460	4" pipe size		100	.080			2.84		2.84	4.23
4480	Single pole thermostat NEMA 4, 30 amp		8	1		515	35.50		550.50	620
4500	NEMA 7, 30 amp		7	1.143		605	40.50		645.50	725
4520	Double pole, NEMA 4, 30 amp		7	1.143		515	40.50		555.50	625
4540	NEMA 7, 30 amp		6	1.333		605	47.50		652.50	735
4560	Thermostat/contactor combination, NEMA 4									
4580	30 amp 4 pole	1 Elec	3.60	2.222	Ea.	540	79		619	715
4600	50 amp 4 pole		3	2.667		545	94.50		639.50	740
4620	75 amp 3 pole		2.50	3.200		825	113		938	1,075
4640	75 amp 4 pole		2.30	3.478		1,175	123		1,298	1,475
4680	Control transformer, 50 VA		4	2		83.50	71		154.50	198
4700	75 VA		3.10	2.581		90.50	91.50		182	237
4720	Expediter fitting		11	.727		20.50	26		46.50	61
5000	Radiant heating ceiling panels, 2' x 4', 500 watt		16	.500		183	17.75		200.75	228
5050	750 watt		16	.500		202	17.75		219.75	249
5200	For recessed plaster frame, add		32	.250		36	8.85		44.85	53
5300	Infra-red quartz heaters, 120 volts, 1000 watts		6.70	1.194		128	42.50		170.50	203
5350	1500 watt		5	1.600		128	56.50		184.50	225
5400	240 volts, 1500 watt		5	1.600		128	56.50		184.50	225
5450	2000 watt		4	2		128	71		199	246
5500	3000 watt		3	2.667		145	94.50		239.50	300
5550	4000 watt		2.60	3.077		257	109		366	445
5570	Modulating control		.80	10		67.50	355		422.50	605
5600	Unit heaters, heavy duty, with fan & mounting bracket									
5650	Single phase, 208-240-277 volt, 3 kW	1 Elec	3.20	2.500	Ea.	350	88.50		438.50	515
5700	4 kW		2.80	2.857		365	101		466	555
5750	5 kW		2.40	3.333		360	118		478	570
5800	7 kW		1.90	4.211		560	149		709	840
5850	10 kW		1.30	6.154		610	218		828	995
5900	13 kW		1	8		980	284		1,264	1,500
5950	15 kW		.90	8.889		1,050	315		1,365	1,625
6000	480 volt, 3 kW		3.30	2.424		460	86		546	635
6020	4 kW		3	2.667		525	94.50		619.50	720
6040	5 kW		2.60	3.077		525	109		634	745
6060	7 kW		2	4		650	142		792	925
6080	10 kW		1.40	5.714		650	203		853	1,025
6100	13 kW		1.10	7.273		1,075	258		1,333	1,550
6120	15 kW		1	8		1,075	284		1,359	1,600
6140	20 kW		.90	8.889		1,375	315		1,690	2,000
6300	3 phase, 208-240 volt, 5 kW		2.40	3.333		355	118		473	565
6320	7 kW		1.90	4.211		545	149		694	825
6340	10 kW		1.30	6.154		590	218		808	975
6360	15 kW		.90	8.889		980	315		1,295	1,550
6380	20 kW		.70	11.429		1,400	405		1,805	2,150
6400	25 kW		.50	16		1,675	565		2,240	2,700
6500	480 volt, 5 kW		2.60	3.077		500	109		609	715
6520	7 kW		2	4		625	142		767	900
6540	10 kW		1.40	5.714		660	203		863	1,025
6560	13 kW		1.10	7.273		1,075	258		1,333	1,550
6580	15 kW		1	8		1,075	284		1,359	1,600
6600	20 kW		.90	8.889		1,400	315		1,715	2,000

15 MECHANICAL

Important: See the Reference Section for critical supporting data - Reference Nos., Crews, & City Cost Indexes

200	15770	Floor-Heating & Snow-Melting Eq.	CREW	DAILY OUTPUT	LABOR-HOURS	UNIT	2002 BARE COSTS MAT.	LABOR	EQUIP.	TOTAL	TOTAL INCL O&P	200
	6620	25 kW	1 Elec	.60	13.333	Ea.	1,675	475		2,150	2,550	
	6630	30 kW		.70	11.429		1,925	405		2,330	2,725	
	6640	40 kW		.60	13.333		2,450	475		2,925	3,400	
	6650	50 kW	↓	.50	16	↓	2,975	565		3,540	4,125	
	6800	Vertical discharge heaters, with fan										
	6820	Single phase, 208-240-277 volt, 10 kW	1 Elec	1.30	6.154	Ea.	1,150	218		1,368	1,575	
	6840	15 kW		.90	8.889		1,400	315		1,715	2,000	
	6900	3 phase, 208-240 volt, 10 kW		1.30	6.154		1,075	218		1,293	1,525	
	6920	15 kW		.90	8.889		1,225	315		1,540	1,825	
	6940	20 kW		.70	11.429		1,450	405		1,855	2,200	
	6960	25 kW		.50	16		1,925	565		2,490	2,975	
	6980	30 kW		.40	20		2,100	710		2,810	3,350	
	7000	40 kW		.36	22.222		2,800	790		3,590	4,250	
	7020	50 kW		.32	25		3,200	885		4,085	4,850	
	7100	480 volt, 10 kW		1.40	5.714		1,275	203		1,478	1,700	
	7120	15 kW		1	8		1,400	284		1,684	1,975	
	7140	20 kW		.90	8.889		1,700	315		2,015	2,350	
	7160	25 kW		.60	13.333		1,950	475		2,425	2,850	
	7180	30 kW		.50	16		2,350	565		2,915	3,425	
	7200	40 kW		.40	20		2,800	710		3,510	4,125	
	7220	50 kW	↓	.35	22.857	↓	3,100	810		3,910	4,600	
	7900	Cabinet convector heaters, 240 volt										
	7920	2' long, 1000 watt	1 Elec	5.30	1.509	Ea.	445	53.50		498.50	570	
	7940	1500 watt		5.30	1.509		470	53.50		523.50	600	
	7960	2000 watt		5.30	1.509		500	53.50		553.50	630	
	7980	3' long, 1500 watt		4.60	1.739		515	61.50		576.50	655	
	8000	2250 watt		4.60	1.739		520	61.50		581.50	660	
	8020	3000 watt		4.60	1.739		575	61.50		636.50	725	
	8040	4' long, 2000 watt		4	2		585	71		656	750	
	8060	3000 watt		4	2		615	71		686	780	
	8080	4000 watt	↓	4	2	↓	645	71		716	815	
	8100	Available also in 208 or 277 volt										
	8200	Cabinet unit heaters, 120 to 277 volt, 1 pole,										
	8220	wall mounted, 2 kW	1 Elec	4.60	1.739	Ea.	1,600	61.50		1,661.50	1,850	
	8230	3 kW		4.60	1.739		1,650	61.50		1,711.50	1,925	
	8240	4 kW		4.40	1.818		1,725	64.50		1,789.50	2,000	
	8250	5 kW		4.40	1.818		1,800	64.50		1,864.50	2,075	
	8260	6 kW		4.20	1.905		1,825	67.50		1,892.50	2,100	
	8270	8 kW		4	2		1,925	71		1,996	2,225	
	8280	10 kW		3.80	2.105		1,950	74.50		2,024.50	2,250	
	8290	12 kW		3.50	2.286		2,100	81		2,181	2,425	
	8300	13.5 kW		2.90	2.759		2,400	98		2,498	2,800	
	8310	16 kW		2.70	2.963		2,400	105		2,505	2,800	
	8320	20 kW		2.30	3.478		2,650	123		2,773	3,100	
	8330	24 kW		1.90	4.211		2,950	149		3,099	3,450	
	8350	Recessed, 2 kW		4.40	1.818		1,800	64.50		1,864.50	2,075	
	8370	3 kW		4.40	1.818		1,800	64.50		1,864.50	2,075	
	8380	4 kW		4.20	1.905		1,850	67.50		1,917.50	2,125	
	8390	5 kW		4.20	1.905		1,925	67.50		1,992.50	2,225	
	8400	6 kW		4	2		1,950	71		2,021	2,225	
	8410	8 kW		3.80	2.105		2,075	74.50		2,149.50	2,375	
	8420	10 kW		3.50	2.286		2,125	81		2,206	2,450	
	8430	12 kW		2.90	2.759		2,250	98		2,348	2,625	
	8440	13.5 kW		2.70	2.963		2,550	105		2,655	2,950	
	8450	16 kW		2.30	3.478		2,575	123		2,698	3,025	
	8460	20 kW	↓	1.90	4.211	↓	2,800	149		2,949	3,300	

MECHANICAL 15

	15770	Floor-Heating & Snow-Melting Eq.	CREW	DAILY OUTPUT	LABOR-HOURS	UNIT	2002 BARE COSTS				TOTAL INCL O&P	
							MAT.	LABOR	EQUIP.	TOTAL		
200	8470	24 kW	1 Elec	1.60	5	Ea.	3,075	177		3,252	3,650	**200**
	8490	Ceiling mounted, 2 kW		3.20	2.500		1,575	88.50		1,663.50	1,875	
	8510	3 kW		3.20	2.500		1,650	88.50		1,738.50	1,925	
	8520	4 kW		3	2.667		1,700	94.50		1,794.50	2,025	
	8530	5 kW		3	2.667		1,775	94.50		1,869.50	2,100	
	8540	6 kW		2.80	2.857		1,800	101		1,901	2,150	
	8550	8 kW		2.40	3.333		1,925	118		2,043	2,275	
	8560	10 kW		2.20	3.636		1,950	129		2,079	2,350	
	8570	12 kW		2	4		2,100	142		2,242	2,500	
	8580	13.5 kW		1.50	5.333		2,225	189		2,414	2,725	
	8590	16 kW		1.30	6.154		2,400	218		2,618	2,975	
	8600	20 kW		.90	8.889		2,650	315		2,965	3,400	
	8610	24 kW	▼	.60	13.333	▼	2,925	475		3,400	3,925	
	8630	208 to 480 V, 3 pole										
	8650	Wall mounted, 2 kW	1 Elec	4.60	1.739	Ea.	1,600	61.50		1,661.50	1,850	
	8670	3 kW		4.60	1.739		1,650	61.50		1,711.50	1,925	
	8680	4 kW		4.40	1.818		1,725	64.50		1,789.50	2,000	
	8690	5 kW		4.40	1.818		1,800	64.50		1,864.50	2,075	
	8700	6 kW		4.20	1.905		1,800	67.50		1,867.50	2,100	
	8710	8 kW		4	2		1,925	71		1,996	2,225	
	8720	10 kW		3.80	2.105		1,975	74.50		2,049.50	2,275	
	8730	12 kW		3.50	2.286		2,100	81		2,181	2,425	
	8740	13.5 kW		2.90	2.759		2,400	98		2,498	2,775	
	8750	16 kW		2.70	2.963		2,400	105		2,505	2,800	
	8760	20 kW		2.30	3.478		2,650	123		2,773	3,100	
	8770	24 kW		1.90	4.211		2,950	149		3,099	3,450	
	8790	Recessed, 2 kW		4.40	1.818		1,200	64.50		1,264.50	1,425	
	8810	3 kW		4.40	1.818		1,775	64.50		1,839.50	2,050	
	8820	4 kW		4.20	1.905		1,850	67.50		1,917.50	2,125	
	8830	5 kW		4.20	1.905		1,925	67.50		1,992.50	2,200	
	8840	6 kW		4	2		1,950	71		2,021	2,225	
	8850	8 kW		3.80	2.105		2,075	74.50		2,149.50	2,375	
	8860	10 kW		3.50	2.286		2,125	81		2,206	2,450	
	8870	12 kW		2.90	2.759		2,225	98		2,323	2,600	
	8880	13.5 kW		2.70	2.963		2,550	105		2,655	2,950	
	8890	16 kW		2.30	3.478		2,575	123		2,698	3,025	
	8900	20 kW		1.90	4.211		2,800	149		2,949	3,300	
	8920	24 kW		1.60	5		3,075	177		3,252	3,650	
	8940	Ceiling mount, 2 kW		3.20	2.500		1,575	88.50		1,663.50	1,875	
	8950	3 kW		3.20	2.500		1,650	88.50		1,738.50	1,925	
	8960	4 kW		3	2.667		1,700	94.50		1,794.50	2,025	
	8970	5 kW		3	2.667		1,775	94.50		1,869.50	2,100	
	8980	6 kW		2.80	2.857		1,800	101		1,901	2,150	
	8990	8 kW		2.40	3.333		1,925	118		2,043	2,275	
	9000	10 kW		2.20	3.636		1,950	129		2,079	2,350	
	9020	13.5 kW		1.50	5.333		2,100	189		2,289	2,575	
	9030	16 kW		1.30	6.154		2,400	218		2,618	2,975	
	9040	20 kW		.90	8.889		2,650	315		2,965	3,400	
	9060	24 kW	▼	.60	13.333	▼	2,950	475		3,425	3,925	
500	0010	**RADIANT FLOOR HEATING**										**500**
	0100	Tubing, PEX (cross-linked polyethylene)										
	0110	Oxygen barrier type for systems with ferrous materials										
	0120	1/2"	Q-5	800	.020	L.F.	.49	.65		1.14	1.52	
	0130	3/4"		535	.030		.62	.97		1.59	2.15	
	0140	1"	▼	400	.040	▼	1.25	1.30		2.55	3.35	

15770	Floor-Heating & Snow-Melting Eq.	CREW	DAILY OUTPUT	LABOR-HOURS	UNIT	2002 BARE COSTS				TOTAL INCL O&P	
						MAT.	LABOR	EQUIP.	TOTAL		
500	0200	Non barrier type for ferrous free systems										**500**
	0210	1/2"	Q-5	800	.020	L.F.	.27	.65		.92	1.28
	0220	3/4"		535	.030		.44	.97		1.41	1.95
	0230	1"	↓	400	.040	↓	.77	1.30		2.07	2.82
	1000	Manifolds									
	1110	Brass									
	1120	Valved									
	1130	1", 2 circuit	Q-5	13	1.231	Ea.	37	40		77	101
	1140	1", 3 circuit		11	1.455		49.50	47.50		97	126
	1150	1", 4 circuit		9	1.778		67.50	58		125.50	162
	1160	1-1/4", 3 circuit		10	1.600		56.50	52		108.50	141
	1170	1-1/4", 4 circuit	↓	8	2	↓	69	65		134	175
	1300	Valveless									
	1310	1", 2 circuit	Q-5	14	1.143	Ea.	18	37		55	76
	1320	1", 3 circuit		12	1.333		25	43.50		68.50	92.50
	1330	1", 4 circuit		10	1.600		31.50	52		83.50	113
	1340	1-1/4", 3 circuit		11	1.455		31	47.50		78.50	106
	1350	1-1/4", 4 circuit	↓	9	1.778	↓	35.50	58		93.50	127
	1400	Manifold connector kit									
	1410	Mounting brackets, couplings, end cap, air vent, valve, plug									
	1420	1"	Q-5	8	2	Ea.	63	65		128	168
	1430	1-1/4"	"	7	2.286	"	76.50	74.50		151	196
	1610	Copper, (cut to size)									
	1620	1" x 1/2" x 72" Lg, 2" sweat drops, 24 circuit	Q-5	3.33	4.805	Ea.	64	157		221	305
	1630	1-1/4" x 1/2" x 72" Lg, 2" sweat drops, 24 circuit		3.20	5		75	163		238	330
	1640	1-1/2" x 1/2" x 72" Lg, 2" sweat drops, 24 circuit		3	5.333		91.50	174		265.50	365
	1650	1-1/4" x 3/4" x 72" Lg, 2" sweat drops, 24 circuit		3.10	5.161		80.50	168		248.50	345
	1660	1-1/2" x 3/4" x 72" Lg, 2" sweat drops, 24 circuit	↓	2.90	5.517	↓	97	180		277	375
	1710	Copper, (modular)									
	1720	1" x 1/2", 2" sweat drops, 2 circuit	Q-5	17.50	.914	Ea.	6	30		36	51.50
	1730	1" x 1/2", 2" sweat drops, 3 circuit		14.50	1.103		9	36		45	64
	1740	1" x 1/2", 2" sweat drops, 4 circuit		12.30	1.301		9.75	42.50		52.25	75
	1750	1-1/4" x 1/2", 2" sweat drops, 2 circuit		17	.941		7	30.50		37.50	53.50
	1760	1-1/4" x 1/2", 2" sweat drops, 3 circuit		14	1.143		10.45	37		47.45	67.50
	1770	1-1/4" x 1/2", 2" sweat drops, 4 circuit	↓	11.80	1.356	↓	13.90	44		57.90	82
	3000	Valves									
	3110	Motorized zone valve operator, 24V	Q-5	30	.533	Ea.	37.50	17.40		54.90	67.50
	3120	Motorized zone valve with operator complete, 24V									
	3130	3/4"	Q-5	35	.457	Ea.	67	14.90		81.90	96
	3140	1"		32	.500		76	16.30		92.30	108
	3150	1-1/4"		29.60	.541		93	17.60		110.60	129
	3200	Flow control/isolation valve, 1/2"	↓	32	.500	↓	21.50	16.30		37.80	48
	3300	Thermostatic mixing valves									
	3310	1/2"	Q-5	26.60	.601	Ea.	70	19.60		89.60	106
	3320	3/4"		25	.640		85	21		106	125
	3330	1"	↓	21.40	.748	↓	91	24.50		115.50	137
	3400	3 Way mixing / diverting valve, manual, brass									
	3410	1/2"	Q-5	21.30	.751	Ea.	88.50	24.50		113	135
	3420	3/4"		20	.800		91.50	26		117.50	141
	3430	1"	↓	17.80	.899	↓	107	29.50		136.50	161
	3500	4 Way mixing / diverting valve, manual, brass									
	3510	1/2"	Q-5	16	1	Ea.	97	32.50		129.50	155
	3520	3/4"		15	1.067		107	35		142	170
	3530	1"		13.30	1.203		119	39		158	190
	3540	1-1/4"		11.40	1.404		131	45.50		176.50	213
	3550	1-1/2"	↓	10.60	1.509	↓	176	49		225	267

MECHANICAL 15

15770		Floor-Heating & Snow-Melting Eq.	CREW	DAILY OUTPUT	LABOR-HOURS	UNIT	2002 BARE COSTS				TOTAL INCL O&P	
							MAT.	LABOR	EQUIP.	TOTAL		
500	3560	2"	Q-5	10	1.600	Ea.	189	52		241	287	**500**
	3570	2-1/2"		8	2		520	65		585	675	
	3800	Motor control, 3 or 4 way, for valves up to 1-1/2"	▼	30	.533	▼	128	17.40		145.40	166	
	5000	Radiant floor heating, zone control panel										
	5110	6 Zone control										
	5120	For zone valves	Q-5	16	1	Ea.	151	32.50		183.50	215	
	5130	For circulators	"	16	1	"	161	32.50		193.50	226	
	7000	PEX tubing fittings, compression type										
	7110	PEX tubing to brass valve conn., 1/2"	Q-5	54	.296	Ea.	2.12	9.65		11.77	16.90	
	7200	PEX x male NPT										
	7210	1/2" x 1/2"	Q-5	54	.296	Ea.	4	9.65		13.65	18.95	
	7220	3/4" x 3/4"		46	.348		6.75	11.35		18.10	24.50	
	7230	1" x 1"	▼	40	.400	▼	15	13.05		28.05	36	
	7300	PEX coupling										
	7310	1/2" x 1/2"	Q-5	50	.320	Ea.	4.88	10.45		15.33	21	
	7320	3/4" x 3/4"		42	.381		14.05	12.40		26.45	34	
	7330	1" x 1"	▼	36	.444	▼	18	14.50		32.50	42	
	7400	PEX x copper sweat										
	7410	1/2" x 1/2"	Q-5	38	.421	Ea.	3.50	13.70		17.20	24.50	
	7420	3/4" x 3/4"		34	.471		7.50	15.35		22.85	31.50	
	7430	1" x 1"	▼	29	.552	▼	16.50	18		34.50	45	

15780		Energy Recovery Equipment										
100	0010	**HEAT RECOVERY PACKAGES**										**100**
	0100	Air to air										
	2000	Kitchen exhaust, commercial, heat pipe exchanger										
	2040	Combined supply/exhaust air volume										
	2080	2.5 to 6.0 MCFM	Q-10	2.80	8.571	MCFM	4,850	281		5,131	5,775	
	2120	6 to 16 MCFM		5	4.800		2,975	157		3,132	3,525	
	2160	16 to 22 MCFM	▼	6	4	▼	2,325	131		2,456	2,750	
	2170	For kitchen ventilation see div. 11420										
	4000	Enthalpy recovery wheel										
	4010	1000 max CFM	Q-9	1.20	13.333	Ea.	3,550	420		3,970	4,550	
	4020	2000 max CFM		1	16		3,900	505		4,405	5,050	
	4030	4000 max CFM		.80	20		4,850	630		5,480	6,325	
	4040	6000 max CFM	▼	.70	22.857		5,725	720		6,445	7,425	
	4050	8000 max CFM	Q-10	1	24		6,375	785		7,160	8,225	
	4060	10,000 max CFM		.90	26.667		7,875	875		8,750	10,000	
	4070	20,000 max CFM		.80	30		13,700	985		14,685	16,500	
	4080	25,000 max CFM		.70	34.286		16,700	1,125		17,825	20,100	
	4090	30,000 max CFM		.50	48		18,400	1,575		19,975	22,700	
	4100	40,000 max CFM		.45	53.333		25,100	1,750		26,850	30,300	
	4110	50,000 max CFM	▼	.40	60	▼	29,900	1,975		31,875	35,900	
	8000	Residential ventilation heat exchanger, not incl. duct										
	8010	70 CFM exchanger	1 Shee	8	1	Ea.	625	35		660	745	
	8020	110 CFM exchanger		7	1.143		945	40		985	1,100	
	8024	210 CFM exchanger	▼	6.60	1.212		945	42.50		987.50	1,125	
	8028	450 CFM exchanger	Q-9	12	1.333		1,925	42		1,967	2,200	
	8032	900 CFM exchanger	"	10	1.600	▼	2,400	50.50		2,450.50	2,725	
	8060	For larger applications										
	8062	Use multiple units										

Important: See the Reference Section for critical supporting data - Reference Nos., Crews, & City Cost Indexes

15810	Ducts		CREW	DAILY OUTPUT	LABOR-HOURS	UNIT	2002 BARE COSTS				TOTAL INCL O&P	
							MAT.	LABOR	EQUIP.	TOTAL		
600	0010	**DUCTWORK** R15810 -050										**600**
	0020	Fabricated rectangular, includes fittings, joints, supports,										
	0030	allowance for flexible connections, no insulation R15810 -070										
	0031	NOTE: Fabrication and installation are combined										
	0040	as LABOR cost. Approx. 25% fittings assumed. R15810 -100										
	0042	Fabrication/Inst. is to commercial quality standards										
	0043	(SMACNA or equiv.) for structure, sealing, leak testing, etc.										
	0050	Add to labor for elevated installation										
	0051	of fabricated ductwork										
	0052	10' to 15' high							6%			
	0053	15' to 20' high							12%			
	0054	20' to 25' high							15%			
	0055	25' to 30' high							21%			
	0056	30' to 35' high							24%			
	0057	35' to 40' high							30%			
	0058	Over 40' high							33%			
	0070	For duct insulation and lining see 15082-200-3000										
	0100	Aluminum, alloy 3003-H14, under 100 lb.		Q-10	75	.320	Lb.	2.09	10.50		12.59	18.45
	0110	100 to 500 lb.			80	.300		1.79	9.85		11.64	17.10
	0120	500 to 1,000 lb.			95	.253		1.59	8.30		9.89	14.50
	0140	1,000 to 2,000 lb.			120	.200		1.49	6.55		8.04	11.75
	0150	2,000 to 5,000 lb.			130	.185		1.45	6.05		7.50	10.95
	0160	Over 5,000 lb.			145	.166		1.41	5.40		6.81	9.90
	0500	Galvanized steel, under 200 lb.			235	.102		.76	3.35		4.11	6
	0520	200 to 500 lb.			245	.098		.46	3.21		3.67	5.45
	0540	500 to 1,000 lb.			255	.094		.44	3.08		3.52	5.25
	0560	1,000 to 2,000 lb.			265	.091		.38	2.97		3.35	5
	0570	2,000 to 5,000 lb.			275	.087		.34	2.86		3.20	4.78
	0580	Over 5,000 lb.		▼	285	.084	▼	.31	2.76		3.07	4.59
	0600	For large quantities special prices available from supplier										
	1000	Stainless steel, type 304, under 100 lb.		Q-10	165	.145	Lb.	1.28	4.77		6.05	8.75
	1020	100 to 500 lb.			175	.137		1.18	4.49		5.67	8.25
	1030	500 to 1,000 lb.			190	.126		1.12	4.14		5.26	7.65
	1040	1,000 to 2,000 lb.			200	.120		1.06	3.93		4.99	7.20
	1050	2,000 to 5,000 lb.			225	.107		.98	3.50		4.48	6.50
	1060	Over 5,000 lb.		▼	235	.102	▼	.96	3.35		4.31	6.20
	1080	Note: Minimum order cost exceeds per lb. cost for min. wt.										
	1100	For medium pressure ductwork, add							15%			
	1200	For high pressure ductwork, add							40%			
	1276	Duct tape, 2" x 60 yd roll					Ea.	5.25			5.25	5.80
	1280	Add to labor for elevated installation										
	1282	of prefabricated (purchased) ductwork										
	1283	10' to 15' high							10%			
	1284	15' to 20' high							20%			
	1285	20' to 25' high							25%			
	1286	25' to 30' high							35%			
	1287	30' to 35' high							40%			
	1288	35' to 40' high							50%			
	1289	Over 40' high							55%			
	1300	Flexible, coated fiberglass fabric on corr. resist. metal helix										
	1400	pressure to 12" (WG) UL-181										
	1500	Non-insulated, 3" diameter		Q-9	400	.040	L.F.	.98	1.26		2.24	3.03
	1520	4" diameter			360	.044		.98	1.40		2.38	3.25
	1540	5" diameter			320	.050		1.14	1.58		2.72	3.69
	1560	6" diameter			280	.057		1.30	1.81		3.11	4.21
	1580	7" diameter		▼	240	.067	▼	1.69	2.11		3.80	5.10

MECHANICAL 15

15810 | Ducts

		CREW	DAILY OUTPUT	LABOR-HOURS	UNIT	2002 BARE COSTS				TOTAL INCL O&P
						MAT.	LABOR	EQUIP.	TOTAL	
600	1600	8" diameter	Q-9	200	.080	L.F.	1.76	2.53		4.29	5.85	600
	1620	9" diameter		180	.089		1.99	2.81		4.80	6.50
	1640	10" diameter		160	.100		2.25	3.16		5.41	7.35
	1660	12" diameter		120	.133		2.67	4.21		6.88	9.45
	1680	14" diameter		80	.200		3.12	6.30		9.42	13.20
	1700	16" diameter		60	.267		4.07	8.45		12.52	17.50
	1800	For adjustable clamps, add				Ea.	.74			.74	.81
	1900	Insulated, 1" thick, PE jacket, 3" diameter	Q-9	380	.042	L.F.	1.07	1.33		2.40	3.23
	1910	4" diameter		340	.047		1.16	1.49		2.65	3.57
	1920	5" diameter		300	.053		1.27	1.69		2.96	4
	1940	6" diameter		260	.062		1.47	1.94		3.41	4.62
	1960	7" diameter		220	.073		1.66	2.30		3.96	5.35
	1980	8" diameter		180	.089		1.84	2.81		4.65	6.35
	2000	9" diameter		160	.100		2.11	3.16		5.27	7.20
	2020	10" diameter		140	.114		2.26	3.61		5.87	8.05
	2040	12" diameter		100	.160		2.78	5.05		7.83	10.85
	2060	14" diameter		80	.200		3.16	6.30		9.46	13.25
	2080	16" diameter		60	.267		3.65	8.45		12.10	17
	2100	18" diameter		45	.356		3.96	11.25		15.21	21.50
	2120	20" diameter	Q-10	65	.369		7.85	12.10		19.95	27.50
	2500	Insulated, heavy duty, coated fiberglass fabric									
	2520	4" diameter	Q-9	340	.047	L.F.	2.47	1.49		3.96	5
	2540	5" diameter		300	.053		2.96	1.69		4.65	5.85
	2560	6" diameter		260	.062		3.32	1.94		5.26	6.65
	2580	7" diameter		220	.073		3.90	2.30		6.20	7.85
	2600	8" diameter		180	.089		4.20	2.81		7.01	8.95
	2620	9" diameter		160	.100		4.85	3.16		8.01	10.20
	2640	10" diameter		140	.114		5.25	3.61		8.86	11.35
	2660	12" diameter		100	.160		6.25	5.05		11.30	14.65
	2680	14" diameter		80	.200		7.45	6.30		13.75	17.95
	2700	16" diameter		60	.267		8.30	8.45		16.75	22
	2720	18" diameter		45	.356		10.35	11.25		21.60	28.50
	2800	Flexible, alum., pressure to 12" (WG) UL-181									
	2820	Non-insulated									
	2830	3" diameter	Q-9	400	.040	L.F.	.97	1.26		2.23	3.02
	2831	4" diameter		360	.044		1.18	1.40		2.58	3.47
	2832	5" diameter		320	.050		1.49	1.58		3.07	4.08
	2833	6" diameter		280	.057		1.82	1.81		3.63	4.78
	2834	7" diameter		240	.067		2.61	2.11		4.72	6.10
	2835	8" diameter		200	.080		2.67	2.53		5.20	6.85
	2836	9" diameter		180	.089		3.12	2.81		5.93	7.75
	2837	10" diameter		160	.100		3.28	3.16		6.44	8.50
	2838	12" diameter		120	.133		4.03	4.21		8.24	10.95
	2839	14" diameter		80	.200		4.84	6.30		11.14	15.05
	2840	15" diameter		70	.229		5.30	7.20		12.50	17
	2841	16" diameter		60	.267		6.50	8.45		14.95	20
	2842	18" diameter		45	.356		7.05	11.25		18.30	25
	2843	20" diameter	Q-10	65	.369		7.55	12.10		19.65	27
	2880	Insulated, 1" thick with 3/4 lb., PE jacket									
	2890	3" diameter	Q-9	380	.042	L.F.	1.78	1.33		3.11	4.01
	2891	4" diameter		340	.047		2.01	1.49		3.50	4.50
	2892	5" diameter		300	.053		2.54	1.69		4.23	5.40
	2893	6" diameter		260	.062		2.89	1.94		4.83	6.20
	2894	7" diameter		220	.073		3.28	2.30		5.58	7.15
	2895	8" diameter		180	.089		3.86	2.81		6.67	8.60
	2896	9" diameter		160	.100		4.36	3.16		7.52	9.65

R15810-050

R15810-070

R15810-100

Important: See the Reference Section for critical supporting data - Reference Nos., Crews, & City Cost Indexes

15800 | Air Distribution

	15810	Ducts		CREW	DAILY OUTPUT	LABOR-HOURS	UNIT	2002 BARE COSTS				TOTAL INCL O&P
								MAT.	LABOR	EQUIP.	TOTAL	
2897	10" diameter	R15810 -050	Q-9	140	.114	L.F.	4.93	3.61		8.54	10.95
2898	12" diameter			100	.160		5.50	5.05		10.55	13.85
2899	14" diameter	R15810 -070		80	.200		6.40	6.30		12.70	16.80
2900	15" diameter			70	.229		7.10	7.20		14.30	18.95
2901	16" diameter	R15810 -100		60	.267		8.40	8.45		16.85	22.50
2902	18" diameter			45	.356		9.45	11.25		20.70	27.50
3000	Transitions, 3" deep, square to round										
3010	6" x 6" to round		Q-9	44	.364	Ea.	7.15	11.50		18.65	25.50
3014	6" x 12" to round			42	.381		13	12.05		25.05	33
3018	8" x 8" to round			40	.400		7.80	12.65		20.45	28
3022	9" x 9" to round			38	.421		9.10	13.30		22.40	30.50
3026	9" x 12" to round			37	.432		13.65	13.65		27.30	36
3030	9" x 15" to round			36	.444		16.90	14.05		30.95	40
3034	10" x 10" to round			34	.471		10.40	14.85		25.25	34.50
3038	10" x 22" to round			32	.500		19.50	15.80		35.30	46
3042	12" x 12" to round			30	.533		11.05	16.85		27.90	38
3046	12" x 18" to round			28	.571		21	18.05		39.05	51
3050	12" x 24" to round			26	.615		23.50	19.45		42.95	55.50
3054	15" x 15" to round			25	.640		14.30	20		34.30	47
3058	15" x 18" to round			24	.667		21	21		42	55.50
3062	18" x 18" to round			22	.727		16.90	23		39.90	54
3066	21" x 21" to round			21	.762		30	24		54	70
3070	22" x 22" to round			20	.800		30	25.50		55.50	72
3074	24" x 24" to round			18	.889		30	28		58	76.50
3490	Rigid fiberglass duct board, foil reinf. kraft facing										
3500	Rectangular, 1" thick, alum. faced, (FRK), std. weight		Q-10	350	.069	SF Surf	1.02	2.25		3.27	4.58
3550	Rigid fiberglass reinforced plastic, FM approved										
3552	for acid fume and smoke exhaust system, nonflammable										
3554	Straight, 4" diameter		Q-9	200	.080	L.F.	14.85	2.53		17.38	20
3555	6" diameter			146	.110		19.45	3.46		22.91	27
3556	8" diameter			106	.151		23.50	4.77		28.27	33.50
3557	10" diameter			85	.188		28	5.95		33.95	40
3558	12" diameter			68	.235		33	7.45		40.45	47.50
3559	14" diameter			50	.320		48.50	10.10		58.60	69
3560	16" diameter			40	.400		54	12.65		66.65	79
3561	18" diameter			31	.516		60.50	16.30		76.80	91.50
3562	20" diameter		Q-10	44	.545		67	17.85		84.85	101
3563	22" diameter			40	.600		72.50	19.65		92.15	110
3564	24" diameter			37	.649		78.50	21.50		100	119
3565	26" diameter			34	.706		84	23		107	128
3566	28" diameter			28	.857		90	28		118	143
3567	30" diameter			24	1		96	33		129	157
3568	32" diameter			22.60	1.062		102	35		137	166
3569	34" diameter			21.60	1.111		108	36.50		144.50	175
3570	36" diameter			20.60	1.165		114	38		152	184
3571	38" diameter			19.80	1.212		176	39.50		215.50	254
3572	42" diameter			18.30	1.311		196	43		239	282
3573	46" diameter			17	1.412		216	46.50		262.50	310
3574	50" diameter			15.90	1.509		236	49.50		285.50	335
3575	54" diameter			14.90	1.611		260	53		313	370
3576	60" diameter			13.60	1.765		296	58		354	415
3577	64" diameter			12.80	1.875		435	61.50		496.50	575
3578	68" diameter			12.20	1.967		465	64.50		529.50	610
3579	72" diameter			11.90	2.017		495	66		561	645
3584	Note: joints are cemented with										
3586	fiberglass resin, included in material cost.										

MECHANICAL 15

	15810	Ducts		CREW	DAILY OUTPUT	LABOR- HOURS	UNIT	2002 BARE COSTS				TOTAL INCL O&P	
								MAT.	LABOR	EQUIP.	TOTAL		
600	3590	Elbow, 90°, 4" diameter	R15810 -050	Q-9	20.30	.788	Ea.	76.50	25		101.50	123	**600**
	3591	6" diameter			13.80	1.159		107	36.50		143.50	175	
	3592	8" diameter	R15810 -070		10	1.600		118	50.50		168.50	208	
	3593	10" diameter			7.80	2.051		156	65		221	271	
	3594	12" diameter	R15810 -100		6.50	2.462		196	78		274	335	
	3595	14" diameter			5.50	2.909		241	92		333	405	
	3596	16" diameter			4.80	3.333		287	105		392	475	
	3597	18" diameter			4.40	3.636		410	115		525	625	
	3598	20" diameter		Q-10	5.90	4.068		435	133		568	685	
	3599	22" diameter			5.40	4.444		535	146		681	815	
	3600	24" diameter			4.90	4.898		610	161		771	920	
	3601	26" diameter			4.60	5.217		620	171		791	945	
	3602	28" diameter			4.20	5.714		780	187		967	1,150	
	3603	30" diameter			3.90	6.154		870	202		1,072	1,275	
	3604	32" diameter			3.70	6.486		965	213		1,178	1,375	
	3605	34" diameter			3.45	6.957		1,125	228		1,353	1,600	
	3606	36" diameter			3.20	7.500		1,300	246		1,546	1,800	
	3607	38" diameter			3.07	7.818		1,850	256		2,106	2,425	
	3608	42" diameter			2.70	8.889		2,200	291		2,491	2,875	
	3609	46" diameter			2.54	9.449		2,600	310		2,910	3,325	
	3610	50" diameter			2.33	10.300		3,000	340		3,340	3,825	
	3611	54" diameter			2.14	11.215		3,450	370		3,820	4,375	
	3612	60" diameter			1.94	12.371		4,575	405		4,980	5,650	
	3613	64" diameter			1.82	13.187		6,675	430		7,105	8,000	
	3614	68" diameter			1.71	14.035		7,400	460		7,860	8,850	
	3615	72" diameter			1.62	14.815		8,175	485		8,660	9,750	
	3626	Elbow, 45°, 4" diameter		Q-9	22.30	.717		73.50	22.50		96	116	
	3627	6" diameter			15.20	1.053		102	33.50		135.50	164	
	3628	8" diameter			11	1.455		109	46		155	190	
	3629	10" diameter			8.60	1.860		115	59		174	217	
	3630	12" diameter			7.20	2.222		126	70		196	246	
	3631	14" diameter			6.10	2.623		154	83		237	297	
	3632	16" diameter			5.28	3.030		183	96		279	350	
	3633	18" diameter			4.84	3.306		215	104		319	395	
	3634	20" diameter		Q-10	6.50	3.692		248	121		369	460	
	3635	22" diameter			5.94	4.040		340	132		472	580	
	3636	24" diameter			5.40	4.444		385	146		531	650	
	3637	26" diameter			5.06	4.743		440	155		595	720	
	3638	28" diameter			4.62	5.195		475	170		645	785	
	3639	30" diameter			4.30	5.581		550	183		733	885	
	3640	32" diameter			4.07	5.897		605	193		798	970	
	3641	34" diameter			3.80	6.316		715	207		922	1,100	
	3642	36" diameter			3.52	6.818		825	223		1,048	1,250	
	3643	38" diameter			3.38	7.101		1,175	233		1,408	1,625	
	3644	42" diameter			2.97	8.081		1,400	265		1,665	1,925	
	3645	46" diameter			2.80	8.571		1,625	281		1,906	2,225	
	3646	50" diameter			2.56	9.375		1,900	305		2,205	2,550	
	3647	54" diameter			2.35	10.213		2,175	335		2,510	2,900	
	3648	60" diameter			2.13	11.268		2,875	370		3,245	3,750	
	3649	64" diameter			2	12		4,175	395		4,570	5,200	
	3650	68" diameter			1.88	12.766		4,625	420		5,045	5,750	
	3651	72" diameter			1.78	13.483		5,125	440		5,565	6,300	
	3660	Tee, 90°, 4" diameter		Q-9	14.06	1.138		42.50	36		78.50	102	
	3661	6" diameter			9.52	1.681		64.50	53		117.50	153	
	3662	8" diameter			6.98	2.292		88	72.50		160.50	209	
	3663	10" diameter			5.45	2.936		115	93		208	269	

Important: See the Reference Section for critical supporting data - Reference Nos., Crews, & City Cost Indexes

	15810	Ducts		CREW	DAILY OUTPUT	LABOR-HOURS	UNIT	2002 BARE COSTS				TOTAL INCL O&P	
								MAT.	LABOR	EQUIP.	TOTAL		
600	3664	12" diameter	R15810 -050	Q-9	4.60	3.478	Ea.	144	110		254	325	**600**
	3665	14" diameter			3.86	4.145		209	131		340	430	
	3666	16" diameter	R15810 -070		3.40	4.706		249	149		398	505	
	3667	18" diameter			3.05	5.246		295	166		461	580	
	3668	20" diameter	R15810 -100	Q-10	4.12	5.825		345	191		536	675	
	3669	22" diameter			3.78	6.349		395	208		603	755	
	3670	24" diameter			3.46	6.936		445	227		672	840	
	3671	26" diameter			3.20	7.500		505	246		751	935	
	3672	28" diameter			2.96	8.108		565	266		831	1,025	
	3673	30" diameter			2.75	8.727		630	286		916	1,125	
	3674	32" diameter			2.54	9.449		695	310		1,005	1,250	
	3675	34" diameter			2.40	10		765	330		1,095	1,350	
	3676	36" diameter			2.27	10.573		840	345		1,185	1,450	
	3677	38" diameter			2.15	11.163		1,250	365		1,615	1,950	
	3678	42" diameter			1.91	12.565		1,500	410		1,910	2,275	
	3679	46" diameter			1.78	13.483		1,775	440		2,215	2,625	
	3680	50" diameter			1.63	14.724		2,075	485		2,560	3,025	
	3681	54" diameter			1.50	16		2,400	525		2,925	3,450	
	3682	60" diameter			1.35	17.778		2,975	585		3,560	4,175	
	3683	64" diameter			1.27	18.898		4,500	620		5,120	5,900	
	3684	68" diameter			1.20	20		5,050	655		5,705	6,550	
	3685	72" diameter			1.13	21.239		5,625	695		6,320	7,275	
	3690	For Y @ 45°, add						22%					
	3700	Blast gate, 4" diameter		Q-9	22.30	.717		204	22.50		226.50	260	
	3701	6" diameter			15.20	1.053		231	33.50		264.50	305	
	3702	8" diameter			11	1.455		310	46		356	415	
	3703	10" diameter			8.60	1.860		355	59		414	480	
	3704	12" diameter			7.20	2.222		435	70		505	585	
	3705	14" diameter			6.10	2.623		480	83		563	660	
	3706	16" diameter			5.28	3.030		535	96		631	735	
	3707	18" diameter			4.84	3.306		585	104		689	800	
	3708	20" diameter		Q-10	6.50	3.692		630	121		751	880	
	3709	22" diameter			5.94	4.040		685	132		817	955	
	3710	24" diameter			5.40	4.444		730	146		876	1,025	
	3711	26" diameter			5.06	4.743		785	155		940	1,100	
	3712	28" diameter			4.62	5.195		845	170		1,015	1,175	
	3713	30" diameter			4.30	5.581		900	183		1,083	1,275	
	3714	32" diameter			4.07	5.897		980	193		1,173	1,375	
	3715	34" diameter			3.80	6.316		1,050	207		1,257	1,500	
	3716	36" diameter			3.52	6.818		1,150	223		1,373	1,625	
	3717	38" diameter			3.38	7.101		1,550	233		1,783	2,050	
	3718	42" diameter			2.97	8.081		1,775	265		2,040	2,375	
	3719	46" diameter			2.80	8.571		2,050	281		2,331	2,675	
	3720	50" diameter			2.56	9.375		2,325	305		2,630	3,025	
	3721	54" diameter			2.35	10.213		2,625	335		2,960	3,400	
	3722	60" diameter			2.13	11.268		3,100	370		3,470	3,975	
	3723	64" diameter			2	12		3,425	395		3,820	4,350	
	3724	68" diameter			1.88	12.766		3,775	420		4,195	4,800	
	3725	72" diameter			1.78	13.483		4,125	440		4,565	5,225	
	3740	Reducers, 4" diameter		Q-9	24.50	.653		51	20.50		71.50	88	
	3741	6" diameter			16.70	.958		76.50	30.50		107	131	
	3742	8" diameter			12	1.333		103	42		145	178	
	3743	10" diameter			9.50	1.684		127	53		180	222	
	3744	12" diameter			7.90	2.025		153	64		217	267	
	3745	14" diameter			6.70	2.388		178	75.50		253.50	310	
	3746	16" diameter			5.80	2.759		203	87		290	360	

15810	Ducts		DAILY OUTPUT	LABOR-HOURS	UNIT	MAT.	LABOR	EQUIP.	TOTAL	TOTAL INCL O&P	
		CREW				2002 BARE COSTS					
600	3747	18" diameter	R15810-050	Q-9	5.30	3.019	Ea.	229	95.50		324.50	400	600
3748	20" diameter		Q-10	7.20	3.333		254	109		363	450
3749	22" diameter	R15810-070		6.50	3.692		410	121		531	635
3750	24" diameter			5.90	4.068		445	133		578	695
3751	26" diameter	R15810-100		5.60	4.286		485	140		625	750
3752	28" diameter			5.10	4.706		520	154		674	815
3753	30" diameter			4.70	5.106		560	167		727	875
3754	32" diameter			4.50	5.333		595	175		770	925
3755	34" diameter			4.20	5.714		635	187		822	985
3756	36" diameter			3.90	6.154		670	202		872	1,050
3757	38" diameter			3.70	6.486		1,025	213		1,238	1,450
3758	42" diameter			3.30	7.273		1,125	238		1,363	1,625
3759	46" diameter			3.10	7.742		1,225	254		1,479	1,750
3760	50" diameter			2.80	8.571		1,350	281		1,631	1,900
3761	54" diameter			2.60	9.231		1,450	300		1,750	2,075
3762	60" diameter			2.34	10.256		1,600	335		1,935	2,300
3763	64" diameter			2.20	10.909		2,525	355		2,880	3,325
3764	68" diameter			2.06	11.650		2,675	380		3,055	3,550
3765	72" diameter		▼	1.96	12.245		2,825	400		3,225	3,750
3780	Flange, per each, 4" diameter		Q-9	40.60	.394		124	12.45		136.45	155
3781	6" diameter			27.70	.578		144	18.25		162.25	187
3782	8" diameter			20.50	.780		180	24.50		204.50	236
3783	10" diameter			16.30	.982		216	31		247	285
3784	12" diameter			13.90	1.151		248	36.50		284.50	330
3785	14" diameter			11.30	1.416		287	44.50		331.50	385
3786	16" diameter			9.70	1.649		325	52		377	435
3787	18" diameter		▼	8.30	1.928		360	61		421	490
3788	20" diameter		Q-10	11.30	2.124		395	69.50		464.50	540
3789	22" diameter			10.40	2.308		430	75.50		505.50	590
3790	24" diameter			9.50	2.526		465	83		548	640
3791	26" diameter			8.80	2.727		500	89.50		589.50	690
3792	28" diameter			8.20	2.927		535	96		631	740
3793	30" diameter			7.60	3.158		570	103		673	790
3794	32" diameter			7.20	3.333		610	109		719	840
3795	34" diameter			6.80	3.529		645	116		761	890
3796	36" diameter			6.40	3.750		680	123		803	935
3797	38" diameter			6.10	3.934		715	129		844	985
3798	42" diameter			5.50	4.364		785	143		928	1,075
3799	46" diameter			5	4.800		860	157		1,017	1,200
3800	50" diameter			4.60	5.217		930	171		1,101	1,300
3801	54" diameter			4.20	5.714		1,000	187		1,187	1,400
3802	60" diameter			3.80	6.316		1,100	207		1,307	1,550
3803	64" diameter			3.60	6.667		1,175	218		1,393	1,625
3804	68" diameter			3.40	7.059		1,250	231		1,481	1,725
3805	72" diameter		▼	3.20	7.500	▼	1,325	246		1,571	1,825
4000	Rigid plastic, corrosive fume resistant PVC										
4020	Straight, 6" diameter		Q-9	220	.073	L.F.	6.50	2.30		8.80	10.70
4030	7" diameter			160	.100		9.60	3.16		12.76	15.40
4040	8" diameter			160	.100		9.60	3.16		12.76	15.40
4050	9" diameter			140	.114		10.85	3.61		14.46	17.50
4060	10" diameter			120	.133		12.10	4.21		16.31	19.85
4070	12" diameter			100	.160		14.15	5.05		19.20	23.50
4080	14" diameter			70	.229		18.35	7.20		25.55	31
4090	16" diameter			62	.258		21	8.15		29.15	36
4100	18" diameter		▼	58	.276		33	8.70		41.70	49.50
4110	20" diameter		Q-10	75	.320	▼	42.50	10.50		53	62.50

Important: See the Reference Section for critical supporting data - Reference Nos., Crews, & City Cost Indexes

			CREW	DAILY OUTPUT	LABOR-HOURS	UNIT	2002 BARE COSTS MAT.	LABOR	EQUIP.	TOTAL	TOTAL INCL O&P	
600	**15810**	**Ducts**										**600**
4120	22" diameter	R15810 -050	Q-10	60	.400	L.F.	48.50	13.10		61.60	73.50	
4130	24" diameter			55	.436		54.50	14.30		68.80	82	
4140	26" diameter	R15810 -070		42	.571		62	18.75		80.75	97	
4150	28" diameter			35	.686		66.50	22.50		89	108	
4160	30" diameter	R15810 -100		28	.857		72	28		100	123	
4170	36" diameter			23	1.043		86	34		120	147	
4180	42" diameter			20	1.200		100	39.50		139.50	171	
4200	48" diameter			18	1.333		114	43.50		157.50	194	
4250	Coupling, 6" diameter		Q-9	88	.182	Ea.	6.75	5.75		12.50	16.30	
4260	7" diameter			82	.195		7.25	6.15		13.40	17.45	
4270	8" diameter			78	.205		7.55	6.50		14.05	18.30	
4280	9" diameter			74	.216		8.25	6.85		15.10	19.60	
4290	10" diameter			70	.229		8.25	7.20		15.45	20	
4300	12" diameter			55	.291		8.80	9.20		18	24	
4310	14" diameter			44	.364		10.75	11.50		22.25	29.50	
4320	16" diameter			40	.400		12.70	12.65		25.35	33.50	
4330	18" diameter			38	.421		14.75	13.30		28.05	36.50	
4340	20" diameter		Q-10	55	.436		25.50	14.30		39.80	50	
4350	22" diameter			50	.480		29.50	15.75		45.25	57	
4360	24" diameter			45	.533		33	17.50		50.50	63.50	
4370	26" diameter			39	.615		35	20		55	69.50	
4380	28" diameter			33	.727		36.50	24		60.50	77	
4390	30" diameter			27	.889		40	29		69	89	
4400	36" diameter			25	.960		63.50	31.50		95	118	
4410	42" diameter			23	1.043		71	34		105	131	
4420	48" diameter			21	1.143		83	37.50		120.50	149	
4470	Elbow, 90°, 6" diameter		Q-9	44	.364		63.50	11.50		75	87	
4480	7" diameter			36	.444		68	14.05		82.05	96	
4490	8" diameter			28	.571		68	18.05		86.05	103	
4500	9" diameter			22	.727		93.50	23		116.50	139	
4510	10" diameter			18	.889		110	28		138	165	
4520	12" diameter			15	1.067		150	33.50		183.50	217	
4530	14" diameter			11	1.455		205	46		251	297	
4540	16" diameter			10	1.600		247	50.50		297.50	350	
4550	18" diameter		Q-10	15	1.600		268	52.50		320.50	375	
4560	20" diameter			14	1.714		284	56		340	400	
4570	22" diameter			13	1.846		360	60.50		420.50	495	
4580	24" diameter			12	2		410	65.50		475.50	550	
4590	26" diameter			11	2.182		465	71.50		536.50	625	
4600	28" diameter			10	2.400		555	78.50		633.50	730	
4610	30" diameter			9	2.667		650	87.50		737.50	850	
4620	36" diameter			7	3.429		895	112		1,007	1,150	
4630	42" diameter			5	4.800		1,225	157		1,382	1,600	
4640	48" diameter			4	6		1,625	197		1,822	2,075	
4750	Elbow 45°, use 90° and deduct						45%					
5400	Spiral preformed, steel, galv., straight lengths, Max 10" spwg.											
5410	4" diameter, 26 ga.		Q-9	360	.044	L.F.	.85	1.40		2.25	3.11	
5416	5" diameter, 26 ga.			320	.050		1.08	1.58		2.66	3.63	
5420	6" diameter, 26 ga.			280	.057		1.23	1.81		3.04	4.13	
5425	7" diameter, 26 ga.			240	.067		1.46	2.11		3.57	4.86	
5430	8" diameter, 26 ga.			200	.080		1.62	2.53		4.15	5.70	
5440	10" diameter, 26 ga.			160	.100		2	3.16		5.16	7.05	
5450	12" diameter, 26 ga.			120	.133		2.46	4.21		6.67	9.20	
5460	14" diameter, 26 ga.			80	.200		2.85	6.30		9.15	12.90	
5480	16" diameter, 24 ga.			60	.267		4.24	8.45		12.69	17.65	
5490	18" diameter, 24 ga.			50	.320		4.70	10.10		14.80	21	

15810 | Ducts

			CREW	DAILY OUTPUT	LABOR-HOURS	UNIT	2002 BARE COSTS				TOTAL INCL O&P
							MAT.	LABOR	EQUIP.	TOTAL	
600 5500	20" diameter, 24 ga.	R15810 -050	Q-10	65	.369	L.F.	5.25	12.10		17.35	24.50	600
5510	22" diameter, 24 ga.			60	.400		5.80	13.10		18.90	26.50
5520	24" diameter, 24 ga.	R15810 -070		55	.436		6.30	14.30		20.60	29
5540	30" diameter, 22 ga.			45	.533		9.55	17.50		27.05	37.50
5600	36" diameter, 22 ga.	R15810 -100		40	.600		11.45	19.65		31.10	43
5800	Connector, 4" diameter		Q-9	100	.160	Ea.	1.58	5.05		6.63	9.55
5810	5" diameter			94	.170		1.69	5.40		7.09	10.15
5820	6" diameter			88	.182		1.82	5.75		7.57	10.85
5840	8" diameter			78	.205		2.35	6.50		8.85	12.60
5860	10" diameter			70	.229		3.71	7.20		10.91	15.25
5880	12" diameter			50	.320		4.47	10.10		14.57	20.50
5900	14" diameter			44	.364		5.20	11.50		16.70	23.50
5920	16" diameter			40	.400		5.80	12.65		18.45	26
5930	18" diameter			37	.432		6.50	13.65		20.15	28
5940	20" diameter			34	.471		7.05	14.85		21.90	31
5950	22" diameter			31	.516		7.70	16.30		24	33.50
5960	24" diameter			28	.571		8.25	18.05		26.30	37
5980	30" diameter			22	.727		10.75	23		33.75	47.50
6000	36" diameter			18	.889		16.10	28		44.10	61.50
6300	Elbow, 45°, 4" diameter			60	.267		1.79	8.45		10.24	14.95
6310	5" diameter			52	.308		2.30	9.70		12	17.55
6320	6" diameter			44	.364		2.51	11.50		14.01	20.50
6340	8" diameter			28	.571		3.31	18.05		21.36	31.50
6360	10" diameter			18	.889		5.05	28		33.05	49
6380	12" diameter			13	1.231		6.50	39		45.50	67
6400	14" diameter			11	1.455		12.20	46		58.20	84.50
6420	16" diameter			10	1.600		20	50.50		70.50	100
6430	18" diameter			9.60	1.667		24.50	52.50		77	108
6440	20" diameter		Q-10	14	1.714		31	56		87	121
6450	22" diameter			13	1.846		31	60.50		91.50	128
6460	24" diameter			12	2		41.50	65.50		107	147
6480	30" diameter			9	2.667		64	87.50		151.50	206
6500	36" diameter			7	3.429		78	112		190	259
6600	Elbow, 90°, 4" diameter		Q-9	60	.267		3.05	8.45		11.50	16.35
6610	5" diameter			52	.308		3.51	9.70		13.21	18.85
6620	6" diameter			44	.364		4.02	11.50		15.52	22
6625	7" diameter			36	.444		4.25	14.05		18.30	26
6630	8" diameter			28	.571		4.93	18.05		22.98	33.50
6640	10" diameter			18	.889		7.05	28		35.05	51.50
6650	12" diameter			13	1.231		9.30	39		48.30	70.50
6660	14" diameter			11	1.455		17.45	46		63.45	90
6670	16" diameter			10	1.600		27.50	50.50		78	109
6676	18" diameter			9.60	1.667		36	52.50		88.50	121
6680	20" diameter		Q-10	14	1.714		48	56		104	139
6684	22" diameter			13	1.846		74.50	60.50		135	176
6690	24" diameter			12	2		83.50	65.50		149	193
6700	30" diameter			9	2.667		128	87.50		215.50	276
6710	36" diameter			7	3.429		156	112		268	345
6720	End cap										
6722	4"		Q-9	110	.145	Ea.	1.58	4.60		6.18	8.85
6724	6"			94	.170		2.10	5.40		7.50	10.60
6725	8"			88	.182		3.01	5.75		8.76	12.15
6726	10"			78	.205		4.37	6.50		10.87	14.80
6727	12"			70	.229		5.80	7.20		13	17.55
6728	14"			50	.320		8.05	10.10		18.15	24.50
6729	16"			44	.364		9.65	11.50		21.15	28.50

Important: See the Reference Section for critical supporting data - Reference Nos., Crews, & City Cost Indexes

	15810	Ducts		CREW	DAILY OUTPUT	LABOR-HOURS	UNIT	MAT.	LABOR	EQUIP.	TOTAL	TOTAL INCL O&P	
600	6730	18″	R15810-050	Q-9	40	.400	Ea.	11.30	12.65		23.95	32	**600**
	6731	20″			37	.432		12.30	13.65		25.95	34.50	
	6732	22″	R15810-070		34	.471		13.70	14.85		28.55	38	
	6733	24″			30	.533		16.10	16.85		32.95	43.50	
	6800	Reducing coupling, 6″ x 4″	R15810-100		46	.348		4.55	11		15.55	22	
	6820	8″ x 6″			40	.400		5.60	12.65		18.25	25.50	
	6840	10″ x 8″			32	.500		7.05	15.80		22.85	32.50	
	6860	12″ x 10″			24	.667		10.10	21		31.10	43.50	
	6880	14″ x 12″			20	.800		16.65	25.50		42.15	57.50	
	6900	16″ x 14″			18	.889		24.50	28		52.50	70	
	6920	18″ x 16″			16	1		30	31.50		61.50	81.50	
	6940	20″ x 18″		Q-10	24	1		36.50	33		69.50	91	
	6950	22″ x 20″			23	1.043		43	34		77	99.50	
	6960	24″ x 22″			22	1.091		49	36		85	109	
	6980	30″ x 28″			18	1.333		73	43.50		116.50	148	
	7000	36″ x 34″			16	1.500		98	49		147	184	
	7100	Tee, 90°, 4″ diameter		Q-9	40	.400		3.99	12.65		16.64	24	
	7110	5″ diameter			34.60	.462		5.30	14.60		19.90	28.50	
	7120	6″ diameter			30	.533		5.75	16.85		22.60	32.50	
	7130	8″ diameter			19	.842		7.50	26.50		34	49.50	
	7140	10″ diameter			12	1.333		10.70	42		52.70	77	
	7150	12″ diameter			9	1.778		13.65	56		69.65	102	
	7160	14″ diameter			7	2.286		26	72		98	140	
	7170	16″ diameter			6.70	2.388		36.50	75.50		112	156	
	7176	18″ diameter			6	2.667		50.50	84.50		135	186	
	7180	20″ diameter		Q-10	8.70	2.759		56	90.50		146.50	201	
	7190	24″ diameter			7	3.429		66.50	112		178.50	246	
	7200	30″ diameter			6	4		82	131		213	292	
	7210	36″ diameter			5	4.800		104	157		261	355	
	7350	Lateral 45°, reducing											
	7354	6″ x 6″ x 4″		Q-9	30	.533	Ea.	7.35	16.85		24.20	34	
	7358	8″ x 8″ x 4″			19	.842		9.05	26.50		35.55	51	
	7362	10″ x 10″ x 4″			12	1.333		12.70	42		54.70	79	
	7366	12″ x 12″ x 4″			9	1.778		14.35	56		70.35	102	
	7370	14″ x 14″ x 4″			7	2.286		20.50	72		92.50	134	
	7374	16″ x 16″ x 4″			6.70	2.388		32.50	75.50		108	152	
	7378	18″ x 18″ x 4″			6	2.667		41	84.50		125.50	175	
	7382	20″ x 20″ x 4″		Q-10	8.70	2.759		44.50	90.50		135	188	
	7386	22″ x 22″ x 4″			8	3		50	98.50		148.50	207	
	7390	24″ x 24″ x 4″			7	3.429		53.50	112		165.50	232	
	7400	Steel, PVC coated both sides, straight lengths											
	7410	4″ diameter, 26 ga.		Q-9	360	.044	L.F.	1.19	1.40		2.59	3.48	
	7416	5″ diameter, 26 ga.			240	.067		1.48	2.11		3.59	4.88	
	7420	6″ diameter, 26 ga.			280	.057		1.65	1.81		3.46	4.60	
	7440	8″ diameter, 26 ga.			200	.080		2.28	2.53		4.81	6.40	
	7460	10″ diameter, 26 ga.			160	.100		3.07	3.16		6.23	8.25	
	7480	12″ diameter, 26 ga.			120	.133		3.80	4.21		8.01	10.70	
	7500	14″ diameter, 26 ga.			80	.200		5.20	6.30		11.50	15.45	
	7520	16″ diameter, 24 ga.			60	.267		7.35	8.45		15.80	21	
	7540	18″ diameter, 24 ga.			45	.356		7.95	11.25		19.20	26	
	7560	20″ diameter, 24 ga.		Q-10	65	.369		9.35	12.10		21.45	29	
	7580	24″ diameter, 24 ga.			55	.436		11.90	14.30		26.20	35	
	7600	30″ diameter, 22 ga.			45	.533		17.25	17.50		34.75	46	
	7620	36″ diameter, 22 ga.			40	.600		21	19.65		40.65	53.50	
	7890	Connector, 4″ diameter		Q-9	100	.160	Ea.	2.49	5.05		7.54	10.55	
	7896	5″ diameter			83	.193		2.66	6.10		8.76	12.35	

MECHANICAL 15

	15810	Ducts	CREW	DAILY OUTPUT	LABOR-HOURS	UNIT	2002 BARE COSTS				TOTAL INCL O&P	
							MAT.	LABOR	EQUIP.	TOTAL		
600	7900	6" diameter `R15810 -050`	Q-9	88	.182	Ea.	2.80	5.75		8.55	11.95	**600**
	7920	8" diameter		78	.205		3.54	6.50		10.04	13.90	
	7940	10" diameter `R15810 -070`		70	.229		4.87	7.20		12.07	16.50	
	7960	12" diameter		50	.320		6.25	10.10		16.35	22.50	
	7980	14" diameter `R15810 -100`		44	.364		8	11.50		19.50	26.50	
	8000	16" diameter		40	.400		8.90	12.65		21.55	29.50	
	8020	18" diameter		37	.432		10	13.65		23.65	32	
	8040	20" diameter		34	.471		10.75	14.85		25.60	35	
	8060	24" diameter		28	.571		12.05	18.05		30.10	41.50	
	8080	30" diameter		22	.727		15.75	23		38.75	53	
	8100	36" diameter		18	.889		21.50	28		49.50	67	
	8390	Elbow, 45°, 4" diameter		60	.267		3.14	8.45		11.59	16.45	
	8396	5" diameter		36	.444		3.79	14.05		17.84	25.50	
	8400	6" diameter		44	.364		4.12	11.50		15.62	22	
	8420	8" diameter		28	.571		5.45	18.05		23.50	34	
	8440	10" diameter		18	.889		7.90	28		35.90	52	
	8460	12" diameter		13	1.231		11.65	39		50.65	73	
	8480	14" diameter		11	1.455		21	46		67	94	
	8500	16" diameter		10	1.600		30	50.50		80.50	111	
	8520	18" diameter	↓	9	1.778		34.50	56		90.50	125	
	8540	20" diameter	Q-10	13	1.846		46.50	60.50		107	145	
	8560	24" diameter		11	2.182		59	71.50		130.50	175	
	8580	30" diameter		9	2.667		90.50	87.50		178	235	
	8600	36" diameter	↓	7	3.429		110	112		222	294	
	8800	Elbow, 90°, 4" diameter	Q-9	60	.267		5.20	8.45		13.65	18.75	
	8804	5" diameter		52	.308		5.95	9.70		15.65	21.50	
	8806	6" diameter		44	.364		6.50	11.50		18	25	
	8808	8" diameter		28	.571		8.10	18.05		26.15	37	
	8810	10" diameter		18	.889		10.95	28		38.95	55.50	
	8812	12" diameter		13	1.231		15.10	39		54.10	76.50	
	8814	14" diameter		11	1.455		27	46		73	101	
	8816	16" diameter		10	1.600		42	50.50		92.50	124	
	8818	18" diameter	↓	9.60	1.667		51.50	52.50		104	138	
	8820	20" diameter	Q-10	14	1.714		65.50	56		121.50	159	
	8822	24" diameter		12	2		118	65.50		183.50	230	
	8824	30" diameter		9	2.667		181	87.50		268.50	335	
	8826	36" diameter	↓	7	3.429		220	112		332	415	
	8850	Tee, 90°, 4" diameter	Q-9	40	.400		5.85	12.65		18.50	26	
	8854	5" diameter		34.60	.462		7.70	14.60		22.30	31	
	8856	6" diameter		30	.533		8.55	16.85		25.40	35.50	
	8858	8" diameter		19	.842		10.95	26.50		37.45	53	
	8860	10" diameter		12	1.333		14.30	42		56.30	81	
	8862	12" diameter		9	1.778		20.50	56		76.50	109	
	8864	14" diameter		7	2.286		35.50	72		107.50	150	
	8866	16" diameter		6.70	2.388		51.50	75.50		127	173	
	8868	18" diameter	↓	6	2.667		74.50	84.50		159	212	
	8870	20" diameter	Q-10	8.70	2.759		80.50	90.50		171	228	
	8872	24" diameter		7	3.429		111	112		223	295	
	8874	30" diameter		6	4		138	131		269	355	
	8876	36" diameter	↓	5	4.800	↓	175	157		332	435	
	9000	Reducing coupling, tapered offset										
	9020	Diameter x any smaller diameter										
	9040	6" x	Q-9	46	.348	Ea.	6.50	11		17.50	24	
	9060	8" x		40	.400		8	12.65		20.65	28.50	
	9080	10" x		32	.500		9.65	15.80		25.45	35	
	9100	12" x	↓	24	.667	↓	13.45	21		34.45	47.50	

Important: See the Reference Section for critical supporting data - Reference Nos., Crews, & City Cost Indexes

15810	Ducts	CREW	DAILY OUTPUT	LABOR- HOURS	UNIT	2002 BARE COSTS				TOTAL INCL O&P
						MAT.	LABOR	EQUIP.	TOTAL	
600											600
9120	14" x	Q-9	20	.800	Ea.	21.50	25.50		47	63
9140	16" x	R15810 -050	18	.889		33	28		61	79.50
9160	18" x		16	1		40	31.50		71.50	92.50
9180	20" x	R15810 -070 Q-10	24	1		49	33		82	105
9200	24" x	R15810 -100	22	1.091		66.50	36		102.50	128
9220	30" x		18	1.333		94	43.50		137.50	171
9240	36" x		16	1.500		122	49		171	210
9400	Lateral 45°, straight barrel									
9410	4" size	Q-9	40	.400	Ea.	6.75	12.65		19.40	27
9416	5" size		34.60	.462		8.85	14.60		23.45	32.50
9420	6" size		30	.533		9.80	16.85		26.65	37
9440	8" size		19	.842		12.60	26.50		39.10	55
9460	10" size		12	1.333		16.45	42		58.45	83
9480	12" size		9	1.778		20.50	56		76.50	109
9500	14" size		7	2.286		27	72		99	141
9520	16" size		6.70	2.388		51	75.50		126.50	172
9540	18" size		6	2.667		72	84.50		156.50	209
9560	20" size	Q-10	8.70	2.759		85.50	90.50		176	233
9580	24" size		7	3.429		119	112		231	305
9600	30" size		6	4		139	131		270	355
9620	36" size		5	4.800		181	157		338	440
9640	For reducing lateral, add					40%				
9650	Lateral 45°, reducing									
9654	6" x 6" x 4"	Q-9	30	.533	Ea.	14.55	16.85		31.40	42
9658	8" x 8" x 4"		19	.842		18.40	26.50		44.90	61.50
9662	10" x 10" x 4"		12	1.333		24	42		66	91.50
9666	12" x 12" x 4"		9	1.778		28	56		84	118
9670	14" x 14" x 4"		7	2.286		46.50	72		118.50	162
9674	16" x 16" x 4"		6.70	2.388		58	75.50		133.50	180
9678	18" x 18" x 4"		6	2.667		91	84.50		175.50	230
9682	20" x 20" x 4"	Q-10	8.70	2.759		116	90.50		206.50	267
9686	22" x 22" x 4"		8	3		132	98.50		230.50	297
9690	24" x 24" x 4"		7	3.429		142	112		254	330
9700	Register boot, elbow or perimeter									
9710	6" duct diameter	Q-9	60	.267	Ea.	24	8.45		32.45	39.50
9720	8" duct diameter		56	.286		29	9.05		38.05	46
9730	10" duct diameter		45	.356		40	11.25		51.25	61
9740	12" duct diameter		36	.444		40	14.05		54.05	65
9800	Steel, stainless, straight lengths									
9810	3" diameter, 24 Ga.	Q-9	400	.040	L.F.	4.28	1.26		5.54	6.65
9820	4" diameter, 24 Ga.		360	.044		5.40	1.40		6.80	8.10
9830	5" diameter, 24 Ga.		320	.050		6.65	1.58		8.23	9.75
9840	6" diameter, 24 Ga.		280	.057		7.80	1.81		9.61	11.40
9850	7" diameter, 24 Ga.		240	.067		8.95	2.11		11.06	13.10
9860	8" diameter, 24 Ga.		200	.080		10.15	2.53		12.68	15.05
9870	9" diameter, 24 Ga.		180	.089		11.30	2.81		14.11	16.80
9880	10" diameter, 24 Ga.		160	.100		12.45	3.16		15.61	18.55
9890	12" diameter, 24 Ga.		120	.133		14.85	4.21		19.06	23
9900	14" diameter, 24 Ga.		80	.200		17.15	6.30		23.45	28.50
9910	16" diameter, 24 Ga.		60	.267		19.50	8.45		27.95	34.50
9920	18" diameter, 24 Ga.		45	.356		22	11.25		33.25	41.50
9930	20" diameter, 24 Ga.	Q-10	65	.369		24	12.10		36.10	45
9940	24" diameter, 24 Ga.		55	.436		29	14.30		43.30	54
9950	30" diameter, 22 Ga.		45	.533		44	17.50		61.50	75.50
9960	36" diameter, 22 Ga.		40	.600		53.50	19.65		73.15	89.50
9970	40" diameter, 20 Ga.		35	.686		71	22.50		93.50	113

MECHANICAL 15

15820	Duct Accessories	CREW	DAILY OUTPUT	LABOR-HOURS	UNIT	2002 BARE COSTS				TOTAL INCL O&P
						MAT.	LABOR	EQUIP.	TOTAL	
300 0010	**DUCT ACCESSORIES**									**300**
0050	Air extractors, 12" x 4"	1 Shee	24	.333	Ea.	15.20	11.70		26.90	35
0100	8" x 6"		22	.364		15.20	12.75		27.95	36.50
0120	12" x 6"		21	.381		21	13.35		34.35	43.50
0140	16" x 6"		20	.400		26.50	14.05		40.55	50.50
0160	24" x 6"		18	.444		44	15.60		59.60	72.50
0180	12" x 8"		20	.400		23.50	14.05		37.55	47.50
0200	20" x 8"		16	.500		34.50	17.55		52.05	65
0240	18" x 10"		14	.571		33.50	20		53.50	68
0260	24" x 10"		12	.667		43	23.50		66.50	83.50
0280	24" x 12"		10	.800		46.50	28		74.50	95
0300	30" x 12"	▼	8	1	▼	58	35		93	118
0350	Duct collar, spin-in type									
0352	Without damper									
0354	Round									
0356	4"	1 Shee	32	.250	Ea.	1.71	8.80		10.51	15.45
0357	5"		30	.267		1.71	9.35		11.06	16.35
0358	6"		28	.286		1.83	10.05		11.88	17.45
0359	7"		26	.308		1.95	10.80		12.75	18.80
0360	8"		24	.333		2.08	11.70		13.78	20.50
0361	9"		22	.364		2.34	12.75		15.09	22
0362	10"		20	.400		2.44	14.05		16.49	24
0363	12"		17	.471		2.87	16.50		19.37	28.50
1000	Duct access door, insulated, 6" x 6"		14	.571		11.60	20		31.60	44
1020	10" x 10"		11	.727		16.25	25.50		41.75	57.50
1040	12" x 12"		10	.800		17.55	28		45.55	63
1050	12" x 18"		9	.889		23	31		54	73.50
1060	16" x 12"		9	.889		24.50	31		55.50	75
1070	18" x 18"		8	1		26	35		61	82.50
1074	24" x 18"		8	1		32.50	35		67.50	89.50
1080	24" x 24"	▼	8	1	▼	34.50	35		69.50	92
1100	Exhaust vent damper, auto., OB with elect. actuator									
1110	6" x 6"	1 Shee	24	.333	Ea.	67	11.70		78.70	92
1114	8" x 6"		23	.348		69	12.20		81.20	95
1116	8" x 8"		22	.364		72	12.75		84.75	98.50
1118	10" x 6"		23	.348		72	12.20		84.20	98
1120	10" x 10"		22	.364		76	12.75		88.75	103
1122	12" x 6"		22	.364		73	12.75		85.75	100
1124	12" x 10"		21	.381		80	13.35		93.35	109
1126	12" x 12"		20	.400		84	14.05		98.05	114
1128	14" x 6"		21	.381		74	13.35		87.35	102
1130	14" x 10"		19	.421		84	14.80		98.80	116
1132	16" x 6"		21	.381		74	13.35		87.35	102
1134	16" x 8"		20	.400		83	14.05		97.05	113
1136	18" x 6"		20	.400		79	14.05		93.05	109
1138	18" x 8"		19	.421	▼	84	14.80		98.80	116
2000	Fabrics for flexible connections, with metal edge		100	.080	L.F.	1.54	2.81		4.35	6
2100	Without metal edge		160	.050	"	1.02	1.76		2.78	3.83
3000	Fire damper, curtain type, 1-1/2 hr rated, vertical, 6" x 6"		24	.333	Ea.	19.85	11.70		31.55	40
3020	8" x 6"		22	.364		19.70	12.75		32.45	41
3021	8" x 8"		22	.364		19.75	12.75		32.50	41
3030	10" x 10"		22	.364		19.75	12.75		32.50	41
3040	12" x 6"		22	.364		19.70	12.75		32.45	41
3044	12" x 12"		20	.400		22.50	14.05		36.55	46.50
3060	20" x 6"		18	.444		23.50	15.60		39.10	49.50
3080	12" x 8"	▼	22	.364	▼	19.50	12.75		32.25	41

Important: See the Reference Section for critical supporting data - Reference Nos., Crews, & City Cost Indexes

15820	Duct Accessories	CREW	DAILY OUTPUT	LABOR-HOURS	UNIT	2002 BARE COSTS MAT.	LABOR	EQUIP.	TOTAL	TOTAL INCL O&P
3100	24" x 8"	1 Shee	16	.500	Ea.	28.50	17.55		46.05	58.50
3120	12" x 10"		21	.381		20.50	13.35		33.85	43
3140	24" x 10"		15	.533		31.50	18.70		50.20	63.50
3160	36" x 10"		12	.667		44.50	23.50		68	85
3180	16" x 12"		20	.400		26	14.05		40.05	50
3200	24" x 12"		13	.615		30.50	21.50		52	67
3220	48" x 12"		10	.800		48	28		76	96.50
3238	14" x 14"		19	.421		27.50	14.80		42.30	53.50
3240	16" x 14"		18	.444		31.50	15.60		47.10	58.50
3260	24" x 14"		12	.667		36.50	23.50		60	76.50
3280	30" x 14"		11	.727		40	25.50		65.50	83.50
3298	16" x 16"		18	.444		28	15.60		43.60	55
3300	18" x 16"		17	.471		32	16.50		48.50	60.50
3320	24" x 16"		11	.727		36.50	25.50		62	80
3340	36" x 16"		10	.800		44.50	28		72.50	92.50
3356	18" x 18"		16	.500		32	17.55		49.55	62
3360	24" x 18"		10	.800		40	28		68	87.50
3380	48" x 18"		8	1		64	35		99	125
3398	20" x 20"		10	.800		33	28		61	80
3400	24" x 20"		8	1		39.50	35		74.50	97.50
3420	36" x 20"		7	1.143		52	40		92	119
3440	24" x 22"		7	1.143		46	40		86	113
3460	30" x 22"		6	1.333		51.50	47		98.50	129
3478	24" x 24"		8	1		38.50	35		73.50	96
3480	26" x 24"		7	1.143		47	40		87	114
3484	32" x 24"	Q-9	13	1.231		48	39		87	113
3500	48" x 24"		12	1.333		67.50	42		109.50	140
3520	28" x 26"		13	1.231		53	39		92	119
3540	30" x 28"		12	1.333		58.50	42		100.50	129
3560	48" x 30"		11	1.455		81.50	46		127.50	161
3580	48" x 36"		10	1.600		88	50.50		138.50	175
3590	32" x 40"		10	1.600		67.50	50.50		118	152
3600	44" x 44"		10	1.600		91.50	50.50		142	179
3620	48" x 48"		8	2		110	63		173	219
3700	U.L. label included in above									
3800	For horizontal operation, add				Ea.	20%				
3900	For cap for blades out of air stream, add					20%				
4000	For 10" 22 ga., U.L. approved sleeve, add					35%				
4100	For 10" 22 ga., U.L. sleeve, 100% free area, add					65%				
4200	For oversize openings group dampers									
4210	Fire damper, round, Type A, 2 Hr									
4214	4" diam.	1 Shee	20	.400	Ea.	37.50	14.05		51.55	63
4216	5" diam.		16	.500		35.50	17.55		53.05	66
4218	6" diam.		14.30	.559		35.50	19.65		55.15	69.50
4220	7" diam.		13.30	.601		35.50	21		56.50	71.50
4222	8" diam.		11.40	.702		35.50	24.50		60	77
4224	9" diam.		10.80	.741		40	26		66	84
4226	10" diam.		10	.800		41	28		69	89
4228	12" diam.		9.50	.842		44	29.50		73.50	93.50
4230	14" diam.		8.20	.976		67.50	34		101.50	127
4232	16" diam.		7.60	1.053		71	37		108	135
4234	18" diam.		7.20	1.111		84.50	39		123.50	153
4236	20" diam.		6.40	1.250		84.50	44		128.50	161
4238	22" diam.		6.10	1.311		101	46		147	182
4240	24" diam.		5.70	1.404		101	49.50		150.50	187
4242	26" diam.		5.50	1.455		120	51		171	211

300

MECHANICAL 15

			DAILY	LABOR-		2002 BARE COSTS				TOTAL	
15820	**Duct Accessories**	CREW	OUTPUT	HOURS	UNIT	MAT.	LABOR	EQUIP.	TOTAL	INCL O&P	
300 4244	28" diam.	1 Shee	5.40	1.481	Ea.	120	52		172	212	**300**
4246	30" diam.	↓	5.30	1.509	↓	120	53		173	214	
4500	Fire/smoke combination damper, louver type, UL										
4506	6" x 6"	1 Shee	24	.333	Ea.	55	11.70		66.70	78.50	
4510	8" x 8"		22	.364		126	12.75		138.75	159	
4520	16" x 8"		20	.400		138	14.05		152.05	174	
4540	18" x 8"		18	.444		138	15.60		153.60	176	
4560	20" x 8"		16	.500		148	17.55		165.55	189	
4580	10" x 10"		21	.381		143	13.35		156.35	178	
4600	24" x 10"		15	.533		168	18.70		186.70	213	
4620	30" x 10"		12	.667		173	23.50		196.50	227	
4640	12" x 12"		20	.400		149	14.05		163.05	186	
4660	18" x 12"		18	.444		152	15.60		167.60	191	
4680	24" x 12"		13	.615		160	21.50		181.50	210	
4700	30" x 12"		11	.727		166	25.50		191.50	223	
4720	14" x 14"		17	.471		160	16.50		176.50	202	
4740	16" x 14"		18	.444		165	15.60		180.60	205	
4760	20" x 14"		14	.571		168	20		188	215	
4780	24" x 14"		12	.667		172	23.50		195.50	225	
4800	30" x 14"		10	.800		182	28		210	244	
4820	16" x 16"		16	.500		177	17.55		194.55	222	
4840	20" x 16"		14	.571		182	20		202	231	
4860	24" x 16"		11	.727		188	25.50		213.50	247	
4880	30" x 16"		8	1		196	35		231	270	
4900	18" x 18"		15	.533		176	18.70		194.70	222	
5000	24" x 18"		10	.800		188	28		216	250	
5020	36" x 18"		7	1.143		205	40		245	288	
5040	20" x 20"		13	.615		196	21.50		217.50	250	
5060	24" x 20"		8	1		201	35		236	275	
5080	30" x 20"		8	1		211	35		246	286	
5100	36" x 20"		7	1.143		222	40		262	305	
5120	24" x 24"		8	1		217	35		252	293	
5130	30" x 24"	↓	7.60	1.053		218	37		255	297	
5140	36" x 24"	Q-9	12	1.333		237	42		279	325	
5141	36" x 30"		10	1.600		208	50.50		258.50	305	
5142	40" x 36"		8	2		445	63		508	590	
5143	48" x 48"	↓	4	4		870	126		996	1,150	
5150	Damper operator motor, 24 or 120 volt	1 Shee	16	.500	↓	191	17.55		208.55	237	
5154	Induction air terminal unit, cooling and heating										
5156	Vertical, cooling w/ electric heating coils										
5158	60 CFM	Q-20	7.60	2.632	Ea.	885	85		970	1,100	
5160	90 CFM		7.20	2.778		1,100	90		1,190	1,350	
5162	120 CFM		6.80	2.941		1,100	95		1,195	1,350	
5164	150 CFM		6.40	3.125		1,250	101		1,351	1,525	
5166	170 CFM	↓	6.20	3.226	↓	1,250	104		1,354	1,525	
5170	Horizontal, cooling w/ electric heating coils										
5172	60 CFM	Q-20	7.60	2.632	Ea.	720	85		805	925	
5174	90 CFM		7.20	2.778		885	90		975	1,125	
5176	120 CFM	↓	6.80	2.941	↓	885	95		980	1,125	
5180	Mixing box, includes electric or pneumatic motor										
5190	Recommend use with attenuator, see 15820-300-9000										
5200	Constant volume, 150 to 270 CFM	Q-9	12	1.333	Ea.	495	42		537	605	
5210	270 to 600 CFM		11	1.455		505	46		551	625	
5230	550 to 1000 CFM		9	1.778		505	56		561	640	
5240	1000 to 1600 CFM		8	2		520	63		583	670	
5250	1300 to 1900 CFM	↓	6	2.667	↓	530	84.50		614.50	715	

15

MECHANICAL

Important: See the Reference Section for critical supporting data - Reference Nos., Crews, & City Cost Indexes

15820	Duct Accessories	CREW	DAILY OUTPUT	LABOR-HOURS	UNIT	2002 BARE COSTS				TOTAL INCL O&P
						MAT.	LABOR	EQUIP.	TOTAL	
300 5260	550 to 2640 CFM	Q-9	5.60	2.857	Ea.	585	90.50		675.50	785 **300**
5270	650 to 3120 CFM		5.20	3.077		620	97		717	835
5500	VAV Cool only, pneum. press indep. 300 to 600 CFM		11	1.455		277	46		323	375
5510	500 to 1000 CFM		9	1.778		288	56		344	400
5520	800 to 1600 CFM		9	1.778		300	56		356	415
5530	1100 to 2000 CFM		8	2		315	63		378	450
5540	1500 to 3000 CFM		7	2.286		340	72		412	480
5550	2000 to 4000 CFM		6	2.667		360	84.50		444.50	525
5560	For electric, w/thermostat, press. dependent, add					20			20	22
5600	VAV Cool & HW coils, damper, actuator and t'stat									
5610	200 CFM	Q-9	11	1.455	Ea.	565	46		611	690
5620	400 CFM		10	1.600		570	50.50		620.50	710
5630	600 CFM		10	1.600		570	50.50		620.50	710
5640	800 CFM		8	2		590	63		653	745
5650	1000 CFM		8	2		590	63		653	745
5660	1250 CFM		6	2.667		645	84.50		729.50	840
5670	1500 CFM		6	2.667		645	84.50		729.50	840
5680	2000 CFM		4	4		700	126		826	965
5684	3000 CFM		3.80	4.211		760	133		893	1,050
5700	VAV Cool only, fan powr'd, damper, actuator, t'stat									
5710	200 CFM	Q-9	10	1.600	Ea.	775	50.50		825.50	935
5720	400 CFM		9	1.778		805	56		861	970
5730	600 CFM		9	1.778		805	56		861	970
5740	800 CFM		7	2.286		860	72		932	1,050
5750	1000 CFM		7	2.286		860	72		932	1,050
5760	1250 CFM		5	3.200		930	101		1,031	1,175
5770	1500 CFM		5	3.200		930	101		1,031	1,175
5780	2000 CFM		4	4		1,025	126		1,151	1,325
5800	VAV Fan powr'd, cool/HW coils, dampers, actuators, t'stat									
5810	200 CFM	Q-9	10	1.600	Ea.	935	50.50		985.50	1,100
5820	400 CFM		9	1.778		960	56		1,016	1,125
5830	600 CFM		9	1.778		960	56		1,016	1,125
5840	800 CFM		7	2.286		1,025	72		1,097	1,225
5850	1000 CFM		7	2.286		1,025	72		1,097	1,225
5860	1250 CFM		5	3.200		1,150	101		1,251	1,425
5870	1500 CFM		5	3.200		1,150	101		1,251	1,425
5880	2000 CFM		3.50	4.571		1,250	144		1,394	1,600
5990	Multi-blade dampers, opposed blade, 8" x 6"	1 Shee	24	.333		18.40	11.70		30.10	38
5994	8" x 8"		22	.364		19.20	12.75		31.95	40.50
5996	10" x 10"		21	.381		21.50	13.35		34.85	44.50
6000	12" x 12"		21	.381		25	13.35		38.35	48
6020	12" x 18"		18	.444		33.50	15.60		49.10	61
6030	14" x 10"		20	.400		21.50	14.05		35.55	45.50
6031	14" x 14"		17	.471		29.50	16.50		46	58
6032	16" x 10"		18	.444		21.50	15.60		37.10	48
6033	16" x 12"		17	.471		29.50	16.50		46	58
6035	16" x 16"		16	.500		37	17.55		54.55	67.50
6036	18" x 16"		15	.533		41	18.70		59.70	74
6037	18" x 14"		16	.500		37	17.55		54.55	67.50
6038	18" x 18"		15	.533		44	18.70		62.70	77.50
6040	18" x 24"		12	.667		57	23.50		80.50	98.50
6060	18" x 28"		10	.800		65.50	28		93.50	116
6068	20" x 6"		18	.444		22.50	15.60		38.10	48.50
6069	20" x 8"		16	.500		25.50	17.55		43.05	55
6070	20" x 16"		14	.571		44	20		64	79.50
6071	20" x 18"		13	.615		48	21.50		69.50	86.50

			DAILY	LABOR-		2002 BARE COSTS				TOTAL	
15820	**Duct Accessories**	CREW	OUTPUT	HOURS	UNIT	MAT.	LABOR	EQUIP.	TOTAL	INCL O&P	
300 6072	20" x 20"	1 Shee	13	.615	Ea.	53	21.50		74.50	91.50	**300**
6073	22" x 6"		20	.400		22.50	14.05		36.55	46	
6074	22" x 18"		14	.571		53	20		73	89	
6075	22" x 20"		12	.667		57.50	23.50		81	99.50	
6076	24" x 16"		11	.727		52	25.50		77.50	96.50	
6077	22" x 22"		10	.800		61.50	28		89.50	112	
6078	24" x 20"		8	1		61.50	35		96.50	122	
6080	24" x 24"		8	1		72	35		107	133	
6100	24" x 28"		6	1.333		83	47		130	164	
6110	26" x 26"		6	1.333		84.50	47		131.50	165	
6120	28" x 28"	Q-9	11	1.455		98	46		144	179	
6130	30" x 18"		10	1.600		69.50	50.50		120	155	
6132	30" x 24"		7	2.286		88	72		160	208	
6133	30" x 30"		6.60	2.424		112	76.50		188.50	242	
6135	32" x 32"		6.40	2.500		128	79		207	263	
6150	34" x 34"		6	2.667		144	84.50		228.50	288	
6151	36" x 12"		10	1.600		59	50.50		109.50	143	
6152	36" x 16"		8	2		74.50	63		137.50	180	
6156	36" x 32"		6.30	2.540		144	80.50		224.50	282	
6157	36" x 34"		6.20	2.581		153	81.50		234.50	294	
6158	36" x 36"		6	2.667		161	84.50		245.50	310	
6160	44" x 28"		5.80	2.759		154	87		241	305	
6180	48" x 36"		5.60	2.857		215	90.50		305.50	375	
6200	56" x 36"		5.40	2.963		250	93.50		343.50	420	
6220	60" x 36"		5.20	3.077		268	97		365	445	
6240	60" x 44"		5	3.200		330	101		431	515	
7000	Splitter damper assembly, self-locking, 1' rod	1 Shee	24	.333		16.55	11.70		28.25	36.50	
7020	3' rod		22	.364		21.50	12.75		34.25	43	
7040	4' rod		20	.400		24	14.05		38.05	48	
7060	6' rod		18	.444		29	15.60		44.60	56	
7500	Variable volume modulating motorized damper, incl. elect. mtr.										
7504	8" x 6"	1 Shee	15	.533	Ea.	101	18.70		119.70	140	
7506	10" x 6"		14	.571		101	20		121	142	
7510	10" x 10"		13	.615		104	21.50		125.50	148	
7520	12" x 12"		12	.667		106	23.50		129.50	153	
7522	12" x 16"		11	.727		110	25.50		135.50	161	
7524	16" x 10"		12	.667		107	23.50		130.50	154	
7526	16" x 14"		10	.800		112	28		140	167	
7528	16" x 18"		9	.889		119	31		150	179	
7540	18" x 12"		10	.800		114	28		142	169	
7542	18" x 18"		8	1		123	35		158	190	
7544	20" x 14"		8	1		126	35		161	192	
7546	20" x 18"		7	1.143		134	40		174	209	
7560	24" x 12"		8	1		127	35		162	194	
7562	24" x 18"		7	1.143		142	40		182	218	
7568	28" x 10"		7	1.143		211	40		251	294	
7580	28" x 16"		6	1.333		224	47		271	320	
7590	30" x 14"		5	1.600		224	56		280	335	
7600	30" x 18"		4	2		238	70		308	370	
7610	30" x 24"		3.80	2.105		258	74		332	395	
7690	48" x 48"	Q-9	6	2.667		665	84.50		749.50	865	
7694	6' x 14' w/2 motors	"	3	5.333		2,600	169		2,769	3,100	
7700	For thermostat, add	1 Shee	8	1		25.50	35		60.50	82	
7800	For transformer 40 VA capacity, add	"	16	.500		20	17.55		37.55	49	
8000	Multi-blade dampers, parallel blade										
8100	8" x 8"	1 Shee	24	.333	Ea.	52.50	11.70		64.20	76	

15 **MECHANICAL**

Important: See the Reference Section for critical supporting data - Reference Nos., Crews, & City Cost Indexes

15820	Duct Accessories	CREW	DAILY OUTPUT	LABOR-HOURS	UNIT	2002 BARE COSTS				TOTAL INCL O&P
						MAT.	LABOR	EQUIP.	TOTAL	
300	8120	12" x 8"	1 Shee	22	.364	Ea.	52.50	12.75		65.25	77.50
8140	16" x 10"		20	.400		55.50	14.05		69.55	82.50
8160	18" x 12"		18	.444		57	15.60		72.60	87
8180	22" x 12"		15	.533		59.50	18.70		78.20	94.50
8200	24" x 16"		11	.727		74	25.50		99.50	121
8220	28" x 16"		10	.800		77	28		105	129
8240	30" x 16"		8	1		79	35		114	141
8260	30" x 18"		7	1.143		86	40		126	157
8300	Relief damper, electronic bypass with tight seal									
8310	8" x 6"	1 Shee	22	.364	Ea.	101	12.75		113.75	131
8314	10" x 6"		22	.364		114	12.75		126.75	145
8318	10" x 10"		21	.381		147	13.35		160.35	183
8322	12" x 12"		20	.400		158	14.05		172.05	195
8326	12" x 16"		19	.421		164	14.80		178.80	203
8330	16" x 10"		20	.400		158	14.05		172.05	196
8334	16" x 14"		18	.444		165	15.60		180.60	206
8338	16" x 18"		17	.471		182	16.50		198.50	226
8342	18" x 12"		18	.444		164	15.60		179.60	205
8346	18" x 18"		15	.533		171	18.70		189.70	217
8350	20" x 14"		14	.571		171	20		191	219
8354	20" x 18"		13	.615		175	21.50		196.50	226
8358	24" x 12"		13	.615		182	21.50		203.50	235
8362	24" x 18"		10	.800		196	28		224	260
8363	24" x 24"		10	.800		204	28		232	268
8364	24" x 36"		9	.889		236	31		267	310
8365	24" x 48"		6	1.333		264	47		311	360
8366	28" x 10"		13	.615		171	21.50		192.50	222
8370	28" x 16"		10	.800		201	28		229	265
8374	30" x 14"		11	.727		197	25.50		222.50	257
8378	30" x 18"		8	1		199	35		234	273
8382	30" x 24"		6	1.333		197	47		244	289
8390	46" x 36"		4	2		277	70		347	415
8394	48" x 48"		3	2.667		320	93.50		413.50	495
8396	54" x 36"		2	4		340	140		480	590
8400	Round damper, butterfly, vol. control w/lever lock reg.									
8410	6" diam.	1 Shee	22	.364	Ea.	21.50	12.75		34.25	43.50
8412	7" diam.		21	.381		22.50	13.35		35.85	45
8414	8" diam.		20	.400		23	14.05		37.05	47
8416	9" diam.		19	.421		24	14.80		38.80	49.50
8418	10" diam.		18	.444		25.50	15.60		41.10	52
8420	12" diam.		16	.500		26.50	17.55		44.05	56
8422	14" diam.		14	.571		33.50	20		53.50	68
8424	16" diam.		13	.615		36	21.50		57.50	73
8426	18" diam.		12	.667		40	23.50		63.50	80
8428	20" diam.		11	.727		43	25.50		68.50	87
8430	24" diam.		10	.800		55	28		83	104
8432	30" diam.		9	.889		76	31		107	132
8434	36" diam.		8	1		102	35		137	166
8500	Round motor operated damper									
8510	6" dia.	1 Shee	22	.364	Ea.	51	12.75		63.75	76
8512	7" dia.		21	.381		53	13.35		66.35	79
8514	8" dia.		20	.400		56	14.05		70.05	83
8516	10" dia.		18	.444		61.50	15.60		77.10	92
8518	12" dia.		16	.500		66.50	17.55		84.05	100
8520	14" dia.		14	.571		75	20		95	114
9000	Silencers, noise control for air flow, duct				MCFM	44.50			44.50	49

15820	Duct Accessories	CREW	DAILY OUTPUT	LABOR-HOURS	UNIT	2002 BARE COSTS				TOTAL INCL O&P
						MAT.	LABOR	EQUIP.	TOTAL	
300											300
9004	Duct sound trap, packaged									
9010	12" x 18" x 36", 2000 CFM	1 Shee	12	.667	Ea.	267	23.50		290.50	330
9014	24" x 36" x 36", 9000 CFM	Q-9	40	.400		600	12.65		612.65	680
9018	36" x 18" x 36", 6300 CFM		42	.381		600	12.05		612.05	680
9022	24" x 24" x 60", 6000 CFM		3.60	4.444		695	140		835	980
9026	24" x 30" x 60", 7000 CFM		3.50	4.571		780	144		924	1,075
9030	24" x 36" x 60", 8000 CFM		3.40	4.706		860	149		1,009	1,175
9034	24" x 48" x 60", 11,000 CFM		3.30	4.848		755	153		908	1,075
9038	36" x 18" x 60", 6300 CFM		3.50	4.571		580	144		724	865
9042	36" x 36" x 60", 12,000 CFM		3.30	4.848		955	153		1,108	1,275
9046	36" x 48" x 60", 17,000 CFM		3.10	5.161		1,225	163		1,388	1,600
9050	36" x 60" x 60", 20,000 CFM		2.80	5.714		1,500	181		1,681	1,900
9054	48" x 48" x 60", 23,000 CFM		2.60	6.154		790	194		984	1,175
9100	Louvers, galvanized				MCFM	60			60	66
9120	Louvers, mill finish aluminum				"	87.50			87.50	96.50
9200	Plenums, measured by panel surface				S.F.	8.25			8.25	9.10
9300	Vent, air transfer				Ea.	139			139	153
9400	Turning vane components									
9410	Turning vane rail	1 Shee	160	.050	L.F.	.82	1.76		2.58	3.61
9420	Double thick, factory fab. vane		300	.027		1.01	.94		1.95	2.55
9428	12" high set		170	.047		5.70	1.65		7.35	8.80
9432	14" high set		160	.050		6.40	1.76		8.16	9.70
9434	16" high set		150	.053		7.05	1.87		8.92	10.65
9436	18" high set		144	.056		7.75	1.95		9.70	11.50
9438	20" high set		138	.058		8.40	2.03		10.43	12.40
9440	22" high set		130	.062		9.10	2.16		11.26	13.35
9442	24" high set		124	.065		9.75	2.26		12.01	14.25
9444	26" high set		116	.069		10.45	2.42		12.87	15.20
9446	30" high set		112	.071		11.80	2.51		14.31	16.80

15830	Fans	CREW	DAILY OUTPUT	LABOR-HOURS	UNIT	MAT.	LABOR	EQUIP.	TOTAL	TOTAL INCL O&P
100											100
0010	**FANS**									
0020	Air conditioning and process air handling	R15700-040								
0030	Axial flow, compact, low sound, 2.5" S.P.	R15800-080								
0050	3,800 CFM, 5 HP	Q-20	3.40	5.882	Ea.	3,600	190		3,790	4,275
0080	6,400 CFM, 5 HP		2.80	7.143		4,025	231		4,256	4,775
0100	10,500 CFM, 7-1/2 HP		2.40	8.333		5,025	270		5,295	5,950
0120	15,600 CFM, 10 HP		1.60	12.500		6,325	405		6,730	7,575
0140	23,000 CFM, 15 HP		.70	28.571		9,700	925		10,625	12,100
0160	28,000 CFM, 20 HP		.40	50		10,800	1,625		12,425	14,400
0200	In-line centrifugal, supply/exhaust booster									
0220	aluminum wheel/hub, disconnect switch, 1/4" S.P.									
0240	500 CFM, 10" diameter connection	Q-20	3	6.667	Ea.	715	216		931	1,125
0260	1,380 CFM, 12" diameter connection		2	10		815	325		1,140	1,400
0280	1,520 CFM, 16" diameter connection		2	10		930	325		1,255	1,525
0300	2,560 CFM, 18" diameter connection		1	20		1,025	645		1,670	2,125
0320	3,480 CFM, 20" diameter connection		.80	25		1,225	810		2,035	2,600
0326	5,080 CFM, 20" diameter connection		.75	26.667		1,350	865		2,215	2,825
0340	7,500 CFM, 22" diameter connection		.70	28.571		1,725	925		2,650	3,325
0350	10,000 CFM, 27" diameter connection		.65	30.769		2,025	995		3,020	3,750
0500	Axial flow, constant speed									
0505	Direct drive, 1/8" S.P.									
0510	12", 1060 CFM, 1/6 HP	Q-20	3	6.667	Ea.	279	216		495	635
0514	12", 2095 CFM, 1/2 HP		3	6.667		305	216		521	665
0518	16", 2490 CFM, 1/3 HP		2.80	7.143		330	231		561	715

15830	Fans			DAILY OUTPUT	LABOR-HOURS	UNIT	2002 BARE COSTS				TOTAL INCL O&P
			CREW				MAT.	LABOR	EQUIP.	TOTAL	
100	0522	20", 4130 CFM, 3/4 HP	R15700 -040	Q-20	2.60	7.692	Ea.	430	249		679	855	100
	0526	22", 4700 CFM, 3/4 HP			2.60	7.692		480	249		729	905
	0530	24", 5850 CFM, 1 HP	R15800 -080		2.50	8		545	259		804	995
	0534	24", 7925 CFM, 1-1/2 HP			2.40	8.333		550	270		820	1,025
	0538	30", 10640 CFM, 1 HP			2.20	9.091		1,000	294		1,294	1,550
	0542	30", 14765 CFM, 2-1/2 HP			2.10	9.524		1,050	310		1,360	1,625
	0546	36", 16780 CFM, 2 HP			2.10	9.524		1,125	310		1,435	1,700
	0550	36", 22920 CFM, 5 HP			1.80	11.111		1,400	360		1,760	2,100
	0560	Belt drive, 1/8" S.P.										
	0562	15", 2800 CFM, 1/3 HP		Q-20	3.20	6.250	Ea.	535	202		737	895
	0564	15", 3400 CFM, 1/2 HP			3	6.667		545	216		761	930
	0568	18", 3280 CFM, 1/3 HP			2.80	7.143		580	231		811	990
	0572	18", 3900 CFM, 1/2 HP			2.80	7.143		585	231		816	1,000
	0576	18", 5250 CFM, 1 HP			2.70	7.407		675	240		915	1,100
	0584	24", 6430 CFM, 1 HP			2.50	8		835	259		1,094	1,325
	0588	24", 8860 CFM, 2 HP			2.40	8.333		840	270		1,110	1,350
	0592	30", 9250 CFM, 1 HP			2.20	9.091		1,025	294		1,319	1,575
	0596	30", 16900 CFM, 5 HP			2	10		1,150	325		1,475	1,750
	0604	36", 14475 CFM, 2 HP			2.30	8.696		1,200	281		1,481	1,725
	0608	36", 20080 CFM, 5 HP			1.80	11.111		1,275	360		1,635	1,950
	0612	36", 14475 CFM, 7-1/2 HP			1.60	12.500		1,400	405		1,805	2,175
	0616	42", 29000 CFM, 7-1/2 HP			1.40	14.286		1,700	460		2,160	2,575
	1000	Return air fan										
	1010	9200 CFM		Q-20	1.40	14.286	Ea.	2,950	460		3,410	3,925
	1020	13,200 CFM			1.30	15.385		4,225	500		4,725	5,400
	1030	16,500 CFM			1.20	16.667		5,275	540		5,815	6,625
	1040	19,500 CFM			1	20		6,250	645		6,895	7,850
	1500	Vaneaxial, low pressure, 2000 CFM, 1/2 HP			3.60	5.556		1,575	180		1,755	2,025
	1520	4,000 CFM, 1 HP			3.20	6.250		1,625	202		1,827	2,075
	1540	8,000 CFM, 2 HP			2.80	7.143		1,900	231		2,131	2,450
	1560	16,000 CFM, 5 HP			2.40	8.333		2,525	270		2,795	3,200
	2000	Blowers, direct drive with motor, complete										
	2020	1045 CFM @ .5" S.P., 1/5 HP		Q-20	18	1.111	Ea.	125	36		161	192
	2040	1385 CFM @ .5" S.P., 1/4 HP			18	1.111		128	36		164	196
	2060	1640 CFM @ .5" S.P., 1/3 HP			18	1.111		152	36		188	222
	2080	1760 CFM @ .5" S.P., 1/2 HP			18	1.111		165	36		201	236
	2090	4 speed										
	2100	1164 to 1739 CFM @ .5" S.P., 1/3 HP		Q-20	16	1.250	Ea.	208	40.50		248.50	290
	2120	1467 to 2218 CFM @ 1.0" S.P., 3/4 HP		"	14	1.429	"	269	46		315	365
	2500	Ceiling fan, right angle, extra quiet, 0.10" S.P.										
	2520	95 CFM		Q-20	20	1	Ea.	116	32.50		148.50	177
	2540	210 CFM			19	1.053		137	34		171	202
	2560	385 CFM			18	1.111		173	36		209	246
	2580	885 CFM			16	1.250		340	40.50		380.50	435
	2600	1,650 CFM			13	1.538		475	50		525	595
	2620	2,960 CFM			11	1.818		630	59		689	785
	2640	For wall or roof cap, add		1 Shee	16	.500		116	17.55		133.55	154
	2660	For straight thru fan, add						10%				
	2680	For speed control switch, add		1 Elec	16	.500		63	17.75		80.75	96
	3000	Paddle blade air circulator, 3 speed switch										
	3020	42", 5,000 CFM high, 3000 CFM low		1 Elec	2.40	3.333	Ea.	77.50	118		195.50	261
	3040	52", 6,500 CFM high, 4000 CFM low		"	2.20	3.636	"	78.50	129		207.50	279
	3100	For antique white motor, same cost										
	3200	For brass plated motor, same cost										
	3300	For light adaptor kit, add					Ea.	29			29	31.50
	3310	Industrial grade, reversable, 4 blade										

15830 | Fans

100			CREW	DAILY OUTPUT	LABOR-HOURS	UNIT	2002 BARE COSTS				TOTAL INCL O&P	100
							MAT.	LABOR	EQUIP.	TOTAL		
3312	5500 CFM	R15700 -040	Q-20	5	4	Ea.	112	129		241	320	
3314	7600 CFM			5	4		122	129		251	330	
3316	21,015 CFM	R15800 -080		4	5		156	162		318	420	
3500	Centrifugal, airfoil, motor and drive, complete											
3520	1000 CFM, 1/2 HP		Q-20	2.50	8	Ea.	870	259		1,129	1,350	
3540	2,000 CFM, 1 HP			2	10		935	325		1,260	1,525	
3560	4,000 CFM, 3 HP			1.80	11.111		1,200	360		1,560	1,875	
3580	8,000 CFM, 7-1/2 HP			1.40	14.286		1,775	460		2,235	2,650	
3600	12,000 CFM, 10 HP			1	20		2,600	645		3,245	3,850	
4000	Single width, belt drive, not incl. motor, capacities											
4020	at 2000 FPM, 2.5" S.P. for indicated motor											
4040	6900 CFM, 5 HP		Q-9	2.40	6.667	Ea.	2,625	211		2,836	3,225	
4060	10,340 CFM, 7-1/2" HP			2.20	7.273		3,675	230		3,905	4,375	
4080	15,320 CFM, 10 HP			2	8		4,325	253		4,578	5,150	
4100	22,780 CFM, 15 HP			1.80	8.889		6,425	281		6,706	7,500	
4120	33,840 CFM, 20 HP			1.60	10		8,525	315		8,840	9,850	
4140	41,400 CFM, 25 HP			1.40	11.429		11,300	360		11,660	13,000	
4160	50,100 CFM, 30 HP			.80	20		14,400	630		15,030	16,800	
4200	Double width wheel, 12,420 CFM, 7.5 HP			2.20	7.273		3,775	230		4,005	4,500	
4220	18,620 CFM, 15 HP			2	8		5,400	253		5,653	6,325	
4240	27,580 CFM, 20 HP			1.80	8.889		6,575	281		6,856	7,675	
4260	40,980 CFM, 25 HP			1.50	10.667		10,300	335		10,635	11,800	
4280	60,920 CFM, 40 HP			1	16		13,900	505		14,405	16,100	
4300	74,520 CFM, 50 HP			.80	20		16,500	630		17,130	19,100	
4320	90,160 CFM, 50 HP			.70	22.857		22,500	720		23,220	25,900	
4340	110,300 CFM, 60 HP			.50	32		30,500	1,000		31,500	35,100	
4360	134,960 CFM, 75 HP			.40	40		40,500	1,275		41,775	46,600	
4500	Corrosive fume resistant, plastic											
4600	roof ventilators, centrifugal, V belt drive, motor											
4620	1/4" S.P., 250 CFM, 1/4 HP		Q-20	6	3.333	Ea.	2,425	108		2,533	2,825	
4640	895 CFM, 1/3 HP			5	4		2,625	129		2,754	3,075	
4660	1630 CFM, 1/2 HP			4	5		3,100	162		3,262	3,650	
4680	2240 CFM, 1 HP			3	6.667		3,225	216		3,441	3,875	
4700	3810 CFM, 2 HP			2	10		3,600	325		3,925	4,450	
4710	5000 CFM, 2 HP			1.80	11.111		7,550	360		7,910	8,850	
4715	8000 CFM, 5 HP			1.40	14.286		8,400	460		8,860	9,950	
4720	11760 CFM, 5 HP			1	20		8,525	645		9,170	10,400	
4740	18810 CFM, 10 HP			.70	28.571		8,850	925		9,775	11,200	
4800	For intermediate capacity, motors may be varied											
4810	For explosion proof motor, add					Ea.	15%					
5000	Utility set, centrifugal, V belt drive, motor											
5020	1/4" S.P., 1200 CFM, 1/4 HP		Q-20	6	3.333	Ea.	2,400	108		2,508	2,825	
5040	1520 CFM, 1/3 HP			5	4		2,400	129		2,529	2,850	
5060	1850 CFM, 1/2 HP			4	5		2,425	162		2,587	2,900	
5080	2180 CFM, 3/4 HP			3	6.667		2,450	216		2,666	3,000	
5100	1/2" S.P., 3600 CFM, 1 HP			2	10		3,550	325		3,875	4,400	
5120	4250 CFM, 1-1/2 HP			1.60	12.500		3,575	405		3,980	4,575	
5140	4800 CFM, 2 HP			1.40	14.286		3,625	460		4,085	4,700	
5160	6920 CFM, 5 HP			1.30	15.385		3,750	500		4,250	4,875	
5180	7700 CFM, 7-1/2 HP			1.20	16.667		3,850	540		4,390	5,075	
5200	For explosion proof motor, add						15%					
5300	Fume exhauster without hose and intake nozzle											
5310	630 CFM, 1-1/2 HP		Q-20	2	10	Ea.	730	325		1,055	1,300	
5320	Hose extension kit 5'		"	20	1	"	54	32.50		86.50	109	
5500	Industrial exhauster, for air which may contain granular material											
5520	1000 CFM, 1-1/2 HP		Q-20	2.50	8	Ea.	2,000	259		2,259	2,600	

15 MECHANICAL

Important: See the Reference Section for critical supporting data - Reference Nos., Crews, & City Cost Indexes

			DAILY	LABOR-		2002 BARE COSTS				TOTAL	
15830	**Fans**	**CREW**	**OUTPUT**	**HOURS**	**UNIT**	**MAT.**	**LABOR**	**EQUIP.**	**TOTAL**	**INCL O&P**	
100	5540	2000 CFM, 3 HP R15700-040	Q-20	2	10	Ea.	2,425	325		2,750	3,175	100
	5560	4000 CFM, 7-1/2 HP		1.80	11.111		3,925	360		4,285	4,850	
	5580	8000 CFM, 15 HP R15800-080		1.40	14.286		6,125	460		6,585	7,425	
	5600	12,000 CFM, 30 HP		1	20		8,500	645		9,145	10,300	
	6000	Propeller exhaust, wall shutter, 1/4" S.P.										
	6020	Direct drive, two speed										
	6100	375 CFM, 1/10 HP	Q-20	10	2	Ea.	207	64.50		271.50	325	
	6120	730 CFM, 1/7 HP		9	2.222		231	72		303	365	
	6140	1000 CFM, 1/8 HP		8	2.500		320	81		401	480	
	6160	1890 CFM, 1/4 HP		7	2.857		287	92.50		379.50	455	
	6180	3275 CFM, 1/2 HP		6	3.333		335	108		443	530	
	6200	4720 CFM, 1 HP		5	4		515	129		644	765	
	6300	V-belt drive, 3 phase										
	6320	6175 CFM, 3/4 HP	Q-20	5	4	Ea.	430	129		559	675	
	6340	7500 CFM, 3/4 HP		5	4		455	129		584	700	
	6360	10,100 CFM, 1 HP		4.50	4.444		555	144		699	830	
	6380	14,300 CFM, 1-1/2 HP		4	5		655	162		817	970	
	6400	19,800 CFM, 2 HP		3	6.667		750	216		966	1,150	
	6420	26,250 CFM, 3 HP		2.60	7.692		895	249		1,144	1,375	
	6440	38,500 CFM, 5 HP		2.20	9.091		1,050	294		1,344	1,625	
	6460	46,000 CFM, 7-1/2 HP		2	10		1,125	325		1,450	1,750	
	6480	51,500 CFM, 10 HP		1.80	11.111		1,175	360		1,535	1,850	
	6490	V-belt drive, 115V., residential, whole house										
	6500	Ceiling-wall, 5200 CFM, 1/4 HP, 30" x 30"	1 Shee	6	1.333	Ea.	226	47		273	320	
	6510	7500 CFM, 1/3 HP, 36"x36"		5	1.600		236	56		292	345	
	6520	10,500 CFM, 1/3 HP, 42"x42"		5	1.600		247	56		303	360	
	6530	13,200 CFM, 1/3 HP, 48" x 48"		4	2		273	70		343	410	
	6540	15,445 CFM, 1/2 HP, 48" x 48"		4	2		284	70		354	420	
	6550	17,025 CFM, 1/2 HP, 54" x 54"		4	2		545	70		615	710	
	6560	For two speed motor, add					20%					
	6570	Shutter, automatic, ceiling/wall										
	6580	30" x 30"	1 Shee	8	1	Ea.	73.50	35		108.50	135	
	6590	36" x 36"		8	1		84	35		119	147	
	6600	42" x 42"		8	1		105	35		140	170	
	6610	48" x 48"		7	1.143		105	40		145	178	
	6620	54" x 54"		6	1.333		150	47		197	237	
	6630	Timer, shut off, to 12 Hr.		20	.400		24	14.05		38.05	48	
	6650	Residential, bath exhaust, grille, back draft damper										
	6660	50 CFM	Q-20	24	.833	Ea.	46	27		73	92	
	6670	110 CFM		22	.909		51.50	29.50		81	102	
	6672	180 CFM		22	.909		81	29.50		110.50	134	
	6673	210 CFM		22	.909		109	29.50		138.50	164	
	6674	260 CFM		22	.909		122	29.50		151.50	179	
	6675	300 CFM		22	.909		166	29.50		195.50	227	
	6680	Light combination, squirrel cage, 100 watt, 70 CFM		24	.833		63.50	27		90.50	112	
	6700	Light/heater combination, ceiling mounted										
	6710	70 CFM, 1450 watt	Q-20	24	.833	Ea.	77	27		104	127	
	6800	Heater combination, recessed, 70 CFM		24	.833		36.50	27		63.50	81.50	
	6820	With 2 infrared bulbs		23	.870		55.50	28		83.50	104	
	6840	Wall mount, 170 CFM		22	.909		109	29.50		138.50	165	
	6846	Ceiling mount, 180 CFM		22	.909		110	29.50		139.50	166	
	6900	Kitchen exhaust, grille, complete, 160 CFM		22	.909		64.50	29.50		94	116	
	6910	180 CFM		20	1		54.50	32.50		87	110	
	6920	270 CFM		18	1.111		99	36		135	164	
	6930	350 CFM		16	1.250		77	40.50		117.50	147	
	6940	Residential roof jacks and wall caps										

MECHANICAL 15

15830	Fans		CREW	DAILY OUTPUT	LABOR-HOURS	UNIT	2002 BARE COSTS				TOTAL INCL O&P
							MAT.	LABOR	EQUIP.	TOTAL	
6944	Wall cap with back draft damper										
6946	3" & 4" dia. round duct	R15700-040	1 Shee	11	.727	Ea.	13.25	25.50		38.75	54
6948	6" dia. round duct		"	11	.727	"	32	25.50		57.50	74.50
6958	Roof jack with bird screen and back draft damper	R15800-080									
6960	3" & 4" dia. round duct		1 Shee	11	.727	Ea.	12.75	25.50		38.25	53.50
6962	3-1/4" x 10" rectangular duct		"	10	.800	"	23.50	28		51.50	69.50
6980	Transition										
6982	3-1/4" x 10" to 6" dia. round		1 Shee	20	.400	Ea.	13.85	14.05		27.90	37
7000	Roof exhauster, centrifugal, aluminum housing, 12" galvanized										
7020	curb, bird screen, back draft damper, 1/4" S.P.										
7100	Direct drive, 320 CFM, 11" sq. damper		Q-20	7	2.857	Ea.	300	92.50		392.50	470
7120	600 CFM, 11" sq. damper			6	3.333		305	108		413	500
7140	815 CFM, 13" sq. damper			5	4		305	129		434	535
7160	1450 CFM, 13" sq. damper			4.20	4.762		390	154		544	665
7180	2050 CFM, 16" sq. damper			4	5		390	162		552	680
7200	V-belt drive, 1650 CFM, 12" sq. damper			6	3.333		740	108		848	980
7220	2750 CFM, 21" sq. damper			5	4		855	129		984	1,150
7230	3500 CFM, 21" sq. damper			4.50	4.444		945	144		1,089	1,275
7240	4910 CFM, 23" sq. damper			4	5		1,175	162		1,337	1,550
7260	8525 CFM, 28" sq. damper			3	6.667		1,475	216		1,691	1,925
7280	13,760 CFM, 35" sq. damper			2	10		2,025	325		2,350	2,725
7300	20,558 CFM, 43" sq. damper			1	20		4,275	645		4,920	5,700
7320	For 2 speed winding, add						15%				
7340	For explosionproof motor, add					Ea.	320			320	350
7360	For belt driven, top discharge, add						15%				
7400	Roof mounted kitchen exhaust, aluminum, centrifugal										
7410	Direct drive, 2 speed, temp to 200°F										
7412	1/12 HP, 9-3/4"		Q-20	8	2.500	Ea.	305	81		386	460
7414	1/3 HP, 12-5/8"			7	2.857		375	92.50		467.50	550
7416	1/2 HP, 13-1/2"			6	3.333		415	108		523	620
7418	3/4 HP, 15"			5	4		590	129		719	850
7424	Belt drive, temp to 250°F										
7426	3/4 HP, 20"		Q-20	5	4	Ea.	1,300	129		1,429	1,625
7428	1-1/2 HP, 24-1/2"			4	5		1,425	162		1,587	1,825
7430	1-1/2 HP, 30"			4	5		1,900	162		2,062	2,325
7450	Upblast, propeller, w/BDD, BS, + C										
7454	30,300 CFM @ 3/8" S.P., 5 HP		Q-20	1.60	12.500	Ea.	1,875	405		2,280	2,700
7458	36,000 CFM @ 1/2" S.P., 15 HP		"	1.60	12.500	"	2,675	405		3,080	3,575
7500	Utility set, steel construction, pedestal, 1/4" S.P.										
7520	Direct drive, 150 CFM, 1/8 HP		Q-20	6.40	3.125	Ea.	605	101		706	820
7540	485 CFM, 1/6 HP			5.80	3.448		760	112		872	1,000
7560	1950 CFM, 1/2 HP			4.80	4.167		895	135		1,030	1,175
7580	2410 CFM, 3/4 HP			4.40	4.545		1,650	147		1,797	2,050
7600	3328 CFM, 1-1/2 HP			3	6.667		1,825	216		2,041	2,350
7680	V-belt drive, drive cover, 3 phase										
7700	800 CFM, 1/4 HP		Q-20	6	3.333	Ea.	435	108		543	645
7720	1,300 CFM, 1/3 HP			5	4		455	129		584	700
7740	2,000 CFM, 1 HP			4.60	4.348		540	141		681	810
7760	2,900 CFM, 3/4 HP			4.20	4.762		730	154		884	1,050
7780	3,600 CFM, 3/4 HP			4	5		900	162		1,062	1,250
7800	4,800 CFM, 1 HP			3.50	5.714		1,050	185		1,235	1,450
7820	6,700 CFM, 1-1/2 HP			3	6.667		1,300	216		1,516	1,775
7830	7,500 CFM, 2 HP			2.50	8		1,300	259		1,559	1,850
7840	11,000 CFM, 3 HP			2	10		2,375	325		2,700	3,100
7860	13,000 CFM, 3 HP			1.60	12.500		2,400	405		2,805	3,275
7880	15,000 CFM, 5 HP			1	20		2,500	645		3,145	3,750

Important: See the Reference Section for critical supporting data - Reference Nos., Crews, & City Cost Indexes

15830 | Fans

			CREW	DAILY OUTPUT	LABOR-HOURS	UNIT	2002 BARE COSTS				TOTAL INCL O&P
							MAT.	LABOR	EQUIP.	TOTAL	
100	7900	17,000 CFM, 7-1/2 HP	Q-20	.80	25	Ea.	2,675	810		3,485	4,175
	7920	20,000 CFM, 7-1/2 HP	↓	.80	25	↓	3,175	810		3,985	4,750
	8000	Ventilation, residential									
	8020	Attic, roof type									
	8030	Aluminum dome, damper & curb									
	8040	6" diameter, 300 CFM	1 Elec	16	.500	Ea.	224	17.75		241.75	273
	8050	7" diameter, 450 CFM		15	.533		245	18.90		263.90	297
	8060	9" diameter, 900 CFM		14	.571		395	20.50		415.50	460
	8080	12" diameter, 1000 CFM (gravity)		10	.800		238	28.50		266.50	305
	8090	16" diameter, 1500 CFM (gravity)		9	.889		288	31.50		319.50	360
	8100	20" diameter, 2500 CFM (gravity)		8	1		355	35.50		390.50	445
	8110	26" diameter, 4000 CFM (gravity)		7	1.143		425	40.50		465.50	530
	8120	32" diameter, 6500 CFM (gravity)		6	1.333		585	47.50		632.50	715
	8130	38" diameter, 8000 CFM (gravity)		5	1.600		870	56.50		926.50	1,050
	8140	50" diameter, 13,000 CFM (gravity)	↓	4	2	↓	1,250	71		1,321	1,475
	8160	Plastic, ABS dome									
	8180	1050 CFM	1 Elec	14	.571	Ea.	70	20.50		90.50	107
	8200	1600 CFM	"	12	.667	"	105	23.50		128.50	152
	8240	Attic, wall type, with shutter, one speed									
	8250	12" diameter, 1000 CFM	1 Elec	14	.571	Ea.	176	20.50		196.50	224
	8260	14" diameter, 1500 CFM		12	.667		191	23.50		214.50	246
	8270	16" diameter, 2000 CFM	↓	9	.889	↓	216	31.50		247.50	285
	8290	Whole house, wall type, with shutter, one speed									
	8300	30" diameter, 4800 CFM	1 Elec	7	1.143	Ea.	465	40.50		505.50	570
	8310	36" diameter, 7000 CFM		6	1.333		505	47.50		552.50	625
	8320	42" diameter, 10,000 CFM		5	1.600		565	56.50		621.50	705
	8330	48" diameter, 16,000 CFM	↓	4	2		700	71		771	880
	8340	For two speed, add				↓	42			42	46
	8350	Whole house, lay-down type, with shutter, one speed									
	8360	30" diameter, 4500 CFM	1 Elec	8	1	Ea.	495	35.50		530.50	600
	8370	36" diameter, 6500 CFM		7	1.143		530	40.50		570.50	645
	8380	42" diameter, 9000 CFM		6	1.333		585	47.50		632.50	710
	8390	48" diameter, 12,000 CFM	↓	5	1.600		660	56.50		716.50	815
	8440	For two speed, add					31.50			31.50	34.50
	8450	For 12 hour timer switch, add	1 Elec	32	.250	↓	31.50	8.85		40.35	48
	8500	Wall exhausters, centrifugal, auto damper, 1/8" S.P.									
	8520	Direct drive, 610 CFM, 1/20 HP	Q-20	14	1.429	Ea.	168	46		214	256
	8540	796 CFM, 1/12 HP		13	1.538		173	50		223	268
	8560	822 CFM, 1/6 HP		12	1.667		274	54		328	385
	8580	1,320 CFM, 1/4 HP		12	1.667		276	54		330	390
	8600	1756 CFM, 1/4 HP		11	1.818		325	59		384	445
	8620	1983 CFM, 1/4 HP		10	2		385	64.50		449.50	525
	8640	2900 CFM, 1/2 HP		9	2.222		465	72		537	625
	8660	3307 CFM, 3/4 HP		8	2.500		515	81		596	690
	8670	5940 CFM, 1 HP	↓	7	2.857	↓	850	92.50		942.50	1,075
	9500	V-belt drive, 3 phase									
	9520	2,800 CFM, 1/4 HP	Q-20	9	2.222	Ea.	760	72		832	945
	9540	3,740 CFM, 1/2 HP		8	2.500		790	81		871	995
	9560	4400 CFM, 3/4 HP		7	2.857		800	92.50		892.50	1,025
	9580	5700 CFM, 1-1/2 HP	↓	6	3.333	↓	835	108		943	1,075

15850 | Air Outlets & Inlets

			CREW	DAILY OUTPUT	LABOR-HOURS	UNIT	MAT.	LABOR	EQUIP.	TOTAL	TOTAL INCL O&P
300	0010	**DIFFUSERS** Aluminum, opposed blade damper unless noted									
	0100	Ceiling, linear, also for sidewall									
	0120	2" wide	1 Shee	32	.250	L.F.	26.50	8.80		35.30	42.50
	0140	3" wide	↓	30	.267	↓	30.50	9.35		39.85	48

				DAILY	LABOR-		2002 BARE COSTS				TOTAL	
15850		**Air Outlets & Inlets**		**OUTPUT**	**HOURS**	**UNIT**	**MAT.**	**LABOR**	**EQUIP.**	**TOTAL**	**INCL O&P**	
			CREW									
300	0160	4" wide	R15800 -080	1 Shee	26	.308	L.F.	34	10.80		44.80	54	300
	0180	6" wide			24	.333		41.50	11.70		53.20	63.50	
	0200	8" wide			22	.364		47	12.75		59.75	71.50	
	0220	10" wide			20	.400		53	14.05		67.05	79.50	
	0240	12" wide			18	.444		67.50	15.60		83.10	98	
	0260	For floor or sill application, add						15%					
	0500	Perforated, 24" x 24" lay-in panel size, 6" x 6"		1 Shee	16	.500	Ea.	68	17.55		85.55	102	
	0520	8" x 8"			15	.533		72	18.70		90.70	108	
	0530	9" x 9"			14	.571		78	20		98	117	
	0540	10" x 10"			14	.571		78.50	20		98.50	117	
	0560	12" x 12"			12	.667		85.50	23.50		109	130	
	0580	15" x 15"			11	.727		108	25.50		133.50	159	
	0590	16" x 16"			11	.727		137	25.50		162.50	191	
	0600	18" x 18"			10	.800		159	28		187	219	
	0610	20" x 20"			10	.800		185	28		213	247	
	0620	24" x 24"			9	.889		226	31		257	296	
	1000	Rectangular, 1 to 4 way blow, 6" x 6"			16	.500		36.50	17.55		54.05	67.50	
	1010	8" x 8"			15	.533		43	18.70		61.70	76	
	1014	9" x 9"			15	.533		46.50	18.70		65.20	80.50	
	1016	10" x 10"			15	.533		52.50	18.70		71.20	86.50	
	1020	12" x 6"			15	.533		58	18.70		76.70	92.50	
	1040	12" x 9"			14	.571		63	20		83	101	
	1060	12" x 12"			12	.667		64.50	23.50		88	107	
	1070	14" x 6"			13	.615		62.50	21.50		84	102	
	1074	14" x 14"			12	.667		80	23.50		103.50	124	
	1080	18" x 12"			11	.727		111	25.50		136.50	162	
	1120	15" x 15"			10	.800		116	28		144	172	
	1140	18" x 15"			9	.889		130	31		161	191	
	1150	18" x 18"			9	.889		118	31		149	177	
	1160	21" x 21"			8	1		149	35		184	218	
	1170	24" x 12"			10	.800		106	28		134	160	
	1180	24" x 24"			7	1.143		242	40		282	330	
	1500	Round, butterfly damper, 6" diameter			18	.444		15.25	15.60		30.85	41	
	1520	8" diameter			16	.500		16.40	17.55		33.95	45	
	1540	10" diameter			14	.571		20.50	20		40.50	53.50	
	1560	12" diameter			12	.667		26.50	23.50		50	65.50	
	1580	14" diameter			10	.800		34	28		62	80.50	
	1600	18" diameter			9	.889		79	31		110	135	
	1610	20" diameter			9	.889		101	31		132	159	
	1620	22" diameter			8	1		112	35		147	178	
	1630	24" diameter			8	1		124	35		159	190	
	1640	28" diameter			7	1.143		138	40		178	214	
	1650	32" diameter			6	1.333		152	47		199	239	
	1700	Round, stl, adjustable core, annular segmented damper											
	1704	6" diameter		1 Shee	18	.444	Ea.	67	15.60		82.60	97.50	
	1708	8" diameter			16	.500		72	17.55		89.55	106	
	1712	10" diameter			14	.571		81	20		101	120	
	1716	12" diameter			12	.667		90	23.50		113.50	135	
	1720	14" diameter			10	.800		100	28		128	154	
	1724	18" diameter			9	.889		153	31		184	216	
	1728	20" diameter			9	.889		177	31		208	243	
	1732	24" diameter			8	1		276	35		311	360	
	1736	30" diameter			7	1.143		480	40		520	590	
	1740	36" diameter			6	1.333		585	47		632	715	
	2000	T bar mounting, 24" x 24" lay-in frame, 6" x 6"			16	.500		76.50	17.55		94.05	112	
	2020	9" x 9"			14	.571		86.50	20		106.50	126	

Important: See the Reference Section for critical supporting data - Reference Nos., Crews, & City Cost Indexes

			DAILY	LABOR-			2002 BARE COSTS				TOTAL	
15850	**Air Outlets & Inlets**	CREW	OUTPUT	HOURS	UNIT	MAT.	LABOR	EQUIP.	TOTAL	INCL O&P	
300	2040	12" x 12"	R15850-060	1 Shee	12	.667	Ea.	113	23.50		136.50	160	**300**
	2060	15" x 15"			11	.727		146	25.50		171.50	200	
	2080	18" x 18"	R15810-070	▼	10	.800	▼	152	28		180	211	
	2500	Combination supply and return											
	2520	21" x 21" supply, 15" x 15" return	R15800-080	Q-9	10	1.600	Ea.	187	50.50		237.50	283	
	2540	24" x 24" supply, 18" x 18" return			9.50	1.684		235	53		288	340	
	2560	27" x 27" supply, 18" x 18" return			9	1.778		287	56		343	400	
	2580	30" x 30" supply, 21" x 21" return			8.50	1.882		365	59.50		424.50	490	
	2600	33" x 33" supply, 24" x 24" return			8	2		425	63		488	565	
	2620	36" x 36" supply, 24" x 24" return		▼	7.50	2.133	▼	500	67.50		567.50	655	
	3000	Baseboard, white enameled steel											
	3100	18" long		1 Shee	20	.400	Ea.	8.20	14.05		22.25	30.50	
	3120	24" long			18	.444		16.05	15.60		31.65	41.50	
	3140	48" long		▼	16	.500		29.50	17.55		47.05	59.50	
	3400	For matching return, deduct					▼	10%					
	4000	Floor, steel, adjustable pattern											
	4100	2" x 10"		1 Shee	34	.235	Ea.	11	8.25		19.25	25	
	4120	2" x 12"			32	.250		12.10	8.80		20.90	27	
	4140	2" x 14"			30	.267		13.75	9.35		23.10	29.50	
	4200	4" x 10"			28	.286		12.75	10.05		22.80	29.50	
	4220	4" x 12"			26	.308		14.10	10.80		24.90	32	
	4240	4" x 14"			25	.320		15.95	11.25		27.20	35	
	4260	6" x 10"			26	.308		19.30	10.80		30.10	37.50	
	4280	6" x 12"			24	.333		20	11.70		31.70	40	
	4300	6" x 14"		▼	22	.364	▼	21	12.75		33.75	42.50	
	4500	Linear bar diffuser in wall, sill or ceiling											
	4510	2" wide		1 Shee	32	.250	L.F.	25.50	8.80		34.30	41.50	
	4512	3" wide			30	.267		30	9.35		39.35	47.50	
	4514	4" wide			26	.308		32	10.80		42.80	52	
	4516	6" wide			24	.333		37.50	11.70		49.20	59.50	
	4520	10" wide			20	.400		46	14.05		60.05	72	
	4522	12" wide		▼	18	.444	▼	60	15.60		75.60	90	
	5000	Sidewall, aluminum, 3 way dispersion											
	5100	8" x 4"		1 Shee	24	.333	Ea.	6.05	11.70		17.75	24.50	
	5110	8" x 6"			24	.333		6.45	11.70		18.15	25	
	5120	10" x 4"			22	.364		6.35	12.75		19.10	26.50	
	5130	10" x 6"			22	.364		6.45	12.75		19.20	27	
	5140	10" x 8"			20	.400		6.80	14.05		20.85	29	
	5160	12" x 4"			18	.444		6.85	15.60		22.45	31.50	
	5170	12" x 6"			17	.471		7.55	16.50		24.05	34	
	5180	12" x 8"			16	.500		9.80	17.55		27.35	38	
	5200	14" x 4"			15	.533		7.60	18.70		26.30	37.50	
	5220	14" x 6"			14	.571		8.40	20		28.40	40.50	
	5240	14" x 8"			13	.615		10.90	21.50		32.40	45.50	
	5260	16" x 6"		▼	12.50	.640		12.05	22.50		34.55	48	
	6000	For steel diffusers instead of aluminum, deduct					▼	10%					
500	0010	**GRILLES**											**500**
	0020	Aluminum											
	0100	Air supply, single deflection, adjustable											
	0120	8" x 4"		1 Shee	30	.267	Ea.	11.30	9.35		20.65	27	
	0140	8" x 8"			28	.286		13.70	10.05		23.75	30.50	
	0160	10" x 4"			24	.333		12.45	11.70		24.15	32	
	0180	10" x 10"			23	.348		15.85	12.20		28.05	36.50	
	0200	12" x 6"			23	.348		14.75	12.20		26.95	35	
	0220	12" x 12"			22	.364		19.35	12.75		32.10	41	
	0230	14" x 6"		▼	23	.348	▼	16.45	12.20		28.65	37	

MECHANICAL 15

		CREW	DAILY OUTPUT	LABOR-HOURS	UNIT	2002 BARE COSTS				TOTAL INCL O&P	
15850	**Air Outlets & Inlets**					MAT.	LABOR	EQUIP.	TOTAL		
500 0240	14″ x 8″	1 Shee	23	.348	Ea.	17.55	12.20		29.75	38	**500**
0260	14″ x 14″		22	.364		24	12.75		36.75	45.50	
0270	18″ x 8″		23	.348		20.50	12.20		32.70	41.50	
0280	18″ x 10″		23	.348		23	12.20		35.20	44.50	
0300	18″ x 18″		21	.381		36	13.35		49.35	60.50	
0320	20″ x 12″		22	.364		26.50	12.75		39.25	48.50	
0340	20″ x 20″		21	.381		44	13.35		57.35	69	
0350	24″ x 8″		20	.400		25.50	14.05		39.55	49.50	
0360	24″ x 14″		18	.444		35.50	15.60		51.10	63	
0380	24″ x 24″		15	.533		62	18.70		80.70	97	
0400	30″ x 8″		20	.400		32.50	14.05		46.55	57.50	
0420	30″ x 10″		19	.421		36	14.80		50.80	62.50	
0440	30″ x 12″		18	.444		39.50	15.60		55.10	67.50	
0460	30″ x 16″		17	.471		51.50	16.50		68	82	
0480	30″ x 18″		17	.471		57.50	16.50		74	89	
0500	30″ x 30″		14	.571		103	20		123	144	
0520	36″ x 12″		17	.471		49	16.50		65.50	79.50	
0540	36″ x 16″		16	.500		62.50	17.55		80.05	96	
0560	36″ x 18″		15	.533		69.50	18.70		88.20	106	
0570	36″ x 20″		15	.533		76	18.70		94.70	113	
0580	36″ x 24″		14	.571		98.50	20		118.50	139	
0600	36″ x 28″		13	.615		121	21.50		142.50	167	
0620	36″ x 30″		12	.667		133	23.50		156.50	182	
0640	36″ x 32″		12	.667		146	23.50		169.50	197	
0660	36″ x 34″		11	.727		159	25.50		184.50	215	
0680	36″ x 36″		11	.727		172	25.50		197.50	230	
0700	For double deflecting, add					70%					
1000	Air return, 6″ x 6″	1 Shee	26	.308		12.35	10.80		23.15	30.50	
1020	10″ x 6″		24	.333		14.95	11.70		26.65	34.50	
1040	14″ x 6″		23	.348		17.55	12.20		29.75	38	
1060	10″ x 8″		23	.348		14.95	12.20		27.15	35.50	
1080	16″ x 8″		22	.364		22	12.75		34.75	44	
1100	12″ x 12″		22	.364		22	12.75		34.75	44	
1120	24″ x 12″		18	.444		39.50	15.60		55.10	67.50	
1140	30″ x 12″		16	.500		47	17.55		64.55	78.50	
1160	14″ x 14″		22	.364		27.50	12.75		40.25	49.50	
1180	16″ x 16″		22	.364		32	12.75		44.75	54.50	
1200	18″ x 18″		21	.381		37	13.35		50.35	61.50	
1220	24″ x 18″		16	.500		47	17.55		64.55	78.50	
1240	36″ x 18″		15	.533		78.50	18.70		97.20	116	
1260	24″ x 24″		15	.533		62	18.70		80.70	97	
1280	36″ x 24″		14	.571		93.50	20		113.50	134	
1300	48″ x 24″		12	.667		153	23.50		176.50	205	
1320	48″ x 30″		11	.727		198	25.50		223.50	257	
1340	36″ x 36″		13	.615		158	21.50		179.50	208	
1360	48″ x 36″		11	.727		222	25.50		247.50	285	
1380	48″ x 48″		8	1		296	35		331	380	
2000	Door grilles, 12″ x 12″		22	.364		31	12.75		43.75	54	
2020	18″ x 12″		22	.364		39	12.75		51.75	62.50	
2040	24″ x 12″		18	.444		46	15.60		61.60	75	
2060	18″ x 18″		18	.444		58.50	15.60		74.10	88.50	
2080	24″ x 18″		16	.500		73	17.55		90.55	107	
2100	24″ x 24″		15	.533		97	18.70		115.70	136	
3000	Filter grille with filter, 12″ x 12″		24	.333		36.50	11.70		48.20	58	
3020	18″ x 12″		20	.400		53.50	14.05		67.55	80	
3040	24″ x 18″		18	.444		77.50	15.60		93.10	109	

Important: See the Reference Section for critical supporting data - Reference Nos., Crews, & City Cost Indexes

15850	Air Outlets & Inlets	CREW	DAILY OUTPUT	LABOR-HOURS	UNIT	2002 BARE COSTS				TOTAL INCL O&P
						MAT.	LABOR	EQUIP.	TOTAL	
500										**500**
3060	24″ x 24″	1 Shee	16	.500	Ea.	101	17.55		118.55	139
3080	30″ x 24″		14	.571		130	20		150	174
3100	30″ x 30″		13	.615		164	21.50		185.50	214
3950	Eggcrate, framed, 6″ x 6″ opening		26	.308		13.45	10.80		24.25	31.50
3954	8″ x 8″ opening		24	.333		13.70	11.70		25.40	33
3960	10″ x 10″ opening		23	.348		22	12.20		34.20	43.50
3970	12″ x 12″ opening		22	.364		18.55	12.75		31.30	40
3980	14″ x 14″ opening		22	.364		23	12.75		35.75	45
3984	16″ x 16″ opening		21	.381		29.50	13.35		42.85	53
3990	18″ x 18″ opening		21	.381		35	13.35		48.35	59
4020	22″ x 22″ opening		17	.471		55.50	16.50		72	86.50
4040	24″ x 24″ opening		15	.533		58.50	18.70		77.20	93.50
4044	28″ x 28″ opening		14	.571		79	20		99	118
4048	36″ x 36″ opening		12	.667		94.50	23.50		118	140
4050	48″ x 24″ opening	▼	12	.667	▼	174	23.50		197.50	227
4060	Eggcrate, lay-in, T-bar system									
4070	48″ x 24″ sheet	1 Shee	40	.200	Ea.	76.50	7		83.50	95.50
5000	Transfer grille, vision proof, 8″ x 4″		30	.267		21	9.35		30.35	37.50
5020	8″ x 6″		28	.286		21	10.05		31.05	38.50
5040	8″ x 8″		26	.308		23.50	10.80		34.30	42
5060	10″ x 6″		24	.333		21.50	11.70		33.20	42
5080	10″ x 10″		23	.348		27.50	12.20		39.70	49
5090	12″ x 6″		23	.348		28	12.20		40.20	49.50
5100	12″ x 10″		22	.364		30	12.75		42.75	52.50
5110	12″ x 12″		22	.364		32.50	12.75		45.25	55
5120	14″ x 10″		22	.364		32.50	12.75		45.25	55
5140	16″ x 8″		21	.381		32.50	13.35		45.85	56
5160	18″ x 12″		20	.400		42.50	14.05		56.55	68
5170	18″ x 18″		21	.381		60	13.35		73.35	86.50
5180	20″ x 12″		20	.400		46	14.05		60.05	72
5200	20″ x 20″		19	.421		72.50	14.80		87.30	103
5220	24″ x 12″		17	.471		51.50	16.50		68	82
5240	24″ x 24″		16	.500		99	17.55		116.55	136
5260	30″ x 6″		15	.533		45.50	18.70		64.20	79.50
5280	30″ x 8″		15	.533		53	18.70		71.70	87.50
5300	30″ x 12″		14	.571		66.50	20		86.50	104
5320	30″ x 16″		13	.615		79	21.50		100.50	121
5340	30″ x 20″		12	.667		100	23.50		123.50	146
5360	30″ x 24″		11	.727		123	25.50		148.50	175
5380	30″ x 30″	▼	10	.800		162	28		190	222
6000	For steel grilles instead of aluminum in above, deduct				▼	10%				
6200	Plastic, eggcrate, lay-in, T-bar system									
6210	48″ x 24″ sheet	1 Shee	50	.160	Ea.	22	5.60		27.60	33
6250	Steel door louver									
6270	With fire link, steel only									
6350	12″ x 18″	1 Shee	22	.364	Ea.	135	12.75		147.75	169
6410	12″ x 24″		21	.381		150	13.35		163.35	186
6470	18″ x 18″		20	.400		150	14.05		164.05	187
6530	18″ x 24″		19	.421		172	14.80		186.80	212
6600	18″ x 36″		17	.471		195	16.50		211.50	241
6660	24″ x 24″		15	.533		181	18.70		199.70	228
6720	24″ x 30″		15	.533		195	18.70		213.70	244
6780	30″ x 30″		13	.615		229	21.50		250.50	286
6840	30″ x 36″		12	.667		265	23.50		288.50	330
6900	36″ x 36″	▼	10	.800	▼	305	28		333	380

MECHANICAL 15

		15850	Air Outlets & Inlets	CREW	DAILY OUTPUT	LABOR-HOURS	UNIT	2002 BARE COSTS				TOTAL INCL O&P	
								MAT.	LABOR	EQUIP.	TOTAL		
600	0010	**LOUVERS**											600
	0100	Aluminum, extruded, with screen, mill finish											
	1000	Brick vent, (see also division 04090)											
	1100	Standard, 4″ deep, 8″ wide, 5″ high	1 Shee	24	.333	Ea.	23	11.70		34.70	43.50	
	1200	Modular, 4″ deep, 7-3/4″ wide, 5″ high		24	.333		24	11.70		35.70	44.50	
	1300	Speed brick, 4″ deep, 11-5/8″ wide, 3-7/8″ high		24	.333		24	11.70		35.70	44.50	
	1400	Fuel oil brick, 4″ deep, 8″ wide, 5″ high		24	.333		42	11.70		53.70	64	
	2000	Cooling tower and mechanical equip., screens, light weight		40	.200	S.F.	9.95	7		16.95	22	
	2020	Standard weight		35	.229		27.50	8		35.50	43	
	2500	Dual combination, automatic, intake or exhaust		20	.400		38	14.05		52.05	63.50	
	2520	Manual operation		20	.400		28	14.05		42.05	52.50	
	2540	Electric or pneumatic operation		20	.400		28	14.05		42.05	52.50	
	2560	Motor, for electric or pneumatic		14	.571	Ea.	330	20		350	395	
	3000	Fixed blade, continuous line										
	3100	Mullion type, stormproof	1 Shee	28	.286	S.F.	28	10.05		38.05	46.50	
	3200	Stormproof		28	.286		28	10.05		38.05	46.50	
	3300	Vertical line		28	.286		37	10.05		47.05	56.50	
	3500	For damper to use with above, add					50%	30%				
	3520	Motor, for damper, electric or pneumatic	1 Shee	14	.571	Ea.	330	20		350	395	
	4000	Operating, 45°, manual, electric or pneumatic		24	.333	S.F.	28	11.70		39.70	49	
	4100	Motor, for electric or pneumatic		14	.571	Ea.	330	20		350	395	
	4200	Penthouse, roof		56	.143	S.F.	16.30	5		21.30	25.50	
	4300	Walls		40	.200		39.50	7		46.50	54	
	5000	Thinline, under 4″ thick, fixed blade		40	.200		15.80	7		22.80	28	
	5010	Finishes, applied by mfr. at additional cost, available in colors										
	5020	Prime coat only, add				S.F.	2.30			2.30	2.53	
	5040	Baked enamel finish coating, add					4.23			4.23	4.65	
	5060	Anodized finish, add					4.59			4.59	5.05	
	5080	Duranodic finish, add					8.30			8.30	9.15	
	5100	Fluoropolymer finish coating, add					13			13	14.30	
	9980	For small orders (under 10 pieces), add					25%					
700	0010	**REGISTERS**											700
	0980	Air supply											
	1000	Ceiling/wall, O.B. damper, anodized aluminum											
	1010	One or two way deflection, adj. curved face bars											
	1014	6″ x 6″	1 Shee	24	.333	Ea.	19.50	11.70		31.20	39.50	
	1020	8″ x 4″		26	.308		19.50	10.80		30.30	38	
	1040	8″ x 8″		24	.333		24.50	11.70		36.20	45	
	1060	10″ x 6″		20	.400		22.50	14.05		36.55	46	
	1080	10″ x 10″		19	.421		30	14.80		44.80	56	
	1100	12″ x 6″		19	.421		21	14.80		35.80	46.50	
	1120	12″ x 12″		18	.444		31	15.60		46.60	58	
	1140	14″ x 8″		17	.471		31	16.50		47.50	59.50	
	1160	14″ x 14″		18	.444		36	15.60		51.60	63.50	
	1170	16″ x 16″		17	.471		40	16.50		56.50	69.50	
	1180	18″ x 8″		18	.444		50	15.60		65.60	79	
	1200	18″ x 18″		17	.471		63.50	16.50		80	95	
	1220	20″ x 4″		19	.421		32	14.80		46.80	58	
	1240	20″ x 6″		18	.444		32	15.60		47.60	59	
	1260	20″ x 8″		18	.444		35	15.60		50.60	62.50	
	1280	20″ x 20″		17	.471		77	16.50		93.50	110	
	1290	22″ x 22″		15	.533		88.50	18.70		107.20	127	
	1300	24″ x 4″		17	.471		39.50	16.50		56	68.50	
	1320	24″ x 6″		16	.500		39.50	17.55		57.05	70	
	1340	24″ x 8″		13	.615		43	21.50		64.50	81	

Important: See the Reference Section for critical supporting data - Reference Nos., Crews, & City Cost Indexes

		DAILY	LABOR-		2002 BARE COSTS				TOTAL	
15850	Air Outlets & Inlets				MAT.	LABOR	EQUIP.	TOTAL	INCL O&P	
		CREW	OUTPUT	HOURS	UNIT					
1350	24" x 18"	1 Shee	12	.667	Ea.	80	23.50		103.50	124
1360	24" x 24"		11	.727		109	25.50		134.50	159
1380	30" x 4"		16	.500		49	17.55		66.55	81
1400	30" x 6"		15	.533		49	18.70		67.70	83
1420	30" x 8"		14	.571		56.50	20		76.50	93
1440	30" x 24"		12	.667		140	23.50		163.50	190
1460	30" x 30"	▼	10	.800	▼	180	28		208	242
1504	4 way deflection, adjustable curved face bars									
1510	6" x 6"	1 Shee	26	.308	Ea.	23.50	10.80		34.30	42
1514	8" x 8"		24	.333		29	11.70		40.70	50
1518	10" x 10"		19	.421		36	14.80		50.80	63
1522	12" x 6"		19	.421		25.50	14.80		40.30	51
1526	12" x 12"		18	.444		37	15.60		52.60	65
1530	14" x 14"		18	.444		43.50	15.60		59.10	71.50
1534	16" x 16"		17	.471		48	16.50		64.50	78.50
1538	18" x 18"		17	.471		76	16.50		92.50	109
1542	22" x 22"	▼	16	.500	▼	106	17.55		123.55	144
1980	One way deflection, adj. vert. or horiz. face bars									
1990	6" x 6"	1 Shee	26	.308	Ea.	20.50	10.80		31.30	39
2000	8" x 4"		26	.308		20	10.80		30.80	38.50
2020	8" x 8"		24	.333		24	11.70		35.70	44
2040	10" x 6"		20	.400		24	14.05		38.05	48
2060	10" x 10"		19	.421		28	14.80		42.80	53.50
2080	12" x 6"		19	.421		26	14.80		40.80	51.50
2100	12" x 12"		18	.444		33.50	15.60		49.10	61
2120	14" x 8"		17	.471		31	16.50		47.50	59.50
2140	14" x 12"		18	.444		37	15.60		52.60	65
2160	14" x 14"		18	.444		42	15.60		57.60	70
2180	16" x 6"		18	.444		31	15.60		46.60	58
2200	16" x 12"		18	.444		41	15.60		56.60	69
2220	16" x 16"		17	.471		51.50	16.50		68	82.50
2240	18" x 8"		18	.444		36	15.60		51.60	63.50
2260	18" x 12"		17	.471		43.50	16.50		60	73.50
2280	18" x 18"		16	.500		74.50	17.55		92.05	109
2300	20" x 10"		18	.444		42.50	15.60		58.10	71
2320	20" x 16"		16	.500		61.50	17.55		79.05	95
2340	20" x 20"		16	.500		77.50	17.55		95.05	113
2360	24" x 12"		15	.533		54	18.70		72.70	88
2380	24" x 16"		14	.571		71.50	20		91.50	110
2400	24" x 20"		12	.667		89.50	23.50		113	135
2420	24" x 24"		10	.800		110	28		138	164
2440	30" x 12"		13	.615		69.50	21.50		91	110
2460	30" x 16"		13	.615		90.50	21.50		112	133
2480	30" x 24"		11	.727		141	25.50		166.50	195
2500	30" x 30"		9	.889		181	31		212	247
2520	36" x 12"		11	.727		86.50	25.50		112	135
2540	36" x 24"		10	.800		174	28		202	236
2560	36" x 36"	▼	8	1		291	35		326	375
2600	For 2 way deflect., adj. vert. or horiz. face bars, add					40%				
2700	Above registers in steel instead of aluminum, deduct				▼	10%				
3000	Baseboard, hand adj. damper, enameled steel									
3012	8" x 6"	1 Shee	26	.308	Ea.	9	10.80		19.80	26.50
3020	10" x 6"		24	.333		10.10	11.70		21.80	29
3040	12" x 5"		23	.348		11.50	12.20		23.70	31.50
3060	12" x 6"		23	.348		10.60	12.20		22.80	30.50
3080	12" x 8"	▼	22	.364	▼	15.40	12.75		28.15	36.50

700 **700**

MECHANICAL **15**

			DAILY	LABOR-		2002 BARE COSTS				TOTAL	
15850	**Air Outlets & Inlets**	CREW	OUTPUT	HOURS	UNIT	MAT.	LABOR	EQUIP.	TOTAL	INCL O&P	
700	3100	14" x 6"	1 Shee	20	.400	Ea.	11.50	14.05		25.55	34	**700**
	4000	Floor, toe operated damper, enameled steel										
	4020	4" x 8"	1 Shee	32	.250	Ea.	16.65	8.80		25.45	32	
	4040	4" x 12"		26	.308		19.55	10.80		30.35	38	
	4060	6" x 8"		28	.286		19.50	10.05		29.55	37	
	4080	6" x 14"		22	.364		22.50	12.75		35.25	44.50	
	4100	8" x 10"		22	.364		20.50	12.75		33.25	42	
	4120	8" x 16"		20	.400		34.50	14.05		48.55	59	
	4140	10" x 10"		20	.400		24	14.05		38.05	48	
	4160	10" x 16"		18	.444		37	15.60		52.60	65	
	4180	12" x 12"		18	.444		30	15.60		45.60	57	
	4200	12" x 24"		16	.500		104	17.55		121.55	141	
	4220	14" x 14"		16	.500		85.50	17.55		103.05	121	
	4240	14" x 20"		15	.533		114	18.70		132.70	154	
	4980	Air return										
	5000	Ceiling or wall, fixed 45° face blades										
	5010	Adjustable O.B. damper, anodized aluminum										
	5020	4" x 8"	1 Shee	26	.308	Ea.	22	10.80		32.80	41	
	5040	6" x 8"		24	.333		22	11.70		33.70	42.50	
	5060	6" x 10"		19	.421		24.50	14.80		39.30	50	
	5070	6" x 12"		19	.421		27.50	14.80		42.30	53	
	5080	6" x 16"		18	.444		30	15.60		45.60	57	
	5100	8" x 10"		19	.421		27.50	14.80		42.30	53	
	5120	8" x 12"		16	.500		30	17.55		47.55	60	
	5140	10" x 10"		18	.444		30	15.60		45.60	57	
	5160	10" x 16"		17	.471		39	16.50		55.50	68.50	
	5170	12" x 12"		18	.444		37	15.60		52.60	65	
	5180	12" x 18"		18	.444		50	15.60		65.60	79	
	5184	12" x 24"		14	.571		62	20		82	99	
	5200	12" x 30"		12	.667		76.50	23.50		100	121	
	5210	12" x 36"		10	.800		96	28		124	150	
	5220	16" x 16"		17	.471		53.50	16.50		70	84	
	5240	18" x 18"		16	.500		64.50	17.55		82.05	98	
	5250	18" x 24"		13	.615		81.50	21.50		103	123	
	5254	18" x 30"		11	.727		101	25.50		126.50	152	
	5260	18" x 36"		10	.800		129	28		157	186	
	5270	20" x 20"		10	.800		76.50	28		104.50	128	
	5274	20" x 24"		12	.667		89.50	23.50		113	135	
	5280	24" x 24"		11	.727		106	25.50		131.50	157	
	5281	24" x 24"		11	.727		106	25.50		131.50	157	
	5290	24" x 30"		10	.800		136	28		164	193	
	5300	24" x 36"		8	1		166	35		201	236	
	5320	24" x 48"		6	1.333		239	47		286	335	
	5340	30" x 30"		6.50	1.231		172	43		215	256	
	5344	30" x 120"	Q-9	6	2.667		730	84.50		814.50	935	
	5360	36" x 36"	1 Shee	6	1.333		278	47		325	375	
	5370	40" x 32"	"	6	1.333		261	47		308	360	
	5400	Ceiling or wall, removable-reversible core, single deflection										
	5402	Adjustable O.B. damper, radiused frame with border, aluminum										
	5410	8" x 4"	1 Shee	26	.308	Ea.	32	10.80		42.80	51.50	
	5414	8" x 6"		24	.333		36	11.70		47.70	57.50	
	5418	10" x 6"		19	.421		39	14.80		53.80	66	
	5422	10" x 10"		18	.444		49	15.60		64.60	78	
	5426	12" x 6"		19	.421		42	14.80		56.80	69	
	5430	12" x 12"		18	.444		63	15.60		78.60	93.50	
	5434	16" x 16"		16	.500		94	17.55		111.55	130	

Important: See the Reference Section for critical supporting data - Reference Nos., Crews, & City Cost Indexes

15850 | Air Outlets & Inlets

		CREW	DAILY OUTPUT	LABOR-HOURS	UNIT	2002 BARE COSTS				TOTAL INCL O&P
						MAT.	LABOR	EQUIP.	TOTAL	
700	5438	18" x 18"	1 Shee	17	.471	Ea.	109	16.50		125.50	146	700
	5442	20" x 20"		16	.500		134	17.55		151.55	174
	5446	24" x 12"		15	.533		92	18.70		110.70	130
	5450	24" x 18"		13	.615		135	21.50		156.50	183
	5454	24" x 24"		11	.727		175	25.50		200.50	233
	5458	30" x 12"		13	.615		110	21.50		131.50	155
	5462	30" x 18"		12.50	.640		158	22.50		180.50	209
	5466	30" x 24"		11	.727		210	25.50		235.50	271
	5470	30" x 30"		9	.889		262	31		293	335
	5474	36" x 12"		11	.727		133	25.50		158.50	186
	5478	36" x 18"		10.50	.762		187	26.50		213.50	247
	5482	36" x 24"		10	.800		259	28		287	330
	5486	36" x 30"		9	.889		340	31		371	420
	5490	36" x 36"		8	1		425	35		460	520
	6000	For steel construction instead of aluminum, deduct					10%				

15854 | Ventilators

		CREW	DAILY OUTPUT	LABOR-HOURS	UNIT	MAT.	LABOR	EQUIP.	TOTAL	TOTAL INCL O&P
300	0010	**VENTILATORS** Base & damper										300
	1280	Rotary ventilators, wind driven, galvanized									
	1300	4" neck diameter	Q-9	20	.800	Ea.	52	25.50		77.50	96.50
	1320	5" neck diameter		18	.889		58	28		86	108
	1340	6" neck diameter		16	1		58	31.50		89.50	113
	1360	8" neck diameter		14	1.143		60.50	36		96.50	122
	1380	10" neck diameter		12	1.333		77.50	42		119.50	151
	1400	12" neck diameter		10	1.600		84.50	50.50		135	171
	1420	14" neck diameter		10	1.600		117	50.50		167.50	207
	1440	16" neck diameter		9	1.778		141	56		197	243
	1460	18" neck diameter		9	1.778		164	56		220	268
	1480	20" neck diameter, 2100 CFM		8	2		182	63		245	298
	1500	24" neck diameter, 3,100 CFM		8	2		245	63		308	370
	1520	30" neck diameter, 4500 CFM		7	2.286		510	72		582	670
	1540	36" neck diameter, 5,500 CFM		6	2.667		720	84.50		804.50	920
	1600	For aluminum, add					300%				
	1620	For stainless steel, add					600%				
	1630	For copper, add					600%				
	2000	Stationary, gravity, syphon, galvanized									
	2100	3" diameter, 40 CFM	Q-9	24	.667	Ea.	47	21		68	84
	2120	4" neck diameter, 50 CFM		20	.800		51.50	25.50		77	96
	2140	5" neck diameter, 58 CFM		18	.889		52	28		80	101
	2160	6" neck diameter, 66 CFM		16	1		45	31.50		76.50	98
	2180	7" neck diameter, 86 CFM		15	1.067		56.50	33.50		90	115
	2200	8" neck diameter, 110 CFM		14	1.143		57	36		93	118
	2220	10" neck diameter, 140 CFM		12	1.333		73.50	42		115.50	146
	2240	12" neck diameter, 160 CFM		10	1.600		94.50	50.50		145	182
	2260	14" neck diameter, 250 CFM		10	1.600		119	50.50		169.50	208
	2280	16" neck diameter, 380 CFM		9	1.778		158	56		214	261
	2300	18" neck diameter, 500 CFM		9	1.778		146	56		202	247
	2320	20" neck diameter, 625 CFM		8	2		223	63		286	345
	2340	24" neck diameter, 900 CFM		8	2		257	63		320	380
	2360	30" neck diameter, 1375 CFM		7	2.286		585	72		657	755
	2380	36" neck diameter, 2,000 CFM		6	2.667		805	84.50		889.50	1,025
	2400	42" neck diameter, 3000 CFM		4	4		1,125	126		1,251	1,450
	2410	48" neck diameter, 4000 CFM		3	5.333		1,525	169		1,694	1,925
	2500	For aluminum, add					300%				
	2520	For stainless steel, add					600%				

MECHANICAL 15

		15854	Ventilators	CREW	DAILY OUTPUT	LABOR-HOURS	UNIT	2002 BARE COSTS				TOTAL INCL O&P	
								MAT.	LABOR	EQUIP.	TOTAL		
300	2530		For copper, add				Ea.	600%					300
	3000		Rotating chimney cap, galvanized, 4" neck diameter	Q-9	20	.800		40.50	25.50		66	83.50	
	3020		5" neck diameter		18	.889		41.50	28		69.50	89	
	3040		6" neck diameter		16	1		42	31.50		73.50	94.50	
	3060		7" neck diameter		15	1.067		45.50	33.50		79	102	
	3080		8" neck diameter		14	1.143		47	36		83	108	
	3100		10" neck diameter		12	1.333		61	42		103	132	
	3600		Stationary chimney rain cap, galvanized, 3" neck diameter		24	.667		35.50	21		56.50	71.50	
	3620		4" neck diameter		20	.800		36	25.50		61.50	78.50	
	3640		6" neck diameter		16	1		37.50	31.50		69	90	
	3680		8" neck diameter		14	1.143		40.50	36		76.50	100	
	3700		10" neck diameter		12	1.333		51.50	42		93.50	122	
	3720		12" neck diameter		10	1.600		85.50	50.50		136	172	
	3740		14" neck diameter		10	1.600		91.50	50.50		142	179	
	3760		16" neck diameter		9	1.778		112	56		168	210	
	3770		18" neck diameter		9	1.778		162	56		218	265	
	3780		20" neck diameter		8	2		216	63		279	335	
	3790		24" neck diameter		8	2		252	63		315	375	
	4200		Stationary mushroom, aluminum, 16" orifice diameter		10	1.600		258	50.50		308.50	360	
	4220		26" orifice diameter		6.15	2.602		380	82		462	545	
	4230		30" orifice diameter		5.71	2.802		560	88.50		648.50	750	
	4240		38" orifice diameter		5	3.200		800	101		901	1,025	
	4250		42" orifice diameter		4.70	3.404		1,050	108		1,158	1,350	
	4260		50" orifice diameter	▼	4.44	3.604	▼	1,250	114		1,364	1,550	
	5000		Relief vent										
	5500		Rectangular, aluminum, galvanized curb										
	5510		intake/exhaust, 0.05" SP										
	5600		500 CFM, 12" x 16"	Q-9	8	2	Ea.	385	63		448	520	
	5620		750 CFM, 12" x 20"		7.20	2.222		420	70		490	575	
	5640		1000 CFM, 12" x 24"		6.60	2.424		455	76.50		531.50	620	
	5660		1500 CFM, 12" x 36"		5.80	2.759		550	87		637	740	
	5680		3000 CFM, 20" x 42"		4	4		790	126		916	1,075	
	5700		6000 CFM, 20" x 84"		2.60	6.154		1,200	194		1,394	1,625	
	5720		8000 CFM, 24" x 96"		2.30	6.957		1,375	220		1,595	1,850	
	5740		10,000 CFM, 48" x 60"		1.80	8.889		1,675	281		1,956	2,275	
	5760		12,500 CFM, 48" x 72"		1.60	10		1,975	315		2,290	2,650	
	5780		15,000 CFM, 48" x 96"		1.30	12.308		2,300	390		2,690	3,125	
	5800		20,000 CFM, 48" x 120"		1.20	13.333		2,650	420		3,070	3,550	
	5820		25,000 CFM, 60" x 120"		.90	17.778		3,500	560		4,060	4,725	
	5840		30,000 CFM, 72" x 120"		.70	22.857		4,600	720		5,320	6,175	
	5860		40,000 CFM, 96" x 120"		.60	26.667		5,925	845		6,770	7,800	
	5880		50,000 CFM, 96" x 144"	▼	.50	32	▼	6,275	1,000		7,275	8,450	
	7000		Note: sizes based on exhaust. Intake, with 0.125" SP										
	7100		loss, approximately twice listed capacity.										
	7500		For power open, spring return damper										
	7600		motor for each S.F. of throat area	Q-9	100	.160	S.F.	40.50	5.05		45.55	53	

		15860	Air Cleaning Devices										
100	0010		AIR FILTERS										100
	0050		Activated charcoal type, full flow				MCFM	600			600	660	
	0500		Chemical media filtration type										
	1100		Industrial air fume & odor scrubber unit w/pump & motor										
	1110		corrosion resistant PVC construction										
	1120		Single pack filter, horizontal type										
	1130		500 CFM	Q-9	14	1.143	Ea.	3,800	36		3,836	4,225	
	1140		1000 CFM	▼	11	1.455	▼	4,450	46		4,496	4,975	

15 MECHANICAL

Important: See the Reference Section for critical supporting data - Reference Nos., Crews, & City Cost Indexes

15860	Air Cleaning Devices	CREW	DAILY OUTPUT	LABOR-HOURS	UNIT	2002 BARE COSTS				TOTAL INCL O&P
						MAT.	LABOR	EQUIP.	TOTAL	
1150	2000 CFM	Q-9	8	2	Ea.	5,400	63		5,463	6,050
1160	3000 CFM		7	2.286		6,250	72		6,322	7,000
1170	5000 CFM		5	3.200		8,150	101		8,251	9,125
1180	8000 CFM		4	4		11,400	126		11,526	12,700
1190	12,000 CFM		3	5.333		15,100	169		15,269	16,900
1200	16,000 CFM		2.50	6.400		18,700	202		18,902	20,800
1210	20,000 CFM		2	8		23,400	253		23,653	26,200
1220	26,000 CFM	↓	1.50	10.667		27,000	335		27,335	30,200
1230	30,000 CFM	Q-10	2	12		30,700	395		31,095	34,300
1240	40,000 CFM	"	1.50	16		40,100	525		40,625	44,900
1250	50,000 CFM	Q-11	2	16		48,800	535		49,335	54,500
1260	55,000 CFM		1.80	17.778		53,500	595		54,095	59,500
1270	60,000 CFM	↓	1.50	21.333	↓	57,500	715		58,215	64,000
1300	Double pack filter, horizontal type									
1310	500 CFM	Q-9	10	1.600	Ea.	4,950	50.50		5,000.50	5,525
1320	1000 CFM		8	2		5,925	63		5,988	6,625
1330	2000 CFM		6	2.667		7,150	84.50		7,234.50	7,975
1340	3000 CFM		5	3.200		8,550	101		8,651	9,550
1350	5000 CFM		4	4		10,900	126		11,026	12,100
1360	8000 CFM		3	5.333		14,300	169		14,469	16,100
1370	12,000 CFM		2.50	6.400		19,500	202		19,702	21,700
1380	16,000 CFM		2	8		25,400	253		25,653	28,300
1390	20,000 CFM		1.50	10.667		30,800	335		31,135	34,400
1400	26,000 CFM	↓	1	16		35,700	505		36,205	40,000
1410	30,000 CFM	Q-10	1.50	16		41,800	525		42,325	46,800
1420	40,000 CFM	"	1.30	18.462		53,500	605		54,105	60,000
1430	50,000 CFM	Q-11	1.50	21.333		67,000	715		67,715	74,500
1440	55,000 CFM		1.30	24.615		73,000	825		73,825	82,000
1450	60,000 CFM	↓	1	32	↓	79,500	1,075		80,575	88,500
1500	Single pack filter, vertical type									
1510	500 CFM	Q-9	24	.667	Ea.	3,625	21		3,646	4,025
1520	1000 CFM		18	.889		4,300	28		4,328	4,775
1530	2000 CFM		12	1.333		5,150	42		5,192	5,725
1540	3000 CFM		9	1.778		6,000	56		6,056	6,700
1550	5000 CFM		6	2.667		8,000	84.50		8,084.50	8,950
1560	8000 CFM		4	4		11,400	126		11,526	12,700
1570	12,000 CFM		3	5.333		15,500	169		15,669	17,400
1580	16,000 CFM		2	8		18,200	253		18,453	20,400
1590	20,000 CFM		1.80	8.889		22,900	281		23,181	25,600
1600	24,000 CFM	↓	1.60	10	↓	25,600	315		25,915	28,600
1650	Double pack filter, vertical type									
1660	500 CFM	Q-9	22	.727	Ea.	4,700	23		4,723	5,200
1670	1000 CFM		16	1		5,625	31.50		5,656.50	6,225
1680	2000 CFM		10	1.600		6,750	50.50		6,800.50	7,500
1690	3000 CFM		7	2.286		7,925	72		7,997	8,825
1700	5000 CFM		5	3.200		10,400	101		10,501	11,600
1710	8000 CFM		3	5.333		14,200	169		14,369	15,900
1720	12,000 CFM		2.50	6.400		19,200	202		19,402	21,400
1730	16,000 CFM		2	8		24,100	253		24,353	27,000
1740	20,000 CFM		1.50	10.667		30,100	335		30,435	33,600
1750	24,000 CFM	↓	1	16	↓	33,600	505		34,105	37,800
1800	Inlet or outlet transition, horizontal									
1810	Single pack to 12,000 CFM, add					4%				
1820	Single pack to 30,000 CFM, add					6%				
1830	Single pack to 60,000 CFM, add					8%				
1840	Double pack to 12,000 CFM, add					3%				

100

15860	Air Cleaning Devices		DAILY	LABOR-		2002 BARE COSTS				TOTAL	
		CREW	OUTPUT	HOURS	UNIT	MAT.	LABOR	EQUIP.	TOTAL	INCL O&P	
100 1850	Double pack to 30,000 CFM, add					5%					**100**
1860	Double pack to 60,000 CFM, add					6%					
1870	Inlet or outlet transition, vertical										
1880	Single pack to 5000 CFM, add					2%					
1890	Single pack to 24,000 CFM, add					3%					
1900	Double pack to 24,000 CFM, add					2%					
2000	Electronic air cleaner, duct mounted										
2150	400 - 1000 CFM	1 Shee	2.30	3.478	Ea.	685	122		807	945	
2200	1000 - 1400 CFM		2.20	3.636		720	128		848	985	
2250	1400 - 2000 CFM		2.10	3.810		795	134		929	1,075	
2260	2000 - 2500 CFM	▼	2	4	▼	815	140		955	1,125	
2950	Mechanical media filtration units										
3000	High efficiency type, with frame, non-supported				MCFM	45			45	49.50	
3100	Supported type					55			55	60.50	
4000	Medium efficiency, extended surface					5			5	5.50	
4500	Permanent washable					20			20	22	
5000	Renewable disposable roll				▼	120			120	132	
5500	Throwaway glass or paper media type				Ea.	4.60			4.60	5.05	
5800	Filter, bag type										
5810	90-95% efficiency										
5820	24" x 12" x 29", .75-1.25 MCFM	1 Shee	1.90	4.211	Ea.	34.50	148		182.50	266	
5830	24" x 24" x 29", 1.5-2.5 MCFM	"	1.90	4.211	"	72	148		220	305	
5850	80-85% efficiency										
5860	24" x 12" x 29", .75-1.25 MCFM	1 Shee	1.90	4.211	Ea.	31	148		179	262	
5870	24" x 24" x 29", 1.5-2.5 MCFM	"	1.90	4.211	"	54.50	148		202.50	288	
6000	HEPA filter complete w/particle board,										
6010	kraft paper frame, separator material										
6020	95% DOP efficiency										
6030	12" x 12" x 6", 150 CFM	1 Shee	3.70	2.162	Ea.	29	76		105	149	
6034	24" x 12" x 6", 375 CFM		1.90	4.211		30.50	148		178.50	262	
6038	24" x 18" x 6", 450 CFM		1.90	4.211		39	148		187	271	
6042	24" x 24" x 6", 700 CFM		1.90	4.211		45.50	148		193.50	278	
6046	12" x 12" x 12", 250 CFM		3.70	2.162		38.50	76		114.50	159	
6050	24" x 12" x 12", 500 CFM		1.90	4.211		45.50	148		193.50	278	
6054	24" x 18" x 12", 875 CFM		1.90	4.211		61.50	148		209.50	296	
6058	24" x 24" x 12", 1000 CFM	▼	1.90	4.211	▼	64	148		212	299	
6100	99% DOP efficiency										
6110	12" x 12" x 6", 150 CFM	1 Shee	3.70	2.162	Ea.	48.50	76		124.50	170	
6114	24" x 12" x 6", 325 CFM		1.90	4.211		53.50	148		201.50	287	
6118	24" x 18" x 6", 550 CFM		1.90	4.211		74.50	148		222.50	310	
6122	24" x 24" x 6", 775 CFM		1.90	4.211		77	148		225	315	
6126	12" x 12" x 12", 250 CFM		3.60	2.222		54.50	78		132.50	180	
6130	24" x 12" x 12", 500 CFM		1.90	4.211		66.50	148		214.50	300	
6134	24" x 18" x 12", 775 CFM		1.90	4.211		106	148		254	345	
6138	24" x 24" x 12", 1100 CFM	▼	1.90	4.211	▼	110	148		258	350	
6500	HEPA filter housing, 14 ga. galv. sheet metal										
6510	12" x 12" x 6"	1 Shee	2.50	3.200	Ea.	435	112		547	655	
6514	24" x 12" x 6"		2	4		485	140		625	745	
6518	12" x 12" x 12"		2.40	3.333		435	117		552	655	
6522	14" x 12" x 12"		2.30	3.478		485	122		607	720	
6526	24" x 18" x 6"		1.90	4.211		560	148		708	850	
6530	24" x 24" x 6"		1.80	4.444		560	156		716	860	
6534	24" x 48" x 6"		1.70	4.706		750	165		915	1,075	
6538	24" x 72" x 6"		1.60	5		965	176		1,141	1,325	
6542	24" x 18" x 12"		1.80	4.444		560	156		716	860	
6546	24" x 24" x 12"	▼	1.70	4.706	▼	560	165		725	875	

Important: See the Reference Section for critical supporting data - Reference Nos., Crews, & City Cost Indexes

			DAILY	LABOR-		2002 BARE COSTS				TOTAL	
15860	Air Cleaning Devices	CREW	OUTPUT	HOURS	UNIT	MAT.	LABOR	EQUIP.	TOTAL	INCL O&P	
100	6550	24" x 48" x 12"	1 Shee	1.60	5	Ea.	750	176		926	1,100	100
	6554	24" x 72" x 12"	↓	1.50	5.333		965	187		1,152	1,350	
	6558	48" x 48" x 6"	Q-9	2.80	5.714		1,075	181		1,256	1,475	
	6562	48" x 72" x 6"		2.60	6.154		1,350	194		1,544	1,800	
	6566	48" x 96" x 6"		2.40	6.667		1,525	211		1,736	2,000	
	6570	48" x 48" x 12"		2.70	5.926		1,075	187		1,262	1,500	
	6574	48" x 72" x 12"		2.50	6.400		1,350	202		1,552	1,800	
	6578	48" x 96" x 12"		2.30	6.957		1,525	220		1,745	2,025	
	6582	114" x 72" x 12"	↓	2	8	↓	2,325	253		2,578	2,950	
500	0010	**EXHAUST SYSTEMS** (D3090 320)										500
	0500	Engine exhaust, garage, in-floor system										
	0510	Single tube outlet assemblies										
	0520	For transite pipe ducting, self-storing tube										
	0530	3" tubing adapter plate	1 Shee	16	.500	Ea.	153	17.55		170.55	195	
	0540	4" tubing adapter plate	↓	16	.500		153	17.55		170.55	195	
	0550	5" tubing adapter plate	↓	16	.500	↓	153	17.55		170.55	195	
	0600	For vitrified tile ducting										
	0610	3" tubing adapter plate, self-storing tube	1 Shee	16	.500	Ea.	153	17.55		170.55	195	
	0620	4" tubing adapter plate, self-storing tube		16	.500		153	17.55		170.55	195	
	0660	5" tubing adapter plate, self-storing tube	↓	16	.500	↓	153	17.55		170.55	195	
	0800	Two tube outlet assemblies										
	0810	For transite pipe ducting, self-storing tube										
	0820	3" tubing, dual exhaust adapter plate	1 Shee	16	.500	Ea.	153	17.55		170.55	195	
	0850	For vitrified tile ducting										
	0860	3" tubing, dual exhaust, self-storing tube	1 Shee	16	.500	Ea.	153	17.55		170.55	195	
	0870	3" tubing, double outlet, non-storing tubes	"	16	.500	"	153	17.55		170.55	195	
	0900	Accessories for metal tubing, (overhead systems also)										
	0910	Adapters, for metal tubing end										
	0920	3" tail pipe type				Ea.	34			34	37.50	
	0930	4" tail pipe type					34			34	37.50	
	0940	5" tail pipe type					34			34	37.50	
	0990	5" diesel stack type					238			238	262	
	1000	6" diesel stack type				↓	238			238	262	
	1100	Bullnose (guide) required for in-floor assemblies										
	1110	3" tubing size				Ea.	17			17	18.70	
	1120	4" tubing size					17			17	18.70	
	1130	5" tubing size				↓	17			17	18.70	
	1150	Plain rings, for tubing end										
	1160	3" tubing size				Ea.	8.10			8.10	8.90	
	1170	4" tubing size				"	10.20			10.20	11.20	
	1200	Tubing, galvanized, flexible, (for overhead systems also)										
	1210	3" ID				L.F.	4.55			4.55	5	
	1220	4" ID					5.50			5.50	6.05	
	1230	5" ID					6.40			6.40	7	
	1240	6" ID				↓	7.30			7.30	8.05	
	1250	Stainless steel, flexible, (for overhead sys, also)										
	1260	3" ID				L.F.	9.45			9.45	10.40	
	1270	4" ID					12.30			12.30	13.50	
	1280	5" ID					14.85			14.85	16.30	
	1290	6" ID				↓	17.70			17.70	19.45	
	1500	Engine exhaust, garage, ovrhd. compon., for neoprene tubing										
	1510	Alternate metal tubing & accessories see above										
	1550	Adapters, for neoprene tubing end										
	1560	3" tail pipe, adjustable, neoprene				Ea.	30			30	32.50	
	1570	3" tail pipe, heavy wall neoprene				↓	38.50			38.50	42	

MECHANICAL 15

			DAILY	LABOR-			2002 BARE COSTS			TOTAL	
15860	**Air Cleaning Devices**	**CREW**	**OUTPUT**	**HOURS**	**UNIT**	**MAT.**	**LABOR**	**EQUIP.**	**TOTAL**	**INCL O&P**	
500	1580	4" tail pipe, heavy wall neoprene	D3090 320			Ea.	41			41	45	**500**
	1590	5" tail pipe, heavy wall neoprene				↓	55.50			55.50	61	
	1650	Connectors, tubing										
	1660	3" interior, aluminum				Ea.	8.10			8.10	8.90	
	1670	4" interior, aluminum					10.20			10.20	11.20	
	1710	5" interior, neoprene					25.50			25.50	28	
	1750	3" spiralock, neoprene					8.10			8.10	8.90	
	1760	4" spiralock, neoprene					10.20			10.20	11.20	
	1780	Y for 3" ID tubing, neoprene, dual exhaust					21.50			21.50	23.50	
	1790	Y for 4" ID tubing, aluminum, dual exhaust				↓	38.50			38.50	42	
	1850	Elbows, aluminum, splice into tubing for strap										
	1860	3" neoprene tubing size				Ea.	25.50			25.50	28	
	1870	4" neoprene tubing size					30			30	32.50	
	1900	Flange assemblies, connect tubing to overhead duct				↓	28			28	31	
	2000	Hardware and accessories										
	2020	Cable, galvanized, 1/8" diameter				L.F.	.30			.30	.33	
	2040	Cleat, tie down cable or rope				Ea.	3.40			3.40	3.74	
	2060	Pulley					4.68			4.68	5.15	
	2080	Pulley hook, universal				↓	3.40			3.40	3.74	
	2100	Rope, nylon, 1/4" diameter				L.F.	.26			.26	.29	
	2120	Winch, 1" diameter				Ea.	72.50			72.50	79.50	
	2150	Lifting strap, mounts on neoprene										
	2160	3" tubing size				Ea.	17			17	18.70	
	2170	4" tubing size					17			17	18.70	
	2180	5" tubing size					17			17	18.70	
	2190	6" tubing size				↓	17			17	18.70	
	2200	Tubing, neoprene, 11' lengths										
	2210	3" ID				L.F.	5.10			5.10	5.60	
	2220	4" ID					6.40			6.40	7	
	2230	5" ID				↓	11.50			11.50	12.65	
	2500	Engine exhaust, thru-door outlet										
	2510	3" tube size	1 Carp	16	.500	Ea.	30	15		45	56	
	2530	4" tube size	"	16	.500	"	30	15		45	56	
	3000	Tubing, exhaust, flex hose, with										
	3010	coupler, damper and tail pipe adapter										
	3020	Neoprene										
	3040	3" x 20'	1 Shee	6	1.333	Ea.	162	47		209	250	
	3050	4" x 15'		5.40	1.481		162	52		214	258	
	3060	4" x 20'		5	1.600		194	56		250	300	
	3070	5" x 15'	↓	4.40	1.818	↓	273	64		337	400	
	3100	Galvanized										
	3110	3" x 20'	1 Shee	6	1.333	Ea.	147	47		194	233	
	3120	4" x 17'		5.60	1.429		153	50		203	246	
	3130	4" x 20'		5	1.600		169	56		225	273	
	3140	5" x 17'	↓	4.60	1.739	↓	188	61		249	300	
	7500	Welding fume elimination accessories for garage exhaust systems										
	7600	Cut off (blast gate)										
	7610	3" tubing size, 3" x 6" opening	1 Shee	24	.333	Ea.	13.60	11.70		25.30	33	
	7620	4" tubing size, 4" x 8" opening		24	.333		15.30	11.70		27	35	
	7630	5" tubing size, 5" x 10" opening		24	.333		20.50	11.70		32.20	40.50	
	7640	6" tubing size		24	.333		24	11.70		35.70	44	
	7650	8" tubing size	↓	24	.333	↓	34	11.70		45.70	55.50	
	7700	Hoods, magnetic, with handle & screen										
	7710	3" tubing size, 3" x 6" opening	1 Shee	24	.333	Ea.	64	11.70		75.70	88	
	7720	4" tubing size, 4" x 8" opening		24	.333		64	11.70		75.70	88	
	7730	5" tubing size, 5" x 10" opening	↓	24	.333	↓	64	11.70		75.70	88	

15 MECHANICAL

Important: See the Reference Section for critical supporting data - Reference Nos., Crews, & City Cost Indexes

15800 | Air Distribution

15860 | Air Cleaning Devices

		CREW	DAILY OUTPUT	LABOR-HOURS	UNIT	2002 BARE COSTS				TOTAL INCL O&P
						MAT.	LABOR	EQUIP.	TOTAL	
500	**8000**	Blower, for tailpipe exhaust system [D3090 320]									**500**
	8010	Direct drive								
	8012	495 CFM, 1/3 HP	Q-9	5	3.200	Ea.	1,025	101		1,126	1,275
	8014	1445 CFM, 3/4 HP		4	4		1,125	126		1,251	1,450
	8016	1840 CFM, 1-1/2 HP	↓	3	5.333	↓	1,550	169		1,719	1,950
	8030	Beltdrive								
	8032	1400 CFM, 1 HP	Q-9	4	4	Ea.	1,225	126		1,351	1,550
	8034	2023 CFM, 1-1/2 HP		3	5.333		1,850	169		2,019	2,300
	8036	2750 CFM, 2 HP		2	8		2,275	253		2,528	2,900
	8038	4400 CFM, 3 HP		1.80	8.889		3,500	281		3,781	4,275
	8040	7060 CFM, 5 HP	↓	1.40	11.429	↓	6,175	360		6,535	7,350

15900 | HVAC Instrumentation & Controls

15905 | HVAC Instrumentation

		CREW	DAILY OUTPUT	LABOR-HOURS	UNIT	2002 BARE COSTS				TOTAL INCL O&P
						MAT.	LABOR	EQUIP.	TOTAL	
960	**0010**	**WATER LEVEL CONTROLS**									**960**
	1000	Electric water feeder	1 Stpi	12	.667	Ea.	194	24		218	251
	2000	Feeder cut-off combination								
	2100	Steam system up to 5000 sq. ft.	1 Stpi	12	.667	Ea.	375	24		399	445
	2200	Steam system above 5000 sq. ft.		12	.667		460	24		484	540
	2300	Steam and hot water, high pressure	↓	10	.800	↓	540	29		569	640
	3000	Low water cut-off for hot water boiler, 50 psi maximum								
	3100	1″ top & bottom equalizing pipes, manual reset	1 Stpi	14	.571	Ea.	204	20.50		224.50	255
	3200	1″ top & bottom equalizing pipes		14	.571		204	20.50		224.50	255
	3300	2-1/2″ side connection for nipple-to-boiler	↓	14	.571	↓	185	20.50		205.50	234
	4000	Low water cut-off for low pressure steam with quick hook-up ftgs.								
	4100	For installation in gauge glass tappings	1 Stpi	16	.500	Ea.	151	18.10		169.10	194
	4200	Built-in type, 2-1/2″ tap - 3-1/8″ insertion		16	.500		127	18.10		145.10	167
	4300	Built-in type, 2-1/2″ tap - 1-3/4″ insertion		16	.500		134	18.10		152.10	176
	4400	Side connection to 2-1/2″ tapping		16	.500		143	18.10		161.10	186
	5000	Pump control, low water cut-off and alarm switch	↓	14	.571	↓	370	20.50		390.50	440
	9000	Water gauges, complete								
	9010	Rough brass, wheel type								
	9020	125 PSI at 350° F								
	9030	3/8″ pipe size	1 Stpi	11	.727	Ea.	30	26.50		56.50	72.50
	9040	1/2″ pipe size	″	10	.800	″	32.50	29		61.50	79.50
	9060	200 PSI at 400° F								
	9070	3/8″ pipe size	1 Stpi	11	.727	Ea.	37	26.50		63.50	80.50
	9080	1/2″ pipe size		10	.800		37	29		66	84.50
	9090	3/4″ pipe size	↓	9	.889	↓	47.50	32		79.50	101
	9130	Rough brass, chain lever type								
	9140	250 PSI at 400° F								
	9200	Polished brass, wheel type								
	9210	200 PSI at 400° F								
	9220	3/8″ pipe size	1 Stpi	11	.727	Ea.	71	26.50		97.50	118
	9230	1/2″ pipe size		10	.800		72	29		101	123
	9240	3/4″ pipe size	↓	9	.889	↓	76.50	32		108.50	133
	9260	Polished brass, chain lever type								
	9270	250 PSI at 400° F								

383

15900 | HVAC Instrumentation & Controls

	15905	HVAC Instrumentation	CREW	DAILY OUTPUT	LABOR-HOURS	UNIT	2002 BARE COSTS				TOTAL INCL O&P	
							MAT.	LABOR	EQUIP.	TOTAL		
960	9280	1/2″ pipe size	1 Stpi	10	.800	Ea.	115	29		144	170	**960**
	9290	3/4″ pipe size	″	9	.889	″	125	32		157	186	
	9400	Bronze, high pressure, ASME										
	9410	1/2″ pipe size	1 Stpi	10	.800	Ea.	123	29		152	179	
	9420	3/4″ pipe size	″	9	.889	″	150	32		182	214	
	9460	Chain lever type										
	9470	1/2″ pipe size	1 Stpi	10	.800	Ea.	186	29		215	249	
	9480	3/4″ pipe size	″	9	.889	″	196	32		228	265	
	9500	316 stainless steel, high pressure, ASME										
	9510	500 PSI at 450° F										
	9520	1/2″ pipe size	1 Stpi	10	.800	Ea.	470	29		499	560	
	9530	3/4″ pipe size	″	9	.889	″	455	32		487	550	

15950 | Testing/Adjusting/Balancing

	15955	HVAC Test/Adjust/Balance	CREW	DAILY OUTPUT	LABOR-HOURS	UNIT	2002 BARE COSTS				TOTAL INCL O&P	
							MAT.	LABOR	EQUIP.	TOTAL		
100	0010	**BALANCING, AIR** (Subcontractor's quote incl. material & labor) R15050 -710										**100**
	0900	Heating and ventilating equipment										
	1000	Centrifugal fans, utility sets				Ea.					259.68	
	1100	Heating and ventilating unit									389.52	
	1200	In-line fan									389.52	
	1300	Propeller and wall fan									73.58	
	1400	Roof exhaust fan									173.12	
	2000	Air conditioning equipment, central station									562.64	
	2100	Built-up low pressure unit									519.36	
	2200	Built-up high pressure unit									605.92	
	2300	Built-up high pressure dual duct									952.16	
	2400	Built-up variable volume									1,125.28	
	2500	Multi-zone A.C. and heating unit									389.52	
	2600	For each zone over one, add									86.56	
	2700	Package A.C. unit									216.40	
	2800	Rooftop heating and cooling unit									302.96	
	3000	Supply, return, exhaust, registers & diffusers, avg. height ceiling									51.94	
	3100	High ceiling									77.90	
	3200	Floor height									43.28	
	3300	Off mixing box									34.62	
	3500	Induction unit									56.26	
	3600	Lab fume hood									259.68	
	3700	Linear supply									129.84	
	3800	Linear supply high									151.48	
	4000	Linear return									43.28	
	4100	Light troffers									51.94	
	4200	Moduline - master									51.94	
	4300	Moduline - slaves									25.97	
	4400	Regenerators									346.24	
	4500	Taps into ceiling plenums									64.92	
	4600	Variable volume boxes									51.94	
700	0010	**PIPING, TESTING**										**700**
	0100	Nondestructive testing										

384   **Important: See the Reference Section for critical supporting data - Reference Nos., Crews, & City Cost Indexes**

15955	HVAC Test/Adjust/Balance	CREW	DAILY OUTPUT	LABOR-HOURS	UNIT	2002 BARE COSTS				TOTAL INCL O&P
						MAT.	LABOR	EQUIP.	TOTAL	
700	0110	Nondestructive hydraulic pressure test, isolate & 1 hr. hold									**700**
	0120	1" - 4" pipe									
	0140	0 - 250 L.F.	1 Stpi	1.33	6.015	Ea.		218		218	330
	0160	250 - 500 L.F.	"	.80	10			360		360	545
	0180	500 - 1000 L.F.	Q-5	1.14	14.035			455		455	690
	0200	1000 - 2000 L.F.	"	.80	20			650		650	985
	0300	6" - 10" pipe									
	0320	0 - 250 L.F.	Q-5	1	16	Ea.		520		520	785
	0340	250 - 500 L.F.		.73	21.918			715		715	1,075
	0360	500 - 1000 L.F.		.53	30.189			985		985	1,475
	0380	1000 - 2000 L.F.		.38	42.105			1,375		1,375	2,075
	1000	Pneumatic pressure test, includes soaping joints									
	1120	1" - 4" pipe									
	1140	0 - 250 L.F.	Q-5	2.67	5.993	Ea.	6.30	195		201.30	300
	1160	250 - 500 L.F.		1.33	12.030		12.60	390		402.60	605
	1180	500 - 1000 L.F.		.80	20		18.90	650		668.90	1,000
	1200	1000 - 2000 L.F.		.50	32		25	1,050		1,075	1,600
	1300	6" - 10" pipe									
	1320	0 - 250 L.F.	Q-5	1.33	12.030	Ea.	6.30	390		396.30	595
	1340	250 - 500 L.F.		.67	23.881		12.60	780		792.60	1,200
	1360	500 - 1000 L.F.		.40	40		25	1,300		1,325	2,000
	1380	1000 - 2000 L.F.		.25	64		31.50	2,075		2,106.50	3,175
	2000	X-Ray of welds									
	2110	2" diam.	1 Stpi	8	1	Ea.	4	36		40	59
	2120	3" diam.		8	1		4	36		40	59
	2130	4" diam.		8	1		6	36		42	61
	2140	6" diam.		8	1		6	36		42	61
	2150	8" diam.		6.60	1.212		6	44		50	72.50
	2160	10" diam.		6	1.333		8	48.50		56.50	82
	3000	Liquid penetration of welds									
	3110	2" diam.	1 Stpi	14	.571	Ea.	1.25	20.50		21.75	32.50
	3120	3" diam.		13.60	.588		1.25	21.50		22.75	33.50
	3130	4" diam.		13.40	.597		1.25	21.50		22.75	34
	3140	6" diam.		13.20	.606		1.25	22		23.25	34.50
	3150	8" diam.		13	.615		1.88	22.50		24.38	35.50
	3160	10" diam.		12.80	.625		1.88	22.50		24.38	36
900	0010	**BALANCING, WATER** (Subcontractor's quote incl. material & labor)									**900**
	0050	Air cooled condenser				Ea.					152.88
	0100	Cabinet unit heater									52.42
	0200	Chiller	R15050 -710								371.28
	0300	Convector									43.68
	0400	Converter									218.40
	0500	Cooling tower									283.92
	0600	Fan coil unit, unit ventilator									78.62
	0700	Fin tube and radiant panels									87.36
	0800	Main and duct re-heat coils									80.81
	0810	Heat exchanger									80.81
	0900	Main balancing cocks									65.52
	1000	Pumps									192.19
	1100	Unit heater									61.15

For information about Means Estimating Seminars, see yellow pages 11 and 12 in back of book

MECHANICAL 15

	CREW	DAILY OUTPUT	LABOR-HOURS	UNIT	2002 BARE COSTS				TOTAL INCL O&P
					MAT.	LABOR	EQUIP.	TOTAL	

Division 16
Electrical

Estimating Tips

16060 Grounding & Bonding
- When taking off grounding system, identify separately the type and size of wire and list each unique type of ground connection.

16100 Wiring Methods
- Conduit should be taken off in three main categories: power distribution, branch power, and branch lighting, so the estimator can concentrate on systems and components, therefore making it easier to ensure all items have been accounted for.
- For cost modifications for elevated conduit installation, add the percentages to labor according to the height of installation and only the quantities exceeding the different height levels, not to the total conduit quantities.
- Remember that aluminum wiring of equal ampacity is larger in diameter than copper and may require larger conduit.
- If more than three wires at a time are being pulled, deduct percentages from the labor hours of that grouping of wires.

- The estimator should take the weights of materials into consideration when completing a takeoff. Topics to consider include: How will the materials be supported? What methods of support are available? How high will the support structure have to reach? Will the final support structure be able to withstand the total burden? Is the support material included or separate from the fixture, equipment and material specified?

16200 Electrical Power
- Do not overlook the costs for equipment used in the installation. If scaffolding or highlifts are available in the field, contractors may use them in lieu of the proposed ladders and rolling staging.

16400 Low-Voltage Distribution
- Supports and concrete pads may be shown on drawings for the larger equipment, or the support system may be just a piece of plywood for the back of a panelboard. In either case, it must be included in the costs.

16500 Lighting
- Fixtures should be taken off room by room, using the fixture schedule, specifications, and the ceiling plan. For large concentrations of lighting fixtures in the same area deduct the percentages from labor hours.

16700 Communications
16800 Sound & Video
- When estimating material costs for special systems, it is always prudent to obtain manufacturers' quotations for equipment prices and special installation requirements which will affect the total costs.

Reference Numbers
Reference numbers are shown in bold squares at the beginning of some major classifications. These numbers refer to related items in the Reference Section. The reference information may be an estimating procedure, an alternate pricing method or technical information.

Note: Not all subdivisions listed here necessarily appear in this publication.

			DAILY	LABOR-			2002 BARE COSTS			TOTAL	
16132	**Conduit & Tubing**	CREW	OUTPUT	HOURS	UNIT	MAT.	LABOR	EQUIP.	TOTAL	INCL O&P	
260	0010	**CUTTING AND DRILLING**										260
	0100	Hole drilling to 10' high, concrete wall										
	0110	8" thick, 1/2" pipe size	1 Elec	12	.667	Ea.		23.50		23.50	35.50	
	0120	3/4" pipe size		12	.667			23.50		23.50	35.50	
	0130	1" pipe size		9.50	.842			30		30	44.50	
	0140	1-1/4" pipe size		9.50	.842			30		30	44.50	
	0150	1-1/2" pipe size		9.50	.842			30		30	44.50	
	0160	2" pipe size		4.40	1.818			64.50		64.50	96	
	0170	2-1/2" pipe size		4.40	1.818			64.50		64.50	96	
	0180	3" pipe size		4.40	1.818			64.50		64.50	96	
	0190	3-1/2" pipe size		3.30	2.424			86		86	128	
	0200	4" pipe size		3.30	2.424			86		86	128	
	0500	12" thick, 1/2" pipe size		9.40	.851			30		30	45	
	0520	3/4" pipe size		9.40	.851			30		30	45	
	0540	1" pipe size		7.30	1.096			39		39	58	
	0560	1-1/4" pipe size		7.30	1.096			39		39	58	
	0570	1-1/2" pipe size		7.30	1.096			39		39	58	
	0580	2" pipe size		3.60	2.222			79		79	118	
	0590	2-1/2" pipe size		3.60	2.222			79		79	118	
	0600	3" pipe size		3.60	2.222			79		79	118	
	0610	3-1/2" pipe size		2.80	2.857			101		101	151	
	0630	4" pipe size		2.50	3.200			113		113	169	
	0650	16" thick, 1/2" pipe size		7.60	1.053			37.50		37.50	55.50	
	0670	3/4" pipe size		7	1.143			40.50		40.50	60.50	
	0690	1" pipe size		6	1.333			47.50		47.50	70.50	
	0710	1-1/4" pipe size		5.50	1.455			51.50		51.50	77	
	0730	1-1/2" pipe size		5.50	1.455			51.50		51.50	77	
	0750	2" pipe size		3	2.667			94.50		94.50	141	
	0770	2-1/2" pipe size		2.70	2.963			105		105	157	
	0790	3" pipe size		2.50	3.200			113		113	169	
	0810	3-1/2" pipe size		2.30	3.478			123		123	184	
	0830	4" pipe size		2	4			142		142	212	
	0850	20" thick, 1/2" pipe size		6.40	1.250			44.50		44.50	66	
	0870	3/4" pipe size		6	1.333			47.50		47.50	70.50	
	0890	1" pipe size		5	1.600			56.50		56.50	84.50	
	0910	1-1/4" pipe size		4.80	1.667			59		59	88	
	0930	1-1/2" pipe size		4.60	1.739			61.50		61.50	92	
	0950	2" pipe size		2.70	2.963			105		105	157	
	0970	2-1/2" pipe size		2.40	3.333			118		118	176	
	0990	3" pipe size		2.20	3.636			129		129	192	
	1010	3-1/2" pipe size		2	4			142		142	212	
	1030	4" pipe size		1.70	4.706			167		167	249	
	1050	24" thick, 1/2" pipe size		5.50	1.455			51.50		51.50	77	
	1070	3/4" pipe size		5.10	1.569			55.50		55.50	83	
	1090	1" pipe size		4.30	1.860			66		66	98.50	
	1110	1-1/4" pipe size		4	2			71		71	106	
	1130	1-1/2" pipe size		4	2			71		71	106	
	1150	2" pipe size		2.40	3.333			118		118	176	
	1170	2-1/2" pipe size		2.20	3.636			129		129	192	
	1190	3" pipe size		2	4			142		142	212	
	1210	3-1/2" pipe size		1.80	4.444			158		158	235	
	1230	4" pipe size		1.50	5.333			189		189	282	
	1500	Brick wall, 8" thick, 1/2" pipe size		18	.444			15.75		15.75	23.50	
	1520	3/4" pipe size		18	.444			15.75		15.75	23.50	
	1540	1" pipe size		13.30	.601			21.50		21.50	32	
	1560	1-1/4" pipe size		13.30	.601			21.50		21.50	32	

Important: See the Reference Section for critical supporting data - Reference Nos., Crews, & City Cost Indexes

			DAILY	LABOR-			2002 BARE COSTS			TOTAL	
16132	**Conduit & Tubing**	CREW	OUTPUT	HOURS	UNIT	MAT.	LABOR	EQUIP.	TOTAL	INCL O&P	
260 1580	1-1/2" pipe size	1 Elec	13.30	.601	Ea.		21.50		21.50	32	260
1600	2" pipe size		5.70	1.404			50		50	74.50	
1620	2-1/2" pipe size		5.70	1.404			50		50	74.50	
1640	3" pipe size		5.70	1.404			50		50	74.50	
1660	3-1/2" pipe size		4.40	1.818			64.50		64.50	96	
1680	4" pipe size		4	2			71		71	106	
1700	12" thick, 1/2" pipe size		14.50	.552			19.55		19.55	29	
1720	3/4" pipe size		14.50	.552			19.55		19.55	29	
1740	1" pipe size		11	.727			26		26	38.50	
1760	1-1/4" pipe size		11	.727			26		26	38.50	
1780	1-1/2" pipe size		11	.727			26		26	38.50	
1800	2" pipe size		5	1.600			56.50		56.50	84.50	
1820	2-1/2" pipe size		5	1.600			56.50		56.50	84.50	
1840	3" pipe size		5	1.600			56.50		56.50	84.50	
1860	3-1/2" pipe size		3.80	2.105			74.50		74.50	111	
1880	4" pipe size		3.30	2.424			86		86	128	
1900	16" thick, 1/2" pipe size		12.30	.650			23		23	34.50	
1920	3/4" pipe size		12.30	.650			23		23	34.50	
1940	1" pipe size		9.30	.860			30.50		30.50	45.50	
1960	1-1/4" pipe size		9.30	.860			30.50		30.50	45.50	
1980	1-1/2" pipe size		9.30	.860			30.50		30.50	45.50	
2000	2" pipe size		4.40	1.818			64.50		64.50	96	
2010	2-1/2" pipe size		4.40	1.818			64.50		64.50	96	
2030	3" pipe size		4.40	1.818			64.50		64.50	96	
2050	3-1/2" pipe size		3.30	2.424			86		86	128	
2070	4" pipe size		3	2.667			94.50		94.50	141	
2090	20" thick, 1/2" pipe size		10.70	.748			26.50		26.50	39.50	
2110	3/4" pipe size		10.70	.748			26.50		26.50	39.50	
2130	1" pipe size		8	1			35.50		35.50	53	
2150	1-1/4" pipe size		8	1			35.50		35.50	53	
2170	1-1/2" pipe size		8	1			35.50		35.50	53	
2190	2" pipe size		4	2			71		71	106	
2210	2-1/2" pipe size		4	2			71		71	106	
2230	3" pipe size		4	2			71		71	106	
2250	3-1/2" pipe size		3	2.667			94.50		94.50	141	
2270	4" pipe size		2.70	2.963			105		105	157	
2290	24" thick, 1/2" pipe size		9.40	.851			30		30	45	
2310	3/4" pipe size		9.40	.851			30		30	45	
2330	1" pipe size		7.10	1.127			40		40	59.50	
2350	1-1/4" pipe size		7.10	1.127			40		40	59.50	
2370	1-1/2" pipe size		7.10	1.127			40		40	59.50	
2390	2" pipe size		3.60	2.222			79		79	118	
2410	2-1/2" pipe size		3.60	2.222			79		79	118	
2430	3" pipe size		3.60	2.222			79		79	118	
2450	3-1/2" pipe size		2.80	2.857			101		101	151	
2470	4" pipe size		2.50	3.200			113		113	169	
3000	Knockouts to 8' high, metal boxes & enclosures										
3020	With hole saw, 1/2" pipe size	1 Elec	53	.151	Ea.		5.35		5.35	8	
3040	3/4" pipe size		47	.170			6.05		6.05	9	
3050	1" pipe size		40	.200			7.10		7.10	10.60	
3060	1-1/4" pipe size		36	.222			7.90		7.90	11.75	
3070	1-1/2" pipe size		32	.250			.8.85		8.85	13.25	
3080	2" pipe size		27	.296			10.50		10.50	15.65	
3090	2-1/2" pipe size		20	.400			14.20		14.20	21	
4010	3" pipe size		16	.500			17.75		17.75	26.50	
4030	3-1/2" pipe size		13	.615			22		22	32.50	

ELECTRICAL 16

16 ELECTRICAL

16132 | Conduit & Tubing

		CREW	DAILY OUTPUT	LABOR-HOURS	UNIT	2002 BARE COSTS				TOTAL INCL O&P
						MAT.	LABOR	EQUIP.	TOTAL	
260	4050	4" pipe size	1 Elec	11	.727	Ea.	26			26	38.50	260
	4070	With hand punch set, 1/2" pipe size		40	.200			7.10		7.10	10.60
	4090	3/4" pipe size		32	.250			8.85		8.85	13.25
	4110	1" pipe size		30	.267			9.45		9.45	14.10
	4130	1-1/4" pipe size		28	.286			10.15		10.15	15.10
	4150	1-1/2" pipe size		26	.308			10.90		10.90	16.30
	4170	2" pipe size		20	.400			14.20		14.20	21
	4190	2-1/2" pipe size		17	.471			16.70		16.70	25
	4200	3" pipe size		15	.533			18.90		18.90	28
	4220	3-1/2" pipe size		12	.667			23.50		23.50	35.50
	4240	4" pipe size		10	.800			28.50		28.50	42.50
	4260	With hydraulic punch, 1/2" pipe size		44	.182			6.45		6.45	9.60
	4280	3/4" pipe size		38	.211			7.45		7.45	11.15
	4300	1" pipe size		38	.211			7.45		7.45	11.15
	4320	1-1/4" pipe size		38	.211			7.45		7.45	11.15
	4340	1-1/2" pipe size		38	.211			7.45		7.45	11.15
	4360	2" pipe size		32	.250			8.85		8.85	13.25
	4380	2-1/2" pipe size		27	.296			10.50		10.50	15.65
	4400	3" pipe size		23	.348			12.35		12.35	18.40
	4420	3-1/2" pipe size		20	.400			14.20		14.20	21
	4440	4" pipe size		18	.444			15.75		15.75	23.50

16139 | Residential Wiring

		CREW	DAILY OUTPUT	LABOR-HOURS	UNIT	MAT.	LABOR	EQUIP.	TOTAL	TOTAL INCL O&P
700	0010	**RESIDENTIAL WIRING**									700
	0020	20' avg. runs and #14/2 wiring incl. unless otherwise noted								
	1000	Service & panel, includes 24' SE-AL cable, service eye, meter,								
	1010	Socket, panel board, main bkr., ground rod, 15 or 20 amp								
	1020	1-pole circuit breakers, and misc. hardware								
	1100	100 amp, with 10 branch breakers	1 Elec	1.19	6.723	Ea.	425	238		663	820
	1110	With PVC conduit and wire		.92	8.696		460	310		770	965
	1120	With RGS conduit and wire		.73	10.959		585	390		975	1,225
	1150	150 amp, with 14 branch breakers		1.03	7.767		650	275		925	1,125
	1170	With PVC conduit and wire		.82	9.756		720	345		1,065	1,300
	1180	With RGS conduit and wire		.67	11.940		940	425		1,365	1,650
	1200	200 amp, with 18 branch breakers	2 Elec	1.80	8.889		855	315		1,170	1,400
	1220	With PVC conduit and wire		1.46	10.959		925	390		1,315	1,600
	1230	With RGS conduit and wire		1.24	12.903		1,225	455		1,680	2,025
	1800	Lightning surge suppressor for above services, add	1 Elec	32	.250		39.50	8.85		48.35	57
	2000	Switch devices								
	2100	Single pole, 15 amp, Ivory, with a 1-gang box, cover plate,								
	2110	Type NM (Romex) cable	1 Elec	17.10	.468	Ea.	6.50	16.60		23.10	32
	2120	Type MC (BX) cable		14.30	.559		17.05	19.85		36.90	48.50
	2130	EMT & wire		5.71	1.401		17.80	49.50		67.30	93.50
	2150	3-way, #14/3, type NM cable		14.55	.550		10	19.50		29.50	40
	2170	Type MC cable		12.31	.650		21.50	23		44.50	58.50
	2180	EMT & wire		5	1.600		20	56.50		76.50	107
	2200	4-way, #14/3, type NM cable		14.55	.550		22.50	19.50		42	54
	2220	Type MC cable		12.31	.650		34	23		57	72
	2230	EMT & wire		5	1.600		32.50	56.50		89	121
	2250	S.P., 20 amp, #12/2, type NM cable		13.33	.600		11.55	21.50		33.05	44.50
	2270	Type MC cable		11.43	.700		21	25		46	60.50
	2280	EMT & wire		4.85	1.649		23.50	58.50		82	114
	2290	S.P. rotary dimmer, 600W, no wiring		17	.471		15.75	16.70		32.45	42.50
	2300	S.P. rotary dimmer, 600W, type NM cable		14.55	.550		18.25	19.50		37.75	49
	2320	Type MC cable		12.31	.650		29	23		52	66

		16139	Residential Wiring	CREW	DAILY OUTPUT	LABOR-HOURS	UNIT	2002 BARE COSTS				TOTAL INCL O&P	
								MAT.	LABOR	EQUIP.	TOTAL		
700	2330		EMT & wire	1 Elec	5	1.600	Ea.	30.50	56.50		87	118	700
	2350		3-way rotary dimmer, type NM cable		13.33	.600		15.65	21.50		37.15	49.50	
	2370		Type MC cable		11.43	.700		26	25		51	66	
	2380		EMT & wire		4.85	1.649		27.50	58.50		86	118	
	2400		Interval timer wall switch, 20 amp, 1-30 min., #12/2										
	2410		Type NM cable	1 Elec	14.55	.550	Ea.	29	19.50		48.50	61	
	2420		Type MC cable		12.31	.650		36	23		59	74.50	
	2430		EMT & wire		5	1.600		41	56.50		97.50	130	
	2500		Decorator style										
	2510		S.P., 15 amp, type NM cable	1 Elec	17.10	.468	Ea.	10.05	16.60		26.65	36	
	2520		Type MC cable		14.30	.559		20.50	19.85		40.35	52	
	2530		EMT & wire		5.71	1.401		21.50	49.50		71	97.50	
	2550		3-way, #14/3, type NM cable		14.55	.550		13.55	19.50		33.05	44	
	2570		Type MC cable		12.31	.650		25	23		48	62	
	2580		EMT & wire		5	1.600		23.50	56.50		80	111	
	2600		4-way, #14/3, type NM cable		14.55	.550		26	19.50		45.50	58	
	2620		Type MC cable		12.31	.650		38	23		61	76	
	2630		EMT & wire		5	1.600		36	56.50		92.50	125	
	2650		S.P., 20 amp, #12/2, type NM cable		13.33	.600		15.10	21.50		36.60	48.50	
	2670		Type MC cable		11.43	.700		25	25		50	64.50	
	2680		EMT & wire		4.85	1.649		27	58.50		85.50	117	
	2700		S.P., slide dimmer, type NM cable		17.10	.468		25.50	16.60		42.10	53	
	2720		Type MC cable		14.30	.559		36	19.85		55.85	69	
	2730		EMT & wire		5.71	1.401		37.50	49.50		87	115	
	2750		S.P., touch dimmer, type NM cable		17.10	.468		21	16.60		37.60	48.50	
	2770		Type MC cable		14.30	.559		32	19.85		51.85	64.50	
	2780		EMT & wire		5.71	1.401		33	49.50		82.50	111	
	2800		3-way touch dimmer, type NM cable		13.33	.600		38.50	21.50		60	74.50	
	2820		Type MC cable		11.43	.700		49.50	25		74.50	91	
	2830		EMT & wire		4.85	1.649		50.50	58.50		109	144	
	3000		Combination devices										
	3100		S.P. switch/15 amp recpt., Ivory, 1-gang box, plate										
	3110		Type NM cable	1 Elec	11.43	.700	Ea.	14.30	25		39.30	53	
	3120		Type MC cable		10	.800		25	28.50		53.50	70	
	3130		EMT & wire		4.40	1.818		26.50	64.50		91	125	
	3150		S.P. switch/pilot light, type NM cable		11.43	.700		14.95	25		39.95	53.50	
	3170		Type MC cable		10	.800		25.50	28.50		54	70.50	
	3180		EMT & wire		4.43	1.806		27	64		91	125	
	3190		2-S.P. switches, 2-#14/2, no wiring		14	.571		5.55	20.50		26.05	36	
	3200		2-S.P. switches, 2-#14/2, type NM cables		10	.800		16.10	28.50		44.60	60.50	
	3220		Type MC cable		8.89	.900		33.50	32		65.50	84.50	
	3230		EMT & wire		4.10	1.951		28	69		97	134	
	3250		3-way switch/15 amp recpt., #14/3, type NM cable		10	.800		20.50	28.50		49	65	
	3270		Type MC cable		8.89	.900		32	32		64	83	
	3280		EMT & wire		4.10	1.951		30.50	69		99.50	137	
	3300		2-3 way switches, 2-#14/3, type NM cables		8.89	.900		27	32		59	77	
	3320		Type MC cable		8	1		46.50	35.50		82	105	
	3330		EMT & wire		4	2		35	71		106	145	
	3350		S.P. switch/20 amp recpt., #12/2, type NM cable		10	.800		25	28.50		53.50	70	
	3370		Type MC cable		8.89	.900		32.50	32		64.50	83	
	3380		EMT & wire		4.10	1.951		37	69		106	144	
	3400		Decorator style										
	3410		S.P. switch/15 amp recpt., type NM cable	1 Elec	11.43	.700	Ea.	17.85	25		42.85	56.50	
	3420		Type MC cable		10	.800		28.50	28.50		57	74	
	3430		EMT & wire		4.40	1.818		30	64.50		94.50	129	
	3450		S.P. switch/pilot light, type NM cable		11.43	.700		18.50	25		43.50	57.50	

ELECTRICAL 16

16139	Residential Wiring	CREW	DAILY OUTPUT	LABOR-HOURS	UNIT	2002 BARE COSTS				TOTAL INCL O&P	
						MAT.	LABOR	EQUIP.	TOTAL		
700	3470	Type MC cable	1 Elec	10	.800	Ea.	29	28.50		57.50	74.50	700
	3480	EMT & wire		4.40	1.818		30.50	64.50		95	130	
	3500	2-S.P. switches, 2-#14/2, type NM cables		10	.800		19.65	28.50		48.15	64	
	3520	Type MC cable		8.89	.900		37	32		69	88.50	
	3530	EMT & wire		4.10	1.951		31.50	69		100.50	138	
	3550	3-way/15 amp recpt., #14/3, type NM cable		10	.800		24	28.50		52.50	69	
	3570	Type MC cable		8.89	.900		35.50	32		67.50	87	
	3580	EMT & wire		4.10	1.951		34	69		103	141	
	3650	2-3 way switches, 2-#14/3, type NM cables		8.89	.900		30.50	32		62.50	81	
	3670	Type MC cable		8	1		50	35.50		85.50	108	
	3680	EMT & wire		4	2		38.50	71		109.50	148	
	3700	S.P. switch/20 amp recpt., #12/2, type NM cable		10	.800		28.50	28.50		57	74	
	3720	Type MC cable		8.89	.900		36	32		68	87	
	3730	EMT & wire		4.10	1.951		40.50	69		109.50	148	
	4000	Receptacle devices										
	4010	Duplex outlet, 15 amp recpt., Ivory, 1-gang box, plate										
	4015	Type NM cable	1 Elec	14.55	.550	Ea.	5.05	19.50		24.55	34.50	
	4020	Type MC cable		12.31	.650		15.60	23		38.60	51.50	
	4030	EMT & wire		5.33	1.501		16.35	53		69.35	97.50	
	4050	With #12/2, type NM cable		12.31	.650		6.20	23		29.20	41.50	
	4070	Type MC cable		10.67	.750		15.90	26.50		42.40	57	
	4080	EMT & wire		4.71	1.699		18.05	60		78.05	110	
	4100	20 amp recpt., #12/2, type NM cable		12.31	.650		11.85	23		34.85	47.50	
	4120	Type MC cable		10.67	.750		21.50	26.50		48	63	
	4130	EMT & wire		4.71	1.699		23.50	60		83.50	116	
	4140	For GFI see line 4300 below										
	4150	Decorator style, 15 amp recpt., type NM cable	1 Elec	14.55	.550	Ea.	8.60	19.50		28.10	38.50	
	4170	Type MC cable		12.31	.650		19.15	23		42.15	55.50	
	4180	EMT & wire		5.33	1.501		19.90	53		72.90	102	
	4200	With #12/2, type NM cable		12.31	.650		9.75	23		32.75	45.50	
	4220	Type MC cable		10.67	.750		19.45	26.50		45.95	61	
	4230	EMT & wire		4.71	1.699		21.50	60		81.50	114	
	4250	20 amp recpt. #12/2, type NM cable		12.31	.650		15.40	23		38.40	51.50	
	4270	Type MC cable		10.67	.750		25	26.50		51.50	67	
	4280	EMT & wire		4.71	1.699		27.50	60		87.50	120	
	4300	GFI, 15 amp recpt., type NM cable		12.31	.650		32.50	23		55.50	70.50	
	4320	Type MC cable		10.67	.750		43.50	26.50		70	87	
	4330	EMT & wire		4.71	1.699		44	60		104	139	
	4350	GFI with #12/2, type NM cable		10.67	.750		34	26.50		60.50	77	
	4370	Type MC cable		9.20	.870		43.50	31		74.50	94	
	4380	EMT & wire		4.21	1.900		45.50	67.50		113	152	
	4400	20 amp recpt., #12/2 type NM cable		10.67	.750		35.50	26.50		62	78.50	
	4420	Type MC cable		9.20	.870		45	31		76	95.50	
	4430	EMT & wire		4.21	1.900		47.50	67.50		115	153	
	4500	Weather-proof cover for above receptacles, add		32	.250		4.40	8.85		13.25	18.10	
	4550	Air conditioner outlet, 20 amp-240 volt recpt.										
	4560	30' of #12/2, 2 pole circuit breaker										
	4570	Type NM cable	1 Elec	10	.800	Ea.	40.50	28.50		69	87	
	4580	Type MC cable		9	.889		53.50	31.50		85	106	
	4590	EMT & wire		4	2		52	71		123	164	
	4600	Decorator style, type NM cable		10	.800		44.50	28.50		73	91.50	
	4620	Type MC cable		9	.889		57	31.50		88.50	110	
	4630	EMT & wire		4	2		56	71		127	168	
	4650	Dryer outlet, 30 amp-240 volt recpt., 20' of #10/3										
	4660	2 pole circuit breaker										
	4670	Type NM cable	1 Elec	6.41	1.248	Ea.	47.50	44		91.50	118	

Important: See the Reference Section for critical supporting data - Reference Nos., Crews, & City Cost Indexes

ELECTRICAL 16

16139	Residential Wiring	CREW	DAILY OUTPUT	LABOR-HOURS	UNIT	MAT.	LABOR	EQUIP.	TOTAL	TOTAL INCL O&P	
						2002 BARE COSTS					
700	4680	Type MC cable	1 Elec	5.71	1.401	Ea.	57.50	49.50		107	137	700
4690	EMT & wire	↓	3.48	2.299	↓	53.50	81.50		135	181	
4700	Range outlet, 50 amp-240 volt recpt., 30' of #8/3										
4710	Type NM cable	1 Elec	4.21	1.900	Ea.	69	67.50		136.50	177	
4720	Type MC cable		4	2		102	71		173	218	
4730	EMT & wire		2.96	2.703		69	96		165	219	
4750	Central vacuum outlet, Type NM cable		6.40	1.250		42.50	44.50		87	113	
4770	Type MC cable		5.71	1.401		60.50	49.50		110	141	
4780	EMT & wire	↓	3.48	2.299	↓	54.50	81.50		136	182	
4800	30 amp-110 volt locking recpt., #10/2 circ. bkr.										
4810	Type NM cable	1 Elec	6.20	1.290	Ea.	50	45.50		95.50	124	
4830	EMT & wire	"	3.20	2.500	"	62.50	88.50		151	201	
4900	Low voltage outlets										
4910	Telephone recpt., 20' of 4/C phone wire	1 Elec	26	.308	Ea.	6.80	10.90		17.70	24	
4920	TV recpt., 20' of RG59U coax wire, F type connector	"	16	.500	"	11.25	17.75		29	39	
4950	Door bell chime, transformer, 2 buttons, 60' of bellwire										
4970	Economy model	1 Elec	11.50	.696	Ea.	49	24.50		73.50	91	
4980	Custom model		11.50	.696		84	24.50		108.50	130	
4990	Luxury model, 3 buttons	↓	9.50	.842	↓	229	30		259	297	
6000	Lighting outlets										
6050	Wire only (for fixture), type NM cable	1 Elec	32	.250	Ea.	3.34	8.85		12.19	16.90	
6070	Type MC cable		24	.333		11.60	11.80		23.40	30.50	
6080	EMT & wire		10	.800		11.40	28.50		39.90	55	
6100	Box (4"), and wire (for fixture), type NM cable		25	.320		7.40	11.35		18.75	25	
6120	Type MC cable		20	.400		15.65	14.20		29.85	38	
6130	EMT & wire	↓	11	.727	↓	15.45	26		41.45	55.50	
6200	Fixtures (use with lines 6050 or 6100 above)										
6210	Canopy style, economy grade	1 Elec	40	.200	Ea.	25.50	7.10		32.60	38.50	
6220	Custom grade		40	.200		46	7.10		53.10	61	
6250	Dining room chandelier, economy grade		19	.421		76	14.95		90.95	106	
6260	Custom grade		19	.421		225	14.95		239.95	271	
6270	Luxury grade		15	.533		495	18.90		513.90	575	
6310	Kitchen fixture (fluorescent), economy grade		30	.267		51.50	9.45		60.95	70.50	
6320	Custom grade		25	.320		161	11.35		172.35	194	
6350	Outdoor, wall mounted, economy grade		30	.267		27	9.45		36.45	43.50	
6360	Custom grade		30	.267		101	9.45		110.45	125	
6370	Luxury grade		25	.320		227	11.35		238.35	267	
6410	Outdoor PAR floodlights, 1 lamp, 150 watt		20	.400		20	14.20		34.20	43	
6420	2 lamp, 150 watt each		20	.400		33.50	14.20		47.70	58	
6430	For infrared security sensor, add		32	.250		87	8.85		95.85	109	
6450	Outdoor, quartz-halogen, 300 watt flood		20	.400		38	14.20		52.20	63	
6600	Recessed downlight, round, pre-wired, 50 or 75 watt trim		30	.267		35	9.45		44.45	52.50	
6610	With shower light trim		30	.267		43	9.45		52.45	61.50	
6620	With wall washer trim		28	.286		52.50	10.15		62.65	73	
6630	With eye-ball trim	↓	28	.286		50	10.15		60.15	70	
6640	For direct contact with insulation, add					1.60			1.60	1.76	
6700	Porcelain lamp holder	1 Elec	40	.200		3.50	7.10		10.60	14.45	
6710	With pull switch		40	.200		3.75	7.10		10.85	14.75	
6750	Fluorescent strip, 1-20 watt tube, wrap around diffuser, 24"		24	.333		51.50	11.80		63.30	74	
6770	2-40 watt tubes, 48"		20	.400		78	14.20		92.20	107	
6780	With residential ballast		20	.400		89.50	14.20		103.70	120	
6800	Bathroom heat lamp, 1-250 watt		28	.286		34.50	10.15		44.65	53	
6810	2-250 watt lamps	↓	28	.286	↓	55	10.15		65.15	75.50	
6820	For timer switch, see line 2400										
6900	Outdoor post lamp, incl. post, fixture, 35' of #14/2										
6910	Type NMC cable	1 Elec	3.50	2.286	Ea.	180	81		261	320	

ELECTRICAL 16

16139	Residential Wiring	CREW	DAILY OUTPUT	LABOR-HOURS	UNIT	2002 BARE COSTS				TOTAL INCL O&P
						MAT.	LABOR	EQUIP.	TOTAL	
700 6920	Photo-eye, add	1 Elec	27	.296	Ea.	29	10.50		39.50	47.50 **700**
6950	Clock dial time switch, 24 hr., w/enclosure, type NM cable		11.43	.700		50.50	25		75.50	92.50
6970	Type MC cable		11	.727		61	26		87	106
6980	EMT & wire		4.85	1.649		61.50	58.50		120	156
7000	Alarm systems									
7050	Smoke detectors, box, #14/3, type NM cable	1 Elec	14.55	.550	Ea.	28	19.50		47.50	59.50
7070	Type MC cable		12.31	.650		37	23		60	75
7080	EMT & wire		5	1.600		35.50	56.50		92	124
7090	For relay output to security system, add					11.75			11.75	12.95
8000	Residential equipment									
8050	Disposal hook-up, incl. switch, outlet box, 3' of flex									
8060	20 amp-1 pole circ. bkr., and 25' of #12/2									
8070	Type NM cable	1 Elec	10	.800	Ea.	19.55	28.50		48.05	64
8080	Type MC cable		8	1		31	35.50		66.50	87
8090	EMT & wire		5	1.600		33	56.50		89.50	121
8100	Trash compactor or dishwasher hook-up, incl. outlet box,									
8110	3' of flex, 15 amp-1 pole circ. bkr., and 25' of #14/2									
8120	Type NM cable	1 Elec	10	.800	Ea.	13.75	28.50		42.25	57.50
8130	Type MC cable		8	1		26.50	35.50		62	82
8140	EMT & wire		5	1.600		27.50	56.50		84	115
8150	Hot water sink dispensor hook-up, use line 8100									
8200	Vent/exhaust fan hook-up, type NM cable	1 Elec	32	.250	Ea.	3.34	8.85		12.19	16.90
8220	Type MC cable		24	.333		11.60	11.80		23.40	30.50
8230	EMT & wire		10	.800		11.40	28.50		39.90	55
8250	Bathroom vent fan, 50 CFM (use with above hook-up)									
8260	Economy model	1 Elec	15	.533	Ea.	21.50	18.90		40.40	51.50
8270	Low noise model		15	.533		30	18.90		48.90	61
8280	Custom model		12	.667		111	23.50		134.50	158
8300	Bathroom or kitchen vent fan, 110 CFM									
8310	Economy model	1 Elec	15	.533	Ea.	57	18.90		75.90	90.50
8320	Low noise model	"	15	.533	"	75	18.90		93.90	111
8350	Paddle fan, variable speed (w/o lights)									
8360	Economy model (AC motor)	1 Elec	10	.800	Ea.	100	28.50		128.50	153
8370	Custom model (AC motor)		10	.800		173	28.50		201.50	233
8380	Luxury model (DC motor)		8	1		340	35.50		375.50	430
8390	Remote speed switch for above, add		12	.667		24	23.50		47.50	62
8500	Whole house exhaust fan, ceiling mount, 36", variable speed									
8510	Remote switch, incl. shutters, 20 amp-1 pole circ. bkr.									
8520	30' of #12/2, type NM cable	1 Elec	4	2	Ea.	575	71		646	740
8530	Type MC cable		3.50	2.286		590	81		671	770
8540	EMT & wire		3	2.667		590	94.50		684.50	790
8600	Whirlpool tub hook-up, incl. timer switch, outlet box									
8610	3' of flex, 20 amp-1 pole GFI circ. bkr.									
8620	30' of #12/2, type NM cable	1 Elec	5	1.600	Ea.	78	56.50		134.50	171
8630	Type MC cable		4.20	1.905		86	67.50		153.50	196
8640	EMT & wire		3.40	2.353		88	83.50		171.50	221
8650	Hot water heater hook-up, incl. 1-2 pole circ. bkr., box;									
8660	3' of flex, 20' of #10/2, type NM cable	1 Elec	5	1.600	Ea.	20.50	56.50		77	107
8670	Type MC cable		4.20	1.905		34.50	67.50		102	139
8680	EMT & wire		3.40	2.353		29	83.50		112.50	156
9000	Heating/air conditioning									
9050	Furnace/boiler hook-up, incl. firestat, local on-off switch									
9060	Emergency switch, and 40' of type NM cable	1 Elec	4	2	Ea.	42.50	71		113.50	153
9070	Type MC cable		3.50	2.286		60	81		141	187
9080	EMT & wire		1.50	5.333		61	189		250	350
9100	Air conditioner hook-up, incl. local 60 amp disc. switch									

Important: See the Reference Section for critical supporting data - Reference Nos., Crews, & City Cost Indexes

16139 | Residential Wiring

		CREW	DAILY OUTPUT	LABOR-HOURS	UNIT	2002 BARE COSTS				TOTAL INCL O&P
						MAT.	LABOR	EQUIP.	TOTAL	
700	9110	3' sealtite, 40 amp, 2 pole circuit breaker									700
	9130	40' of #8/2, type NM cable	1 Elec	3.50	2.286	Ea.	137	81		218	272
	9140	Type MC cable		3	2.667		188	94.50		282.50	350
	9150	EMT & wire	↓	1.30	6.154	↓	150	218		368	490
	9200	Heat pump hook-up, 1-40 & 1-100 amp 2 pole circ. bkr.									
	9210	Local disconnect switch, 3' sealtite									
	9220	40' of #8/2 & 30' of #3/2									
	9230	Type NM cable	1 Elec	1.30	6.154	Ea.	325	218		543	680
	9240	Type MC cable		1.08	7.407		465	263		728	900
	9250	EMT & wire	↓	.94	8.511	↓	350	300		650	830
	9500	Thermostat hook-up, using low voltage wire									
	9520	Heating only	1 Elec	24	.333	Ea.	5.80	11.80		17.60	24
	9530	Heating/cooling	"	20	.400	"	7.15	14.20		21.35	29

16150 | Wiring Connections

		CREW	DAILY OUTPUT	LABOR-HOURS	UNIT	MAT.	LABOR	EQUIP.	TOTAL	TOTAL INCL O&P	
275	0010	**MOTOR CONNECTIONS**										275
	0020	Flexible conduit and fittings, 115 volt, 1 phase, up to 1 HP motor	1 Elec	8	1	Ea.	4.38	35.50		39.88	58
	0050	2 HP motor		6.50	1.231		4.48	43.50		47.98	70
	0100	3 HP motor		5.50	1.455		7.45	51.50		58.95	85
	0120	230 volt, 10 HP motor, 3 phase		4.20	1.905		8.05	67.50		75.55	110
	0150	15 HP motor		3.30	2.424		15.50	86		101.50	145
	0200	25 HP motor		2.70	2.963		17.65	105		122.65	176
	0400	50 HP motor		2.20	3.636		48	129		177	245
	0600	100 HP motor		1.50	5.333		122	189		311	415
	1500	460 volt, 5 HP motor, 3 phase		8	1		4.88	35.50		40.38	58.50
	1520	10 HP motor		8	1		4.88	35.50		40.38	58.50
	1530	25 HP motor		6	1.333		9.20	47.50		56.70	80.50
	1540	30 HP motor		6	1.333		9.20	47.50		56.70	80.50
	1550	40 HP motor		5	1.600		14.55	56.50		71.05	101
	1560	50 HP motor		5	1.600		16.50	56.50		73	103
	1570	60 HP motor		3.80	2.105		22	74.50		96.50	135
	1580	75 HP motor		3.50	2.286		29.50	81		110.50	154
	1590	100 HP motor		2.50	3.200		41	113		154	214
	1600	125 HP motor		2	4		54	142		196	271
	1610	150 HP motor		1.80	4.444		56	158		214	297
	1620	200 HP motor		1.50	5.333		97	189		286	390
	2005	460 Volt, 5 HP motor, 3 Phase, w/sealtite		8	1		9	35.50		44.50	63
	2010	10 HP motor		8	1		9	35.50		44.50	63
	2015	25 HP motor		6	1.333		16.50	47.50		64	88.50
	2020	30 HP motor		6	1.333		16.50	47.50		64	88.50
	2025	40 HP motor		5	1.600		30	56.50		86.50	118
	2030	50 HP motor		5	1.600		28.50	56.50		85	116
	2035	60 HP motor		3.80	2.105		42.50	74.50		117	158
	2040	75 HP motor		3.50	2.286		46	81		127	172
	2045	100 HP motor		2.50	3.200		62.50	113		175.50	238
	2055	150 HP motor		1.80	4.444		67.50	158		225.50	310
	2060	200 HP motor	↓	1.50	5.333	↓	410	189		599	730

ELECTRICAL 16

16220	Motors & Generators	CREW	DAILY OUTPUT	LABOR-HOURS	UNIT	2002 BARE COSTS				TOTAL INCL O&P	
						MAT.	LABOR	EQUIP.	TOTAL		
900	**0010**	**VARIABLE FREQUENCY DRIVES/ADJUSTABLE FREQUENCY DRIVES**									**900**
0100	Enclosed (NEMA 1), 460 volt, for 3 HP motor size	1 Elec	.80	10	Ea.	1,525	355		1,880	2,200	
0110	5 HP motor size		.80	10		1,750	355		2,105	2,450	
0120	7.5 HP motor size		.67	11.940		2,475	425		2,900	3,350	
0130	10 HP motor size		.67	11.940		2,475	425		2,900	3,350	
0140	15 HP motor size	2 Elec	.89	17.978		2,675	635		3,310	3,900	
0150	20 HP motor size		.89	17.978		3,950	635		4,585	5,275	
0160	25 HP motor size		.67	23.881		4,350	845		5,195	6,075	
0170	30 HP motor size		.67	23.881		5,025	845		5,870	6,800	
0180	40 HP motor size		.67	23.881		7,350	845		8,195	9,375	
0190	50 HP motor size		.53	30.189		8,025	1,075		9,100	10,400	
0200	60 HP motor size	R-3	.56	35.714		8,975	1,250	282	10,507	12,100	
0210	75 HP motor size		.56	35.714		12,200	1,250	282	13,732	15,600	
0220	100 HP motor size		.50	40		14,000	1,400	315	15,715	17,800	
0230	125 HP motor size		.50	40		16,500	1,400	315	18,215	20,500	
0240	150 HP motor size		.50	40		18,900	1,400	315	20,615	23,200	
0250	200 HP motor size		.42	47.619		21,200	1,675	375	23,250	26,200	
1100	Custom-engineered, 460 volt, for 3 HP motor size	1 Elec	.56	14.286		2,450	505		2,955	3,450	
1110	5 HP motor size		.56	14.286		2,675	505		3,180	3,700	
1120	7.5 HP motor size		.47	17.021		3,600	605		4,205	4,875	
1130	10 HP motor size		.47	17.021		3,825	605		4,430	5,100	
1140	15 HP motor size	2 Elec	.62	25.806		4,700	915		5,615	6,550	
1150	20 HP motor size		.62	25.806		5,550	915		6,465	7,475	
1160	25 HP motor size		.47	34.043		6,400	1,200		7,600	8,825	
1170	30 HP motor size		.47	34.043		7,650	1,200		8,850	10,200	
1180	40 HP motor size		.47	34.043		8,175	1,200		9,375	10,800	
1190	50 HP motor size		.37	43.243		10,300	1,525		11,825	13,700	
1200	60 HP motor size	R-3	.39	51.282		14,200	1,800	405	16,405	18,800	
1210	75 HP motor size		.39	51.282		15,500	1,800	405	17,705	20,200	
1220	100 HP motor size		.35	57.143		16,800	2,000	450	19,250	21,900	
1230	125 HP motor size		.35	57.143		19,200	2,000	450	21,650	24,600	
1240	150 HP motor size		.35	57.143		21,300	2,000	450	23,750	27,000	
1250	200 HP motor size		.29	68.966		24,900	2,425	545	27,870	31,600	
2000	For complex & special design systems to meet specific										
2010	requirements, obtain quote from vendor.										

16420	Enclosed Controllers	CREW	DAILY OUTPUT	LABOR-HOURS	UNIT	2002 BARE COSTS				TOTAL INCL O&P	
						MAT.	LABOR	EQUIP.	TOTAL		
220	**0010**	**CONTROL STATIONS**									**220**
0050	NEMA 1, heavy duty, stop/start	1 Elec	8	1	Ea.	116	35.50		151.50	181	
0100	Stop/start, pilot light		6.20	1.290		156	45.50		201.50	241	
0200	Hand/off/automatic		6.20	1.290		85.50	45.50		131	163	
0400	Stop/start/reverse		5.30	1.509		156	53.50		209.50	252	
0500	NEMA 7, heavy duty, stop/start		6	1.333		188	47.50		235.50	278	
0600	Stop/start, pilot light		4	2		230	71		301	360	

Important: See the Reference Section for critical supporting data - Reference Nos., Crews, & City Cost Indexes

16440	Swbds, Panels & Cont Centers	CREW	DAILY OUTPUT	LABOR-HOURS	UNIT	2002 BARE COSTS				TOTAL INCL O&P
						MAT.	LABOR	EQUIP.	TOTAL	
660	0010	**MOTOR STARTERS & CONTROLS**									660
0050	Magnetic, FVNR, with enclosure and heaters, 480 volt									
0080	2 HP, size 00	1 Elec	3.50	2.286	Ea.	171	81		252	310
0100	5 HP, size 0		2.30	3.478		207	123		330	410
0200	10 HP, size 1	↓	1.60	5		232	177		409	520
0300	25 HP, size 2	2 Elec	2.20	7.273		435	258		693	865
0400	50 HP, size 3		1.80	8.889		710	315		1,025	1,250
0500	100 HP, size 4		1.20	13.333		1,575	475		2,050	2,425
0600	200 HP, size 5		.90	17.778		3,700	630		4,330	5,000
0610	400 HP, size 6	↓	.80	20		10,400	710		11,110	12,500
0620	NEMA 7, 5 HP, size 0	1 Elec	1.60	5		865	177		1,042	1,225
0630	10 HP, size 1	"	1.10	7.273		905	258		1,163	1,375
0640	25 HP, size 2	2 Elec	1.80	8.889		1,450	315		1,765	2,075
0650	50 HP, size 3		1.20	13.333		2,150	475		2,625	3,075
0660	100 HP, size 4		.90	17.778		3,525	630		4,155	4,825
0670	200 HP, size 5	↓	.50	32		8,250	1,125		9,375	10,800
0700	Combination, with motor circuit protectors, 5 HP, size 0	1 Elec	1.80	4.444		670	158		828	975
0800	10 HP, size 1	"	1.30	6.154		695	218		913	1,100
0900	25 HP, size 2	2 Elec	2	8		975	284		1,259	1,500
1000	50 HP, size 3		1.32	12.121		1,400	430		1,830	2,200
1200	100 HP, size 4	↓	.80	20		3,075	710		3,785	4,425
1220	NEMA 7, 5 HP, size 0	1 Elec	1.30	6.154		1,525	218		1,743	2,000
1230	10 HP, size 1	"	1	8		1,575	284		1,859	2,150
1240	25 HP, size 2	2 Elec	1.32	12.121		2,075	430		2,505	2,950
1250	50 HP, size 3		.80	20		3,425	710		4,135	4,825
1260	100 HP, size 4		.60	26.667		5,325	945		6,270	7,250
1270	200 HP, size 5	↓	.40	40		11,500	1,425		12,925	14,800
1400	Combination, with fused switch, 5 HP, size 0	1 Elec	1.80	4.444		515	158		673	800
1600	10 HP, size 1	"	1.30	6.154		550	218		768	930
1800	25 HP, size 2	2 Elec	2	8		890	284		1,174	1,400
2000	50 HP, size 3		1.32	12.121		1,500	430		1,930	2,300
2200	100 HP, size 4	↓	.80	20	↓	2,625	710		3,335	3,950
3500	Magnetic FVNR with NEMA 12, enclosure & heaters, 480 volt									
3600	5 HP, size 0	1 Elec	2.20	3.636	Ea.	195	129		324	405
3700	10 HP, size 1	"	1.50	5.333		293	189		482	605
3800	25 HP, size 2	2 Elec	2	8		550	284		834	1,025
3900	50 HP, size 3		1.60	10		845	355		1,200	1,450
4000	100 HP, size 4		1	16		2,025	565		2,590	3,075
4100	200 HP, size 5	↓	.80	20		4,850	710		5,560	6,375
4200	Combination, with motor circuit protectors, 5 HP, size 0	1 Elec	1.70	4.706		645	167		812	960
4300	10 HP, size 1	"	1.20	6.667		670	236		906	1,100
4400	25 HP, size 2	2 Elec	1.80	8.889		1,000	315		1,315	1,575
4500	50 HP, size 3		1.20	13.333		1,625	475		2,100	2,500
4600	100 HP, size 4	↓	.74	21.622		3,700	765		4,465	5,200
4700	Combination, with fused switch, 5 HP, size 0	1 Elec	1.70	4.706		625	167		792	935
4800	10 HP, size 1	"	1.20	6.667		650	236		886	1,075
4900	25 HP, size 2	2 Elec	1.80	8.889		990	315		1,305	1,575
5000	50 HP, size 3		1.20	13.333		1,600	475		2,075	2,450
5100	100 HP, size 4	↓	.74	21.622	↓	3,225	765		3,990	4,700
5200	Factory installed controls, adders to size 0 thru 5									
5300	Start-stop push button	1 Elec	32	.250	Ea.	41	8.85		49.85	58.50
5400	Hand-off-auto-selector switch		32	.250		41	8.85		49.85	58.50
5500	Pilot light		32	.250		76.50	8.85		85.35	97.50
5600	Start-stop-pilot		32	.250		117	8.85		125.85	142
5700	Auxiliary contact, NO or NC		32	.250		56	8.85		64.85	75
5800	NO-NC	↓	32	.250	↓	112	8.85		120.85	136

ELECTRICAL 16

Division Notes

	CREW	DAILY OUTPUT	LABOR-HOURS	UNIT	2002 BARE COSTS				TOTAL INCL O&P
					MAT.	LABOR	EQUIP.	TOTAL	

Division 17
Square Foot & Cubic Foot Costs

Estimating Tips

- The cost figures in Division 17 were derived from approximately 10,400 projects contained in the Means database of completed construction projects, and include the contractor's overhead and profit, but do not generally include architectural fees or land costs. The figures have been adjusted to January of the current year. New projects are added to our files each year, and outdated projects are discarded. For this reason, certain costs may not show a uniform annual progression. In no case are all subdivisions of a project listed.

- These projects were located throughout the U.S. and reflect a tremendous variation in square foot (S.F.) and cubic foot (C.F.) costs. This is due to differences, not only in labor and material costs, but also in individual owners' requirements. For instance, a bank in a large city would have different features than one in a rural area. This is true of all the different types of buildings analyzed. Therefore, caution should be exercised when using Division 17 costs. For example, for court houses, costs in the database are local court house costs and will not apply to the larger, more elaborate federal court houses. As a general rule, the projects in the 1/4 column do not include any site work or equipment, while the projects in the 3/4 column may include both equipment and site work. The median figures do not generally include site work.

- None of the figures "go with" any others. All individual cost items were computed and tabulated separately. Thus the sum of the median figures for Plumbing, HVAC and Electrical will not normally total up to the total Mechanical and Electrical costs arrived at by separate analysis and tabulation of the projects.

- Each building was analyzed as to total and component costs and percentages. The figures were arranged in ascending order with the results tabulated as shown. The 1/4 column shows that 25% of the projects had lower costs, 75% higher. The 3/4 column shows that 75% of the projects had lower costs, 25% had higher. The median column shows that 50% of the projects had lower costs, 50% had higher.

- There are two times when square foot costs are useful. The first is in the conceptual stage when no details are available. Then square foot costs make a useful starting point. The second is after the bids are in and the costs can be worked back into their appropriate units for information purposes. As soon as details become available in the project design, the square foot approach should be discontinued and the project priced as to its particular components. When more precision is required or for estimating the replacement cost of specific buildings, the current edition of *Means Square Foot Costs* should be used.

- In using the figures in Division 17, it is recommended that the median column be used for preliminary figures if no additional information is available. The median figures, when multiplied by the total city construction cost index figures (see City Cost Indexes) and then multiplied by the project size modifier in Reference Number R17100-100, should present a fairly accurate base figure, which would then have to be adjusted in view of the estimator's experience, local economic conditions, code requirements and the owner's particular requirements. There is no need to factor the percentage figures, as these should remain constant from city to city. All tabulations mentioning air conditioning had at least partial air conditioning.

- The editors of this book would greatly appreciate receiving cost figures on one or more of your recent projects which would then be included in the averages for next year. All cost figures received will be kept confidential except that they will be averaged with other similar projects to arrive at S.F. and C.F. cost figures for next year's book. See the last page of the book for details and the discount available for submitting one or more of your projects.

17100 | S.F. & C.F. Costs			UNIT	UNIT COSTS			% OF TOTAL		
				1/4	MEDIAN	3/4	1/4	MEDIAN	3/4
010	0010	**APARTMENTS** Low Rise (1 to 3 story) R17100-100	S.F.	45.50	57	76				**010**
	0020	Total project cost	C.F.	4.08	5.40	6.70			
	2720	Plumbing	S.F.	3.50	4.55	5.75	6.70%	9%	10.10%
	2770	Heating, ventilating, air conditioning		2.26	2.78	4.09	4.20%	5.80%	7.70%
	2900	Electrical		2.62	3.43	4.68	5.20%	6.70%	8.40%
	3100	Total: Mechanical & Electrical		9.05	11.55	14.50	16%	18.20%	23%
	9000	Per apartment unit, total cost	Apt.	42,000	64,000	95,500			
	9500	Total: Mechanical & Electrical	"	7,900	12,600	16,500			
020	0010	**APARTMENTS** Mid Rise (4 to 7 story) R17100-100	S.F.	59	72	88				**020**
	0020	Total project costs	C.F.	4.71	6.50	8.90			
	2720	Plumbing	S.F.	3.54	5.65	6.30	6.30%	7.40%	10%
	2900	Electrical		4.16	5.70	6.60	6.70%	7.50%	8.90%
	3100	Total: Mechanical & Electrical		12	16.25	19.95	18.80%	21.60%	25.10%
	9000	Per apartment unit, total cost	Apt.	68,000	80,500	133,500			
	9500	Total: Mechanical & Electrical	"	12,600	15,100	19,100			
030	0010	**APARTMENTS** High Rise (8 to 24 story) R17100-100	S.F.	68.50	82.50	101				**030**
	0020	Total project costs	C.F.	5.80	8.15	9.90			
	2720	Plumbing	S.F.	5.05	5.95	7.30	7%	9.10%	10.60%
	2900	Electrical		4.70	5.95	8	6.40%	7.70%	8.90%
	3100	Total: Mechanical & Electrical		14.10	17.95	21.50	18%	22.30%	24.40%
	9000	Per apartment unit, total cost	Apt.	71,000	77,000	108,500			
	9500	Total: Mechanical & Electrical	"	15,400	17,600	18,600			
040	0010	**AUDITORIUMS** R17100-100	S.F.	70.50	95.50	128				**040**
	0020	Total project costs	C.F.	4.45	6.20	9.30			
	2720	Plumbing	S.F.	4.50	6.25	7.85	6.30%	7.20%	9.05%
	2900	Electrical		5.75	7.90	9.90	6.80%	9%	10.60%
	3100	Total: Mechanical & Electrical		11	14.70	25.50	14.40%	18.50%	23.60%
050	0010	**AUTOMOTIVE SALES** R17100-100	S.F.	51.50	59	87				**050**
	0020	Total project costs	C.F.	3.41	4.16	5.40			
	2720	Plumbing	S.F.	2.56	4.16	4.54	4.70%	6.50%	7%
	2900	Electrical		4.08	6.35	7.15	5.60%	8.20%	12.20%
	3100	Total: Mechanical & Electrical		9.25	13.30	16.60	17%	20.40%	26%
060	0010	**BANKS** R17100-100	S.F.	102	128	162				**060**
	0020	Total project costs	C.F.	7.35	10	13.10			
	2720	Plumbing	S.F.	3.25	4.65	6.75	2.80%	3.90%	4.90%
	2770	Heating, ventilating, air conditioning		6.15	8.10	10.65	5%	7.20%	8.50%
	2900	Electrical		9.90	13.05	17	8.30%	10.30%	12.20%
	3100	Total: Mechanical & Electrical		23.50	31.50	38.50	16.60%	20.40%	24.40%
130	0010	**CHURCHES** R17100-100	S.F.	68.50	86	112				**130**
	0020	Total project costs	C.F.	4.31	5.45	7.15			
	2720	Plumbing	S.F.	2.70	3.78	5.55	3.60%	5%	6.30%
	2770	Heating, ventilating, air conditioning		6.35	8.25	12.10	7.60%	10%	12.20%
	2900	Electrical		5.60	7.85	10.20	7.30%	8.80%	10.90%
	3100	Total: Mechanical & Electrical		14.80	22	32.50	19.60%	22.90%	25.40%
150	0010	**CLUBS, COUNTRY** R17100-100	S.F.	70	86	113				**150**
	0020	Total project costs	C.F.	6	7.30	10.10			
	2720	Plumbing	S.F.	4.78	6.70	15.20	6.40%	9.60%	10%
	2900	Electrical		5.85	8.40	11.05	7%	9.30%	11.40%
	3100	Total: Mechanical & Electrical		12.65	16.30	31	16.10%	22.30%	30.90%
170	0010	**CLUBS, SOCIAL** Fraternal R17100-100	S.F.	58.50	83.50	109				**170**
	0020	Total project costs	C.F.	3.51	5.65	6.55			
	2720	Plumbing	S.F.	3.36	4.65	5.65	5.20%	6.80%	8.30%
	2770	Heating, ventilating, air conditioning		5.40	6.50	8.40	8.70%	11%	14.40%

17 SQUARE FOOT

See R17100-100 and City Cost Indexes in the Reference Section

		17100 | S.F. & C.F. Costs		UNIT	UNIT COSTS			% OF TOTAL			
					1/4	MEDIAN	3/4	1/4	MEDIAN	3/4	
170	2900	Electrical		S.F.	4.73	7.35	8.80	7.30%	9.50%	11.40%	170
	3100	Total: Mechanical & Electrical		↓	12.65	17.10	25	19.30%	23%	33.10%	
180	0010	**CLUBS, Y.M.C.A.**	R17100 -100	S.F.	68	83.50	115				180
	0020	Total project costs		C.F.	3.44	5.75	8.60				
	2720	Plumbing		S.F.	4.96	9.40	10.55	6%	9.70%	16%	
	2900	Electrical			5.65	7.45	11	6.40%	9.20%	10.80%	
	3100	Total: Mechanical & Electrical		↓	12.40	17.70	27.50	16.20%	21.90%	28.50%	
190	0010	**COLLEGES** Classrooms & Administration	R17100 -100	S.F.	85	112	149				190
	0020	Total project costs		C.F.	6.20	8.65	14				
	2720	Plumbing		S.F.	4.18	8.20	15.20	4.80%	8.90%	11.80%	
	2900	Electrical			7	10.75	13.25	8.40%	10%	12.10%	
	3100	Total: Mechanical & Electrical		↓	14.85	27.50	42.50	14.20%	24.70%	32.50%	
210	0010	**COLLEGES** Science, Engineering, Laboratories	R17100 -100	S.F.	135	164	199				210
	0020	Total project costs		C.F.	8.05	11.75	13.30				
	2900	Electrical		S.F.	11.60	16.75	25.50	7.10%	9.60%	15.40%	
	3100	Total: Mechanical & Electrical		"	46	51	79	28%	31.60%	39.80%	
230	0010	**COLLEGES** Student Unions	R17100 -100	S.F.	89.50	125	147				230
	0020	Total project costs		C.F.	4.99	6.55	8.45				
	3100	Total: Mechanical & Electrical		S.F.	23.50	33.50	36.50	17.40%	23.90%	28.90%	
250	0010	**COMMUNITY CENTERS**	R17100 -100	"	67	89.50	119				250
	0020	Total project costs		C.F.	4.82	6.90	8.90				
	2720	Plumbing		S.F.	3.58	6.10	8.90	4.90%	7.10%	9.50%	
	2770	Heating, ventilating, air conditioning			5.90	8.50	11.70	7.70%	10.80%	13%	
	2900	Electrical			5.90	7.80	12.15	7.40%	9.40%	11.50%	
	3100	Total: Mechanical & Electrical		↓	22	26	37.50	25.10%	27.90%	34.40%	
280	0010	**COURT HOUSES**	R17100 -100	S.F.	106	122	141				280
	0020	Total project costs		C.F.	8.05	9.80	13.55				
	2720	Plumbing		S.F.	5.05	7.05	10.15	5.90%	7.40%	9.60%	
	2900	Electrical			9.30	11.35	14.90	8.40%	9.80%	11.60%	
	3100	Total: Mechanical & Electrical		↓	21	28.50	37	20.10%	25.70%	28.70%	
300	0010	**DEPARTMENT STORES**	R17100 -100	S.F.	39.50	53.50	67.50				300
	0020	Total project costs		C.F.	2.07	2.74	3.72				
	2720	Plumbing		S.F.	1.28	1.55	2.35	3.10%	4.20%	5.90%	
	2770	Heating, ventilating, air conditioning			3.58	5.50	8.30	8.30%	12.40%	14.80%	
	2900	Electrical			4.52	6.20	7.30	9.10%	12.20%	14.90%	
	3100	Total: Mechanical & Electrical		↓	7.95	10.30	16.65	20%	19%	26.50%	
310	0010	**DORMITORIES** Low Rise (1 to 3 story)	R17100 -100	S.F.	64.50	92	111				310
	0020	Total project costs		C.F.	4.46	6.80	10.20				
	2720	Plumbing		S.F.	4.49	6	7.60	8%	9%	9.60%	
	2770	Heating, ventilating, air conditioning			4.75	5.70	7.60	4.60%	8%	10%	
	2900	Electrical			4.75	7.25	9.10	6.60%	8.90%	9.60%	
	3100	Total: Mechanical & Electrical		↓	25	26	27.50	22.50%	24.10%	29.10%	
	9000	Per bed, total cost		Bed	31,700	35,100	75,500				
320	0010	**DORMITORIES** Mid Rise (4 to 8 story)	R17100 -100	S.F.	92	120	148				320
	0020	Total project costs		C.F.	10.10	11.10	13.30				
	2900	Electrical		S.F.	6.80	10.65	13.65	7.40%	8.90%	10.40%	
	3100	Total: Mechanical & Electrical		"	20.50	27.50	35.50	18.10%	19.50%	30.60%	
	9000	Per bed, total cost		Bed	13,100	29,800	61,500				
340	0010	**FACTORIES**	R17100 -100	S.F.	35	51.50	80				340
	0020	Total project costs		C.F.	2.23	3.30	5.50				
	2720	Plumbing		S.F.	2.02	3.48	5.90	4.60%	6.20%	8.70%	
	2770	Heating, ventilating, air conditioning		↓	3.64	5.20	7.05	5.20%	8.50%	11.40%	

SQUARE FOOT 17

		17100 | S.F. & C.F. Costs		UNIT	UNIT COSTS 1/4	MEDIAN	3/4	% OF TOTAL 1/4	MEDIAN	3/4	
340	2900	Electrical		S.F.	4.25	6.80	10.40	8.10%	10.50%	14.20%	340
	3100	Total: Mechanical & Electrical		↓	10.25	16.45	25	20.40%	28.10%	34.50%	
360	0010	**FIRE STATIONS**	R17100 -100	S.F.	67	90	116				360
	0020	Total project costs		C.F.	3.92	5.45	7.30				
	2720	Plumbing		S.F.	4.36	6.35	9.35	6.20%	7.70%	9.70%	
	2770	Heating, ventilating, air conditioning			3.76	6.20	9.45	4.90%	7.50%	9.30%	
	2900	Electrical			4.77	8.20	10.90	7%	9.20%	11%	
	3100	Total: Mechanical & Electrical		↓	21	26	31	18.80%	23.80%	27.30%	
370	0010	**FRATERNITY HOUSES** and Sorority Houses	R17100 -100	S.F.	69	88.50	125				370
	0020	Total project costs		C.F.	6.60	7.15	9.15				
	2720	Plumbing		S.F.	5.15	5.95	10.90	5.20%	8%	10.90%	
	2900	Electrical			4.54	9.80	12.05	6.60%	9.90%	10.70%	
	3100	Total: Mechanical & Electrical		↓	12.55	17.85	21.50	14.60%	19.90%	20.70%	
380	0010	**FUNERAL HOMES**	R17100 -100	S.F.	72.50	99	180				380
	0020	Total project costs		C.F.	7.40	8.25	15.85				
	2900	Electrical		S.F.	3.21	5.90	6.95	3.60%	4.40%	11%	
	3100	Total: Mechanical & Electrical		"	11.55	17.25	22.50	12.90%	17.80%	18.80%	
390	0010	**GARAGES, COMMERCIAL** (Service)	R17100 -100	S.F.	39.50	63	86.50				390
	0020	Total project costs		C.F.	2.52	3.77	5.40				
	2720	Plumbing		S.F.	2.68	4.01	7.85	4.70%	7.40%	11%	
	2730	Heating & ventilating			3.74	5.10	7.05	5.30%	6.90%	9.60%	
	2900	Electrical			3.78	5.80	8	7.20%	9%	11.20%	
	3100	Total: Mechanical & Electrical		↓	8.90	16.55	24.50	14.10%	22.50%	26.90%	
400	0010	**GARAGES, MUNICIPAL** (Repair)	R17100 -100	S.F.	60.50	80.50	115				400
	0020	Total project costs		C.F.	3.78	4.79	7.80				
	2720	Plumbing		S.F.	2.71	5.20	9.80	3.60%	6.70%	8%	
	2730	Heating & ventilating			4.64	6.70	12.95	6.20%	7.40%	13.50%	
	2900	Electrical			4.46	6.60	10.10	6.30%	8%	11.70%	
	3100	Total: Mechanical & Electrical		↓	11.30	20.50	33.50	15.20%	25.50%	32.70%	
410	0010	**GARAGES, PARKING**	R17100 -100	S.F.	23	33	59				410
	0020	Total project costs		C.F.	2.21	3	4.37				
	2720	Plumbing		S.F.	.61	1.03	1.59	2.60%	3.40%	3.90%	
	2900	Electrical			.96	1.46	2.14	4.30%	5.20%	6.30%	
	3100	Total: Mechanical & Electrical		↓	1.28	3.77	4.83	6.50%	9.40%	12.80%	
	9000	Per car, total cost		Car	9,900	12,500	16,100				
430	0010	**GYMNASIUMS**	R17100 -100	S.F.	65.50	83.50	106				430
	0020	Total project costs		C.F.	3.26	4.17	5.40				
	2720	Plumbing		S.F.	4.14	5.10	6.10	5.40%	7.30%	7.90%	
	2770	Heating, ventilating, air conditioning		"	4.45	6.80	13.60	9%	11.10%	22.60%	
	2900	Electrical		S.F.	4.95	6.15	8.15	6.60%	8.30%	10.70%	
	3100	Total: Mechanical & Electrical		↓	15.35	21	28.50	20.60%	26.20%	29.40%	
460	0010	**HOSPITALS**	R17100 -100	↓	135	157	233				460
	0020	Total project costs		C.F.	9.50	11.75	16.95				
	2720	Plumbing		S.F.	11.25	15.10	19.45	7.80%	9.40%	11.80%	
	2770	Heating, ventilating, air conditioning			15.85	21	28.50	8.40%	14.60%	17%	
	2900	Electrical			13.50	17.55	27.50	9.90%	12%	14.50%	
	3100	Total: Mechanical & Electrical		↓	38.50	51	83.50	26.60%	33.10%	39%	
	9000	Per bed or person, total cost		Bed	48,400	103,000	162,000				
480	0010	**HOUSING** For the Elderly	R17100 -100	S.F.	62	78	95.50				480
	0020	Total project costs		C.F.	4.38	6.10	7.80				
	2720	Plumbing		S.F.	4.58	5.90	7.85	8.10%	9.70%	10.90%	
	2730	Heating, ventilating, air conditioning			2.35	3.33	4.98	3.30%	5.60%	7.20%	
	2900	Electrical			4.60	6.20	7.95	7.30%	8.50%	10.50%	
	3100	Total: Mechanical & Electrical		↓	15.85	18.60	24.50	18.10%	22%	29.10%	
	9000	Per rental unit, total cost		Unit	57,500	67,000	74,500				
	9500	Total: Mechanical & Electrical		"	12,300	14,700	17,200				

See R17100-100 and City Cost Indexes in the Reference Section

		17100 | S.F. & C.F. Costs		UNIT	UNIT COSTS			% OF TOTAL			
					1/4	MEDIAN	3/4	1/4	MEDIAN	3/4	
500	0010	HOUSING Public (Low Rise)	R17100 -100	S.F.	52	72	94				**500**
	0020	Total project costs		C.F.	4.12	5.75	7.15				
	2720	Plumbing		S.F.	3.53	4.95	6.25	6.80%	9%	11.60%	
	2730	Heating, ventilating, air conditioning			1.88	3.65	4	4.20%	6%	6.40%	
	2900	Electrical			3.14	4.68	6.50	5.10%	6.60%	8.30%	
	3100	Total: Mechanical & Electrical		↓	14.90	19.20	21.50	14.50%	17.60%	26.50%	
	9000	Per apartment, total cost		Apt.	57,000	65,000	81,500				
	9500	Total: Mechanical & Electrical		"	12,200	15,000	16,600				
510	0010	ICE SKATING RINKS	R17100 -100	S.F.	46	80	114				**510**
	0020	Total project costs		C.F.	3.26	3.34	3.85				
	2720	Plumbing		S.F.	1.66	3.11	3.18	3.10%	5.60%	6.70%	
	2900	Electrical			4.75	7.30	7.70	6.70%	15%	15.80%	
	3100	Total: Mechanical & Electrical		↓	8	11.50	14.35	9.90%	25.90%	29.80%	
520	0010	JAILS	R17100 -100	S.F.	136	174	225				**520**
	0020	Total project costs		C.F.	12.75	15.65	19.90				
	2720	Plumbing		S.F.	13.75	17.45	23	7%	8.90%	13.80%	
	2770	Heating, ventilating, air conditioning			12.20	16.25	31.50	7.50%	9.40%	17.70%	
	2900	Electrical			14.85	19.10	23.50	8.10%	11.40%	12.40%	
	3100	Total: Mechanical & Electrical		↓	38	67.50	79.50	29.20%	31.10%	34.10%	
530	0010	LIBRARIES	R17100 -100	S.F.	80.50	103	131				**530**
	0020	Total project costs		C.F.	5.65	7.05	9.15				
	2720	Plumbing		S.F.	3.18	4.60	6.20	3.60%	4.90%	5.70%	
	2770	Heating, ventilating, air conditioning			7	11.80	15.40	8%	11%	14.60%	
	2900	Electrical			8.30	10.40	13.30	8.30%	11%	12.10%	
	3100	Total: Mechanical & Electrical		↓	24	32.50	41	18.90%	25.30%	27.60%	
540	0010	LIVING, ASSISTED		S.F.	81.50	93.50	110				**540**
	0020	Total project costs		C.F.	6.80	7.80	9.15				
	2720	Plumbing		S.F.	6.60	8.80	9.35	6%	8.10%	10.60%	
	2770	Heating, ventilating, air conditioning			7.80	8.15	9	7.90%	8.80%	9.70%	
	2900	Electrical			7.75	8.80	9.80	9.30%	10.40%	10.70%	
	3100	Total: Mechanical & Electrical		↓	22.50	26.50	29	27.90%	30.10%	32%	
550	0010	MEDICAL CLINICS	R17100 -100	S.F.	79.50	98	123				**550**
	0020	Total project costs		C.F.	5.90	7.65	10.20				
	2720	Plumbing		S.F.	5.35	7.50	10.05	6.10%	8.40%	10%	
	2770	Heating, ventilating, air conditioning			6.35	8.35	12.25	6.70%	9%	11.30%	
	2900	Electrical			6.75	9.60	12.75	8.10%	10%	12.20%	
	3100	Total: Mechanical & Electrical		↓	21	30	41.50	22%	27.60%	34.30%	
570	0010	MEDICAL OFFICES	R17100 -100	S.F.	74.50	92.50	114				**570**
	0020	Total project costs		C.F.	5.55	7.60	10.40				
	2720	Plumbing		S.F.	4.18	6.45	8.75	5.70%	6.80%	8.60%	
	2770	Heating, ventilating, air conditioning			5.05	7.40	9.60	6.20%	8%	9.70%	
	2900	Electrical			5.90	8.60	12	7.60%	9.80%	11.40%	
	3100	Total: Mechanical & Electrical		↓	14.70	21	31.50	18.50%	22%	24.90%	
590	0010	MOTELS	R17100 -100	S.F.	47.50	70.50	91				**590**
	0020	Total project costs		C.F.	4.16	5.85	9.60				
	2720	Plumbing		S.F.	4.83	6.15	7.35	9.40%	10.60%	12.50%	
	2770	Heating, ventilating, air conditioning			2.95	4.39	7.90	5.60%	5.60%	10%	
	2900	Electrical			4.50	5.75	7.50	7.10%	8.20%	10.40%	
	3100	Total: Mechanical & Electrical		↓	15.30	19.20	33	18.50%	21%	24.40%	
	9000	Per rental unit, total cost		Unit	24,300	46,200	49,900				
	9500	Total: Mechanical & Electrical		"	4,700	7,200	8,300				
600	0010	NURSING HOMES	R17100 -100	S.F.	71.50	92	114				**600**
	0020	Total project costs		C.F.	5.75	7.30	10.05				
	2720	Plumbing		S.F.	6.75	8.60	11.90	9.40%	10.70%	14.20%	
	2770	Heating, ventilating, air conditioning			6.70	10.15	11.90	9.30%	11.40%	11.80%	
	2900	Electrical			7.40	9.30	12.45	9.70%	11%	13%	
	3100	Total: Mechanical & Electrical		↓	17.65	24.50	36	26%	29.90%	30.50%	

SQUARE FOOT 17

		17100 | S.F. & C.F. Costs		UNIT	UNIT COSTS			% OF TOTAL			
					1/4	MEDIAN	3/4	1/4	MEDIAN	3/4	
600	9000	Per bed or person, total cost		Bed	31,000	38,200	51,000				600
610	0010	OFFICES Low Rise (1 to 4 story)	R17100 -100	S.F.	60.50	77	102				610
	0020	Total project costs		C.F.	4.37	6.10	8.30				
	2720	Plumbing		S.F.	2.30	3.48	4.92	3.70%	4.50%	6.10%	
	2770	Heating, ventilating, air conditioning			4.97	6.85	10.15	7.20%	10.50%	11.90%	
	2900	Electrical			5.10	7.10	9.90	7.50%	9.60%	11.10%	
	3100	Total: Mechanical & Electrical		↓	12.05	16.70	24.50	18%	21.80%	26.50%	
620	0010	OFFICES Mid Rise (5 to 10 story)	R17100 -100	S.F.	66.50	81	110				620
	0020	Total project costs		C.F.	4.66	5.90	8.55				
	2720	Plumbing		S.F.	2.02	3.12	4.49	2.80%	3.70%	4.50%	
	2770	Heating, ventilating, air conditioning			5.05	7.25	11.55	7.60%	9.40%	11%	
	3100	Total: Mechanical & Electrical		↓	12.65	16.15	32	17.90%	22.30%	29.90%	
630	0010	OFFICES High Rise (11 to 20 story)	R17100 -100	S.F.	80.50	103	127				630
	0020	Total project costs		C.F.	5.15	7.15	10.25				
	2900	Electrical		S.F.	4.12	5.95	9	5.80%	7%	10.50%	
	3100	Total: Mechanical & Electrical		"	15.05	19.25	35.50	15.40%	17%	34.10%	
640	0010	POLICE STATIONS	R17100 -100	S.F.	96.50	129	163				640
	0020	Total project costs		C.F.	7.85	9.45	13.15				
	2720	Plumbing		S.F.	5.50	8.90	12.05	5.60%	6.90%	11.30%	
	2770	Heating, ventilating, air conditioning			8.60	11.40	15.50	7%	10.70%	12%	
	2900	Electrical			10.70	16.05	20.50	9.70%	11.90%	14.90%	
	3100	Total: Mechanical & Electrical		↓	34	41	57.50	25.10%	31.30%	36.40%	
650	0010	POST OFFICES	R17100 -100	S.F.	75.50	95.50	119				650
	0020	Total project costs		C.F.	4.60	5.90	7				
	2720	Plumbing		S.F.	3.49	4.48	5.45	4.40%	5.60%	5.60%	
	2770	Heating, ventilating, air conditioning			5.45	6.75	7.50	7%	8.40%	10.20%	
	2900	Electrical			6.35	8.85	10.30	7.20%	9.60%	11%	
	3100	Total: Mechanical & Electrical		↓	11.25	21.50	27.50	16.40%	20.40%	22.30%	
660	0010	POWER PLANTS	R17100 -100	S.F.	560	715	1,300				660
	0020	Total project costs		C.F.	14.80	30	69				
	2900	Electrical		S.F.	38	80.50	120	9.50%	12.80%	18.40%	
	8100	Total: Mechanical & Electrical		"	94.50	310	690	32.50%	32.60%	52.60%	
670	0010	RELIGIOUS EDUCATION	R17100 -100	S.F.	61	79	89.50				670
	0020	Total project costs		C.F.	3.47	4.97	6.45				
	2720	Plumbing		S.F.	2.60	3.69	5.20	4.30%	4.90%	8.40%	
	2770	Heating, ventilating, air conditioning			6.10	6.90	7.60	9.80%	10%	11.80%	
	2900	Electrical			4.96	6.55	8.65	7.90%	9.10%	10.30%	
	3100	Total: Mechanical & Electrical		↓	12.80	22.50	26.50	17.60%	20.50%	22%	
690	0010	RESEARCH Laboratories and facilities	R17100 -100	S.F.	93.50	134	195				690
	0020	Total project costs		C.F.	6.95	11.70	15.75				
	2720	Plumbing		S.F.	9.40	12	19.30	7.30%	9.20%	13.30%	
	2770	Heating, ventilating, air conditioning			8.40	28.50	33.50	7.20%	16.50%	17.50%	
	2900	Electrical			10.75	18.05	31	9.60%	12%	15.80%	
	3100	Total: Mechanical & Electrical		↓	25	47	90	23.90%	35.60%	41.70%	
700	0010	RESTAURANTS	R17100 -100	S.F.	90	116	151				700
	0020	Total project costs		C.F.	7.55	9.95	13.10				
	2720	Plumbing		S.F.	7.15	8.70	12.40	6.10%	8.30%	9.10%	
	2770	Heating, ventilating, air conditioning			9.45	12.60	16.45	9.40%	12.20%	13%	
	2900	Electrical			9.60	11.85	15.40	8.50%	10.60%	11.60%	
	3100	Total: Mechanical & Electrical		↓	30	31.50	39.50	21.10%	24%	29.50%	
	9000	Per seat unit, total cost		Seat	3,300	4,400	5,600				
	9500	Total: Mechanical & Electrical		"	700	995	1,300				
720	0010	RETAIL STORES	R17100 -100	S.F.	42	56.50	74.50				720
	0020	Total project costs		C.F.	2.86	4.08	5.65				
	2720	Plumbing		S.F.	1.54	2.56	4.38	3.20%	4.60%	6.80%	
	2770	Heating, ventilating, air conditioning		↓	3.32	4.54	6.85	6.70%	8.70%	10.10%	

17 SQUARE FOOT

See R17100-100 and City Cost Indexes in the Reference Section

17100 | S.F. & C.F. Costs			UNIT	UNIT COSTS			% OF TOTAL		
				1/4	MEDIAN	3/4	1/4	MEDIAN	3/4
720	2900	Electrical	S.F.	3.79	5.15	7.45	7.30%	9.90%	11.60%	**720**
	3100	Total: Mechanical & Electrical	↓	10.15	13	16.85	17.10%	21.40%	23.80%	
740	0010	**SCHOOLS** Elementary R17100 -100	S.F.	67	82.50	99.50				**740**
	0020	Total project costs	C.F.	4.53	5.80	7.50				
	2720	Plumbing	S.F.	3.87	5.60	7.50	5.70%	7.10%	9.40%	
	2730	Heating, ventilating, air conditioning		5.95	9.50	13.25	8.10%	10.80%	15.20%	
	2900	Electrical		6.25	7.95	10.15	8.40%	10%	11.60%	
	3100	Total: Mechanical & Electrical	↓	21.50	24	33.50	22.20%	28%	30.40%	
	9000	Per pupil, total cost	Ea.	6,700	11,000	36,000				
	9500	Total: Mechanical & Electrical	"	2,250	2,825	11,300				
760	0010	**SCHOOLS** Junior High & Middle R17100 -100	S.F.	68	84	98.50				**760**
	0020	Total project costs	C.F.	4.47	5.85	6.55				
	2720	Plumbing	S.F.	4.52	5	6.65	5.60%	6.90%	8.10%	
	2770	Heating, ventilating, air conditioning		5.20	10.05	14	8.70%	12.70%	17.40%	
	2900	Electrical		6.55	8.25	9.95	7.80%	9.40%	10.60%	
	3100	Total: Mechanical & Electrical	↓	18.95	24	31.50	23.30%	25.70%	27.30%	
	9000	Per pupil, total cost	Ea.	8,500	9,800	12,200				
780	0010	**SCHOOLS** Senior High R17100 -100	S.F.	73.50	84	118				**780**
	0020	Total project costs	C.F.	4.70	6.10	8.60				
	2720	Plumbing	S.F.	3.79	6.85	11.40	5%	6.90%	12%	
	2770	Heating, ventilating, air conditioning		8.50	9.65	18.55	8.90%	11%	15%	
	2900	Electrical		7.20	9.50	15.80	8.30%	10.10%	12.60%	
	3100	Total: Mechanical & Electrical	↓	19.70	26	48.50	19.80%	23.10%	27.80%	
	9000	Per pupil, total cost	Ea.	7,000	11,900	17,800				
800	0010	**SCHOOLS** Vocational R17100 -100	S.F.	60	83.50	106				**800**
	0020	Total project costs	C.F.	3.67	5.30	7.40				
	2720	Plumbing	S.F.	3.84	5.85	8.40	5.40%	7%	8.50%	
	2770	Heating, ventilating, air conditioning		5.35	10	16.70	8.80%	11.90%	14.60%	
	2900	Electrical		6.15	8.55	12.05	8.40%	11.20%	13.80%	
	3100	Total: Mechanical & Electrical	↓	16.25	21.50	41.50	21.70%	27.30%	33.10%	
	9000	Per pupil, total cost	Ea.	8,400	22,400	33,400				
830	0010	**SPORTS ARENAS** R17100 -100	S.F.	51	70	103				**830**
	0020	Total project costs	C.F.	2.85	5.05	6.60				
	2720	Plumbing	S.F.	2.71	4.51	8.50	4.50%	6.30%	9.40%	
	2770	Heating, ventilating, air conditioning		5.55	7.75	10.10	5.80%	10.20%	13.50%	
	2900	Electrical		4.51	7.10	8.90	7.70%	9.80%	12.30%	
	3100	Total: Mechanical & Electrical	↓	13.25	23.50	30.50	13.40%	22.50%	30.80%	
850	0010	**SUPERMARKETS** R17100 -100	S.F.	48.50	57	66				**850**
	0020	Total project costs	C.F.	2.70	3.26	4.94				
	2720	Plumbing	S.F.	2.71	3.41	3.97	5.40%	6%	7.50%	
	2770	Heating, ventilating, air conditioning		3.98	4.89	5.80	8.60%	8.60%	9.60%	
	2900	Electrical		5.65	7	8.25	10.40%	12.40%	13.60%	
	3100	Total: Mechanical & Electrical	↓	15.85	17.10	23.50	17.40%	20.40%	28.40%	
860	0010	**SWIMMING POOLS** R17100 -100	S.F.	85.50	131	183				**860**
	0020	Total project costs	C.F.	6.30	7.85	8.55				
	2720	Plumbing	S.F.	7.30	8.30	11.55	9.70%	12.30%	12.30%	
	2900	Electrical		5.90	9.15	11.95	7.50%	8%	8%	
	3100	Total: Mechanical & Electrical	↓	12.55	26.50	49	17.70%	24.90%	31.60%	
870	0010	**TELEPHONE EXCHANGES** R17100 -100	S.F.	104	153	195				**870**
	0020	Total project costs	C.F.	6.45	9.90	14				
	2720	Plumbing	S.F.	4.40	6.80	9.95	4.50%	5.80%	6.90%	
	2770	Heating, ventilating, air conditioning		8.95	20	25.50	11.80%	11.80%	18.40%	
	2900	Electrical		10.60	16.45	30	10.90%	14%	17.80%	
	3100	Total: Mechanical & Electrical	↓	23	31.50	77.50	27.30%	33.40%	44.40%	

SQUARE FOOT 17

For expanded coverage of these items see *Means Square Foot Costs 2002*

		17100 | S.F. & C.F. Costs		UNIT	UNIT COSTS			% OF TOTAL			
					1/4	MEDIAN	3/4	1/4	MEDIAN	3/4	
910	0010	**THEATERS**	R17100	S.F.	65.50	84	124				910
	0020	Total project costs	-100	C.F.	3.03	4.47	6.60				
	2720	Plumbing		S.F.	2.04	2.37	7.10	2.90%	4.70%	6.80%	
	2770	Heating, ventilating, air conditioning			6.35	7.70	9.55	8%	12.20%	13.40%	
	2900	Electrical			5.75	7.75	15.75	8%	10%	12.40%	
	3100	Total: Mechanical & Electrical		↓	14.75	22	45.50	22.90%	26.60%	27.50%	
940	0010	**TOWN HALLS** City Halls & Municipal Buildings	R17100	S.F.	73	92.50	121				940
	0020	Total project costs	-100	C.F.	5.50	7.65	10.25				
	2720	Plumbing		S.F.	2.60	5.35	8.90	4.20%	6.10%	7.90%	
	2770	Heating, ventilating, air conditioning			5.50	10.95	16	7%	9%	13.50%	
	2900	Electrical			6.45	8.85	12.55	8.20%	9.50%	11.70%	
	3100	Total: Mechanical & Electrical		↓	23	24	30.50	22%	25.10%	30.40%	
970	0010	**WAREHOUSES** And Storage Buildings	R17100	S.F.	27.50	38	58.50				970
	0020	Total project costs	-100	C.F.	1.37	2.24	3.71				
	2720	Plumbing		S.F.	.89	1.63	3.05	2.90%	4.80%	6.60%	
	2730	Heating, ventilating, air conditioning			1.04	2.93	3.93	2.40%	5%	8.90%	
	2900	Electrical			1.62	3.04	5.05	5%	7.30%	10%	
	3100	Total: Mechanical & Electrical		↓	4.50	6.65	15.10	12.80%	18.40%	26.20%	
990	0010	**WAREHOUSE & OFFICES** Combination	R17100	S.F.	33	44	58.50				990
	0020	Total project costs	-100	C.F.	1.70	2.49	3.68				
	2720	Plumbing		S.F.	1.27	2.19	3.45	3.60%	4.70%	6.30%	
	2770	Heating, ventilating, air conditioning			2.02	3.19	4.48	5%	5.60%	9.60%	
	2900	Electrical		S.F.	2.22	3.28	5.15	5.70%	7.70%	10%	
	3100	Total: Mechanical & Electrical		↓	5.25	8.70	10.85	13.80%	18.60%	23%	

For information about Means Estimating Seminars, see yellow pages 11 and 12 in back of book

17 SQUARE FOOT

See R17100-100 and City Cost Indexes in the Reference Section

Assemblies Section

Table of Contents

Table No.	Page
A SUBSTRUCTURE	411
A2020 Basement Walls	
A2020 100　Basement Wall Construction	412
D SERVICES	413
D3010 Energy Supply	
D3010 500　Hot Water Supply System	414
D3010 700　Solar Energy System	417
D3020 Heat Generating Systems	
D3020 100　Boilers	438
D3020 300　Auxiliary Equipment	448
D3030 Cooling Generating Systems	
D3030 100　Chilled Water Systems	449
D3030 200　Direct Expansion Systems	456
D3030 300　Other Cooling Systems Components	469
D3040 Distribution Systems	
D3040 100　Air Distribution Systems	470
D3040 200　Exhaust Ventilation Systems	480
D3040 600　Change-over Distribution Systems	482

Table No.	Page
D3050 Terminal & Package Units	
D3050 100　Terminal Self-contained Units	483
D3050 200　Package Units	485
D3060 Controls & Instrumentation	
D3060 400　Exhaust/Ventilating Sytems	490
D3060 500　Hoods & Exhaust Systems	492
D3090 Other HVAC Sys. & Equip.	
D3090 300　Dust/Fume Collectors	493
G BUILDING SITEWORK	495
G1030 Site Earthwork	
G1030 800　Utilities Trenching	496
G3010 Water Supply	
G3010 100　Potable Water Distribution/Storage	501

Table No.	Page
G3020 Sanitary Sewer	
G3020 100　Piping	502
G3030 Storm Sewer	
G3030 200　Manholes	504
G3060 Fuel Distribution	
G3060 100　Fuel Piping	506

How to Use the
Assemblies Cost Tables

The following is a detailed explanation of a sample Assemblies Cost Table. Most Assembly Tables are separated into three parts: 1) an illustration of the system to be estimated; 2) the components and related costs of a typical system; and 3) the costs for similar systems with dimensional and/or size variations. For costs of the components that comprise these systems or "assemblies" refer to the Unit Price Section. Next to each bold number below is the item being described with the appropriate component of the sample entry following in parenthesis. In most cases, if the work is to be subcontracted, the general contractor will need to add an additional markup (R.S. Means suggests using 10%) to the "Total" figures.

System/Line Numbers (D3010 510 1760)

Each Assemblies Cost Line has been assigned a unique identification number based on the UniFormat classification sytem.

UNIFORMAT II Major Group

D3010 **510** **1760**

UNIFORMAT II Level 3

Means Major Classification

Means Individual Line Number

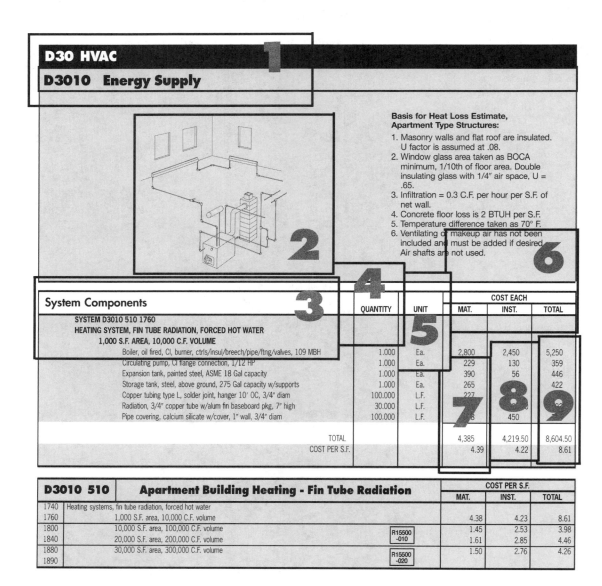

D30 HVAC
D3010 Energy Supply

Basis for Heat Loss Estimate, Apartment Type Structures:

1. Masonry walls and flat roof are insulated. U factor is assumed at .08.
2. Window glass area taken as BOCA minimum, 1/10th of floor area. Double insulating glass with 1/4" air space, U = .65.
3. Infiltration = 0.3 C.F. per hour per S.F. of net wall.
4. Concrete floor loss is 2 BTUH per S.F.
5. Temperature difference taken as 70° F.
6. Ventilating of makeup air has not been included and must be added if desired. Air shafts are not used.

System Components	QUANTITY	UNIT	COST EACH		
			MAT.	INST.	TOTAL
SYSTEM D3010 510 1760					
HEATING SYSTEM, FIN TUBE RADIATION, FORCED HOT WATER					
1,000 S.F. AREA, 10,000 C.F. VOLUME					
Boiler, oil fired, CI, burner, ctrls/insul/breech/pipe/ftng/valves, 109 MBH	1.000	Ea.	2,800	2,450	5,250
Circulating pump, CI flange connection, 1/12 HP	1.000	Ea.	229	130	359
Expansion tank, painted steel, ASME 18 Gal capacity	1.000	Ea.	390	56	446
Storage tank, steel, above ground, 275 Gal capacity w/supports	1.000	Ea.	265		422
Copper tubing type L, solder joint, hanger 10' OC, 3/4" diam	100.000	L.F.	227		
Radiation, 3/4" copper tube w/alum fin baseboard pkg, 7" high	30.000	L.F.		450	
Pipe covering, calcium silicate w/cover, 1" wall, 3/4" diam	100.000	L.F.			
TOTAL			4,385	4,219.50	8,604.50
COST PER S.F.			4.39	4.22	8.61

D3010 510	Apartment Building Heating - Fin Tube Radiation		COST PER S.F.		
			MAT.	INST.	TOTAL
1740	Heating systems, fin tube radiation, forced hot water				
1760	1,000 S.F. area, 10,000 C.F. volume		4.38	4.23	8.61
1800	10,000 S.F. area, 100,000 C.F. volume	R15500 -010	1.45	2.53	3.98
1840	20,000 S.F. area, 200,000 C.F. volume		1.61	2.85	4.46
1880	30,000 S.F. area, 300,000 C.F. volume	R15500 -020	1.50	2.76	4.26
1890					

2 Illustration

At the top of most assembly pages is an illustration, a brief description, and the design criteria used to develop the cost.

3 System Components

The components of a typical system are listed separately to show what has been included in the development of the total system price. The table below contains prices for other similar systems with dimensional and/or size variations.

4 Quantity

This is the number of line item units required for one system unit. For example, we assume that it will take 30 linear feet of radiation for the 1000 S.F. area shown.

5 Unit of Measure for Each Item

The abbreviated designation indicates the unit of measure, as defined by industry standards, upon which the price of the component is based. For example, baseboard radiation is priced by the linear foot. For a complete listing of abbreviations, see the Reference Section.

6 Unit of Measure for Each System (Each)

Costs shown in the three right hand columns have been adjusted by the component quantity and unit of measure for the entire system. In this example, "Cost Each" is the unit of measure for this system or "assembly."

7 Materials (4,385.00)

This column contains the Materials Cost of each component. These cost figures are bare costs plus 10% for profit.

8 Installation (4,219.50)

Installation includes labor and equipment plus the installing contractor's overhead and profit. Equipment costs are the bare rental costs plus 10% for profit. The labor overhead and profit is defined on the inside back cover of this book.

9 Total (8,604.50)

The figure in this column is the sum of the material and installation costs.

Material Cost	+	Installation Cost	=	Total
$4,385	+	$4,219.50	=	$8,604.50

A SUBSTRUCTURE

A2020 180	Subdrainage Piping	COST PER L.F.		
		MAT.	INST.	TOTAL
2000	Piping, excavation & backfill excluded, PVC, perforated			
2110	3" diameter	1.57	2.65	4.22
2130	4" diameter	1.57	2.65	4.22
2140	5" diameter	2.73	2.84	5.57
2150	6" diameter	2.73	2.84	5.57
3000	Metal alum. or steel, perforated asphalt coated			
3150	6" diameter	2.88	5.40	8.28
3160	8" diameter	4.19	5.50	9.69
3170	10" diameter	5.25	5.70	10.95
3180	12" diameter	5.85	7.15	13
3220	18" diameter	8.80	9.95	18.75
4000	Porous wall concrete			
4130	4" diameter	2.02	2.97	4.99
4150	6" diameter	2.63	3.16	5.79
4160	8" diameter	3.25	4.40	7.65
4180	12" diameter	6.90	4.79	11.69
4200	15" diameter	7.90	5.95	13.85
4220	18" diameter	10.40	8.25	18.65
5000	Vitrified clay C-211, perforated			
5130	4" diameter	2.20	5.10	7.30
5150	6" diameter	3.64	6.45	10.09
5160	8" diameter	5.20	7.05	12.25
5180	12" diameter	10.90	7.40	18.30

For information about Means Estimating Seminars, see yellow pages 11 and 12 in back of book

SUBSTRUCTURE A

D SERVICES

No part of this publication may be reproduced, stored in a retrieval system, or transmitted in any form or by any means without prior written permission of R.S. Means Company, Inc.

413

Basis for Heat Loss Estimate, Apartment Type Structures:

1. Masonry walls and flat roof are insulated. U factor is assumed at .08.
2. Window glass area taken as BOCA minimum, 1/10th of floor area. Double insulating glass with 1/4″ air space, U = .65.
3. Infiltration = 0.3 C.F. per hour per S.F. of net wall.
4. Concrete floor loss is 2 BTUH per S.F.
5. Temperature difference taken as 70° F.
6. Ventilating or makeup air has not been included and must be added if desired. Air shafts are not used.

System Components	QUANTITY	UNIT	COST EACH		
			MAT.	INST.	TOTAL
SYSTEM D3010 510 1760					
HEATING SYSTEM, FIN TUBE RADIATION, FORCED HOT WATER					
1,000 S.F. AREA, 10,000 C.F. VOLUME					
Boiler, oil fired, CI, burner, ctrls/insul/breech/pipe/ftng/valves, 109 MBH	1.000	Ea.	2,800	2,450	5,250
Circulating pump, CI flange connection, 1/12 HP	1.000	Ea.	229	130	359
Expansion tank, painted steel, ASME 18 Gal capacity	1.000	Ea.	390	56	446
Storage tank, steel, above ground, 275 Gal capacity w/supports	1.000	Ea.	265	157	422
Copper tubing type L, solder joint, hanger 10′ OC, 3/4″ diam	100.000	L.F.	227	570	797
Radiation, 3/4″ copper tube w/alum fin baseboard pkg, 7″ high	30.000	L.F.	186	406.50	592.50
Pipe covering, calcium silicate w/cover, 1″ wall, 3/4″ diam	100.000	L.F.	288	450	738
TOTAL			4,385	4,219.50	8,604.50
COST PER S.F.			4.39	4.22	8.61

D3010 510	Apartment Building Heating - Fin Tube Radiation		COST PER S.F.		
			MAT.	INST.	TOTAL
1740	Heating systems, fin tube radiation, forced hot water				
1760	1,000 S.F. area, 10,000 C.F. volume		4.38	4.23	8.61
1800	10,000 S.F. area, 100,000 C.F. volume	R15500 -010	1.45	2.53	3.98
1840	20,000 S.F. area, 200,000 C.F. volume		1.61	2.85	4.46
1880	30,000 S.F. area, 300,000 C.F. volume	R15500 -020	1.50	2.76	4.26
1890					

SERVICES D

D3010 Energy Supply

Fin Tube Radiator

Basis for Heat Loss Estimate, Factory or Commercial Type Structures:

1. Walls and flat roof are of lightly insulated concrete block or metal. U factor is assumed at .17.
2. Windows and doors are figured at 1-1/2 S.F. per linear foot of building perimeter. The U factor for flat single glass is 1.13.
3. Infiltration is approximately 5 C.F. per hour per S.F. of wall.
4. Concrete floor loss is 2 BTUH per S.F. of floor.
5. Temperature difference is assumed at 70° F.
6. Ventilation or makeup air has not been included and must be added if desired.

System Components	QUANTITY	UNIT	COST EACH		
			MAT.	INST.	TOTAL
SYSTEM D3010 520 1960					
HEATING SYSTEM, FIN TUBE RADIATION, FORCED HOT WATER					
1,000 S.F. BLDG., ONE FLOOR					
Boiler, oil fired, CI, burner/ctrls/insul/breech/pipe/ftngs/valves, 109 MBH	1.000	Ea.	2,800	2,450	5,250
Expansion tank, painted steel, ASME 18 Gal capacity	1.000	Ea.	1,400	65.50	1,465.50
Storage tank, steel, above ground, 550 Gal capacity w/supports	1.000	Ea.	1,250	291	1,541
Circulating pump, CI flanged, 1/8 HP	1.000	Ea.	385	130	515
Pipe, steel, black, sch. 40, threaded, cplg & hngr 10'OC, 1-1/2" diam.	260.000	L.F.	717.60	2,535	3,252.60
Pipe covering, calcium silicate w/cover , 1" wall, 1-1/2" diam	260.000	L.F.	733.20	1,203.80	1,937
Radiation, steel 1-1/4" tube & 4-1/4" fin w/cover & damper, wall hung	42.000	L.F.	1,134	924	2,058
Rough in, steel fin tube radiation w/supply & balance valves	4.000	Set	628	1,920	2,548
TOTAL			9,047.80	9,519.30	18,567.10
COST PER S.F.			9.05	9.52	18.57

SERVICES

D

D3010 520	Commercial Building Heating - Fin Tube Radiation		COST PER S.F.		
			MAT.	INST.	TOTAL
1940	Heating systems, fin tube radiation, forced hot water				
1960	1,000 S.F. bldg, one floor		9.05	9.53	18.58
2000	10,000 S.F., 100,000 C.F., total two floors	R15500 -010	2.47	3.63	6.10
2040	100,000 S.F., 1,000,000 C.F., total three floors		1.13	1.65	2.78
2080	1,000,000 S.F., 10,000,000 C.F., total five floors	R15500 -020	.54	.87	1.41
2090					

D30 HVAC

D3010 Energy Supply

Unit Heater

Basis for Heat Loss Estimate, Factory or Commercial Type Structures:

1. Walls and flat roof are of lightly insulated concrete block or metal. U factor is assumed at .17.
2. Windows and doors are figured at 1-1/2 S.F. per linear foot of building perimeter. The U factor for flat single glass is 1.13.
3. Infiltration is approximately 5 C.F. per hour per S.F. of wall.
4. Concrete floor loss is 2 BTUH per S.F. of floor.
5. Temperature difference is assumed at 70° F.
6. Ventilation or makeup air has not been included and must be added if desired.

SERVICES D

System Components	QUANTITY	UNIT	COST EACH MAT.	COST EACH INST.	COST EACH TOTAL
SYSTEM D3010 530 1880					
HEATING SYSTEM, TERMINAL UNIT HEATERS, FORCED HOT WATER					
1,000 S.F. BLDG., ONE FLOOR					
Boiler oil fired, CI, burner/ctrls/insul/breech/pipe/ftngs/valves, 109 MBH	1.000	Ea.	2,800	2,450	5,250
Expansion tank, painted steel, ASME 18 Gal capacity	1.000	Ea.	1,400	65.50	1,465.50
Storage tank, steel, above ground, 550 Gal capacity w/supports	1.000	Ea.	1,250	291	1,541
Circulating pump, CI, flanged, 1/8 HP	1.000	Ea.	385	130	515
Pipe, steel, black, schedule 40, threaded, cplg & hngr 10'OC 1-1/2" diam	260.000	L.F.	717.60	2,535	3,252.60
Pipe covering, calcium silicate w/cover, 1" wall, 1-1/2" diam	260.000	L.F.	733.20	1,203.80	1,937
Unit heater, 1 speed propeller, horizontal, 200° EWT, 26.9 MBH	2.000	Ea.	880	197	1,077
Unit heater piping hookup with controls	2.000	Set	720	1,860	2,580
TOTAL			8,885.80	8,732.30	17,618.10
COST PER S.F.			8.89	8.73	17.62

D3010 530	Commercial Bldg. Heating - Terminal Unit Heaters		COST PER S.F. MAT.	COST PER S.F. INST.	COST PER S.F. TOTAL
1860	Heating systems, terminal unit heaters, forced hot water				
1880	1,000 S.F. bldg., one floor		8.88	8.73	17.61
1920	10,000 S.F. bldg., 100,000 C.F. total two floors	R15500 -010	2.21	3.01	5.22
1960	100,000 S.F. bldg., 1,000,000 C.F. total three floors		1.17	1.52	2.69
2000	1,000,000 S.F. bldg., 10,000,000 C.F. total five floors	R15500 -020	.72	.97	1.69
2010					

Important: See the Reference Section for critical supporting data - Reference Numbers and City Cost Indexes

Many styles of active solar energy systems exist. Those shown on the following page represent the majority of systems now being installed in different regions of the country.

The five active domestic hot water (DHW) systems which follow are typically specified as two or three panel systems with additional variations being the type of glazing and size of the storage tanks. Various combinations have been costed for the user's evaluation and comparison. The basic specifications from which the following systems were developed satisfy the construction detail requirements specified in the HUD Intermediate Minimum Property Standards (IMPS). If these standards are not complied with, the renewable energy system's costs could be significantly lower than shown.

To develop the system's specifications and costs it was necessary to make a number of assumptions about the systems. Certain systems are more appropriate to one climatic region than another or the systems may require modifications to be usable in particular locations. Specific instances in which the systems are not appropriate throughout the country as specified, or in which modification will be needed include the following:

- The freeze protection mechanisms provided in the DHW systems vary greatly. In harsh climates, where freeze is a major concern, a closed-loop indirect collection system may be more appropriate than a direct collection system.

- The thermosyphon water heater system described cannot be used when temperatures drop below 32° F.

- In warm climates it may be necessary to modify the systems installed to prevent overheating.

For each renewable resource (solar) system a schematic diagram and descriptive summary of the system is provided along with a list of all the components priced as part of the system. The costs were developed based on these specifications.

Considerations affecting costs which may increase or decrease beyond the estimates presented here include the following:

- Special structural qualities (allowance for earthquake, future expansion, high winds, and unusual spans or shapes);

- Isolated building site or rough terrain that would affect the transportation of personnel, material, or equipment;

- Unusual climatic conditions during the construction process;

- Substitution of other materials or system components for those used in the system specifications.

SERVICES

D

In this closed-loop indirect collection system, fluid with a low freezing temperature, such as propylene glycol, transports heat from the collectors to water storage. The transfer fluid is contained in a closed-loop consisting of collectors, supply and return piping, and a remote heat exchanger. The heat exchanger transfers heat energy from the fluid in the collector loop to potable water circulated in a storage loop. A typical two-or-three panel system contains 5 to 6 gallons of heat transfer fluid.

When the collectors become approximately 20° F warmer than the storage temperature, a controller activates the circulator on the collector and storage loops. The circulators will move the fluid and potable water through the heat exchanger until heat collection no longer occurs. At that point, the system shuts down. Since the heat transfer medium is a fluid with a very low freezing temperature, there is no need for it to be drained from the system between periods of collection.

SERVICES D

D3010 Energy Supply

System Components	QUANTITY	UNIT	COST EACH		
			MAT.	INST.	TOTAL
SYSTEM D3010 715 2760					
SOLAR, CLOSED LOOP, ADD-ON HOT WATER SYS., EXTERNAL HEAT EXCHANGER					
3/4″ TUBING, TWO 3′X7′ BLACK CHROME COLLECTORS					
A,B,G,L,K,M Heat exchanger fluid-fluid pkg incl 2 circulators, expansion tank,					
Check valve, relief valve, controller, hi temp cutoff, & 2 sensors	1.000	Ea.	765	310	1,075
C Thermometer, 2″ dial	3.000	Ea.	63	81	144
D, T Fill & drain valve, brass, 3/4″ connection	1.000	Ea.	5.45	18.05	23.50
E Air vent, manual, 1/8″ fitting	2.000	Ea.	4.26	27.10	31.36
F Air purger	1.000	Ea.	45	36	81
H Strainer, Y type, bronze body, 3/4″ IPS	1.000	Ea.	19.15	23	42.15
I Valve, gate, bronze, NRS, soldered 3/4″ diam	6.000	Ea.	117	129	246
J Neoprene vent flashing	2.000	Ea.	19.60	43	62.60
N-1, N Relief valve temp & press, 150 psi 210°F self-closing 3/4″ IPS	1.000	Ea.	8.65	14.45	23.10
O Pipe covering, urethane, ultraviolet cover, 1″ wall 3/4″ diam	20.000	L.F.	32.20	83.20	115.40
P Pipe covering, fiberglass, all service jacket, 1″ wall, 3/4″ diam	50.000	L.F.	73.50	166	239.50
Q Collector panel solar energy blk chrome on copper, 1/8″ temp glass 3′x7′	2.000	Ea.	1,200	164	1,364
Roof clamps for solar energy collector panels	2.000	Set	3.54	22.30	25.84
R Valve, swing check, bronze, regrinding disc, 3/4″ diam	2.000	Ea.	69	43	112
S Pressure gauge, 60 psi, 2″ dial	1.000	Ea.	23	13.55	36.55
U Valve, water tempering, bronze, sweat connections, 3/4″ diam	1.000	Ea.	56.50	21.50	78
W-2, V Tank water storage w/heating element, drain, relief valve, existing	1.000	Ea.			
Copper tubing type L, solder joint, hanger 10′ OC 3/4″ diam	20.000	L.F.	45.40	114	159.40
Copper tubing, type M, solder joint, hanger 10′ OC 3/4″ diam	70.000	L.F.	128.80	388.50	517.30
Sensor wire, #22-2 conductor multistranded	.500	C.L.F.	4.33	21.25	25.58
Solar energy heat transfer fluid, propylene glycol anti-freeze	6.000	Gal.	58.20	93	151.20
Wrought copper fittings & solder, 3/4″ diam	76.000	Ea.	67.64	1,748	1,815.64
Total			2,809.22	3,559.90	6,369.12

D3010 715	Solar, Closed Loop, Add-On Hot Water Systems	COST EACH		
		MAT.	INST.	TOTAL
2550	Solar, closed loop, add-on hot water system, external heat exchanger			
2570	3/8″ tubing, 3 ea. 4′ x 4′-4″ vacuum tube collectors	4,325	3,325	7,650
2580	1/2″ tubing, 4 ea 4 x 4′-4″ vacuum tube collectors	5,200	3,575	8,775
2600	2 ea 3′x7′ black chrome collectors	2,675	3,375	6,050
2620	3 ea 3′x7′ black chrome collectors	3,275	3,475	6,750
2640	2 ea 3′x7′ flat black collectors	2,425	3,375	5,800
2660	3 ea 3′x7′ flat black collectors	2,900	3,475	6,375
2700	3/4″ tubing, 3 ea 3′x7′ black chrome collectors	3,400	3,650	7,050
2720	3 ea 3′x7′ flat black absorber plate collectors	3,025	3,675	6,700
2740	2 ea 4′x9′ flat black w/plastic glazing collectors	2,700	3,675	6,375
2760	2 ea 3′x7′ black chrome collectors	2,800	3,550	6,350
2780	1″ tubing,4 ea 2′x9′ plastic absorber & glazing collectors	4,050	4,125	8,175
2800	4 ea 3′x7′ black chrome absorber collectors	4,275	4,125	8,400
2820	4 ea 3′x7′ flat black absorber collectors	3,775	4,150	7,925

Note: row 2580/2600 area shows reference box: R13600 -610

SERVICES D

In the drainback indirect-collection system, the heat transfer fluid is distilled water contained in a loop consisting of collectors, supply and return piping, and an unpressurized holding tank. A large heat exchanger containing incoming potable water is immersed in the holding tank. When a controller activates solar collection, the distilled water is pumped through the collectors and heated and pumped back down to the holding tank. When the temperature differential between the water in the collectors and water in storage is such that collection no longer occurs, the pump turns off and gravity causes the distilled water in the collector loop to drain back to the holding tank. All the loop piping is pitched so that the water can drain out of the collectors and piping and not freeze there. As hot water is needed in the home, incoming water first flows through the holding tank with the immersed heat exchanger and is warmed and then flows through a conventional heater for any supplemental heating that is necessary.

D3010 Energy Supply

System Components	QUANTITY	UNIT	COST EACH		
			MAT.	INST.	TOTAL
SYSTEM D3010 720 2760					
SOLAR, DRAINBACK, ADD ON, HOT WATER, IMMERSED HEAT EXCHANGER					
3/4″ TUBING, THREE EA 3′X7′ BLACK CHROME COLLECTOR					
A, B Differential controller 2 sensors, thermostat, solar energy system	1.000	Ea.	85.50	36	121.50
C Thermometer 2″ dial	3.000	Ea.	63	81	144
D, T Fill & drain valve, brass, 3/4″ connection	1.000	Ea.	5.45	18.05	23.50
E-1 Automatic air vent 1/8″ fitting	1.000	Ea.	9.70	13.55	23.25
H Strainer, Y type, bronze body, 3/4″ IPS	1.000	Ea.	19.15	23	42.15
I Valve, gate, bronze, NRS, soldered 3/4″ diam	2.000	Ea.	39	43	82
J Neoprene vent flashing	2.000	Ea.	19.60	43	62.60
L Circulator, solar heated liquid, 1/20 HP	1.000	Ea.	161	65	226
N Relief valve temp. & press. 150 psi 210°F self-closing 3/4″ IPS	1.000	Ea.	8.65	14.45	23.10
O Pipe covering, urethane, ultraviolet cover, 1″ wall, 3/4″ diam	20.000	L.F.	32.20	83.20	115.40
P Pipe covering, fiberglass, all service jacket, 1″ wall, 3/4″ diam	50.000	L.F.	73.50	166	239.50
Q Collector panel solar energy blk chrome on copper, 1/8″ temp glas 3′x7′	3.000	Ea.	1,800	246	2,046
Roof clamps for solar energy collector panels	3.000	Set	5.31	33.45	38.76
R Valve, swing check, bronze, regrinding disc, 3/4″ diam	1.000	Ea.	34.50	21.50	56
U Valve, water tempering, bronze sweat connections, 3/4″ diam	1.000	Ea.	56.50	21.50	78
W Tank, water storage immersed heat exchr elec 2″x1/2# insul 120 gal	1.000	Ea.	925	310	1,235
V Tank, water storage w/heating element, drain, relief valve, existing	1.000	Ea.			
X Valve, globe, bronze, rising stem, 3/4″ diam, soldered	3.000	Ea.	148.50	64.50	213
Y Flow control valve	1.000	Ea.	54.50	19.70	74.20
Z Valve, ball, bronze, solder 3/4″ diam, solar loop flow control	1.000	Ea.	12.20	21.50	33.70
Copper tubing, type L, solder joint, hanger 10′ OC 3/4″ diam	20.000	L.F.	45.40	114	159.40
Copper tubing, type M, solder joint, hanger 10′ OC 3/4″ diam	70.000	L.F.	128.80	388.50	517.30
Sensor wire, #22-2 conductor, multistranded	.500	C.L.F.	4.33	21.25	25.58
Wrought copper fittings & solder, 3/4″ diam	76.000	Ea.	67.64	1,748	1,815.64
Total			3,799.43	3,596.15	7,395.58

D3010 720	Solar, Drainback, Hot Water Systems		COST EACH		
			MAT.	INST.	TOTAL
2550	Solar, drainback, hot water, immersed heat exchanger				
2560	3/8″ tubing, 3 ea. 4′ x 4′-4″ vacuum tube collectors	R13600 -610	4,575	3,225	7,800
2580	1/2″ tubing, 4 ea 4′ x 4′-4″ vacuum tube collectors, 80 gal tank		5,450	3,475	8,925
2600	120 gal tank		5,550	3,525	9,075
2640	2 ea. 3′x7′ blk chrome collectors, 80 gal tank		2,925	3,275	6,200
2660	3 ea. 3′x7′ blk chrome collectors, 120 gal tank		3,625	3,400	7,025
2700	2 ea. 3′x7′ flat blk collectors, 120 gal tank		2,775	3,325	6,100
2720	3 ea. 3′x7′ flat blk collectors, 120 gal tank		3,250	3,425	6,675
2760	3/4″ tubing, 3 ea 3′x7′ black chrome collectors, 120 gal tank		3,800	3,600	7,400
2780	3 ea. 3′x7′ flat black absorber collectors, 120 gal tank		3,425	3,600	7,025
2800	2 ea. 4′x9′ flat blk w/plastic glazing collectors 120 gal tank		3,075	3,625	6,700
2840	1″ tubing, 4 ea. 2′x9′ plastic absorber & glazing collectors, 120 gal tank		4,475	4,050	8,525
2860	4 ea. 3′x7′ black chrome absorber collectors, 120 gal tank		4,675	4,075	8,750
2880	4 ea. 3′x7′ flat black absorber collectors, 120 gal tank		4,175	4,100	8,275

In the draindown direct-collection system, incoming domestic water is heated in the collectors. When the controller activates solar collection, domestic water is first heated as it flows through the collectors and is then pumped to storage. When conditions are no longer suitable for heat collection, the pump shuts off and the water in the loop drains down and out of the system by means of solenoid valves and properly pitched piping.

D3010 Energy Supply

System Components	QUANTITY	UNIT	COST EACH		
			MAT.	INST.	TOTAL
SYSTEM D3010 730 2760					
SOLAR, DRAINDOWN, HOT WATER, DIRECT COLLECTION					
3/4" TUBING, THREE 3'X7' BLACK CHROME COLLECTORS					
A, B Differential controller, 2 sensors, thermostat, solar energy system	1.000	Ea.	85.50	36	121.50
A-1 Solenoid valve, solar heating loop, brass, 3/4" diam, 24 volts	3.000	Ea.	699	144	843
B-1 Solar energy sensor, freeze prevention	1.000	Ea.	22	13.55	35.55
C Thermometer, 2" dial	3.000	Ea.	63	81	144
E-1 Vacuum relief valve, 3/4" diam	1.000	Ea.	27.50	13.55	41.05
F-1 Air vent, automatic, 1/8" fitting	1.000	Ea.	9.70	13.55	23.25
H Strainer, Y type, bronze body, 3/4" IPS	1.000	Ea.	19.15	23	42.15
I Valve, gate, bronze, NRS, soldered, 3/4" diam	2.000	Ea.	39	43	82
J Vent flashing neoprene	2.000	Ea.	19.60	43	62.60
K Circulator, solar heated liquid, 1/25 HP	1.000	Ea.	153	56	209
N Relief valve temp & press 150 psi 210°F self-closing 3/4" IPS	1.000	Ea.	8.65	14.45	23.10
O Pipe covering, urethane, ultraviolet cover, 1" wall, 3/4" diam	20.000	L.F.	32.20	83.20	115.40
P Pipe covering, fiberglass, all service jacket, 1" wall, 3/4" diam	50.000	L.F.	73.50	166	239.50
Roof clamps for solar energy collector panels	3.000	Set	5.31	33.45	38.76
Q Collector panel solar energy blk chrome on copper, 1/8" temp glass 3'x7'	3.000	Ea.	1,800	246	2,046
R Valve, swing check, bronze, regrinding disc, 3/4" diam, soldered	2.000	Ea.	69	43	112
T Drain valve, brass, 3/4" connection	2.000	Ea.	10.90	36.10	47
U Valve, water tempering, bronze, sweat connections, 3/4" diam	1.000	Ea.	56.50	21.50	78
W-2, W Tank, water storage elec elem 2"x1/2# insul 120 gal	1.000	Ea.	925	310	1,235
X Valve, globe, bronze, rising stem, 3/4" diam, soldered	1.000	Ea.	49.50	21.50	71
Copper tubing, type L, solder joints, hangers 10' OC 3/4" diam	20.000	L.F.	45.40	114	159.40
Copper tubing, type M, solder joints, hangers 10' OC 3/4" diam	70.000	L.F.	128.80	388.50	517.30
Sensor wire, #22-2 conductor, multistranded	.500	C.L.F.	4.33	21.25	25.58
Wrought copper fittings & solder, 3/4" diam	76.000	Ea.	67.64	1,748	1,815.64
Total			4,414.18	3,713.60	8,127.78

D3010 730	Solar, Draindown, Hot Water Systems		COST EACH		
			MAT.	INST.	TOTAL
2550	Solar, draindown, hot water				
2560	3/8" tubing, 3 ea. 4' x 4'-4" vacuum tube collectors, 80 gal tank		5,150	3,500	8,650
2580	1/2" tubing, 4 ea. 4' x 4'-4" vacuum tube collectors, 80 gal tank	R13600 -610	6,100	3,600	9,700
2600	120 gal tank		6,200	3,650	9,850
2640	2 ea 3'x7' black chrome collectors, 80 gal tank		3,575	3,400	6,975
2660	3 ea 3'x7' black chrome collectors, 120 gal tank		4,275	3,525	7,800
2700	2 ea 3'x7' flat black collectors, 120 gal tank		3,425	3,450	6,875
2720	3 ea 3'x7' flat black collectors, 120 gal tank		3,875	3,525	7,400
2760	3/4" tubing, 3 ea 3'x7' black chrome collectors, 120 gal tank		4,425	3,725	8,150
2780	3 ea 3'x7' flat collectors, 120 gal tank		4,050	3,725	7,775
2800	2 ea. 4'x9' flat black & plastic glazing collectors, 120 gal tank		3,700	3,750	7,450
2840	1" tubing, 4 ea. 2'x9' plastic absorber & glazing collectors, 120 gal tank		5,050	4,175	9,225
2860	4 ea 3'x7' black chrome absorber collectors, 120 gal tank		5,250	4,200	9,450
2880	4 ea 3'x7' flat black absorber collectors, 120 gal tank		4,750	4,200	8,950

D SERVICES

SERVICES D

In the recirculation system, (a direct-collection system), incoming domestic water is heated in the collectors. When the controller activates solar collection, domestic water is heated as it flows through the collectors and then it flows back to storage. When conditions are not suitable for heat collection, the pump shuts off and the flow of the water stops. In this type of system, water remains in the collector loop at all times. A "frost sensor" at the collector activates the circulation of warm water from storage through the collectors when protection from freezing is required.

D30 HVAC

D3010 Energy Supply

System Components	QUANTITY	UNIT	COST EACH MAT.	INST.	TOTAL
SYSTEM D3010 735 2820					
SOLAR, RECIRCULATION, HOT WATER					
3/4" TUBING, TWO 3'X7' BLACK CHROME COLLECTORS					
A, B Differential controller 2 sensors, thermostat, for solar energy system	1.000	Ea.	85.50	36	121.50
A-1 Solenoid valve, solar heating loop, brass, 3/4" IPS, 24 volts	2.000	Ea.	466	96	562
B-1 Solar energy sensor freeze prevention	1.000	Ea.	22	13.55	35.55
C Thermometer, 2" dial	3.000	Ea.	63	81	144
D Drain valve, brass, 3/4" connection	1.000	Ea.	5.45	18.05	23.50
F-1 Air vent, automatic, 1/8" fitting	1.000	Ea.	9.70	13.55	23.25
H Strainer, Y type, bronze body, 3/4" IPS	1.000	Ea.	19.15	23	42.15
I Valve, gate, bronze, 125 lb, soldered 3/4" diam	3.000	Ea.	58.50	64.50	123
J Vent flashing, neoprene	2.000	Ea.	19.60	43	62.60
L Circulator, solar heated liquid, 1/20 HP	1.000	Ea.	161	65	226
N Relief valve, temp & press 150 psi 210°F self-closing 3/4" IPS	2.000	Ea.	17.30	28.90	46.20
O Pipe covering, urethane, ultraviolet cover, 1" wall, 3/4" diam	20.000	L.F.	32.20	83.20	115.40
P Pipe covering, fiberglass, all service jacket, 1" wall 3/4" diam	50.000	L.F.	73.50	166	239.50
Q Collector panel solar energy blk chrome on copper, 1/8" temp glass 3'x7'	2.000	Ea.	1,200	164	1,364
Roof clamps for solar energy collector panels	2.000	Set	3.54	22.30	25.84
R Valve, swing check, bronze, 125 lb, regrinding disc, soldered 3/4" diam	2.000	Ea.	69	43	112
U Valve, water tempering, bronze, sweat connections, 3/4" diam	1.000	Ea.	56.50	21.50	78
V Tank, water storage, w/heating element, drain, relief valve, existing	1.000	Ea.			
X Valve, globe, bronze, 125 lb, 3/4" diam	2.000	Ea.	99	43	142
Copper tubing, type L, solder joints, hangers 10' OC 3/4" diam	20.000	L.F.	45.40	114	159.40
Copper tubing, type M, solder joints, hangers 10' OC 3/4" diam	70.000	L.F.	128.80	388.50	517.30
Wrought copper fittings & solder, 3/4" diam	76.000	Ea.	67.64	1,748	1,815.64
Sensor wire, #22-2 conductor, multistranded	.500	C.L.F.	4.33	21.25	25.58
Total			2,707.11	3,297.30	6,004.41

D3010 735	Solar Recirculation, Domestic Hot Water Systems	COST EACH MAT.	INST.	TOTAL
2550	Solar recirculation, hot water			
2560	3/8" tubing, 3 ea. 4' x 4'-4" vacuum tube collectors	4,200	3,050	7,250
2580	1/2" tubing, 4 ea. 4' x 4'-4" vacuum tube collectors	5,075	3,325	8,400
2640	2 ea. 3'x7' black chrome collectors	2,550	3,100	5,650
2660	3 ea. 3'x7' black chrome collectors	3,150	3,200	6,350
2700	2 ea. 3'x7' flat black collectors	2,300	3,125	5,425
2720	3 ea. 3'x7' flat black collectors	2,775	3,225	6,000
2760	3/4" tubing, 3 ea. 3'x7' black chrome collectors	3,300	3,400	6,700
2780	3 ea. 3'x7' flat black absorber plate collectors	2,925	3,400	6,325
2800	2 ea. 4'x9' flat black w/plastic glazing collectors	2,575	3,425	6,000
2820	2 ea. 3'x7' black chrome collectors	2,700	3,300	6,000
2840	1" tubing, 4 ea. 2'x9' black plastic absorber & glazing collectors	3,975	3,850	7,825
2860	4 ea. 3'x7' black chrome absorber collectors	4,200	3,875	8,075
2880	4 ea. 3'x7' flat black absorber collectors	3,700	3,900	7,600

R13600 -610

SERVICES D

SERVICES D

RIDGE

WIRE 220V

The thermosyphon domestic hot water system, a direct collection system, operates under city water pressure and does not require pumps for system operation. An insulated water storage tank is located above the collectors. As the sun heats the collectors, warm water in them rises by means of natural convection; the colder water in the storage tank flows into the collectors by means of gravity. As long as the sun is shining the water continues to flow through the collectors and to become warmer.

To prevent freezing, the system must be drained or the collectors covered with an insulated lid when the temperature drops below 32°F.

D3010 Energy Supply

System Components	QUANTITY	UNIT	COST EACH		
			MAT.	INST.	TOTAL
SYSTEM D3010 745 0960					
SOLAR, THERMOSYPHON, WATER HEATER					
3/4" TUBING, TWO 3'X7' BLACK CHROME COLLECTORS					
D-1 Framing lumber, fir, 2" x 6" x 8', tank cradle	.008	M.B.F.	4.96	4.80	9.76
F-1 Framing lumber, fir, 2" x 4" x 24', sleepers	.016	M.B.F.	9.36	14.40	23.76
I Valve, gate, bronze, 125 lb, soldered 1/2" diam	2.000	Ea.	34.10	36.10	70.20
J Vent flashing, neoprene	4.000	Ea.	39.20	86	125.20
O Pipe covering, urethane, ultraviolet cover, 1" wall, 1/2" diam	40.000	L.F.	57.60	163.20	220.80
P Pipe covering fiberglass all service jacket 1" wall 1/2" diam	160.000	L.F.	206.40	510.40	716.80
Q Collector panel solar, blk chrome on copper, 3/16" temp glass 3'-8" x 6'	2.000	Ea.	950	173	1,123
Y Flow control valve, globe, bronze, 125#, soldered, 1/2" diam	1.000	Ea.	36	18.05	54.05
U Valve, water tempering, bronze, sweat connections, 1/2" diam	1.000	Ea.	46.50	18.05	64.55
W Tank, water storage, solar energy system, 80 Gal, 2" x 1/2 lb insul	1.000	Ea.	820	271	1,091
Copper tubing type L, solder joints, hangers 10' OC 1/2" diam	150.000	L.F.	253.50	802.50	1,056
Copper tubing type M, solder joints, hangers 10' OC 1/2" diam	50.000	L.F.	69.50	257.50	327
Sensor wire, #22-2 gauge multistranded	.500	C.L.F.	4.33	21.25	25.58
Wrought copper fittings & solder, 1/2" diam	75.000	Ea.	30.75	1,612.50	1,643.25
Total			2,562.20	3,988.75	6,550.95

D3010 745	Thermosyphon, Hot Water		MAT.	INST.	TOTAL
0960	Solar, thermosyphon, hot water, two collector system	R13600 -610	2,550	4,000	6,550
0970					

SERVICES D

In this closed-loop indirect collection system, fluid with a low freezing temperature, propylene glycol, transports heat from the collectors to water storage. The transfer fluid is contained in a closed loop consisting of collectors, supply and return piping, and a heat exchanger immersed in the storage tank.

When the collectors become approximately 20° F warmer than the storage temperature, the controller activates the circulator. The circulator moves the fluid continuously until the temperature difference between fluid in the collectors and storage is such that the collection will no longer occur and then the circulator turns off. Since the heat transfer fluid has a very low freezing temperature, there is no need for it to be drained from the collectors between periods of collection.

D3010 Energy Supply

System Components	QUANTITY	UNIT	COST EACH		
			MAT.	INST.	TOTAL
SYSTEM D3010 750 2750					
SOLAR, CLOSED LOOP, SPACE/HOT WATER					
1" TUBING, TEN 3'X7' BLK CHROME ON COPPER ABSORBER COLLECTORS					
A, B Differential controller 2 sensors, thermostat, solar energy system	2.000	Ea.	171	72	243
C Thermometer, 2" dial	10.000	Ea.	210	270	480
C-1 Heat exchanger, solar energy system, fluid to air, up flow, 80 MBH	1.000	Ea.	460	260	720
D, T Fill & drain valves, brass, 3/4" connection	5.000	Ea.	27.25	90.25	117.50
D-1 Fan center	1.000	Ea.	119	94	213
E Air vent, manual, 1/8" fitting	1.000	Ea.	2.13	13.55	15.68
E-2 Thermostat, 2 stage for sensing room temperature	1.000	Ea.	122	54	176
F Air purger	2.000	Ea.	90	72	162
F-1 Controller, liquid temperature, solar energy system	1.000	Ea.	101	86.50	187.50
G Expansion tank	2.000	Ea.	128	27.10	155.10
I Valve, gate, bronze, 125 lb, soldered, 1" diam	5.000	Ea.	140	115	255
J Vent flashing, neoprene	2.000	Ea.	19.60	43	62.60
K Circulator, solar heated liquid, 1/25 HP	2.000	Ea.	306	112	418
L Circulator, solar heated liquid, 1/20 HP	1.000	Ea.	161	65	226
N Relief valve, temp & pressure 150 psi 210°F self-closing	4.000	Ea.	34.60	57.80	92.40
N-1 Relief valve, pressure poppet, bronze, 30 psi, 3/4" IPS	3.000	Ea.	148.50	46.50	195
O Pipe covering, urethane, ultraviolet cover, 1" wall, 1" diam	50.000	L.F.	80.50	208	288.50
P Pipe covering, fiberglass, all service jacket, 1" wall, 1" diam	60.000	L.F.	88.20	199.20	287.40
Q Collector panel solar energy blk chrome on copper 1/8" temp glass 3'x7'	10.000	Ea.	6,000	820	6,820
Roof clamps for solar energy collector panel	10.000	Set	17.70	111.50	129.20
R Valve, swing check, bronze, 125 lb, regrinding disc, 3/4" & 1" diam	4.000	Ea.	208	92	300
S Pressure gage, 0-60 psi, for solar energy system	2.000	Ea.	46	27.10	73.10
U Valve, water tempering, bronze, sweat connections, 3/4" diam	1.000	Ea.	56.50	21.50	78
W-2, W-1, W Tank, water storage immersed heat xchr elec elem 2"x1/2# ins 120 gal	4.000	Ea.	3,700	1,240	4,940
X Valve, globe, bronze, 125 lb, rising stem, 1" diam	3.000	Ea.	229.50	69	298.50
Y Valve, flow control	1.000	Ea.	54.50	19.70	74.20
Copper tubing type M, solder joint, hanger 10' OC, 1" diam	110.000	L.F.	269.50	682	951.50
Copper tubing, type L, solder joint, hanger 10' OC 3/4" diam	20.000	L.F.	45.40	114	159.40
Wrought copper fittings & solder, 3/4" & 1" diam	121.000	Ea.	266.20	3,267	3,533.20
Sensor, wire, #22-2 conductor, multistranded	.700	C.L.F.	6.06	29.75	35.81
Ductwork, galvanized steel, for heat exchanger	8.000	Lb.	6.72	41.20	47.92
Solar energy heat transfer fluid propylene glycol, anti-freeze	25.000	Gal.	242.50	387.50	630
TOTAL			13,557.36	8,808.15	22,365.51

D3010 750	Solar, Closed Loop, Space/Hot Water Systems		COST EACH		
			MAT.	INST.	TOTAL
2540	Solar, closed loop, space/hot water				
2550	1/2" tubing, 12 ea. 4'x4'4" vacuum tube collectors		18,000	8,200	26,200
2600	3/4" tubing, 12 ea. 4'x4'4" vacuum tube collectors	R13600	18,400	8,475	26,875
2650	10 ea. 3' x 7' black chrome absorber collectors	-610	13,200	8,250	21,450
2700	10 ea. 3' x 7' flat black absorber collectors		12,000	8,300	20,300
2750	1" tubing, 10 ea. 3' x 7' black chrome absorber collectors		13,600	8,800	22,400
2800	10 ea. 3' x 7' flat black absorber collectors		12,300	8,875	21,175
2850	6 ea. 4' x 9' flat black w/plastic glazing collectors		10,800	8,800	19,600
2900	12 ea. 2' x 9' plastic absorber and glazing collectors		14,100	8,950	23,050

D SERVICES

SERVICES D

This draindown pool system uses a differential thermostat similar to those used in solar domestic hot water and space heating applications. To heat the pool, the pool water passes through the conventional pump-filter loop and then flows through the collectors. When collection is not possible, or when the pool temperature is reached, all water drains from the solar loop back to the pool through the existing piping. The modes are controlled by solenoid valves or other automatic valves in conjunction with a vacuum breaker relief valve, which facilitates draindown.

Important: See the Reference Section for critical supporting data - Reference Numbers and City Cost Indexes

D3010 Energy Supply

System Components	QUANTITY	UNIT	COST EACH		
			MAT.	INST.	TOTAL
SYSTEM D3010 760 2640					
SOLAR SWIMMING POOL HEATER, ROOF MOUNTED COLLECTORS					
TEN 4' X 10' FULLY WETTED UNGLAZED PLASTIC ABSORBERS					
A Differential thermostat/controller, 110V, adj pool pump system	1.000	Ea.	282	217	499
A-1 Solenoid valve, PVC, normally 1 open 1 closed (included)	2.000	Ea.			
B Sensor, thermistor type (included)	2.000	Ea.			
E-1 Valve, vacuum relief	1.000	Ea.	27.50	13.55	41.05
Q Collector panel, solar energy, plastic, liquid full wetted, 4' x 10'	10.000	Ea.	2,150	1,560	3,710
R Valve, ball check, PVC, socket, 1-1/2″ diam	1.000	Ea.	75	21.50	96.50
Z Valve, ball, PVC, socket, 1-1/2″ diam	3.000	Ea.	250.50	64.50	315
Pipe, PVC, sch 40, 1-1/2″ diam	80.000	L.F.	243.20	964	1,207.20
Pipe fittings, PVC sch 40, socket joint, 1-1/2″ diam	10.000	Ea.	11	215	226
Sensor wire, #22-2 conductor, multistranded	.500	C.L.F.	4.33	21.25	25.58
Roof clamps for solar energy collector panels	10.000	Set	17.70	111.50	129.20
Roof strap, teflon for solar energy collector panels	26.000	L.F.	271.70	54.86	326.56
TOTAL			3,332.93	3,243.16	6,576.09

D3010 760	Solar Swimming Pool Heater Systems		COST EACH		
			MAT.	INST.	TOTAL
2530	Solar swimming pool heater systems, roof mounted collectors				
2540	10 ea. 3'x7' black chrome absorber, 1/8″ temp. glass		7,175	2,500	9,675
2560	10 ea. 4'x8' black chrome absorber, 3/16″ temp. glass	R13600	8,275	2,975	11,250
2580	10 ea. 3'8″x6' flat black absorber, 3/16″ temp. glass	-610	5,925	2,550	8,475
2600	10 ea. 4'x9' flat black absorber, plastic glazing		6,575	3,100	9,675
2620	10 ea. 2'x9' rubber absorber, plastic glazing		12,100	3,350	15,450
2640	10 ea. 4'x10' fully wetted unglazed plastic absorber		3,325	3,250	6,575
2660	Ground mounted collectors				
2680	10 ea. 3'x7' black chrome absorber, 1/8″ temp. glass		7,375	2,900	10,275
2700	10 ea. 4'x8' black chrome absorber, 3/16″ temp glass		8,475	3,375	11,850
2720	10 ea. 3'8″x6' flat blk absorber, 3/16″ temp. glass		6,125	2,950	9,075
2740	10 ea. 4'x9' flat blk absorber, plastic glazing		6,775	3,500	10,275
2760	10 ea. 2'x9' rubber absorber, plastic glazing		12,300	3,650	15,950
2780	10 ea. 4'x10' fully wetted unglazed plastic absorber		3,525	3,650	7,175

D SERVICES

ATTIC FLOOR

CWS

HWS

SERVICES D

This domestic hot water pre-heat system includes heat exchanger with a circulating pump, blower, air-to-water, coil and controls, mounted in the upper collector manifold. Heat from the hot air coming out of the collectors is transferred through the heat exchanger. For each degree of DHW preheating gained, one degree less heating is needed from the fuel fired water heater. The system is simple, inexpensive to operate and can provide a substantial portion of DHW requirements for modest additional cost.

D3010 Energy Supply

System Components	QUANTITY	UNIT	COST EACH		
			MAT.	INST.	TOTAL
SYSTEM D3010 765 2560					
SOLAR, HOT WATER, AIR TO WATER HEAT EXCHANGE					
THREE COLLECTORS, OPTICAL BLACK ON ALUMINUM, 10'X2',80GAL TANK					
A, B Differential controller, 2 sensors, thermostat, solar energy system	1.000	Ea.	85.50	36	121.50
C Thermometer, 2″ dial	2.000	Ea.	42	54	96
C-1 Heat exchanger, air to fluid, up flow 70 MBH	1.000	Ea.	345	223	568
E Air vent, manual, for solar energy system 1/8″ fitting	1.000	Ea.	2.13	13.55	15.68
F Air purger	1.000	Ea.	45	36	81
G Expansion tank	1.000	Ea.	64	13.55	77.55
H Strainer, Y type, bronze body, 1/2″ IPS	1.000	Ea.	17.50	22	39.50
I Valve, gate, bronze, 125 lb, NRS, soldered 1/2″ diam	5.000	Ea.	85.25	90.25	175.50
N Relief valve, temp & pressure solar 150 psi 210°F self-closing	1.000	Ea.	8.65	14.45	23.10
P Pipe covering, fiberglass, all service jacket, 1″ wall, 1/2″ diam	60.000	L.F.	77.40	191.40	268.80
Q Collector panel solar energy, air, black on alum plate, flush mount, 10'x2'	30.000	L.F.	4,830	243	5,073
B-1, R-1 Shutter damper for solar heater circulator	1.000	Ea.	49.50	65	114.50
B-1, R-1 Shutter motor for solar heater circulator	1.000	Ea.	107	48.50	155.50
R Backflow preventer for solar energy system, 1/2″ diam	2.000	Ea.	128	54	182
T Drain valve, brass, 3/4″ connection	1.000	Ea.	5.45	18.05	23.50
U Valve, water tempering, bronze, sweat connections, 1/2″ diam	1.000	Ea.	46.50	18.05	64.55
W Tank, water storage, solar energy system, 80 Gal, 2″ x 2 lb insul	1.000	Ea.	820	271	1,091
V Tank, water storage, w/heating element, drain, relief valve, existing	1.000	System			
X Valve, globe, bronze, 125 lb, rising stem, 1/2″ diam	1.000	Ea.	36	18.05	54.05
Copper tubing type M, solder joints, hangers 10' OC 1/2″ diam	50.000	L.F.	69.50	257.50	327
Copper tubing type L, solder joints, hangers 10' OC 1/2″ diam	10.000	L.F.	16.90	53.50	70.40
Wrought copper fittings & solder, 1/2″ diam	10.000	Ea.	4.10	215	219.10
Sensor wire, #22-2 conductor multistranded	.500	C.L.F.	4.33	21.25	25.58
Q-1, Q-2 Ductwork, fiberglass, aluminzed jacket, 1-1/2″ thick, 8″ diam	32.000	S.F.	147.84	138.56	286.40
Q-3 Manifold for flush mount solar energy collector panels, air	6.000	L.F.	177	29.22	206.22
TOTAL			7,214.55	2,144.88	9,359.43

D3010 765	Solar Hot Water, Air To Water Heat Exchange		COST EACH		
			MAT.	INST.	TOTAL
2550	Solar hot water, air to water heat exchange				
2560	Three collectors, optical black on aluminum, 10' x 2', 80 Gal tank		7,225	2,150	9,375
2580	Four collectors, optical black on aluminum, 10' x 2', 80 Gal tank	R13600 -610	8,925	2,275	11,200
2600	Four collectors, optical black on aluminum, 10' x 2', 120 Gal tank		9,025	2,300	11,325

SERVICES

D

433

SERVICES D

The complete Solar Air Heating System provides maximum savings of conventional fuel with both space heating and year-round domestic hot water heating. It allows for the home air conditioning to operate simultaneously and independently from the solar domestic water heating in summer. The system's modes of operation are:

Mode 1: The building is heated directly from the collectors with air circulated by the Solar Air Mover.

Mode 2: When heat is not needed in the building, the dampers change within the air mover to circulate the air from the collectors to the rock storage bin.

Mode 3: When heat is not available from the collector array and is available in rock storage, the air mover draws heated air from rock storage and directs it into the building. When heat is not available from the collectors or the rock storage bin, the auxiliary heating unit will provide heat for the building. The size of the collector array is typically 25% the size of the main floor area.

Important: See the Reference Section for critical supporting data - Reference Numbers and City Cost Indexes

D3010 Energy Supply

System Components	QUANTITY	UNIT	COST EACH		
			MAT.	INST.	TOTAL
SYSTEM D3010 775 1210					
SOLAR, SPACE/HOT WATER, AIR TO WATER HEAT EXCHANGE					
A, B Differential controller 2 sensors thermos., solar energy sys liquid loop	1.000	Ea.	85.50	36	121.50
A-1, B Differential controller 2 sensors 6 station solar energy sys air loop	1.000	Ea.	222	145	367
B-1 Solar energy sensor, freeze prevention	1.000	Ea.	22	13.55	35.55
C Thermometer for solar energy system, 2" dial	2.000	Ea.	42	54	96
C-1 Heat exchanger, solar energy system, air to fluid, up flow, 70 MBH	1.000	Ea.	345	223	568
D Drain valve, brass, 3/4" connection	2.000	Ea.	10.90	36.10	47
E Air vent, manual, for solar energy system 1/8" fitting	1.000	Ea.	2.13	13.55	15.68
E-1 Thermostat, 2 stage for sensing room temperature	1.000	Ea.	122	54	176
F Air purger	1.000	Ea.	45	36	81
G Expansion tank, for solar energy system	1.000	Ea.	64	13.55	77.55
I Valve, gate, bronze, 125 lb, NRS, soldered 3/4" diam	2.000	Ea.	39	43	82
K Circulator, solar heated liquid, 1/25 HP	1.000	Ea.	153	56	209
N Relief valve temp & press 150 psi 210°F self-closing, 3/4" IPS	1.000	Ea.	8.65	14.45	23.10
N-1 Relief valve, pressure, poppet, bronze, 30 psi, 3/4" IPS	1.000	Ea.	49.50	15.50	65
P Pipe covering, fiberglass, all service jacket, 1" wall, 3/4" diam	60.000	L.F.	88.20	199.20	287.40
Q-3 Manifold for flush mount solar energy collector panels	20.000	L.F.	590	97.40	687.40
Q Collector panel solar energy, air, black on alum. plate, flush mount 10'x2'	100.000	L.F.	16,100	810	16,910
R Valve, swing check, bronze, 125 lb, regrinding disc, 3/4" diam	2.000	Ea.	69	43	112
S Pressure gage, 2" dial, for solar energy system	1.000	Ea.	23	13.55	36.55
U Valve, water tempering, bronze, sweat connections, 3/4" diam	1.000	Ea.	56.50	21.50	78
W-2, W Tank, water storage, solar, elec element 2"x1/2# insul, 80 Gal	1.000	Ea.	820	271	1,091
X Valve, globe, bronze, 125 lb, soldered 3/4" diam	1.000	Ea.	49.50	21.50	71
Copper tubing type L, solder joints, hangers 10' OC 3/4" diam	10.000	L.F.	22.70	57	79.70
Copper tubing type M, solder joints, hangers 10' OC 3/4" diam	60.000	L.F.	110.40	333	443.40
Wrought copper fittings & solder, 3/4" diam	26.000	Ea.	23.14	598	621.14
Sensor wire, #22-2 conductor multistranded	1.200	C.L.F.	10.38	51	61.38
Q-1 Duct work, rigid fiberglass, rectangular	400.000	S.F.	448	1,384	1,832
Ductwork, transition pieces	8.000	Ea.	118.40	260	378.40
R-2 Shutter/damper for solar heater circulator	9.000	Ea.	445.50	585	1,030.50
K-1 Shutter motor for solar heater circulator blower	9.000	Ea.	963	436.50	1,399.50
K-1 Fan, solar energy heated air circulator, space & DHW system	1.000	Ea.	1,550	1,550	3,100
Z-1 Tank, solar energy air storage, 6'-3"H 7'x7' = 306 CF/2000 Gal	1.000	Ea.	8,950	1,000	9,950
Z-2 Crushed stone 1-1/2"	11.000	C.Y.	171.60	56.43	228.03
C-2 Thermometer, remote probe, 2" dial	1.000	Ea.	32	54	86
R-1 Solenoid valve	1.000	Ea.	233	48	281
V, R-4, R-3 Furnace, supply diffusers, return grilles, existing	1.000	Ea.	8,950	1,000	9,950
Total			41,035	9,643.78	50,678.78

D3010 775	Air To Water Heat Exchange		COST EACH		
			MAT.	INST.	TOTAL
1210 1220	Solar, air to water heat exchange, for space / hot water heating	R13600 -610	41,000	9,625	50,625

SERVICES

D

SERVICES D

In this closed-loop indirect collection system, fluid with a low freezing temperature, such as propylene glycol, transports heat from the collectors to water storage. The transfer fluid is contained in a closed-loop consisting of collectors, supply and return piping, and a heat exchanger immersed in the storage tank. A typical two-or-three panel system contains 5 to 6 gallons of heat transfer fluid.

When the collectors become approximately 20° F warmer than the storage temperature, a controller activates the circulator. The circulator moves the fluid continuously through the collectors until the temperature difference between the collectors and storage is such that heat collection no longer occurs; at that point, the circulator shuts off. Since the heat transfer fluid has a very low freezing temperature, there is no need for it to be drained from the collectors between periods of collection.

D3010 Energy Supply

System Components	QUANTITY	UNIT	COST EACH		
			MAT.	INST.	TOTAL
SYSTEM D3010 780 2760					
SOLAR, CLOSED LOOP, HOT WATER SYSTEM, IMMERSED HEAT EXCHANGER					
3/4″ TUBING, THREE 3′ X 7′ BLACK CHROME COLLECTORS					
A, B Differential controller, 2 sensors, thermostat, solar energy system	1.000	Ea.	85.50	36	121.50
C Thermometer 2″ dial	3.000	Ea.	63	81	144
D, T Fill & drain valves, brass, 3/4″ connection	3.000	Ea.	16.35	54.15	70.50
E Air vent, manual, 1/8″ fitting	1.000	Ea.	2.13	13.55	15.68
F Air purger	1.000	Ea.	45	36	81
G Expansion tank	1.000	Ea.	64	13.55	77.55
I Valve, gate, bronze, NRS, soldered 3/4″ diam	3.000	Ea.	58.50	64.50	123
J Neoprene vent flashing	2.000	Ea.	19.60	43	62.60
K Circulator, solar heated liquid, 1/25 HP	1.000	Ea.	153	56	209
N-1, N Relief valve, temp & press 150 psi 210°F self-closing 3/4″ IPS	2.000	Ea.	17.30	28.90	46.20
O Pipe covering, urethane ultraviolet cover, 1″ wall, 3/4″ diam	20.000	L.F.	32.20	83.20	115.40
P Pipe covering, fiberglass, all service jacket, 1″ wall, 3/4″ diam	50.000	L.F.	73.50	166	239.50
Roof clamps for solar energy collector panel	3.000	Set	5.31	33.45	38.76
Q Collector panel solar blk chrome on copper, 1/8″ temp glass, 3′x7′	3.000	Ea.	1,800	246	2,046
R-1 Valve, swing check, bronze, regrinding disc, 3/4″ diam, soldered	1.000	Ea.	34.50	21.50	56
S Pressure gauge, 60 psi, 2-1/2″ dial	1.000	Ea.	23	13.55	36.55
U Valve, water tempering, bronze, sweat connections, 3/4″ diam	1.000	Ea.	56.50	21.50	78
W-2, W Tank, water storage immersed heat exchr elec elem 2″x2# insul 120 Gal	1.000	Ea.	925	310	1,235
X Valve, globe, bronze, rising stem, 3/4″ diam, soldered	1.000	Ea.	49.50	21.50	71
Copper tubing type L, solder joint, hanger 10′ OC 3/4″ diam	20.000	L.F.	45.40	114	159.40
Copper tubing, type M, solder joint, hanger 10′ OC 3/4″ diam	70.000	L.F.	128.80	388.50	517.30
Sensor wire, #22-2 conductor multistranded	.500	C.L.F.	4.33	21.25	25.58
Solar energy heat transfer fluid, propylene glycol, anti-freeze	6.000	Gal.	58.20	93	151.20
Wrought copper fittings & solder, 3/4″ diam	76.000	Ea.	67.64	1,748	1,815.64
TOTAL			3,828.26	3,708.10	7,536.36

D3010 780	Solar, Closed Loop, Hot Water Systems		COST EACH		
			MAT.	INST.	TOTAL
2550	Solar, closed loop, hot water system, immersed heat exchanger				
2560	3/8″ tubing, 3 ea. 4′ x 4′-4″ vacuum tube collectors, 80 gal. tank		4,650	3,350	8,000
2580	1/2″ tubing, 4 ea. 4′ x 4′-4″ vacuum tube collectors, 80 gal. tank		5,525	3,600	9,125
2600	120 gal. tank		5,625	3,650	9,275
2640	2 ea. 3′x7′ black chrome collectors, 80 gal. tank	R13600 -610	3,000	3,450	6,450
2660	120 gal. tank		3,100	3,425	6,525
2700	2 ea. 3′x7′ flat black collectors, 120 gal. tank		2,850	3,450	6,300
2720	3 ea. 3′x7′ flat black collectors, 120 gal. tank		3,325	3,550	6,875
2760	3/4″ tubing, 3 ea. 3′x7′ black chrome collectors, 120 gal. tank		3,825	3,700	7,525
2780	3 ea. 3′x7′ flat black collectors, 120 gal. tank		3,450	3,725	7,175
2800	2 ea. 4′x9′ flat black w/plastic glazing collectors 120 gal. tank		3,100	3,725	6,825
2840	1″ tubing, 4 ea. 2′x9′ plastic absorber & glazing collectors 120 gal. tank		4,425	4,150	8,575
2860	4 ea. 3′x7′ black chrome collectors, 120 gal. tank		4,650	4,175	8,825
2880	4 ea. 3′x7′ flat black absorber collectors, 120 gal. tank		4,150	4,175	8,325

D3020 Heat Generating Systems

Boiler   **Baseboard Radiation**

Small Electric Boiler
System Considerations:
1. Terminal units are fin tube baseboard radiation rated at 720 BTU/hr with 200° water temperature or 820 BTU/hr steam.
2. Primary use being for residential or smaller supplementary areas, the floor levels are based on 7-1/2′ ceiling heights.
3. All distribution piping is copper for boilers through 205 MBH. All piping for larger systems is steel pipe.

System Components			COST EACH		
	QUANTITY	UNIT	MAT.	INST.	TOTAL
SYSTEM D3020 102 1120					
SMALL HEATING SYSTEM, HYDRONIC, ELECTRIC BOILER					
1,480 S.F., 61 MBH, STEAM, 1 FLOOR					
Boiler, electric steam, std cntrls, trim, ftngs and valves, 18 KW, 61.4 MBH	1.000	Ea.	10,065	1,100	11,165
Copper tubing type L, solder joint, hanger 10′OC, 1-1/4″ diam	160.000	L.F.	611.20	1,200	1,811.20
Radiation, 3/4″ copper tube w/alum fin baseboard pkg 7″ high	60.000	L.F.	372	813	1,185
Rough in baseboard panel or fin tube with valves & traps	10.000	Set	1,140	4,100	5,240
Pipe covering, calcium silicate w/cover 1″ wall 1-1/4″ diam	160.000	L.F.	441.60	740.80	1,182.40
Low water cut-off, quick hookup, in gage glass tappings	1.000	Ea.	166	27.50	193.50
TOTAL			12,795.80	7,981.30	20,777.10
COST PER S.F.			8.65	5.39	14.04

D3020 102	Small Heating Systems, Hydronic, Electric Boilers		COST PER S.F.		
			MAT.	INST.	TOTAL
1100	Small heating systems, hydronic, electric boilers				
1120	Steam, 1 floor, 1480 S.F., 61 M.B.H.		8.65	5.39	14.04
1160	3,000 S.F., 123 M.B.H.	R15500 -010	5.10	4.72	9.82
1200	5,000 S.F., 205 M.B.H.		3.86	4.35	8.21
1240	2 floors, 12,400 S.F., 512 M.B.H.	R15500 -020	2.75	4.33	7.08
1280	3 floors, 24,800 S.F., 1023 M.B.H.		2.40	4.27	6.67
1320	34,750 S.F., 1,433 M.B.H.	R15500 -030	2.12	4.15	6.27
1360	Hot water, 1 floor, 1,000 S.F., 41 M.B.H.		7.85	3.02	10.87
1400	2,500 S.F., 103 M.B.H.		4.92	5.40	10.32
1440	2 floors, 4,850 S.F., 205 M.B.H.		4.52	6.50	11.02
1480	3 floors, 9,700 S.F., 410 M.B.H.		4.39	6.70	11.09

D3020   Heat Generating Systems

Boiler

Unit Heater

Large Electric Boiler System Considerations:

1. Terminal units are all unit heaters of the same size. Quantities are varied to accommodate total requirements.
2. All air is circulated through the heaters a minimum of three times per hour.
3. As the capacities are adequate for commercial use, floor levels are based on 10′ ceiling heights.
4. All distribution piping is black steel pipe.

System Components	QUANTITY	UNIT	COST EACH		
			MAT.	INST.	TOTAL
SYSTEM D3020 104 1240					
LARGE HEATING SYSTEM, HYDRONIC, ELECTRIC BOILER					
9,280 S.F., 150 KW, 510 MBH, 1 FLOOR					
Boiler, electric hot water, std ctrls, trim, ftngs, valves, 150 KW, 510 MBH	1.000	Ea.	14,137.50	2,400	16,537.50
Expansion tank, painted steel, 60 Gal capacity ASME	1.000	Ea.	2,250	131	2,381
Circulating pump, CI, close cpld, 50 GPM, 2 HP, 2″ pipe conn	1.000	Ea.	1,225	260	1,485
Unit heater, 1 speed propeller, horizontal, 200° EWT, 72.7 MBH	7.000	Ea.	4,410	1,001	5,411
Unit heater piping hookup with controls	7.000	Set	2,520	6,510	9,030
Pipe, steel, black, schedule 40, welded, 2-1/2″ diam	380.000	L.F.	2,166	7,018.60	9,184.60
Pipe covering, calcium silicate w/cover, 1″ wall, 2-1/2″ diam	380.000	L.F.	1,352.80	1,816.40	3,169.20
TOTAL			28,061.30	19,137	47,198.30
COST PER S.F.			3.02	2.06	5.08

SERVICES

D

D3020 104	Large Heating Systems, Hydronic, Electric Boilers		COST PER S.F.		
			MAT.	INST.	TOTAL
1230	Large heating systems, hydronic, electric boilers				
1240	9,280 S.F., 150 K.W., 510 M.B.H., 1 floor		3.03	2.06	5.09
1280	14,900 S.F., 240 K.W., 820 M.B.H., 2 floors	R15500 -010	3.15	3.39	6.54
1320	18,600 S.F., 296 K.W., 1,010 M.B.H., 3 floors		3.05	3.68	6.73
1360	26,100 S.F., 420 K.W., 1,432 M.B.H., 4 floors	R15500 -020	3.18	3.62	6.80
1400	39,100 S.F., 666 K.W., 2,273 M.B.H., 4 floors		2.81	3.01	5.82
1440	57,700 S.F., 900 K.W., 3,071 M.B.H., 5 floors	R15500 -030	2.63	2.97	5.60
1480	111,700 S.F., 1,800 K.W., 6,148 M.B.H., 6 floors		2.31	2.56	4.87
1520	149,000 S.F., 2,400 K.W., 8,191 M.B.H., 8 floors		2.26	2.56	4.82
1560	223,300 S.F., 3,600 K.W., 12,283 M.B.H., 14 floors		2.29	2.89	5.18

D3020 Heat Generating Systems

Unit Heater

Fossil Fuel Boiler System Considerations:

1. Terminal units are horizontal unit heaters. Quantities are varied to accommodate total heat loss per building.
2. Unit heater selection was determined by their capacity to circulate the building volume a minimum of three times per hour in addition to the BTU output.

3. Systems shown are forced hot water. Steam boilers cost slightly more than hot water boilers. However, this is compensated for by the smaller size or fewer terminal units required with steam.
4. Floor levels are based on 10' story heights.
5. MBH requirements are gross boiler output.

SERVICES D

System Components	QUANTITY	UNIT	COST EACH		
			MAT.	INST.	TOTAL
SYSTEM D3020 108 1280					
HEATING SYSTEM, HYDRONIC, FOSSIL FUEL, TERMINAL UNIT HEATERS					
CAST IRON BOILER, GAS, 80 MBH, 1,070 S.F. BUILDING					
Boiler, gas, hot water, CI, burner, controls, insulation, breeching, 80 MBH	1.000	Ea.	1,181.25	1,207.50	2,388.75
Pipe, steel, black, schedule 40, threaded, cplg & hngr 10'OC, 2" diam	200.000	L.F.	700	2,440	3,140
Unit heater, 1 speed propeller, horizontal, 200° EWT, 72.7 MBH	2.000	Ea.	1,260	286	1,546
Unit heater piping hookup with controls	2.000	Set	720	1,860	2,580
Expansion tank, painted steel, ASME, 18 Gal capacity	1.000	Ea.	1,400	65.50	1,465.50
Circulating pump, CI, flange connection, 1/12 HP	1.000	Ea.	229	130	359
Pipe covering, calcium silicate w/cover, 1" wall, 2" diam	200.000	L.F.	666	956	1,622
TOTAL			6,156.25	6,945	13,101.25
COST PER S.F.			5.75	6.49	12.24

D3020 108	Heating Systems, Unit Heaters		COST PER S.F.		
			MAT.	INST.	TOTAL
1260	Heating systems, hydronic, fossil fuel, terminal unit heaters,				
1280	Cast iron boiler, gas, 80 M.B.H., 1,070 S.F. bldg.		5.75	6.50	12.25
1320	163 M.B.H., 2,140 S.F. bldg.	R15500 -010	3.88	4.40	8.28
1360	544 M.B.H., 7,250 S.F. bldg.		2.71	3.17	5.88
1400	1,088 M.B.H., 14,500 S.F. bldg.	R15500 -020	2.38	3	5.38
1440	3,264 M.B.H., 43,500 S.F. bldg.		1.95	2.28	4.23
1480	5,032 M.B.H., 67,100 S.F. bldg.	R15500 -030	2.39	2.31	4.70
1520	Oil, 109 M.B.H., 1,420 S.F. bldg.		5.90	5.75	11.65
1560	235 M.B.H., 3,150 S.F. bldg.		4.12	4.20	8.32
1600	940 M.B.H., 12,500 S.F. bldg.		2.87	2.75	5.62
1640	1,600 M.B.H., 21,300 S.F. bldg.		2.79	2.62	5.41
1680	2,480 M.B.H., 33,100 S.F. bldg.		2.77	2.37	5.14
1720	3,350 M.B.H., 44,500 S.F. bldg.		2.38	2.42	4.80
1760	Coal, 148 M.B.H., 1,975 S.F. bldg.		4.90	3.76	8.66
1800	300 M.B.H., 4,000 S.F. bldg.		3.79	2.95	6.74
1840	2,360 M.B.H., 31,500 S.F. bldg.		2.66	2.43	5.09
1880	Steel boiler, gas, 72 M.B.H., 1,020 S.F. bldg.		5.70	4.78	10.48
1920	240 M.B.H., 3,200 S.F. bldg.		4.02	3.84	7.86
1960	480 M.B.H., 6,400 S.F. bldg.		3.26	2.85	6.11
2000	800 M.B.H., 10,700 S.F. bldg.		2.80	2.54	5.34

D30 HVAC

D3020 Heat Generating Systems

D3020 108	Heating Systems, Unit Heaters	COST PER S.F.		
		MAT.	INST.	TOTAL
2040	1,960 M.B.H., 26,100 S.F. bldg.	2.48	2.27	4.75
2080	3,000 M.B.H., 40,000 S.F. bldg.	2.45	2.32	4.77
2120	Oil, 97 M.B.H., 1,300 S.F. bldg.	5.45	5.45	10.90
2160	315 M.B.H., 4,550 S.F. bldg.	3.39	2.78	6.17
2200	525 M.B.H., 7,000 S.F. bldg.	3.60	2.84	6.44
2240	1,050 M.B.H., 14,000 S.F. bldg.	3.06	2.86	5.92
2280	2,310 M.B.H. 30,800 S.F. bldg.	2.68	2.47	5.15
2320	3,150 M.B.H., 42,000 S.F. bldg.	2.48	2.53	5.01

D3020 Heat Generating Systems

Fin Tube Radiator

Fossil Fuel Boiler System Considerations:

1. Terminal units are commercial steel fin tube radiation. Quantities are varied to accommodate total heat loss per building.
2. Systems shown are forced hot water. Steam boilers cost slightly more than hot water boilers. However, this is compensated for by the smaller size or fewer terminal units required with steam.
3. Floor levels are based on 10' story heights.
4. MBH requirements are gross boiler output.

SERVICES D

System Components	QUANTITY	UNIT	COST EACH MAT.	COST EACH INST.	COST EACH TOTAL
SYSTEM D3020 110 3240					
HEATING SYSTEM, HYDRONIC, FOSSIL FUEL, FIN TUBE RADIATION					
CAST IRON BOILER, GAS, 80 MBH, 1,070 S.F. BLDG.					
Boiler, gas, hot water, CI, burner, controls, insulation, breeching, 80 MBH	1.000	Ea.	1,181.25	1,207.50	2,388.75
Pipe, steel, black, schedule 40, threaded, 2″ diam	200.000	L.F.	700	2,440	3,140
Radiation, steel 1-1/4″ tube & 4-1/4″ fin w/cover & damper, wall hung	80.000	L.F.	2,160	1,760	3,920
Rough-in, wall hung steel fin radiation with supply & balance valves	8.000	Set	1,256	3,840	5,096
Expansion tank, painted steel, ASME, 18 Gal capacity	1.000	Ea.	1,400	65.50	1,465.50
Circulating pump, CI flanged, 1/12 HP	1.000	Ea.	229	130	359
Pipe covering, calcium silicate w/cover, 1″ wall, 2″ diam	200.000	L.F.	666	956	1,622
TOTAL			7,592.25	10,399	17,991.25
COST PER S.F.			7.10	9.72	16.82

D3020 110	Heating System, Fin Tube Radiation		COST PER S.F. MAT.	COST PER S.F. INST.	COST PER S.F. TOTAL
3230	Heating systems, hydronic, fossil fuel, fin tube radiation				
3240	Cast iron boiler, gas, 80 MBH, 1,070 S.F. bldg.		7.10	9.72	16.82
3280	169 M.B.H., 2,140 S.F. bldg.		4.48	6.15	10.63
3320	544 M.B.H., 7,250 S.F. bldg.	R15500 -010	3.69	5.35	9.04
3360	1,088 M.B.H., 14,500 S.F. bldg.		3.44	5.25	8.69
3400	3,264 M.B.H., 43,500 S.F. bldg.	R15500 -020	3.09	4.59	7.68
3440	5,032 M.B.H., 67,100 S.F. bldg.		3.55	4.65	8.20
3480	Oil, 109 M.B.H., 1,420 S.F. bldg.	R15500 -030	8.35	10.60	18.95
3520	235 M.B.H., 3,150 S.F. bldg.		5.10	6.30	11.40
3560	940 M.B.H., 12,500 S.F. bldg.		3.95	4.99	8.94
3600	1,600 M.B.H., 21,300 S.F. bldg.		3.94	4.96	8.90
3640	2,480 M.B.H., 33,100 S.F. bldg.		3.94	4.71	8.65
3680	3,350 M.B.H., 44,500 S.F. bldg.		3.53	4.74	8.27
3720	Coal, 148 M.B.H., 1,975 S.F. bldg.		5.85	5.85	11.70
3760	300 M.B.H., 4,000 S.F. bldg.		4.74	5.05	9.79
3800	2,360 M.B.H., 31,500 S.F. bldg.		3.78	4.72	8.50
3840	Steel boiler, gas, 72 M.B.H., 1,020 S.F. bldg.		8.10	9.20	17.30
3880	240 M.B.H., 3,200 S.F. bldg.		5.05	6.10	11.15
3920	480 M.B.H., 6,400 S.F. bldg.		4.27	5.05	9.32
3960	800 M.B.H., 10,700 S.F. bldg.		4.19	5.35	9.54
4000	1,960 M.B.H., 26,100 S.F. bldg.		3.60	4.56	8.16
4040	3,000 M.B.H., 40,000 S.F. bldg.		3.54	4.59	8.13

D3020 Heat Generating Systems

D3020 110	Heating System, Fin Tube Radiation	COST PER S.F.		
		MAT.	INST.	TOTAL
4080	Oil, 97 M.B.H., 1,300 S.F. bldg.	6.30	8.70	15
4120	315 M.B.H., 4,550 S.F. bldg.	4.31	4.75	9.06
4160	525 M.B.H., 7,000 S.F. bldg.	4.70	5.10	9.80
4200	1,050 M.B.H., 14,000 S.F. bldg.	4.16	5.10	9.26
4240	2,310 M.B.H., 30,800 S.F. bldg.	3.79	4.74	8.53
4280	3,150 M.B.H., 42,000 S.F. bldg.	3.59	4.81	8.40

D **SERVICES**

Pressure Relief Valve

Supply Piping

Circulating Pump

Return Piping

Expansion Tank

Electric Boiler, Hot Water

System Components	QUANTITY	UNIT	COST EACH		
			MAT.	INST.	TOTAL
SYSTEM D3020 126 1010					
BOILER, ELECTRIC, HOT WATER, 15 kW, 52 MBH					
Boilers, electric, ASME, hot water, 15 KW, 52 MBH	1.000	Ea.	3,750	930	4,680
Pipe, black steel, Sch 40, threaded, W/coupling & hangers, 10' OC, 3/4" dia	10.000	L.F.	18.63	81.65	100.28
Pipe, black steel, Sch 40, threaded, W/coupling & hangers, 10' OC, 1" dia	20.000	L.F.	44.51	176.30	220.81
Elbow, 90°, black, straight, 3/4" dia.	6.000	Ea.	9.12	186	195.12
Elbow, 90°, black, straight, 1" dia.	6.000	Ea.	16.20	201	217.20
Tee, black, straight, 3/4" dia.	1.000	Ea.	2.46	48	50.46
Tee, black, straight, 1" dia.	2.000	Ea.	8.40	108	116.40
Tee, black, reducing, 1" dia.	2.000	Ea.	9.60	108	117.60
Union, black with brass seat, 3/4" dia.	1.000	Ea.	6.90	33.50	40.40
Union, black with brass seat, 1" dia.	2.000	Ea.	19.80	72	91.80
Pipe nipples, 3/4" diam.	3.000	Ea.	2.43	10.65	13.08
Pipe nipples, 1" diam.	3.000	Ea.	3.11	12.30	15.41
Valves, bronze, gate, N.R.S., threaded, class 150, 3/4" size	2.000	Ea.	59	43	102
Valves, bronze, gate, N.R.S., threaded, class 150, 1" size	2.000	Ea.	77	46	123
Thermometer, stem type, 9" case, 8" stem, 3/4" NPT	1.000	Ea.	266	31.20	297.20
Tank, steel, liquid expansion, ASME, painted, 15 gallon capacity	1.000	Ea.	385	46	431
Pump, circulating, bronze, flange connection, 3/4" to 1-1/2" size, 1/8 HP	1.000	Ea.	605	130	735
Insulation, fiberglass pipe covering, 1" wall, 1" IPS	20.000	L.F.	30	69.60	99.60
TOTAL			5,313.16	2,333.20	7,646.36

D3020 126	Electric Boiler, Hot Water	COST EACH		
		MAT.	INST.	TOTAL
1010	Boiler, electric, hot water, 15 kW, 52 MBH	5,325	2,325	7,650
1020	30 kW, 103 MBH	5,725	2,450	8,175
1030	60 kW, 205 MBH	5,800	2,525	8,325
1040	120 kW, 410 MBH	8,700	3,050	11,750
1050	150 kW, 510 MBH	14,700	5,975	20,675
1060	210 kW, 716 MBH	17,500	6,600	24,100
1070	296 kW, 1010 MBH	18,400	6,775	25,175
1080	444 kW, 1515 MBH	25,600	9,100	34,700
1090	666 kW, 2273 MBH	33,000	9,825	42,825
1100	900 kW, 3071 MBH	42,600	11,400	54,000
1110	1320 kW, 4505 MBH	53,500	12,000	65,500
1120	2100 kW, 7167 MBH	69,500	14,500	84,000

D3020 Heat Generating Systems

D3020 126	Electric Boiler, Hot Water	COST EACH		
		MAT.	INST.	TOTAL
1130	2610 kW, 8905 MBH	83,000	17,500	100,500
1140	3600 kW, 12,283 MBH	96,000	21,000	117,000

D3020 128	Electric Boiler, Steam	COST EACH		
		MAT.	INST.	TOTAL
1010	Boiler, electric, steam, 18 kW, 61.4 MBH	9,775	2,225	12,000
1020	36 kW, 123 MBH	10,100	2,375	12,475
1030	60 kW, 205 MBH	13,700	4,800	18,500
1040	120 kW, 409 MBH	15,400	5,275	20,675
1050	150 kW, 512 MBH	18,800	8,125	26,925
1060	210 kW, 716 MBH	22,100	8,725	30,825
1070	300 kW, 1023 MBH	24,700	9,550	34,250
1080	510 kW, 1740 MBH	29,000	12,400	41,400
1090	720 kW, 2456 MBH	35,000	13,000	48,000
1100	1080 kW, 3685 MBH	42,400	15,400	57,800
1110	1260 kW, 4300 MBH	52,500	16,400	68,900
1120	1620 kW, 5527 MBH	59,000	18,500	77,500
1130	2070 kW, 7063 MBH	71,500	19,400	90,900
1140	2340 kW, 7984 MBH	78,500	22,100	100,600

SERVICES

D

D3020 Heat Generating Systems

Cast Iron Boiler, Hot Water, Gas Fired

System Components	QUANTITY	UNIT	COST EACH		
			MAT.	INST.	TOTAL
SYSTEM D3020 130 1010					
BOILER, CAST IRON, GAS, HOT WATER, 100 MBH					
Boilers, gas fired, std controls, CI, insulated, HW, gross output 100 MBH	1.000	Ea.	1,300	1,225	2,525
Pipe, black steel, Sch 40, threaded, W/coupling & hangers, 10' OC, 3/4" dia	20.000	L.F.	36.45	159.75	196.20
Pipe, black steel, Sch 40, threaded, W/coupling & hangers, 10' OC, 1" dia.	20.000	L.F.	44.51	176.30	220.81
Elbow, 90°, black, straight, 3/4" dia.	9.000	Ea.	13.68	279	292.68
Elbow, 90°, black, straight, 1" dia.	6.000	Ea.	16.20	201	217.20
Tee, black, straight, 3/4" dia.	2.000	Ea.	4.92	96	100.92
Tee, black, straight, 1" dia.	2.000	Ea.	8.40	108	116.40
Tee, black, reducing, 1" dia.	2.000	Ea.	9.60	108	117.60
Pipe cap, black, 3/4" dia.	1.000	Ea.	1.75	13.55	15.30
Union, black with brass seat, 3/4" dia.	2.000	Ea.	13.80	67	80.80
Union, black with brass seat, 1" dia.	2.000	Ea.	19.80	72	91.80
Pipe nipples, black, 3/4" dia	5.000	Ea.	4.05	17.75	21.80
Pipe nipples, black, 1" dia	3.000	Ea.	3.11	12.30	15.41
Valves, bronze, gate, N.R.S., threaded, class 150, 3/4" size	2.000	Ea.	59	43	102
Valves, bronze, gate, N.R.S., threaded, class 150, 1" size	2.000	Ea.	77	46	123
Gas cock, brass, 3/4" size	1.000	Ea.	11.35	19.70	31.05
Thermometer, stem type, 9" case, 8" stem, 3/4" NPT	2.000	Ea.	266	31.20	297.20
Tank, steel, liquid expansion, ASME, painted, 15 gallon capacity	1.000	Ea.	385	46	431
Pump, circulating, bronze, flange connection, 3/4" to 1-1/2" size, 1/8 HP	1.000	Ea.	605	130	735
Vent chimney, all fuel, pressure tight, double wall, SS, 6" dia	20.000	L.F.	780	260	1,040
Vent chimney, elbow, 90° fixed, 6" dia	2.000	Ea.	440	52	492
Vent chimney, Tee, 6" dia	2.000	Ea.	290	65	355
Vent chimney, ventilated roof thimble, 6" dia	1.000	Ea.	210	30	240
Vent chimney, adjustable roof flashing, 6" dia	1.000	Ea.	53.50	26	79.50
Vent chimney, stack cap, 6" diameter	1.000	Ea.	154	16.95	170.95
Insulation, fiberglass pipe covering, 1" wall, 1" IPS	20.000	L.F.	30	69.60	99.60
TOTAL			4,837.12	3,371.10	8,208.22

D3020 130	Boiler, Cast Iron, Hot Water, Gas	COST EACH		
		MAT.	INST.	TOTAL
1010	Boiler, cast iron, gas, hot water, 100 MBH	4,825	3,375	8,200
1020	200 MBH	5,575	3,875	9,450
1030	320 MBH	5,950	4,200	10,150
1040	440 MBH	7,975	5,400	13,375

Important: See the Reference Section for critical supporting data - Reference Numbers and City Cost Indexes

D3020 Heat Generating Systems

D3020 130	Boiler, Cast Iron, Hot Water, Gas	COST EACH		
		MAT.	INST.	TOTAL
1050	544 MBH	12,000	8,625	20,625
1060	765 MBH	15,100	9,250	24,350
1070	1088 MBH	16,900	9,875	26,775
1080	1530 MBH	20,800	13,100	33,900
1090	2312 MBH	25,200	15,400	40,600
1100	2856 MBH	31,000	16,500	47,500
1110	3808 MBH	34,900	19,400	54,300
1120	4720 MBH	66,000	21,700	87,700
1130	6100 MBH	74,000	24,100	98,100
1140	6970 MBH	81,000	29,600	110,600

D3020 134	Boiler, Cast Iron, Steam, Gas	COST EACH		
		MAT.	INST.	TOTAL
1010	Boiler, cast iron, gas, steam, 100 MBH	4,075	3,275	7,350
1020	200 MBH	7,150	6,150	13,300
1030	320 MBH	8,400	6,900	15,300
1040	544 MBH	11,800	11,000	22,800
1050	765 MBH	13,800	11,400	25,200
1060	1275 MBH	17,000	12,500	29,500
1070	2675 MBH	25,600	18,300	43,900
1080	3570 MBH	30,800	20,500	51,300
1090	4720 MBH	34,900	21,800	56,700
1100	6100 MBH	69,500	25,300	94,800
1110	6970 MBH	77,000	29,600	106,600

D3020 136	Boiler, Cast Iron, Hot Water, Gas/Oil	COST EACH		
		MAT.	INST.	TOTAL
1010	Boiler, cast iron, gas & oil, hot water, 584 MBH	14,300	9,550	23,850
1020	876 MBH	18,700	10,200	28,900
1030	1168 MBH	19,600	11,600	31,200
1040	1460 MBH	22,400	14,000	36,400
1050	2044 MBH	25,800	14,400	40,200
1060	2628 MBH	33,300	17,500	50,800
1070	3210 MBH	37,600	18,800	56,400
1080	3796 MBH	40,300	19,300	59,600
1090	4672 MBH	46,500	20,300	66,800
1100	5256 MBH	59,000	24,400	83,400
1110	6000 MBH	67,000	24,800	91,800
1120	9800 MBH	83,500	42,300	125,800
1130	12,200 MBH	103,500	49,200	152,700
1140	13,500 MBH	112,000	52,000	164,000

D3020 138	Boiler, Cast Iron, Steam, Gas/Oil	COST EACH		
		MAT.	INST.	TOTAL
1010	Boiler, cast iron, gas & oil, steam, 810 MBH	15,700	12,400	28,100
1020	1360 MBH	18,600	13,200	31,800
1030	2040 MBH	24,300	17,000	41,300
1040	2700 MBH	26,600	18,800	45,400
1050	3270 MBH	31,200	21,100	52,300
1060	3770 MBH	56,500	21,500	78,000
1070	4650 MBH	63,000	24,300	87,300
1080	5520 MBH	73,500	25,500	99,000
1090	6100 MBH	74,500	26,800	101,300
1100	6970 MBH	81,500	32,700	114,200

SERVICES

D

D3020 Heat Generating Systems

Base Mounted End—Suction Pump

System Components	QUANTITY	UNIT	COST EACH		
			MAT.	INST.	TOTAL
SYSTEM D3020 330 1010					
PUMP, BASE MTD. WITH MOTOR, END-SUCTION, 2-1/2" SIZE, 3 HP, TO 150 GPM					
Pump, circulating, CI, base mounted, 2-1/2" size, 3 HP, to 150 GPM	1.000	Ea.	2,075	435	2,510
Pipe, black steel, Sch. 40, on yoke & roll hangers, 10' O.C., 2-1/2" dia	12.000	L.F.	68.40	221.64	290.04
Elbow, 90°, weld joint, steel, 2-1/2" pipe size	1.000	Ea.	10.20	108.50	118.70
Flange, weld neck, 150 LB, 2-1/2" pipe size	8.000	Ea.	176	436	612
Valve, iron body, gate, 125 lb., N.R.S., flanged, 2-1/2" size	1.000	Ea.	320	156	476
Strainer, Y type, iron body, flanged, 125 lb., 2-1/2" pipe size	1.000	Ea.	80	157	237
Multipurpose valve, CI body, 2" size	1.000	Ea.	385	54.50	439.50
Expansion joint, flanged spool, 6" F to F, 2-1/2" dia	2.000	Ea.	374	127	501
T-O-L, weld joint, socket, 1/4" pipe size, nozzle	2.000	Ea.	9.48	75.66	85.14
T-O-L, weld joint, socket, 1/2" pipe size, nozzle	2.000	Ea.	9.48	79	88.48
Control gauges, pressure or vacuum, 3-1/2" diameter dial	2.000	Ea.	46	27.30	73.30
Pressure/temperature relief plug, 316 SS, 3/4" OD, 7-1/2" insertion	2.000	Ea.	99	27.30	126.30
Insulation, fiberglass pipe covering, 1-1/2" wall, 2-1/2" IPS	12.000	L.F.	46.92	51	97.92
Pump control system	1.000	Ea.	945	405	1,350
Pump balancing	1.000	Ea.		192	192
TOTAL			4,644.48	2,552.90	7,197.38

D3020 330	Circulating Pump Systems, End Suction	COST EACH		
		MAT.	INST.	TOTAL
1010	Pump, base mtd with motor, end-suction, 2-1/2" size, 3 HP, to 150 GPM	4,650	2,550	7,200
1020	3" size, 5 HP, to 225 GPM	5,150	2,925	8,075
1030	4" size, 7-1/2 HP, to 350 GPM	6,325	3,500	9,825
1040	5" size, 15 HP, to 1000 GPM	8,825	5,325	14,150
1050	6" size, 25 HP, to 1550 GPM	11,800	6,450	18,250

D3020 340	Circulating Pump Systems, Double Suction	COST EACH		
		MAT.	INST.	TOTAL
1010	Pump, base mtd w/ motor, double suction, 6" size, 50 HP, to 1200 GPM	15,500	9,250	24,750
1020	8" size, 75 HP, to 2500 GPM	22,600	12,500	35,100
1030	8" size, 100 HP, to 3000 GPM	33,200	14,000	47,200
1040	10" size, 150 HP, to 4000 GPM	47,500	15,900	63,400

D3030 Cooling Generating Systems

Chilled Water Supply & Return Piping — Air Cooled Water Chiller Unit — Roof — Insulate — Return — Fan Coil Unit — Supply — Finish Ceiling

Design Assumptions: The chilled water, air cooled systems priced, utilize reciprocating hermetic compressors and propeller-type condenser fans. Piping with pumps and expansion tanks is included based on a two pipe system. No ducting is included and the fan-coil units are cooling only. Water treatment and balancing are not included. Chilled water piping is insulated. Area distribution is through the use of multiple fan coil units. Fewer but larger fan coil units with duct distribution would be approximately the same S.F. cost.

System Components	QUANTITY	UNIT	COST EACH		
			MAT.	INST.	TOTAL
SYSTEM D3030 110 1200					
PACKAGED CHILLER, AIR COOLED, WITH FAN COIL UNIT					
APARTMENT CORRIDORS, 3,000 S.F., 5.50 TON					
Fan coil air conditioning unit, cabinet mounted & filters chilled water	1.000	Ea.	4,399.20	361.10	4,760.30
Water chiller,air conditioning unit, reciprocating, air cooled,	1.000	Ea.	5,775	1,361.25	7,136.25
Chilled water unit coil connections	1.000	Ea.	550	985	1,535
Chilled water distribution piping	440.000	L.F.	4,642	13,200	17,842
TOTAL			15,366.20	15,907.35	31,273.55
COST PER S.F.			5.12	5.30	10.42

*Cooling requirements would lead to choosing a water cooled unit

D3030 110	Chilled Water, Air Cooled Condenser Systems	COST PER S.F.		
		MAT.	INST.	TOTAL
1180	Packaged chiller, air cooled, with fan coil unit			
1200	Apartment corridors, 3,000 S.F., 5.50 ton	5.13	5.30	10.43
1240	6,000 S.F., 11.00 ton	3.84	4.24	8.08
1280	10,000 S.F., 18.33 ton	3.43	3.23	6.66
1320	20,000 S.F., 36.66 ton	2.86	2.34	5.20
1360	40,000 S.F., 73.33 ton	3.05	2.37	5.42
1440	Banks and libraries, 3,000 S.F., 12.50 ton	7.45	6.25	13.70
1480	6,000 S.F., 25.00 ton	7.05	5.10	12.15
1520	10,000 S.F., 41.66 ton	5.90	3.56	9.46
1560	20,000 S.F., 83.33 ton	5.60	3.24	8.84
1600	40,000 S.F., 167 ton*			
1680	Bars and taverns, 3,000 S.F., 33.25 ton	14.95	7.40	22.35
1720	6,000 S.F., 66.50 ton	12.40	5.75	18.15
1760	10,000 S.F., 110.83 ton	11.90	1.84	13.74
1800	20,000 S.F., 220 ton*			
1840	40,000 S.F., 440 ton*			
1920	Bowling alleys, 3,000 S.F., 17.00 ton	9.60	7.05	16.65
1960	6,000 S.F., 34.00 ton	8.05	4.93	12.98
2000	10,000 S.F., 56.66 ton	6.55	3.59	10.14
2040	20,000 S.F., 113.33 ton	7.05	3.28	10.33
2080	40,000 S.F., 227 ton*			
2160	Department stores, 3,000 S.F., 8.75 ton	7.45	5.90	13.35
2200	6,000 S.F., 17.50 ton	5.25	4.57	9.82
2240	10,000 S.F., 29.17 ton	4.50	3.31	7.81

D SERVICES

D3030 Cooling Generating Systems

D3030 110	Chilled Water, Air Cooled Condenser Systems	COST PER S.F.		
		MAT.	INST.	TOTAL
2280	20,000 S.F., 58.33 ton	3.61	2.45	6.06
2320	40,000 S.F., 116.66 ton	4.30	2.45	6.75
2400	Drug stores, 3,000 S.F., 20.00 ton	11.05	7.25	18.30
2440	6,000 S.F., 40.00 ton	9.50	5.40	14.90
2480	10,000 S.F., 66.66 ton	8.70	4.64	13.34
2520	20,000 S.F., 133.33 ton	8.55	3.44	11.99
2560	40,000 S.F., 267 ton*			
2640	Factories, 2,000 S.F., 10.00 ton	6.40	6	12.40
2680	6,000 S.F., 20.00 ton	5.95	4.83	10.78
2720	10,000 S.F., 33.33 ton	4.98	3.39	8.37
2760	20,000 S.F., 66.66 ton	4.79	3.08	7.87
2800	40,000 S.F., 133.33 ton	4.76	2.52	7.28
2880	Food supermarkets, 3,000 S.F., 8.50 ton	7.30	5.90	13.20
2920	6,000 S.F., 17.00 ton	5.15	4.54	9.69
2960	10,000 S.F., 28.33 ton	4.31	3.21	7.52
3000	20,000 S.F., 56.66 ton	3.48	2.39	5.87
3040	40,000 S.F., 113.33 ton	4.19	2.45	6.64
3120	Medical centers, 3,000 S.F., 7.00 ton	6.40	5.70	12.10
3160	6,000 S.F., 14.00 ton	4.48	4.40	8.88
3200	10,000 S.F., 23.33 ton	4.13	3.37	7.50
3240	20,000 S.F., 46.66 ton	3.50	2.47	5.97
3280	40,000 S.F., 93.33 ton	3.71	2.41	6.12
3360	Offices, 3,000 S.F., 9.50 ton	6.15	5.85	12
3400	6,000 S.F., 19.00 ton	5.75	4.77	10.52
3440	10,000 S.F., 31.66 ton	4.79	3.36	8.15
3480	20,000 S.F., 63.33 ton	4.74	3.13	7.87
3520	40,000 S.F., 126.66 ton	4.64	2.54	7.18
3600	Restaurants, 3,000 S.F., 15.00 ton	8.55	6.50	15.05
3640	6,000 S.F., 30.00 ton	7.45	4.92	12.37
3680	10,000 S.F., 50.00 ton	6.10	3.63	9.73
3720	20,000 S.F., 100.00 ton	7	3.42	10.42
3760	40,000 S.F., 200 ton*			
3840	Schools and colleges, 3,000 S.F., 11.50 ton	7.05	6.15	13.20
3880	6,000 S.F., 23.00 ton	6.60	4.98	11.58
3920	10,000 S.F., 38.33 ton	5.55	3.49	9.04
3960	20,000 S.F., 76.66 ton	5.40	3.26	8.66
4000	40,000 S.F., 153 ton*			

D3030 Cooling Generating Systems

General: Water cooled chillers are available in the same sizes as air cooled units. They are also available in larger capacities.

Design Assumptions: The chilled water systems with water cooled condenser include reciprocating hermetic compressors, water cooling tower, pumps, piping and expansion tanks and are based on a two pipe system. Chilled water piping is insulated. No ducts are included and fan-coil units are cooling only. Area distribution is through use of multiple fan coil units. Fewer but larger fan coil units with duct distribution would be approximately the same S.F. cost. Water treatment and balancing are not included.

System Components	QUANTITY	UNIT	COST EACH MAT.	COST EACH INST.	COST EACH TOTAL
SYSTEM D3030 115 1320					
PACKAGED CHILLER, WATER COOLED, WITH FAN COIL UNIT					
APARTMENT CORRIDORS, 4,000 S.F., 7.33 TON					
Fan coil air conditioner unit, cabinet mounted & filters, chilled water	2.000	Ea.	5,863.20	481.27	6,344.47
Water chiller, reciprocating, water cooled, 1 compressor semihermetic	1.000	Ea.	9,892.80	2,335.80	12,228.60
Cooling tower, draw thru single flow, belt drive	1.000	Ea.	648.71	99.69	748.40
Cooling tower pumps & piping	1.000	System	326.19	234.56	560.75
Chilled water unit coil connections	2.000	Ea.	1,100	1,970	3,070
Chilled water distribution piping	520.000	L.F.	5,486	15,600	21,086
TOTAL			23,316.90	20,721.32	44,038.22
COST PER S.F.			5.83	5.18	11.01

*Cooling requirements would lead to choosing a water cooled unit

SERVICES

D

D3030 115	Chilled Water, Cooling Tower Systems		COST PER S.F. MAT.	COST PER S.F. INST.	COST PER S.F. TOTAL
1300	Packaged chiller, water cooled, with fan coil unit				
1320	Apartment corridors, 4,000 S.F., 7.33 ton		5.83	5.18	11.01
1360	6,000 S.F., 11.00 ton	R15700 -020	4.39	4.47	8.86
1400	10,000 S.F., 18.33 ton		3.99	3.31	7.30
1440	20,000 S.F., 26.66 ton		3.14	2.47	5.61
1480	40,000 S.F., 73.33 ton		3.18	2.55	5.73
1520	60,000 S.F., 110.00 ton		3.09	2.61	5.70
1600	Banks and libraries, 4,000 S.F., 16.66 ton		9.30	5.70	15
1640	6,000 S.F., 25.00 ton		7.60	5	12.60
1680	10,000 S.F., 41.66 ton		6.30	3.75	10.05
1720	20,000 S.F., 83.33 ton		5.90	3.60	9.50
1760	40,000 S.F., 166.66 ton		5.60	4.44	10.04
1800	60,000 S.F., 250.00 ton		5.50	4.69	10.19
1880	Bars and taverns, 4,000 S.F., 44.33 ton		15.55	7.20	22.75
1920	6,000 S.F., 66.50 ton		15.35	7.45	22.80
1960	10,000 S.F., 110.83 ton		12.85	5.95	18.80
2000	20,000 S.F., 221.66 ton		11.85	6.35	18.20
2040	40,000 S.F., 440 ton*				
2080	60,000 S.F., 660 ton*				
2160	Bowling alleys, 4,000 S.F., 22.66 ton		10.80	6.25	17.05
2200	6,000 S.F., 34.00 ton		9	5.50	14.50
2240	10,000 S.F., 56.66 ton		7.75	4.05	11.80

D3030 Cooling Generating Systems

D3030 115	Chilled Water, Cooling Tower Systems	COST PER S.F.		
		MAT.	INST.	TOTAL
2280	20,000 S.F., 113.33 ton	7	3.80	10.80
2320	40,000 S.F., 226.66 ton	6.60	4.57	11.17
2360	60,000 S.F., 340 ton			
2440	Department stores, 4,000 S.F., 11.66 ton	6.30	5.65	11.95
2480	6,000 S.F., 17.50 ton	6.70	4.63	11.33
2520	10,000 S.F., 29.17 ton	4.81	3.47	8.28
2560	20,000 S.F., 58.33 ton	3.95	2.58	6.53
2600	40,000 S.F., 116.66 ton	4.01	2.68	6.69
2640	60,000 S.F., 175.00 ton	4.42	4.38	8.80
2720	Drug stores, 4,000 S.F., 26.66 ton	11.55	6.45	18
2760	6,000 S.F., 40.00 ton	10	5.50	15.50
2800	10,000 S.F., 66.66 ton	9.65	5.05	14.70
2840	20,000 S.F., 133.33 ton	8	3.95	11.95
2880	40,000 S.F., 266.67 ton	7.70	5	12.70
2920	60,000 S.F., 400 ton*			
3000	Factories, 4,000 S.F., 13.33 ton	7.80	5.45	13.25
3040	6,000 S.F., 20.00 ton	6.70	4.68	11.38
3080	10,000 S.F., 33.33 ton	5.35	3.59	8.94
3120	20,000 S.F., 66.66 ton	5.20	3.24	8.44
3160	40,000 S.F., 133.33 ton	4.47	2.78	7.25
3200	60,000 S.F., 200.00 ton	4.80	4.51	9.31
3280	Food supermarkets, 4,000 S.F., 11.33 ton	6.20	5.60	11.80
3320	6,000 S.F., 17.00 ton	5.85	4.54	10.39
3360	10,000 S.F., 28.33 ton	4.71	3.44	8.15
3400	20,000 S.F., 56.66 ton	4.03	2.59	6.62
3440	40,000 S.F., 113.33 ton	3.96	2.69	6.65
3480	60,000 S.F., 170.00 ton	4.36	4.38	8.74
3560	Medical centers, 4.000 S.F., 9.33 ton	5.25	5.15	10.40
3600	6,000 S.F., 14.00 ton	5.65	4.44	10.09
3640	10,000 S.F., 23.33 ton	4.44	3.33	7.77
3680	20,000 S.F., 46.66 ton	3.52	2.52	6.04
3720	40,000 S.F., 93.33 ton	3.61	2.63	6.24
3760	60,000 S.F., 140.00 ton	3.99	4.33	8.32
3840	Offices, 4,000 S.F., 12.66 ton	7.50	5.35	12.85
3880	6,000 S.F., 19.00 ton	6.55	4.77	11.32
3920	10,000 S.F., 31.66 ton	5.25	3.61	8.86
3960	20,000 S.F., 63.33 ton	5.05	3.24	8.29
4000	40,000 S.F., 126.66 ton	4.65	4.24	8.89
4040	60,000 S.F., 190.00 ton	4.62	4.45	9.07
4120	Restaurants, 4,000 S.F., 20.00 ton	9.60	5.80	15.40
4160	6,000 S.F., 30.00 ton	8.05	5.15	13.20
4200	10,000 S.F., 50.00 ton	7.05	3.88	10.93
4240	20,000 S.F., 100.00 ton	6.70	3.78	10.48
4280	40,000 S.F., 200.00 ton	5.85	4.30	10.15
4320	60,000 S.F., 300.00 ton	6.20	4.85	11.05
4400	Schools and colleges, 4,000 S.F., 15.33 ton	8.70	5.60	14.30
4440	6,000 S.F., 23.00 ton	7.10	4.92	12.02
4480	10,000 S.F., 38.33 ton	5.90	3.66	9.56
4520	20,000 S.F., 76.66 ton	5.60	3.56	9.16
4560	40,000 S.F., 153.33 ton	5.25	4.33	9.58
4600	60,000 S.F., 230.00 ton	5.05	4.52	9.57

Important: See the Reference Section for critical supporting data - Reference Numbers and City Cost Indexes

D3030 Cooling Generating Systems

Chilled Water Piping   Reciprocating Chiller, Water Cooled

Condenser Water Piping

System Components	QUANTITY	UNIT	COST EACH		
			MAT.	INST.	TOTAL
SYSTEM D3030 130 1010					
CHILLER, RECIPROCATING, WATER COOLED, STD. CONTROLS, 60 TON					
Water chiller, reciprocating, water cooled, 60 ton cooling	1.000	Ea.	41,900	6,550	48,450
Pipe, black steel, welded, Sch. 40, on yoke & roll hgrs, 10' O.C., 4" dia	40.000	L.F.	396	935.20	1,331.20
Elbow, 90°, weld joint, butt, 4" pipe size	8.000	Ea.	164	1,388.80	1,552.80
Flange, welding neck, 150 lb., 4" pipe size	16.000	Ea.	456	1,388.80	1,844.80
Valves, iron body, butterfly, lug type, lever actuator, 4" size	4.000	Ea.	460	624	1,084
Expansion joints, bellows type, flange spool, 6" F to F, 4" dia	4.000	Ea.	908	374	1,282
T-O-L socket, weld joint, 1/4" pipe size, nozzle	4.000	Ea.	18.96	151.32	170.28
T-O-L socket, weld joint, 3/4" pipe size, nozzle	4.000	Ea.	22.20	164.76	186.96
Thermometers, stem type, 9" case, 8" stem, 3/4" NPT	4.000	Ea.	532	62.40	594.40
Gauges, pressure or vacuum, 3-1/2" diameter dial	4.000	Ea.	92	54.60	146.60
Insulation, fiberglass pipe covering, 1-1/2" wall, 4" IPS	20.000	L.F.	92	109	201
Chiller balancing	1.000	Ea.		370	370
TOTAL			45,041.16	12,172.88	57,214.04

D3030 130	Reciprocating Chiller, Water Cooled	COST EACH		
		MAT.	INST.	TOTAL
1010	Chiller, reciprocating, water cooled, std. controls, 60 Ton	45,000	12,200	57,200
1020	100 Ton	63,000	18,300	81,300
1030	150 Ton	84,000	20,100	104,100
1040	200 Ton	101,500	21,100	122,600

D3030 135	Reciprocating Chiller, Air Cooled	COST EACH		
		MAT.	INST.	TOTAL
1010	Chiller, reciprocating, air cooled, std. controls, 20 Ton	22,300	7,175	29,475
1020	30 Ton	27,000	7,525	34,525
1030	40 Ton	36,800	8,025	44,825
1040	60 Ton	53,000	9,050	62,050
1050	70 Ton	60,500	9,300	69,800
1060	80 Ton	65,000	9,450	74,450
1070	90 Ton	70,500	9,600	80,100
1080	100 Ton	78,500	9,775	88,275
1090	110 Ton	86,000	10,500	96,500
1100	130 Ton	94,500	10,700	105,200
1110	150 Ton	109,000	10,800	119,800
1120	175 Ton	116,000	11,000	127,000
1130	210 Ton	143,500	11,900	155,400

SERVICES

D

Chilled Water Piping

Centrifugal Chiller, Water Cooled, Hermetic

Condenser Water Piping

System Components	QUANTITY	UNIT	COST EACH		
			MAT.	INST.	TOTAL
SYSTEM D3030 140 1010					
CHILLER, CENTRIF., WATER COOLED, PKGD. HERMETIC, STD. CONTROLS, 200 TON					
Water chiller, centrifugal, water cooled, not incl tower, 200 ton	1.000	Ea.	77,500	13,100	90,600
Pipe, black steel, welded, Sch. 40, on yoke & roll hgrs, 10' o.c., 6" dia	40.000	L.F.	690	1,437.60	2,127.60
Elbow, 90°, weld joint, 6" pipe size	8.000	Ea.	384	2,084.40	2,468.40
Flange, welding neck, 150 lb., 6" pipe size	16.000	Ea.	704	2,076.80	2,780.80
Valves, iron body, butterfly, lug type, lever actuator, 6" size	4.000	Ea.	764	972	1,736
Expansion joints, bellows type, flange spool, 6" F to F, 6" dia	4.000	Ea.	1,148	464	1,612
T-O-L socket, weld jiont, 1/4" pipe size, nozzle	4.000	Ea.	18.96	151.32	170.28
T-O-L socket, weld joint, 3/4" pipe size, nozzle	4.000	Ea.	22.20	164.76	186.96
Thermometers, stem type, 9" case, 8" stem, 3/4" NPT	4.000	Ea.	532	62.40	594.40
Gauges, pressure or vacuum, 3-1/2" diameter dial	4.000	Ea.	92	54.60	146.60
Insulation, fiberglass pipe covering, 1-1/2" wall, 6" IPS	40.000	L.F.	210	278	488
Chiller balancing	1.000	Ea.		370	370
TOTAL			82,065.16	21,215.88	103,281.04

D3030 140	Centrifugal Chiller, Water Cooled	COST EACH		
		MAT.	INST.	TOTAL
1010	Chiller, centrif., water cooled, pkgd. hermetic, std controls, 200 Ton	82,000	21,200	103,200
1020	400 Ton	147,000	29,200	176,200
1030	1,000 Ton	373,000	44,000	417,000
1040	1,500 Ton	549,500	54,000	603,500

SERVICES D

D3030 Cooling Generating Systems

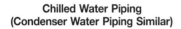

Chilled Water Piping
(Condenser Water Piping Similar)

Gas Absorption Chiller, Water Cooled

Gas Piping

System Components	QUANTITY	UNIT	COST EACH		
			MAT.	INST.	TOTAL
SYSTEM D3030 145 1010					
CHILLER, GAS FIRED ABSORPTION, WATER COOLED, STD. CONTROLS, 800 TON					
Absorption water chiller, gas fired, water cooled, 800 ton	1.000	Ea.	549,500	19,800	569,300
Pipe, black steel, weld, Sch. 40, on yoke & roll hangers, 10' O.C., 10" dia	20.000	L.F.	950	1,083.20	2,033.20
Pipe, black steel, welded, Sch. 40, (no hgrs incl) 14" dia.	44.000	L.F.	1,936	3,821.40	5,757.40
Pipe, blacksteel, welded, Sch. 40, on yoke & roll hangers, 10' O.C., 4" dia	20.000	L.F.	198	467.60	665.60
Elbow, 90°, weld joint, 10" pipe size	4.000	Ea.	664	1,738	2,402
Elbow, 90°, weld joint, 14" pipe size	6.000	Ea.	1,980	3,894	5,874
Elbow, 90°, weld joint, 4" pipe size	3.000	Ea.	61.50	520.80	582.30
Flange, welding neck, 150 lb., 10" pipe size	8.000	Ea.	992	1,733.20	2,725.20
Flange, welding neck, 150 lb., 14" pipe size	14.000	Ea.	3,640	4,053	7,693
Flange, welding neck, 150 lb., 4" pipe size	3.000	Ea.	85.50	260.40	345.90
Valves, iron body, butterfly, lug type, gear operated, 10" size	2.000	Ea.	1,070	610	1,680
Valves, iron body, butterfly, lug type, gear operated, 14" size	5.000	Ea.	4,475	2,650	7,125
Valves, semi-steel, lubricated plug valve, flanged, 4" pipe size	1.000	Ea.	365	260	625
Expansion joints, bellows type, flange spool, 6" F to F, 10" dia	2.000	Ea.	880	314	1,194
Expansion joints, bellows type, flange spool, 10" F to F, 14" dia	2.000	Ea.	1,430	414	1,844
T-O-L socket, weld joint, 1/4" pipe size, nozzle	4.000	Ea.	18.96	151.32	170.28
T-O-L socket, weld joint, 3/4" pipe size, nozzle	4.000	Ea.	22.20	164.76	186.96
Thermometers, stem type, 9" case, 8" stem, 3/4" NPT	4.000	Ea.	532	62.40	594.40
Gauges, pressure or vacuum, 3-1/2" diameter dial	4.000	Ea.	92	54.60	146.60
Insulation, fiberglass pipe covering, 2" wall, 10" IPS	20.000	L.F.	239	218	457
Vent chimney, all fuel, pressure tight, double wall, SS, 12" dia	20.000	L.F.	1,220	354	1,574
Chiller balancing	1.000	Ea.		370	370
TOTAL			570,351.16	42,994.68	613,345.84

D3030 145	Gas Absorption Chiller, Water Cooled	MAT.	INST.	TOTAL
1010	Chiller, gas fired absorption, water cooled, std. ctrls, 800 Ton	570,500	43,000	613,500
1020	1000 Ton	721,500	55,500	777,000

D3030 150	Steam Absorption Chiller, Water Cooled	COST EACH		
		MAT.	INST.	TOTAL
1010	Chiller, steam absorption, water cooled, std. ctrls, 750 Ton	383,500	51,000	434,500
1020	955 Ton	445,500	61,000	506,500
1030	1465 Ton	643,000	70,000	713,000
1040	1660 Ton	768,500	79,500	848,000

SERVICES

D

455

D3030 Cooling Generating Systems

System Description: Rooftop single zone units are electric cooling and gas heat. Duct systems are low velocity, galvanized steel supply and return. Price variations between sizes are due to several factors. Jumps in the cost of the rooftop unit occur when the manufacturer shifts from the largest capacity unit on a small frame to the smallest capacity on the next larger frame, or changes from one compressor to two. As the unit capacity increases for larger areas the duct distribution grows in proportion. For most applications there is a tradeoff point where it is less expensive and more efficient to utilize smaller units with short simple distribution systems. Larger units also require larger initial supply and return ducts which can create a space problem. Supplemental heat may be desired in colder locations. The table below is based on one unit supplying the area listed. The 10,000 S.F. unit for bars and taverns is not listed because a nominal 110 ton unit would be required and this is above the normal single zone rooftop capacity.

System Components				COST EACH		
	QUANTITY	UNIT		MAT.	INST.	TOTAL
SYSTEM D3030 202 1280						
ROOFTOP, SINGLE ZONE, AIR CONDITIONER						
APARTMENT CORRIDORS, 500 S.F., .92 TON						
Rooftop air-conditioner, 1 zone, electric cool, standard controls, curb	1.000	Ea.		2,058.50	517.50	2,576
Ductwork package for rooftop single zone units	1.000	System		229.08	745.20	974.28
TOTAL				2,287.58	1,262.70	3,550.28
COST PER S.F.				4.58	2.53	7.11
*Size would suggest multiple units						

D3030 202	Rooftop Single Zone Unit Systems		COST PER S.F.		
			MAT.	INST.	TOTAL
1260	Rooftop, single zone, air conditioner				
1280	Apartment corridors, 500 S.F., .92 ton		4.60	2.55	7.15
1320	1,000 S.F., 1.83 ton	R15700 -020	4.55	2.51	7.06
1360	1500 S.F., 2.75 ton		3.05	2.08	5.13
1400	3,000 S.F., 5.50 ton		2.81	2	4.81
1440	5,000 S.F., 9.17 ton		2.97	1.82	4.79
1480	10,000 S.F., 18.33 ton		2.59	1.71	4.30
1560	Banks or libraries, 500 S.F., 2.08 ton		10.35	5.70	16.05
1600	1,000 S.F., 4.17 ton		6.95	4.73	11.68
1640	1,500 S.F., 6.25 ton		6.35	4.55	10.90
1680	3,000 S.F., 12.50 ton		6.75	4.14	10.89
1720	5,000 S.F., 20.80 ton		5.85	3.89	9.74
1760	10,000 S.F., 41.67 ton		5.85	3.87	9.72
1840	Bars and taverns, 500 S.F. 5.54 ton		15.55	7.60	23.15
1880	1,000 S.F., 11.08 ton		16.55	6.50	23.05
1920	1,500 S.F., 16.62 ton		12.45	6.10	18.55
1960	3,000 S.F., 33.25 ton		14.65	5.80	20.45
2000	5,000 S.F., 55.42 ton		13.70	5.80	19.50
2040	10,000 S.F., 110.83 ton*				
2080	Bowling alleys, 500 S.F., 2.83 ton		9.40	6.40	15.80
2120	1,000 S.F., 5.67 ton		8.65	6.20	14.85
2160	1,500 S.F., 8.50 ton		9.15	5.60	14.75
2200	3,000 S.F., 17.00 ton		7.10	5.40	12.50
2240	5,000 S.F., 28.33 ton		8.20	5.25	13.45

Important: See the Reference Section for critical supporting data - Reference Numbers and City Cost Indexes

D3030 Cooling Generating Systems

D3030 202	Rooftop Single Zone Unit Systems	COST PER S.F.		
		MAT.	INST.	TOTAL
2280	10,000 S.F., 56.67 ton	7.70	5.25	12.95
2360	Department stores, 500 S.F., 1.46 ton	7.25	4.01	11.26
2400	1,000 S.F., 2.92 ton	4.85	3.32	8.17
2440	1,500 S.F., 4.37 ton	4.46	3.18	7.64
2480	3,000 S.F., 8.75 ton	4.73	2.89	7.62
2520	5,000 S.F., 14.58 ton	3.65	2.78	6.43
2560	10,000 S.F., 29.17 ton	4.22	2.71	6.93
2640	Drug stores, 500 S.F., 3.33 ton	11.05	7.55	18.60
2680	1,000 S.F., 6.67 ton	10.20	7.25	17.45
2720	1,500 S.F., 10.00 ton	10.80	6.60	17.40
2760	3,000 S.F., 20.00 ton	9.40	6.25	15.65
2800	5,000 S.F., 33.33 ton	9.65	6.20	15.85
2840	10,000 S.F., 66.67 ton	9.05	6.20	15.25
2920	Factories, 500 S.F., 1.67 ton	8.25	4.58	12.83
2960	1,000 S.F., 3.33 ton	5.55	3.78	9.33
3000	1,500 S.F., 5.00 ton	5.10	3.63	8.73
3040	3,000 S.F., 10.00 ton	5.40	3.31	8.71
3080	5,000 S.F., 16.67 ton	4.16	3.18	7.34
3120	10,000 S.F., 33.33 ton	4.82	3.09	7.91
3200	Food supermarkets, 500 S.F., 1.42 ton	7.05	3.89	10.94
3240	1,000 S.F., 2.83 ton	4.66	3.20	7.86
3280	1,500 S.F., 4.25 ton	4.34	3.09	7.43
3320	3,000 S.F., 8.50 ton	4.59	2.82	7.41
3360	5,000 S.F., 14.17 ton	3.54	2.71	6.25
3400	10,000 S.F., 28.33 ton	4.10	2.63	6.73
3480	Medical centers, 500 S.F., 1.17 ton	5.80	3.21	9.01
3520	1,000 S.F., 2.33 ton	5.80	3.20	9
3560	1,500 S.F., 3.50 ton	3.88	2.65	6.53
3600	3,000 S.F., 7.00 ton	3.57	2.54	6.11
3640	5,000 S.F., 11.67 ton	3.77	2.32	6.09
3680	10,000 S.F., 23.33 ton	3.29	2.18	5.47
3760	Offices, 500 S.F., 1.58 ton	7.90	4.34	12.24
3800	1,000 S.F., 3.17 ton	5.25	3.60	8.85
3840	1,500 S.F., 4.75 ton	4.84	3.46	8.30
3880	3,000 S.F., 9.50 ton	5.15	3.15	8.30
3920	5,000 S.F., 15.83 ton	3.96	3.01	6.97
3960	10,000 S.F., 31.67 ton	4.58	2.94	7.52
4000	Restaurants, 500 S.F., 2.50 ton	12.45	6.85	19.30
4040	1,000 S.F., 5.00 ton	7.65	5.45	13.10
4080	1,500 S.F., 7.50 ton	8.10	4.96	13.06
4120	3,000 S.F., 15.00 ton	6.25	4.77	11.02
4160	5,000 S.F., 25.00 ton	7.05	4.68	11.73
4200	10,000 S.F., 50.00 ton	6.80	4.64	11.44
4240	Schools and colleges, 500 S.F., 1.92 ton	9.55	5.25	14.80
4280	1,000 S.F., 3.83 ton	6.35	4.35	10.70
4320	1,500 S.F., 5.75 ton	5.85	4.18	10.03
4360	3,000 S.F., 11.50 ton	6.20	3.81	10.01
4400	5,000 S.F., 19.17 ton	5.40	3.59	8.99
4440	10,000 S.F., 38.33 ton	5.40	3.56	8.96
5000				
5100				
5200				
9000	Components of ductwork packages for above systems, per ton of cooling:			
9010	Ductwork; galvanized steel, 120 pounds			
9020	Insulation; fiberglass 2" thick, FRK faced, 52 SF			
9030	Diffusers; aluminum, 24" x 12", one			
9040	Registers; aluminum, return with dampers, one			

SERVICES

D

D3030 Cooling Generating Systems

System Description: Rooftop units are multizone with up to 12 zones, and include electric cooling, gas heat, thermostats, filters, supply and return fans complete. Duct systems are low velocity, galvanized steel supply and return with insulated supplies.

Multizone units cost more per ton of cooling than single zone. However, they offer flexibility where load conditions are varied due to heat generating areas or exposure to radiational heating. For example, perimeter offices on the "sunny side" may require cooling at the same time "shady side" or central offices may require heating. It is possible to accomplish similar results using duct heaters in branches of the single zone unit. However, heater location could be a problem and total system operating energy efficiency could be lower.

System Components	QUANTITY	UNIT	COST EACH		
			MAT.	INST.	TOTAL
SYSTEM D3030 206 1280					
ROOFTOP, MULTIZONE, AIR CONDITIONER					
APARTMENT CORRIDORS, 3,000 S.F., 5.50 TON					
Rooftop multizone unit, standard controls, curb	1.000	Ea.	23,980	1,199	25,179
Ductwork package for rooftop multizone units	1.000	System	1,815	8,525	10,340
TOTAL			25,795	9,724	35,519
COST PER S.F.			8.60	3.24	11.84

Note A: Small single zone unit recommended
Note B: A combination of multizone units recommended

D3030 206	Rooftop Multizone Unit Systems		COST PER S.F.		
			MAT.	INST.	TOTAL
1240	Rooftop, multizone, air conditioner				
1260	Apartment corridors, 1,500 S.F., 2.75 ton. See Note A.				
1280	3,000 S.F., 5.50 ton	R15700	8.60	3.24	11.84
1320	10,000 S.F., 18.30 ton	-020	6.35	3.13	9.48
1360	15,000 S.F., 27.50 ton		5.50	3.11	8.61
1400	20,000 S.F., 36.70 ton		5.55	3.11	8.66
1440	25,000 S.F., 45.80 ton		5.55	3.13	8.68
1520	Banks or libraries, 1,500 S.F., 6.25 ton		19.55	7.35	26.90
1560	3,000 S.F., 12.50 ton		16.50	7.20	23.70
1600	10,000 S.F., 41.67 ton		12.65	7.15	19.80
1640	15,000 S.F., 62.50 ton		8.45	7.05	15.50
1680	20,000 S.F., 83.33 ton		8.45	7.10	15.55
1720	25,000 S.F., 104.00 ton		7.95	7.05	15
1800	Bars and taverns, 1,500 S.F., 16.62 ton		42	10.60	52.60
1840	3,000 S.F., 33.24 ton		32	10.25	42.25
1880	10,000 S.F., 110.83 ton		19.40	10.20	29.60
1920	15,000 S.F., 165 ton, See Note B				
1960	20,000 S.F., 220 ton, See Note B				
2000	25,000 S.F., 275 ton, See Note B				
2080	Bowling alleys, 1,500 S.F., 8.50 ton		26.50	10	36.50
2120	3,000 S.F., 17.00 ton		22.50	9.80	32.30
2160	10,000 S.F., 56.70 ton		17.20	9.70	26.90
2200	15,000 S.F., 85.00 ton		11.50	9.60	21.10
2240	20,000 S.F., 113.00 ton		10.80	9.55	20.35

D3030 Cooling Generating Systems

D3030 206	Rooftop Multizone Unit Systems	COST PER S.F.		
		MAT.	INST.	TOTAL
2280	25,000 S.F., 140.00 ton see Note B			
2360	Department stores, 1,500 S.F., 4.37 ton, See Note A.			
2400	3,000 S.F., 8.75 ton	13.70	5.15	18.85
2440	10,000 S.F., 29.17 ton	8.85	4.95	13.80
2480	15,000 S.F., 43.75 ton	8.85	4.99	13.84
2520	20,000 S.F., 58.33 ton	5.90	4.95	10.85
2560	25,000 S.F., 72.92 ton	5.90	4.95	10.85
2640	Drug stores, 1,500 S.F., 10.00 ton	31.50	11.80	43.30
2680	3,000 S.F., 20.00 ton	23	11.35	34.35
2720	10,000 S.F., 66.66 ton	13.55	11.30	24.85
2760	15,000 S.F., 100.00 ton	12.75	11.30	24.05
2800	20,000 S.F., 135 ton, See Note B			
2840	25,000 S.F., 165 ton, See Note B			
2920	Factories, 1,500 S.F., 5 ton, See Note A			
2960	3,000 S.F., 10.00 ton	15.65	5.90	21.55
3000	10,000 S.F., 33.33 ton	10.10	5.65	15.75
3040	15,000 S.F., 50.00 ton	10.10	5.70	15.80
3080	20,000 S.F., 66.66 ton	6.75	5.65	12.40
3120	25,000 S.F., 83.33 ton	6.75	5.65	12.40
3200	Food supermarkets, 1,500 S.F., 4.25 ton, See Note A			
3240	3,000 S.F., 8.50 ton	13.30	5	18.30
3280	10,000 S.F., 28.33 ton	8.45	4.81	13.26
3320	15,000 S.F., 42.50 ton	8.60	4.84	13.44
3360	20,000 S.F., 56.67 ton	5.75	4.81	10.56
3400	25,000 S.F., 70.83 ton	5.75	4.81	10.56
3480	Medical centers, 1,500 S.F., 3.5 ton, See Note A			
3520	3,000 S.F., 7.00 ton	10.95	4.13	15.08
3560	10,000 S.F., 23.33 ton	7.40	3.97	11.37
3600	15,000 S.F., 35.00 ton	7.05	3.97	11.02
3640	20,000 S.F., 46.66 ton	7.10	3.99	11.09
3680	25,000 S.F., 58.33 ton	4.74	3.96	8.70
3760	Offices, 1,500 S.F., 4.75 ton, See Note A			
3800	3,000 S.F., 9.50 ton	14.85	5.60	20.45
3840	10,000 S.F., 31.66 ton	9.60	5.40	15
3880	15,000 S.F., 47.50 ton	9.60	5.40	15
3920	20,000 S.F., 63.33 ton	6.40	5.40	11.80
3960	25,000 S.F., 79.16 ton	6.40	5.40	11.80
4000	Restaurants, 1,500 S.F., 7.50 ton	23.50	8.85	32.35
4040	3,000 S.F., 15.00 ton	19.80	8.65	28.45
4080	10,000 S.F., 50.00 ton	15.15	8.55	23.70
4120	15,000 S.F., 75.00 ton	10.15	8.50	18.65
4160	20,000 S.F., 100.00 ton	9.60	8.45	18.05
4200	25,000 S.F., 125 ton, See Note B			
4240	Schools and colleges, 1,500 S.F., 5.75 ton	18	6.80	24.80
4280	3,000 S.F., 11.50 ton	15.20	6.65	21.85
4320	10,000 S.F., 38.33 ton	11.60	6.55	18.15
4360	15,000 S.F., 57.50 ton	7.80	6.50	14.30
4400	20,000 S.F., 76.66 ton	7.75	6.50	14.25
4440	25,000 S.F., 95.83 ton	7.35	6.50	13.85
4450				
9000	Components of ductwork packages for above systems, per ton of cooling:			
9010	Ductwork; galvanized steel, 240 pounds			
9020	Insulation; fiberglass, 2" thick, FRK faced, 104 SF			
9030	Diffusers; aluminum, 24" x 12", two			
9040	Registers; aluminum, return with dampers, one			

SERVICES

D

System Description: Self-contained, single package water cooled units include cooling tower, pump, piping allowance. Systems for 1000 S.F. and up include duct and diffusers to provide for even distribution of air. Smaller units distribute air through a supply air plenum, which is integral with the unit.

Returns are not ducted and supplies are not insulated.

Hot water or steam heating coils are included but piping to boiler and the boiler itself is not included.

Where local codes or conditions permit single pass cooling for the smaller units, deduct 10%.

System Components	QUANTITY	UNIT	COST EACH		
			MAT.	INST.	TOTAL
SYSTEM D3030 210 1300					
SELF-CONTAINED, WATER COOLED UNIT					
APARTMENT CORRIDORS, 500 S.F., .92 TON					
Self contained, water cooled, single package air conditioner unit	1.000	Ea.	1,251.60	411.60	1,663.20
Ductwork package for water or air cooled packaged units	1.000	System	50.14	570.40	620.54
Cooling tower, draw thru single flow, belt drive	1.000	Ea.	81.42	12.51	93.93
Cooling tower pumps & piping	1.000	System	40.94	29.44	70.38
TOTAL			1,424.10	1,023.95	2,448.05
COST PER S.F.			2.85	2.05	4.90

D3030 210	Self-contained, Water Cooled Unit Systems	COST PER S.F.		
		MAT.	INST.	TOTAL
1280	Self-contained, water cooled unit	2.84	2.05	4.89
1300	Apartment corridors, 500 S.F., .92 ton	2.85	2.05	4.90
1320	1,000 S.F., 1.83 ton	2.84	2.03	4.87
1360	3,000 S.F., 5.50 ton	2.26	1.71	3.97
1400	5,000 S.F., 9.17 ton	2.33	1.59	3.92
1440	10,000 S.F., 18.33 ton	1.96	1.43	3.39
1520	Banks or libraries, 500 S.F., 2.08 ton	6.25	2.06	8.31
1560	1,000 S.F., 4.17 ton	5.15	3.89	9.04
1600	3,000 S.F., 12.50 ton	5.30	3.62	8.92
1640	5,000 S.F., 20.80 ton	4.47	3.24	7.71
1680	10,000 S.F., 41.66 ton	3.88	3.25	7.13
1760	Bars and taverns, 500 S.F., 5.54 ton	12.95	3.46	16.41
1800	1,000 S.F., 11.08 ton	13.70	6.15	19.85
1840	3,000 S.F., 33.25 ton	10.35	4.90	15.25
1880	5,000 S.F., 55.42 ton	9.75	5.10	14.85
1920	10,000 S.F., 110.00 ton	9.55	5.05	14.60
2000	Bowling alleys, 500 S.F., 2.83 ton	8.50	2.80	11.30
2040	1,000 S.F., 5.66 ton	7	5.30	12.30
2080	3,000 S.F., 17.00 ton	6.10	4.42	10.52
2120	5,000 S.F., 28.33 ton	5.50	4.26	9.76
2160	10,000 S.F., 56.66 ton	5.15	4.38	9.53
2200	Department stores, 500 S.F., 1.46 ton	4.38	1.44	5.82
2240	1,000 S.F., 2.92 ton	3.60	2.73	6.33
2280	3,000 S.F., 8.75 ton	3.71	2.53	6.24

R15700 -020

SERVICES D

D3030 Cooling Generating Systems

D3030 210	Self-contained, Water Cooled Unit Systems	COST PER S.F.		
		MAT.	INST.	TOTAL
2320	5,000 S.F., 14.58 ton	3.13	2.27	5.40
2360	10,000 S.F., 29.17 ton	2.83	2.19	5.02
2440	Drug stores, 500 S.F., 3.33 ton	10	3.29	13.29
2480	1,000 S.F., 6.66 ton	8.20	6.25	14.45
2520	3,000 S.F., 20.00 ton	7.15	5.20	12.35
2560	5,000 S.F., 33.33 ton	7.50	5.15	12.65
2600	10,000 S.F., 66.66 ton	6.05	5.15	11.20
2680	Factories, 500 S.F., 1.66 ton	4.98	1.65	6.63
2720	1,000 S.F. 3.37 ton	4.15	3.16	7.31
2760	3,000 S.F., 10.00 ton	4.24	2.91	7.15
2800	5,000 S.F., 16.66 ton	3.57	2.61	6.18
2840	10,000 S.F., 33.33 ton	3.22	2.52	5.74
2920	Food supermarkets, 500 S.F., 1.42 ton	4.25	1.40	5.65
2960	1,000 S.F., 2.83 ton	4.40	3.16	7.56
3000	3,000 S.F., 8.50 ton	3.49	2.65	6.14
3040	5,000 S.F., 14.17 ton	3.60	2.47	6.07
3080	10,000 S.F., 28.33 ton	2.74	2.14	4.88
3160	Medical centers, 500 S.F., 1.17 ton	3.49	1.15	4.64
3200	1,000 S.F., 2.33 ton	3.62	2.60	6.22
3240	3,000 S.F., 7.00 ton	2.88	2.18	5.06
3280	5,000 S.F., 11.66 ton	2.97	2.02	4.99
3320	10,000 S.F., 23.33 ton	2.51	1.82	4.33
3400	Offices, 500 S.F., 1.58 ton	4.74	1.56	6.30
3440	1,000 S.F., 3.17 ton	4.91	3.52	8.43
3480	3,000 S.F., 9.50 ton	4.02	2.74	6.76
3520	5,000 S.F., 15.83 ton	3.39	2.46	5.85
3560	10,000 S.F., 31.67 ton	3.06	2.38	5.44
3640	Restaurants, 500 S.F., 2.50 ton	7.50	2.47	9.97
3680	1,000 S.F., 5.00 ton	6.15	4.68	10.83
3720	3,000 S.F., 15.00 ton	6.35	4.35	10.70
3760	5,000 S.F., 25.00 ton	5.35	3.90	9.25
3800	10,000 S.F., 50.00 ton	3.27	3.59	6.86
3880	Schools and colleges, 500 S.F., 1.92 ton	5.75	1.89	7.64
3920	1,000 S.F., 3.83 ton	4.72	3.58	8.30
3960	3,000 S.F., 11.50 ton	4.87	3.33	8.20
4000	5,000 S.F., 19.17 ton	4.11	2.99	7.10
4040	10,000 S.F., 38.33 ton	3.57	2.99	6.56
4050				
9000	Components of ductwork packages for above systems, per ton of cooling:			
9010	Ductwork; galvanized steel, 108 pounds			
9020	Diffusers, aluminum, 24" x 12", two			

D3030 Cooling Generating Systems

System Description: Self-contained air cooled units with remote air cooled condenser and interconnecting tubing. Systems for 1000 S.F. and up include duct and diffusers. Smaller units distribute air directly.

Returns are not ducted and supplies are not insulated.

Potential savings may be realized by using a single zone rooftop system or through-the-wall unit, especially in the smaller capacities, if the application permits.

Hot water or steam heating coils are included but piping to boiler and the boiler itself is not included.

Condenserless models are available for 15% less where remote refrigerant source is available.

System Components	QUANTITY	UNIT	COST EACH		
			MAT.	INST.	TOTAL
SYSTEM D3030 214 1320					
SELF-CONTAINED, AIR COOLED UNIT					
APARTMENT CORRIDORS, 500 S.F., .92 TON					
Air cooled, package unit	1.000	Ea.	1,802	266.90	2,068.90
Ductwork package for water or air cooled packaged units	1.000	System	50.14	570.40	620.54
Refrigerant piping	1.000	System	128.80	363.40	492.20
Air cooled condenser, direct drive, propeller fan	1.000	Ea.	252.65	102.30	354.95
TOTAL			2,233.59	1,303	3,536.59
COST PER S.F.			4.47	2.61	7.08

D3030 214	Self-contained, Air Cooled Unit Systems	COST PER S.F.		
		MAT.	INST.	TOTAL
1300	Self-contained, air cooled unit			
1320	Apartment corridors, 500 S.F., .92 ton	4.45	2.60	7.05
1360	1,000 S.F., 1.83 ton	4.41	2.58	6.99
1400	3,000 S.F., 5.50 ton	3.60	2.41	6.01
1440	5,000 S.F., 9.17 ton	3.08	2.30	5.38
1480	10,000 S.F., 18.33 ton	2.55	2.12	4.67
1560	Banks or libraries, 500 S.F., 2.08 ton	9.75	3.28	13.03
1600	1,000 S.F., 4.17 ton	8.15	5.50	13.65
1640	3,000 S.F., 12.50 ton	7	5.25	12.25
1680	5,000 S.F., 20.80 ton	5.80	4.80	10.60
1720	10,000 S.F., 41.66 ton	5.30	4.71	10.01
1800	Bars and taverns, 500 S.F., 5.54 ton	17.80	7.15	24.95
1840	1,000 S.F., 11.08 ton	18.30	10.45	28.75
1880	3,000 S.F., 33.25 ton	13.95	9.10	23.05
1920	5,000 S.F., 55.42 ton	13.40	9.10	22.50
1960	10,000 S.F., 110.00 ton	13.55	9.05	22.60
2040	Bowling alleys, 500 S.F., 2.83 ton	13.35	4.49	17.84
2080	1,000 S.F., 5.66 ton	11.10	7.50	18.60
2120	3,000 S.F., 17.00 ton	7.90	6.55	14.45
2160	5,000 S.F., 28.33 ton	7.25	6.40	13.65
2200	10,000 S.F., 56.66 ton	7	6.40	13.40
2240	Department stores, 500 S.F., 1.46 ton	6.85	2.31	9.16

R15700-020

SERVICES D

D3030 Cooling Generating Systems

D3030 214	Self-contained, Air Cooled Unit Systems	COST PER S.F.		
		MAT.	INST.	TOTAL
2280	1,000 S.F., 2.92 ton	5.75	3.85	9.60
2320	3,000 S.F., 8.75 ton	4.90	3.66	8.56
2360	5,000 S.F., 14.58 ton	4.90	3.66	8.56
2400	10,000 S.F., 29.17 ton	3.75	3.29	7.04
2480	Drug stores, 500 S.F., 3.33 ton	15.65	5.30	20.95
2520	1,000 S.F., 6.66 ton	13.05	8.80	21.85
2560	3,000 S.F., 20.00 ton	9.30	7.70	17
2600	5,000 S.F., 33.33 ton	8.55	7.50	16.05
2640	10,000 S.F., 66.66 ton	8.30	7.55	15.85
2720	Factories, 500 S.F., 1.66 ton	7.95	2.66	10.61
2760	1,000 S.F., 3.33 ton	6.60	4.41	11.01
2800	3,000 S.F., 10.00 ton	5.60	4.19	9.79
2840	5,000 S.F., 16.66 ton	4.61	3.85	8.46
2880	10,000 S.F., 33.33 ton	4.27	3.76	8.03
2960	Food supermarkets, 500 S.F., 1.42 ton	6.70	2.25	8.95
3000	1,000 S.F., 2.83 ton	6.80	4	10.80
3040	3,000 S.F., 8.50 ton	5.55	3.74	9.29
3080	5,000 S.F., 14.17 ton	4.76	3.56	8.32
3120	10,000 S.F., 28.33 ton	3.64	3.21	6.85
3200	Medical centers, 500 S.F., 1.17 ton	5.55	1.86	7.41
3240	1,000 S.F., 2.33 ton	5.65	3.31	8.96
3280	3,000 S.F., 7.00 ton	4.59	3.08	7.67
3320	5,000 S.F., 16.66 ton	3.94	2.93	6.87
3360	10,000 S.F., 23.33 ton	3.26	2.70	5.96
3440	Offices, 500 S.F., 1.58 ton	7.45	2.51	9.96
3480	1,000 S.F., 3.16 ton	7.65	4.48	12.13
3520	3,000 S.F., 9.50 ton	5.35	3.98	9.33
3560	5,000 S.F., 15.83 ton	4.42	3.65	8.07
3600	10,000 S.F., 31.66 ton	4.07	3.57	7.64
3680	Restaurants, 500 S.F., 2.50 ton	11.70	3.96	15.66
3720	1,000 S.F., 5.00 ton	9.80	6.60	16.40
3760	3,000 S.F., 15.00 ton	8.40	6.30	14.70
3800	5,000 S.F., 25.00 ton	6.40	5.65	12.05
3840	10,000 S.F., 50.00 ton	6.40	5.65	12.05
3920	Schools and colleges, 500 S.F., 1.92 ton	9	3.03	12.03
3960	1,000 S.F., 3.83 ton	7.60	5.05	12.65
4000	3,000 S.F., 11.50 ton	6.45	4.81	11.26
4040	5,000 S.F., 19.17 ton	5.35	4.42	9.77
4080	10,000 S.F., 38.33 ton	4.88	4.33	9.21
4090				
9000	Components of ductwork packages for above systems, per ton of cooling:			
9010	Ductwork; galvanized steel, 102 pounds			
9020	Diffusers; aluminum, 24" x 12", two			

SERVICES

D

D3030 Cooling Generating Systems

General: Split systems offer several important advantages which should be evaluated when a selection is to be made. They provide a greater degree of flexibility in component selection which permits an accurate match-up of the proper equipment size and type with the particular needs of the building. This allows for maximum use of modern energy saving concepts in heating and cooling. Outdoor installation of the air cooled condensing unit allows space savings in the building and also isolates the equipment operating sounds from building occupants.

Design Assumptions: The systems below are comprised of a direct expansion air handling unit and air cooled condensing unit with interconnecting copper tubing. Ducts and diffusers are also included for distribution of air. Systems are priced for cooling only. Heat can be added as desired either by putting hot water/steam coils into the air unit or into the duct supplying the particular area of need. Gas fired duct furnaces are also available. Refrigerant liquid line is insulated.

System Components	QUANTITY	UNIT	COST EACH		
			MAT.	INST.	TOTAL
SYSTEM D3030 218 1280					
SPLIT SYSTEM, AIR COOLED CONDENSING UNIT					
APARTMENT CORRIDORS, 1,000 S.F., 1.80 TON					
Fan coil AC unit, cabinet mntd & filters direct expansion air cool	1.000	Ea.	411.75	95.89	507.64
Ductwork package, for split system, remote condensing unit	1.000	System	47.12	539.85	586.97
Refrigeration piping	1.000	System	143.66	603.90	747.56
Condensing unit, air cooled, incls compressor & standard controls	1.000	Ea.	1,061.40	384.30	1,445.70
TOTAL			1,663.93	1,623.94	3,287.87
COST PER S.F.			1.66	1.62	3.28
*Cooling requirements would lead to choosing a water cooled unit					

D3030 218	Split Systems With Air Cooled Condensing Units	COST PER S.F.		
		MAT.	INST.	TOTAL
1260	Split system, air cooled condensing unit			
1280	Apartment corridors, 1,000 S.F., 1.83 ton	1.68	1.63	3.31
1320	2,000 S.F., 3.66 ton	1.35	1.64	2.99
1360	5,000 S.F., 9.17 ton	1.80	2.06	3.86
1400	10,000 S.F., 18.33 ton	2.18	2.24	4.42
1440	20,000 S.F., 36.66 ton	1.93	2.28	4.21
1520	Banks and libraries, 1,000 S.F., 4.17 ton	3.10	3.73	6.83
1560	2,000 S.F., 8.33 ton	4.12	4.69	8.81
1600	5,000 S.F., 20.80 ton	4.96	5.10	10.06
1640	10,000 S.F., 41.66 ton	4.40	5.20	9.60
1680	20,000 S.F., 83.32 ton	4.11	5.40	9.51
1760	Bars and taverns, 1,000 S.F., 11.08 ton	10.25	7.35	17.60
1800	2,000 S.F., 22.16 ton	15.30	8.80	24.10
1840	5,000 S.F., 55.42 ton	10.70	8.25	18.95
1880	10,000 S.F., 110.84 ton	9.75	8.55	18.30
1920	20,000 S.F., 220 ton*			
2000	Bowling alleys, 1,000 S.F., 5.66 ton	4.18	7	11.18
2040	2,000 S.F., 11.33 ton	5.60	6.35	11.95
2080	5,000 S.F., 28.33 ton	6.80	6.95	13.75
2120	10,000 S.F., 56.66 ton	6	7.05	13.05
2160	20,000 S.F., 113.32 ton	5.75	7.60	13.35
2320	Department stores, 1,000 S.F., 2.92 ton	2.26	2.56	4.82
2360	2,000 S.F., 5.83 ton	2.16	3.59	5.75
2400	5,000 S.F., 14.58 ton	2.89	3.28	6.17

R15700 -020

D30 HVAC

D3030 Cooling Generating Systems

D3030 218	Split Systems With Air Cooled Condensing Units	COST PER S.F.		
		MAT.	INST.	TOTAL
2440	10,000 S.F., 29.17 ton	3.49	3.57	7.06
2480	20,000 S.F., 58.33 ton	3.08	3.64	6.72
2560	Drug stores, 1,000 S.F., 6.66 ton	4.93	8.20	13.13
2600	2,000 S.F., 13.32 ton	6.60	7.50	14.10
2640	5,000 S.F., 33.33 ton	7.05	8.30	15.35
2680	10,000 S.F., 66.66 ton	6.65	8.65	15.30
2720	20,000 S.F., 133.32 ton*			
2800	Factories, 1,000 S.F., 3.33 ton	2.57	2.92	5.49
2840	2,000 S.F., 6.66 ton	2.46	4.10	6.56
2880	5,000 S.F., 16.66 ton	3.98	4.09	8.07
2920	10,000 S.F., 33.33 ton	3.51	4.15	7.66
2960	20,000 S.F., 66.66 ton	3.34	4.34	7.68
3040	Food supermarkets, 1,000 S.F., 2.83 ton	2.18	2.49	4.67
3080	2,000 S.F., 5.66 ton	2.10	3.49	5.59
3120	5,000 S.F., 14.66 ton	2.81	3.18	5.99
3160	10,000 S.F., 28.33 ton	3.39	3.47	6.86
3200	20,000 S.F., 56.66 ton	2.99	3.53	6.52
3280	Medical centers, 1,000 S.F., 2.33 ton	1.84	2.01	3.85
3320	2,000 S.F., 4.66 ton	1.72	2.88	4.60
3360	5,000 S.F., 11.66 ton	2.31	2.63	4.94
3400	10,000 S.F., 23.33 ton	2.79	2.86	5.65
3440	20,000 S.F., 46.66 ton	2.46	2.91	5.37
3520	Offices, 1,000 S.F., 3.17 ton	2.44	2.78	5.22
3560	2,000 S.F., 6.33 ton	2.34	3.90	6.24
3600	5,000 S.F., 15.83 ton	3.14	3.56	6.70
3640	10,000 S.F., 31.66 ton	3.78	3.88	7.66
3680	20,000 S.F., 63.32 ton	3.16	4.12	7.28
3760	Restaurants, 1,000 S.F., 5.00 ton	3.71	6.15	9.86
3800	2,000 S.F., 10.00 ton	4.95	5.65	10.60
3840	5,000 S.F., 25.00 ton	6	6.15	12.15
3880	10,000 S.F., 50.00 ton	5.30	6.25	11.55
3920	20,000 S.F., 100.00 ton	5.10	6.75	11.85
4000	Schools and colleges, 1,000 S.F., 3.83 ton	2.84	3.43	6.27
4040	2,000 S.F., 7.66 ton	3.80	4.30	8.10
4080	5,000 S.F., 19.17 ton	4.58	4.69	9.27
4120	10,000 S.F., 38.33 ton	4.05	4.78	8.83
4160	20,000 S.F., 76.66 ton	3.86	4.96	8.82
5000				
5100				
5200				
9000	Components of ductwork packages for above systems, per ton of cooling:			
9010	Ductwork; galvanized steel, 102 pounds			
9020	Diffusers; aluminum, 24" x 12", two			

SERVICES

D

D3030 Cooling Generating Systems

Computer rooms impose special requirements on air conditioning systems. A prime requirement is reliability, due to the potential monetary loss that could be incurred by a system failure. A second basic requirement is the tolerance of control with which temperature and humidity are regulated, and dust eliminated. As the air conditioning system reliability is so vital, the additional cost of reserve capacity and redundant components is often justified.

System Descriptions: Computer areas may be environmentally controlled by one of three methods as follows:

1. Self-contained Units
   These are units built to higher standards of performance and reliability. They

usually contain alarms and controls to indicate component operation failure, filter change, etc. It should be remembered that these units in the room will occupy space that is relatively expensive to build and that all alterations and service of the equipment will also have to be accomplished within the computer area.

2. Decentralized Air Handling Units In operation these are similar to the self-contained units except that their cooling capability comes from remotely located refrigeration equipment as refrigerant or chilled water. As no compressors or refrigerating equipment are required in the air units, they are smaller and require less service than

self-contained units. An added plus for this type of system occurs if some of the computer components themselves also require chilled water for cooling.

3. Central System Supply Cooling is obtained from a central source which, since it is not located within the computer room, may have excess capacity and permit greater flexibility without interfering with the computer components. System performance criteria must still be met.

Note: The costs shown below do not include an allowance for ductwork or piping.

D3030 222	Computer Room Cooling Units	COST EACH		
		MAT.	INST.	TOTAL
0560	Computer room unit, air cooled, includes remote condenser			
0580	3 ton	8,275	1,575	9,850
0600	5 ton	11,100	1,750	12,850
0620	8 ton	23,100	2,900	26,000
0640	10 ton	24,000	3,150	27,150
0660	15 ton	26,800	3,575	30,375
0680	20 ton	32,000	4,225	36,225
0700	23 ton	33,600	4,375	37,975
0800	Chilled water, for connection to existing chiller system			
0820	5 ton	9,325	1,050	10,375
0840	8 ton	13,300	1,575	14,875
0860	10 ton	14,800	1,600	16,400
0880	15 ton	15,900	1,650	17,550
0900	20 ton	16,300	1,700	18,000
0920	23 ton	17,100	1,875	18,975
1000	Glycol system, complete except for interconnecting tubing			
1020	3 ton	10,100	1,975	12,075
1040	5 ton	13,200	2,075	15,275
1060	8 ton	28,500	3,425	31,925
1080	10 ton	30,900	3,750	34,650
1100	15 ton	37,500	4,700	42,200
1120	20 ton	40,100	5,100	45,200
1140	23 ton	43,700	5,550	49,250
1240	Water cooled, not including condenser water supply or cooling tower			
1260	3 ton	7,475	1,275	8,750
1280	5 ton	10,500	1,450	11,950
1300	8 ton	21,700	2,375	24,075
1320	15 ton	26,500	2,900	29,400
1340	20 ton	31,300	3,225	34,525
1360	23 ton	33,200	3,600	36,800

SERVICES D

D3030 Cooling Generating Systems

Gas Cock

Roof Curb

Gas Piping

Rooftop Air Conditioner Unit

System Components	QUANTITY	UNIT	COST EACH		
			MAT.	INST.	TOTAL
SYSTEM D3030 228 1010					
A/C, ROOFTOP, DX COOL, GAS HEAT, CURB, ECONOMIZER, FILTERS., 5 TON					
Roof top A/C, curb, economizer, sgl zone, elec cool, gas ht, 5 ton, 112 MBH	1.000	Ea.	6,400	1,400	7,800
Pipe, black steel, Sch 40, threaded, W/cplgs & hangers, 10' OC, 1" dia	20.000	L.F.	43.47	172.20	215.67
Elbow, 90°, black, straight, 3/4" dia.	3.000	Ea.	4.56	93	97.56
Elbow, 90°, black, straight, 1" dia.	3.000	Ea.	8.10	100.50	108.60
Tee, black, reducing, 1" dia.	1.000	Ea.	4.80	54	58.80
Union, black with brass seat, 1" dia.	1.000	Ea.	9.90	36	45.90
Pipe nipple, black, 3/4" dia	2.000	Ea.	3.24	14.20	17.44
Pipe nipple, black, 1" dia	2.000	Ea.	2.07	8.20	10.27
Cap, black, 1" dia.	1.000	Ea.	2.12	14.45	16.57
Gas cock, brass, 1" size	1.000	Ea.	13.85	23	36.85
Control system, pneumatic, Rooftop A/C unit	1.000	Ea.	2,050	1,550	3,600
Rooftop unit heat/cool balancing	1.000	Ea.		305	305
TOTAL			8,542.11	3,770.55	12,312.66

D3030 228	Rooftop Air Conditioner, Const. Volume	COST EACH		
		MAT.	INST.	TOTAL
1010	A/C, Rooftop, DX cool, gas heat, curb, economizer, fltrs, 5 Ton	8,550	3,775	12,325
1020	7-1/2 Ton	12,300	3,925	16,225
1030	12-1/2 Ton	17,500	4,300	21,800
1040	18 Ton	20,000	4,725	24,725
1050	25 Ton	29,700	5,375	35,075
1060	40 Ton	48,400	7,175	55,575

D3030 230	Rooftop Air Conditioner, Variable Air Volume	COST EACH		
		MAT.	INST.	TOTAL
1010	A/C, Rooftop, DX cool, gas heat, curb, ecmizr, fltrs, VAV, 12-1/2 Ton	20,200	4,600	24,800
1020	18 Ton	23,000	5,050	28,050
1030	25 Ton	34,100	5,825	39,925
1040	40 Ton	56,500	7,875	64,375
1050	60 Ton	79,000	10,700	89,700
1060	80 Ton	122,500	13,400	135,900

SERVICES

D

D3030 Cooling Generating Systems

D3030 232	Packaged A/C, Elec. Ht., Const. Volume	COST EACH		
		MAT.	INST.	TOTAL
1010	A/C Packaged, DX, air cooled, electric heat, constant vol., 5 Ton	9,500	2,750	12,250
1020	10 Ton	15,600	3,775	19,375
1030	20 Ton	24,700	4,950	29,650
1040	30 Ton	33,500	5,475	38,975
1050	40 Ton	43,200	6,300	49,500
1060	50 Ton	55,000	7,050	62,050

D3030 234	Packaged Air Conditioner, Elect. Heat, VAV	COST EACH		
		MAT.	INST.	TOTAL
1010	A/C Packaged, DX, air cooled, electric heat, VAV, 10 Ton	18,800	4,725	23,525
1020	20 Ton	29,400	6,250	35,650
1030	30 Ton	39,500	6,775	46,275
1040	40 Ton	52,000	7,775	59,775
1050	50 Ton	64,000	8,650	72,650
1060	60 Ton	74,000	10,200	84,200

D3030 236	Packaged A/C, Hot Wtr. Heat, Const. Volume	COST EACH		
		MAT.	INST.	TOTAL
1010	A/C Packaged, DX, air cooled, H/W heat, constant vol., 5 Ton	8,700	4,025	12,725
1020	10 Ton	13,800	5,275	19,075
1030	20 Ton	22,200	7,100	29,300
1040	30 Ton	28,400	8,900	37,300
1050	40 Ton	38,600	10,000	48,600
1060	50 Ton	47,200	10,900	58,100

D3030 238	Packaged A/C, Hot Wtr. Heat, VAV	COST EACH		
		MAT.	INST.	TOTAL
1010	A/C Packaged, DX, air cooled, H/W heat, VAV, 10 Ton	16,800	6,125	22,925
1020	20 Ton	27,000	8,050	35,050
1030	30 Ton	34,400	10,100	44,500
1040	40 Ton	47,200	10,900	58,100
1050	50 Ton	56,500	12,700	69,200
1060	60 Ton	68,500	14,300	82,800

Important: See the Reference Section for critical supporting data - Reference Numbers and City Cost Indexes

D3030 Cooling Generating Systems

Cold Water Fill

Drain To Waste Piping

Cooling Tower

Condenser Water Piping

System Components	QUANTITY	UNIT	COST EACH		
			MAT.	INST.	TOTAL
SYSTEM D3030 310 1010					
COOLING TOWER, GALVANIZED STEEL, PACKAGED UNIT, DRAW THRU, 60 TON					
Cooling towers, draw thru, single flow, belt drive, 60 tons	1.000	Ea.	5,310	816	6,126
Pipe, black steel, welded, Sch. 40, on yoke & roll hangers, 10' OC, 4" dia.	40.000	L.F.	396	935.20	1,331.20
Tubing, copper, Type L, couplings & hangers 10' OC, 1-1/2" dia.	20.000	L.F.	93.60	167	260.60
Elbow, 90°, copper, cu x cu, 1-1/2" dia.	3.000	Ea.	15.60	100.50	116.10
Elbow, 90°, weld joint, 4" pipe size	6.000	Ea.	123	1,041.60	1,164.60
Flange, welding neck, 150 lb., 4" pipe size	6.000	Ea.	171	520.80	691.80
Valves, bronze, gate, N.R.S., soldered, 125 psi, 1-1/2" size	1.000	Ea.	49.50	33.50	83
Valves, iron body, butterfly, lug type, lever actuator, 4" size	2.000	Ea.	230	312	542
Electric heat trace system, 400 degree, 115v, 5 watts per L.F.	160.000	L.F.	1,040	128	1,168
Insulation, fiberglass pipe covering, , 1-1/2" wall, 4" IPS	40.000	L.F.	184	218	402
Insulation, fiberglass pipe covering, 1" wall, 1-1/2" IPS	20.000	L.F.	36.60	72.80	109.40
Cooling tower condenser control system	1.000	Ea.	4,100	1,800	5,900
Cooling tower balancing	1.000	Ea.		284	284
TOTAL			11,749.30	6,429.40	18,178.70

D3030 310	Galvanized Draw Through Cooling Tower	COST EACH		
		MAT.	INST.	TOTAL
1010	Cooling tower, galvanized steel, packaged unit, draw thru, 60 Ton	11,700	6,425	18,125
1020	110 Ton	15,400	7,575	22,975
1030	300 Ton	29,300	11,300	40,600
1040	600 Ton	52,000	16,900	68,900
1050	1000 Ton	66,500	24,900	91,400

D3030 320	Fiberglass Draw Through Cooling Tower	COST EACH		
		MAT.	INST.	TOTAL
1010	Cooling tower, fiberglass, packaged unit, draw thru, 60 Ton	9,500	6,425	15,925
1020	125 Ton	13,900	7,575	21,475
1030	300 Ton	24,400	11,300	35,700
1040	600 Ton	41,600	16,900	58,500
1050	1000 Ton	67,000	24,900	91,900

D3030 330	Stainless Steel Draw Through Cooling Tower	COST EACH		
		MAT.	INST.	TOTAL
1010	Cooling tower, stainless steel, packaged unit, draw thru, 60 Ton	15,000	6,425	21,425
1020	110 Ton	20,700	7,575	28,275
1030	300 Ton	46,300	11,300	57,600
1040	600 Ton	71,500	16,900	88,400
1050	1000 Ton	112,500	24,900	137,400

SERVICES

D

D3040 Distribution Systems

Two-Way Valve Piping

Three-Way Valve Piping

Built-Up Air Handling Unit

System Components	QUANTITY	UNIT	COST EACH		
			MAT.	INST.	TOTAL
SYSTEM D3040 106 1010					
AHU, FIELD FAB., BLT. UP, COOL/HEAT COILS, FLTRS., CONST. VOL., 40,000 CFM					
AHU, Built-up, cool/heat coils, filters, mix box, constant volume	1.000	Ea.	20,400	4,075	24,475
Pipe, black steel, welded, Sch. 40, on yoke & roll hgrs, 10' OC, 3" dia.	26.000	L.F.	171.60	525.20	696.80
Pipe, black steel, welded, Sch. 40, on yoke & roll hgrs, 10' OC, 4" dia	20.000	L.F.	198	467.60	665.60
Elbow, 90°, weld joint, 3" pipe size	5.000	Ea.	59.75	622.75	682.50
Elbow, 90°, weld joint, 4" pipe size	4.000	Ea.	82	694.40	776.40
Tee, welded, reducing on outlet, 3" by 2-1/2" pipe	3.000	Ea.	109.50	651	760.50
Tee, welded, reducing on outlet, 4" by 3" pipe	2.000	Ea.	93	579	672
Reducer, eccentric, weld joint, 3" pipe size	4.000	Ea.	84	434	518
Reducer, eccentric, weld joint, 4" pipe size	4.000	Ea.	110	578.60	688.60
Flange, weld neck, 150 LB, 2-1/2" pipe size	3.000	Ea.	66	163.50	229.50
Flange, weld neck, 150 lb., 3" pipe size	14.000	Ea.	336	872.20	1,208.20
Flange, weld neck, 150 lb., 4" pipe size	10.000	Ea.	285	868	1,153
Valves, iron body, butterfly, lug type, lever actuator, 3" size	2.000	Ea.	188	195	383
Valves, iron body, butterfly, lug type, lever actuator, 4" size	2.000	Ea.	230	312	542
Strainer, Y type, iron body, flanged, 125 lb., 3" pipe size	1.000	Ea.	93	175	268
Strainer, Y type, iron body, flanged, 125 lb., 4" pipe size	1.000	Ea.	170	262	432
Valve, electric motor actuated, iron body, 3 way, flanged, 2-1/2" pipe	1.000	Ea.	850	215	1,065
Valve, electric motor actuated, iron body, two way, flanged, 3" pipe size	1.000	Ea.	515	215	730
Expansion joints, bellows type, flange spool, 6" F to F, 3" dia	2.000	Ea.	424	138	562
Expansion joints, bellows type, flange spool, 6" F to F, 4" dia	2.000	Ea.	454	187	641
T-O-L socket, weld joint, 1/4" pipe size, nozzle	4.000	Ea.	18.96	151.32	170.28
T-O-L socket, weld joint, 1/2" pipe size, nozzle	12.000	Ea.	56.88	474	530.88
Thermometers, stem type, 9" case, 8" stem, 3/4" NPT	4.000	Ea.	532	62.40	594.40
Pressure/temperature relief plug, 316 SS, 3/4"OD, 7-1/2" insertion	4.000	Ea.	198	54.60	252.60
Insulation, fiberglass pipe covering, 1-1/2" wall, 3" IPS	26.000	L.F.	104.52	117	221.52
Insulation, fiberglass pipe covering, 1-1/2" wall, 4" IPS	20.000	L.F.	92	109	201
Fan coil control system	1.000	Ea.	925	405	1,330
Re-heat coil balancing	2.000	Ea.		162	162
Air conditioner balancing	1.000	Ea.		520	520
TOTAL			26,846.21	14,285.57	41,131.78

D3040 106	Field Fabricated Air Handling Unit	COST EACH		
		MAT.	INST.	TOTAL
1010	AHU, Field fab, blt up, cool/heat coils, fltrs, const vol, 40,000 CFM	26,800	14,300	41,100
1020	60,000 CFM	63,000	22,000	85,000
1030	75,000 CFM	76,500	23,100	99,600

Important: See the Reference Section for critical supporting data - Reference Numbers and City Cost Indexes

D3040 108	Field Fabricated VAV Air Handling Unit	COST EACH		
		MAT.	INST.	TOTAL
1010	AHU, Field fab, blt up, cool/heat coils, fltrs, VAV, 75,000 CFM	78,000	26,500	104,500
1020	100,000 CFM	126,000	35,000	161,000
1030	150,000 CFM	205,500	56,000	261,500

SERVICES

D

471

D3040 Distribution Systems

Chilled Water Piping   Central Station Air Handling Unit   Hot Water Piping

System Components	QUANTITY	UNIT	COST EACH MAT.	COST EACH INST.	COST EACH TOTAL
SYSTEM D3040 110 1010					
AHU, CENTRAL STATION, COOL/HEAT COILS, CONST. VOL., FLTRS., 2,000 CFM					
Central station air handling unit, chilled water, 1900 CFM	1.000	Ea.	3,622.50	822.25	4,444.75
Pipe, black steel, Sch 40, threaded, W/cplgs & hangers, 10' OC, 1" dia.	26.000	L.F.	67.28	266.50	333.78
Pipe, black steel, Sch 40, threaded, W/cplg & hangers, 10' OC, 1-1/4" dia.	20.000	L.F.	63.96	227.50	291.46
Elbow, 90°, black, straight, 1" dia.	4.000	Ea.	10.80	134	144.80
Elbow, 90°, black, straight, 1-1/4" dia.	4.000	Ea.	17.68	142	159.68
Tee, black, reducing, 1" dia.	8.000	Ea.	38.40	432	470.40
Tee, black, reducing, 1-1/4" dia.	8.000	Ea.	66.40	448	514.40
Reducer, concentric, black, 1" dia.	4.000	Ea.	14.84	116	130.84
Reducer, concentric, black, 1-1/4" dia.	4.000	Ea.	21.80	120	141.80
Union, black with brass seat, 1" dia.	4.000	Ea.	39.60	144	183.60
Union, black with brass seat, 1-1/4" dia.	2.000	Ea.	26.40	74	100.40
Pipe nipple, black, 1" dia	13.000	Ea.	13.46	53.30	66.76
Pipe nipple, black, 1-1/4" dia	12.000	Ea.	14.76	52.50	67.26
Valves, bronze, gate, N.R.S., threaded, class 150, 1" size	2.000	Ea.	77	46	123
Valves, bronze, gate, N.R.S., threaded, class 150, 1-1/4" size	2.000	Ea.	103	58	161
Strainer, Y type, bronze body, screwed, 150 lb., 1" pipe size	1.000	Ea.	20.50	25.50	46
Strainer, Y type, bronze body, screwed, 150 lb., 1-1/4" pipe size	1.000	Ea.	46	29	75
Valve, electric motor actuated, brass, 3 way, screwed, 3/4" pipe	1.000	Ea.	261	24	285
Valve, electric motor actuated, brass, 2 way, screwed, 1" pipe size	1.000	Ea.	365	23	388
Thermometers, stem type, 9" case, 8" stem, 3/4" NPT	4.000	Ea.	532	62.40	594.40
Pressure/temperature relief plug, 316 SS, 3/4"OD, 7-1/2" insertion	4.000	Ea.	198	54.60	252.60
Insulation, fiberglass pipe covering, 1" wall, 1" IPS	26.000	L.F.	39	90.48	129.48
Insulation, fiberglass pipe covering, 1" wall, 1-1/4" IPS	20.000	L.F.	34.20	72.80	107
Fan coil control system	1.000	Ea.	925	405	1,330
Re-heat coil balancing	2.000	Ea.		162	162
Multizone A/C balancing	1.000	Ea.		260	260
TOTAL			6,618.58	4,344.83	10,963.41

D3040 110	Central Station Air Handling Unit	COST EACH MAT.	COST EACH INST.	COST EACH TOTAL
1010	AHU, Central station, cool/heat coils, constant vol, fltrs, 2,000 CFM	6,625	4,350	10,975
1020	5,000 CFM	10,000	5,700	15,700
1030	10,000 CFM	13,300	8,175	21,475
1040	15,000 CFM	18,700	11,200	29,900
1050	20,000 CFM	22,300	13,100	35,400

Important: See the Reference Section for critical supporting data - Reference Numbers and City Cost Indexes

D3040 112	Central Station Air Handling Unit, VAV	COST EACH		
		MAT.	INST.	TOTAL
1010	AHU, Central station, cool/heat coils, VAV, fltrs, 5,000 CFM	12,200	5,700	17,900
1020	10,000 CFM	16,400	8,000	24,400
1030	15,000 CFM	21,900	10,600	32,500
1040	20,000 CFM	27,300	12,800	40,100
1050	30,000 CFM	40,300	14,700	55,000

SERVICES

D

Rooftop Air Handling Unit

System Components	QUANTITY	UNIT	COST EACH		
			MAT.	INST.	TOTAL
SYSTEM D3040 114 1010					
AHU, ROOFTOP, COOL/HEAT COILS, FLTRS., CONST. VOL., 2,000 CFM					
AHU, Built-up, cool/heat coils, filter, mix box, rooftop, constant volume	1.000	Ea.	4,075	785	4,860
Pipe, black steel, Sch 40, threaded, W/cplgs & hangers, 10' OC, 1" dia.	26.000	L.F.	67.28	266.50	333.78
Pipe, black steel, Sch 40, threaded, W/cplg & hangers, 10' OC, 1-1/4" dia.	20.000	L.F.	63.96	227.50	291.46
Elbow, 90°, black, straight, 1" dia.	4.000	Ea.	10.80	134	144.80
Elbow, 90°, black, straight, 1-1/4" dia.	4.000	Ea.	17.68	142	159.68
Tee, black, reducing, 1" dia.	8.000	Ea.	38.40	432	470.40
Tee, black, reducing, 1-1/4" dia.	8.000	Ea.	66.40	448	514.40
Reducer, black, concentric, 1" dia.	4.000	Ea.	14.84	116	130.84
Reducer, black, concentric, 1-1/4" dia.	4.000	Ea.	21.80	120	141.80
Union, black with brass seat, 1" dia.	4.000	Ea.	39.60	144	183.60
Union, black with brass seat, 1-1/4" dia.	2.000	Ea.	26.40	74	100.40
Pipe nipples, 1" dia	13.000	Ea.	13.46	53.30	66.76
Pipe nipples, 1-1/4" dia	12.000	Ea.	14.76	52.50	67.26
Valves, bronze, gate, N.R.S., threaded, class 150, 1" size	2.000	Ea.	77	46	123
Valves, bronze, gate, N.R.S., threaded, class 150, 1-1/4" size	2.000	Ea.	103	58	161
Strainer, Y type, bronze body, screwed, 150 lb., 1" pipe size	1.000	Ea.	20.50	25.50	46
Strainer, Y type, bronze body, screwed, 150 lb., 1-1/4" pipe size	1.000	Ea.	46	29	75
Valve, electric motor actuated, brass, 3 way, screwed, 3/4" pipe	1.000	Ea.	261	24	285
Valve, electric motor actuated, brass, 2 way, screwed, 1" pipe size	1.000	Ea.	365	23	388
Thermometers, stem type, 9" case, 8" stem, 3/4" NPT	4.000	Ea.	532	62.40	594.40
Pressure/temperature relief plug, 316 SS, 3/4"OD, 7-1/2" insertion	4.000	Ea.	198	54.60	252.60
Insulation, fiberglass pipe covering, 1" wall, 1" IPS	26.000	L.F.	39	90.48	129.48
Insulation, fiberglass pipe covering, 1" wall, 1-1/4" IPS	20.000	L.F.	34.20	72.80	107
Insulation pipe covering .016" aluminum jacket finish, add	45.000	S.F.	20.25	171.90	192.15
Fan coil control system	1.000	Ea.	925	405	1,330
Re-heat coil balancing	2.000	Ea.		162	162
Rooftop unit heat/cool balancing	1.000	Ea.		305	305
TOTAL			7,091.33	4,524.48	11,615.81

D3040 114	Rooftop Air Handling Unit	COST EACH		
		MAT.	INST.	TOTAL
1010	AHU, Rooftop, cool/heat coils, constant vol., fltrs, 2,000 CFM	7,100	4,525	11,625
1020	5,000 CFM	10,900	5,775	16,675

D30 HVAC

D3040 Distribution Systems

D3040 114	Rooftop Air Handling Unit	COST EACH		
		MAT.	INST.	TOTAL
1030	10,000 CFM	14,400	8,125	22,525
1040	15,000 CFM	20,300	11,000	31,300
1050	20,000 CFM	24,300	13,100	37,400

D3040 116	Rooftop Air Handling Unit, VAV	COST EACH		
		MAT.	INST.	TOTAL
1010	AHU, Rooftop, cool/heat coils, VAV, fltrs, 5,000 CFM	12,800	5,975	18,775
1020	10,000 CFM	17,200	8,225	25,425
1030	15,000 CFM	23,000	10,900	33,900
1040	20,000 CFM	28,600	13,200	41,800
1050	30,000 CFM	42,700	15,100	57,800

SERVICES

D

D3040   Distribution Systems

Galvanized Steel Duct

Fiberglass Duct Insulation

Flexible Fiberglass Duct

Supply Piping

Return Piping

Horizontal Fan Air Conditioning System

SERVICES D

System Components	QUANTITY	UNIT	COST EACH		
			MAT.	INST.	TOTAL
SYSTEM D3040 124 1010					
FAN COIL A/C SYSTEM, HORIZONTAL, W/HOUSING, CONTROLS, 2 PIPE, 1/2 TON					
Fan coil A/C, horizontal housing, filters, chilled water, 1/2 ton cooling	1.000	Ea.	945	98.50	1,043.50
Pipe, black steel, Sch 40, threaded, W/cplgs & hangers, 10' OC, 3/4" dia.	20.000	L.F.	34.83	152.65	187.48
Elbow, 90°, black, straight, 3/4" dia.	6.000	Ea.	9.12	186	195.12
Tee, black, straight, 3/4" dia.	2.000	Ea.	4.92	96	100.92
Union, black with brass seat, 3/4" dia.	2.000	Ea.	13.80	67	80.80
Pipe nipples, 3/4" diam	3.000	Ea.	2.43	10.65	13.08
Valves, bronze, gate, N.R.S., threaded, class 150, 3/4" size	2.000	Ea.	59	43	102
Circuit setter, balance valve, bronze body, threaded, 3/4" pipe size	1.000	Ea.	52	22	74
Insulation, fiberglass pipe covering, 1" wall, 3/4" IPS	20.000	L.F.	29.40	66.40	95.80
Ductwork, 12" x 8" fabricated, galvanized steel, 12 LF	55.000	Lb.	46.20	283.25	329.45
Insulation, ductwork, blanket type, fiberglass, 1" thk, 1-1/2 LB density	40.000	S.F.	18.80	87.20	106
Diffusers, aluminum, OB damper, ceiling, perf, 24"x24" panel size, 6"x6"	2.000	Ea.	150	54	204
Ductwork, flexible, fiberglass fabric, insulated, 1"thk, PE jacket, 6" dia	16.000	L.F.	25.92	48	73.92
Round volume control damper 6" dia.	2.000	Ea.	48	39.30	87.30
Fan coil unit balancing	1.000	Ea.		56	56
Re-heat coil balancing	1.000	Ea.		81	81
Diffuser/register, high, balancing	2.000	Ea.		156	156
TOTAL			1,439.42	1,546.95	2,986.37

D3040 118	Fan Coil A/C Unit, Two Pipe	COST EACH		
		MAT.	INST.	TOTAL
1010	Fan coil A/C system, cabinet mounted, controls, 2 pipe, 1/2 Ton	1,050	820	1,870
1020	1 Ton	1,275	920	2,195
1030	1-1/2 Ton	1,425	930	2,355
1040	2 Ton	2,125	1,125	3,250
1050	3 Ton	3,050	1,175	4,225

D3040 120	Fan Coil A/C Unit, Two Pipe, Electric Heat	COST EACH		
		MAT.	INST.	TOTAL
1010	Fan coil A/C system, cabinet mntd, elect. ht, controls, 2 pipe, 1/2 Ton	1,225	820	2,045
1020	1 Ton	1,475	920	2,395
1030	1-1/2 Ton	1,750	930	2,680
1040	2 Ton	2,600	1,150	3,750
1050	3 Ton	4,425	1,175	5,600

D3040 Distribution Systems

D3040 122	Fan Coil A/C Unit, Four Pipe	COST EACH		
		MAT.	INST.	TOTAL
1010	Fan coil A/C system, cabinet mounted, controls, 4 pipe, 1/2 Ton	1,600	1,600	3,200
1020	1 Ton	1,875	1,700	3,575
1030	1-1/2 Ton	2,075	1,700	3,775
1040	2 Ton	2,625	1,775	4,400
1050	3 Ton	3,975	1,900	5,875

D3040 124	Fan Coil A/C, Horizontal, Duct Mount, 2 Pipe	COST EACH		
		MAT.	INST.	TOTAL
1010	Fan coil A/C system, horizontal w/housing, controls, 2 pipe, 1/2 Ton	1,450	1,550	3,000
1020	1 Ton	1,875	2,250	4,125
1030	1-1/2 Ton	2,350	2,925	5,275
1040	2 Ton	3,000	3,525	6,525
1050	3 Ton	3,600	4,675	8,275
1060	3-1/2 Ton	3,800	4,850	8,650
1070	4 Ton	3,900	5,500	9,400
1080	5 Ton	4,600	7,000	11,600
1090	6 Ton	4,650	7,500	12,150
1100	7 Ton	4,675	7,800	12,475
1110	8 Ton	4,675	7,950	12,625
1120	10 Ton	5,125	8,250	13,375

D3040 126	Fan Coil A/C, Horiz., Duct Mount, 2 Pipe, Elec. Ht.	COST EACH		
		MAT.	INST.	TOTAL
1010	Fan coil A/C system, horiz. hsng, elect. ht ctrls, 2 pipe, 1/2 Ton	1,650	1,550	3,200
1020	1 Ton	2,125	2,250	4,375
1030	1-1/2 Ton	2,525	2,925	5,450
1040	2 Ton	3,300	3,525	6,825
1050	3 Ton	5,325	4,850	10,175
1060	3-1/2 Ton	6,025	4,850	10,875
1070	4 Ton	6,800	5,525	12,325
1080	5 Ton	7,775	7,000	14,775
1090	6 Ton	7,825	7,525	15,350
1100	7 Ton	7,925	7,475	15,400
1110	8 Ton	8,675	7,550	16,225
1120	10 Ton	9,250	7,775	17,025

D3040 128	Fan Coil A/C, Horizontal, Duct Mount, 4 Pipe	COST EACH		
		MAT.	INST.	TOTAL
1010	Fan coil A/C system, horiz. w/cabinet, controls, 4 pipe, 1/2 Ton	2,025	2,300	4,325
1020	1 Ton	2,550	3,100	5,650
1030	1-1/2 Ton	3,100	3,725	6,825
1040	2 Ton	3,575	4,175	7,750
1050	3 Ton	4,350	5,575	9,925
1060	3.5 Ton	5,000	6,800	11,800
1070	4 Ton	5,000	6,875	11,875
1080	5 Ton	5,975	9,475	15,450
1090	6 Ton	6,350	11,200	17,550
1100	7 Ton	7,175	12,600	19,775
1110	8 Ton	7,275	13,300	20,575
1120	10 Ton	8,750	16,300	25,050

SERVICES

D

D3040 Distribution Systems

VAV Terminal   Fan Powered VAV Terminal

System Components

System Components	QUANTITY	UNIT	COST EACH MAT.	COST EACH INST.	COST EACH TOTAL
SYSTEM D3040 132 1010					
VAV TERMINAL, COOLING ONLY, WITH ACTUATOR / CONTROLS, 200 CFM					
Mixing box variable volume 300 to 600CFM, cool only	1.000	Ea.	335.50	78.10	413.60
Ductwork, 12″ x 8″ fabricated, galvanized steel, 12 LF	55.000	Lb.	46.20	283.25	329.45
Insulation, ductwork, blanket type, fiberglass, 1″ thk, 1-1/2 LB density	40.000	S.F.	18.80	87.20	106
Diffusers, aluminum, OB damper, ceiling, perf, 24″x24″ panel size, 6″x6″	2.000	Ea.	150	54	204
Ductwork, flexible, fiberglass fabric, insulated, 1″thk, PE jacket, 6″ dia	16.000	L.F.	25.92	48	73.92
Round volume control damper 6″ dia.	2.000	Ea.	48	39.30	87.30
VAV box balancing	1.000	Ea.		52	52
Diffuser/register, high, balancing	2.000	Ea.		156	156
TOTAL			624.42	797.85	1,422.27

D3040 132	VAV Terminal, Cooling Only	COST EACH MAT.	COST EACH INST.	COST EACH TOTAL
1010	VAV Terminal, cooling only, with actuator / controls, 200 CFM	625	800	1,425
1020	400 CFM	800	1,400	2,200
1030	600 CFM	1,050	2,100	3,150
1040	800 CFM	1,125	2,550	3,675
1050	1000 CFM	1,325	2,950	4,275
1060	1250 CFM	1,425	3,750	5,175
1070	1500 CFM	1,550	4,425	5,975
1080	2000 CFM	2,025	6,275	8,300

D3040 134	VAV Terminal, Hot Water Reheat	COST EACH MAT.	COST EACH INST.	COST EACH TOTAL
1010	VAV Terminal, cooling, HW reheat, with actuator / controls, 200 CFM	1,450	1,475	2,925
1020	400 CFM	1,625	2,025	3,650
1030	600 CFM	1,875	2,750	4,625
1040	800 CFM	1,975	3,225	5,200
1050	1000 CFM	2,125	3,625	5,750
1060	1250 CFM	2,300	4,475	6,775
1070	1500 CFM	2,550	5,300	7,850
1080	2000 CFM	3,050	7,175	10,225

SERVICES D

D3040 Distribution Systems

D3040 136	Fan Powered VAV Terminal, Cooling Only	COST EACH		
		MAT.	INST.	TOTAL
1010	VAV Terminal, cooling, fan powrd, with actuator / controls, 200 CFM	1,150	800	1,950
1020	400 CFM	1,300	1,325	2,625
1030	600 CFM	1,600	2,075	3,675
1040	800 CFM	1,650	2,500	4,150
1050	1000 CFM	1,800	2,900	4,700
1060	1250 CFM	2,100	3,800	5,900
1070	1500 CFM	2,175	4,450	6,625
1080	2000 CFM	2,725	6,475	9,200

D3040 138	Fan Powered VAV Terminal, Hot Water Reheat	COST EACH		
		MAT.	INST.	TOTAL
1010	VAV Terminal, cool, HW reht, fan powrd, with actuator/ctrls, 200 CFM	1,850	1,475	3,325
1020	400 CFM	2,050	2,025	4,075
1030	600 CFM	2,300	2,775	5,075
1040	800 CFM	2,350	3,200	5,550
1050	1000 CFM	2,525	3,600	6,125
1060	1250 CFM	2,875	4,500	7,375
1070	1500 CFM	3,100	5,350	8,450
1080	2000 CFM	3,650	7,375	11,025

SERVICES

D

D3040   Distribution Systems

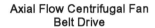

Axial Flow Centrifugal Fan Belt Drive

Belt Drive Utility Set

Centrifugal Roof Exhaust Fan Direct Drive

System Components	QUANTITY	UNIT	COST EACH		
			MAT.	INST.	TOTAL
SYSTEM D3040 220 1010					
FAN SYSTEM, IN-LINE CENTRIFUGAL, 500 CFM					
Fans, in-line centrifugal, supply/exhaust, 500 CFM, 10″ dia conn	1.000	Ea.	785	330	1,115
Louver, alum., W/screen, damper, mill fin, fxd blade, cont line, stormproof	4.050	S.F.	125.55	62.57	188.12
Motor for louver damper, electric or pneumatic	1.000	Ea.	365	31	396
Ductwork, 12″ x 8″ fabricated, galvanized steel, 60 LF	200.000	Lb.	132.60	1,287	1,419.60
Grilles, aluminum, air supply, single deflection, adjustable, 12″ x 6″	1.000	Ea.	48.60	56.40	105
Fire damper, curtain type, vertical, 12″ x 8″	1.000	Ea.	21.50	19.65	41.15
Duct access door, insulated, 10″ x 10″	1.000	Ea.	17.90	39.50	57.40
Duct access door, insulated, 16″ x 12″	1.000	Ea.	27	48	75
Pressure controller/switch, air flow controller	1.000	Ea.	121	36.50	157.50
Fan control switch and mounting base	1.000	Ea.	70.50	27.50	98
Fan balancing	1.000	Ea.		260	260
Diffuser/register balancing	3.000	Ea.		156	156
TOTAL			1,714.65	2,354.12	4,068.77

D3040 220	Centrifugal In Line Fan Systems	COST EACH		
		MAT.	INST.	TOTAL
1010	Fan system, in-line centrifugal, 500 CFM	1,725	2,350	4,075
1020	1300 CFM	2,125	4,075	6,200
1030	1500 CFM	2,450	4,700	7,150
1040	2500 CFM	3,550	14,400	17,950
1050	3500 CFM	4,450	19,400	23,850
1060	5000 CFM	5,600	28,700	34,300
1070	7500 CFM	6,800	29,400	36,200
1080	10,000 CFM	8,150	34,400	42,550

D3040 230	Utility Set Fan Systems	COST EACH		
		MAT.	INST.	TOTAL
1010	Utility fan set system, belt drive, 2000 CFM	1,525	7,625	9,150
1020	3500 CFM	2,475	13,500	15,975
1030	5000 CFM	3,125	18,700	21,825
1040	7500 CFM	3,825	21,700	25,525
1050	10,000 CFM	5,625	28,500	34,125
1060	15,000 CFM	6,175	29,100	35,275
1070	20,000 CFM	6,825	32,900	39,725

SERVICES D

D3040   Distribution Systems

D3040 240	Roof Exhaust Fan Systems	COST EACH		
		MAT.	INST.	TOTAL
1010	Roof vent. system, centrifugal, alum., galv curb, BDD, 500 CFM	665	1,550	2,215
1020	800 CFM	830	2,975	3,805
1030	1500 CFM	1,050	4,425	5,475
1040	2750 CFM	2,150	10,000	12,150
1050	3500 CFM	2,550	12,900	15,450
1060	5000 CFM	3,525	21,200	24,725
1070	8500 CFM	4,350	27,200	31,550
1080	13,800 CFM	6,125	39,500	45,625

D3040 Distribution Systems

Plate Heat Exchanger   Shell and Tube Heat Exchanger

System Components	QUANTITY	UNIT	COST EACH		
			MAT.	INST.	TOTAL
SYSTEM D3040 620 1010					
SHELL & TUBE HEAT EXCHANGER, 40 GPM					
Heat exchanger, 4 pass, 3/4" O.D. copper tubes, by steam at 10 psi, 40 GPM	1.000	Ea.	2,375	197	2,572
Pipe, black steel, Sch 40, threaded, W/couplings & hangers, 10' OC, 2" dia.	40.000	L.F.	176.75	616.10	792.85
Elbow, 90°, straight, 2" dia.	16.000	Ea.	160.80	696	856.80
Reducer, black steel, concentric, 2" dia.	2.000	Ea.	19.60	74	93.60
Valves, bronze, gate, rising stem, threaded, class 150, 2" size	6.000	Ea.	507	237	744
Valves, bronze, globe, class 150, rising stem, threaded, 2" size	2.000	Ea.	442	79	521
Strainers, Y type, bronze body, screwed, 150 lb., 2" pipe size	2.000	Ea.	118	67	185
Valve, electric motor actuated, brass, 2 way, screwed, 1-1/2" pipe size	1.000	Ea.	390	34	424
Union, black with brass seat, 2" dia.	8.000	Ea.	164	368	532
Tee, black, straight, 2" dia.	9.000	Ea.	106.20	639	745.20
Tee, black, reducing run and outlet, 2" dia.	4.000	Ea.	73.20	284	357.20
Pipe nipple, black, 2" dia.	21.000	Ea.	1.75	6.10	7.85
Thermometers, stem type, 9" case, 8" stem, 3/4" NPT	2.000	Ea.	266	31.20	297.20
Gauges, pressure or vacuum, 3-1/2" diameter dial	2.000	Ea.	46	27.30	73.30
Insulation, fiberglass pipe covering, 1" wall, 2" IPS	40.000	L.F.	79.60	152.80	232.40
Heat exchanger control system	1.000	Ea.	2,000	1,400	3,400
Coil balancing	1.000	Ea.		80.50	80.50
TOTAL			6,925.90	4,989	11,914.90

D3040 610	Heat Exchanger, Plate Type	COST EACH		
		MAT.	INST.	TOTAL
1010	Plate heat exchanger, 400 GPM	25,700	10,300	36,000
1020	800 GPM	40,400	13,100	53,500
1030	1200 GPM	52,500	17,600	70,100
1040	1800 GPM	77,000	21,900	98,900

D3040 620	Heat Exchanger, Shell & Tube	COST EACH		
		MAT.	INST.	TOTAL
1010	Shell & tube heat exchanger, 40 GPM	6,925	5,000	11,925
1020	96 GPM	12,100	9,100	21,200
1030	240 GPM	21,700	12,300	34,000
1040	600 GPM	40,600	17,200	57,800

SERVICES D

D3050 Terminal & Package Units

Cabinet Unit Heater | Gas Unit Heater | Hydronic Unit Heater

System Components	QUANTITY	UNIT	COST EACH		
			MAT.	INST.	TOTAL
SYSTEM D3050 120 1010					
SPACE HEATER, SUSPENDED, GAS FIRED, PROPELLER FAN, 20 MBH					
Space heater, propeller fan, 20 MBH output	1.000	Ea.	460	92.50	552.50
Pipe, black steel, Sch 40, threaded, W/coupling & hangers, 10' OC, 3/4" dia	20.000	L.F.	34.02	149.10	183.12
Elbow, 90°, black steel, straight, 1/2" dia.	3.000	Ea.	3.84	87	90.84
Elbow, 90°, black steel, straight, 3/4" dia.	3.000	Ea.	4.56	93	97.56
Tee, black steel, reducing, 3/4" dia.	1.000	Ea.	4.68	48	52.68
Union, black with brass seat, 3/4" dia.	1.000	Ea.	6.90	33.50	40.40
Pipe nipples, black, 1/2" dia	2.000	Ea.	2.90	13.80	16.70
Pipe nipples, black, 3/4" dia	2.000	Ea.	1.62	7.10	8.72
Pipe cap, black, 3/4" dia.	1.000	Ea.	1.75	13.55	15.30
Gas cock, brass, 3/4" size	1.000	Ea.	11.35	19.70	31.05
Thermostat, 1 set back, electric, timed	1.000	Ea.	89	54	143
Wiring, thermostat hook-up	1.000	Ea.	6.35	17.65	24
Vent chimney, prefab metal, U.L. listed, gas, double wall, galv. st, 4" dia	12.000	L.F.	57.48	137.40	194.88
Vent chimney, gas, double wall, galv. steel, elbow 90°, 4" dia.	2.000	Ea.	38.50	46	84.50
Vent chimney, gas, double wall, galv. steel, Tee, 4" dia.	1.000	Ea.	24.50	30	54.50
Vent chimney, gas, double wall, galv. steel, T cap, 4" dia.	1.000	Ea.	1.86	18.55	20.41
Vent chimney, gas, double wall, galv. steel, roof flashing, 4" dia.	1.000	Ea.	6.45	23	29.45
Vent chimney, gas, double wall, galv. steel, top, 4" dia.	1.000	Ea.	9.25	17.70	26.95
TOTAL			765.01	901.55	1,666.56

D3050 120	Unit Heaters, Gas	COST EACH		
		MAT.	INST.	TOTAL
1010	Space heater, suspended, gas fired, propeller fan, 20 MBH	765	900	1,665
1020	60 MBH	895	935	1,830
1030	100 MBH	1,125	995	2,120
1040	160 MBH	1,325	1,100	2,425
1050	200 MBH	1,600	1,200	2,800
1060	280 MBH	2,025	1,275	3,300
1070	320 MBH	2,750	1,450	4,200

D3050 130	Unit Heaters, Hydronic	COST EACH		
		MAT.	INST.	TOTAL
1010	Space heater, suspended, horiz. mount, HW, prop. fan, 20 MBH	1,325	965	2,290
1020	60 MBH	1,700	1,075	2,775

SERVICES

D

D3050 Terminal & Package Units

D3050 130	Unit Heaters, Hydronic	COST EACH		
		MAT.	INST.	TOTAL
1030	100 MBH	2,050	1,175	3,225
1040	150 MBH	2,550	1,300	3,850
1050	200 MBH	2,700	1,475	4,175
1060	300 MBH	2,950	1,725	4,675

D3050 140	Cabinet Unit Heaters, Hydronic	COST EACH		
		MAT.	INST.	TOTAL
1010	Unit heater, cabinet type, horizontal blower, hot water, 20 MBH	1,625	945	2,570
1020	60 MBH	2,350	1,050	3,400
1030	100 MBH	2,550	1,150	3,700
1040	120 MBH	2,625	1,175	3,800

SERVICES D

Important: See the Reference Section for critical supporting data - Reference Numbers and City Cost Indexes

D3050   Terminal & Package Units

Condenser Supply Water Piping

Condenser Return Water Piping

Self—Contained Air Conditioner Unit, Water Cooled

System Components	QUANTITY	UNIT	COST EACH		
			MAT.	INST.	TOTAL
SYSTEM D3050 210 1010					
A/C, SELF CONTAINED, SINGLE PKG., WATER COOLED, ELECT. HEAT, 5 TON					
Self-contained, water cooled, elect. heat, not inc tower, 5 ton, const vol	1.000	Ea.	7,400	1,025	8,425
Pipe, black steel, Sch 40, threaded, W/cplg & hangers, 10' OC, 1-1/4" dia.	20.000	L.F.	52.89	188.13	241.02
Elbow, 90°, black, straight, 1-1/4" dia.	6.000	Ea.	26.52	213	239.52
Tee, black, straight, 1-1/4" dia.	2.000	Ea.	13.60	112	125.60
Tee, black, reducing, 1-1/4" dia.	2.000	Ea.	16.60	112	128.60
Thermometers, stem type, 9" case, 8" stem, 3/4" NPT	2.000	Ea.	266	31.20	297.20
Union, black with brass seat, 1-1/4" dia.	2.000	Ea.	26.40	74	100.40
Pipe nipple, black, 1-1/4" dia	3.000	Ea.	3.69	13.13	16.82
Valves, bronze, gate, N.R.S., threaded, class 150, 1-1/4" size	2.000	Ea.	103	58	161
Circuit setter, bal valve, bronze body, threaded, 1-1/4" pipe size	1.000	Ea.	99.50	29	128.50
Insulation, fiberglass pipe covering, 1" wall, 1-1/4" IPS	20.000	L.F.	34.20	72.80	107
Control system, pneumatic, A/C Unit with heat	1.000	Ea.	2,050	1,550	3,600
Re-heat coil balancing	1.000	Ea.		81	81
Rooftop unit heat/cool balancing	1.000	Ea.		305	305
TOTAL			10,092.40	3,864.26	13,956.66

D3050 210	AC Unit, Package, Elec. Ht., Water Cooled	COST EACH		
		MAT.	INST.	TOTAL
1010	A/C, Self contained, single pkg., water cooled, elect. heat, 5 Ton	10,100	3,875	13,975
1020	10 Ton	16,800	4,750	21,550
1030	20 Ton	22,500	6,150	28,650
1040	30 Ton	36,400	6,675	43,075
1050	40 Ton	46,100	7,375	53,475
1060	50 Ton	58,000	8,825	66,825

D3050 215	AC Unit, Package, Elec. Ht., Water Cooled, VAV	COST EACH		
		MAT.	INST.	TOTAL
1010	A/C, Self contained, single pkg., water cooled, elect. ht, VAV, 10 Ton	19,500	4,750	24,250
1020	20 Ton	26,900	6,150	33,050
1030	30 Ton	41,900	6,675	48,575
1040	40 Ton	54,500	7,375	61,875
1050	50 Ton	66,500	8,825	75,325
1060	60 Ton	77,000	10,900	87,900

D SERVICES

D3050 Terminal & Package Units

D3050 220	AC Unit, Package, Hot Water Coil, Water Cooled	COST EACH		
		MAT.	INST.	TOTAL
1010	A/C, Self contn'd, single pkg., water cool, H/W ht, const. vol, 5 Ton	8,875	5,350	14,225
1020	10 Ton	14,500	6,375	20,875
1030	20 Ton	22,500	8,075	30,575
1040	30 Ton	29,200	10,400	39,600
1050	40 Ton	35,800	12,500	48,300
1060	50 Ton	44,100	15,400	59,500

D3050 225	AC Unit, Package, HW Coil, Water Cooled, VAV	COST EACH		
		MAT.	INST.	TOTAL
1010	A/C, Self contn'd, single pkg., water cool, H/W ht, VAV, 10 Ton	17,200	6,000	23,200
1020	20 Ton	23,800	7,850	31,650
1030	30 Ton	37,000	9,925	46,925
1040	40 Ton	49,700	10,800	60,500
1050	50 Ton	59,500	12,800	72,300
1060	60 Ton	71,500	13,900	85,400

Important: See the Reference Section for critical supporting data - Reference Numbers and City Cost Indexes

D3050 Terminal & Package Units

Heat Pump, Air/Air, Rooftop w/ Economizer

System Components	QUANTITY	UNIT	COST EACH		
			MAT.	INST.	TOTAL
SYSTEM D3050 245 1010					
HEAT PUMP, ROOF TOP, AIR/AIR, CURB., ECONOMIZER., SUP. ELECT. HEAT, 10 TON					
Heat pump, air to air, rooftop, curb, economizer, elect ht, 10 ton cooling	1.000	Ea.	39,800	3,050	42,850
Ductwork, 32" x 12" fabricated, galvanized steel, 16 LF	176.000	Lb.	147.84	906.40	1,054.24
Insulation, ductwork, blanket type, fiberglass, 1" thk, 1-1/2 LB density	117.000	S.F.	54.99	255.06	310.05
Rooftop unit heat/cool balancing	1.000	Ea.		305	305
TOTAL			40,002.83	4,516.46	44,519.29

D3050 230	Heat Pump, Water Source, Central Station	COST EACH		
		MAT.	INST.	TOTAL
1010	Heat pump, central station, water source, constant vol., 5 Ton	3,025	2,175	5,200
1020	10 Ton	9,325	3,000	12,325
1030	20 Ton	10,300	5,725	16,025
1040	30 Ton	18,900	8,250	27,150
1050	40 Ton	19,400	9,025	28,425
1060	50 Ton	28,300	12,100	40,400

D3050 235	Heat Pump, Water Source, Console	COST EACH		
		MAT.	INST.	TOTAL
1020	Heat pump, console, water source, 1 Ton	1,350	1,175	2,525
1030	1-1/2 Ton	1,550	1,225	2,775
1040	2 Ton	2,000	1,450	3,450
1050	3 Ton	2,150	1,550	3,700
1060	3-1/2 Ton	2,250	1,600	3,850

D3050 240	Heat Pump, Horizontal, Duct Mounted	COST EACH		
		MAT.	INST.	TOTAL
1020	Heat pump, horizontal, ducted, water source, 1 Ton	1,800	2,550	4,350
1030	1-1/2 Ton	2,200	3,225	5,425
1040	2 Ton	2,650	3,850	6,500
1050	3 Ton	3,175	5,050	8,225
1060	3-1/2 Ton	3,275	5,225	8,500

D SERVICES

D3050 245	Heat Pump, Roof Top, Air/Air	COST EACH		
		MAT.	INST.	TOTAL
1010	Heat pump, roof top, air/air, curb, economizer, sup. elect heat, 10 Ton	40,000	4,525	44,525
1020	20 Ton	57,000	7,700	64,700
1030	30 Ton	76,500	10,700	87,200
1040	40 Ton	85,500	12,400	97,900
1050	50 Ton	100,000	13,900	113,900
1060	60 Ton	105,500	15,800	121,300

SERVICES D

Important: See the Reference Section for critical supporting data - Reference Numbers and City Cost Indexes

D3050 Terminal & Package Units

Thru-Wall Heat Pump

Thru-Wall Air Conditioner

System Components	QUANTITY	UNIT	COST EACH		
			MAT.	INST.	TOTAL
SYSTEM D3050 255 1010 **A/C UNIT, THRU-THE-WALL, SUP. ELECT. HEAT, CABINET, LOUVER, 1/2 TON** A/C unit, thru-wall, electric heat., cabinet, louver, 1/2 ton	1.000	Ea.	1,150	131	1,281
TOTAL			1,150	131	1,281

D3050 255	Thru-Wall A/C Unit	COST EACH		
		MAT.	INST.	TOTAL
1010	A/C Unit, thru-the-wall, sup. elect. heat, cabinet, louver, 1/2 Ton	1,150	131	1,281
1020	3/4 Ton	1,150	157	1,307
1030	1 Ton	1,200	197	1,397
1040	1-1/2 Ton	1,500	330	1,830
1050	2 Ton	1,500	415	1,915

D3050 260	Thru-Wall Heat Pump	COST EACH		
		MAT.	INST.	TOTAL
1010	Heat pump, thru-the-wall, cabinet, louver, 1/2 Ton	1,450	98.50	1,548.50
1020	3/4 Ton	1,500	131	1,631
1030	1 Ton	1,625	197	1,822
1040	Sup. elect. heat, 1-1/2 Ton	3,400	505	3,905
1050	2 Ton	3,475	525	4,000

D SERVICES

D3060 Controls & Instrumentation

Commercial/Industrial Dust Collection System

System Components	QUANTITY	UNIT	COST EACH		
			MAT.	INST.	TOTAL
SYSTEM D3060 410 1010					
COMMERCIAL/INDUSTRIAL VACUUM DUST COLLECTION SYSTEM, 500 CFM					
Dust collection central vac unit inc. stand, filter & shaker, 500 CFM 2 HP	1.000	Ea.	3,575	415	3,990
Galvanized tubing, 16 ga., 4″ OD	80.000	L.F.	331.20	173.60	504.80
Galvanized tubing, 90° ell, 4″ dia.	3.000	Ea.	70.50	46.80	117.30
Galvanized tubing, 45° ell, 4″ dia.	4.000	Ea.	82	62.40	144.40
Galvanized tubing, TY slip fit, 4″ dia.	4.000	Ea.	200	104	304
Flexible rubber hose, 4″ dia.	32.000	L.F.	264	69.44	333.44
Galvanized air gate valve, 4″ dia.	4.000	Ea.	600	136	736
Galvanized compression coupling, 4″ dia.	36.000	Ea.	707.40	846	1,553.40
Pipe hangers, steel, 4″ pipe size	16.000	Ea.	38.24	113.60	151.84
TOTAL			5,868.34	1,966.84	7,835.18

D3060 410	Commercial/Industrial Vacuum Dust Collection	COST EACH		
		MAT.	INST.	TOTAL
1010	Commercial / industrial, vacuum dust collection system, 500 CFM	5,875	1,975	7,850
1020	1000 CFM	7,450	2,225	9,675
1030	1500 CFM	9,075	2,475	11,550
1040	3000 CFM	18,800	3,475	22,275
1050	5000 CFM	28,000	5,225	33,225

D3060 Controls & Instrumentation

Gas Fired Make-up Air Unit

Roof Fan

Grease Filters

Removable Grease Trough

Make-up Air

Exhaust Air

Commercial Kitchen Exhaust/Make-up Air Rooftop System

System Components	QUANTITY	UNIT	COST EACH		
			MAT.	INST.	TOTAL
SYSTEM D3060 420 1010					
COMMERCIAL KITCHEN EXHAUST/MAKE-UP AIR SYSTEM, ROOFTOP, GAS, 2000 CFM					
Make-up air rooftop unit, gas, 2090 CFM, 159 MBH	1.000	Ea.	5,130	442.80	5,572.80
Pipe, black steel, Sch 40, thrded, nipples, W/cplngs/hngrs, 10′ OC, 1″ dia.	20.000	L.F.	45.54	180.40	225.94
Elbow, 90°, black steel, straight, 3/4″ dia.	2.000	Ea.	3.04	62	65.04
Elbow, 90°, black steel, straight, 1″ dia.	3.000	Ea.	8.10	100.50	108.60
Tee, black steel, reducing, 1″ dia.	1.000	Ea.	4.80	54	58.80
Union, black with brass seat, 1″ dia.	1.000	Ea.	9.90	36	45.90
Pipe nipples, black, 3/4″ dia	2.000	Ea.	3.24	14.20	17.44
Pipe cap, malleable iron, black, 1″ dia.	1.000	Ea.	2.12	14.45	16.57
Gas cock, brass, 1″ size	1.000	Ea.	13.85	23	36.85
Ductwork, 20″ x 14″ fabricated, galv. steel, 30 LF	220.000	Lb.	105.60	1,045	1,150.60
Ductwork, 18″ x 10″ fabricated, galv. steel, 12 LF	72.000	Lb.	34.56	342	376.56
Ductwork, 30″ x 10″ fabricated, galv. steel, 30 LF	260.000	Lb.	124.80	1,235	1,359.80
Ductwork, 16″ x 10″ fabricated, galv. steel, 12 LF	70.000	Lb.	33.60	332.50	366.10
Kitchen ventilation, island style, water wash, Stainless Steel	5.000	L.F.	8,250	855	9,105
Control system, Make-up Air Unit / Exhauster	1.000	Ea.	2,050	1,550	3,600
Rooftop unit heat/cool balancing	1.000	Ea.		305	305
Roof fan balancing	1.000	Ea.		173	173
TOTAL			15,819.15	6,764.85	22,584

D3060 420	Kitchen Exhaust/Make-Up Air	COST EACH		
		MAT.	INST.	TOTAL
1010	Commercial kitchen exhaust/make-up air system, rooftop, gas, 2000 CFM	15,800	6,775	22,575
1020	3000 CFM	16,600	9,225	25,825
1030	5000 CFM	20,900	11,200	32,100
1040	8000 CFM	33,100	16,600	49,700
1050	12,000 CFM	49,000	20,200	69,200
1060	16,000 CFM	60,000	32,100	92,100

D3060   Controls & Instrumentation

Roof Exhaust Fan

10″ Dia. Fiberglass 90° Elbows →

10″ Dia. Fiberglass Reinforced Duct

Fume Hood →

Fume Hood Exhaust

System Components	QUANTITY	UNIT	COST EACH		
			MAT.	INST.	TOTAL
SYSTEM D3060 510 1010					
FUME HOOD EXHAUST SYSTEM, 3′ LONG, 1000 CFM					
Fume hood	3.000	L.F.	3,900	930	4,830
Ductwork, plastic, FM approved for acid fume & smoke, straight, 10″ dia	60.000	L.F.	1,860	549	2,409
Ductwork, plastic, FM approved for acid fume & smoke, elbow 90°, 10″ dia	3.000	Ea.	513	300	813
Fan, corrosive fume resistant plastic, 1630 CFM, 1/2 HP	1.000	Ea.	3,400	248	3,648
Fan control switch & mounting base	1.000	Ea.	70.50	27.50	98
Roof fan balancing	1.000	Ea.		173	173
TOTAL			9,743.50	2,227.50	11,971

D3060 510	Fume Hood Exhaust Systems	COST EACH		
		MAT.	INST.	TOTAL
1010	Fume hood exhaust system, 3′ long, 1000 CFM	9,750	2,225	11,975
1020	4′ long, 2000 CFM	11,600	2,825	14,425
1030	6′ long, 3500 CFM	16,300	4,200	20,500
1040	6′ long, 5000 CFM	22,300	4,950	27,250
1050	10′ long, 8000 CFM	29,800	6,800	36,600

SERVICES D

MECHANICAL SYSTEMS | D3090-000 | Sample Worksheet

Heating & Air Conditioning Estimate Worksheet

Equipment	Type	Table Number	Quantity	Unit Cost	Total Costs	Comments
Heat Source				$	$	
				$	$	
				$	$	
				$	$	
Pipe				$	$	
Duct				$	$	
Terminals				$	$	
Other				$	$	
				$	$	
				$	$	
				$	$	
Cold Source				$	$	
				$	$	
				$	$	
				$	$	
Pipe				$	$	
Duct				$	$	
Terminals				$	$	
Other				$	$	
SUBTOTAL				$	$	
Quality Complexity			%		$	
TOTAL					$	

D3090   Other HVAC Systems & Equipment

Vitrified Clay Garage Exhaust System

Dual Exhaust System

System Components	QUANTITY	UNIT	COST EACH MAT.	COST EACH INST.	COST EACH TOTAL
SYSTEM D3090 320 1040					
GARAGE, EXHAUST, SINGLE OUTLET, 3" EXHAUST, CARS & LIGHT TRUCKS					
A Outlet top assy. for engine exhaust system with adapters and ftngs, 3" diam	1.000	Ea.	168	27	195
F Bullnose (guide) for engine exhaust system, 3" diam	1.000	Ea.	18.70		18.70
G Galvanized flexible tubing for engine exhaust system, 3" diam	8.000	L.F.	40		40
H Adapter for metal tubing end of engine exhaust system, 3" tail pipe	1.000	Ea.	37.50		37.50
J Pipe, sewer, vitrified clay, premium joint, 8" diam	18.000	L.F.	78.48	122.76	201.24
Excavate utility trench w/chain trencher, 12 H.P., oper. walking	18.000	L.F.		23.40	23.40
Backfill utility trench by hand, incl. compaction, 8" wide 24" deep	18.000	L.F.		32.22	32.22
Stand for blower, concrete over polystyrene core, 6" high	1.000	Ea.	7.50	13.55	21.05
AC&V duct spiral reducer 10"x8"	1.000	Ea.	7.80	24.50	32.30
AC&V duct spiral reducer 12"x10"	1.000	Ea.	11.10	32.50	43.60
AC&V duct spiral preformed 45° elbow 8" diam	1.000	Ea.	3.64	28	31.64
AC&V utility fan, belt drive, 3 phase, 2000 CFM, 1 HP	2.000	Ea.	1,190	430	1,620
Safety switch, heavy duty fused, 240V, 3 pole, 30 amp	1.000	Ea.	123	132	255
Total			1,685.72	865.93	2,551.65

D3090 320	Garage Exhaust Systems	COST PER BAY MAT.	COST PER BAY INST.	COST PER BAY TOTAL
1040	Garage, single exhaust, 3" outlet, cars & light trucks, one bay	1,675	865	2,540
1060	Additional bays up to seven bays	310	126	436
1500	4" outlet, trucks, one bay	1,700	865	2,565
1520	Additional bays up to six bays	315	126	441
1600	5" outlet, diesel trucks, one bay	1,925	865	2,790
1650	Additional single bays up to six	555	146	701
1700	Two adjoining bays	1,925	865	2,790
2000	Dual exhaust, 3" outlets, pair of adjoining bays	1,975	965	2,940
2100	Additional pairs of adjoining bays	575	146	721

For information about Means Estimating Seminars, see yellow pages 11 and 12 in back of book

Important: See the Reference Section for critical supporting data - Reference Numbers and City Cost Indexes

G BUILDING SITEWORK

Trenching Systems are shown on a cost per linear foot basis. The systems include: excavation; backfill and removal of spoil; and compaction for various depths and trench bottom widths. The backfill has been reduced to accommodate a pipe of suitable diameter and bedding.

The slope for trench sides varies from 0:1 to 2:1.

The Expanded System Listing shows Trenching Systems that range from 2' to 12' in width. Depths range from 2' to 25'.

System Components			COST PER L.F.		
	QUANTITY	UNIT	EQUIP.	LABOR	TOTAL
SYSTEM G1030 805 1310					
TRENCHING, BACKHOE, 0 TO 1 SLOPE, 2' WIDE, 2' DP, 3/8 C.Y. BUCKET					
Excavation, trench, hyd. backhoe, track mtd., 3/8 C.Y. bucket	.174	C.Y.	.23	.78	1.01
Backfill and load spoil, from stockpile	.174	C.Y.	.10	.23	.33
Compaction by rammer tamper, 8" lifts, 4 passes	.014	C.Y.	.01	.03	.04
Remove excess spoil, 6 C.Y. dump truck, 2 mile roundtrip	.160	C.Y.	.62	.49	1.11
Total			.96	1.53	2.49

G1030 805	Trenching	COST PER L.F.		
		EQUIP.	LABOR	TOTAL
1310	Trenching, backhoe, 0 to 1 slope, 2' wide, 2' deep, 3/8 C.Y. bucket	.96	1.53	2.49
1320	3' deep, 3/8 C.Y. bucket	1.16	2.22	3.38
1330	4' deep, 3/8 C.Y. bucket	1.39	2.93	4.32
1340	6' deep, 3/8 C.Y. bucket	1.73	3.73	5.46
1350	8' deep, 1/2 C.Y. bucket	1.98	4.73	6.71
1360	10' deep, 1 C.Y. bucket	3.20	5.70	8.90
1400	4' wide, 2' deep, 3/8 C.Y. bucket	2.02	3.07	5.09
1410	3' deep, 3/8 C.Y. bucket	2.85	4.55	7.40
1420	4' deep, 1/2 C.Y. bucket	2.71	4.55	7.26
1430	6' deep, 1/2 C.Y. bucket	3.50	6.95	10.45
1440	8' deep, 1/2 C.Y. bucket	5.75	9.10	14.85
1450	10' deep, 1 C.Y. bucket	6.70	11.30	18
1460	12' deep, 1 C.Y. bucket	8.45	14.50	22.95
1470	15' deep, 1-1/2 C.Y. bucket	7.55	12.30	19.85
1480	18' deep, 2-1/2 C.Y. bucket	10.30	17.65	27.95
1520	6' wide, 6' deep, 5/8 C.Y. bucket	7.40	9.80	17.20
1530	8' deep, 3/4 C.Y. bucket	8.80	11.85	20.65
1540	10' deep, 1 C.Y. bucket	10	14.25	24.25
1550	12' deep, 1-1/4 C.Y. bucket	9.05	11.15	20.20
1560	16' deep, 2 C.Y. bucket	14.55	12.70	27.25
1570	20' deep, 3-1/2 C.Y. bucket	16	23	39
1580	24' deep, 3-1/2 C.Y. bucket	25.50	38	63.50
1640	8' wide, 12' deep, 1-1/4 C.Y. bucket	13.60	15.25	28.85
1650	15' deep, 1-1/2 C.Y. bucket	15.90	18.80	34.70
1660	18' deep, 2-1/2 C.Y. bucket	22.50	18.95	41.45
1680	24' deep, 3-1/2 C.Y. bucket	35.50	51	86.50
1730	10' wide, 20' deep, 3-1/2 C.Y. bucket	29	38.50	67.50
1740	24' deep, 3-1/2 C.Y. bucket	45	64	109
1780	12' wide, 20' deep, 3-1/2 C.Y. bucket	35.50	46.50	82
1790	25' deep, bucket	60	85.50	145.50
1800	1/2 to 1 slope, 2' wide, 2' deep, 3/8 C.Y. bucket	1.28	1.96	3.24
1810	3' deep, 3/8 C.Y. bucket	1.76	3.21	4.97
1820	4' deep, 3/8 C.Y. bucket	2.22	4.69	6.91
1840	6' deep, 3/8 C.Y. bucket	3.25	7.25	10.50

BUILDING SITEWORK G

G1030 Site Earthwork

G1030 805	Trenching	COST PER L.F.		
		EQUIP.	LABOR	TOTAL
1860	8' deep, 1/2 C.Y. bucket	4.41	11	15.41
1880	10' deep, 1 C.Y. bucket	8.75	15.50	24.25
2300	4' wide, 2' deep, 3/8 C.Y. bucket	2.14	3.26	5.40
2310	3' deep, 3/8 C.Y. bucket	3.53	5.40	8.93
2320	4' deep, 1/2 C.Y. bucket	3.53	5.60	9.13
2340	6' deep, 1/2 C.Y. bucket	4.95	9.70	14.65
2360	8' deep, 1/2 C.Y. bucket	8.75	13.45	22.20
2380	10' deep, 1 C.Y. bucket	11.90	19.85	31.75
2400	12' deep, 1 C.Y. bucket	13.15	21	34.15
2430	15' deep, 1-1/2 C.Y. bucket	16.50	26.50	43
2460	18' deep, 2-1/2 C.Y. bucket	28.50	31.50	60
2840	6' wide, 6' deep, 5/8 C.Y. bucket	8.40	10.85	19.25
2860	8' deep, 3/4 C.Y. bucket	12.20	16.25	28.45
2880	10' deep, 1 C.Y. bucket	14.90	21.50	36.40
2900	12' deep, 1-1/4 C.Y. bucket	14.40	17.75	32.15
2940	16' deep, 2 C.Y. bucket	22	28.50	50.50
2980	20' deep, 3-1/2 C.Y. bucket	34.50	49	83.50
3020	24' deep, 3-1/2 C.Y. bucket	62	91.50	153.50
3100	8' wide, 12' deep, 1-1/4 C.Y. bucket	19.70	22	41.70
3120	15' deep, 1-1/2 C.Y. bucket	25	29.50	54.50
3140	18' deep, 2-1/2 C.Y. bucket	38.50	32	70.50
3180	24' deep, 3-1/2 C.Y. bucket	71.50	102	173.50
3270	10' wide, 20' deep, 3-1/2 C.Y. bucket	44.50	61	105.50
3280	24' deep, 3-1/2 C.Y. bucket	81.50	116	197.50
3370	12' wide, 20' deep, 3-1/2 C.Y. bucket	51	68	119
3380	25' deep, 3-1/2 C.Y. bucket	95.50	134	229.50
3500	1 to 1 slope, 2' wide, 2' deep, 3/8 C.Y. bucket	1.50	2.29	3.79
3520	3' deep, 3/8 C.Y. bucket	2.29	4	6.29
3540	4' deep, 3/8 C.Y. bucket	2.91	6	8.91
3560	6' deep, 3/8 C.Y. bucket	4.38	9.55	13.93
3580	8' deep, 1/2 C.Y. bucket	6.15	15	21.15
3600	10' deep, 1 C.Y. bucket	14.35	24.50	38.85
3800	4' wide, 2' deep, 3/8 C.Y. bucket	2.55	3.90	6.45
3820	3' deep, 3/8 C.Y. bucket	4.51	6.85	11.36
3840	4' deep, 1/2 C.Y. bucket	4.26	6.25	10.51
3860	6' deep, 1/2 C.Y. bucket	6.15	11.60	17.75
3880	8' deep, 1/2 C.Y. bucket	11.75	17.70	29.45
3900	10' deep, 1 C.Y. bucket	15.80	25.50	41.30
3920	12' deep, 1 C.Y. bucket	22.50	37	59.50
3940	15' deep, 1-1/2 C.Y. bucket	23	34	57
3960	18' deep, 2-1/2 C.Y. bucket	40.50	41	81.50
4030	6' wide, 6' deep, 5/8 C.Y. bucket	10.20	13.15	23.35
4040	8' deep, 3/4 C.Y. bucket	15.35	21	36.35
4050	10' deep, 1 C.Y. bucket	19.45	29	48.45
4060	12' deep, 1-1/4 C.Y. bucket	19.45	26	45.45
4070	16' deep, 2 C.Y. bucket	38	37	75
4080	20' deep, 3-1/2 C.Y. bucket	50	77.50	127.50
4090	24' deep, 3-1/2 C.Y. bucket	93.50	150	243.50
4500	8' wide, 12' deep, 1-1/4 C.Y. bucket	25.50	29.50	55
4550	15' deep, 1-1/2 C.Y. bucket	33	42	75
4600	18' deep, 2-1/2 C.Y. bucket	52.50	49	101.50
4650	24' deep, 3-1/2 C.Y. bucket	103	160	263
4800	10' wide, 20' deep, 3-1/2 C.Y. bucket	63	89.50	152.50
4850	24' deep, 3-1/2 C.Y. bucket	112	170	282
4950	12' wide, 20' deep, 3-1/2 C.Y. bucket	69	95.50	164.50
4980	25' deep, 3-1/2 C.Y. bucket	136	207	343
5000	1-1/2 to 1 slope, 2' wide, 2' deep, 3/8 C.Y. bucket	1.74	2.65	4.39
5020	3' deep, 3/8 C.Y. bucket	2.79	4.73	7.52

BUILDING SITEWORK

G

G1030 Site Earthwork

G1030 805	Trenching	COST PER L.F.		
		EQUIP.	LABOR	TOTAL
5040	4' deep, 3/8 C.Y. bucket	3.53	7.05	10.58
5060	6' deep, 3/8 C.Y. bucket	5.35	11.20	16.55
5080	8' deep, 1/2 C.Y. bucket	7.50	17.70	25.20
5100	10' deep, 1 C.Y. bucket	15.80	25.50	41.30
5300	4' wide, 2' deep, 3/8 C.Y. bucket	2.41	3.67	6.08
5320	3' deep, 3/8 C.Y. bucket	4.40	6.70	11.10
5340	4' deep, 1/2 C.Y. bucket	5.10	7.35	12.45
5360	6' deep, 1/2 C.Y. bucket	7.25	13.20	20.45
5380	8' deep, 1/2 C.Y. bucket	14	20	34
5400	10' deep, 1 C.Y. bucket	18.95	29	47.95
5420	12' deep, 1 C.Y. bucket	27.50	43	70.50
5450	15' deep, 1-1/2 C.Y. bucket	27.50	38	65.50
5480	18' deep, 2-1/2 C.Y. bucket	50	45	95
5660	6' wide, 6' deep, 5/8 C.Y. bucket	11.80	14.70	26.50
5680	8' deep, 3/4 C.Y. bucket	17.85	23.50	41.35
5700	10' deep, 1 C.Y. bucket	23	32.50	55.50
5720	12' deep, 1-1/4 C.Y. bucket	22.50	28	50.50
5760	16' deep, 2 C.Y. bucket	46	39.50	85.50
5800	20' deep, 3-1/2 C.Y. bucket	60.50	88	148.50
5840	24' deep, 3-1/2 C.Y. bucket	115	171	286
6020	8' wide, 12' deep, 1-1/4 C.Y. bucket	30	32.50	62.50
6050	15' deep, 1-1/2 C.Y. bucket	39	46	85
6080	18' deep, 2-1/2 C.Y. bucket	63	52.50	115.50
6140	24' deep, 3-1/2 C.Y. bucket	125	181	306
6300	10' wide, 20' deep, 3-1/2 C.Y. bucket	75	100	175
6350	24' deep, 3-1/2 C.Y. bucket	133	190	323
6450	12' wide, 20' deep, 3-1/2 C.Y. bucket	82	106	188
6480	25' deep, 3-1/2 C.Y. bucket	162	228	390
6600	2 to 1 slope, 2' wide, 2' deep, 3/8 C.Y. bucket	2.65	2.47	5.12
6620	3' deep, 3/8 C.Y. bucket	3.21	5.25	8.46
6640	4' deep, 3/8 C.Y. bucket	4.05	7.85	11.90
6660	6' deep, 3/8 C.Y. bucket	5.55	11.85	17.40
6680	8' deep, 1/2 C.Y. bucket	8.55	19.95	28.50
6700	10' deep, 1 C.Y. bucket	18	29	47
6900	4' wide, 2' deep, 3/8 C.Y. bucket	2.41	3.67	6.08
6920	3' deep, 3/8 C.Y. bucket	4.51	6.85	11.36
6940	4' deep, 1/2 C.Y. bucket	5.70	7.85	13.55
6960	6' deep, 1/2 C.Y. bucket	8.05	14.30	22.35
6980	8' deep, 1/2 C.Y. bucket	15.55	22	37.55
7000	10' deep, 1 C.Y. bucket	21	32	53
7020	12' deep, 1 C.Y. bucket	31	48	79
7050	15' deep, 1-1/2 C.Y. bucket	31.50	42.50	74
7080	18' deep, 2-1/2 C.Y. bucket	57	50.50	107.50
7260	6' wide, 6' deep, 5/8 C.Y. bucket	13	15.75	28.75
7280	8' deep, 3/4 C.Y. bucket	19.60	25.50	45.10
7300	10' deep, 3/4 C.Y. bucket	26	36.50	62.50
7320	12' deep, 1-1/4 C.Y. bucket	25.50	31.50	57
7360	16' deep, 2 C.Y. bucket	51	44	95
7400	20' deep, 3-1/2 C.Y. bucket	68	98.50	166.50
7440	24' deep, 3-1/2 C.Y. bucket	129	192	321
7620	8' wide, 12' deep, 1-1/4 C.Y. bucket	33.50	35.50	69
7650	15' deep, 1-1/2 C.Y. bucket	43.50	50.50	94
7680	18' deep, 2-1/2 C.Y. bucket	70	58	128
7740	24' deep, 3-1/2 C.Y. bucket	140	201	341
7920	10' wide, 20' deep, 3-1/2 C.Y. bucket	83	110	193
7940	24' deep, 3-1/2 C.Y. bucket	148	210	358
8060	12' wide, 20' deep, 3-1/2 C.Y. bucket	90.50	115	205.50
8080	25' deep, 3-1/2 C.Y. bucket	179	251	430

The Pipe Bedding System is shown for various pipe diameters. Compacted bank sand is used for pipe bedding and to fill 12″ over the pipe. No backfill is included. Various side slopes are shown to accommodate different soil conditions. Pipe sizes vary from 6″ to 84″ diameter.

System Components	QUANTITY	UNIT	COST PER L.F.		
			MAT.	INST.	TOTAL
SYSTEM G1030 815 1440					
PIPE BEDDING, SIDE SLOPE 0 TO 1, 1′ WIDE, PIPE SIZE 6″ DIAMETER					
Borrow, bank sand, 2 mile haul, machine spread	.067	C.Y.	.29	.35	.64
Compaction, vibrating plate	.067	C.Y.		.12	.12
TOTAL			.29	.47	.76

G1030 815	Pipe Bedding	COST PER L.F.		
		MAT.	INST.	TOTAL
1440	Pipe bedding, side slope 0 to 1, 1′ wide, pipe size 6″ diameter	.29	.47	.76
1460	2′ wide, pipe size 8″ diameter	.63	1.03	1.66
1480	Pipe size 10″ diameter	.65	1.05	1.70
1500	Pipe size 12″ diameter	.67	1.08	1.75
1520	3′ wide, pipe size 14″ diameter	1.10	1.78	2.88
1540	Pipe size 15″ diameter	1.11	1.80	2.91
1560	Pipe size 16″ diameter	1.12	1.82	2.94
1580	Pipe size 18″ diameter	1.15	1.86	3.01
1600	4′ wide, pipe size 20″ diameter	1.66	2.69	4.35
1620	Pipe size 21″ diameter	1.67	2.71	4.38
1640	Pipe size 24″ diameter	1.72	2.78	4.50
1660	Pipe size 30″ diameter	1.75	2.85	4.60
1680	6′ wide, pipe size 32″ diameter	3.07	4.98	8.05
1700	Pipe size 36″ diameter	3.15	5.10	8.25
1720	7′ wide, pipe size 48″ diameter	4.10	6.65	10.75
1740	8′ wide, pipe size 60″ diameter	5.10	8.30	13.40
1760	10′ wide, pipe size 72″ diameter	7.40	12	19.40
1780	12′ wide, pipe size 84″ diameter	10.05	16.30	26.35
2140	Side slope 1/2 to 1, 1′ wide, pipe size 6″ diameter	.60	.99	1.59
2160	2′ wide, pipe size 8″ diameter	1	1.62	2.62
2180	Pipe size 10″ diameter	1.08	1.75	2.83
2200	Pipe size 12″ diameter	1.16	1.88	3.04
2220	3′ wide, pipe size 14″ diameter	1.66	2.70	4.36
2240	Pipe size 15″ diameter	1.70	2.75	4.45
2260	Pipe size 16″ diameter	1.75	2.85	4.60
2280	Pipe size 18″ diameter	1.86	3.01	4.87
2300	4′ wide, pipe size 20″ diameter	2.45	3.97	6.42
2320	Pipe size 21″ diameter	2.51	4.07	6.58
2340	Pipe size 24″ diameter	2.69	4.36	7.05
2360	Pipe size 30″ diameter	3.02	4.89	7.91
2380	6′ wide, pipe size 32″ diameter	4.45	7.20	11.65
2400	Pipe size 36″ diameter	4.76	7.70	12.46
2420	7′ wide, pipe size 48″ diameter	6.50	10.55	17.05
2440	8′ wide, pipe size 60″ diameter	8.45	13.70	22.15
2460	10′ wide, pipe size 72″ diameter	11.85	19.20	31.05
2480	12′ wide, pipe size 84″ diameter	15.75	25.50	41.25
2620	Side slope 1 to 1, 1′ wide, pipe size 6″ diameter	.92	1.49	2.41
2640	2′ wide, pipe size 8″ diameter	1.38	2.24	3.62

BUILDING SITEWORK

G

G1030   Site Earthwork

G1030 815	Pipe Bedding	COST PER L.F.		
		MAT.	INST.	TOTAL
2660	Pipe size 10" diameter	1.51	2.46	3.97
2680	Pipe size 12" diameter	1.66	2.70	4.36
2700	3' wide, pipe size 14" diameter	2.22	3.60	5.82
2720	Pipe size 15" diameter	2.30	3.74	6.04
2740	Pipe size 16" diameter	2.39	3.87	6.26
2760	Pipe size 18" diameter	2.57	4.17	6.74
2780	4' wide, pipe size 20" diameter	3.25	5.25	8.50
2800	Pipe size 21" diameter	3.34	5.40	8.74
2820	Pipe size 24" diameter	3.66	5.90	9.56
2840	Pipe size 30" diameter	4.27	6.90	11.17
2860	6' wide, pipe size 32" diameter	5.80	9.45	15.25
2880	Pipe size 36" diameter	6.35	10.30	16.65
2900	7' wide, pipe size 48" diameter	8.90	14.45	23.35
2920	8' wide, pipe size 60" diameter	11.80	19.10	30.90
2940	10' wide, pipe size 72" diameter	16.30	26.50	42.80
2960	12' wide, pipe size 84" diameter	21.50	35	56.50
3000	Side slope 1-1/2 to 1, 1' wide, pipe size 6" diameter	1.23	2.01	3.24
3020	2' wide, pipe size 8" diameter	1.75	2.84	4.59
3040	Pipe size 10" diameter	1.94	3.14	5.08
3060	Pipe size 12" diameter	2.15	3.49	5.64
3080	3' wide, pipe size 14" diameter	2.78	4.51	7.29
3100	Pipe size 15" diameter	2.90	4.69	7.59
3120	Pipe size 16" diameter	3.02	4.89	7.91
3140	Pipe size 18" diameter	3.28	5.30	8.58
3160	4' wide, pipe size 20" diameter	4.05	6.55	10.60
3180	Pipe size 21" diameter	4.18	6.75	10.93
3200	Pipe size 24" diameter	4.62	7.50	12.12
3220	Pipe size 30" diameter	5.55	9	14.55
3240	6' wide, pipe size 32" diameter	7.20	11.65	18.85
3260	Pipe size 36" diameter	7.95	12.90	20.85
3280	7' wide, pipe size 48" diameter	11.30	18.30	29.60
3300	8' wide, pipe size 60" diameter	15.15	24.50	39.65
3320	10' wide, pipe size 72" diameter	20.50	33.50	54
3340	12' wide, pipe size 84" diameter	27	44	71
3400	Side slope 2 to 1, 1' wide, pipe size 6" diameter	1.58	2.56	4.14
3420	2' wide, pipe size 8" diameter	2.12	3.44	5.56
3440	Pipe size 10" diameter	2.37	3.84	6.21
3460	Pipe size 12" diameter	2.64	4.28	6.92
3480	3' wide, pipe size 14" diameter	3.34	5.40	8.74
3500	Pipe size 15" diameter	3.48	5.65	9.13
3520	Pipe size 16" diameter	3.65	5.90	9.55
3540	Pipe size 18" diameter	3.99	6.50	10.49
3560	4' wide, pipe size 20" diameter	4.84	7.85	12.69
3580	Pipe size 21" diameter	5	8.10	13.10
3600	Pipe size 24" diameter	5.60	9.05	14.65
3620	Pipe size 30" diameter	6.80	11.05	17.85
3640	6' wide, pipe size 32" diameter	8.55	13.85	22.40
3660	Pipe size 36" diameter	9.55	15.50	25.05
3680	7' wide, pipe size 48" diameter	13.65	22	35.65
3700	8' wide, pipe size 60" diameter	18.50	30	48.50
3720	10' wide, pipe size 72" diameter	25	41	66
3740	12' wide, pipe size 84" diameter	33	53.50	86.50

BUILDING SITEWORK **G**

G30 Site Mechanical Utilities

G3010 Water Supply

G3010 110	Water Distribution Piping	COST PER L.F.		
		MAT.	INST.	TOTAL
2000	Piping, excav. & backfill excl., ductile iron class 250, mech. joint			
2130	4" diameter	10.20	6.90	17.10
2150	6" diameter	10.50	7.90	18.40
2160	8" diameter	13.45	9.20	22.65
2170	10" diameter	17.50	11.05	28.55
2180	12" diameter	22.50	18.95	41.45
2210	16" diameter	30	30	60
2220	18" diameter	39.50	32.50	72
3000	Tyton joint			
3130	4" diameter	7.45	6.40	13.85
3150	6" diameter	8.50	7.35	15.85
3160	8" diameter	12.10	8.65	20.75
3170	10" diameter	18.40	10.15	28.55
3180	12" diameter	19.35	12.75	32.10
3210	16" diameter	29.50	26.50	56
3220	18" diameter	33	31.50	64.50
3230	20" diameter	36	33	69
3250	24" diameter	47.50	34.50	82
4000	Copper tubing, type K			
4050	3/4" diameter	2.15	1.95	4.10
4060	1" diameter	2.73	2.44	5.17
4080	1-1/2" diameter	4.37	2.95	7.32
4090	2" diameter	7	3.39	10.39
4110	3" diameter	14	5.80	19.80
4130	4" diameter	22	8.20	30.20
4150	6" diameter	69	12.80	81.80
5000	Polyvinyl chloride class 160, S.D.R. 26			
5130	1-1/2" diameter	.65	3.32	3.97
5150	2" diameter	.98	3.98	4.96
5160	4" diameter	3.40	4.98	8.38
5170	6" diameter	7.35	5.55	12.90
5180	8" diameter	12.40	8.55	20.95
6000	Polyethylene 160 psi, S.D.R. 7			
6050	3/4" diameter	.23	1.90	2.13
6060	1" diameter	.39	2.05	2.44
6080	1-1/2" diameter	.89	2.21	3.10
6090	2" diameter	1.52	2.73	4.25

BUILDING SITEWORK

501

G3020 110	Drainage & Sewage Piping	COST PER L.F.		
		MAT.	INST.	TOTAL
2000	Piping, excavation & backfill excluded, PVC, plain			
2130	4" diameter	1.57	2.65	4.22
2150	6" diameter	2.73	2.84	5.57
2160	8" diameter	4.84	2.97	7.81
2900	Box culvert, precast, 8' long			
3000	6' x 3'	172	24.50	196.50
3020	6' x 7'	267	27	294
3040	8' x 3'	197	25.50	222.50
3060	8' x 8'	325	34	359
3080	10' x 3'	420	31	451
3100	10' x 8'	405	42.50	447.50
3120	12' x 3'	335	34	369
3140	12' x 8'	585	51	636
4000	Concrete, nonreinforced			
4150	6" diameter	3.85	7.70	11.55
4160	8" diameter	4.24	9.15	13.39
4170	10" diameter	4.70	9.45	14.15
4180	12" diameter	5.80	10.25	16.05
4200	15" diameter	6.75	11.35	18.10
4220	18" diameter	8.30	14.15	22.45
4250	24" diameter	12.50	20.50	33
4400	Reinforced, no gasket			
4580	12" diameter	11.30	9.75	21.05
4600	15" diameter	13.95	11.65	25.60
4620	18" diameter	14.05	15.70	29.75
4650	24" diameter	20.50	20.50	41
4670	30" diameter	28.50	34	62.50
4680	36" diameter	40.50	41.50	82
4690	42" diameter	55	45.50	100.50
4700	48" diameter	68.50	51	119.50
4720	60" diameter	110	68.50	178.50
4730	72" diameter	155	81.50	236.50
4740	84" diameter	282	102	384
4800	With gasket			
4980	12" diameter	12.45	8.15	20.60
5000	15" diameter	14.90	8.55	23.45
5020	18" diameter	18.65	9	27.65
5050	24" diameter	28	10.05	38.05
5070	30" diameter	37.50	34	71.50
5080	36" diameter	56	41.50	97.50
5090	42" diameter	73	41	114
5100	48" diameter	91.50	51	142.50
5120	60" diameter	124	65	189
5130	72" diameter	236	81.50	317.50
5140	84" diameter	340	136	476
5700	Corrugated metal, alum. or galv. bit. coated			
5760	8" diameter	6.45	6.20	12.65
5770	10" diameter	7.75	7.85	15.60
5780	12" diameter	9.35	9.75	19.10
5800	15" diameter	11.30	10.25	21.55
5820	18" diameter	14.50	10.75	25.25
5850	24" diameter	17.80	12.80	30.60
5870	30" diameter	23.50	25	48.50
5880	36" diameter	34.50	25	59.50
5900	48" diameter	52	30	82
5920	60" diameter	67.50	44	111.50
5930	72" diameter	103	73	176
6000	Plain			

G3020 Sanitary Sewer

G3020 110	Drainage & Sewage Piping	COST PER L.F.		
		MAT.	INST.	TOTAL
6060	8" diameter	5.10	5.75	10.85
6070	10" diameter	5.60	7.30	12.90
6080	12" diameter	6.35	9.30	15.65
6100	15" diameter	8.60	9.30	17.90
6120	18" diameter	11.35	9.95	21.30
6140	24" diameter	15.55	11.65	27.20
6170	30" diameter	19.80	23	42.80
6180	36" diameter	32.50	23	55.50
6200	48" diameter	43.50	27.50	71
6220	60" diameter	68.50	41.50	110
6230	72" diameter	101	95	196
6300	Steel or alum. oval arch, coated & paved invert			
6400	15" equivalent diam	19.15	10.25	29.40
6420	18" equivalent diam	24.50	13.60	38.10
6450	24" equivalent diam	35.50	16.30	51.80
6470	30" equivalent diam	44	20.50	64.50
6480	36" equivalent diam	64.50	30	94.50
6490	42" equivalent diam	77.50	33.50	111
6500	48" equivalent diam	87.50	40	127.50
6600	Plain			
6700	15" equivalent diam	10.50	9.05	19.55
6720	18" equivalent diam	12.45	11.65	24.10
6750	24" equivalent diam	20	13.60	33.60
6770	30" equivalent diam	25	28	53
6780	36" equivalent diam	41.50	28	69.50
6790	42" equivalent diam	49	32.50	81.50
6800	48" equivalent diam	43	40	83
8000	Polyvinyl chloride SDR 35			
8130	4" diameter	1.57	2.65	4.22
8150	6" diameter	2.73	2.84	5.57
8160	8" diameter	4.84	2.97	7.81
8170	10" diameter	7.20	4.14	11.34
8180	12" diameter	8.10	4.26	12.36
8200	15" diameter	12.30	7.15	19.45
9000	Vitrified clay, C-2000, premium joint			
9030	4" diameter	1.83	3.76	5.59
9050	6" diameter	3.06	4.98	8.04
9060	8" diameter	4.36	6.80	11.16
9070	10" diameter	7.30	7.15	14.45
9080	12" diameter	9.70	9.10	18.80
9100	15" diameter	17	12.45	29.45
9120	18" diameter	25.50	15.50	41
9150	24" diameter	46.50	30.50	77
9170	30" diameter	82	50	132
9180	36" diameter	135	77.50	212.50

BUILDING SITEWORK

G

Manhole   Catch Basin

The Manhole and Catch Basin System includes: excavation with a backhoe; a formed concrete footing; frame and cover; cast iron steps and compacted backfill.

The Expanded System Listing shows manholes that have a 4', 5' and 6' inside diameter riser. Depths range from 4' to 14'. Construction material shown is either concrete, concrete block, precast concrete, or brick.

BUILDING SITEWORK G

System Components	QUANTITY	UNIT	COST PER EACH		
			MAT.	INST.	TOTAL
SYSTEM G3030 210 1920					
MANHOLE/CATCH BASIN, BRICK, 4' I.D. RISER, 4' DEEP					
Excavation, hydraulic backhoe, 3/8 C.Y. bucket	14.815	C.Y.		80	80
Trim sides and bottom of excavation	64.000	S.F.		40.96	40.96
Forms in place, manhole base, 4 uses	20.000	SFCA	12.20	68.20	80.40
Reinforcing in place footings, #4 to #7	.019	Ton	11.12	16.72	27.84
Concrete, 3000 psi	.925	C.Y.	70.30		70.30
Place and vibrate concrete, footing, direct chute	.925	C.Y.		31.84	31.84
Catch basin or MH, brick, 4' ID, 4' deep	1.000	Ea.	315	665	980
Catch basin or MH steps; heavy galvanized cast iron	1.000	Ea.	10.40	9.35	19.75
Catch basin or MH frame and cover	1.000	Ea.	177	146	323
Fill, granular	12.954	C.Y.	112.70		112.70
Backfill, spread with wheeled front end loader	12.954	C.Y.		20.59	20.59
Backfill compaction, 12" lifts, air tamp	12.954	C.Y.		76.17	76.17
TOTAL			708.72	1,154.83	1,863.55

G3030 210	Manholes & Catch Basins	COST PER EACH		
		MAT.	INST.	TOTAL
1920	Manhole/catch basin, brick, 4' I.D. riser, 4' deep	710	1,150	1,860
1940	6' deep	950	1,600	2,550
1960	8' deep	1,225	2,200	3,425
1980	10' deep	1,550	2,725	4,275
3000	12' deep	2,000	2,950	4,950
3020	14' deep	2,475	4,150	6,625
3200	Block, 4' I.D. riser, 4' deep	650	930	1,580
3220	6' deep	850	1,325	2,175
3240	8' deep	1,075	1,825	2,900
3260	10' deep	1,275	2,250	3,525
3280	12' deep	1,550	2,875	4,425
3300	14' deep	1,900	3,500	5,400
4620	Concrete, cast-in-place, 4' I.D. riser, 4' deep	785	1,625	2,410
4640	6' deep	1,075	2,175	3,250
4660	8' deep	1,475	3,150	4,625
4680	10' deep	1,750	3,925	5,675
4700	12' deep	2,150	4,875	7,025
4720	14' deep	2,600	5,825	8,425
5820	Concrete, precast, 4' I.D. riser, 4' deep	855	865	1,720
5840	6' deep	1,100	1,175	2,275

G3030 Storm Sewer

G3030 210	Manholes & Catch Basins	COST PER EACH		
		MAT.	INST.	TOTAL
5860	8' deep	1,350	1,650	3,000
5880	10' deep	1,650	2,025	3,675
5900	12' deep	2,050	2,475	4,525
5920	14' deep	2,500	3,175	5,675
6000	5' I.D. riser, 4' deep	925	935	1,860
6020	6' deep	1,225	1,325	2,550
6040	8' deep	1,550	1,750	3,300
6060	10' deep	1,975	2,200	4,175
6080	12' deep	2,425	2,825	5,250
6100	14' deep	2,925	3,475	6,400
6200	6' I.D. riser, 4' deep	1,325	1,200	2,525
6220	6' deep	1,700	1,625	3,325
6240	8' deep	2,100	2,275	4,375
6260	10' deep	2,675	2,875	5,550
6280	12' deep	3,275	3,625	6,900
6300	14' deep	3,925	4,425	8,350

BUILDING SITEWORK

G

G3060 Fuel Distribution

G3060 110	Gas Service Piping	COST PER L.F.		
		MAT.	INST.	TOTAL
2000	Piping, excavation & backfill excluded, polyethylene			
2070	1-1/4" diam, SDR 10	.76	2.55	3.31
2090	2" diam, SDR 11	.94	2.85	3.79
2110	3" diam, SDR 11.5	1.98	3.40	5.38
2130	4" diam, SDR 11	4.55	6.55	11.10
2150	6" diam, SDR 21	14.15	7.05	21.20
2160	8" diam, SDR 21	19.30	8.45	27.75
3000	Steel, schedule 40, plain end, tarred & wrapped			
3060	1" diameter	2.33	5.80	8.13
3090	2" diameter	3.66	6.20	9.86
3110	3" diameter	6.05	6.70	12.75
3130	4" diameter	7.90	10.70	18.60
3140	5" diameter	11.45	12.45	23.90
3150	6" diameter	14	15.20	29.20
3160	8" diameter	22	19.50	41.50

Important: See the Reference Section for critical supporting data - Reference Numbers and City Cost Indexes

BUILDING SITEWORK G

Fiberglass Underground Fuel Storage Tank, Single Wall

Labels on figure:
- 3" Fill Box
- 2" Vent Cap
- Supply
- Manway Extension
- Return
- Hold Down Straps
- Pea Gravel Backfill
- Concrete Slab w/ Reinforcing Steel

System Components	QUANTITY	UNIT	COST EACH		
			MAT.	INST.	TOTAL
SYSTEM G3060 305 1010					
STORAGE TANK, FUEL, UNDERGROUND, SINGLE WALL FIBERGLASS, 550 GAL.					
Tank, fiberglass, underground, single wall, U.L. listed, 550 gal	1.000	Ea.	1,600	294	1,894
Tank, fiberglass, manways, add	1.000	Ea.	925		925
Tank, for hold-downs 500-4000 gal, add	1.000	Ea.	178	104	282
Foot valve, single poppet, 3/4" dia	1.000	Ea.	41.50	24.50	66
Tubing, copper, Type L, 3/4" dia	120.000	L.F.	272.40	684	956.40
Elbow 90°, copper, wrought, cu x cu, 3/4" dia.	8.000	Ea.	7.12	184	191.12
Fuel oil specialties, valve, ball chk, globe type, fusible, 3/4" dia	2.000	Ea.	64	44	108
Elbow 90°, black steel, straight, 2" dia.	3.000	Ea.	30.15	130.50	160.65
Union, black steel, with brass seat, 2" dia.	1.000	Ea.	20.50	46	66.50
Pipe and nipples, steel, Sch. 40, threaded, black, 2" dia	2.000	Ea.	133	463.60	596.60
Vent protector / breather, 2" dia.	1.000	Ea.	14.10	13.65	27.75
Pipe, black steel, welded, Sch. 40, 3" dia.	30.000	L.F.	198	606	804
Elbow 90°, steel, weld joint, butt, 3" dia.	3.000	Ea.	35.85	373.65	409.50
Flange, weld neck, 150 lb., 3" pipe size	1.000	Ea.	24	62.30	86.30
Fuel fill box, locking inner cover, 3" dia.	1.000	Ea.	112	87.50	199.50
Remote tank gauging system, 30', 5" pointer travel	1.000	Ea.	4,075	218	4,293
Tank leak detector system, 8 channel, external monitoring	1.000	Ea.	1,750		1,750
Tank leak detection system, probes, well monitoring, liquid phase detection	2.000	Ea.	1,570		1,570
Excavating, trench, NO sheeting or dewatering, 6'-10' D,3/4 CY hyd backhoe	18.000	C.Y.		97.20	97.20
Concrete hold down pad, 6' x 5' x 8" thick	30.000	C.Y.	60.30	29.70	90
Reinforcing in hold down pad, #3 to #7	120.000	Lb.	37.20	52.80	90
Stone back-fill, hand spread, pea gravel	9.000	C.Y.	301.50	288	589.50
Corrosion resistance, wrap & coat, small diam pipe, 1" diam, add	120.000	L.F.	134.40		134.40
Corrosion resistance, wrap & coat, small diameter pipe, 2" diam, add	36.000	L.F.	44.28		44.28
Corrosion resistance wrap & coat, 4" diam, add	30.000	L.F.	44.70		44.70
TOTAL			11,673	3,803.40	15,476.40

G3060 305	Fiberglass Fuel Tank, Single Wall	COST EACH		
		MAT.	INST.	TOTAL
1010	Storage tank, fuel, underground, single wall fiberglass, 550 Gal.	11,700	3,800	15,500
1020	2000 Gal.	13,900	5,175	19,075
1030	4000 Gal.	18,000	7,075	25,075
1040	6000 Gal.	19,100	7,900	27,000

G3060 Fuel Distribution

G3060 305	Fiberglass Fuel Tank, Single Wall	COST EACH		
		MAT.	INST.	TOTAL
1050	8000 Gal.	23,300	9,675	32,975
1060	10,000 Gal.	25,400	10,400	35,800
1070	15,000 Gal.	29,200	13,000	42,200
1080	20,000 Gal.	36,700	15,700	52,400
1090	25,000 Gal.	46,500	18,400	64,900
1100	30,000 Gal.	53,500	19,800	73,300
1110	40,000 Gal.	67,500	25,000	92,500
1120	48,000 Gal.	93,000	28,800	121,800

G3060 310	Fiberglass Fuel Tank, Double Wall	COST EACH		
		MAT.	INST.	TOTAL
1010	Storage tank, fuel, underground, double wall fiberglass, 550 Gal.	13,900	4,350	18,250
1020	2500 Gal.	18,500	6,050	24,550
1030	4000 Gal.	22,600	7,500	30,100
1040	6000 Gal.	24,300	8,450	32,750
1050	8000 Gal.	30,200	10,200	40,400
1060	10,000 Gal.	32,400	11,800	44,200
1070	15,000 Gal.	42,100	13,600	55,700
1080	20,000 Gal.	51,000	16,400	67,400
1090	25,000 Gal.	60,500	19,100	79,600
1100	30,000 Gal.	69,000	20,300	89,300

G3060 Fuel Distribution

Steel Underground Fuel Storage Tank, Single Wall

System Components	QUANTITY	UNIT	COST EACH		
			MAT.	INST.	TOTAL
SYSTEM G3060 315 1010					
STORAGE TANK, FUEL, UNDERGROUND, SINGLE WALL STEEL, 550 GAL.					
Tank, steel, underground, sti-P3, set in place, 500 gal, 7 Ga. shell	1.000	Ea.	905	291	1,196
Tank, for manways, add	1.000	Ea.	650		650
Tanks, for hold-downs, 500-2000 gal, add	1.000	Ea.	310	104	414
Foot valve, single poppet, 3/4″ dia	1.000	Ea.	41.50	24.50	66
Tubing, copper, Type L, 3/4″ dia.	120.000	L.F.	272.40	684	956.40
Elbow 90°, copper, wrought, cu x cu, 3/4″ dia.	8.000	Ea.	7.12	184	191.12
Fuel oil valve, ball chk, globe type, fusible, 3/4″ dia.	2.000	Ea.	64	44	108
Elbow 90°, black steel, straight, 2″ dia.	3.000	Ea.	30.15	130.50	160.65
Union, black steel, with brass seat, 3/4″ dia.	1.000	Ea.	6.90	33.50	40.40
Pipe and nipples, steel, Sch. 40, threaded, black, 2″ dia	2.000	Ea.	133	463.60	596.60
Vent protector / breather, 2″ dia.	1.000	Ea.	14.10	13.65	27.75
Pipe, black steel, welded, Sch. 40, 3″ dia.	30.000	L.F.	198	606	804
Elbow 90°, steel, weld joint, butt, 90° elbow, 3″ dia.	3.000	Ea.	35.85	373.65	409.50
Flange, steel, weld neck, 150 lb., 3″ pipe size	1.000	Ea.	24	62.30	86.30
Fuel fill box, locking inner cover, 3″ dia.	1.000	Ea.	112	87.50	199.50
Remote tank gauging system, 30′, 5″ pointer travel	1.000	Ea.	4,075	218	4,293
Tank leak detector system, 8 channel, external monitoring	1.000	Ea.	1,750		1,750
Tank leak detection system, probes, well monitoring, liquid phase detection	2.000	Ea.	1,570		1,570
Excavation, NO sheeting or dewatering,6′-10′ D,3/4 CY hyd backhoe	18.000	C.Y.		97.20	97.20
Concrete hold down pad, 6′ x 5′ x 8″ thick	30.000	C.Y.	60.30	29.70	90
Reinforcing in hold down pad, #3 to #7	120.000	Lb.	37.20	52.80	90
Stone back-fill, hand spread, pea gravel	9.000	C.Y.	301.50	288	589.50
Corrosion resistance, wrap & coat, small diam pipe, 1″ dia., add	120.000	L.F.	134.40		134.40
Corrosion resistance, wrap & coat, small diameter pipe, 2″ dia., add	36.000	L.F.	44.28		44.28
Corrosion resistance wrap & coat, 4″ dia., add	30.000	L.F.	44.70		44.70
TOTAL			10,821.40	3,787.90	14,609.30

G3060 315	Steel Fuel Tank, Single Wall	COST EACH		
		MAT.	INST.	TOTAL
1010	Storage tank, fuel, underground, single wall steel, 550 Gal.	10,800	3,800	14,600
1020	2000 Gal.	13,800	5,175	18,975
1030	5000 Gal.	19,500	7,300	26,800
1040	10,000 Gal.	28,400	11,200	39,600

BUILDING SITEWORK

G

G3060 Fuel Distribution

G3060 315	Steel Fuel Tank, Single Wall	COST EACH		
		MAT.	INST.	TOTAL
1050	15,000 Gal.	33,400	13,000	46,400
1060	20,000 Gal.	40,600	15,600	56,200
1070	25,000 Gal.	48,700	18,300	67,000
1080	30,000 Gal.	56,000	19,600	75,600
1090	40,000 Gal.	69,000	24,700	93,700
1100	50,000 Gal.	83,000	29,900	112,900

G3060 320	Steel Fuel Tank, Double Wall	COST EACH		
		MAT.	INST.	TOTAL
1010	Storage tank, fuel, underground, double wall steel, 500 Gal.	13,600	4,450	18,050
1020	2000 Gal.	17,300	5,825	23,125
1030	4000 Gal.	22,800	7,600	30,400
1040	6000 Gal.	27,200	8,375	35,575
1050	8000 Gal.	31,900	10,400	42,300
1060	10,000 Gal.	35,100	11,900	47,000
1070	15,000 Gal.	41,200	13,900	55,100
1080	20,000 Gal.	49,200	16,600	65,800
1090	25,000 Gal.	66,000	19,100	85,100
1100	30,000 Gal.	75,500	20,400	95,900
1110	40,000 Gal.	92,000	25,600	117,600
1120	50,000 Gal.	109,000	30,700	139,700

G3060 325	Steel Tank, Above Ground, Single Wall	COST EACH		
		MAT.	INST.	TOTAL
1010	Storage tank, fuel, above ground, single wall steel, 550 Gal.	8,200	3,200	11,400
1020	2000 Gal.	10,300	3,250	13,550
1030	5000 Gal.	12,600	4,975	17,575
1040	10,000 Gal.	20,500	5,300	25,800
1050	15,000 Gal.	23,100	5,450	28,550
1060	20,000 Gal.	27,000	5,600	32,600
1070	25,000 Gal.	29,100	5,725	34,825
1080	30,000 Gal.	34,300	5,975	40,275

G3060 330	Steel Tank, Above Ground, Double Wall	COST EACH		
		MAT.	INST.	TOTAL
1010	Storage tank, fuel, above ground, double wall steel, 500 Gal.	12,000	3,225	15,225
1020	2000 Gal.	14,600	3,300	17,900
1030	4000 Gal.	19,300	3,375	22,675
1040	6000 Gal.	22,100	5,150	27,250
1050	8000 Gal.	24,300	5,300	29,600
1060	10,000 Gal.	25,600	5,375	30,975
1070	15,000 Gal.	32,900	5,550	38,450
1080	20,000 Gal.	36,000	5,725	41,725
1090	25,000 Gal.	41,500	5,900	47,400
1100	30,000 Gal.	44,400	6,125	50,525

For information about Means Estimating Seminars, see yellow pages 11 and 12 in back of book

Reference Section

All the reference information is in one section, making it easy to find what you need to know . . . and easy to use the book on a daily basis. This section is visually identified by a vertical gray bar on the edge of pages.

In the reference number information that follows, you'll see the background that relates to the "reference numbers" that appeared in the Unit Price Section. You'll find reference tables, explanations and estimating information that support how we arrived at the unit price data. Also included are alternate pricing methods, technical data and estimating procedures along with information on design and economy in construction.

Also in this Reference Section, we've included Change Orders, information on pricing changes to contract documents; Crew Listings, a full listing of all the crews, equipment and their costs; Historical Cost Indexes for cost comparisons over time; City Cost Indexes and Location Factors for adjusting costs to the region you are in, and an explanation of all abbreviations used in the book.

Table of Contents

Reference Numbers

R011 Overhead & Misc. Data
R01100-005 Tips for Accurate Estimating.... 512
R01100-040 Builder's Risk Insurance........... 512
R01100-060 Workers' Compensation 513
R01100-070 Contractor's Overhead
& Profit 516
R01100-080 Performance Bond 517
R01100-090 Sales Tax by State 517
R01100-100 Unemployment &
Social Security Taxes 517
R01100-110 Overtime.................................. 518
R01100-710 Unit Gross Area Require. 518
R01100-720 Floor Area Ratios 519
R01100-730 Occupancy Determinations..... 519
R01107-030 Engineering Fees 520

R012 Price & Payment Procedures
R01250-010 Repair & Remodeling.............. 521

R015 Construction Aids
R01540-100 Steel Tubular Scaffolding.......... 522
R01590-100 Contractor Equipment 523

R021 Subsurface Invest. & Demolition
R02115-200 Underground Storage Tank
Removal 524

R022 Site Prep. & Excav. Support
R02240-900 Wellpoints................................ 524
R02250-400 Wood Sheet Piling.................... 525
R02250-450 Steel Sheet Piling...................... 526

R132 Hazardous Material Remediation
R13281-120 Asbestos Removal Process....... 527

R136 Solar & Wind Energy Equip.
R13600-610 Solar Heating (Space & H.W.).. 527

R150 Basic Materials & Methods
R15050-710 Subcontractors 528
R15050-720 Demolition (Selective vs.
Removal for Replacement) 528

R151 Plumbing
R15100-050 Pipe Material Considerations... 528
R15100-070 Piping to 10' High 529
R15100-080 Valve Materials......................... 530
R15100-090 Valve Select. Considerations 530

Reference Numbers (cont.)

R155 Heating
R15500-010 Heating Systems & Heat
Transmission............................. 533
R15500-020 Heating Approximations for
Quick Estimating 533
R15500-030 Basics of a Heating System 534
R15500-050 Factor for Determining
Heat Loss 535
R15500-070 Transmission of Heat............... 536
R15500-080 Transmission of Heat
(Low Rate)................................ 536

R157 Air Conditioning & Ventilation
R15700-010 Air Conditioning...................... 537
R15700-020 Air Conditioning
Requirements 537
R15700-030 Psychrometric Table 538
R15700-040 Recommended Ventilation
Air Changes 539
R15700-090 Quality/Complexity
Adjust. for A/C Systems............ 539

R158 Air Distribution
R15800-080 Loudness Levels for Moving
Air thru Fans, Diffusers,
Registers, etc............................. 540
R15810-050 Ductwork (Duct Weight
in Pounds per L.F.
Straight Runs).......................... 541
R15810-070 Duct Fabrication/Installation ... 541
R15810-100 Sheet Metal Calculator
(Weight in Lb./Ft. of Length)... 542
R15810-110 Ductwork Packages
(per Ton of Cooling) 543
R15850-060 Diffuser Evaluation.................. 543

R171 Information
R17100-100 Square Foot Project Size
Modifier 544

Change Orders 545

Crew Listings 549

Historical Cost Indexes 574

City Cost Indexes 577

Location Factors 595

Abbreviations 601

R01100-005 Tips for Accurate Estimating

1. Use pre-printed or columnar forms for orderly sequence of dimensions and locations and for recording telephone quotations.

2. Use only the front side of each paper or form except for certain pre-printed summary forms.

3. Be consistent in listing dimensions: For example, length x width x height. This helps in rechecking to ensure that, the total length of partitions is appropriate for the building area.

4. Use printed (rather than measured) dimensions where given.

5. Add up multiple printed dimensions for a single entry where possible.

6. Measure all other dimensions carefully.

7. Use each set of dimensions to calculate multiple related quantities.

8. Convert foot and inch measurements to decimal feet when listing. Memorize decimal equivalents to .01 parts of a foot (1/8″ equals approximately .01′).

9. Do not "round off" quantities until the final summary.

10. Mark drawings with different colors as items are taken off.

11. Keep similar items together, different items separate.

12. Identify location and drawing numbers to aid in future checking for completeness.

13. Measure or list everything on the drawings or mentioned in the specifications.

14. It may be necessary to list items not called for to make the job complete.

15. Be alert for: Notes on plans such as N.T.S. (not to scale); changes in scale throughout the drawings; reduced size drawings; discrepancies between the specifications and the drawings.

16. Develop a consistent pattern of performing an estimate.
    For example:
    a. Start the quantity takeoff at the lower floor and move to the next higher floor.
    b. Proceed from the main section of the building to the wings.
    c. Proceed from south to north or vice versa, clockwise or counterclockwise.
    d. Take off floor plan quantities first, elevations next, then detail drawings.

17. List all gross dimensions that can be either used again for different quantities, or used as a rough check of other quantities for verification (exterior perimeter, gross floor area, individual floor areas, etc.).

18. Utilize design symmetry or repetition (repetitive floors, repetitive wings, symmetrical design around a center line, similar room layouts, etc.). Note: Extreme caution is needed here so as not to omit or duplicate an area.

19. Do not convert units until the final total is obtained. For instance, when estimating concrete work, keep all units to the nearest cubic foot, then summarize and convert to cubic yards.

20. When figuring alternatives, it is best to total all items involved in the basic system, then total all items involved in the alternates. Therefore you work with positive numbers in all cases. When adds and deducts are used, it is often confusing whether to add or subtract a portion of an item; especially on a complicated or involved alternate.

R01100-040 Builder's Risk Insurance

Builder's Risk Insurance is insurance on a building during construction. Premiums are paid by the owner or the contractor. Blasting, collapse and underground insurance would raise total insurance costs above those listed. Floater policy for materials delivered to the job runs $.75 to $1.25 per $100 value. Contractor equipment insurance runs $.50 to $1.50 per $100 value. Insurance for miscellaneous tools to $1,500 value runs from $3.00 to $7.50 per $100 value.

Tabulated below are New England Builder's Risk insurance rates in dollars per $100 value for $1,000 deductible. For $25,000 deductible, rates can be reduced 13% to 34%. On contracts over $1,000,000, rates may be lower than those tabulated. Policies are written annually for the total completed value in place. For "all risk" insurance (excluding flood, earthquake and certain other perils) add $.025 to total rates below.

Coverage	Frame Construction (Class 1) Range			Frame Construction (Class 1) Average	Brick Construction (Class 4) Range			Brick Construction (Class 4) Average	Fire Resistive (Class 6) Range			Fire Resistive (Class 6) Average
Fire Insurance	$.300	to	$.420	$.394	$.132	to	$.189	$.174	$.052	to	$.080	$.070
Extended Coverage	.115	to	.150	.144	.080	to	.105	.101	.081	to	.105	.100
Vandalism	.012	to	.016	.015	.008	to	.011	.011	.008	to	.011	.010
Total Annual Rate	$.427	to	$.586	$.553	$.220	to	$.305	$.286	$.141	to	$.196	$.180

R01100-060 Workers' Compensation Insurance Rates by Trade

The table below tabulates the national averages for Workers' Compensation insurance rates by trade and type of building. The average "Insurance Rate" is multiplied by the "% of Building Cost" for each trade. This produces the "Workers' Compensation Cost" by % of total labor cost, to be added for each trade by building type to determine the weighted average Workers' Compensation rate for the building types analyzed.

Trade	Insurance Rate (% Labor Cost) Range	Insurance Rate (% Labor Cost) Average	% of Building Cost Office Bldgs.	% of Building Cost Schools & Apts.	% of Building Cost Mfg.	Workers' Compensation Office Bldgs.	Workers' Compensation Schools & Apts.	Workers' Compensation Mfg.
Excavation, Grading, etc.	3.8 % to 24.7%	10.5%	4.8%	4.9%	4.5%	.50%	.51%	.47%
Piles & Foundations	5.9 to 62.5	24.8	7.1	5.2	8.7	1.76	1.29	2.16
Concrete	5.9 to 36.8	16.9	5.0	14.8	3.7	.85	2.50	.63
Masonry	5.3 to 31.1	15.9	6.9	7.5	1.9	1.10	1.19	.30
Structural Steel	5.9 to 112.0	39.5	10.7	3.9	17.6	4.23	1.54	6.95
Miscellaneous & Ornamental Metals	4.9 to 25.3	12.8	2.8	4.0	3.6	.36	.51	.46
Carpentry & Millwork	5.9 to 43.6	18.0	3.7	4.0	0.5	.67	.72	.09
Metal or Composition Siding	5.9 to 32.2	16.5	2.3	0.3	4.3	.38	.05	.71
Roofing	5.9 to 85.9	32.3	2.3	2.6	3.1	.74	.84	1.00
Doors & Hardware	4.1 to 25.5	10.9	0.9	1.4	0.4	.10	.15	.04
Sash & Glazing	5.7 to 38.8	13.7	3.5	4.0	1.0	.48	.55	.14
Lath & Plaster	5.2 to 53.5	15.7	3.3	6.9	0.8	.52	1.08	.13
Tile, Marble & Floors	3.7 to 22.6	9.8	2.6	3.0	0.5	.25	.29	.05
Acoustical Ceilings	3.5 to 26.7	11.0	2.4	0.2	0.3	.26	.02	.03
Painting	5.9 to 27.6	13.7	1.5	1.6	1.6	.21	.22	.22
Interior Partitions	5.9 to 43.6	18.0	3.9	4.3	4.4	.70	.77	.79
Miscellaneous Items	1.1 to 128.0	17.6	5.2	3.7	9.7	.91	.65	1.70
Elevators	2.1 to 17.0	7.7	2.1	1.1	2.2	.16	.08	.17
Sprinklers	2.5 to 19.5	8.7	0.5	—	2.0	.04	—	.17
Plumbing	3.0 to 16.9	8.2	4.9	7.2	5.2	.40	.59	.43
Heat., Vent., Air Conditioning	3.3 to 32.5	11.6	13.5	11.0	12.9	1.57	1.28	1.50
Electrical	2.4 to 10.5	6.7	10.1	8.4	11.1	.68	.56	.74
Total	2.0 % to 132.9%	—	100.0%	100.0%	100.0%	16.87%	15.39%	18.88%
	Overall Weighted Average 17.05%							

Workers' Compensation Insurance Rates by States

The table below lists the weighted average Workers' Compensation base rate for each state with a factor comparing this with the national average of 16.8%.

State	Weighted Average	Factor	State	Weighted Average	Factor	State	Weighted Average	Factor
Alabama	31.9%	190	Kentucky	19.7%	117	North Dakota	14.8%	88
Alaska	11.0	65	Louisiana	27.2	162	Ohio	19.1	114
Arizona	11.0	65	Maine	21.7	129	Oklahoma	21.8	130
Arkansas	14.2	85	Maryland	11.8	70	Oregon	13.6	81
California	18.5	110	Massachusetts	20.4	121	Pennsylvania	22.8	136
Colorado	26.4	157	Michigan	18.5	110	Rhode Island	20.9	124
Connecticut	20.0	119	Minnesota	27.7	165	South Carolina	13.7	82
Delaware	11.5	68	Mississippi	20.4	121	South Dakota	14.5	86
District of Columbia	21.9	130	Missouri	17.6	105	Tennessee	15.8	94
Florida	27.8	165	Montana	17.6	105	Texas	15.8	94
Georgia	24.8	148	Nebraska	15.7	93	Utah	12.9	77
Hawaii	16.7	99	Nevada	17.3	103	Vermont	16.4	98
Idaho	10.7	64	New Hampshire	24.9	148	Virginia	10.9	65
Illinois	23.2	138	New Jersey	9.0	54	Washington	9.2	55
Indiana	7.2	43	New Mexico	16.7	99	West Virginia	13.0	77
Iowa	13.4	80	New York	15.3	91	Wisconsin	13.0	77
Kansas	10.1	60	North Carolina	13.0	77	Wyoming	6.5	39
			Weighted Average for U.S. is	17.0% of payroll = 100%				

Rates in the following table are the base or manual costs per $100 of payroll for Workers' Compensation in each state. Rates are usually applied to straight time wages only and not to premium time wages and bonuses.

The weighted average skilled worker rate for 35 trades is 16.8%. For bidding purposes, apply the full value of Workers' Compensation directly to total labor costs, or if labor is 38%, materials 42% and overhead and profit 20% of total cost, carry 38/80 x 16.8% = 8.0% of cost (before overhead and profit) into overhead. Rates vary not only from state to state but also with the experience rating of the contractor.

Rates are the most current available at the time of publication.

R01100-060 Workers' Compensation Insurance Rates by Trade and State (cont.)

State	Carpentry — 3 stories or less 5651	Carpentry — interior cab. work 5437	Carpentry — general 5403	Concrete Work — NOC 5213	Concrete Work — flat (flr., sdwk.) 5221	Electrical Wiring — inside 5190	Excavation — earth NOC 6217	Excavation — rock 6217	Glaziers 5462	Insulation Work 5479	Lathing 5443	Masonry 5022	Painting & Decorating 5474	Pile Driving 6003	Plastering 5480	Plumbing 5183	Roofing 5551	Sheet Metal Work (HVAC) 5538	Steel Erection — door & sash 5102	Steel Erection — inter., ornam. 5102	Steel Erection — structure 5040	Steel Erection — NOC 5057	Tile Work — (interior ceramic) 5348	Waterproofing 9014	Wrecking 5701
AL	26.11	14.28	27.87	21.48	16.84	10.36	22.37	22.37	38.82	20.88	14.51	30.14	23.55	44.38	53.45	12.19	85.89	28.75	20.74	20.74	60.88	63.07	18.28	5.98	60.88
AK	8.81	8.42	9.30	9.30	5.59	5.67	8.82	8.82	9.10	10.01	5.64	10.99	8.49	29.93	9.97	5.25	17.33	6.01	9.40	9.40	17.20	17.20	6.23	6.38	25.07
AZ	10.70	6.49	19.81	10.24	6.84	5.94	6.75	6.75	10.18	16.00	6.83	10.93	7.55	14.38	10.72	5.67	17.04	8.57	8.41	8.41	21.63	15.24	4.58	3.93	50.31
AR	14.01	9.01	15.49	14.63	6.89	7.42	9.31	9.31	12.71	17.62	14.48	9.93	11.08	17.41	14.18	6.87	26.87	10.39	8.68	8.68	27.84	27.73	7.84	4.40	27.84
CA	28.28	9.07	28.28	13.18	13.18	9.93	7.88	7.88	16.37	23.34	10.90	14.55	19.45	21.18	18.12	11.59	39.79	15.33	13.55	13.55	23.93	21.91	7.74	19.45	21.91
CO	32.22	13.54	20.95	20.96	15.13	9.20	15.99	15.99	15.98	21.31	14.68	31.06	21.85	41.53	37.40	14.50	59.10	12.54	11.83	11.83	73.46	43.49	15.04	12.26	73.46
CT	19.61	13.45	27.11	19.35	12.19	7.00	10.28	10.28	19.33	28.37	8.73	23.82	14.96	20.24	18.86	9.39	30.72	10.04	11.49	11.49	68.73	28.67	11.59	5.85	50.86
DE	11.47	11.47	9.21	9.66	7.08	4.77	7.98	7.98	9.66	9.21	9.21	11.55	11.55	15.68	10.11	5.74	21.77	8.75	10.55	10.55	23.18	10.55	7.79	9.26	23.18
DC	16.82	9.38	15.57	36.80	20.66	8.85	11.48	11.48	27.21	17.44	10.18	24.94	12.47	27.38	20.34	12.15	29.23	10.14	18.73	18.73	63.23	32.24	22.61	4.48	66.43
FL	30.50	25.48	28.80	33.02	15.37	10.50	14.27	14.27	20.62	27.82	16.71	26.13	23.62	62.53	36.48	10.75	53.17	16.89	15.61	15.61	51.50	42.80	12.40	9.83	51.50
GA	31.04	16.02	32.71	20.67	15.12	9.82	19.55	19.55	21.97	19.15	26.66	23.03	21.20	40.88	19.50	11.14	40.72	18.65	13.74	13.74	36.03	53.09	13.08	10.16	36.03
HI	17.38	11.62	28.20	13.74	12.27	7.10	7.73	7.73	20.55	21.19	10.94	17.87	11.33	20.15	15.61	6.06	33.24	8.02	11.11	11.11	31.71	21.70	9.78	11.54	31.71
ID	8.79	6.95	14.19	10.24	10.45	4.56	6.41	6.41	6.96	10.12	7.12	7.92	7.51	14.30	8.97	3.95	20.53	5.87	8.05	8.05	27.60	20.85	4.78	6.59	26.68
IL	20.76	12.31	21.64	35.48	12.83	8.86	8.69	8.69	23.09	24.90	14.24	20.34	13.15	37.63	13.78	11.68	35.88	14.58	21.53	21.53	58.63	52.47	13.87	6.26	58.63
IN	7.08	4.06	7.07	6.68	3.23	3.00	3.79	3.79	5.72	7.01	3.48	5.32	5.94	9.68	5.16	3.00	11.87	4.29	4.92	4.92	28.84	13.32	3.69	3.26	27.19
IA	10.92	5.52	13.64	13.21	7.34	4.07	6.62	6.62	11.38	13.86	6.44	11.20	8.76	16.66	9.76	4.88	20.62	7.78	11.86	11.86	50.80	29.35	6.77	4.79	50.80
KS	14.05	8.56	10.18	10.45	7.04	4.39	5.62	5.62	7.79	12.99	5.65	10.08	8.05	12.08	9.29	5.45	20.85	7.09	6.51	6.51	21.14	15.34	4.84	3.84	16.49
KY	16.47	15.26	24.34	22.12	9.53	8.95	13.67	13.67	15.96	14.68	10.62	22.00	19.83	30.44	13.94	10.38	32.83	15.92	13.21	13.21	45.21	33.73	10.60	8.85	45.21
LA	29.63	22.95	26.01	20.86	14.80	10.48	24.73	24.73	23.88	20.82	14.00	21.27	27.63	54.39	21.42	13.78	49.11	21.35	15.51	15.51	66.41	36.94	12.82	10.26	66.41
ME	13.43	10.42	43.59	24.29	11.12	10.02	14.75	14.75	13.35	15.39	14.03	17.26	16.07	31.44	17.95	11.12	32.07	17.70	15.08	15.08	37.84	61.81	11.66	9.37	37.84
MD	10.55	5.95	10.55	11.35	5.05	5.15	9.25	9.25	13.20	13.05	6.45	11.35	6.75	27.45	6.65	5.55	22.80	7.00	9.15	9.15	26.80	18.50	6.35	3.50	26.80
MA	11.05	7.61	13.61	22.12	9.81	3.69	5.91	5.91	9.92	18.24	8.84	17.55	9.04	17.59	7.39	5.69	39.96	7.65	11.96	11.96	50.78	46.56	10.48	3.89	34.86
MI	19.40	11.07	18.72	21.94	10.81	5.88	10.90	10.90	10.46	15.25	11.61	20.40	14.92	38.97	15.76	8.34	35.54	9.78	12.69	12.69	38.97	34.62	11.74	6.17	38.97
MN	18.94	20.38	41.30	20.33	14.86	6.79	15.72	15.72	18.94	22.75	23.83	25.13	18.89	29.83	23.83	10.56	65.54	10.92	16.66	16.66	112.03	36.34	17.83	6.43	132.92
MS	21.02	15.11	22.05	17.14	9.43	9.14	12.39	12.39	14.99	19.16	9.24	18.34	14.80	45.04	18.23	9.45	28.88	21.00	11.13	11.13	53.38	36.68	12.11	8.95	37.42
MO	20.97	8.42	13.26	14.87	11.94	6.48	12.38	12.38	10.05	22.39	10.83	15.20	13.18	18.91	15.93	7.95	28.83	10.93	15.48	15.48	53.74	35.80	6.53	5.96	53.74
MT	17.20	9.33	17.59	12.16	12.16	5.19	13.34	13.34	13.87	16.21	9.35	14.03	16.64	37.35	15.06	7.40	52.56	9.42	11.92	11.92	37.61	22.44	6.61	6.06	37.61
NE	16.19	9.23	15.94	17.31	12.73	5.43	10.19	10.19	12.36	16.79	9.34	18.07	11.58	22.01	14.62	9.06	33.91	13.86	10.75	10.75	25.03	26.96	7.88	5.47	25.03
NV	19.48	9.43	14.46	12.79	10.94	10.06	10.36	10.36	13.43	19.86	10.26	12.00	14.34	19.15	15.91	11.64	26.00	17.26	15.98	15.98	44.29	28.75	9.64	9.14	44.29
NH	17.78	12.10	21.93	24.63	10.87	8.22	16.18	16.18	15.21	36.78	10.52	27.78	22.59	25.01	20.25	12.04	63.35	14.19	14.90	14.90	55.86	67.49	12.19	8.71	55.86
NJ	9.98	6.90	9.98	9.63	6.39	3.32	5.32	5.32	7.23	8.69	8.00	10.06	8.24	11.24	8.00	4.66	25.36	5.65	8.03	8.03	12.98	7.00	5.03	4.41	20.62
NM	16.50	8.23	14.95	12.65	9.18	6.26	7.72	7.72	13.81	18.99	7.40	15.55	9.63	21.98	10.41	7.05	31.76	16.01	19.41	19.41	58.27	21.01	6.70	7.48	58.27
NY	13.08	5.99	13.73	20.80	12.36	6.22	8.91	8.91	12.93	9.97	16.41	19.61	13.21	22.21	12.85	8.23	31.40	14.91	12.42	12.42	22.45	22.62	10.57	5.76	32.53
NC	13.20	10.11	16.28	13.67	6.49	7.71	8.80	8.80	8.40	11.75	5.73	10.04	10.66	16.13	15.67	5.71	28.64	9.64	9.33	9.33	28.82	19.64	7.35	3.88	39.03
ND	11.64	11.64	11.64	7.53	7.53	4.71	7.02	7.02	6.69	11.64	11.18	7.53	7.47	25.26	11.18	6.68	27.33	6.68	25.26	25.26	25.26	25.26	8.56	27.33	17.90
OH	17.06	17.06	17.06	15.16	15.16	9.12	15.16	15.16	22.67	17.90	17.90	20.44	22.67	15.16	17.90	10.47	33.45	32.47	15.16	15.16	15.16	15.16	13.22	15.16	15.16
OK	26.16	11.79	19.33	15.51	11.50	7.41	18.45	18.45	13.55	18.00	12.05	16.80	17.10	36.26	18.29	8.72	42.26	12.45	11.17	11.17	69.45	46.76	11.42	8.20	69.45
OR	18.86	8.20	17.03	11.06	9.20	4.58	9.96	9.96	14.54	10.23	7.59	11.11	12.25	14.70	11.26	5.50	25.26	11.19	8.09	8.09	37.79	19.03	8.54	5.55	37.79
PA	15.80	15.80	19.85	28.18	11.84	7.80	12.74	12.74	15.69	19.85	18.94	18.96	22.05	32.12	18.94	11.57	43.30	12.58	24.71	24.71	59.05	24.71	11.83	18.96	81.34
RI	19.53	11.65	18.07	18.23	16.24	4.43	10.38	10.38	12.85	22.78	11.97	25.11	24.13	37.66	17.25	8.52	33.92	10.29	14.07	14.07	59.49	37.50	14.36	7.70	78.79
SC	19.10	13.03	19.24	12.88	6.19	7.02	8.32	8.32	12.22	10.69	7.27	9.55	11.19	16.64	18.89	6.88	28.98	11.42	9.49	9.49	21.80	23.66	6.23	4.65	21.80
SD	13.70	6.67	17.59	16.61	6.07	5.60	9.23	9.23	11.50	15.30	7.77	10.02	12.97	31.49	12.79	8.27	20.70	7.55	9.64	9.64	45.63	21.23	6.90	4.75	46.57
TN	15.52	11.50	14.25	14.34	8.85	5.88	10.07	10.07	9.46	17.08	9.49	13.44	12.46	16.80	14.75	16.86	32.77	10.46	9.28	9.28	53.04	19.67	7.22	6.14	53.04
TX	16.89	11.74	15.14	17.55	10.64	8.28	11.28	11.28	11.25	16.15	11.21	15.42	10.91	19.24	20.99	8.55	27.75	14.40	10.87	10.87	29.38	21.35	8.64	8.09	27.22
UT	11.17	11.17	11.17	17.68	7.99	7.05	6.38	6.38	9.55	10.95	12.32	13.75	15.29	17.24	10.60	6.41	26.27	6.30	8.99	8.99	23.51	23.51	6.06	5.83	26.53
VT	16.27	10.49	15.14	30.87	8.27	5.84	9.03	9.03	15.93	16.11	8.60	16.58	9.08	21.46	15.25	7.22	25.15	10.54	12.62	12.62	35.79	31.64	8.14	10.00	35.79
VA	9.26	6.36	8.82	9.86	4.74	3.53	7.10	7.10	7.64	7.24	8.39	7.13	8.20	18.78	6.65	5.68	18.86	6.84	15.65	15.65	22.67	23.32	8.20	3.11	22.67
WA	7.57	7.57	6.89	6.22	5.94	2.41	6.79	6.79	9.41	6.43	6.89	8.97	7.91	13.73	9.99	4.12	17.12	3.34	12.96	12.96	10.06	10.06	7.91	10.23	10.06
WV	12.67	12.67	12.67	25.84	25.84	5.06	8.01	8.01	6.00	6.00	16.88	11.83	13.35	11.28	13.35	5.10	12.21	6.00	14.98	14.98	10.68	14.50	16.02	4.16	13.27
WI	9.94	7.47	17.64	9.67	8.55	4.57	6.14	6.14	9.74	10.23	14.32	13.69	8.90	14.88	10.54	5.34	31.67	6.78	13.03	13.03	32.90	15.65	12.11	4.09	36.51
WY	5.90	5.90	5.90	5.90	5.90	5.90	5.90	5.90	5.90	5.90	5.90	5.90	5.90	5.90	5.90	5.90	5.90	5.90	5.90	5.90	5.90	5.90	5.90	5.90	5.90
AVG.	16.48	10.88	17.96	16.88	10.53	6.66	10.51	10.51	13.73	16.17	11.03	15.87	13.73	24.78	15.69	8.25	32.28	11.61	12.79	12.79	39.50	28.33	9.77	7.69	40.91

R01100-060 Workers' Compensation (cont.) (Canada in Canadian dollars)

Province		Alberta	British Columbia	Manitoba	Ontario	New Brunswick	Newfndld. & Labrador	Northwest Territories	Nova Scotia	Prince Edward Island	Quebec	Saskat-chewan	Yukon
Carpentry—3 stories or less	Rate	2.72	4.65	4.76	4.80	3.48	6.00	2.50	6.91	6.81	12.69	4.83	3.25
	Code	40400	721028	40102	723	422	403	4-41	4022	401	80110	B12-02	202
Carpentry—interior cab. work	Rate	1.17	4.85	4.76	4.80	3.48	6.00	2.50	6.91	3.82	12.69	3.32	3.25
	Code	42133	721021	40102	723	427	403	4-41	4022	402	80110	B11-27	202
CARPENTRY—general	Rate	2.72	4.65	4.76	4.80	3.48	6.00	2.50	6.91	6.81	12.69	4.83	3.25
	Code	40400	721028	40102	723	422	403	4-41	4022	401	80110	B12-02	202
CONCRETE WORK—NOC	Rate	4.69	6.18	5.36	18.00	3.48	6.00	2.50	5.36	6.81	13.89	6.45	3.25
	Code	42104	721010	40110	748	422	403	4-41	4224	401	80100	B13-14	203
CONCRETE WORK—flat (flr. sidewalk)	Rate	4.69	6.18	5.36	18.00	3.48	6.00	2.50	5.36	6.81	13.89	6.45	3.25
	Code	42104	721010	40110	748	422	403	4-41	4224	401	80100	B13-14	203
ELECTRICAL Wiring—inside	Rate	1.99	3.36	2.35	3.13	1.59	4.83	2.00	2.71	3.82	5.69	3.32	2.35
	Code	42124	721019	40203	704	426	400	4-46	4261	402	80170	B11-05	206
EXCAVATION—earth NOC	Rate	2.57	3.40	4.90	3.99	2.57	6.00	2.50	3.74	3.95	7.71	3.62	3.25
	Code	40604	721031	40706	711	421	403	4-43	4214	404	80030	R11-06	207
EXCAVATION—rock	Rate	2.57	3.40	4.90	3.99	2.57	6.00	2.50	3.74	3.95	7.71	3.62	3.25
	Code	40604	721031	40706	711	421	403	4-43	4214	404	80030	R11-06	207
GLAZIERS	Rate	2.79	2.92	3.61	8.22	4.98	4.55	2.50	8.59	3.82	14.40	6.45	2.35
	Code	42121	715020	40109	751	423	402	4-41	4233	402	80150	B13-04	212
INSULATION WORK	Rate	2.89	5.80	4.76	8.22	4.98	4.55	2.50	8.59	6.81	14.15	4.83	3.25
	Code	42184	721029	40102	751	423	402	4-41	4234	401	80120	B12-07	202
LATHING	Rate	5.76	7.29	4.76	4.80	3.48	4.55	2.50	5.70	3.82	14.15	6.45	3.25
	Code	42135	721033	40102	723	427	402	4-41	4271	402	80120	B13-16	202
MASONRY	Rate	4.69	7.29	4.76	11.66	4.98	6.00	2.50	8.59	6.81	13.89	6.45	3.25
	Code	42102	721037	40102	741	423	403	4-41	4231	401	80100	B13-18	202
PAINTING & DECORATING	Rate	4.07	5.80	5.27	7.60	3.48	4.55	2.50	5.70	3.82	12.69	4.83	3.25
	Code	42111	721041	40105	719	427	402	4-41	4275	402	80110	B12-01	202
PILE DRIVING	Rate	4.69	14.31	4.90	6.26	3.48	9.60	2.50	5.36	6.81	7.71	6.45	3.25
	Code	42159	722004	40706	732	422	404	4-43	4221	401	80030	B13-10	202
PLASTERING	Rate	5.76	7.29	5.27	7.60	3.48	4.55	2.50	5.70	3.82	12.69	4.83	3.25
	Code	42135	721042	40108	719	427	402	4-41	4271	402	80110	B12-21	202
PLUMBING	Rate	1.99	3.65	2.80	3.89	1.95	3.88	2.00	3.35	3.82	7.13	3.32	1.40
	Code	42122	721043	40204	707	424	401	4-46	4241	402	80160	B11-01	214
ROOFING	Rate	9.70	9.42	6.41	12.15	7.82	6.00	2.50	8.59	6.81	22.17	6.45	3.25
	Code	42118	721036	40403	728	430	403	4-41	4236	401	80130	B13-20	202
SHEET METAL WORK (HVAC)	Rate	1.99	3.65	4.76	3.89	1.95	3.88	2.00	3.35	3.82	7.13	3.32	2.35
	Code	42117	721043	40402	707	424	401	4-46	4244	402	80160	B11-07	208
STEEL ERECTION—door & sash	Rate	2.89	14.31	6.07	18.00	3.48	9.60	2.50	8.59	6.81	32.47	6.45	3.25
	Code	42106	722005	40502	748	422	404	4-41	4227	401	80080	B13-22	202
STEEL ERECTION—inter., ornam.	Rate	2.89	14.31	6.07	18.00	3.48	6.00	2.50	8.59	6.81	32.47	6.45	3.25
	Code	42106	722005	40502	748	422	403	4-41	4227	401	80080	B13-22	202
STEEL ERECTION—structure	Rate	2.89	14.31	6.07	18.00	3.48	9.60	2.50	8.59	6.81	32.47	6.45	3.25
	Code	42106	722005	40502	748	422	404	4-41	4227	401	80080	B13-22	202
STEEL ERECTION—NOC	Rate	2.89	14.31	6.07	18.00	3.48	9.60	2.50	8.59	6.81	32.47	6.45	3.25
	Code	42106	722005	40502	748	422	404	4-41	4227	401	80080	B13-22	202
TILE WORK—inter. (ceramic)	Rate	2.94	3.27	2.35	7.60	3.48	4.55	2.50	5.70	3.82	12.69	6.45	3.25
	Code	42113	721054	40103	719	427	402	4-41	4276	402	80110	B13-01	202
WATERPROOFING	Rate	4.07	5.80	4.76	4.80	4.98	6.00	2.50	8.59	3.82	22.17	4.83	3.25
	Code	42139	721016	40102	723	423	403	4-41	4239	402	80130	B12-17	202
WRECKING	Rate	2.57	5.90	5.27	18.00	2.57	6.00	2.50	3.74	6.81	12.69	6.45	3.25
	Code	40604	721005	40106	748	421	403	4-43	4211	401	80110	B13-09	202

R01100-070 Contractor's Overhead & Profit

Below are the **average** installing contractor's percentage mark-ups applied to base labor rates to arrive at typical billing rates.

Column A: Labor rates are based on union wages averaged for 30 major U.S. cities. Base rates including fringe benefits are listed hourly and daily. These figures are the sum of the wage rate and employer-paid fringe benefits such as vacation pay, employer-paid health and welfare costs, pension costs, plus appropriate training and industry advancement funds costs.

Column B: Workers' Compensation rates are the national average of state rates established for each trade.

Column C: Column C lists average fixed overhead figures for all trades. Included are Federal and State Unemployment costs set at 7.0%; Social Security Taxes (FICA) set at 7.65%; Builder's Risk Insurance costs set at 0.34%; and Public Liability costs set at 1.55%. All the percentages except those for Social Security Taxes vary from state to state as well as from company to company.

Columns D and E: Percentages in Columns D and E are based on the presumption that the installing contractor has annual billing of $1,500,000 and up. Overhead percentages may increase with smaller annual billing. The overhead percentages for any given contractor may vary greatly and depend on a number of factors, such as the contractor's annual volume, engineering and logistical support costs, and staff requirements. The figures for overhead and profit will also vary depending on the type of job, the job location, and the prevailing economic conditions. All factors should be examined very carefully for each job.

Column F: Column F lists the total of Columns B, C, D, and E.

Column G: Column G is Column A (hourly base labor rate) multiplied by the percentage in Column F (O&P percentage).

Column H: Column H is the total of Column A (hourly base labor rate) plus Column G (Total O&P).

Column I: Column I is Column H multiplied by eight hours.

Abbr.	Trade	A Base Rate Incl. Fringes Hourly	A Base Rate Incl. Fringes Daily	B Workers' Comp. Ins.	C Average Fixed Overhead	D Overhead	E Profit	F Total Overhead & Profit %	F Total Overhead & Profit Amount	H Rate with O & P Hourly	I Rate with O & P Daily
Skwk	Skilled Workers Average (35 trades)	$30.95	$247.60	16.8%	16.5%	13.0%	10.0%	56.3%	$17.40	$48.35	$386.80
	Helpers Average (5 trades)	22.75	182.00	18.5		11.0		56.0	$12.75	35.50	284.00
	Foreman Average, Inside ($0.50 over trade)	31.45	251.60	16.8		13.0		56.3	17.70	49.15	393.20
	Foreman Average, Outside ($2.00 over trade)	32.95	263.60	16.8		13.0		56.3	18.55	51.50	412.00
Clab	Common Building Laborers	23.45	187.60	18.1		11.0		55.6	13.05	36.50	292.00
Asbe	Asbestos/Insulation Workers/Pipe Coverers	33.45	267.60	16.2		16.0		58.7	19.65	53.10	424.80
Boil	Boilermakers	36.25	290.00	14.7		16.0		57.2	20.75	57.00	456.00
Bric	Bricklayers	30.50	244.00	16.0		11.0		53.5	16.30	46.80	374.40
Brhe	Bricklayer Helpers	23.50	188.00	16.0		11.0		53.5	12.55	36.05	288.40
Carp	Carpenters	30.00	240.00	18.1		11.0		55.6	16.70	46.70	373.60
Cefi	Cement Finishers	28.70	229.60	10.6		11.0		48.1	13.80	42.50	340.00
Elec	Electricians	35.45	283.60	6.7		16.0		49.2	17.45	52.90	423.20
Elev	Elevator Constructors	37.10	296.80	7.7		16.0		50.2	18.60	55.70	445.60
Eqhv	Equipment Operators, Crane or Shovel	32.35	258.80	10.6		14.0		51.1	16.55	48.90	391.20
Eqmd	Equipment Operators, Medium Equipment	31.20	249.60	10.6		14.0		51.1	15.95	47.15	377.20
Eqlt	Equipment Operators, Light Equipment	29.80	238.40	10.6		14.0		51.1	15.25	45.05	360.40
Eqol	Equipment Operators, Oilers	26.65	213.20	10.6		14.0		51.1	13.60	40.25	322.00
Eqmm	Equipment Operators, Master Mechanics	32.80	262.40	10.6		14.0		51.1	16.75	49.55	396.40
Glaz	Glaziers	30.00	240.00	13.8		11.0		51.3	15.40	45.40	363.20
Lath	Lathers	28.75	230.00	11.1		11.0		48.6	13.95	42.70	341.60
Marb	Marble Setters	30.10	240.80	16.0		11.0		53.5	16.10	46.20	369.60
Mill	Millwrights	31.75	254.00	10.6		11.0		48.1	15.25	47.00	376.00
Mstz	Mosaic and Terrazzo Workers	29.25	234.00	9.8		11.0		47.3	13.85	43.10	344.80
Pord	Painters, Ordinary	27.15	217.20	13.8		11.0		51.3	13.95	41.10	328.80
Psst	Painters, Structural Steel	27.90	223.20	48.4		11.0		85.9	23.95	51.85	414.80
Pape	Paper Hangers	27.10	216.80	13.8		11.0		51.3	13.90	41.00	328.00
Pile	Pile Drivers	29.80	238.40	24.9		16.0		67.4	20.10	49.90	399.20
Plas	Plasterers	28.10	224.80	15.8		11.0		53.3	15.00	43.10	344.80
Plah	Plasterer Helpers	23.70	189.60	15.8		11.0		53.3	12.65	36.35	290.80
Plum	Plumbers	35.95	287.60	8.3		16.0		50.8	18.25	54.20	433.60
Rodm	Rodmen (Reinforcing)	34.25	274.00	28.3		14.0		68.8	23.55	57.80	462.40
Rofc	Roofers, Composition	26.60	212.80	32.6		11.0		70.1	18.65	45.25	362.00
Rots	Roofers, Tile and Slate	26.75	214.00	32.6		11.0		70.1	18.75	45.50	364.00
Rohe	Roofer Helpers (Composition)	19.80	158.40	32.6		11.0		70.1	13.90	33.70	269.60
Shee	Sheet Metal Workers	35.10	280.80	11.7		16.0		54.2	19.00	54.10	432.80
Spri	Sprinkler Installers	36.20	289.60	8.7		16.0		51.2	18.55	54.75	438.00
Stpi	Steamfitters or Pipefitters	36.20	289.60	8.3		16.0		50.8	18.40	54.60	436.80
Ston	Stone Masons	30.65	245.20	16.0		11.0		53.5	16.40	47.05	376.40
Sswk	Structural Steel Workers	34.25	274.00	39.8		14.0		80.3	27.50	61.75	494.00
Tilf	Tile Layers	29.15	233.20	9.8		11.0		47.3	13.80	42.95	343.60
Tilh	Tile Layer Helpers	23.35	186.80	9.8		11.0		47.3	11.05	34.40	275.20
Trlt	Truck Drivers, Light	24.30	194.40	14.9		11.0		52.4	12.75	37.05	296.40
Trhv	Truck Drivers, Heavy	25.00	200.00	14.9		11.0		52.4	13.10	38.10	304.80
Sswl	Welders, Structural Steel	34.25	274.00	39.8		14.0		80.3	27.50	61.75	494.00
Wrck	*Wrecking	23.45	187.60	41.2		11.0		78.7	18.45	41.90	335.20

*Not included in Averages.

R01100-080 Performance Bond

This table shows the cost of a Performance Bond for a construction job scheduled to be completed in 12 months. Add 1% of the premium cost per month for jobs requiring more than 12 months to complete. The rates are "standard" rates offered to contractors that the bonding company considers financially sound and capable of doing the work. Preferred rates are offered by some bonding companies based upon financial strength of the contractor. Actual rates vary from contractor to contractor and from bonding company to bonding company. Contractors should prequalify through a bonding agency before submitting a bid on a contract that requires a bond.

Contract Amount	Building Construction Class B Projects			Highways & Bridges					
				Class A New Construction			Class A-1 Highway Resurfacing		
First $ 100,000 bid	$25.00 per M			$15.00 per M			$9.40 per M		
Next 400,000 bid	$ 2,500	plus	$15.00 per M	$ 1,500	plus	$10.00 per M	$ 940	plus	$7.20 per M
Next 2,000,000 bid	8,500	plus	10.00 per M	5,500	plus	7.00 per M	3,820	plus	5.00 per M
Next 2,500,000 bid	28,500	plus	7.50 per M	19,500	plus	5.50 per M	15,820	plus	4.50 per M
Next 2,500,000 bid	47,250	plus	7.00 per M	33,250	plus	5.00 per M	28,320	plus	4.50 per M
Over 7,500,000 bid	64,750	plus	6.00 per M	45,750	plus	4.50 per M	39,570	plus	4.00 per M

R01100-090 Sales Tax by State

State sales tax on materials is tabulated below (5 states have no sales tax). Many states allow local jurisdictions, such as a county or city, to levy additional sales tax.

Some projects may be sales tax exempt, particularly those constructed with public funds.

State	Tax (%)	State	Tax (%)	State	Tax (%)	State	Tax (%)
Alabama	4	Illinois	6.25	Montana	0	Rhode Island	7
Alaska	0	Indiana	5	Nebraska	5	South Carolina	5
Arizona	5	Iowa	5	Nevada	6.5	South Dakota	4
Arkansas	4.625	Kansas	4.9	New Hampshire	0	Tennessee	6
California	7	Kentucky	6	New Jersey	6	Texas	6.25
Colorado	2.9	Louisiana	4	New Mexico	5	Utah	4.75
Connecticut	6	Maine	5.5	New York	4	Vermont	5
Delaware	0	Maryland	5	North Carolina	4	Virginia	4.5
District of Columbia	5.75	Massachusetts	5	North Dakota	5	Washington	6.5
Florida	6	Michigan	6	Ohio	5	West Virginia	6
Georgia	4	Minnesota	6.5	Oklahoma	4.5	Wisconsin	5
Hawaii	4	Mississippi	7	Oregon	0	Wyoming	4
Idaho	5	Missouri	4.225	Pennsylvania	6	Average	4.70 %

R01100-090 Sales Tax by Province (Canada)

GST - a value-added tax, which the government imposes on most goods and services provided in or imported into Canada.
PST - a retail sales tax, which five of the provinces impose on the price of most goods and some services.

QST - a value-added tax, similar to the federal GST, which Quebec imposes.
HST - Three provinces have combined their retail sales tax with the federal GST into one harmonized tax.

Province	PST (%)	QST (%)	GST(%)	HST(%)
Alberta	0	0	7	0
British Columbia	7	0	7	0
Manitoba	7	0	7	0
New Brunswick	0	0	0	15
Newfoundland	0	0	0	15
Northwest Territories	0	0	7	0
Nova Scotia	0	0	0	15
Ontario	8	0	7	0
Prince Edward Island	10.7	0	7	0
Quebec	0	8.02	7	0
Saskatchewan	6	0	7	0
Yukon	0	0	7	0

R01100-100 Unemployment Taxes and Social Security Taxes

Mass. State Unemployment tax ranges from 1.325% to 7.225% plus an experience rating assessment the following year, on the first $10,800 of wages. Federal Unemployment tax is 6.2% of the first $7,000 of wages. This is reduced by a credit for payment to the state. The minimum Federal Unemployment tax is .8% after all credits.

Combined rates in Mass. thus vary from 2.125% to 8.025% of the first $10,800 of wages. Combined average U.S. rate is about 7.0% of the first $7,000. Contractors with permanent workers will pay less since the average annual wages for skilled workers is $30.95 x 2,000 hours or about $61,900 per year. The average combined rate for U.S. would thus be 7.0% x $7,000 ÷ $61,900 = 0.8% of total wages for permanent employees.

Rates vary not only from state to state but also with the experience rating of the contractor.

Social Security (FICA) for 2002 is estimated at time of publication to be 7.65% of wages up to $80,400.

517

R01100-110 Overtime

One way to improve the completion date of a project or eliminate negative float from a schedule is to compress activity duration times. This can be achieved by increasing the crew size or working overtime with the proposed crew.

To determine the costs of working overtime to compress activity duration times, consider the following examples. Below is an overtime efficiency and cost chart based on a five, six, or seven day week with an eight through twelve hour day. Payroll percentage increases for time and one half and double time are shown for the various working days.

Days per Week	Hours per Day	Production Efficiency					Payroll Cost Factors	
		1 Week	2 Weeks	3 Weeks	4 Weeks	Average 4 Weeks	@ 1-1/2 Times	@ 2 Times
5	8	100%	100%	100%	100%	100 %	100 %	100 %
	9	100	100	95	90	96.25	105.6	111.1
	10	100	95	90	85	91.25	110.0	120.0
	11	95	90	75	65	81.25	113.6	127.3
	12	90	85	70	60	76.25	116.7	133.3
6	8	100	100	95	90	96.25	108.3	116.7
	9	100	95	90	85	92.50	113.0	125.9
	10	95	90	85	80	87.50	116.7	133.3
	11	95	85	70	65	78.75	119.7	139.4
	12	90	80	65	60	73.75	122.2	144.4
7	8	100	95	85	75	88.75	114.3	128.6
	9	95	90	80	70	83.75	118.3	136.5
	10	90	85	75	65	78.75	121.4	142.9
	11	85	80	65	60	72.50	124.0	148.1
	12	85	75	60	55	68.75	126.2	152.4

R01100-710 Unit Gross Area Requirements

The figures in the table below indicate typical ranges in square feet as a function of the "occupant" unit. This table is best used in the preliminary design stages to help determine the probable size requirement for the total project. See R17100-100 for the typical total size ranges for various types of buildings.

Building Type	Unit	Gross Area in S.F.		
		1/4	Median	3/4
Apartments	Unit	660	860	1,100
Auditorium & Play Theaters	Seat	18	25	38
Bowling Alleys	Lane		940	
Churches & Synagogues	Seat	20	28	39
Dormitories	Bed	200	230	275
Fraternity & Sorority Houses	Bed	220	315	370
Garages, Parking	Car	325	355	385
Hospitals	Bed	685	850	1,075
Hotels	Rental Unit	475	600	710
Housing for the elderly	Unit	515	635	755
Housing, Public	Unit	700	875	1,030
Ice Skating Rinks	Total	27,000	30,000	36,000
Motels	Rental Unit	360	465	620
Nursing Homes	Bed	290	350	450
Restaurants	Seat	23	29	39
Schools, Elementary	Pupil	65	77	90
Junior High & Middle		85	110	129
Senior High		102	130	145
Vocational		110	135	195
Shooting Ranges	Point		450	
Theaters & Movies	Seat		15	

R01100-720 Floor Area Ratios

Table below lists commonly used gross to net area and net to gross area ratios expressed in % for various building types.

Building Type	Gross to Net Ratio	Net to Gross Ratio	Building Type	Gross to Net Ratio	Net to Gross Ratio
Apartment	156	64	School Buildings (campus type)		
Bank	140	72	Administrative	150	67
Church	142	70	Auditorium	142	70
Courthouse	162	61	Biology	161	62
Department Store	123	81	Chemistry	170	59
Garage	118	85	Classroom	152	66
Hospital	183	55	Dining Hall	138	72
Hotel	158	63	Dormitory	154	65
Laboratory	171	58	Engineering	164	61
Library	132	76	Fraternity	160	63
Office	135	75	Gymnasium	142	70
Restaurant	141	70	Science	167	60
Warehouse	108	93	Service	120	83
			Student Union	172	59

The gross area of a building is the total floor area based on outside dimensions.

The net area of a building is the usable floor area for the function intended and excludes such items as stairways, corridors and mechanical rooms. In the case of a commercial building, it might be considered as the "leasable area."

R01100-730 Occupancy Determinations

Description		S.F. Required per Person		
		BOCA	SBC	UBC
Assembly Areas	Fixed Seats	**	6	7
	Movable Seats		15	15
	Concentrated	7		
	Unconcentrated	15		
	Standing Space	3		
Educational	Unclassified			
	Classrooms	20	40	20
	Shop Areas	50	100	50
Institutional	Unclassified		125	
	In-Patient Areas	240		
	Sleeping Areas	120		
Mercantile	Basement	30	30	20
	Ground Floor	30	30	30
	Upper Floors	60	60	50
Office		100	100	100

BOCA=Building Officials & Code Administrators
SBC=Southern Building Code
UBC=Uniform Building Code

** The occupancy load for assembly area with fixed seats shall be determined by the number of fixed seats installed.

519

R01107-030 Engineering Fees

Typical **Structural Engineering Fees** based on type of construction and total project size. These fees are included in Architectural Fees.

Type of Construction	Total Project Size (in thousands of dollars)			
	$500	$500-$1,000	$1,000-$5,000	Over $5000
Industrial buildings, factories & warehouses	Technical payroll times 2.0 to 2.5	1.60%	1.25%	1.00%
Hotels, apartments, offices, dormitories, hospitals, public buildings, food stores		2.00%	1.70%	1.20%
Museums, banks, churches and cathedrals		2.00%	1.75%	1.25%
Thin shells, prestressed concrete, earthquake resistive		2.00%	1.75%	1.50%
Parking ramps, auditoriums, stadiums, convention halls, hangars & boiler houses		2.50%	2.00%	1.75%
Special buildings, major alterations, underpinning & future expansion		Add to above 0.5%	Add to above 0.5%	Add to above 0.5%

For complex reinforced concrete or unusually complicated structures, add 20% to 50%.

Typical **Mechanical and Electrical Engineering Fees** are based on the size of the subcontract. The fee structure for Mechanical Engineering is shown below; Electrical Engineering fees range from 2 to 5%. These fees are included in Architectural Fees.

Type of Construction	Subcontract Size							
	$25,000	$50,000	$100,000	$225,000	$350,000	$500,000	$750,000	$1,000,000
Simple structures	6.4%	5.7%	4.8%	4.5%	4.4%	4.3%	4.2%	4.1%
Intermediate structures	8.0	7.3	6.5	5.6	5.1	5.0	4.9	4.8
Complex structures	12.0	9.0	9.0	8.0	7.5	7.5	7.0	7.0

For renovations, add 15% to 25% to applicable fee.

R01250-010　Repair and Remodeling

Cost figures are based on new construction utilizing the most cost-effective combination of labor, equipment and material with the work scheduled in proper sequence to allow the various trades to accomplish their work in an efficient manner.

The costs for repair and remodeling work must be modified due to the following factors that may be present in any given repair and remodeling project.

1. Equipment usage curtailment due to the physical limitations of the project, with only hand-operated equipment being used.

2. Increased requirement for shoring and bracing to hold up the building while structural changes are being made and to allow for temporary storage of construction materials on above-grade floors.

3. Material handling becomes more costly due to having to move within the confines of an enclosed building. For multi-story construction, low capacity elevators and stairwells may be the only access to the upper floors.

4. Large amount of cutting and patching and attempting to match the existing construction is required. It is often more economical to remove entire walls rather than create many new door and window openings. This sort of trade-off has to be carefully analyzed.

5. Cost of protection of completed work is increased since the usual sequence of construction usually cannot be accomplished.

6. Economies of scale usually associated with new construction may not be present. If small quantities of components must be custom fabricated due to job requirements, unit costs will naturally increase. Also, if only small work areas are available at a given time, job scheduling between trades becomes difficult and subcontractor quotations may reflect the excessive start-up and shut-down phases of the job.

7. Work may have to be done on other than normal shifts and may have to be done around an existing production facility which has to stay in production during the course of the repair and remodeling.

8. Dust and noise protection of adjoining non-construction areas can involve substantial special protection and alter usual construction methods.

9. Job may be delayed due to unexpected conditions discovered during demolition or removal. These delays ultimately increase construction costs.

10. Piping and ductwork runs may not be as simple as for new construction. Wiring may have to be snaked through walls and floors.

11. Matching "existing construction" may be impossible because materials may no longer be manufactured. Substitutions may be expensive.

12. Weather protection of existing structure requires additional temporary structures to protect building at openings.

13. On small projects, because of local conditions, it may be necessary to pay a tradesman for a minimum of four hours for a task that is completed in one hour.

All of the above areas can contribute to increased costs for a repair and remodeling project. Each of the above factors should be considered in the planning, bidding and construction stage in order to minimize the increased costs associated with repair and remodeling jobs.

R01540-100   Steel Tubular Scaffolding

On new construction, tubular scaffolding is efficient up to 60' high or five stories. Above this it is usually better to use a hung scaffolding if construction permits. Swing scaffolding operations may interfere with tenants. In this case, the tubular is more practical at all heights.

In repairing or cleaning the front of an existing building the cost of tubular scaffolding per S.F. of building front increases as the height increases above the first tier. The first tier cost is relatively high due to leveling and alignment.

The minimum efficient crew for erection is three workers. For heights over 50', a crew of four is more efficient. Use two or more on top and two at the bottom for handing up or hoisting. Four workers can erect and dismantle about nine frames per hour up to five stories. From five to eight stories they will average six frames per hour. With 7' horizontal spacing this will run about 400 S.F. and 265 S.F. of wall surface, respectively. Time for placing planks must be added to the above. On heights above 50', five planks can be placed per labor-hour.

The table below shows the number of pieces required to erect tubular steel scaffolding for 1000 S.F. of building frontage. This area is made up of a scaffolding system that is 12 frames (11 bays) long by 2 frames high.

For jobs under twenty-five frames, add 50% to rental cost. Rental rates will be lower for jobs over three months duration. Large quantities for long periods can reduce rental rates by 20%.

Description of Component	CSI Line Item	Number of pieces for 1000 S.F. of Building Front	Unit
5' Wide Standard Frame, 6'-4" High	01540-750-2200	24	Ea.
Leveling Jack & Plate	01540-750-2650	24	
Cross Brace	01540-750-2500	44	
Side Arm Bracket, 21"	01540-750-2700	12	
Guardrail Post	01540-750-2550	12	
Guardrail, 7' section	01540-750-2600	22	
Stairway Section	01540-750-2900	2	
Stairway Starter Bar	01540-750-2910	1	
Stairway Inside Handrail	01540-750-2920	2	
Stairway Outside Handrail	01540-750-2930	2	
Walk-Thru Frame Guardrail	01540-750-2940	2	

Scaffolding is often used as falsework over 15' high during construction of cast-in-place concrete beams and slabs. Two foot wide scaffolding is generally used for heavy beam construction. The span between frames depends upon the load to be carried with a maximum span of 5'.

Heavy duty shoring frames with a capacity of 10,000#/leg can be spaced up to 10' O. C. depending upon form support design and loading.

Scaffolding used as horizontal shoring requires less than half the material required with conventional shoring.

On new construction, erection is done by carpenters.

Rolling towers supporting horizontal shores can reduce labor and speed the job. For maintenance work, catwalks with spans up to 70' can be supported by the rolling towers.

R01590-100   Contractor Equipment

Rental Rates shown in the front of the book pertain to late model high quality machines in excellent working condition, rented from equipment dealers. Rental rates from contractors may be substantially lower than the rental rates from equipment dealers depending upon economic conditions. For older, less productive machines, reduce rates by a maximum of 15%. Any overtime must be added to the base rates. For shift work, rates are lower. Usual rule of thumb is 150% of one shift rate for two shifts; 200% for three shifts.

For periods of less than one week, operated equipment is usually more economical to rent than renting bare equipment and hiring an operator.

Equipment moving and mobilization costs must be added to rental rates where applicable. A large crane, for instance, may take two days to erect and two days to dismantle.

Rental rates vary throughout the country with larger cities generally having lower rates. Lease plans for new equipment are available for periods in excess of six months with a percentage of payments applying toward purchase.

Monthly rental rates vary from 2% to 5% of the cost of the equipment depending on the anticipated life of the equipment and its wearing parts. Weekly rates are about 1/3 the monthly rates and daily rental rates about 1/3 the weekly rate.

The hourly operating costs for each piece of equipment include costs to the user such as fuel, oil, lubrication, normal expendables for the equipment, and a percentage of mechanic's wages chargeable to maintenance. The hourly operating costs listed do not include the operator's wages.

The daily cost for equipment used in the standard crews is figured by dividing the weekly rate by five, then adding eight times the hourly operating cost to give the total daily equipment cost, not including the operator. This figure is in the right hand column of Division 01590 under Crew Equipment Cost/Day.

Pile Driving rates shown for pile hammer and extractor do not include leads, crane, boiler or compressor. Vibratory pile driving requires an added field specialist during set-up and pile driving operation for the electric model. The hydraulic model requires a field specialist for set-up only. Up to 125 reuses of sheet piling are possible using vibratory drivers. For normal conditions, crane capacity for hammer type and size are as follows.

Crane Capacity	Hammer Type and Size		
	Air or Steam	Diesel	Vibratory
25 ton	to 8,750 ft.-lb.		70 H.P.
40 ton	15,000 ft.-lb.	to 32,000 ft.-lb.	170 H.P.
60 ton	25,000 ft.-lb.		300 H.P.
100 ton		112,000 ft.-lb.	

Cranes should be specified for the job by size, building and site characteristics, availability, performance characteristics, and duration of time required.

Backhoes & Shovels rent for about the same as equivalent size cranes but maintenance and operating expense is higher. Crane operators rate must be adjusted for high boom heights. Average adjustments: for 150' boom add 2% per hour; over 185', add 4% per hour; over 210', add 6% per hour; over 250', add 8% per hour and over 295', add 12% per hour.

Tower Cranes of the climbing or static type have jibs from 50' to 200' and capacities at maximum reach range from 4,000 to 14,000 pounds. Lifting capacities increase up to maximum load as the hook radius decreases.

Typical rental rates, based on purchase price are about 2% to 3% per month.

Erection and dismantling runs between 500 and 2000 labor hours. Climbing operation takes 10 labor hours per 20' climb. Crane dead time is about 5 hours per 40' climb. If crane is bolted to side of the building add cost of ties and extra mast sections. Climbing cranes have from 80' to 180' of mast while static cranes have 80' to 800' of mast.

Truck Cranes can be converted to tower cranes by using tower attachments. Mast heights over 400' have been used. See Division 01590-600 for rental rates of high boom cranes.

A single 100' high material **Hoist and Tower** can be erected and dismantled in about 400 labor hours; a double 100' high hoist and tower in about 600 labor hours. Erection times for additional heights are 3 and 4 labor hours per vertical foot respectively up to 150', and 4 to 5 labor hours per vertical foot over 150' high. A 40' high portable Buck hoist takes about 160 labor hours to erect and dismantle. Additional heights take 2 labor hours per vertical foot to 80' and 3 labor hours per vertical foot for the next 100'. Most material hoists do not meet local code requirements for carrying personnel.

A 150' high **Personnel Hoist** requires about 500 to 800 labor hours to erect and dismantle. Budget erection time at 5 labor hours per vertical foot for all trades. Local code requirements or labor scarcity requiring overtime can add up to 50% to any of the above erection costs.

Earthmoving Equipment: The selection of earthmoving equipment depends upon the type and quantity of material, moisture content, haul distance, haul road, time available, and equipment available. Short haul cut and fill operations may require dozers only, while another operation may require excavators, a fleet of trucks, and spreading and compaction equipment. Stockpiled material and granular material are easily excavated with front end loaders. Scrapers are most economically used with hauls between 300' and 1-1/2 miles if adequate haul roads can be maintained. Shovels are often used for blasted rock and any material where a vertical face of 8' or more can be excavated. Special conditions may dictate the use of draglines, clamshells, or backhoes. Spreading and compaction equipment must be matched to the soil characteristics, the compaction required and the rate the fill is being supplied.

2 SITE CONSTRUCTION

R02115-200 Underground Storage Tank Removal

Underground Storage Tank Removal can be divided into two categories: Non-Leaking and Leaking. Prior to removing an underground storage tank, tests should be made, with the proper authorities present, to determine whether a tank has been leaking or the surrounding soil has been contaminated.

To safely remove Liquid Underground Storage Tanks:
1. Excavate to the top of the tank.
2. Disconnect all piping.
3. Open all tank vents and access ports.
4. Remove all liquids and/or sludge.
5. Purge the tank with an inert gas.
6. Provide access to the inside of the tank and clean out the interior using proper personal protective equipment (PPE).
7. Excavate soil surrounding the tank using proper PPE for on-site personnel.
8. Pull and properly dispose of the tank.
9. Clean up the site of all contaminated material.
10. Install new tanks or close the excavation.

R02240-900 Wellpoints

A single stage wellpoint system is usually limited to dewatering an average 15' depth below normal ground water level. Multi-stage systems are employed for greater depth with the pumping equipment installed only at the lowest header level. Ejectors, with unlimited lift capacity, can be economical when two or more stages of wellpoints can be replaced or when horizontal clearance is restricted, such as in deep trenches or tunneling projects, and where low water flows are expected. Wellpoints are usually spaced on 2-1/2' to 10' centers along a header pipe. Wellpoint spacing, header size, and pump size are all determined by the expected flow, as dictated by soil conditions.

In almost all soils encountered in wellpoint dewatering, the wellpoints may be jetted into place. Cemented soils and stiff clays may require sand wicks about 12" in diameter around each wellpoint to increase efficiency and eliminate weeping into the excavation. These sand wicks require 1/2 to 3 C.Y. of washed filter sand and are installed by using a 12" diameter steel casing and hole puncher jetted into the ground 2' deeper than the wellpoint. Rock may require predrilled holes.

Labor required for the complete installation and removal of a single stage wellpoint system is in the range of 3/4 to 2 labor-hours per linear foot of header, depending upon jetting conditions, wellpoint spacing, etc.

Continuous pumping is necessary except in some free draining soil where temporary flooding is permissible (as in trenches which are backfilled after each day's work). Good practice requires provision of a stand-by pump during the continuous pumping operation.

Systems for continuous trenching below the water table should be installed three to four times the length of expected daily progress to insure uninterrupted digging, and header pipe size should not be changed during the job.

For pervious free draining soils, deep wells in place of wellpoints may be economical because of lower installation and maintenance costs. Daily production ranges between two to three wells per day, for 25' to 40' depths, to one well per day for depths over 50'.

Detailed analysis and estimating for any dewatering problem is available at no cost from wellpoint manufacturers. Major firms will quote "sufficient equipment" quotes or their affiliates offer lump sum proposals to cover complete dewatering responsibility.

	Description for 200' System with 8" Header	Quantities	1st Month		Thereafter	
			Unit	Total	Unit	Total
Equipment & Material	Wellpoints 25' long, 2" diameter @ 5' O.C.	40 Each	$ 29.00	$ 1,160.00	$ 21.75	$ 870.00
	Header pipe, 8" diameter	200 L.F.	5.50	1,100.00	2.75	550.00
	Discharge pipe, 8" diameter	100 L.F.	3.72	372.00	2.42	242.00
	8" valves	3 Each	143.00	429.00	92.95	278.85
	Combination Jetting & Wellpoint pump (standby)	1 Each	2,300.00	2,300.00	1,610.00	1,610.00
	Wellpoint pump, 8" diameter	1 Each	2,275.00	2,275.00	1,592.50	1,592.50
	Transportation to and from site	1 Day	670.60	670.60	—	—
	Fuel 30 days x 60 gal./day	1800 Gallons	1.05	1,890.00	1.05	1,890.00
	Lubricants for 30 days x 16 lbs./day	480 Lbs.	1.10	528.00	1.10	528.00
	Sand for points	40 C.Y.	18.00	720.00	—	—
	Equipment & Materials Subtotal			$11,444.60		$ 7,561.35
Labor	Technician to supervise installation	1 Week	$ 32.80	$ 1,312.00	—	—
	Labor for installation and removal of system	300 Labor-hours	23.45	7,035.00	—	—
	4 Operators straight time 40 hrs./wk. for 4.33 wks.	693 Hrs.	29.80	20,651.40	$ 29.80	$20,651.40
	4 Operators overtime 2 hrs./wk. for 4.33 wks.	35 Hrs.	59.60	2,086.00	59.60	2,086.00
	Bare Labor Subtotal			$31,084.40		$22,737.40
	Total Bare Cost			$42,529.00		$30,298.75
	Monthly Bare Cost/L.F. Header			$ 212.65		$ 151.49

R02250-400   Wood Sheet Piling

Wood sheet piling may be used for depths to 20' where there is no ground water. If moderate ground water is encountered Tongue & Groove sheeting will help to keep it out. When considerable ground water is present, steel sheeting must be used.

For estimating purposes on trench excavation, sizes are as follows:

Depth	Sheeting	Wales	Braces	B.F. per S.F.
To 8'	3 x 12's	6 x 8's, 2 line	6 x 8's, @ 10'	4.0 @ 8'
8' x 12'	3 x 12's	10 x 10's, 2 line	10 x 10's, @ 9'	5.0 average
12' to 20'	3 x 12's	12 x 12's, 3 line	12 x 12's, @ 8'	7.0 average

Sheeting to be toed in at least 2' depending upon soil conditions. A five person crew with an air compressor and sheeting driver can drive and brace 440 SF/day at 8' deep, 360 SF/day at 12' deep, and 320 SF/day at 16' deep. For normal soils, piling can be pulled in 1/3 the time to install. Pulling difficulty increases with the time in the ground. Production can be increased by high pressure jetting. Figures below assume 50% of lumber is salvaged and includes pulling costs. Some jurisdictions require an equipment operator in addition to Crew B-31.

Sheeting Pulled

Daily Cost Crew B-31	L.H./ Day	Hourly Cost	Daily Cost	8' Depth, 440 S.F./Day		16' Depth, 320 S.F./Day	
				To Drive (1 Day)	To Pull (1/3 Day)	To Drive (1 Day)	To Pull (1/3 Day)
1 Foreman	8	$25.45	$ 203.60	$ 203.60	$ 67.80	$ 203.60	$ 67.80
3 Laborers	24	23.45	562.80	$ 562.80	187.41	$ 562.80	187.41
1 Carpenter	8	30.00	240.00	$ 240.00	79.92	$ 240.00	79.92
1 Air Compressor			135.00	$ 135.00	44.96	$ 135.00	44.96
1 Sheeting Driver			10.20	$ 10.20	3.40	$ 10.20	3.40
2 -50 Ft. Air Hoses, 1-1/2" Diam.			18.80	$ 18.80	6.26	$ 18.80	6.26
Lumber (50% salvage)				1.76 MBF 642.40		2.24 MBF 817.60	
Bare Total			$1,170.40	$1,812.80	$389.75	$1,988.00	$389.75
Bare Total/S.F.				$ 4.12	$ .89	$ 4.52	$ 1.22
Bare Total (Drive and Pull)/S.F.				$ 5.01		$ 5.74	

Sheeting Left in Place

Daily Cost	8' Depth 440 S.F./Day		10' Depth 400 S.F./Day		12' Depth 360 S.F./Day		16' Depth 320 S.F./Day		18' Depth 305 S.F./Day		20' Depth 280 S.F./Day	
Crew B-31		$1,161.00		$1,161.00		$1,161.00		$1,161.00		$1,161.00		$1,161.00
Lumber	1.76 MBF	1,284.80	1.8 MBF	1,314.00	1.8 MBF	1,314.00	2.24 MBF	1,635.20	2.1 MBF	1,533.00	1.9 MBF	1,387.00
Bare Total in Place		$2,445.80		$2,475.00		$2,475.00		$2,796.20		$2,694.00		$2,548.00
Bare Total/S.F.		$ 5.56		$ 6.19		$ 6.88		$ 8.74		$ 8.83		$ 9.10
Bare Total/M.B.F.		$1,389.66		$1,375.00		$1,375.00		$1,248.30		$1,282.86		$1,341.05

R02250-450 Steel Sheet Piling

Limiting weights are 22 to 38#/S.F. of wall surface with 27#/S.F. average for usual types and sizes. (Weights of piles themselves are from 30.7#/L.F. to 57#/L.F. but they are 15" to 21" wide.) Lightweight sections 12" to 28" wide from 3 ga. to 12 ga. thick are also available for shallow excavations. Piles may be driven two at a time with an impact or vibratory hammer (use vibratory to pull) hung from a crane without leads. A reasonable estimate of the life of steel sheet piling is 10 uses with up to 125 uses possible if a vibratory hammer is used. Used piling costs from 50% to 80% of new piling depending on location and market conditions. Sheet piling and H piles can be rented for about 30% of the delivered mill price for the first month and 5% per month thereafter. Allow 1 labor-hour per pile for cleaning and trimming after driving. These costs increase with depth and hydrostatic head. Vibratory drivers are faster in wet granular soils and are excellent for pile extraction. Pulling difficulty increases with the time in the ground and may cost more than driving. It is often economical to abandon the sheet piling, especially if it can be used as the outer wall form. Allow about 1/3 additional length or more, for toeing into ground. Add bracing, waler and strut costs. Waler costs can equal the cost per ton of sheeting.

Cost of Sheet Piling & Production Rate by Ton & S.F.

Depth of Excavation	15' Depth		20' Depth		25' Depth	
Description of Pile	22 psf = 90.9 S.F./Ton		27 psf = 74 S.F./Ton		38 psf = 52.6 S.F./Ton	
Type of operation:	Drive &	Drive &	Drive &	Drive &	Drive &	Drive &
Left in Place or Removed	Left	Extract*	Left	Extract*	Left	Extract*
Labor & Equip. to Drive 1 ton	$ 382.89	$689.83	$ 319.61	$631.91	$ 217.84	$394.19
Piling (75% Salvage)	845.00	211.25	845.00	211.25	845.00	211.25
Bare Cost/Ton in Place	$1,227.89	$901.08	$1,164.61	$843.16	$1,062.84	$605.44
Production Rate, Tons/Day	10.81	6.00	12.95	6.55	19.00	10.50
Bare Cost, S.F. in Place (incl. 33% toe in)	$ 13.51	$ 9.91	$ 15.74	$ 11.39	$ 20.21	$ 11.51
Production Rate, S.F./Day (incl. 33% toe in)	983	545	959	485	1000	553

Crew B-40	Bare Costs		Inc. Subs O&P		Installation Cost per Ton
	Hr.	Daily	Hr.	Daily	
1 Pile Driver Foreman (Out)	$31.80	$ 254.40	$53.25	$ 426.00	for 15' Deep Excavation plus 33% toe-in
4 Pile Drivers	29.80	953.60	49.90	1596.80	$4,139 per day/10.81 tons = $382.88 per ton
2 Equip. Oper. (crane)	32.35	517.60	48.90	782.40	Installation Cost per S.F.
1 Equip. Oper. Oiler	26.65	213.20	40.25	322.00	$\frac{2000\#}{22\#/S.F.}$ = 90.9 S.F. x 10.81 tons = 983 S.F. per day
1 Crane, 40 Ton		809.20		890.10	
Vibratory Hammer & Gen.		1391.00		1530.10	$4,139 / 983 S.F. per day = $4.21 per S.F.
64 L.H., Daily Totals		$4,139.00		$5,547.40	

*For driving & extracting, two mobilizations & demobilizations will be necessary

R13281-120 Asbestos Removal Process

Asbestos removal is accomplished by a specialty contractor who understands the federal and state regulations regarding the handling and disposal of the material. The process of asbestos removal is divided into many individual steps. An accurate estimate can be calculated only after all the steps have been priced.

The steps are generally as follows:

1. Obtain an asbestos abatement plan from an industrial hygienist.
2. Monitor the air quality in and around the removal area and along the path of travel between the removal area and transport area. This establishes the background contamination.
3. Construct a two part decontamination chamber at entrance to removal area.
4. Install a HEPA filter to create a negative pressure in the removal area.
5. Install wall, floor and ceiling protection as required by the plan, usually 2 layers of fireproof 6 mil polyethylene.

6. Industrial hygienist visually inspects work area to verify compliance with plan.
7. Provide temporary supports for conduit and piping affected by the removal process.
8. Proceed with asbestos removal and bagging process. Monitor air quality as described in Step #2. Discontinue operations when contaminate levels exceed applicable standards.
9. Document the legal disposal of materials in accordance with EPA standards.
10. Thoroughly clean removal area including all ledges, crevices and surfaces.
11. Post abatement inspection by industrial hygienist to verify plan compliance.
12. Provide a certificate from a licensed industrial hygienist attesting that contaminate levels are within acceptable standards before returning area to regular use.

R13600-610 Solar Heating (Space and Hot Water)

Collectors should face as close to due South as possible, however, variations of up to 20 degrees on either side of true South are acceptable. Local climate and collector type may influence the choice between east or west deviations. Obviously they should be located so they are not shaded from the sun's rays. Incline collectors at a slope of latitude minus 5 degrees for domestic hot water and latitude plus 15 degrees for space heating.

Flat plate collectors consist of a number of components as follows: Insulation to reduce heat loss through the bottom and sides of the collector. The enclosure which contains all the components in this assembly is usually weatherproof and prevents dust, wind and water from coming in contact with the absorber plate. The cover plate usually consists of one or more layers of a variety of glass or plastic and reduces the reradiation by creating an air space which traps the heat between the cover and the absorber plates.

The absorber plate must have a good thermal bond with the fluid passages. The absorber plate is usually metallic and treated with a surface coating which improves absorptivity. Black or dark paints or selective coatings are used for this purpose, and the design of this passage and plate combination helps determine a solar system's effectiveness.

Heat transfer fluid passage tubes are attached above and below or integral with an absorber plate for the purpose of transferring thermal energy from the absorber plate to a heat transfer medium. The heat exchanger is a device for transferring thermal energy from one fluid to another.

Piping and storage tanks should be well insulated to minimize heat losses.

Size domestic water heating storage tanks to hold 20 gallons of water per user, minimum, plus 10 gallons per dishwasher or washing machine. For domestic water heating an optimum collector size is approximately 3/4 square foot of area per gallon of water storage. For space heating of residences and small commercial applications the collector is commonly sized between 30% and 50% of the internal floor area. For space heating of large commercial applications, collector areas less than 30% of the internal floor area can still provide significant heat reductions.

A supplementary heat source is recommended for Northern states for December through February.

The solar energy transmission per square foot of collector surface varies greatly with the material used. Initial cost, heat transmittance and useful life are obviously interrelated.

SPECIAL CONSTRUCTION 13

REFERENCE NOS.

R15050-710 Subcontractors

On the unit cost pages of the R.S. Means Cost Data books, the last column is entitled "Total Incl. O&P". This is normally the cost of the installing contractor. In Division 15, this is the cost of the mechanical contractor. If the particular work being estimated is to be performed by a sub to the mechanical contractor, the mechanical's profit and handling charge (usually 10%) is added to the total of the last column.

R15050-720 Demolition (Selective vs. Removal for Replacement)

Demolition can be divided into two basic categories.

One type of demolition involves the removal of material with no concern for its replacement. The labor-hours to estimate this work are found in Div. 15055 under "Selective Demolition". It is selective in that individual items or all the material installed as a system or trade grouping such as plumbing or heating systems are removed. This may be accomplished by the easiest way possible, such as sawing, torch cutting, or sledge hammer as well as simple unbolting.

The second type of demolition is the removal of some item for repair or replacement. This removal may involve careful draining, opening of unions, disconnecting and tagging of electrical connections, capping of pipes/ducts to prevent entry of debris or leakage of the material contained as well as transport of the item away from its in-place location to a truck/dumpster. An approximation of the time required to accomplish this type of demolition is to use half of the time indicated as necessary to install a new unit. For example; installation of a new pump might be listed as requiring 6 labor-hours so if we had to estimate the removal of the old pump we would allow an additional 3 hours for a total of 9 hours. That is, the complete replacement of a defective pump with a new pump would be estimated to take 9 labor-hours.

R15100-050 Pipe Material Considerations

1. Malleable fittings should be used for gas service.
2. Malleable fittings are used where there are stresses/strains due to expansion and vibration.
3. Cast fittings may be broken as an aid to disassembling of heating lines frozen by long use, temperature and minerals.
4. Cast iron pipe is extensively used for underground and submerged service.
5. Type M (light wall) copper tubing is available in hard temper only and is used for nonpressure and less severe applications than K and L.
6. Type L (medium wall) copper tubing, available hard or soft for interior service.
7. Type K (heavy wall) copper tubing, available in hard or soft temper for use where conditions are severe. For underground and interior service.
8. Hard drawn tubing requires fewer hangers or supports but should not be bent. Silver brazed fittings are recommended, however soft solder is normally used.
9. Type DMV (very light wall) copper tubing designed for drainage, waste and vent plus other non-critical pressure services.

Domestic/Imported Pipe and Fittings Cost

The prices shown in this publication for steel/cast iron pipe and steel, cast iron, malleable iron fittings are based on domestic production sold at the normal trade discounts. The above listed items of foreign manufacture may be available at prices of 1/3 to 1/2 those shown. Some imported items after minor machining or finishing operations are being sold as domestic to further complicate the system.

Caution: Most pipe prices in this book also include a coupling and pipe hangers which for the larger sizes can add significantly to the per foot cost and should be taken into account when comparing "book cost" with quoted supplier's cost.

R15100-070 Piping to 10' High

When taking off pipe, it is important to identify the different material types and joining procedures, as well as distances between supports and components required for proper support.

During the takeoff, measure through all fittings. Do not subtract the lengths of the fittings, valves, or strainers, etc. This added length plus the final rounding of the totals will compensate for nipples and waste.

When rounding off totals always increase the actual amount to correspond with manufacturer's shipping lengths.

A. Both red brass and yellow brass pipe are normally furnished in 12' lengths, plain end. Division 15107-220 includes in the linear foot costs two field threads and one coupling per 10' length. A carbon steel clevis type hanger assembly every 10' is also prorated into the linear foot costs, including both material and labor.

B. Cast iron soil pipe is furnished in either 5' or 10' lengths. For pricing purposes, Division 15107-320 in *Means Plumbing Cost Data* features 10' lengths with a joint and a carbon steel clevis hanger assembly every 5' prorated into the per foot costs of both material and labor.

Three methods of joining are considered in Division 15107-320, lead and oakum poured joints, or push-on gasket type joints for the bell and spigot pipe, and a joint clamp for the no-hub soil pipe. The labor and material costs for each of these individual joining procedures are also prorated into the linear costs per foot.

C. Copper tubing in Division 15107-420 covers types K, L, M, and DWV which are furnished in 20' lengths. Means pricing data is based on a tubing cut each length and a coupling and two soft soldered joints every 10'. A carbon steel, clevis type hanger assembly every 10' is also prorated into the per foot costs. The prices for refrigeration tubing are for materials only. Labor for full lengths may be based on the type L labor but short cut measures in tight areas can increase the installation labor-hours from 20 to 40%.

D. Corrosion-resistant piping does not lend itself to one particular standard of hanging or support assembly due to its diversity of application and placement. The several varieties of corrosion-resistant piping do not include any material or labor costs for hanger assemblies (See Division 15060-300 for appropriate selection). The estimator should refer to Division 15107-560 in *Means Plumbing Cost Data* for the compatible corrosion-resistant fittings and piping.

E. Glass pipe, as shown in *Means Plumbing Cost Data* Division 15106-120, is furnished in standard lengths either 5' or 10' long, beaded on one end. Special orders for diverse lengths beaded on both ends are also available. For pricing purposes, R.S. Means features 10' lengths with a coupling and a carbon steel band hanger assembly every 10' prorated into the per foot linear costs.

Glass pipe is also available with conical ends and standard lengths ranging from 6" through 3' in 6" increments, then up to 10' in 12" increments. Special lengths can be customized for particular installation requirements.

For pricing purposes, Means has based the labor and material pricing on 10' lengths. Included in these costs per linear foot are the prorated costs for a flanged assembly every 10' consisting of two flanges, a gasket, two insertable seals, and the required number of bolts and nuts. A carbon steel band hanger assembly based on 10' center lines has also been prorated into the costs per foot for labor and materials.

F. Plastic pipe of several compositions and joining methods are considered in Division 15108-520. Fiberglass reinforced pipe (FRP) is priced, based on 10' lengths (20' lengths are also available), with coupling and epoxy joints every 10'. FRP is furnished in both "General Service" and "High Strength." A carbon steel clevis hanger assembly, 3 for every 10', is built into the prorated labor and material costs on a per foot basis.

The PVC and CPVC pipe schedules 40, 80 and 120, plus SDR ratings are all based on 20' lengths with a coupling installed every 10', as well as a carbon steel clevis hanger assembly every 3'. The PVC and ABS type DWV piping is based on 10' lengths with solvent weld couplings every 10', and with carbon steel clevis hanger assemblies, 3 for every 10'. The rest of the plastic piping in this section is based on flexible 100' coils and does not include any coupling or supports.

This section ends with PVC drain and sewer piping based on 10' lengths with bell and spigot ends and O-ring type, push-on joints. Material costs only are given, but the complete installation costs should be similar to those shown in line number 02620-270.

G. Stainless steel piping, Division 15107-900, includes both weld end and threaded piping, both in the type 304 and 316 specification and in the following schedules, 5, 10, 40, 80, and 160. Although this piping is usually furnished in 20' lengths, this cost grouping has a joint (either heli-arc butt-welded or threads and coupling) every 10'. A carbon steel clevis type hanger assembly is also included at 10' intervals and prorated into the linear foot costs.

H. Division 15107-620 covers carbon steel pipe, both black and galvanized. This section encompasses schedules 40 (standard) and 80 (extra heavy).

Several common methods of joining steel pipe — such as thread and coupled, butt welded, and flanged (150 lb. weld neck flanges) are also included.

For estimating purposes, it is assumed that the piping is purchased in 20' lengths and that a compatible joint is made up every 10'. These joints are prorated into the labor and material costs per linear foot. The following hanger and support assemblies every 10' are also included: carbon steel clevis for the T & C pipe, and single rod roll type for both the welded and flanged piping. All of these hangers are oversized to accommodate pipe insulation 3/4" thick through 5" pipe size and 1-1/2" thick from 6" through 12" pipe size.

I. Division 15107-690 prices grooved joint steel pipe, both black and galvanized, in schedules 10, 40, and 80, furnished in 20' lengths. This section describes two joining methods: cut groove and roll groove. The schedule 10 piping is roll-grooved, while the heavier schedules are cut-grooved. The labor and material costs are prorated into per linear foot prices, including a coupled joint every 10', as well as a carbon steel clevis hanger assembly.

Notes:

The pipe hanger assemblies mentioned in the preceding paragraphs include the described hanger; appropriately sized steel, box-type insert and nut; plus a threaded hanger rod. On average, the distance from the pipe center line to the insert face is 2'.

C clamps are used when the pipe is to be supported from steel shapes rather than anchored in the slab. C clamps are slightly less costly than inserts. However, to save time in estimating, it is advisable to use the given line number cost, rather than substituting a C clamp for the insert.

Add to piping labor for elevated installation:

10' to 15' high	10%	30' to 35' high	40%
15' to 20' high	20%	35' to 40' high	50%
20' to 25' high	25%	Over 40' high	55%
25' to 30' high	35%		

When using the percentage adds for elevated piping installations as shown above, bear in mind that the given heights are for the pipe supports, even though the insert, anchor, or clamp may be several feet higher than the pipe itself.

An allowance has been included in the piping installation time for testing and minor tightening of leaking joints, fittings, stuffing boxes, packing glands, etc. For extraordinary test requirements such as x-rays, prolonged pressure or demonstration tests, a percentage of the piping labor, based on the estimator's experience, must be added to the labor total. A testing service specializing in weld x-rays should be consulted for pricing if it is an estimate requirement. Equipment installation time includes start-up with associated adjustments.

R15100-080 Valve Materials

VALVE MATERIALS

Bronze:
Bronze is one of the oldest materials used to make valves. It is most commonly used in hot and cold water systems and other non-corrosive services. It is often used as a seating surface in larger iron body valves to ensure tight closure.

Carbon Steel:
Carbon steel is a high strength material. Therefore, valves made from this metal are used in higher pressure services, such as steam lines up to 600 psi at 850°F. Many steel valves are available with butt-weld ends for economy and are generally used in high pressure steam service as well as other higher pressure non-corrosive services.

Forged Steel:
Valves from tough carbon steel are used in service up to 2000 psi and temperatures up to 1000°F in Gate, Globe and Check valves.

Iron:
Valves are normally used in medium to large pipe lines to control non-corrosive fluid and gases, where pressures do not exceed 250 psi at 450° or 500 psi cold water, oil or gas.

Stainless Steel:
Developed steel alloys can be used in over 90% corrosive services.

Plastic PVC:
This is used in a great variety of valves generally in high corrosive service with lower temperatures and pressures.

VALVE SERVICE PRESSURES

Pressure ratings on valves provide an indication of the safe operating pressure for a valve at some elevated temperature. This temperature is dependent upon the materials used and the fabrication of the valve. When specific data is not available, a good "rule-of-thumb" to follow is the temperature of saturated steam on the primary rating indicated on the valve body. Example: The valve has the number 150S printed on the side indicating 150 psi and hence, a maximum operating temperature of 367°F (temperature of saturated steam and 150 psi).

DEFINITIONS

1. "WOG" – Water, oil, gas (cold working pressures).
2. "SWP" – Steam working pressure.
3. 100% area (full port) – means the area through the valve is equal to or greater than the area of standard pipe.
4. "Standard Opening" – means that the area through the valve is less than the area of standard pipe and therefore these valves should be used only where restriction of flow is unimportant.
5. "Round Port" – means the valve has a full round opening through the plug and body, of the same size and area as standard pipe.
6. "Rectangular Port" – valves have rectangular shaped ports through the plug body. The area of the port is either equal to 100% of the area of standard pipe, or restricted (standard opening). In either case it is clearly marked.
7. "ANSI" – American National Standards Institute.

R15100-090 Valve Selection Considerations

INTRODUCTION: In any piping application, valve performance is critical. Valves should be selected to give the best performance at the lowest cost.

The following is a list of performance characteristics generally expected of valves.

1. Stopping flow or starting it.
2. Throttling flow (Modulation).
3. Flow direction changing.
4. Checking backflow (Permitting flow in only one direction).
5. Relieving or regulating pressure.

In order to properly select the right valve, some facts must be determined.

A. What liquid or gas will flow through the valve?
B. Does the fluid contain suspended particles?
C. Does the fluid remain in liquid form at all times?
D. Which metals does fluid corrode?
E. What are the pressure and temperature limits? (As temperature and pressure rise, so will the price of the valve.)
F. Is there constant line pressure?
G. Is the valve merely an on-off valve?
H. Will checking of backflow be required?
I. Will the valve operate frequently or infrequently?

Valves are classified by design type into such classifications as Gate, Globe, Angle, Check, Ball, Butterfly and Plug. They are also classified by end connection, stem, pressure restrictions and material such as bronze, cast iron, etc. Each valve has a specific use. A quality valve used correctly will provide a lifetime of trouble-free service, but a high quality valve installed in the wrong service may require frequent attention.

STEM TYPES
(OS & Y)—Rising Stem-Outside Screw and Yoke

Offers a visual indication of whether the valve is open or closed. Recommended where high temperatures, corrosives, and solids in the line might cause damage to inside-valve stem threads. The stem threads are engaged by the yoke bushing so the stem rises through the hand wheel as it is turned.

(R.S.)—Rising Stem-Inside Screw

Adequate clearance for operation must be provided because both the hand wheel and the stem rise.
The valve wedge position is indicated by the position of the stem and hand wheel.

(N.R.S.)—Non-Rising Stem-Inside Screw

A minimum clearance is required for operating this type of valve. Excessive wear or damage to stem threads inside the valve may be caused by heat, corrosion, and solids. Because the hand wheel and stem do not rise, wedge position cannot be visually determined.

VALVE TYPES
Gate Valves

Provide full flow, minute pressure drop, minimum turbulence and minimum fluid trapped in the line.
They are normally used where operation is infrequent.

Globe Valves

Globe valves are designed for throttling and/or frequent operation with positive shut-off. Particular attention must be paid to the several types of seating materials available to avoid unnecessary wear. The seats must be compatible with the fluid in service and may be composition or metal. The configuration of the Globe valve opening causes turbulence which results in increased resistance. Most bronze Globe valves are rising stem-inside screw, but they are also available on O.S. & Y.

Angle Valves

The fundamental difference between the Angle valve and the Globe valve is the fluid flow through the Angle valve. It makes a 90° turn and offers less resistance to flow than the Globe valve while replacing an elbow. An Angle valve thus reduces the number of joints and installation time.

MECHANICAL **15**

REFERENCE NOS.

Check Valves

Check valves are designed to prevent backflow by automatically seating when the direction of fluid is reversed.
Swing Check valves are generally installed with Gate-valves, as they provide comparable full flow. Usually recommended for lines where flow velocities are low and should not be used on lines with pulsating flow. Recommended for horizontal installation, or in vertical lines only where flow is upward.

Lift Check Valves

These are commonly used with Globe and Angle valves since they have similar diaphragm seating arrangements and are recommended for preventing backflow of steam, air, gas and water, and on vapor lines with high flow velocities. For horizontal lines, horizontal lift checks should be used and vertical lift checks for vertical lines.

Ball Valves

Ball valves are light and easily installed, yet because of modern elastomeric seats, provide tight closure. Flow is controlled by rotating up to 90° a drilled ball which fits tightly against resilient seals. This ball seats with flow in either direction, and valve handle indicates the degree of opening. Recommended for frequent operation readily adaptable to automation, ideal for installation where space is limited.

Butterfly Valves

Butterfly valves provide bubble-tight closure with excellent throttling characteristics. They can be used for full-open, closed and for throttling applications.

The Butterfly valve consists of a disc within the valve body which is controlled by a shaft. In its closed position, the valve disc seals against a resilient seat. The disc position throughout the full 90° rotation is visually indicated by the position of the operator.

A Butterfly valve is only a fraction of the weight of a Gate valve and requires no gaskets between flanges in most cases. Recommended for frequent operation and adaptable to automation where space is limited.

Wafer and Lug type bodies when installed between two pipe flanges, can be easily removed from the line. The pressure of the bolted flanges holds the valve in place.
Locating lugs makes installation easier.

Plug Valves

Lubricated plug valves, because of the wide range of service to which they are adapted, may be classified as all purpose valves. They can be safely used at all pressure and vacuums, and at all temperatures up to the limits of available lubricants. They are the most satisfactory valves for the handling of gritty suspensions and many other destructive, erosive, corrosive and chemical solutions.

R15500-010   Heating Systems

Heating Systems

The basic function of a heating system is to bring an enclosed volume up to a desired temperature and then maintain that temperature within a reasonable range. To accomplish this, the selected system must have sufficient capacity to offset transmission losses resulting from the temperature difference on the interior and exterior of the enclosing walls in addition to losses due to cold air infiltration through cracks, crevices and around doors and windows. The amount of heat to be furnished is dependent upon the building size, construction, temperature difference, air leakage, use, shape, orientation and exposure. Air circulation is also an important consideration. Circulation will prevent stratification which could result in heat losses through uneven temperatures at various levels. For example, the most

Heat Transmission

Heat transfer is an important parameter to be considered during selection of the exterior wall style, material and window area. A high rate of transfer will permit greater heat loss during the wintertime with the resultant increase in heating energy costs and a greater rate of heat gain in the summer with proportionally greater cooling cost. Several terms are used to describe various aspects of heat transfer. However, for general estimating purposes this book lists U values for systems of construction materials. U is the "overall heat transfer coefficient." It is defined as the heat flow per hour through one square foot when the temperature difference in the air on either side of the structure wall, roof, ceiling or floor is one degree Fahrenheit. The structural segment may be a single homogeneous material or a composite.

efficient use of unit heaters can usually be achieved by circulating the space volume through the total number of units once every 20 minutes or 3 times an hour. This general rule must, of course, be adapted for special cases such as large buildings with low ratios of heat transmitting surface to cubical volume. The type of occupancy of a building will have considerable bearing on the number of heat transmitting units and the location selected. It is axiomatic, however, that the basis of any successful heating system is to provide the maximum amount of heat at the points of maximum heat loss such as exposed walls, windows, and doors. Large roof areas, wind direction, and wide doorways create problems of excessive heat loss and require special consideration and treatment.

Total heat transfer is found using the following equation:

$Q = AU(T_2 - T_1)$ where
   Q = Heat flow, BTU per hour
   A = Area, square feet
   U = Overall heat transfer coefficient
$(T_2 - T_1)$ = Difference in temperature of
      air on each side of the construction
      component. (Also abbreviated TD)

Note that heat can flow through all surfaces of any building and this flow is in addition to heat gain or loss due to ventilation, infiltration and generation (appliances, machinery, people).

R15500-020   Heating Approximations for Quick Estimating

Oil Piping & Boiler Room Piping
   Small System . . . . 20 to 30% of Boiler
   Complex System
   with Pumps, Headers, Etc. . . . . 80 to 110% of Boiler
Breeching With Insulation:
   Small . . . . 10 to 15% of Boiler
   Large . . . . 15 to 25% of Boiler
Coils: . . . . 15 to 30% of Containing Unit
Balancing (Independent) . . . . 1/2% of H.V.A.C. Estimate

Quality/Complexity Adjustment: For all heating installations add these adjustments to the estimate to more closely allow for the equipment and conditions of the particular job under consideration.
   Economy installation, add . . . . 0 to 5% of System
   Good quality, medium complexity, add . . . . 5 to 15% of System
   Above average quality and complexity, add . . . . 15 to 25% of System

R15500-030 The Basics of a Heating System

The function of a heating system is to achieve and maintain a desired temperature in a room or building by replacing the amount of heat being dissipated. There are four kinds of heating systems: hot-water, steam, warm-air and electric resistance. Each has certain essential and similar elements with the exception of electric resistance heating which is covered in Section 16.

The basic elements of a heating system are:

A. A **combustion chamber** in which fuel is burned and heat transferred to a conveying medium.

B. The **"fluid"** used for conveying the heat (water, steam or air).

C. **Conductors** or pipes for transporting the fluid to specific desired locations.

D. A means of disseminating the heat, sometimes called **terminal units.**

A. The **combustion chamber** in a furnace heats air which is then distributed. This is called a warm-air system.

The combustion chamber in a boiler heats water which is either distributed as hot water or steam and this is termed a hydronic system.

The maximum allowable working pressures are limited by ASME "Code for Heating Boilers" to 15 PSI for steam and 160 PSI for hot water heating boilers, with a maximum temperature limitation of 250°F. Hot water boilers are generally rated for a working pressure of 30 PSI. High pressure boilers are governed by the ASME "Code for Power Boilers" which is used almost universally for boilers operating over 15 PSIG. High pressure boilers used for a combination of heating/process loads are usually designed for 150 PSIG.

Boiler ratings are usually indicated as either Gross or Net Output. The Gross Load is equal to the Net Load plus a piping and pickup allowance. When this allowance cannot be determined, divide the gross output rating by 1.25 for a value equal to or greater than the net heat loss requirement of the building.

B. Of the three **fluids** used, steam carries the greatest amount of heat per unit volume. This is due to the fact that it gives up its latent heat of vaporization at a temperature considerably above room temperature. Another advantage is that the pressure to produce a positive circulation is readily available. Piping conducts the steam to terminal units and returns condensate to the boiler.

The **steam system** is well adapted to large buildings because of its positive circulation, its comparatively economical installation and its ability to deliver large quantities of heat. Nearly all large office buildings, stores, hotels, and industrial buildings are so heated, in addition to many residences.

Hot water, when used as the heat carrying fluid, gives up a portion of its sensible heat and then returns to the boiler or heating apparatus for reheating. As the heat conveyed by each pound of water is about one-fiftieth of the heat conveyed by a pound of steam, it is necessary to circulate about fifty times as much water as steam by weight (although only one-thirtieth as much by volume). The hot water system is usually, although not necessarily, designed to operate at temperatures below that of the ordinary steam system and so the amount of heat transfer surface must be correspondingly greater. A temperature of 190°F to 200°F is normally the maximum. Circulation in small buildings may depend on the difference in density between hot water and the cool water returning to the boiler; circulating pumps are normally used to maintain a desired rate of flow. Pumps permit a greater degree of flexibility and better control.

In **warm-air** furnace systems, cool air is taken from one or more points in the building, passed over the combustion chamber and flue gas passages and then distributed through a duct system. A disadvantage of this system is that the ducts take up much more building volume than steam or hot water pipes. Advantages of this system are the relative ease with which humidification can be accomplished by the evaporation of water as the air circulates through the heater, and the lack of need for expensive disseminating units as the warm air simply becomes part of the interior atmosphere of the building.

C. Conductors (pipes and ducts) have been lightly treated in the discussion of conveying fluids. For more detailed information such as sizing and distribution methods, the reader is referred to technical publications such as the American Society of Heating, Refrigerating and Air-Conditioning Engineers "Handbook of Fundamentals."

D. Terminal units come in an almost infinite variety of sizes and styles, but the basic principles of operation are very limited. As previously mentioned, warm-air systems require only a simple register or diffuser to mix heated air with that present in the room. Special application items such as radiant coils and infrared heaters are available to meet particular conditions but are not usually considered for general heating needs. Most heating is accomplished by having air flow over coils or pipes containing the heat transporting medium (steam, hot-water, electricity). These units, while varied, may be separated into two general types, (1) radiator/convectors and (2) unit heaters.

Radiator/convectors may be cast, fin-tube or pipe assemblies. They may be direct, indirect, exposed, concealed or mounted within a cabinet enclosure, upright or baseboard style. These units are often collectively referred to as "radiators" or "radiation" although none gives off heat either entirely by radiation or by convection but rather a combination of both. The air flows over the units as a gravity "current." It is necessary to have one or more heat-emitting units in each room. The most efficient placement is low along an outside wall or under a window to counteract the cold coming into the room and achieve an even distribution.

In contrast to radiator/convectors which operate most effectively against the walls of smaller rooms, **unit heaters** utilize a fan to move air over heating coils and are very effective in locations of relatively large volume. Unit heaters, while usually suspended overhead, may be floor mounted. They also may take in fresh outside air for ventilation. The heat distributed by unit heaters may be from a remote source and conveyed by a fluid or it may be from the combustion of fuel in each individual heater. In the latter case the only piping required would be for fuel, however, a vent for the products of combustion would be necessary.

The following list gives may of the advantages of unit heaters for applications other than office or residential:

a. Large capacity so smaller number of units are required,
b. Piping system simplified, **c.** Space saved where they are located overhead out of the way, **d.** Rapid heating directed where needed with effective wide distribution, **e.** Difference between floor and ceiling temperature reduced, **f.** Circulation of air obtained, and ventilation with introduction of fresh air possible, **g.** Heat output flexible and easily controlled.

15 MECHANICAL REFERENCE NOS.

R15500-050 Factor for Determining Heat Loss for Various Types of Buildings

General: While the most accurate estimates of heating requirements would naturally be based on detailed information about the building being considered, it is possible to arrive at a reasonable approximation using the following procedure:

1. Calculate the cubic volume of the room or building.
2. Select the appropriate factor from Table 1 below. Note that the factors apply only to inside temperatures listed in the first column and to 0°F outside temperature.

3. If the building has bad north and west exposures, multiply the heat loss factor by 1.1.
4. If the outside design temperature is other than 0°F, multiply the factor from Table 1 by the factor from Table 2.
5. Multiply the cubic volume by the factor selected from Table 1. This will give the estimated BTUH heat loss which must be made up to maintain inside temperature.

Table 1 — Building Type	Conditions	Qualifications	Loss Factor*
Factories & Industrial Plants General Office Areas at 70°F	One Story	Skylight in Roof	6.2
		No Skylight in Roof	5.7
	Multiple Story	Two Story	4.6
		Three Story	4.3
		Four Story	4.1
		Five Story	3.9
		Six Story	3.6
	All Walls Exposed	Flat Roof	6.9
		Heated Space Above	5.2
	One Long Warm Common Wall	Flat Roof	6.3
		Heated Space Above	4.7
	Warm Common Walls on Both Long Sides	Flat Roof	5.8
		Heated Space Above	4.1
Warehouses at 60°F	All Walls Exposed	Skylights in Roof	5.5
		No Skylight in Roof	5.1
		Heated Space Above	4.0
	One Long Warm Common Wall	Skylight in Roof	5.0
		No Skylight in Roof	4.9
		Heated Space Above	3.4
	Warm Common Walls on Both Long Sides	Skylight in Roof	4.7
		No Skylight in Roof	4.4
		Heated Space Above	3.0

*Note: This table tends to be conservative particularly for new buildings designed for minimum energy consumption.

Table 2 — Outside Design Temperature Correction Factor (for Degrees Fahrenheit)									
Outside Design Temperature	50	40	30	20	10	0	-10	-20	-30
Correction Factor	.29	.43	.57	.72	.86	1.00	1.14	1.28	1.43

R15500-070 Transmission of Heat

R15500-070 and R15500-080 provide a way to calculate heat transmission of various construction materials from their U values and the TD (Temperature Difference).

1. From the Exterior Enclosure Division or elsewhere, determine U values for the construction desired.
2. Determine the coldest design temperature. The difference between this temperature and the desired interior temperature is the TD (temperature difference).

3. Enter R15500-070 or R15500-080 at correct U Value. Cross horizontally to the intersection with appropriate TD. Read transmission per square foot from bottom of figure.
4. Multiply this value of BTU per hour transmission per square foot of area by the total surface area of that type of construction.

R15500-080 Transmission of Heat (Low Rate)

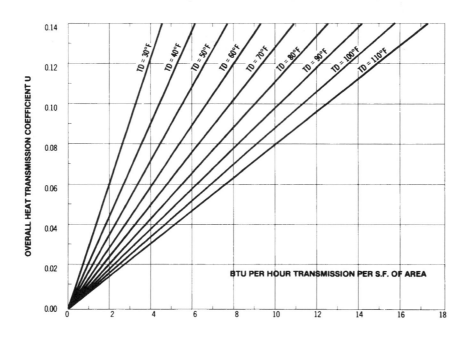

15

MECHANICAL

REFERENCE NOS.

R15700-010 Air Conditioning

General: The purpose of air conditioning is to control the environment of a space so that comfort is provided for the occupants and/or conditions are suitable for the processes or equipment contained therein. The several items which should be evaluated to define system objectives are:

Temperature Control
Humidity Control
Cleanliness
Odor, smoke and fumes
Ventilation

Efforts to control the above parameters must also include consideration of the degree or tolerance of variation, the noise level introduced, the velocity of air motion and the energy requirements to accomplish the desired results.

The variation in **temperature** and **humidity** is a function of the sensor and the controller. The controller reacts to a signal from the sensor and produces the appropriate suitable response in either the terminal unit, the conductor of the transporting medium (air, steam, chilled water, etc.), or the source (boiler, evaporating coils, etc.).

The **noise level** is a by-product of the energy supplied to moving components of the system. Those items which usually contribute the most noise are pumps, blowers, fans, compressors and diffusers. The level of noise can be partially controlled through use of vibration pads, isolators, proper sizing, shields, baffles and sound absorbing liners.

Some **air motion** is necessary to prevent stagnation and stratification. The maximum acceptable velocity varies with the degree of heating or cooling which is taking place. Most people feel air moving past them at velocities in excess of 25 FPM as an annoying draft. However, velocities up to 45 FPM may be acceptable in certain cases. Ventilation, expressed as air changes per hour and percentage of fresh air, is usually an item regulated by local codes.

Selection of the system to be used for a particular application is usually a trade-off. In some cases the building size, style, or room available for mechanical use limits the range of possibilities. Prime factors influencing the decision are first cost and total life (operating, maintenance and replacement costs). The accuracy with which each parameter is determined will be an important measure of the reliability of the decision and subsequent satisfactory operation of the installed system.

Heat delivery may be desired from an air conditioning system. Heating capability usually is added as follows: A gas fired burner or hot water/steam/electric coils may be added to the air handling unit directly and heat all air equally. For limited or localized heat requirements the water/steam/electric coils may be inserted into the duct branch supplying the cold areas. Gas fired duct furnaces are also available.

Note: When water or steam coils are used the cost of the piping and boiler must also be added. For a rough estimate use the cost per square foot of the appropriate sized hydronic system with unit heaters. This will provide a cost for the boiler and piping, and the unit heaters of the system would equate to the approximate cost of the heating coils.

R15700-020 Air Conditioning Requirements

BTU's per hour per S.F. of floor area and S.F. per ton of air conditioning.

Type of Building	BTU per S.F.	S.F. per Ton	Type of Building	BTU per S.F.	S.F. per Ton	Type of Building	BTU per S.F.	S.F. per Ton
Apartments, Individual	26	450	Dormitory, Rooms	40	300	Libraries	50	240
Corridors	22	550	Corridors	30	400	Low Rise Office, Exterior	38	320
Auditoriums & Theaters	40	300/18*	Dress Shops	43	280	Interior	33	360
Banks	50	240	Drug Stores	80	150	Medical Centers	28	425
Barber Shops	48	250	Factories	40	300	Motels	28	425
Bars & Taverns	133	90	High Rise Office—Ext. Rms.	46	263	Office (small suite)	43	280
Beauty Parlors	66	180	Interior Rooms	37	325	Post Office, Individual Office	42	285
Bowling Alleys	68	175	Hospitals, Core	43	280	Central Area	46	260
Churches	36	330/20*	Perimeter	46	260	Residences	20	600
Cocktail Lounges	68	175	Hotel, Guest Rooms	44	275	Restaurants	60	200
Computer Rooms	141	85	Corridors	30	400	Schools & Colleges	46	260
Dental Offices	52	230	Public Spaces	55	220	Shoe Stores	55	220
Dept. Stores, Basement	34	350	Industrial Plants, Offices	38	320	Shop'g. Ctrs., Supermarkets	34	350
Main Floor	40	300	General Offices	34	350	Retail Stores	48	250
Upper Floor	30	400	Plant Areas	40	300	Specialty	60	200

*Persons per ton
12,000 BTU = 1 ton of air conditioning

R15700-030 Psychrometric Table

Dewpoint or Saturation Temperature (F)

Relative humidity (%)															
100	32	35	40	45	50	55	60	65	70	75	80	85	90	95	100
90	30	33	37	42	47	52	57	62	67	72	77	82	87	92	97
80	27	30	34	39	44	49	54	58	64	68	73	78	83	88	93
70	24	27	31	36	40	45	50	55	60	64	69	74	79	84	88
60	20	24	28	32	36	41	46	51	55	60	65	69	74	79	83
50	16	20	24	28	33	36	41	46	50	55	60	64	69	73	78
40	12	15	18	23	27	31	35	40	45	49	53	58	62	67	71
30	8	10	14	18	21	25	29	33	37	42	46	50	54	59	62
20	6	7	8	9	13	16	20	24	28	31	35	40	43	48	52
10	4	4	5	5	6	8	9	10	13	17	20	24	27	30	34
	32	35	40	45	50	55	60	65	70	75	80	85	90	95	100

Dry bulb temperature (F)

This table shows the relationship between RELATIVE HUMIDITY, DRY BULB TEMPERATURE AND DEWPOINT.

As an example, assume that the thermometer in a room reads 75°F, and we know that the relative humidity is 50%. The chart shows the dewpoint temperature to be 55°. That is, any surface colder than 55°F will "sweat" or collect condensing moisture. This surface could be the outside of an uninsulated chilled water pipe in the summertime, or the inside surface of a wall or deck in the wintertime. After determining the extreme ambient parameters, the table at the left is useful in determining which surfaces need insulation or vapor barrier protection.

TEMPERATURE DEGREES FAHRENHEIT
TOTAL PRESSURE = 14.696 LB. PER SQ. IN. ABS.

Psychrometric chart showing different variables based on one pound of dry air. Space marked A B C D is temperature-humidity range which is most comfortable for majority of people.

R15700-040   Recommended Ventilation Air Changes

Table below lists range of time in minutes per change for various types of facilities.

Assembly Halls	2-10	Dance Halls	2-10	Laundries	1-3
Auditoriums	2-10	Dining Rooms	3-10	Markets	2-10
Bakeries	2-3	Dry Cleaners	1-5	Offices	2-10
Banks	3-10	Factories	2-5	Pool Rooms	2-5
Bars	2-5	Garages	2-10	Recreation Rooms	2-10
Beauty Parlors	2-5	Generator Rooms	2-5	Sales Rooms	2-10
Boiler Rooms	1-5	Gymnasiums	2-10	Theaters	2-8
Bowling Alleys	2-10	Kitchens-Hospitals	2-5	Toilets	2-5
Churches	5-10	Kitchens-Restaurant	1-3	Transformer Rooms	1-5

CFM air required for changes = Volume of room in cubic feet ÷ Minutes per change.

R15700-090   Quality/Complexity Adjustment for Air Conditioning Systems

Economy installation, add .... 0 to 5%
Good quality, medium complexity, add .... 5 to 15%
Above average quality and complexity, add .... 15 to 25%

Add the above adjustments to the estimate to more closely allow for the equipment and conditions of the particular job under consideration.

R15800-080 Loudness Levels for Moving Air thru Fans, Diffusers, Register, Etc. (Measured in Sones)

Area	Very Quiet	Quiet	Noisy	Area	Very Quiet	Quiet	Noisy	Area	Very Quiet	Quiet	Noisy
Auditoriums				**Hotels**				**Offices**			
Auditorium lobbies	3	4	6	Banquet Rooms	1.5	3	6	Conference rooms	1	1.7	3
Concert and opera halls	0.8	1	1.5	Individual rooms, suites	1	2	4	Drafting	2	4	8
Courtrooms	2	3	4	Kitchens and laundries	4	7	10	General open offices	2	4	8
Lecture halls	1.5	2	3	Lobbies	2	4	8	Halls and corridors	2.5	5	10
Movie theaters	1.5	2	3	**Indoor Sports**				Professional offices	1.5	3	6
Churches and Schools				Bowling alleys	3	4	6	Reception room	1.5	3	6
Kitchens	4	6	8	Gymnasiums	2	4	6	Tabulation & computation	3	6	12
Laboratories	2	4	6	Swimming pools	4	7	10	**Public Buildings**			
Libraries	1.5	2	3	**Manufacturing Areas**				Banks	2	4	6
Recreation halls	2	4	8	Assembly lines	5	12	30	Court houses	2.5	4	6
Sanctuaries	1	1.7	3	Foreman's office	3	5	8	Museums	2	3	4
Schools and classrooms	1.5	2.5	4	Foundries	10	20	40	Planetariums	1.5	2	3
Hospital and Clinics				General storage	5	10	20	Post offices	2.5	4	6
Laboratories	2	4	6	Heavy machinery	10	25	60	Public libraries	1.5	2	4
Lobbies, waiting rooms	2	4	6	Light machinery	5	12	30	Waiting rooms	3	5	8
Halls and corridors	2	4	6	Tool maintenance	4	7	10	**Restaurants**			
Operating rooms	1.5	2.5	4	**Stores**				Cafeterias	3	6	10
Private rooms	1	1.7	3	Department stores	3	6	8	Night clubs	2.5	4	6
Wards	1.5	2.5	4	Supermarkets	4	7	10	Restaurants	2	4	8

R15810-050 Ductwork

Duct Weight in Pounds per L.F., Straight Runs

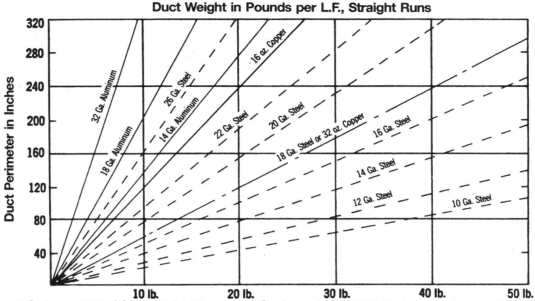

Add to the above for fittings; 90° elbow is 3 L.F.; 45° elbow is 2.5 L.F.; offset is 4 L.F.; transition offset is 6 L.F.; square-to-round transition is 4 L.F.; 90° reducing elbow is 5 L.F. For bracing and waste, add 20% to aluminum and copper, 15% to steel.

R15810-070 Duct Fabrication/Installation

The labor cost for sheet metal duct using lines 15810-600-0070 thru 1060 includes both the cost of fabrication and installation of the duct. The split is approximately 40% for fabrication, 60% for installation. It is for this reason that the percentage add for elevated installation is less than the percentage add for prefabricated duct as found starting on line 15810-600-1283.

Example: assume a piece of duct cost $100 installed (labor only)

Sheet Metal Fabrication = 40% = $40

Installation = 60% = $60

The add for elevated installation is:

$$\text{Based on total labor (fabrication \& installation)} = \$100 \times 6\% = \$6.00$$

$$\text{Based on installation cost only (Material purchased prefabricated)} = \$60 \times 10\% = \$6.00$$

The $6.00 markup (10' to 15' high) is the same.

R15810-100 Sheet Metal Calculator (Weight in Lb./Ft. of Length)

Gauge	26	24	22	20	18	16	Gauge	26	24	22	20	18	16
Wt.-Lb./S.F.	.906	1.156	1.406	1.656	2.156	2.656	Wt.-Lb./S.F.	.906	1.156	1.406	1.656	2.156	2.656
SMACNA Max. Dimension – Long Side		30"	54"	84"	85" Up		SMACNA Max. Dimension – Long Side		30"	54"	84"	85" Up	
Sum-2 sides							**Sum-2 Sides**						
2	.3	.40	.50	.60	.80	.90	56	9.3	12.0	14.0	16.2	21.3	25.2
3	.5	.65	.80	.90	1.1	1.4	57	9.5	12.3	14.3	16.5	21.7	25.7
4	.7	.85	1.0	1.2	1.5	1.8	58	9.7	12.5	14.5	16.8	22.0	26.1
5	.8	1.1	1.3	1.5	1.9	2.3	59	9.8	12.7	14.8	17.1	22.4	26.6
6	1.0	1.3	1.5	1.7	2.3	2.7	60	10.0	12.9	15.0	17.4	22.8	27.0
7	1.2	1.5	1.8	2.0	2.7	3.2	61	10.2	13.1	15.3	17.7	23.2	27.5
8	1.3	1.7	2.0	2.3	3.0	3.6	62	10.3	13.3	15.5	18.0	23.6	27.9
9	1.5	1.9	2.3	2.6	3.4	4.1	63	10.5	13.5	15.8	18.3	24.0	28.4
10	1.7	2.2	2.5	2.9	3.8	4.5	64	10.7	13.7	16.0	18.6	24.3	28.8
11	1.8	2.4	2.8	3.2	4.2	5.0	65	10.8	13.9	16.3	18.9	24.7	29.3
12	2.0	2.6	3.0	3.5	4.6	5.4	66	11.0	14.1	16.5	19.1	25.1	29.7
13	2.2	2.8	3.3	3.8	4.9	5.9	67	11.2	14.3	16.8	19.4	25.5	30.2
14	2.3	3.0	3.5	4.1	5.3	6.3	68	11.3	14.6	17.0	19.7	25.8	30.6
15	2.5	3.2	3.8	4.4	5.7	6.8	69	11.5	14.8	17.3	20.0	26.2	31.1
16	2.7	3.4	4.0	4.6	6.1	7.2	70	11.7	15.0	17.5	20.3	26.6	31.5
17	2.8	3.7	4.3	4.9	6.5	7.7	71	11.8	15.2	17.8	20.6	27.0	32.0
18	3.0	3.9	4.5	5.2	6.8	8.1	72	12.0	15.4	18.0	20.9	27.4	32.4
19	3.2	4.1	4.8	5.5	7.2	8.6	73	12.2	15.6	18.3	21.2	27.7	32.9
20	3.3	4.3	5.0	5.8	7.6	9.0	74	12.3	15.8	18.5	21.5	28.1	33.3
21	3.5	4.5	5.3	6.1	8.0	9.5	75	12.5	16.1	18.8	21.8	28.5	33.8
22	3.7	4.7	5.5	6.4	8.4	9.9	76	12.7	16.3	19.0	22.0	28.9	34.2
23	3.8	5.0	5.8	6.7	8.7	10.4	77	12.8	16.5	19.3	22.3	29.3	34.7
24	4.0	5.2	6.0	7.0	9.1	10.8	78	13.0	16.7	19.5	22.6	29.6	35.1
25	4.2	5.4	6.3	7.3	9.5	11.3	79	13.2	16.9	19.8	22.9	30.0	35.6
26	4.3	5.6	6.5	7.5	9.9	11.7	80	13.3	17.1	20.0	23.2	30.4	36.0
27	4.5	5.8	6.8	7.8	10.3	12.2	81	13.5	17.3	20.3	23.5	30.8	36.5
28	4.7	6.0	7.0	8.1	10.6	12.6	82	13.7	17.5	20.5	23.8	31.2	36.9
29	4.8	6.2	7.3	8.4	11.0	13.1	83	13.8	17.8	20.8	24.1	31.5	37.4
30	5.0	6.5	7.5	8.7	11.4	13.5	84	14.0	18.0	21.0	24.4	31.9	37.8
31	5.2	6.7	7.8	9.0	11.8	14.0	85	14.2	18.2	21.3	24.7	32.3	38.3
32	5.3	6.9	8.0	9.3	12.2	14.4	86	14.3	18.4	21.5	24.9	32.7	38.7
33	5.5	7.1	8.3	9.6	12.5	14.9	87	14.5	18.6	21.8	25.2	33.1	39.2
34	5.7	7.3	8.5	9.9	12.9	15.3	88	14.7	18.8	22.0	25.5	33.4	39.6
35	5.8	7.5	8.8	10.2	13.3	15.8	89	14.8	19.0	22.3	25.8	33.8	40.1
36	6.0	7.8	9.0	10.4	13.7	16.2	90	15.0	19.3	22.5	26.1	34.2	40.5
37	6.2	8.0	9.3	10.7	14.1	16.7	91	15.2	19.5	22.8	26.4	34.6	41.0
38	6.3	8.2	9.5	11.0	14.4	17.1	92	15.3	19.7	23.0	26.7	35.0	41.4
39	6.5	8.4	9.8	11.3	14.8	17.6	93	15.5	19.9	23.3	27.0	35.3	41.9
40	6.7	8.6	10.0	11.6	15.2	18.0	94	15.7	20.1	23.5	27.3	35.7	42.3
41	6.8	8.8	10.3	11.9	15.6	18.5	95	15.8	20.3	23.8	27.6	36.1	42.8
42	7.0	9.0	10.5	12.2	16.0	18.9	96	16.0	20.5	24.0	27.8	36.5	43.2
43	7.2	9.2	10.8	12.5	16.3	19.4	97	16.2	20.8	24.3	28.1	36.9	43.7
44	7.3	9.5	11.0	12.8	16.7	19.8	98	16.3	21.0	24.5	28.4	37.2	44.1
45	7.5	9.7	11.3	13.1	17.1	20.3	99	16.5	21.2	24.8	28.7	37.6	44.6
46	7.7	9.9	11.5	13.3	17.5	20.7	100	16.7	21.4	25.0	29.0	38.0	45.0
47	7.8	10.1	11.8	13.6	17.9	21.2	101	16.8	21.6	25.3	29.3	38.4	45.5
48	8.0	10.3	12.0	13.9	18.2	21.6	102	17.0	21.8	25.5	29.6	38.8	45.9
49	8.2	10.5	12.3	14.2	18.6	22.1	103	17.2	22.0	25.8	29.9	39.1	46.4
50	8.3	10.7	12.5	14.5	19.0	22.5	104	17.3	22.3	26.0	30.2	39.5	46.8
51	8.5	11.0	12.8	14.8	19.4	23.0	105	17.5	22.5	26.3	30.5	39.9	47.3
52	8.7	11.2	13.0	15.1	19.8	23.4	106	17.7	22.7	26.5	30.7	40.3	47.7
53	8.8	11.4	13.3	15.4	20.1	23.9	107	17.8	22.9	26.8	31.0	40.7	48.2
54	9.0	11.6	13.5	15.7	20.5	24.3	108	18.0	23.1	27.0	31.3	41.0	48.6
55	9.2	11.8	13.8	16.0	20.9	24.8	109	18.2	23.3	27.3	31.6	41.4	49.1
							110	18.3	23.5	27.5	31.9	41.8	49.5

Example: If duct is 34" x 20" x 15' long, 34" is greater than 30" maximum, for 24 ga. so must be 22 ga. 34" + 20" = 54" going across from 54" find 13.5 lb. per foot. 13.5 x 15' = 202.5 lbs. For S.F. of surface area 202.5 ÷

1.406 = 144 S.F.
Note: Figures include an allowance for scrap.

R15810-110 Ductwork Packages (per Ton of Cooling)

System	Sheet Metal	Insulation	Diffusers	Return Register
Roof Top Unit Single Zone	120 Lbs.	52 S.F.	1	1
Roof Top Unit Multizone	240 Lbs.	104 S.F.	2	1
Self-contained Air or Water Cooled	108 Lbs.	—	2	—
Split System Air Cooled	102 Lbs.	—	2	—

Systems reflect most common usage.
Refer to system graphics for duct layout.

R15850-060 Diffuser Evaluation

$CFM = V \times A_n \times K$ where V = Outlet velocity in feet per minute. A_n = Neck area in square feet and K = Diffuser delivery factor. An undersized diffuser for a desired CFM will produce a high velocity and noise level. When air moves past people at a velocity in excess of 25 FPM, an annoying draft is felt. An oversized diffuser will result in low velocity with poor mixing. Consideration must be given to avoid vertical stratification or horizontal areas of stagnation.

MECHANICAL 15

REFERENCE NOS.

R17100-100   Square Foot Project Size Modifier

One factor that affects the S.F. cost of a particular building is the size. In general, for buildings built to the same specifications in the same locality, the larger building will have the lower S.F. cost. This is due mainly to the decreasing contribution of the exterior walls plus the economy of scale usually achievable in larger buildings. The Area Conversion Scale shown below will give a factor to convert costs for the typical size building to an adjusted cost for the particular project.

The Square Foot Base Size lists the median costs, most typical project size in our accumulated data and the range in size of the projects.

The Size Factor for your project is determined by dividing your project area in S.F. by the typical project size for the particular Building Type. With this factor, enter the Area Conversion Scale at the appropriate Size Factor and determine the appropriate cost multiplier for your building size.

Example: Determine the cost per S.F. for a 100,000 S.F. Mid-rise apartment building.

$$\frac{\text{Proposed building area} = 100,000 \text{ S.F.}}{\text{Typical size from below} = 50,000 \text{ S.F.}} = 2.00$$

Enter Area Conversion scale at 2.0, intersect curve, read horizontally the appropriate cost multiplier of .94. Size adjusted cost becomes .94 x $72.00 = $67.70 based on national average costs.

Note: For Size Factors less than .50, the Cost Multiplier is 1.1
    For Size Factors greater than 3.5, the Cost Multiplier is .90

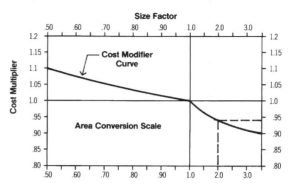

Square Foot Base Size							
Building Type	Median Cost per S.F.	Typical Size Gross S.F.	Typical Range Gross S.F.	Building Type	Median Cost per S.F.	Typical Size Gross S.F.	Typical Range Gross S.F.
Apartments, Low Rise	$ 57.00	21,000	9,700 - 37,200	Jails	$174.00	40,000	5,500 - 145,000
Apartments, Mid Rise	72.00	50,000	32,000 - 100,000	Libraries	103.00	12,000	7,000 - 31,000
Apartments, High Rise	82.55	145,000	95,000 - 600,000	Medical Clinics	98.20	7,200	4,200 - 15,700
Auditoriums	95.30	25,000	7,600 - 39,000	Medical Offices	92.25	6,000	4,000 - 15,000
Auto Sales	58.95	20,000	10,800 - 28,600	Motels	70.65	40,000	15,800 - 120,000
Banks	128.00	4,200	2,500 - 7,500	Nursing Homes	92.10	23,000	15,000 - 37,000
Churches	86.05	17,000	2,000 - 42,000	Offices, Low Rise	76.95	20,000	5,000 - 80,000
Clubs, Country	85.80	6,500	4,500 - 15,000	Offices, Mid Rise	80.80	120,000	20,000 - 300,000
Clubs, Social	83.45	10,000	6,000 - 13,500	Offices, High Rise	103.00	260,000	120,000 - 800,000
Clubs, YMCA	83.70	28,300	12,800 - 39,400	Police Stations	129.00	10,500	4,000 - 19,000
Colleges (Class)	112.00	50,000	15,000 - 150,000	Post Offices	95.30	12,400	6,800 - 30,000
Colleges (Science Lab)	164.00	45,600	16,600 - 80,000	Power Plants	716.00	7,500	1,000 - 20,000
College (Student Union)	125.00	33,400	16,000 - 85,000	Religious Education	78.95	9,000	6,000 - 12,000
Community Center	89.70	9,400	5,300 - 16,700	Research	134.00	19,000	6,300 - 45,000
Court Houses	122.00	32,400	17,800 - 106,000	Restaurants	116.00	4,400	2,800 - 6,000
Dept. Stores	53.25	90,000	44,000 - 122,000	Retail Stores	56.60	7,200	4,000 - 17,600
Dormitories, Low Rise	91.95	25,000	10,000 - 95,000	Schools, Elementary	82.45	41,000	24,500 - 55,000
Dormitories, Mid Rise	120.00	85,000	20,000 - 200,000	Schools, Jr. High	84.00	92,000	52,000 - 119,000
Factories	51.60	26,400	12,900 - 50,000	Schools, Sr. High	83.95	101,000	50,500 - 175,000
Fire Stations	90.15	5,800	4,000 - 8,700	Schools, Vocational	83.65	37,000	20,500 - 82,000
Fraternity Houses	88.65	12,500	8,200 - 14,800	Sports Arenas	70.10	15,000	5,000 - 40,000
Funeral Homes	99.10	10,000	4,000 - 20,000	Supermarkets	56.80	44,000	12,000 - 60,000
Garages, Commercial	62.95	9,300	5,000 - 13,600	Swimming Pools	131.00	20,000	10,000 - 32,000
Garages, Municipal	80.55	8,300	4,500 - 12,600	Telephone Exchange	153.00	4,500	1,200 - 10,600
Garages, Parking	33.00	163,000	76,400 - 225,300	Theaters	84.05	10,500	8,800 - 17,500
Gymnasiums	83.30	19,200	11,600 - 41,000	Town Halls	92.45	10,800	4,800 - 23,400
Hospitals	157.00	55,000	27,200 - 125,000	Warehouses	38.10	25,000	8,000 - 72,000
House (Elderly)	77.95	37,000	21,000 - 66,000	Warehouse & Office	44.00	25,000	8,000 - 72,000
Housing (Public)	72.15	36,000	14,400 - 74,400				
Ice Rinks	80.20	29,000	27,200 - 33,600				

Change Orders

Change Order Considerations

A Change Order is a written document, usually prepared by the design professional, and signed by the owner, the architect/engineer and the contractor. A change order states the agreement of the parties to: an addition, deletion, or revision in the work; an adjustment in the contract sum, if any; or an adjustment in the contract time, if any. Change orders, or "extras" in the construction process occur after execution of the construction contract and impact architects/engineers, contractors and owners.

Change orders that are properly recognized and managed can ensure orderly, professional and profitable progress for all who are involved in the project. There are many causes for change orders and change order requests. In all cases, change orders or change order requests should be addressed promptly and in a precise and prescribed manner. The following paragraphs include information regarding change order pricing and procedures.

The Causes of Change Orders

Reasons for issuing change orders include:

- Unforeseen field conditions that require a change in the work
- Correction of design discrepancies, errors or omissions in the contract documents
- Owner-requested changes, either by design criteria, scope of work, or project objectives
- Completion date changes for reasons unrelated to the construction process
- Changes in building code interpretations, or other public authority requirements that require a change in the work
- Changes in availability of existing or new materials and products

Procedures

Properly written contract documents must include the correct change order procedures for all parties—owners, design professionals and contractors—to follow in order to avoid costly delays and litigation.

Being "in the right" is not always a sufficient or acceptable defense. The contract provisions requiring notification and documentation must be adhered to within a defined or reasonable time frame.

The appropriate method of handling change orders is by a written proposal and acceptance by all parties involved. Prior to starting work on a project, all parties should identify their authorized agents who may sign and accept change orders, as well as any limits placed on their authority.

Time may be a critical factor when the need for a change arises. For such cases, the contractor might be directed to proceed on a "time and materials" basis, rather than wait for all paperwork to be processed—a delay that could impede progress. In this situation, the contractor must still follow the prescribed change order procedures, including but not limited to, notification and documentation.

All forms used for change orders should be dated and signed by the proper authority. Lack of documentation can be very costly, especially if legal judgments are to be made and if certain field personnel are no longer available. For time and material change orders, the contractor should keep accurate daily records of all labor and material allocated to the change. Forms that can be used to document change order work are available in *Means Forms for Building Construction Professionals.*

Owners or awarding authorities who do considerable and continual building construction (such as the federal government) realize the inevitability of change orders for numerous reasons, both predictable and unpredictable. As a result, the federal government, the American Institute of Architects (AIA), the Engineers Joint Contract Documents Committee (EJCDC) and other contractor, legal and technical organizations have developed standards and procedures to be followed by all parties to achieve contract continuance and timely completion, while being financially fair to all concerned.

In addition to the change order standards put forth by industry associations, there are also many books available on the subject.

Pricing Change Orders

When pricing change orders, regardless of their cause, the most significant factor is *when* the change occurs. The need for a change may be perceived in the field or requested by the architect/engineer *before* any of the actual installation has begun, or may evolve or appear *during* construction when the item of work in question is partially installed. In the latter cases, the original sequence of construction is disrupted, along with all contiguous and supporting systems. Change orders cause the greatest impact when they occur *after* the installation has been completed and must be uncovered, or even replaced. Post-completion changes may be caused by necessary design changes, product failure, or changes in the owner's requirements that are not discovered until the building or the systems begin to function.

Specified procedures of notification and record keeping must be adhered to and enforced regardless of the stage of construction: *before, during,* or *after* installation. Some bidding documents anticipate change orders by requiring that unit prices including overhead and profit percentages—for additional as well as deductible changes—be listed. Generally these unit prices do not fully take into account the ripple effect, or impact on other trades, and should be used for general guidance only.

When pricing change orders, it is important to classify the time frame in which the change occurs. There are two basic time frames for change orders: *pre-installation change orders*, which occur before the start of construction, and *post-installation change orders*, which involve reworking after the original installation. Change orders that occur between these stages may be priced according to the extent of work completed using a combination of techniques developed for pricing *pre-* and *post-installation* changes.

The following factors are the basis for a check list to use when preparing a change order estimate.

Factors To Consider When Pricing Change Orders

As an estimator begins to prepare a change order, the following questions should be reviewed to determine their impact on the final price.

General

- Is the change order work *pre-installation* or *post-installation*?

  Change order work costs vary according to how much of the installation has been completed. Once workers have the project scoped in their mind, even though they have not started, it can be difficult to refocus.

Consequently they may spend more than the normal amount of time understanding the change. Also, modifications to work in place such as trimming or refitting usually take more time than was initially estimated. The greater the amount of work in place, the more reluctant workers are to change it. Psychologically they may resent the change and as a result the rework takes longer than normal. Post-installation change order estimates must include demolition of existing work as required to accomplish the change. If the work is performed at a later time, additional obstacles such as building finishes may be present which must be

protected. Regardless of whether the change occurs pre-installation or post-installation, attempt to isolate the identifiable factors and price them separately. For example, add shipping costs that may be required pre-installation or any demolition required post-installation. Then analyze the potential impact on productivity of psychological and/ or learning curve factors and adjust the output rates accordingly. One approach is to break down the typical workday into segments and quantify the impact on each segment. The following chart may be useful as a guide:

Activities (Productivity) Expressed as Percentages of a Workday			
Task	Means Mechanical Cost Data (for New Construction)	Pre-Installation Change Orders	Post-Installation Change Orders
1. Study plans	3%	6%	6%
2. Material procurement	3%	3%	3%
3. Receiving and storing	3%	3%	3%
4. Mobilization	5%	5%	5%
5. Site movement	5%	5%	8%
6. Layout and marking	8%	10%	12%
7. Actual installation	64%	59%	54%
8. Clean-up	3%	3%	3%
9. Breaks—non-productive	6%	6%	6%
Total	100%	100%	100%

Change Order Installation Efficiency

The labor-hours expressed (for new construction) are based on average installation time, using an efficiency level of approximately 60-65%. For change order situations, adjustments to this efficiency level should reflect the daily labor-hour allocation for that particular occurrence.

If any of the specific percentages expressed in the above chart do not apply to a particular project situation, then those percentage points should be reallocated to the appropriate task(s). Example: Using data for new construction, assume there is no new material being utilized. The percentages for Tasks 2 and 3 would therefore be reallocated to other tasks. If the time required for Tasks 2 and 3 can now be applied to installation, we can add the time allocated for *Material Procurement* and *Receiving and Storing* to the *Actual Installation* time for new construction, thereby increasing the Actual Installation percentage.

This chart shows that, due to reduced productivity, labor costs will be higher than those for new construction by 5% to 15% for pre-installation change orders and by 15% to 25% for post-installation change orders. Each job and change order is unique and must be examined individually. Many factors, covered elsewhere in this section, can each have a significant impact on productivity and change order costs. All such factors should be considered in every case.

- Will the change substantially delay the original completion date?

  A significant change in the project may cause the original completion date to be extended. The extended schedule may subject the contractor to new wage rates dictated by relevant labor contracts. Project supervision and other project overhead must also be extended beyond the original completion date. The schedule extension may also put installation into a new weather season. For example, underground piping scheduled for October installation was delayed until January. As a result, frost penetrated the trench area, thereby changing the degree of difficulty of the task. Changes and delays may have a ripple effect throughout the project. This effect must be analyzed and negotiated with the owner.

- What is the net effect of a deduct change order?

  In most cases, change orders resulting in a deduction or credit reflect only bare costs. The contractor may retain the overhead and profit based on the original bid.

Materials

- Will you have to pay more or less for the new material, required by the change order, than you paid for the original purchase?

  The same material prices or discounts will usually apply to materials purchased for change orders as new construction. In some instances, however, the contractor may forfeit the advantages of competitive pricing for change orders. Consider the following example:

  A contractor purchased over $20,000 worth of fan coil units for an installation, and obtained the maximum discount. Some time later it was determined the project required an additional matching unit. The contractor has to purchase this unit from the original supplier to ensure a match. The supplier at this time may not discount the unit because of the small quantity, and the fact that he is no longer in a competitive situation. The impact of quantity on purchase can add between 0% and 25% to material prices and/or subcontractor quotes.

- If materials have been ordered or delivered to the job site, will they be subject to a cancellation charge or restocking fee?

  Check with the supplier to determine if ordered materials are subject to a cancellation charge. Delivered materials not used as result of a change order may be subject to a restocking fee if returned to the supplier. Common restocking charges run between 20% and 40%. Also, delivery charges to return the goods to the supplier must be added.

Labor

- How efficient is the existing crew at the actual installation?

  Is the same crew that performed the initial work going to do the change order? Possibly the change consists of the installation of a unit identical to one already installed; therefore the change should take less time. Be sure to consider this potential productivity increase and modify the productivity rates accordingly.

- If the crew size is increased, what impact will that have on supervision requirements?

  Under most bargaining agreements or management practices, there is a point at which a working foreman is replaced by a nonworking foreman. This replacement increases project overhead by adding a nonproductive worker. If additional workers are added to accelerate the project or to perform changes while maintaining the schedule, be sure to add additional supervision time if warranted. Calculate the hours involved and the additional cost directly if possible.

- What are the other impacts of increased crew size?

  The larger the crew, the greater the potential for productivity to decrease. Some of the factors that cause this productivity loss are: overcrowding (producing restrictive conditions in the working space), and possibly a shortage of any special tools and equipment required. Such factors affect not only the crew working on the elements directly involved in the change order, but other crews whose movement may also be hampered.

As the crew increases, check its basic composition for changes by the addition or deletion of apprentices or nonworking foreman and quantify the potential effects of equipment shortages or other logistical factors.

- As new crews, unfamiliar with the project, are brought onto the site, how long will it take them to become oriented to the project requirements?

  The orientation time for a new crew to become 100% effective varies with the site and type of project. Orientation is easiest at a new construction site, and most difficult at existing, very restrictive renovation sites. The type of work also affects orientation time. When all elements of the work are exposed, such as concrete or masonry work, orientation is decreased. When the work is concealed or less visible, such as existing electrical systems, orientation takes longer. Usually orientation can be accomplished in one day or less. Costs for added orientation should be itemized and added to the total estimated cost.

- How much actual production can be gained by working overtime?

  Short term overtime can be used effectively to accomplish more work in a day. However, as overtime is scheduled to run beyond several weeks, studies have shown marked decreases in output. The following chart shows the effect of long term overtime on worker efficiency. If the anticipated change requires extended overtime to keep the job on schedule, these factors can be used as a guide to predict the impact on time and cost. Add project overhead, particularly supervision, that may also be incurred.

Days per Week	Hours per Day	Production Efficiency					Payroll Cost Factors	
		1 Week	2 Weeks	3 Weeks	4 Weeks	Average 4 Weeks	@ 1-1/2 Times	@ 2 Times
5	8	100%	100%	100%	100%	100%	100%	100%
	9	100	100	95	90	96.25	105.6	111.1
	10	100	95	90	85	91.25	110.0	120.0
	11	95	90	75	65	81.25	113.6	127.3
	12	90	85	70	60	76.25	116.7	133.3
6	8	100	100	95	90	96.25	108.3	116.7
	9	100	95	90	85	92.50	113.0	125.9
	10	95	90	85	80	87.50	116.7	133.3
	11	95	85	70	65	78.75	119.7	139.4
	12	90	80	65	60	73.75	122.2	144.4
7	8	100	95	85	75	88.75	114.3	128.6
	9	95	90	80	70	83.75	118.3	136.5
	10	90	85	75	65	78.75	121.4	142.9
	11	85	80	65	60	72.50	124.0	148.1
	12	85	75	60	55	68.75	126.2	152.4

Effects of Overtime

Caution: Under many labor agreements, Sundays and holidays are paid at a higher premium than the normal overtime rate.

The use of long-term overtime is counterproductive on almost any construction job; that is, the longer the period of overtime, the lower the actual production rate. Numerous studies have been conducted, and while they have resulted in slightly different numbers, all reach the same conclusion. The figure above tabulates the effects of overtime work on efficiency.

As illustrated, there can be a difference between the *actual* payroll cost per hour and the *effective* cost per hour for overtime work. This is due to the reduced production efficiency with the increase in weekly hours beyond 40. This difference between actual and effective cost results from overtime work over a prolonged period. Short-term overtime work does not result in as great a reduction in efficiency, and in such cases, effective cost may not vary significantly from the actual payroll cost. As the total hours per week are increased on a regular basis, more time is lost because of fatigue, lowered morale, and an increased accident rate.

As an example, assume a project where workers are working 6 days a week, 10 hours per day. From the figure above (based on productivity studies), the average effective productive hours over a four-week period are:

$$0.875 \times 60 = 52.5$$

Depending upon the locale and day of week, overtime hours may be paid at time and a half or double time. For time and a half, the overall (average) *actual* payroll cost (including regular and overtime hours) is determined as follows:

$$\frac{40 \text{ reg. hrs.} + (20 \text{ overtime hrs.} \times 1.5)}{60 \text{ hrs.}} = 1.167$$

Based on 60 hours, the payroll cost per hour will be 116.7% of the normal rate at 40 hours per week. However, because the effective production (efficiency) for 60 hours is reduced to the equivalent of 52.5 hours, the effective cost of overtime is calculated as follows:

For time and a half:

$$\frac{40 \text{ reg. hrs.} + (20 \text{ overtime hrs.} \times 1.5)}{52.5 \text{ hrs.}} = 1.33$$

Installed cost will be 133% of the normal rate (for labor).

Thus, when figuring overtime, the actual cost per unit of work will be higher than the apparent overtime payroll dollar increase, due to the reduced productivity of the longer workweek. These efficiency calculations are true only for those cost factors determined by hours worked. Costs that are applied weekly or monthly, such as equipment rentals, will not be similarly affected.

Equipment

• What equipment is required to complete the change order?

Change orders may require extending the rental period of equipment already on the job site, or the addition of special equipment brought in to accomplish the change work. In either case, the additional rental charges and operator labor charges must be added.

Summary

The preceding considerations and others you deem appropriate should be analyzed and applied to a change order estimate. The impact of each should be quantified and listed on the estimate to form an audit trail.

Change orders that are properly identified, documented, and managed help to ensure the orderly, professional and profitable progress of the work. They also minimize potential claims or disputes at the end of the project.

Crew A-1 – A-6

Crew No.	Bare Costs Hr.	Daily	Incl. Subs O & P Hr.	Daily	Cost Per Labor-Hour Bare Costs	Incl. O&P
Crew A-1						
1 Building Laborer	$23.45	$187.60	$36.50	$292.00	$23.45	$36.50
1 Gas Eng. Power Tool		60.60		66.65	7.58	8.33
8 L.H., Daily Totals		$248.20		$358.65	$31.03	$44.83
Crew A-1A						
1 Skilled Worker	$30.95	$247.60	$48.35	$386.80	$30.95	$48.35
1 Shot Blaster, 20"		173.40		190.75	21.68	23.84
8 L.H., Daily Totals		$421.00		$577.55	$52.63	$72.19
Crew A-2						
2 Laborers	$23.45	$375.20	$36.50	$584.00	$23.73	$36.68
1 Truck Driver (light)	24.30	194.40	37.05	296.40		
1 Light Truck, 1.5 Ton		155.60		171.15	6.48	7.13
24 L.H., Daily Totals		$725.20		$1051.55	$30.21	$43.81
Crew A-2A						
2 Laborers	$23.45	$375.20	$36.50	$584.00	$23.73	$36.68
1 Truck Driver (light)	24.30	194.40	37.05	296.40		
1 Light Truck, 1.5 Ton		155.60		171.15		
1 Concrete Saw		103.40		113.75	10.79	11.87
24 L.H., Daily Totals		$828.60		$1165.30	$34.52	$48.55
Crew A-3						
1 Truck Driver (heavy)	$25.00	$200.00	$38.10	$304.80	$25.00	$38.10
1 Dump Truck, 12 Ton		353.20		388.50	44.15	48.57
8 L.H., Daily Totals		$553.20		$693.30	$69.15	$86.67
Crew A-3A						
1 Truck Driver (light)	$24.30	$194.40	$37.05	$296.40	$24.30	$37.05
1 Pickup Truck (4x4)		90.20		99.20	11.28	12.40
8 L.H., Daily Totals		$284.60		$395.60	$35.58	$49.45
Crew A-3B						
1 Equip. Oper. (medium)	$31.20	$249.60	$47.15	$377.20	$28.10	$42.63
1 Truck Driver (heavy)	25.00	200.00	38.10	304.80		
1 Dump Truck, 16 Ton		470.60		517.65		
1 F.E. Loader, 3 C.Y.		370.40		407.45	52.56	57.82
16 L.H., Daily Totals		$1290.60		$1607.10	$80.66	$100.45
Crew A-3C						
1 Equip. Oper. (light)	$29.80	$238.40	$45.05	$360.40	$29.80	$45.05
1 Wheeled Skid Steer Loader		194.80		214.30	24.35	26.79
8 L.H., Daily Totals		$433.20		$574.70	$54.15	$71.84
Crew A-4						
2 Carpenters	$30.00	$480.00	$46.70	$747.20	$29.05	$44.83
1 Painter, Ordinary	27.15	217.20	41.10	328.80		
24 L.H., Daily Totals		$697.20		$1076.00	$29.05	$44.83
Crew A-5						
2 Laborers	$23.45	$375.20	$36.50	$584.00	$23.54	$36.56
.25 Truck Driver (light)	24.30	48.60	37.05	74.10		
.25 Light Truck, 1.5 Ton		38.90		42.80	2.16	2.38
18 L.H., Daily Totals		$462.70		$700.90	$25.70	$38.94
Crew A-6						
1 Chief Of Party	$36.25	$290.00	$57.00	$456.00	$33.60	$52.67
1 Instrument Man	30.95	247.60	48.35	386.80		
16 L.H., Daily Totals		$537.60		$842.80	$33.60	$52.67

Crew A-7 – A-13

Crew No.	Bare Costs Hr.	Daily	Incl. Subs O & P Hr.	Daily	Cost Per Labor-Hour Bare Costs	Incl. O&P
Crew A-7						
1 Chief Of Party	$36.25	$290.00	$57.00	$456.00	$32.40	$50.25
1 Instrument Man	30.95	247.60	48.35	386.80		
1 Rodman/Chainman	30.00	240.00	45.40	363.20		
24 L.H., Daily Totals		$777.60		$1206.00	$32.40	$50.25
Crew A-8						
1 Chief Of Party	$36.25	$290.00	$57.00	$456.00	$31.80	$49.04
1 Instrument Man	30.95	247.60	48.35	386.80		
2 Rodmen/Chainmen	30.00	480.00	45.40	726.40		
32 L.H., Daily Totals		$1017.60		$1569.20	$31.80	$49.04
Crew A-9						
1 Asbestos Foreman	$33.95	$271.60	$53.90	$431.20	$33.51	$53.20
7 Asbestos Workers	33.45	1873.20	53.10	2973.60		
64 L.H., Daily Totals		$2144.80		$3404.80	$33.51	$53.20
Crew A-10						
1 Asbestos Foreman	$33.95	$271.60	$53.90	$431.20	$33.51	$53.20
7 Asbestos Workers	33.45	1873.20	53.10	2973.60		
64 L.H., Daily Totals		$2144.80		$3404.80	$33.51	$53.20
Crew A-10A						
1 Asbestos Foreman	$33.95	$271.60	$53.90	$431.20	$33.62	$53.37
2 Asbestos Workers	33.45	535.20	53.10	849.60		
24 L.H., Daily Totals		$806.80		$1280.80	$33.62	$53.37
Crew A-10B						
1 Asbestos Foreman	$33.95	$271.60	$53.90	$431.20	$33.58	$53.30
3 Asbestos Workers	33.45	802.80	53.10	1274.40		
32 L.H., Daily Totals		$1074.40		$1705.60	$33.58	$53.30
Crew A-10C						
3 Asbestos Workers	$33.45	$802.80	$53.10	$1274.40	$33.45	$53.10
1 Flatbed Truck		155.60		171.15	6.48	7.13
24 L.H., Daily Totals		$958.40		$1445.55	$39.93	$60.23
Crew A-10D						
2 Asbestos Workers	$33.45	$535.20	$53.10	$849.60	$31.48	$48.84
1 Equip. Oper. (crane)	32.35	258.80	48.90	391.20		
1 Equip. Oper. Oiler	26.65	213.20	40.25	322.00		
1 Hydraulic Crane, 33 Ton		857.40		943.15	26.79	29.47
32 L.H., Daily Totals		$1864.60		$2505.95	$58.27	$78.31
Crew A-11						
1 Asbestos Foreman	$33.95	$271.60	$53.90	$431.20	$33.51	$53.20
7 Asbestos Workers	33.45	1873.20	53.10	2973.60		
2 Chipping Hammers		29.60		32.55	.46	.51
64 L.H., Daily Totals		$2174.40		$3437.35	$33.97	$53.71
Crew A-12						
1 Asbestos Foreman	$33.95	$271.60	$53.90	$431.20	$33.51	$53.20
7 Asbestos Workers	33.45	1873.20	53.10	2973.60		
1 Large Prod. Vac. Loader		490.25		539.30	7.66	8.43
64 L.H., Daily Totals		$2635.05		$3944.10	$41.17	$61.63
Crew A-13						
1 Equip. Oper. (light)	$29.80	$238.40	$45.05	$360.40	$29.80	$45.05
1 Large Prod. Vac. Loader		490.25		539.30	61.28	67.41
8 L.H., Daily Totals		$728.65		$899.70	$91.08	$112.46

CREWS

Crew No.	Bare Costs		Incl. Subs O & P		Cost Per Labor-Hour	

Crew B-1

	Hr.	Daily	Hr.	Daily	Bare Costs	Incl. O&P
1 Labor Foreman (outside)	$25.45	$203.60	$39.60	$316.80	$24.12	$37.53
2 Laborers	23.45	375.20	36.50	584.00		
24 L.H., Daily Totals		$578.80		$900.80	$24.12	$37.53

Crew B-2

	Hr.	Daily	Hr.	Daily	Bare Costs	Incl. O&P
1 Labor Foreman (outside)	$25.45	$203.60	$39.60	$316.80	$23.85	$37.12
4 Laborers	23.45	750.40	36.50	1168.00		
40 L.H., Daily Totals		$954.00		$1484.80	$23.85	$37.12

Crew B-3

	Hr.	Daily	Hr.	Daily	Bare Costs	Incl. O&P
1 Labor Foreman (outside)	$25.45	$203.60	$39.60	$316.80	$25.59	$39.33
2 Laborers	23.45	375.20	36.50	584.00		
1 Equip. Oper. (med.)	31.20	249.60	47.15	377.20		
2 Truck Drivers (heavy)	25.00	400.00	38.10	609.60		
1 F.E. Loader, T.M., 2.5 C.Y.		690.60		759.65		
2 Dump Trucks, 16 Ton		941.20		1035.30	34.00	37.40
48 L.H., Daily Totals		$2860.20		$3682.55	$59.59	$76.73

Crew B-3A

	Hr.	Daily	Hr.	Daily	Bare Costs	Incl. O&P
4 Laborers	$23.45	$750.40	$36.50	$1168.00	$25.00	$38.63
1 Equip. Oper. (med.)	31.20	249.60	47.15	377.20		
1 Hyd. Excavator, 1.5 C.Y.		705.00		775.50	17.63	19.39
40 L.H., Daily Totals		$1705.00		$2320.70	$42.63	$58.02

Crew B-3B

	Hr.	Daily	Hr.	Daily	Bare Costs	Incl. O&P
2 Laborers	$23.45	$375.20	$36.50	$584.00	$25.78	$39.56
1 Equip. Oper. (med.)	31.20	249.60	47.15	377.20		
1 Truck Drivers (heavy)	25.00	200.00	38.10	304.80		
1 Backhoe Loader, 80 H.P.		215.60		237.15		
1 Dump Trucks, 16 Ton		470.60		517.65	21.44	23.59
32 L.H., Daily Totals		$1511.00		$2020.80	$47.22	$63.15

Crew B-3C

	Hr.	Daily	Hr.	Daily	Bare Costs	Incl. O&P
3 Laborers	$23.45	$562.80	$36.50	$876.00	$25.39	$39.16
1 Equip. Oper. (med.)	31.20	249.60	47.15	377.20		
1 F.E. Crawler Ldr, 4 C.Y.		971.20		1068.30	30.35	33.39
32 L.H., Daily Totals		$1783.60		$2321.50	$55.74	$72.55

Crew B-4

	Hr.	Daily	Hr.	Daily	Bare Costs	Incl. O&P
1 Labor Foreman (outside)	$25.45	$203.60	$39.60	$316.80	$24.04	$37.28
4 Laborers	23.45	750.40	36.50	1168.00		
1 Truck Driver (heavy)	25.00	200.00	38.10	304.80		
1 Tractor, 4 x 2, 195 H.P.		291.80		321.00		
1 Platform Trailer		150.80		165.90	9.22	10.14
48 L.H., Daily Totals		$1596.60		$2276.50	$33.26	$47.42

Crew B-5

	Hr.	Daily	Hr.	Daily	Bare Costs	Incl. O&P
1 Labor Foreman (outside)	$25.45	$203.60	$39.60	$316.80	$25.95	$39.99
4 Laborers	23.45	750.40	36.50	1168.00		
2 Equip. Oper. (med.)	31.20	499.20	47.15	754.40		
1 Air Compr., 250 C.F.M.		135.00		148.50		
2 Air Tools & Accessories		33.20		36.50		
2-50 Ft. Air Hoses, 1.5" Dia.		9.40		10.35		
1 F.E. Loader, T.M., 2.5 C.Y.		690.60		759.65	15.50	17.05
56 L.H., Daily Totals		$2321.40		$3194.20	$41.45	$57.04

Crew B-5A

	Hr.	Daily	Hr.	Daily	Bare Costs	Incl. O&P
1 Foreman	$25.45	$203.60	$39.60	$316.80	$25.70	$39.51
6 Laborers	23.45	1125.60	36.50	1752.00		
2 Equip. Oper. (med.)	31.20	499.20	47.15	754.40		
1 Equip. Oper. (light)	29.80	238.40	45.05	360.40		
2 Truck Drivers (heavy)	25.00	400.00	38.10	609.60		
1 Air Compr. 365 C.F.M.		163.00		179.30		
2 Pavement Breakers		33.20		36.50		
8 Air Hoses w/Coup.,1"		27.60		30.35		
2 Dump Trucks, 12 Ton		706.40		777.05	9.69	10.66
96 L.H., Daily Totals		$3397.00		$4816.40	$35.39	$50.17

Crew B-5B

	Hr.	Daily	Hr.	Daily	Bare Costs	Incl. O&P
1 Powderman	$30.95	$247.60	$48.35	$386.80	$28.06	$42.83
2 Equip. Oper. (med.)	31.20	499.20	47.15	754.40		
3 Truck Drivers (heavy)	25.00	600.00	38.10	914.40		
1 F.E. Ldr. 2-1/2 CY		370.40		407.45		
3 Dump Trucks, 16 Ton		1411.80		1553.00		
1 Air Compr. 365 C.F.M.		163.00		179.30	40.53	44.58
48 L.H., Daily Totals		$3292.00		$4195.35	$68.59	$87.41

Crew B-5C

	Hr.	Daily	Hr.	Daily	Bare Costs	Incl. O&P
3 Laborers	$23.45	$562.80	$36.50	$876.00	$26.32	$40.25
1 Equip. Oper. (medium)	31.20	249.60	47.15	377.20		
2 Truck Drivers (heavy)	25.00	400.00	38.10	609.60		
1 Equip. Oper. (crane)	32.35	258.80	48.90	391.20		
1 Equip. Oper. Oiler	26.65	213.20	40.25	322.00		
2 Dump Truck, 16 Ton		941.20		1035.30		
1 F.E. Crawler Ldr, 4 C.Y.		971.20		1068.30		
1 Hyd. Crane, 25 Ton		629.40		692.35	39.72	43.69
64 L.H., Daily Totals		$4226.20		$5371.95	$66.04	$83.94

Crew B-6

	Hr.	Daily	Hr.	Daily	Bare Costs	Incl. O&P
2 Laborers	$23.45	$375.20	$36.50	$584.00	$25.57	$39.35
1 Equip. Oper. (light)	29.80	238.40	45.05	360.40		
1 Backhoe Loader, 48 H.P.		178.60		196.45	7.44	8.19
24 L.H., Daily Totals		$792.20		$1140.85	$33.01	$47.54

Crew B-6A

	Hr.	Daily	Hr.	Daily	Bare Costs	Incl. O&P
.5 Labor Foreman (outside)	$25.45	$101.80	$39.60	$158.40	$26.95	$41.38
1 Laborers	23.45	187.60	36.50	292.00		
1 Equip. Oper. (med.)	31.20	249.60	47.15	377.20		
1 Vacuum Trk.,5000 Gal.		302.20		332.40	15.11	16.62
20 L.H., Daily Totals		$841.20		$1160.00	$42.06	$58.00

Crew B-7

	Hr.	Daily	Hr.	Daily	Bare Costs	Incl. O&P
1 Labor Foreman (outside)	$25.45	$203.60	$39.60	$316.80	$25.08	$38.79
4 Laborers	23.45	750.40	36.50	1168.00		
1 Equip. Oper. (med.)	31.20	249.60	47.15	377.20		
1 Chipping Machine		182.60		200.85		
1 F.E. Loader, T.M., 2.5 C.Y.		690.60		759.65		
2 Chain Saws, 36"		79.20		87.10	19.84	21.83
48 L.H., Daily Totals		$2156.00		$2909.60	$44.92	$60.62

Crew B-7A

	Hr.	Daily	Hr.	Daily	Bare Costs	Incl. O&P
2 Laborers	$23.45	$375.20	$36.50	$584.00	$25.57	$39.35
1 Equip. Oper. (light)	29.80	238.40	45.05	360.40		
1 Rake w/Tractor		123.30		135.65		
2 Chain Saws, 18"		67.70		74.45	7.96	8.75
24 L.H., Daily Totals		$804.60		$1154.50	$33.53	$48.10

Crews

Crew No.	Bare Costs		Incl. Subs O & P		Cost Per Labor-Hour	
Crew B-8	Hr.	Daily	Hr.	Daily	Bare Costs	Incl. O&P
1 Labor Foreman (outside)	$25.45	$203.60	$39.60	$316.80	$26.43	$40.42
2 Laborers	23.45	375.20	36.50	584.00		
2 Equip. Oper. (med.)	31.20	499.20	47.15	754.40		
1 Equip. Oper. Oiler	26.65	213.20	40.25	322.00		
2 Truck Drivers (heavy)	25.00	400.00	38.10	609.60		
1 Hyd. Crane, 25 Ton		724.80		797.30		
1 F.E. Loader, T.M., 2.5 C.Y.		690.60		759.65		
2 Dump Trucks, 16 Ton		941.20		1035.30	36.82	40.50
64 L.H., Daily Totals		$4047.80		$5179.05	$63.25	$80.92
Crew B-9	Hr.	Daily	Hr.	Daily	Bare Costs	Incl. O&P
1 Labor Foreman (outside)	$25.45	$203.60	$39.60	$316.80	$23.85	$37.12
4 Laborers	23.45	750.40	36.50	1168.00		
1 Air Compr., 250 C.F.M.		135.00		148.50		
2 Air Tools & Accessories		33.20		36.50		
2-50 Ft. Air Hoses, 1.5" Dia.		9.40		10.35	4.44	4.88
40 L.H., Daily Totals		$1131.60		$1680.15	$28.29	$42.00
Crew B-9A	Hr.	Daily	Hr.	Daily	Bare Costs	Incl. O&P
2 Laborers	$23.45	$375.20	$36.50	$584.00	$23.97	$37.03
1 Truck Driver (heavy)	25.00	200.00	38.10	304.80		
1 Water Tanker		163.40		179.75		
1 Tractor		291.80		321.00		
2-50 Ft. Disch. Hoses		10.70		11.75	19.41	21.35
24 L.H., Daily Totals		$1041.10		$1401.30	$43.38	$58.38
Crew B-9B	Hr.	Daily	Hr.	Daily	Bare Costs	Incl. O&P
2 Laborers	$23.45	$375.20	$36.50	$584.00	$23.97	$37.03
1 Truck Driver (heavy)	25.00	200.00	38.10	304.80		
2-50 Ft. Disch. Hoses		10.70		11.75		
1 Water Tanker		163.40		179.75		
1 Tractor		291.80		321.00		
1 Pressure Washer		53.60		58.95	21.65	23.81
24 L.H., Daily Totals		$1094.70		$1460.25	$45.62	$60.84
Crew B-9C	Hr.	Daily	Hr.	Daily	Bare Costs	Incl. O&P
1 Labor Foreman (outside)	$25.45	$203.60	$39.60	$316.80	$23.85	$37.12
4 Laborers	23.45	750.40	36.50	1168.00		
1 Air Compr., 250 C.F.M.		135.00		148.50		
2-50 Ft. Air Hoses, 1.5" Dia.		9.40		10.35		
2 Breaker, Pavement, 60 lb.		33.20		36.50	4.44	4.88
40 L.H., Daily Totals		$1131.60		$1680.15	$28.29	$42.00
Crew B-10	Hr.	Daily	Hr.	Daily	Bare Costs	Incl. O&P
1 Equip. Oper. (med.)	$31.20	$249.60	$47.15	$377.20	$28.62	$43.60
.5 Laborer	23.45	93.80	36.50	146.00		
12 L.H., Daily Totals		$343.40		$523.20	$28.62	$43.60
Crew B-10A	Hr.	Daily	Hr.	Daily	Bare Costs	Incl. O&P
1 Equip. Oper. (med.)	$31.20	$249.60	$47.15	$377.20	$28.62	$43.60
.5 Laborer	23.45	93.80	36.50	146.00		
1 Roll. Compact., 2K Lbs.		123.20		135.50	10.27	11.29
12 L.H., Daily Totals		$466.60		$658.70	$38.89	$54.89
Crew B-10B	Hr.	Daily	Hr.	Daily	Bare Costs	Incl. O&P
1 Equip. Oper. (med.)	$31.20	$249.60	$47.15	$377.20	$28.62	$43.60
.5 Laborer	23.45	93.80	36.50	146.00		
1 Dozer, 200 H.P.		823.40		905.75	68.62	75.48
12 L.H., Daily Totals		$1166.80		$1428.95	$97.24	$119.08

Crew No.	Bare Costs		Incl. Subs O & P		Cost Per Labor-Hour	
Crew B-10C	Hr.	Daily	Hr.	Daily	Bare Costs	Incl. O&P
1 Equip. Oper. (med.)	$31.20	$249.60	$47.15	$377.20	$28.62	$43.60
.5 Laborer	23.45	93.80	36.50	146.00		
1 Dozer, 200 H.P.		823.40		905.75		
1 Vibratory Roller, Towed		174.60		192.05	83.17	91.48
12 L.H., Daily Totals		$1341.40		$1621.00	$111.79	$135.08
Crew B-10D	Hr.	Daily	Hr.	Daily	Bare Costs	Incl. O&P
1 Equip. Oper. (med.)	$31.20	$249.60	$47.15	$377.20	$28.62	$43.60
.5 Laborer	23.45	93.80	36.50	146.00		
1 Dozer, 200 H.P.		823.40		905.75		
1 Sheepsft. Roller, Towed		72.80		80.10	74.68	82.15
12 L.H., Daily Totals		$1239.60		$1509.05	$103.30	$125.75
Crew B-10E	Hr.	Daily	Hr.	Daily	Bare Costs	Incl. O&P
1 Equip. Oper. (med.)	$31.20	$249.60	$47.15	$377.20	$28.62	$43.60
.5 Laborer	23.45	93.80	36.50	146.00		
1 Tandem Roller, 5 Ton		117.20		128.90	9.77	10.74
12 L.H., Daily Totals		$460.60		$652.10	$38.39	$54.34
Crew B-10F	Hr.	Daily	Hr.	Daily	Bare Costs	Incl. O&P
1 Equip. Oper. (med.)	$31.20	$249.60	$47.15	$377.20	$28.62	$43.60
.5 Laborer	23.45	93.80	36.50	146.00		
1 Tandem Roller, 10 Ton		168.40		185.25	14.03	15.44
12 L.H., Daily Totals		$511.80		$708.45	$42.65	$59.04
Crew B-10G	Hr.	Daily	Hr.	Daily	Bare Costs	Incl. O&P
1 Equip. Oper. (med.)	$31.20	$249.60	$47.15	$377.20	$28.62	$43.60
.5 Laborer	23.45	93.80	36.50	146.00		
1 Sheepsft. Roll., 130 H.P.		764.40		840.85	63.70	70.07
12 L.H., Daily Totals		$1107.80		$1364.05	$92.32	$113.67
Crew B-10H	Hr.	Daily	Hr.	Daily	Bare Costs	Incl. O&P
1 Equip. Oper. (med.)	$31.20	$249.60	$47.15	$377.20	$28.62	$43.60
.5 Laborer	23.45	93.80	36.50	146.00		
1 Diaphr. Water Pump, 2"		32.60		35.85		
1-20 Ft. Suction Hose, 2"		5.10		5.60		
2-50 Ft. Disch. Hoses, 2"		8.60		9.45	3.86	4.24
12 L.H., Daily Totals		$389.70		$574.10	$32.48	$47.84
Crew B-10I	Hr.	Daily	Hr.	Daily	Bare Costs	Incl. O&P
1 Equip. Oper. (med.)	$31.20	$249.60	$47.15	$377.20	$28.62	$43.60
.5 Laborer	23.45	93.80	36.50	146.00		
1 Diaphr. Water Pump, 4"		68.80		75.70		
1-20 Ft. Suction Hose, 4"		10.25		11.30		
2-50 Ft. Disch. Hoses, 4"		14.50		15.95	7.80	8.58
12 L.H., Daily Totals		$436.95		$626.15	$36.42	$52.18
Crew B-10J	Hr.	Daily	Hr.	Daily	Bare Costs	Incl. O&P
1 Equip. Oper. (med.)	$31.20	$249.60	$47.15	$377.20	$28.62	$43.60
.5 Laborer	23.45	93.80	36.50	146.00		
1 Centr. Water Pump, 3"		39.00		42.90		
1-20 Ft. Suction Hose, 3"		7.75		8.55		
2-50 Ft. Disch. Hoses, 3"		10.70		11.75	4.79	5.27
12 L.H., Daily Totals		$400.85		$586.40	$33.41	$48.87

CREWS

551

Crews

Crew No.	Bare Costs Hr.	Daily	Incl. Subs O & P Hr.	Daily	Cost Per Labor-Hour Bare Costs	Incl. O&P
Crew B-10K						
1 Equip. Oper. (med.)	$31.20	$249.60	$47.15	$377.20	$28.62	$43.60
.5 Laborer	23.45	93.80	36.50	146.00		
1 Centr. Water Pump, 6"		198.20		218.00		
1-20 Ft. Suction Hose, 6"		19.50		21.45		
2-50 Ft. Disch. Hoses, 6"		34.80		38.30	21.04	23.15
12 L.H., Daily Totals		$595.90		$800.95	$49.66	$66.75
Crew B-10L						
1 Equip. Oper. (med.)	$31.20	$249.60	$47.15	$377.20	$28.62	$43.60
.5 Laborer	23.45	93.80	36.50	146.00		
1 Dozer, 75 H.P.		291.20		320.30	24.27	26.69
12 L.H., Daily Totals		$634.60		$843.50	$52.89	$70.29
Crew B-10M						
1 Equip. Oper. (med.)	$31.20	$249.60	$47.15	$377.20	$28.62	$43.60
.5 Laborer	23.45	93.80	36.50	146.00		
1 Dozer, 300 H.P.		1034.00		1137.40	86.17	94.78
12 L.H., Daily Totals		$1377.40		$1660.60	$114.79	$138.38
Crew B-10N						
1 Equip. Oper. (med.)	$31.20	$249.60	$47.15	$377.20	$28.62	$43.60
.5 Laborer	23.45	93.80	36.50	146.00		
1 F.E. Loader, T.M., 1.5 C.Y		263.60		289.95	21.97	24.16
12 L.H., Daily Totals		$607.00		$813.15	$50.59	$67.76
Crew B-100						
1 Equip. Oper. (med.)	$31.20	$249.60	$47.15	$377.20	$28.62	$43.60
.5 Laborer	23.45	93.80	36.50	146.00		
1 F.E. Loader, T.M., 2.25 C.Y.		498.40		548.25	41.53	45.69
12 L.H., Daily Totals		$841.80		$1071.45	$70.15	$89.29
Crew B-10P						
1 Equip. Oper. (med.)	$31.20	$249.60	$47.15	$377.20	$28.62	$43.60
.5 Laborer	23.45	93.80	36.50	146.00		
1 F.E. Loader, T.M., 2.5 C.Y.		690.60		759.65	57.55	63.31
12 L.H., Daily Totals		$1034.00		$1282.85	$86.17	$106.91
Crew B-10Q						
1 Equip. Oper. (med.)	$31.20	$249.60	$47.15	$377.20	$28.62	$43.60
.5 Laborer	23.45	93.80	36.50	146.00		
1 F.E. Loader, T.M., 5 C.Y.		971.20		1068.30	80.93	89.03
12 L.H., Daily Totals		$1314.60		$1591.50	$109.55	$132.63
Crew B-10R						
1 Equip. Oper. (med.)	$31.20	$249.60	$47.15	$377.20	$28.62	$43.60
.5 Laborer	23.45	93.80	36.50	146.00		
1 F.E. Loader, W.M., 1 C.Y.		214.40		235.85	17.87	19.65
12 L.H., Daily Totals		$557.80		$759.05	$46.49	$63.25
Crew B-10S						
1 Equip. Oper. (med.)	$31.20	$249.60	$47.15	$377.20	$28.62	$43.60
.5 Laborer	23.45	93.80	36.50	146.00		
1 F.E. Loader, W.M., 1.5 C.Y.		276.40		304.05	23.03	25.34
12 L.H., Daily Totals		$619.80		$827.25	$51.65	$68.94
Crew B-10T						
1 Equip. Oper. (med.)	$31.20	$249.60	$47.15	$377.20	$28.62	$43.60
.5 Laborer	23.45	93.80	36.50	146.00		
1 F.E. Loader, W.M.,2.5 C.Y.		370.40		407.45	30.87	33.95
12 L.H., Daily Totals		$713.80		$930.65	$59.49	$77.55

Crew No.	Bare Costs Hr.	Daily	Incl. Subs O & P Hr.	Daily	Cost Per Labor-Hour Bare Costs	Incl. O&P
Crew B-10U						
1 Equip. Oper. (med.)	$31.20	$249.60	$47.15	$377.20	$28.62	$43.60
.5 Laborer	23.45	93.80	36.50	146.00		
1 F.E. Loader, W.M., 5.5 C.Y.		777.60		855.35	64.80	71.28
12 L.H., Daily Totals		$1121.00		$1378.55	$93.42	$114.88
Crew B-10V						
1 Equip. Oper. (med.)	$31.20	$249.60	$47.15	$377.20	$28.62	$43.60
.5 Laborer	23.45	93.80	36.50	146.00		
1 Dozer, 700 H.P.		2924.00		3216.40	243.67	268.03
12 L.H., Daily Totals		$3267.40		$3739.60	$272.29	$311.63
Crew B-10W						
1 Equip. Oper. (med.)	$31.20	$249.60	$47.15	$377.20	$28.62	$43.60
.5 Laborer	23.45	93.80	36.50	146.00		
1 Dozer, 105 H.P.		432.60		475.85	36.05	39.66
12 L.H., Daily Totals		$776.00		$999.05	$64.67	$83.26
Crew B-10X						
1 Equip. Oper. (med.)	$31.20	$249.60	$47.15	$377.20	$28.62	$43.60
.5 Laborer	23.45	93.80	36.50	146.00		
1 Dozer, 410 H.P.		1374.00		1511.40	114.50	125.95
12 L.H., Daily Totals		$1717.40		$2034.60	$143.12	$169.55
Crew B-10Y						
1 Equip. Oper. (med.)	$31.20	$249.60	$47.15	$377.20	$28.62	$43.60
.5 Laborer	23.45	93.80	36.50	146.00		
1 Vibratory Drum Roller		303.40		333.75	25.28	27.81
12 L.H., Daily Totals		$646.80		$856.95	$53.90	$71.41
Crew B-11						
1 Equipment Oper. (med.)	$31.20	$249.60	$47.15	$377.20	$27.33	$41.83
1 Laborer	23.45	187.60	36.50	292.00		
16 L.H., Daily Totals		$437.20		$669.20	$27.33	$41.83
Crew B-11A						
1 Equipment Oper. (med.)	$31.20	$249.60	$47.15	$377.20	$27.33	$41.83
1 Laborer	23.45	187.60	36.50	292.00		
1 Dozer, 200 H.P.		823.40		905.75	51.46	56.61
16 L.H., Daily Totals		$1260.60		$1574.95	$78.79	$98.44
Crew B-11B						
1 Equipment Oper. (med.)	$31.20	$249.60	$47.15	$377.20	$27.33	$41.83
1 Laborer	23.45	187.60	36.50	292.00		
1 Dozer, 200 H.P.		823.40		905.75		
1 Air Powered Tamper		21.10		23.20		
1 Air Compr. 365 C.F.M.		163.00		179.30		
2-50 Ft. Air Hoses, 1.5" Dia.		9.40		10.35	63.56	69.91
16 L.H., Daily Totals		$1454.10		$1787.80	$90.89	$111.74
Crew B-11C						
1 Equipment Oper. (med.)	$31.20	$249.60	$47.15	$377.20	$27.33	$41.83
1 Laborer	23.45	187.60	36.50	292.00		
1 Backhoe Loader, 48 H.P.		178.60		196.45	11.16	12.28
16 L.H., Daily Totals		$615.80		$865.65	$38.49	$54.11
Crew B-11K						
1 Equipment Oper. (med.)	$31.20	$249.60	$47.15	$377.20	$27.33	$41.83
1 Laborer	23.45	187.60	36.50	292.00		
1 Trencher, 8' D., 16" W.		849.60		934.55	53.10	58.41
16 L.H., Daily Totals		$1286.80		$1603.75	$80.43	$100.24

Crews

Crew B-11L

Crew No.	Bare Costs Hr.	Daily	Incl. Subs O&P Hr.	Daily	Bare Costs	Incl. O&P
Crew B-11L						
1 Equipment Oper. (med.)	$31.20	$249.60	$47.15	$377.20	$27.33	$41.83
1 Laborer	23.45	187.60	36.50	292.00		
1 Grader, 30,000 Lbs.		431.60		474.75	26.98	29.67
16 L.H., Daily Totals		$868.80		$1143.95	$54.31	$71.50

Crew B-11M

Crew No.	Bare Costs Hr.	Daily	Incl. Subs O&P Hr.	Daily	Bare Costs	Incl. O&P
Crew B-11M						
1 Equipment Oper. (med.)	$31.20	$249.60	$47.15	$377.20	$27.33	$41.83
1 Laborer	23.45	187.60	36.50	292.00		
1 Backhoe Loader, 80 H.P.		215.60		237.15	13.48	14.82
16 L.H., Daily Totals		$652.80		$906.35	$40.81	$56.65

Crew B-11N

Crew No.	Bare Costs Hr.	Daily	Incl. Subs O&P Hr.	Daily	Bare Costs	Incl. O&P
Crew B-11N						
1 Labor Foreman	$25.45	$203.60	$39.60	$316.80	$26.43	$40.28
2 Equipment Operators (med.)	31.20	499.20	47.15	754.40		
6 Truck Drivers (hvy.)	25.00	1200.00	38.10	1828.80		
1 F.E. Loader, 5.5 C.Y.		777.60		855.35		
1 Dozer, 400 H.P.		1374.00		1511.40		
6 Off Hwy. Tks. 50 Ton		8184.00		9002.40	143.55	157.91
72 L.H., Daily Totals		$12238.40		$14269.15	$169.98	$198.19

Crew B-11Q

Crew No.	Bare Costs Hr.	Daily	Incl. Subs O&P Hr.	Daily	Bare Costs	Incl. O&P
Crew B-11Q						
1 Equipment Operator (med.)	$31.20	$249.60	$47.15	$377.20	$28.62	$43.60
.5 Laborer	23.45	93.80	36.50	146.00		
1 Dozer, 140 H.P.		464.40		510.85	38.70	42.57
12 L.H., Daily Totals		$807.80		$1034.05	$67.32	$86.17

Crew B-11R

Crew No.	Bare Costs Hr.	Daily	Incl. Subs O&P Hr.	Daily	Bare Costs	Incl. O&P
Crew B-11R						
1 Equipment Operator (med.)	$31.20	$249.60	$47.15	$377.20	$28.62	$43.60
.5 Laborer	23.45	93.80	36.50	146.00		
1 Dozer, 215 H.P.		823.40		905.75	68.62	75.48
12 L.H., Daily Totals		$1166.80		$1428.95	$97.24	$119.08

Crew B-11S

Crew No.	Bare Costs Hr.	Daily	Incl. Subs O&P Hr.	Daily	Bare Costs	Incl. O&P
Crew B-11S						
1 Equipment Operator	$31.20	$249.60	$47.15	$377.20	$28.62	$43.60
.5 Laborer	23.45	93.80	36.50	146.00		
1 Dozer, 285 H.P.		1034.00		1137.40	86.17	94.78
12 L.H., Daily Totals		$1377.40		$1660.60	$114.79	$138.38

Crew B-11T

Crew No.	Bare Costs Hr.	Daily	Incl. Subs O&P Hr.	Daily	Bare Costs	Incl. O&P
Crew B-11T						
1 Equipment Operator (med.)	$31.20	$249.60	$47.15	$377.20	$28.62	$43.60
.5 Laborer	23.45	93.80	36.50	146.00		
1 Dozer, 370 H.P.		1374.00		1511.40	114.50	125.95
12 L.H., Daily Totals		$1717.40		$2034.60	$143.12	$169.55

Crew B-11U

Crew No.	Bare Costs Hr.	Daily	Incl. Subs O&P Hr.	Daily	Bare Costs	Incl. O&P
Crew B-11U						
1 Equipment Operator (med.)	$31.20	$249.60	$47.15	$377.20	$28.62	$43.60
.5 Laborer	23.45	93.80	36.50	146.00		
1 Dozer, 520 H.P.		1832.00		2015.20	152.67	167.93
12 L.H., Daily Totals		$2175.40		$2538.40	$181.29	$211.53

Crew B-11V

Crew No.	Bare Costs Hr.	Daily	Incl. Subs O&P Hr.	Daily	Bare Costs	Incl. O&P
Crew B-11V						
3 Laborer	$23.45	$562.80	$36.50	$876.00	$23.45	$36.50
1 Roll. Compact., 2K Lbs.		123.20		135.50	5.13	5.65
24 L.H., Daily Totals		$686.00		$1011.50	$28.58	$42.15

Crew B-12

Crew No.	Bare Costs Hr.	Daily	Incl. Subs O&P Hr.	Daily	Bare Costs	Incl. O&P
Crew B-12						
1 Equip. Oper. (crane)	$32.35	$258.80	$48.90	$391.20	$29.50	$44.58
1 Equip. Oper. Oiler	26.65	213.20	40.25	322.00		
16 L.H., Daily Totals		$472.00		$713.20	$29.50	$44.58

Crew B-12A

Crew No.	Bare Costs Hr.	Daily	Incl. Subs O&P Hr.	Daily	Bare Costs	Incl. O&P
Crew B-12A						
1 Equip. Oper. (crane)	$32.35	$258.80	$48.90	$391.20	$29.50	$44.58
1 Equip. Oper. Oiler	26.65	213.20	40.25	322.00		
1 Hyd. Excavator, 1 C.Y.		563.40		619.75	35.21	38.73
16 L.H., Daily Totals		$1035.40		$1332.95	$64.71	$83.31

Crew B-12B

Crew No.	Bare Costs Hr.	Daily	Incl. Subs O&P Hr.	Daily	Bare Costs	Incl. O&P
Crew B-12B						
1 Equip. Oper. (crane)	$32.35	$258.80	$48.90	$391.20	$29.50	$44.58
1 Equip. Oper. Oiler	26.65	213.20	40.25	322.00		
1 Hyd. Excavator, 1.5 C.Y.		705.00		775.50	44.06	48.47
16 L.H., Daily Totals		$1177.00		$1488.70	$73.56	$93.05

Crew B-12C

Crew No.	Bare Costs Hr.	Daily	Incl. Subs O&P Hr.	Daily	Bare Costs	Incl. O&P
Crew B-12C						
1 Equip. Oper. (crane)	$32.35	$258.80	$48.90	$391.20	$29.50	$44.58
1 Equip. Oper. Oiler	26.65	213.20	40.25	322.00		
1 Hyd. Excavator, 2 C.Y.		962.20		1058.40	60.14	66.15
16 L.H., Daily Totals		$1434.20		$1771.60	$89.64	$110.73

Crew B-12D

Crew No.	Bare Costs Hr.	Daily	Incl. Subs O&P Hr.	Daily	Bare Costs	Incl. O&P
Crew B-12D						
1 Equip. Oper. (crane)	$32.35	$258.80	$48.90	$391.20	$29.50	$44.58
1 Equip. Oper. Oiler	26.65	213.20	40.25	322.00		
1 Hyd. Excavator, 3.5 C.Y.		2094.00		2303.40	130.88	143.96
16 L.H., Daily Totals		$2566.00		$3016.60	$160.38	$188.54

Crew B-12E

Crew No.	Bare Costs Hr.	Daily	Incl. Subs O&P Hr.	Daily	Bare Costs	Incl. O&P
Crew B-12E						
1 Equip. Oper. (crane)	$32.35	$258.80	$48.90	$391.20	$29.50	$44.58
1 Equip. Oper. Oiler	26.65	213.20	40.25	322.00		
1 Hyd. Excavator, .5 C.Y.		319.80		351.80	19.99	21.99
16 L.H., Daily Totals		$791.80		$1065.00	$49.49	$66.57

Crew B-12F

Crew No.	Bare Costs Hr.	Daily	Incl. Subs O&P Hr.	Daily	Bare Costs	Incl. O&P
Crew B-12F						
1 Equip. Oper. (crane)	$32.35	$258.80	$48.90	$391.20	$29.50	$44.58
1 Equip. Oper. Oiler	26.65	213.20	40.25	322.00		
1 Hyd. Excavator, .75 C.Y.		457.00		502.70	28.56	31.42
16 L.H., Daily Totals		$929.00		$1215.90	$58.06	$76.00

Crew B-12G

Crew No.	Bare Costs Hr.	Daily	Incl. Subs O&P Hr.	Daily	Bare Costs	Incl. O&P
Crew B-12G						
1 Equip. Oper. (crane)	$32.35	$258.80	$48.90	$391.20	$29.50	$44.58
1 Equip. Oper. Oiler	26.65	213.20	40.25	322.00		
1 Power Shovel, .5 C.Y.		311.75		342.95		
1 Clamshell Bucket, .5 C.Y.		42.40		46.65	22.13	24.35
16 L.H., Daily Totals		$826.15		$1102.80	$51.63	$68.93

Crew B-12H

Crew No.	Bare Costs Hr.	Daily	Incl. Subs O&P Hr.	Daily	Bare Costs	Incl. O&P
Crew B-12H						
1 Equip. Oper. (crane)	$32.35	$258.80	$48.90	$391.20	$29.50	$44.58
1 Equip. Oper. Oiler	26.65	213.20	40.25	322.00		
1 Power Shovel, 1 C.Y.		640.80		704.90		
1 Clamshell Bucket, 1 C.Y.		61.60		67.75	43.90	48.29
16 L.H., Daily Totals		$1174.40		$1485.85	$73.40	$92.87

Crew B-12I

Crew No.	Bare Costs Hr.	Daily	Incl. Subs O&P Hr.	Daily	Bare Costs	Incl. O&P
Crew B-12I						
1 Equip. Oper. (crane)	$32.35	$258.80	$48.90	$391.20	$29.50	$44.58
1 Equip. Oper. Oiler	26.65	213.20	40.25	322.00		
1 Power Shovel, .75 C.Y.		502.10		552.30		
1 Dragline Bucket, .75 C.Y.		18.20		20.00	32.52	35.77
16 L.H., Daily Totals		$992.30		$1285.50	$62.02	$80.35

CREWS

Crew No.	Bare Costs Hr.	Daily	Incl. Subs O & P Hr.	Daily	Cost Per Labor-Hour Bare Costs	Incl. O&P
Crew B-12J	Hr.	Daily	Hr.	Daily	Bare Costs	Incl. O&P
1 Equip. Oper. (crane)	$32.35	$258.80	$48.90	$391.20	$29.50	$44.58
1 Equip. Oper. Oiler	26.65	213.20	40.25	322.00		
1 Gradall, 3 Ton, .5 C.Y.		736.40		810.05	46.03	50.63
16 L.H., Daily Totals		$1208.40		$1523.25	$75.53	$95.21
Crew B-12K	Hr.	Daily	Hr.	Daily	Bare Costs	Incl. O&P
1 Equip. Oper. (crane)	$32.35	$258.80	$48.90	$391.20	$29.50	$44.58
1 Equip. Oper. Oiler	26.65	213.20	40.25	322.00		
1 Gradall, 3 Ton, 1 C.Y.		868.60		955.45	54.29	59.72
16 L.H., Daily Totals		$1340.60		$1668.65	$83.79	$104.30
Crew B-12L	Hr.	Daily	Hr.	Daily	Bare Costs	Incl. O&P
1 Equip. Oper. (crane)	$32.35	$258.80	$48.90	$391.20	$29.50	$44.58
1 Equip. Oper. Oiler	26.65	213.20	40.25	322.00		
1 Power Shovel, .5 C.Y.		311.75		342.95		
1 F.E. Attachment, .5 C.Y.		38.80		42.70	21.91	24.10
16 L.H., Daily Totals		$822.55		$1098.85	$51.41	$68.68
Crew B-12M	Hr.	Daily	Hr.	Daily	Bare Costs	Incl. O&P
1 Equip. Oper. (crane)	$32.35	$258.80	$48.90	$391.20	$29.50	$44.58
1 Equip. Oper. Oiler	26.65	213.20	40.25	322.00		
1 Power Shovel, .75 C.Y.		502.10		552.30		
1 F.E. Attachment, .75 C.Y.		43.80		48.20	34.12	37.53
16 L.H., Daily Totals		$1017.90		$1313.70	$63.62	$82.11
Crew B-12N	Hr.	Daily	Hr.	Daily	Bare Costs	Incl. O&P
1 Equip. Oper. (crane)	$32.35	$258.80	$48.90	$391.20	$29.50	$44.58
1 Equip. Oper. Oiler	26.65	213.20	40.25	322.00		
1 Power Shovel, 1 C.Y.		640.80		704.90		
1 F.E. Attachment, 1 C.Y.		50.20		55.20	43.19	47.51
16 L.H., Daily Totals		$1163.00		$1473.30	$72.69	$92.09
Crew B-12O	Hr.	Daily	Hr.	Daily	Bare Costs	Incl. O&P
1 Equip. Oper. (crane)	$32.35	$258.80	$48.90	$391.20	$29.50	$44.58
1 Equip. Oper. Oiler	26.65	213.20	40.25	322.00		
1 Power Shovel, 1.5 C.Y.		809.20		890.10		
1 F.E. Attachment, 1.5 C.Y.		110.00		121.00	57.45	63.20
16 L.H., Daily Totals		$1391.20		$1724.30	$86.95	$107.78
Crew B-12P	Hr.	Daily	Hr.	Daily	Bare Costs	Incl. O&P
1 Equip. Oper. (crane)	$32.35	$258.80	$48.90	$391.20	$29.50	$44.58
1 Equip. Oper. Oiler	26.65	213.20	40.25	322.00		
1 Crawler Crane, 40 Ton		809.20		890.10		
1 Dragline Bucket, 1.5 C.Y.		29.60		32.55	52.43	57.67
16 L.H., Daily Totals		$1310.80		$1635.85	$81.93	$102.25
Crew B-12Q	Hr.	Daily	Hr.	Daily	Bare Costs	Incl. O&P
1 Equip. Oper. (crane)	$32.35	$258.80	$48.90	$391.20	$29.50	$44.58
1 Equip. Oper. Oiler	26.65	213.20	40.25	322.00		
1 Hyd. Excavator, 5/8 C.Y.		393.20		432.50	24.58	27.03
16 L.H., Daily Totals		$865.20		$1145.70	$54.08	$71.61
Crew B-12R	Hr.	Daily	Hr.	Daily	Bare Costs	Incl. O&P
1 Equip. Oper. (crane)	$32.35	$258.80	$48.90	$391.20	$29.50	$44.58
1 Equip. Oper. Oiler	26.65	213.20	40.25	322.00		
1 Hyd. Excavator, 1.5 C.Y.		705.00		775.50	44.06	48.47
16 L.H., Daily Totals		$1177.00		$1488.70	$73.56	$93.05

Crew No.	Bare Costs Hr.	Daily	Incl. Subs O & P Hr.	Daily	Cost Per Labor-Hour Bare Costs	Incl. O&P
Crew B-12S	Hr.	Daily	Hr.	Daily	Bare Costs	Incl. O&P
1 Equip. Oper. (crane)	$32.35	$258.80	$48.90	$391.20	$29.50	$44.58
1 Equip. Oper. Oiler	26.65	213.20	40.25	322.00		
1 Hyd. Excavator, 2.5 C.Y.		1700.00		1870.00	106.25	116.88
16 L.H., Daily Totals		$2172.00		$2583.20	$135.75	$161.46
Crew B-12T	Hr.	Daily	Hr.	Daily	Bare Costs	Incl. O&P
1 Equip. Oper. (crane)	$32.35	$258.80	$48.90	$391.20	$29.50	$44.58
1 Equip. Oper. Oiler	26.65	213.20	40.25	322.00		
1 Crawler Crane, 75 Ton		1103.00		1213.30		
1 F.E. Attachment, 3 C.Y.		193.20		212.50	81.01	89.11
16 L.H., Daily Totals		$1768.20		$2139.00	$110.51	$133.69
Crew B-12V	Hr.	Daily	Hr.	Daily	Bare Costs	Incl. O&P
1 Equip. Oper. (crane)	$32.35	$258.80	$48.90	$391.20	$29.50	$44.58
1 Equip. Oper. Oiler	26.65	213.20	40.25	322.00		
1 Crawler Crane, 75 Ton		1103.00		1213.30		
1 Dragline Bucket, 3 C.Y.		64.80		71.30	72.99	80.29
16 L.H., Daily Totals		$1639.80		$1997.80	$102.49	$124.87
Crew B-13	Hr.	Daily	Hr.	Daily	Bare Costs	Incl. O&P
1 Labor Foreman (outside)	$25.45	$203.60	$39.60	$316.80	$25.46	$39.25
4 Laborers	23.45	750.40	36.50	1168.00		
1 Equip. Oper. (crane)	32.35	258.80	48.90	391.20		
1 Equip. Oper. Oiler	26.65	213.20	40.25	322.00		
1 Hyd. Crane, 25 Ton		724.80		797.30	12.94	14.24
56 L.H., Daily Totals		$2150.80		$2995.30	$38.40	$53.49
Crew B-13A	Hr.	Daily	Hr.	Daily	Bare Costs	Incl. O&P
1 Foreman	$25.45	$203.60	$39.60	$316.80	$26.39	$40.44
2 Laborers	23.45	375.20	36.50	584.00		
2 Equipment Operator	31.20	499.20	47.15	754.40		
2 Truck Drivers (heavy)	25.00	400.00	38.10	609.60		
1 Crane, 75 Ton		1103.00		1213.30		
1 F.E. Lder, 3.75 C.Y.		971.20		1068.30		
2 Dump Trucks, 12 Ton		706.40		777.05	49.65	54.62
56 L.H., Daily Totals		$4258.60		$5323.45	$76.04	$95.06
Crew B-13B	Hr.	Daily	Hr.	Daily	Bare Costs	Incl. O&P
1 Labor Foreman (outside)	$25.45	$203.60	$39.60	$316.80	$25.46	$39.25
4 Laborers	23.45	750.40	36.50	1168.00		
1 Equip. Oper. (crane)	32.35	258.80	48.90	391.20		
1 Equip. Oper. Oiler	26.65	213.20	40.25	322.00		
1 Hyd. Crane, 55 Ton		972.40		1069.65	17.36	19.10
56 L.H., Daily Totals		$2398.40		$3267.65	$42.82	$58.35
Crew B-13C	Hr.	Daily	Hr.	Daily	Bare Costs	Incl. O&P
1 Labor Foreman (outside)	$25.45	$203.60	$39.60	$316.80	$25.46	$39.25
4 Laborers	23.45	750.40	36.50	1168.00		
1 Equip. Oper. (crane)	32.35	258.80	48.90	391.20		
1 Equip. Oper. Oiler	26.65	213.20	40.25	322.00		
1 Crawler Crane, 100 Ton		1383.00		1521.30	24.70	27.17
56 L.H., Daily Totals		$2809.00		$3719.30	$50.16	$66.42
Crew B-14	Hr.	Daily	Hr.	Daily	Bare Costs	Incl. O&P
1 Labor Foreman (outside)	$25.45	$203.60	$39.60	$316.80	$24.84	$38.44
4 Laborers	23.45	750.40	36.50	1168.00		
1 Equip. Oper. (light)	29.80	238.40	45.05	360.40		
1 Backhoe Loader, 48 H.P.		178.60		196.45	3.72	4.09
48 L.H., Daily Totals		$1371.00		$2041.65	$28.56	$42.53

CREWS

Crews

Crew B-15

Crew No.	Bare Costs Hr.	Daily	Incl. Subs O & P Hr.	Daily	Cost Per Labor-Hour Bare Costs	Incl. O&P
1 Equipment Oper. (med)	$31.20	$249.60	$47.15	$377.20	$26.55	$40.46
.5 Laborer	23.45	93.80	36.50	146.00		
2 Truck Drivers (heavy)	25.00	400.00	38.10	609.60		
2 Dump Trucks, 16 Ton		941.20		1035.30		
1 Dozer, 200 H.P.		823.40		905.75	63.02	69.32
28 L.H., Daily Totals		$2508.00		$3073.85	$89.57	$109.78

Crew B-16

Crew No.	Bare Costs Hr.	Daily	Incl. Subs O & P Hr.	Daily	Cost Per Labor-Hour Bare Costs	Incl. O&P
1 Labor Foreman (outside)	$25.45	$203.60	$39.60	$316.80	$24.34	$37.67
2 Laborers	23.45	375.20	36.50	584.00		
1 Truck Driver (heavy)	25.00	200.00	38.10	304.80		
1 Dump Truck, 16 Ton		470.60		517.65	14.71	16.18
32 L.H., Daily Totals		$1249.40		$1723.25	$39.05	$53.85

Crew B-17

Crew No.	Bare Costs Hr.	Daily	Incl. Subs O & P Hr.	Daily	Cost Per Labor-Hour Bare Costs	Incl. O&P
2 Laborers	$23.45	$375.20	$36.50	$584.00	$25.43	$39.04
1 Equip. Oper. (light)	29.80	238.40	45.05	360.40		
1 Truck Driver (heavy)	25.00	200.00	38.10	304.80		
1 Backhoe Loader, 48 H.P.		178.60		196.45		
1 Dump Truck, 12 Ton		353.20		388.50	16.62	18.28
32 L.H., Daily Totals		$1345.40		$1834.15	$42.05	$57.32

Crew B-18

Crew No.	Bare Costs Hr.	Daily	Incl. Subs O & P Hr.	Daily	Cost Per Labor-Hour Bare Costs	Incl. O&P
1 Labor Foreman (outside)	$25.45	$203.60	$39.60	$316.80	$24.12	$37.53
2 Laborers	23.45	375.20	36.50	584.00		
1 Vibrating Compactor		55.60		61.15	2.32	2.55
24 L.H., Daily Totals		$634.40		$961.95	$26.44	$40.08

Crew B-19

Crew No.	Bare Costs Hr.	Daily	Incl. Subs O & P Hr.	Daily	Cost Per Labor-Hour Bare Costs	Incl. O&P
1 Pile Driver Foreman	$31.80	$254.40	$53.25	$426.00	$30.29	$48.86
4 Pile Drivers	29.80	953.60	49.90	1596.80		
2 Equip. Oper. (crane)	32.35	517.60	48.90	782.40		
1 Equip. Oper. Oiler	26.65	213.20	40.25	322.00		
1 Crane, 40 Ton & Access.		809.20		890.10		
60 L.F. Leads, 15K Ft. Lbs.		456.00		501.60		
1 Hammer, 15K Ft. Lbs.		320.60		352.65		
1 Air Compr., 600 C.F.M.		263.60		289.95		
2-50 Ft. Air Hoses, 3" Dia.		35.40		38.95	29.45	32.40
64 L.H., Daily Totals		$3823.60		$5200.45	$59.74	$81.26

Crew B-19A

Crew No.	Bare Costs Hr.	Daily	Incl. Subs O & P Hr.	Daily	Cost Per Labor-Hour Bare Costs	Incl. O&P
1 Pile Driver Foreman	$31.80	$254.40	$53.25	$426.00	$30.29	$48.86
4 Pile Drivers	29.80	953.60	49.90	1596.80		
2 Equip. Oper. (crane)	32.35	517.60	48.90	782.40		
1 Equip. Oper. Oiler	26.65	213.20	40.25	322.00		
1 Crawler Crane, 75 Ton		1103.00		1213.30		
60 L.F. Leads, 25K Ft. Lbs.		696.00		765.60		
1 Air Compressor, 750 CFM		326.00		358.60		
4-50 Ft. Air Hose, 3" Dia.		70.80		77.90	34.31	37.74
64 L.H., Daily Totals		$4134.60		$5542.60	$64.60	$86.60

Crew B-20

Crew No.	Bare Costs Hr.	Daily	Incl. Subs O & P Hr.	Daily	Cost Per Labor-Hour Bare Costs	Incl. O&P
1 Labor Foreman (out)	$25.45	$203.60	$39.60	$316.80	$26.62	$41.48
1 Skilled Worker	30.95	247.60	48.35	386.80		
1 Laborer	23.45	187.60	36.50	292.00		
24 L.H., Daily Totals		$638.80		$995.60	$26.62	$41.48

Crew B-20A

Crew No.	Bare Costs Hr.	Daily	Incl. Subs O & P Hr.	Daily	Cost Per Labor-Hour Bare Costs	Incl. O&P
1 Labor Foreman	$25.45	$203.60	$39.60	$316.80	$28.40	$43.41
1 Laborer	23.45	187.60	36.50	292.00		
1 Plumber	35.95	287.60	54.20	433.60		
1 Plumber Apprentice	28.75	230.00	43.35	346.80		
32 L.H., Daily Totals		$908.80		$1389.20	$28.40	$43.41

Crew B-21

Crew No.	Bare Costs Hr.	Daily	Incl. Subs O & P Hr.	Daily	Cost Per Labor-Hour Bare Costs	Incl. O&P
1 Labor Foreman (out)	$25.45	$203.60	$39.60	$316.80	$27.44	$42.54
1 Skilled Worker	30.95	247.60	48.35	386.80		
1 Laborer	23.45	187.60	36.50	292.00		
.5 Equip. Oper. (crane)	32.35	129.40	48.90	195.60		
.5 S.P. Crane, 5 Ton		157.70		173.45	5.63	6.20
28 L.H., Daily Totals		$925.90		$1364.65	$33.07	$48.74

Crew B-21A

Crew No.	Bare Costs Hr.	Daily	Incl. Subs O & P Hr.	Daily	Cost Per Labor-Hour Bare Costs	Incl. O&P
1 Labor Foreman	$25.45	$203.60	$39.60	$316.80	$29.19	$44.51
1 Laborer	23.45	187.60	36.50	292.00		
1 Plumber	35.95	287.60	54.20	433.60		
1 Plumber Apprentice	28.75	230.00	43.35	346.80		
1 Equip. Oper. (crane)	32.35	258.80	48.90	391.20		
1 S.P. Crane, 12 Ton		473.20		520.50	11.83	13.01
40 L.H., Daily Totals		$1640.80		$2300.90	$41.02	$57.52

Crew B-22

Crew No.	Bare Costs Hr.	Daily	Incl. Subs O & P Hr.	Daily	Cost Per Labor-Hour Bare Costs	Incl. O&P
1 Labor Foreman (out)	$25.45	$203.60	$39.60	$316.80	$27.76	$42.97
1 Skilled Worker	30.95	247.60	48.35	386.80		
1 Laborer	23.45	187.60	36.50	292.00		
.75 Equip. Oper. (crane)	32.35	194.10	48.90	293.40		
.75 S.P. Crane, 5 Ton		236.55		260.20	7.89	8.67
30 L.H., Daily Totals		$1069.45		$1549.20	$35.65	$51.64

Crew B-22A

Crew No.	Bare Costs Hr.	Daily	Incl. Subs O & P Hr.	Daily	Cost Per Labor-Hour Bare Costs	Incl. O&P
1 Labor Foreman (out)	$25.45	$203.60	$39.60	$316.80	$26.86	$41.61
1 Skilled Worker	30.95	247.60	48.35	386.80		
2 Laborers	23.45	375.20	36.50	584.00		
.75 Equipment Oper. (crane)	32.35	194.10	48.90	293.40		
.75 Crane, 5 Ton		236.55		260.20		
1 Generator, 5 KW		45.20		49.70		
1 Butt Fusion Machine		361.20		397.30	16.92	18.61
38 L.H., Daily Totals		$1663.45		$2288.20	$43.78	$60.22

Crew B-22B

Crew No.	Bare Costs Hr.	Daily	Incl. Subs O & P Hr.	Daily	Cost Per Labor-Hour Bare Costs	Incl. O&P
1 Skilled Worker	$30.95	$247.60	$48.35	$386.80	$27.20	$42.42
1 Laborer	23.45	187.60	36.50	292.00		
1 Electro Fusion Machine		151.00		166.10	9.44	10.38
16 L.H., Daily Totals		$586.20		$844.90	$36.64	$52.80

Crew B-23

Crew No.	Bare Costs Hr.	Daily	Incl. Subs O & P Hr.	Daily	Cost Per Labor-Hour Bare Costs	Incl. O&P
1 Labor Foreman (outside)	$25.45	$203.60	$39.60	$316.80	$23.85	$37.12
4 Laborers	23.45	750.40	36.50	1168.00		
1 Drill Rig, Wells		1993.00		2192.30		
1 Light Truck, 3 Ton		182.40		200.65	54.39	59.82
40 L.H., Daily Totals		$3129.40		$3877.75	$78.24	$96.94

Crew B-23A

Crew No.	Bare Costs Hr.	Daily	Incl. Subs O & P Hr.	Daily	Cost Per Labor-Hour Bare Costs	Incl. O&P
1 Labor Foreman (outside)	$25.45	$203.60	$39.60	$316.80	$26.70	$41.08
1 Laborers	23.45	187.60	36.50	292.00		
1 Equip. Operator (medium)	31.20	249.60	47.15	377.20		
1 Drill Rig, Wells		1993.00		2192.30		
1 Pickup Truck, 3/4 Ton		84.00		92.40	86.54	95.20
24 L.H., Daily Totals		$2717.80		$3270.70	$113.24	$136.28

Crew No.	Bare Costs		Incl. Subs O & P		Cost Per Labor-Hour	
Crew B-23B	Hr.	Daily	Hr.	Daily	Bare Costs	Incl. O&P
1 Labor Foreman (outside)	$25.45	$203.60	$39.60	$316.80	$26.70	$41.08
1 Laborer	23.45	187.60	36.50	292.00		
1 Equip. Operator (medium)	31.20	249.60	47.15	377.20		
1 Drill Rig, Wells		1993.00		2192.30		
1 Pickup Truck, 3/4 Ton		84.00		92.40		
1 Pump, Cntfgl, 6"		198.20		218.00	94.80	104.28
24 L.H., Daily Totals		$2916.00		$3488.70	$121.50	$145.36
Crew B-24	Hr.	Daily	Hr.	Daily	Bare Costs	Incl. O&P
1 Cement Finisher	$28.70	$229.60	$42.50	$340.00	$27.38	$41.90
1 Laborer	23.45	187.60	36.50	292.00		
1 Carpenter	30.00	240.00	46.70	373.60		
24 L.H., Daily Totals		$657.20		$1005.60	$27.38	$41.90
Crew B-25	Hr.	Daily	Hr.	Daily	Bare Costs	Incl. O&P
1 Labor Foreman	$25.45	$203.60	$39.60	$316.80	$25.75	$39.69
7 Laborers	23.45	1313.20	36.50	2044.00		
3 Equip. Oper. (med.)	31.20	748.80	47.15	1131.60		
1 Asphalt Paver, 130 H.P		1290.00		1419.00		
1 Tandem Roller, 10 Ton		168.40		185.25		
1 Roller, Pneumatic Wheel		237.80		261.60	19.28	21.20
88 L.H., Daily Totals		$3961.80		$5358.25	$45.03	$60.89
Crew B-25B	Hr.	Daily	Hr.	Daily	Bare Costs	Incl. O&P
1 Labor Foreman	$25.45	$203.60	$39.60	$316.80	$26.20	$40.31
7 Laborers	23.45	1313.20	36.50	2044.00		
4 Equip. Oper. (medium)	31.20	998.40	47.15	1508.80		
1 Asphalt Paver, 130 H.P.		1290.00		1419.00		
2 Rollers, Steel Wheel		336.80		370.50		
1 Roller, Pneumatic Wheel		237.80		261.60	19.42	21.37
96 L.H., Daily Totals		$4379.80		$5920.70	$45.62	$61.68
Crew B-25C	Hr.	Daily	Hr.	Daily	Bare Costs	Incl. O&P
1 Labor Foreman	$25.45	$203.60	$39.60	$316.80	$26.37	$40.57
3 Laborers	23.45	562.80	36.50	876.00		
2 Equip. Oper. (medium)	31.20	499.20	47.15	754.40		
1 Asphalt Paver, 130 H.P.		1290.00		1419.00		
1 Rollers, Steel Wheel		168.40		185.25	30.38	33.42
48 L.H., Daily Totals		$2724.00		$3551.45	$56.75	$73.99
Crew B-26	Hr.	Daily	Hr.	Daily	Bare Costs	Incl. O&P
1 Labor Foreman (outside)	$25.45	$203.60	$39.60	$316.80	$26.50	$41.20
6 Laborers	23.45	1125.60	36.50	1752.00		
2 Equip. Oper. (med.)	31.20	499.20	47.15	754.40		
1 Rodman (reinf.)	34.25	274.00	57.80	462.40		
1 Cement Finisher	28.70	229.60	42.50	340.00		
1 Grader, 30,000 Lbs.		431.60		474.75		
1 Paving Mach. & Equip.		1451.00		1596.10	21.39	23.53
88 L.H., Daily Totals		$4214.60		$5696.45	$47.89	$64.73
Crew B-27	Hr.	Daily	Hr.	Daily	Bare Costs	Incl. O&P
1 Labor Foreman (outside)	$25.45	$203.60	$39.60	$316.80	$23.95	$37.28
3 Laborers	23.45	562.80	36.50	876.00		
1 Berm Machine		174.40		191.85	5.45	6.00
32 L.H., Daily Totals		$940.80		$1384.65	$29.40	$43.28
Crew B-28	Hr.	Daily	Hr.	Daily	Bare Costs	Incl. O&P
2 Carpenters	$30.00	$480.00	$46.70	$747.20	$27.82	$43.30
1 Laborer	23.45	187.60	36.50	292.00		
24 L.H., Daily Totals		$667.60		$1039.20	$27.82	$43.30

Crew No.	Bare Costs		Incl. Subs O & P		Cost Per Labor-Hour	
Crew B-29	Hr.	Daily	Hr.	Daily	Bare Costs	Incl. O&P
1 Labor Foreman (outside)	$25.45	$203.60	$39.60	$316.80	$25.46	$39.25
4 Laborers	23.45	750.40	36.50	1168.00		
1 Equip. Oper. (crane)	32.35	258.80	48.90	391.20		
1 Equip. Oper. Oiler	26.65	213.20	40.25	322.00		
1 Gradall, 3 Ton, 1/2 C.Y.		736.40		810.05	13.15	14.47
56 L.H., Daily Totals		$2162.40		$3008.05	$38.61	$53.72
Crew B-30	Hr.	Daily	Hr.	Daily	Bare Costs	Incl. O&P
1 Equip. Oper. (med.)	$31.20	$249.60	$47.15	$377.20	$27.07	$41.12
2 Truck Drivers (heavy)	25.00	400.00	38.10	609.60		
1 Hyd. Excavator, 1.5 C.Y.		705.00		775.50		
2 Dump Trucks, 16 Ton		941.20		1035.30	68.59	75.45
24 L.H., Daily Totals		$2295.80		$2797.60	$95.66	$116.57
Crew B-31	Hr.	Daily	Hr.	Daily	Bare Costs	Incl. O&P
1 Labor Foreman (outside)	$25.45	$203.60	$39.60	$316.80	$25.16	$39.16
3 Laborers	23.45	562.80	36.50	876.00		
1 Carpenter	30.00	240.00	46.70	373.60		
1 Air Compr., 250 C.F.M.		135.00		148.50		
1 Sheeting Driver		10.20		11.20		
2-50 Ft. Air Hoses, 1.5" Dia.		9.40		10.35	3.87	4.25
40 L.H., Daily Totals		$1161.00		$1736.45	$29.03	$43.41
Crew B-32	Hr.	Daily	Hr.	Daily	Bare Costs	Incl. O&P
1 Laborer	$23.45	$187.60	$36.50	$292.00	$29.26	$44.49
3 Equip. Oper. (med.)	31.20	748.80	47.15	1131.60		
1 Grader, 30,000 Lbs.		431.60		474.75		
1 Tandem Roller, 10 Ton		168.40		185.25		
1 Dozer, 200 H.P.		823.40		905.75	44.48	48.93
32 L.H., Daily Totals		$2359.80		$2989.35	$73.74	$93.42
Crew B-32A	Hr.	Daily	Hr.	Daily	Bare Costs	Incl. O&P
1 Laborer	$23.45	$187.60	$36.50	$292.00	$28.62	$43.60
2 Equip. Oper. (medium)	31.20	499.20	47.15	754.40		
1 Grader, 30,000 Lbs.		431.60		474.75		
1 Roller, Vibratory, 29,000 Lbs.		434.60		478.05	36.09	39.70
24 L.H., Daily Totals		$1553.00		$1999.20	$64.71	$83.30
Crew B-32B	Hr.	Daily	Hr.	Daily	Bare Costs	Incl. O&P
1 Laborer	$23.45	$187.60	$36.50	$292.00	$28.62	$43.60
2 Equip. Oper. (medium)	31.20	499.20	47.15	754.40		
1 Dozer, 200 H.P.		823.40		905.75		
1 Roller, Vibratory, 29,000 Lbs.		434.60		478.05	52.42	57.66
24 L.H., Daily Totals		$1944.80		$2430.20	$81.04	$101.26
Crew B-32C	Hr.	Daily	Hr.	Daily	Bare Costs	Incl. O&P
1 Labor Foreman	$25.45	$203.60	$39.60	$316.80	$27.66	$42.34
2 Laborers	23.45	375.20	36.50	584.00		
3 Equip. Oper. (medium)	31.20	748.80	47.15	1131.60		
1 Grader, 30,000 Lbs.		431.60		474.75		
1 Roller, Steel Wheel		168.40		185.25		
1 Dozer, 200 H.P.		823.40		905.75	29.65	32.62
48 L.H., Daily Totals		$2751.00		$3598.15	$57.31	$74.96
Crew B-33	Hr.	Daily	Hr.	Daily	Bare Costs	Incl. O&P
1 Equip. Oper. (med.)	$31.20	$249.60	$47.15	$377.20	$28.99	$44.11
.5 Laborer	23.45	93.80	36.50	146.00		
.25 Equip. Oper. (med.)	31.20	62.40	47.15	94.30		
14 L.H., Daily Totals		$405.80		$617.50	$28.99	$44.11

Crews

Crew No.	Bare Costs Hr.	Daily	Incl. Subs O & P Hr.	Daily	Cost Per Labor-Hour Bare Costs	Incl. O&P
Crew B-33A	Hr.	Daily	Hr.	Daily	Bare Costs	Incl. O&P
1 Equip. Oper. (med.)	$31.20	$249.60	$47.15	$377.20	$28.99	$44.11
.5 Laborer	23.45	93.80	36.50	146.00		
.25 Equip. Oper. (med.)	31.20	62.40	47.15	94.30		
1 Scraper, Towed, 7 C.Y.		123.35		135.70		
1.25 Dozer, 300 H.P.		1292.50		1421.75	101.13	111.25
14 L.H., Daily Totals		$1821.65		$2174.95	$130.12	$155.36

Crew B-33B	Hr.	Daily	Hr.	Daily	Bare Costs	Incl. O&P
1 Equip. Oper. (med.)	$31.20	$249.60	$47.15	$377.20	$28.99	$44.11
.5 Laborer	23.45	93.80	36.50	146.00		
.25 Equip. Oper. (med.)	31.20	62.40	47.15	94.30		
1 Scraper, Towed, 10 C.Y.		157.25		173.00		
1.25 Dozer, 300 H.P.		1292.50		1421.75	103.55	113.91
14 L.H., Daily Totals		$1855.55		$2212.25	$132.54	$158.02

Crew B-33C	Hr.	Daily	Hr.	Daily	Bare Costs	Incl. O&P
1 Equip. Oper. (med.)	$31.20	$249.60	$47.15	$377.20	$28.99	$44.11
.5 Laborer	23.45	93.80	36.50	146.00		
.25 Equip. Oper. (med.)	31.20	62.40	47.15	94.30		
1 Scraper, Towed, 12 C.Y.		157.25		173.00		
1.25 Dozer, 300 H.P.		1292.50		1421.75	103.55	113.91
14 L.H., Daily Totals		$1855.55		$2212.25	$132.54	$158.02

Crew B-33D	Hr.	Daily	Hr.	Daily	Bare Costs	Incl. O&P
1 Equip. Oper. (med.)	$31.20	$249.60	$47.15	$377.20	$28.99	$44.11
.5 Laborer	23.45	93.80	36.50	146.00		
.25 Equip. Oper. (med.)	31.20	62.40	47.15	94.30		
1 S.P. Scraper, 14 C.Y.		1419.00		1560.90		
.25 Dozer, 300 H.P.		258.50		284.35	119.82	131.80
14 L.H., Daily Totals		$2083.30		$2462.75	$148.81	$175.91

Crew B-33E	Hr.	Daily	Hr.	Daily	Bare Costs	Incl. O&P
1 Equip. Oper. (med.)	$31.20	$249.60	$47.15	$377.20	$28.99	$44.11
.5 Laborer	23.45	93.80	36.50	146.00		
.25 Equip. Oper. (med.)	31.20	62.40	47.15	94.30		
1 S.P. Scraper, 24 C.Y.		2174.00		2391.40		
.25 Dozer, 300 H.P.		258.50		284.35	173.75	191.13
14 L.H., Daily Totals		$2838.30		$3293.25	$202.74	$235.24

Crew B-33F	Hr.	Daily	Hr.	Daily	Bare Costs	Incl. O&P
1 Equip. Oper. (med.)	$31.20	$249.60	$47.15	$377.20	$28.99	$44.11
.5 Laborer	23.45	93.80	36.50	146.00		
.25 Equip. Oper. (med.)	31.20	62.40	47.15	94.30		
1 Elev. Scraper, 11 C.Y.		744.80		819.30		
.25 Dozer, 300 H.P.		258.50		284.35	71.66	78.83
14 L.H., Daily Totals		$1409.10		$1721.15	$100.65	$122.94

Crew B-33G	Hr.	Daily	Hr.	Daily	Bare Costs	Incl. O&P
1 Equip. Oper. (med.)	$31.20	$249.60	$47.15	$377.20	$28.99	$44.11
.5 Laborer	23.45	93.80	36.50	146.00		
.25 Equip. Oper. (med.)	31.20	62.40	47.15	94.30		
1 Elev. Scraper, 20 C.Y.		1373.00		1510.30		
.25 Dozer, 300 H.P.		258.50		284.35	116.54	128.19
14 L.H., Daily Totals		$2037.30		$2412.15	$145.53	$172.30

Crew B-33H	Hr.	Daily	Hr.	Daily	Bare Costs	Incl. O&P
.25 Laborer	$23.45	$46.90	$36.50	$73.00	$29.86	$45.31
1 Equipment Operator (med.)	31.20	249.60	47.15	377.20		
.2 Equipment Operator (med.)	31.20	49.92	47.15	75.44		
1 Scraper, 32-44 C.Y.		2691.00		2960.10		
.2 Dozer, 298 kw		274.80		302.30	255.67	281.24
11 L.H., Daily Totals		$3312.22		$3788.04	$285.53	$326.55

Crew B-33J	Hr.	Daily	Hr.	Daily	Bare Costs	Incl. O&P
1 Equipment Operator (med.)	$31.20	$249.60	$47.15	$377.20	$31.20	$47.15
1 Scraper 17 C.Y.		1419.00		1560.90	177.38	195.11
8 L.H., Daily Totals		$1668.60		$1938.10	$208.58	$242.26

Crew B-34A	Hr.	Daily	Hr.	Daily	Bare Costs	Incl. O&P
1 Truck Driver (heavy)	$25.00	$200.00	$38.10	$304.80	$25.00	$38.10
1 Dump Truck, 12 Ton		353.20		388.50	44.15	48.57
8 L.H., Daily Totals		$553.20		$693.30	$69.15	$86.67

Crew B-34B	Hr.	Daily	Hr.	Daily	Bare Costs	Incl. O&P
1 Truck Driver (heavy)	$25.00	$200.00	$38.10	$304.80	$25.00	$38.10
1 Dump Truck, 16 Ton		470.60		517.65	58.83	64.71
8 L.H., Daily Totals		$670.60		$822.45	$83.83	$102.81

Crew B-34C	Hr.	Daily	Hr.	Daily	Bare Costs	Incl. O&P
1 Truck Driver (heavy)	$25.00	$200.00	$38.10	$304.80	$25.00	$38.10
1 Truck Tractor, 40 Ton		383.60		421.95		
1 Dump Trailer, 16.5 C.Y.		119.40		131.35	62.88	69.16
8 L.H., Daily Totals		$703.00		$858.10	$87.88	$107.26

Crew B-34D	Hr.	Daily	Hr.	Daily	Bare Costs	Incl. O&P
1 Truck Driver (heavy)	$25.00	$200.00	$38.10	$304.80	$25.00	$38.10
1 Truck Tractor, 40 Ton		383.60		421.95		
1 Dump Trailer, 20 C.Y.		129.00		141.90	64.08	70.48
8 L.H., Daily Totals		$712.60		$868.65	$89.08	$108.58

Crew B-34E	Hr.	Daily	Hr.	Daily	Bare Costs	Incl. O&P
1 Truck Driver (heavy)	$25.00	$200.00	$38.10	$304.80	$25.00	$38.10
1 Truck, Off Hwy., 25 Ton		927.60		1020.35	115.95	127.55
8 L.H., Daily Totals		$1127.60		$1325.15	$140.95	$165.65

Crew B-34F	Hr.	Daily	Hr.	Daily	Bare Costs	Incl. O&P
1 Truck Driver (heavy)	$25.00	$200.00	$38.10	$304.80	$25.00	$38.10
1 Truck, Off Hwy., 22 C.Y.		965.00		1061.50	120.63	132.69
8 L.H., Daily Totals		$1165.00		$1366.30	$145.63	$170.79

Crew B-34G	Hr.	Daily	Hr.	Daily	Bare Costs	Incl. O&P
1 Truck Driver (heavy)	$25.00	$200.00	$38.10	$304.80	$25.00	$38.10
1 Truck, Off Hwy., 34 C.Y.		1364.00		1500.40	170.50	187.55
8 L.H., Daily Totals		$1564.00		$1805.20	$195.50	$225.65

Crew B-34H	Hr.	Daily	Hr.	Daily	Bare Costs	Incl. O&P
1 Truck Driver (heavy)	$25.00	$200.00	$38.10	$304.80	$25.00	$38.10
1 Truck, Off Hwy., 42 C.Y.		1453.00		1598.30	181.63	199.79
8 L.H., Daily Totals		$1653.00		$1903.10	$206.63	$237.89

Crew B-34J	Hr.	Daily	Hr.	Daily	Bare Costs	Incl. O&P
1 Truck Driver (heavy)	$25.00	$200.00	$38.10	$304.80	$25.00	$38.10
1 Truck, Off Hwy., 60 C.Y.		2072.00		2279.20	259.00	284.90
8 L.H., Daily Totals		$2272.00		$2584.00	$284.00	$323.00

CREWS

Crews

Crew B-34K

	Bare Costs Hr.	Bare Costs Daily	Incl. Subs O & P Hr.	Incl. Subs O & P Daily	Cost Per Labor-Hour Bare Costs	Cost Per Labor-Hour Incl. O&P
1 Truck Driver (heavy)	$25.00	$200.00	$38.10	$304.80	$25.00	$38.10
1 Truck Tractor, 240 H.P.		455.20		500.70		
1 Low Bed Trailer		248.80		273.70	88.00	96.80
8 L.H., Daily Totals		$904.00		$1079.20	$113.00	$134.90

Crew B-35

	Bare Costs Hr.	Bare Costs Daily	Incl. Subs O & P Hr.	Incl. Subs O & P Daily	Cost Per Labor-Hour Bare Costs	Cost Per Labor-Hour Incl. O&P
1 Laborer Foreman (out)	$25.45	$203.60	$39.60	$316.80	$29.13	$44.63
1 Skilled Worker	30.95	247.60	48.35	386.80		
1 Welder (plumber)	35.95	287.60	54.20	433.60		
1 Laborer	23.45	187.60	36.50	292.00		
1 Equip. Oper. (crane)	32.35	258.80	48.90	391.20		
1 Equip. Oper. Oiler	26.65	213.20	40.25	322.00		
1 Electric Welding Mach.		79.95		87.95		
1 Hyd. Excavator, .75 C.Y.		457.00		502.70	11.19	12.31
48 L.H., Daily Totals		$1935.35		$2733.05	$40.32	$56.94

Crew B-35A

	Bare Costs Hr.	Bare Costs Daily	Incl. Subs O & P Hr.	Incl. Subs O & P Daily	Cost Per Labor-Hour Bare Costs	Cost Per Labor-Hour Incl. O&P
1 Laborer Foreman (out)	$25.45	$203.60	$39.60	$316.80	$28.32	$43.47
2 Laborers	23.45	375.20	36.50	584.00		
1 Skilled Worker	30.95	247.60	48.35	386.80		
1 Welder (plumber)	35.95	287.60	54.20	433.60		
1 Equip. Oper. (crane)	32.35	258.80	48.90	391.20		
1 Equip. Oper. Oiler	26.65	213.20	40.25	322.00		
1 Welder, 300 amp		88.60		97.45		
1 Crane, 75 Ton		1103.00		1213.30	21.28	23.41
56 L.H., Daily Totals		$2777.60		$3745.15	$49.60	$66.88

Crew B-36

	Bare Costs Hr.	Bare Costs Daily	Incl. Subs O & P Hr.	Incl. Subs O & P Daily	Cost Per Labor-Hour Bare Costs	Cost Per Labor-Hour Incl. O&P
1 Labor Foreman (outside)	$25.45	$203.60	$39.60	$316.80	$26.95	$41.38
2 Laborers	23.45	375.20	36.50	584.00		
2 Equip. Oper. (med.)	31.20	499.20	47.15	754.40		
1 Dozer, 200 H.P.		823.40		905.75		
1 Aggregate Spreader		62.00		68.20		
1 Tandem Roller, 10 Ton		168.40		185.25	26.35	28.98
40 L.H., Daily Totals		$2131.80		$2814.40	$53.30	$70.36

Crew B-36A

	Bare Costs Hr.	Bare Costs Daily	Incl. Subs O & P Hr.	Incl. Subs O & P Daily	Cost Per Labor-Hour Bare Costs	Cost Per Labor-Hour Incl. O&P
1 Labor Foreman (outside)	$25.45	$203.60	$39.60	$316.80	$28.16	$43.03
2 Laborers	23.45	375.20	36.50	584.00		
4 Equip. Oper. (med.)	31.20	998.40	47.15	1508.80		
1 Dozer, 200 H.P.		823.40		905.75		
1 Aggregate Spreader		62.00		68.20		
1 Roller, Steel Wheel		168.40		185.25		
1 Roller, Pneumatic Wheel		237.80		261.60	23.06	25.37
56 L.H., Daily Totals		$2868.80		$3830.40	$51.22	$68.40

Crew B-36B

	Bare Costs Hr.	Bare Costs Daily	Incl. Subs O & P Hr.	Incl. Subs O & P Daily	Cost Per Labor-Hour Bare Costs	Cost Per Labor-Hour Incl. O&P
1 Labor Foreman (outside)	$25.45	$203.60	$39.60	$316.80	$27.77	$42.41
2 Laborers	23.45	375.20	36.50	584.00		
4 Equip. Oper. (medium)	31.20	998.40	47.15	1508.80		
1 Truck Driver, Heavy	25.00	200.00	38.10	304.80		
1 Grader, 30,000 Lbs.		431.60		474.75		
1 F.E. Loader, crl. 1.5 C.Y.		322.60		354.85		
1 Dozer, 300 H.P.		1034.00		1137.40		
1 Roller, Vibratory		434.60		478.05		
1 Truck, Tractor, 240 H.P.		455.20		500.70		
1 Water Tanker, 5000 Gal.		163.40		179.75	44.40	48.84
64 L.H., Daily Totals		$4618.60		$5839.90	$72.17	$91.25

Crew B-36C

	Bare Costs Hr.	Bare Costs Daily	Incl. Subs O & P Hr.	Incl. Subs O & P Daily	Cost Per Labor-Hour Bare Costs	Cost Per Labor-Hour Incl. O&P
1 Labor Foreman (outside)	$25.45	$203.60	$39.60	$316.80	$28.81	$43.83
3 Equip. Oper. (medium)	31.20	748.80	47.15	1131.60		
1 Truck Driver, Heavy	25.00	200.00	38.10	304.80		
1 Grader, 30,000 Lbs.		431.60		474.75		
1 Dozer, 300 H.P.		1034.00		1137.40		
1 Roller, Vibratory		434.60		478.05		
1 Truck, Tractor, 240 H.P.		455.20		500.70		
1 Water Tanker, 5000 Gal.		163.40		179.75	62.97	69.27
40 L.H., Daily Totals		$3671.20		$4523.85	$91.78	$113.10

Crew B-37

	Bare Costs Hr.	Bare Costs Daily	Incl. Subs O & P Hr.	Incl. Subs O & P Daily	Cost Per Labor-Hour Bare Costs	Cost Per Labor-Hour Incl. O&P
1 Labor Foreman (outside)	$25.45	$203.60	$39.60	$316.80	$24.84	$38.44
4 Laborers	23.45	750.40	36.50	1168.00		
1 Equip. Oper. (light)	29.80	238.40	45.05	360.40		
1 Tandem Roller, 5 Ton		117.20		128.90	2.44	2.69
48 L.H., Daily Totals		$1309.60		$1974.10	$27.28	$41.13

Crew B-38

	Bare Costs Hr.	Bare Costs Daily	Incl. Subs O & P Hr.	Incl. Subs O & P Daily	Cost Per Labor-Hour Bare Costs	Cost Per Labor-Hour Incl. O&P
1 Labor Foreman (outside)	$25.45	$203.60	$39.60	$316.80	$26.67	$40.96
2 Laborers	23.45	375.20	36.50	584.00		
1 Equip. Oper. (light)	29.80	238.40	45.05	360.40		
1 Equip. Oper. (medium)	31.20	249.60	47.15	377.20		
1 Backhoe Loader, 48 H.P.		178.60		196.45		
1 Hyd. Hammer, (1200 lb)		162.20		178.40		
1 F.E. Loader (170 H.P.)		484.80		533.30		
1 Pavt. Rem. Bucket		45.00		49.50	21.77	23.94
40 L.H., Daily Totals		$1937.40		$2596.05	$48.44	$64.90

Crew B-39

	Bare Costs Hr.	Bare Costs Daily	Incl. Subs O & P Hr.	Incl. Subs O & P Daily	Cost Per Labor-Hour Bare Costs	Cost Per Labor-Hour Incl. O&P
1 Labor Foreman (outside)	$25.45	$203.60	$39.60	$316.80	$24.84	$38.44
4 Laborers	23.45	750.40	36.50	1168.00		
1 Equip. Oper. (light)	29.80	238.40	45.05	360.40		
1 Air Compr., 250 C.F.M.		135.00		148.50		
2 Air Tools & Accessories		33.20		36.50		
2-50 Ft. Air Hoses, 1.5" Dia.		9.40		10.35	3.70	4.07
48 L.H., Daily Totals		$1370.00		$2040.55	$28.54	$42.51

Crew B-40

	Bare Costs Hr.	Bare Costs Daily	Incl. Subs O & P Hr.	Incl. Subs O & P Daily	Cost Per Labor-Hour Bare Costs	Cost Per Labor-Hour Incl. O&P
1 Pile Driver Foreman (out)	$31.80	$254.40	$53.25	$426.00	$30.29	$48.86
4 Pile Drivers	29.80	953.60	49.90	1596.80		
2 Equip. Oper. (crane)	32.35	517.60	48.90	782.40		
1 Equip. Oper. Oiler	26.65	213.20	40.25	322.00		
1 Crane, 40 Ton		809.20		890.10		
1 Vibratory Hammer & Gen.		1391.00		1530.10	34.38	37.82
64 L.H., Daily Totals		$4139.00		$5547.40	$64.67	$86.68

Crew B-41

	Bare Costs Hr.	Bare Costs Daily	Incl. Subs O & P Hr.	Incl. Subs O & P Daily	Cost Per Labor-Hour Bare Costs	Cost Per Labor-Hour Incl. O&P
1 Labor Foreman (outside)	$25.45	$203.60	$39.60	$316.80	$24.36	$37.80
4 Laborers	23.45	750.40	36.50	1168.00		
.25 Equip. Oper. (crane)	32.35	64.70	48.90	97.80		
.25 Equip. Oper. Oiler	26.65	53.30	40.25	80.50		
.25 Crawler Crane, 40 Ton		202.30		222.55	4.60	5.06
44 L.H., Daily Totals		$1274.30		$1885.65	$28.96	$42.86

Crew B-42

Crew No.	Bare Costs Hr.	Daily	Incl. Subs O & P Hr.	Daily	Cost Per Labor-Hour Bare Costs	Incl. O&P
1 Labor Foreman (outside)	$25.45	$203.60	$39.60	$316.80	$26.56	$42.06
4 Laborers	23.45	750.40	36.50	1168.00		
1 Equip. Oper. (crane)	32.35	258.80	48.90	391.20		
1 Equip. Oper. Oiler	26.65	213.20	40.25	322.00		
1 Welder	34.25	274.00	61.75	494.00		
1 Hyd. Crane, 25 Ton		724.80		797.30		
1 Gas Welding Machine		88.60		97.45		
1 Horz. Boring Csg. Mch.		465.00		511.50	19.98	21.97
64 L.H., Daily Totals		$2978.40		$4098.25	$46.54	$64.03

Crew B-43

Crew No.	Bare Costs Hr.	Daily	Incl. Subs O & P Hr.	Daily	Cost Per Labor-Hour Bare Costs	Incl. O&P
1 Labor Foreman (outside)	$25.45	$203.60	$39.60	$316.80	$25.80	$39.71
3 Laborers	23.45	562.80	36.50	876.00		
1 Equip. Oper. (crane)	32.35	258.80	48.90	391.20		
1 Equip. Oper. Oiler	26.65	213.20	40.25	322.00		
1 Drill Rig & Augers		1993.00		2192.30	41.52	45.67
48 L.H., Daily Totals		$3231.40		$4098.30	$67.32	$85.38

Crew B-44

Crew No.	Bare Costs Hr.	Daily	Incl. Subs O & P Hr.	Daily	Cost Per Labor-Hour Bare Costs	Incl. O&P
1 Pile Driver Foreman	$31.80	$254.40	$53.25	$426.00	$29.89	$48.39
4 Pile Drivers	29.80	953.60	49.90	1596.80		
2 Equip. Oper. (crane)	32.35	517.60	48.90	782.40		
1 Laborer	23.45	187.60	36.50	292.00		
1 Crane, 40 Ton, & Access.		809.20		890.10		
45 L.F. Leads, 15K Ft. Lbs.		342.00		376.20	17.99	19.79
64 L.H., Daily Totals		$3064.40		$4363.50	$47.88	$68.18

Crew B-45

Crew No.	Bare Costs Hr.	Daily	Incl. Subs O & P Hr.	Daily	Cost Per Labor-Hour Bare Costs	Incl. O&P
1 Equip. Oper. (med.)	$31.20	$249.60	$47.15	$377.20	$28.10	$42.63
1 Truck Driver (heavy)	25.00	200.00	38.10	304.80		
1 Dist. Tank Truck, 3K Gal.		287.60		316.35		
1 Tractor, 4 x 2, 250 H.P.		378.80		416.70	41.65	45.82
16 L.H., Daily Totals		$1116.00		$1415.05	$69.75	$88.45

Crew B-46

Crew No.	Bare Costs Hr.	Daily	Incl. Subs O & P Hr.	Daily	Cost Per Labor-Hour Bare Costs	Incl. O&P
1 Pile Driver Foreman	$31.80	$254.40	$53.25	$426.00	$26.96	$43.76
2 Pile Drivers	29.80	476.80	49.90	798.40		
3 Laborers	23.45	562.80	36.50	876.00		
1 Chain Saw, 36" Long		39.60		43.55	.83	.91
48 L.H., Daily Totals		$1333.60		$2143.95	$27.79	$44.67

Crew B-47

Crew No.	Bare Costs Hr.	Daily	Incl. Subs O & P Hr.	Daily	Cost Per Labor-Hour Bare Costs	Incl. O&P
1 Blast Foreman	$25.45	$203.60	$39.60	$316.80	$26.23	$40.38
1 Driller	23.45	187.60	36.50	292.00		
1 Equip. Oper. (light)	29.80	238.40	45.05	360.40		
1 Crawler Type Drill, 4"		530.40		583.45		
1 Air Compr., 600 C.F.M.		263.60		289.95		
2-50 Ft. Air Hoses, 3" Dia.		35.40		38.95	34.56	38.01
24 L.H., Daily Totals		$1459.00		$1881.55	$60.79	$78.39

Crew B-47A

Crew No.	Bare Costs Hr.	Daily	Incl. Subs O & P Hr.	Daily	Cost Per Labor-Hour Bare Costs	Incl. O&P
1 Drilling Foreman	$25.45	$203.60	$39.60	$316.80	$28.15	$42.92
1 Equip. Oper. (heavy)	32.35	258.80	48.90	391.20		
1 Oiler	26.65	213.20	40.25	322.00		
1 Quarry Drill		793.80		873.20	33.08	36.38
24 L.H., Daily Totals		$1469.40		$1903.20	$61.23	$79.30

Crew B-47C

Crew No.	Bare Costs Hr.	Daily	Incl. Subs O & P Hr.	Daily	Cost Per Labor-Hour Bare Costs	Incl. O&P
1 Laborer	$23.45	$187.60	$36.50	$292.00	$26.63	$40.78
1 Equip. Oper. (light)	29.80	238.40	45.05	360.40		
1 Air Compressor, 750 CFM		326.00		358.60		
2-50' Air Hose, 3"		35.40		38.95		
1 Air Track Drill, 4"		530.40		583.45	55.74	61.31
16 L.H., Daily Totals		$1317.80		$1633.40	$82.37	$102.09

Crew B-47E

Crew No.	Bare Costs Hr.	Daily	Incl. Subs O & P Hr.	Daily	Cost Per Labor-Hour Bare Costs	Incl. O&P
1 Laborer Foreman	$25.45	$203.60	$39.60	$316.80	$23.95	$37.28
3 Laborers	23.45	562.80	36.50	876.00		
1 Truck, Flatbed, 3 Ton		182.40		200.65	5.70	6.27
32 L.H., Daily Totals		$948.80		$1393.45	$29.65	$43.55

Crew B-48

Crew No.	Bare Costs Hr.	Daily	Incl. Subs O & P Hr.	Daily	Cost Per Labor-Hour Bare Costs	Incl. O&P
1 Labor Foreman (outside)	$25.45	$203.60	$39.60	$316.80	$26.37	$40.47
3 Laborers	23.45	562.80	36.50	876.00		
1 Equip. Oper. (crane)	32.35	258.80	48.90	391.20		
1 Equip. Oper. Oiler	26.65	213.20	40.25	322.00		
1 Equip. Oper. (light)	29.80	238.40	45.05	360.40		
1 Centr. Water Pump, 6"		198.20		218.00		
1-20 Ft. Suction Hose, 6"		19.50		21.45		
1-50 Ft. Disch. Hose, 6"		17.40		19.15		
1 Drill Rig & Augers		1993.00		2192.30	39.79	43.77
56 L.H., Daily Totals		$3704.90		$4717.30	$66.16	$84.24

Crew B-49

Crew No.	Bare Costs Hr.	Daily	Incl. Subs O & P Hr.	Daily	Cost Per Labor-Hour Bare Costs	Incl. O&P
1 Labor Foreman (outside)	$25.45	$203.60	$39.60	$316.80	$27.56	$42.93
3 Laborers	23.45	562.80	36.50	876.00		
2 Equip. Oper. (crane)	32.35	517.60	48.90	782.40		
2 Equip. Oper. Oilers	26.65	426.40	40.25	644.00		
1 Equip. Oper. (light)	29.80	238.40	45.05	360.40		
2 Pile Drivers	29.80	476.80	49.90	798.40		
1 Hyd. Crane, 25 Ton		724.80		797.30		
1 Centr. Water Pump, 6"		198.20		218.00		
1-20 Ft. Suction Hose, 6"		19.50		21.45		
1-50 Ft. Disch. Hose, 6"		17.40		19.15		
1 Drill Rig & Augers		1993.00		2192.30	33.56	36.91
88 L.H., Daily Totals		$5378.50		$7026.20	$61.12	$79.84

Crew B-50

Crew No.	Bare Costs Hr.	Daily	Incl. Subs O & P Hr.	Daily	Cost Per Labor-Hour Bare Costs	Incl. O&P
2 Pile Driver Foremen	$31.80	$508.80	$53.25	$852.00	$28.86	$46.67
6 Pile Drivers	29.80	1430.40	49.90	2395.20		
2 Equip. Oper. (crane)	32.35	517.60	48.90	782.40		
1 Equip. Oper. Oiler	26.65	213.20	40.25	322.00		
3 Laborers	23.45	562.80	36.50	876.00		
1 Crane, 40 Ton		809.20		890.10		
60 L.F. Leads, 15K Ft. Lbs.		456.00		501.60		
1 Hammer, 15K Ft. Lbs.		320.60		352.65		
1 Air Compr., 600 C.F.M.		263.60		289.95		
2-50 Ft. Air Hoses, 3" Dia.		35.40		38.95		
1 Chain Saw, 36" Long		39.60		43.55	17.18	18.90
112 L.H., Daily Totals		$5157.20		$7344.40	$46.04	$65.57

Crew B-51

Crew No.	Bare Costs Hr.	Daily	Incl. Subs O & P Hr.	Daily	Cost Per Labor-Hour Bare Costs	Incl. O&P
1 Labor Foreman (outside)	$25.45	$203.60	$39.60	$316.80	$23.93	$37.11
4 Laborers	23.45	750.40	36.50	1168.00		
1 Truck Driver (light)	24.30	194.40	37.05	296.40		
1 Light Truck, 1.5 Ton		155.60		171.15	3.24	3.57
48 L.H., Daily Totals		$1304.00		$1952.35	$27.17	$40.68

CREWS

Crew No.	Bare Costs		Incl. Subs O & P		Cost Per Labor-Hour	

Crew B-52

Crew B-52	Hr.	Daily	Hr.	Daily	Bare Costs	Incl. O&P
1 Carpenter Foreman	$32.00	$256.00	$49.80	$398.40	$27.68	$43.00
1 Carpenter	30.00	240.00	46.70	373.60		
3 Laborers	23.45	562.80	36.50	876.00		
1 Cement Finisher	28.70	229.60	42.50	340.00		
.5 Rodman (reinf.)	34.25	137.00	57.80	231.20		
.5 Equip. Oper. (med.)	31.20	124.80	47.15	188.60		
.5 F.E. Ldr., T.M., 2.5 C.Y.		345.30		379.85	6.17	6.78
56 L.H., Daily Totals		$1895.50		$2787.65	$33.85	$49.78

Crew B-53

Crew B-53	Hr.	Daily	Hr.	Daily	Bare Costs	Incl. O&P
1 Equip. Oper. (light)	$29.80	$238.40	$45.05	$360.40	$29.80	$45.05
1 Trencher, Chain, 12 H.P.		86.00		94.60	10.75	11.83
8 L.H., Daily Totals		$324.40		$455.00	$40.55	$56.88

Crew B-54

Crew B-54	Hr.	Daily	Hr.	Daily	Bare Costs	Incl. O&P
1 Equip. Oper. (light)	$29.80	$238.40	$45.05	$360.40	$29.80	$45.05
1 Trencher, Chain, 40 H.P.		193.00		212.30	24.13	26.54
8 L.H., Daily Totals		$431.40		$572.70	$53.93	$71.59

Crew B-54A

Crew B-54A	Hr.	Daily	Hr.	Daily	Bare Costs	Incl. O&P
.17 Labor Foreman (outside)	$25.45	$34.61	$39.60	$53.86	$30.36	$46.05
1 Equipment Operator (med.)	31.20	249.60	47.15	377.20		
1 Wheel Trencher, 67 H.P.		488.40		537.25	52.18	57.40
9.36 L.H., Daily Totals		$772.61		$968.31	$82.54	$103.45

Crew B-54B

Crew B-54B	Hr.	Daily	Hr.	Daily	Bare Costs	Incl. O&P
.25 Labor Foreman (outside)	$25.45	$50.90	$39.60	$79.20	$30.05	$45.64
1 Equipment Operator (med.)	31.20	249.60	47.15	377.20		
1 Wheel Trencher, 150 H.P.		753.60		828.95	75.36	82.90
10 L.H., Daily Totals		$1054.10		$1285.35	$105.41	$128.54

Crew B-55

Crew B-55	Hr.	Daily	Hr.	Daily	Bare Costs	Incl. O&P
2 Laborers	$23.45	$375.20	$36.50	$584.00	$23.73	$36.68
1 Truck Driver (light)	24.30	194.40	37.05	296.40		
1 Auger, 4" to 36" Dia.		552.40		607.65		
1 Flatbed 3 Ton Truck		182.40		200.65	30.62	33.68
24 L.H., Daily Totals		$1304.40		$1688.70	$54.35	$70.36

Crew B-56

Crew B-56	Hr.	Daily	Hr.	Daily	Bare Costs	Incl. O&P
1 Laborer	$23.45	$187.60	$36.50	$292.00	$26.63	$40.78
1 Equip. Oper. (light)	29.80	238.40	45.05	360.40		
1 Crawler Type Drill, 4"		530.40		583.45		
1 Air Compr., 600 C.F.M.		263.60		289.95		
1-50 Ft. Air Hose, 3" Dia.		17.70		19.45	50.73	55.80
16 L.H., Daily Totals		$1237.70		$1545.25	$77.36	$96.58

Crew B-57

Crew B-57	Hr.	Daily	Hr.	Daily	Bare Costs	Incl. O&P
1 Labor Foreman (outside)	$25.45	$203.60	$39.60	$316.80	$26.86	$41.13
2 Laborers	23.45	375.20	36.50	584.00		
1 Equip. Oper. (crane)	32.35	258.80	48.90	391.20		
1 Equip. Oper. (light)	29.80	238.40	45.05	360.40		
1 Equip. Oper. Oiler	26.65	213.20	40.25	322.00		
1 Power Shovel, 1 C.Y.		640.80		704.90		
1 Clamshell Bucket, 1 C.Y.		61.60		67.75		
1 Centr. Water Pump, 6"		198.20		218.00		
1-20 Ft. Suction Hose, 6"		19.50		21.45		
20-50 Ft. Disch. Hoses, 6"		348.00		382.80	26.42	29.06
48 L.H., Daily Totals		$2557.30		$3369.30	$53.28	$70.19

Crew No.	Bare Costs		Incl. Subs O & P		Cost Per Labor-Hour	

Crew B-58

Crew B-58	Hr.	Daily	Hr.	Daily	Bare Costs	Incl. O&P
2 Laborers	$23.45	$375.20	$36.50	$584.00	$25.57	$39.35
1 Equip. Oper. (light)	29.80	238.40	45.05	360.40		
1 Backhoe Loader, 48 H.P.		178.60		196.45		
1 Small Helicopter, w/pilot		2031.00		2234.10	92.07	101.27
24 L.H., Daily Totals		$2823.20		$3374.95	$117.64	$140.62

Crew B-59

Crew B-59	Hr.	Daily	Hr.	Daily	Bare Costs	Incl. O&P
1 Truck Driver (heavy)	$25.00	$200.00	$38.10	$304.80	$25.00	$38.10
1 Truck, 30 Ton		291.80		321.00		
1 Water tank, 5000 Gal.		163.40		179.75	56.90	62.59
8 L.H., Daily Totals		$655.20		$805.55	$81.90	$100.69

Crew B-59A

Crew B-59A	Hr.	Daily	Hr.	Daily	Bare Costs	Incl. O&P
2 Laborers	$23.45	$375.20	$36.50	$584.00	$23.97	$37.03
1 Truck Driver (heavy)	25.00	200.00	38.10	304.80		
1 Water tank, 5K w/pum		163.40		179.75		
1 Truck, 30 Ton		291.80		321.00	18.97	20.86
24 L.H., Daily Totals		$1030.40		$1389.55	$42.94	$57.89

Crew B-60

Crew B-60	Hr.	Daily	Hr.	Daily	Bare Costs	Incl. O&P
1 Labor Foreman (outside)	$25.45	$203.60	$39.60	$316.80	$27.28	$41.69
2 Laborers	23.45	375.20	36.50	584.00		
1 Equip. Oper. (crane)	32.35	258.80	48.90	391.20		
2 Equip. Oper. (light)	29.80	476.80	45.05	720.80		
1 Equip. Oper. Oiler	26.65	213.20	40.25	322.00		
1 Crawler Crane, 40 Ton		809.20		890.10		
45 L.F. Leads, 15K Ft. Lbs.		342.00		376.20		
1 Backhoe Loader, 48 H.P.		178.60		196.45	23.75	26.12
56 L.H., Daily Totals		$2857.40		$3797.55	$51.03	$67.81

Crew B-61

Crew B-61	Hr.	Daily	Hr.	Daily	Bare Costs	Incl. O&P
1 Labor Foreman (outside)	$25.45	$203.60	$39.60	$316.80	$25.12	$38.83
3 Laborers	23.45	562.80	36.50	876.00		
1 Equip. Oper. (light)	29.80	238.40	45.05	360.40		
1 Cement Mixer, 2 C.Y.		165.60		182.15		
1 Air Compr., 160 C.F.M.		97.20		106.90	6.57	7.23
40 L.H., Daily Totals		$1267.60		$1842.25	$31.69	$46.06

Crew B-62

Crew B-62	Hr.	Daily	Hr.	Daily	Bare Costs	Incl. O&P
2 Laborers	$23.45	$375.20	$36.50	$584.00	$25.57	$39.35
1 Equip. Oper. (light)	29.80	238.40	45.05	360.40		
1 Loader, Skid Steer		130.80		143.90	5.45	6.00
24 L.H., Daily Totals		$744.40		$1088.30	$31.02	$45.35

Crew B-63

Crew B-63	Hr.	Daily	Hr.	Daily	Bare Costs	Incl. O&P
4 Laborers	$23.45	$750.40	$36.50	$1168.00	$24.72	$38.21
1 Equip. Oper. (light)	29.80	238.40	45.05	360.40		
1 Loader, Skid Steer		130.80		143.90	3.27	3.60
40 L.H., Daily Totals		$1119.60		$1672.30	$27.99	$41.81

Crew B-64

Crew B-64	Hr.	Daily	Hr.	Daily	Bare Costs	Incl. O&P
1 Laborer	$23.45	$187.60	$36.50	$292.00	$23.88	$36.78
1 Truck Driver (light)	24.30	194.40	37.05	296.40		
1 Power Mulcher (small)		114.40		125.85		
1 Light Truck, 1.5 Ton		155.60		171.15	16.88	18.56
16 L.H., Daily Totals		$652.00		$885.40	$40.76	$55.34

Crews

Crew No.	Bare Costs			Incl. Subs O & P		Cost Per Labor-Hour	

Crew B-65	Hr.	Daily	Hr.	Daily	Bare Costs	Incl. O&P
1 Laborer	$23.45	$187.60	$36.50	$292.00	$23.88	$36.78
1 Truck Driver (light)	24.30	194.40	37.05	296.40		
1 Power Mulcher (large)		230.60		253.65		
1 Light Truck, 1.5 Ton		155.60		171.15	24.14	26.55
16 L.H., Daily Totals		$768.20		$1013.20	$48.02	$63.33

Crew B-66	Hr.	Daily	Hr.	Daily	Bare Costs	Incl. O&P
1 Equip. Oper. (light)	$29.80	$238.40	$45.05	$360.40	$29.80	$45.05
1 Backhoe Ldr. w/Attchmt.		155.80		171.40	19.48	21.42
8 L.H., Daily Totals		$394.20		$531.80	$49.28	$66.47

Crew B-67	Hr.	Daily	Hr.	Daily	Bare Costs	Incl. O&P
1 Millwright	$31.75	$254.00	$47.00	$376.00	$30.77	$46.03
1 Equip. Oper. (light)	29.80	238.40	45.05	360.40		
1 Forklift		232.20		255.40	14.51	15.96
16 L.H., Daily Totals		$724.60		$991.80	$45.28	$61.99

Crew B-68	Hr.	Daily	Hr.	Daily	Bare Costs	Incl. O&P
2 Millwrights	$31.75	$508.00	$47.00	$752.00	$31.10	$46.35
1 Equip. Oper. (light)	29.80	238.40	45.05	360.40		
1 Forklift		232.20		255.40	9.68	10.64
24 L.H., Daily Totals		$978.60		$1367.80	$40.78	$56.99

Crew B-69	Hr.	Daily	Hr.	Daily	Bare Costs	Incl. O&P
1 Labor Foreman (outside)	$25.45	$203.60	$39.60	$316.80	$25.80	$39.71
3 Laborers	23.45	562.80	36.50	876.00		
1 Equip. Oper. (crane)	32.35	258.80	48.90	391.20		
1 Equip. Oper. Oiler	26.65	213.20	40.25	322.00		
1 Truck Crane, 80 Ton		1356.00		1491.60	28.25	31.08
48 L.H., Daily Totals		$2594.40		$3397.60	$54.05	$70.79

Crew B-69A	Hr.	Daily	Hr.	Daily	Bare Costs	Incl. O&P
1 Labor Foreman	$25.45	$203.60	$39.60	$316.80	$25.95	$39.79
3 Laborers	23.45	562.80	36.50	876.00		
1 Equip. Oper. (medium)	31.20	249.60	47.15	377.20		
1 Concrete Finisher	28.70	229.60	42.50	340.00		
1 Curb Paver		491.40		540.55	10.24	11.26
48 L.H., Daily Totals		$1737.00		$2450.55	$36.19	$51.05

Crew B-69B	Hr.	Daily	Hr.	Daily	Bare Costs	Incl. O&P
1 Labor Foreman	$25.45	$203.60	$39.60	$316.80	$25.95	$39.79
3 Laborers	23.45	562.80	36.50	876.00		
1 Equip. Oper. (medium)	31.20	249.60	47.15	377.20		
1 Cement Finisher	28.70	229.60	42.50	340.00		
1 Curb/Gutter Paver		816.80		898.50	17.02	18.72
48 L.H., Daily Totals		$2062.40		$2808.50	$42.97	$58.51

Crew B-70	Hr.	Daily	Hr.	Daily	Bare Costs	Incl. O&P
1 Labor Foreman (outside)	$25.45	$203.60	$39.60	$316.80	$27.06	$41.51
3 Laborers	23.45	562.80	36.50	876.00		
3 Equip. Oper. (med.)	31.20	748.80	47.15	1131.60		
1 Motor Grader, 30,000 Lb.		431.60		474.75		
1 Grader Attach., Ripper		59.00		64.90		
1 Road Sweeper, S.P.		310.20		341.20		
1 F.E. Loader, 1-3/4 C.Y.		276.40		304.05	19.24	21.16
56 L.H., Daily Totals		$2592.40		$3509.30	$46.30	$62.67

Crew B-71	Hr.	Daily	Hr.	Daily	Bare Costs	Incl. O&P
1 Labor Foreman (outside)	$25.45	$203.60	$39.60	$316.80	$27.06	$41.51
3 Laborers	23.45	562.80	36.50	876.00		
3 Equip. Oper. (med.)	31.20	748.80	47.15	1131.60		
1 Pvmt. Profiler, 450 H.P.		3748.00		4122.80		
1 Road Sweeper, S.P.		310.20		341.20		
1 F.E. Loader, 1-3/4 C.Y.		276.40		304.05	77.40	85.14
56 L.H., Daily Totals		$5849.80		$7092.45	$104.46	$126.65

Crew B-72	Hr.	Daily	Hr.	Daily	Bare Costs	Incl. O&P
1 Labor Foreman (outside)	$25.45	$203.60	$39.60	$316.80	$27.58	$42.21
3 Laborers	23.45	562.80	36.50	876.00		
4 Equip. Oper. (med.)	31.20	998.40	47.15	1508.80		
1 Pvmt. Profiler, 450 H.P.		3748.00		4122.80		
1 Hammermill, 250 H.P.		1267.00		1393.70		
1 Windrow Loader		793.20		872.50		
1 Mix Paver 165 H.P.		1471.00		1618.10		
1 Roller, Pneu. Tire, 12 T.		237.80		261.60	117.45	129.20
64 L.H., Daily Totals		$9281.80		$10970.30	$145.03	$171.41

Crew B-73	Hr.	Daily	Hr.	Daily	Bare Costs	Incl. O&P
1 Labor Foreman (outside)	$25.45	$203.60	$39.60	$316.80	$28.54	$43.54
2 Laborers	23.45	375.20	36.50	584.00		
5 Equip. Oper. (med.)	31.20	1248.00	47.15	1886.00		
1 Road Mixer, 310 H.P.		1316.00		1447.60		
1 Roller, Tandem, 12 Ton		168.40		185.25		
1 Hammermill, 250 H.P.		1267.00		1393.70		
1 Motor Grader, 30,000 Lb.		431.60		474.75		
.5 F.E. Loader, 1-3/4 C.Y.		138.20		152.00		
.5 Truck, 30 Ton		145.90		160.50		
.5 Water Tank 5000 Gal.		81.70		89.85	55.45	61.00
64 L.H., Daily Totals		$5375.60		$6690.45	$83.99	$104.54

Crew B-74	Hr.	Daily	Hr.	Daily	Bare Costs	Incl. O&P
1 Labor Foreman (outside)	$25.45	$203.60	$39.60	$316.80	$27.96	$42.61
1 Laborer	23.45	187.60	36.50	292.00		
4 Equip. Oper. (med.)	31.20	998.40	47.15	1508.80		
2 Truck Drivers (heavy)	25.00	400.00	38.10	609.60		
1 Motor Grader, 30,000 Lb.		431.60		474.75		
1 Grader Attach., Ripper		59.00		64.90		
2 Stabilizers, 310 H.P.		1869.60		2056.55		
1 Flatbed Truck, 3 Ton		182.40		200.65		
1 Chem. Spreader, Towed		82.00		90.20		
1 Vibr. Roller, 29,000 Lb.		434.60		478.05		
1 Water Tank 5000 Gal.		163.40		179.75		
1 Truck, 30 Ton		291.80		321.00	54.91	60.40
64 L.H., Daily Totals		$5304.00		$6593.05	$82.87	$103.01

Crew B-75	Hr.	Daily	Hr.	Daily	Bare Costs	Incl. O&P
1 Labor Foreman (outside)	$25.45	$203.60	$39.60	$316.80	$28.39	$43.26
1 Laborer	23.45	187.60	36.50	292.00		
4 Equip. Oper. (med.)	31.20	998.40	47.15	1508.80		
1 Truck Driver (heavy)	25.00	200.00	38.10	304.80		
1 Motor Grader, 30,000 Lb.		431.60		474.75		
1 Grader Attach., Ripper		59.00		64.90		
2 Stabilizers, 310 H.P.		1869.60		2056.55		
1 Dist. Truck, 3000 Gal.		287.60		316.35		
1 Vibr. Roller, 29,000 Lb.		434.60		478.05	55.04	60.55
56 L.H., Daily Totals		$4672.00		$5813.00	$83.43	$103.81

Crew No.	Bare Costs		Incl. Subs O & P		Cost Per Labor-Hour	
Crew B-76	Hr.	Daily	Hr.	Daily	Bare Costs	Incl. O&P
1 Dock Builder Foreman	$31.80	$254.40	$53.25	$426.00	$30.24	$48.98
5 Dock Builders	29.80	1192.00	49.90	1996.00		
2 Equip. Oper. (crane)	32.35	517.60	48.90	782.40		
1 Equip. Oper. Oiler	26.65	213.20	40.25	322.00		
1 Crawler Crane, 50 Ton		1027.00		1129.70		
1 Barge, 400 Ton		374.80		412.30		
1 Hammer, 15K Ft. Lbs.		320.60		352.65		
60 L.F. Leads, 15K Ft. Lbs.		456.00		501.60		
1 Air Compr., 600 C.F.M.		263.60		289.95		
2-50 Ft. Air Hoses, 3" Dia.		35.40		38.95	34.41	37.85
72 L.H., Daily Totals		$4654.60		$6251.55	$64.65	$86.83

Crew B-77	Hr.	Daily	Hr.	Daily	Bare Costs	Incl. O&P
1 Labor Foreman	$25.45	$203.60	$39.60	$316.80	$24.02	$37.23
3 Laborers	23.45	562.80	36.50	876.00		
1 Truck Driver (light)	24.30	194.40	37.05	296.40		
1 Crack Cleaner, 25 H.P.		64.40		70.85		
1 Crack Filler, Trailer Mtd.		138.20		152.00		
1 Flatbed Truck, 3 Ton		182.40		200.65	9.63	10.59
40 L.H., Daily Totals		$1345.80		$1912.70	$33.65	$47.82

Crew B-78	Hr.	Daily	Hr.	Daily	Bare Costs	Incl. O&P
1 Labor Foreman	$25.45	$203.60	$39.60	$316.80	$23.93	$37.11
4 Laborers	23.45	750.40	36.50	1168.00		
1 Truck Driver (light)	24.30	194.40	37.05	296.40		
1 Paint Striper, S.P.		177.40		195.15		
1 Flatbed Truck, 3 Ton		182.40		200.65		
1 Pickup Truck, 3/4 Ton		84.00		92.40	9.25	10.17
48 L.H., Daily Totals		$1592.20		$2269.40	$33.18	$47.28

Crew B-79	Hr.	Daily	Hr.	Daily	Bare Costs	Incl. O&P
1 Labor Foreman	$25.45	$203.60	$39.60	$316.80	$24.02	$37.23
3 Laborers	23.45	562.80	36.50	876.00		
1 Truck Driver (light)	24.30	194.40	37.05	296.40		
1 Thermo. Striper, T.M.		475.40		522.95		
1 Flatbed Truck, 3 Ton		182.40		200.65		
2 Pickup Truck, 3/4 Ton		168.00		184.80	20.65	22.71
40 L.H., Daily Totals		$1786.60		$2397.60	$44.67	$59.94

Crew B-80	Hr.	Daily	Hr.	Daily	Bare Costs	Incl. O&P
1 Labor Foreman	$25.45	$203.60	$39.60	$316.80	$25.75	$39.55
1 Laborer	23.45	187.60	36.50	292.00		
1 Truck Driver (light)	24.30	194.40	37.05	296.40		
1 Equip. Oper. (light)	29.80	238.40	45.05	360.40		
1 Flatbed Truck, 3 Ton		182.40		200.65		
1 Fence Post Auger, T.M.		377.80		415.60	17.51	19.26
32 L.H., Daily Totals		$1384.20		$1881.85	$43.26	$58.81

Crew B-80A	Hr.	Daily	Hr.	Daily	Bare Costs	Incl. O&P
3 Laborers	$23.45	$562.80	$36.50	$876.00	$23.45	$36.50
1 Flatbed Truck, 3 Ton		182.40		200.65	7.60	8.36
24 L.H., Daily Totals		$745.20		$1076.65	$31.05	$44.86

Crew B-80B	Hr.	Daily	Hr.	Daily	Bare Costs	Incl. O&P
3 Laborers	$23.45	$562.80	$36.50	$876.00	$25.04	$38.64
1 Equip. Oper. (light)	29.80	238.40	45.05	360.40		
1 Crane, Flatbed Mnt.		205.60		226.15	6.43	7.07
32 L.H., Daily Totals		$1006.80		$1462.55	$31.47	$45.71

Crew No.	Bare Costs		Incl. Subs O & P		Cost Per Labor-Hour	
Crew B-81	Hr.	Daily	Hr.	Daily	Bare Costs	Incl. O&P
1 Laborer	$23.45	$187.60	$36.50	$292.00	$26.55	$40.58
1 Equip. Oper. (med.)	31.20	249.60	47.15	377.20		
1 Truck Driver (heavy)	25.00	200.00	38.10	304.80		
1 Hydromulcher, T.M.		249.40		274.35		
1 Tractor Truck, 4x2		291.80		321.00	22.55	24.81
24 L.H., Daily Totals		$1178.40		$1569.35	$49.10	$65.39

Crew B-82	Hr.	Daily	Hr.	Daily	Bare Costs	Incl. O&P
1 Laborer	$23.45	$187.60	$36.50	$292.00	$26.63	$40.78
1 Equip. Oper. (light)	29.80	238.40	45.05	360.40		
1 Horiz. Borer, 6 H.P.		51.60		56.75	3.23	3.55
16 L.H., Daily Totals		$477.60		$709.15	$29.86	$44.33

Crew B-83	Hr.	Daily	Hr.	Daily	Bare Costs	Incl. O&P
1 Tugboat Captain	$31.20	$249.60	$47.15	$377.20	$27.33	$41.83
1 Tugboat Hand	23.45	187.60	36.50	292.00		
1 Tugboat, 250 H.P.		485.60		534.15	30.35	33.39
16 L.H., Daily Totals		$922.80		$1203.35	$57.68	$75.22

Crew B-84	Hr.	Daily	Hr.	Daily	Bare Costs	Incl. O&P
1 Equip. Oper. (med.)	$31.20	$249.60	$47.15	$377.20	$31.20	$47.15
1 Rotary Mower/Tractor		199.20		219.10	24.90	27.39
8 L.H., Daily Totals		$448.80		$596.30	$56.10	$74.54

Crew B-85	Hr.	Daily	Hr.	Daily	Bare Costs	Incl. O&P
3 Laborers	$23.45	$562.80	$36.50	$876.00	$25.31	$38.95
1 Equip. Oper. (med.)	31.20	249.60	47.15	377.20		
1 Truck Driver (heavy)	25.00	200.00	38.10	304.80		
1 Aerial Lift Truck, 80'		493.60		542.95		
1 Brush Chipper, 130 H.P.		182.60		200.85		
1 Pruning Saw, Rotary		17.90		19.70	17.35	19.09
40 L.H., Daily Totals		$1706.50		$2321.50	$42.66	$58.04

Crew B-86	Hr.	Daily	Hr.	Daily	Bare Costs	Incl. O&P
1 Equip. Oper. (med.)	$31.20	$249.60	$47.15	$377.20	$31.20	$47.15
1 Stump Chipper, S.P.		135.65		149.20	16.96	18.65
8 L.H., Daily Totals		$385.25		$526.40	$48.16	$65.80

Crew B-86A	Hr.	Daily	Hr.	Daily	Bare Costs	Incl. O&P
1 Equip. Oper. (medium)	$31.20	$249.60	$47.15	$377.20	$31.20	$47.15
1 Grader, 30,000 Lbs.		431.60		474.75	53.95	59.35
8 L.H., Daily Totals		$681.20		$851.95	$85.15	$106.50

Crew B-86B	Hr.	Daily	Hr.	Daily	Bare Costs	Incl. O&P
1 Equip. Oper. (medium)	$31.20	$249.60	$47.15	$377.20	$31.20	$47.15
1 Dozer, 200 H.P.		823.40		905.75	102.93	113.22
8 L.H., Daily Totals		$1073.00		$1282.95	$134.13	$160.37

Crew B-87	Hr.	Daily	Hr.	Daily	Bare Costs	Incl. O&P
1 Laborer	$23.45	$187.60	$36.50	$292.00	$29.65	$45.02
4 Equip. Oper. (med.)	31.20	998.40	47.15	1508.80		
2 Feller Bunchers, 50 H.P.		806.40		887.05		
1 Log Chipper, 22" Tree		1543.00		1697.30		
1 Dozer, 105 H.P.		291.20		320.30		
1 Chainsaw, Gas, 36" Long		39.60		43.55	67.01	73.71
40 L.H., Daily Totals		$3866.20		$4749.00	$96.66	$118.73

Crew No.	Bare Costs		Incl. Subs O & P		Cost Per Labor-Hour	

Crew B-88

	Hr.	Daily	Hr.	Daily	Bare Costs	Incl. O&P
1 Laborer	$23.45	$187.60	$36.50	$292.00	$30.09	$45.63
6 Equip. Oper. (med.)	31.20	1497.60	47.15	2263.20		
2 Feller Bunchers, 50 H.P.		806.40		887.05		
1 Log Chipper, 22" Tree		1543.00		1697.30		
2 Log Skidders, 50 H.P.		1200.80		1320.90		
1 Dozer, 105 H.P.		291.20		320.30		
1 Chainsaw, Gas, 36" Long		39.60		43.55	69.30	76.23
56 L.H., Daily Totals		$5566.20		$6824.30	$99.39	$121.86

Crew B-89

	Hr.	Daily	Hr.	Daily	Bare Costs	Incl. O&P
1 Equip. Oper. (light)	$29.80	$238.40	$45.05	$360.40	$27.05	$41.05
1 Truck Driver (light)	24.30	194.40	37.05	296.40		
1 Truck, Stake Body, 3 Ton		182.40		200.65		
1 Concrete Saw		103.40		113.75		
1 Water Tank, 65 Gal.		13.50		14.85	18.71	20.58
16 L.H., Daily Totals		$732.10		$986.05	$45.76	$61.63

Crew B-89A

	Hr.	Daily	Hr.	Daily	Bare Costs	Incl. O&P
1 Skilled Worker	$30.95	$247.60	$48.35	$386.80	$27.20	$42.42
1 Laborer	23.45	187.60	36.50	292.00		
1 Core Drill (large)		93.05		102.35	5.82	6.40
16 L.H., Daily Totals		$528.25		$781.15	$33.02	$48.82

Crew B-89B

	Hr.	Daily	Hr.	Daily	Bare Costs	Incl. O&P
1 Equip. Oper. (light)	$29.80	$238.40	$45.05	$360.40	$27.05	$41.05
1 Truck Driver, Light	24.30	194.40	37.05	296.40		
1 Wall Saw, Hydraulic, 10 H.P.		93.05		102.35		
1 Generator, Diesel, 100 KW		206.00		226.60		
1 Water Tank, 65 Gal.		13.50		14.85		
1 Flatbed Truck, 3 Ton		182.40		200.65	30.93	34.03
16 L.H., Daily Totals		$927.75		$1201.25	$57.98	$75.08

Crew B-90

	Hr.	Daily	Hr.	Daily	Bare Costs	Incl. O&P
1 Labor Foreman (outside)	$25.45	$203.60	$39.60	$316.80	$25.68	$39.42
3 Laborers	23.45	562.80	36.50	876.00		
2 Equip. Oper. (light)	29.80	476.80	45.05	720.80		
2 Truck Drivers (heavy)	25.00	400.00	38.10	609.60		
1 Road Mixer, 310 H.P.		1316.00		1447.60		
1 Dist. Truck, 2000 Gal.		271.20		298.30	24.80	27.28
64 L.H., Daily Totals		$3230.40		$4269.10	$50.48	$66.70

Crew B-90A

	Hr.	Daily	Hr.	Daily	Bare Costs	Incl. O&P
1 Labor Foreman	$25.45	$203.60	$39.60	$316.80	$28.16	$43.03
2 Laborers	23.45	375.20	36.50	584.00		
4 Equip. Oper. (medium)	31.20	998.40	47.15	1508.80		
2 Graders, 30,000 Lbs.		863.20		949.50		
1 Roller, Steel Wheel		168.40		185.25		
1 Roller, Pneumatic Wheel		237.80		261.60	22.67	24.93
56 L.H., Daily Totals		$2846.60		$3805.95	$50.83	$67.96

Crew B-90B

	Hr.	Daily	Hr.	Daily	Bare Costs	Incl. O&P
1 Labor Foreman	$25.45	$203.60	$39.60	$316.80	$27.66	$42.34
2 Laborers	23.45	375.20	36.50	584.00		
3 Equip. Oper. (medium)	31.20	748.80	47.15	1131.60		
1 Roller, Steel Wheel		168.40		185.25		
1 Roller, Pneumatic Wheel		237.80		261.60		
1 Road Mixer, 310 H.P.		1316.00		1447.60	35.88	39.47
48 L.H., Daily Totals		$3049.80		$3926.85	$63.54	$81.81

Crew B-91

	Hr.	Daily	Hr.	Daily	Bare Costs	Incl. O&P
1 Labor Foreman (outside)	$25.45	$203.60	$39.60	$316.80	$27.77	$42.41
2 Laborers	23.45	375.20	36.50	584.00		
4 Equip. Oper. (med.)	31.20	998.40	47.15	1508.80		
1 Truck Driver (heavy)	25.00	200.00	38.10	304.80		
1 Dist. Truck, 3000 Gal.		287.60		316.35		
1 Aggreg. Spreader, S.P.		678.20		746.00		
1 Roller, Pneu. Tire, 12 Ton		237.80		261.60		
1 Roller, Steel, 10 Ton		168.40		185.25	21.44	23.58
64 L.H., Daily Totals		$3149.20		$4223.60	$49.21	$65.99

Crew B-92

	Hr.	Daily	Hr.	Daily	Bare Costs	Incl. O&P
1 Labor Foreman (outside)	$25.45	$203.60	$39.60	$316.80	$23.95	$37.28
3 Laborers	23.45	562.80	36.50	876.00		
1 Crack Cleaner, 25 H.P.		64.40		70.85		
1 Air Compressor		63.60		69.95		
1 Tar Kettle, T.M.		42.30		46.55		
1 Flatbed Truck, 3 Ton		182.40		200.65	11.02	12.12
32 L.H., Daily Totals		$1119.10		$1580.80	$34.97	$49.40

Crew B-93

	Hr.	Daily	Hr.	Daily	Bare Costs	Incl. O&P
1 Equip. Oper. (med.)	$31.20	$249.60	$47.15	$377.20	$31.20	$47.15
1 Feller Buncher, 50 H.P.		403.20		443.50	50.40	55.44
8 L.H., Daily Totals		$652.80		$820.70	$81.60	$102.59

Crew B-94A

	Hr.	Daily	Hr.	Daily	Bare Costs	Incl. O&P
1 Laborer	$23.45	$187.60	$36.50	$292.00	$23.45	$36.50
1 Diaph. Water Pump, 2"		32.60		35.85		
1-20 Ft. Suction Hose, 2"		5.10		5.60		
2-50 Ft. Disch. Hoses, 2"		8.60		9.45	5.79	6.37
8 L.H., Daily Totals		$233.90		$342.90	$29.24	$42.87

Crew B-94B

	Hr.	Daily	Hr.	Daily	Bare Costs	Incl. O&P
1 Laborer	$23.45	$187.60	$36.50	$292.00	$23.45	$36.50
1 Diaph. Water Pump, 4"		68.80		75.70		
1-20 Ft. Suction Hose, 4"		10.25		11.30		
2-50 Ft. Disch. Hoses, 4"		14.50		15.95	11.69	12.86
8 L.H., Daily Totals		$281.15		$394.95	$35.14	$49.36

Crew B-94C

	Hr.	Daily	Hr.	Daily	Bare Costs	Incl. O&P
1 Laborer	$23.45	$187.60	$36.50	$292.00	$23.45	$36.50
1 Centr. Water Pump, 3"		39.00		42.90		
1-20 Ft. Suction Hose, 3"		7.75		8.55		
2-50 Ft. Disch. Hoses, 3"		10.70		11.75	7.18	7.90
8 L.H., Daily Totals		$245.05		$355.20	$30.63	$44.40

Crew B-94D

	Hr.	Daily	Hr.	Daily	Bare Costs	Incl. O&P
1 Laborer	$23.45	$187.60	$36.50	$292.00	$23.45	$36.50
1 Centr. Water Pump, 6"		198.20		218.00		
1-20 Ft. Suction Hose, 6"		19.50		21.45		
2-50 Ft. Disch. Hoses, 6"		34.80		38.30	31.56	34.72
8 L.H., Daily Totals		$440.10		$569.75	$55.01	$71.22

Crew B-95

	Hr.	Daily	Hr.	Daily	Bare Costs	Incl. O&P
1 Equip. Oper. (crane)	$32.35	$258.80	$48.90	$391.20	$27.90	$42.70
1 Laborer	23.45	187.60	36.50	292.00		
16 L.H., Daily Totals		$446.40		$683.20	$27.90	$42.70

Crew No.		Bare Costs		Incl. Subs O & P		Cost Per Labor-Hour	
Crew B-95A	Hr.	Daily	Hr.	Daily	Bare Costs	Incl. O&P
1 Equip. Oper. (crane)	$32.35	$258.80	$48.90	$391.20	$27.90	$42.70
1 Laborer	23.45	187.60	36.50	292.00		
1 Hyd. Excavator, 5/8 C.Y.		393.20		432.50	24.58	27.03
16 L.H., Daily Totals		$839.60		$1115.70	$52.48	$69.73
Crew B-95B	Hr.	Daily	Hr.	Daily	Bare Costs	Incl. O&P
1 Equip. Oper. (crane)	$32.35	$258.80	$48.90	$391.20	$27.90	$42.70
1 Laborer	23.45	187.60	36.50	292.00		
1 Hyd. Excavator, 1.5 C.Y.		705.00		775.50	44.06	48.47
16 L.H., Daily Totals		$1151.40		$1458.70	$71.96	$91.17
Crew B-95C	Hr.	Daily	Hr.	Daily	Bare Costs	Incl. O&P
1 Equip. Oper. (crane)	$32.35	$258.80	$48.90	$391.20	$27.90	$42.70
1 Laborer	23.45	187.60	36.50	292.00		
1 Hyd. Excavator, 2.5 C.Y.		1700.00		1870.00	106.25	116.88
16 L.H., Daily Totals		$2146.40		$2553.20	$134.15	$159.58
Crew C-1	Hr.	Daily	Hr.	Daily	Bare Costs	Incl. O&P
3 Carpenters	$30.00	$720.00	$46.70	$1120.80	$28.36	$44.15
1 Laborer	23.45	187.60	36.50	292.00		
32 L.H., Daily Totals		$907.60		$1412.80	$28.36	$44.15
Crew C-2	Hr.	Daily	Hr.	Daily	Bare Costs	Incl. O&P
1 Carpenter Foreman (out)	$32.00	$256.00	$49.80	$398.40	$29.24	$45.52
4 Carpenters	30.00	960.00	46.70	1494.40		
1 Laborer	23.45	187.60	36.50	292.00		
48 L.H., Daily Totals		$1403.60		$2184.80	$29.24	$45.52
Crew C-2A	Hr.	Daily	Hr.	Daily	Bare Costs	Incl. O&P
1 Carpenter Foreman (out)	$32.00	$256.00	$49.80	$398.40	$29.03	$44.82
3 Carpenters	30.00	720.00	46.70	1120.80		
1 Cement Finisher	28.70	229.60	42.50	340.00		
1 Laborer	23.45	187.60	36.50	292.00		
48 L.H., Daily Totals		$1393.20		$2151.20	$29.03	$44.82
Crew C-3	Hr.	Daily	Hr.	Daily	Bare Costs	Incl. O&P
1 Rodman Foreman	$36.25	$290.00	$61.20	$489.60	$31.24	$51.31
4 Rodmen (reinf.)	34.25	1096.00	57.80	1849.60		
1 Equip. Oper. (light)	29.80	238.40	45.05	360.40		
2 Laborers	23.45	375.20	36.50	584.00		
3 Stressing Equipment		66.60		73.25		
.5 Grouting Equipment		97.60		107.35	2.57	2.82
64 L.H., Daily Totals		$2163.80		$3464.20	$33.81	$54.13
Crew C-4	Hr.	Daily	Hr.	Daily	Bare Costs	Incl. O&P
1 Rodman Foreman	$36.25	$290.00	$61.20	$489.60	$34.75	$58.65
3 Rodmen (reinf.)	34.25	822.00	57.80	1387.20		
3 Stressing Equipment		66.60		73.25	2.08	2.29
32 L.H., Daily Totals		$1178.60		$1950.05	$36.83	$60.94
Crew C-5	Hr.	Daily	Hr.	Daily	Bare Costs	Incl. O&P
1 Rodman Foreman	$36.25	$290.00	$61.20	$489.60	$33.18	$54.51
4 Rodmen (reinf.)	34.25	1096.00	57.80	1849.60		
1 Equip. Oper. (crane)	32.35	258.80	48.90	391.20		
1 Equip. Oper. Oiler	26.65	213.20	40.25	322.00		
1 Hyd. Crane, 25 Ton		724.80		797.30	12.94	14.24
56 L.H., Daily Totals		$2582.80		$3849.70	$46.12	$68.75

Crew No.		Bare Costs		Incl. Subs O & P		Cost Per Labor-Hour	
Crew C-6	Hr.	Daily	Hr.	Daily	Bare Costs	Incl. O&P
1 Labor Foreman (outside)	$25.45	$203.60	$39.60	$316.80	$24.66	$38.02
4 Laborers	23.45	750.40	36.50	1168.00		
1 Cement Finisher	28.70	229.60	42.50	340.00		
2 Gas Engine Vibrators		71.20		78.30	1.48	1.63
48 L.H., Daily Totals		$1254.80		$1903.10	$26.14	$39.65
Crew C-7	Hr.	Daily	Hr.	Daily	Bare Costs	Incl. O&P
1 Labor Foreman (outside)	$25.45	$203.60	$39.60	$316.80	$25.47	$39.11
5 Laborers	23.45	938.00	36.50	1460.00		
1 Cement Finisher	28.70	229.60	42.50	340.00		
1 Equip. Oper. (med.)	31.20	249.60	47.15	377.20		
1 Equip. Oper. (oiler)	26.65	213.20	40.25	322.00		
2 Gas Engine Vibrators		71.20		78.30		
1 Concrete Bucket, 1 C.Y.		22.60		24.85		
1 Hyd. Crane, 55 Ton		972.40		1069.65	14.81	16.29
72 L.H., Daily Totals		$2900.20		$3988.80	$40.28	$55.40
Crew C-7A	Hr.	Daily	Hr.	Daily	Bare Costs	Incl. O&P
1 Labor Foreman (outside)	$25.45	$203.60	$39.60	$316.80	$24.09	$37.29
5 Laborers	23.45	938.00	36.50	1460.00		
2 Truck Driver (Heavy)	25.00	400.00	38.10	609.60		
2 Conc. Transit Mixers		1607.20		1767.90	25.11	27.62
64 L.H., Daily Totals		$3148.80		$4154.30	$49.20	$64.91
Crew C-7B	Hr.	Daily	Hr.	Daily	Bare Costs	Incl. O&P
1 Labor Foreman (outside)	$25.45	$203.60	$39.60	$316.80	$25.21	$38.91
5 Laborers	23.45	938.00	36.50	1460.00		
1 Equipment Operator (heavy)	32.35	258.80	48.90	391.20		
1 Equipment Oiler	26.65	213.20	40.25	322.00		
1 Conc. Bucket, 2 C.Y.		31.40		34.55		
1 Truck Crane, 165 Ton		1958.00		2153.80	31.08	34.19
64 L.H., Daily Totals		$3603.00		$4678.35	$56.29	$73.10
Crew C-7C	Hr.	Daily	Hr.	Daily	Bare Costs	Incl. O&P
1 Labor Foreman (outside)	$25.45	$203.60	$39.60	$316.80	$25.64	$39.55
5 Laborers	23.45	938.00	36.50	1460.00		
2 Equipment Operator (medium)	31.20	499.20	47.15	754.40		
2 Wheel Loader, 4 C.Y.		969.60		1066.55	15.15	16.67
64 L.H., Daily Totals		$2610.40		$3597.75	$40.79	$56.22
Crew C-7D	Hr.	Daily	Hr.	Daily	Bare Costs	Incl. O&P
1 Labor Foreman (outside)	$25.45	$203.60	$39.60	$316.80	$24.84	$38.46
5 Laborers	23.45	938.00	36.50	1460.00		
1 Equipment Operator (med.)	31.20	249.60	47.15	377.20		
1 Concrete Conveyer		169.40		186.35	3.03	3.33
56 L.H., Daily Totals		$1560.60		$2340.35	$27.87	$41.79
Crew C-8	Hr.	Daily	Hr.	Daily	Bare Costs	Incl. O&P
1 Labor Foreman (outside)	$25.45	$203.60	$39.60	$316.80	$26.34	$40.18
3 Laborers	23.45	562.80	36.50	876.00		
2 Cement Finishers	28.70	459.20	42.50	680.00		
1 Equip. Oper. (med.)	31.20	249.60	47.15	377.20		
1 Concrete Pump (small)		681.40		749.55	12.17	13.38
56 L.H., Daily Totals		$2156.60		$2999.55	$38.51	$53.56
Crew C-8A	Hr.	Daily	Hr.	Daily	Bare Costs	Incl. O&P
1 Labor Foreman (outside)	$25.45	$203.60	$39.60	$316.80	$25.53	$39.02
3 Laborers	23.45	562.80	36.50	876.00		
2 Cement Finishers	28.70	459.20	42.50	680.00		
48 L.H., Daily Totals		$1225.60		$1872.80	$25.53	$39.02

Crew No.	Bare Costs		Incl. Subs O & P		Cost Per Labor-Hour	

Crew C-8B

	Hr.	Daily	Hr.	Daily	Bare Costs	Incl. O&P
1 Labor Foreman (outside)	$25.45	$203.60	$39.60	$316.80	$25.40	$39.25
3 Laborers	23.45	562.80	36.50	876.00		
1 Equipment Operator	31.20	249.60	47.15	377.20		
1 Vibrating Screed		53.00		58.30		
1 Vibratory Roller		434.60		478.05		
1 Dozer, 200 H.P.		823.40		905.75	32.78	36.05
40 L.H., Daily Totals		$2327.00		$3012.10	$58.18	$75.30

Crew C-8C

	Hr.	Daily	Hr.	Daily	Bare Costs	Incl. O&P
1 Labor Foreman (outside)	$25.45	$203.60	$39.60	$316.80	$25.95	$39.79
3 Laborers	23.45	562.80	36.50	876.00		
1 Cement Finisher	28.70	229.60	42.50	340.00		
1 Equipment Operator (med.)	31.20	249.60	47.15	377.20		
1 Shotcrete Rig, 12 CY/hr		298.20		328.00	6.21	6.83
48 L.H., Daily Totals		$1543.80		$2238.00	$32.16	$46.62

Crew C-8D

	Hr.	Daily	Hr.	Daily	Bare Costs	Incl. O&P
1 Labor Foreman (outside)	$25.45	$203.60	$39.60	$316.80	$26.85	$40.91
1 Laborers	23.45	187.60	36.50	292.00		
1 Cement Finisher	28.70	229.60	42.50	340.00		
1 Equipment Operator (light)	29.80	238.40	45.05	360.40		
1 Compressor, 250 CFM		135.00		148.50		
2 Hoses, 1", 50'		6.90		7.60	4.43	4.88
32 L.H., Daily Totals		$1001.10		$1465.30	$31.28	$45.79

Crew C-8E

	Hr.	Daily	Hr.	Daily	Bare Costs	Incl. O&P
1 Labor Foreman (outside)	$25.45	$203.60	$39.60	$316.80	$26.85	$40.91
1 Laborers	23.45	187.60	36.50	292.00		
1 Cement Finisher	28.70	229.60	42.50	340.00		
1 Equipment Operator (light)	29.80	238.40	45.05	360.40		
1 Compressor, 250 CFM		135.00		148.50		
2 Hoses, 1", 50'		6.90		7.60		
1 Concrete Pump (small)		681.40		749.55	25.73	28.30
32 L.H., Daily Totals		$1682.50		$2214.85	$52.58	$69.21

Crew C-10

	Hr.	Daily	Hr.	Daily	Bare Costs	Incl. O&P
1 Laborer	$23.45	$187.60	$36.50	$292.00	$26.95	$40.50
2 Cement Finishers	28.70	459.20	42.50	680.00		
24 L.H., Daily Totals		$646.80		$972.00	$26.95	$40.50

Crew C-11

	Hr.	Daily	Hr.	Daily	Bare Costs	Incl. O&P
1 Struc. Steel Foreman	$36.25	$290.00	$65.35	$522.80	$33.42	$58.33
6 Struc. Steel Workers	34.25	1644.00	61.75	2964.00		
1 Equip. Oper. (crane)	32.35	258.80	48.90	391.20		
1 Equip. Oper. Oiler	26.65	213.20	40.25	322.00		
1 Truck Crane, 150 Ton		1690.00		1859.00	23.47	25.82
72 L.H., Daily Totals		$4096.00		$6059.00	$56.89	$84.15

Crew C-12

	Hr.	Daily	Hr.	Daily	Bare Costs	Incl. O&P
1 Carpenter Foreman (out)	$32.00	$256.00	$49.80	$398.40	$29.63	$45.88
3 Carpenters	30.00	720.00	46.70	1120.80		
1 Laborer	23.45	187.60	36.50	292.00		
1 Equip. Oper. (crane)	32.35	258.80	48.90	391.20		
1 Hyd. Crane, 12 Ton		607.60		668.35	12.66	13.92
48 L.H., Daily Totals		$2030.00		$2870.75	$42.29	$59.80

Crew C-13

	Hr.	Daily	Hr.	Daily	Bare Costs	Incl. O&P
1 Struc. Steel Worker	$34.25	$274.00	$61.75	$494.00	$32.83	$56.73
1 Welder	34.25	274.00	61.75	494.00		
1 Carpenter	30.00	240.00	46.70	373.60		
1 Gas Welding Machine		88.60		97.45	3.69	4.06
24 L.H., Daily Totals		$876.60		$1459.05	$36.52	$60.79

Crew C-14

	Hr.	Daily	Hr.	Daily	Bare Costs	Incl. O&P
1 Carpenter Foreman (out)	$32.00	$256.00	$49.80	$398.40	$29.40	$46.37
5 Carpenters	30.00	1200.00	46.70	1868.00		
4 Laborers	23.45	750.40	36.50	1168.00		
4 Rodmen (reinf.)	34.25	1096.00	57.80	1849.60		
2 Cement Finishers	28.70	459.20	42.50	680.00		
1 Equip. Oper. (crane)	32.35	258.80	48.90	391.20		
1 Equip. Oper. Oiler	26.65	213.20	40.25	322.00		
1 Crane, 80 Ton, & Tools		1356.00		1491.60	9.42	10.36
144 L.H., Daily Totals		$5589.60		$8168.80	$38.82	$56.73

Crew C-14A

	Hr.	Daily	Hr.	Daily	Bare Costs	Incl. O&P
1 Carpenter Foreman (out)	$32.00	$256.00	$49.80	$398.40	$30.23	$47.63
16 Carpenters	30.00	3840.00	46.70	5977.60		
4 Rodmen (reinf.)	34.25	1096.00	57.80	1849.60		
2 Laborers	23.45	375.20	36.50	584.00		
1 Cement Finisher	28.70	229.60	42.50	340.00		
1 Equip. Oper. (med.)	31.20	249.60	47.15	377.20		
1 Gas Engine Vibrator		35.60		39.15		
1 Concrete Pump (small)		681.40		749.55	3.59	3.94
200 L.H., Daily Totals		$6763.40		$10315.50	$33.82	$51.57

Crew C-14B

	Hr.	Daily	Hr.	Daily	Bare Costs	Incl. O&P
1 Carpenter Foreman (out)	$32.00	$256.00	$49.80	$398.40	$30.17	$47.44
16 Carpenters	30.00	3840.00	46.70	5977.60		
4 Rodmen (reinf.)	34.25	1096.00	57.80	1849.60		
2 Laborers	23.45	375.20	36.50	584.00		
2 Cement Finishers	28.70	459.20	42.50	680.00		
1 Equip. Oper. (med.)	31.20	249.60	47.15	377.20		
1 Gas Engine Vibrator		35.60		39.15		
1 Concrete Pump (small)		681.40		749.55	3.45	3.79
208 L.H., Daily Totals		$6993.00		$10655.50	$33.62	$51.23

Crew C-14C

	Hr.	Daily	Hr.	Daily	Bare Costs	Incl. O&P
1 Carpenter Foreman (out)	$32.00	$256.00	$49.80	$398.40	$28.79	$45.29
6 Carpenters	30.00	1440.00	46.70	2241.60		
2 Rodmen (reinf.)	34.25	548.00	57.80	924.80		
4 Laborers	23.45	750.40	36.50	1168.00		
1 Cement Finisher	28.70	229.60	42.50	340.00		
1 Gas Engine Vibrator		35.60		39.15	.32	.35
112 L.H., Daily Totals		$3259.60		$5111.95	$29.11	$45.64

Crew C-14D

	Hr.	Daily	Hr.	Daily	Bare Costs	Incl. O&P
1 Carpenter Foreman (out)	$32.00	$256.00	$49.80	$398.40	$29.89	$46.75
18 Carpenters	30.00	4320.00	46.70	6724.80		
2 Rodmen (reinf.)	34.25	548.00	57.80	924.80		
2 Laborers	23.45	375.20	36.50	584.00		
1 Cement Finisher	28.70	229.60	42.50	340.00		
1 Equip. Oper. (med.)	31.20	249.60	47.15	377.20		
1 Gas Engine Vibrator		35.60		39.15		
1 Concrete Pump (small)		681.40		749.55	3.59	3.94
200 L.H., Daily Totals		$6695.40		$10137.90	$33.48	$50.69

CREWS

Crew C-14E	Hr.	Daily	Hr.	Daily	Bare Costs	Incl. O&P
1 Carpenter Foreman (out)	$32.00	$256.00	$49.80	$398.40	$29.82	$47.85
2 Carpenters	30.00	480.00	46.70	747.20		
4 Rodmen (reinf.)	34.25	1096.00	57.80	1849.60		
3 Laborers	23.45	562.80	36.50	876.00		
1 Cement Finisher	28.70	229.60	42.50	340.00		
1 Gas Engine Vibrator		35.60		39.15	.40	.45
88 L.H., Daily Totals		$2660.00		$4250.35	$30.22	$48.30

Crew C-14F	Hr.	Daily	Hr.	Daily	Bare Costs	Incl. O&P
1 Laborer Foreman (out)	$25.45	$203.60	$39.60	$316.80	$27.17	$40.84
2 Laborers	23.45	375.20	36.50	584.00		
6 Cement Finishers	28.70	1377.60	42.50	2040.00		
1 Gas Engine Vibrator		35.60		39.15	.49	.54
72 L.H., Daily Totals		$1992.00		$2979.95	$27.66	$41.38

Crew C-14G	Hr.	Daily	Hr.	Daily	Bare Costs	Incl. O&P
1 Laborer Foreman (out)	$25.45	$203.60	$39.60	$316.80	$26.74	$40.37
2 Laborers	23.45	375.20	36.50	584.00		
4 Cement Finishers	28.70	918.40	42.50	1360.00		
1 Gas Engine Vibrator		35.60		39.15	.64	.70
56 L.H., Daily Totals		$1532.80		$2299.95	$27.38	$41.07

Crew C-14H	Hr.	Daily	Hr.	Daily	Bare Costs	Incl. O&P
1 Carpenter Foreman (out)	$32.00	$256.00	$49.80	$398.40	$29.73	$46.67
2 Carpenters	30.00	480.00	46.70	747.20		
1 Rodman (reinf.)	34.25	274.00	57.80	462.40		
1 Laborer	23.45	187.60	36.50	292.00		
1 Cement Finisher	28.70	229.60	42.50	340.00		
1 Gas Engine Vibrator		35.60		39.15	.74	.82
48 L.H., Daily Totals		$1462.80		$2279.15	$30.47	$47.49

Crew C-15	Hr.	Daily	Hr.	Daily	Bare Costs	Incl. O&P
1 Carpenter Foreman (out)	$32.00	$256.00	$49.80	$398.40	$28.22	$43.94
2 Carpenters	30.00	480.00	46.70	747.20		
3 Laborers	23.45	562.80	36.50	876.00		
2 Cement Finishers	28.70	459.20	42.50	680.00		
1 Rodman (reinf.)	34.25	274.00	57.80	462.40		
72 L.H., Daily Totals		$2032.00		$3164.00	$28.22	$43.94

Crew C-16	Hr.	Daily	Hr.	Daily	Bare Costs	Incl. O&P
1 Labor Foreman (outside)	$25.45	$203.60	$39.60	$316.80	$28.10	$44.09
3 Laborers	23.45	562.80	36.50	876.00		
2 Cement Finishers	28.70	459.20	42.50	680.00		
1 Equip. Oper. (med.)	31.20	249.60	47.15	377.20		
2 Rodmen (reinf.)	34.25	548.00	57.80	924.80		
1 Concrete Pump (small)		681.40		749.55	9.46	10.41
72 L.H., Daily Totals		$2704.60		$3924.35	$37.56	$54.50

Crew C-17	Hr.	Daily	Hr.	Daily	Bare Costs	Incl. O&P
2 Skilled Worker Foremen	$32.95	$527.20	$51.50	$824.00	$31.35	$48.98
8 Skilled Workers	30.95	1980.80	48.35	3094.40		
80 L.H., Daily Totals		$2508.00		$3918.40	$31.35	$48.98

Crew C-17A	Hr.	Daily	Hr.	Daily	Bare Costs	Incl. O&P
2 Skilled Worker Foremen	$32.95	$527.20	$51.50	$824.00	$31.36	$48.98
8 Skilled Workers	30.95	1980.80	48.35	3094.40		
.125 Equip. Oper. (crane)	32.35	32.35	48.90	48.90		
.125 Crane, 80 Ton, & Tools		169.50		186.45	2.09	2.30
81 L.H., Daily Totals		$2709.85		$4153.75	$33.45	$51.28

Crew C-17B	Hr.	Daily	Hr.	Daily	Bare Costs	Incl. O&P
2 Skilled Worker Foremen	$32.95	$527.20	$51.50	$824.00	$31.37	$48.98
8 Skilled Workers	30.95	1980.80	48.35	3094.40		
.25 Equip. Oper. (crane)	32.35	64.70	48.90	97.80		
.25 Crane, 80 Ton, & Tools		339.00		372.90		
.25 Walk Behind Power Tools		9.85		10.85	4.25	4.68
82 L.H., Daily Totals		$2921.55		$4399.95	$35.62	$53.66

Crew C-17C	Hr.	Daily	Hr.	Daily	Bare Costs	Incl. O&P
2 Skilled Worker Foremen	$32.95	$527.20	$51.50	$824.00	$31.39	$48.98
8 Skilled Workers	30.95	1980.80	48.35	3094.40		
.375 Equip. Oper. (crane)	32.35	97.05	48.90	146.70		
.375 Crane, 80 Ton & Tools		508.50		559.35	6.13	6.74
83 L.H., Daily Totals		$3113.55		$4624.45	$37.52	$55.72

Crew C-17D	Hr.	Daily	Hr.	Daily	Bare Costs	Incl. O&P
2 Skilled Worker Foremen	$32.95	$527.20	$51.50	$824.00	$31.40	$48.98
8 Skilled Workers	30.95	1980.80	48.35	3094.40		
.5 Equip. Oper. (crane)	32.35	129.40	48.90	195.60		
.5 Crane, 80 Ton & Tools		678.00		745.80	8.07	8.88
84 L.H., Daily Totals		$3315.40		$4859.80	$39.47	$57.86

Crew C-17E	Hr.	Daily	Hr.	Daily	Bare Costs	Incl. O&P
2 Skilled Worker Foremen	$32.95	$527.20	$51.50	$824.00	$31.35	$48.98
8 Skilled Workers	30.95	1980.80	48.35	3094.40		
1 Hyd. Jack with Rods		74.70		82.15	.93	1.03
80 L.H., Daily Totals		$2582.70		$4000.55	$32.28	$50.01

Crew C-18	Hr.	Daily	Hr.	Daily	Bare Costs	Incl. O&P
.125 Labor Foreman (out)	$25.45	$25.45	$39.60	$39.60	$23.67	$36.84
1 Laborer	23.45	187.60	36.50	292.00		
1 Concrete Cart, 10 C.F.		46.40		51.05	5.16	5.67
9 L.H., Daily Totals		$259.45		$382.65	$28.83	$42.51

Crew C-19	Hr.	Daily	Hr.	Daily	Bare Costs	Incl. O&P
.125 Labor Foreman (out)	$25.45	$25.45	$39.60	$39.60	$23.67	$36.84
1 Laborer	23.45	187.60	36.50	292.00		
1 Concrete Cart, 18 C.F.		67.40		74.15	7.49	8.24
9 L.H., Daily Totals		$280.45		$405.75	$31.16	$45.08

Crew C-20	Hr.	Daily	Hr.	Daily	Bare Costs	Incl. O&P
1 Labor Foreman (outside)	$25.45	$203.60	$39.60	$316.80	$25.33	$38.97
5 Laborers	23.45	938.00	36.50	1460.00		
1 Cement Finisher	28.70	229.60	42.50	340.00		
1 Equip. Oper. (med.)	31.20	249.60	47.15	377.20		
2 Gas Engine Vibrators		71.20		78.30		
1 Concrete Pump (small)		681.40		749.55	11.76	12.94
64 L.H., Daily Totals		$2373.40		$3321.85	$37.09	$51.91

Crew C-21	Hr.	Daily	Hr.	Daily	Bare Costs	Incl. O&P
1 Labor Foreman (outside)	$25.45	$203.60	$39.60	$316.80	$25.33	$38.97
5 Laborers	23.45	938.00	36.50	1460.00		
1 Cement Finisher	28.70	229.60	42.50	340.00		
1 Equip. Oper. (med.)	31.20	249.60	47.15	377.20		
2 Gas Engine Vibrators		71.20		78.30		
1 Concrete Conveyer		169.40		186.35	3.76	4.14
64 L.H., Daily Totals		$1861.40		$2758.65	$29.09	$43.11

Crews

Crew No.	Bare Costs Hr.	Daily	Incl. Subs O & P Hr.	Daily	Cost Per Labor-Hour Bare Costs	Incl. O&P
Crew C-22	Hr.	Daily	Hr.	Daily	Bare Costs	Incl. O&P
1 Rodman Foreman	$36.25	$290.00	$61.20	$489.60	$34.40	$57.82
4 Rodmen (reinf.)	34.25	1096.00	57.80	1849.60		
.125 Equip. Oper. (crane)	32.35	32.35	48.90	48.90		
.125 Equip. Oper. Oiler	26.65	26.65	40.25	40.25		
.125 Hyd. Crane, 25 Ton		90.60		99.65	2.16	2.37
42 L.H., Daily Totals		$1535.60		$2528.00	$36.56	$60.19

Crew C-23	Hr.	Daily	Hr.	Daily	Bare Costs	Incl. O&P
2 Skilled Worker Foremen	$32.95	$527.20	$51.50	$824.00	$31.06	$48.22
6 Skilled Workers	30.95	1485.60	48.35	2320.80		
1 Equip. Oper. (crane)	32.35	258.80	48.90	391.20		
1 Equip. Oper. Oiler	26.65	213.20	40.25	322.00		
1 Crane, 90 Ton		1421.00		1563.10	17.76	19.54
80 L.H., Daily Totals		$3905.80		$5421.10	$48.82	$67.76

Crew C-24	Hr.	Daily	Hr.	Daily	Bare Costs	Incl. O&P
2 Skilled Worker Foremen	$32.95	$527.20	$51.50	$824.00	$31.06	$48.22
6 Skilled Workers	30.95	1485.60	48.35	2320.80		
1 Equip. Oper. (crane)	32.35	258.80	48.90	391.20		
1 Equip. Oper. Oiler	26.65	213.20	40.25	322.00		
1 Truck Crane, 150 Ton		1690.00		1859.00	21.13	23.24
80 L.H., Daily Totals		$4174.80		$5717.00	$52.19	$71.46

Crew C-25	Hr.	Daily	Hr.	Daily	Bare Costs	Incl. O&P
2 Rodmen (reinf.)	$34.25	$548.00	$57.80	$924.80	$27.02	$45.75
2 Rodmen Helpers	19.80	316.80	33.70	539.20		
32 L.H., Daily Totals		$864.80		$1464.00	$27.02	$45.75

Crew D-1	Hr.	Daily	Hr.	Daily	Bare Costs	Incl. O&P
1 Bricklayer	$30.50	$244.00	$46.80	$374.40	$27.00	$41.42
1 Bricklayer Helper	23.50	188.00	36.05	288.40		
16 L.H., Daily Totals		$432.00		$662.80	$27.00	$41.42

Crew D-2	Hr.	Daily	Hr.	Daily	Bare Costs	Incl. O&P
3 Bricklayers	$30.50	$732.00	$46.80	$1123.20	$27.91	$42.88
2 Bricklayer Helpers	23.50	376.00	36.05	576.80		
.5 Carpenter	30.00	120.00	46.70	186.80		
44 L.H., Daily Totals		$1228.00		$1886.80	$27.91	$42.88

Crew D-3	Hr.	Daily	Hr.	Daily	Bare Costs	Incl. O&P
3 Bricklayers	$30.50	$732.00	$46.80	$1123.20	$27.81	$42.70
2 Bricklayer Helpers	23.50	376.00	36.05	576.80		
.25 Carpenter	30.00	60.00	46.70	93.40		
42 L.H., Daily Totals		$1168.00		$1793.40	$27.81	$42.70

Crew D-4	Hr.	Daily	Hr.	Daily	Bare Costs	Incl. O&P
1 Bricklayer	$30.50	$244.00	$46.80	$374.40	$26.83	$40.99
2 Bricklayer Helpers	23.50	376.00	36.05	576.80		
1 Equip. Oper. (light)	29.80	238.40	45.05	360.40		
1 Grout Pump, 50 C.F./hr		104.95		115.45		
1 Hoses & Hopper		26.80		29.50		
1 Accessories		11.20		12.30	4.47	4.91
32 L.H., Daily Totals		$1001.35		$1468.85	$31.30	$45.90

Crew D-5	Hr.	Daily	Hr.	Daily	Bare Costs	Incl. O&P
1 Bricklayer	$30.50	$244.00	$46.80	$374.40	$30.50	$46.80
8 L.H., Daily Totals		$244.00		$374.40	$30.50	$46.80

Crew D-6	Hr.	Daily	Hr.	Daily	Bare Costs	Incl. O&P
3 Bricklayers	$30.50	$732.00	$46.80	$1123.20	$27.12	$41.64
3 Bricklayer Helpers	23.50	564.00	36.05	865.20		
.25 Carpenter	30.00	60.00	46.70	93.40		
50 L.H., Daily Totals		$1356.00		$2081.80	$27.12	$41.64

Crew D-7	Hr.	Daily	Hr.	Daily	Bare Costs	Incl. O&P
1 Tile Layer	$29.15	$233.20	$42.95	$343.60	$26.25	$38.68
1 Tile Layer Helper	23.35	186.80	34.40	275.20		
16 L.H., Daily Totals		$420.00		$618.80	$26.25	$38.68

Crew D-8	Hr.	Daily	Hr.	Daily	Bare Costs	Incl. O&P
3 Bricklayers	$30.50	$732.00	$46.80	$1123.20	$27.70	$42.50
2 Bricklayer Helpers	23.50	376.00	36.05	576.80		
40 L.H., Daily Totals		$1108.00		$1700.00	$27.70	$42.50

Crew D-9	Hr.	Daily	Hr.	Daily	Bare Costs	Incl. O&P
3 Bricklayers	$30.50	$732.00	$46.80	$1123.20	$27.00	$41.42
3 Bricklayer Helpers	23.50	564.00	36.05	865.20		
48 L.H., Daily Totals		$1296.00		$1988.40	$27.00	$41.42

Crew D-10	Hr.	Daily	Hr.	Daily	Bare Costs	Incl. O&P
1 Bricklayer Foreman	$32.50	$260.00	$49.90	$399.20	$28.47	$43.54
1 Bricklayer	30.50	244.00	46.80	374.40		
2 Bricklayer Helpers	23.50	376.00	36.05	576.80		
1 Equip. Oper. (crane)	32.35	258.80	48.90	391.20		
1 Truck Crane, 12.5 Ton		473.20		520.50	11.83	13.01
40 L.H., Daily Totals		$1612.00		$2262.10	$40.30	$56.55

Crew D-11	Hr.	Daily	Hr.	Daily	Bare Costs	Incl. O&P
1 Bricklayer Foreman	$32.50	$260.00	$49.90	$399.20	$28.83	$44.25
1 Bricklayer	30.50	244.00	46.80	374.40		
1 Bricklayer Helper	23.50	188.00	36.05	288.40		
24 L.H., Daily Totals		$692.00		$1062.00	$28.83	$44.25

Crew D-12	Hr.	Daily	Hr.	Daily	Bare Costs	Incl. O&P
1 Bricklayer Foreman	$32.50	$260.00	$49.90	$399.20	$27.50	$42.20
1 Bricklayer	30.50	244.00	46.80	374.40		
2 Bricklayer Helpers	23.50	376.00	36.05	576.80		
32 L.H., Daily Totals		$880.00		$1350.40	$27.50	$42.20

Crew D-13	Hr.	Daily	Hr.	Daily	Bare Costs	Incl. O&P
1 Bricklayer Foreman	$32.50	$260.00	$49.90	$399.20	$28.73	$44.07
1 Bricklayer	30.50	244.00	46.80	374.40		
2 Bricklayer Helpers	23.50	376.00	36.05	576.80		
1 Carpenter	30.00	240.00	46.70	373.60		
1 Equip. Oper. (crane)	32.35	258.80	48.90	391.20		
1 Truck Crane, 12.5 Ton		473.20		520.50	9.86	10.84
48 L.H., Daily Totals		$1852.00		$2635.70	$38.59	$54.91

Crew E-1	Hr.	Daily	Hr.	Daily	Bare Costs	Incl. O&P
1 Welder Foreman	$36.25	$290.00	$65.35	$522.80	$33.43	$57.38
1 Welder	34.25	274.00	61.75	494.00		
1 Equip. Oper. (light)	29.80	238.40	45.05	360.40		
1 Gas Welding Machine		88.60		97.45	3.69	4.06
24 L.H., Daily Totals		$891.00		$1474.65	$37.12	$61.44

Crew No.	Bare Costs		Incl. Subs O & P		Cost Per Labor-Hour	

Crew E-2	Hr.	Daily	Hr.	Daily	Bare Costs	Incl. O&P
1 Struc. Steel Foreman	$36.25	$290.00	$65.35	$522.80	$33.18	$57.36
4 Struc. Steel Workers	34.25	1096.00	61.75	1976.00		
1 Equip. Oper. (crane)	32.35	258.80	48.90	391.20		
1 Equip. Oper. Oiler	26.65	213.20	40.25	322.00		
1 Crane, 90 Ton		1421.00		1563.10	25.38	27.91
56 L.H., Daily Totals		$3279.00		$4775.10	$58.56	$85.27

Crew E-3	Hr.	Daily	Hr.	Daily	Bare Costs	Incl. O&P
1 Struc. Steel Foreman	$36.25	$290.00	$65.35	$522.80	$34.92	$62.95
1 Struc. Steel Worker	34.25	274.00	61.75	494.00		
1 Welder	34.25	274.00	61.75	494.00		
1 Gas Welding Machine		88.60		97.45	3.69	4.06
24 L.H., Daily Totals		$926.60		$1608.25	$38.61	$67.01

Crew E-4	Hr.	Daily	Hr.	Daily	Bare Costs	Incl. O&P
1 Struc. Steel Foreman	$36.25	$290.00	$65.35	$522.80	$34.75	$62.65
3 Struc. Steel Workers	34.25	822.00	61.75	1482.00		
1 Gas Welding Machine		88.60		97.45	2.77	3.05
32 L.H., Daily Totals		$1200.60		$2102.25	$37.52	$65.70

Crew E-5	Hr.	Daily	Hr.	Daily	Bare Costs	Incl. O&P
2 Struc. Steel Foremen	$36.25	$580.00	$65.35	$1045.60	$33.70	$59.03
5 Struc. Steel Workers	34.25	1370.00	61.75	2470.00		
1 Equip. Oper. (crane)	32.35	258.80	48.90	391.20		
1 Welder	34.25	274.00	61.75	494.00		
1 Equip. Oper. Oiler	26.65	213.20	40.25	322.00		
1 Crane, 90 Ton		1421.00		1563.10		
1 Gas Welding Machine		88.60		97.45	18.87	20.76
80 L.H., Daily Totals		$4205.60		$6383.35	$52.57	$79.79

Crew E-6	Hr.	Daily	Hr.	Daily	Bare Costs	Incl. O&P
3 Struc. Steel Foremen	$36.25	$870.00	$65.35	$1568.40	$33.75	$59.23
9 Struc. Steel Workers	34.25	2466.00	61.75	4446.00		
1 Equip. Oper. (crane)	32.35	258.80	48.90	391.20		
1 Welder	34.25	274.00	61.75	494.00		
1 Equip. Oper. Oiler	26.65	213.20	40.25	322.00		
1 Equip. Oper. (light)	29.80	238.40	45.05	360.40		
1 Crane, 90 Ton		1421.00		1563.10		
1 Gas Welding Machine		88.60		97.45		
1 Air Compr., 160 C.F.M.		97.20		106.90		
2 Impact Wrenches		42.40		46.65	12.88	14.17
128 L.H., Daily Totals		$5969.60		$9396.10	$46.63	$73.40

Crew E-7	Hr.	Daily	Hr.	Daily	Bare Costs	Incl. O&P
1 Struc. Steel Foreman	$36.25	$290.00	$65.35	$522.80	$33.70	$59.03
4 Struc. Steel Workers	34.25	1096.00	61.75	1976.00		
1 Equip. Oper. (crane)	32.35	258.80	48.90	391.20		
1 Equip. Oper. Oiler	26.65	213.20	40.25	322.00		
1 Welder Foreman	36.25	290.00	65.35	522.80		
2 Welders	34.25	548.00	61.75	988.00		
1 Crane, 90 Ton		1421.00		1563.10		
2 Gas Welding Machines		177.20		194.90	19.98	21.98
80 L.H., Daily Totals		$4294.20		$6480.80	$53.68	$81.01

Crew E-8	Hr.	Daily	Hr.	Daily	Bare Costs	Incl. O&P
1 Struc. Steel Foreman	$36.25	$290.00	$65.35	$522.80	$33.48	$58.38
4 Struc. Steel Workers	34.25	1096.00	61.75	1976.00		
1 Welder Foreman	36.25	290.00	65.35	522.80		
4 Welders	34.25	1096.00	61.75	1976.00		
1 Equip. Oper. (crane)	32.35	258.80	48.90	391.20		
1 Equip. Oper. Oiler	26.65	213.20	40.25	322.00		
1 Equip. Oper. (light)	29.80	238.40	45.05	360.40		
1 Crane, 90 Ton		1421.00		1563.10		
4 Gas Welding Machines		354.40		389.85	17.07	18.78
104 L.H., Daily Totals		$5257.80		$8024.15	$50.55	$77.16

Crew E-9	Hr.	Daily	Hr.	Daily	Bare Costs	Incl. O&P
2 Struc. Steel Foremen	$36.25	$580.00	$65.35	$1045.60	$33.75	$59.23
5 Struc. Steel Workers	34.25	1370.00	61.75	2470.00		
1 Welder Foreman	36.25	290.00	65.35	522.80		
5 Welders	34.25	1370.00	61.75	2470.00		
1 Equip. Oper. (crane)	32.35	258.80	48.90	391.20		
1 Equip. Oper. Oiler	26.65	213.20	40.25	322.00		
1 Equip. Oper. (light)	29.80	238.40	45.05	360.40		
1 Crane, 90 Ton		1421.00		1563.10		
5 Gas Welding Machines		443.00		487.30	14.56	16.02
128 L.H., Daily Totals		$6184.40		$9632.40	$48.31	$75.25

Crew E-10	Hr.	Daily	Hr.	Daily	Bare Costs	Incl. O&P
1 Welder Foreman	$36.25	$290.00	$65.35	$522.80	$35.25	$63.55
1 Welder	34.25	274.00	61.75	494.00		
4 Gas Welding Machines		354.40		389.85		
1 Truck, 3 Ton		182.40		200.65	33.55	36.91
16 L.H., Daily Totals		$1100.80		$1607.30	$68.80	$100.46

Crew E-11	Hr.	Daily	Hr.	Daily	Bare Costs	Incl. O&P
2 Painters, Struc. Steel	$27.90	$446.40	$51.85	$829.60	$27.26	$46.31
1 Building Laborer	23.45	187.60	36.50	292.00		
1 Equip. Oper. (light)	29.80	238.40	45.05	360.40		
1 Air Compressor 250 C.F.M.		135.00		148.50		
1 Sand Blaster		26.80		29.50		
1 Sand Blasting Accessories		11.20		12.30	5.41	5.95
32 L.H., Daily Totals		$1045.40		$1672.30	$32.67	$52.26

Crew E-12	Hr.	Daily	Hr.	Daily	Bare Costs	Incl. O&P
1 Welder Foreman	$36.25	$290.00	$65.35	$522.80	$33.03	$55.20
1 Equip. Oper. (light)	29.80	238.40	45.05	360.40		
1 Gas Welding Machine		88.60		97.45	5.54	6.09
16 L.H., Daily Totals		$617.00		$980.65	$38.57	$61.29

Crew E-13	Hr.	Daily	Hr.	Daily	Bare Costs	Incl. O&P
1 Welder Foreman	$36.25	$290.00	$65.35	$522.80	$34.10	$58.58
.5 Equip. Oper. (light)	29.80	119.20	45.05	180.20		
1 Gas Welding Machine		88.60		97.45	7.38	8.12
12 L.H., Daily Totals		$497.80		$800.45	$41.48	$66.70

Crew E-14	Hr.	Daily	Hr.	Daily	Bare Costs	Incl. O&P
1 Welder Foreman	$36.25	$290.00	$65.35	$522.80	$36.25	$65.35
1 Gas Welding Machine		88.60		97.45	11.08	12.18
8 L.H., Daily Totals		$378.60		$620.25	$47.33	$77.53

Crew E-16	Hr.	Daily	Hr.	Daily	Bare Costs	Incl. O&P
1 Welder Foreman	$36.25	$290.00	$65.35	$522.80	$35.25	$63.55
1 Welder	34.25	274.00	61.75	494.00		
1 Gas Welding Machine		88.60		97.45	5.54	6.09
16 L.H., Daily Totals		$652.60		$1114.25	$40.79	$69.64

Crews

Crew E-17

Crew E-17	Hr.	Daily	Hr.	Daily	Bare Costs	Incl. O&P
1 Structural Steel Foreman	$36.25	$290.00	$65.35	$522.80	$35.25	$63.55
1 Structural Steel Worker	34.25	274.00	61.75	494.00		
1 Power Tool		3.40		3.75	.21	.23
16 L.H., Daily Totals		$567.40		$1020.55	$35.46	$63.78

Crew E-18

Crew E-18	Hr.	Daily	Hr.	Daily	Bare Costs	Incl. O&P
1 Structural Steel Foreman	$36.25	$290.00	$65.35	$522.80	$34.04	$59.55
3 Structural Steel Workers	34.25	822.00	61.75	1482.00		
1 Equipment Operator (med.)	31.20	249.60	47.15	377.20		
1 Crane, 20 Ton		588.55		647.40	14.71	16.19
40 L.H., Daily Totals		$1950.15		$3029.40	$48.75	$75.74

Crew E-19

Crew E-19	Hr.	Daily	Hr.	Daily	Bare Costs	Incl. O&P
1 Structural Steel Worker	$34.25	$274.00	$61.75	$494.00	$33.43	$57.38
1 Structural Steel Foreman	36.25	290.00	65.35	522.80		
1 Equip. Oper. (light)	29.80	238.40	45.05	360.40		
1 Power Tool		3.40		3.75		
1 Crane, 20 Ton		588.55		647.40	24.66	27.13
24 L.H., Daily Totals		$1394.35		$2028.35	$58.09	$84.51

Crew E-20

Crew E-20	Hr.	Daily	Hr.	Daily	Bare Costs	Incl. O&P
1 Structural Steel Foreman	$36.25	$290.00	$65.35	$522.80	$33.31	$57.91
5 Structural Steel Workers	34.25	1370.00	61.75	2470.00		
1 Equip. Oper. (crane)	32.35	258.80	48.90	391.20		
1 Oiler	26.65	213.20	40.25	322.00		
1 Power Tool		3.40		3.75		
1 Crane, 40 Ton		847.60		932.35	13.30	14.63
64 L.H., Daily Totals		$2983.00		$4642.10	$46.61	$72.54

Crew E-22

Crew E-22	Hr.	Daily	Hr.	Daily	Bare Costs	Incl. O&P
1 Skilled Worker Foreman	$32.95	$263.60	$51.50	$412.00	$31.62	$49.40
2 Skilled Worker	30.95	495.20	48.35	773.60		
24 L.H., Daily Totals		$758.80		$1185.60	$31.62	$49.40

Crew E-24

Crew E-24	Hr.	Daily	Hr.	Daily	Bare Costs	Incl. O&P
3 Structural Steel Worker	$34.25	$822.00	$61.75	$1482.00	$33.49	$58.10
1 Equipment Operator (medium)	31.20	249.60	47.15	377.20		
1-25 Ton Crane		724.80		797.30	22.65	24.92
32 L.H., Daily Totals		$1796.40		$2656.50	$56.14	$83.02

Crew E-25

Crew E-25	Hr.	Daily	Hr.	Daily	Bare Costs	Incl. O&P
1 Welder Foreman	$36.25	$290.00	$65.35	$522.80	$36.25	$65.35
1 Cutting Torch		28.00		30.80		
1 Gases		34.00		37.40	7.75	8.53
8 L.H., Daily Totals		$352.00		$591.00	$44.00	$73.88

Crew F-3

Crew F-3	Hr.	Daily	Hr.	Daily	Bare Costs	Incl. O&P
4 Carpenters	$30.00	$960.00	$46.70	$1494.40	$30.47	$47.14
1 Equip. Oper. (crane)	32.35	258.80	48.90	391.20		
1 Hyd. Crane, 12 Ton		607.60		668.35	15.19	16.71
40 L.H., Daily Totals		$1826.40		$2553.95	$45.66	$63.85

Crew F-4

Crew F-4	Hr.	Daily	Hr.	Daily	Bare Costs	Incl. O&P
4 Carpenters	$30.00	$960.00	$46.70	$1494.40	$29.83	$45.99
1 Equip. Oper. (crane)	32.35	258.80	48.90	391.20		
1 Equip. Oper. Oiler	26.65	213.20	40.25	322.00		
1 Hyd. Crane, 55 Ton		972.40		1069.65	20.26	22.28
48 L.H., Daily Totals		$2404.40		$3277.25	$50.09	$68.27

Crew F-5

Crew F-5	Hr.	Daily	Hr.	Daily	Bare Costs	Incl. O&P
1 Carpenter Foreman	$32.00	$256.00	$49.80	$398.40	$30.50	$47.48
3 Carpenters	30.00	720.00	46.70	1120.80		
32 L.H., Daily Totals		$976.00		$1519.20	$30.50	$47.48

Crew F-6

Crew F-6	Hr.	Daily	Hr.	Daily	Bare Costs	Incl. O&P
2 Carpenters	$30.00	$480.00	$46.70	$747.20	$27.85	$43.06
2 Building Laborers	23.45	375.20	36.50	584.00		
1 Equip. Oper. (crane)	32.35	258.80	48.90	391.20		
1 Hyd. Crane, 12 Ton		607.60		668.35	15.19	16.71
40 L.H., Daily Totals		$1721.60		$2390.75	$43.04	$59.77

Crew F-7

Crew F-7	Hr.	Daily	Hr.	Daily	Bare Costs	Incl. O&P
2 Carpenters	$30.00	$480.00	$46.70	$747.20	$26.73	$41.60
2 Building Laborers	23.45	375.20	36.50	584.00		
32 L.H., Daily Totals		$855.20		$1331.20	$26.73	$41.60

Crew G-1

Crew G-1	Hr.	Daily	Hr.	Daily	Bare Costs	Incl. O&P
1 Roofer Foreman	$28.60	$228.80	$48.65	$389.20	$24.94	$42.44
4 Roofers, Composition	26.60	851.20	45.25	1448.00		
2 Roofer Helpers	19.80	316.80	33.70	539.20		
1 Application Equipment		159.60		175.55		
1 Tar Kettle/Pot		61.75		67.95		
1 Crew Truck		81.40		89.55	5.41	5.95
56 L.H., Daily Totals		$1699.55		$2709.45	$30.35	$48.39

Crew G-2

Crew G-2	Hr.	Daily	Hr.	Daily	Bare Costs	Incl. O&P
1 Plasterer	$28.10	$224.80	$43.10	$344.80	$25.08	$38.65
1 Plasterer Helper	23.70	189.60	36.35	290.80		
1 Building Laborer	23.45	187.60	36.50	292.00		
1 Grouting Equipment		104.95		115.45	4.37	4.81
24 L.H., Daily Totals		$706.95		$1043.05	$29.45	$43.46

Crew G-3

Crew G-3	Hr.	Daily	Hr.	Daily	Bare Costs	Incl. O&P
2 Sheet Metal Workers	$35.10	$561.60	$54.10	$865.60	$29.27	$45.30
2 Building Laborers	23.45	375.20	36.50	584.00		
32 L.H., Daily Totals		$936.80		$1449.60	$29.27	$45.30

Crew G-4

Crew G-4	Hr.	Daily	Hr.	Daily	Bare Costs	Incl. O&P
1 Labor Foreman (outside)	$25.45	$203.60	$39.60	$316.80	$24.12	$37.53
2 Building Laborers	23.45	375.20	36.50	584.00		
1 Light Truck, 1.5 Ton		155.60		171.15		
1 Air Compr., 160 C.F.M.		97.20		106.90	10.53	11.59
24 L.H., Daily Totals		$831.60		$1178.85	$34.65	$49.12

Crew G-5

Crew G-5	Hr.	Daily	Hr.	Daily	Bare Costs	Incl. O&P
1 Roofer Foreman	$28.60	$228.80	$48.65	$389.20	$24.28	$41.31
2 Roofers, Composition	26.60	425.60	45.25	724.00		
2 Roofer Helpers	19.80	316.80	33.70	539.20		
1 Application Equipment		159.60		175.55	3.99	4.39
40 L.H., Daily Totals		$1130.80		$1827.95	$28.27	$45.70

Crew G-6A

Crew G-6A	Hr.	Daily	Hr.	Daily	Bare Costs	Incl. O&P
2 Roofers Composition	$26.60	$425.60	$45.25	$724.00	$26.60	$45.25
1 Small Compressor		19.15		21.05		
2 Pneumatic Nailers		40.90		45.00	3.75	4.13
16 L.H., Daily Totals		$485.65		$790.05	$30.35	$49.38

CREWS

Crew No.	Bare Costs				Incl. Subs O & P		Cost Per Labor-Hour	
	Hr.	Daily	Hr.	Daily			Bare Costs	Incl. O&P
Crew G-7	Hr.	Daily	Hr.	Daily			Bare Costs	Incl. O&P
1 Carpenter	$30.00	$240.00	$46.70	$373.60			$30.00	$46.70
1 Small Compressor		19.15		21.05				
1 Pneumatic Nailer		20.45		22.50			4.95	5.45
8 L.H., Daily Totals		$279.60		$417.15			$34.95	$52.15
Crew H-1	Hr.	Daily	Hr.	Daily			Bare Costs	Incl. O&P
2 Glaziers	$30.00	$480.00	$45.40	$726.40			$32.13	$53.58
2 Struc. Steel Workers	34.25	548.00	61.75	988.00				
32 L.H., Daily Totals		$1028.00		$1714.40			$32.13	$53.58
Crew H-2	Hr.	Daily	Hr.	Daily			Bare Costs	Incl. O&P
2 Glaziers	$30.00	$480.00	$45.40	$726.40			$27.82	$42.43
1 Building Laborer	23.45	187.60	36.50	292.00				
24 L.H., Daily Totals		$667.60		$1018.40			$27.82	$42.43
Crew H-3	Hr.	Daily	Hr.	Daily			Bare Costs	Incl. O&P
1 Glazier	$30.00	$240.00	$45.40	$363.20			$26.38	$40.45
1 Helper	22.75	182.00	35.50	284.00				
16 L.H., Daily Totals		$422.00		$647.20			$26.38	$40.45
Crew J-1	Hr.	Daily	Hr.	Daily			Bare Costs	Incl. O&P
3 Plasterers	$28.10	$674.40	$43.10	$1034.40			$26.34	$40.40
2 Plasterer Helpers	23.70	379.20	36.35	581.60				
1 Mixing Machine, 6 C.F.		71.40		78.55			1.79	1.96
40 L.H., Daily Totals		$1125.00		$1694.55			$28.13	$42.36
Crew J-2	Hr.	Daily	Hr.	Daily			Bare Costs	Incl. O&P
3 Plasterers	$28.10	$674.40	$43.10	$1034.40			$26.74	$40.78
2 Plasterer Helpers	23.70	379.20	36.35	581.60				
1 Lather	28.75	230.00	42.70	341.60				
1 Mixing Machine, 6 C.F.		71.40		78.55			1.49	1.64
48 L.H., Daily Totals		$1355.00		$2036.15			$28.23	$42.42
Crew J-3	Hr.	Daily	Hr.	Daily			Bare Costs	Incl. O&P
1 Terrazzo Worker	$29.25	$234.00	$43.10	$344.80			$26.60	$39.20
1 Terrazzo Helper	23.95	191.60	35.30	282.40				
1 Terrazzo Grinder, Electric		61.85		68.05				
1 Terrazzo Mixer		94.80		104.30			9.79	10.77
16 L.H., Daily Totals		$582.25		$799.55			$36.39	$49.97
Crew J-4	Hr.	Daily	Hr.	Daily			Bare Costs	Incl. O&P
1 Tile Layer	$29.15	$233.20	$42.95	$343.60			$26.25	$38.68
1 Tile Layer Helper	23.35	186.80	34.40	275.20				
16 L.H., Daily Totals		$420.00		$618.80			$26.25	$38.68
Crew K-1	Hr.	Daily	Hr.	Daily			Bare Costs	Incl. O&P
1 Carpenter	$30.00	$240.00	$46.70	$373.60			$27.15	$41.88
1 Truck Driver (light)	24.30	194.40	37.05	296.40				
1 Truck w/Power Equip.		182.40		200.65			11.40	12.54
16 L.H., Daily Totals		$616.80		$870.65			$38.55	$54.42
Crew K-2	Hr.	Daily	Hr.	Daily			Bare Costs	Incl. O&P
1 Struc. Steel Foreman	$36.25	$290.00	$65.35	$522.80			$31.60	$54.72
1 Struc. Steel Worker	34.25	274.00	61.75	494.00				
1 Truck Driver (light)	24.30	194.40	37.05	296.40				
1 Truck w/Power Equip.		182.40		200.65			7.60	8.36
24 L.H., Daily Totals		$940.80		$1513.85			$39.20	$63.08

Crew No.	Bare Costs				Incl. Subs O & P		Cost Per Labor-Hour	
Crew L-1	Hr.	Daily	Hr.	Daily			Bare Costs	Incl. O&P
1 Electrician	$35.45	$283.60	$52.90	$423.20			$35.70	$53.55
1 Plumber	35.95	287.60	54.20	433.60				
16 L.H., Daily Totals		$571.20		$856.80			$35.70	$53.55
Crew L-2	Hr.	Daily	Hr.	Daily			Bare Costs	Incl. O&P
1 Carpenter	$30.00	$240.00	$46.70	$373.60			$26.38	$41.10
1 Carpenter Helper	22.75	182.00	35.50	284.00				
16 L.H., Daily Totals		$422.00		$657.60			$26.38	$41.10
Crew L-3	Hr.	Daily	Hr.	Daily			Bare Costs	Incl. O&P
1 Carpenter	$30.00	$240.00	$46.70	$373.60			$32.64	$50.10
.5 Electrician	35.45	141.80	52.90	211.60				
.5 Sheet Metal Worker	35.10	140.40	54.10	216.40				
16 L.H., Daily Totals		$522.20		$801.60			$32.64	$50.10
Crew L-3A	Hr.	Daily	Hr.	Daily			Bare Costs	Incl. O&P
1 Carpenter Foreman (outside)	$32.00	$256.00	$49.80	$398.40			$33.03	$51.23
.5 Sheet Metal Worker	35.10	140.40	54.10	216.40				
12 L.H., Daily Totals		$396.40		$614.80			$33.03	$51.23
Crew L-4	Hr.	Daily	Hr.	Daily			Bare Costs	Incl. O&P
2 Skilled Workers	$30.95	$495.20	$48.35	$773.60			$28.22	$44.07
1 Helper	22.75	182.00	35.50	284.00				
24 L.H., Daily Totals		$677.20		$1057.60			$28.22	$44.07
Crew L-5	Hr.	Daily	Hr.	Daily			Bare Costs	Incl. O&P
1 Struc. Steel Foreman	$36.25	$290.00	$65.35	$522.80			$34.26	$60.43
5 Struc. Steel Workers	34.25	1370.00	61.75	2470.00				
1 Equip. Oper. (crane)	32.35	258.80	48.90	391.20				
1 Hyd. Crane, 25 Ton		724.80		797.30			12.94	14.24
56 L.H., Daily Totals		$2643.60		$4181.30			$47.20	$74.67
Crew L-5A	Hr.	Daily	Hr.	Daily			Bare Costs	Incl. O&P
1 Structural Steel Foreman	$36.25	$290.00	$65.35	$522.80			$34.28	$59.44
2 Structural Steel Workers	34.25	548.00	61.75	988.00				
1 Equip. Oper. (crane)	32.35	258.80	48.90	391.20				
1 Crane, SP, 25 Ton		629.40		692.35			19.67	21.64
32 L.H., Daily Totals		$1726.20		$2594.35			$53.95	$81.08
Crew L-6	Hr.	Daily	Hr.	Daily			Bare Costs	Incl. O&P
1 Plumber	$35.95	$287.60	$54.20	$433.60			$35.78	$53.77
.5 Electrician	35.45	141.80	52.90	211.60				
12 L.H., Daily Totals		$429.40		$645.20			$35.78	$53.77
Crew L-7	Hr.	Daily	Hr.	Daily			Bare Costs	Incl. O&P
2 Carpenters	$30.00	$480.00	$46.70	$747.20			$28.91	$44.67
1 Building Laborer	23.45	187.60	36.50	292.00				
.5 Electrician	35.45	141.80	52.90	211.60				
28 L.H., Daily Totals		$809.40		$1250.80			$28.91	$44.67
Crew L-8	Hr.	Daily	Hr.	Daily			Bare Costs	Incl. O&P
2 Carpenters	$30.00	$480.00	$46.70	$747.20			$31.19	$48.20
.5 Plumber	35.95	143.80	54.20	216.80				
20 L.H., Daily Totals		$623.80		$964.00			$31.19	$48.20

Crew L-9

Crew No.	Bare Costs Hr.	Daily	Incl. Subs O & P Hr.	Daily	Cost Per Labor-Hour Bare Costs	Incl. O&P
1 Labor Foreman (inside)	$23.95	$191.60	$37.25	$298.00	$27.29	$44.10
2 Building Laborers	23.45	375.20	36.50	584.00		
1 Struc. Steel Worker	34.25	274.00	61.75	494.00		
.5 Electrician	35.45	141.80	52.90	211.60		
36 L.H., Daily Totals		$982.60		$1587.60	$27.29	$44.10

Crew L-10

Crew No.	Bare Costs Hr.	Daily	Incl. Subs O & P Hr.	Daily	Cost Per Labor-Hour Bare Costs	Incl. O&P
1 Structural Steel Foreman	$36.25	$290.00	$65.35	$522.80	$34.28	$58.67
1 Structural Steel Worker	34.25	274.00	61.75	494.00		
1 Equip. Oper. (crane)	32.35	258.80	48.90	391.20		
1 Hyd. Crane, 12 Ton		607.60		668.35	25.32	27.85
24 L.H., Daily Totals		$1430.40		$2076.35	$59.60	$86.52

Crew M-1

Crew No.	Bare Costs Hr.	Daily	Incl. Subs O & P Hr.	Daily	Cost Per Labor-Hour Bare Costs	Incl. O&P
3 Elevator Constructors	$37.10	$890.40	$55.70	$1336.80	$35.25	$52.93
1 Elevator Apprentice	29.70	237.60	44.60	356.80		
5 Hand Tools		83.00		91.30	2.59	2.85
32 L.H., Daily Totals		$1211.00		$1784.90	$37.84	$55.78

Crew M-3

Crew No.	Bare Costs Hr.	Daily	Incl. Subs O & P Hr.	Daily	Cost Per Labor-Hour Bare Costs	Incl. O&P
1 Electrician Foreman (out)	$37.45	$299.60	$55.90	$447.20	$31.88	$48.11
1 Common Laborer	23.45	187.60	36.50	292.00		
.25 Equipment Operator, Medium	31.20	62.40	47.15	94.30		
1 Elevator Constructor	37.10	296.80	55.70	445.60		
1 Elevator Apprentice	29.70	237.60	44.60	356.80		
.25 Crane, SP, 4 x 4, 20 ton		142.10		156.30	4.18	4.60
34 L.H., Daily Totals		$1226.10		$1792.20	$36.06	$52.71

Crew M-4

Crew No.	Bare Costs Hr.	Daily	Incl. Subs O & P Hr.	Daily	Cost Per Labor-Hour Bare Costs	Incl. O&P
1 Electrician Foreman (out)	$37.45	$299.60	$55.90	$447.20	$31.66	$47.78
1 Common Laborer	23.45	187.60	36.50	292.00		
.25 Equipment Operator, Crane	32.35	64.70	48.90	97.80		
.25 Equipment Operator, Oiler	26.65	53.30	40.25	80.50		
1 Elevator Constructor	37.10	296.80	55.70	445.60		
1 Elevator Apprentice	29.70	237.60	44.60	356.80		
.25 Crane, Hyd, SP, 4WD, 40 Ton		245.20		269.70	6.81	7.49
36 L.H., Daily Totals		$1384.80		$1989.60	$38.47	$55.27

Crew Q-1

Crew No.	Bare Costs Hr.	Daily	Incl. Subs O & P Hr.	Daily	Cost Per Labor-Hour Bare Costs	Incl. O&P
1 Plumber	$35.95	$287.60	$54.20	$433.60	$32.35	$48.78
1 Plumber Apprentice	28.75	230.00	43.35	346.80		
16 L.H., Daily Totals		$517.60		$780.40	$32.35	$48.78

Crew Q-1C

Crew No.	Bare Costs Hr.	Daily	Incl. Subs O & P Hr.	Daily	Cost Per Labor-Hour Bare Costs	Incl. O&P
1 Plumber	$35.95	$287.60	$54.20	$433.60	$31.97	$48.23
1 Plumber Apprentice	28.75	230.00	43.35	346.80		
1 Equip. Oper. (medium)	31.20	249.60	47.15	377.20		
1 Trencher, Chain		849.60		934.55	35.40	38.94
24 L.H., Daily Totals		$1616.80		$2092.15	$67.37	$87.17

Crew Q-2

Crew No.	Bare Costs Hr.	Daily	Incl. Subs O & P Hr.	Daily	Cost Per Labor-Hour Bare Costs	Incl. O&P
2 Plumbers	$35.95	$575.20	$54.20	$867.20	$33.55	$50.58
1 Plumber Apprentice	28.75	230.00	43.35	346.80		
24 L.H., Daily Totals		$805.20		$1214.00	$33.55	$50.58

Crew Q-3

Crew No.	Bare Costs Hr.	Daily	Incl. Subs O & P Hr.	Daily	Cost Per Labor-Hour Bare Costs	Incl. O&P
1 Plumber Foreman (inside)	$36.45	$291.60	$54.95	$439.60	$34.28	$51.67
2 Plumbers	35.95	575.20	54.20	867.20		
1 Plumber Apprentice	28.75	230.00	43.35	346.80		
32 L.H., Daily Totals		$1096.80		$1653.60	$34.28	$51.67

Crew Q-4

Crew No.	Bare Costs Hr.	Daily	Incl. Subs O & P Hr.	Daily	Cost Per Labor-Hour Bare Costs	Incl. O&P
1 Plumber Foreman (inside)	$36.45	$291.60	$54.95	$439.60	$34.28	$51.67
1 Plumber	35.95	287.60	54.20	433.60		
1 Welder (plumber)	35.95	287.60	54.20	433.60		
1 Plumber Apprentice	28.75	230.00	43.35	346.80		
1 Electric Welding Mach.		79.95		87.95	2.50	2.75
32 L.H., Daily Totals		$1176.75		$1741.55	$36.78	$54.42

Crew Q-5

Crew No.	Bare Costs Hr.	Daily	Incl. Subs O & P Hr.	Daily	Cost Per Labor-Hour Bare Costs	Incl. O&P
1 Steamfitter	$36.20	$289.60	$54.60	$436.80	$32.58	$49.13
1 Steamfitter Apprentice	28.95	231.60	43.65	349.20		
16 L.H., Daily Totals		$521.20		$786.00	$32.58	$49.13

Crew Q-6

Crew No.	Bare Costs Hr.	Daily	Incl. Subs O & P Hr.	Daily	Cost Per Labor-Hour Bare Costs	Incl. O&P
2 Steamfitters	$36.20	$579.20	$54.60	$873.60	$33.78	$50.95
1 Steamfitter Apprentice	28.95	231.60	43.65	349.20		
24 L.H., Daily Totals		$810.80		$1222.80	$33.78	$50.95

Crew Q-7

Crew No.	Bare Costs Hr.	Daily	Incl. Subs O & P Hr.	Daily	Cost Per Labor-Hour Bare Costs	Incl. O&P
1 Steamfitter Foreman (inside)	$36.70	$293.60	$55.35	$442.80	$34.51	$52.05
2 Steamfitters	36.20	579.20	54.60	873.60		
1 Steamfitter Apprentice	28.95	231.60	43.65	349.20		
32 L.H., Daily Totals		$1104.40		$1665.60	$34.51	$52.05

Crew Q-8

Crew No.	Bare Costs Hr.	Daily	Incl. Subs O & P Hr.	Daily	Cost Per Labor-Hour Bare Costs	Incl. O&P
1 Steamfitter Foreman (inside)	$36.70	$293.60	$55.35	$442.80	$34.51	$52.05
1 Steamfitter	36.20	289.60	54.60	436.80		
1 Welder (steamfitter)	36.20	289.60	54.60	436.80		
1 Steamfitter Apprentice	28.95	231.60	43.65	349.20		
1 Electric Welding Mach.		79.95		87.95	2.50	2.75
32 L.H., Daily Totals		$1184.35		$1753.55	$37.01	$54.80

Crew Q-9

Crew No.	Bare Costs Hr.	Daily	Incl. Subs O & P Hr.	Daily	Cost Per Labor-Hour Bare Costs	Incl. O&P
1 Sheet Metal Worker	$35.10	$280.80	$54.10	$432.80	$31.60	$48.72
1 Sheet Metal Apprentice	28.10	224.80	43.35	346.80		
16 L.H., Daily Totals		$505.60		$779.60	$31.60	$48.72

Crew Q-10

Crew No.	Bare Costs Hr.	Daily	Incl. Subs O & P Hr.	Daily	Cost Per Labor-Hour Bare Costs	Incl. O&P
2 Sheet Metal Workers	$35.10	$561.60	$54.10	$865.60	$32.77	$50.52
1 Sheet Metal Apprentice	28.10	224.80	43.35	346.80		
24 L.H., Daily Totals		$786.40		$1212.40	$32.77	$50.52

Crew Q-11

Crew No.	Bare Costs Hr.	Daily	Incl. Subs O & P Hr.	Daily	Cost Per Labor-Hour Bare Costs	Incl. O&P
1 Sheet Metal Foreman (inside)	$35.60	$284.80	$54.90	$439.20	$33.47	$51.61
2 Sheet Metal Workers	35.10	561.60	54.10	865.60		
1 Sheet Metal Apprentice	28.10	224.80	43.35	346.80		
32 L.H., Daily Totals		$1071.20		$1651.60	$33.47	$51.61

Crew Q-12

Crew No.	Bare Costs Hr.	Daily	Incl. Subs O & P Hr.	Daily	Cost Per Labor-Hour Bare Costs	Incl. O&P
1 Sprinkler Installer	$36.20	$289.60	$54.75	$438.00	$32.58	$49.25
1 Sprinkler Apprentice	28.95	231.60	43.75	350.00		
16 L.H., Daily Totals		$521.20		$788.00	$32.58	$49.25

Crew Q-13

Crew No.	Bare Costs Hr.	Daily	Incl. Subs O & P Hr.	Daily	Cost Per Labor-Hour Bare Costs	Incl. O&P
1 Sprinkler Foreman (inside)	$36.70	$293.60	$55.50	$444.00	$34.51	$52.19
2 Sprinkler Installers	36.20	579.20	54.75	876.00		
1 Sprinkler Apprentice	28.95	231.60	43.75	350.00		
32 L.H., Daily Totals		$1104.40		$1670.00	$34.51	$52.19

Crew No.	Bare Costs		Incl. Subs O & P		Cost Per Labor-Hour	

Crew Q-14	Hr.	Daily	Hr.	Daily	Bare Costs	Incl. O&P
1 Asbestos Worker	$33.45	$267.60	$53.10	$424.80	$30.10	$47.78
1 Asbestos Apprentice	26.75	214.00	42.45	339.60		
16 L.H., Daily Totals		$481.60		$764.40	$30.10	$47.78

Crew Q-15	Hr.	Daily	Hr.	Daily	Bare Costs	Incl. O&P
1 Plumber	$35.95	$287.60	$54.20	$433.60	$32.35	$48.78
1 Plumber Apprentice	28.75	230.00	43.35	346.80		
1 Electric Welding Mach.		79.95		87.95	5.00	5.50
16 L.H., Daily Totals		$597.55		$868.35	$37.35	$54.28

Crew Q-16	Hr.	Daily	Hr.	Daily	Bare Costs	Incl. O&P
2 Plumbers	$35.95	$575.20	$54.20	$867.20	$33.55	$50.58
1 Plumber Apprentice	28.75	230.00	43.35	346.80		
1 Electric Welding Mach.		79.95		87.95	3.33	3.66
24 L.H., Daily Totals		$885.15		$1301.95	$36.88	$54.24

Crew Q-17	Hr.	Daily	Hr.	Daily	Bare Costs	Incl. O&P
1 Steamfitter	$36.20	$289.60	$54.60	$436.80	$32.58	$49.13
1 Steamfitter Apprentice	28.95	231.60	43.65	349.20		
1 Electric Welding Mach.		79.95		87.95	5.00	5.50
16 L.H., Daily Totals		$601.15		$873.95	$37.58	$54.63

Crew Q-17A	Hr.	Daily	Hr.	Daily	Bare Costs	Incl. O&P
1 Steamfitter	$36.20	$289.60	$54.60	$436.80	$32.50	$49.05
1 Steamfitter Apprentice	28.95	231.60	43.65	349.20		
1 Equip. Oper. (crane)	32.35	258.80	48.90	391.20		
1 Truck Crane, 12 Ton		607.60		668.35		
1 Electric Welding Mach.		79.95		87.95	28.65	31.51
24 L.H., Daily Totals		$1467.55		$1933.50	$61.15	$80.56

Crew Q-18	Hr.	Daily	Hr.	Daily	Bare Costs	Incl. O&P
2 Steamfitters	$36.20	$579.20	$54.60	$873.60	$33.78	$50.95
1 Steamfitter Apprentice	28.95	231.60	43.65	349.20		
1 Electric Welding Mach.		79.95		87.95	3.33	3.66
24 L.H., Daily Totals		$890.75		$1310.75	$37.11	$54.61

Crew Q-19	Hr.	Daily	Hr.	Daily	Bare Costs	Incl. O&P
1 Steamfitter	$36.20	$289.60	$54.60	$436.80	$33.53	$50.38
1 Steamfitter Apprentice	28.95	231.60	43.65	349.20		
1 Electrician	35.45	283.60	52.90	423.20		
24 L.H., Daily Totals		$804.80		$1209.20	$33.53	$50.38

Crew Q-20	Hr.	Daily	Hr.	Daily	Bare Costs	Incl. O&P
1 Sheet Metal Worker	$35.10	$280.80	$54.10	$432.80	$32.37	$49.56
1 Sheet Metal Apprentice	28.10	224.80	43.35	346.80		
.5 Electrician	35.45	141.80	52.90	211.60		
20 L.H., Daily Totals		$647.40		$991.20	$32.37	$49.56

Crew Q-21	Hr.	Daily	Hr.	Daily	Bare Costs	Incl. O&P
2 Steamfitters	$36.20	$579.20	$54.60	$873.60	$34.20	$51.44
1 Steamfitter Apprentice	28.95	231.60	43.65	349.20		
1 Electrician	35.45	283.60	52.90	423.20		
32 L.H., Daily Totals		$1094.40		$1646.00	$34.20	$51.44

Crew Q-22	Hr.	Daily	Hr.	Daily	Bare Costs	Incl. O&P
1 Plumber	$35.95	$287.60	$54.20	$433.60	$32.35	$48.78
1 Plumber Apprentice	28.75	230.00	43.35	346.80		
1 Truck Crane, 12 Ton		607.60		668.35	37.98	41.77
16 L.H., Daily Totals		$1125.20		$1448.75	$70.33	$90.55

Crew Q-22A	Hr.	Daily	Hr.	Daily	Bare Costs	Incl. O&P
1 Plumber	$35.95	$287.60	$54.20	$433.60	$30.13	$45.74
1 Plumber Apprentice	28.75	230.00	43.35	346.80		
1 Laborer	23.45	187.60	36.50	292.00		
1 Equip. Oper. (crane)	32.35	258.80	48.90	391.20		
1 Truck Crane, 12 Ton		607.60		668.35	18.99	20.89
32 L.H., Daily Totals		$1571.60		$2131.95	$49.12	$66.63

Crew Q-23	Hr.	Daily	Hr.	Daily	Bare Costs	Incl. O&P
1 Plumber Foreman	$37.95	$303.60	$57.25	$458.00	$35.03	$52.87
1 Plumber	35.95	287.60	54.20	433.60		
1 Equip. Oper. (medium)	31.20	249.60	47.15	377.20		
1 Power Tools		3.40		3.75		
1 Crane, 20 Ton		588.55		647.40	24.66	27.13
24 L.H., Daily Totals		$1432.75		$1919.95	$59.69	$80.00

Crew R-1	Hr.	Daily	Hr.	Daily	Bare Costs	Incl. O&P
1 Electrician Foreman	$35.95	$287.60	$53.65	$429.20	$31.30	$47.23
3 Electricians	35.45	850.80	52.90	1269.60		
2 Helpers	22.75	364.00	35.50	568.00		
48 L.H., Daily Totals		$1502.40		$2266.80	$31.30	$47.23

Crew R-1A	Hr.	Daily	Hr.	Daily	Bare Costs	Incl. O&P
1 Electrician	$35.45	$283.60	$52.90	$423.20	$29.10	$44.20
1 Helper	22.75	182.00	35.50	284.00		
16 L.H., Daily Totals		$465.60		$707.20	$29.10	$44.20

Crew R-2	Hr.	Daily	Hr.	Daily	Bare Costs	Incl. O&P
1 Electrician Foreman	$35.95	$287.60	$53.65	$429.20	$31.45	$47.46
3 Electricians	35.45	850.80	52.90	1269.60		
2 Helpers	22.75	364.00	35.50	568.00		
1 Equip. Oper. (crane)	32.35	258.80	48.90	391.20		
1 S.P. Crane, 5 Ton		315.40		346.95	5.63	6.20
56 L.H., Daily Totals		$2076.60		$3004.95	$37.08	$53.66

Crew R-3	Hr.	Daily	Hr.	Daily	Bare Costs	Incl. O&P
1 Electrician Foreman	$35.95	$287.60	$53.65	$429.20	$35.03	$52.40
1 Electrician	35.45	283.60	52.90	423.20		
.5 Equip. Oper. (crane)	32.35	129.40	48.90	195.60		
.5 S.P. Crane, 5 Ton		157.70		173.45	7.89	8.67
20 L.H., Daily Totals		$858.30		$1221.45	$42.92	$61.07

Crew R-4	Hr.	Daily	Hr.	Daily	Bare Costs	Incl. O&P
1 Struc. Steel Foreman	$36.25	$290.00	$65.35	$522.80	$34.89	$60.70
3 Struc. Steel Workers	34.25	822.00	61.75	1482.00		
1 Electrician	35.45	283.60	52.90	423.20		
1 Gas Welding Machine		88.60		97.45	2.22	2.44
40 L.H., Daily Totals		$1484.20		$2525.45	$37.11	$63.14

Crew R-5	Hr.	Daily	Hr.	Daily	Bare Costs	Incl. O&P
1 Electrician Foreman	$35.95	$287.60	$53.65	$429.20	$30.88	$46.64
4 Electrician Linemen	35.45	1134.40	52.90	1692.80		
2 Electrician Operators	35.45	567.20	52.90	846.40		
4 Electrician Groundmen	22.75	728.00	35.50	1136.00		
1 Crew Truck		81.40		89.55		
1 Tool Van		113.90		125.30		
1 Pickup Truck, 3/4 Ton		84.00		92.40		
.2 Crane, 55 Ton		194.48		213.95		
.2 Crane, 12 Ton		121.52		133.65		
.2 Auger, Truck Mtd.		398.60		438.45		
1 Tractor w/Winch		248.60		273.45	14.12	15.53
88 L.H., Daily Totals		$3959.70		$5471.15	$45.00	$62.17

Crew No.	Bare Costs		Incl. Subs O & P		Cost Per Labor-Hour	
Crew R-6	Hr.	Daily	Hr.	Daily	Bare Costs	Incl. O&P
1 Electrician Foreman	$35.95	$287.60	$53.65	$429.20	$30.88	$46.64
4 Electrician Linemen	35.45	1134.40	52.90	1692.80		
2 Electrician Operators	35.45	567.20	52.90	846.40		
4 Electrician Groundmen	22.75	728.00	35.50	1136.00		
1 Crew Truck		81.40		89.55		
1 Tool Van		113.90		125.30		
1 Pickup Truck, 3/4 Ton		84.00		92.40		
.2 Crane, 55 Ton		194.48		213.95		
.2 Crane, 12 Ton		121.52		133.65		
.2 Auger, Truck Mtd.		398.60		438.45		
1 Tractor w/Winch		248.60		273.45		
3 Cable Trailers		462.30		508.55		
.5 Tensioning Rig		152.32		167.55		
.5 Cable Pulling Rig		881.50		969.65	31.12	34.23
88 L.H., Daily Totals		$5455.82		$7116.90	$62.00	$80.87
Crew R-7	Hr.	Daily	Hr.	Daily	Bare Costs	Incl. O&P
1 Electrician Foreman	$35.95	$287.60	$53.65	$429.20	$24.95	$38.53
5 Electrician Groundmen	22.75	910.00	35.50	1420.00		
1 Crew Truck		81.40		89.55	1.70	1.87
48 L.H., Daily Totals		$1279.00		$1938.75	$26.65	$40.40
Crew R-8	Hr.	Daily	Hr.	Daily	Bare Costs	Incl. O&P
1 Electrician Foreman	$35.95	$287.60	$53.65	$429.20	$31.30	$47.23
3 Electrician Linemen	35.45	850.80	52.90	1269.60		
2 Electrician Groundmen	22.75	364.00	35.50	568.00		
1 Pickup Truck, 3/4 Ton		84.00		92.40		
1 Crew Truck		81.40		89.55	3.45	3.79
48 L.H., Daily Totals		$1667.80		$2448.75	$34.75	$51.02
Crew R-9	Hr.	Daily	Hr.	Daily	Bare Costs	Incl. O&P
1 Electrician Foreman	$35.95	$287.60	$53.65	$429.20	$29.16	$44.29
1 Electrician Lineman	35.45	283.60	52.90	423.20		
2 Electrician Operators	35.45	567.20	52.90	846.40		
4 Electrician Groundmen	22.75	728.00	35.50	1136.00		
1 Pickup Truck, 3/4 Ton		84.00		92.40		
1 Crew Truck		81.40		89.55	2.58	2.84
64 L.H., Daily Totals		$2031.80		$3016.75	$31.74	$47.13
Crew R-10	Hr.	Daily	Hr.	Daily	Bare Costs	Incl. O&P
1 Electrician Foreman	$35.95	$287.60	$53.65	$429.20	$33.42	$50.13
4 Electrician Linemen	35.45	1134.40	52.90	1692.80		
1 Electrician Groundman	22.75	182.00	35.50	284.00		
1 Crew Truck		81.40		89.55		
3 Tram Cars		329.55		362.50	8.56	9.42
48 L.H., Daily Totals		$2014.95		$2858.05	$41.98	$59.55
Crew R-11	Hr.	Daily	Hr.	Daily	Bare Costs	Incl. O&P
1 Electrician Foreman	$35.95	$287.60	$53.65	$429.20	$31.99	$48.18
4 Electricians	35.45	1134.40	52.90	1692.80		
1 Helper	22.75	182.00	35.50	284.00		
1 Common Laborer	23.45	187.60	36.50	292.00		
1 Crew Truck		81.40		89.55		
1 Crane, 12 Ton		607.60		668.35	12.30	13.53
56 L.H., Daily Totals		$2480.60		$3455.90	$44.29	$61.71

Crew No.	Bare Costs		Incl. Subs O & P		Cost Per Labor-Hour	
Crew R-12	Hr.	Daily	Hr.	Daily	Bare Costs	Incl. O&P
1 Carpenter Foreman	$30.50	$244.00	$47.45	$379.60	$28.16	$44.47
4 Carpenters	30.00	960.00	46.70	1494.40		
4 Common Laborers	23.45	750.40	36.50	1168.00		
1 Equip. Oper. (med.)	31.20	249.60	47.15	377.20		
1 Steel Worker	34.25	274.00	61.75	494.00		
1 Dozer, 200 H.P.		823.40		905.75		
1 Pickup Truck, 3/4 Ton		84.00		92.40	10.31	11.34
88 L.H., Daily Totals		$3385.40		$4911.35	$38.47	$55.81
Crew R-13	Hr.	Daily	Hr.	Daily	Bare Costs	Incl. O&P
1 Electrician Foreman	$35.95	$287.60	$53.65	$429.20	$33.72	$50.44
3 Electricians	35.45	850.80	52.90	1269.60		
.25 Equip. Oper. (crane)	32.35	64.70	48.90	97.80		
1 Equipment Oiler	26.65	213.20	40.25	322.00		
.25-1 Hyd. Crane, 33 Ton		214.35		235.80	5.10	5.61
42 L.H., Daily Totals		$1630.65		$2354.40	$38.82	$56.05
Crew R-15	Hr.	Daily	Hr.	Daily	Bare Costs	Incl. O&P
1 Electrician Foreman	$35.95	$287.60	$53.65	$429.20	$34.59	$51.72
4 Electricians	35.45	1134.40	52.90	1692.80		
1 Equipment Operator	29.80	238.40	45.05	360.40		
1 Aerial Lift Truck		214.00		235.40	4.46	4.90
48 L.H., Daily Totals		$1874.40		$2717.80	$39.05	$56.62
Crew R-18	Hr.	Daily	Hr.	Daily	Bare Costs	Incl. O&P
.25 Electrician Foreman	$35.95	$71.90	$53.65	$107.30	$27.67	$42.25
1 Electrician	35.45	283.60	52.90	423.20		
2 Helpers	22.75	364.00	35.50	568.00		
26 L.H., Daily Totals		$719.50		$1098.50	$27.67	$42.25
Crew R-19	Hr.	Daily	Hr.	Daily	Bare Costs	Incl. O&P
.5 Electrician Foreman	$35.95	$143.80	$53.65	$214.60	$35.55	$53.05
2 Electricians	35.45	567.20	52.90	846.40		
20 L.H., Daily Totals		$711.00		$1061.00	$35.55	$53.05
Crew R-21	Hr.	Daily	Hr.	Daily	Bare Costs	Incl. O&P
1 Electrician Foreman	$35.95	$287.60	$53.65	$429.20	$35.47	$52.94
3 Electricians	35.45	850.80	52.90	1269.60		
.1 Equip. Oper. (med.)	31.20	24.96	47.15	37.72		
.1 Hyd. Crane 25 Ton		62.94		69.25	1.92	2.11
32. L.H., Daily Totals		$1226.30		$1805.77	$37.39	$55.05
Crew R-22	Hr.	Daily	Hr.	Daily	Bare Costs	Incl. O&P
.66 Electrician Foreman	$35.95	$189.82	$53.65	$283.27	$30.07	$45.54
2 Helpers	22.75	364.00	35.50	568.00		
2 Electricians	35.45	567.20	52.90	846.40		
37.28 L.H., Daily Totals		$1121.02		$1697.67	$30.07	$45.54
Crew R-30	Hr.	Daily	Hr.	Daily	Bare Costs	Incl. O&P
.25 Electricians	$37.45	$74.90	$55.90	$111.80	$28.22	$43.04
1 Electricians	35.45	283.60	52.90	423.20		
2 Laborers, (Semi-Skilled)	23.45	375.20	36.50	584.00		
26 L.H., Daily Totals		$733.70		$1119.00	$28.22	$43.04

Historical Cost Indexes

The table below lists both the Means Historical Cost Index based on Jan. 1, 1993 = 100 as well as the computed value of an index based on Jan. 1, 2002 costs. Since the Jan. 1, 2002 figure is estimated, space is left to write in the actual index figures as they become available through either the quarterly "Means Construction Cost Indexes" or as printed in the "Engineering News-Record." To compute the actual index based on Jan. 1, 2002 = 100, divide the Historical Cost Index for a particular year by the actual Jan. 1, 2002 Construction Cost Index. Space has been left to advance the index figures as the year progresses.

Year	Historical Cost Index Jan. 1, 1993 = 100		Current Index Based on Jan. 1, 2002 = 100		Year	Historical Cost Index Jan. 1, 1993 = 100	Current Index Based on Jan. 1, 2002 = 100		Year	Historical Cost Index Jan. 1, 1993 = 100	Current Index Based on Jan. 1, 2002 = 100	
	Est.	Actual	Est.	Actual		Actual	Est.	Actual		Actual	Est.	Actual
Oct 2002					July 1987	87.7	69.5		July 1969	26.9	21.3	
July 2002					1986	84.2	66.8		1968	24.9	19.7	
April 2002					1985	82.6	65.5		1967	23.5	18.6	
Jan 2002	126.1		100.0	100.0	1984	82.0	65.0		1966	22.7	18.0	
July 2001		125.1	99.2		1983	80.2	63.6		1965	21.7	17.2	
2000		120.9	95.9		1982	76.1	60.4		1964	21.2	16.8	
1999		117.6	93.3		1981	70.0	55.5		1963	20.7	16.4	
1998		115.1	91.3		1980	62.9	49.9		1962	20.2	16.0	
1997		112.8	89.5		1979	57.8	45.8		1961	19.8	15.7	
1996		110.2	87.4		1978	53.5	42.4		1960	19.7	15.6	
1995		107.6	85.3		1977	49.5	39.3		1959	19.3	15.3	
1994		104.4	82.8		1976	46.9	37.2		1958	18.8	14.9	
1993		101.7	80.7		1975	44.8	35.5		1957	18.4	14.6	
1992		99.4	78.9		1974	41.4	32.8		1956	17.6	14.0	
1991		96.8	76.8		1973	37.7	29.9		1955	16.6	13.2	
1990		94.3	74.8		1972	34.8	27.6		1954	16.0	12.7	
1989		92.1	73.1		1971	32.1	25.5		1953	15.8	12.5	
1988		89.9	71.3		1970	28.7	22.8		1952	15.4	12.2	

Adjustments to Costs

The Historical Cost Index can be used to convert National Average building costs at a particular time to the approximate building costs for some other time.

Example:

Estimate and compare construction costs for different years in the same city.

To estimate the National Average construction cost of a building in 1970, knowing that it cost $900,000 in 2002:

INDEX in 1970 = 28.7

INDEX in 2002 = 126.1

Note: The City Cost Indexes for Canada can be used to convert U.S. National averages to local costs in Canadian dollars.

Time Adjustment using the Historical Cost Indexes:

$$\frac{\text{Index for Year A}}{\text{Index for Year B}} \times \text{Cost in Year B} = \text{Cost in Year A}$$

$$\frac{\text{INDEX 1970}}{\text{INDEX 2002}} \times \text{Cost 2002} = \text{Cost 1970}$$

$$\frac{28.7}{126.1} \times \$900,000 = .228 \times \$900,000 = \$205,200$$

The construction cost of the building in 1970 is $205,200.

How to Use the City Cost Indexes

What you should know before you begin

Means City Cost Indexes (CCI) are an extremely useful tool to use when you want to compare costs from city to city and region to region.

This publication contains average construction cost indexes for 316 major U.S. and Canadian cities and Location Factors covering over 930 three-digit zip code locations.

Keep in mind that a City Cost Index number is a *percentage ratio* of a specific city's cost to the national average cost of the same item at a stated time period.

In other words, these index figures represent relative construction *factors* (or, if you prefer, multipliers) for Material and Installation costs, as well as the weighted average for Total In Place costs for each CSI MasterFormat division. Installation costs include both labor and equipment rental costs.

The 30 City Average Index is the average of 30 major U.S. cities and serves as a National Average.

Index figures for both material and installation are based on the 30 major city average of 100 and represent the cost relationship as of July 1, 2001. The index for each division is computed from representative material and labor quantities for that division. The weighted average for each city is a weighted total of the components listed above it, but does not include relative productivity between trades or cities.

As changes occur in local material prices, labor rates and equipment rental rates, the impact of these changes should be accurately measured by the change in the City Cost Index for each particular city (as compared to the 30 City Average).

> *Therefore, if you know (or have estimated) building costs in one city today, you can easily convert those costs to expected building costs in another city.*
>
> *In addition, by using the Historical Cost Index, you can easily convert National Average building costs at a particular time to the approximate building costs for some other time. The City Cost Indexes can then be applied to calculate the costs for a particular city.*

Quick Calculations

Location Adjustment Using the City Cost Indexes:

$$\frac{\text{Index for City A}}{\text{Index for City B}} \times \text{Cost in City B} = \text{Cost in City A}$$

**Time Adjustment for the National Average
Using the Historical Cost Index:**

$$\frac{\text{Index for Year A}}{\text{Index for Year B}} \times \text{Cost in Year B} = \text{Cost in Year A}$$

Adjustment from the National Average:

$$\frac{\text{Index for City A}}{100} \times \text{National Average Cost} = \text{Cost in City A}$$

Since each of the other R.S. Means publications contains many different items, any *one* item multiplied by the particular city index may give incorrect results. However, the larger the number of items compiled, the closer the results should be to actual costs for that particular city.

The City Cost Indexes for Canadian cities are calculated using Canadian material and equipment prices and labor rates, in Canadian dollars. Therefore, indexes for Canadian cities can be used to convert U.S. National Average prices to local costs in Canadian dollars.

How to use this section

1. Compare costs from city to city.

In using the Means Indexes, remember that an index number is not a fixed number but a *ratio:* It's a percentage ratio of a building component's cost at any stated time to the National Average cost of that same component at the same time period. Put in the form of an equation:

$$\frac{\text{Specific City Cost}}{\text{National Average Cost}} \times 100 = \text{City Index Number}$$

Therefore, when making cost comparisons between cities, do not subtract one city's index number from the index number of another city and read the result as a percentage difference. Instead, divide one city's index number by that of the other city. The resulting number may then be used as a multiplier to calculate cost differences from city to city.

The formula used to find cost differences between cities for the purpose of comparison is as follows:

$$\frac{\text{City A Index}}{\text{City B Index}} \times \text{City B Cost (Known)} = \text{City A Cost (Unknown)}$$

In addition, you can use *Means CCI* to calculate and compare costs division by division between cities using the same basic formula. (Just be sure that you're comparing similar divisions.)

2. Compare a specific city's construction costs with the National Average.

When you're studying construction location feasibility, it's advisable to compare a prospective project's cost index with an index of the National Average cost.

For example, divide the weighted average index of construction costs of a specific city by that of the 30 City Average, which = 100.

$$\frac{\text{City Index}}{100} = \% \text{ of National Average}$$

As a result, you get a ratio that indicates the relative cost of construction in that city in comparison with the National Average.

3. Convert U.S. National Average to actual costs in Canadian City

$$\frac{\text{Index for Canadian City}}{100} \times \text{National Average Cost} = \text{Cost in Canadian City in \$ CAN}$$

4. Adjust construction cost data based on a National Average.

When you use a source of construction cost data which is based on a National Average (such as *Means cost data publications*), it is necessary to adjust those costs to a specific location.

$$\frac{\text{City Index}}{100} \quad x \quad \frac{\text{"Book" Cost Based on}}{\text{National Average Costs}} \quad = \quad \frac{\text{City Cost}}{\text{(Unknown)}}$$

5. When applying the City Cost Indexes to demolition projects, use the appropriate division index. For example, for removal of existing doors and windows, use the Division 8 index.

What you might like to know about how we developed the Indexes

To create a reliable index, R.S. Means researched the building type most often constructed in the United States and Canada. Because it was concluded that no one type of building completely represented the building construction industry, nine different types of buildings were combined to create a composite model.

The exact material, labor and equipment quantities are based on detailed analysis of these nine building types, then each quantity is weighted in proportion to expected usage. These various material items, labor hours, and equipment rental rates are thus combined to form a composite building representing as closely as possible the actual usage of materials, labor and equipment used in the North American Building Construction Industry.

The following structures were chosen to make up that composite model:

1. Factory, 1 story
2. Office, 2–4 story
3. Store, Retail
4. Town Hall, 2–3 story
5. High School, 2–3 story
6. Hospital, 4–8 story
7. Garage, Parking
8. Apartment, 1–3 story
9. Hotel/Motel, 2–3 story

For the purposes of ensuring the timeliness of the data, the components of the index for the composite model have been streamlined. They currently consist of:

- specific quantities of 66 commonly used construction materials;
- specific labor-hours for 21 building construction trades; and
- specific days of equipment rental for 6 types of construction equipment (normally used to install the 66 material items by the 21 trades.)

A sophisticated computer program handles the updating of all costs for each city on a quarterly basis. Material and equipment price quotations are gathered quarterly from over 316 cities in the United States and Canada. These prices and the latest negotiated labor wage rates for 21 different building trades are used to compile the quarterly update of the City Cost Index.

The 30 major U.S. cities used to calculate the National Average are:

Atlanta, GA	Memphis, TN
Baltimore, MD	Milwaukee, WI
Boston, MA	Minneapolis, MN
Buffalo, NY	Nashville, TN
Chicago, IL	New Orleans, LA
Cincinnati, OH	New York, NY
Cleveland, OH	Philadelphia, PA
Columbus, OH	Phoenix, AZ
Dallas, TX	Pittsburgh, PA
Denver, CO	St. Louis, MO
Detroit, MI	San Antonio, TX
Houston, TX	San Diego, CA
Indianapolis, IN	San Francisco, CA
Kansas City, MO	Seattle, WA
Los Angeles, CA	Washington, DC

F.Y.I.: The CSI MasterFormat Divisions

1. General Requirements
2. Site Work
3. Concrete
4. Masonry
5. Metals
6. Wood & Plastics
7. Thermal & Moisture Protection
8. Doors & Windows
9. Finishes
10. Specialties
11. Equipment
12. Furnishings
13. Special Construction
14. Conveying Systems
15. Mechanical
16. Electrical

The information presented in the CCI is organized according to the Construction Specifications Institute (CSI) MasterFormat.

What the CCI does not indicate

The weighted average for each city is a total of the components listed above weighted to reflect typical usage, but it does *not* include the productivity variations between trades or cities.

In addition, the CCI does not take into consideration factors such as the following:

- managerial efficiency
- competitive conditions
- automation
- restrictive union practices
- Unique local requirements
- regional variations due to specific building codes

City Cost Indexes

DIVISION		UNITED STATES 30 CITY AVERAGE			ALABAMA BIRMINGHAM			HUNTSVILLE			MOBILE			MONTGOMERY			TUSCALOOSA		
		MAT.	INST.	TOTAL	MAT.	INST.	TOTAL	MAT.	INST.	TOTAL	MAT.	INST.	TOTAL	MAT.	INST.	TOTAL	MAT.	INST.	TOTAL
01590	EQUIPMENT RENTAL	.0	100.0	100.0	.0	101.6	101.6	.0	101.5	101.5	.0	98.0	98.0	.0	98.0	98.0	.0	101.5	101.5
02	SITE CONSTRUCTION	100.0	100.0	100.0	87.3	93.8	92.2	85.3	92.9	91.1	96.4	86.7	89.0	96.9	87.2	89.6	85.8	92.6	91.0
03100	CONCRETE FORMS & ACCESSORIES	100.0	100.0	100.0	91.5	76.7	78.6	93.3	55.6	60.5	93.3	58.7	63.3	91.7	53.1	58.2	93.2	43.7	50.2
03200	CONCRETE REINFORCEMENT	100.0	100.0	100.0	92.9	85.5	88.6	92.9	85.9	88.8	95.9	63.2	76.8	95.9	84.5	89.2	92.9	84.8	88.2
03300	CAST-IN-PLACE CONCRETE	100.0	100.0	100.0	93.8	66.8	82.9	88.7	63.5	78.5	93.7	61.6	80.7	95.3	55.1	79.3	92.4	53.0	76.5
03	CONCRETE	100.0	100.0	100.0	89.2	76.1	82.7	86.8	65.8	76.3	89.7	62.2	76.0	90.4	61.7	76.1	88.6	56.8	72.8
04	MASONRY	100.0	100.0	100.0	85.6	72.8	77.8	85.5	57.7	68.5	86.1	58.2	69.1	86.6	40.4	58.4	85.8	42.9	59.6
05	METALS	100.0	100.0	100.0	98.5	95.6	97.5	98.2	94.4	96.8	96.8	85.6	92.7	96.8	93.5	95.6	97.2	94.0	96.1
06	WOOD & PLASTICS	100.0	100.0	100.0	92.1	76.9	84.3	91.3	53.9	72.1	91.3	58.8	74.6	89.5	53.9	71.2	91.3	42.8	66.4
07	THERMAL & MOISTURE PROTECTION	100.0	100.0	100.0	96.1	80.2	88.6	95.8	72.3	84.7	95.8	73.5	85.3	95.5	66.5	81.8	95.8	65.2	81.3
08	DOORS & WINDOWS	100.0	100.0	100.0	96.4	78.5	92.1	96.4	60.0	87.6	96.4	59.2	87.4	96.4	61.1	87.9	96.4	60.2	87.7
09200	PLASTER & GYPSUM BOARD	100.0	100.0	100.0	110.1	76.7	88.5	106.9	53.0	72.1	106.9	58.1	75.3	106.9	53.0	72.1	106.9	41.5	64.7
095,098	CEILINGS & ACOUSTICAL TREATMENT	100.0	100.0	100.0	95.7	76.7	82.9	95.7	53.0	67.0	95.7	58.1	70.4	95.7	53.0	67.0	95.7	41.5	59.3
09600	FLOORING	100.0	100.0	100.0	101.2	58.0	90.6	101.2	50.2	88.6	109.8	62.9	98.2	109.8	32.5	90.7	101.2	47.7	88.0
097,099	WALL FINISHES, PAINTS & COATINGS	100.0	100.0	100.0	92.6	54.7	70.4	92.6	53.7	69.8	92.6	62.1	74.7	92.6	61.0	74.1	92.6		69.1
09	FINISHES	100.0	100.0	100.0	98.4	70.6	83.9	97.8	53.2	74.6	101.8	59.4	79.7	101.9	48.9	74.3	97.8	43.9	69.7
10-14	TOTAL DIV. 10000 - 14000	100.0	100.0	100.0	100.0	81.5	96.1	100.0	74.9	94.8	100.0	71.1	94.0	100.0	72.6	94.3	100.0	69.7	93.7
15	MECHANICAL	100.0	100.0	100.0	99.9	69.8	86.5	99.9	52.0	78.6	99.9	65.2	84.5	99.9	46.9	76.3	99.9	41.2	73.8
16	ELECTRICAL	100.0	100.0	100.0	96.3	64.8	74.4	96.7	72.5	79.8	96.7	59.5	70.8	96.3	72.2	79.6	96.7	64.8	74.5
01-16	WEIGHTED AVERAGE	100.0	100.0	100.0	96.2	75.8	86.4	95.8	67.2	82.0	96.7	66.5	82.1	96.8	62.7	80.3	95.9	59.7	78.4

DIVISION		ALASKA ANCHORAGE			FAIRBANKS			JUNEAU			ARIZONA FLAGSTAFF			MESA/TEMPE			PHOENIX		
		MAT.	INST.	TOTAL	MAT.	INST.	TOTAL	MAT.	INST.	TOTAL	MAT.	INST.	TOTAL	MAT.	INST.	TOTAL	MAT.	INST.	TOTAL
01590	EQUIPMENT RENTAL	.0	118.9	118.9	.0	118.9	118.9	.0	118.9	118.9	.0	93.5	93.5	.0	94.0	94.0	.0	94.6	94.6
02	SITE CONSTRUCTION	135.4	135.4	135.4	119.2	135.4	131.5	131.2	135.4	134.4	86.4	99.4	96.3	83.1	99.0	95.2	83.5	99.0	95.9
03100	CONCRETE FORMS & ACCESSORIES	131.1	115.5	117.6	132.7	122.4	123.7	132.5	115.5	117.8	106.0	71.2	75.8	103.0	70.6	74.9	104.3	76.7	80.4
03200	CONCRETE REINFORCEMENT	141.5	102.7	118.9	118.9	102.7	109.5	105.2	102.7	103.7	103.7	78.6	89.1	103.9	71.2	84.9	102.1	79.2	88.7
03300	CAST-IN-PLACE CONCRETE	189.8	117.6	160.7	158.3	118.2	142.1	190.6	117.6	161.2	97.3	84.7	92.2	99.2	76.6	90.1	99.3	84.7	93.4
03	CONCRETE	153.7	113.1	133.5	130.0	116.3	123.2	149.8	113.1	131.6	122.5	77.0	99.9	100.1	72.6	86.4	99.7	79.6	89.7
04	MASONRY	195.4	122.8	151.0	182.4	122.8	146.0	184.6	122.8	146.9	101.4	63.1	78.0	109.3	57.6	77.7	96.2	72.6	81.8
05	METALS	130.0	98.4	118.4	130.0	98.8	118.6	130.3	98.4	118.6	98.0	70.9	88.1	98.3	68.9	87.6	99.6	73.5	90.1
06	WOOD & PLASTICS	118.0	113.5	115.7	118.3	122.3	120.3	118.0	113.5	115.7	107.0	70.7	88.3	101.0	77.5	89.0	102.2	77.7	89.6
07	THERMAL & MOISTURE PROTECTION	199.7	115.4	159.7	195.7	118.9	159.3	196.4	115.4	158.0	111.2	73.5	93.4	108.6	68.5	89.6	108.5	75.9	93.1
08	DOORS & WINDOWS	128.0	105.9	122.7	125.1	111.0	121.7	125.1	105.9	120.5	101.1	72.2	94.1	98.0	70.5	91.4	99.2	75.9	93.6
09200	PLASTER & GYPSUM BOARD	129.7	113.9	119.4	129.7	122.9	125.3	129.7	113.9	119.4	90.9	69.9	77.3	91.5	76.8	82.0	92.0	76.9	82.3
095,098	CEILINGS & ACOUSTICAL TREATMENT	129.4	113.9	118.9	129.4	122.9	125.0	129.4	113.9	118.9	102.9	69.9	80.7	105.6	76.8	86.2	105.6	76.9	86.3
09600	FLOORING	127.3	126.3	127.1	127.5	126.3	127.2	127.3	126.3	127.1	98.9	58.7	89.0	102.1	73.2	95.0	102.4	76.4	96.0
097,099	WALL FINISHES, PAINTS & COATINGS	116.2	107.8	111.3	116.2	117.9	117.2	116.2	107.8	111.3	98.7	55.7	73.5	109.9	59.9	80.6	109.9	69.0	85.9
09	FINISHES	140.2	116.9	128.1	138.0	123.2	130.3	138.7	116.9	127.4	97.1	66.6	81.2	98.5	70.0	83.6	98.7	75.6	86.7
10-14	TOTAL DIV. 10000 - 14000	100.0	114.1	102.9	100.0	115.2	103.2	100.0	114.1	102.9	100.0	82.1	96.3	100.0	77.5	95.3	100.0	83.2	96.5
15	MECHANICAL	108.2	108.0	108.1	108.2	117.3	112.2	108.2	108.0	108.1	100.2	83.1	92.6	100.1	73.0	88.0	100.1	83.2	92.6
16	ELECTRICAL	161.7	110.8	126.3	164.6	110.8	127.2	164.6	110.8	127.2	100.4	55.8	64.9	98.1	37.7	56.1	107.2	66.9	79.2
01-16	WEIGHTED AVERAGE	134.1	113.7	124.3	129.5	117.2	123.6	132.6	113.7	123.5	102.5	73.4	88.4	99.6	67.2	83.9	99.8	78.2	89.4

DIVISION		ARIZONA PRESCOTT			TUCSON			ARKANSAS FORT SMITH			JONESBORO			LITTLE ROCK			PINE BLUFF		
		MAT.	INST.	TOTAL	MAT.	INST.	TOTAL	MAT.	INST.	TOTAL	MAT.	INST.	TOTAL	MAT.	INST.	TOTAL	MAT.	INST.	TOTAL
01590	EQUIPMENT RENTAL	.0	93.5	93.5	.0	94.0	94.0	.0	85.1	85.1	.0	106.6	106.6	.0	85.1	85.1	.0	85.1	85.1
02	SITE CONSTRUCTION	73.6	99.0	92.9	80.3	99.6	94.9	77.9	83.7	82.3	101.8	98.8	99.5	77.7	83.7	82.3	79.9	83.7	82.8
03100	CONCRETE FORMS & ACCESSORIES	100.9	65.3	70.0	103.7	76.2	79.8	101.5	47.4	54.6	85.7	54.4	58.5	95.6	52.0	57.8	77.5	51.9	55.3
03200	CONCRETE REINFORCEMENT	103.7	75.2	87.1	101.0	78.6	87.9	95.6	73.7	82.8	91.5	52.8	68.9	95.8	69.7	80.6	95.8	69.7	80.6
03300	CAST-IN-PLACE CONCRETE	97.2	71.9	87.0	99.3	84.5	93.3	91.2	68.8	82.2	86.9	62.8	77.2	91.2	68.8	82.2	83.7	68.8	77.7
03	CONCRETE	107.0	69.3	88.2	99.5	79.2	89.4	87.8	60.7	74.3	84.4	58.7	71.6	87.5	61.9	74.8	85.1	61.9	73.5
04	MASONRY	101.9	68.7	81.6	96.7	63.0	76.1	96.8	58.5	73.4	90.8	53.8	68.2	95.0	58.5	72.7	114.4	58.5	80.2
05	METALS	98.1	68.2	87.2	98.9	71.6	88.9	96.7	72.8	87.9	90.9	80.6	87.2	96.3	71.5	87.2	95.4	71.4	86.6
06	WOOD & PLASTICS	102.2	64.2	82.7	102.7	77.7	89.9	104.8	46.4	74.8	89.1	55.8	72.0	101.5	52.5	76.3	81.5	52.5	66.6
07	THERMAL & MOISTURE PROTECTION	109.4	69.8	90.7	109.2	70.1	90.7	99.2	53.2	77.4	109.1	57.5	84.7	98.1	53.8	77.1	97.8	53.8	76.9
08	DOORS & WINDOWS	101.1	64.6	92.3	96.0	75.9	91.1	96.6	49.4	85.2	98.1	53.3	87.3	96.6	52.6	86.0	92.0	52.6	82.5
09200	PLASTER & GYPSUM BOARD	88.9	63.1	72.3	92.3	76.9	82.4	96.0	45.7	63.5	102.1	55.0	71.6	96.0	51.9	67.5	89.3	51.9	65.1
095,098	CEILINGS & ACOUSTICAL TREATMENT	102.9	63.1	76.2	107.0	76.9	86.8	101.9	45.7	64.1	102.5	55.0	70.5	101.9	51.9	68.3	97.6	51.9	66.9
09600	FLOORING	97.2	58.5	87.7	101.5	58.7	91.0	118.1	74.4	107.3	79.1	51.6	72.3	119.4	74.4	108.3	106.8	74.4	98.8
097,099	WALL FINISHES, PAINTS & COATINGS	98.7	55.7	73.5	107.4	55.7	77.2	95.8	65.3	78.0	84.7	58.7	69.5	95.8	50.0	69.0	95.8	50.0	69.0
09	FINISHES	95.1	62.6	78.1	98.4	70.7	84.0	101.9	54.1	77.0	92.5	54.5	72.7	102.3	55.8	78.1	96.6	55.8	75.4
10-14	TOTAL DIV. 10000 - 14000	100.0	80.8	96.0	100.0	83.2	96.5	100.0	70.0	93.8	100.0	63.0	92.3	100.0	70.8	93.9	100.0	70.8	93.9
15	MECHANICAL	100.2	79.8	91.1	100.0	74.5	88.7	100.1	48.8	77.3	100.2	48.8	77.4	100.1	58.3	81.5	100.1	51.6	78.5
16	ELECTRICAL	100.0	53.7	67.8	101.6	64.9	76.1	95.6	69.5	77.5	100.7	55.0	68.9	96.7	69.9	78.1	94.5	69.9	77.4
01-16	WEIGHTED AVERAGE	99.9	70.7	85.8	98.8	74.3	87.0	96.7	61.6	79.8	95.5	60.9	78.8	96.6	64.0	80.8	95.8	62.7	79.8

COST INDEXES

City Cost Indexes

| | | ARKANSAS | | | CALIFORNIA | | | | | | | | | | | | | | | | |
| | | TEXARKANA | | | ANAHEIM | | | BAKERSFIELD | | | FRESNO | | | LOS ANGELES | | | OAKLAND | | |
DIVISION		MAT.	INST.	TOTAL	MAT.	INST.	TOTAL	MAT.	INST.	TOTAL	MAT.	INST.	TOTAL	MAT.	INST.	TOTAL	MAT.	INST.	TOTAL
01590	EQUIPMENT RENTAL	.0	85.8	85.8	.0	102.4	102.4	.0	99.7	99.7	.0	99.7	99.7	.0	98.2	98.2	.0	101.6	101.6
02	SITE CONSTRUCTION	97.1	84.5	87.5	95.8	110.2	106.7	100.3	107.1	105.5	101.8	107.3	105.9	90.8	109.0	104.6	130.1	104.3	110.5
03100	CONCRETE FORMS & ACCESSORIES	86.6	45.1	50.5	103.7	120.2	118.1	96.1	120.0	116.8	100.4	122.8	119.8	105.4	119.9	118.0	110.2	138.7	134.9
03200	CONCRETE REINFORCEMENT	95.3	51.1	69.5	106.9	114.4	111.3	106.5	114.2	111.0	107.0	114.3	111.3	108.4	114.4	111.9	98.9	115.3	108.4
03300	CAST-IN-PLACE CONCRETE	91.2	49.0	74.2	104.0	120.3	110.6	99.3	119.1	107.3	108.7	114.4	111.0	86.2	117.6	98.8	129.6	118.3	125.0
03	CONCRETE	84.4	48.6	66.6	111.9	118.2	115.0	109.0	117.6	113.3	114.1	117.2	115.7	105.1	117.0	111.0	126.2	125.9	126.0
04	MASONRY	96.8	37.7	60.7	87.7	116.6	105.4	107.3	113.9	111.3	109.8	113.3	112.0	94.8	118.6	109.3	150.4	128.3	136.9
05	METALS	87.7	62.9	78.7	110.5	100.9	107.0	104.9	100.0	103.1	106.9	100.8	104.6	111.3	99.0	106.8	101.6	106.6	103.4
06	WOOD & PLASTICS	91.2	48.4	69.2	94.3	119.0	107.0	86.1	119.1	103.1	96.8	122.1	109.8	91.9	118.5	105.6	107.6	140.6	124.6
07	THERMAL & MOISTURE PROTECTION	98.7	47.0	74.2	120.9	117.3	119.2	105.5	109.0	107.1	101.7	111.2	106.2	111.8	116.6	114.1	111.4	131.7	121.0
08	DOORS & WINDOWS	97.0	45.7	84.7	103.2	115.1	106.0	102.0	113.3	104.7	102.0	114.9	105.1	98.3	114.8	102.3	106.2	130.0	112.0
09200	PLASTER & GYPSUM BOARD	92.9	47.7	63.7	93.6	119.7	110.4	93.6	119.7	110.4	92.6	122.7	112.1	85.2	119.7	107.5	98.4	141.6	126.3
095,098	CEILINGS & ACOUSTICAL TREATMENT	103.3	47.7	65.9	118.1	119.7	119.1	118.1	119.7	119.1	118.1	122.7	121.2	116.5	119.7	118.6	115.4	141.6	133.0
09600	FLOORING	109.7	49.2	94.7	124.3	107.0	120.0	119.7	71.0	107.7	120.5	142.9	126.0	116.8	107.0	114.3	112.4	121.9	114.8
097,099	WALL FINISHES, PAINTS & COATINGS	95.8	41.2	63.8	110.7	112.8	111.9	113.8	102.3	107.1	113.2	98.2	104.4	107.4	112.8	110.6	116.9	131.5	125.4
09	FINISHES	99.6	45.6	71.5	113.5	116.8	115.2	114.7	109.5	112.0	114.8	124.4	119.8	108.8	116.5	112.8	115.4	135.8	126.0
10 - 14	TOTAL DIV. 10000 - 14000	100.0	45.0	88.5	100.0	115.4	103.2	100.0	132.1	106.7	100.0	132.4	106.8	100.0	114.3	103.0	100.0	136.2	107.6
15	MECHANICAL	100.1	43.1	74.8	100.2	111.8	105.4	100.2	109.4	104.3	100.2	112.0	105.5	100.2	111.7	105.3	100.1	137.8	116.8
16	ELECTRICAL	97.7	45.7	61.6	95.0	109.2	104.9	95.4	97.3	96.7	94.1	99.5	97.9	109.8	116.5	114.5	113.6	133.9	127.7
01 - 16	WEIGHTED AVERAGE	95.3	50.2	73.5	104.2	112.6	108.2	103.7	108.6	106.1	104.8	111.4	108.0	103.4	113.4	108.3	110.9	127.9	119.1

| | | CALIFORNIA | | | | | | | | | | | | | | | | | | | |
| | | OXNARD | | | REDDING | | | RIVERSIDE | | | SACRAMENTO | | | SAN DIEGO | | | SAN FRANCISCO | | |
DIVISION		MAT.	INST.	TOTAL	MAT.	INST.	TOTAL	MAT.	INST.	TOTAL	MAT.	INST.	TOTAL	MAT.	INST.	TOTAL	MAT.	INST.	TOTAL
01590	EQUIPMENT RENTAL	.0	98.3	98.3	.0	99.3	99.3	.0	100.9	100.9	.0	101.2	101.2	.0	97.4	97.4	.0	106.8	106.8
02	SITE CONSTRUCTION	101.3	104.8	104.0	106.7	106.4	106.5	93.6	107.8	104.4	97.7	110.6	107.5	102.5	102.1	102.2	132.2	110.7	115.9
03100	CONCRETE FORMS & ACCESSORIES	102.2	120.3	117.9	101.6	122.8	120.0	104.9	120.2	118.1	108.6	123.1	121.2	103.8	112.3	111.2	110.5	139.5	135.7
03200	CONCRETE REINFORCEMENT	106.5	114.2	111.0	103.2	114.4	109.7	105.5	114.2	110.6	100.0	114.5	108.4	110.5	114.1	112.6	112.6	115.6	114.4
03300	CAST-IN-PLACE CONCRETE	105.4	119.4	111.0	119.4	114.3	117.3	103.0	120.3	110.0	110.9	114.5	112.4	107.8	104.2	106.4	129.5	119.8	125.6
03	CONCRETE	112.5	117.8	115.1	123.0	117.2	120.1	111.3	118.1	114.7	116.8	117.3	117.1	116.2	109.1	112.7	127.8	126.8	127.3
04	MASONRY	112.4	113.0	112.8	113.8	115.2	114.7	85.4	114.2	103.0	120.7	115.3	117.4	102.4	111.5	108.0	150.6	132.4	139.5
05	METALS	104.4	100.3	102.9	108.8	100.6	105.8	110.5	100.6	106.9	97.7	100.2	98.6	110.5	99.7	106.6	106.8	107.8	107.2
06	WOOD & PLASTICS	93.9	119.1	106.9	94.7	122.1	108.7	94.3	119.0	107.0	101.7	122.1	112.2	97.5	108.8	103.3	107.6	140.8	124.7
07	THERMAL & MOISTURE PROTECTION	110.8	114.4	112.5	111.5	114.3	112.8	119.9	115.0	117.6	122.1	114.0	118.3	113.9	106.9	110.6	111.4	133.7	122.0
08	DOORS & WINDOWS	100.8	115.1	104.3	103.6	117.1	106.9	103.2	115.1	106.0	117.8	117.2	117.7	104.2	108.9	105.3	110.7	130.1	115.3
09200	PLASTER & GYPSUM BOARD	93.6	119.7	110.4	93.1	122.7	112.2	93.3	119.7	110.3	96.6	122.7	113.5	89.1	109.0	102.0	100.1	141.6	126.9
095,098	CEILINGS & ACOUSTICAL TREATMENT	118.1	119.7	119.1	125.1	122.7	123.5	116.7	119.7	118.7	122.5	122.7	122.6	115.1	109.0	111.0	120.7	141.6	134.8
09600	FLOORING	119.7	107.0	116.6	119.0	112.8	117.5	124.2	107.0	120.0	115.2	112.8	114.6	118.5	116.2	118.0	112.4	121.9	114.8
097,099	WALL FINISHES, PAINTS & COATINGS	113.2	105.5	108.7	113.2	126.1	120.8	110.7	112.8	111.9	115.3	126.1	121.6	111.9	112.8	112.4	116.9	140.6	130.8
09	FINISHES	114.5	116.1	115.3	115.8	122.3	119.2	112.9	116.8	115.0	115.6	122.4	119.2	111.4	113.0	112.3	116.7	137.0	127.3
10 - 14	TOTAL DIV. 10000 - 14000	100.0	115.8	103.3	100.0	132.4	106.8	100.0	115.4	103.2	100.0	132.5	106.8	100.0	114.4	103.0	100.0	136.9	107.7
15	MECHANICAL	100.2	111.8	105.4	100.2	111.4	105.2	100.2	111.8	105.3	100.0	117.4	107.8	100.1	111.8	105.3	100.1	163.1	128.1
16	ELECTRICAL	99.2	110.7	107.2	104.4	100.9	102.0	95.1	103.6	101.0	103.9	100.9	101.8	92.8	94.0	93.6	109.3	151.8	138.8
01 - 16	WEIGHTED AVERAGE	104.7	111.7	108.1	107.8	111.5	109.6	103.8	111.0	107.3	107.5	113.1	110.2	105.3	106.3	105.8	112.3	137.3	124.4

| | | CALIFORNIA | | | | | | | | | | | | COLORADO | | | | | |
| | | SAN JOSE | | | SANTA BARBARA | | | STOCKTON | | | VALLEJO | | | COLORADO SPRINGS | | | DENVER | | |
DIVISION		MAT.	INST.	TOTAL	MAT.	INST.	TOTAL	MAT.	INST.	TOTAL	MAT.	INST.	TOTAL	MAT.	INST.	TOTAL	MAT.	INST.	TOTAL
01590	EQUIPMENT RENTAL	.0	100.0	100.0	.0	99.7	99.7	.0	99.3	99.3	.0	101.9	101.9	.0	96.5	96.5	.0	101.7	101.7
02	SITE CONSTRUCTION	137.9	100.6	109.6	101.7	107.1	105.8	99.7	106.6	104.9	94.6	110.8	106.9	106.6	98.4	100.4	105.0	107.8	107.1
03100	CONCRETE FORMS & ACCESSORIES	104.4	138.8	134.3	102.9	120.3	118.0	96.9	122.8	119.4	110.6	137.4	133.9	90.6	79.6	81.1	100.2	80.0	82.6
03200	CONCRETE REINFORCEMENT	103.5	115.3	110.4	106.5	114.3	111.0	107.0	114.5	111.4	98.6	115.4	108.4	99.0	83.3	89.8	98.4	84.1	90.1
03300	CAST-IN-PLACE CONCRETE	123.6	118.5	121.5	105.0	119.2	110.7	104.9	114.4	108.8	116.2	115.9	116.1	109.2	84.8	99.4	102.1	84.4	95.0
03	CONCRETE	121.4	126.1	123.8	112.3	117.8	115.0	111.9	117.3	114.6	119.4	124.3	121.8	103.8	82.4	93.2	108.4	82.6	95.5
04	MASONRY	149.1	128.9	136.7	107.8	113.9	111.5	112.0	117.3	115.2	86.1	130.6	113.3	106.1	69.1	83.5	105.6	84.2	92.5
05	METALS	111.1	108.0	110.0	104.2	100.4	102.8	106.5	101.2	104.6	100.9	102.3	101.4	101.5	87.1	96.3	105.1	88.3	98.9
06	WOOD & PLASTICS	102.0	140.5	121.8	93.9	119.1	106.9	89.9	122.1	106.4	99.8	140.3	120.6	90.9	80.6	85.6	99.0	80.5	89.5
07	THERMAL & MOISTURE PROTECTION	107.3	133.1	119.5	106.8	113.2	109.8	110.9	113.2	112.0	125.1	130.7	127.8	106.5	81.0	94.4	105.9	84.6	95.8
08	DOORS & WINDOWS	93.1	129.9	102.0	102.0	115.1	105.2	101.3	117.1	105.1	119.2	129.8	121.7	97.3	86.3	94.7	98.7	86.2	95.7
09200	PLASTER & GYPSUM BOARD	94.1	141.6	124.8	93.6	119.7	110.4	94.6	122.7	112.8	97.1	141.6	125.9	86.5	80.0	82.3	95.4	80.1	85.5
095,098	CEILINGS & ACOUSTICAL TREATMENT	109.6	141.6	131.1	118.1	119.7	119.1	118.1	122.7	121.2	122.5	141.6	135.3	99.8	80.0	86.5	98.4	80.1	86.1
09600	FLOORING	116.6	121.9	117.9	119.7	112.8	118.0	119.7	112.9	118.0	119.5	121.9	120.1	110.9	95.6	107.1	111.9	95.6	107.9
097,099	WALL FINISHES, PAINTS & COATINGS	114.9	129.7	123.6	113.2	105.5	108.7	113.2	100.7	105.9	113.3	131.5	124.0	110.6	60.2	81.1	110.6	76.8	90.8
09	FINISHES	113.7	135.6	125.1	114.7	117.1	116.0	114.6	119.4	117.1	114.2	135.2	125.1	100.2	79.5	89.4	101.3	82.3	91.4
10 - 14	TOTAL DIV. 10000 - 14000	100.0	136.1	107.5	100.0	115.8	103.3	100.0	132.4	106.8	100.0	134.6	107.2	100.0	85.8	97.0	100.0	85.3	96.9
15	MECHANICAL	100.2	146.6	120.8	100.2	111.8	105.4	100.2	112.0	105.5	100.0	128.7	112.8	100.2	81.3	91.8	100.1	84.5	93.2
16	ELECTRICAL	109.9	145.2	134.4	91.9	109.5	104.2	103.4	110.5	108.3	99.8	114.0	109.6	95.2	86.5	89.2	97.3	92.4	93.9
01 - 16	WEIGHTED AVERAGE	109.7	131.4	120.2	104.0	111.9	107.8	105.2	113.2	109.1	106.0	122.8	114.1	100.7	83.4	92.4	102.2	87.9	95.3

COST INDEXES

City Cost Indexes

		COLORADO												CONNECTICUT					
		FORT COLLINS			GRAND JUNCTION			GREELEY			PUEBLO			BRIDGEPORT			BRISTOL		
DIVISION		MAT.	INST.	TOTAL	MAT.	INST.	TOTAL	MAT.	INST.	TOTAL	MAT.	INST.	TOTAL	MAT.	INST.	TOTAL	MAT.	INST.	TOTAL
01590	EQUIPMENT RENTAL	.0	98.0	98.0	.0	98.6	98.6	.0	98.0	98.0	.0	95.7	95.7	.0	103.3	103.3	.0	103.3	103.3
02	SITE CONSTRUCTION	117.8	101.3	105.3	129.6	99.9	107.1	102.9	100.1	100.8	120.9	95.0	101.2	106.0	105.3	105.5	105.1	105.3	105.3
03100	CONCRETE FORMS & ACCESSORIES	99.6	77.9	80.8	110.7	50.5	58.4	96.9	54.7	60.3	105.0	80.0	83.3	104.6	104.7	104.7	104.6	110.1	109.3
03200	CONCRETE REINFORCEMENT	103.1	82.3	91.0	109.0	82.0	93.3	102.7	55.2	75.1	101.6	83.2	90.5	112.0	116.4	114.6	112.0	116.4	114.5
03300	CAST-IN-PLACE CONCRETE	118.6	82.5	104.0	120.0	50.3	91.9	99.1	67.4	86.3	107.5	85.6	98.7	106.4	116.7	110.6	99.7	116.6	106.5
03	CONCRETE	123.3	80.6	102.0	120.4	57.4	89.1	105.8	59.8	82.9	108.9	82.9	95.9	110.1	111.0	110.5	106.7	113.2	110.0
04	MASONRY	119.7	67.4	87.8	145.9	55.1	90.4	113.3	46.5	72.5	106.3	67.0	82.2	106.6	117.6	113.4	96.4	117.6	109.4
05	METALS	100.9	84.4	94.9	101.4	82.3	94.4	100.9	65.3	87.9	102.9	88.1	97.5	99.1	112.4	104.0	99.1	112.1	103.9
06	WOOD & PLASTICS	98.7	80.4	89.3	106.6	51.9	78.5	96.1	55.6	75.3	102.1	81.0	91.2	103.0	101.0	101.9	103.0	108.6	105.9
07	THERMAL & MOISTURE PROTECTION	106.5	75.5	91.8	106.9	61.4	85.4	105.6	65.0	86.4	106.3	79.7	93.7	101.2	113.5	107.0	101.3	107.8	104.4
08	DOORS & WINDOWS	94.7	76.9	90.5	99.6	58.1	89.6	94.7	48.1	83.5	94.2	86.4	92.3	108.1	111.0	108.8	108.1	110.7	108.8
09200	PLASTER & GYPSUM BOARD	94.2	80.0	85.0	101.1	50.5	68.4	93.2	54.4	68.1	86.3	80.0	82.3	98.8	99.8	99.4	98.8	107.7	104.5
095,098	CEILINGS & ACOUSTICAL TREATMENT	92.7	80.0	84.2	97.3	50.5	65.8	92.7	54.4	67.0	107.2	80.0	88.9	104.9	99.8	101.4	104.9	107.7	106.7
09600	FLOORING	111.7	80.4	104.0	120.4	80.4	110.5	110.3	80.4	102.9	116.7	95.6	111.5	99.8	106.3	101.4	99.8	106.3	101.4
097,099	WALL FINISHES, PAINTS & COATINGS	110.6	59.7	80.8	121.8	35.0	71.0	110.6	37.2	67.6	121.8	55.8	83.1	95.2	106.0	101.5	95.2	105.2	101.1
09	FINISHES	100.7	76.7	88.2	107.2	54.3	79.7	99.4	57.3	77.4	105.4	80.2	92.2	101.6	104.3	103.0	101.7	108.6	105.3
10-14	TOTAL DIV. 10000-14000	100.0	70.2	93.8	100.0	61.9	92.1	100.0	63.1	92.3	100.0	86.7	97.2	100.0	112.9	102.7	100.0	113.9	102.9
15	MECHANICAL	100.1	82.9	92.4	100.0	38.7	72.7	100.1	77.0	89.8	100.0	76.9	89.7	99.9	104.5	101.9	99.9	104.4	101.9
16	ELECTRICAL	90.1	68.1	74.8	94.0	75.4	81.1	90.1	68.1	74.8	95.3	83.0	86.8	104.0	101.5	102.2	104.0	102.5	102.9
01-16	WEIGHTED AVERAGE	103.6	79.0	91.7	106.7	62.8	85.5	100.4	67.1	84.3	102.2	81.7	92.3	103.1	107.5	105.2	102.1	108.4	105.1

		CONNECTICUT																	
		HARTFORD			NEW BRITAIN			NEW HAVEN			NORWALK			STAMFORD			WATERBURY		
DIVISION		MAT.	INST.	TOTAL	MAT.	INST.	TOTAL	MAT.	INST.	TOTAL	MAT.	INST.	TOTAL	MAT.	INST.	TOTAL	MAT.	INST.	TOTAL
01590	EQUIPMENT RENTAL	.0	103.3	103.3	.0	103.3	103.3	.0	103.8	103.8	.0	103.3	103.3	.0	103.3	103.3	.0	103.3	103.3
02	SITE CONSTRUCTION	105.3	105.3	105.3	105.3	105.3	105.3	105.2	106.2	106.0	105.3	106.2	106.0	106.5	105.4	105.6	105.5	105.3	105.4
03100	CONCRETE FORMS & ACCESSORIES	103.5	110.1	109.2	104.9	110.1	109.4	104.4	110.2	109.4	104.6	104.8	104.8	104.5	105.0	105.0	104.6	110.2	109.5
03200	CONCRETE REINFORCEMENT	112.0	116.4	114.5	112.0	116.4	114.6	112.0	116.4	114.6	112.0	116.5	114.6	112.0	116.6	114.7	112.0	116.4	114.6
03300	CAST-IN-PLACE CONCRETE	99.4	116.6	106.3	101.3	116.6	107.5	103.0	116.6	108.5	104.7	120.7	111.1	106.4	120.8	112.2	106.4	116.7	110.5
03	CONCRETE	106.5	113.2	109.9	107.6	113.3	110.4	109.3	113.3	111.3	109.3	112.4	110.8	110.1	112.5	111.3	110.2	113.3	111.7
04	MASONRY	96.3	117.6	109.4	96.5	117.6	109.4	96.6	117.6	109.5	96.7	119.8	110.8	96.9	119.8	110.9	96.9	117.6	109.6
05	METALS	99.7	112.1	104.2	95.9	112.2	101.8	96.1	112.2	102.0	99.1	112.6	104.1	99.1	112.8	104.1	99.1	112.3	103.9
06	WOOD & PLASTICS	103.0	108.6	105.9	103.0	108.6	105.9	103.0	108.6	105.9	103.0	101.0	101.9	103.0	101.0	101.9	103.0	108.6	105.9
07	THERMAL & MOISTURE PROTECTION	100.0	107.8	103.7	101.3	110.8	105.8	101.5	110.8	105.9	101.3	114.6	107.6	101.3	114.6	107.6	101.3	110.8	105.8
08	DOORS & WINDOWS	108.1	110.7	108.8	108.1	110.7	108.8	108.1	115.1	109.8	108.1	111.0	108.8	108.1	111.0	108.8	108.1	115.1	109.8
09200	PLASTER & GYPSUM BOARD	98.8	107.7	104.5	98.8	107.7	104.5	98.8	107.7	104.5	98.8	99.8	99.4	98.8	99.8	99.4	98.8	107.7	104.5
095,098	CEILINGS & ACOUSTICAL TREATMENT	104.9	107.7	106.7	104.9	107.7	106.7	104.9	107.7	106.7	104.9	99.8	101.4	104.9	99.8	101.4	104.9	107.7	106.7
09600	FLOORING	99.8	106.3	101.4	99.8	106.3	101.4	99.8	106.3	101.4	99.8	106.3	101.4	99.8	106.3	101.4	99.8	106.3	101.4
097,099	WALL FINISHES, PAINTS & COATINGS	95.2	105.2	101.1	95.2	105.2	101.1	95.2	99.5	97.7	95.2	106.0	101.5	95.2	106.0	101.5	95.2	104.1	100.4
09	FINISHES	101.6	108.6	105.3	101.7	108.6	105.3	101.7	108.0	105.0	101.7	104.3	103.0	101.7	104.3	103.0	101.5	108.5	105.2
10-14	TOTAL DIV. 10000-14000	100.0	113.9	102.9	100.0	113.9	102.9	100.0	113.9	102.9	100.0	112.9	102.7	100.0	113.0	102.7	100.0	113.9	102.9
15	MECHANICAL	99.9	104.4	101.9	99.9	104.4	101.9	99.9	104.4	101.9	99.9	104.5	101.9	99.9	104.5	101.9	99.9	104.5	101.9
16	ELECTRICAL	103.2	102.8	102.9	104.0	102.5	103.0	104.0	102.5	102.9	104.0	99.0	100.5	104.0	130.7	122.5	103.2	101.5	102.0
01-16	WEIGHTED AVERAGE	102.1	108.5	105.2	101.7	108.5	105.3	102.0	108.7	105.2	102.4	107.5	104.9	102.6	113.0	107.6	102.5	108.5	105.4

		D.C.			DELAWARE			FLORIDA											
		WASHINGTON			WILMINGTON			DAYTONA BEACH			FORT LAUDERDALE			JACKSONVILLE			MELBOURNE		
DIVISION		MAT.	INST.	TOTAL	MAT.	INST.	TOTAL	MAT.	INST.	TOTAL	MAT.	INST.	TOTAL	MAT.	INST.	TOTAL	MAT.	INST.	TOTAL
01590	EQUIPMENT RENTAL	.0	103.8	103.8	.0	120.0	120.0	.0	98.0	98.0	.0	89.5	89.5	.0	98.0	98.0	.0	98.0	98.0
02	SITE CONSTRUCTION	102.8	90.2	93.2	87.7	113.8	107.5	118.4	86.9	94.5	104.0	73.6	80.9	118.6	87.3	94.8	126.4	87.2	96.6
03100	CONCRETE FORMS & ACCESSORIES	98.6	77.1	79.9	97.5	102.4	101.7	93.8	69.0	72.2	91.6	63.6	67.3	93.5	57.8	62.5	88.0	71.2	73.5
03200	CONCRETE REINFORCEMENT	99.3	89.8	93.8	98.5	94.9	96.4	95.9	84.3	89.2	95.9	72.8	82.4	95.9	56.4	72.9	96.9	84.4	89.6
03300	CAST-IN-PLACE CONCRETE	113.4	85.3	102.1	75.9	92.0	82.4	90.9	72.3	83.4	95.3	68.2	84.4	91.8	57.3	77.9	105.6	75.2	93.3
03	CONCRETE	105.2	84.0	94.6	95.6	98.4	97.0	88.3	74.4	81.4	90.5	68.6	79.6	88.8	59.7	74.6	97.9	76.5	87.2
04	MASONRY	89.3	79.4	83.2	106.3	88.4	95.4	86.4	67.5	74.9	86.7	61.0	71.0	85.9	54.5	66.7	82.9	72.6	76.6
05	METALS	96.9	110.0	101.7	97.7	114.3	103.8	99.1	96.3	98.1	98.5	92.8	96.4	98.7	84.1	93.4	107.1	96.6	103.3
06	WOOD & PLASTICS	93.3	77.3	85.1	99.0	103.8	101.4	91.9	70.2	80.7	87.7	65.3	76.2	91.9	58.4	74.7	86.4	70.2	78.1
07	THERMAL & MOISTURE PROTECTION	95.2	83.2	89.5	98.3	102.4	100.2	95.8	71.8	84.4	95.8	66.8	82.1	96.1	62.3	80.1	96.0	74.9	86.0
08	DOORS & WINDOWS	101.6	87.6	98.2	95.1	104.6	97.4	98.7	69.6	91.7	96.4	64.1	88.6	98.7	55.4	88.3	97.9	72.9	91.9
09200	PLASTER & GYPSUM BOARD	102.7	76.4	85.7	106.6	103.7	104.7	106.9	69.8	82.9	106.4	64.7	79.5	106.9	57.7	75.1	103.6	69.8	81.8
095,098	CEILINGS & ACOUSTICAL TREATMENT	101.9	76.4	84.7	100.5	103.7	102.6	95.7	69.8	78.3	95.7	64.7	74.9	95.7	57.7	70.1	91.5	69.8	76.9
09600	FLOORING	96.4	88.9	94.5	79.6	90.4	82.3	115.4	71.7	104.6	115.4	68.5	103.8	115.4	56.2	100.8	112.1	71.7	102.1
097,099	WALL FINISHES, PAINTS & COATINGS	105.7	92.4	97.9	86.3	98.1	93.2	108.6	54.0	81.9	105.0	56.1	76.3	108.6	54.0	76.6	108.6	94.8	100.5
09	FINISHES	96.7	80.4	88.2	97.2	99.6	98.5	106.7	70.8	88.0	104.7	63.3	83.1	106.7	56.8	80.7	104.9	73.7	88.6
10-14	TOTAL DIV. 10000-14000	100.0	89.0	97.7	100.0	107.5	101.6	100.0	78.8	95.6	100.0	80.7	96.0	100.0	70.7	93.9	100.0	80.8	96.0
15	MECHANICAL	100.1	89.8	95.5	100.2	108.0	103.7	99.9	68.9	86.1	99.9	65.7	84.7	99.9	56.0	80.4	99.9	71.8	87.4
16	ELECTRICAL	97.4	96.8	97.0	99.8	103.5	102.4	96.7	64.3	74.2	96.7	80.8	85.6	96.2	69.8	77.8	96.4	73.3	80.4
01-16	WEIGHTED AVERAGE	99.1	89.5	94.5	98.2	103.9	101.0	98.3	73.5	86.3	97.5	71.5	84.9	98.2	64.8	82.1	100.3	77.0	89.1

COST INDEXES

City Cost Indexes

FLORIDA

DIVISION		MIAMI MAT.	INST.	TOTAL	ORLANDO MAT.	INST.	TOTAL	PANAMA CITY MAT.	INST.	TOTAL	PENSACOLA MAT.	INST.	TOTAL	ST. PETERSBURG MAT.	INST.	TOTAL	TALLAHASSEE MAT.	INST.	TOTAL
01590	EQUIPMENT RENTAL	.0	89.5	89.5	.0	98.0	98.0	.0	98.0	98.0	.0	98.0	98.0	.0	98.0	98.0	.0	98.0	98.0
02	SITE CONSTRUCTION	103.3	73.3	80.6	119.1	86.6	94.4	133.4	84.4	96.2	130.9	86.6	97.3	119.4	86.4	94.3	119.9	86.0	94.2
03100	CONCRETE FORMS & ACCESSORIES	91.4	63.6	67.2	93.6	60.5	64.8	92.5	31.8	39.8	82.9	58.3	61.6	91.1	54.2	59.1	93.5	44.6	51.0
03200	CONCRETE REINFORCEMENT	95.9	72.8	82.5	95.9	80.3	86.8	100.0	55.2	73.9	102.6	55.6	75.2	99.3	63.6	78.5	95.9	55.8	72.5
03300	CAST-IN-PLACE CONCRETE	92.8	68.4	83.0	98.8	72.1	88.0	96.4	38.8	73.2	96.4	61.0	82.2	102.8	62.0	86.3	95.1	53.8	78.5
03	CONCRETE	89.2	68.6	78.9	92.3	69.9	81.2	96.9	40.6	68.9	95.8	60.5	78.2	94.7	60.5	77.7	90.5	52.0	71.3
04	MASONRY	84.2	61.4	70.3	87.0	67.5	75.1	90.6	31.7	54.6	87.8	57.2	69.1	132.3	56.6	86.1	87.4	44.5	61.2
05	METALS	99.0	92.3	96.5	108.3	94.4	103.2	97.3	70.2	87.4	97.3	83.7	92.3	101.1	86.4	95.7	99.0	82.6	93.0
06	WOOD & PLASTICS	87.7	65.3	76.2	91.9	59.2	75.1	90.7	32.1	60.6	81.5	59.2	70.1	89.4	54.4	71.4	91.9	43.1	66.8
07	THERMAL & MOISTURE PROTECTION	98.8	68.3	84.3	96.2	70.7	84.1	96.4	35.1	67.4	96.1	59.0	78.5	95.7	55.5	76.6	96.3	52.0	75.3
08	DOORS & WINDOWS	96.4	64.8	88.8	98.7	61.8	89.8	96.4	30.2	80.5	96.4	56.8	86.9	97.4	52.6	86.6	98.7	46.5	86.1
09200	PLASTER & GYPSUM BOARD	106.4	64.7	79.5	110.1	58.5	76.7	105.3	30.5	56.9	101.3	58.5	73.6	104.8	53.5	71.6	106.9	41.8	64.8
095,098	CEILINGS & ACOUSTICAL TREATMENT	95.7	64.7	74.9	95.7	58.5	70.7	90.1	30.5	50.0	90.1	58.5	68.8	90.1	53.5	65.5	95.7	41.8	59.5
09600	FLOORING	122.7	69.3	109.5	115.4	71.7	104.6	114.8	21.6	91.8	109.0	59.8	96.9	113.8	59.6	100.4	115.4	43.7	97.7
097,099	WALL FINISHES, PAINTS & COATINGS	105.0	56.1	76.3	108.6	62.0	81.3	108.6	27.6	61.2	108.6	62.8	81.7	108.6	52.3	75.6	108.6	44.5	71.0
09	FINISHES	107.1	63.4	84.3	107.2	62.2	83.8	106.4	29.0	66.1	103.8	59.1	80.5	104.9	54.8	78.8	106.9	43.6	73.9
10-14	TOTAL DIV. 10000-14000	100.0	80.7	96.0	100.0	77.3	95.3	100.0	53.4	90.3	100.0	59.7	91.6	100.0	62.5	92.2	100.0	69.8	93.7
15	MECHANICAL	99.9	68.5	86.0	99.9	61.7	82.9	99.9	28.7	68.2	99.9	56.9	80.8	99.9	56.7	80.7	99.9	45.3	75.7
16	ELECTRICAL	97.2	78.5	84.2	97.3	51.3	65.4	94.9	38.4	55.7	100.4	62.1	73.8	97.1	55.5	68.2	97.3	47.5	62.7
01-16	WEIGHTED AVERAGE	97.6	71.7	85.1	100.3	67.6	84.5	99.3	42.3	71.7	98.9	64.0	82.1	101.6	62.2	82.6	98.7	54.4	77.3

FLORIDA / GEORGIA

DIVISION		TAMPA (FLORIDA) MAT.	INST.	TOTAL	ALBANY MAT.	INST.	TOTAL	ATLANTA MAT.	INST.	TOTAL	AUGUSTA MAT.	INST.	TOTAL	COLUMBUS MAT.	INST.	TOTAL	MACON MAT.	INST.	TOTAL
01590	EQUIPMENT RENTAL	.0	98.0	98.0	.0	90.3	90.3	.0	93.1	93.1	.0	92.4	92.4	.0	90.3	90.3	.0	102.8	102.8
02	SITE CONSTRUCTION	119.7	86.4	94.4	104.2	76.2	83.0	107.3	95.2	98.1	103.6	92.5	95.2	104.2	76.5	83.2	105.0	94.3	96.9
03100	CONCRETE FORMS & ACCESSORIES	95.0	54.5	59.8	93.2	48.8	54.7	93.2	81.7	83.2	90.2	58.8	63.0	94.5	62.5	66.7	92.2	62.3	67.0
03200	CONCRETE REINFORCEMENT	95.9	63.6	77.1	95.9	91.0	93.0	103.0	92.7	97.0	108.8	81.0	92.6	95.9	91.2	93.2	98.2	91.3	94.2
03300	CAST-IN-PLACE CONCRETE	100.5	62.1	85.0	97.1	46.6	76.7	97.7	76.6	89.2	92.3	54.7	77.2	96.7	47.3	76.8	95.4	51.0	77.5
03	CONCRETE	93.3	60.7	77.1	91.4	57.8	74.7	93.2	81.9	87.5	89.5	62.1	75.9	91.3	64.2	77.8	90.7	65.7	78.3
04	MASONRY	87.4	56.6	68.6	88.7	36.7	56.9	88.9	73.1	79.2	89.1	46.3	62.9	88.7	40.4	59.2	102.9	44.1	67.0
05	METALS	102.0	86.7	96.4	96.4	91.6	94.6	92.4	81.7	88.5	91.1	73.1	84.5	96.8	92.8	95.3	91.7	93.2	92.2
06	WOOD & PLASTICS	93.2	54.4	73.2	91.3	49.0	69.6	91.3	84.7	87.9	88.6	61.5	74.7	92.8	67.5	79.8	99.0	66.3	82.2
07	THERMAL & MOISTURE PROTECTION	96.1	56.5	77.3	95.9	56.0	77.0	91.9	78.3	85.4	91.3	55.7	74.5	95.5	58.8	78.1	94.4	62.9	79.5
08	DOORS & WINDOWS	98.7	52.9	87.7	96.4	54.6	86.3	98.2	79.9	93.8	92.4	59.3	84.5	96.4	64.7	88.8	95.0	65.0	87.8
09200	PLASTER & GYPSUM BOARD	106.9	53.5	72.4	106.9	48.0	68.8	113.5	84.6	94.8	112.5	60.6	79.0	106.9	67.1	81.2	115.4	65.8	83.3
095,098	CEILINGS & ACOUSTICAL TREATMENT	95.7	53.5	67.3	95.7	48.0	63.6	101.7	84.6	90.2	103.3	60.6	74.6	95.7	67.1	76.5	87.8	65.8	73.0
09600	FLOORING	115.4	59.6	101.6	115.4	36.5	95.9	86.7	82.2	85.6	85.6	46.5	75.9	115.4	42.6	97.4	90.2	42.9	78.5
097,099	WALL FINISHES, PAINTS & COATINGS	108.6	52.3	75.6	105.0	46.5	70.7	93.2	85.4	88.6	93.2	44.2	64.5	105.0	44.5	69.6	106.7	54.4	76.1
09	FINISHES	106.8	54.8	79.7	104.7	45.1	73.7	93.3	82.5	87.6	92.8	55.1	73.1	104.6	56.9	79.8	91.1	58.3	74.0
10-14	TOTAL DIV. 10000-14000	100.0	75.1	94.8	100.0	72.9	94.4	100.0	82.4	96.3	100.0	67.6	93.3	100.0	75.2	94.8	100.0	76.7	95.1
15	MECHANICAL	99.9	57.0	80.9	99.9	51.3	78.3	100.1	80.3	91.3	100.1	48.6	77.2	99.9	42.0	74.2	99.9	47.3	76.5
16	ELECTRICAL	96.2	55.5	67.9	90.7	63.8	72.0	94.0	86.7	88.9	96.1	52.3	65.7	92.0	45.1	59.4	90.1	68.7	75.3
01-16	WEIGHTED AVERAGE	99.4	62.7	81.7	97.1	59.0	78.7	96.0	82.8	89.6	94.6	59.3	77.5	97.2	57.5	78.0	95.6	65.2	80.9

GEORGIA / HAWAII / IDAHO

DIVISION		SAVANNAH (GEORGIA) MAT.	INST.	TOTAL	VALDOSTA MAT.	INST.	TOTAL	HONOLULU (HAWAII) MAT.	INST.	TOTAL	BOISE (IDAHO) MAT.	INST.	TOTAL	LEWISTON MAT.	INST.	TOTAL	POCATELLO MAT.	INST.	TOTAL
01590	EQUIPMENT RENTAL	.0	91.4	91.4	.0	90.3	90.3	.0	99.5	99.5	.0	100.3	100.3	.0	93.2	93.2	.0	100.3	100.3
02	SITE CONSTRUCTION	104.8	77.4	84.0	114.7	76.7	85.9	130.7	108.2	113.6	82.7	101.4	96.9	87.8	95.2	93.4	83.7	101.4	97.1
03100	CONCRETE FORMS & ACCESSORIES	93.0	59.0	63.5	79.8	50.9	54.7	105.7	153.6	147.3	99.6	86.0	87.8	111.7	75.3	80.1	99.5	85.6	87.5
03200	CONCRETE REINFORCEMENT	101.6	81.3	89.8	101.7	48.5	70.7	106.5	124.7	117.1	100.3	79.6	88.3	113.4	97.0	103.8	100.7	79.2	88.1
03300	CAST-IN-PLACE CONCRETE	93.7	55.7	78.4	95.0	56.4	79.4	193.7	133.2	169.3	101.6	95.5	99.2	110.2	91.6	102.7	100.8	95.3	98.6
03	CONCRETE	90.3	63.6	77.0	95.4	54.2	74.9	157.3	139.3	148.4	104.5	87.9	96.3	116.5	85.3	100.9	104.2	87.6	95.9
04	MASONRY	91.9	58.2	71.3	94.9	52.6	69.0	132.5	137.2	135.4	135.9	82.2	103.1	136.3	89.4	107.7	130.9	74.2	96.3
05	METALS	97.1	88.8	94.1	96.6	77.5	89.6	116.3	111.1	114.4	113.1	80.4	101.2	96.7	89.4	94.0	113.4	79.5	101.0
06	WOOD & PLASTICS	91.4	57.9	74.1	78.5	47.8	62.7	97.5	158.6	128.9	94.9	84.8	89.7	103.1	69.5	85.8	94.9	84.8	89.7
07	THERMAL & MOISTURE PROTECTION	96.0	57.8	77.9	95.7	62.0	79.7	113.3	136.1	124.1	98.9	84.4	92.1	171.5	83.7	129.9	99.2	76.4	88.4
08	DOORS & WINDOWS	96.4	56.7	86.8	92.0	43.8	80.4	106.3	145.6	115.8	93.7	79.2	90.2	113.7	73.7	104.1	93.7	73.8	88.9
09200	PLASTER & GYPSUM BOARD	106.9	57.1	74.7	100.3	46.7	65.7	112.9	160.4	143.6	84.9	84.1	84.4	145.1	68.5	95.6	84.9	84.1	84.4
095,098	CEILINGS & ACOUSTICAL TREATMENT	95.7	57.1	69.8	90.1	46.7	60.9	118.1	160.4	146.6	107.2	84.1	91.7	155.4	68.5	97.0	107.2	84.1	91.7
09600	FLOORING	115.4	54.8	100.5	107.3	43.8	91.6	120.6	137.2	124.7	103.6	63.2	93.7	143.1	92.9	130.7	104.0	63.2	93.9
097,099	WALL FINISHES, PAINTS & COATINGS	105.0	55.2	75.8	105.0	40.3	67.1	113.2	157.1	138.9	108.1	58.2	78.9	135.4	77.9	101.7	108.1	61.0	80.5
09	FINISHES	104.9	57.7	80.3	100.8	47.8	73.2	119.4	152.6	136.7	98.6	78.9	88.3	166.7	77.8	120.4	98.7	79.2	88.5
10-14	TOTAL DIV. 10000-14000	100.0	69.5	93.6	100.0	68.1	93.4	100.0	128.8	106.0	100.0	90.6	98.0	100.0	90.2	98.0	100.0	90.6	98.0
15	MECHANICAL	99.9	52.4	78.8	99.9	45.9	75.9	100.2	124.8	111.2	100.0	82.6	92.3	101.2	91.0	96.7	100.0	82.6	92.2
16	ELECTRICAL	93.4	60.6	70.6	89.0	34.8	51.3	114.3	128.9	124.4	81.4	79.8	80.3	83.6	90.4	88.3	84.5	78.9	80.6
01-16	WEIGHTED AVERAGE	97.5	63.0	80.8	97.1	52.7	75.6	116.3	130.7	123.2	101.7	84.1	93.2	112.9	87.5	100.7	101.7	82.6	92.5

COST INDEXES

DIVISION		IDAHO			ILLINOIS													
		TWIN FALLS			CHICAGO			DECATUR			EAST ST. LOUIS			JOLIET			PEORIA		
		MAT.	INST.	TOTAL	MAT.	INST.	TOTAL	MAT.	INST.	TOTAL	MAT.	INST.	TOTAL	MAT.	INST.	TOTAL	MAT.	INST.	TOTAL
01590	EQUIPMENT RENTAL	.0	100.3	100.3	.0	93.8	93.8	.0	103.0	103.0	.0	109.7	109.7	.0	91.5	91.5	.0	102.3	102.3
02	SITE CONSTRUCTION	90.8	99.4	97.3	89.5	91.0	90.6	83.7	95.4	92.6	100.6	96.1	97.2	89.5	90.2	90.0	91.5	94.7	93.9
03100	CONCRETE FORMS & ACCESSORIES	102.7	40.3	48.5	103.0	131.9	128.1	97.8	108.1	106.7	90.8	111.9	109.1	104.0	132.5	128.7	95.8	113.3	111.0
03200	CONCRETE REINFORCEMENT	109.4	54.2	77.2	95.0	162.3	134.2	96.5	108.8	103.7	102.1	116.9	110.8	95.0	142.9	122.9	96.5	116.7	108.3
03300	CAST-IN-PLACE CONCRETE	103.5	50.0	81.9	113.3	130.3	120.1	99.6	91.7	96.4	93.2	114.7	101.9	113.2	125.8	118.3	97.3	112.1	103.3
03	CONCRETE	113.4	47.4	80.5	107.4	136.0	121.7	95.6	102.7	99.1	86.8	114.6	100.7	107.4	131.0	119.2	94.3	113.6	103.9
04	MASONRY	134.3	44.8	79.7	94.5	132.7	117.8	69.2	98.9	87.3	73.1	118.2	100.7	95.9	123.3	112.6	113.6	115.6	114.8
05	METALS	113.3	66.9	96.3	96.2	125.9	107.1	95.5	108.0	100.1	93.0	124.1	104.3	94.3	115.6	102.1	95.5	113.3	102.0
06	WOOD & PLASTICS	97.9	40.3	68.3	102.3	131.2	117.1	99.9	108.8	104.5	96.1	110.4	103.4	104.1	133.8	119.4	100.0	111.2	105.8
07	THERMAL & MOISTURE PROTECTION	100.3	52.2	77.5	99.9	128.4	113.4	97.6	97.2	97.4	92.4	108.4	100.0	99.7	124.2	111.3	97.4	107.9	102.3
08	DOORS & WINDOWS	96.9	40.7	83.4	104.0	138.7	112.3	96.7	104.2	98.5	86.3	117.7	93.9	101.8	135.4	109.9	96.7	109.0	99.7
09200	PLASTER & GYPSUM BOARD	84.7	38.3	54.7	99.7	131.8	120.4	106.5	108.7	107.9	103.4	110.4	107.9	97.2	134.6	121.4	106.5	111.2	109.6
095,098	CEILINGS & ACOUSTICAL TREATMENT	101.5	38.3	59.0	95.7	131.8	119.9	98.5	108.7	105.4	92.9	110.4	104.7	95.7	134.6	121.8	98.5	111.2	107.1
09600	FLOORING	105.3	63.2	94.9	84.9	130.1	96.0	100.9	93.6	99.1	110.5	59.6	97.9	84.4	118.2	92.8	100.9	98.8	100.4
097,099	WALL FINISHES, PAINTS & COATINGS	108.1	35.9	65.8	79.9	134.6	112.0	90.7	106.2	99.8	98.8	102.8	101.1	77.7	131.3	109.1	90.7	92.8	91.9
09	FINISHES	98.7	43.8	70.1	88.5	131.8	111.1	97.5	106.0	101.9	97.4	100.7	99.1	87.9	130.3	110.0	97.5	108.4	103.2
10 - 14	TOTAL DIV. 10000 - 14000	100.0	55.9	90.8	100.0	122.7	104.7	100.0	101.1	100.2	100.0	108.8	101.8	100.0	122.0	104.6	100.0	104.4	100.9
15	MECHANICAL	100.0	41.3	73.9	100.0	120.9	109.3	100.2	99.4	99.8	100.1	100.5	100.3	100.2	122.0	109.9	100.2	106.0	102.8
16	ELECTRICAL	87.0	36.3	51.8	101.1	131.9	122.5	100.6	93.4	95.6	96.9	105.1	102.6	100.7	121.8	115.4	99.5	98.2	98.6
01 - 16	WEIGHTED AVERAGE	103.9	50.4	78.0	99.1	125.8	112.0	95.9	100.2	98.0	93.5	107.8	100.4	98.6	121.3	109.6	98.4	106.6	102.4

DIVISION		ILLINOIS						INDIANA											
		ROCKFORD			SPRINGFIELD			ANDERSON			BLOOMINGTON			EVANSVILLE			FORT WAYNE		
		MAT.	INST.	TOTAL	MAT.	INST.	TOTAL	MAT.	INST.	TOTAL	MAT.	INST.	TOTAL	MAT.	INST.	TOTAL	MAT.	INST.	TOTAL
01590	EQUIPMENT RENTAL	.0	102.3	102.3	.0	103.0	103.0	.0	96.6	96.6	.0	86.5	86.5	.0	122.2	122.2	.0	96.6	96.6
02	SITE CONSTRUCTION	91.7	94.6	93.9	90.0	95.4	94.1	89.1	95.4	93.9	78.2	96.1	91.8	83.6	130.8	119.4	90.0	95.5	94.2
03100	CONCRETE FORMS & ACCESSORIES	100.8	110.5	109.2	100.2	108.1	107.0	91.5	85.4	86.2	104.9	84.3	87.0	92.9	85.4	86.4	90.9	82.3	83.4
03200	CONCRETE REINFORCEMENT	96.5	141.2	122.5	96.5	108.9	103.7	96.8	87.2	91.2	88.9	87.1	87.8	97.3	84.1	89.6	96.8	81.6	87.9
03300	CAST-IN-PLACE CONCRETE	99.6	104.8	101.7	92.5	99.0	95.1	104.9	86.1	97.3	103.1	82.5	94.8	98.4	96.3	97.6	111.4	82.6	99.8
03	CONCRETE	95.8	114.5	105.1	92.1	105.2	98.7	98.4	86.5	92.5	105.5	83.8	94.7	108.3	89.0	98.7	101.7	82.8	92.3
04	MASONRY	84.2	115.0	103.0	67.9	98.0	86.3	92.8	85.4	88.3	92.6	83.2	86.9	88.6	87.4	87.9	89.9	88.2	88.9
05	METALS	95.5	124.2	106.0	95.5	108.5	100.2	102.0	93.4	98.9	97.4	78.7	90.6	90.7	89.4	90.2	102.0	91.3	98.1
06	WOOD & PLASTICS	99.9	107.8	104.0	99.8	108.8	104.4	104.8	85.4	94.8	121.6	83.4	102.0	96.6	83.2	89.7	104.3	80.9	92.3
07	THERMAL & MOISTURE PROTECTION	97.2	113.5	105.0	97.1	103.2	100.0	101.0	82.9	92.5	92.1	87.0	89.7	95.4	89.2	92.5	100.4	88.8	94.9
08	DOORS & WINDOWS	96.7	115.3	101.2	96.7	104.2	98.5	100.5	86.7	97.2	103.6	85.6	99.3	96.1	82.5	92.8	100.5	80.5	95.7
09200	PLASTER & GYPSUM BOARD	106.5	107.6	107.2	106.5	108.7	107.9	106.6	85.5	93.0	104.1	83.7	90.9	100.2	81.6	88.2	101.7	80.9	88.3
095,098	CEILINGS & ACOUSTICAL TREATMENT	98.5	107.6	104.7	98.5	108.7	105.4	94.3	85.5	88.4	82.5	83.7	83.3	94.3	81.6	85.8	94.3	80.9	85.3
09600	FLOORING	100.9	103.2	101.5	101.1	93.7	99.3	86.9	93.3	88.5	102.3	90.8	99.5	96.0	90.9	94.8	86.9	80.7	85.4
097,099	WALL FINISHES, PAINTS & COATINGS	90.7	107.9	100.8	90.7	102.6	97.6	91.2	76.0	82.3	90.5	93.5	92.3	97.4	92.0	94.3	91.2	79.9	84.6
09	FINISHES	97.5	108.2	103.1	97.6	105.3	101.6	91.6	85.9	88.6	96.5	86.4	91.2	95.9	86.9	91.2	90.7	81.4	85.9
10 - 14	TOTAL DIV. 10000 - 14000	100.0	110.8	102.3	100.0	100.7	100.1	100.0	89.9	97.9	100.0	89.3	97.8	100.0	92.8	98.5	100.0	94.0	98.7
15	MECHANICAL	100.2	108.4	103.8	100.2	95.5	98.1	99.9	85.7	93.6	99.8	88.7	94.9	100.0	90.4	95.7	99.9	86.3	93.8
16	ELECTRICAL	99.5	114.8	110.1	100.5	94.9	96.6	81.3	95.1	90.9	101.0	90.1	93.4	95.7	86.6	89.4	81.8	89.0	86.8
01 - 16	WEIGHTED AVERAGE	97.0	111.5	104.0	95.6	100.1	97.8	97.3	89.1	93.4	99.3	87.1	93.4	97.2	92.3	94.8	97.5	87.2	92.5

DIVISION		INDIANA															IOWA		
		GARY			INDIANAPOLIS			MUNCIE			SOUTH BEND			TERRE HAUTE			CEDAR RAPIDS		
		MAT.	INST.	TOTAL	MAT.	INST.	TOTAL	MAT.	INST.	TOTAL	MAT.	INST.	TOTAL	MAT.	INST.	TOTAL	MAT.	INST.	TOTAL
01590	EQUIPMENT RENTAL	.0	96.6	96.6	.0	91.3	91.3	.0	96.7	96.7	.0	106.8	106.8	.0	122.2	122.2	.0	92.4	92.4
02	SITE CONSTRUCTION	89.8	97.1	95.4	89.2	99.0	96.6	78.0	96.3	91.9	89.2	96.2	94.5	84.7	130.8	119.7	85.8	94.2	92.2
03100	CONCRETE FORMS & ACCESSORIES	91.9	107.5	105.5	91.9	90.7	90.8	90.4	85.2	85.9	96.9	85.3	86.8	94.4	88.0	88.8	102.7	83.1	85.7
03200	CONCRETE REINFORCEMENT	96.8	105.8	102.0	96.2	86.9	90.8	98.3	87.2	91.8	96.8	78.6	86.2	97.3	83.4	89.2	100.2	79.7	88.3
03300	CAST-IN-PLACE CONCRETE	109.5	109.9	109.7	100.9	90.4	96.7	108.3	85.6	99.1	102.0	86.4	95.7	95.3	89.4	92.9	104.7	85.1	96.8
03	CONCRETE	100.8	108.0	104.4	100.6	89.3	95.0	103.5	86.2	94.9	92.8	85.9	89.4	108.0	87.7	97.9	97.7	83.5	90.6
04	MASONRY	94.2	104.0	100.2	99.8	87.3	92.2	94.8	85.4	89.0	87.2	83.5	85.0	95.7	86.9	90.3	107.9	84.1	93.4
05	METALS	102.0	106.4	103.6	104.0	81.1	95.6	99.0	93.5	97.0	102.0	104.9	103.1	91.4	89.5	90.7	89.5	86.4	88.3
06	WOOD & PLASTICS	103.9	107.5	105.7	101.8	90.9	96.2	110.0	85.2	97.2	105.1	85.2	94.8	98.9	87.8	93.2	108.5	82.4	95.1
07	THERMAL & MOISTURE PROTECTION	100.1	103.6	101.8	99.8	89.1	94.7	93.7	83.2	88.7	99.9	88.2	94.3	95.5	86.8	91.4	99.0	83.7	91.7
08	DOORS & WINDOWS	100.5	105.4	101.7	108.4	89.5	103.8	98.5	86.6	95.6	93.7	82.8	91.1	96.7	86.9	94.3	100.5	82.3	96.1
09200	PLASTER & GYPSUM BOARD	102.7	108.3	106.3	102.6	90.9	95.1	99.0	85.5	90.3	106.1	85.3	92.6	100.2	86.4	91.2	103.6	82.4	89.9
095,098	CEILINGS & ACOUSTICAL TREATMENT	94.3	108.3	103.7	92.9	90.9	91.6	83.9	85.5	85.0	94.3	85.3	88.2	94.3	86.4	89.0	114.0	82.4	92.7
09600	FLOORING	86.9	111.7	93.0	86.7	96.5	89.1	95.1	93.3	94.7	86.9	76.2	84.3	96.0	93.3	95.4	126.3	65.7	111.3
097,099	WALL FINISHES, PAINTS & COATINGS	91.2	110.2	102.3	91.2	93.5	92.6	90.5	76.0	82.0	91.2	88.3	89.5	97.4	87.3	91.5	111.4	84.1	95.4
09	FINISHES	91.0	108.7	102.0	91.1	92.4	91.7	93.4	85.8	89.4	91.5	83.9	87.5	95.9	89.0	92.3	113.6	79.3	95.7
10 - 14	TOTAL DIV. 10000 - 14000	100.0	108.2	101.7	100.0	92.0	98.3	100.0	89.4	97.8	100.0	87.2	97.3	100.0	94.5	98.9	100.0	87.5	97.4
15	MECHANICAL	99.9	100.5	100.1	99.9	91.2	96.0	99.7	85.6	93.5	99.9	79.5	90.8	100.0	88.3	94.8	100.3	83.8	93.0
16	ELECTRICAL	88.4	112.6	105.2	99.0	95.1	96.3	89.5	86.9	87.7	94.2	84.7	87.6	92.9	90.1	91.0	96.2	80.9	85.5
01 - 16	WEIGHTED AVERAGE	98.1	105.6	101.7	100.3	91.1	95.9	97.6	87.7	92.8	96.4	86.7	91.7	97.6	92.8	95.3	99.5	84.0	92.0

COST INDEXES

IOWA

DIVISION		COUNCIL BLUFFS MAT.	INST.	TOTAL	DAVENPORT MAT.	INST.	TOTAL	DES MOINES MAT.	INST.	TOTAL	DUBUQUE MAT.	INST.	TOTAL	SIOUX CITY MAT.	INST.	TOTAL	WATERLOO MAT.	INST.	TOTAL
01590	EQUIPMENT RENTAL	.0	92.5	92.5	.0	96.7	96.7	.0	98.6	98.6	.0	91.2	91.2	.0	96.7	96.7	.0	96.7	96.7
02	SITE CONSTRUCTION	89.5	91.9	91.3	84.5	99.4	95.8	77.8	101.1	95.5	84.0	92.4	90.3	93.7	96.5	95.8	85.0	95.7	93.1
03100	CONCRETE FORMS & ACCESSORIES	78.4	65.7	67.4	102.3	96.1	96.9	104.4	82.2	85.1	80.3	69.1	70.6	102.7	70.0	74.3	103.3	55.7	62.0
03200	CONCRETE REINFORCEMENT	102.3	77.5	87.9	100.2	95.7	97.6	100.2	79.0	87.8	98.8	79.4	87.5	100.2	68.5	81.7	100.2	79.1	87.9
03300	CAST-IN-PLACE CONCRETE	108.9	74.2	94.9	100.9	105.1	102.5	105.0	85.3	97.1	102.5	67.9	88.6	103.8	62.1	87.0	104.7	52.9	83.8
03	CONCRETE	99.5	71.7	85.6	95.7	99.2	97.5	98.0	83.1	90.5	94.0	71.4	82.7	97.3	67.7	82.5	97.8	60.3	79.1
04	MASONRY	107.7	77.7	89.4	107.1	83.3	92.6	101.7	90.0	94.6	108.9	76.0	88.8	102.1	67.8	81.1	102.1	61.4	77.2
05	METALS	94.8	84.4	91.0	89.5	96.3	92.0	89.4	88.1	88.9	88.0	85.4	87.1	89.5	80.4	86.2	89.5	84.4	87.6
06	WOOD & PLASTICS	84.6	62.4	73.2	108.5	95.8	101.9	110.0	81.1	95.1	86.7	66.9	76.5	108.5	69.5	88.4	109.2	56.5	82.1
07	THERMAL & MOISTURE PROTECTION	98.1	69.6	84.6	98.3	88.2	93.5	99.2	82.2	91.1	98.5	71.4	85.6	98.3	64.2	82.1	98.0	57.6	78.8
08	DOORS & WINDOWS	99.5	65.8	91.4	100.5	94.6	99.1	100.5	79.9	95.5	99.5	74.0	93.4	100.5	67.5	92.5	95.9	65.0	88.5
09200	PLASTER & GYPSUM BOARD	95.3	61.8	73.7	103.6	95.9	98.6	100.1	80.8	87.6	95.8	66.4	76.8	103.6	68.8	81.1	103.6	55.4	72.4
095,098	CEILINGS & ACOUSTICAL TREATMENT	109.8	61.8	77.5	114.0	95.9	101.8	112.6	80.8	91.2	109.8	66.4	80.6	114.0	68.8	83.6	114.0	55.4	74.6
09600	FLOORING	101.0	53.7	89.3	112.0	90.0	106.6	111.8	49.0	96.3	115.5	45.7	98.3	112.6	69.2	101.9	113.7	69.6	102.8
097,099	WALL FINISHES, PAINTS & COATINGS	102.6	56.7	75.7	107.0	99.6	102.7	107.0	81.7	92.2	110.1	80.9	93.0	108.3	69.4	85.5	108.3	41.1	69.0
09	FINISHES	102.6	61.4	81.1	108.5	95.4	101.7	106.9	75.2	90.4	108.0	65.0	85.6	109.8	69.7	88.9	109.2	56.3	81.6
10-14	TOTAL DIV. 10000 - 14000	100.0	82.8	96.4	100.0	91.8	98.3	100.0	88.1	97.5	100.0	83.6	96.6	100.0	84.1	96.7	100.0	78.1	95.4
15	MECHANICAL	100.3	82.8	92.5	100.3	96.5	98.6	100.3	86.3	94.1	100.3	77.5	90.2	100.3	82.9	92.6	100.3	52.8	79.2
16	ELECTRICAL	104.5	85.7	91.5	93.6	85.7	88.1	96.9	87.3	90.2	102.2	80.8	87.3	96.2	79.2	84.4	96.2	66.3	75.4
01-16	WEIGHTED AVERAGE	99.7	78.6	89.5	98.5	93.5	96.0	98.3	86.3	92.5	98.3	77.5	88.3	99.0	77.1	88.4	98.2	65.6	82.4

KANSAS / KENTUCKY

DIVISION		DODGE CITY MAT.	INST.	TOTAL	KANSAS CITY MAT.	INST.	TOTAL	SALINA MAT.	INST.	TOTAL	TOPEKA MAT.	INST.	TOTAL	WICHITA MAT.	INST.	TOTAL	BOWLING GREEN MAT.	INST.	TOTAL
01590	EQUIPMENT RENTAL	.0	104.2	104.2	.0	100.8	100.8	.0	104.2	104.2	.0	102.3	102.3	.0	104.2	104.2	.0	96.8	96.8
02	SITE CONSTRUCTION	107.3	93.0	96.4	87.1	90.4	89.6	96.9	93.3	94.1	89.4	90.6	90.3	90.8	93.5	92.8	70.7	100.8	93.5
03100	CONCRETE FORMS & ACCESSORIES	94.0	49.9	55.7	99.9	90.5	91.7	89.4	52.7	57.5	99.2	51.2	57.5	96.1	62.8	67.2	83.9	86.0	85.7
03200	CONCRETE REINFORCEMENT	104.5	65.4	81.7	99.3	87.2	92.2	103.9	83.9	92.2	96.5	90.6	93.1	96.5	84.9	89.8	88.2	72.0	78.7
03300	CAST-IN-PLACE CONCRETE	115.6	64.7	95.1	90.6	93.5	91.8	100.3	55.1	82.1	91.6	59.9	78.9	87.4	63.4	77.7	89.9	86.2	88.4
03	CONCRETE	112.4	59.7	86.2	93.5	91.5	92.5	99.4	61.1	80.4	91.7	63.4	77.6	89.3	68.6	79.0	96.5	83.5	90.0
04	MASONRY	103.4	49.3	70.3	103.0	90.0	95.1	118.2	47.4	74.9	96.7	59.3	73.9	91.3	70.2	78.4	96.0	84.1	88.7
05	METALS	92.7	84.8	89.9	97.6	99.0	98.1	92.6	92.9	92.7	95.5	97.4	96.2	95.5	94.8	95.3	94.8	80.3	89.5
06	WOOD & PLASTICS	92.8	50.1	70.8	99.6	90.3	94.8	88.9	54.0	70.9	96.7	47.7	71.5	94.6	62.0	77.9	92.6	84.3	88.3
07	THERMAL & MOISTURE PROTECTION	98.5	55.5	78.1	96.4	91.5	94.1	97.9	61.1	80.5	97.7	72.3	85.7	97.4	71.2	85.0	85.8	82.9	84.4
08	DOORS & WINDOWS	96.6	53.6	86.2	95.5	90.7	94.3	96.6	58.4	87.4	96.7	64.3	88.9	96.7	67.5	89.7	95.8	77.5	91.4
09200	PLASTER & GYPSUM BOARD	104.1	48.2	68.0	100.0	89.7	93.3	103.1	52.2	70.2	106.5	45.8	67.2	106.5	60.5	76.8	95.9	84.1	88.2
095,098	CEILINGS & ACOUSTICAL TREATMENT	91.5	48.2	62.4	98.5	89.7	92.6	91.5	52.2	65.1	98.5	45.8	63.1	98.5	60.5	73.0	90.1	84.1	86.0
09600	FLOORING	100.2	57.8	89.8	90.4	61.7	83.3	98.3	42.2	84.4	101.8	53.7	89.9	100.9	82.0	96.3	92.1	70.8	86.8
097,099	WALL FINISHES, PAINTS & COATINGS	90.7	57.0	70.9	98.8	95.7	97.0	90.7	44.9	63.9	90.7	66.5	76.5	90.7	66.8	76.7	97.4	78.4	86.3
09	FINISHES	96.7	51.7	73.3	95.8	85.4	90.4	95.3	49.1	71.3	97.8	51.3	73.6	97.6	66.5	81.4	93.3	82.3	87.6
10-14	TOTAL DIV. 10000 - 14000	100.0	58.5	91.3	100.0	93.7	98.7	100.0	72.8	94.3	100.0	77.4	95.3	100.0	76.8	95.2	100.0	72.6	94.3
15	MECHANICAL	100.2	54.9	80.0	100.0	95.6	98.1	100.2	41.4	74.1	100.2	74.2	88.6	100.2	71.3	87.3	100.0	87.4	94.4
16	ELECTRICAL	95.2	58.2	69.5	104.5	76.4	85.0	94.7	73.8	80.2	101.4	78.5	85.5	99.3	73.8	81.6	93.2	87.4	89.2
01-16	WEIGHTED AVERAGE	99.7	61.6	81.3	97.9	89.5	93.8	98.4	62.8	81.2	97.2	72.4	85.2	96.5	74.9	86.0	95.6	85.7	90.8

KENTUCKY / LOUISIANA

DIVISION		LEXINGTON MAT.	INST.	TOTAL	LOUISVILLE MAT.	INST.	TOTAL	OWENSBORO MAT.	INST.	TOTAL	ALEXANDRIA MAT.	INST.	TOTAL	BATON ROUGE MAT.	INST.	TOTAL	LAKE CHARLES MAT.	INST.	TOTAL
01590	EQUIPMENT RENTAL	.0	104.3	104.3	.0	96.8	96.8	.0	122.2	122.2	.0	85.8	85.8	.0	87.0	87.0	.0	87.0	87.0
02	SITE CONSTRUCTION	78.0	102.7	96.8	67.9	100.4	92.5	82.7	130.6	119.0	104.4	85.2	89.8	119.2	86.8	94.6	121.0	86.5	94.8
03100	CONCRETE FORMS & ACCESSORIES	94.8	74.1	76.8	93.3	84.2	85.4	88.5	73.9	75.9	80.8	49.8	53.9	106.6	54.9	61.7	106.4	55.8	62.5
03200	CONCRETE REINFORCEMENT	97.3	96.3	96.8	97.3	97.3	97.3	88.3	97.3	93.5	97.1	66.5	79.3	98.1	56.2	73.7	98.1	56.1	73.6
03300	CAST-IN-PLACE CONCRETE	97.1	74.6	88.1	94.0	82.8	89.5	92.7	86.2	90.1	95.4	48.0	76.3	86.3	57.7	74.8	91.4	59.3	78.5
03	CONCRETE	98.4	78.9	88.7	96.7	86.3	91.5	106.2	83.0	94.6	89.4	53.5	71.5	94.4	57.0	75.8	97.0	57.9	77.6
04	MASONRY	92.8	55.2	69.8	93.1	83.2	87.1	93.0	59.8	72.7	114.0	49.5	74.6	101.4	59.1	75.6	100.2	59.5	75.4
05	METALS	96.3	90.3	94.1	96.3	90.7	94.2	86.2	91.3	88.1	86.6	75.6	82.6	101.1	70.6	90.0	100.6	70.6	89.7
06	WOOD & PLASTICS	98.9	77.1	87.7	101.7	84.3	92.8	92.6	73.4	82.7	85.5	50.3	67.4	116.7	55.4	85.2	114.4	56.3	84.5
07	THERMAL & MOISTURE PROTECTION	95.9	67.7	82.5	86.0	83.2	84.6	95.4	74.1	85.3	99.2	54.1	77.8	98.0	60.7	80.3	100.9	60.6	81.8
08	DOORS & WINDOWS	96.7	83.4	93.5	96.7	87.2	94.4	94.1	81.5	91.0	98.7	55.6	88.3	103.1	55.0	91.5	103.1	57.2	92.0
09200	PLASTER & GYPSUM BOARD	100.2	75.3	84.1	103.4	84.1	90.9	95.6	71.5	80.0	90.3	49.7	64.1	104.7	54.9	72.5	104.7	55.7	73.1
095,098	CEILINGS & ACOUSTICAL TREATMENT	94.3	75.3	81.5	94.3	84.1	87.4	81.6	71.5	74.8	100.5	49.7	66.3	100.1	54.9	69.7	100.1	55.7	70.3
09600	FLOORING	96.9	49.2	85.1	95.4	75.4	90.5	94.3	70.8	88.5	107.0	56.4	94.5	110.2	68.0	99.8	110.4	59.4	97.8
097,099	WALL FINISHES, PAINTS & COATINGS	97.4	67.1	79.7	97.4	80.9	87.7	97.4	104.1	101.3	95.8	47.3	67.4	97.4	54.2	72.1	97.4	50.0	69.6
09	FINISHES	96.3	68.4	81.8	95.9	81.9	88.6	92.5	76.1	84.0	98.3	50.3	73.3	104.2	57.3	79.7	104.2	55.8	79.0
10-14	TOTAL DIV. 10000 - 14000	100.0	91.9	98.3	100.0	93.6	98.7	100.0	93.3	98.6	100.0	82.3	96.3	100.0	71.2	94.0	100.0	71.5	94.0
15	MECHANICAL	100.0	55.1	80.1	100.0	84.2	93.0	100.0	60.6	82.5	100.1	43.5	74.9	100.2	49.6	77.7	100.2	61.0	82.8
16	ELECTRICAL	94.0	58.5	69.3	94.0	87.4	89.4	92.4	86.0	87.9	95.5	58.2	69.6	96.4	58.1	69.8	96.4	61.0	71.8
01-16	WEIGHTED AVERAGE	96.9	70.8	84.3	96.1	87.2	91.8	95.6	81.3	88.7	96.9	57.5	77.8	100.8	60.5	81.3	101.1	63.2	82.8

LOUISIANA / MAINE

DIVISION		MONROE MAT.	INST.	TOTAL	NEW ORLEANS MAT.	INST.	TOTAL	SHREVEPORT MAT.	INST.	TOTAL	AUGUSTA MAT.	INST.	TOTAL	BANGOR MAT.	INST.	TOTAL	LEWISTON MAT.	INST.	TOTAL
01590	EQUIPMENT RENTAL	.0	85.8	85.8	.0	88.2	88.2	.0	85.8	85.8	.0	103.3	103.3	.0	103.3	103.3	.0	103.3	103.3
02	SITE CONSTRUCTION	104.4	85.0	89.7	123.7	88.9	97.3	103.0	84.9	89.3	86.5	103.4	99.3	86.4	103.0	99.0	84.7	103.0	98.6
03100	CONCRETE FORMS & ACCESSORIES	80.0	50.6	54.5	105.0	62.9	68.4	102.8	53.7	60.1	103.0	73.9	77.7	96.5	91.8	92.4	103.2	91.8	93.3
03200	CONCRETE REINFORCEMENT	96.1	66.6	78.9	98.1	67.7	80.4	95.6	66.4	78.6	91.1	116.1	105.7	91.1	116.4	105.9	112.0	116.4	114.6
03300	CAST-IN-PLACE CONCRETE	95.4	52.7	78.2	90.5	61.3	78.8	94.0	56.9	79.0	82.5	70.8	77.8	82.5	73.1	78.7	91.9	73.1	84.3
03	CONCRETE	89.2	55.5	72.4	96.4	64.0	80.3	89.3	58.2	73.8	100.1	80.8	90.5	99.2	89.3	94.3	102.7	89.3	96.1
04	MASONRY	108.6	51.3	73.6	102.3	61.1	77.1	102.0	46.3	68.0	93.2	60.2	73.0	110.4	69.2	85.2	94.2	69.2	78.9
05	METALS	86.6	75.2	82.4	103.4	77.2	93.8	87.5	74.6	82.8	95.1	88.4	92.7	95.0	87.0	92.1	98.2	87.0	94.1
06	WOOD & PLASTICS	84.7	50.8	67.3	110.1	63.9	86.3	106.9	55.3	80.4	101.9	73.0	87.1	95.6	95.2	95.4	101.9	95.2	98.4
07	THERMAL & MOISTURE PROTECTION	99.2	56.4	78.9	101.6	65.3	84.4	98.2	55.6	78.0	101.6	59.9	81.8	101.5	67.5	85.4	101.3	67.5	85.3
08	DOORS & WINDOWS	98.7	60.6	89.5	103.8	67.0	95.0	96.6	58.3	87.4	105.0	68.7	96.2	104.9	82.5	99.5	108.1	82.5	102.0
09200	PLASTER & GYPSUM BOARD	89.9	50.2	64.2	103.8	63.5	77.8	96.0	54.8	69.4	97.5	71.0	80.4	94.8	93.8	94.1	99.3	93.8	95.7
095,098	CEILINGS & ACOUSTICAL TREATMENT	100.5	50.2	66.7	98.5	63.5	75.0	101.9	54.8	70.2	96.4	71.0	79.3	95.0	93.8	94.2	104.9	93.8	97.4
09600	FLOORING	106.6	49.0	92.4	110.9	58.2	97.9	117.9	51.5	101.5	99.8	49.3	87.3	97.4	61.5	88.5	99.8	61.5	90.3
097,099	WALL FINISHES, PAINTS & COATINGS	95.8	48.8	68.3	99.2	63.6	78.4	95.8	47.3	67.4	95.2	40.9	63.4	95.2	37.7	61.5	95.2	37.7	61.5
09	FINISHES	98.1	49.5	72.8	104.4	61.9	82.3	102.9	52.2	76.5	98.8	65.2	81.3	97.4	81.0	88.8	100.2	81.0	90.2
10-14	TOTAL DIV. 10000 - 14000	100.0	70.8	93.9	100.0	73.3	94.4	100.0	73.3	94.4	100.0	74.1	94.6	100.0	85.5	97.0	100.0	85.6	97.0
15	MECHANICAL	100.1	53.7	79.4	100.2	64.6	84.3	100.1	55.9	80.5	99.9	62.2	83.1	99.9	82.9	92.3	99.9	82.9	92.3
16	ELECTRICAL	98.6	58.1	70.5	96.7	66.8	75.9	96.6	76.7	82.8	104.0	88.8	93.4	100.6	87.2	91.3	104.0	90.1	94.3
01-16	WEIGHTED AVERAGE	96.7	59.7	78.8	101.8	68.0	85.5	96.7	63.4	80.6	99.3	76.3	88.2	99.7	85.0	92.6	100.6	85.5	93.3

MAINE / MARYLAND / MASSACHUSETTS

DIVISION		PORTLAND MAT.	INST.	TOTAL	BALTIMORE MAT.	INST.	TOTAL	HAGERSTOWN MAT.	INST.	TOTAL	BOSTON MAT.	INST.	TOTAL	BROCKTON MAT.	INST.	TOTAL	FALL RIVER MAT.	INST.	TOTAL
01590	EQUIPMENT RENTAL	.0	103.3	103.3	.0	101.9	101.9	.0	97.9	97.9	.0	110.2	110.2	.0	105.5	105.5	.0	106.7	106.7
02	SITE CONSTRUCTION	84.1	103.0	98.5	99.5	92.2	93.6	87.6	88.0	87.9	91.9	110.3	105.9	89.5	106.2	102.2	88.5	106.3	102.0
03100	CONCRETE FORMS & ACCESSORIES	102.3	91.8	93.2	99.5	76.6	79.6	87.5	78.0	79.2	109.3	133.1	130.0	108.9	116.7	115.7	108.9	111.5	111.2
03200	CONCRETE REINFORCEMENT	112.0	116.4	114.6	97.3	91.1	93.7	86.0	77.5	81.0	110.8	139.5	127.5	112.0	139.1	127.8	112.0	125.1	119.6
03300	CAST-IN-PLACE CONCRETE	85.5	73.1	80.5	94.9	80.9	89.2	80.7	65.2	74.5	106.4	143.6	121.4	101.3	134.1	114.5	98.0	140.3	115.1
03	CONCRETE	99.4	89.3	94.4	98.8	82.1	90.5	84.4	74.7	79.5	113.3	137.1	125.2	110.4	126.4	118.3	108.7	123.6	116.1
04	MASONRY	92.2	69.2	78.1	88.1	75.9	80.7	92.2	77.8	83.4	113.3	144.3	132.2	108.9	133.6	124.0	108.7	139.8	127.7
05	METALS	98.3	87.0	94.2	96.0	101.7	98.1	95.0	94.4	94.8	101.0	123.9	109.4	98.1	120.4	106.2	98.1	115.4	104.4
06	WOOD & PLASTICS	101.9	95.2	98.4	101.3	77.8	89.2	89.7	77.7	83.5	106.0	132.8	119.8	104.9	113.7	109.4	104.9	106.9	106.0
07	THERMAL & MOISTURE PROTECTION	101.2	67.5	85.2	95.6	80.2	88.3	94.8	75.8	85.8	101.9	137.9	119.0	101.7	129.5	114.9	101.6	125.0	112.7
08	DOORS & WINDOWS	108.1	82.5	102.0	93.8	87.2	92.2	91.0	79.2	88.2	104.0	131.2	110.5	101.9	121.0	106.5	101.9	111.9	104.3
09200	PLASTER & GYPSUM BOARD	99.3	93.8	95.7	104.7	77.8	87.3	99.8	77.8	85.5	96.0	132.7	119.7	96.0	112.9	107.0	96.0	105.5	102.2
095,098	CEILINGS & ACOUSTICAL TREATMENT	104.9	93.8	97.4	90.1	77.8	81.8	91.7	77.8	82.3	103.3	132.7	123.1	106.4	112.9	110.8	106.4	105.5	105.8
09600	FLOORING	99.8	61.5	90.3	91.1	79.1	88.1	86.2	79.3	84.5	100.4	147.4	112.0	100.9	147.4	112.4	100.7	147.4	112.2
097,099	WALL FINISHES, PAINTS & COATINGS	95.2	37.7	61.5	97.2	77.4	85.6	97.2	46.2	67.3	98.4	137.0	121.0	97.8	124.1	113.2	97.8	124.1	113.2
09	FINISHES	100.2	81.0	90.2	91.5	76.7	83.8	88.9	74.7	81.5	99.7	136.5	118.9	100.4	122.9	112.1	100.3	119.0	110.1
10-14	TOTAL DIV. 10000 - 14000	100.0	85.6	97.0	100.0	84.5	96.8	100.0	86.7	97.2	100.0	126.0	105.4	100.0	121.9	104.6	100.0	122.0	104.6
15	MECHANICAL	99.9	82.9	92.3	99.9	83.3	92.6	99.9	86.1	93.8	99.9	121.2	109.3	99.9	102.9	101.2	99.9	102.8	101.2
16	ELECTRICAL	104.2	90.1	94.4	101.9	91.7	94.8	98.0	82.8	87.4	99.1	123.6	116.1	98.7	98.6	98.7	98.1	98.6	98.5
01-16	WEIGHTED AVERAGE	100.1	85.5	93.0	96.9	85.7	91.5	93.9	82.2	88.3	102.8	128.0	115.0	101.4	114.5	107.7	101.1	113.2	107.0

MASSACHUSETTS

DIVISION		HYANNIS MAT.	INST.	TOTAL	LAWRENCE MAT.	INST.	TOTAL	LOWELL MAT.	INST.	TOTAL	NEW BEDFORD MAT.	INST.	TOTAL	PITTSFIELD MAT.	INST.	TOTAL	SPRINGFIELD MAT.	INST.	TOTAL
01590	EQUIPMENT RENTAL	.0	105.5	105.5	.0	105.5	105.5	.0	103.3	103.3	.0	106.7	106.7	.0	103.3	103.3	.0	103.3	103.3
02	SITE CONSTRUCTION	86.2	106.2	101.4	90.1	106.2	102.3	89.1	106.1	102.0	88.1	106.3	101.9	90.2	105.1	101.5	89.4	105.1	101.3
03100	CONCRETE FORMS & ACCESSORIES	98.3	111.0	109.4	108.8	115.3	114.4	105.0	115.4	114.0	108.9	111.5	111.2	104.9	97.9	98.8	105.1	104.2	104.3
03200	CONCRETE REINFORCEMENT	89.8	121.9	108.5	111.0	125.1	119.2	112.0	124.7	119.4	112.0	125.1	119.6	92.9	110.4	103.1	112.0	117.4	115.2
03300	CAST-IN-PLACE CONCRETE	92.4	139.5	111.4	102.1	139.7	117.3	92.9	139.8	111.8	95.1	140.3	113.3	101.3	112.2	105.7	95.1	114.3	102.9
03	CONCRETE	100.3	122.5	111.4	110.6	125.9	117.8	100.8	124.8	112.8	107.2	123.6	115.4	102.8	104.9	103.8	102.0	109.7	105.8
04	MASONRY	107.9	139.9	127.5	108.1	136.4	125.4	94.3	127.7	114.7	108.4	139.8	127.6	94.9	107.7	102.7	94.5	110.2	104.1
05	METALS	94.7	113.5	101.6	95.7	115.1	102.8	95.7	111.3	101.4	98.1	115.4	104.4	95.5	101.0	97.5	98.0	104.3	100.3
06	WOOD & PLASTICS	95.1	106.5	101.0	104.9	111.6	108.4	104.0	111.6	107.9	104.9	106.9	106.0	104.0	95.6	99.7	104.0	103.2	103.5
07	THERMAL & MOISTURE PROTECTION	101.1	126.1	112.9	101.6	130.7	115.4	101.4	128.5	114.2	101.6	125.0	112.7	101.4	107.4	104.3	101.4	109.1	105.0
08	DOORS & WINDOWS	98.0	110.8	101.1	101.9	116.0	105.3	108.1	116.0	110.0	101.9	111.9	104.3	108.1	103.0	106.9	108.1	108.9	108.3
09200	PLASTER & GYPSUM BOARD	90.6	105.5	100.2	99.3	110.8	106.7	99.3	110.8	106.7	96.0	105.5	102.2	99.3	94.2	96.0	99.3	102.0	101.1
095,098	CEILINGS & ACOUSTICAL TREATMENT	96.6	105.5	102.6	104.9	110.8	108.8	104.9	110.8	108.8	106.4	105.5	105.8	104.9	94.2	97.7	104.9	102.0	103.0
09600	FLOORING	97.1	147.4	109.5	99.8	147.4	111.5	99.8	147.4	111.5	100.7	147.4	112.2	100.0	113.2	103.3	99.8	116.1	103.8
097,099	WALL FINISHES, PAINTS & COATINGS	97.8	122.8	112.5	95.3	124.1	112.1	95.2	122.8	111.4	97.8	122.8	112.5	95.2	96.2	95.8	97.1	96.2	96.6
09	FINISHES	96.4	118.6	108.0	100.0	121.7	111.3	100.0	121.5	111.2	100.3	118.9	110.0	100.0	100.2	100.1	100.0	105.4	102.8
10-14	TOTAL DIV. 10000 - 14000	100.0	120.9	104.4	100.0	121.6	104.5	100.0	121.6	104.5	100.0	122.0	104.6	100.0	107.6	101.6	100.0	108.9	101.9
15	MECHANICAL	99.9	102.6	101.1	99.9	108.1	103.5	99.9	118.0	107.9	99.9	102.8	101.2	99.9	87.0	94.1	99.9	98.9	99.4
16	ELECTRICAL	93.9	98.6	97.2	103.2	108.1	106.6	104.0	108.1	106.9	99.1	98.6	98.8	104.0	85.4	91.1	104.0	92.0	95.7
01-16	WEIGHTED AVERAGE	98.2	112.8	105.3	101.4	116.4	108.6	100.1	117.0	108.3	101.0	113.2	106.9	100.4	97.4	99.0	100.6	103.1	101.8

COST INDEXES

583

	DIVISION	MASSACHUSETTS			MICHIGAN														
		WORCESTER			ANN ARBOR			DEARBORN			DETROIT			FLINT			GRAND RAPIDS		
		MAT.	INST.	TOTAL	MAT.	INST.	TOTAL	MAT.	INST.	TOTAL	MAT.	INST.	TOTAL	MAT.	INST.	TOTAL	MAT.	INST.	TOTAL
01590	EQUIPMENT RENTAL	.0	103.3	103.3	.0	112.3	112.3	.0	112.3	112.3	.0	97.9	97.9	.0	112.3	112.3	.0	106.6	106.6
02	SITE CONSTRUCTION	89.1	106.1	102.0	81.0	93.9	90.8	80.7	94.0	90.8	97.4	95.8	96.2	69.8	93.1	87.5	83.3	88.0	86.8
03100	CONCRETE FORMS & ACCESSORIES	105.6	114.5	113.4	95.2	115.9	113.2	95.1	122.8	119.1	96.5	122.8	119.4	96.4	97.6	97.4	97.8	77.7	80.4
03200	CONCRETE REINFORCEMENT	112.0	138.0	127.2	93.9	127.1	113.2	93.9	127.5	113.4	93.3	127.4	113.2	93.9	126.7	113.0	97.3	92.3	94.4
03300	CAST-IN-PLACE CONCRETE	92.9	131.0	108.3	95.3	122.9	106.4	93.1	123.3	105.3	102.2	123.3	110.7	95.9	105.3	99.7	97.3	87.6	93.3
03	CONCRETE	100.9	124.0	112.4	91.3	120.8	106.0	90.2	124.0	107.0	94.8	122.5	108.6	91.7	106.6	99.1	98.7	83.4	91.1
04	MASONRY	94.3	130.5	116.4	101.1	117.1	110.9	100.9	123.7	114.8	99.9	123.7	114.4	101.1	105.1	103.5	97.3	58.0	73.3
05	METALS	98.1	116.0	104.6	100.1	126.3	109.7	100.2	127.3	110.1	100.7	103.8	101.8	100.2	124.1	108.9	97.3	85.0	92.8
06	WOOD & PLASTICS	104.5	111.6	108.1	95.4	115.4	105.7	95.4	123.2	109.7	96.3	123.2	110.1	95.4	96.4	95.9	100.1	78.5	89.0
07	THERMAL & MOISTURE PROTECTION	101.4	122.5	111.4	94.9	114.1	104.0	94.0	123.5	108.0	92.4	123.5	107.1	93.2	96.9	94.9	91.9	64.7	79.0
08	DOORS & WINDOWS	108.1	119.6	110.9	96.4	112.6	100.3	96.4	116.7	101.3	98.0	117.6	102.7	96.4	100.9	97.5	91.4	73.0	87.0
09200	PLASTER & GYPSUM BOARD	99.3	110.8	106.7	103.2	114.2	110.3	103.2	122.1	115.4	103.2	122.1	115.4	103.2	94.5	97.6	100.2	71.9	81.9
095,098	CEILINGS & ACOUSTICAL TREATMENT	104.9	110.8	108.8	100.5	114.2	109.7	100.5	122.1	115.0	102.0	122.1	115.6	100.5	94.5	96.5	94.3	71.9	79.2
09600	FLOORING	99.8	139.3	109.5	90.9	118.7	97.8	90.5	120.5	97.9	90.6	120.5	98.0	90.7	84.5	89.2	95.6	49.9	84.3
097,099	WALL FINISHES, PAINTS & COATINGS	95.2	116.8	107.8	92.2	114.4	105.2	92.2	114.4	105.2	94.1	114.4	106.0	92.2	92.3	92.3	97.4	48.8	68.9
09	FINISHES	100.0	119.2	110.0	96.4	116.1	106.6	96.2	121.9	109.6	97.4	121.9	110.1	95.7	94.0	94.8	95.8	68.9	81.8
10-14	TOTAL DIV. 10000 - 14000	100.0	112.8	102.7	100.0	116.3	103.4	100.0	117.9	103.7	100.0	117.9	103.7	100.0	105.8	101.2	100.0	109.0	101.9
15	MECHANICAL	99.9	105.1	102.2	100.0	102.1	100.9	100.0	114.5	106.4	100.0	116.2	107.2	100.0	103.0	101.3	100.0	61.4	82.8
16	ELECTRICAL	104.0	97.9	99.8	95.2	100.8	99.1	95.2	113.0	107.6	96.5	113.0	107.9	95.2	101.9	99.9	93.5	70.3	77.3
01-16	WEIGHTED AVERAGE	100.5	112.9	106.5	97.1	110.1	103.4	96.9	116.9	106.6	98.4	115.1	106.5	96.8	103.1	99.8	96.7	73.0	85.2

	DIVISION	MICHIGAN												MINNESOTA					
		KALAMAZOO			LANSING			MUSKEGON			SAGINAW			DULUTH			MINNEAPOLIS		
		MAT.	INST.	TOTAL	MAT.	INST.	TOTAL	MAT.	INST.	TOTAL	MAT.	INST.	TOTAL	MAT.	INST.	TOTAL	MAT.	INST.	TOTAL
01590	EQUIPMENT RENTAL	.0	106.7	106.7	.0	112.3	112.3	.0	106.6	106.6	.0	112.3	112.3	.0	97.3	97.3	.0	101.6	101.6
02	SITE CONSTRUCTION	84.0	88.4	87.3	88.1	93.1	91.9	82.1	88.4	86.9	72.1	93.0	87.9	82.8	103.1	98.2	82.5	109.1	102.7
03100	CONCRETE FORMS & ACCESSORIES	96.9	93.9	94.3	98.5	96.4	96.7	97.8	91.1	92.0	95.2	95.1	95.1	97.7	118.7	115.9	98.4	129.6	125.5
03200	CONCRETE REINFORCEMENT	97.3	91.0	93.6	93.9	126.4	112.8	97.9	93.5	95.4	93.9	126.3	112.8	95.6	105.9	101.6	95.8	117.0	108.1
03300	CAST-IN-PLACE CONCRETE	98.9	109.1	103.0	95.3	103.6	98.6	96.7	109.0	101.6	94.1	103.4	97.8	106.6	107.3	106.9	105.0	118.6	110.5
03	CONCRETE	102.0	97.7	99.9	91.5	105.5	98.5	96.6	96.9	96.8	90.7	104.8	97.7	95.0	112.5	103.7	95.7	123.3	109.5
04	MASONRY	100.0	95.5	97.2	94.8	99.0	97.3	97.1	83.8	89.0	102.6	97.1	99.2	109.4	118.6	115.0	109.5	127.5	120.5
05	METALS	97.5	87.9	94.0	99.4	123.2	108.1	95.5	88.1	92.8	100.2	123.0	108.5	96.7	116.8	104.0	98.7	123.9	107.9
06	WOOD & PLASTICS	98.9	93.9	96.3	98.2	95.5	96.8	95.9	90.7	93.2	95.4	94.4	94.9	108.2	118.9	113.7	108.6	129.2	119.2
07	THERMAL & MOISTURE PROTECTION	91.5	90.2	90.9	94.6	95.8	95.2	90.7	79.6	85.4	93.7	94.1	93.9	99.0	118.7	108.3	98.8	129.4	113.3
08	DOORS & WINDOWS	91.4	88.0	90.6	96.4	100.4	97.3	90.6	88.6	90.1	96.4	99.9	97.2	98.7	120.6	104.0	101.9	134.4	109.7
09200	PLASTER & GYPSUM BOARD	100.2	87.7	92.1	106.4	93.6	98.1	90.2	84.5	86.5	103.2	92.5	96.3	104.8	120.3	114.8	104.8	130.7	121.5
095,098	CEILINGS & ACOUSTICAL TREATMENT	94.3	87.7	89.9	100.5	93.6	95.8	100.0	84.5	89.6	100.5	92.5	95.1	89.9	120.3	110.4	89.9	130.7	117.4
09600	FLOORING	95.6	57.8	86.3	90.9	73.9	86.7	94.8	72.0	89.1	90.9	67.2	85.1	111.8	113.9	112.3	109.0	113.9	110.2
097,099	WALL FINISHES, PAINTS & COATINGS	97.4	84.3	89.7	92.2	103.1	98.6	96.6	70.7	81.5	92.2	89.1	90.4	99.4	111.2	106.3	106.1	121.7	115.2
09	FINISHES	95.8	85.6	90.5	97.3	92.3	94.7	94.6	84.9	89.6	96.1	88.4	92.1	102.8	117.7	110.6	102.3	126.6	115.0
10-14	TOTAL DIV. 10000 - 14000	100.0	112.1	102.5	100.0	105.5	101.2	100.0	113.4	102.8	100.0	104.9	101.0	100.0	104.8	101.0	100.0	109.1	101.9
15	MECHANICAL	100.0	92.9	96.9	100.0	91.7	96.3	99.9	95.3	97.8	100.0	96.0	98.2	99.9	103.4	101.4	99.8	112.7	105.5
16	ELECTRICAL	93.1	82.3	85.6	93.9	97.5	96.4	93.4	77.2	82.2	97.6	93.3	94.6	102.2	104.3	103.7	103.4	96.6	98.7
01-16	WEIGHTED AVERAGE	97.2	90.4	93.9	96.9	99.1	98.0	95.8	88.5	92.2	97.0	98.3	97.7	99.3	110.4	104.6	100.0	116.5	108.0

	DIVISION	MINNESOTA									MISSISSIPPI								
		ROCHESTER			SAINT PAUL			ST. CLOUD			BILOXI			GREENVILLE			JACKSON		
		MAT.	INST.	TOTAL	MAT.	INST.	TOTAL	MAT.	INST.	TOTAL	MAT.	INST.	TOTAL	MAT.	INST.	TOTAL	MAT.	INST.	TOTAL
01590	EQUIPMENT RENTAL	.0	97.3	97.3	.0	97.3	97.3	.0	99.0	99.0	.0	98.5	98.5	.0	98.5	98.5	.0	98.5	98.5
02	SITE CONSTRUCTION	81.8	102.1	97.2	84.8	103.4	98.9	78.1	105.5	98.9	106.2	86.8	91.5	110.4	86.6	92.3	102.6	86.6	90.5
03100	CONCRETE FORMS & ACCESSORIES	98.3	101.6	101.2	88.7	122.4	118.0	81.0	104.4	101.3	95.9	50.0	56.0	75.9	41.3	45.9	90.9	47.2	52.9
03200	CONCRETE REINFORCEMENT	95.6	115.9	107.4	92.3	116.9	106.6	103.5	116.6	111.1	95.9	66.9	79.0	103.9	50.3	72.7	95.9	54.6	71.8
03300	CAST-IN-PLACE CONCRETE	103.0	100.2	101.9	105.9	117.8	110.7	95.1	117.7	104.2	105.6	52.4	84.2	106.1	46.0	81.8	103.6	48.8	81.5
03	CONCRETE	93.2	100.6	96.9	97.2	119.3	108.5	88.1	112.0	100.0	95.9	56.0	76.0	98.4	46.8	72.7	94.5	51.2	72.9
04	MASONRY	109.1	98.2	102.4	121.5	127.5	125.2	111.0	121.5	117.4	89.2	45.0	62.2	131.5	42.7	77.3	92.0	42.7	61.9
05	METALS	96.6	121.0	105.5	95.3	123.7	105.7	96.4	123.4	106.3	98.1	87.1	94.1	96.4	79.6	90.3	98.0	81.5	92.0
06	WOOD & PLASTICS	108.6	99.9	104.1	99.1	119.6	109.6	91.2	95.5	93.4	95.4	50.3	72.2	75.5	40.8	57.7	90.2	48.5	68.8
07	THERMAL & MOISTURE PROTECTION	98.8	100.4	99.5	98.7	128.2	112.7	97.8	119.0	107.8	95.7	53.1	75.5	95.8	47.3	72.8	95.4	48.8	73.3
08	DOORS & WINDOWS	98.7	118.7	103.6	95.9	129.2	103.9	92.5	116.4	98.3	96.4	54.5	86.3	95.8	44.0	83.4	96.8	49.3	85.3
09200	PLASTER & GYPSUM BOARD	104.8	100.7	102.2	100.3	121.0	113.7	90.0	96.2	94.0	108.4	49.3	70.2	100.3	39.5	61.0	108.4	47.4	69.0
095,098	CEILINGS & ACOUSTICAL TREATMENT	89.9	100.7	97.2	87.1	121.0	109.9	70.2	96.2	87.7	95.7	49.3	64.5	90.1	39.5	56.1	95.7	47.4	63.2
09600	FLOORING	111.6	81.7	104.2	104.5	113.9	106.8	100.3	113.9	103.6	115.4	47.8	98.7	105.4	40.8	89.8	115.4	47.2	98.6
097,099	WALL FINISHES, PAINTS & COATINGS	101.9	102.8	102.4	106.1	120.7	114.7	107.0	121.7	115.6	105.0	41.5	67.8	105.0	40.9	67.5	105.0	40.9	67.5
09	FINISHES	102.9	98.5	100.6	100.0	120.9	110.9	94.4	107.0	101.0	104.9	48.3	75.4	100.0	41.0	69.2	104.9	46.6	74.5
10-14	TOTAL DIV. 10000 - 14000	100.0	99.8	99.9	100.0	107.5	101.6	100.0	102.2	100.5	100.0	65.2	92.7	100.0	63.2	92.3	100.0	64.2	92.5
15	MECHANICAL	99.8	92.0	96.3	99.8	112.8	105.5	99.7	112.6	105.4	99.9	63.1	83.5	99.9	41.5	74.0	99.9	41.2	73.8
16	ELECTRICAL	102.2	90.1	93.8	101.3	99.3	99.9	98.9	99.3	99.2	96.7	53.8	66.9	95.6	44.8	60.3	97.2	44.8	60.8
01-16	WEIGHTED AVERAGE	99.0	99.9	99.4	99.2	114.9	106.8	96.3	110.6	103.2	98.4	60.9	80.3	100.2	51.4	76.6	98.3	53.1	76.5

City Cost Indexes

		MISSISSIPPI			MISSOURI															
DIVISION		MERIDIAN			CAPE GIRARDEAU			COLUMBIA			JOPLIN			KANSAS CITY			SPRINGFIELD		
		MAT.	INST.	TOTAL	MAT.	INST.	TOTAL	MAT.	INST.	TOTAL	MAT.	INST.	TOTAL	MAT.	INST.	TOTAL	MAT.	INST.	TOTAL
01590	EQUIPMENT RENTAL	.0	98.5	98.5	.0	109.6	109.6	.0	109.7	109.7	.0	105.0	105.0	.0	103.2	103.2	.0	103.4	103.4
02	SITE CONSTRUCTION	101.5	86.6	90.2	93.3	93.3	93.3	98.6	94.4	95.5	100.3	95.8	96.9	92.6	94.6	94.1	91.9	93.5	93.1
03100	CONCRETE FORMS & ACCESSORIES	74.4	40.2	44.7	83.1	83.9	83.8	84.2	75.7	76.8	103.0	68.9	73.4	102.4	103.9	103.7	96.5	106.5	102.3
03200	CONCRETE REINFORCEMENT	102.6	54.1	74.3	106.1	92.9	98.4	100.4	106.5	103.9	104.0	78.2	89.0	98.9	107.7	104.0	96.5	106.5	102.3
03300	CAST-IN-PLACE CONCRETE	100.0	48.8	79.4	84.6	92.2	87.7	89.1	74.9	83.4	103.3	76.7	92.6	96.0	99.7	97.5	100.7	71.2	88.8
03	CONCRETE	91.1	48.0	69.7	85.0	90.2	87.6	83.3	83.3	83.3	97.3	74.8	86.1	93.4	103.6	98.5	96.4	73.5	85.0
04	MASONRY	88.9	34.8	55.8	117.4	82.7	96.2	131.2	75.8	97.3	93.5	61.1	73.7	98.4	104.1	101.9	86.8	73.0	78.4
05	METALS	96.4	80.7	90.6	93.8	115.4	101.7	93.0	121.0	103.2	95.8	95.4	95.7	99.3	112.2	104.0	96.3	106.7	100.1
06	WOOD & PLASTICS	74.2	39.2	56.2	83.0	83.4	83.2	89.8	72.0	80.7	100.9	69.3	84.6	100.9	104.0	102.4	100.0	54.4	76.5
07	THERMAL & MOISTURE PROTECTION	95.2	45.8	71.8	96.2	85.9	91.4	92.6	79.1	86.2	98.3	67.2	83.6	97.5	103.2	100.2	96.8	72.9	85.5
08	DOORS & WINDOWS	95.8	43.4	83.2	94.2	84.0	91.7	91.2	93.0	91.6	92.2	77.5	88.7	98.5	105.5	100.2	96.7	70.8	90.5
09200	PLASTER & GYPSUM BOARD	99.9	37.8	59.8	97.4	82.5	87.8	100.9	70.8	81.5	109.8	68.0	82.8	104.5	103.7	104.0	106.5	52.6	71.7
095,098	CEILINGS & ACOUSTICAL TREATMENT	90.1	37.8	54.9	93.1	82.5	86.0	92.9	70.8	78.1	97.8	68.0	77.8	103.5	103.7	103.6	98.5	52.6	67.7
09600	FLOORING	104.8	33.2	87.1	98.6	87.8	96.0	107.5	73.1	99.0	118.4	59.1	103.8	96.5	95.7	96.3	109.9	59.1	97.4
097,099	WALL FINISHES, PAINTS & COATINGS	105.0	40.4	67.2	99.7	88.2	93.0	98.8	69.9	81.8	92.9	52.1	69.0	97.7	105.3	102.1	93.2	74.5	82.3
09	FINISHES	98.9	38.2	67.3	93.8	84.4	88.9	95.9	72.4	83.7	105.8	64.8	84.4	100.4	102.3	101.4	100.8	58.5	78.8
10 - 14	TOTAL DIV. 10000 - 14000	100.0	63.1	92.3	100.0	79.8	95.8	100.0	96.0	99.2	100.0	88.5	97.6	100.0	99.5	99.9	100.0	86.3	97.1
15	MECHANICAL	99.9	41.2	73.8	100.1	94.3	97.5	100.1	95.5	98.0	100.2	61.4	83.0	100.1	101.9	100.9	100.2	72.6	87.9
16	ELECTRICAL	95.0	59.1	70.1	95.3	114.3	108.5	93.8	83.6	86.7	89.2	70.3	76.1	103.4	101.6	102.1	100.2	75.7	83.2
01 - 16	WEIGHTED AVERAGE	96.4	52.9	75.4	96.1	95.6	95.9	96.5	88.6	92.7	97.6	73.0	85.7	98.8	102.7	100.7	97.7	76.8	87.6

		MISSOURI						MONTANA												
DIVISION		ST. JOSEPH			ST. LOUIS			BILLINGS			BUTTE			GREAT FALLS			HELENA		
		MAT.	INST.	TOTAL	MAT.	INST.	TOTAL	MAT.	INST.	TOTAL	MAT.	INST.	TOTAL	MAT.	INST.	TOTAL	MAT.	INST.	TOTAL
01590	EQUIPMENT RENTAL	.0	101.8	101.8	.0	110.8	110.8	.0	95.7	95.7	.0	95.4	95.4	.0	95.4	95.4	.0	95.4	95.4
02	SITE CONSTRUCTION	94.0	91.6	92.2	92.9	96.9	95.9	83.7	97.5	94.1	89.7	96.1	94.6	93.2	96.7	95.9	95.0	96.6	96.2
03100	CONCRETE FORMS & ACCESSORIES	102.3	77.0	80.3	99.3	109.0	107.7	96.3	73.0	76.1	82.3	70.2	71.8	102.6	71.3	75.5	102.7	71.7	75.8
03200	CONCRETE REINFORCEMENT	97.7	102.0	100.2	97.0	110.4	104.8	100.2	81.1	89.0	108.7	81.0	92.5	100.2	81.0	89.0	103.7	75.0	87.0
03300	CAST-IN-PLACE CONCRETE	95.9	96.7	96.2	84.6	114.1	96.5	115.6	75.2	99.3	119.0	75.3	101.4	125.7	62.7	100.3	128.1	73.5	106.1
03	CONCRETE	93.2	89.7	91.5	85.0	112.1	98.5	102.8	75.8	89.4	102.5	74.6	88.6	108.3	70.8	89.6	110.0	73.5	91.8
04	MASONRY	97.8	93.1	94.9	98.6	110.6	105.9	123.9	76.0	94.6	122.2	73.7	92.6	126.0	75.8	95.3	123.2	73.5	92.8
05	METALS	97.9	108.8	101.9	96.3	126.3	107.2	98.2	85.2	93.5	97.6	84.6	92.9	98.7	84.8	93.6	97.8	81.6	91.9
06	WOOD & PLASTICS	101.5	71.1	85.9	98.1	106.9	102.6	100.5	73.3	86.5	88.4	71.6	79.8	108.3	71.4	89.3	108.5	71.4	89.4
07	THERMAL & MOISTURE PROTECTION	98.1	93.1	95.7	96.2	105.7	100.7	98.9	74.6	87.4	98.1	71.7	85.6	99.1	72.2	86.4	99.1	73.4	86.9
08	DOORS & WINDOWS	97.2	86.5	94.6	93.2	113.2	98.0	100.2	69.7	92.9	96.9	68.7	90.1	100.5	68.7	92.8	99.9	67.3	92.1
09200	PLASTER & GYPSUM BOARD	110.6	69.8	84.3	104.2	106.7	105.8	103.6	73.0	83.8	96.1	71.3	80.1	103.6	71.0	82.5	103.0	71.0	82.3
095,098	CEILINGS & ACOUSTICAL TREATMENT	102.0	69.8	80.4	97.3	106.7	103.6	114.0	73.0	86.4	111.2	71.3	84.4	114.0	71.0	85.1	111.2	71.0	84.2
09600	FLOORING	99.1	94.6	98.0	106.2	99.1	104.5	110.4	57.5	97.4	102.1	45.8	88.2	110.4	62.1	98.5	110.4	60.1	98.0
097,099	WALL FINISHES, PAINTS & COATINGS	93.5	90.1	91.5	99.7	109.6	105.5	102.6	70.9	84.0	102.6	48.3	70.8	102.6	54.3	74.4	102.6	60.3	77.8
09	FINISHES	101.6	80.3	90.5	98.1	106.9	102.6	107.6	69.7	87.8	103.1	62.9	82.2	107.7	67.5	86.7	107.2	68.1	86.8
10 - 14	TOTAL DIV. 10000 - 14000	100.0	93.1	98.6	100.0	106.4	101.3	100.0	75.4	94.9	100.0	74.1	94.6	100.0	75.2	94.8	100.0	66.8	93.1
15	MECHANICAL	100.2	88.9	95.2	100.1	107.9	103.6	100.3	81.1	91.8	100.3	76.3	89.6	100.3	81.2	91.8	100.3	79.9	91.3
16	ELECTRICAL	102.8	79.3	86.4	99.0	114.3	109.7	95.0	82.9	86.6	106.5	73.9	83.9	94.9	78.1	83.2	94.9	78.0	83.2
01 - 16	WEIGHTED AVERAGE	98.6	88.7	93.8	96.1	110.4	103.0	101.5	80.0	91.1	101.2	76.0	89.0	102.7	78.0	90.8	102.6	77.4	90.4

		MONTANA			NEBRASKA												NEVADA		
DIVISION		MISSOULA			GRAND ISLAND			LINCOLN			NORTH PLATTE			OMAHA			CARSON CITY		
		MAT.	INST.	TOTAL	MAT.	INST.	TOTAL	MAT.	INST.	TOTAL	MAT.	INST.	TOTAL	MAT.	INST.	TOTAL	MAT.	INST.	TOTAL
01590	EQUIPMENT RENTAL	.0	95.4	95.4	.0	102.3	102.3	.0	102.3	102.3	.0	102.3	102.3	.0	91.3	91.3	.0	100.3	100.3
02	SITE CONSTRUCTION	73.7	96.0	90.6	94.8	90.7	91.7	85.9	90.7	89.6	95.7	90.2	91.5	77.4	89.4	86.5	64.4	102.1	93.0
03100	CONCRETE FORMS & ACCESSORIES	86.7	66.1	68.8	97.4	57.6	62.9	102.5	50.7	57.5	97.3	54.1	59.8	97.1	74.5	77.5	99.3	106.4	105.5
03200	CONCRETE REINFORCEMENT	110.6	85.3	95.8	105.1	75.0	87.6	96.5	74.9	83.9	106.5	74.8	88.1	101.0	76.5	86.7	108.5	115.1	112.3
03300	CAST-IN-PLACE CONCRETE	87.6	72.3	81.5	116.3	65.4	95.8	105.1	68.0	90.1	116.3	58.9	93.2	109.0	76.6	95.9	118.3	92.4	107.9
03	CONCRETE	81.1	72.6	76.8	106.5	65.0	85.9	98.6	62.8	80.8	106.7	61.4	84.1	100.0	75.8	88.0	113.9	102.8	108.4
04	MASONRY	148.7	69.3	100.2	105.1	57.3	75.9	94.9	54.7	70.4	90.4	46.5	63.6	100.7	83.9	90.4	133.8	81.7	102.0
05	METALS	97.7	85.7	93.3	92.9	86.1	90.4	95.8	85.7	92.1	93.1	89.7	91.8	98.4	77.9	90.9	103.8	101.7	103.1
06	WOOD & PLASTICS	92.9	67.4	79.8	96.7	52.4	73.9	101.2	43.3	71.4	96.5	52.4	73.8	95.2	74.0	84.3	94.1	110.2	102.4
07	THERMAL & MOISTURE PROTECTION	97.3	77.2	87.8	97.3	63.1	81.1	97.7	55.8	77.8	97.2	53.1	76.3	92.9	75.7	84.7	105.4	91.2	98.6
08	DOORS & WINDOWS	96.9	68.3	90.1	90.4	55.0	81.9	95.9	52.2	85.4	89.7	60.2	82.6	99.3	68.6	91.9	93.4	112.0	97.8
09200	PLASTER & GYPSUM BOARD	98.1	66.9	77.9	103.1	50.6	69.2	106.5	41.2	64.3	103.4	50.6	69.3	112.6	73.6	87.4	83.7	110.3	100.9
095,098	CEILINGS & ACOUSTICAL TREATMENT	111.2	66.9	81.4	91.5	50.6	64.0	98.5	41.2	60.0	92.9	50.6	64.5	135.3	73.6	93.9	101.5	110.3	107.4
09600	FLOORING	104.1	71.5	96.0	99.2	40.8	84.7	100.9	42.5	86.5	99.1	42.5	85.1	129.6	51.0	110.2	105.1	75.0	97.7
097,099	WALL FINISHES, PAINTS & COATINGS	102.6	54.3	74.4	90.7	48.1	65.8	90.7	48.6	66.0	90.7	58.2	71.7	145.1	58.2	94.2	108.1	89.1	97.0
09	FINISHES	103.2	65.4	83.5	95.8	52.2	73.1	97.9	47.3	71.6	96.1	51.7	73.0	123.2	67.9	94.4	97.1	99.0	98.1
10 - 14	TOTAL DIV. 10000 - 14000	100.0	63.2	92.3	100.0	83.5	96.6	100.0	82.3	96.3	100.0	64.1	92.5	100.0	84.5	96.8	100.0	118.7	103.9
15	MECHANICAL	100.3	74.0	88.6	100.2	83.4	92.7	100.2	83.3	92.7	100.2	78.8	90.7	99.9	85.6	93.5	100.0	94.0	97.3
16	ELECTRICAL	103.1	78.2	85.8	91.0	50.8	63.0	100.5	50.8	66.0	94.3	50.7	64.0	88.3	86.8	87.3	92.7	89.1	90.2
01 - 16	WEIGHTED AVERAGE	99.4	75.8	88.0	97.6	68.2	83.4	97.8	66.5	82.7	97.0	65.4	81.8	100.4	80.8	90.9	101.7	96.7	99.3

COST INDEXES

585

DIVISION		NEVADA						NEW HAMPSHIRE						NEW JERSEY		
		LAS VEGAS			RENO			MANCHESTER			NASHUA			PORTSMOUTH			CAMDEN
		MAT.	INST.	TOTAL	MAT.	INST.	TOTAL	MAT.	INST.	TOTAL	MAT.	INST.	TOTAL	MAT.	INST.	TOTAL	MAT.	INST.	TOTAL
01590	EQUIPMENT RENTAL	.0	100.3	100.3	.0	100.3	100.3	.0	103.3	103.3	.0	103.3	103.3	.0	103.3	103.3	.0	101.1	101.1
02	SITE CONSTRUCTION	64.3	103.9	94.3	64.7	102.1	93.1	89.8	102.1	99.2	91.4	102.1	99.5	85.5	101.5	97.6	94.6	106.0	103.2
03100	CONCRETE FORMS & ACCESSORIES	95.7	112.4	110.2	99.7	106.4	105.6	104.6	73.6	77.7	105.1	73.6	77.8	90.7	69.5	72.3	105.1	120.8	118.7
03200	CONCRETE REINFORCEMENT	100.2	119.6	111.5	100.2	119.2	111.2	112.0	105.4	108.2	112.0	105.4	108.2	89.5	105.3	98.7	112.0	106.3	108.7
03300	CAST-IN-PLACE CONCRETE	110.8	110.5	110.7	120.5	92.4	109.2	100.7	90.6	96.6	93.8	90.6	92.5	89.0	85.0	87.4	81.6	128.7	100.6
03	CONCRETE	108.9	112.6	110.8	114.1	103.6	108.9	107.2	85.4	96.4	103.8	85.4	94.6	95.2	81.6	88.4	97.7	119.6	108.8
04	MASONRY	127.8	96.4	108.6	133.9	81.7	102.0	96.3	102.9	100.3	96.0	102.9	100.2	90.4	93.5	92.3	88.1	123.7	109.8
05	METALS	104.5	106.3	105.2	104.4	103.3	104.0	98.1	88.3	94.5	98.0	88.3	94.5	94.7	87.0	91.9	97.9	98.7	98.2
06	WOOD & PLASTICS	89.8	112.1	101.2	94.5	110.2	102.6	104.0	66.8	84.9	104.0	66.8	84.9	89.8	66.8	78.0	104.0	120.4	112.4
07	THERMAL & MOISTURE PROTECTION	105.0	101.1	103.2	105.4	91.2	98.6	101.2	98.5	99.9	101.6	98.5	100.1	101.1	97.9	99.6	101.0	119.3	109.7
08	DOORS & WINDOWS	93.7	114.1	98.6	93.7	111.0	97.9	108.1	72.9	99.7	108.1	72.9	99.7	109.2	66.1	98.8	108.1	111.6	109.0
09200	PLASTER & GYPSUM BOARD	83.9	112.2	102.2	84.9	110.3	101.3	99.3	64.6	76.8	99.3	64.6	76.8	91.3	64.6	74.0	99.3	119.8	112.5
095,098	CEILINGS & ACOUSTICAL TREATMENT	107.2	112.2	110.6	107.2	110.3	109.3	104.9	64.6	77.8	104.9	64.6	77.8	95.0	64.6	74.5	104.9	119.8	114.9
09600	FLOORING	105.1	89.3	101.2	105.1	75.0	97.7	100.0	107.5	101.8	99.8	107.5	101.7	94.6	107.5	97.8	99.8	131.9	107.7
097,099	WALL FINISHES, PAINTS & COATINGS	108.1	107.7	107.8	108.1	89.1	97.0	95.2	96.8	96.1	95.2	96.8	96.1	95.2	47.4	67.2	95.2	135.1	118.5
09	FINISHES	98.1	107.4	103.0	98.3	99.0	98.7	100.2	81.3	90.3	100.4	81.3	90.4	95.5	73.3	83.9	100.7	124.8	113.2
10-14	TOTAL DIV. 10000 - 14000	100.0	109.8	102.1	100.0	118.7	103.9	100.0	92.6	98.4	100.0	92.6	98.4	100.0	88.8	97.7	100.0	115.7	103.3
15	MECHANICAL	100.0	111.5	105.1	100.0	94.0	97.3	99.9	91.5	96.1	99.9	91.5	96.1	99.9	86.5	93.9	99.9	112.8	105.6
16	ELECTRICAL	94.8	106.9	103.2	92.7	89.1	90.2	104.2	81.8	88.7	104.0	81.8	88.6	100.7	81.8	87.6	104.0	123.4	117.4
01-16	WEIGHTED AVERAGE	101.0	107.5	104.2	101.9	96.9	99.5	101.4	88.9	95.4	101.0	88.9	95.2	98.2	85.0	91.8	99.9	116.3	107.8

DIVISION		NEW JERSEY															NEW MEXICO		
		ELIZABETH			JERSEY CITY			NEWARK			PATERSON			TRENTON			ALBUQUERQUE		
		MAT.	INST.	TOTAL	MAT.	INST.	TOTAL	MAT.	INST.	TOTAL	MAT.	INST.	TOTAL	MAT.	INST.	TOTAL	MAT.	INST.	TOTAL
01590	EQUIPMENT RENTAL	.0	103.3	103.3	.0	101.1	101.1	.0	103.3	103.3	.0	103.3	103.3	.0	100.6	100.6	.0	114.8	114.8
02	SITE CONSTRUCTION	109.3	106.7	107.3	94.6	106.4	103.6	113.0	106.4	108.0	107.6	106.4	106.7	95.8	106.9	104.2	83.4	109.8	103.5
03100	CONCRETE FORMS & ACCESSORIES	116.7	122.9	122.1	105.1	123.7	121.3	103.3	123.9	121.2	103.9	123.9	121.3	104.0	121.4	119.1	97.6	72.9	76.1
03200	CONCRETE REINFORCEMENT	86.2	122.4	107.3	112.0	122.5	118.1	112.0	122.5	118.1	112.0	122.5	118.1	112.0	116.7	114.7	100.9	66.2	80.7
03300	CAST-IN-PLACE CONCRETE	89.9	129.2	105.7	81.6	129.7	101.0	91.9	129.6	107.1	104.0	129.6	114.3	91.9	122.0	104.0	110.7	78.2	97.6
03	CONCRETE	101.3	123.9	112.6	97.7	124.3	110.9	102.7	124.5	113.6	108.9	124.6	116.7	102.8	119.6	111.1	109.0	74.5	91.8
04	MASONRY	109.4	125.2	119.0	88.1	125.2	110.8	97.5	125.2	114.4	92.8	125.2	112.6	89.0	120.8	108.4	109.4	67.9	84.0
05	METALS	94.4	107.3	99.1	98.0	105.2	100.6	97.9	106.9	101.2	98.0	108.3	101.7	98.7	102.6	100.1	105.5	89.1	99.5
06	WOOD & PLASTICS	119.0	123.4	121.3	104.0	123.1	113.8	105.9	123.4	114.9	105.9	123.4	114.9	104.0	120.1	112.3	94.9	75.0	84.7
07	THERMAL & MOISTURE PROTECTION	101.9	118.8	109.9	101.0	119.8	109.9	101.3	118.8	109.6	101.8	119.5	110.2	99.7	117.9	108.3	104.2	76.3	91.0
08	DOORS & WINDOWS	109.3	121.3	112.2	108.1	121.2	111.3	108.1	121.2	111.3	114.4	120.4	115.8	114.4	121.3	116.0	93.7	74.7	89.1
09200	PLASTER & GYPSUM BOARD	102.2	122.9	115.6	99.3	122.6	114.4	99.3	122.9	114.6	99.3	122.9	114.6	99.3	119.6	112.4	88.1	73.6	78.7
095,098	CEILINGS & ACOUSTICAL TREATMENT	95.0	122.9	113.8	104.9	122.6	116.8	104.9	122.9	117.0	104.9	122.9	117.0	104.9	119.6	114.8	107.2	73.6	84.6
09600	FLOORING	104.3	63.8	94.3	99.8	128.1	106.8	100.0	128.1	106.9	99.8	128.1	106.8	100.0	118.6	104.6	105.1	68.4	96.0
097,099	WALL FINISHES, PAINTS & COATINGS	95.1	135.1	118.5	95.2	135.1	118.6	95.1	135.1	118.5	95.1	135.1	118.5	95.2	135.1	118.6	108.1	62.4	81.3
09	FINISHES	101.5	112.7	107.3	100.7	126.0	113.9	101.2	126.1	114.2	100.8	126.1	114.0	100.8	122.2	111.9	99.2	71.0	84.5
10-14	TOTAL DIV. 10000 - 14000	100.0	96.4	99.3	100.0	97.3	99.4	100.0	97.4	99.5	100.0	97.4	99.4	100.0	116.1	103.4	100.0	80.3	95.9
15	MECHANICAL	99.9	111.8	105.2	99.9	122.6	110.0	99.9	123.8	110.5	99.9	122.7	110.0	99.9	123.2	110.3	100.0	75.0	88.9
16	ELECTRICAL	98.8	110.8	107.1	105.7	132.1	124.0	105.4	132.1	124.0	105.7	127.7	121.0	104.2	129.0	121.4	87.2	75.2	78.9
01-16	WEIGHTED AVERAGE	101.5	114.0	107.6	100.0	121.2	110.3	102.5	121.6	111.7	102.9	120.8	111.5	100.7	119.4	109.7	100.3	78.6	89.8

DIVISION		NEW MEXICO												NEW YORK					
		FARMINGTON			LAS CRUCES			ROSWELL			SANTA FE			ALBANY			BINGHAMTON		
		MAT.	INST.	TOTAL	MAT.	INST.	TOTAL	MAT.	INST.	TOTAL	MAT.	INST.	TOTAL	MAT.	INST.	TOTAL	MAT.	INST.	TOTAL
01590	EQUIPMENT RENTAL	.0	114.8	114.8	.0	85.7	85.7	.0	114.8	114.8	.0	114.8	114.8	.0	116.3	116.3	.0	115.8	115.8
02	SITE CONSTRUCTION	89.1	109.8	104.8	95.1	85.3	87.7	93.9	109.8	106.0	82.6	109.8	103.3	72.5	107.7	99.2	95.0	90.3	91.4
03100	CONCRETE FORMS & ACCESSORIES	97.5	72.9	76.1	95.0	70.9	74.0	97.5	72.8	76.0	97.5	72.9	76.1	97.5	90.7	91.6	103.0	78.9	82.1
03200	CONCRETE REINFORCEMENT	110.2	66.2	84.6	104.1	59.9	78.3	109.4	60.6	80.9	108.2	66.2	83.7	100.8	92.7	96.1	99.8	91.2	94.8
03300	CAST-IN-PLACE CONCRETE	111.3	78.2	98.0	96.9	67.6	85.1	102.5	76.6	92.1	104.6	78.2	94.0	87.5	101.3	93.1	104.5	96.2	101.2
03	CONCRETE	112.6	74.5	93.6	88.8	68.3	78.6	112.9	72.8	92.9	106.9	74.5	90.7	101.2	96.0	98.6	100.6	89.6	95.2
04	MASONRY	124.9	67.9	90.1	110.0	58.7	78.7	125.7	67.9	90.4	114.4	67.9	86.0	91.0	97.2	94.8	107.1	90.1	96.7
05	METALS	104.7	89.1	99.0	98.9	76.6	90.8	104.7	86.4	98.0	104.7	89.1	99.0	100.1	110.4	103.9	95.2	121.6	104.8
06	WOOD & PLASTICS	94.9	75.0	84.7	86.7	73.4	79.8	94.9	75.0	84.7	94.9	75.0	84.7	99.5	88.8	94.0	108.8	75.0	91.4
07	THERMAL & MOISTURE PROTECTION	105.1	76.3	91.5	90.4	67.7	79.7	105.4	76.3	91.6	104.5	76.3	91.1	89.3	95.0	92.0	99.7	87.9	94.1
08	DOORS & WINDOWS	96.5	74.7	91.3	86.0	72.1	82.6	92.6	72.9	87.9	92.8	74.7	88.4	95.9	84.3	93.1	91.1	76.9	87.7
09200	PLASTER & GYPSUM BOARD	83.1	73.6	77.0	85.2	73.6	77.7	83.1	73.6	77.0	83.1	73.6	77.0	99.0	88.1	92.0	105.4	73.6	84.9
095,098	CEILINGS & ACOUSTICAL TREATMENT	98.7	73.6	81.8	100.7	73.6	82.5	98.7	73.6	81.8	98.7	73.6	81.8	97.1	88.1	91.1	97.1	73.6	81.3
09600	FLOORING	105.1	68.4	96.0	138.0	59.0	118.5	105.1	68.4	96.0	105.1	68.4	96.0	84.4	96.0	87.2	94.5	90.8	93.6
097,099	WALL FINISHES, PAINTS & COATINGS	108.1	62.4	81.3	101.6	62.4	78.6	108.1	62.4	81.3	108.1	62.4	81.3	81.7	85.0	83.6	85.2	82.2	83.5
09	FINISHES	97.3	71.0	83.6	114.8	67.9	90.4	97.8	71.0	83.8	97.1	71.0	83.5	94.5	90.7	92.5	95.3	80.7	87.7
10-14	TOTAL DIV. 10000 - 14000	100.0	80.3	95.9	100.0	75.8	95.0	100.0	80.3	95.9	100.0	80.3	95.9	100.0	95.5	99.1	100.0	94.2	98.8
15	MECHANICAL	100.0	75.0	88.9	100.2	74.4	88.8	100.0	74.9	88.8	100.0	75.0	88.9	100.2	95.1	97.9	100.5	90.9	96.2
16	ELECTRICAL	84.5	75.2	78.0	85.7	63.4	70.2	85.7	75.2	78.4	87.2	75.2	78.9	101.8	95.2	97.2	102.8	82.6	88.8
01-16	WEIGHTED AVERAGE	101.7	78.6	90.5	97.4	70.5	84.4	101.6	78.1	90.2	99.9	78.6	89.6	97.7	97.0	97.4	98.5	90.0	94.4

COST INDEXES

City Cost Indexes

NEW YORK

DIVISION		BUFFALO MAT.	INST.	TOTAL	HICKSVILLE MAT.	INST.	TOTAL	NEW YORK MAT.	INST.	TOTAL	RIVERHEAD MAT.	INST.	TOTAL	ROCHESTER MAT.	INST.	TOTAL	SCHENECTADY MAT.	INST.	TOTAL
01590	EQUIPMENT RENTAL	.0	93.3	93.3	.0	118.4	118.4	.0	117.0	117.0	.0	118.4	118.4	.0	116.2	116.2	.0	116.3	116.3
02	SITE CONSTRUCTION	98.3	93.3	94.5	114.2	131.7	127.5	137.0	128.8	130.8	114.5	131.7	127.5	75.1	107.5	99.7	72.0	107.8	99.2
03100	CONCRETE FORMS & ACCESSORIES	102.5	112.2	111.0	88.4	160.2	150.7	109.6	176.1	167.3	92.8	160.2	151.4	102.3	99.6	99.9	102.5	91.5	92.9
03200	CONCRETE REINFORCEMENT	101.7	104.1	103.1	101.2	182.8	148.8	105.2	183.9	151.0	103.1	182.9	149.6	102.5	91.8	95.3	99.5	92.7	95.5
03300	CAST-IN-PLACE CONCRETE	116.2	118.9	117.3	97.7	152.1	119.6	126.5	166.1	142.5	95.9	152.1	118.6	110.5	107.5	109.3	95.9	102.4	98.6
03	CONCRETE	106.2	112.2	109.2	104.5	160.1	132.2	125.4	172.1	148.6	103.4	160.2	131.7	111.6	101.8	106.7	105.6	96.7	101.2
04	MASONRY	103.6	116.6	111.5	109.6	159.8	140.3	109.8	169.7	146.4	116.4	159.8	142.9	99.7	105.6	103.3	91.6	99.3	96.3
05	METALS	100.2	95.9	98.6	109.7	141.3	121.3	109.6	142.7	121.7	109.8	141.4	121.3	103.5	110.4	106.0	100.1	110.4	103.8
06	WOOD & PLASTICS	106.7	112.0	109.4	90.2	163.4	127.8	111.2	178.2	145.6	94.2	163.4	129.8	103.1	99.7	101.3	105.1	88.8	96.7
07	THERMAL & MOISTURE PROTECTION	99.2	106.5	102.6	103.1	153.8	127.1	104.2	162.0	131.6	103.2	153.8	127.2	95.8	100.0	97.8	89.4	95.9	92.5
08	DOORS & WINDOWS	94.2	103.8	96.5	88.6	164.2	106.8	97.1	172.3	115.2	88.6	164.2	106.8	97.8	93.6	96.8	95.9	84.3	93.1
09200	PLASTER & GYPSUM BOARD	96.5	111.9	106.4	94.1	165.4	140.2	103.5	180.4	153.2	95.1	165.4	140.5	98.1	99.5	99.0	99.0	88.1	92.0
095,098	CEILINGS & ACOUSTICAL TREATMENT	95.6	111.9	106.5	81.6	165.4	137.9	104.7	180.4	155.6	81.6	165.4	137.9	100.0	99.5	99.7	97.1	88.1	91.1
09600	FLOORING	90.0	116.7	96.6	94.2	102.1	96.1	98.6	164.7	114.9	95.5	102.1	97.1	80.0	97.2	84.2	84.4	96.0	87.2
097,099	WALL FINISHES, PAINTS & COATINGS	86.3	119.0	105.4	107.4	158.1	137.1	97.8	158.1	133.1	107.4	158.1	137.1	83.6	101.3	94.0	81.7	85.0	83.6
09	FINISHES	93.0	114.4	104.2	105.3	148.4	127.8	107.8	172.7	141.6	105.8	148.4	128.0	94.3	99.5	97.0	94.2	91.2	92.6
10-14	TOTAL DIV. 10000-14000	100.0	105.5	101.2	100.0	135.5	107.4	100.0	150.5	110.5	100.0	135.5	107.4	100.0	99.5	99.9	100.0	96.3	99.2
15	MECHANICAL	99.9	96.8	98.5	99.8	150.7	122.4	100.2	164.1	128.6	99.8	150.7	122.4	99.9	93.3	97.0	100.2	96.1	98.4
16	ELECTRICAL	99.5	99.0	99.1	102.4	156.3	139.8	111.1	175.2	155.7	103.6	163.7	145.3	104.5	91.7	95.6	102.3	95.2	97.3
01-16	WEIGHTED AVERAGE	99.5	103.7	101.5	102.2	151.2	125.9	107.7	163.3	134.6	102.7	152.4	126.7	100.5	99.4	100.0	98.3	97.6	98.0

NEW YORK / NORTH CAROLINA

DIVISION		SYRACUSE MAT.	INST.	TOTAL	UTICA MAT.	INST.	TOTAL	WATERTOWN MAT.	INST.	TOTAL	WHITE PLAINS MAT.	INST.	TOTAL	YONKERS MAT.	INST.	TOTAL	ASHEVILLE MAT.	INST.	TOTAL
01590	EQUIPMENT RENTAL	.0	116.3	116.3	.0	116.3	116.3	.0	116.3	116.3	.0	116.8	116.8	.0	116.8	116.8	.0	93.2	93.2
02	SITE CONSTRUCTION	93.7	107.6	104.3	70.1	106.7	97.8	78.4	108.4	101.2	126.3	126.0	126.1	136.3	125.9	128.4	103.2	72.7	80.1
03100	CONCRETE FORMS & ACCESSORIES	101.8	88.2	90.0	103.0	81.0	83.9	83.5	81.9	82.1	110.3	140.8	136.8	109.3	141.5	137.3	93.1	44.8	51.2
03200	CONCRETE REINFORCEMENT	100.8	91.5	95.4	100.8	87.1	92.8	101.4	65.0	80.2	98.5	181.5	146.8	102.4	182.3	148.9	92.7	42.8	63.6
03300	CAST-IN-PLACE CONCRETE	97.0	96.7	96.9	88.7	93.4	90.6	103.3	93.9	99.5	111.5	126.4	117.5	124.8	127.5	125.9	94.2	52.1	77.2
03	CONCRETE	104.2	93.0	98.6	102.2	87.9	95.1	115.8	84.6	100.2	112.5	142.4	127.4	124.3	143.3	133.8	96.1	49.1	72.7
04	MASONRY	97.1	95.0	95.8	90.2	88.4	89.1	91.4	97.7	95.2	104.3	125.7	117.4	109.5	127.7	120.6	77.0	40.6	54.8
05	METALS	100.0	108.8	103.2	98.2	106.8	101.4	98.3	104.5	100.6	99.6	136.7	113.2	106.5	137.1	117.6	93.9	78.7	88.4
06	WOOD & PLASTICS	105.1	85.7	95.1	105.1	79.0	91.7	86.2	77.6	81.8	113.3	143.5	128.9	112.0	143.5	128.2	94.0	45.6	69.1
07	THERMAL & MOISTURE PROTECTION	97.6	95.8	96.7	89.4	92.5	90.9	89.7	93.4	91.5	105.1	137.5	120.4	105.1	138.4	121.1	96.0	46.2	72.4
08	DOORS & WINDOWS	93.8	83.2	91.3	95.9	78.6	91.8	96.0	76.1	91.2	91.9	152.0	106.3	95.2	153.8	109.3	91.7	42.9	80.0
09200	PLASTER & GYPSUM BOARD	99.0	84.9	89.9	99.0	78.1	85.5	92.1	76.6	82.1	99.9	144.7	128.9	103.2	144.7	130.0	101.7	43.8	64.3
095,098	CEILINGS & ACOUSTICAL TREATMENT	97.1	84.9	88.9	97.1	78.1	84.3	97.1	76.6	83.3	84.9	144.7	125.1	103.3	144.7	131.1	87.1	43.8	58.0
09600	FLOORING	85.9	85.7	85.8	84.4	81.9	83.8	77.0	83.8	78.7	94.8	164.7	112.0	94.3	164.7	111.7	92.3	46.6	81.0
097,099	WALL FINISHES, PAINTS & COATINGS	86.9	90.3	88.9	81.7	88.7	85.8	81.7	82.7	82.3	95.3	151.5	128.2	95.3	151.5	128.2	105.9	44.2	69.8
09	FINISHES	95.7	87.4	91.4	94.2	81.3	87.5	91.4	81.4	86.2	101.7	146.9	125.2	105.9	147.4	127.5	92.6	44.9	67.8
10-14	TOTAL DIV. 10000-14000	100.0	98.5	99.7	100.0	92.3	98.4	100.0	93.2	98.6	100.0	129.3	106.1	100.0	138.8	108.1	100.0	64.0	92.5
15	MECHANICAL	100.2	92.6	96.9	100.2	86.3	94.1	100.2	82.4	92.3	100.5	126.4	112.0	100.5	127.5	112.5	100.1	48.1	77.0
16	ELECTRICAL	102.3	86.6	91.4	102.2	85.7	90.8	102.2	91.1	94.5	95.5	145.5	130.2	108.6	145.5	134.2	99.6	46.2	62.5
01-16	WEIGHTED AVERAGE	99.2	94.0	96.7	97.5	89.8	93.8	99.0	90.3	94.8	101.8	136.6	118.6	106.6	137.5	121.5	95.5	52.1	74.5

NORTH CAROLINA

DIVISION		CHARLOTTE MAT.	INST.	TOTAL	DURHAM MAT.	INST.	TOTAL	FAYETTEVILLE MAT.	INST.	TOTAL	GREENSBORO MAT.	INST.	TOTAL	RALEIGH MAT.	INST.	TOTAL	WILMINGTON MAT.	INST.	TOTAL
01590	EQUIPMENT RENTAL	.0	93.2	93.2	.0	99.6	99.6	.0	99.6	99.6	.0	99.6	99.6	.0	99.6	99.6	.0	93.2	93.2
02	SITE CONSTRUCTION	103.4	72.7	80.1	103.1	82.8	87.7	101.2	82.8	87.2	103.0	82.8	87.6	104.3	82.8	87.9	104.2	72.7	80.3
03100	CONCRETE FORMS & ACCESSORIES	101.2	44.9	52.3	95.3	45.2	51.8	90.5	45.2	51.1	95.3	45.1	51.7	98.2	45.2	52.2	94.7	45.2	51.7
03200	CONCRETE REINFORCEMENT	93.1	42.8	63.8	93.1	56.7	71.9	92.2	56.7	71.5	93.1	56.7	71.9	93.1	56.7	71.9	93.4	56.7	72.0
03300	CAST-IN-PLACE CONCRETE	96.6	52.2	78.7	94.4	52.3	77.4	91.1	52.3	75.5	93.6	52.3	76.9	99.9	52.3	80.7	93.7	52.3	77.0
03	CONCRETE	97.3	49.1	73.3	95.8	52.0	74.0	93.4	52.0	72.8	95.4	51.9	73.8	98.8	52.0	75.5	96.0	52.0	74.1
04	MASONRY	78.5	40.6	55.4	77.1	40.6	54.8	80.8	40.6	56.2	76.9	40.6	54.7	83.6	40.6	57.4	67.6	40.6	51.1
05	METALS	95.2	78.8	89.2	95.3	84.4	91.3	94.2	84.4	90.6	96.0	84.3	91.8	95.3	84.4	91.3	94.3	84.4	90.7
06	WOOD & PLASTICS	103.3	45.6	73.7	96.5	45.6	70.4	91.1	45.6	67.7	96.5	45.6	70.4	99.8	45.6	71.9	95.6	45.6	69.9
07	THERMAL & MOISTURE PROTECTION	96.0	47.3	72.9	96.6	46.8	73.0	96.3	46.8	72.8	96.6	46.8	73.0	96.4	46.9	73.0	96.0	46.8	72.7
08	DOORS & WINDOWS	95.7	42.9	83.0	95.7	48.5	84.3	91.8	48.5	81.4	95.7	48.5	84.3	92.5	48.5	81.9	91.9	48.5	81.4
09200	PLASTER & GYPSUM BOARD	108.5	43.8	66.7	108.5	43.8	66.7	102.0	43.8	64.4	108.5	43.8	66.7	111.7	43.8	67.8	102.5	43.8	64.6
095,098	CEILINGS & ACOUSTICAL TREATMENT	91.3	43.8	59.4	91.3	43.8	59.4	88.5	43.8	58.4	91.3	43.8	59.4	91.3	43.8	59.4	88.5	43.8	58.4
09600	FLOORING	95.7	46.6	83.5	95.9	46.6	83.7	92.4	46.6	81.1	95.9	46.6	83.7	93.2	46.6	81.7	93.2	46.6	81.7
097,099	WALL FINISHES, PAINTS & COATINGS	105.9	44.2	69.8	105.9	44.2	69.8	105.9	44.2	69.8	105.9	44.2	69.8	105.9	44.2	69.8	105.9	44.2	69.8
09	FINISHES	95.3	44.9	69.1	95.5	44.9	69.2	92.9	44.9	67.9	95.5	44.9	69.2	96.0	44.9	69.4	93.2	44.9	68.1
10-14	TOTAL DIV. 10000-14000	100.0	64.0	92.5	100.0	72.7	94.3	100.0	72.7	94.3	100.0	64.0	92.5	100.0	72.7	94.3	100.0	72.7	94.3
15	MECHANICAL	100.1	48.1	77.0	100.1	48.3	77.1	100.1	48.3	77.1	100.1	48.3	77.1	100.1	48.3	77.1	100.1	48.3	77.1
16	ELECTRICAL	99.1	41.5	59.1	97.5	44.2	60.5	92.7	44.2	59.0	98.6	44.2	60.8	98.2	44.2	60.7	99.3	44.2	61.0
01-16	WEIGHTED AVERAGE	96.7	51.3	74.8	96.3	54.1	75.9	94.9	54.1	75.2	96.4	53.9	75.9	96.8	54.1	76.2	95.1	53.1	74.8

COST INDEXES

587

		NORTH CAROLINA			NORTH DAKOTA													OHIO		
	DIVISION	WINSTON-SALEM			BISMARCK			FARGO			GRAND FORKS			MINOT			AKRON		
		MAT.	INST.	TOTAL	MAT.	INST.	TOTAL	MAT.	INST.	TOTAL	MAT.	INST.	TOTAL	MAT.	INST.	TOTAL	MAT.	INST.	TOTAL
01590	EQUIPMENT RENTAL	.0	99.6	99.6	.0	95.4	95.4	.0	95.4	95.4	.0	95.4	95.4	.0	95.4	95.4	.0	100.1	100.1
02	SITE CONSTRUCTION	103.4	82.8	87.7	86.6	96.1	93.8	86.1	96.2	93.7	94.1	95.5	95.2	91.8	96.2	95.1	100.5	109.3	107.2
03100	CONCRETE FORMS & ACCESSORIES	96.9	44.9	51.7	95.1	56.9	61.9	95.8	57.8	62.8	91.6	50.4	55.8	86.1	65.1	67.9	104.5	100.8	101.3
03200	CONCRETE REINFORCEMENT	93.1	42.8	63.8	108.7	75.3	89.2	100.2	75.3	85.7	107.0	58.3	78.6	110.7	75.8	90.4	99.1	98.0	98.5
03300	CAST-IN-PLACE CONCRETE	96.6	52.2	78.7	101.9	61.6	85.6	106.8	64.2	89.6	104.3	59.8	86.4	104.3	60.1	86.5	102.8	108.5	105.1
03	CONCRETE	97.0	49.2	73.2	96.8	62.9	79.9	98.3	64.1	81.3	102.8	56.1	79.6	101.3	66.0	83.7	102.7	101.8	102.3
04	MASONRY	77.1	40.6	54.8	107.3	57.8	77.1	107.6	57.4	76.9	107.5	59.1	77.9	108.3	66.9	83.0	92.2	105.0	100.0
05	METALS	95.2	78.8	89.2	95.4	77.0	88.7	95.3	77.2	88.7	95.4	69.8	86.1	95.8	77.9	89.2	91.7	83.5	88.7
06	WOOD & PLASTICS	96.5	45.6	70.4	87.4	52.9	69.7	87.4	53.7	70.1	84.2	48.7	65.9	79.2	63.6	71.2	99.2	99.4	99.3
07	THERMAL & MOISTURE PROTECTION	96.6	46.8	73.0	99.6	57.2	79.5	100.2	58.4	80.4	100.4	57.7	80.2	100.0	60.4	81.3	106.0	102.8	104.5
08	DOORS & WINDOWS	95.7	44.8	83.5	100.7	55.1	89.7	100.7	55.5	89.8	100.7	45.7	87.5	100.9	60.8	91.2	104.5	100.0	103.4
09200	PLASTER & GYPSUM BOARD	108.5	43.8	66.7	119.7	52.0	75.9	119.7	52.8	76.5	118.7	47.6	72.8	117.3	63.1	82.2	105.6	98.6	101.1
095,098	CEILINGS & ACOUSTICAL TREATMENT	91.3	43.8	59.4	147.2	52.0	83.2	147.2	52.8	83.7	147.2	47.6	80.2	147.2	63.1	90.6	98.9	98.6	98.7
09600	FLOORING	95.9	46.6	83.7	111.0	53.5	96.8	110.8	52.6	96.4	109.3	52.6	95.3	106.6	70.1	97.6	107.9	99.5	105.8
097,099	WALL FINISHES, PAINTS & COATINGS	105.9	44.2	69.8	102.6	43.7	68.1	102.6	82.7	91.0	102.6	33.4	62.1	102.6	36.6	64.0	105.3	117.8	112.6
09	FINISHES	95.5	44.9	69.2	116.1	54.0	83.8	115.9	58.8	86.2	116.0	48.5	80.8	114.7	62.9	87.7	104.3	102.2	103.2
10 - 14	TOTAL DIV. 10000 - 14000	100.0	64.0	92.5	100.0	85.2	96.9	100.0	85.2	96.9	100.0	55.2	90.7	100.0	86.6	97.2	100.0	101.8	100.4
15	MECHANICAL	100.1	48.2	77.0	100.6	66.5	85.4	100.6	72.8	88.2	100.6	53.0	79.4	100.6	66.5	85.4	100.1	104.6	102.1
16	ELECTRICAL	98.6	44.2	60.8	96.9	67.9	76.8	96.8	48.8	63.5	101.7	66.6	77.3	106.8	67.9	79.8	98.0	101.8	100.6
01 - 16	WEIGHTED AVERAGE	96.6	52.8	75.4	100.5	67.4	84.5	100.6	66.1	83.9	101.7	61.2	82.2	101.7	70.3	86.5	99.7	101.7	100.7

		OHIO																		
	DIVISION	CANTON			CINCINNATI			CLEVELAND			COLUMBUS			DAYTON			LORAIN		
		MAT.	INST.	TOTAL	MAT.	INST.	TOTAL	MAT.	INST.	TOTAL	MAT.	INST.	TOTAL	MAT.	INST.	TOTAL	MAT.	INST.	TOTAL
01590	EQUIPMENT RENTAL	.0	100.1	100.1	.0	104.2	104.2	.0	100.4	100.4	.0	95.7	95.7	.0	98.4	98.4	.0	100.1	100.1
02	SITE CONSTRUCTION	100.6	108.2	106.4	77.1	110.8	102.7	100.5	109.0	107.0	84.9	100.1	96.5	75.9	110.7	102.3	99.8	108.6	106.5
03100	CONCRETE FORMS & ACCESSORIES	104.5	89.3	91.3	102.0	88.8	90.5	104.7	107.2	106.9	96.0	93.1	93.5	101.9	87.6	89.5	104.6	90.5	92.4
03200	CONCRETE REINFORCEMENT	99.1	82.6	89.5	93.4	85.8	89.0	99.6	98.2	98.8	103.2	85.2	92.7	93.4	84.2	88.0	99.1	98.1	98.5
03300	CAST-IN-PLACE CONCRETE	103.8	105.4	104.4	86.4	90.9	88.2	98.0	116.0	107.0	87.8	96.5	91.3	78.7	92.5	84.2	97.8	111.5	103.4
03	CONCRETE	103.2	92.8	98.0	92.9	89.0	91.0	101.8	107.4	104.6	95.1	92.2	93.6	89.1	88.1	88.6	100.2	98.4	99.3
04	MASONRY	93.1	94.7	94.1	86.4	96.1	92.3	96.8	111.5	105.8	97.0	95.3	95.9	85.8	93.0	90.2	89.1	105.0	98.8
05	METALS	91.7	76.5	86.1	94.9	89.4	92.9	93.1	86.5	90.7	96.3	81.3	90.8	94.2	78.6	88.5	92.2	84.2	89.3
06	WOOD & PLASTICS	99.2	87.9	93.4	102.7	86.4	94.3	98.4	104.4	101.5	99.4	93.0	96.1	105.1	86.8	95.7	99.2	86.0	92.4
07	THERMAL & MOISTURE PROTECTION	106.6	98.2	102.6	90.1	97.8	93.8	104.7	114.6	109.4	100.7	97.9	99.3	95.5	93.6	94.6	106.5	106.5	106.5
08	DOORS & WINDOWS	98.9	81.1	94.6	96.7	85.2	94.0	95.3	102.7	97.1	99.8	88.7	97.1	95.0	84.4	92.4	98.9	92.9	97.4
09200	PLASTER & GYPSUM BOARD	105.6	86.8	93.5	102.6	86.3	92.0	104.9	103.8	104.2	102.2	92.5	95.9	102.6	86.7	92.4	105.6	84.8	92.2
095,098	CEILINGS & ACOUSTICAL TREATMENT	98.9	86.8	90.7	91.3	86.3	87.9	97.5	103.8	101.7	95.7	92.5	93.6	92.9	86.7	88.8	98.9	84.8	89.4
09600	FLOORING	108.1	88.8	103.3	93.1	101.8	95.2	107.7	114.4	109.3	97.4	92.6	96.2	95.5	90.9	94.4	108.1	111.6	109.0
097,099	WALL FINISHES, PAINTS & COATINGS	105.3	91.7	97.3	102.3	97.4	99.4	105.3	121.0	114.5	99.8	106.0	103.4	102.3	95.9	98.6	105.3	121.0	114.5
09	FINISHES	104.4	88.6	96.2	93.6	91.9	92.7	103.9	109.7	107.0	97.8	94.4	96.0	94.7	88.9	91.7	104.4	97.0	100.5
10 - 14	TOTAL DIV. 10000 - 14000	100.0	90.0	97.9	100.0	90.4	98.0	100.0	106.1	101.3	100.0	94.0	98.8	100.0	89.2	97.7	100.0	101.0	100.2
15	MECHANICAL	100.1	89.5	95.4	99.8	98.5	99.2	100.1	111.0	104.9	100.1	95.1	97.9	100.8	89.0	95.5	100.1	95.0	97.8
16	ELECTRICAL	96.5	97.4	97.1	95.8	84.9	88.3	97.1	106.5	103.7	102.1	89.4	93.3	94.5	90.1	91.4	96.5	99.1	98.3
01 - 16	WEIGHTED AVERAGE	99.1	92.2	95.8	95.3	93.4	94.4	98.8	106.8	102.7	98.2	92.7	95.5	94.9	90.5	92.8	98.5	98.0	98.3

		OHIO									OKLAHOMA									
	DIVISION	SPRINGFIELD			TOLEDO			YOUNGSTOWN			ENID			LAWTON			MUSKOGEE		
		MAT.	INST.	TOTAL	MAT.	INST.	TOTAL	MAT.	INST.	TOTAL	MAT.	INST.	TOTAL	MAT.	INST.	TOTAL	MAT.	INST.	TOTAL
01590	EQUIPMENT RENTAL	.0	98.4	98.4	.0	98.7	98.7	.0	100.1	100.1	.0	76.1	76.1	.0	77.2	77.2	.0	85.8	85.8
02	SITE CONSTRUCTION	76.4	109.8	101.7	84.1	101.4	97.2	100.4	109.6	107.4	110.8	89.0	94.2	105.9	90.6	94.3	96.1	85.0	87.7
03100	CONCRETE FORMS & ACCESSORIES	101.9	92.1	93.3	96.0	107.5	106.0	104.5	94.0	95.4	96.9	45.7	52.4	101.4	62.8	67.9	103.1	41.3	49.4
03200	CONCRETE REINFORCEMENT	93.4	84.2	88.0	103.2	92.1	96.7	99.1	91.9	94.9	95.4	84.3	88.9	95.6	84.4	89.0	95.4	43.3	65.0
03300	CAST-IN-PLACE CONCRETE	82.5	93.0	86.7	87.8	110.9	97.1	101.8	109.9	105.1	94.1	56.8	79.1	91.0	56.9	77.2	84.6	43.5	68.1
03	CONCRETE	91.0	90.2	90.6	95.1	104.9	100.0	102.3	98.2	100.2	91.2	57.1	74.2	87.7	64.6	76.2	83.2	43.7	63.5
04	MASONRY	86.1	94.1	91.0	107.4	110.0	109.0	92.3	103.1	98.9	101.3	58.6	75.2	96.8	58.6	73.5	110.0	49.7	73.2
05	METALS	94.2	78.5	88.5	96.1	88.7	93.4	91.8	81.8	88.1	93.4	68.5	84.3	95.4	68.6	85.6	93.3	63.0	82.2
06	WOOD & PLASTICS	105.1	92.4	98.6	99.4	108.1	103.9	99.2	91.3	95.1	101.1	43.0	71.2	104.8	66.2	85.0	106.9	42.5	73.8
07	THERMAL & MOISTURE PROTECTION	95.5	94.7	95.1	102.5	109.2	105.6	106.7	102.6	104.8	99.6	64.0	82.7	99.3	66.3	83.7	99.1	51.1	76.4
08	DOORS & WINDOWS	95.0	87.3	93.1	97.7	100.0	98.2	98.9	96.3	98.3	95.1	56.5	85.8	96.6	68.9	89.9	95.1	41.4	82.1
09200	PLASTER & GYPSUM BOARD	102.6	92.5	96.1	102.2	108.1	106.0	105.6	90.2	95.7	93.3	42.3	60.4	96.0	66.2	76.7	95.3	41.7	60.6
095,098	CEILINGS & ACOUSTICAL TREATMENT	92.9	92.5	92.6	95.7	108.1	104.0	98.9	90.2	93.1	93.4	42.3	59.1	101.9	66.2	77.9	93.4	41.7	58.6
09600	FLOORING	95.5	90.9	94.4	96.6	106.9	99.2	108.1	104.0	107.1	115.1	60.3	101.6	118.1	60.3	103.9	119.4	48.6	101.9
097,099	WALL FINISHES, PAINTS & COATINGS	102.3	95.9	98.6	99.8	115.4	108.9	105.3	107.0	106.3	95.8	64.2	77.3	95.8	64.2	77.3	95.8	40.8	63.6
09	FINISHES	94.7	92.4	93.5	97.5	108.7	103.3	104.4	96.7	100.4	100.9	48.7	73.7	103.3	62.1	81.9	101.5	42.9	70.9
10 - 14	TOTAL DIV. 10000 - 14000	100.0	90.0	97.9	100.0	97.7	99.5	100.0	99.9	100.0	100.0	73.8	94.5	100.0	76.8	95.2	100.0	38.0	87.1
15	MECHANICAL	100.8	89.6	95.8	100.1	105.1	102.3	100.1	95.7	98.1	100.1	68.2	85.9	100.1	68.3	85.9	100.1	34.2	70.8
16	ELECTRICAL	94.5	89.4	90.9	102.8	106.8	105.6	96.5	92.6	93.8	95.0	72.4	79.3	97.7	63.7	74.1	94.4	40.5	56.9
01 - 16	WEIGHTED AVERAGE	95.2	91.5	93.4	98.5	104.2	101.3	98.9	96.7	97.8	97.5	65.5	82.1	97.6	67.7	83.2	96.7	47.5	72.9

City Cost Indexes

DIVISION		OKLAHOMA						OREGON												
		OKLAHOMA CITY			TULSA			EUGENE			MEDFORD			PORTLAND			SALEM		
		MAT.	INST.	TOTAL	MAT.	INST.	TOTAL	MAT.	INST.	TOTAL	MAT.	INST.	TOTAL	MAT.	INST.	TOTAL	MAT.	INST.	TOTAL
01590	EQUIPMENT RENTAL	.0	77.4	77.4	.0	85.8	85.8	.0	99.7	99.7	.0	99.7	99.7	.0	99.7	99.7	.0	99.7	99.7
02	SITE CONSTRUCTION	107.3	91.2	95.1	102.7	86.0	90.0	101.0	106.4	105.1	110.2	106.3	107.3	102.2	106.4	105.4	101.2	106.4	105.1
03100	CONCRETE FORMS & ACCESSORIES	102.8	54.0	60.4	102.7	54.6	60.9	102.3	110.8	109.7	96.9	106.6	105.4	103.7	111.2	110.2	103.7	111.0	110.1
03200	CONCRETE REINFORCEMENT	95.6	84.3	89.0	95.6	84.3	89.0	109.3	98.7	103.2	100.9	98.7	99.6	104.5	99.2	101.4	110.5	99.1	103.9
03300	CAST-IN-PLACE CONCRETE	98.4	56.0	81.3	92.2	55.5	77.4	102.9	109.7	105.6	106.2	109.6	107.6	110.4	109.9	110.2	106.7	109.8	107.9
03	CONCRETE	91.6	60.4	76.1	88.4	61.6	75.1	111.5	107.6	109.6	118.7	105.7	112.3	114.9	107.9	111.4	113.7	107.8	110.8
04	MASONRY	99.1	61.1	75.9	96.0	61.2	74.7	112.2	104.2	107.3	109.0	104.2	106.1	113.5	111.8	112.5	114.2	111.8	112.8
05	METALS	95.6	68.6	85.7	96.0	84.3	91.7	95.1	96.6	95.7	94.7	96.4	95.3	93.8	97.5	95.2	95.7	97.4	96.3
06	WOOD & PLASTICS	106.9	53.0	79.2	105.6	54.6	79.4	94.0	110.2	102.3	88.5	104.8	96.9	95.6	110.2	103.1	95.6	110.2	103.1
07	THERMAL & MOISTURE PROTECTION	98.4	66.1	83.1	99.1	63.8	82.4	110.5	97.9	104.5	111.4	95.0	103.6	110.4	105.3	107.9	110.3	105.0	107.8
08	DOORS & WINDOWS	96.6	61.8	88.2	96.6	62.1	88.3	99.2	108.4	101.4	101.9	105.5	102.8	96.7	108.4	99.5	98.5	108.4	100.9
09200	PLASTER & GYPSUM BOARD	96.0	52.6	68.0	96.0	54.1	68.9	93.6	110.4	104.5	90.3	105.0	99.8	92.6	110.4	104.1	94.3	110.4	104.7
095,098	CEILINGS & ACOUSTICAL TREATMENT	101.9	52.6	68.7	101.9	54.1	69.7	118.1	110.4	112.9	116.7	105.0	108.8	118.1	110.4	112.9	126.5	110.4	115.7
09600	FLOORING	118.1	60.3	103.9	117.9	62.8	104.3	118.2	102.5	114.3	115.8	102.5	112.5	118.2	102.5	114.3	118.2	102.5	114.3
097,099	WALL FINISHES, PAINTS & COATINGS	95.8	64.2	77.3	95.8	60.2	75.0	118.8	75.3	93.4	118.8	63.4	86.4	118.8	75.3	93.4	118.8	75.3	93.4
09	FINISHES	103.4	55.1	78.2	102.8	56.1	78.5	114.4	105.6	109.8	113.6	101.1	107.1	114.3	105.6	109.8	115.9	105.6	110.5
10-14	TOTAL DIV. 10000 - 14000	100.0	76.1	95.0	100.0	74.2	94.6	100.0	108.3	101.7	100.0	107.6	101.6	100.0	108.4	101.7	100.0	108.3	101.7
15	MECHANICAL	100.1	69.6	86.5	100.1	60.4	82.4	100.2	110.2	104.7	100.2	110.1	104.6	100.2	112.2	105.6	100.2	110.2	104.7
16	ELECTRICAL	96.5	72.4	79.7	97.7	52.8	66.5	104.2	101.1	102.0	109.8	92.2	97.6	104.0	111.5	109.2	103.8	101.1	101.9
01-16	WEIGHTED AVERAGE	98.2	67.9	83.6	97.6	63.9	81.3	103.4	105.1	104.2	104.9	102.5	103.7	103.4	108.3	105.8	103.9	106.1	105.0

DIVISION		PENNSYLVANIA																	
		ALLENTOWN			ALTOONA			ERIE			HARRISBURG			PHILADELPHIA			PITTSBURGH		
		MAT.	INST.	TOTAL	MAT.	INST.	TOTAL	MAT.	INST.	TOTAL	MAT.	INST.	TOTAL	MAT.	INST.	TOTAL	MAT.	INST.	TOTAL
01590	EQUIPMENT RENTAL	.0	116.3	116.3	.0	116.3	116.3	.0	116.3	116.3	.0	115.6	115.6	.0	95.8	95.8	.0	117.2	117.2
02	SITE CONSTRUCTION	92.5	107.6	103.9	96.9	107.6	105.0	93.4	108.4	104.8	82.9	106.2	100.6	102.2	95.6	97.1	99.0	110.3	107.6
03100	CONCRETE FORMS & ACCESSORIES	101.7	108.8	107.8	81.5	91.6	90.3	100.2	101.8	101.6	93.5	90.0	90.5	100.8	129.0	125.3	98.8	104.1	103.4
03200	CONCRETE REINFORCEMENT	100.8	106.6	104.2	97.8	97.3	97.5	99.8	99.2	99.5	100.8	99.2	99.9	100.2	130.1	117.6	102.6	110.4	107.2
03300	CAST-IN-PLACE CONCRETE	87.8	99.5	92.5	97.9	91.8	95.4	96.3	93.7	95.2	97.0	92.4	95.1	93.6	122.3	105.2	94.9	101.9	97.7
03	CONCRETE	98.5	106.3	102.3	93.3	94.3	93.8	92.2	99.8	96.0	102.3	94.2	98.3	105.0	126.4	115.7	94.3	105.8	100.0
04	MASONRY	95.2	96.1	95.7	97.4	89.4	92.5	87.8	96.1	92.9	93.4	88.3	90.3	95.2	123.4	112.4	90.4	105.3	99.5
05	METALS	98.1	124.3	107.7	91.8	118.8	101.7	91.9	121.6	102.8	99.9	120.5	107.4	100.0	122.4	108.2	98.5	127.9	109.2
06	WOOD & PLASTICS	105.1	112.3	108.8	82.1	93.2	87.8	100.5	102.4	101.5	98.0	89.8	93.8	101.8	128.9	115.8	99.3	103.9	101.7
07	THERMAL & MOISTURE PROTECTION	97.6	112.0	104.4	96.6	95.6	96.1	96.7	100.7	98.6	101.4	105.8	103.4	98.8	128.2	112.7	95.1	104.0	99.3
08	DOORS & WINDOWS	93.8	107.8	97.2	88.3	100.4	91.2	88.5	99.8	91.2	93.8	99.2	95.1	97.5	133.4	106.1	93.1	113.0	97.8
09200	PLASTER & GYPSUM BOARD	99.0	112.3	107.6	87.5	92.6	90.8	95.8	102.2	99.9	99.0	89.1	92.6	98.6	129.6	118.6	91.9	103.6	99.4
095,098	CEILINGS & ACOUSTICAL TREATMENT	97.1	112.3	107.4	92.9	92.6	92.7	97.1	102.2	100.5	97.1	89.1	91.7	99.2	129.6	119.7	96.1	103.6	101.1
09600	FLOORING	85.9	93.7	87.8	80.3	66.9	77.0	87.4	87.6	87.5	86.1	88.1	86.6	81.6	132.3	94.1	96.2	106.1	98.6
097,099	WALL FINISHES, PAINTS & COATINGS	86.9	98.7	93.8	82.5	107.6	97.2	91.9	90.2	90.9	86.9	85.8	86.3	88.1	134.0	115.0	92.3	111.9	103.8
09	FINISHES	95.7	105.1	100.6	91.7	88.2	89.9	96.5	97.9	97.2	94.6	88.8	91.5	96.5	130.1	114.0	98.0	105.0	101.7
10-14	TOTAL DIV. 10000 - 14000	100.0	103.4	100.7	100.0	100.4	100.1	100.0	102.5	100.5	100.0	97.6	99.5	100.0	123.4	104.9	100.0	105.2	101.1
15	MECHANICAL	100.2	100.3	100.3	99.8	86.6	93.9	99.8	96.8	98.5	100.2	92.1	96.6	99.9	124.3	110.8	99.9	100.7	100.2
16	ELECTRICAL	102.2	86.1	91.0	87.6	102.9	98.2	90.0	85.5	86.8	101.3	82.6	88.3	97.5	129.6	119.8	95.0	102.9	100.5
01-16	WEIGHTED AVERAGE	98.1	102.6	100.2	94.3	97.0	95.6	94.4	99.1	96.7	98.3	94.7	96.6	99.5	123.6	111.2	97.0	106.8	101.7

DIVISION		PENNSYLVANIA									PUERTO RICO			RHODE ISLAND			SOUTH CAROLINA		
		READING			SCRANTON			YORK			SAN JUAN			PROVIDENCE			CHARLESTON		
		MAT.	INST.	TOTAL	MAT.	INST.	TOTAL	MAT.	INST.	TOTAL	MAT.	INST.	TOTAL	MAT.	INST.	TOTAL	MAT.	INST.	TOTAL
01590	EQUIPMENT RENTAL	.0	119.8	119.8	.0	116.3	116.3	.0	115.6	115.6	.0	90.7	90.7	.0	105.1	105.1	.0	99.3	99.3
02	SITE CONSTRUCTION	99.7	113.2	110.0	93.0	107.6	104.1	82.1	106.3	100.5	122.3	94.3	101.0	83.9	105.4	100.2	96.1	81.8	85.3
03100	CONCRETE FORMS & ACCESSORIES	100.3	92.0	93.1	101.8	92.4	93.6	81.1	90.5	89.3	100.7	23.1	33.3	107.8	110.5	110.1	95.3	43.7	50.5
03200	CONCRETE REINFORCEMENT	98.5	103.2	101.3	100.8	103.9	102.6	99.8	99.1	99.4	201.3	13.8	92.1	112.0	116.5	114.6	93.1	62.6	75.4
03300	CAST-IN-PLACE CONCRETE	68.4	97.0	80.0	100.5	97.1	99.1	86.4	93.1	89.1	106.4	35.5	77.8	87.1	119.2	100.0	80.2	54.6	69.9
03	CONCRETE	92.0	97.5	94.7	100.5	97.1	98.8	98.8	94.6	96.7	111.9	26.3	69.3	103.1	114.2	108.6	88.6	53.1	70.9
04	MASONRY	96.8	94.2	95.3	95.5	96.9	96.4	94.5	88.0	90.5	213.1	20.0	95.1	103.8	117.4	112.1	84.1	41.5	58.1
05	METALS	97.0	122.9	106.5	99.9	122.8	108.3	97.3	120.4	105.7	129.2	33.7	94.3	98.1	109.4	102.2	95.2	83.8	91.1
06	WOOD & PLASTICS	101.9	89.9	95.7	105.1	90.0	97.4	89.1	89.8	89.4	93.4	23.5	57.5	104.8	108.4	106.7	96.5	43.2	69.1
07	THERMAL & MOISTURE PROTECTION	98.8	108.3	103.3	97.5	106.2	101.6	96.2	105.9	100.8	167.9	27.0	101.1	100.2	110.9	105.3	96.2	49.2	73.9
08	DOORS & WINDOWS	95.1	94.9	95.0	93.8	100.8	95.5	90.7	99.2	92.8	155.6	19.4	122.8	101.9	110.7	104.0	95.7	45.4	83.6
09200	PLASTER & GYPSUM BOARD	106.6	89.4	95.5	99.0	89.4	92.8	94.1	89.1	90.9	245.7	20.8	100.3	96.0	107.5	103.5	108.5	41.3	65.1
095,098	CEILINGS & ACOUSTICAL TREATMENT	100.5	89.4	93.0	97.1	89.4	91.9	90.1	89.1	89.4	377.3	20.8	137.6	106.4	107.5	107.2	91.3	41.3	57.7
09600	FLOORING	79.8	95.0	83.6	85.9	92.3	87.5	81.9	88.1	83.5	202.1	19.2	157.0	100.7	125.6	106.8	95.9	46.1	83.6
097,099	WALL FINISHES, PAINTS & COATINGS	86.3	99.8	94.2	86.9	99.8	94.5	86.9	85.8	86.3	207.0	17.7	96.2	97.8	117.0	109.0	105.9	44.4	69.9
09	FINISHES	97.1	92.8	94.9	95.6	92.5	94.0	91.4	89.1	90.2	246.3	22.1	129.5	100.2	114.2	107.5	95.8	43.7	68.7
10-14	TOTAL DIV. 10000 - 14000	100.0	100.2	100.0	100.0	101.0	100.2	100.0	98.1	99.6	100.0	24.5	84.2	100.0	110.5	102.2	100.0	71.4	94.0
15	MECHANICAL	100.2	100.2	100.2	100.2	91.9	96.5	100.2	92.7	96.9	101.3	18.1	64.3	99.9	102.6	101.1	100.1	50.5	78.1
16	ELECTRICAL	100.1	79.6	85.9	102.3	89.7	93.5	91.9	82.6	85.4	144.7	18.1	56.8	98.6	98.4	98.5	98.5	38.2	56.6
01-16	WEIGHTED AVERAGE	97.5	98.1	97.8	98.6	98.0	98.3	95.9	94.9	95.4	139.7	29.1	86.3	100.0	107.9	103.8	95.7	53.4	75.2

COST INDEXES

DIVISION		SOUTH CAROLINA											SOUTH DAKOTA					
		COLUMBIA			FLORENCE			GREENVILLE			SPARTANBURG			ABERDEEN			PIERRE		
		MAT.	INST.	TOTAL	MAT.	INST.	TOTAL	MAT.	INST.	TOTAL	MAT.	INST.	TOTAL	MAT.	INST.	TOTAL	MAT.	INST.	TOTAL
01590	EQUIPMENT RENTAL	.0	99.3	99.3	.0	99.3	99.3	.0	99.3	99.3	.0	99.3	99.3	.0	95.4	95.4	.0	95.4	95.4
02	SITE CONSTRUCTION	95.8	81.8	85.2	107.4	81.8	88.0	102.2	81.4	86.4	102.0	81.4	86.4	82.4	95.0	92.0	80.8	95.0	91.6
03100	CONCRETE FORMS & ACCESSORIES	100.3	46.2	53.3	79.6	46.2	50.6	95.0	46.1	52.5	99.1	46.1	53.0	95.4	43.4	50.3	93.6	45.7	52.0
03200	CONCRETE REINFORCEMENT	93.1	62.3	75.2	92.8	62.6	75.2	92.7	44.1	64.4	92.7	44.1	64.4	107.3	54.6	76.6	106.7	63.5	81.5
03300	CAST-IN-PLACE CONCRETE	78.6	50.5	67.3	70.9	54.5	64.3	70.9	54.4	64.3	70.9	54.4	64.3	99.4	54.9	81.5	96.6	50.8	78.1
03	CONCRETE	88.1	52.7	70.5	90.4	54.1	72.3	89.1	50.6	69.9	89.3	50.6	70.0	96.1	50.8	73.5	93.8	52.1	73.1
04	MASONRY	83.1	36.3	54.5	70.9	41.5	53.0	69.0	41.5	52.2	70.9	41.5	53.0	109.3	56.5	77.0	106.0	47.6	70.3
05	METALS	95.2	83.2	90.8	94.2	83.6	90.3	94.0	77.0	87.8	94.1	77.0	87.8	107.2	71.1	94.0	107.1	74.3	95.1
06	WOOD & PLASTICS	102.3	47.0	73.9	80.3	47.0	63.2	95.8	47.0	70.7	100.1	47.0	72.8	100.3	41.2	69.9	98.5	44.6	70.8
07	THERMAL & MOISTURE PROTECTION	96.1	47.1	72.9	96.5	49.5	74.2	96.5	49.5	74.2	96.5	49.5	74.3	97.9	51.1	75.7	98.1	49.0	74.8
08	DOORS & WINDOWS	95.7	47.5	84.1	91.8	47.5	81.1	91.7	43.0	80.0	91.8	43.0	80.0	96.1	44.4	83.6	99.5	48.9	87.3
09200	PLASTER & GYPSUM BOARD	108.5	45.2	67.6	97.1	45.2	63.6	102.7	45.2	65.5	104.7	45.2	66.2	104.8	39.9	62.9	103.6	43.4	64.7
095,098	CEILINGS & ACOUSTICAL TREATMENT	91.3	45.2	60.3	88.5	45.2	59.4	87.1	45.2	58.9	87.1	45.2	58.9	110.7	39.9	63.1	109.3	43.4	65.0
09600	FLOORING	95.7	46.1	83.4	87.0	46.1	76.9	93.5	47.0	82.0	95.1	47.0	83.2	111.3	66.2	100.2	110.4	47.2	94.8
097,099	WALL FINISHES, PAINTS & COATINGS	105.9	44.4	69.9	105.9	44.4	69.9	105.9	44.4	69.9	105.9	44.4	69.9	102.6	39.5	65.7	102.6	39.5	65.7
09	FINISHES	95.7	45.9	69.8	91.8	45.9	67.9	94.1	46.1	69.1	94.9	46.1	69.5	107.2	47.0	75.8	106.4	45.0	74.4
10-14	TOTAL DIV. 10000 - 14000	100.0	71.9	94.1	100.0	71.8	94.1	100.0	71.9	94.1	100.0	71.9	94.1	100.0	65.8	92.9	100.0	66.3	93.0
15	MECHANICAL	100.1	42.6	74.5	100.1	42.6	74.5	100.1	42.6	74.5	100.1	42.6	74.5	100.2	39.9	73.4	100.2	40.0	73.4
16	ELECTRICAL	99.6	40.3	58.4	95.4	23.0	45.1	98.9	36.7	55.7	98.9	36.7	55.7	102.5	46.4	63.6	96.1	55.7	68.1
01-16	WEIGHTED AVERAGE	95.7	51.9	74.5	94.1	49.8	72.7	94.3	50.8	73.3	94.6	50.8	73.4	101.0	54.2	78.4	100.4	55.3	78.6

DIVISION		SOUTH DAKOTA						TENNESSEE										
		RAPID CITY			SIOUX FALLS			CHATTANOOGA			JACKSON			JOHNSON CITY			KNOXVILLE		
		MAT.	INST.	TOTAL	MAT.	INST.	TOTAL	MAT.	INST.	TOTAL	MAT.	INST.	TOTAL	MAT.	INST.	TOTAL	MAT.	INST.	TOTAL
01590	EQUIPMENT RENTAL	.0	95.4	95.4	.0	96.7	96.7	.0	104.7	104.7	.0	105.1	105.1	.0	97.6	97.6	.0	97.6	97.6
02	SITE CONSTRUCTION	81.0	94.7	91.4	82.0	97.0	93.4	104.1	97.4	99.0	103.4	95.4	97.3	113.5	86.1	92.7	91.2	86.1	87.4
03100	CONCRETE FORMS & ACCESSORIES	106.8	42.0	50.5	93.7	44.0	50.6	93.6	52.4	57.8	84.3	37.9	44.0	78.2	54.6	57.7	92.5	54.7	59.7
03200	CONCRETE REINFORCEMENT	100.2	64.3	79.3	100.2	64.5	79.4	94.0	49.2	67.9	99.7	48.6	70.0	98.0	49.4	69.7	94.0	49.4	68.0
03300	CAST-IN-PLACE CONCRETE	95.8	47.9	76.5	99.9	55.1	81.8	103.5	54.1	83.6	101.6	38.4	76.1	83.2	60.4	74.0	97.2	55.4	80.3
03	CONCRETE	93.5	49.7	71.7	94.7	53.1	74.0	94.0	54.3	74.2	93.6	42.4	68.1	98.8	57.5	78.2	91.0	55.8	73.5
04	MASONRY	106.0	56.3	75.7	103.2	65.1	80.0	97.4	51.1	69.1	110.6	30.7	61.8	108.4	50.8	73.2	75.7	50.8	60.5
05	METALS	109.2	75.7	97.0	109.8	76.3	97.5	97.0	82.5	91.7	95.1	79.1	89.3	94.7	82.2	90.1	97.2	82.3	91.8
06	WOOD & PLASTICS	108.3	41.7	74.1	98.5	41.2	69.0	95.7	53.4	74.0	82.3	40.1	60.6	71.8	56.7	64.0	85.1	56.7	70.5
07	THERMAL & MOISTURE PROTECTION	98.6	50.8	76.0	98.0	55.2	77.7	102.0	54.3	79.4	100.0	40.1	71.6	96.5	55.1	76.9	94.1	54.5	75.4
08	DOORS & WINDOWS	100.5	47.3	87.7	100.5	47.1	87.6	99.0	52.5	87.8	99.8	43.9	86.3	95.1	55.8	85.7	91.8	55.8	83.2
09200	PLASTER & GYPSUM BOARD	104.7	40.4	63.2	104.7	39.9	62.9	93.2	52.6	66.9	96.8	38.8	59.3	104.5	55.9	73.1	111.1	55.9	75.4
095,098	CEILINGS & ACOUSTICAL TREATMENT	114.9	40.4	64.8	114.9	39.9	64.5	95.9	52.6	66.8	96.4	38.8	57.7	87.1	55.9	66.1	92.7	55.9	68.0
09600	FLOORING	110.4	75.7	101.9	110.4	81.5	103.3	99.3	58.0	89.1	90.4	28.1	75.0	92.6	59.6	84.4	98.1	59.6	88.6
097,099	WALL FINISHES, PAINTS & COATINGS	102.6	39.5	65.7	102.6	39.5	65.7	105.0	48.9	72.1	94.1	31.9	57.7	102.9	62.5	79.3	102.9	62.5	79.3
09	FINISHES	107.7	47.8	76.5	107.6	50.1	77.7	97.5	53.1	74.4	95.5	36.0	64.5	98.2	56.6	76.5	92.8	56.6	73.9
10-14	TOTAL DIV. 10000 - 14000	100.0	63.7	92.4	100.0	65.9	92.9	100.0	64.0	92.5	100.0	57.7	91.2	100.0	69.6	93.7	100.0	69.7	93.7
15	MECHANICAL	100.2	37.1	72.1	100.2	40.1	73.5	100.2	52.9	79.1	100.1	46.9	76.5	99.9	58.6	81.5	99.9	62.5	83.2
16	ELECTRICAL	96.9	55.7	68.3	95.8	69.7	77.6	103.5	62.3	74.9	99.3	35.7	55.1	90.9	38.1	54.2	100.4	61.2	73.1
01-16	WEIGHTED AVERAGE	101.1	55.6	79.1	101.0	60.5	81.4	98.7	61.8	80.9	98.5	49.1	74.6	98.1	58.9	79.1	94.8	63.3	79.6

DIVISION		TENNESSEE						TEXAS										
		MEMPHIS			NASHVILLE			ABILENE			AMARILLO			AUSTIN			BEAUMONT		
		MAT.	INST.	TOTAL	MAT.	INST.	TOTAL	MAT.	INST.	TOTAL	MAT.	INST.	TOTAL	MAT.	INST.	TOTAL	MAT.	INST.	TOTAL
01590	EQUIPMENT RENTAL	.0	102.8	102.8	.0	105.4	105.4	.0	85.8	85.8	.0	85.8	85.8	.0	86.1	86.1	.0	88.0	88.0
02	SITE CONSTRUCTION	97.5	92.0	93.3	95.2	100.1	98.9	103.2	85.0	89.4	103.5	85.9	90.2	86.1	85.7	85.8	94.9	87.2	89.1
03100	CONCRETE FORMS & ACCESSORIES	94.6	62.6	66.8	92.0	68.3	71.4	99.3	49.9	56.4	103.0	59.2	65.0	97.4	64.8	69.0	107.0	62.2	68.1
03200	CONCRETE REINFORCEMENT	92.3	65.6	76.7	95.1	64.5	77.2	95.6	58.1	73.8	95.6	57.8	73.6	94.1	60.3	74.4	94.9	51.4	69.6
03300	CAST-IN-PLACE CONCRETE	91.9	65.9	81.4	96.2	65.0	83.6	95.6	47.8	76.4	100.5	54.9	82.1	91.5	56.7	77.5	87.8	62.8	77.7
03	CONCRETE	87.5	66.2	76.9	91.7	68.0	79.9	89.9	51.8	70.9	92.6	58.2	75.5	82.0	61.6	71.9	86.7	61.1	73.9
04	MASONRY	85.4	61.5	70.8	85.7	64.0	72.4	99.3	53.1	71.1	103.1	46.1	68.3	93.8	52.0	68.2	105.6	64.6	80.6
05	METALS	96.5	94.2	95.7	99.7	92.3	97.0	96.7	69.3	86.7	96.7	69.3	86.7	95.7	69.8	86.2	95.7	69.6	86.2
06	WOOD & PLASTICS	94.3	64.3	78.9	88.4	70.4	79.1	101.8	50.1	75.2	104.8	62.6	83.1	98.3	69.2	83.3	109.9	63.8	86.2
07	THERMAL & MOISTURE PROTECTION	98.0	64.5	82.2	97.7	66.5	82.9	99.2	55.1	78.3	101.8	52.1	78.3	96.5	58.3	78.4	102.4	65.1	84.7
08	DOORS & WINDOWS	98.8	67.3	91.2	96.5	69.9	90.1	92.3	51.8	82.5	92.3	55.9	83.5	94.6	66.5	87.9	95.8	57.6	86.6
09200	PLASTER & GYPSUM BOARD	102.4	63.5	77.3	101.9	70.0	81.3	96.0	49.4	65.9	96.0	62.4	74.3	99.2	69.1	79.7	98.0	63.5	75.7
095,098	CEILINGS & ACOUSTICAL TREATMENT	89.6	63.5	72.1	89.6	70.0	76.4	101.9	49.4	66.6	101.9	62.4	75.3	91.0	69.1	76.3	106.6	63.5	77.6
09600	FLOORING	91.7	52.9	82.1	103.2	72.0	95.5	118.1	62.3	104.4	117.9	50.5	101.3	97.9	58.1	88.1	116.8	76.3	106.8
097,099	WALL FINISHES, PAINTS & COATINGS	93.8	63.4	76.0	109.1	58.9	79.7	94.7	63.8	76.6	94.7	44.2	65.1	95.3	49.6	68.5	91.4	59.6	72.8
09	FINISHES	92.8	60.5	76.0	101.9	68.1	84.3	102.8	53.5	77.1	102.9	56.0	78.5	93.8	62.1	77.3	97.9	64.7	80.6
10-14	TOTAL DIV. 10000 - 14000	100.0	76.4	95.1	100.0	71.7	94.1	100.0	72.1	94.2	100.0	73.7	94.5	100.0	74.7	94.7	100.0	77.1	95.2
15	MECHANICAL	100.0	69.7	86.5	100.0	69.3	86.3	100.1	53.3	79.3	100.1	59.1	81.9	99.8	64.9	84.2	100.0	64.7	84.3
16	ELECTRICAL	99.1	72.5	80.7	101.6	69.9	79.6	97.7	57.7	69.9	98.7	65.1	75.3	97.1	73.5	80.7	96.6	75.4	81.9
01-16	WEIGHTED AVERAGE	96.0	72.1	84.4	97.7	73.7	86.1	97.6	58.8	78.8	98.3	62.0	80.8	94.8	67.0	81.3	97.1	68.7	83.4

COST INDEXES

TEXAS

DIVISION		CORPUS CHRISTI			DALLAS			EL PASO			FORT WORTH			HOUSTON			LAREDO		
		MAT.	INST.	TOTAL	MAT.	INST.	TOTAL	MAT.	INST.	TOTAL	MAT.	INST.	TOTAL	MAT.	INST.	TOTAL	MAT.	INST.	TOTAL
01590	EQUIPMENT RENTAL	.0	94.8	94.8	.0	93.8	93.8	.0	85.8	85.8	.0	85.8	85.8	.0	98.3	98.3	.0	86.1	86.1
02	SITE CONSTRUCTION	120.1	80.9	90.3	132.7	82.3	94.5	103.6	84.9	89.4	103.0	85.2	89.5	122.3	85.3	94.2	86.2	85.5	85.6
03100	CONCRETE FORMS & ACCESSORIES	100.9	45.5	52.7	98.3	60.6	65.6	100.0	55.6	61.4	102.1	60.2	65.7	94.0	73.6	76.3	89.6	45.6	51.4
03200	CONCRETE REINFORCEMENT	92.3	54.1	70.1	94.5	58.6	73.6	95.6	56.9	73.1	95.6	58.3	73.9	97.6	62.4	77.1	94.1	54.3	70.9
03300	CAST-IN-PLACE CONCRETE	108.5	52.4	85.9	99.2	61.1	83.9	98.0	42.1	75.5	94.0	56.4	78.9	90.5	74.3	84.0	83.5	60.5	74.2
03	CONCRETE	92.5	51.6	72.1	88.8	62.0	75.5	91.2	51.9	71.6	89.3	59.2	74.3	86.1	73.3	79.7	81.5	53.3	67.5
04	MASONRY	85.0	51.0	64.2	96.2	61.8	75.2	98.1	50.5	69.0	94.3	58.1	72.2	100.5	70.0	81.9	93.4	55.0	69.9
05	METALS	95.2	82.5	90.6	96.7	85.1	92.4	96.5	67.4	85.8	96.4	70.2	86.8	100.0	92.4	97.2	96.5	64.2	84.7
06	WOOD & PLASTICS	111.8	45.4	77.7	103.9	61.6	82.2	103.0	60.2	81.0	110.5	61.6	85.4	94.3	74.6	84.2	88.8	45.3	66.4
07	THERMAL & MOISTURE PROTECTION	99.3	52.4	77.1	95.4	62.8	79.9	98.8	57.0	79.0	99.7	58.1	80.0	93.9	73.0	84.0	95.3	56.1	76.7
08	DOORS & WINDOWS	101.7	45.5	88.2	106.5	58.0	94.8	92.3	53.6	83.0	86.3	58.0	79.5	105.7	70.1	97.1	95.3	44.4	83.0
09200	PLASTER & GYPSUM BOARD	100.3	44.4	64.2	98.4	61.4	74.5	96.0	59.9	72.7	96.0	61.4	73.6	96.1	74.5	82.2	96.7	44.4	62.9
095,098	CEILINGS & ACOUSTICAL TREATMENT	96.6	44.4	61.5	101.5	61.4	74.5	101.9	59.9	73.6	101.9	61.4	74.6	106.3	74.5	84.9	91.0	44.4	59.7
09600	FLOORING	109.9	57.4	97.0	110.0	70.4	100.3	118.1	64.3	104.9	153.1	56.2	129.2	104.3	69.8	95.8	94.8	57.4	85.6
097,099	WALL FINISHES, PAINTS & COATINGS	109.1	44.3	71.2	101.9	60.4	77.6	94.7	42.5	64.1	95.8	63.0	76.6	99.8	69.8	82.2	95.3	54.0	71.1
09	FINISHES	100.4	47.0	72.6	105.4	62.4	83.0	102.8	56.4	78.6	114.8	59.8	86.1	103.3	72.5	87.2	92.6	48.0	69.4
10 - 14	TOTAL DIV. 10000 - 14000	100.0	73.0	94.4	100.0	75.2	94.8	100.0	72.0	94.2	100.0	75.2	94.8	100.0	81.6	96.2	100.0	70.0	93.7
15	MECHANICAL	99.8	54.4	79.6	99.9	68.3	85.9	100.1	40.6	73.6	100.1	63.8	84.0	99.9	75.8	89.2	99.7	48.9	77.1
16	ELECTRICAL	94.0	56.6	68.0	98.9	73.2	81.0	96.5	62.0	72.5	96.5	71.9	79.4	98.5	73.0	80.8	97.1	69.2	77.7
01 - 16	WEIGHTED AVERAGE	98.0	58.3	78.8	100.0	69.4	85.2	97.5	57.2	78.1	97.7	66.1	82.4	99.6	76.3	88.3	94.6	58.8	77.3

TEXAS / UTAH

DIVISION		LUBBOCK			ODESSA			SAN ANTONIO			WACO			WICHITA FALLS			LOGAN		
		MAT.	INST.	TOTAL	MAT.	INST.	TOTAL	MAT.	INST.	TOTAL	MAT.	INST.	TOTAL	MAT.	INST.	TOTAL	MAT.	INST.	TOTAL
01590	EQUIPMENT RENTAL	.0	96.0	96.0	.0	85.8	85.8	.0	89.1	89.1	.0	85.8	85.8	.0	85.8	85.8	.0	99.3	99.3
02	SITE CONSTRUCTION	134.8	83.1	95.6	103.2	85.4	89.7	85.8	90.0	89.0	101.9	85.2	89.2	102.7	85.0	89.3	95.1	99.6	98.5
03100	CONCRETE FORMS & ACCESSORIES	101.2	49.1	55.9	99.3	46.9	53.8	89.6	62.9	66.5	101.5	46.8	54.0	101.5	50.1	56.8	102.5	69.7	74.0
03200	CONCRETE REINFORCEMENT	94.9	57.7	73.2	97.7	57.6	74.3	94.1	57.5	72.7	95.6	60.1	74.9	95.6	58.2	73.8	100.7	74.6	85.5
03300	CAST-IN-PLACE CONCRETE	95.8	55.0	79.4	95.6	50.4	77.4	81.9	66.3	75.7	86.3	58.3	75.0	92.2	54.5	77.0	93.5	77.1	86.9
03	CONCRETE	88.8	54.9	71.9	90.2	51.2	70.8	80.7	63.7	72.2	85.4	54.4	69.9	88.3	54.1	71.3	112.4	73.5	93.0
04	MASONRY	98.8	42.6	64.5	99.3	42.6	64.7	93.3	62.6	74.5	96.4	54.3	70.7	96.8	54.3	70.9	120.1	71.6	90.5
05	METALS	100.7	85.3	95.1	96.8	68.5	86.5	97.1	71.0	87.6	96.6	70.1	86.9	96.5	70.0	86.8	105.2	77.5	95.1
06	WOOD & PLASTICS	104.2	50.2	76.4	101.8	47.6	73.9	88.8	63.4	75.7	109.3	44.2	75.8	109.3	50.0	78.8	87.4	68.6	77.7
07	THERMAL & MOISTURE PROTECTION	91.2	54.5	73.8	99.2	49.4	75.6	95.3	68.9	82.8	100.0	53.8	78.1	100.0	56.4	79.3	104.6	75.3	90.8
08	DOORS & WINDOWS	103.9	49.7	90.8	92.3	47.9	81.6	97.3	61.4	88.6	86.3	45.2	76.4	86.3	51.7	77.9	88.4	66.5	83.1
09200	PLASTER & GYPSUM BOARD	96.3	49.4	66.0	96.0	46.8	64.2	96.7	63.0	74.9	96.0	43.4	62.0	96.0	49.4	65.9	84.9	67.5	73.6
095,098	CEILINGS & ACOUSTICAL TREATMENT	103.3	49.4	67.0	101.9	46.8	64.9	91.0	63.0	72.2	101.9	43.4	62.5	101.9	49.4	66.6	107.2	67.5	80.5
09600	FLOORING	109.3	47.7	94.1	118.1	47.1	100.6	94.8	61.6	86.6	153.3	43.8	126.3	153.8	64.3	131.7	105.1	70.0	96.4
097,099	WALL FINISHES, PAINTS & COATINGS	106.5	39.7	67.3	94.7	39.7	62.5	95.3	54.0	71.1	95.8	40.9	63.7	97.7	64.8	78.5	108.1	55.7	77.4
09	FINISHES	104.4	47.2	74.6	102.8	45.7	73.1	92.6	61.4	76.3	114.8	44.5	78.2	115.1	54.0	83.2	100.0	67.9	83.2
10 - 14	TOTAL DIV. 10000 - 14000	100.0	71.7	94.1	100.0	71.1	94.0	100.0	75.6	94.9	100.0	72.9	94.4	100.0	72.2	94.2	100.0	76.8	95.2
15	MECHANICAL	99.8	55.5	80.1	100.1	44.5	75.4	99.7	73.0	87.8	100.1	61.7	83.0	100.1	51.7	78.6	100.0	72.0	87.5
16	ELECTRICAL	96.2	54.2	67.0	97.7	50.4	64.9	97.1	69.2	77.7	97.7	66.2	75.8	102.6	66.3	77.3	92.7	79.4	83.5
01 - 16	WEIGHTED AVERAGE	100.0	58.4	79.9	97.6	53.4	76.3	94.8	69.6	82.6	97.4	61.0	79.8	98.2	60.5	80.0	101.4	76.0	89.1

UTAH / VERMONT / VIRGINIA

DIVISION		OGDEN			PROVO			SALT LAKE CITY			BURLINGTON			RUTLAND			ALEXANDRIA		
		MAT.	INST.	TOTAL	MAT.	INST.	TOTAL	MAT.	INST.	TOTAL	MAT.	INST.	TOTAL	MAT.	INST.	TOTAL	MAT.	INST.	TOTAL
01590	EQUIPMENT RENTAL	.0	99.3	99.3	.0	98.1	98.1	.0	99.3	99.3	.0	103.3	103.3	.0	103.3	103.3	.0	99.6	99.6
02	SITE CONSTRUCTION	82.7	99.6	95.5	91.4	97.6	96.1	82.3	99.5	95.4	82.8	101.3	96.8	82.7	101.3	96.8	114.1	83.6	91.0
03100	CONCRETE FORMS & ACCESSORIES	102.5	69.7	74.0	104.2	69.7	74.3	101.0	69.8	73.9	93.5	57.5	62.3	105.6	57.6	63.9	91.7	72.4	74.9
03200	CONCRETE REINFORCEMENT	100.7	74.6	85.5	108.9	74.6	88.9	98.5	74.6	84.6	112.0	61.6	82.7	112.0	61.6	82.7	82.8	83.5	83.2
03300	CAST-IN-PLACE CONCRETE	94.9	77.1	87.7	93.6	77.1	86.9	102.7	77.1	92.4	98.3	72.0	87.7	93.7	72.0	85.0	97.4	79.2	90.1
03	CONCRETE	101.4	73.5	87.5	111.8	73.5	92.7	104.9	73.5	89.3	105.3	63.8	84.7	103.8	63.9	83.9	95.9	78.1	87.0
04	MASONRY	113.1	71.6	87.7	125.5	71.6	92.6	131.6	71.6	94.9	95.7	69.6	79.7	83.6	69.6	75.0	83.4	70.4	75.4
05	METALS	105.7	77.5	95.4	102.6	77.6	93.5	104.4	77.6	94.6	98.7	72.1	89.0	98.0	72.2	88.6	95.3	97.4	96.0
06	WOOD & PLASTICS	87.4	68.6	77.7	89.0	68.6	78.5	87.5	68.6	77.8	91.1	55.8	72.9	104.5	55.8	79.4	95.5	72.9	83.9
07	THERMAL & MOISTURE PROTECTION	103.1	75.3	89.9	107.1	75.3	92.0	106.4	75.3	91.7	101.0	62.9	83.0	100.8	64.2	83.5	96.4	79.3	88.3
08	DOORS & WINDOWS	88.4	66.5	83.1	92.7	66.5	86.4	88.4	66.5	83.1	108.1	51.6	94.5	108.1	51.6	94.5	95.7	75.8	90.9
09200	PLASTER & GYPSUM BOARD	84.9	67.5	73.6	83.9	67.5	73.3	84.9	67.5	73.6	99.3	53.2	69.5	99.3	53.2	69.5	108.5	71.9	84.9
095,098	CEILINGS & ACOUSTICAL TREATMENT	107.2	67.5	80.5	100.1	67.5	78.2	107.2	67.5	80.5	104.9	53.2	70.1	104.9	53.2	70.1	91.3	71.9	78.3
09600	FLOORING	105.1	70.0	96.4	105.9	70.0	97.0	104.6	70.0	96.1	99.8	80.6	95.0	99.8	80.6	95.0	95.9	78.5	91.6
097,099	WALL FINISHES, PAINTS & COATINGS	108.1	55.7	77.4	108.1	64.7	82.7	108.1	64.7	82.7	95.2	45.8	66.3	95.2	45.8	66.3	116.7	88.9	100.4
09	FINISHES	98.8	67.9	82.7	99.2	68.9	83.4	99.2	68.9	83.4	99.4	60.1	78.9	99.3	60.1	78.8	96.2	75.0	85.1
10 - 14	TOTAL DIV. 10000 - 14000	100.0	76.8	95.2	100.0	76.8	95.2	100.0	76.8	95.2	100.0	95.6	99.1	100.0	95.6	99.1	100.0	82.6	96.4
15	MECHANICAL	100.0	72.0	87.5	100.0	72.0	87.5	100.2	72.0	87.6	99.9	72.0	87.5	99.9	72.0	87.5	100.1	84.4	93.1
16	ELECTRICAL	92.7	79.4	83.5	93.2	79.4	83.6	93.6	83.6	86.7	105.2	69.1	80.1	104.0	69.1	79.7	98.7	92.0	94.1
01 - 16	WEIGHTED AVERAGE	99.2	76.0	88.0	101.7	76.0	89.3	100.7	76.9	89.2	100.9	71.0	86.4	100.0	71.0	86.0	97.1	82.9	90.2

COST INDEXES

VIRGINIA

DIVISION		ARLINGTON			NEWPORT NEWS			NORFOLK			PORTSMOUTH			RICHMOND			ROANOKE		
		MAT.	INST.	TOTAL	MAT.	INST.	TOTAL	MAT.	INST.	TOTAL	MAT.	INST.	TOTAL	MAT.	INST.	TOTAL	MAT.	INST.	TOTAL
01590	EQUIPMENT RENTAL	.0	99.6	99.6	.0	104.7	104.7	.0	105.4	105.4	.0	104.6	104.6	.0	104.6	104.6	.0	99.6	99.6
02	SITE CONSTRUCTION	125.1	83.2	93.3	107.1	84.8	90.2	106.4	85.9	90.9	105.6	84.4	89.5	107.7	85.5	90.9	104.8	81.5	87.1
03100	CONCRETE FORMS & ACCESSORIES	88.7	71.2	73.5	95.3	52.6	58.2	99.9	52.7	58.9	82.2	52.5	56.4	96.1	63.6	67.9	95.1	40.9	48.0
03200	CONCRETE REINFORCEMENT	93.4	70.5	80.0	93.1	69.0	79.1	93.1	69.0	79.1	92.8	71.1	80.2	93.1	72.6	81.2	93.1	70.1	79.7
03300	CAST-IN-PLACE CONCRETE	94.8	77.6	87.9	95.4	58.6	80.6	98.2	58.6	82.3	94.5	60.5	80.8	101.5	61.1	85.2	107.2	52.1	85.0
03	CONCRETE	103.3	74.6	89.0	96.3	59.6	78.0	98.1	59.7	79.0	95.0	60.6	77.9	99.5	66.0	82.8	102.3	52.4	77.5
04	MASONRY	94.8	68.0	78.4	85.4	56.8	67.9	93.0	56.8	70.9	90.1	56.7	69.7	81.8	57.2	66.8	85.7	38.6	56.9
05	METALS	94.2	91.3	93.1	95.3	90.2	93.4	94.5	90.3	93.0	94.5	90.2	92.9	95.3	92.4	94.2	95.1	85.8	91.7
06	WOOD & PLASTICS	91.6	72.9	82.0	96.5	53.0	74.2	102.0	53.0	76.8	83.0	53.0	67.6	97.7	66.5	81.7	96.5	40.7	67.8
07	THERMAL & MOISTURE PROTECTION	97.1	75.1	86.7	96.2	52.4	75.4	95.9	52.4	75.3	96.1	52.4	75.4	95.7	57.4	77.6	96.1	46.3	72.5
08	DOORS & WINDOWS	93.7	72.1	88.5	95.7	53.6	85.6	95.7	56.6	86.3	95.8	57.1	86.5	95.7	60.6	87.3	95.7	46.3	83.8
09200	PLASTER & GYPSUM BOARD	104.0	71.9	83.3	108.5	50.5	71.0	108.5	50.5	71.0	99.2	50.5	67.7	108.5	64.4	80.0	108.5	38.7	63.4
095,098	CEILINGS & ACOUSTICAL TREATMENT	88.5	71.9	77.4	91.3	50.5	63.9	91.3	50.5	63.9	91.3	50.5	63.9	91.3	64.4	73.2	91.3	38.7	55.9
09600	FLOORING	94.4	78.5	90.5	95.9	45.8	83.5	95.7	45.8	83.3	88.1	45.8	77.6	95.7	72.7	90.0	95.9	41.1	82.3
097,099	WALL FINISHES, PAINTS & COATINGS	116.7	88.9	100.4	105.9	48.1	72.0	105.9	62.1	80.3	105.9	42.1	68.6	105.9	62.8	80.7	105.9	37.5	65.9
09	FINISHES	95.6	74.5	84.6	95.5	50.3	72.0	95.5	51.9	72.8	91.7	49.6	69.8	95.4	65.6	79.9	95.4	40.1	66.6
10 - 14	TOTAL DIV. 10000 - 14000	100.0	81.7	96.2	100.0	73.7	94.5	100.0	73.8	94.5	100.0	73.8	94.5	100.0	81.4	96.1	100.0	67.8	93.3
15	MECHANICAL	100.1	82.5	92.3	100.1	65.5	84.7	100.1	62.6	83.4	100.1	62.6	83.4	100.1	70.0	86.7	100.1	45.8	75.9
16	ELECTRICAL	95.1	92.0	93.0	98.6	60.7	72.3	98.6	62.6	73.6	95.7	62.6	72.7	99.6	69.5	78.7	98.6	40.0	57.9
01 - 16	WEIGHTED AVERAGE	98.3	80.9	89.9	97.0	64.6	81.4	97.6	64.8	81.8	96.2	64.5	80.9	97.3	70.7	84.4	97.7	51.9	75.6

WASHINGTON

DIVISION		EVERETT			RICHLAND			SEATTLE			SPOKANE			TACOMA			VANCOUVER		
		MAT.	INST.	TOTAL	MAT.	INST.	TOTAL	MAT.	INST.	TOTAL	MAT.	INST.	TOTAL	MAT.	INST.	TOTAL	MAT.	INST.	TOTAL
01590	EQUIPMENT RENTAL	.0	106.7	106.7	.0	89.8	89.8	.0	106.4	106.4	.0	89.8	89.8	.0	106.7	106.7	.0	99.5	99.5
02	SITE CONSTRUCTION	94.7	118.1	112.4	98.5	90.7	92.5	98.9	116.1	111.9	99.3	90.7	92.7	97.5	118.4	113.3	110.7	103.5	105.2
03100	CONCRETE FORMS & ACCESSORIES	108.3	104.2	104.7	113.3	86.6	90.1	92.2	105.3	103.6	121.1	86.6	91.2	92.2	105.0	103.4	96.5	101.9	101.2
03200	CONCRETE REINFORCEMENT	107.2	94.1	99.6	101.2	88.9	94.0	101.2	94.3	97.2	101.8	88.7	94.2	101.2	94.2	97.1	104.1	94.1	98.3
03300	CAST-IN-PLACE CONCRETE	98.8	107.7	102.4	114.0	88.0	103.5	113.6	112.1	107.1	118.0	88.0	105.9	101.5	112.0	105.7	113.4	105.5	110.2
03	CONCRETE	102.2	102.7	102.5	109.5	87.3	98.4	104.5	104.9	104.7	112.0	87.3	99.7	103.5	104.6	104.0	115.3	101.3	108.3
04	MASONRY	129.2	105.2	114.5	113.6	91.0	99.8	123.4	108.8	114.5	115.3	87.7	98.4	123.2	106.9	113.2	124.6	106.7	113.7
05	METALS	99.2	88.2	95.2	93.1	84.3	89.9	100.7	91.4	97.3	92.9	84.1	89.7	100.2	89.2	96.2	98.7	91.5	96.1
06	WOOD & PLASTICS	100.9	103.6	102.3	99.9	86.3	92.9	85.8	103.6	95.0	108.9	86.3	97.3	84.6	103.6	94.4	82.1	100.7	91.7
07	THERMAL & MOISTURE PROTECTION	108.7	101.6	105.3	173.0	85.0	131.3	108.5	103.9	106.3	170.3	84.4	129.6	108.3	102.8	105.7	111.3	96.7	104.4
08	DOORS & WINDOWS	98.7	99.3	98.9	112.7	80.6	105.0	100.9	100.4	100.8	112.5	79.7	104.6	99.3	100.4	99.5	95.8	97.1	96.1
09200	PLASTER & GYPSUM BOARD	100.8	103.3	102.4	123.0	85.9	99.0	95.3	103.3	100.5	126.4	85.9	100.2	98.3	103.3	101.6	99.7	100.9	100.5
095,098	CEILINGS & ACOUSTICAL TREATMENT	110.7	103.3	105.8	120.9	85.9	97.3	111.8	103.3	106.1	120.9	85.9	97.3	113.5	103.3	106.7	122.1	100.9	107.8
09600	FLOORING	123.3	103.6	118.4	113.3	50.9	97.9	115.0	103.6	112.2	116.9	81.3	108.1	115.7	103.6	112.7	121.3	93.6	114.4
097,099	WALL FINISHES, PAINTS & COATINGS	113.7	87.6	98.4	116.2	78.2	93.9	113.7	94.7	102.6	116.2	78.2	93.9	113.7	94.7	102.6	121.7	72.4	92.8
09	FINISHES	116.6	102.3	109.1	129.1	78.3	102.7	113.5	103.5	108.3	130.8	84.6	106.7	114.3	103.5	108.7	114.8	96.9	105.5
10 - 14	TOTAL DIV. 10000 - 14000	100.0	104.4	100.9	100.0	90.1	97.9	100.0	106.8	101.4	100.0	90.1	97.9	100.0	106.7	101.4	100.0	97.5	99.5
15	MECHANICAL	100.1	101.6	100.8	100.6	99.9	100.3	100.1	110.0	104.5	100.6	89.8	95.8	100.2	102.3	101.1	100.3	106.7	103.2
16	ELECTRICAL	106.3	97.6	100.3	104.2	95.2	98.0	106.0	111.2	109.6	94.8	88.5	90.4	106.0	95.4	98.6	119.1	109.2	112.2
01 - 16	WEIGHTED AVERAGE	103.9	101.9	102.9	107.9	90.2	99.3	104.0	106.8	105.4	107.8	87.4	98.0	103.6	102.5	103.1	106.0	102.6	104.3

DIVISION		WASHINGTON			WEST VIRGINIA												WISCONSIN		
		YAKIMA			CHARLESTON			HUNTINGTON			PARKERSBURG			WHEELING			EAU CLAIRE		
		MAT.	INST.	TOTAL	MAT.	INST.	TOTAL	MAT.	INST.	TOTAL	MAT.	INST.	TOTAL	MAT.	INST.	TOTAL	MAT.	INST.	TOTAL
01590	EQUIPMENT RENTAL	.0	106.7	106.7	.0	99.6	99.6	.0	99.6	99.6	.0	99.6	99.6	.0	99.6	99.6	.0	97.7	97.7
02	SITE CONSTRUCTION	100.9	117.3	113.3	102.6	85.8	89.8	104.0	86.2	90.5	108.8	85.8	91.3	109.4	85.8	91.5	83.1	102.2	97.6
03100	CONCRETE FORMS & ACCESSORIES	94.6	99.8	99.1	103.2	91.7	93.2	96.1	93.9	94.2	85.0	90.4	89.7	87.0	90.0	89.6	97.8	96.0	96.3
03200	CONCRETE REINFORCEMENT	103.6	93.3	97.6	93.1	80.6	85.9	93.1	89.3	90.9	91.8	90.6	91.1	91.2	90.9	91.0	103.8	99.0	101.0
03300	CAST-IN-PLACE CONCRETE	108.5	89.0	100.6	93.4	106.0	98.4	100.7	106.8	103.2	94.1	95.4	94.6	94.1	100.7	96.7	97.2	99.0	98.0
03	CONCRETE	109.2	94.1	101.7	95.8	95.4	95.6	99.1	98.3	98.7	99.0	93.0	96.0	99.0	94.7	96.9	94.9	97.7	96.3
04	MASONRY	115.3	84.3	96.4	83.8	93.6	89.8	85.2	97.4	92.7	73.9	89.4	83.4	95.8	92.4	93.7	93.7	99.2	97.1
05	METALS	99.1	84.9	93.9	95.3	99.3	96.7	95.3	102.8	98.1	94.1	102.5	97.2	94.2	103.2	97.5	93.9	94.8	95.3
06	WOOD & PLASTICS	86.9	103.6	95.5	104.9	91.1	97.9	96.5	92.9	94.7	85.4	88.9	87.2	87.3	88.2	87.8	112.1	95.8	103.7
07	THERMAL & MOISTURE PROTECTION	108.5	83.9	96.9	96.0	89.4	92.9	96.4	91.0	93.8	96.6	88.1	92.6	96.7	89.3	93.2	97.4	91.3	94.5
08	DOORS & WINDOWS	98.9	87.6	96.2	96.9	81.9	93.3	95.7	84.8	93.1	96.4	82.0	92.9	97.3	88.1	95.1	101.9	91.6	99.4
09200	PLASTER & GYPSUM BOARD	98.4	103.3	101.6	108.5	90.7	97.0	108.5	92.6	98.2	101.0	88.4	92.9	101.5	87.7	92.6	114.6	96.2	102.7
095,098	CEILINGS & ACOUSTICAL TREATMENT	109.3	103.3	105.3	91.3	90.7	90.9	91.3	92.6	92.2	88.5	88.4	88.5	88.5	87.7	88.0	109.8	96.2	100.6
09600	FLOORING	117.0	76.5	107.0	95.7	101.3	97.1	95.7	110.6	99.4	91.0	94.9	91.9	91.9	104.1	94.9	92.7	71.1	87.4
097,099	WALL FINISHES, PAINTS & COATINGS	113.7	78.2	92.9	105.9	93.8	98.8	105.9	89.1	96.1	105.9	98.1	101.4	105.9	89.5	96.3	92.0	77.1	83.3
09	FINISHES	114.2	93.6	103.5	95.5	94.1	94.8	95.3	96.8	96.1	92.7	92.1	92.4	93.1	92.4	92.7	101.5	89.1	95.0
10 - 14	TOTAL DIV. 10000 - 14000	100.0	102.8	100.6	100.0	95.8	99.1	100.0	96.6	99.3	100.0	95.9	99.1	100.0	93.4	98.6	100.0	91.8	98.3
15	MECHANICAL	100.2	100.2	100.2	100.1	84.3	93.1	100.1	88.0	94.7	100.1	90.7	95.9	100.1	92.8	96.9	100.2	90.2	95.7
16	ELECTRICAL	110.8	95.2	100.0	98.6	75.3	82.4	98.6	87.8	91.1	99.4	93.9	95.6	94.6	94.3	94.4	105.0	88.0	93.2
01 - 16	WEIGHTED AVERAGE	104.1	95.7	100.0	97.0	88.2	92.7	97.3	92.6	95.0	96.3	91.9	94.2	97.4	93.2	95.4	98.4	93.6	96.1

COST INDEXES

WISCONSIN

DIVISION		GREEN BAY			KENOSHA			LA CROSSE			MADISON			MILWAUKEE			RACINE		
		MAT.	INST.	TOTAL	MAT.	INST.	TOTAL	MAT.	INST.	TOTAL	MAT.	INST.	TOTAL	MAT.	INST.	TOTAL	MAT.	INST.	TOTAL
01590	EQUIPMENT RENTAL	.0	95.4	95.4	.0	97.7	97.7	.0	97.7	97.7	.0	99.9	99.9	.0	86.1	86.1	.0	99.9	99.9
02	SITE CONSTRUCTION	85.4	98.0	94.9	90.5	101.7	99.0	77.4	102.1	96.1	85.6	104.8	100.2	86.8	93.7	92.0	85.9	105.5	100.8
03100	CONCRETE FORMS & ACCESSORIES	117.3	95.7	98.5	113.4	104.4	105.6	80.5	96.0	93.9	100.6	97.2	97.7	103.6	108.5	107.9	100.3	104.4	103.9
03200	CONCRETE REINFORCEMENT	100.2	90.3	94.4	111.7	99.4	104.5	102.2	89.0	94.5	104.7	89.3	95.7	102.9	99.6	101.0	104.7	99.4	101.6
03300	CAST-IN-PLACE CONCRETE	100.6	98.8	99.9	111.7	101.6	107.6	87.4	95.8	90.8	100.6	98.4	99.7	100.6	107.8	103.5	100.6	101.3	100.9
03	CONCRETE	96.6	95.8	96.2	105.9	102.4	104.2	86.1	94.6	90.3	98.6	96.2	97.4	98.8	105.7	102.3	98.6	102.3	100.4
04	MASONRY	115.3	98.2	104.8	100.4	109.8	106.1	92.9	99.3	96.8	104.2	102.6	103.2	104.2	114.5	110.5	104.2	109.7	107.6
05	METALS	96.4	94.2	95.6	96.5	100.8	98.1	93.8	93.6	93.7	96.8	95.3	96.3	98.4	91.4	95.8	96.8	100.8	98.3
06	WOOD & PLASTICS	124.2	95.8	109.6	120.5	103.1	111.6	94.7	95.8	95.3	113.9	96.5	104.9	116.9	107.4	112.0	114.1	103.1	108.5
07	THERMAL & MOISTURE PROTECTION	100.1	89.3	95.0	97.9	102.4	100.0	96.7	89.9	93.5	96.9	97.6	97.2	96.2	107.3	101.5	97.6	101.8	99.6
08	DOORS & WINDOWS	100.3	88.1	97.3	97.2	103.9	98.8	101.8	85.6	97.9	102.1	96.1	100.6	104.4	106.2	104.8	102.1	103.9	102.5
09200	PLASTER & GYPSUM BOARD	99.1	96.2	97.2	96.4	103.7	101.2	107.6	96.2	100.2	107.7	96.9	100.7	107.7	107.9	107.8	107.7	103.7	105.1
095,098	CEILINGS & ACOUSTICAL TREATMENT	104.5	96.2	98.9	87.8	103.7	98.5	104.1	96.2	98.8	97.1	96.9	97.0	97.1	107.0	104.4	97.1	103.7	101.6
09600	FLOORING	111.5	106.3	110.2	108.7	109.5	108.9	85.5	106.3	90.6	92.2	101.3	94.4	94.9	112.4	99.3	92.2	109.5	96.5
097,099	WALL FINISHES, PAINTS & COATINGS	102.6	69.6	83.3	103.6	92.0	96.8	92.0	77.1	83.3	92.9	92.5	92.7	95.4	104.6	100.8	92.9	97.1	95.4
09	FINISHES	105.9	95.3	100.3	101.4	104.3	102.9	96.8	96.4	96.6	98.3	97.6	97.9	99.4	109.1	104.5	98.3	104.9	101.7
10 - 14	TOTAL DIV. 10000 - 14000	100.0	90.7	98.1	100.0	99.1	99.8	100.0	91.8	98.3	100.0	92.0	98.3	100.0	100.8	100.2	100.0	99.1	99.8
15	MECHANICAL	100.6	89.2	95.5	100.4	97.6	99.1	100.2	87.6	94.6	100.1	94.1	97.4	100.1	99.3	99.7	100.1	97.6	99.0
16	ELECTRICAL	96.1	87.8	90.3	93.3	99.1	97.3	105.4	88.0	93.3	92.1	94.8	94.0	92.9	103.7	100.4	91.0	98.9	96.5
01 - 16	WEIGHTED AVERAGE	100.2	92.8	96.6	99.5	101.6	100.5	96.5	92.9	94.7	98.8	97.0	97.9	99.5	102.9	101.1	98.7	102.0	100.3

WYOMING / CANADA

DIVISION		CASPER			CHEYENNE			ROCK SPRINGS			CALGARY, ALBERTA			EDMONTON, ALBERTA			HALIFAX, NOVA SCOTIA		
		MAT.	INST.	TOTAL	MAT.	INST.	TOTAL	MAT.	INST.	TOTAL	MAT.	INST.	TOTAL	MAT.	INST.	TOTAL	MAT.	INST.	TOTAL
01590	EQUIPMENT RENTAL	.0	100.3	100.3	.0	100.3	100.3	.0	100.3	100.3	.0	98.6	98.6	.0	98.6	98.6	.0	98.6	98.6
02	SITE CONSTRUCTION	83.6	99.0	95.3	83.4	99.0	95.2	87.1	98.6	95.8	108.9	97.2	100.0	110.3	97.2	100.4	97.8	96.6	96.9
03100	CONCRETE FORMS & ACCESSORIES	100.1	49.1	55.8	100.2	49.1	55.8	105.7	34.7	44.0	123.2	78.7	84.6	120.6	78.7	84.2	89.2	75.0	76.8
03200	CONCRETE REINFORCEMENT	107.3	52.5	75.4	100.7	52.6	72.7	109.4	52.2	76.1	161.2	61.5	103.1	161.2	61.5	103.1	147.9	56.1	94.4
03300	CAST-IN-PLACE CONCRETE	105.3	63.8	88.6	105.3	63.8	88.6	105.8	48.0	82.5	162.8	81.8	130.1	163.6	81.8	130.6	178.4	76.8	137.5
03	CONCRETE	107.3	55.6	81.6	106.5	55.7	81.2	108.1	43.9	76.1	146.7	77.1	112.0	146.9	77.1	112.1	152.3	72.7	112.6
04	MASONRY	107.0	44.8	69.0	106.0	47.2	70.1	172.6	36.6	89.5	177.8	80.0	118.1	177.8	80.0	118.1	160.2	79.4	110.8
05	METALS	105.0	65.6	90.6	106.3	65.7	91.5	105.0	64.3	90.2	100.8	81.4	93.7	100.8	81.4	93.7	100.7	78.3	92.5
06	WOOD & PLASTICS	95.4	47.6	70.9	94.6	47.6	70.5	101.5	33.8	66.7	114.8	78.1	95.9	111.1	78.1	94.1	83.9	74.3	78.9
07	THERMAL & MOISTURE PROTECTION	105.1	60.3	83.9	105.1	60.9	84.2	105.8	45.9	77.4	104.6	79.4	92.7	104.5	79.4	92.6	103.4	76.2	90.5
08	DOORS & WINDOWS	92.5	48.7	82.0	93.3	48.7	82.5	99.5	41.3	85.5	90.5	73.5	86.4	90.5	73.5	86.4	81.3	68.8	78.3
09200	PLASTER & GYPSUM BOARD	83.4	45.8	59.1	84.9	45.8	59.6	87.3	31.5	51.3	151.9	77.5	103.8	149.4	77.5	102.9	188.1	73.5	114.1
095,098	CEILINGS & ACOUSTICAL TREATMENT	100.1	45.8	63.6	107.2	45.8	65.9	100.1	31.5	54.0	101.9	77.5	85.5	101.9	77.5	85.5	101.9	73.5	82.8
09600	FLOORING	105.1	48.0	91.0	105.1	73.8	97.4	107.4	47.1	92.5	132.5	80.9	119.8	132.5	80.9	119.8	110.2	70.6	100.5
097,099	WALL FINISHES, PAINTS & COATINGS	108.1	63.4	81.9	108.0	63.4	81.9	108.1	38.5	67.3	109.3	76.4	90.0	109.4	76.4	90.1	109.3	68.1	85.2
09	FINISHES	97.6	49.6	72.6	99.0	54.9	76.0	98.9	36.5	66.4	120.6	79.3	99.1	120.4	79.3	99.0	117.1	73.5	94.4
10 - 14	TOTAL DIV. 10000 - 14000	100.0	77.3	95.3	100.0	77.2	95.3	100.0	57.9	91.2	100.0	86.3	97.1	100.0	82.1	96.3	100.0	71.6	94.1
15	MECHANICAL	100.0	66.6	85.1	100.0	50.4	78.0	100.0	35.6	71.4	99.2	70.6	86.5	99.2	70.6	86.5	98.6	78.6	89.7
16	ELECTRICAL	92.6	65.1	73.5	92.6	60.1	70.0	91.6	51.8	64.0	114.0	77.9	89.0	113.9	77.9	88.9	123.7	81.1	94.2
01 - 16	WEIGHTED AVERAGE	100.0	63.0	82.1	100.3	59.9	80.8	104.9	49.4	78.1	112.6	79.0	96.4	112.6	78.9	96.3	110.7	78.8	95.2

CANADA

DIVISION		HAMILTON, ONTARIO			KITCHENER, ONTARIO			LAVAL, QUEBEC			LONDON, ONTARIO			MONTREAL, QUEBEC			OSHAWA, ONTARIO		
		MAT.	INST.	TOTAL	MAT.	INST.	TOTAL	MAT.	INST.	TOTAL	MAT.	INST.	TOTAL	MAT.	INST.	TOTAL	MAT.	INST.	TOTAL
01590	EQUIPMENT RENTAL	.0	98.6	98.6	.0	98.6	98.6	.0	98.6	98.6	.0	98.6	98.6	.0	98.6	98.6	.0	98.6	98.6
02	SITE CONSTRUCTION	112.2	97.4	101.0	99.5	97.0	97.6	92.6	97.6	96.4	112.2	97.1	100.8	92.4	97.6	96.3	113.8	97.2	101.2
03100	CONCRETE FORMS & ACCESSORIES	122.6	102.4	105.0	116.4	93.7	96.7	132.3	94.0	99.0	122.4	97.1	100.4	132.3	94.0	99.0	124.2	98.2	101.6
03200	CONCRETE REINFORCEMENT	174.5	100.8	131.6	110.2	100.7	104.6	156.6	92.7	119.4	123.0	99.0	109.0	159.3	92.7	120.5	174.5	101.5	131.9
03300	CAST-IN-PLACE CONCRETE	157.2	104.8	136.1	146.7	84.7	121.7	140.1	100.7	124.2	157.2	102.1	135.0	138.6	100.7	123.3	169.7	90.7	137.8
03	CONCRETE	145.4	102.7	124.1	125.4	91.6	108.6	135.3	96.0	115.7	139.2	99.1	119.2	134.8	96.0	115.5	151.8	96.1	124.1
04	MASONRY	161.8	108.9	129.5	159.9	104.0	125.8	158.8	95.2	119.9	161.8	105.6	127.5	158.7	95.2	119.9	162.7	108.0	129.3
05	METALS	100.9	97.0	99.5	100.7	90.5	96.9	101.0	91.7	97.6	99.2	96.5	98.2	101.0	91.7	97.6	100.7	96.7	99.3
06	WOOD & PLASTICS	115.6	101.6	108.4	109.3	92.4	100.6	128.3	94.3	110.8	115.6	95.7	105.3	128.3	94.3	110.8	116.6	96.7	106.3
07	THERMAL & MOISTURE PROTECTION	104.6	100.6	102.7	103.1	95.6	99.5	104.8	96.1	100.7	104.6	98.1	101.5	104.8	96.7	101.0	104.6	96.2	100.6
08	DOORS & WINDOWS	90.5	99.6	92.7	82.0	88.9	83.7	90.5	82.6	88.6	91.4	95.3	92.3	90.5	82.6	88.6	89.7	96.6	91.3
09200	PLASTER & GYPSUM BOARD	209.7	101.7	139.9	153.2	92.2	113.8	148.3	94.2	113.3	209.7	95.6	136.0	148.3	94.2	113.3	157.1	96.6	118.0
095,098	CEILINGS & ACOUSTICAL TREATMENT	101.9	101.7	101.8	101.9	92.2	95.4	101.9	94.2	96.7	101.9	95.6	97.7	101.9	94.2	96.7	101.9	96.6	98.3
09600	FLOORING	132.5	105.4	125.8	128.2	105.4	122.5	132.5	106.2	126.0	132.5	105.4	125.8	132.5	106.2	126.0	132.5	108.1	126.5
097,099	WALL FINISHES, PAINTS & COATINGS	109.3	113.7	111.9	109.3	104.2	106.3	109.3	99.0	103.3	109.3	113.7	111.9	109.3	99.0	103.3	109.3	119.5	115.3
09	FINISHES	128.0	104.5	115.8	118.2	97.2	107.3	119.1	97.4	107.8	128.0	100.6	113.8	119.1	97.4	107.8	120.9	102.7	111.4
10 - 14	TOTAL DIV. 10000 - 14000	100.0	106.7	101.4	100.0	104.1	100.8	100.0	91.5	98.2	100.0	105.1	101.1	100.0	91.5	98.2	100.0	105.7	101.2
15	MECHANICAL	99.2	98.3	98.8	98.6	93.9	96.5	98.6	85.5	92.8	99.2	94.4	97.0	99.2	85.5	93.1	98.6	96.0	97.4
16	ELECTRICAL	127.1	108.8	114.4	127.1	105.0	111.8	123.7	85.9	97.5	127.2	105.2	111.9	123.7	85.9	97.5	127.2	105.8	112.4
01 - 16	WEIGHTED AVERAGE	113.3	102.6	108.1	108.1	96.9	102.7	110.2	91.5	101.1	112.4	99.5	106.2	110.3	91.5	101.2	113.2	100.1	106.9

City Cost Indexes

		CANADA																
	DIVISION	OTTAWA, ONTARIO			QUEBEC, QUEBEC			REGINA, SASKATCHEWAN			SASKATOON, SASKATCHEWAN			ST CATHARINES, ONTARIO			ST JOHNS, NEWFOUNDLAND		
		MAT.	INST.	TOTAL	MAT.	INST.	TOTAL	MAT.	INST.	TOTAL	MAT.	INST.	TOTAL	MAT.	INST.	TOTAL	MAT.	INST.	TOTAL
01590	EQUIPMENT RENTAL	.0	98.6	98.6	.0	98.6	98.6	.0	98.6	98.6	.0	98.6	98.6	.0	98.6	98.6	.0	98.6	98.6
02	SITE CONSTRUCTION	112.5	97.2	100.9	94.1	97.6	96.7	110.8	95.6	99.3	105.8	95.7	98.1	100.0	97.0	97.7	113.7	95.3	99.7
03100	CONCRETE FORMS & ACCESSORIES	120.9	98.2	101.2	132.3	94.1	99.1	101.8	64.4	69.3	101.8	64.3	69.2	113.7	95.7	98.1	99.9	64.5	69.2
03200	CONCRETE REINFORCEMENT	174.5	99.0	130.5	147.9	92.7	115.7	113.8	73.6	90.4	118.8	73.6	92.5	111.2	100.7	105.1	147.9	56.2	94.5
03300	CAST-IN-PLACE CONCRETE	159.6	102.1	136.4	151.5	100.8	131.1	162.8	73.8	126.9	149.1	73.7	118.7	140.1	89.4	119.6	179.6	71.6	136.1
03	CONCRETE	146.5	99.6	123.1	140.0	96.1	118.1	132.8	70.0	101.5	126.5	70.0	98.4	122.0	94.1	108.1	165.9	66.2	116.3
04	MASONRY	162.0	105.1	127.2	159.7	95.2	120.3	167.9	70.5	108.4	166.9	70.5	108.0	159.4	101.0	123.7	160.3	68.3	104.1
05	METALS	100.9	96.5	99.3	101.0	91.8	97.7	98.2	78.5	91.0	98.4	78.4	91.1	97.9	90.6	95.2	102.1	74.9	92.2
06	WOOD & PLASTICS	114.7	97.6	105.9	128.3	94.3	110.8	93.9	62.9	78.0	93.9	62.9	78.0	106.2	97.4	101.7	93.3	63.7	78.1
07	THERMAL & MOISTURE PROTECTION	104.6	97.0	101.0	104.8	96.7	101.0	103.9	69.9	87.7	103.7	68.8	87.2	103.1	96.9	100.2	106.2	66.1	87.2
08	DOORS & WINDOWS	90.5	96.3	91.9	90.5	87.6	89.8	85.5	61.3	79.7	84.1	61.3	78.6	81.6	92.0	84.1	96.7	60.3	88.0
09200	PLASTER & GYPSUM BOARD	259.6	97.6	154.9	216.1	94.2	137.3	125.1	61.9	84.2	125.1	61.9	84.2	149.6	97.4	115.8	194.1	62.6	109.1
095,098	CEILINGS & ACOUSTICAL TREATMENT	101.9	97.6	99.0	101.9	94.2	96.7	99.1	61.9	74.0	99.1	61.9	74.0	99.1	97.4	97.9	104.7	62.6	76.4
09600	FLOORING	132.5	103.9	125.4	132.5	106.2	126.0	120.2	67.1	107.0	120.2	67.1	107.0	126.5	105.4	121.3	115.1	59.5	101.3
097,099	WALL FINISHES, PAINTS & COATINGS	109.3	106.4	107.6	109.3	99.0	103.3	109.3	70.9	86.8	109.3	60.5	80.7	109.3	106.5	107.6	109.3	65.0	83.3
09	FINISHES	134.8	100.4	116.9	128.3	97.4	112.2	112.3	65.3	87.8	112.3	64.1	87.2	116.7	99.3	107.6	121.1	63.6	91.2
10 - 14	TOTAL DIV. 10000 - 14000	100.0	102.9	100.6	100.0	91.5	98.2	100.0	69.4	93.6	100.0	69.4	93.6	100.0	79.3	95.7	100.0	70.0	93.7
15	MECHANICAL	99.2	94.7	97.2	99.2	85.5	93.1	98.6	70.9	86.3	98.6	70.9	86.3	98.6	92.0	95.7	98.6	66.2	84.2
16	ELECTRICAL	123.7	106.4	111.7	123.7	85.9	97.5	128.9	73.5	90.4	129.0	73.5	90.5	127.1	106.5	112.8	123.8	67.8	84.9
01 - 16	WEIGHTED AVERAGE	113.9	99.7	107.1	111.9	91.7	102.2	109.2	73.1	91.7	108.1	72.9	91.1	107.0	96.7	102.0	115.4	69.8	93.4

		CANADA																	
	DIVISION	THUNDER BAY, ONTARIO			TORONTO, ONTARIO			VANCOUVER, B C			WINDSOR, ONTARIO			WINNIPEG, MANITOBA					
		MAT.	INST.	TOTAL	MAT.	INST.	TOTAL	MAT.	INST.	TOTAL	MAT.	INST.	TOTAL	MAT.	INST.	TOTAL	MAT.	INST.	TOTAL
01590	EQUIPMENT RENTAL	.0	98.6	98.6	.0	98.6	98.6	.0	98.6	98.6	.0	98.6	98.6	.0	98.6	98.6	.0	.0	.0
02	SITE CONSTRUCTION	105.9	97.1	99.2	113.1	97.8	101.5	111.3	98.7	101.7	95.7	97.3	96.9	110.5	96.1	99.6	.0	.0	.0
03100	CONCRETE FORMS & ACCESSORIES	124.0	97.1	100.6	124.2	110.0	111.9	115.4	98.5	100.7	124.0	97.8	101.2	123.3	77.5	83.5	.0	.0	.0
03200	CONCRETE REINFORCEMENT	99.4	100.0	99.7	170.8	101.9	130.6	161.2	93.0	121.4	108.8	99.2	103.2	161.2	61.7	103.2	.0	.0	.0
03300	CAST-IN-PLACE CONCRETE	155.4	102.2	133.9	164.5	112.7	143.6	159.6	102.5	136.6	143.5	103.5	127.4	151.2	77.8	121.6	.0	.0	.0
03	CONCRETE	132.4	99.2	115.9	148.8	109.0	129.0	150.7	98.7	124.8	124.1	99.9	112.1	133.9	75.1	104.6	.0	.0	.0
04	MASONRY	160.4	102.0	124.7	184.9	118.4	144.3	175.3	102.0	130.5	159.7	109.6	129.1	165.7	73.8	109.6	.0	.0	.0
05	METALS	97.3	95.5	96.6	100.9	99.0	100.2	101.1	93.0	98.2	97.4	96.8	97.2	100.8	80.2	93.3	.0	.0	.0
06	WOOD & PLASTICS	116.6	97.8	106.9	116.6	108.9	112.6	103.7	96.5	100.0	116.6	96.1	106.1	114.6	78.8	96.2	.0	.0	.0
07	THERMAL & MOISTURE PROTECTION	103.6	96.1	100.0	104.9	108.2	106.5	104.4	98.2	101.5	103.2	100.4	101.8	104.6	77.6	91.8	.0	.0	.0
08	DOORS & WINDOWS	80.6	95.4	84.2	89.7	105.6	93.5	90.5	93.1	91.1	80.3	95.1	83.9	90.5	69.4	85.5	.0	.0	.0
09200	PLASTER & GYPSUM BOARD	156.2	97.8	118.5	147.4	109.2	122.7	154.2	96.5	116.9	156.2	96.1	117.3	138.5	78.2	99.5	.0	.0	.0
095,098	CEILINGS & ACOUSTICAL TREATMENT	97.6	97.8	97.8	101.9	109.2	106.8	101.9	96.5	98.3	97.6	96.1	96.6	101.9	78.2	85.9	.0	.0	.0
09600	FLOORING	132.5	60.7	114.8	132.5	111.7	127.4	132.5	103.7	125.4	132.5	106.2	126.0	132.5	73.7	118.0	.0	.0	.0
097,099	WALL FINISHES, PAINTS & COATINGS	109.3	108.8	109.0	112.2	119.5	116.5	109.3	112.0	110.8	109.3	108.3	108.7	109.4	63.6	82.6	.0	.0	.0
09	FINISHES	119.7	92.7	105.6	119.9	111.8	115.7	121.5	100.6	110.6	119.4	100.7	109.6	118.7	75.7	96.3	.0	.0	.0
10 - 14	TOTAL DIV. 10000 - 14000	100.0	80.6	95.9	100.0	109.4	102.0	100.0	102.0	100.4	100.0	82.0	96.3	100.0	72.3	94.2	.0	.0	.0
15	MECHANICAL	98.6	93.1	96.1	99.2	105.1	101.8	99.2	96.3	97.9	98.6	96.1	97.5	99.2	78.5	90.0	.0	.0	.0
16	ELECTRICAL	127.1	104.4	111.3	127.5	109.6	115.0	131.2	97.0	107.4	127.2	106.4	112.8	129.0	82.2	96.5	.0	.0	.0
01 - 16	WEIGHTED AVERAGE	108.7	97.1	103.1	114.2	107.4	110.9	114.2	97.8	106.3	107.3	100.1	103.8	111.2	79.2	95.8	.0	.0	.0

Location Factors

Costs shown in *Means cost data publications* are based on National Averages for materials and installation. To adjust these costs to a specific location, simply multiply the base cost by the factor and divide by 100 for that city. The data is arranged alphabetically by state and postal zip code numbers. For a city not listed, use the factor for a nearby city with similar economic characteristics.

STATE/ZIP	CITY	MAT.	INST.	TOTAL
ALABAMA				
350-352	Birmingham	96.2	75.8	86.4
354	Tuscaloosa	95.9	59.7	78.4
355	Jasper	97.3	56.1	77.4
356	Decatur	95.9	63.8	80.4
357-358	Huntsville	95.8	67.2	82.0
359	Gadsden	96.9	63.4	80.7
360-361	Montgomery	96.8	62.7	80.3
362	Anniston	95.5	51.3	74.1
363	Dothan	96.4	55.8	76.7
364	Evergreen	95.7	57.1	77.0
365-366	Mobile	96.7	66.5	82.1
367	Selma	96.0	57.2	77.2
368	Phenix City	96.7	62.8	80.3
369	Butler	96.1	55.7	76.6
ALASKA				
995-996	Anchorage	134.1	113.7	124.3
997	Fairbanks	129.5	117.2	123.6
998	Juneau	132.6	113.7	123.5
999	Ketchikan	143.9	113.7	129.3
ARIZONA				
850,853	Phoenix	99.8	78.2	89.4
852	Mesa/Tempe	99.6	67.2	83.9
855	Globe	100.2	68.3	84.8
856-857	Tucson	98.8	74.3	87.0
859	Show Low	100.4	68.5	85.0
860	Flagstaff	102.5	73.4	88.4
863	Prescott	99.9	70.7	85.8
864	Kingman	98.4	71.0	85.2
865	Chambers	98.4	68.3	83.8
ARKANSAS				
716	Pine Bluff	95.8	62.7	79.8
717	Camden	94.0	44.7	70.1
718	Texarkana	95.3	50.2	73.5
719	Hot Springs	93.1	44.1	69.4
720-722	Little Rock	96.6	64.0	80.8
723	West Memphis	95.4	60.9	78.7
724	Jonesboro	95.5	60.9	78.8
725	Batesville	94.0	55.4	75.4
726	Harrison	95.5	55.4	76.1
727	Fayetteville	92.6	38.4	66.4
728	Russellville	94.1	53.4	74.4
729	Fort Smith	96.7	61.6	79.8
CALIFORNIA				
900-902	Los Angeles	103.4	113.4	108.3
903-905	Inglewood	99.4	112.8	105.9
906-908	Long Beach	101.2	112.4	106.6
910-912	Pasadena	100.7	113.0	106.6
913-916	Van Nuys	104.6	113.0	108.6
917-918	Alhambra	103.2	113.0	107.9
919-921	San Diego	105.3	106.3	105.8
922	Palm Springs	101.6	109.4	105.4
923-924	San Bernardino	99.1	110.3	104.5
925	Riverside	103.8	111.0	107.3
926-927	Santa Ana	101.4	111.8	106.4
928	Anaheim	104.2	112.6	108.2
930	Oxnard	104.7	111.7	108.1
931	Santa Barbara	104.0	111.9	107.8
932-933	Bakersfield	103.7	108.6	106.1
934	San Luis Obispo	105.5	111.7	108.5
935	Mojave	102.1	108.4	105.1
936-938	Fresno	104.8	111.4	108.0
939	Salinas	106.4	117.5	111.8
940-941	San Francisco	112.3	137.3	124.4
942,956-958	Sacramento	107.5	113.1	110.2
943	Palo Alto	105.5	131.6	118.1
944	San Mateo	108.7	129.4	118.7
945	Vallejo	106.0	122.8	114.1
946	Oakland	110.9	127.9	119.1
947	Berkeley	110.2	127.3	118.5
948	Richmond	110.0	125.2	117.3
949	San Rafael	111.7	129.1	120.1
950	Santa Cruz	110.5	117.6	113.9

STATE/ZIP	CITY	MAT.	INST.	TOTAL
CALIFORNIA (CONT'D)				
951	San Jose	109.7	131.4	120.2
952	Stockton	105.2	113.2	109.1
953	Modesto	105.4	113.0	109.0
954	Santa Rosa	106.5	128.5	117.1
955	Eureka	107.7	109.9	108.8
959	Marysville	106.6	111.4	108.9
960	Redding	107.8	111.5	109.6
961	Susanville	107.7	111.5	109.5
COLORADO				
800-802	Denver	102.2	87.9	95.3
803	Boulder	100.0	67.1	84.1
804	Golden	102.6	82.6	92.9
805	Fort Collins	103.6	79.0	91.7
806	Greeley	100.4	67.1	84.3
807	Fort Morgan	100.8	81.0	91.3
808-809	Colorado Springs	100.7	83.4	92.4
810	Pueblo	102.2	81.7	92.3
811	Alamosa	104.6	68.1	87.0
812	Salida	104.6	68.9	87.3
813	Durango	105.2	65.1	85.8
814	Montrose	103.5	62.5	83.7
815	Grand Junction	106.7	62.8	85.5
816	Glenwood Springs	104.4	75.6	90.5
CONNECTICUT				
060	New Britain	101.7	108.5	105.0
061	Hartford	102.1	108.5	105.2
062	Willimantic	102.4	106.4	104.3
063	New London	98.9	108.5	103.5
064	Meriden	101.5	106.9	104.1
065	New Haven	102.0	108.7	105.2
066	Bridgeport	103.1	107.5	105.2
067	Waterbury	102.5	108.5	105.4
068	Norwalk	102.4	107.5	104.9
069	Stamford	102.6	113.0	107.6
D.C.				
200-205	Washington	99.1	89.5	94.5
DELAWARE				
197	Newark	98.9	103.9	101.3
198	Wilmington	98.2	103.9	101.0
199	Dover	98.8	103.9	101.3
FLORIDA				
320,322	Jacksonville	98.2	64.8	82.1
321	Daytona Beach	98.3	73.5	86.3
323	Tallahassee	98.7	54.4	77.3
324	Panama City	99.3	42.3	71.7
325	Pensacola	98.9	64.0	82.1
326,344	Gainesville	99.8	60.9	81.0
327-328,347	Orlando	100.3	67.6	84.5
329	Melbourne	100.3	77.0	89.1
330-332,340	Miami	97.6	71.7	85.1
333	Fort Lauderdale	97.5	71.5	84.9
334,349	West Palm Beach	96.3	68.1	82.7
335-336,346	Tampa	99.4	62.7	81.7
337	St. Petersburg	101.6	62.2	82.6
338	Lakeland	98.3	62.4	80.9
339,341	Fort Myers	97.7	59.6	79.3
342	Sarasota	99.5	59.9	80.4
GEORGIA				
300-303,399	Atlanta	96.0	82.8	89.6
304	Statesboro	95.9	49.9	73.6
305	Gainesville	94.7	64.3	80.0
306	Athens	93.9	68.9	81.8
307	Dalton	95.7	35.8	66.8
308-309	Augusta	94.6	59.3	77.5
310-312	Macon	95.6	65.2	80.9
313-314	Savannah	97.5	63.0	80.8
315	Waycross	97.4	48.3	73.7
316	Valdosta	97.1	52.7	75.6
317	Albany	97.1	59.0	78.7
318-319	Columbus	97.2	57.5	78.0

STATE/ZIP	CITY	MAT.	INST.	TOTAL
HAWAII				
967	Hilo	115.6	130.7	122.9
968	Honolulu	116.3	130.7	123.2
STATES & POSS.				
969	Guam	201.7	59.1	132.8
IDAHO				
832	Pocatello	101.7	82.6	92.5
833	Twin Falls	103.9	50.4	78.0
834	Idaho Falls	101.1	61.5	82.0
835	Lewiston	112.9	87.5	100.7
836-837	Boise	101.7	84.1	93.2
838	Coeur d'Alene	112.7	61.6	88.0
ILLINOIS				
600-603	North Suburban	98.6	121.0	109.5
604	Joliet	98.6	121.3	109.6
605	South Suburban	98.6	120.3	109.1
606	Chicago	99.1	125.8	112.0
609	Kankakee	94.9	105.9	100.2
610-611	Rockford	97.0	111.5	104.0
612	Rock Island	94.8	100.2	97.4
613	La Salle	96.2	100.1	98.1
614	Galesburg	95.9	106.1	100.8
615-616	Peoria	98.4	106.6	102.4
617	Bloomington	95.1	106.2	100.5
618-619	Champaign	99.1	102.4	100.7
620-622	East St. Louis	93.5	107.8	100.4
623	Quincy	95.2	98.4	96.8
624	Effingham	94.5	102.8	98.5
625	Decatur	95.9	100.2	98.0
626-627	Springfield	95.6	100.1	97.8
628	Centralia	92.6	106.0	99.1
629	Carbondale	92.3	101.6	96.8
INDIANA				
460	Anderson	97.3	89.1	93.4
461-462	Indianapolis	100.3	91.1	95.9
463-464	Gary	98.1	105.6	101.7
465-466	South Bend	96.4	86.7	91.7
467-468	Fort Wayne	97.5	87.2	92.5
469	Kokomo	95.6	88.4	92.1
470	Lawrenceburg	93.3	86.2	89.9
471	New Albany	95.0	81.8	88.6
472	Columbus	97.3	87.8	92.7
473	Muncie	97.6	87.7	92.8
474	Bloomington	99.3	87.1	93.4
475	Washington	95.6	89.6	92.7
476-477	Evansville	97.2	92.3	94.8
478	Terre Haute	97.6	92.8	95.3
479	Lafayette	97.1	86.9	92.2
IOWA				
500-503,509	Des Moines	98.3	86.3	92.5
504	Mason City	96.7	63.8	80.8
505	Fort Dodge	96.9	58.2	78.2
506-507	Waterloo	98.2	65.6	82.4
508	Creston	97.5	70.2	84.3
510-511	Sioux City	99.0	77.1	88.4
512	Sibley	97.8	57.6	78.3
513	Spencer	100.0	54.9	78.2
514	Carroll	96.4	61.6	79.6
515	Council Bluffs	99.7	78.6	89.5
516	Shenandoah	96.5	56.2	77.0
520	Dubuque	98.3	77.5	88.3
521	Decorah	97.6	59.6	79.3
522-524	Cedar Rapids	99.5	84.0	92.0
525	Ottumwa	97.7	74.1	86.3
526	Burlington	96.5	74.8	86.0
527-528	Davenport	98.5	93.5	96.0
KANSAS				
660-662	Kansas City	97.9	89.5	93.8
664-666	Topeka	97.2	72.4	85.2
667	Fort Scott	97.0	69.0	83.4
668	Emporia	96.8	63.2	80.5
669	Belleville	98.8	61.6	80.8
670-672	Wichita	96.5	74.9	86.0
673	Independence	98.5	58.1	79.0
674	Salina	98.4	62.8	81.2
675	Hutchinson	93.5	57.0	75.9
676	Hays	98.0	61.6	80.4
677	Colby	98.7	61.6	80.8

STATE/ZIP	CITY	MAT.	INST.	TOTAL
KANSAS (CONT'D)				
678	Dodge City	99.7	61.6	81.3
679	Liberal	97.7	51.1	75.2
KENTUCKY				
400-402	Louisville	96.1	87.2	91.8
403-405	Lexington	96.9	70.8	84.3
406	Frankfort	96.2	74.4	85.7
407-409	Corbin	94.6	49.3	72.7
410	Covington	95.5	93.6	94.6
411-412	Ashland	93.7	99.9	96.7
413-414	Campton	95.5	49.8	73.4
415-416	Pikeville	96.6	69.1	83.3
417-418	Hazard	94.8	49.5	72.9
420	Paducah	93.6	91.0	92.3
421-422	Bowling Green	95.6	85.7	90.8
423	Owensboro	95.6	81.3	88.7
424	Henderson	93.1	91.6	92.4
425-426	Somerset	92.5	49.7	71.8
427	Elizabethtown	92.1	87.1	89.7
LOUISIANA				
700-701	New Orleans	101.8	68.0	85.5
703	Thibodaux	101.9	67.9	85.4
704	Hammond	98.6	66.4	83.0
705	Lafayette	100.9	59.9	81.1
706	Lake Charles	101.1	63.2	82.8
707-708	Baton Rouge	100.8	60.5	81.3
710-711	Shreveport	96.7	63.4	80.6
712	Monroe	96.7	59.7	78.8
713-714	Alexandria	96.9	57.5	77.8
MAINE				
039	Kittery	97.2	77.2	87.5
040-041	Portland	100.1	85.5	93.0
042	Lewiston	100.6	85.5	93.3
043	Augusta	99.3	76.3	88.2
044	Bangor	99.7	85.0	92.6
045	Bath	98.5	79.0	89.1
046	Machias	98.0	75.9	87.3
047	Houlton	98.1	80.0	89.4
048	Rockland	97.0	74.6	86.2
049	Waterville	98.5	71.0	85.2
MARYLAND				
206	Waldorf	97.0	77.3	87.5
207-208	College Park	96.9	83.6	90.5
209	Silver Spring	96.2	82.0	89.3
210-212	Baltimore	96.9	85.7	91.5
214	Annapolis	96.4	82.7	89.8
215	Cumberland	93.6	81.7	87.8
216	Easton	95.2	50.0	73.4
217	Hagerstown	93.9	82.2	88.3
218	Salisbury	95.7	56.9	76.9
219	Elkton	92.6	73.3	83.3
MASSACHUSETTS				
010-011	Springfield	100.6	103.1	101.8
012	Pittsfield	100.4	97.4	99.0
013	Greenfield	98.5	101.8	100.1
014	Fitchburg	96.9	111.7	104.0
015-016	Worcester	100.5	112.9	106.5
017	Framingham	96.5	117.2	106.5
018	Lowell	100.1	117.0	108.3
019	Lawrence	101.4	116.4	108.6
020-022, 024	Boston	102.8	128.0	115.0
023	Brockton	101.4	114.5	107.7
025	Buzzards Bay	95.7	112.8	103.9
026	Hyannis	98.2	112.8	105.3
027	New Bedford	101.0	113.2	106.9
MICHIGAN				
480,483	Royal Oak	94.6	109.2	101.7
481	Ann Arbor	97.1	110.1	103.4
482	Detroit	98.4	115.1	106.5
484-485	Flint	96.8	103.1	99.8
486	Saginaw	97.0	98.3	97.7
487	Bay City	96.6	98.2	97.3
488-489	Lansing	96.9	99.1	98.0
490	Battle Creek	96.8	92.2	94.6
491	Kalamazoo	97.2	90.4	93.9
492	Jackson	95.3	97.5	96.4
493,495	Grand Rapids	96.7	73.0	85.2
494	Muskegon	95.8	88.5	92.2

STATE/ZIP	CITY	MAT.	INST.	TOTAL
MICHIGAN (CONT'D)				
496	Traverse City	94.7	73.3	84.4
497	Gaylord	96.2	79.3	88.1
498-499	Iron Mountain	98.2	92.1	95.3
MINNESOTA				
550-551	Saint Paul	99.2	114.9	106.8
553-555	Minneapolis	100.0	116.5	108.0
556-558	Duluth	99.3	110.4	104.6
559	Rochester	99.0	99.9	99.4
560	Mankato	97.2	100.1	98.6
561	Windom	96.0	81.7	89.1
562	Willmar	95.2	89.4	92.4
563	St. Cloud	96.3	110.6	103.2
564	Brainerd	97.0	101.0	98.9
565	Detroit Lakes	99.3	89.9	94.8
566	Bemidji	98.5	98.1	98.4
567	Thief River Falls	97.6	89.9	93.9
MISSISSIPPI				
386	Clarksdale	96.5	33.8	66.2
387	Greenville	100.2	51.4	76.6
388	Tupelo	98.0	43.0	71.4
389	Greenwood	97.9	36.3	68.1
390-392	Jackson	98.3	53.1	76.5
393	Meridian	96.4	52.9	75.4
394	Laurel	97.9	37.5	68.7
395	Biloxi	98.4	60.9	80.3
396	McComb	96.3	41.9	70.0
397	Columbus	97.9	42.5	71.1
MISSOURI				
630-631	St. Louis	96.1	110.4	103.0
633	Bowling Green	95.8	93.0	94.4
634	Hannibal	94.5	91.0	92.8
635	Kirksville	97.0	82.4	89.9
636	Flat River	96.8	97.5	97.1
637	Cape Girardeau	96.1	95.6	95.9
638	Sikeston	94.6	89.9	92.3
639	Poplar Bluff	94.1	88.8	91.5
640-641	Kansas City	98.8	102.7	100.7
644-645	St. Joseph	98.6	88.7	93.8
646	Chillicothe	95.8	74.6	85.5
647	Harrisonville	95.3	97.6	96.4
648	Joplin	97.6	73.0	85.7
650-651	Jefferson City	95.0	88.4	91.8
652	Columbia	96.5	88.6	92.7
653	Sedalia	95.8	87.7	91.9
654-655	Rolla	94.5	83.6	89.2
656-658	Springfield	97.7	76.8	87.6
MONTANA				
590-591	Billings	101.5	80.0	91.1
592	Wolf Point	101.9	76.6	89.7
593	Miles City	99.8	77.2	88.9
594	Great Falls	102.7	78.0	90.8
595	Havre	100.6	76.8	89.1
596	Helena	102.6	77.4	90.4
597	Butte	101.2	76.0	89.0
598	Missoula	99.4	75.8	88.0
599	Kalispell	98.5	75.5	87.4
NEBRASKA				
680-681	Omaha	100.4	80.8	90.9
683-685	Lincoln	97.8	66.5	82.7
686	Columbus	96.5	47.2	72.7
687	Norfolk	98.9	65.1	82.6
688	Grand Island	97.6	68.2	83.4
689	Hastings	97.2	56.9	77.7
690	Mccook	96.9	52.2	75.3
691	North Platte	97.0	65.4	81.8
692	Valentine	100.5	44.7	73.5
693	Alliance	99.8	40.7	71.2
NEVADA				
889-891	Las Vegas	101.0	107.5	104.2
893	Ely	102.0	85.6	94.1
894-895	Reno	101.9	96.9	99.5
897	Carson City	101.7	96.7	99.3
898	Elko	100.5	85.3	93.2
NEW HAMPSHIRE				
030	Nashua	101.0	88.9	95.2
031	Manchester	101.4	88.9	95.4

STATE/ZIP	CITY	MAT.	INST.	TOTAL
NEW HAMPSHIRE (CONT'D)				
032-033	Concord	98.9	88.9	94.1
034	Keene	98.0	59.3	79.3
035	Littleton	98.0	66.3	82.7
036	Charleston	97.4	55.9	77.3
037	Claremont	96.5	55.9	76.9
038	Portsmouth	98.2	85.0	91.8
NEW JERSEY				
070-071	Newark	102.5	121.6	111.7
072	Elizabeth	101.5	114.0	107.6
073	Jersey City	100.0	121.2	110.3
074-075	Paterson	102.9	120.8	111.5
076	Hackensack	100.0	119.8	109.6
077	Long Branch	99.5	117.1	108.0
078	Dover	100.3	118.5	109.1
079	Summit	100.4	112.7	106.3
080,083	Vineland	97.9	117.8	107.5
081	Camden	99.9	116.3	107.8
082,084	Atlantic City	98.8	117.2	107.7
085-086	Trenton	100.7	119.4	109.7
087	Point Pleasant	100.2	117.2	108.4
088-089	New Brunswick	100.6	119.3	109.7
NEW MEXICO				
870-872	Albuquerque	100.3	78.6	89.8
873	Gallup	101.6	78.6	90.5
874	Farmington	101.7	78.6	90.5
875	Santa Fe	99.9	78.6	89.6
877	Las Vegas	99.8	78.6	89.6
878	Socorro	99.3	78.6	89.3
879	Truth/Consequences	99.2	73.9	86.9
880	Las Cruces	97.4	70.5	84.4
881	Clovis	100.0	78.0	89.4
882	Roswell	101.6	78.1	90.2
883	Carrizozo	102.3	78.6	90.9
884	Tucumcari	101.0	78.1	89.9
NEW YORK				
100-102	New York	107.7	163.3	134.6
103	Staten Island	103.9	159.6	130.8
104	Bronx	101.2	159.8	129.5
105	Mount Vernon	101.9	139.0	119.8
106	White Plains	101.8	136.6	118.6
107	Yonkers	106.6	137.5	121.5
108	New Rochelle	102.5	139.0	120.2
109	Suffern	102.5	126.8	114.2
110	Queens	102.4	160.0	130.2
111	Long Island City	104.3	160.0	131.2
112	Brooklyn	104.5	159.4	131.0
113	Flushing	104.9	160.0	131.5
114	Jamaica	102.9	160.0	130.5
115,117,118	Hicksville	102.2	151.2	125.9
116	Far Rockaway	105.1	159.8	131.6
119	Riverhead	102.7	152.4	126.7
120-122	Albany	97.7	97.0	97.4
123	Schenectady	98.3	97.6	98.0
124	Kingston	102.6	116.1	109.1
125-126	Poughkeepsie	101.7	121.3	111.2
127	Monticello	100.9	113.3	106.9
128	Glens Falls	92.7	92.8	92.8
129	Plattsburgh	98.3	88.0	93.4
130-132	Syracuse	99.2	94.0	96.7
133-135	Utica	97.5	89.8	93.8
136	Watertown	99.0	90.3	94.8
137-139	Binghamton	98.5	90.0	94.4
140-142	Buffalo	99.5	103.7	101.5
143	Niagara Falls	97.6	108.2	102.8
144-146	Rochester	100.5	99.4	100.0
147	Jamestown	96.8	91.8	94.4
148-149	Elmira	96.2	89.9	93.1
NORTH CAROLINA				
270,272-274	Greensboro	96.4	53.9	75.9
271	Winston-Salem	96.6	52.8	75.4
275-276	Raleigh	96.8	54.1	76.2
277	Durham	96.3	54.1	75.9
278	Rocky Mount	94.9	39.8	68.3
279	Elizabeth City	95.5	42.2	69.8
280	Gastonia	96.5	51.0	74.5
281-282	Charlotte	96.7	51.3	74.8
283	Fayetteville	94.9	54.1	75.2
284	Wilmington	95.1	53.1	74.8
285	Kinston	93.4	39.4	67.3

597

STATE/ZIP	CITY	MAT.	INST.	TOTAL
NORTH CAROLINA (CONT'D)				
286	Hickory	93.7	38.3	66.9
287-288	Asheville	95.5	52.1	74.5
289	Murphy	95.0	37.4	67.1
NORTH DAKOTA				
580-581	Fargo	100.6	66.1	83.9
582	Grand Forks	101.7	61.2	82.2
583	Devils Lake	101.7	58.2	80.7
584	Jamestown	101.6	59.1	81.1
585	Bismarck	100.5	67.4	84.5
586	Dickinson	102.3	66.9	85.2
587	Minot	101.7	70.3	86.5
588	Williston	100.7	58.2	80.2
OHIO				
430-432	Columbus	98.2	92.7	95.5
433	Marion	95.2	91.3	93.3
434-436	Toledo	98.5	104.2	101.3
437-438	Zanesville	95.3	88.0	91.8
439	Steubenville	97.1	97.2	97.1
440	Lorain	98.5	98.0	98.3
441	Cleveland	98.8	106.8	102.7
442-443	Akron	99.7	101.7	100.7
444-445	Youngstown	98.9	96.7	97.8
446-447	Canton	99.1	92.2	95.8
448-449	Mansfield	96.4	92.4	94.4
450	Hamilton	95.1	92.6	93.9
451-452	Cincinnati	95.3	93.4	94.4
453-454	Dayton	94.9	90.5	92.8
455	Springfield	95.2	91.5	93.4
456	Chillicothe	94.4	97.8	96.1
457	Athens	97.8	83.0	90.6
458	Lima	98.4	91.9	95.3
OKLAHOMA				
730-731	Oklahoma City	98.2	67.9	83.6
734	Ardmore	95.8	67.4	82.0
735	Lawton	97.6	67.7	83.2
736	Clinton	97.3	65.5	81.9
737	Enid	97.5	65.5	82.1
738	Woodward	95.9	65.6	81.2
739	Guymon	97.2	37.2	68.2
740-741	Tulsa	97.6	63.9	81.3
743	Miami	94.6	70.2	82.8
744	Muskogee	96.7	47.5	72.9
745	Mcalester	94.1	59.7	77.5
746	Ponca City	94.7	66.3	81.0
747	Durant	94.7	66.1	80.9
748	Shawnee	96.4	64.4	81.0
749	Poteau	93.7	67.3	80.9
OREGON				
970-972	Portland	103.4	108.3	105.8
973	Salem	103.9	106.1	105.0
974	Eugene	103.4	105.1	104.2
975	Medford	104.9	102.5	103.7
976	Klamath Falls	105.7	102.2	104.0
977	Bend	104.5	105.1	104.8
978	Pendleton	98.4	104.0	101.1
979	Vale	95.8	95.6	95.7
PENNSYLVANIA				
150-152	Pittsburgh	97.0	106.8	101.7
153	Washington	94.2	106.1	99.9
154	Uniontown	94.4	103.7	98.9
155	Bedford	95.4	96.7	96.0
156	Greensburg	95.6	103.8	99.6
157	Indiana	94.3	101.8	97.9
158	Dubois	95.8	98.7	97.2
159	Johnstown	95.4	99.7	97.5
160	Butler	92.3	103.5	97.7
161	New Castle	92.3	103.6	97.8
162	Kittanning	92.9	105.9	99.2
163	Oil City	92.2	98.7	95.4
164-165	Erie	94.4	99.1	96.7
166	Altoona	94.3	97.0	95.6
167	Bradford	95.8	98.3	97.0
168	State College	95.5	97.3	96.4
169	Wellsboro	96.6	92.0	94.3
170-171	Harrisburg	98.3	94.7	96.6
172	Chambersburg	96.1	93.0	94.6
173-174	York	95.9	94.9	95.4
175-176	Lancaster	94.7	91.9	93.4

STATE/ZIP	CITY	MAT.	INST.	TOTAL
PENNSYLVANIA (CONT'D)				
177	Williamsport	93.0	86.3	89.8
178	Sunbury	95.4	92.8	94.1
179	Pottsville	94.5	92.8	93.7
180	Lehigh Valley	96.2	109.3	102.5
181	Allentown	98.1	102.6	100.2
182	Hazleton	95.4	96.2	95.8
183	Stroudsburg	95.4	104.8	99.9
184-185	Scranton	98.6	98.0	98.3
186-187	Wilkes-Barre	95.1	97.3	96.1
188	Montrose	94.6	97.4	96.0
189	Doylestown	94.9	117.8	106.0
190-191	Philadelphia	99.5	123.6	111.2
193	Westchester	96.4	116.3	106.0
194	Norristown	95.2	120.0	107.2
195-196	Reading	97.5	98.1	97.8
PUERTO RICO				
009	San Juan	139.7	29.1	86.3
RHODE ISLAND				
028	Newport	99.7	107.9	103.6
029	Providence	100.0	107.9	103.8
SOUTH CAROLINA				
290-292	Columbia	95.7	51.9	74.5
293	Spartanburg	94.6	50.8	73.4
294	Charleston	95.7	53.4	75.2
295	Florence	94.1	49.8	72.7
296	Greenville	94.3	50.8	73.3
297	Rock Hill	94.2	37.2	66.7
298	Aiken	95.3	69.4	82.7
299	Beaufort	95.9	41.3	69.5
SOUTH DAKOTA				
570-571	Sioux Falls	101.0	60.5	81.4
572	Watertown	99.8	54.2	77.7
573	Mitchell	98.6	53.9	77.0
574	Aberdeen	101.0	54.2	78.4
575	Pierre	100.4	55.3	78.6
576	Mobridge	99.3	54.2	77.5
577	Rapid City	101.1	55.6	79.1
TENNESSEE				
370-372	Nashville	97.7	73.7	86.1
373-374	Chattanooga	98.7	61.8	80.9
375,380-381	Memphis	96.0	72.1	84.4
376	Johnson City	98.1	58.9	79.1
377-379	Knoxville	94.8	63.3	79.6
382	Mckenzie	96.9	38.5	68.7
383	Jackson	98.5	49.1	74.6
384	Columbia	95.3	54.8	75.7
385	Cookeville	96.6	37.3	68.0
TEXAS				
750	Mckinney	99.3	61.7	81.1
751	Waxahackie	99.3	63.3	81.9
752-753	Dallas	100.0	69.4	85.2
754	Greenville	99.5	43.6	72.5
755	Texarkana	98.6	53.2	76.7
756	Longview	99.1	46.5	73.7
757	Tyler	99.7	58.7	79.9
758	Palestine	95.6	47.3	72.2
759	Lufkin	96.6	53.4	75.7
760-761	Fort Worth	97.7	66.1	82.4
762	Denton	97.7	57.5	78.3
763	Wichita Falls	98.2	60.5	80.0
764	Eastland	97.1	46.2	72.5
765	Temple	95.6	55.4	76.2
766-767	Waco	97.4	61.0	79.8
768	Brownwood	98.0	43.0	71.4
769	San Angelo	97.6	51.9	75.5
770-772	Houston	99.6	76.3	88.3
773	Huntsville	98.2	46.3	73.1
774	Wharton	99.6	50.9	76.1
775	Galveston	97.5	74.8	86.5
776-777	Beaumont	97.1	68.7	83.4
778	Bryan	94.6	69.3	82.4
779	Victoria	99.8	55.1	78.2
780	Laredo	94.6	58.8	77.3
781-782	San Antonio	94.8	69.6	82.6
783-784	Corpus Christi	98.0	58.3	78.8
785	Mc Allen	98.3	51.7	75.8
786-787	Austin	94.8	67.0	81.3

Location Factors

STATE/ZIP	CITY	MAT.	INST.	TOTAL
TEXAS (CONT'D)				
788	Del Rio	97.6	35.8	67.8
789	Giddings	94.6	46.3	71.3
790-791	Amarillo	98.3	62.0	80.8
792	Childress	97.6	57.2	78.0
793-794	Lubbock	100.0	58.4	79.9
795-796	Abilene	97.6	58.8	78.8
797	Midland	100.7	55.0	78.6
798-799,885	El Paso	97.5	57.2	78.1
UTAH				
840-841	Salt Lake City	100.7	76.9	89.2
842,844	Ogden	99.2	76.0	88.0
843	Logan	101.4	76.0	89.1
845	Price	101.9	56.2	79.8
846-847	Provo	101.7	76.0	89.3
VERMONT				
050	White River Jct.	99.5	42.0	71.7
051	Bellows Falls	97.9	45.6	72.6
052	Bennington	98.2	41.7	70.9
053	Brattleboro	98.7	45.8	73.2
054	Burlington	100.9	71.0	86.4
056	Montpelier	97.9	71.0	84.9
057	Rutland	100.0	71.0	86.0
058	St. Johnsbury	99.7	48.5	75.0
059	Guildhall	98.1	48.0	73.9
VIRGINIA				
220-221	Fairfax	97.4	81.3	89.6
222	Arlington	98.3	80.9	89.9
223	Alexandria	97.1	82.9	90.2
224-225	Fredericksburg	95.8	71.4	84.0
226	Winchester	96.7	59.5	78.7
227	Culpeper	96.4	60.3	79.0
228	Harrisonburg	96.8	52.6	75.5
229	Charlottesville	97.0	65.0	81.5
230-232	Richmond	97.3	70.7	84.4
233-235	Norfolk	97.6	64.8	81.8
236	Newport News	97.0	64.6	81.4
237	Portsmouth	96.2	64.5	80.9
238	Petersburg	96.8	70.7	84.2
239	Farmville	96.2	47.0	72.4
240-241	Roanoke	97.7	51.9	75.6
242	Bristol	95.9	51.1	74.2
243	Pulaski	95.6	46.3	71.8
244	Staunton	96.4	51.0	74.4
245	Lynchburg	96.4	55.2	76.5
246	Grundy	95.8	46.3	71.9
WASHINGTON				
980-981,987	Seattle	104.0	106.8	105.4
982	Everett	103.9	101.9	102.9
983-984	Tacoma	103.6	102.5	103.1
985	Olympia	103.5	102.4	103.0
986	Vancouver	106.0	102.6	104.3
988	Wenatchee	104.6	89.9	97.5
989	Yakima	104.1	95.7	100.0
990-992	Spokane	107.8	87.4	98.0
993	Richland	107.9	90.2	99.3
994	Clarkston	106.9	88.3	97.9
WEST VIRGINIA				
247-248	Bluefield	94.7	82.0	88.5
249	Lewisburg	96.2	82.9	89.8
250-253	Charleston	97.0	88.2	92.7
254	Martinsburg	96.2	54.1	75.8
255-257	Huntington	97.3	92.6	95.0
258-259	Beckley	94.5	85.9	90.3
260	Wheeling	97.4	93.2	95.4
261	Parkersburg	96.3	91.9	94.2
262	Buckhannon	96.1	91.1	93.7
263-264	Clarksburg	96.6	91.4	94.1
265	Morgantown	96.6	91.4	94.1
266	Gassaway	95.9	89.9	93.0
267	Romney	95.9	84.6	90.4
268	Petersburg	95.9	87.9	92.0
WISCONSIN				
530,532	Milwaukee	99.5	102.9	101.1
531	Kenosha	99.5	101.6	100.5
534	Racine	98.7	102.0	100.3
535	Beloit	98.7	97.6	98.1
537	Madison	98.8	97.0	97.9

STATE/ZIP	CITY	MAT.	INST.	TOTAL
WISCONSIN (CONT'D)				
538	Lancaster	96.9	80.3	88.9
539	Portage	95.2	97.0	96.1
540	New Richmond	96.8	93.8	95.3
541-543	Green Bay	100.2	92.8	96.6
544	Wausau	95.8	92.8	94.3
545	Rhinelander	99.3	89.0	94.3
546	La Crosse	96.5	92.9	94.7
547	Eau Claire	98.4	93.6	96.1
548	Superior	96.6	98.0	97.3
549	Oshkosh	96.4	90.7	93.6
WYOMING				
820	Cheyenne	100.3	59.9	80.8
821	Yellowstone Nat'l Park	99.6	53.6	77.4
822	Wheatland	101.4	52.9	77.9
823	Rawlins	102.8	49.0	76.8
824	Worland	100.4	49.3	75.7
825	Riverton	101.6	52.4	77.8
826	Casper	100.0	63.0	82.1
827	Newcastle	100.2	49.0	75.4
828	Sheridan	101.3	57.7	80.2
829-831	Rock Springs	104.9	49.4	78.1
CANADIAN FACTORS (reflect Canadian currency)				
ALBERTA	Calgary	112.6	79.0	96.4
	Edmonton	112.6	78.9	96.3
	Fort McMurray	110.5	78.9	95.2
	Lethbridge	110.6	79.0	95.3
	Lloydminster	110.5	78.9	95.2
	Medicine Hat	110.6	79.0	95.3
	Red Deer	110.3	79.0	95.2
BRITISH COLUMBIA	Kamloops	110.6	94.7	102.9
	Prince George	114.2	94.7	104.7
	Vancouver	114.2	97.8	106.3
	Victoria	114.6	94.7	105.0
MANITOBA	Brandon	112.0	79.2	96.2
	Portage la Prairie	111.2	79.2	95.7
	Winnipeg	111.2	79.2	95.8
NEW BRUNSWICK	Bathurst	109.8	70.1	90.6
	Dalhousie	109.8	70.1	90.6
	Fredericton	110.4	75.1	93.4
	Moncton	109.5	70.1	90.4
	Newcastle	109.8	70.1	90.6
	Saint John	111.9	75.1	94.1
NEWFOUNDLAND	Corner Brook	115.8	69.8	93.6
	St. John's	115.4	69.8	93.4
NORTHWEST TERRITORIES	Yellowknife	108.3	71.3	90.4
NOVA SCOTIA	Dartmouth	111.1	77.4	94.8
	Halifax	110.7	78.8	95.2
	New Glasgow	111.1	77.4	94.8
	Sydney	108.2	77.4	93.3
	Yarmouth	110.8	77.4	94.7
ONTARIO	Barrie	115.6	96.8	106.5
	Brantford	113.5	101.6	107.8
	Cornwall	113.7	97.1	105.7
	Hamilton	113.3	102.6	108.1
	Kingston	113.7	98.8	106.5
	Kitchener	108.1	96.9	102.7
	London	112.4	99.5	106.2
	North Bay	113.5	96.7	105.4
	Oshawa	113.2	100.1	106.9
	Ottawa	113.9	99.7	107.1
	Owen Sound	115.7	96.7	106.5
	Peterborough	113.5	96.9	105.5
	Sarnia	112.6	103.1	108.0
	St. Catharines	107.0	96.7	102.0
	Sudbury	107.3	97.0	102.4

Location Factors

STATE/ZIP	CITY	MAT.	INST.	TOTAL
ONTARIO (CONT'D)				
	Thunder Bay	108.7	97.1	103.1
	Toronto	114.2	107.4	110.9
	Windsor	107.3	100.1	103.8
PRINCE EDWARD ISLAND				
	Charlottetown	113.1	65.5	90.1
	Summerside	113.2	65.5	90.1
QUEBEC				
	Cap-de-la-Madeleine	111.2	91.7	101.8
	Charlesbourg	111.2	91.7	101.8
	Chicoutimi	109.9	91.4	100.9
	Gatineau	108.4	91.5	100.2
	Laval	110.2	91.5	101.1
	Montreal	110.3	91.5	101.2
	Quebec	111.9	91.7	102.2
	Sherbrooke	110.6	91.5	101.3
	Trois Rivieres	111.2	91.7	101.8
SASKATCHEWAN				
	Moose Jaw	108.5	73.0	91.3
	Prince Albert	107.8	72.8	90.9
	Regina	109.2	73.1	91.7
	Saskatoon	108.1	72.9	91.1
YUKON				
	Whitehorse	108.3	71.3	90.4

A	Area Square Feet; Ampere
ABS	Acrylonitrile Butadiene Stryrene; Asbestos Bonded Steel
A.C.	Alternating Current; Air-Conditioning; Asbestos Cement; Plywood Grade A & C
A.C.I.	American Concrete Institute
AD	Plywood, Grade A & D
Addit.	Additional
Adj.	Adjustable
af	Audio-frequency
A.G.A.	American Gas Association
Agg.	Aggregate
A.H.	Ampere Hours
A hr.	Ampere-hour
A.H.U.	Air Handling Unit
A.I.A.	American Institute of Architects
AIC	Ampere Interrupting Capacity
Allow.	Allowance
alt.	Altitude
Alum.	Aluminum
a.m.	Ante Meridiem
Amp.	Ampere
Anod.	Anodized
Approx.	Approximate
Apt.	Apartment
Asb.	Asbestos
A.S.B.C.	American Standard Building Code
Asbe.	Asbestos Worker
A.S.H.R.A.E.	American Society of Heating, Refrig. & AC Engineers
A.S.M.E.	American Society of Mechanical Engineers
A.S.T.M.	American Society for Testing and Materials
Attchmt.	Attachment
Avg.	Average
A.W.G.	American Wire Gauge
AWWA	American Water Works Assoc.
Bbl.	Barrel
B. & B.	Grade B and Better; Balled & Burlapped
B. & S.	Bell and Spigot
B. & W.	Black and White
b.c.c.	Body-centered Cubic
B.C.Y.	Bank Cubic Yards
BE	Bevel End
B.F.	Board Feet
Bg. cem.	Bag of Cement
BHP	Boiler Horsepower; Brake Horsepower
B.I.	Black Iron
Bit.; Bitum.	Bituminous
Bk.	Backed
Bkrs.	Breakers
Bldg.	Building
Blk.	Block
Bm.	Beam
Boil.	Boilermaker
B.P.M.	Blows per Minute
BR	Bedroom
Brg.	Bearing
Brhe.	Bricklayer Helper
Bric.	Bricklayer
Brk.	Brick
Brng.	Bearing
Brs.	Brass
Brz.	Bronze
Bsn.	Basin
Btr.	Better
BTU	British Thermal Unit
BTUH	BTU per Hour
B.U.R.	Built-up Roofing
BX	Interlocked Armored Cable
c	Conductivity, Copper Sweat
C	Hundred; Centigrade
C/C	Center to Center, Cedar on Cedar

Cab.	Cabinet
Cair.	Air Tool Laborer
Calc	Calculated
Cap.	Capacity
Carp.	Carpenter
C.B.	Circuit Breaker
C.C.A.	Chromate Copper Arsenate
C.C.F.	Hundred Cubic Feet
cd	Candela
cd/sf	Candela per Square Foot
CD	Grade of Plywood Face & Back
CDX	Plywood, Grade C & D, exterior glue
Cefi.	Cement Finisher
Cem.	Cement
CF	Hundred Feet
C.F.	Cubic Feet
CFM	Cubic Feet per Minute
c.g.	Center of Gravity
CHW	Chilled Water; Commercial Hot Water
C.I.	Cast Iron
C.I.P.	Cast in Place
Circ.	Circuit
C.L.	Carload Lot
Clab.	Common Laborer
C.L.F.	Hundred Linear Feet
CLF	Current Limiting Fuse
CLP	Cross Linked Polyethylene
cm	Centimeter
CMP	Corr. Metal Pipe
C.M.U.	Concrete Masonry Unit
CN	Change Notice
Col.	Column
CO_2	Carbon Dioxide
Comb.	Combination
Compr.	Compressor
Conc.	Concrete
Cont.	Continuous; Continued
Corr.	Corrugated
Cos	Cosine
Cot	Cotangent
Cov.	Cover
C/P	Cedar on Paneling
CPA	Control Point Adjustment
Cplg.	Coupling
C.P.M.	Critical Path Method
CPVC	Chlorinated Polyvinyl Chloride
C.Pr.	Hundred Pair
CRC	Cold Rolled Channel
Creos.	Creosote
Crpt.	Carpet & Linoleum Layer
CRT	Cathode-ray Tube
CS	Carbon Steel, Constant Shear Bar Joist
Csc	Cosecant
C.S.F.	Hundred Square Feet
CSI	Construction Specifications Institute
C.T.	Current Transformer
CTS	Copper Tube Size
Cu	Copper, Cubic
Cu. Ft.	Cubic Foot
cw	Continuous Wave
C.W.	Cool White; Cold Water
Cwt.	100 Pounds
C.W.X.	Cool White Deluxe
C.Y.	Cubic Yard (27 cubic feet)
C.Y./Hr.	Cubic Yard per Hour
Cyl.	Cylinder
d	Penny (nail size)
D	Deep; Depth; Discharge
Dis.;Disch.	Discharge
Db.	Decibel
Dbl.	Double
DC	Direct Current
DDC	Direct Digital Control
Demob.	Demobilization

d.f.u.	Drainage Fixture Units
D.H.	Double Hung
DHW	Domestic Hot Water
Diag.	Diagonal
Diam.	Diameter
Distrib.	Distribution
Dk.	Deck
D.L.	Dead Load; Diesel
DLH	Deep Long Span Bar Joist
Do.	Ditto
Dp.	Depth
D.P.S.T.	Double Pole, Single Throw
Dr.	Driver
Drink.	Drinking
D.S.	Double Strength
D.S.A.	Double Strength A Grade
D.S.B.	Double Strength B Grade
Dty.	Duty
DWV	Drain Waste Vent
DX	Deluxe White, Direct Expansion
dyn	Dyne
e	Eccentricity
E	Equipment Only; East
Ea.	Each
E.B.	Encased Burial
Econ.	Economy
EDP	Electronic Data Processing
EIFS	Exterior Insulation Finish System
E.D.R.	Equiv. Direct Radiation
Eq.	Equation
Elec.	Electrician; Electrical
Elev.	Elevator; Elevating
EMT	Electrical Metallic Conduit; Thin Wall Conduit
Eng.	Engine, Engineered
EPDM	Ethylene Propylene Diene Monomer
EPS	Expanded Polystyrene
Eqhv.	Equip. Oper., Heavy
Eqlt.	Equip. Oper., Light
Eqmd.	Equip. Oper., Medium
Eqmm.	Equip. Oper., Master Mechanic
Eqol.	Equip. Oper., Oilers
Equip.	Equipment
ERW	Electric Resistance Welded
E.S.	Energy Saver
Est.	Estimated
esu	Electrostatic Units
E.W.	Each Way
EWT	Entering Water Temperature
Excav.	Excavation
Exp.	Expansion, Exposure
Ext.	Exterior
Extru.	Extrusion
f.	Fiber stress
F	Fahrenheit; Female; Fill
Fab.	Fabricated
FBGS	Fiberglass
F.C.	Footcandles
f.c.c.	Face-centered Cubic
f'c.	Compressive Stress in Concrete; Extreme Compressive Stress
F.E.	Front End
FEP	Fluorinated Ethylene Propylene (Teflon)
F.G.	Flat Grain
F.H.A.	Federal Housing Administration
Fig.	Figure
Fin.	Finished
Fixt.	Fixture
Fl. Oz.	Fluid Ounces
Flr.	Floor
F.M.	Frequency Modulation; Factory Mutual
Fmg.	Framing
Fndtn.	Foundation
Fori.	Foreman, Inside
Foro.	Foreman, Outside

Fount.	Fountain	J.I.C.	Joint Industrial Council	M.C.P.	Motor Circuit Protector
FPM	Feet per Minute	K	Thousand; Thousand Pounds;	MD	Medium Duty
FPT	Female Pipe Thread		Heavy Wall Copper Tubing, Kelvin	M.D.O.	Medium Density Overlaid
Fr.	Frame	K.A.H.	Thousand Amp. Hours	Med.	Medium
F.R.	Fire Rating	KCMIL	Thousand Circular Mils	MF	Thousand Feet
FRK	Foil Reinforced Kraft	KD	Knock Down	M.F.B.M.	Thousand Feet Board Measure
FRP	Fiberglass Reinforced Plastic	K.D.A.T.	Kiln Dried After Treatment	Mfg.	Manufacturing
FS	Forged Steel	kg	Kilogram	Mfrs.	Manufacturers
FSC	Cast Body; Cast Switch Box	kG	Kilogauss	mg	Milligram
Ft.	Foot; Feet	kgf	Kilogram Force	MGD	Million Gallons per Day
Ftng.	Fitting	kHz	Kilohertz	MGPH	Thousand Gallons per Hour
Ftg.	Footing	Kip.	1000 Pounds	MH, M.H.	Manhole; Metal Halide; Man-Hour
Ft. Lb.	Foot Pound	KJ	Kiljoule	MHz	Megahertz
Furn.	Furniture	K.L.	Effective Length Factor	Mi.	Mile
FVNR	Full Voltage Non-Reversing	K.L.F.	Kips per Linear Foot	MI	Malleable Iron; Mineral Insulated
FXM	Female by Male	Km	Kilometer	mm	Millimeter
Fy.	Minimum Yield Stress of Steel	K.S.F.	Kips per Square Foot	Mill.	Millwright
g	Gram	K.S.I.	Kips per Square Inch	Min., min.	Minimum, minute
G	Gauss	kV	Kilovolt	Misc.	Miscellaneous
Ga.	Gauge	kVA	Kilovolt Ampere	ml	Milliliter, Mainline
Gal.	Gallon	K.V.A.R.	Kilovar (Reactance)	M.L.F.	Thousand Linear Feet
Gal./Min.	Gallon per Minute	KW	Kilowatt	Mo.	Month
Galv.	Galvanized	KWh	Kilowatt-hour	Mobil.	Mobilization
Gen.	General	L	Labor Only; Length; Long;	Mog.	Mogul Base
G.F.I.	Ground Fault Interrupter		Medium Wall Copper Tubing	MPH	Miles per Hour
Glaz.	Glazier	Lab.	Labor	MPT	Male Pipe Thread
GPD	Gallons per Day	lat	Latitude	MRT	Mile Round Trip
GPH	Gallons per Hour	Lath.	Lather	ms	Millisecond
GPM	Gallons per Minute	Lav.	Lavatory	M.S.F.	Thousand Square Feet
GR	Grade	lb.; #	Pound	Mstz.	Mosaic & Terrazzo Worker
Gran.	Granular	L.B.	Load Bearing; L Conduit Body	M.S.Y.	Thousand Square Yards
Grnd.	Ground	L. & E.	Labor & Equipment	Mtd.	Mounted
H	High; High Strength Bar Joist;	lb./hr.	Pounds per Hour	Mthe.	Mosaic & Terrazzo Helper
	Henry	lb./L.F.	Pounds per Linear Foot	Mtng.	Mounting
H.C.	High Capacity	lbf/sq.in.	Pound-force per Square Inch	Mult.	Multi; Multiply
H.D.	Heavy Duty; High Density	L.C.L.	Less than Carload Lot	M.V.A.	Million Volt Amperes
H.D.O.	High Density Overlaid	Ld.	Load	M.V.A.R.	Million Volt Amperes Reactance
Hdr.	Header	LE	Lead Equivalent	MV	Megavolt
Hdwe.	Hardware	LED	Light Emitting Diode	MW	Megawatt
Help.	Helpers Average	L.F.	Linear Foot	MXM	Male by Male
HEPA	High Efficiency Particulate Air	Lg.	Long; Length; Large	MYD	Thousand Yards
	Filter	L & H	Light and Heat	N	Natural; North
Hg	Mercury	LH	Long Span Bar Joist	nA	Nanoampere
HIC	High Interrupting Capacity	L.H.	Labor Hours	NA	Not Available; Not Applicable
HM	Hollow Metal	L.L.	Live Load	N.B.C.	National Building Code
H.O.	High Output	L.L.D.	Lamp Lumen Depreciation	NC	Normally Closed
Horiz.	Horizontal	L-O-L	Lateralolet	N.E.M.A.	National Electrical Manufacturers
H.P.	Horsepower; High Pressure	lm	Lumen		Assoc.
H.P.F.	High Power Factor	lm/sf	Lumen per Square Foot	NEHB	Bolted Circuit Breaker to 600V.
Hr.	Hour	lm/W	Lumen per Watt	N.L.B.	Non-Load-Bearing
Hrs./Day	Hours per Day	L.O.A.	Length Over All	NM	Non-Metallic Cable
HSC	High Short Circuit	log	Logarithm	nm	Nanometer
Ht.	Height	L.P.	Liquefied Petroleum; Low Pressure	No.	Number
Htg.	Heating	L.P.F.	Low Power Factor	NO	Normally Open
Htrs.	Heaters	LR	Long Radius	N.O.C.	Not Otherwise Classified
HVAC	Heating, Ventilation & Air-	L.S.	Lump Sum	Nose.	Nosing
	Conditioning	Lt.	Light	N.P.T.	National Pipe Thread
Hvy.	Heavy	Lt. Ga.	Light Gauge	NQOD	Combination Plug-on/Bolt on
HW	Hot Water	L.T.L.	Less than Truckload Lot		Circuit Breaker to 240V.
Hyd.;Hydr.	Hydraulic	Lt. Wt.	Lightweight	N.R.C.	Noise Reduction Coefficient
Hz.	Hertz (cycles)	L.V.	Low Voltage	N.R.S.	Non Rising Stem
I.	Moment of Inertia	M	Thousand; Material; Male;	ns	Nanosecond
I.C.	Interrupting Capacity		Light Wall Copper Tubing	nW	Nanowatt
ID	Inside Diameter	M²CA	Meters Squared Contact Area	OB	Opposing Blade
I.D.	Inside Dimension; Identification	m/hr; M.H.	Man-hour	OC	On Center
I.F.	Inside Frosted	mA	Milliampere	OD	Outside Diameter
I.M.C.	Intermediate Metal Conduit	Mach.	Machine	O.D.	Outside Dimension
In.	Inch	Mag. Str.	Magnetic Starter	ODS	Overhead Distribution System
Incan.	Incandescent	Maint.	Maintenance	O.G.	Ogee
Incl.	Included; Including	Marb.	Marble Setter	O.H.	Overhead
Int.	Interior	Mat; Mat'l.	Material	O & P	Overhead and Profit
Inst.	Installation	Max.	Maximum	Oper.	Operator
Insul.	Insulation/Insulated	MBF	Thousand Board Feet	Opng.	Opening
I.P.	Iron Pipe	MBH	Thousand BTU's per hr.	Orna.	Ornamental
I.P.S.	Iron Pipe Size	MC	Metal Clad Cable	OSB	Oriented Strand Board
I.P.T.	Iron Pipe Threaded	M.C.F.	Thousand Cubic Feet	O. S. & Y.	Outside Screw and Yoke
I.W.	Indirect Waste	M.C.F.M.	Thousand Cubic Feet per Minute	Ovhd.	Overhead
J	Joule	M.C.M.	Thousand Circular Mils	OWG	Oil, Water or Gas

Abbreviations

Oz.	Ounce	SCFM	Standard Cubic Feet per Minute	THW.	Insulated Strand Wire
P.	Pole; Applied Load; Projection	Scaf.	Scaffold	THWN;	Nylon Jacketed Wire
p.	Page	Sch.; Sched.	Schedule	T.L.	Truckload
Pape.	Paperhanger	S.C.R.	Modular Brick	T.M.	Track Mounted
P.A.P.R.	Powered Air Purifying Respirator	S.D.	Sound Deadening	Tot.	Total
PAR	Parabolic Reflector	S.D.R.	Standard Dimension Ratio	T-O-L	Threadolet
Pc., Pcs.	Piece, Pieces	S.E.	Surfaced Edge	T.S.	Trigger Start
P.C.	Portland Cement; Power Connector	Sel.	Select	Tr.	Trade
P.C.F.	Pounds per Cubic Foot	S.E.R.;	Service Entrance Cable	Transf.	Transformer
P.C.M.	Phase Contract Microscopy	S.E.U.	Service Entrance Cable	Trhv.	Truck Driver, Heavy
P.E.	Professional Engineer;	S.F.	Square Foot	Trlr	Trailer
	Porcelain Enamel;	S.F.C.A.	Square Foot Contact Area	Trlt.	Truck Driver, Light
	Polyethylene; Plain End	S.F.G.	Square Foot of Ground	TV	Television
Perf.	Perforated	S.F. Hor.	Square Foot Horizontal	T.W.	Thermoplastic Water Resistant
Ph.	Phase	S.F.R.	Square Feet of Radiation		Wire
P.I.	Pressure Injected	S.F. Shlf.	Square Foot of Shelf	UCI	Uniform Construction Index
Pile.	Pile Driver	S4S	Surface 4 Sides	UF	Underground Feeder
Pkg.	Package	Shee.	Sheet Metal Worker	UGND	Underground Feeder
Pl.	Plate	Sin.	Sine	U.H.F.	Ultra High Frequency
Plah.	Plasterer Helper	Skwk.	Skilled Worker	U.L.	Underwriters Laboratory
Plas.	Plasterer	SL	Saran Lined	Unfin.	Unfinished
Pluh.	Plumbers Helper	S.L.	Slimline	URD	Underground Residential
Plum.	Plumber	Sldr.	Solder		Distribution
Ply.	Plywood	SLH	Super Long Span Bar Joist	US	United States
p.m.	Post Meridiem	S.N.	Solid Neutral	USP	United States Primed
Pntd.	Painted	S-O-L	Socketolet	UTP	Unshielded Twisted Pair
Pord.	Painter, Ordinary	sp	Standpipe	V	Volt
pp	Pages	S.P.	Static Pressure; Single Pole; Self-	V.A.	Volt Amperes
PP; PPL	Polypropylene		Propelled	V.C.T.	Vinyl Composition Tile
P.P.M.	Parts per Million	Spri.	Sprinkler Installer	VAV	Variable Air Volume
Pr.	Pair	spwg	Static Pressure Water Gauge	VC	Veneer Core
P.E.S.B.	Pre-engineered Steel Building	Sq.	Square; 100 Square Feet	Vent.	Ventilation
Prefab.	Prefabricated	S.P.D.T.	Single Pole, Double Throw	Vert.	Vertical
Prefin.	Prefinished	SPF	Spruce Pine Fir	V.F.	Vinyl Faced
Prop.	Propelled	S.P.S.T.	Single Pole, Single Throw	V.G.	Vertical Grain
PSF; psf	Pounds per Square Foot	SPT	Standard Pipe Thread	V.H.F.	Very High Frequency
PSI; psi	Pounds per Square Inch	Sq. Hd.	Square Head	VHO	Very High Output
PSIG	Pounds per Square Inch Gauge	Sq. In.	Square Inch	Vib.	Vibrating
PSP	Plastic Sewer Pipe	S.S.	Single Strength; Stainless Steel	V.L.F.	Vertical Linear Foot
Pspr.	Painter, Spray	S.S.B.	Single Strength B Grade	Vol.	Volume
Psst.	Painter, Structural Steel	sst	Stainless Steel	VRP	Vinyl Reinforced Polyester
P.T.	Potential Transformer	Sswk.	Structural Steel Worker	W	Wire; Watt; Wide; West
P. & T.	Pressure & Temperature	Sswl.	Structural Steel Welder	w/	With
Ptd.	Painted	St.; Stl.	Steel	W.C.	Water Column; Water Closet
Ptns.	Partitions	S.T.C.	Sound Transmission Coefficient	W.F.	Wide Flange
Pu	Ultimate Load	Std.	Standard	W.G.	Water Gauge
PVC	Polyvinyl Chloride	STK	Select Tight Knot	Wldg.	Welding
Pvmt.	Pavement	STP	Standard Temperature & Pressure	W. Mile	Wire Mile
Pwr.	Power	Stpi.	Steamfitter, Pipefitter	W-O-L	Weldolet
Q	Quantity Heat Flow	Str.	Strength; Starter; Straight	W.R.	Water Resistant
Quan.; Qty.	Quantity	Strd.	Stranded	Wrck.	Wrecker
Q.C.	Quick Coupling	Struct.	Structural	W.S.P.	Water, Steam, Petroleum
r	Radius of Gyration	Sty.	Story	WT., Wt.	Weight
R	Resistance	Subj.	Subject	WWF	Welded Wire Fabric
R.C.P.	Reinforced Concrete Pipe	Subs.	Subcontractors	XFER	Transfer
Rect.	Rectangle	Surf.	Surface	XFMR	Transformer
Reg.	Regular	Sw.	Switch	XHD	Extra Heavy Duty
Reinf.	Reinforced	Swbd.	Switchboard	XHHW; XLPE	Cross-Linked Polyethylene Wire
Req'd.	Required	S.Y.	Square Yard		Insulation
Res.	Resistant	Syn.	Synthetic	XLP	Cross-linked Polyethylene
Resi.	Residential	S.Y.P.	Southern Yellow Pine	Y	Wye
Rgh.	Rough	Sys.	System	yd	Yard
RGS	Rigid Galvanized Steel	t.	Thickness	yr	Year
R.H.W.	Rubber, Heat & Water Resistant;	T	Temperature; Ton	Δ	Delta
	Residential Hot Water	Tan	Tangent	%	Percent
rms	Root Mean Square	T.C.	Terra Cotta	~	Approximately
Rnd.	Round	T & C	Threaded and Coupled	Ø	Phase
Rodm.	Rodman	T.D.	Temperature Difference	@	At
Rofc.	Roofer, Composition	T.E.M.	Transmission Electron Microscopy	#	Pound; Number
Rofp.	Roofer, Precast	TFE	Tetrafluoroethylene (Teflon)	<	Less Than
Rohe.	Roofer Helpers (Composition)	T. & G.	Tongue & Groove;	>	Greater Than
Rots.	Roofer, Tile & Slate		Tar & Gravel		
R.O.W.	Right of Way	Th.; Thk.	Thick		
RPM	Revolutions per Minute	Thn.	Thin		
R.S.	Rapid Start	Thrded	Threaded		
Rsr	Riser	Tilf.	Tile Layer, Floor		
RT	Round Trip	Tilh.	Tile Layer, Helper		
S.	Suction; Single Entrance; South	THHN	Nylon Jacketed Wire		

A

Abandon catch basin 22
Abatement asbestos 101, 102
ABS DWV pipe 207
Absorber vibration 129
Absorption chiller 455
Absorption water chillers 312
A/C coils furnace 296
  fan coil system 476, 477
  packaged DX 468
  packaged DXVAV 468
  rooftop DX 467
  rooftop DXVAV 467
  self contained 485, 486
  self contained VAV 485, 486
Access door duct 358
  floor 78
  road and parking area 11
Accessories duct 358
  fuel oil 273
Accessory formwork 50
  masonry 54
Accumulator refrigerant 271
Acoustical sealant 68
Adapter fill 273
  polybutylene 217, 218
Adjustable frequency drives 396
Adsorptive desiccant 279
  filter 279
Aerator 44
Aerial lift 14
Aftercooler air-cooled 277
  compressor 277
  separator 277, 278
  water-cooled 278
Ahu built up 470
  built up VAV 471
  central station 472
  central station VAV 473
  rooftop 474
  rooftop VAV 475
Air & dirt eliminator 244
  balancing 384
  cell pipe-covering 132
  cleaner electronic 380
  cock 248
  compressor 14
  compressor control system . . . 108
  compressor package 276, 277
  condition and ventilation . 539, 540
  conditioner central station 322
  conditioner cooling & heating . 325
  conditioner direct expansion . . 337
  conditioner fan coil 336
  conditioner gas heat 325
  conditioner packaged terminal 324
  conditioner portable 327
  conditioner receptacle 392
  conditioner removal 120
  conditioner roof top 326
  conditioner rooftop 325
  conditioner self-contained 327
  conditioner thru-wall 325
  conditioner ventilating 347
  conditioner window 327
  conditioner wiring 394
  conditioning 537, 541, 543
  conditioning and ventilation . . 542,
                                         543
  conditioning computer 324
  conditioning fan 364
  conditioning pump 272
  conditioning ventilating . . 108, 314,
  315, 318 , 319, 322, 324, 325, 327, 352,
  353, 355 , 357, 359, 362, 365, 368-370,

  372 , 373, 375, 376, 378, 381, 382
  control 244
  cooled belt drive condenser . . 315
  cooled condensing unit 317
  cooled direct drive condenser . 316
  curtain 72
  curtain door 72
  dryer adsorption 279
  dryer compressor 278
  dryer deliquescent 278
  dryer refrigeration 278
  dryer regenerative 279
  eliminator 244
  extractor 358
  filter 279, 378
  filter industrial 378, 379
  filter regulator lubricator 280
  filter roll type 380
  filtration 102
  fitting boiler 245
  fitting compression tank 245
  handler heater 297
  handler modular 297
  handling fan 364
  handling unit 321, 322
  hose 14
  line couplers 280
  make-up unit 323
  register 374
  return grille 372
  sampling 102
  scoop 245
  supply grille 371
  supply pneumatic control 108
  supply register 374
  tool 14
  tube system 118
  unit induction 360
  vent automatic 245
Air-conditioner air cooled . . . 452, 462,
                                   463-466
  chilled water . . . 449, 450, 452, 466
  direct expansion 456, 457, 459-465
  fan coil 449, 450, 452, 465
  multizone 326, 459
  penthouse 326
  self-contained 460-463
  single zone 456, 457
  split system 464, 465
  water cooled 460, 461, 466
Air-conditioning air cooled 450
  computer 466
  drug store 459
Airfoil fan centrifugal 366
Airless sprayer 102
Alarm burglar 115
  detection 115
  residential 394
  sprinkler 115
  standpipe 115
  vent 274
All fuel chimney 305-308
  service jacket insulation 137
Aluminum corrugated pipe 503
  diffuser 369, 370
  diffuser perforated 370
  downspout 66
  ductwork 347
  flashing 66, 269
  flex duct 348
  grille 371
  handrail 78
  louver 78, 374
  nail 62
  operating louver 374
  pipe 40, 41

  register 374
  rivet 58
  salvage 121
Anchor chemical 56
  epoxy 56
  expansion 56
  hollow wall 56
  jute fiber 57
  lead screw 57
  machinery 58
  pipe conduit casing 38
  plastic screw 57
  selfdrilling 56
  wedge 57
Anchors drilling 56
Angle framing steel 58
  valve 236
Anti-freeze inhibited 270
Anti-siphon device 267
  water 267
Apartment S.F. & C.F. 400
Appliance 86
  plumbing 285, 286
  residential 82, 394
Approach ramp 78
Aquastats 108
Architectural woodwork 63
Area clean-up 102
Arena sport S.F. & C.F. 405
Asbestos abatement 101, 102
  demolition 101
  disposal 103
  removal 100, 101
  removal process 527
As-built drawings 18
Ash conveyor 84
Asphalt flashing 66
Asphaltic binder 43
  concrete 20, 43
  pavement 43
  wearing course 43
Aspirator dental 91
Atomizer water 102
Attic ventilation fan 369
Auditorium S.F. & C.F. 400
Auger 13
Automatic timed thermostat 110
  washing machine 87
Automotive equipment 83
  exhaust system 494
  lift 83
  sales S.F. & C.F. 400
Autopsy equipment 90
Axial flow fan 364

B

Backer rod 68
Backfill 25, 26
  trench 27
Backflow preventer 266
  preventer solar 105
Backhoe excavation 26
  trenching 496
Bag disposable 102
  glove 101
Balance valve circuit setter 247
Balancing air 384
  hydronic 264
  water 385
Baler 84
Ball check valve 237
  joint weld end 265
  valve 228, 236, 240
  valve grooved joint 188

Band vent-chimney 310
Bank air-conditioning . 449, 451, 456,
                        458 , 460, 462, 464
  S.F. & C.F. 400
Barometric damper 301
Barricade 12
  tape 12
Barrier separation 101
Barriers and enclosures 12
Baseboard diffuser 371
  expansion coupling 250
  heat 338
  heat electric 340
  register 375
Basket type strainer 261
Bath steam 95
  whirlpool 91
Bathroom exhaust fan 367
  heater & fan 367
Bathtub removal 120
Bead board insulation 94
Beam attachment welded 121
  clamp 122
  soldier 25
Bedding pipe 26
Bellow expansion joint pipe 250
Bellows expansion joints 251
Belt driven fan 366
Bibb hose 248
Binder asphaltic 43
Bituminous coating 47
Blanket insulation 131
Blasting 26
Block manhole 42
  manhole/catch basin 504
Blowdown boiler 286
Blower 365
  pneumatic tube 118
Boiler 286
  air fitting 245
  blowdown 286
  coal fired 295
  coal-fired hydronic 442
  control system 108
  cut off low water 383
  demolition 120
  drain 248
  electric 287, 288, 439
Boiler electric HW 444
  electric steam 287
Boiler electric steam 445
Boiler feed pump 280
  fire tube 293
  fossil fuel 291
  fossil fuel hydronic 443
  gas fired 288-290
  gas pulse 290
  gas/oil combination 290, 291
  general 120
  hot water 288-294
  hot-water electric 439
  hydronic fossil fuel 441, 443
  insulation 131
  oil fired 292, 293
  oil-fired hydronic 441, 442
  scotch marine 293
  solid fuel 294
  steam 289, 290, 292, 293
  steam electric 438
  stoker fired 295
  wall hung 290
  water gauge 383
  water tube 294
Bolt & gasket set 172
  toggle 57
Bond performance 7, 517

Boom lift 14
Booster coil 334, 335
  fan 364
Boot spiral duct register 357
Booth painting 83
Boring horizontal 28
  service 28
Bowling alley air-conditioning . . 449,
    451 , 456, 458, 460, 462, 464
Box buffalo 32
  culvert precast 502
  storage 9
  vent 54
Boxes VAV 361
Brace cross 59
Bracket pipe 121
Braided bronze hose 252
  steel hose 253
Brass pipe 141
  pipe fitting 141
  salvage 121
  screw 63
  valve 227, 228
Break glass station 115
Breaker vacuum 267
Breather cap fuel 274
Breeching draft inducer 311
  insulation 131
Brick catch basin 42
  chimney simulated 79
  vent 374
  vent box 54
Bridge sidewalk 10
Bronze body strainer 263
  butterfly valve 228
  valve 228
Broom finish concrete 51
Bucket crane 13
Buffalo box 32
Builder's risk insurance 512
Building air conditioning 449
  air-conditioning . . 449-452, 456-466
  permit 7
  portable 9
  temporary 9
Built-in range 86
Bulldozer 25
Burglar alarm 115
Burner gas conversion 286
  gas/oil 286
  gun type 286
  industrial 287
  jet 287
  oil 287
  residential 286
Butterfly valve 20, 30, 232
  valve bronze 228
  valve grooved joint 189
Butyl caulking 69

C

Cabinet convector heater 343
  laboratory 90
  unit heater 340, 343, 344
Cabinet unit heater 484
Cable heating 340
  jack 16
Calcium silicate insulation 131
Cap vent 274
Carbon monoxide detector 108
Carousel compactor 84
Carpentry rough 63
Carpet computer room 78
Carrier pipe 122, 126

Case refrigerated 82
Cased evaporator coils 297
Casing anchor pipe conduit . . . 38
Cast in place concrete 50
  iron fitting 170
  iron manhole cover 42
  iron pipe fitting 160
  iron radiator 338
  iron trap 268
Catch basin 42
  basin and manhole 505
  basin brick 42
  basin frame and cover 43
  basin masonry 42
  basin precast 42
  basin removal 22
  basin/manhole 504
Catwalk 11
Caulking 68, 69
  polyurethane 69
  sealant 68
Ceiling diffuser 369, 370
  drilling 56
  fan 365
  heater 87
  polystyrene 94
  register 374
Cellular glass insulation 133
Cement plaster 94
Central station air handling unit . 322
Central station heat pump 487
Central vacuum 82
Centrifugal airfoil fan 366
Centrifugal chiller 454
Centrifugal fan 364, 366
Centrifugal fan system 480
Centrifugal in-line motor 281
  liquid chiller 312
  pump 14, 281
Chain hoist 16
  link fence 12
  saw 15
  trencher 13
Chair dental 91
  hydraulic 91
Chamber decontamination 101
Channel framing steel 59
  slotted framing 59
Charge powder 58
Charges disposal 103
Chart recorders 111
Chase formwork 50
Check lift valve 229
  swing valve 228, 234
  valve 20, 30, 238, 240-242
  valve ball 237
  valve grooved joint 189
  valve silent 231, 234, 240, 243
  valve wafer 231, 235, 243
Chemical anchor 56
  feeder shot 296
  media filter 378
  toilet 15
Chiller reciprocating 453
Chiller water 312
Chimney 54
  accessories 79
  all fuel 305-308
  cap ventilator 378
  concrete 54
  draft inducer 311
  foundation 51
  high temperature vent 309
  metal 54, 79, 309-311
  positive pressure 305-308
  prefabricated metal 305-308

rain cap 378
  simulated brick 79
  vent 302
  vent fitting 308
  vent oval 305
Chlorination system 96
Church S.F. & C.F. 400
Chute linen 118
  package 118
  refuse 118
  rubbish 23
Chutes 118
C.I.P. concrete 50
Circuit sensor 246, 247
  setter 247
Circular saw 15
Circulating pump 270
City hall S.F. & C.F. 406
Clamp beam 122
  pipe 45, 46, 121, 129
  riser 122
Clamshell bucket 13
Classroom heating & cooling 338
Clay pipe 35, 41
Clean room 94
  tank 21
Cleaning internal pipe 45
  up 18
Clean-up area 102
Clevis pipe hanger 124
Climbing crane 15
  jack 16
Clock timer 394
Clothes dryer commercial 82
  dryer residential 87
Club country S.F. & C.F. 400
Coal burning furnace 299
  fired boiler 295
  stoker 287
Coalescent filter 279
Coating bituminous 47
  epoxy 47
  polyethylene 47
Cock air 248
  gauge 249
  pet 248
  steam 248
Cocks gas 227
Coil booster 334, 335
  condenser 332-334
  cooling 332
  heating 322
  water 332-334
Cold storage door 72
Collar spin-in duct 358
  storm 308
Collection systems dust 88
Collector solar energy system . . . 103,
                                    104
College S.F. & C.F. 401
Combination device 391
Comfort station 95
Commercial air-conditioner 449
  air-conditioning . 450-452, 456-464,
                                    466
  fin tube 338
  fin-tube heating 415
  heater electric 341
Commercial kitchen exhaust 491
Commercial kitchen vent 85
  refrigeration 94
  unit-heater heating 416
Commissioning 18
Common nail 62
Community center S.F. & C.F. . . . 401
Compact fill 26

Compaction soil 25
Compactor 84
  earth 13
  waste 84
Companion flange 172
Compensation workers' 7
Component control 106-108
  pneumatic 112, 113
Components furnace 296
Compression joint fitting 149
  tank 251
Compressor accessory 277
  aftercooler 277
  air 14
  air package 277
  filter 279
  reciprocating 276
  reciprocating hermetic 276
  refrigerant 276
  refrigeration 276
  tank mounted 108
Computer air conditioning 324
  floor 78
  room carpet 78
Concrete asphaltic 20, 43
  broom finish 51
  cast in place 50
  catch basin 42
  chimney 54
  C.I.P. 50
  core 22
  demo 23
  drill 22, 388
  drilling 56
  finish 51
  float finish 51
  footing 51
  formwork 50
  foundation 51
  hole cutting 389
  hole drilling 388
  insert hanger 123
  manhole 42
  monolithic finish 51
  patching 51
  pipe 33, 34, 40
  placing 51
  ready mix 50
  reinforcement 50
  removal 22
  structural 50
  tank 97
Condensate pump 271
  steam meter 273
Condenser 315
  air cooled belt drive 315
  air cooled direct drive 316
  coil 332-334
  evaporative 316, 317
  pad 320
Condensing unit air cooled 317
  unit water cooled 317
Conditioner portable air 327
  roof top air 326
Conditioning ventilating air . 318, 319,
322 , 324, 325, 327, 352, 353, 355, 357,
370 , 372, 373, 375, 376, 381, 382
Conduit & fitting flexible 395
  fitting pipe 37
  pipe 36
  pipe prefabricated 39
  prefabricated pipe . . 36, 37, 39, 40
Connection motor 395
Connector flexible 252
  water copper tubing 252
Console heat pump 487

Construction aids 9
  cost index 7
  management fee 6
  photography 8
  sump hole 24
  temporary 9
Contained pipe 223
Contaminated soil 21
Contingencies 6
Contractor equipment 13
  overhead 18
  pump 15
Control & motor starter 397
  air 244
  component 106-108, 113
  component electronic 114
  component relay 110
  component sensor 110
  component transmitter 111
  draft 301
  electronic 114
  flow check 254
  hand-off-automatic 396
  pneumatic 109, 112, 113
  radiator supply 254
  station 396
  station stop-start 396
  system 106, 107
  system air compressor 108
  system boiler 108
  system cooling 108
  system electronic 107
  system split system 108
  system VAV box 108
  system ventilator 108
  valve heating 111, 254
  water feeder 383
  water level 383
Control-component 111
Controller pneumatic 113
Controls D.C. 107
  DDC 106
Convector cabinet heater 343
  cover 62
  heating unit 338
Conveyor ash 84
  door 73
Cooking equipment 85
  range 86
Cooler 94
  beverage 85
  door 72
  evaporator 318
Cooling & heating A/C 325
  coil 332
  computer 466
  control system 108
  electric furnace 299
  tower 315
  tower louver 374
  towers fiberglass 315
  towers stainless 315
Copper ACR tubing 144
  drum trap 268
  DWV tubing 143
  fitting 145
  flashing 66, 270
  grooved joint 153
  pipe 142, 143
  rivet 58
  salvage 121
  tubing oxygen class 143
Core drill 22
Cork insulation 94
Corridor air-conditioning . . . 449, 451,
            456 , 458, 460, 462, 464

Corrosion resist. bkflo. preven . . 266
  resistant duct fitting 353
  resistant fan 366
  resistant tank 251
Corrosive fume resistant duct 349,
                    350–352
Corrugated metal pipe 40, 41
  pipe 40
Cost mark-up 8
Country club S.F. & C.F. 400
Couplers air line 280
Coupling expansion 250
  mechanical 182
  plastic 216
Course wearing 43
Courthouse S.F. & C.F. 401
Cover convector 62
  manhole 43
  stair tread 9
Covering calcium silicate pipe . . . 134,
                    135
  protection saddle 127
CPVC pipe 207
  valve 236, 237
Crane bucket 13
  climbing 15
  crawler 15
  hydraulic 16
  material handling 15
  tower 15
Crawler crane 15
Crematory 84
Crew 6
Cross brace 59
Crosses 28
Crossflow fuel heater 275
Cubic foot building cost 400
Culvert box precast 502
  end 34
  reinforced 33
Curb prefabricated 320
Curtain air 72
  damper fire 358
Cut in sleeve 29
  in valve 29
  off low water boiler 383
  pipe groove 189
Cutoff feeder combination 383
Cutting steel 23
  torch 15

D

Dairy case 82
Damper 359, 361, 362
  barometric 301
  fire round 359
  foundation vent 54
  louver type fire 360
  multi-blade 361
  relief 363
  round 363
  vent 301
Darkroom equipment 87
Dasher hockey 96
DDC controls 106, 107
Deck roof 63
Decontamination chamber 101
  enclosure 101, 102
  equipment 102
Decorator device 391
  switch 391
Deep well pump 33
Dehumidifier 86
  selfcontained 330

Delicatessen case 82
Deliquescent desiccant 278
Demo concrete 23
Demolition 20, 120
  asbestos 101
  boiler 120
  ductwork 120
  enclosure 102
  furnace 120
  HVAC 120
  pavement 22
  plumbing 120
  selective 23
  site 22
Dental equipment 91
Department store A/C 452, 457
  store S.F. & C.F. 401
Derrick 16
  guyed 16
  stiffleg 16
Desiccant dryer 278, 279
Detection alarm 115
  leak 116
  probe leak 115
  system 115
  tape 20
Detector carbon monoxide 108
  check valve 30
  meter water supply 265
  smoke 115
  temperature rise 115
Device anti-siphon 267
  combination 391
  decorator 391
  GFI 392
  receptacle 392
  residential 390
Dewater 17, 23, 24
  pumping 24
Dewatering 23
Diaphragm pump 15
Dielectric union 249
Diffuser aluminum 369, 370
  baseboard 371
  ceiling 369, 370
  floor 371
  linear 369
  opposed blade damper 369
  perforated aluminum 370
  pump suction 264
  rectangular 370
  sidewall 371
  steel 371
  supply/return combination . . . 371
  T-bar mount 370
Dimmer switch 390
Direct expansion A/C 337
Discharge hose 14
Dishwasher 85, 86
Dispenser product 83
Disposable bag 102
Disposal asbestos 103
  charges 103
  garbage 87
  waste 84
Distiller water 90
Ditching 23
Diverter valve 233, 239, 242
Door air curtain 72
  bell residential 393
  cold storage 72
  conveyor 73
  cooler 72
  duct access 358
  fire 115
  grille 372

refrigerator 72
  special 72
  steel louver 373
Dormitory S.F. & C.F. 401
Double wall pipe 35, 219
Downspout 66
  aluminum 66
  painting 76
  steel 66
Dozer 25
Draft control 301
  damper vent 301
  inducer breeching 311
  inducer chimney 311
Drain boiler 248
  pipe 24
Drainage pipe 33, 34, 41, 208
  site 42
  trap 268
Drainers liquid 255, 256
Drawings as-built 18
Drill & tap main 31
  concrete 22, 388
  core 22
  main 31
  rock 26
Drilling anchors 56
  ceiling 56
  concrete 56
  horizontal 28
  plaster 56
Driveway removal 22
Drug store air-conditioning . 450, 452,
           457 , 461, 463, 465
Drum trap copper 268
Dryer adsorption air 279
  clothes 87
  commercial clothes 82
  compressor air 278
  deliquescent air 278
  desiccant 278, 279
  receptacle 392
  refrigeration air 278
  regenerative air 279
  residential 82
  system 279
  vent 87
Drywall nail 62
Dual wall hdpe pipe 223
Duck tarpaulin 11
Duct access door 358
  accessories 358
  collar spin-in 358
  connection fabric flexible 358
  corrosive fume resistant . . . 349-352
  fiberglass 349-352
  fitting plastic 353
  fitting spiral 354-356
  flexible insulated 348
  flexible noninsulated 347
  furnace 299
  grille 78
  heater electric 330-332
  humidifier 330
  HVAC 347
  insulation 131
  liner 132
  mechanical 347
  mixing box 360
  plastic 352
  preformed spiral 353, 355
  rigid fiberglass 349-352
  rigid plastic 352
  silencer 363
  smoke exhaust 349-352
  sound trap 364

Ducted evaporative cooler 322
Ductile iron fitting 31
   iron pipe 31
   iron pipe cement lined 30
Ductless split system 328
Ductwork .. 347, 348, 353, 354, 356,
                  541
   aluminum 347
   demolition 120
   fabric coated flexible 347
   fabricated 347
   flexible 348
   galvanized 347
   laboratory 90
   rectangular 347
   removal 120
   rigid 347
Dump charge 23
   truck 13, 27
Dumpster 23
Duplex style strainer 262, 263
Dust collection systems 88
   partition 23
DWV pipe ABS 207
   PVC pipe 207
   tubing copper 143
DX coil evaporator 317-319

E

Earth compactor 13
   vibrator 25, 26
Earthwork 25
   equipment 13, 25
Education religious S.F. & C.F. ... 404
Eggcrate framed grille 373
   grille lay-in 373
   T-bar grille 373
Ejector pump 284
Elbow pipe 185
Elderly housing S.F. & C.F. 402
Electric baseboard heater 340
   boiler 287, 288, 438
   cabinet heater 343
   duct heater 330-332
   fuel oil heater 274
   furnace 297
   generator 14
   heater 87
   heating 340
   hot air heater 343
   induction unit 360
   infra-red heater 340
   log 79
   pool heater 296
   unit heater 341-343
Electrical fee 6
   installation drilling 388
   knockout 389
   laboratory 90
Electronic air cleaner 380
   control 114
   control component 114
   control system 107
Electrostatic precipitator 85
Elementary school S.F. & C.F. ... 405
Elevated water tank 97
Elevator construction 16
   fee 6
Eliminator air 244
   air & dirt 244
Employer liability 7
Encapsulation 103
   pipe 103
Enclosure decontamination . 101, 102

demolition 102
End culvert 34
Energy circulator air solar 104
   recovery wheel 346
   system air purger solar 105
   system air vent solar 105
   system balancing valve solar .. 105
   system control valve solar ... 105
   system controller solar 104
   system expansion tank solar .. 105
   system gauge pressure solar .. 105
   system heat exchanger solar .. 104
   system storage tank solar 105
   system vacuum relief solar ... 106
Engine exhaust elimination 381
Engineering fee 6
   fees 520, 521
Enthalpy control 109
   recovery wheel 346
Entrainment eliminator 268
Epoxy anchor 56
   coating 47
   grout 51
   resin fitting 209
Equipment 13, 82
   darkroom 87
   dental 91
   earthwork 13, 25
   feedwater 296
   formwork 50
   foundation 50
   general 8
   hospital 90
   insurance 7
   laboratory 90
   laundry 82
   medical 90
   rental 13, 15, 523
   service station 83
   swimming pool 96
   waste handling 83
Escutcheon 259
Estimate electrical heating 340
Estimating 512
Ethylene glycol 270
Evaporative condenser 316, 317
   cooler 322
   cooler ducted 322
Evaporator coils cased 297
   cooler 318
   DX coil 317-319
   freezer 318
Excavation 25, 27
   backhoe 26
   hand 25, 26
   hauling 27
   machine 25
   structural 26
   trench 26, 27, 496-498
Excavator rental 13
Exchanger heat 320
Exhaust elimination engine 381
Exhaust fume hood 492
Exhaust granular fan 366
   hood 87, 90
   hood kitchen 86
   registers 376
   system 381
   vent 377
   vent damper 358
Exhaust-elimination garage 381
Exhauster industrial 366
   roof 368
Expansion anchor 56
   coupling 250
   coupling baseboard 250

grooved joint 188
joint 250
joints bellows 251
shield 56
tank 251
tank steel 251
Exterior sprinkler 44
Extra work 8
Extractor air 358
   fitting 273
   industrial 82
Eye rod linked 125
   rod welded 125

F

Fabricated ductwork 347
Factor 6
Factory air-conditioning 450, 452, 457,
           459, 463, 465
   S.F. & C.F. 401
Fan 364, 366
   air conditioning 364
   air handling 364
   attic ventilation 369
   axial flow 364
   bathroom exhaust 367
   belt driven 366
   booster 364
   ceiling 365
   centrifugal 364, 366
   coil air conditioning 337
   corrosion resistant 366
   exhaust granular 366
   house ventilation 369
   induced draft 311
   in-line 364
   kitchen exhaust 367
   low sound 364
   paddle 394
   propeller exhaust 367
   residential 367, 394
   roof 368
   utility set 366, 368
   vane-axial 365
   ventilation 394
   ventilator 364
   wall exhaust 369
   wiring 394
Fastener timber 62
   wood 63
Fastenings metal 56
Faucet & fitting 282
Fee engineering 6
Feeder control water 383
   cut-off combination 383
Feedwater equipment 296
Fence chain link 12
   plywood 12
   temporary 12
   wire 12
Fibergalss tank 98
Fiberglass cooling tower 469
Fiberglass cooling towers 315
   duct 349-352
   insulation 131
   panel 9
   pipe covering 136, 137
   reinforced ceiling 94
Fiberglass storage tank 507, 508
Fiberglass tank 99
Field office 9
   office expense 9
Fill 25, 26
   adapter 273

box fuel 273
cap fuel 273
gravel 26
Film equipment 87
Filter air 279, 378
   chemical media 378
   compressor 279
   dryer refrigerant 271
   grille 372
   mechanical media 380
   odor 279
   oil 273
   regulator lubricator air 280
   swimming pool 96
Filtration air 102
   equipment 96
Fin tube commercial 338
   tube radiation 338
   tube radition 338
Finish concrete 51
   nail 62
   refrigeration 94
Fire alarm 115
   call pullbox 115
   damper curtain type 358
   damper louver type 360
   damper round 359
   door 115
   horn 115
   protection kitchen 85
   signal bell 115
   station S.F. & C.F. 402
   tube boiler 293
Fired burner gas 287
   pool heater oil 296
Fireplace free standing 79
   prefabricated 78
Fireproofing 67
Firestopping 67
Fitting adapter pipe 147
   cast iron 170
   cast iron pipe 162, 163
   compression joint 149
   compression tank air 245
   copper 145
   copper pipe 147
   epoxy resin 209
   extractor 273
   flanged 171, 172
   flare joint 150
   forged steel pipe 177, 179, 180
   gas vent 302, 304
   grooved joint ... 153, 185, 187, 188
   grooved joint pipe 185
   grooved unit 153, 186
   insert pipe 218
   malleable iron 164
   malleable iron pipe   167, 169, 180,
              182
   mechanical joint pipe 180, 182
   monoflow 256
   pipe 160, 161, 165, 171, 176
   pipe conduit 37
   plastic 209
   plastic pipe 213, 217
   polybutylene pipe 217, 218
   PVC 211
   PVC pipe 219
   refrigeration 150
   stainless steel 195
   stainless steel pipe .. 195-197, 199,
            201 -204
   steel 160, 170
   steel carbon pipe 175
   steel pipe 177, 179-181
   tee 31

vent chimney . . 302, 304, 308, 310
water pipe 31
weld 176
weld joint pipe 177
welded steel pipe 181
wye 31
Fittings hdpe 220, 221
Fixed blade louver 374
roof tank 98
Fixture plumbing 33, 270, 280
removal 121
residential 393
Flange companion 172
gasket 172
grooved joint 187
stainless steel 200
steel 187
steel weld-on 179
threaded 172, 173
Flanged fitting 171, 172
pipe 155, 158
Flare joint fitting 150
Flashing 66
aluminum 66, 269
asphalt 66
copper 66, 270
galvanized 270
masonry 66
neoprene 270
stainless 66
vent 269
vent chimney 303, 309
Flatbed truck 13
Flex duct aluminum 348
Flexible conduit & fitting 395
connector 252
duct connection fabric 358
ductwork 348
ductwork fabric coated 347
insulated duct 348
metal hose 252
Float finish concrete 51
valve 254
Floater equipment 7
Floating roof tank 98
Floodlight 14
Floor access 78
cleaning 18
diffuser 371
heating radiant 344-346
nail 62
patching concrete 51
pedestal 78
refrigeration 94
register 376
Flow check control 254
fitting hydronic 256
measurement 246, 247
meter 246, 247, 265
meter circuit sensor 246
Flue chimney metal 302
prefab metal 302
Fluid heat transfer 105
Foam pipe covering 139
urethane 69
Food service equipment 85
Foot valve 237, 273
valve oil 273
Footing concrete 51
Forged valves 240
Formwork accessory 50
chase 50
concrete 50
equipment 50
sleeve 50
structural 50

Fossil fuel boiler . . 289, 291-293, 295
fuel burner 287
fuel hydronic heating 440-442
Foundation chimney 51
concrete 51
equipment 50
mat 51
vent 54
Fountain 44
lighting 44
Framing lightweight 58
metal 58
steel angle 58
steel channel 59
Fraternal club S.F. & C.F. 400
Fraternity house S.F. & C.F. 402
Freezer 82, 85, 94
door 72
evaporator 318
Freezestat 109
Fuel breather cap 274
burner fossil 287
fill box 273
fill cap 273
heater crossflow 275
heater safety type 275
heater steam 275
heater suction bell 275
oil gauge 274
oil heater 274, 275
oil heater electric 274
oil pump 274, 282
oil pump & motor set 274
oil specialties 273
oil valve 274
spill container 273
storage tank 98
system components 273
tank 98, 99
Fume elimination welding 382
hood 90
Funeral home S.F. & C.F. 402
Furnace A/C coils 296
coal burning 299
components 296
demolition 120
duct 299
electric 297
gas 299
gas fired 297, 299
heating & cooling 299
hot air 297
oil fired 298, 299
type humidifier 330
wall 301
wood coal/oil combination 299
Fused pipe 31
Fusible oil valve 274

G

Gage stick wood 273
Galvanized ductwork 347
flashing 270
Garage exhaust system 494
exhaust-elimination 381
S.F. & C.F. 402
Garbage disposal 87
Gas boiler hw 446
boiler steam 447
Gas cocks 227
connector 252
conversion burner 286
fired boiler 289, 290
fired burner 287

fired furnace 297, 299
fired infra-red heater 340
fired pool heater 295
fired space heater 300, 301
furnace 299
heat air conditioner 325
incinerator 84
log 79
meter residential 275
pipe 36
pressure regulators 257
pulse combustion 290
regulator 257
safety valve 287
station piping 35
station tank 98, 99
stop valve 228
vent 302
vent fitting 302, 304
Gasket & bolt set 172
flange 172
neoprene 69
Gas/oil boiler hw 447
boiler steam 447
Gas/oil burner 286
combination boiler 291
steam boiler 291
Gasoline pump 83
tank 98
Gas-service polyethlene pipe 506
Gate box 29
valve 20, 29, 233, 239, 241, 242
valve grooved-joint 189
valve soldered 229
Gauge cock 249
fuel oil 274
pressure 109
pressure/vacuum 109
vacuum 109
water 383
General equipment 8
Generator construction 14
steam 90
GFI receptacle 392
Gland seal pipe conduit 36
Glass insulation cellular 133
lined water heater 87
Glassware washer 90
Globe valve . . 21, 230, 233, 239, 241, 243
Glove bag 101
Glycol ethylene 270
Gradall 13
Grading 28
Grass sprinkler 44
Gravel fill 26
pack well 32
Gravity syphon ventilator 377
Grille air return 372
aluminum 371
convector cover 62
door 372
duct 78
eggcrate framed 373
filter 372
lay-in T-bar 373
painting 76
plastic 373
transfer 373
Grinder pump system 284
Groove cut labor steel 189
roll labor steel 190
Grooved joint check valve 189
joint copper 153
joint expansion 188
joint fitting 153, 185, 187, 188

joint flange 187
joint pipe 183
joint plug valve 189
joint strainer 188
joint suction diffuser 188
reducer 186
unit fitting 153, 186
Ground box hydrant 283
post hydrant 283, 284
Grout epoxy 51
topping 51
underlayment 51
Guardrail temporary 12
Guide pipe 123
Gutter swimming pool 96
Guyed derrick 16
Gymnasium S.F. & C.F. 402
Gypsum lath nail 62

H

Hand excavation 25-27
Handling rubbish 23
unit air 321, 322
waste 83
Hand-off-automatic control 396
Handrail aluminum 78
Hanger & support pipe . 121-123, 125, 128, 129
assembly pipe 126
clevis 124
pipe 121
plastic pipe 129
rod coupling 125
rod pipe 125
rod socket 128
roll 126
split ring 124
trapeze 126
U-bolt pipe 129
U-hook 129
vibration absorbing 129
Hanging lintel 59
Hauling excavation 27
truck 28
Hazardous waste handling 84
Hdpe fittings 220, 221
pipe 31, 219
pipe dual wall 223
Header pipe 17
Heat baseboard 338
coil steam 335, 336
electric baseboard 340
exchanger 320
exchanger plate 320
exchanger residential 346
loss 535
pump 329
Heat pump central station 487
pump console 487
Heat pump residential 395
pump rooftop 329
radiant 340
recovery 84, 85
recovery air to air 346
recovery ventilation 346
recovery wheel 346
steam 336
temporary 8
trace system 341
transfer fluid 105
transfer package 286
Heater & fan bathroom 367
air handler 297
cabinet convector 343

Index

ceiling mount unit 344
contractor 14
electric 87
electric commercial 341
electric hot air 343
electric residential 341
electric wall 341
floor mounted space 300
fuel oil 274, 275
infra-red 340
infra-red quartz 342
raceway 341
sauna 95
swimming pool 295
swimming pool solar 431
terminal 339, 340
unit 300, 338, 340
valance 340
vertical discharge 343
wall mounted 343
warm air 299
water 87
Heating .. 97, 104, 105, 131, 135-137,
139 , 274, 287, 288, 290, 292, 293, 296,
300 , 302, 304, 309, 310, 320, 338, 340,
534
& cooling classroom 338
cable 340
coal hydronic 440
coil 322
coil tank 275
commercial bldg 416
control transformer 342
control valve 254
electric 340
estimate electrical 340
fin-tube 443
fin-tube residential 414
hot air 297, 299, 338-340
hot water solar 421
hydronic 287-289, 292, 338
hydronic fossil fuel .. 440, 442, 443
industrial 287
insulation 131
oil hydronic 440, 442
oil-fired hydronic 441
panel radiant 342
solar-energy 431, 433
subsoil 96
systems 533
terminal unit-heater 440, 441
unit convector 338
Heavy duty shoring 9, 10
High impact pipe 205
school S.F. & C.F. 405
temperature vent chimney 310
Historical Cost Index 7, 574
Hockey dasher 96
Hoist and tower 16
automotive 83
contractor 16
lift equipment 15
personnel 16
Hole cutting electrical 389
drilling electrical 388
Hollow wall anchor 56
Hood exhaust 90
fire protection 85
fume 90
range 87
Hook type pipe outlet 182
Hopper refuse 118
Horizontal boring 28
drilling 28
Horizontal heat pump 487
Horn fire 115

Hose adapter vapor 273
air 14
bibb 248
bibb sillcock 282
braided bronze 252
braided stainless steel 253
braided steel 253
discharge 14
metal flexible 252
suction 14
water 14
Hospital equipment 90
S.F. & C.F. 402
Hot air furnace 297
air heater electric 343
air heating 297, 299, 338-340
water boiler 288-294
water coil 334, 335
water heating 288, 338
water-steam exchange ... 286, 320
water-water exchange 320
Hot-water heating forced 414
solar-energy ... 419, 421, 423, 425,
427 , 429, 437
House fan whole 367
ventilation fan 369
Housing elderly S.F. & C.F. 402
public S.F. & C.F. 403
HTRW 100
Humidifier 87
duct 330
furnace type 330
room 330
Humidifiers 330
HVAC demolition 120
duct 347
piping specialties .. 250, 252, 256,
261 , 263, 264, 268, 383
removal 120
Hydrant ground box 283
ground post 283, 284
removal 22
wall 282
water 282
Hydraulic chair 91
crane 16
jack 16
oil pump 282
pressure test 385
Hydrocumulator 284
Hydronic balancing 264
expansion coupling 250
flow fitting 256
heating 287-289, 292, 338

I

Ice rink 96
skating equipment 96
skating rink S.F. & C.F. 403
Icemaker 85, 87
Impact wrench 14
Incinerator gas 84
municipal 84
waste 84
Increaser vent chimney 310
Index construction cost 7
Indicator thermoflo 264
Induced draft fan 311
Induction air unit 360
unit electric 360
unit vertical 360
Industrial air filter 378, 379
burner 287
equipment 88

exhauster 366
heating 287
Inert gas 21
Infra-red heater 340
heater gas fired 340
quartz heater 342
In-line fan 364
Insert fitting PVC 218
hanger concrete 123
Insulated protectors ada 131
Insulation 131, 132, 136, 137, 139
all service jacket 137
blanket 131
boiler 131
breeching 131
calcium silicate 131
duct 131
fiberglass 131
pipe 132, 139
polyethylene 132, 138
polyolefin 139
refrigeration 94
removal 100
subsoil 97
vapor proof 133
vaporproof 139
water heater 131
Insurance 7, 512
builder risk 7
equipment 7
public liability 7
Intake-exhaust louver 374
vent 378
Interval timer 391
Intrusion system 115
Invert manhole 43
Inverted bucket steam trap 259
Investigation subsurface 22
Iron body valve 232, 234
Irrigation system 44

J

J hook clamp w/nail 129
Jack cable 16
hydraulic 16
Jacking 28
Jail S.F. & C.F. 403
Jet burner 287
water system 33
Jetting pump 17
Job condition 6
Joint expansion 250
sealer 68
weld end ball 265
Jute fiber anchor 57

K

Kettle 85
Kitchen equipment 85
equipment fee 6
exhaust fan 367
exhaust hood 86
heat recovery 346
smoke control 86
vent commercial 85
ventilation 85
Knockout electrical 389

L

Labels pipe & valve 130

Labor index 7
pipe groove roll 190
pipe threading st steel 194
pipe threading steel 159
pipe weld stainless steel 194
pipe weld steel 159
Laboratory cabinet 90
equipment 90
research S.F. & C.F. 404
sink 90
Ladder 14
rolling 11
Lag screw 58
screw shield 57
Lagging 25
Landfill fees 23
Lateral-o-let 179
Latex caulking 69
Laundry equipment 82
Lavatory removal 121
Lead flashing 66
salvage 121
screw anchor 57
Leak detection 116
detection probe 115
detection tank 115
Liability employer 7
Library air-conditioning 449, 451, 456,
460, 462
S.F. & C.F. 403
Lift aerial 14
automotive 83
Light dental 91
stand 102
temporary 8
tower 14
underwater 96
Lighting fountain 44
outdoor 14
outlet 393
residential 393
Lightning suppressor 390
Lightweight framing 58
Linear diffuser 369
Linen chute 118
collector 118
Liner duct 132
Link seals pipe 259
Linked eye rod 125
Lintel hanging 59
Liquid chiller centrifugal 312
drainers 255, 256
Living assisted 403
Loadcenter residential 390
Loader vacuum 102
Locking receptacle 393
Log electric 79
gas 79
Louver 78, 374
aluminum 78, 374
aluminum operating 374
coating 374
cooling tower 374
door steel 373
fixed blade 374
intake-exhaust 374
midget 78
mullion type 374
redwood 63
silencer 364
ventilation 63
Lube equipment 83
Lubricated plug valve 238
Lubricating oil 317

INDEX

M

Machine excavation 25
  screw 58
  welding 15
Machinery anchor 58
Magnetic motor starter 397
Maintenance 18
Make-up air unit 323
  air unit rooftop 323
Malleable iron fitting 164
  iron pipe fitting 164, 165
Management fee construction 6
Manhole 42
  concrete 42
  cover 43
  invert 43
  precast 42
  raise 43
  removal 22
  step 43
Marker pipe 130
Mark-up cost 8
Masonry accessory 54
  catch basin 42
  flashing 66
  manhole 42
  nail 62
  saw 15
  ventilator 54
Masons scaffolding 11
Mat foundation 51
Material handling system 118
  hoist 16
  index 7
Measurement flow 246, 247
Meat case 82
Mechanical 528
  coupling 182
  duct 347
  equipment demolition 120
  fee 6
  media filter 380
Medical center air-conditioner . . . 465
  center air-conditioning . . . 450, 452,
                457 , 459, 461, 463
  clean copper tubing 143
  clinic S.F. & C.F. 403
  equipment 90
  office S.F. & C.F. 403
Melting snow 340
Metal chimney 54, 79, 309-311
  ductwork 347
  fastenings 56
  flexible hose 252
  flue chimney 302
  framing 58
  pipe 30, 41
Meter flow 246, 247
  steam condensate 273
  venturi flow 265
  water supply 265
  water supply detector 265
  water supply domestic 265
Meters 275
Microtunneling 28
Miscellaneous pumps 281
Mixing box duct 360
  boxes VAV 361
  valve 256
Modification to cost 6
Modular air handler 297
Modulating damper motorized . . 362
Monoflow fitting 256
Monolithic finish concrete 51
Monoxide detector carbon 108

N

Nail 62
Needle valve 237
Neoprene flashing 270
  gasket 69
Nondestructive pipe testing 384
Nonfusible oil valve 274
Nursing home S.F. & C.F. 403
Nut threaded 125

O

Observation well 32
Occupancy determinations 519
Odor filter 279
Off highway truck 14
Office air-conditioner 465
  air-conditioning . 450, 452, 457, 459,
                 461, 463
  field 9
  floor 78
  medical S.F. & C.F. 403
  S.F. & C.F. 404
  trailer 9
Oil burner 287
  filter 273
  fired boiler 292, 293
  fired furnace 298, 299
  fired heating 292
  fired infra-red heater 340
  fired pool heater 296
  fired space heater 301
  heater temporary 14
  lubricating 317
  pump fuel 282
  pump hydraulic 282
  refrigeration 317
  shut-off valve 274
  specialties fuel 273
  storage tank 98
  valve fuel 274
Oils 317
Omitted work 8
Opening roof frame 59
Operating cost equipment 13
OSHA testing 103

O

Outdoor lighting 14
Outlet heater tank 275
  lighting 393
Oval chimney vent 305
Oven 85, 86
Overhaul 23, 28
Overhead & profit 8
  and profit 516
  contractor 18
Overtime 7, 8, 518
Oxygen lance cutting 23

P

P trap 269
  trap refrigeration 272
Package chute 118
Packaged ventilation 85
Packaging waste 103
Pad condenser 320
  prefabricated 320
  vibration absorbing 130
Paddle blade air circulator 365
  fan 394
Painting booth 83
  downspout 76
  grille 76
  pipe 76
  swimming pool 96
Paints & coatings 76
Pan shower 66
Panel fiberglass 9
  fire 115
  radiant heat 342
Parking garage S.F. & C.F. 402
  lots paving 44
Partition dust 23
  refrigeration 94
Patching asphalt 20
  concrete 51
  concrete floor 51
  concrete wall 51
Pavement asphaltic 43
  demolition 22
  replacement 44
Paving parking lots 44
Pedestal floor 78
Penthouse air-conditioner 326
  roof louver 374
Perforated aluminum pipe 40
  pipe 40, 41
  PVC pipe 41
Performance bond 7, 517
Permit building 7
Personal respirator 102
Personnel hoist 16
Pet cock 248
Pex tubing 344, 345
Photography 8
  construction 8
  time lapse 8
Pickup truck 15
Pigtail steam syphon 249
Pile driving 523
  H 25
  steel sheet 24
  wood sheet 24
Piling sheet 24, 526
Pillow tank 98
Pin powder 58
Pipe 42
  & fittings . 36, 38, 39, 121, 122, 124,
125, 127, 129, 142, 143, 145, 147,
154-156, 158-161, 163, 165, 167, 170,
173, 175, 179-181, 183-185, 190, 191,

193-197, 199, 201-208, 213, 216, 218,
228, 230, 233, 234, 236-240, 242, 250,
            265, 267, 269, 283
  & valve labels 130
  aluminum 40, 41
  and fittings 528
  bedding 26, 499, 500
  bedding trench 27
  bellow expansion joint 250
  bituminous coated metal 502
  bituminous fiber 502
  bituminous fiber perforated . . . 412
  brass 141
  carrier 126
  clamp 45, 46, 121
  clamp plastic 129
  clay 35, 41
  cleaning internal 45
  concrete 33, 34, 40
  concrete nonreinforced 502
  concrete porous wall 412
  concrete reinforced 502
  conduit 36
  conduit gland seal 36
  conduit prefabricated . . 36, 37, 39,
                      40
  conduit system 36, 37
  contained 223
  copper 142, 143
  copper in trench 501
  corrugated 40
  corrugated metal 40, 41
  corrugated metal coated 502
  corrugated metal plain 502
  corrugated oval arch coated . . . 503
  corrugated oval arch plain 503
  covering 132, 136, 137
  covering calcium silicate . 134, 135
  covering fiberglass 136, 137
  covering foam 139
  CPVC 207
  double wall 35, 219
  drain 24
  drainage 33, 34, 41
  dual wall hdpe 223
  ductile iron 31
  ductile iron in trench 501
  DWV ABS 207
  DWV PVC 207
  elevated installation 120
  encapsulation 103
  fitting 160, 161, 165, 171, 176
  fitting adapter 147
  fitting brass 141
  fitting cast iron 160, 162, 163
  fitting copper 145, 147
  fitting DWV 213
  fitting flanged cast iron 170
  fitting forged steel . . . 177, 179, 180
  fitting grooved joint 185
  fitting insert 218
  fitting malleable iron 165, 167, 169
                    , 180, 182
  fitting mechanical joint . . . 180, 182
  fitting plastic 211, 213, 216, 217
  fitting polybutylene 217, 218
  fitting PVC 219
  fitting stainless steel . 195-197, 199,
                  201 -204
  fitting steel 173, 177, 179-181
  fitting steel carbon 173, 175
  fitting weld joint 177
  fitting weld steel 173
  fitting welded steel 181
  flanged 155, 158
  fused 31

gas 36
groove roll labor 190
grooved joint 183
hanger 121
hanger & support . . . 121-123, 125,
128 , 129
hanger assembly 126
hanger clevis 124
hanger plastic 129
hanger rod 125
hanger strap 128
hanger U-bolt 129
HDPE 31, 219
header 17
high impact 205
in trench 412, 501-503, 506
inspection internal 45
insulation 132, 139
insulation removal 101
iron ductile mech joint 501
iron ductile tyton joint 501
link seals 259
marker 130
metal 41
metal perforated coated 412
painting 76
perforated 41
perforated aluminum 40
plastic 205-207, 209
plastic, FRP 204
polyethylene 36
polyethylene gas-service 506
polyethylene in trench 501
polyvinyl chloride 503
PVC 34, 207
PVC in trench 501
PVC pressure 206
quick coupling 17
reinforced concrete 33
relay 24
removal 121
repair 45, 46
ring hanger 123
rodding 45
roll single 126
sewage 33, 34, 41
shock absorber 268
specialties 256, 266
stainless steel 190-192, 195
steel 40, 41, 153, 155-158, 160
steel gas-service 506
steel tarred & wrapped 506
subdrainage 40
subdrainage plastic 41
support framing 59
support saddle 127
testing 384
testing nondestructive 384
threading st steel labor 194
threading steel labor 159
vitrified clay 34, 41
vitrified clay perforated 412
vitrified clay premium joint 503
weld joint . . . 154, 157, 190, 191, 193,
194
weld stainless steel labor 194
weld steel labor 159
wrapping 47
X-ray 385
Piping drainage & sewage 502
excavation 496
gas station 35
gas-service 506
specialties 383
specialties HVAC 250, 252, 256, 261,
263, 264, 268, 383

subdrainage 40
to 10' high 529
valve 20
water distribution 501
Pit excavation 26
sump 23
Placing concrete 51
Plank scaffolding 10
Plant power S.F. & C.F. 404
screening 13
Plaster cement 94
drilling 56
Plastic convector cover 62
coupling 216
duct 352
duct fitting 353
fitting 209
Plastic, FRP pipe 204
Plastic grille 373
pipe 205-207, 209
pipe clamp 129
pipe fitting 209, 211, 213, 216
pipe hanger 129
pipe polyethylene 219
pipe weld 222
roof ventilator 366
screw anchor 57
trap 269
valve 236
Plate heat exchanger 320, 482
Platform trailer 15
Plating zinc 62
Plenum silencer 364
Plug valve 238
valve grooved joint 189
Plumbing 266
appliance 285, 286
demolition 120
fixture 33, 270, 280
fixtures removal 120
laboratory 90
Plywood fence 12
mounting board 63
sheathing, roof and walls 63
sidewalk 12
Pneumatic air operated valve . . . 112
component 112, 113
control 109, 112, 113
control system 107
controller 113
ejector pumps 84
stud driver 58
thermostat 111
tube 113, 114
tube fitting 113, 114
tube system 118
Police station S.F. & C.F. 404
Polybutylene adapter 217, 218
Polyethylene coating 47
insulation 132, 138
pipe 36, 208
plastic pipe 219
tarpaulin 11
Polyolefin insulation 139
Polypropylene trap 269
valve 236
Polystyrene ceiling 94
insulation 94
Polysulfide caulking 69
Polyurethane caulking 69
Polyvinyl tarpaulin 11
Pool heater electric 296
heater gas fired 295
swimming 96
swimming S.F. & C.F. 405
Porous wall concrete pipe 412

Portable air compressor 14
building 9
heater 14
Positive pressure chimney . . . 305-308
Post office S.F. & C.F. 404
Powder actuated tool 58
charge 58
pin 58
Power plant S.F. & C.F. 404
Power roof vent 481
Power temporary 8
Precast catch basin 42
manhole 42
Precipitator electrostatic 85
Pre-engineered structure 95
Prefabricated comfort station 95
fireplace 78
metal chimney 305-308
Pressure booster pump 284
gauge 109
pipe 206
pipe PVC 206
regulator 257, 258
regulator oil 257
regulator steam 258
regulators gas 257
relief valve 230
test 385
test hydraulic 385
valve relief 230
Pressure/temperature safety plug 258
Pressure/vacuum gauge 109
Prison S.F. & C.F. 403
Prisons work in 6
Process air handling fan 364
Produce case 82
Product dispenser 83
piping 35
Project overhead 8
sign 12
size modifier 544
Propeller exhaust fan 367
Propylene glycol 270
Protection winter 9
worker 102
Protectors ada insulated 131
Psychrometric table 538
P&T relief valve 230
Public housing S.F. & C.F. 403
Pulse boiler gas 290
Pump 33, 280, 284
air conditioning 272
centrifugal 281
circulating 270
condensate 271
condensate return 271
contractor 23
diaphragm 15
Pump double suction 448
end suction 448
Pump fuel oil 274
gasoline 83
heat 329
jetting 17
operator 24
pressure booster 284
rental 15
rotary 282
sewage ejector 284
shallow well 33
submersible 15, 285
sump 87, 285
trash 15
turbine 282
vacuum 272
water 14, 17, 32

wellpoint 17
Pumping 23
dewater 24
Pumps miscellaneous 281
pneumatic ejector 84
vacuum 280
Putlog scaffolding 11
PVC DWV pipe 207
fitting 211
insert fitting 218
pipe 31, 34, 205, 207
pipe perforated 41
SDR35 pipe 503
trap 269
valve 236, 237

Q

Quartz heater 342

R

Raceway 388-390
for heater cable 341
Radiant floor heating 344-346
heat 340
heating panel 342
Radiation fin tube 338
Radiator air vent 246
cast iron 338
supply control 254
supply valve 254
thermostat control sys. 108
Radition fin tube 338
Rain cap chimney 378
Raise manhole 43
manhole frame 43
Raised floor 78
Ramp approach 78
temporary 11
Range cooking 86
hood 87
receptacle 393
Reach-in cooler evaporator 317
Ready mix concrete 50
Receptacle air conditioner 392
device 392
dryer 392
GFI 392
locking 393
range 393
telephone 393
television 393
weatherproof 392
Reciprocating chiller 453
Reciprocating compressor 276
hermetic compressor 276
water chiller 313
Recorders chart 111
Recovery heat 85
wheel energy 346
Rectangular diffuser 370
ductwork 347
Reducer grooved 186
Redwood louver 63
tank 98
Refrigerant accumulator 271
compressor 276
filter dryer 271
removal 103
Refrigerated case 82
Refrigeration 94
commercial 82, 85
compressor 276

Index

equipment	96
fitting	150
floor	94
insulation	94
mortuary	90
oil	317
P trap	272
panel fiberglass	94
partition	94
sightglass	272
specialties	271
tubing	144
valves	272
walk-in	94
Refrigerator door	72
Refuse chute	118
hopper	118
Register air supply	374
baseboard	375
boot spiral duct	357
return	376
steel	375
wall	376
Registers exhaust	376
Regulator gas	257
oil pressure	257
pressure	257, 258
steam pressure	258
Reinforced concrete pipe	33, 34
culvert	33
Reinforcement concrete	50
Reinforcing steel	50
Relay control component	110
pipe	24
Relief damper	363
vent ventilator	378
water vacuum	267
Religious education S.F. & C.F.	404
Remote oil gauge	274
Removal air conditioner	120
asbestos	100, 101
catch basin	22
concrete	22
driveway	22
hydrant	22
insulation	100
lavatory	121
pipe	121
pipe insulation	101
plumbing fixtures	120
refrigerant	103
sidewalk	23
tank	21
urinal	121
water closet	121
water fountain	121
Rental equipment	13, 15
equipment rate	13
generator	14
Repair pipe	45, 46
Replacement pavement	44
Residential alarm	394
appliance	82, 394
burner	286
device	390
door bell	393
dryer	82
fan	367, 394
fin-tube heating	414
fixture	393
gas meter	275
heat exchanger	346
heat pump	395
heater electric	341
lighting	393
loadcenter	390

roof jack	368
service	390
smoke detector	394
switch	390
transition	368
ventilation	369
wall cap	368
washer	82
water heater	394
wiring	390, 394
Resistant duct fitting corrosion	353
Respirator	102
personal	102
Restaurant air-conditioner	465
air-conditioning	450, 452, 457, 459, 461, 463
S.F. & C.F.	404
Retail store S.F. & C.F.	404
Return register	376
Ridge vent	78
Rigid fiberglass duct	349-352
plastic duct	352
Ring hanger pipe	123
Rink ice	96
Riser clamp	122
Rivet	58
aluminum	58
copper	58
steel	58
tool	58
Road temporary	11
Rock drill	26
Rod backer	68
coupling hanger	125
pipe hanger	125
socket	128
threaded	125
tie	59
Roll hanger	126
type air filter	380
Roller compaction	25
sheepsfoot	25
Rolling ladder	11
tower	11
Roof exhauster	368
flashing vent chimney	309
frame opening	59
jack residential	368
sheathing	63
specialties, prefab	66
ventilator plastic	366
Rooftop air conditioner	325
air-conditioner	456-459
heat pump	329
Rooftop heat pump	488
Rooftop make-up air unit	323
multizone	458, 459
multizone air cond.	325
single zone	457
space heater	300
Room clean	94
humidifier	330
Rotary pump	282
Rotating chimney cap	378
Rough carpentry	63
Round damper	363
damper fire	359
diffuser	370
fire damper	359
volume damper	363
Rubber pipe insulation	139
Rubbish chute	23
handling	23

S

Saddle covering protection	127
pipe support	127
Safety equipment laboratory	90
plug pressure/temperature	258
type fuel heater	275
valve gas	287
Salamander	15
Sales tax	7, 517
Sampling air	102
Sand backfill	499
Sauna	95
Saw	15
chain	15
circular	15
masonry	15
Scaffold specialties	10
Scaffolding	522
masons	11
plank	10
putlog	11
tubular	9
School air-conditioner	465
air-conditioning	450, 452, 457, 459, 461, 463
elementary S.F. & C.F.	405
S.F. & C.F.	405
Scoop air	245
Scotch marine boiler	293
Screening plant	13
Screw brass	63
lag	58
machine	58
sheet metal	62
steel	63
wood	63
Sealant caulking	68
tape	69
Sealer joint	68
Sediment strainer Y valve	238
Selective demolition	23
Self-closing relief valve	230
Self-contained air conditioner	327
Selfcontained dehumidifier	330
Selfdrilling anchor	56
Semi-steel valve	238
Sensor circuit	246, 247
control component	110
Separation barrier	101
Separator	268
aftercooler	277, 278
Service boring	28
residential	390
station equipment	83
Setter circuit	247
Sewage ejector pump	284
pipe	33, 34, 41
piping	502
Sheathing	63
roof	63
Sheathing, roof & walls plywood	63
Sheathing wall	63
Sheepsfoot roller	25
Sheet metal screw	62
metal transitions	349
piling	24, 526
Sheeting	24
open	25
tie back	25
wale	24
wood	24, 525
Shell/tube heat exchanger	482
Shelving refrigeration	94
Shield expansion	56
lag screw	57

Shift work	7
Shipping door	72
Shock absorber	268
absorber pipe	268
Shoring	24
heavy duty	9, 10
slab	10
temporary	25
Shot chemical feeder	296
Shower pan	66
Shredder compactor	84
Shut-off valve oil	274
Sidewalk bridge	10
removal	23
temporary	12
Sidewall diffuser	371
Sightglass refrigeration	272
Sign	12
project	12
Signal bell fire	115
Silencer duct	363
louver	364
plenum	364
Silent check valve	231, 234, 240, 243
Silicone caulking	69
Sillcock hose bibb	282
Simplex style strainer	261
Single pipe roll	126
zone rooftop unit	325
Sink darkroom	88
laboratory	90
removal	121
Site demolition	22
drainage	42
improvement	44
preparation	22
utility	499
Sitework manhole	504
Skating rink S.F. & C.F.	403
Slab on grade	51
shoring	10
Sleeve	29, 259
cut in	29
formwork	50
wall	259
Slotted framing channel	59
pipe	34
Small tools	11
Smoke control kitchen	86
detector	115
exhaust duct	349-352
vent chimney	302
Smokestack	54
Snow melting	340
Snubber valve	249
Social club S.F. & C.F.	400
Socket hanger rod	128
Socket-o-let	179
Softener water	87
Soil compaction	25
compactor	13
decontamination	21
tamping	25
Solar backflow preventer	105
drain back hot water	420
draindown hot-water	423
energy	103-106
energy circulator air	104
energy system	419, 425, 427, 429, 437
energy system air purger	105
energy system air vent	105
energy system balancing valve	105
energy system collector	104
energy system control valve	105
energy system controller	104

Index

energy system expansion tank . 105
energy system gauge pressure 105
energy system heat exchanger 104
energy system storage tank ... 105
energy system thermometer .. 106
energy system vacuum relief .. 106
heating 527
system solenoid valve 105
Solar-energy pool heater 431
water-heater 433
Solder 142
Soldier beam 25
Solid fuel boiler 294
fuel furnace 299
Sorority house S.F. & C.F. 402
Sound trap duct 364
Space heater floor mounted 300
heater rental 14
heater rooftop 300
heater suspended 301
Special door 72
purpose room 94
systems 341-344
Specialties & accessories, roof 66
piping 383
piping HVAC .. 250, 252, 256, 261,
263 , 268
refrigeration 271
scaffold 10
Spill container fuel 273
Spinner ventilator 377
Spiral duct fitting 354-356
duct preformed 353, 355
duct register boot 357
Split ring hanger 124
system control system 107
system ductless 328
Splitter damper assembly 362
Sport arena S.F. & C.F. 405
Spotter 28
Spray substrate 102
Sprayer airless 102
Spread footing 51
Sprinkler alarm 115
grass 44
system 44
Square foot building cost 400
Stainless cooling tower 469
Stainless cooling towers 315
duct 347
flashing 66
steel fitting 195
steel flange 200
steel pipe 190-195
steel spiral duct 357
steel valve 242, 243
Stair temporary protection 9
Stand heat pump 320
Standpipe alarm 115
steel 97
Starter board & switch 397
motor 397
Station control 396
Stationary ventilator 377
ventilator mushroom 378
Steam bath 95
bath residential 95
boiler 289, 290, 292, 293
boiler electric 287
cock 248
condensate meter 273
fuel heater 275
heat 336
heat coil 335, 336
heating commercial 415
humidifier 330

jacketed kettle 85
pressure reducing valve 256
pressure valve 254
radiator valve 254
regulator pressure 258
syphon pigtail 249
trap 260, 261
Steel cooling tower 469
Steel cutting 23
diffuser 371
downspout 66
expansion tank 251
fitting 160
flange 179, 187
oval arch corrugated pipe 503
pipe 40, 41, 153, 155-158, 160
pipe downspout 67
pipe fitting 160, 173
pipe stainless 193, 194
register 375
reinforcing 50
rivet 58
salvage 121
screw 63
sheet pile 24
spiral duct 353, 355
spiral duct stainless 357
standpipe 97
Steel storage tank 509, 510
Steel structural 58
tank 97, 251
valve 238, 241
well screen 32
Step manhole 43
Sterilizer dental 91
medical 90
Stiffleg derrick 16
Stoker coal 287
fired boiler 295
Stop-start control station 396
Storage box 9
building S.F. & C.F. 406
door cold 72
tank 97-99
Store air-conditioning . 449, 459, 460,
463 , 464
department S.F. & C.F. 401
retail S.F. & C.F. 405
Storm collar 308
Stove 79
wood burning 79
Strainer basket type 261, 262
bronze body 263
duplex style 262, 263
grooved joint 188
simplex style 261
Y type 263
Y type iron body 263
Strap pipe hanger 128
Strapless outlet 182
Strip footing 51
Structural concrete 50
excavation 26
formwork 50
steel 58
Structure pre-engineered 95
Stud driver pneumatic 58
Subcontractor O & P 8
Subcontractors 528
Subdrainage pipe 40
piping 40, 412
plastic pipe 41
system 40
Submersible pump . 15, 32, 281, 285
Subsoil heating 96
Subsurface investigation 22

Suction bell fuel heater 275
diffuser grooved joint 188
diffuser pump 264
hose 14
Sump hole construction 24
pit 23
pump 87, 285
Supermarket air-conditioning .. 450,
452 , 457, 461, 463, 465
air-condtioning 459
S.F. & C.F. 405
Supply/return combination diff. .. 371
Support framing pipe 59
pipe 121
Suppressor lightning 390
Surfactant 102
Suspended space heater 301
Suspended space heater 483
Suspension mounted heater 300
Swimming pool 96
pool equipment 96
pool heater 295
pool painting 96
pool S.F. & C.F. 405
Swing check valve 228
Switch decorator 391
dimmer 390
residential 390
Switchboard electric 396
System boiler fossil fuel 441
boiler hot-water 439
boiler oil fired 443
boiler steam 438
components fuel 273
control 106, 107
dryer 279
exhaust 381
fin-tube radiation 443
grinder pump 284
heat trace 341
heating hydronic 441, 443
hydronic fossil fuel 440, 442
irrigation 44
pipe conduit 36, 37, 39
sitework catch basin 504
sitework trenching 496
solar-energy 425, 431
solenoid valve solar 105
sprinkler 44
subdrainage 40
tube 118
VAV box control 108

T

Tag valve 131
Tamper 14, 26
Tamping soil 25
Tank clean 21
compression 251
corrosion resistant 251
darkroom 87
disposal 21
expansion 251
fibergalss 98
fixed roof 98
gasoline 98
heating coil 275
leak detection 115
outlet heater 275
pillow 98
removal 21
steel 97, 251
storage 97-99
water 97-99

water storage solar 106
water-heater storage 285
Tape barricade 12
detection 20
sealant 69
underground 20
Tapping main 31
Tarpaulin 9, 11
duck 11
mylar 11
polyethylene 11
polyvinyl 11
Tavern air-conditioning 449, 451, 456,
458, 460, 462, 464
Tax 7
sales 7
social security 7
unemployment 7
Taxes 517
T-bar grille eggcrate 373
grille lay-in 373
mount diffuser 370
Tee fitting 31
pipe 185
Telephone exchange S.F. & C.F. 405
receptacle 393
Television receptacle 393
Temperature relief valve 230
rise detector 115
Tempering valve 231
valve water 256
Temporary building 9
construction 8, 9, 12
facility 12
fence 12
guardrail 12
heat 8
light 8
oil heater 14
shoring 25
toilet 15
utility 8
Terminal A/C packaged 324
heater 339, 340
Terne coated flashing 66
Test pressure 385
well 32
Testing OSHA 103
pipe 384
Theater S.F. & C.F. 406
Thermoflo indicator 264
Thermometer 110
solar energy system 106
well 182
Thermometers 110
Thermostat 110
automatic timed 110
contactor combination 342
control sys. radiator 108
integral 341
pneumatic 111
wire 395
Threaded flange 172, 173
nut 125
rod 125
Threading labor st steel 194
labor steel 159
Thread-o-let 178
Thru-the-wall A/C 489
heat pump 489
Thru-wall air conditioner 325
Tie back sheeting 25
rod 59
Timber fastener 62
Time lapse photography 8
Timer clock 394

613

INDEX

interval 391
   switch ventilator 369
Toggle bolt 57
Toilet chemical 15
   temporary 15
   trailer 15
Tool air 14
   powder actuated 58
   rivet 58
Tools small 11
Topping grout 51
Torch cutting 15, 23
Tower cooling 315
   crane 15, 523
   hoist 16
   light 14
   rolling 11
Town hall S.F. & C.F. 406
Tractor 25, 26
   truck 15
Trailer office 9
   platform 15
   toilet 15
   truck 13, 15
Training 18
Transfer grille 373
Transformer heating control 342
Transition residential 368
Transitions sheet metal 349
Transmission of heat 536
Transmitter control component . . 111
Trap cast iron 268
   drainage 268
   inverted bucket steam 259
   plastic 269
   polypropylene 269
   PVC 269
   steam 260, 261
Trapeze hanger 126
Trash pump 15
Trench backfill 27, 499, 500
   excavation 26, 27
   utility 26
Trencher 13
   chain 13
Trenching 23, 497, 498
Truck dump 13
   flatbed 13
   hauling 28
   mounted crane 16
   off highway 14
   pickup 15
   rental 13
   tractor 15
   trailer 13, 15
   vacuum 15
Trucking 28
Tube fitting pneumatic 113, 114
   pneumatic 113, 114
   system 118
   system air 118
   system pneumatic 118
Tubing copper 142
   copper type K 501
   pex 344, 345
Tubular scaffolding 9
Tunneling 28
Turbine pump 282
   water meter 265
Turnbuckle 125
Turning vanes 364

U

U-bolt pipe hanger 129

U-hook hanger 129
Underdrain 42
Undereave vent 78
Underground piping 499
   storage tank removal 524
   tape 20
Underlayment grout 51
Underwater light 96
Unemployment tax 7
Union dielectric 249
Unit gross area requirements 518
   heater 300, 338, 340
   heater cabinet 340, 343, 344
   heater electric 341-343
   water cooled condensing 317
Urethane foam 69
   insulation 94
Urinal removal 121
Utensil washer medical 90
Utility excavation 496
Utility fan system 480
Utility pump 280
   set fan 366, 368
   structure 504
   temporary 8
   trench 26

V

Vacuum breaker 267
   central 82
   cleaning 82
Vacuum dust collection 490
Vacuum gauge 109
   loader 102
   pump 272
   pumps 280
   relief water 267
   truck 15
Valance heater 340
Valve 227-230, 234, 236-238, 242,
                  243, 270
   angle 236
   balancing & shut off 254
   ball 228, 236, 240
   brass 227
   bronze 228
   bronze butterfly 228
   butterfly 20, 232
   check 20, 30, 238, 240-242
   check lift 229
   check swing 234
   circuit setter balance 247
   control solar energy system . . . 105
   CPVC 236
   cut in 29
   diverter 233, 239, 242
   float 254
   foot 237, 273
   foot oil 273
   fusible oil 274
   gas stop 228
   gate 20, 29, 233, 239, 241, 242
   globe 21, 230, 233, 239, 241, 243
   grooved joint butterfly 189
   grooved-joint gate 189
   heating control 111, 254
   hot water radiator 254
   iron body 232, 234
   lubricated plug 238
   materials 530-532
   mixing 256
   motor actuated 111
   motor controlled 111
   motorized zone 111

multipurpose 235
needle 237
nonfusible oil 274
oil shut-off 274
piping 20
plastic 236
plug 238
pneumatic air operated 112
polypropylene 236
PVC 236
radiator supply 254
relief pressure 230
selection considerations 530
semi-steel 238
snubber 249
soldered gate 229
stainless steel 242, 243
steam pressure 254
steam radiator 254
steel 238, 241
swing check 228
tag 131
tempering 231
water pressure 230
water tempering 256
Y sediment strainer 238
Y-check 238
zone 266
Valves forged 240
   refrigeration 272
Vane-axial fan 365
Vanes turning 364
Vapor barrier 94
   hose adapter 273
   proof insulation 133
Vaporproof insulation 139
Variable air volume 326
   frequency drives 396
   volume damper 362
VAV boxes 361
   mixing boxes 361
   terminal 478, 479
Vent air automatic 246
   alarm 274
   automatic air 245
   box 54
   brick 374
   cap 274
   chimney 302-304, 309, 310
   chimney all fuel 308
   chimney fitting . 302, 304, 308, 310
   chimney flashing 303, 309
   chimney high temperature 309,
                      310
   chimney increaser 310
   damper 301
   damper exhaust 358
   draft damper 301
   dryer 87
   exhaust 377
   flashing 269
   foundation 54
   intake-exhaust 378
   metal chimney 302
   radiator air 246
   relief 378
   ridge 78
   ridge strip 78
   silencer 364
Vent-chimney band 310
Ventilating air conditioning . 108, 314,
315, 327 , 347, 359, 362, 365, 368, 369,
                  378
Ventilation 539
   fan 394
   garage 381

heat recovery 346
kitchen 85
louver 63
packaged 85
residential 369
Ventilator chimney cap 378
   control system 108
   fan 364
   gravity syphon 377
   masonry 54
   mushroom stationary 378
   relief vent 378
   spinner 377
   timer switch 369
Venturi flow meter 265
Vertical induction unit 360
Vibration absorber 129
   absorbing pad 130
Vibrator earth 25, 26
   plate 25
Vibratory equipment 13
Vitrified clay pipe 34, 41
Vocational school S.F. & C.F. 405
Volume control damper 362
   damper round 363

W

Wafer check valve 231, 235, 243
Wale sheeting 24
Wales 24
Walk-in cooler evaporator 318
   refrigeration 94
Wall cap residential 368
   exhaust fan 369
   furnace 301
   heater 87
   heater electric 341
   hung boiler 290
   hydrant 282
   patching concrete 51
   register 376
   sheathing 63
   sleeve 259
Warehouse S.F. & C.F. 406
Warm air heater 299
Washer commercial 82
   darkroom 87
   residential 82, 87
Washers 124
Washing machine automatic 87
Waste compactor 84
   disposal 84
   handling 83
   handling equipment 83
   incinerator 84
   packaging 103
Water anti-siphon 267
   atomizer 102
   balancing 385
   chiller 312
   chillers absorption 312
   closet removal 121
   coil 332-334
   coil hot 334, 335
   copper tubing connector 252
   distiller 90
   distribution system 30
   fountain removal 121
   gauge 383
   gauge boiler 383
   heater 87, 394
   heater insulation 131
   heater removal 121
   heater wrap kit 131

heating hot 288, 338
hose 14
hydrant 282
level control 383
meter turbine 265
piping polyethylene 501
pressure relief valve 230
pressure valve 230
pump 14, 17, 32, 87, 270
pumping 23
softener 87
storage solar tank 106
supply detector meter 265
supply domestic meter 265
supply meter 265
tank 97-99
tank elevated 97
tempering valve 231, 256
tube boiler 294
vacuum relief 267
well 32
Water-heater solar-energy .. 421, 423,
                  425 , 427, 429, 437
storage tank 285
Wearing course 43
Weatherproof receptacle 392
Wedge anchor 57
Weld fitting 176
joint pipe . 154, 157, 190, 191, 193,
                                194
plastic pipe 222
X-ray 385
Welded beam attachment 121
eye rod 125
Welding fume elimination 382
labor stainless-steel 194
labor steel 159
machine 15
Weld-o-let 178
Weld-on flange steel 179
Well 24, 32
gravel pack 32
pump deep 33
pump shallow 33
screen steel 32
thermometer 182
water 32
Wellpoint 24
equipment 17
header pipe 17
pump 17
Wellpoints 524
Wheelbarrow 15
Whirlpool bath 91
Whole house fan 367
Window air conditioner 327
Winter protection 9
Wire fence 12
thermostat 395
Wiring air conditioner 394
fan 394
methods 388
residential 390, 394
Wood burning furnace 299
coal/oil combination furnace .. 299
fastener 63
gage stick 273
screw 63
sheet piling 525
sheeting 24, 525
tank 98
Woodwork architectural 63
Work extra 8
in prisons 6
Worker protection 102
Workers' compensation ...7, 513-515

Wrapping pipe 47
Wrench impact 14
Wye fitting 31

X

X-ray pipe 385
weld 385

Y

Y sediment strainer valve 238
type iron body strainer 263
Y-check valve 238
YMCA club S.F. & C.F. 401

Z

Zinc plating 62
Zone valve 266

Notes

Notes

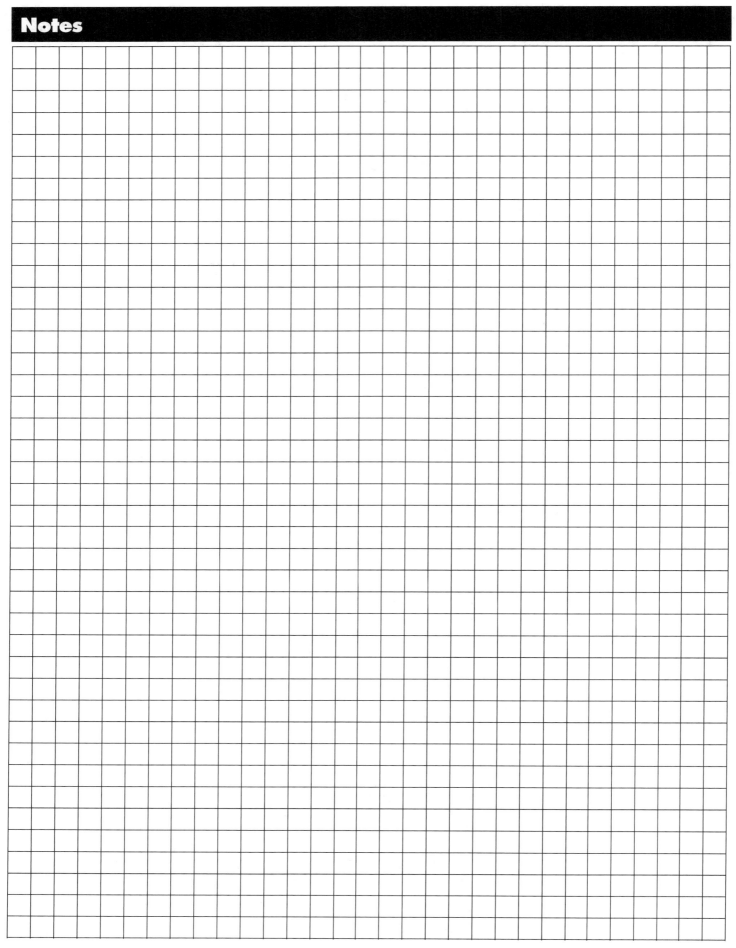

CMD . . .

R.S. Means Company, Inc., a CMD company, is the leading provider of construction cost data in North America and supplies comprehensive construction cost guides, related technical publications and education services.

CMD, a leading worldwide provider of total construction information solutions, is comprised of three synergistic product groups designed specifically to help construction professionals advance their businesses with timely, accurate and actionable project, product and cost data. CMD is a division of Cahners Business Information, a member of the Reed Elsevier plc group of companies.

The *Project, Product, and Cost & Estimating* divisions offer a variety of innovative products designed for the full spectrum of design, construction and manufacturing professionals. Together with Cahners, CMD created *Buildingteam.com,* a valuable Internet portal of the construction community. Through it's *International* companies, CMD's reputation for quality construction market data is growing worldwide.

Project Data
CMD provides complete, accurate and relevant project information through all stages of construction. Customers are supplied industry data through leads, project reports, contact lists, plans and specifications surveys, market penetration analyses and sales evaluation reports. Any of these products can pinpoint a county, look at a state, or cover the country. Data is delivered via paper, e-mail, CD-ROM or the Internet.

Building Product Information
The First Source suite of products is the only integrated building product information system offered to the commercial construction industry for comparing and specifying building products. These print and online resources include *First Source for Products,* SPEC-DATA™, MANU-SPEC™, CADBlocks, First Source Exchange (www.firstsourceexchange.com), and Manufacturer Catalogs. Written by industry professionals and organized using CSI's MasterFormat™, construction professionals use this information to make better design decisions.

Cost Information
R.S. Means, the undisputed market leader and authority on construction costs, publishes current cost and estimating information in annual cost books and on the CostWorks CD-ROM. R.S. Means furnishes the construction industry with a rich library of complementary reference books and a series of professional seminars that are designed to sharpen professional skills and maximize the effective use of cost estimating and management tools. R.S. Means also provides construction cost consulting for Owners, Manufacturers Designers and Contractors.

Buildingteam.com
Combining CMD's project, product and cost data with news and information from Cahners' *Building Design & Construction* and *Consulting-Specifying Engineer,* this industry-focused site offers easy and unlimited access to vital information for all construction professionals.

International
BIMSA/Mexico provides construction project news, product information, cost-data, seminars and consulting services to construction professionals in Mexico. Its subsidiary, PRISMA, provides job costing software.

Byggfakta Scandinavia AB, founded in 1936, is the parent company for the leaders of customized construction market data for Denmark, Estonia, Finland, Norway and Sweden. Each company fully covers the local construction market and provides information across several platforms including subscription, ad-hoc basis, electronically and on paper.

CMD Canada serves the Canadian construction market with reliable and comprehensive project and product information services that cover all facets of construction. Core services include: Buildcore, product selection and specification tools available in print and on the Internet; CMD Building Reports, a national construction project lead service; CanaData, statistical and forecasting information; *Daily Commercial News,* a construction newspaper reporting on news and projects in Ontario; and *Journal of Commerce,* reporting news in British Columbia and Alberta.

Cordell Building Information Services, with its complete range of project and cost and estimating services, is Australia's specialist in the construction information industry. Cordell provides in-depth and historical information on all aspects of construction projects and estimation, including several customized reports, construction and sales leads, and detailed cost information among others.

For more information, please visit our website at www.cmdg.com.

CMD Corporate Office
30 Technology Parkway South
Norcross, GA 30092-2912
(800) 793-0304
(700) 417-4002 (fax)
info@cmdg.com
www.cmdg.com

RSMeans
CMD

Means Project Cost Report

By filling out and returning the Project Description, you can receive a discount of $20.00 off any one of the Means products advertised in the following pages. The cost information required includes all items marked (✔) except those where no costs occurred. The sum of all major items should equal the Total Project Cost.

$20.00 Discount per product for each report you submit.

DISCOUNT PRODUCTS AVAILABLE—FOR U.S. CUSTOMERS ONLY—STRICTLY CONFIDENTIAL

Project Description (No remodeling projects, please.)

✔ Type Building _____

✔ Location _____

Capacity _____

✔ Frame _____

✔ Exterior _____

✔ Basement: full ☐ partial ☐ none ☐ crawl ☐

✔ Height in Stories _____

✔ Total Floor Area _____

Ground Floor Area _____

✔ Volume in C.F. _____

% Air Conditioned _____ Tons _____

Comments _____

Owner _____

Architect _____

General Contractor _____

✔ Bid Date _____

Typical Bay Size _____

✔ Labor Force: _____ % Union _____ % Non-Union

✔ Project Description (Circle one number in each line)
   1. Economy 2. Average 3. Custom 4. Luxury
   1. Square 2. Rectangular 3. Irregular 4. Very Irregular

	✔ Total Project Cost				$
A	**✔ General Conditions**				$
B	**✔ Site Work**				$
BS	Site Clearing & Improvement				
BE	Excavation	(		C.Y.)	
BF	Caissons & Piling	(		L.F.)	
BU	Site Utilities				
BP	Roads & Walks Exterior Paving	(		S.Y.)	
C	**✔ Concrete**				$
C	Cast in Place	(		C.Y.)	
CP	Precast	(		S.F.)	
D	**✔ Masonry**				$
DB	Brick	(		M)	
DC	Block	(		M)	
DT	Tile	(		S.F.)	
DS	Stone	(		S.F.)	
E	**✔ Metals**				$
ES	Structural Steel	(		Tons)	
EM	Misc. & Ornamental Metals				
F	**✔ Wood & Plastics**				$
FR	Rough Carpentry	(		MBF)	
FF	Finish Carpentry				
FM	Architectural Millwork				
G	**✔ Thermal & Moisture Protection**				$
GW	Waterproofing-Dampproofing	(		S.F.)	
GN	Insulation	(		S.F.)	
GR	Roofing & Flashing	(		S.F.)	
GM	Metal Siding/Curtain Wall	(		S.F.)	
H	**✔ Doors and Windows**				$
HD	Doors	(		Ea.)	
HW	Windows	(		S.F.)	
HH	Finish Hardware				
HG	Glass & Glazing	(		S.F.)	
HS	Storefronts	(		S.F.)	

J	**✔ Finishes**				$
JL	Lath & Plaster	(		S.Y.)	
JD	Drywall	(		S.F.)	
JM	Tile & Marble	(		S.F.)	
JT	Terrazzo	(		S.F.)	
JA	Acoustical Treatment	(		S.F.)	
JC	Carpet	(		S.Y.)	
JF	Hard Surface Flooring	(		S.F.)	
JP	Painting & Wall Covering	(		S.F.)	
K	**✔ Specialties**				$
KB	Bathroom Partitions & Access.	(		S.F.)	
KF	Other Partitions	(		S.F.)	
KL	Lockers	(		Ea.)	
L	**✔ Equipment**				$
LK	Kitchen				
LS	School				
LO	Other				
M	**✔ Furnishings**				$
MW	Window Treatment				
MS	Seating	(		Ea.)	
N	**✔ Special Construction**				$
NA	Acoustical	(		S.F.)	
NB	Prefab. Bldgs.	(		S.F.)	
NO	Other				
P	**✔ Conveying Systems**				$
PE	Elevators	(		Ea.)	
PS	Escalators	(		Ea.)	
PM	Material Handling				
Q	**✔ Mechanical**				$
QP	Plumbing	(No. of fixtures		)	
QS	Fire Protection (Sprinklers)				
QF	Fire Protection (Hose Standpipes)				
QB	Heating, Ventilating & A.C.				
QH	Heating & Ventilating	(BTU Output		)	
QA	Air Conditioning	(		Tons)	
R	**✔ Electrical**				$
RL	Lighting	(		S.F.)	
RP	Power Service				
RD	Power Distribution				
RA	Alarms				
RG	Special Systems				
S	**✔ Mech./Elec. Combined**				$

Product Name _____

Product Number _____

Your Name _____

Title _____

Company _____

☐ Company

☐ Home   Street Address _____

City, State, Zip _____

☐ Please send _____ forms.

Please specify the Means product you wish to receive. Complete the address information as requested and return this form with your check (product cost less $20.00) to address below.

R.S. Means Company, Inc.,
Square Foot Costs Department
P.O. Box 800
Kingston, MA 02364-9988

R.S. Means Company, Inc...
a tradition of excellence in Construction Cost Information and Services since 1942.

For more information
visit Means Web Site
at www.rsmeans.com

Table of Contents

Annual Cost Guides, Page 2
Reference Books, Page 6
Seminars, Page 11
Consulting Services, Page 13
Electronic Data, Page 14
New Titles, Page 15
Order Form, Page 16

Book Selection Guide

The following table provides definitive information on the content of each cost data publication. The number of lines of data provided in each unit price or assemblies division, as well as the number of reference tables and crews is listed for each book. The presence of other elements such as an historical cost index, city cost indexes, square foot models or cross-referenced index is also indicated. You can use the table to help select the Means' book that has the quantity and type of information you most need in your work.

Unit Cost Divisions	Building Construction Costs	Mechanical	Electrical	Repair & Remodel.	Square Foot	Site Work Landsc.	Assemblies	Interior	Concrete Masonry	Open Shop	Heavy Construc.	Light Commercial	Facil. Construc.	Plumbing	Western Construction Costs	Residential
1	1103	590	668	866		1091		492	991	1100	1133	625	1617	663	1095	420
2	3141	1462	508	1881		9148		851	1655	3091	5620	984	5014	1825	3114	1065
3	1511	91	71	767		1315		198	1932	1505	1489	227	1379	43	1510	266
4	905	27	0	695		753		624	1164	882	651	428	1111	0	877	345
5	1928	178	176	978		790		989	708	1895	1070	810	1877	95	1911	780
6	1356	96	83	1290		491		1299	337	1347	610	1456	1353	59	1734	1529
7	1365	180	91	1329		490		546	451	1364	354	1036	1357	188	1364	803
8	1902	60	0	1907		337		1821	723	1883	9	1187	2035	0	1903	1134
9	1690	50	0	1513		188		1788	324	1638	114	1434	1887	50	1685	1293
10	992	58	31	565		218		823	196	996	0	467	997	250	992	250
11	1157	434	212	622		151		916	34	1051	95	237	1220	370	1050	116
12	352	0	0	48		228		1537	30	344	0	68	1538	0	343	72
13	1343	1181	460	567		448		1029	106	1317	307	561	2053	1080	1298	236
14	364	43	0	261		36		330	0	364	40	18	362	17	362	9
15	2514	14716	763	2254		1828		1466	80	2516	2030	1563	12320	10884	2546	1100
16	1521	540	10211	1202		879		1290	62	1538	916	1312	9896	481	1463	705
17	429	354	429	0		0		0	0	429	0	0	429	356	429	0
Totals	23573	20060	13703	16745		18391		15999	8793	23260	14438	12413	46445	16361	23676	10123

Assembly Divisions	Building Construction Costs	Mechanical	Electrical	Repair & Remodel.	Square Foot	Site Work Landsc.	Assemblies	Interior	Concrete Masonry	Open Shop	Heavy Construc.	Light Commercial	Facil. Construc.	Plumbing	Western Construction Costs	Asm Div	Residential
A		28	0	333	169	787	863	0	777		765	168	95	2		1	841
B		0	0	1331	2716	0	6707	495	2515		0	2218	265	3		2	584
C		0	0	1150	954	0	1490	1809	182		0	850	343	3		3	1544
D		2269	1546	1513	2172	0	4422	1402	0		0	1608	1511	2106		4	2049
E		0	0	129	295	0	348	153	0		0	297	153	1		5	1081
F		0	0	0	148	0	150	0	0		0	149	52	0		6	1077
G		622	163	631	104	2436	715	0	692		554	104	125	680		7	709
																8	1686
																9	238
																10	0
																11	0
																12	0
Totals		2919	1709	5087	6558	3223	14695	3859	4166		1319	5394	2544	2795			9809

Reference Section	Building Construction Costs	Mechanical	Electrical	Repair & Remodel.	Square Foot	Site Work Landsc.	Assemblies	Interior	Concrete Masonry	Open Shop	Heavy Construc.	Light Commercial	Facil. Construc.	Plumbing	Western Construction Costs	Residential
Tables	129	45	85	69	4	80	219	46	71	130	61	57	104	51	131	42
Models					102							43				32
Crews	410	410	410	391		410		410	410	393	410	393	391	410	410	393
City Cost Indexes	yes	yes	yes	yes	yes	yes	yes	yes	yes	yes	yes	yes	yes	yes	yes	yes
Historical Cost Indexes	yes	yes	yes	yes	yes	yes	yes	yes	yes	yes	yes	yes	yes	yes	yes	no
Index	yes	yes	yes	yes	no	yes	yes	yes	yes	yes	yes	yes	yes	yes	yes	yes

Annual Cost Guides

For more information
visit Means Web Site
at www.rsmeans.com

Means Building Construction Cost Data 2002

Available in Both Softbound and Looseleaf Editions

The "Bible" of the industry comes in the standard softcover edition or the looseleaf edition.

Many customers enjoy the convenience and flexibility of the looseleaf binder, which increases the usefulness of *Means Building Construction Cost Data 2002* by making it easy to add and remove pages. You can insert your own cost information pages, so everything is in one place. Copying pages for faxing is easier also. Whichever edition you prefer, softbound or the convenient looseleaf edition, you'll be eligible to receive a FREE e-mail subscription to *The Change Notice* quarterly newsletter.

$99.95 per copy, Softbound
Catalog No. 60012

$124.95 per copy, Looseleaf
Catalog No. 61012

Means Building Construction Cost Data 2002

Offers you unchallenged unit price reliability in an easy-to-use arrangement. Whether used for complete, finished estimates or for periodic checks, it supplies more cost facts better and faster than any comparable source. Over 23,000 unit prices for 2002. The City Cost Indexes cover over 930 areas, for indexing to any project location in North America. Order and get *The Change Notice* e-mail quarterly newsletter sent to you FREE. You'll have year-long access to the Means Estimating **Hotline** FREE with your subscription. Expert assistance when using Means data is just a phone call away.

$99.95 per copy
Over 650 pages, illustrated, available Oct. 2001
Catalog No. 60012

Means Building Construction Cost Data 2002

Metric Version

The Federal Government has stated that all federal construction projects must now use metric documentation. The *Metric Version* of *Means Building Construction Cost Data 2002* is presented in metric measurements covering all construction areas. Don't miss out on these billion dollar opportunities. Make the switch to metric today.

$99.95 per copy
Over 650 pages, illus., available Nov. 2001
Catalog No. 63012

For more information
visit Means Web Site
at www.rsmeans.com

Annual Cost Guides

Means Mechanical Cost Data 2002

• HVAC • Controls

Total unit and systems price guidance for mechanical construction...materials, parts, fittings, and complete labor cost information. Includes prices for piping, heating, air conditioning, ventilation, and all related construction.

Plus new 2002 unit costs for:
• Over 2500 installed HVAC/controls assemblies
• "On Site" Location Factors for over 930 cities and towns in the U.S. and Canada
• Crews, labor and equipment

$99.95 per copy
Over 600 pages, illustrated, available Oct. 2001
Catalog No. 60022

Means Plumbing Cost Data 2002

Comprehensive unit prices and assemblies for plumbing, irrigation systems, commercial and residential fire protection, point-of-use water heaters, and the latest approved materials. This publication and its companion, *Means Mechanical Cost Data*, provide full-range cost estimating coverage for all the mechanical trades.

$99.95 per copy
Over 500 pages, illustrated, available Nov. 2001
Catalog No. 60212

Means Electrical Cost Data 2002

Pricing information for every part of electrical cost planning: More than 15,000 unit and systems costs with design tables; clear specifications and drawings; engineering guides and illustrated estimating procedures; complete labor-hour and materials costs for better scheduling and procurement; the latest electrical products and construction methods.
• A Variety of Special Electrical Systems including Cathodic Protection
• Costs for maintenance, demolition, HVAC/ mechanical, specialties, equipment, and more

$99.95 per copy
Over 450 pages, illustrated, available Oct. 2001
Catalog No. 60032

Means Electrical Change Order Cost Data 2002

You are provided with electrical unit prices exclusively for pricing change orders based on the recent, direct experience of contractors and suppliers. Analyze and check your own change order estimates against the experience others have had doing the same work. It also covers productivity analysis and change order cost justifications. With useful information for calculating the effects of change orders and dealing with their administration.

$99.95 per copy
Over 450 pages, available Oct. 2001
Catalog No. 60232

Means Facilities Maintenance & Repair Cost Data 2002

Published in a looseleaf format, *Means Facilities Maintenance & Repair Cost Data* gives you a complete system to manage and plan your facility repair and maintenance costs and budget efficiently. Guidelines for auditing a facility and developing an annual maintenance plan. Budgeting is included, along with reference tables on cost and management and information on frequency and productivity of maintenance operations.

The only nationally recognized source of maintenance and repair costs. Developed in cooperation with the Army Corps of Engineers.

$219.95 per copy
Over 600 pages, illustrated, available Dec. 2001
Catalog No. 60302

Means Square Foot Costs 2002

It's Accurate and Easy To Use!

• **Updated 2002 price information,** based on nationwide figures from suppliers, estimators, labor experts and contractors.
• "How-to-Use" Sections, with **clear examples** of commercial, residential, industrial, and institutional structures.
• Realistic graphics, offering true-to-life illustrations of building projects.
• Extensive information on using square foot cost data, including **sample estimates** and **alternate pricing methods.**

$109.95 per copy
Over 450 pages, illustrated, available Oct. 2001
Catalog No. 60052

Annual Cost Guides

For more information
visit Means Web Site
at www.rsmeans.com

Means Repair & Remodeling Cost Data 2002

Commercial/Residential

You can use this valuable tool to estimate commercial and residential renovation and remodeling.

Includes: New costs for hundreds of unique methods, materials and conditions that only come up in repair and remodeling. PLUS:

- Unit costs for over 16,000 construction components
- Installed costs for over 90 assemblies
- Costs for 300+ construction crews
- Over 930 "On Site" localization factors for the U.S. and Canada.

$87.95 per copy
Over 600 pages, illustrated, available Nov. 2001
Catalog No. 60042

Means Facilities Construction Cost Data 2002

For the maintenance and construction of commercial, industrial, municipal, and institutional properties. Costs are shown for new and remodeling construction and are broken down into materials, labor, equipment, overhead, and profit. Special emphasis is given to sections on mechanical, electrical, furnishings, site work, building maintenance, finish work, and demolition. More than 45,000 unit costs, plus assemblies and reference sections are included.

$241.95 per copy
Over 1150 pages, illustrated, available Nov. 2001
Catalog No. 60202

Means Residential Cost Data 2002

Contains square foot costs for 30 basic home models with the look of today, plus hundreds of custom additions and modifications you can quote right off the page. With costs for the 100 residential systems you're most likely to use in the year ahead. Complete with blank estimating forms, sample estimates and step-by-step instructions.

$87.95 per copy
Over 550 pages, illustrated, available Dec. 2001
Catalog No. 60172

Means Light Commercial Cost Data 2002

Specifically addresses the light commercial market, which is an increasingly specialized niche in the industry. Aids you, the owner/designer/contractor, in preparing all types of estimates, from budgets to detailed bids. Includes new advances in methods and materials. Assemblies section allows you to evaluate alternatives in the early stages of design/planning.

Over 13,000 unit costs for 2002 ensure you have the prices you need...when you need them.

$87.95 per copy
Over 600 pages, illustrated, available Dec. 2001
Catalog No. 60182

Means Assemblies Cost Data 2002

Means Assemblies Cost Data 2002 takes the guesswork out of preliminary or conceptual estimates. Now you don't have to try to calculate the assembled cost by working up individual components costs. We've done all the work for you.

Presents detailed illustrations, descriptions, specifications and costs for every conceivable building assembly—240 types in all—arranged in the easy-to-use UNIFORMAT II system. Each illustrated "assembled" cost includes a complete grouping of materials and associated installation costs including the installing contractor's overhead and profit.

$164.95 per copy
Over 550 pages, illustrated, available Oct. 2001
Catalog No. 60062

Means Site Work & Landscape Cost Data 2002

Means Site Work & Landscape Cost Data 2002 is organized to assist you in all your estimating needs. Hundreds of fact-filled pages help you make accurate cost estimates efficiently.

Updated for 2002!

- Demolition features—including ceilings, doors, electrical, flooring, HVAC, millwork, plumbing, roofing, walls and windows
- State-of-the-art segmental retaining walls
- Flywheel trenching costs and details
- Updated Wells section
- Landscape materials, flowers, shrubs and trees

$99.95 per copy
Over 600 pages, illustrated, available Oct. 2001
Catalog No. 60282

For more information
visit Means Web Site
at www.rsmeans.com

Annual Cost Guides

Means Open Shop Building Construction Cost Data 2002

The latest costs for accurate budgeting and estimating of new commercial and residential construction... renovation work... change orders... cost engineering. *Means Open Shop BCCD* will assist you to...
- Develop benchmark prices for change orders
- Plug gaps in preliminary estimates, budgets
- Estimate complex projects
- Substantiate invoices on contracts
- Price ADA-related renovations

$99.95 per copy
Over 650 pages, illustrated, available Nov. 2001
Catalog No. 60152

Means Heavy Construction Cost Data 2002

A comprehensive guide to heavy construction costs. Includes costs for highly specialized projects such as tunnels, dams, highways, airports, and waterways. Information on different labor rates, equipment, and material costs is included. Has unit price costs, systems costs, and numerous reference tables for costs and design. Valuable not only to contractors and civil engineers, but also to government agencies and city/town engineers.

$99.95 per copy
Over 450 pages, illustrated, available Nov. 2001
Catalog No. 60162

Means Building Construction Cost Data 2002
Western Edition

This regional edition provides more precise cost information for western North America. Labor rates are based on union rates from 13 western states and western Canada. Included are western practices and materials not found in our national edition: tilt-up concrete walls, glu-lam structural systems, specialized timber construction, seismic restraints, landscape and irrigation systems.

$99.95 per copy
Over 600 pages, illustrated, available Dec. 2001
Catalog No. 60222

Means Heavy Construction Cost Data 2002
Metric Version

Make sure you have the Means industry standard metric costs for the federal, state, municipal and private marketplace. With thousands of up-to-date metric unit prices in tables by CSI standard divisions. Supplies you with assemblies costs using the metric standard for reliable cost projections in the design stage of your project. Helps you determine sizes, material amounts, and has tips for handling metric estimates.

$99.95 per copy
Over 450 pages, illustrated, available Nov. 2001
Catalog No. 63162

Means Construction Cost Indexes 2002

Who knows what 2002 holds? What materials and labor costs will change unexpectedly? By how much?
- Breakdowns for 305 major cities.
- National averages for 30 key cities.
- Expanded five major city indexes.
- Historical construction cost indexes.

$218.00 per year
$54.50 individual quarters
Catalog No. 60142 A,B,C,D

Means Interior Cost Data 2002

Provides you with prices and guidance needed to make accurate interior work estimates. Contains costs on materials, equipment, hardware, custom installations, furnishings, labor costs . . . every cost factor for new and remodel commercial and industrial interior construction, including updated information on office furnishings, plus more than 50 reference tables. For contractors, facility managers, owners.

$99.95 per copy
Over 550 pages, illustrated, available Oct. 2001
Catalog No. 60092

Means Concrete & Masonry Cost Data 2002

Provides you with cost facts for virtually all concrete/masonry estimating needs, from complicated formwork to various sizes and face finishes of brick and block, all in great detail. The comprehensive unit cost section contains more than 8,500 selected entries. Also contains an assemblies cost section, and a detailed reference section which supplements the cost data.

$92.95 per copy
Over 450 pages, illustrated, available Nov. 2001
Catalog No. 60112

Means Labor Rates for the Construction Industry 2002

Complete information for estimating labor costs, making comparisons and negotiating wage rates by trade for over 300 cities (United States and Canada). With 46 construction trades listed by local union number in each city, and historical wage rates included for comparison. No similar book is available through the trade.

Each city chart lists the county and is alphabetically arranged with handy visual flip tabs for quick reference.

$219.95 per copy
Over 300 pages, available Dec. 2001
Catalog No. 60122

Reference Books

For more information visit Means Web Site at www.rsmeans.com

Preventive Maintenance Guidelines for School Facilities

NEW!

By John C. Maciha

This new publication is a complete PM program for K-12 schools that ensures sustained security, safety, property integrity, user satisfaction, and reasonable ongoing expenditures.

Includes schedules for weekly, monthly, semiannual, and annual maintenance. Comes as a 3-ring binder with hard copy and electronic forms available at Means website, and a laminated wall chart.

$149.95 per copy
Over 225 pages, Hardcover
Catalog No. 67326

Means Spanish/English Construction Dictionary

By R.S. Means, The International Conference of Building Officials (ICBO), and Rolf Jensen & Associates (RJA)

Designed to facilitate communication among Spanish- and English-speaking construction personnel—improving performance and job-site safety. Features the most common words and phrases used in the construction industry, with easy-to-follow pronunciations. Includes extensive building systems and tools illustrations.

$22.95 per copy
250 pages, illustrated, Softcover
Catalog No. 67327

Project Scheduling & Management for Construction

New 2nd Edition

By David R. Pierce, Jr.

A comprehensive yet easy-to-follow guide to construction project scheduling and control—from vital project management principles through the latest scheduling, tracking, and controlling techniques. The author is a leading authority on scheduling with years of field and teaching experience at leading academic institutions. Spend a few hours with this book and come away with a solid understanding of this essential management topic.

$64.95 per copy
Over 250 pages; illustrated, Hardcover Catalog No. 67247A

Residential & Light Commercial Construction Standards

A Unique Collection of Industry Standards That Define Quality in Construction

For Contractors & Subcontractors, Owners, Developers, Architects & Engineers, Attorneys & Insurance Personnel

Compiled from the nation's major building codes, professional associations, and product manufacturer institutes, this unique resource enables you to:
• Set a standard for subcontractors and employees
• Protect yourself against defect claims
• Resolve disputes
• Answer client questions with authority

$59.95 per copy
Over 500 pages, illustrated, Softcover Catalog No. 67322

The Building Professional's Guide to Contract Documents

New 3rd Edition

By Waller S. Poage, AIA, CSI, CVS

This comprehensive treatment of Contract Documents is an important reference for owners, design professionals, contractors, and students.

• Structure your Documents for Maximum Efficiency
• Effectively communicate construction requirements to all concerned
• Understand the Roles and Responsibilities of Construction Professionals
• Improve Methods of Project Delivery

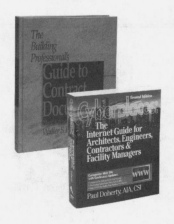

$64.95 per copy, 400 pages
Diagrams and construction forms, Hardcover
Catalog No. 67261A

Cyberplaces: The Internet Guide for AECs & Facility Managers

2nd Edition

By Paul Doherty, AIA, CSI

Completely updated with a section on the New Economy and e-commerce—such as AEC arbitrage, aggregate purchasing, e-supply chains, and more. Plus the best Knowledge Management and Internet tools, including PDAs and the newest Web software for scheduling and project management.

The **Web site** keeps you current on the best Internet tools for building design, construction, and management.

$59.95 per copy
Over 700 pages, illustrated, Softcover
Catalog No. 67317A

For more information
visit Means Web Site
at www.rsmeans.com

Reference Books

Value Engineering: Practical Applications

. . . For Design, Construction, Maintenance & Operations
By Alphonse Dell'Isola, PE

A tool for immediate application—for engineers, architects, facility managers, owners, and contractors. Includes: Making the Case for VE—The Management Briefing, Integrating VE into Planning and Budgeting, Conducting Life Cycle Costing, Integrating VE into the Design Process, Using VE Methodology in Design Review and Consultant Selection, Case Studies, VE Workbook, and a Life Cycle Costing program on disk.

$79.95 per copy
Over 450 pages, illustrated, Hardcover
Catalog No. 67319

Facilities Operations & Engineering Reference

By the Association for Facilities Engineering and R.S. Means

An all-in-one technical referance for planning and managing facility projects and solving day-to-day operations problems. Selected as the official Certified Plant Engineer reference, this handbook covers financial analysis, maintenance, HVAC and energy efficiency, and more.

$99.95 per copy
Over 700 pages, illustrated, Hardcover
Catalog No. 67318

Concrete Repair and Maintenance Illustrated

By Peter H. Emmons

Illustrated guide to analyzing, repairing, and maintaining concrete structures for optimal performance and cost–effectiveness. From parking garages, to roads and bridges, to structural concrete, this book describes the causes, effects, and remedies for concrete failure.

$69.95 per copy
Catalog No. 67146

HVAC: Design Criteria, Options, Selection

Expanded 2nd Edition
By William H. Rowe III, AIA, PE

Includes Indoor Air Quality, CFC Removal, Energy Efficient Systems, and Special Systems by Building Type. Helps you solve a wide range of HVAC system design and selection problems effectively and economically. Gives you clear explanations of the latest ASHRAE standards.

$84.95 per copy
Over 600 pages, illustrated, Hardcover
Catalog No. 67306

Total Productive Facilities Management

By Richard W. Sievert, Jr.

This book provides a comprehensive program to:
• Achieve business goals by optimizing facility resources
• Implement best practices through benchmarking, evaluation & project management
• Increase your value to the organization

$39.98 per copy
Over 270 pages, illustrated, Hardcover
Catalog No. 67321

Facilities Maintenance Management

By Gregory H. Magee, PE

Now you can get successful management methods and techniques for all aspects of facilities maintenance. This comprehensive reference explains and demonstrates successful management techniques for all aspects of maintenance, repair, and improvements for buildings, machinery, equipment, and grounds. Plus, guidance for outsourcing and managing internal staffs.

$86.95 per copy
Over 280 pages with illustrations, Hardcover
Catalog No. 67249

Cost Planning & Estimating for Facilities Maintenance

In this unique book, a team of facilities management authorities shares their expertise at:
• Evaluating and budgeting maintenance operations
• Maintaining & repairing key building components
• Applying *Means Facilities Maintenance & Repair Cost Data* to your estimating

Covers special maintenance requirements of the 10 major building types.

$82.95 per copy
Over 475 pages, Hardcover
Catalog No. 67314

Maintenance Management Audit

This annual audit program is essential for every organization in need of a proper assessment of its maintenance operation. Downloadable electronic forms from Means Web Site allow managers to identify and correct problems and enhance productivity.

Now $32.48 per copy, limited quantity
125 pages, spiral bound, illustrated, Hardcover
Catalog No. 67299

Reference Books

For more information
visit Means Web Site
at www.rsmeans.com

Builder's Essentials: Plan Reading & Material Takeoff
By Wayne J. DelPico

For Residential and Light Commercial Construction

A valuable tool for understanding plans and specs, and accurately calculating material quantities.
Step-by-step instructions and takeoff procedures based on a full set of working drawings.

$35.95 per copy
Over 420 pages, Softcover
Catalog No. 67307

Planning & Managing Interior Projects
New 2nd Edition
By Carol E. Farren, CFM

Addresses changes in technology and business, guiding you through commercial design and construction from initial client meetings to post-project administration. Includes: evaluating space requirements, alternative work models, telecommunications and data management, and environmental issues.

$69.95 per copy
Over 400 pages, illustrated, Hardcover
Catalog No. 67245A

Builder's Essentials: Best Business Practices for Builders & Remodelers:
An Easy-to-Use Checklist System
By Thomas N. Frisby

A comprehensive guide covering all aspects of running a construction business, with more than 40 user–friendly checklists. This book provides expert guidance on: increasing your revenue and keeping more of your profit; planning for long-term growth; keeping good employees and managing subcontractors.

$29.95 per copy
Over 220 pages, Softcover
Catalog No. 67329

Builder's Essentials: Framing & Rough Carpentry 2nd Edition
By Scot Simpson

A complete, training manual for apprentice and experienced carpenters. Develop and improve your skills with "framer-friendly," easy-to-follow instructions, and step-by-step illustrations. Learn proven techniques for framing walls, floors, roofs, stairs, doors, and windows. Updated guidance on standards, building codes, safety requirements, and more.

$24.95 per copy
Over 150 pages, Softcover
Catalog No. 67298A

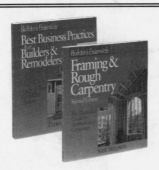

How to Estimate with Means Data & CostWorks
By R.S. Means and Saleh Mubarak New 2nd Edition!

Learn estimating techniques using Means cost data. Includes an instructional version of Means CostWorks CD–ROM with Sample Building Plans.
The step-by-step guide takes you through all the major construction items. Over 300 sample estimating problems are included.

$59.95 per copy
Over 190 pages, Softcover
Catalog No. 67324

Means Heavy Construction Handbook

Informed guidance for planning, estimating and performing today's heavy construction projects. Provides expert advice on every aspect of heavy construction work including hazardous waste remediation and estimating. To assist planning, estimating, performing, or overseeing work.

$74.95 per copy
Over 430 pages, illustrated, Hardcover
Catalog No. 67148

Superintending for Contractors:
How to Bring Jobs in On-Time, On-Budget
By Paul J. Cook

This book examines the complex role of the superintendent/field project manager, and provides guidelines for the efficient organization of this job. Includes administration of contracts, change orders, purchase orders, and more.

$35.95 per copy
Over 220 pages, illustrated, Softcover
Catalog No. 67233

HVAC Systems Evaluation
By Harold R. Colen, PE

You get direct comparisons of how each type of system works, with the relative costs of installation, operation, maintenance and applications by type of building. With requirements for hooking up electrical power to HVAC components. Contains experienced advice for repairing operational problems in existing HVAC systems, ductwork, fans, cooling coils, and much more!

$84.95 per copy
Over 500 pages, illustrated, Hardcover
Catalog No. 67281

Means Landscape Estimating Methods
3rd Edition
By Sylvia H. Fee

This revised edition offers expert guidance for preparing accurate estimates for new landscape construction and grounds maintenance. Includes a complete project estimate featuring the latest equipment and methods, and **two totally new chapters—Life Cycle Costing, and Landscape Maintenance Estimating.**

$62.95 per copy
Over 300 pages, illustrated, Hardcover
Catalog No. 67295A

Means Environmental Remediation Estimating Methods
By Richard R. Rast

Guidelines for estimating 50 standard remediation technologies. Use it to prepare preliminary budgets, develop estimates, compare costs and solutions, estimate liability, review quotes, negotiate settlements.

A valuable support tool for *Means Environmental Remediation Unit Price* and *Assemblies books.*

$99.95 per copy
Over 600 pages, illustrated, Hardcover
Catalog No. 64777

For more information
visit Means Web Site
at www.rsmeans.com

Reference Books

Means Estimating Handbook

This comprehensive reference covers a full spectrum of technical data for estimating, with information on sizing, productivity, equipment requirements, codes, design standards, and engineering factors.
Means Estimating Handbook will help you: evaluate architectural plans and specifications, prepare accurate quantity takeoffs, prepare estimates from conceptual to detailed, and evaluate change orders.

$99.95 per copy
Over 900 pages, Hardcover
Catalog No. 67276

Means Repair and Remodeling Estimating
3rd Edition
By Edward B. Wetherill & R.S. Means

Focuses on the unique problems of estimating renovations of existing structures. It helps you determine the true costs of remodeling through careful evaluation of architectural details and a site visit.
New section on disaster restoration costs.

$69.95 per copy
Over 450 pages, illustrated, Hardcover
Catalog No. 67265A

Facilities Planning & Relocation
New, lower price and user-friendly format.
By David D. Owen

A complete system for planning space needs and managing relocations. Includes step-by-step manual, over 50 forms, and extensive reference section on materials and furnishings.

$89.95 per copy
Over 450 pages, Softcover
Catalog No. 67301

Means Square Foot & Assemblies Estimating Methods
New 3rd Edition!

Develop realistic Square Foot and Assemblies Costs for budgeting and construction funding. The new edition features updated guidance on square foot and assemblies estimating using UNIFORMAT II. An essential reference for anyone who performs conceptual estimates.

$69.95 per copy
Over 300 pages, illustrated, Hardcover
Catalog No. 67145B

Means Electrical Estimating Methods
2nd Edition

Expanded version includes sample estimates and cost information in keeping with the latest version of the CSI MasterFormat. Contains new coverage of Fiber Optic and Uninterruptible Power Supply electrical systems, broken down by components and explained in detail. A practical companion to *Means Electrical Cost Data.*

$64.95 per copy
Over 325 pages, Hardcover
Catalog No. 67230A

Means Mechanical Estimating Methods
2nd Edition

This guide assists you in making a review of plans, specs, and bid packages with suggestions for takeoff procedures, listings, substitutions and pre-bid scheduling. Includes suggestions for budgeting labor and equipment usage. Compares materials and construction methods to allow you to select the best option.

$64.95 per copy
Over 350 pages, illustrated, Hardcover
Catalog No. 67294

Successful Estimating Methods:
From Concept to Bid
By John D. Bledsoe, PhD, PE

A highly practical, all-in-one guide to the tips and practices of today's successful estimator. Presents techniques for all types of estimates, and advanced topics such as life cycle cost analysis, value engineering, and automated estimating. Estimate spreadsheets available at Means Web site.

$64.95 per copy
Over 300 pages, illustrated, Hardcover
Catalog No. 67287

The ADA in Practice
By Deborah S. Kearney, PhD

Helps you meet and budget for the requirements of the Americans with Disabilities Act. Shows how to do the job right by understanding what the law requires, allows, and enforces. Includes a "buyers guide" for 70 ADA-compliant products.

*Winner of IFMA's "Distinguished Author of the Year" award.

$59.98 per copy
Over 600 pages, illustrated, Softcover
Catalog No. 67147A

Means Productivity Standards for Construction
Expanded Edition (*Formerly Man-Hour Standards*)

Here is the working encyclopedia of labor productivity information for construction professionals, with labor requirements for thousands of construction functions in CSI MasterFormat.
Completely updated, with over 3,000 new work items.

$69.98 per copy
Over 800 pages, Hardcover
Catalog No. 67236A

Means ADA Compliance Pricing Guide

Gives you detailed cost estimates for budgeting modification projects, including estimates for each of 260 alternates. You get the 75 most commonly needed modifications for ADA compliance, each an assembly estimate with detailed cost breakdown.

$59.98 per copy
Over 350 pages, illustrated, Softcover
Catalog No. 67310

Reference Books

For more information
visit Means Web Site
at www.rsmeans.com

Means Unit Price Estimating Methods

2nd Edition
$59.95 per copy
Catalog No. 67303

Means Plumbing Estimating Methods

New 2nd Edition
By Joseph J. Galeno &
Sheldon T. Greene
$59.95 per copy
Catalog No. 67283A

Business Management for Contractors

How to Make Profits in Today's Market
By Paul J. Cook
Now $17.98 per copy, limited quantity
Catalog No. 67250

Risk Management for Building Professionals

By Thomas E. Papageorge, RA
Now $15.98 per copy, limited quantity
Catalog No. 67254

Contractor's Business Handbook

By Michael S. Milliner
Now $21.48 per copy, limited quantity
Catalog No. 67255

Basics for Builders: How to Survive and Prosper in Construction

By Thomas N. Frisby
$34.95 per copy
Catalog No. 67273

Means Scheduling Manual

3rd Edition
$32.48 per copy
Catalog No. 67291

Successful Interior Projects Through Effective Contract Documents

By Joel Downey & Patricia K. Gilbert
Now $24.98 per copy, limited quantity
Catalog No. 67313

Estimating for Contractors:

How to Make Estimates that Win Jobs
By Paul J. Cook
$35.95 per copy
Catalog No. 67160

Means Forms for Contractors

For general and specialty contractors
$49.98 per copy
Catalog No. 67288

Understanding Building Automation Systems

By Reinhold A. Carlson, PE &
Robert Di Giandomenico
$29.98 per copy, limited quantity
Catalog No. 67284

Illustrated Construction Dictionary, Condensed

$59.95 per copy
Catalog No. 67282

Managing Construction Purchasing

By John G. McConville, CCC, CPE
Now $19.98 per copy, limited quantity
Catalog No. 67302

Hazardous Material & Hazardous Waste

By Francis J. Hopcroft, PE,
David L. Vitale, M. Ed., &
Donald L. Anglehart, Esq.
Now $44.98 per copy, limited quantity
Catalog No. 67258

Fundamentals of the Construction Process

By Kweku K. Bentil, AIC
Now $34.98 per copy, limited quantity
Catalog No. 67260

Construction Delays

By Theodore J. Trauner, Jr., PE, PP
$29.48 per copy
Catalog No. 67278

Interior Home Improvement Costs

New 7th Edition
$19.95 per copy
Catalog No. 67308C

Exterior Home Improvement Costs

New 7th Edition
$19.95 per copy
Catalog No. 67309C

Means Facilities Maintenance Standards

By Roger W. Liska, PE, AIC
Now $69.95 per copy
Catalog No. 67246

How to Estimate with Metric Units

Now $9.98 per copy, limited quantity
Catalog No. 67304

Quantity Takeoff for Contractors

How to Get Accurate Material Counts
By Paul J. Cook
Now $17.98 per copy, limited quantity
Catalog No. 67262

For more information
visit Means Web Site
at www.rsmeans.com

Seminars

Developing and Managing Facility Assessment Programs

This two-day program instructs attendees on organizing, staffing, conducting and reporting the results of a facilities audit as the basis for sound capital planning and budgeting.

Some Of What You'll Learn:
- Conducting regular condition inspections to identify deferred maintenance problems, cost estimating, formulating accurate budgets, and planning staff workloads
- The management process required for planning, conducting and documenting the physical condition and functional adequacy of buildings or other facilities
- Gathering reliable data to identify and gauge deferred maintenance requirements
- Essential steps for conducting facilities inspection programs

Who Should Attend: Plant Managers, Facility Executives, Real Estate Planners and Developers, Estimators... and others who plan for capital renewal and replacement and/or whose financial decisions and budgets are based on the conditions of facilities.

Repair and Remodeling Estimating

This two-day seminar emphasizes all the underlying considerations unique to repair/remodeling estimating and presents the correct methods for generating accurate, reliable R&R project costs using the unit price and assemblies methods.

Some Of What You'll Learn:
- Estimating considerations like labor hours, building code compliance, working within existing structures, purchasing materials in smaller quantities, and unforeseen deficiencies
- Identifies problems and provides solutions to estimating building alterations
- Rules for factoring in minimum labor costs, accurate productivity estimates, and allowances for project contingencies
- R&R estimating examples are calculated using unit price and assemblies data

Who Should Attend: Facilities Managers, Plant Engineers, Architects, Contractors, Estimators, Builders... and others who are concerned with the proper preparation and/or evaluation of repair and remodeling estimates.

Mechanical and Electrical Estimating

This two-day course teaches attendees how to prepare more accurate and complete mechanical/ electrical estimates, avoiding the pitfalls of omission and double-counting, while understanding the composition and rationale within the Means mechanical/electrical database.

Some Of What You'll Learn:
- The unique way mechanical and electrical systems are interrelated
- M&E estimates, conceptual, planning, budgeting and bidding stages
- Order of magnitude, square foot, assemblies and unit price estimating
- Comparative cost analysis of equipment and design alternatives

Who Should Attend: Architects, Engineers, Facilities Managers, Mechanical and Electrical Contractors... and others needing a highly reliable method for developing, understanding and evaluating mechanical and electrical contracts.

Unit Price Estimating

This interactive two-day seminar teaches attendees how to interpret project information and process it into final, detailed estimates with the greatest accuracy level. The single most important credential an estimator can take to the job is the ability to visualize construction in the mind's eye, and thereby estimate accurately.

Some Of What You'll Learn:
- Interpreting the design in terms of cost
- The most detailed, time-tested methodology for accurate "pricing"
- Key cost drivers: material, labor, equipment, staging, and subcontracts
- Understanding direct and indirect costs for accurate job cost accounting and change order management

Who Should Attend: Corporate and Government Estimators and Purchasers, Architects, Engineers... and others needing to produce accurate project estimates.

Square Foot and Assemblies Cost Estimating

This two-day course teaches attendees how to quickly deliver accurate square foot estimates using limited budget and design information.

Some Of What You'll Learn:
- Using square foot costing to get the estimate done faster
- Taking advantage of a "systems" or "assemblies" format
- The Means "building assemblies/square foot cost approach"
- How to create a very reliable preliminary systems estimate using bare-bones design information

Who Should Attend: Facilities Managers, Facilities Engineers, Estimators, Planners, Developers, Construction Finance Professionals... and others needing to make quick, accurate construction cost estimates at commercial, government, educational and medical facilities.

Scheduling and Project Management

This two-day course teaches attendees the most current and proven scheduling and management techniques needed to bring projects in on time and on budget.

Some Of What You'll Learn:
- Crucial phases of planning and scheduling
- How to establish project priorities, develop realistic schedules and management techniques
- Critical Path and Precedence Methods
- Special emphasis on cost control

Who Should Attend: Construction Project Managers, Supervisors, Engineers, Estimators, Contractors... and others who want to improve their project planning, scheduling and management skills.

Facilities Maintenance and Repair Estimating

This two-day course teaches attendees how to plan, budget, and estimate the cost of ongoing and preventive maintenance and repair for existing buildings and grounds.

Some Of What You'll Learn:
- The most financially favorable maintenance, repair and replacement scheduling and estimating
- Auditing and value engineering facilities
- Preventive planning and facilities upgrading
- Determining both in-house and contract-out service costs
- An annual, asset-protecting M&R plan

Who Should Attend: Facility Managers, Maintenance Supervisors, Buildings and Grounds Superintendents, Plant Managers, Planners, Estimators... and others involved in facilities planning and budgeting.

Managing Facilities Construction and Maintenance

This two-day program teaches attendees the management techniques needed to effectively plan, organize, control and get the most out of limited facilities resources.

Some Of What You'll Learn:
- Helping an organization allocate physical resources
- Evaluating the impact of today's facility decisions on tomorrow's budgets.
- Developing budgets, reducing expenditures and assessing productivity
- Common problems and challenges encountered during new construction, renovation or maintenance projects

Who Should Attend: Facility Managers, Maintenance Managers, Plant Managers, Facility Project Managers, Facilities Foremen and Supervisors... and others involved in or aspiring to a role in facilities management.

Advanced Project Management

This two-day seminar will teach you how to effectively manage and control the entire design-build process and allow you to take home tangible skills that will be immediately applicable on existing projects.

Some Of What You'll Learn:
- Value engineering, bonding, fast-tracking and bid package creation
- How estimates and schedules can be integrated to provide advanced project management tools
- Cost engineering, quality control, productivity measurement and improvement
- Front-loading a project and predicting its cash flow

Who Should Attend: Owners, Project Managers, Architectural and Engineering Managers, Construction Managers, Contractors... and anyone else who is responsible for the timely design and completion of construction projects.

Seminars

For more information
visit Means Web Site
at www.rsmeans.com

2002 Means Seminar Schedule

Location	Dates
Las Vegas, NV	March 11-14
Washington, DC	April 22-25
Denver, CO	May 20-23
San Francisco, CA	June 10-13
Cape Cod, MA	September 9-12
Washington, DC	September 23-26
San Diego, CA	October 21-24
Orlando, FL	November 4-7
Atlantic City, NJ	November 18-21
Las Vegas, NV	December 2-5

Note: Call for exact dates and details.

Registration Information

Register Early... Save up to $150! Register 30 days before the start date of a seminar and save $150 off your total fee. *Note: This discount can be applied only once per order.*

How to Register Register by phone today! Means toll-free number for making reservations is: **1-800-448-8182.**

Individual Seminar Registration Fee $875 To register by mail, complete the registration form and return with your full fee to: Seminar Division, R.S. Means Company, Inc., 63 Smiths Lane, Kingston, MA 02364.

Federal Government Pricing All Federal Government employees save 25% off regular seminar price. Other promotional discounts cannot be combined with Federal Government discount.

Team Discount Program Two to four seminar registrations: $760 per person– Five or more seminar registrations: $710 per person–Ten or more seminar registrations: Call for pricing.

Consecutive Seminar Offer One individual signing up for two separate courses at the same location during the designated time period pays only $1,400. You get the second course for only $525 **(a 40% discount).** Payment must be received at least ten days prior to seminar dates to confirm attendance.

Refunds Cancellations will be accepted up to ten days prior to the seminar start. There are no refunds for cancellations postmarked later than ten working days prior to the first day of the seminar. A $150 processing fee will be charged for all cancellations. Written notice or telegram is required for all cancellations. Substitutions can be made at anytime before the session starts. **No-shows are subject to the full seminar fee.**

AACE Approved Courses The R.S. Means Construction Estimating and Management Seminars described and offered to you here have each been approved for 14 hours (1.4 recertification credits) of credit by the AACE International Certification Board toward meeting the continuing education requirements for re-certification as a Certified Cost Engineer/Certified Cost Consultant.

AIA Continuing Education R.S. Means is registered with the AIA Continuing Education System (AIA/CES) and is committed to developing quality learning activities in accordance with the CES criteria. R.S. Means seminars meet the AIA/ CES criteria for Quality Level 2. AIA members will receive (28) learning units (LUs) for each two day R.S. Means Course.

Daily Course Schedule The first day of each seminar session begins at 8:30 A.M. and ends at 4:30 P.M. The second day is 8:00 A.M.–4:00 P.M. Participants are urged to bring a hand-held calculator since many actual problems will be worked out in each session.

Continental Breakfast Your registration includes the cost of a continental breakfast, a morning coffee break, and an afternoon break. These informal segments will allow you to discuss topics of mutual interest with other members of the seminar. (You are free to make your own lunch and dinner arrangements.)

Hotel/Transportation Arrangements R.S. Means has arranged to hold a block of rooms at each hotel hosting a seminar. To take advantage of special group rates when making your reservation be sure to mention that you are attending the Means Seminar. You are of course free to stay at the lodging place of your choice. **(Hotel reservations and transportation arrangements should be made directly by seminar attendees.)**

Important Class sizes are limited, so please register as soon as possible.

Registration Form   Call 1-800-448-8182 x5115 to register or FAX 1-800-632-6732. Visit our Web site www.rsmeans.com

Please register the following people for the Means Construction Seminars as shown here. Full payment or deposit is enclosed, and we understand that we must make our own hotel reservations if overnight stays are necessary.

☐ Full payment of $ _____ enclosed.

☐ Bill me

Name of Registrant(s)
(To appear on certificate of completion)

_____

_____

_____

P.O. #: _____
GOVERNMENT AGENCIES MUST SUPPLY PURCHASE ORDER NUMBER

Firm Name _____

Address _____

City/State/Zip _____

Telephone No. Fax No. _____

E-Mail Address _____

Charge our registration(s) to: ☐ MasterCard ☐ VISA ☐ American Express ☐ Discover

Account No. _____ Exp. Date _____

Cardholder's Signature _____

Seminar Name	City	Dates

Please mail check to: R.S. Means Company, Inc., 63 Smiths Lane, P.O.Box 800, Kingston, MA 02364 USA

Consulting Services Group

Solutions For Your Construction and Facilities Cost Management Problems

We are leaders in the cost engineering field, supporting the unique construction and facilities management costing challenges of clients from the Federal Government, Fortune 1000 Corporations, Building Product Manufacturers, and some of the World's Largest Design and Construction Firms.

Research...

- **Custom Database Development**—Means expertise in construction cost engineering and database management can be put to work creating customized cost databases and applications.
- **Data Licensing & Integration**—To enhance applications dealing with construction, any segment of Means vast database can be licensed for use, and harnessed in a format compatible with a previously developed proprietary system.
- **Cost Modeling**—Pre-built custom cost models provide organizations that expend countless hours estimating repetitive work with a systematic time-saving estimating solution.
- **Database Auditing & Maintenance**—For clients with in-house data, Means can help organize it, and by linking it with Means database, fill in any gaps that exist and provide necessary updates to maintain current and relevant proprietary cost data.

Estimating...

- **Estimating Service**—Means expertise is available to perform construction cost estimates, as well as to develop baseline schedules and establish management plans for projects of all sizes and types. Conceptual, budget and detailed estimates are available.
- **Benchmarking**—Means can run baseline estimates on existing project estimates. Gauging estimating accuracy and identifying inefficiencies improves the success ratio, precision and productivity of estimates.
- **Project Feasibility Studies**—The Consulting Services Group can assist in the review and clarification of the most sound and practical construction approach in terms of time, cost and use.
- **Litigation Support**—Means is available to provide opinions of value and to supply expert interpretations or testimony in the resolution of construction cost claims, litigation and mediation.

Training...

- **Core Curriculum**—Means educational programs, delivered on-site, are designed to sharpen professional skills and to maximize effective use of cost estimating and management tools. On-site training cuts down on travel expenses and time away from the office.
- **Custom Curriculum**—Means can custom-tailor courses to meet the specific needs and requirements of clients. The goal is to simultaneously boost skills and broaden cost estimating and management knowledge while focusing on applications that bring immediate benefits to unique operations, challenges, or markets.
- **Staff Assessments and Development Programs**—In addition to custom curricula, Means can work with a client's Human Resources Department or with individual operating units to create programs consistent with long-term employee development objectives.

For more information and a copy of our capabilities brochure, please call 1-800-448-8182 and ask for the Consulting Services Group, or reach us at www.rsmeans.com

MeansData™

CONSTRUCTION COSTS FOR SOFTWARE APPLICATIONS

Your construction estimating software is only as good as your cost data.

Software Integration

A proven construction cost database is a mandatory part of any estimating package. We have linked MeansData™ directly into the industry's leading software applications. The following list of software providers can offer you MeansData™ as an added feature for their estimating systems. Visit them on-line at *www.rsmeans.com/demo/* for more information and free demos. Or call their numbers listed below.

3D International
713-871-7000 venegas@3di.com

4Clicks-Solutions, LLC
719-574-7721
mbrown@4clicks-solutions.com

ACT
Applied Computer Technologies
Facility Management Software
919-859-1335 info@srs.net

AEPCO, Inc.
301-670-4642 blueworks@aepco.com

**American Contractor/
Maxwell Systems**
800-333-8435 info@amercon.com

ArenaSoft Estimating
888-370-8806 info@arenasoft.com

Ares Corporation
650-401-7100
sales@arescorporation.com

**AssetWork
CSI-Maximus**
Facility Management Software
800-659-9001 info@assetworks.com

Benchmark, Inc.
800-393-9193 sales@benchmark-inc.com

BSD
Building Systems Design, Inc.
888-273-7638 bsd@bsdsoftlink.com

CProjects, Inc.
203-262-6248 sales@cprojects.com

cManagement
800-945-7093 sales@cmanagement.com

CDCI
Construction Data Controls, Inc.
800-285-3929 sales@cdci.com

CMS
Computerized Micro Solutions
800-255-7407 cms@proest.com

Conac Group
800-663-2338 sales@conac.com

Eagle Point Software
800-678-6565 sales@eaglepoint.com

Estimating Systems, Inc.
800-967-8572 pulsar@capecod.net

G2 Estimator
A Div. of Valli Info. Syst., Inc.
800-657-6312 info@g2estimator.com

G/C EMUNI, Inc.
514-953-5148 rpa@gcei.ca

Geac Commercial Systems, Inc.
800-554-9865 info@geac.com

Hard Dollar
800-637-7496 sales@harddollar.com

IQ Beneco
801-565-1122 mdover@beneco.com

Luqs International
888-682-5573 info@luqs.com

MC²
Management Computer Controls
800-225-5622 vkeys@mc2-ice.com

Prism Computer Corporation
Facility Management Software
800-774-7622 famis@prismcc.com

Quest Solutions, Inc.
800-452-2342 info@questsolutions.com

Sanders Software, Inc.
800-280-9760 hsander@vallnet.com

Sinisoft, Inc.
877-669-4949 info.usa@sinistre.com

Timberline Software Corp.
800-628-6583
product.info@timberline.com

TMA Systems, Inc.
Facility Management Software
800-862-1130 sales@tmasys.com

US Cost, Inc.
800-372-4003
sales@uscost.com

Vertigraph, Inc.
800-989-4243
info-request@vertigraph.com

Wendlware
714-895-7222 sales@corecon.com

Winestimator, Inc.
800-950-2374 sales@winest.com

DemoSource™

One-stop shopping for the latest cost estimating software for just $19.95. This evaluation tool includes product literature and demo diskettes for ten or more estimating systems, all of which link to MeansData™. **Call 1-800-334-3509 to order.**

FOR MORE INFORMATION ON ELECTRONIC PRODUCTS CALL
1-800-448-8182 OR FAX 1-800-632-6732.

MeansData™ is a registered trademark of R.S. Means Co., Inc., *CMD*.

14

For more information
visit Means Web Site
at www.rsmeans.com

New Titles

From Model Codes to the IBC: A Transitional Guide

By Rolf Jensen & Associates, Inc.

NEW!

A time–saving resource for Architects, Engineers, Building Officials and Authorities Having Jurisdiction (AHJs), Contractors, Manufacturers, Building Owners, and Facility Managers.

Provides comprehensive, user-friendly guidance on making the transition to the International Building Code® from the model codes you're familiar with. Includes side-by-side code comparison of the IBC to the UBC, NBC, SBC, and NFPA 101®. Also features professional code commentary, quick-find indexes, and a Web site with regular code updates.

Also contains illustrations, abbreviations key, and an extensive resource section.

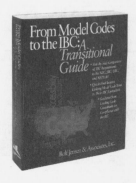

$114.95 per copy
880 pages, Softcover
Catalog No. 67328

Historic Preservation: Project Planning & Estimating

By Swanke Hayden Connell Architects

• *Managing Historic Restoration, Rehabilitation, and Preservation Building Projects and*
• *Determining and Controlling Their Costs*

The authors explain:
• How to determine whether a structure qualifies as historic
• Where to obtain funding and other assistance
• How to evaluate and repair more than 75 historic building materials
• How to properly research, document, and manage the project to meet code, agency, and other special requirements
• How to approach the upgrade of major building systems

$99.95 per copy
Over 675 pages, Hardcover
Catalog No. 67323

Means Illustrated Construction Dictionary, 3rd Edition

New Updated Edition with interactive CD-ROM

Long regarded as the Industry's finest, the Means Illustrated Construction Dictionary is now even better. With the addition of over 1,000 new terms and hundreds of new illustrations, it is the clear choice for the most comprehensive and current information.

The companion CD-ROM that comes with this new edition adds many extra features: larger graphics, expanded definitions, and links to both CSI MasterFormat numbers and product information.

• 19,000 construction words, terms, phrases, symbols, weights, measures, and equivalents
• 1,000 new entries
• 1,200 helpful illustrations
• Easy-to-use format, with thumbtabs

$99.95 per copy
Over 790 pages, illustrated Hardcover
Catalog No. 67292A

2002 Order Form

Qty.	Book No.	COST ESTIMATING BOOKS	Unit Price	Total
	60062	Assemblies Cost Data 2002	$164.95	
	60012	Building Construction Cost Data 2002	99.95	
	61012	Building Const. Cost Data–Looseleaf Ed. 2002	124.95	
	63012	Building Const. Cost Data–Metric Version 2002	99.95	
	60222	Building Const. Cost Data–Western Ed. 2002	99.95	
	60112	Concrete & Masonry Cost Data 2002	92.95	
	60142	Construction Cost Indexes 2002	218.00	
	60142A	Construction Cost Index–January 2002	54.50	
	60142B	Construction Cost Index–April 2002	54.50	
	60142C	Construction Cost Index–July 2002	54.50	
	60142D	Construction Cost Index–October 2002	54.50	
	60341	Contr. Pricing Guide: Resid. R & R Costs 2002	36.95	
	60332	Contr. Pricing Guide: Resid. Detailed 2002	36.95	
	60322	Contr. Pricing Guide: Resid. Sq. Ft. 2002	39.95	
	64022	ECHOS Assemblies Cost Book 2002	164.95	
	64012	ECHOS Unit Cost Book 2002	109.95	
	54002	ECHOS (Combo set of both books)	229.95	
	60232	Electrical Change Order Cost Data 2002	99.95	
	60032	Electrical Cost Data 2002	99.95	
	60202	Facilities Construction Cost Data 2002	241.95	
	60302	Facilities Maintenance & Repair Cost Data 2002	219.95	
	60162	Heavy Construction Cost Data 2002	99.95	
	63162	Heavy Const. Cost Data–Metric Version 2002	99.95	
	60092	Interior Cost Data 2002	99.95	
	60122	Labor Rates for the Const. Industry 2002	219.95	
	60182	Light Commercial Cost Data 2002	87.95	
	60022	Mechanical Cost Data 2002	99.95	
	60152	Open Shop Building Const. Cost Data 2002	99.95	
	60212	Plumbing Cost Data 2002	99.95	
	60042	Repair and Remodeling Cost Data 2002	87.95	
	60172	Residential Cost Data 2002	87.95	
	60282	Site Work & Landscape Cost Data 2002	99.95	
	60052	Square Foot Costs 2002	109.95	
		REFERENCE BOOKS		
	67147A	ADA in Practice	59.98	
	67310	ADA Pricing Guide	59.98	
	67273	Basics for Builders: How to Survive and Prosper	34.95	
	67330	Bldrs Essentials: Adv. Framing Techniques	24.95	
	67329	Bldrs Essentials: Best Bus. Practices for Bldrs	29.95	
	67298A	Bldrs Essentials: Framing/Carpentry 2nd Ed.	24.95	
	67298AS	Bldrs Essentials: Framing/Carpentry Spanish	24.95	
	67307	Bldrs Essentials: Plan Reading & Takeoff	35.95	
	67261A	Bldg. Prof. Guide to Contract Documents–3rd Ed.	64.95	
	67312	Building Spec Homes Profitably	29.95	
	67250	Business Management for Contractors	17.98	
	67146	Concrete Repair & Maintenance Illustrated	69.95	
	67278	Construction Delays	29.48	
	67255	Contractor's Business Handbook	21.48	
	67314	Cost Planning & Est. for Facil. Maint.	82.95	
	67317A	Cyberplaces: The Internet Guide–2nd Ed.	59.95	
	67230A	Electrical Estimating Methods–2nd Ed.	64.95	
	64777	Environmental Remediation Est. Methods	99.95	
	67160	Estimating for Contractors	35.95	
	67276	Estimating Handbook	99.95	

Qty.	Book No.	REFERENCE BOOKS (Cont.)	Unit Price	Total
	67249	Facilities Maintenance Management	$ 86.95	
	67246	Facilities Maintenance Standards	69.95	
	67318	Facilities Operations & Engineering Reference	99.95	
	67301	Facilities Planning & Relocation	89.95	
	67231	Forms for Building Const. Professional	47.48	
	67288	Forms for Contractors	49.98	
	67328	From Model Codes to IBC: Transitional Guide	114.95	
	67260	Fundamentals of the Construction Process	34.98	
	67148	Heavy Construction Handbook	74.95	
	67323	Historic Preservation: Proj. Planning & Est.	99.95	
	67308C	Home Improvement Costs–Int. Projects 7th Ed.	19.95	
	67309C	Home Improvement Costs–Ext. Projects 7th Ed.	19.95	
	67324	How to Estimate w/Means Data & CostWorks	59.95	
	67304	How to Estimate with Metric Units	9.98	
	67306	HVAC: Design Criteria, Options, Select.–2nd Ed.	84.95	
	67281	HVAC Systems Evaluation	84.95	
	67282	Illustrated Construction Dictionary, Condensed	59.95	
	67292A	Illustrated Construction Dictionary, w/CD-ROM	99.95	
	67295A	Landscape Estimating–3rd Ed.	62.95	
	67299	Maintenance Management Audit	32.48	
	67302	Managing Construction Purchasing	19.98	
	67294	Mechanical Estimating–2nd Ed.	64.95	
	67245A	Planning and Managing Interior Projects–2nd Ed.	69.95	
	67283A	Plumbing Estimating Methods–2nd Ed.	59.95	
	67326	Preventive Maint. Guidelines for School Facil.	149.95	
	67236A	Productivity Standards for Constr.–3rd Ed.	69.98	
	67247A	Project Scheduling & Management for Constr.	64.95	
	67262	Quantity Takeoff for Contractors	17.98	
	67265A	Repair & Remodeling Estimating–3rd Ed.	69.95	
	67322	Residential & Light Commercial Const. Stds.	59.95	
	67254	Risk Management for Building Professionals	15.98	
	67291	Scheduling Manual–3rd Ed.	32.48	
	67327	Spanish/English Construction Dictionary	22.95	
	67145B	Sq. Ft. & Assem. Estimating Methods–3rd Ed.	69.95	
	67287	Successful Estimating Methods	64.95	
	67313	Successful Interior Projects	24.98	
	67233	Superintending for Contractors	35.95	
	67321	Total Productive Facilities Management	39.98	
	67284	Understanding Building Automation Systems	29.98	
	67303	Unit Price Estimating Methods–2nd Ed.	59.95	
	67319	Value Engineering: Practical Applications	79.95	

MA residents add 5% state sales tax

Shipping & Handling**

Total (U.S. Funds)*

Prices are subject to change and are for U.S. delivery only. *Canadian customers may call for current prices. **Shipping & handling charges: Add 7% of total order for check and credit card payments. Add 9% of total order for invoiced orders.

Send Order To: **ADDV-1001**

Name (Please Print) _____

Company _____

☐ **Company**
☐ **Home**   Address _____

City/State/Zip _____

Phone # _____ P.O. # _____